The series Lecture Notes in Bioinformatics (LNBI) was established in 2003 as a topical subseries of LNCS devoted to bioinformatics and computational biology.

The series publishes state-of-the-art research results at a high level. As with the LNCS mother series, the mission of the series is to serve the international R & D community by providing an invaluable service, mainly focused on the publication of conference and workshop proceedings and postproceedings.

Aurelio López Fernández ·
Alejandro Rodríguez-González ·
Raquel Leirós-Rodríguez ·
Christian Mata Miquel ·
Víctor Manuel González Suárez

Editors

Artificial Intelligence in Biomedicine

First Conference of the Spanish Society
of Artificial Intelligence in Biomedicine, CIABiomed 2025
Seville, Spain, October 23–24, 2025
Proceedings

 Springer

Editors
Aurelio López Fernández
Universidad Pablo de Olavide
Seville, Spain

Raquel Leirós-Rodríguez
University of León
Ponferrada, Spain

Víctor Manuel González Suárez
University of Oviedo
Gijón, Spain

Alejandro Rodríguez-González
Universidad Politecnica de Madrid
Madrid, Spain

Christian Mata Miquel
Universitat Politècnica de Catalunya
Barcelona, Spain

ISSN 0302-9743　　　　　ISSN 1611-3349　(electronic)
Lecture Notes in Bioinformatics
ISBN 978-3-032-10660-5　　　ISBN 978-3-032-10661-2　(eBook)
https://doi.org/10.1007/978-3-032-10661-2

LNCS Sublibrary: SL8 – Bioinformatics

Preface

We are delighted to present the accepted contributions of the 1st Conference of the Spanish Society of Artificial Intelligence in Biomedicine (CIABiomed 2025), held in Seville, Spain, on October 23–24, 2025, at the School of Computer Engineering of the University of Seville.

This inaugural edition of CIABiomed was conceived with the aim of establishing an international reference point for the exchange of knowledge and experiences in the ethical, rigorous, and responsible use of artificial intelligence in the biomedical domain. The event brought together researchers, professionals from the healthcare and technology sectors, and students from diverse disciplines to discuss the latest scientific advances, current challenges, and future perspectives in this rapidly evolving field.

The scientific programme of CIABiomed 2025 included keynote lectures, oral presentations, and special tracks, which provided a broad view of the state of the art and emerging trends in AI applied to biomedicine. Topics ranged from the design of models and algorithms to their integration in real clinical contexts, addressing issues of ethics, explainability, interoperability, and technology transfer, among others.

For the sake of consistency and readability of these proceedings, the accepted papers have been organised into nine chapters, aligned with the official thematic areas of CIABiomed 2025:

1. **AI for multi-omic integration and analysis of heterogeneous biomedical data.** This chapter gathers contributions that explore advanced computational methods for integrating genomics, transcriptomics, proteomics, metabolomics, and other omic layers with heterogeneous clinical data. Emphasis is placed on data harmonization, fusion strategies, and innovative AI-driven models that enable a deeper understanding of complex biological processes and disease mechanisms.
2. **AI-based clinical decision support systems.** This section includes works devoted to the design, development, and evaluation of intelligent systems that assist healthcare professionals in decision-making. The focus is on predictive and prescriptive models capable of providing actionable recommendations, improving diagnostic accuracy, and personalising treatments in real-world clinical contexts.
3. **Biomedical signal processing.** Contributions in this chapter address the processing, analysis, and interpretation of biomedical signals, including electrophysiological recordings, biosensor outputs, and multimodal time-series data. AI-based approaches for noise reduction, feature extraction, and disease-related pattern recognition are particularly emphasised.
4. **Ethics, privacy, and security in AI applications in healthcare.** This chapter brings together research on the ethical, legal, and social challenges arising from the application of AI in healthcare. Topics include privacy preservation, data protection, algorithmic fairness, bias mitigation, and the development of regulatory and governance frameworks to ensure safe and equitable adoption of AI technologies.

5. **Explainable and interpretable AI in clinical and biomedical contexts.** Papers in this section discuss methods and frameworks aimed at making AI systems transparent, interpretable, and trustworthy for biomedical applications. The works highlight novel strategies for ante-hoc and post-hoc explainability, visualisation techniques, and human-in-the-loop paradigms that enhance user confidence and accountability in AI-assisted medical decision-making.

6. **Medical image processing and computer vision in clinical environments.** This chapter includes contributions that leverage AI for the analysis, segmentation, classification, and reconstruction of medical images. Special attention is given to computer vision techniques applied to radiology, pathology, and multimodal imaging, as well as to the integration of image-based biomarkers into diagnostic and prognostic workflows.

7. **Natural language processing (NLP) in biomedicine.** Contributions here focus on the use of NLP to process and analyse biomedical texts, electronic health records, and clinical narratives. Research includes language models, text mining, clinical concept extraction, and conversational agents, with applications ranging from information retrieval to decision support in patient care.

8. **Prediction of protein structures and molecular interactions with AI techniques.** This chapter highlights advances in the prediction of protein folding, structural modelling, and the characterisation of molecular interactions through AI methods. Works cover the development of models for drug design, protein–protein and protein–ligand interactions, and the application of generative and deep learning techniques in structural bioinformatics.

9. **Predictive modelling and personalised medicine using artificial intelligence.** The final chapter is devoted to contributions that apply AI to predictive modelling in healthcare, enabling early disease detection, prognosis, and patient stratification. Papers also address personalised medicine, integrating patient-specific data to design tailored therapeutic strategies and improve health outcomes.

Submissions were handled via Springer's workflow and underwent single-blind peer review by at least three members of the Program Committee. Following this rigorous evaluation process, 47 articles were accepted from 55 submissions, according to the recommendations of the reviewers and the preferences of the authors, to be included in the LNBI proceedings.

To complement the general programme with high-impact and emerging themes, CIABiomed 2025 hosted a set of Special Tracks that addressed specific topics of great relevance for the community. These sessions were carefully curated by expert organisers and provided a platform for in-depth exploration of cutting-edge research questions, novel methodologies, and application-driven case studies that extend beyond the scope of the general submissions. The Special Tracks served not only to highlight areas where artificial intelligence is making transformative contributions to biomedicine, but also to foster interdisciplinary collaboration between computer scientists, clinicians, biologists, and industry stakeholders. By combining rigorous methodological exchange with a strong emphasis on real-world challenges, these tracks created a unique opportunity to bridge basic research and clinical translation, and to stimulate dialogue among participants from diverse backgrounds who share a common interest in advancing trustworthy and impactful AI solutions for healthcare, and in the first edition of CIABiomed 2025 the following were received:

– **Special track on Generative AI in Healthcare: Models, Agents, and Clinical Decision Support (IAGAM).**

Generative artificial intelligence (Generative AI) is rapidly transforming the healthcare sector, offering new opportunities for clinical knowledge generation, medical task automation, and improved decision support. This special session brought together recent research and innovative applications that explore the role of generative models and intelligent agents in clinical settings, with a particular focus on their ability to provide reliable, ethical, and effective support in medical practice.

The session explored the state of the art in the use of generative models in the clinical setting, shared experiences on the implementation of agents in healthcare systems, and analysed evaluation and validation frameworks for generative AI in clinical decision-making. In addition, it fostered interdisciplinary collaboration between AI, medicine, bioethics, and healthcare policy. The topics of interest in this special track were as follows:

- Generative models applied to structured and unstructured clinical data.
- Conversational agents in clinical contexts (e.g., automated medical history, triage).
- Automatic generation of medical texts (medical record summaries, discharge notes, etc.).
- Evaluation of generative models in real or simulated healthcare settings.
- Ethical and regulatory considerations in the use of generative AI in healthcare.
- Integration of generative AI into medical record systems.

Organizers: **Cristina Rubio-Escudero**, *Universidad de Sevilla, Spain.*
Miguel Ángel Armengol de la Hoz, *Fundación Progreso y Salud, Spain.*
Belén Vega Márquez, *Universidad de Sevilla, Spain.*
David Gutiérrez Avilés, *Universidad de Sevilla, Spain.*
Laura Melgar García, *Universidad Politécnica de Madrid, Spain.*
Isabel Amaya Rodríguez, *Fundación Progreso y Salud, Spain.*

– **Special track on Computational Multiomics-Clinical Integration for the Discovery and Validation of Predictive Biomarkers (ICMBIO)**

Modern medicine is on the cusp of a revolution, driven by the ability to integrate vast amounts of biological and clinical data. This session showcased concepts, techniques and best practices for multiomics–clinical integration, combining omics layers (genomics, transcriptomics, proteomics, metabolomics, epigenomics, and microbiomics) with clinical data (medical history, laboratory tests, medical imaging, electronic health records), adopting a bidirectional laboratory–clinical approach that spanned bench-to-bedside and bedside-to-bench translation.

The goal was to promote a multidisciplinary exchange between laboratory researchers, clinicians, and data scientists, catalyzing the rapid translation of biomarker innovations into clinical practice and enabling the discovery of new predictive markers.

Multidisciplinary exchange between bench researchers, clinicians, and data scientists is catalyzing the rapid translation of biomarker innovations into clinical practice and sparking the discovery of entirely new predictive markers. Topics of interest for this special session included (but were not limited to) the following:

- Multiomic-Clinical Integration.
- AI for Omics Data (Omics AI).
- Predictive Modelling and Digital Twins.
- Translational Research.
- Computer-Aided Modelling of Biological Processes and Gene Networks.

Organizers: **Juan Antonio Ortega Ramírez**, *Universidad de Sevilla, Spain.*
Aurea Simón-Soro, *Universidad de Sevilla, Spain.*
Francisco Antonio Gómez Vela, *Universidad Pablo de Olavide, Spain.*

- **Special track on Explainable, Robust, and Trustworthy Artificial Intelligence (ERTAI)**

As data scientists, we are accustomed to validating our models according to metrics of correctness. In the medical domain, though, alternative measures of quality are preferred, particularly those that assess robustness, trustworthiness, and explainability. The development of explainable, robust, and trustworthy artificial intelligence (AI) approaches is essential to foster transparency, accountability, and user confidence in the increasingly pervasive role of these technologies in health and biomedicine. By ensuring that AI systems can be understood, monitored, and relied upon to make decisions that comply with regulations and are consistent with ethical principles, risks and biases are mitigated while domain users are empowered to engage with these technologies with greater assurance, fostering a harmonious integration of AI into the field of biomedicine.

The aim of this special session was to discuss the principles that govern the behaviour of AI technology in the light of current regulatory efforts, as well as of the operators, users, and other stakeholders who are impacted by decisions informed by such technologies in the medical domain. Moreover, it also explored how clear, understandable, and interpretable explanations of AI decisions could be made to enhance transparency and foster user trust.

The session welcomed contributions that addressed either theoretical developments or practical applications, presenting innovative approaches and technological advancements within Explainable, Robust, and Trustworthy AI. Topics of interest for this special session included (but were not limited to) the following:

- Explainable and Interpretable AI in (bio)medicine.
- AI in safety-critical medical systems.
- Uncertainty management in medical applications.
- Compliance with regulations of medical AI-based systems.
- Human-in-the-loop in medical AI-based systems.
- Intrinsically interpretable methods applied in medicine.
- Ante-hoc vs post-hoc methods applied in medicine.
- Data visualization in biomedicine.
- Application domain specificity of methods.

- Trustworthy AI in medicine.
- Bias in AI methods applied to medical problems.
- Fairness of AI methods applied to medical problems.
- Accountability of AI methods applied to medical problems.
- Impact of AI methods on the human workforce.
- AI for federated learning in healthcare.
- AI ethics in medical application domains.
- Legal implications of the use of AI in healthcare.
- Regulatory frameworks for the use of AI in healthcare.

Organizers: **Caroline König**, *Universitat Politècnica de Catalunya, Spain.*
Alfredo Vellido, *Universitat Politècnica de Catalunya, Spain.*
José M. Juarez, *Universidad de Murcia, Spain.*

- **Special track on Intelligent Systems for the Assessment and Treatment of Early Stages of Alzheimer's Disease and Other Dementias (SIETAD)**

Advances in intelligent systems are opening up new possibilities for the early detection, clinical assessment, and personalised treatment of neurodegenerative diseases such as Alzheimer's and other dementias. The integration of tools based on artificial intelligence (AI), machine learning, smart sensors, multimodal data analysis, and interactive digital platforms is revolutionising early diagnosis, therapeutic strategies, and continuous monitoring.

This special track brought together researchers, clinicians, and technology developers in a multidisciplinary forum to present and analyse innovative approaches that incorporate intelligent technologies into the comprehensive management of the early stages of these diseases. Topics of interest for this special session included (but were not limited to) the following:

- Machine learning and deep learning models for early diagnosis.
- Digital biomarkers and neuroimaging analysis using AI.
- Automated cognitive assessment and intelligent tools.
- Clinical decision support systems.
- AI applications in personalised cognitive interventions.
- Remote monitoring and disease progression prediction.
- Integration of multimodal data (clinical, genomic, behavioural, etc.).
- Ethics and explainability in AI applied to neurodegeneration.

Organizers: **Rafael Martínez Tomás**, *Universidad Nacional de Educación a Distancia, Spain.*
Mariano Rincón Zamorano, *Universidad Nacional de Educación a Distancia, Spain.*

CIABiomed 2025 was organized by the Spanish Society of Artificial Intelligence in Biomedicine (IABiomed). We thank the University of Seville, in particular the School of Computer Engineering, for hosting the event. We gratefully acknowledge the support of our sponsors MSD Spain, MINSAIT, i+Med, and GSK. Our sincere thanks go to the members of the committees, the reviewers, and the organizers of the special tracks for their dedication and excellent work, to the Springer Meteor support team for their assistance during the submission and review process, and to Springer for their continuous cooperation and support.

October 2025

Aurelio López-Fernández
Alejandro Rodríguez-González
Raquel Leirós Rodríguez
Christian Mata Miquel
Victor Manuel González Suárez

Organization

General Chairs

Aurelio López-Fernández   Pablo de Olavide University, Spain
Alejandro Rodríguez-González   Polytechnic University of Madrid, Spain

Program Committee Chairs

Victor Manuel González Suárez   University of Oviedo, Spain
Raquel Leirós Rodríguez   University of León, Spain
Christian Mata Miquel   Polytechnic University of Catalonia, Spain

Steering Committee

Isabel Amaya Rodríguez   Fundación Progreso y Salud, Spain
Miguel Ángel Armengol de la Hoz   Fundación Progreso y Salud, Spain
José Alberto Benítez Andrades   University of León, Spain
Francisco Antonio Gómez Vela   Pablo de Olavide University, Spain
David Gutiérrez Avilés   University of Seville, Spain
José Manuel Juárez Herrero   University of Murcia, Spain
Caroline König   Polytechnic University of Catalonia, Spain
Laura Melgar García   Polytechnic University of Madrid, Spain
Rafael Martínez Tomás   National University of Distance Education, Spain
Juan Antonio Ortega Ramírez   University of Seville, Spain
Beatriz Pontes Balanza   University of Seville, Spain
Mariano Rincón Zamorano   National University of Distance Education, Spain
Cristina Rubio-Escudero   University of Seville, Spain
Aurea Simón-Soro   University of Seville, Spain
Belén Vega Márquez   University of Seville, Spain
Alfredo Vellido   Polytechnic University of Catalonia, Spain

Program Committee

Joao Rafael Almeida   University of Aveiro, Portugal
Andrea Alvarez Perez   Polytechnic University of Madrid, Spain

Gualberto Asencio Cortés	Pablo de Olavide University, Spain
Martín Bayón Gutiérrez	University of León, Spain
Fernando M. Delgado Chaves	University of Hamburg, Germany
Antonio Jesus Díaz Honrubia	Polytechnic University of Madrid, Spain
Norberto Díaz Díaz	Pablo de Olavide University, Spain
Federico Divina	Pablo de Olavide University, Spain
Jorge González Domínguez	Universidade da Coruña, Spain
María Teresa García Ordás	University of León, Spain
Miguel García Torres	Pablo de Olavide University, Spain
Valerio Guarrasi	University Campus Bio-Medico di Roma, Italy
Alicia Merayo Corcoba	University of León, Spain
Ernestina Menasalvas Ruiz	Polytechnic University of Madrid, Spain
Isabel Nepomuceno Chamorro	University of Seville, Spain
Juan A. Nepomuceno Chamorro	Pablo de Olavide University, Spain
Dulcenombre M. del Saz Navarro	Pablo de Olavide University, Spain
José Antonio Aveleira Mata	University of León, Spain
Mónica López Lacort	FISABIO, Spain
Francisco Martínez Álvarez	Pablo de Olavide University, Spain
Lucia Prieto Santamaria	Polytechnic University of Madrid, Spain
Pablo Reina Jiménez	University of Seville, Spain
Marc Ríos-Cadenas	Pablo de Olavide University, Spain
Domingo S. Rodríguez Baena	Pablo de Olavide University, Spain
Sara Rubio Lanchas	Valladolid University Clinical Hospital, Spain
José Enrique Sánchez López	University of Seville, Spain
Paloma Tejera Nevado	Polytechnic University of Madrid, Spain
José F. Torres-Maldonado	Pablo de Olavide University, Spain
Ángela Troncoso García	Pablo de Olavide University, Spain

Contents

AI-based Clinical Decision Support Systems

Biomedical Signal Processing

Natural Language Processing (NLP) in Biomedicine

Medical Image Processing and Computer Vision in Clinical Environments

Prediction of Protein Structures and Molecular Interactions with AI Techniques

**AI for Multi-omic Integration and Analysis of Heterogeneous
biomedical data**

Predictive Modelling and Personalised Medicine Using Artificial Intelligence

Explainable and Interpretable AI in Clinical and Biomedical Contexts

An Iterative Random Forest Framework for Statistical Feature Selection in High-Dimensional Biomedical Data: A Case Study on Alzheimer's Diagnosis

Pablo Zubasti[(✉)] [iD], Miguel A. Patricio [iD], Antonio Berlanga [iD], and José M. Molina [iD]

Universidad Carlos III de Madrid, ROR: https://ror.org/03ths8210, Computer Science and Engineering Department, Applied Artificial Intelligence Group, Madrid, Spain

pzubasti@pa.uc3m.es, mpatrici@inf.uc3m.es, {aberlan,molina}@ia.uc3m.es
https://giaa.uc3m.es/

Abstract. This study presents a robust methodology to address the challenge of high-dimensionality in clinical datasets, a pervasive issue in biomedical informatics. Leveraging the predictive capabilities of Machine Learning—particularly in healthcare applications—has shown considerable promise. However, the presence of redundant or low-relevance features in training data often hampers model efficiency and inflates both computational and clinical costs. These concerns are especially critical in medical scenarios, where data acquisition may involve expensive or invasive procedures. To mitigate these challenges, we introduce an iterative framework that employs multiple *Random Forest* models and a Gaussian modeling approach to statistically assess and rank attribute importance. Through repeated training cycles, our approach identifies non-contributory and statistically insignificant features, enabling their systematic exclusion.

Keywords: Random Forest · Machine Learning · Feature selection · Alzheimer · Dementia

1 Introduction

The early diagnosis of Alzheimer's Disease (AD) remains a critical challenge in the biomedical and computational sciences. Recent advances in machine learning (ML), particularly in explainable artificial intelligence (XAI) and feature selection methodologies, have enabled the extraction of clinically relevant insights from large-scale biomedical datasets such as the *National Alzheimer's Coordinating Center* (NACC) *Uniform Data Set* (UDS). Medical datasets, especially those related to neurodegenerative diseases like Alzheimer's, often include hundreds or even thousands of features collected through diverse procedures—ranging from

A. López Fernández et al. (Eds.): CIABiomed 2025, LNBI 16148, pp. 3–16, 2026.
https://doi.org/10.1007/978-3-032-10661-2_1

cognitive assessments and blood tests to advanced neuroimaging. Many of these features, however, are either redundant or weakly correlated with the outcome of interest. Their presence not only adds noise to the learning process but also inflates computational costs and, crucially, can lead to unnecessary or even invasive diagnostic procedures. Given the high dimensionality of clinical data, discerning the most informative features is essential not only to improve predictive accuracy but also to streamline the diagnostic process by eliminating superfluous or non-contributive medical examinations.

To address this issue, we propose a robust and statistically grounded methodology designed to identify and eliminate low-relevance attributes from high-dimensional clinical datasets. The core idea is to treat feature importance as an intrinsic property of the dataset, but one that can vary depending on the hyperparameter configuration of the model. Our framework employs multiple iterations of *Random Forest* models trained under varying hyperparameter settings to obtain a stable estimate of feature relevance. Importances are then aggregated, statistically analyzed, and thresholded using a Gaussian modeling approach to identify and discard non-contributory variables. We apply this methodology to real-world data from the *National Alzheimer's Coordinating Center* (NACC), demonstrating that it is possible to achieve a substantial reduction in feature space without compromising diagnostic performance. In doing so, we enable the construction of more parsimonious and interpretable models that are better suited to support early-stage diagnosis, while simultaneously reducing the clinical burden associated with high-cost or low-utility data acquisition.

2   Previous Work

A growing body of literature has focused on developing interpretable models that provide insight into which clinical variables are most significant in predicting cognitive decline. The authors of [1] introduced an explainable machine learning framework employing *SHapley Additive exPlanations* (SHAP) in conjunction with tree-based classifiers to analyze the NACC dataset. Their approach revealed that a relatively small subset of clinical features could account for the majority of predictive power in AD classification, enabling the removal of redundant tests without compromising model performance. Similarly, the authors of [15] proposed an XGBoost-SHAP integrated diagnostic model which not only demonstrated high accuracy but also emphasized interpretability and clinical validation through feature impact visualization. Their findings corroborated that certain frequently administered tests in the NACC dataset, such as general physical metrics, had limited diagnostic value compared to cognitive and neuropsychological assessments.

In line with the objective of minimizing diagnostic costs while preserving accuracy, the authors of [11] developed a cost-effective feature selection strategy. Their approach filtered features not only based on statistical contribution but also on economic and logistical constraints, making it particularly relevant for resource-limited healthcare settings. Furthermore, in their comprehensive review of ML methods for Alzheimer's detection, [14] emphasized that attribute reduction techniques such as recursive feature elimination (RFE) and LASSO have become standard for preprocessing datasets like NACC and ADNI. They also noted the increasing use of ensemble methods to cross-validate feature importance rankings across different cohorts and modalities.

The integration of deep learning into feature selection has also been explored. In [6] the authors utilized convolutional neural networks (CNNs) for structural MRI analysis but included NACC-based metadata for external validation, thereby highlighting the utility of hybrid models in verifying feature relevance across modalities. Their work supports the notion that feature harmonization between imaging and clinical records enhances model generalizability. In [4] the authors proposed a multimodal classification framework using ensemble deep random vector functional link (RVFL) networks. Their model, trained on a fusion of cognitive scores and clinical demographics from NACC, leveraged dimensionality reduction and late fusion techniques to isolate critical diagnostic factors. A separate study [10], performed a comparative evaluation of various ML models using an augmented NACC dataset that included derived features. They applied wrapper-based feature selection and reported that up to 40% of input variables could be safely excluded without statistically significant degradation in classification performance.

In addition to model-specific studies, broader architectural strategies have been explored. The authors of [13] applied filter and wrapper methods in tandem, noting that features related to socioeconomic status and basic vital signs were consistently filtered out as irrelevant, reinforcing the need for domain-informed feature curation. In [7], the authors introduced a deep learning model with an embedded feature selection module that autonomously learned to suppress low-impact features. Their embedded selection framework enabled real-time feature pruning, a particularly advantageous characteristic in clinical deployment scenarios. Finally, the authors of [17] proposed a novel dimensionality reduction methodology grounded in distance-correlation-based feature spaces [3, 12], which preserves high-order dependencies often missed by linear correlation methods. Their study demonstrated that this approach, applied to dementia diagnosis, not only improves model interpretability but also enhances generalization by filtering out attributes with minimal non-linear contribution. In a related work [16], the same authors combined graph-based embeddings with unsupervised learning to evaluate latent relationships among clinical attributes, thereby identifying clusters of interdependent features. This approach offers a new paradigm for feature evaluation in dementia studies—moving beyond individual significance and toward structural interplay among attributes.

In summary, not only do these methodologies reduce computational complexity and overfitting, but they also align with clinical objectives by eliminating unnecessary medical tests. ultimately facilitating more ethical, cost-effective, and scalable diagnostic pathways. The contribution of the current study builds upon this body of work by leveraging the NACC dataset to further investigate the role of feature significance and its implications for the rational exclusion of redundant clinical attributes.

3   Proposed Methodology

The following section provides a detailed and mathematically rigorous exposition of the methodology proposed in this work, which was briefly outlined in the abstract and the introduction.

Multi-random Forest Iterative Feature Importance

In Random Forest models, feature importance can be estimated by quantifying the contribution of each feature to the reduction of node impurity during the construction of decision trees. This method, also known as *Mean Decrease in Impurity* (MDI), is one of the most commonly used techniques for interpreting ensemble models.

Let $\mathcal{T}$ be a decision tree constructed during the training of a Random Forest. At each internal node t of the tree, a split is performed using a feature X_j selected among the available attributes. The quality of the split is evaluated based on a decrease in impurity, which depends on the learning task:

- For **classification tasks**, the impurity $i(t)$ is typically measured using the *Gini impurity* or *entropy*.
- For **regression tasks**, the impurity is usually measured by the *variance* or *mean squared error*.

Let $i(t)$ be the impurity at node t, and let t_L and t_R be the left and right child nodes resulting from a split at node t. The decrease in impurity $\Delta i(t)$ produced by a split using feature X_j is defined as:

$$\Delta i(t) = i(t) - p_L \cdot i(t_L) - p_R \cdot i(t_R) \tag{1}$$

where p_L and p_R are the proportions of samples that go to the left and right child nodes, respectively.

Let $\mathcal{F} = \{\mathcal{T}_1, \mathcal{T}_2, \ldots, \mathcal{T}_M\}$ be the set of M trees in the Random Forest. The importance of feature X_j, denoted as $\mathrm{Imp}(X_j)$, is computed as the sum of the impurity decreases over all nodes where X_j is used to split, averaged over all trees:

$$\mathrm{Imp}(X_j) = \frac{1}{M} \sum_{m=1}^{M} \sum_{t \in \mathcal{T}_m : v(t)=j} \Delta i(t) \tag{2}$$

where $v(t) = j$ indicates that the feature X_j is used for the split at node t in tree $\mathcal{T}_m$.

To facilitate interpretability, the raw importance values are typically normalized so that they sum to 1:

$$\widehat{\text{Imp}}(X_j) = \frac{\text{Imp}(X_j)}{\sum_{k=1}^{d} \text{Imp}(X_k)} \tag{3}$$

where d is the total number of features.

One of the fundamental characteristics of the impurity-based method for estimating feature importance in supervised learning problems is its computational efficiency and methodological simplicity. While more advanced approaches based on feature subset selection exist and can provide alternative importance estimates, the impurity-based method remains robust and yields reliable results. Therefore, it is important to note that, throughout this work, any reference to the computation of feature importances from the trees of a Random Forest model shall refer specifically to the procedure described above.

The core idea of the proposed methodology is that feature importance in a learning problem constitutes an intrinsic and implicit property of the problem itself. However, since it is estimated from trained models that are configurable through a variety of hyperparameters, it may exhibit some degree of variability. Therefore, in order to support a reliable estimation of feature importance, the methodology establishes the necessity of training multiple models iteratively— one distinct hyperparameter configuration per iteration—to obtain a set of feature importance vectors, each associated with a specific iteration. Formally, the methodology is defined as follows:

1. Define a set of hyperparameter configurations $\mathcal{H} = \{h_1, h_2, \ldots h_T\} = \{h_t\}_{t=1}^{T}$, where each h_t represents a distinct configuration to be applied during the iterative training process, stored as a K-dimensional column vector $h_t \in \mathbb{O}^K$ (each dimension represents the value of a hyperparameter: numerical, discrete, categorical, etc.). Accordingly, we use the notation $\mathbb{O}$ to indicate that each element of the vector corresponds, in an abstract manner, to an object that may take on values of any of the types described above. It is important to emphasize that, although the term iterative is used, a new model is trained from scratch for each configuration; no pre-existing model is reused or incrementally retrained.

2. For each given configuration $h_t \in \mathcal{H}$, a Random Forest model $\mathcal{M}_t(\mathbf{X}; h_t)$ is trained and a feature importance vector is obtained:

$$\Phi\left[\mathcal{M}_t(\mathbf{X}; h_t)\right] = \begin{bmatrix} \widehat{\text{Imp}}(X_1) \\ \widehat{\text{Imp}}(X_2) \\ \vdots \\ \widehat{\text{Imp}}(X_N) \end{bmatrix} \tag{4}$$

Where $\text{Imp}(X_i)$ is computed as described in Eq. (2) and its normalized form is obtained from the definition in Eq. (3). The value N determines the number of attributes in the input space (the features of the problem).

3. An estimated importance value for each feature is computed by applying the statistical mean operator over all values obtained for that feature across iterations. Mathematically:

$$\hat{\Omega}(X_i) = \frac{1}{T} \sum_{t=1}^{T} \Phi_i \left[\mathcal{M}_t(\mathbf{X}; h_t) \right] \tag{5}$$

As a result, the vector $\widehat{\boldsymbol{\Omega}} \in \mathbb{R}^N$ of estimated feature importances is obtained, defined as:

$$\widehat{\boldsymbol{\Omega}} = \begin{bmatrix} \hat{\Omega}(X_1) \\ \hat{\Omega}(X_2) \\ \vdots \\ \hat{\Omega}(X_N) \end{bmatrix} \tag{6}$$

4. Based on the values collected in the vector $\widehat{\boldsymbol{\Omega}}$, the sample mean μ and standard deviation σ are computed in order to fit a Gaussian distribution over the estimated importances (see Eq. (7)). From this fitted distribution, statistical inference is performed to determine the z-value such that the cumulative probability to its left equals a predefined threshold α, i.e., the z-score that corresponds to the lower α proportion of importances (representing the least relevant features).

$$f(x) = \frac{1}{\sigma \sqrt{2\pi}} \exp \left[-\frac{1}{2} \left(\frac{x - \mu}{\sigma} \right)^2 \right] \tag{7}$$

$$\text{Compute } z_\alpha \text{ such that:} \quad \int_{-\infty}^{z_\alpha} f(x)dx = \alpha$$

$$\text{Where } \alpha \in [0,1] \text{ since} \quad \int_{-\infty}^{+\infty} f(x)dx = 1 \tag{8}$$

5. Using the previously computed threshold value z_α as a selection criterion, all features with an importance value less than or equal to z_α are discarded. It is important to note that, in order to determine z_α, the data (i.e., the importance scores) must first be standardized using the transformation:

$$\tilde{x}_i = \frac{x_i - \mu}{\sigma} \tag{9}$$

So after the z_α value is identified in the standardized domain, it must be de-standardized back to the original scale using:

$$z_\alpha = \tilde{z}_\alpha \cdot \sigma + \mu \tag{10}$$

The previously described and formulated methodology thus provides a systematic approach for eliminating low-relevance features in supervised learning problems, leveraging trained models and statistically sound inference procedures.

4   Experimental Results

Basic Dataset: OASIS-2

Although the core of this work focuses on applying the proposed methodology to the dataset generously provided by the *National Alzheimer's Coordinating Center* (NACC), an initial experimental round on the OASIS-2 dataset [8,9] is conducted to observe the behavior and outcomes of the methodology in the context of a substantially smaller classification problem (14 features and 373 patient instances). This smaller scale facilitates the visualization of the various steps of the methodology, which would be considerably more complex to illustrate using the NACC's UDS dataset [2,5,15].

A fundamental consideration before applying the methodology is that the OASIS-2 dataset contains several attributes that are irrelevant to the learning task (such as the patient ID and MRI session ID), as well as certain categorical features that, in this specific context, are empirically found to be uninformative for predictive modeling. Therefore, prior to any processing, the dataset is systematically cleaned, resulting in a total of 10 input features and one target variable. Although the target variable originally takes three possible values—"Demented", "Nondemented", and "Converted"—the "Converted" label is reclassified as "Demented", based on the clinical definition that converted patients are those who were initially diagnosed as non-demented but subsequently progressed to a diagnosis of dementia in follow-up evaluations. The first step in the experimental pipeline (for both the OASIS-2 and UDS datasets) was the definition of the set of hyperparameter vectors that determine both the successive iterations and the distinct Random Forest models to be trained. The selected configurations are detailed in Tables 1 and 2.

Table 1. Hyperparameters used in the experimental evaluation.

NE	MD	MSS	MSL	MF	Criterion
10	5	2	1	None	gini
100	10	3	2	None	gini
200	15	4	3	sqrt	gini
300	20	5	4	sqrt	gini
400	25	6	5	log2	gini
500	30	7	6	log2	gini
1000	35	8	7	log2	gini

Applying the proposed methodology, feature importance estimates are first computed, with the results for each attribute presented in Fig. 1. It can be observed that the variation in importance across the different iterations is not particularly high. Additionally, Fig. 2 illustrates the evolution of the importance

Table 2. Description of acronyms used in Table 1.

Acronym	Full Hyperparameter Name
NE	`n_estimators`
MD	`max_depth`
MSS	`min_samples_split`
MSL	`min_samples_leaf`
MF	`max_features`
Criterion	`criterion`

values for each attribute in the OASIS-2 dataset throughout the iterations. Notably, from configuration 3 onward, the importance of the "CDR" attribute decreases significantly, while the importance of the "MMSE" attribute increases.

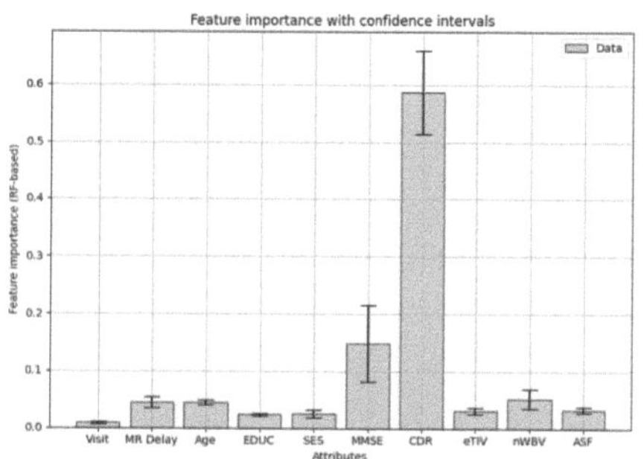

Fig. 1. Feature importance of each attribute in the OASIS-2 dataset.

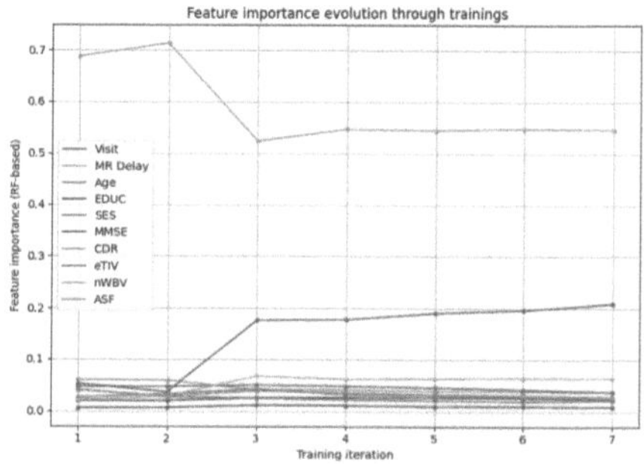

Fig. 2. Feature importance evolution through the different iterations of the methodology (using the OASIS-2 data).

After computing the estimated importance values for each attribute, a histogram of the raw values and their cumulative distribution was constructed, as shown in Figs. 3 and 4. These visualizations illustrate the distribution of the importance scores and the fitted normal distribution, which will subsequently be used to compute the threshold value z_α.

Following the remaining steps of the proposed methodology, a threshold value of $z_\alpha = 0.071$ ($\alpha = 5\%$) was selected, resulting in the elimination of 8 out of 10 variables and retaining only "MMSE" and "CDR". The classification accuracy obtained by training a Random Forest model on the original dataset (prior to applying the methodology) was **0.9643**, while the accuracy obtained after applying the methodology to the reduced dataset was **0.9554**. Figures 5 and 6 present a detailed comparison of the results before and after feature reduction, including additional classification metrics such as *precision, recall,* and *f1-score.* The results indicate a nearly perfect preservation of model performance when

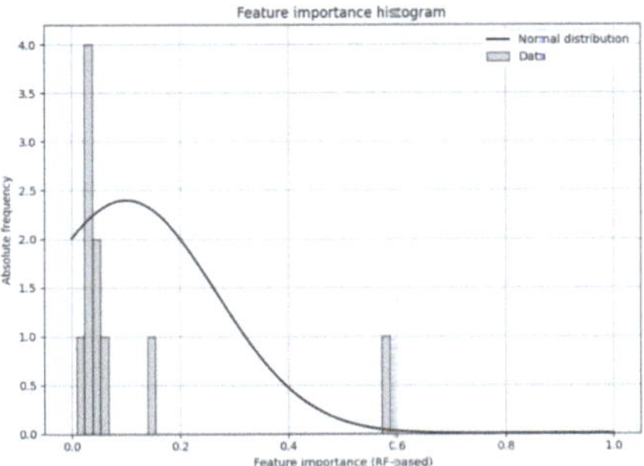
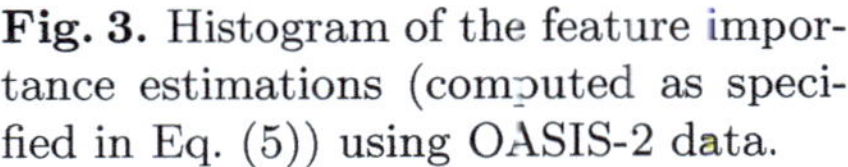
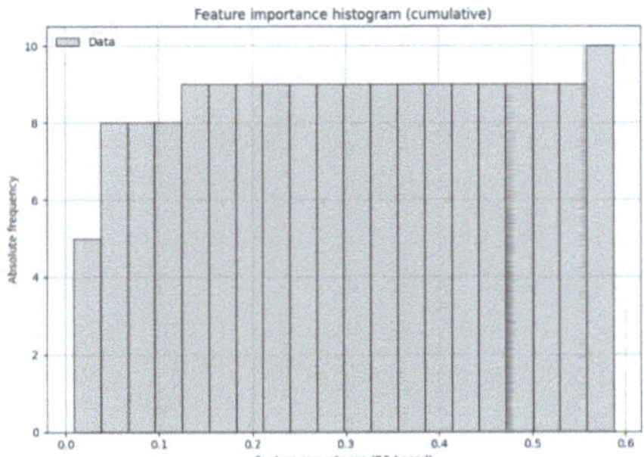

Fig. 3. Histogram of the feature importance estimations (computed as specified in Eq. (5)) using OASIS-2 data.

Fig. 4. Cumulative histogram distribution of the data displayed in Fig. 3.

trained on the reduced feature set, suggesting that the most relevant attributes are those directly associated with in-clinic assessments, such as memory and cognitive function tests, rather than features derived from MRI-based imaging. It is also noteworthy that the attributes related to age ("Age") and years of education ("EDUC") were discarded, reinforcing the principle that correlation does not imply causation, and indicating that age alone is not a reliable predictor for diagnosing dementia through machine learning models.

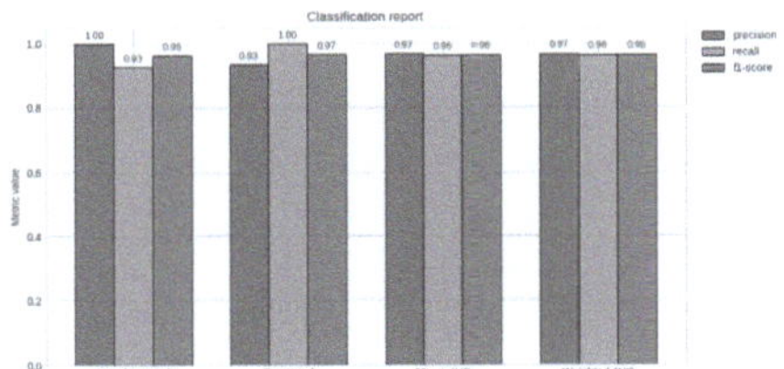
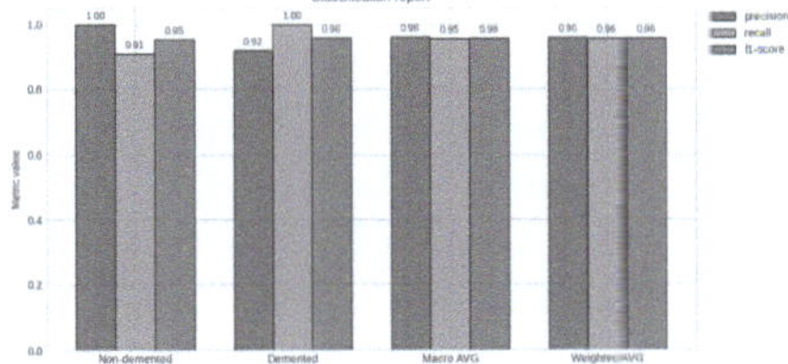

Fig. 5. Classification report for the OASIS-2 dataset before applying the proposed methodology.

Fig. 6. Classification report for the OASIS-2 dataset after applying the proposed methodology.

Complex Dataset: UDS from NACC

The second block of experiments is conducted on the UDS-NACC dataset. The scale of this dataset far exceeds that of OASIS-2, containing nearly 200,000 instances and 1,936 attributes. As with the OASIS-2 dataset—and in order to operate within a more controlled environment—features with excessive missing values were filtered out, allowing at most 10% of missing instances per attribute.

If missing values were present within this tolerance, they were imputed using the k-Nearest Neighbors method (with $k = 3$). Additionally, categorical attributes lacking meaningful semantic content were discarded to simplify the complexity of the original high-dimensional space.

After preprocessing, the number of attributes was reduced to 846, which still represents a considerable dimensionality. The first results, which can be visualized graphically, are presented in Figs. 7 and 8, showing the histogram and cumulative histogram of the feature importance values obtained from the 846 variables, respectively. It is important to highlight that the importance of most variables is close to zero. This begins to support the hypothesis that a vast majority of the features are irrelevant to the problem. However, this observation must be interpreted with caution, as the importance values are normalized to sum to one; thus, as the number of features increases, the magnitude of individual normalized values naturally tends to diminish or even vanish.

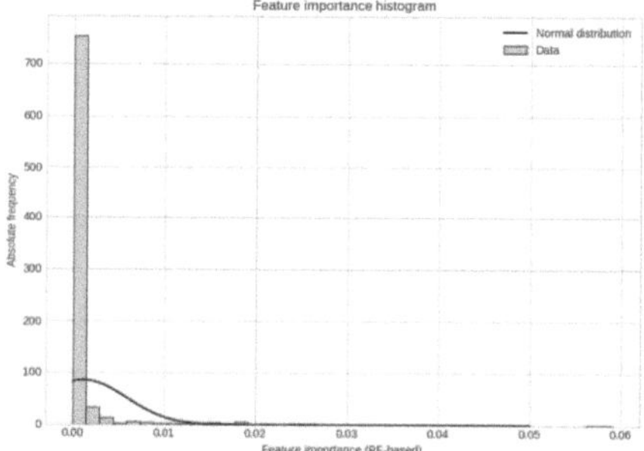

Fig. 7. Histogram of the feature importance estimations (computed as specified in Eq. (5)) using UDS-NACC data.

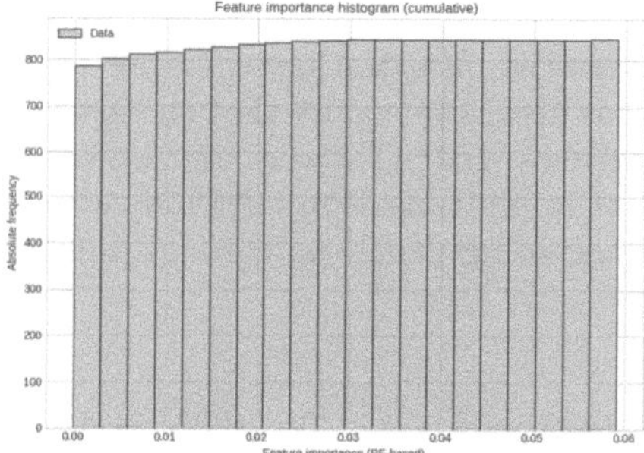

Fig. 8. Cumulative histogram distribution of the data displayed in Fig. 7.

Furthermore, Fig. 9 displays the evolution of feature importance values throughout the iterative process. It is particularly noteworthy how many attributes begin with an importance value effectively equal to zero (bearing in mind that, although not visually distinguishable in the figure, the input space consists of a total of 846 features). It is also striking that some variables initially exhibit relatively significant importance compared to others, but later stabilize in the final configurations. This observation is important, as it reinforces the notion that overly optimistic or pessimistic importance estimates can arise from specific hyperparameter configurations, highlighting the value of aggregating across multiple model instances.

Finally, in the comparison phase between the model trained on the full feature set and the one trained on the reduced set, a dimensionality reduction from 846 to 107 features is achieved using a threshold of $z_\alpha = 0.00115$, computed from

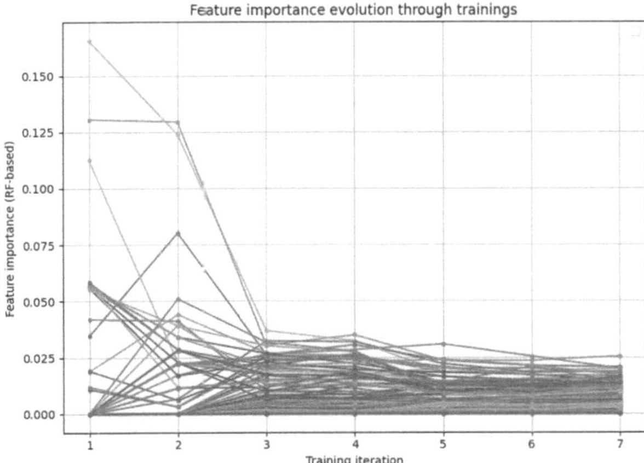

Fig. 9. Feature importance evolution through the different iterations of the methodology (using the UDS-NACC data).

a significance level of $\alpha = 5\%$. The task remains a classification problem, but
unlike OASIS-2—which was converted into a binary classification setting—the
target variable in UDS-NACC spans four distinct classes, representing varying
degrees of Mild Cognitive Impairment (MCI), ranging from cognitively normal
individuals to diagnosed dementia cases. The reader is referred to [17] for a
detailed explanation of the semantic meaning of each of these diagnostic categories. The comparative results between both models are identical. All classification metrics (*accuracy*, *precision*, *recall*, and *f1-score*) fall within the range of
0.98 to **1.0** for both models, rendering graphical representation of these metrics of limited practical value. It is likely that these highly optimistic results are
due to the presence of one or more input variables that implicitly encode information closely correlated with the target variable, thus making the predictions
somewhat "unfair". Nonetheless, the findings effectively demonstrate the utility
of the proposed methodology, which successfully **eliminated a total of 739
features** that were entirely irrelevant to the task of early Alzheimer's diagnosis.

To provide insight into the nature of the removed variables, we present the
following as representative examples of **features that were excluded** during
the dimensionality reduction process:

- **NACCDAYS**: Days from initial visit to most recent visit.
- **NACCREAS**: Primary reason for attending the ADC (Alzheimer's Disease
  Center).
- **THYROID**: History of thyroid disease.
- **APNEA**: History of sleep apnea reported at the initial visit.
- **INSOMN**: History of hyposomnia or insomnia reported at the initial visit.
- **ALCOHOL**: History of clinically significant alcohol abuse over a 12-month
  period, with manifestations in work, driving, legal, or social domains.
- **DIABETES**: Diagnosis of diabetes.

- **DIABTYPE**: Type of diabetes.
- **HYPERTEN**: Diagnosis of hypertension.

Conversely, examples of **retained variables that were preserved** in the model after dimensionality reduction include:

- **MEMORY**: Level of memory impairment.
- **INDEPEND**: Level of functional independence.
- **ORIENT**: Level of orientation.
- **HUNTIF**: Huntington's disease as a primary, contributing, or noncontributing cause of cognitive impairment.
- **ANXIETIF**: Anxiety as a primary, contributing, or noncontributing cause of cognitive impairment.
- **CORTIF**: Corticobasal degeneration (CBD) as a primary, contributing, or noncontributing cause of cognitive impairment.

These examples illustrate how the proposed methodology prioritizes variables with strong cognitive or functional relevance while discarding those with limited direct diagnostic value in the context of Alzheimer's disease.

5   Conclusions

This work has introduced an innovative methodology that leverages advanced learning models—specifically Random Forests—to obtain a statistically robust estimation of feature importance in high-dimensional clinical datasets. The initial hypothesis is confirmed: most clinical tests and patient data collected in the context of Alzheimer's disease monitoring are superfluous and do not play a significant role in training models aimed at supporting early diagnosis. In the case of the UDS-NACC dataset, this hypothesis is strongly validated: a vast number of clinical measurements and assessments are found to be of minimal relevance to the predictive task. Therefore, this study addresses a classical problem in Computer Science and Artificial Intelligence: feature selection and management in machine learning settings. The proposed methodology contributes directly to improving the models and techniques currently applied to the medical domain, particularly those focused on the early detection of Alzheimer's disease.

6   Future Work

Future work includes the exploration of alternative learning models that rely on different strategies for estimating feature importance, thereby extending the present methodology, which is exclusively based on Random Forests. Given that the current framework employs a Gaussian distribution to determine the z-value threshold for discarding low-relevance attributes, a critical extension would involve conducting normality tests to assess the goodness-of-fit of the normal distribution to the estimated importance values. This would provide stronger statistical justification for assuming Gaussianity in the elimination process. Additionally, we propose the incorporation of reliability indicators associated with the importance estimates produced by each model, based on the predictive quality achieved during training. Poorly fitted models may yield misleading or unreliable importance scores. Introducing a reliability layer to weight or qualify these estimates could pave the way for more robust and refined versions of the proposed methodology.

Acknowledgements. This study was funded by public research projects of the Spanish Ministry of Science and Innovation PID2023-151605OB-C22 and the project under the call PEICTI 2021-2023 with the identifier TED2021-131520B-C22. The NACC database is funded by NIA/NIH Grant U24 AG072122. NACC data are contributed by the NIA-funded ADRCs: P30 AG062429 (PI James Brewer, MD, PhD), P30 AG066468 (PI Oscar Lopez, MD), P30 AG062421 (PI Bradley Hyman, MD, PhD), P30 AG066509 (PI Thomas Grabowski, MD), P30 AG066514 (PI Mary Sano, PhD), P30 AG066530 (PI Helena Chui, MD), P30 AG066507 (PI Marilyn Albert, PhD), P30 AG063444 (PI David Holtzman, MD), P30 AG066518 (PI Lisa Silbert, MD, MCR), P30 AG066512 (PI Thomas Wisniewski, MD), P30 AG066462 (PI Scott Small, MD), P30 AG072979 (PI David Wolk, MD), P30 AG072972 (PI Charles DeCarli, MD), P30 AG072976 (PI Andrew Saykin, PsyD), P30 AG072975 (PI Julie A. Schneider, MD, MS), P30 AG072978 (PI Ann McKee, MD), P30 AG072977 (PI Robert Vassar, PhD), P30 AG066519 (PI Frank LaFerla, PhD), P30 AG062677 (PI Ronald Petersen, MD, PhD), P30 AG079280 (PI Jessica Langbaum, PhD), P30 AG062422 (PI Gil Rabinovici, MD), P30 AG066511 (PI Allan Levey, MD, PhD), P30 AG072946 (PI Linda Van Eldik, PhD), P30 AG062715 (PI Sanjay Asthana, MD, FRCP), P30 AG072973 (PI Russell Swerdlow, MD), P30 AG066506 (PI Glenn Smith, PhD, ABPP), P30 AG063508 (PI Stephen Strittmatter, MD, PhD), P30 AG066515 (PI Victor Henderson, MD, MS), P30 AG072947 (PI Suzanne Craft, PhD), P30 AG072931 (PI Henry Paulson, MD, PhD), P30 AG066546 (PI Sudha Seshadri, MD), P30 AG086401 (PI Erik Roberson, MD, PhD), P30 AG086404 (PI Gary Rosenberg, MD), P20 AG068082 (PI Angela Jefferson, PhD), P30 AG072958 (PI Heather Whitson, MD), P30 AG072959 (PI James Leverenz, MD).

References

1. Alatrany, A.S., Khan, W., Hussain, A., Kolivand, H., Al-Jumeily, D.: An explainable machine learning approach for Alzheimer's disease classification. Sci. Rep. **14**(1) (2024). https://doi.org/10.1038/s41598-024-51985-w
2. Barnes, J., Dickerson, B.C. Frost, C., Jiskoot, L.C., Wolk, D., Van Der Flier, W.M.: Alzheimer's disease first symptoms are age dependent: evidence from the NACC dataset. Alzheimer's Dementia **11**(11) (2015). https://doi.org/10.1016/j.jalz.2014.12.007
3. Chaudhuri, A., Hu, W.: A fast algorithm for computing distance correlation. Comput. Stat. Data Anal. **135** (2019). https://doi.org/10.1016/j.csda.2019.01.016
4. Henríquez, P.A., Araya, N.: Multimodal Alzheimer's disease classification through ensemble deep random vector functional link neural network. PeerJ Comput. Sci. **10**, e2590 (2024). https://doi.org/10.7717/peerj-cs.2590
5. Lin, M., Gong, P., Yang, T., Ye, J., Albin, R.L., Dodge, H.H.: Big data analytical approaches to the NACC dataset: aiding preclinical trial enrichment. Alzheimer Dis. Assoc. Disord. **32**(1) (2018). https://doi.org/10.1097/WAD.0000000000000228

6. Liu, S., et al.: Generalizable deep learning model for early Alzheimer's disease detection from structural MRIs. Sci. Rep. **12**(1) (2022). https://doi.org/10.1038/s41598-022-20674-x

7. Mahendran, N., PM, D.R.V.: A deep learning framework with an embedded-based feature selection approach for the early detection of the Alzheimer's disease. Comput. Biol. Med. **141** (2022). https://doi.org/10.1016/j.compbiomed.2021.105056

8. Marcus, D.S., Fotenos, A.F., Csernansky, J.G., Morris, J.C., Buckner, R.L.: Open access series of imaging studies: longitudinal MRI data in nondemented and demented older adults. J. Cogn. Neurosci. **22**(12) (2010). https://doi.org/10.1162/jocn.2009.21407

9. Marcus, D.S., Wang, T.H., Parker, J., Csernansky, J.G., Morris, J.C., Buckner, R.L.: Open access series of imaging studies (OASIS): cross-sectional MRI data in young, middle aged, nondemented, and demented older adults. J. Cogn. Neurosci. **19**(9) (2007). https://doi.org/10.1162/jocn.2007.19.9.1498

10. Meteumba, L.G., Ojha, V.P., Yarahmadian, S.: Comprehensive evaluation of machine learning models for predicting the cognitive status of Alzheimer's disease subjects and susceptible. Discov. Data **3**(1), 28 (2025). https://doi.org/10.1007/s44248-025-00068-w

11. Shubar, A., Ramakrishnan, K., Ho, C.K.: Optimizing machine learning models for accessible early cognitive impairment prediction: a novel cost-effective model selection algorithm. IEEE Access (2024). https://doi.org/10.1109/ACCESS.2024.3505038

12. Székely, G.J., Rizzo, M.L., Bakirov, N.K.: Measuring and testing dependence by correlation of distances. Ann. Stat. **35**(6) (2007). https://doi.org/10.1214/009053607000000505

13. Thapa, S., Singh, P., Jain, D.K., Bharill, N., Gupta, A., Prasad, M.: Data-driven approach based on feature selection technique for early diagnosis of Alzheimer's disease. In: Proceedings of the International Joint Conference on Neural Networks (2020). https://doi.org/10.1109/IJCNN48605.2020.9207359

14. Thulasimani, V., Shanmugavadivel, K., Cho, J., Veerappampalayam Easwaramoorthy, S.: A review of datasets, optimization strategies, and learning algorithms for analyzing Alzheimer's dementia detection. Neuropsychiatr. Dis. Treat. **20**, 2203–2225 (2024). https://doi.org/10.2147/NDT.S496307

15. Yi, F., et al.: XGBoost-SHAP-based interpretable diagnostic framework for Alzheimer's disease. BMC Med. Inform. Decis. Mak. **23**(1) (2023). https://doi.org/10.1186/s12911-023-02238-9

16. Zubasti, P., Berlanga, A., Patricio, M.A., Molina, J.M.: Assessing the interplay of attributes in dementia prediction through the integration of graph embeddings and unsupervised learning. In: Ferrández Vicente, J.M., Val Calvo, M., Adeli, H. (eds.) Artificial Intelligence for Neuroscience and Emotional Systems, pp. 371–380. Springer, Cham (2024)

17. Zubasti, P., Patricio, M.A., Berlanga, A., Molina, J.: Optimizing dementia diagnosis through distance-correlation feature space and dimensionality reduction. Int. J. Neural Syst. (2025). https://doi.org/10.1142/S012906572550042X

Towards an Explainability Agent: Leveraging LLMs to Interpret LIME Outputs

Belén Vega-Márquez[(✉)] [iD], Cristina Rubio-Escudero[iD], and Beatriz Pontes-Balanza[iD]

Department of Computer Languages and Systems, University of Seville, 41012 Seville, Spain
bvega@us.es

Abstract. In clinical decision-making, the adoption of machine learning models requires not only high predictive performance but also transparent, trustworthy explanations. This work presents a hybrid explainability framework that combines Local Interpretable Model-agnostic Explanations (LIME) with a Large Language Model (LLM) to generate natural language explanations of classification outputs in healthcare applications. While LIME provides local feature attributions for individual predictions, these are often difficult to interpret by non-technical clinical staff. Our system leverages an LLM to translate LIME outputs into fluent, domain-aware explanations that align with the reasoning needs of healthcare professionals. We evaluate the method on two medical datasets involving patient risk classification tasks and obtain promising results in terms of interpretability and consistency with the model's decision logic. However, the current evaluation lacks validation by external healthcare professionals, which we identify as an essential next step. Despite this, our approach represents a strong foundation for the development of adaptive explainability agents in clinical contexts. We discuss its potential impact on future decision support systems and propose directions for evolving toward interactive, trustworthy, and user-aligned AI tools in medicine.

Keywords: Explainable AI (XAI) · Large Language Models (LLMs) · LIME · Clinical Decision Support · Natural Language Explanations

1 Introduction

In recent years, artificial intelligence (AI) has significantly influenced medicine and healthcare [8], finding applications across a wide range of specialties including radiology, dermatology, cardiology, ophthalmology, gastroenterology, and mental health. AI models are increasingly used in medical diagnostics due to their ability to detect complex patterns in data. However, their adoption is often hindered by the lack of transparency in decision-making processes [5].

A. López Fernández et al. (Eds.): CIABiomed 2025, LNBI 16148, pp. 17–28, 2026.
https://doi.org/10.1007/978-3-032-10661-2_2

Black box models refer to artificial intelligence systems whose predictions cannot be fully understood based only on the input data [1]. Incorporating expert involvement alongside advanced interpretability techniques in a collaborative approach can enhance the reliability and credibility of the predictions made by these models.

Despite recent progress in explainable AI (XAI), many interpretability methods remain either too technical for clinical users or insufficiently informative for real-world decision-making [9]. One common approach to improve interpretability involves feature attribution methods, such as SHAP or LIME, which provide local explanations for individual predictions by estimating the contribution of each feature [14]. However, these explanations are often presented in a purely numerical form, making them difficult to interpret for non-technical users, such as clinicians or patients.

To address this communication gap, recent studies have explored the use of natural language generation to translate model explanations into more accessible textual formats [3]. Large language models (LLMs), such as GPT-based architectures, offer promising capabilities in generating human-like explanations and summaries. By leveraging these models, it becomes possible to convert raw interpretability outputs into clearer narratives that can assist in clinical understanding and foster trust in AI-assisted decisions. This new methodology could be a push towards more sophisticated models, such as AI agents [7]. This new paradigm uses an LLM as its main component, which is responsible for a specific task.

In this work, we propose a novel framework that combines feature attribution methods (specifically LIME) with large language models to generate natural language explanations of predictions made by black-box classifiers in the healthcare domain. We evaluate our approach on two publicly available medical datasets, and we assess the quality of the explanations quantitatively using semantic similarity metrics. All code and materials are made publicly available in our GitHub repository to ensure reproducibility and foster future research.

The rest of this paper is organized as follows: Sect. 2 describes the methodology used; Sect. 3 presents experimental results; and Sect. 4 concludes the paper.

2 Methodology

This work presents a hybrid methodology for producing understandable, faithful, and context-aware explanations of classification model predictions applied to healthcare data. The approach is motivated by the need for explainable artificial intelligence (XAI) tools in clinical environments, where decisions must be transparent and trustworthy. Our framework combines a machine learning classifier trained on structured health data, a post-hoc explainability method (LIME), and a large language model (LLM) that reformulates technical explanations into natural language understandable by healthcare professionals. The core idea is to bridge the gap between complex machine learning models and clinical users by offering explanations that are both technically accurate and human-readable. The process followed can be seen graphically in Fig. 1.

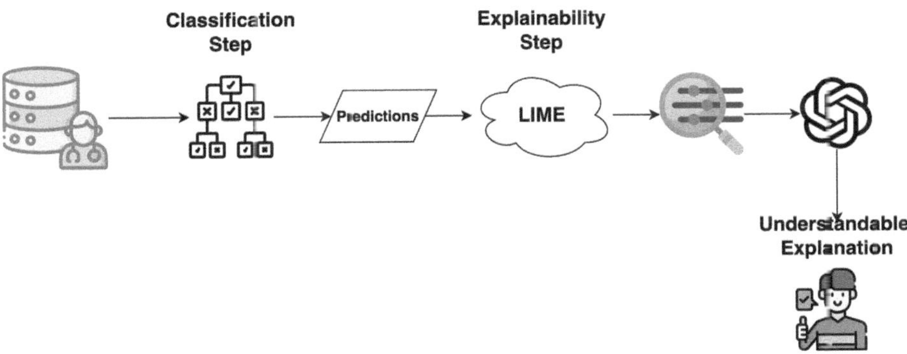

Fig. 1. Overview of the methodological pipeline proposed in this work. First, a classification model is trained and used to generate predictions on patient data. Subsequently, LIME is applied to each prediction to identify and quantify the most influential features. Finally, the explanations provided by LIME are incorporated into structured prompts that are processed by a large language model (LLM), which generates human-readable, context-aware explanations aimed at clinical interpretation.

The following subsections will explain in more detail the main steps taken and the decisions made. First, the classification of the samples will be explained, with special emphasis on the classification model used and the experimental framework. Finally, the process for making explainable predictions will be described, describing both the model used for explainability and the language model [16].

2.1 Datasets Description

In order to carry out the study, two medical datasets widely used in the literature were used. The characteristics of each are presented in Table 1. The first dataset is Breast Cancer [19]; this dataset was generated using interactive image processing techniques and a linear-programming-based classifier to support breast tumor diagnosis. A digitized portion of fine needle aspirate slides was analyzed to extract morphological features of cell nuclei. For each nucleus, ten features related to size, shape, and texture were computed, and their mean, worst, and standard error values were aggregated, resulting in 30 features per sample. The second dataset is Diabetes [6]; its target variable is a quantitative measure of disease progression one year after baseline.

With respect to the diabetes dataset, it should be noted that the aim of this work is to make predictions in classification models explainable in a simple way. In order to perform classification in this dataset, a discretization of the target variable has been used, thus converting it into a binary variable suitable for classification tasks, assigning 1 to samples with values above the median and 0 to those below or equal to it.

Table 1. Summary of the `breast_cancer` and `diabetes` datasets from `scikit-learn`.

Feature	Breast Cancer	Diabetes
Problem type	Classification	Regression
Number of samples	569	442
Number of features	30	10
Feature type	Continuous	Continuous
Target variable	Binary (0,1)	Continuous
Number of classes	2	-
Class names	malignant, benign	-

2.2   Classification Step

The model used to make the predictions was Random Forest. Random Forest (RF) is an ensemble method based on the combinations of decision trees. It mitigates overfitting by aggregating the outputs of multiple overfitted decision trees, thereby increasing generalization. Each individual tree in the forest independently produces a classification result. The final prediction is determined through a majority voting mechanism, where the class receiving the most votes across all trees is selected as the output label for a given input sample [18].

For model training, the default parameters of scikit-learn were used. `scikit-learn`, which include: `n_estimators=100`, `criterion='gini'`, `min_samples_split=2`, `min_samples_leaf=1`, and `bootstrap=True`, among others. A train-test split was performed, with 30% of the data used for validation and the remaining 70% for training.

2.3   Explainability Step

To improve the interpretability of the model's predictions and support human-understandable reasoning, an explainability pipeline was implemented. This pipeline aims to bridge the gap between complex machine learning outputs and domain-level insights by combining local model explanations with natural language generation.

As a first step, a Local Interpretable Model-agnostic Explanations (LIME) model was applied. LIME is a widely used technique that approximates the behavior of a black-box classifier using an interpretable linear model in the vicinity of each individual prediction. LIME assigns local importance scores to the input features, highlighting which features contributed most significantly (positively or negatively) to a given prediction. These feature attributions offer a quantitative explanation of the decision process on a per-sample basis [14]. The output shown in the extract below corresponds to the explanation generated by LIME for a single prediction:

```
"lime_explanation": [
    ["515.30 < worst area <= 686.50", 0.0598],
    ["84.11 < worst perimeter <= 97.66", 0.0449],
    ["13.01 < worst radius <= 14.97", 0.0386],
    ["0.23 < worst concavity <= 0.38", -0.0294],
    ["21.08 < worst texture <= 25.41", 0.0245]
]
```

Each entry in the explanation represents a condition on a feature's value for the given instance, along with the corresponding weight assigned by LIME. These weights are coefficients of a locally fitted linear model that approximates the behavior of the original classifier around this specific prediction. Positive weights indicate that the feature range contributes to a prediction in favor of the positive class (e.g., malignant), while negative weights suggest a contribution toward the negative class (e.g., benign). For example, the feature `worst area` falling between 515.30 and 686.50 contributes positively to the prediction with a weight of 0.0598, whereas `worst concavity` in the range 0.23 to 0.38 contributes negatively. This localized explanation helps identify which features and value ranges were most influential in the model's decision for that specific case.

This previous output will serve as input to the language model used. In this case, the LLM chosen was GPT-Neo 1.3B [4]. GPT-Neo 1.3B is a transformer-based language model developed by EleutherAI, replicating the GPT-3 architecture. It contains 1.3 billion parameters and was trained on The Pile, a large curated dataset designed specifically for language model training. The model was trained over 380 billion tokens across 362,000 steps, using a masked autoregressive approach with cross-entropy loss.

The scores produced by LIME, are formatted into a structured prompt as shown in the code snippet. This prompt includes detailed information about the patient's features, the model's predicted class, and the key feature contributions identified by LIME. This comprehensive prompt is then provided as input to the large language model (LLM), enabling it to generate clear, human-readable explanations that contextualize the prediction and highlight the most influential factors in a way that is accessible to non-expert users.

```python
for feature, impact in lime:
    sign = "positive" if impact > 0 else "negative"
    lime_lines.append(f"- '{feature}' between {impact: 4f}
                        impact: {sign} ({impact:+.
                        4f})")


prompt = (
    f"You have a sample from the dataset '{dataset_name}'.\n\
                        n"
    f"Patient features:\n"
    + "\n".join([f"- {k}: {v}" for k, v in sample.items()])
    + f"\n\nThe model predicted class {prediction}.\n"
    f"LIME explanation indicates the following features had
                        the most impact:\n"
```

```
    + "\n".join(lime_lines)
    + "\n\nPlease provide a clear and simple explanation for
                        a patient, describing why
                        the model predicted this
                        class."
)
return prompt
```

2.4 Evaluation Step

To assess the quality of the language model's generated explanations, we employed semantic similarity metrics [2]. Instead of relying solely on exact lexical matches, semantic similarity measures evaluate how closely the meaning of the generated text aligns with reference explanations or expected outputs. This approach allows for a more flexible and meaningful evaluation by capturing paraphrases and variations in wording that convey the same information. Concretely, we computed embeddings of both the generated explanations and the reference features using a pre-trained sentence transformer model, and measured their cosine similarity. This embedding-based approach ensures that the evaluation reflects the true semantic alignment between explanations and references, which is particularly important when dealing with natural language generation tasks in the medical domain.

2.5 Code Reproducibility

All the code used for model training, evaluation, and interpretability analysis has been made publicly available in our GitHub [17] repository to ensure transparency and reproducibility.

The implementation is in Python [15], using the following open-source libraries:

- Pandas [10].
- Scikit-Learn [12]
- lime [13]
- Pytorch [11].
- Transformers [20].

All experiments were conducted on a computer with the following specifications: Ubuntu 22.04.2 LTS, AMD Ryzen Threadripper PRO 3955WX 16-Cores CPU, NVIDIA RTX A5000 24 GB GPU, and 8X16 GB (128GB) DDR4 RAM.

3 Results

The following section presents the key findings derived from the experimental data and analyses described in Sect. 2. Results are organized according to the main objectives of the study. All data are reported with appropriate measures; relevant figures and tables are included to support the interpretation of the results. We first assess the overall accuracy of the models to understand their predictive capabilities. Then, we delve into the explainability of these predictions by analyzing feature importance using LIME, which sheds light on how the models make decisions. Lastly, we evaluate the quality of the explanations through semantic similarity measure, providing insight into how well the generated explanations capture the model's reasoning.

Figure 2 shows the classification accuracy achieved by the model on the two datasets used in this study. The breast cancer dataset obtained a high accuracy of approximately 97.1%, indicating excellent performance in distinguishing benign from malignant cases. In contrast, the diabetes dataset showed a moderate accuracy of about 75.2%, which may be due in part to the discretization applied to the target variable, making the classification task more challenging. These results highlight the model's strong predictive capability on breast cancer data, while also pointing to potential limitations when working with transformed or more complex datasets like diabetes.

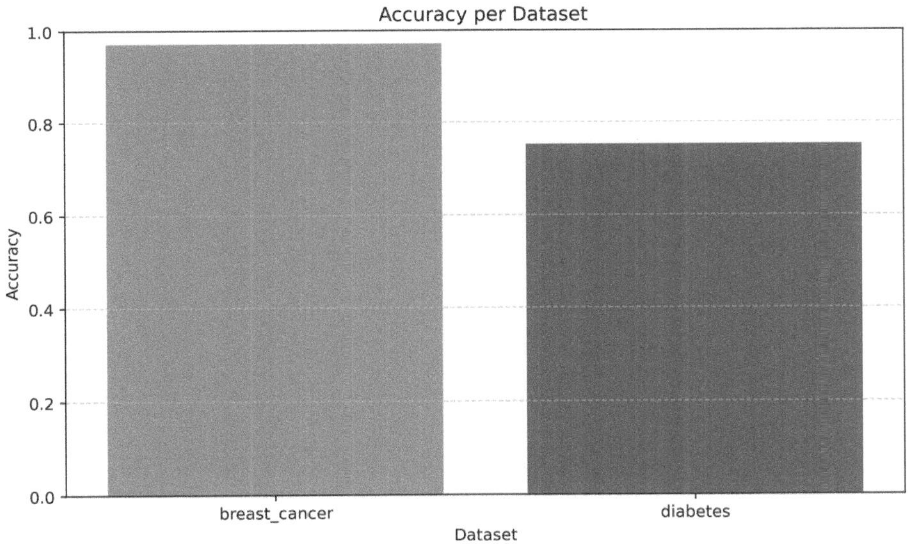

Fig. 2. Accuracy obtained by the classification model for each of the datasets used.

The analysis of feature importance using LIME (Fig. 3) revealed the top 20 features that most positively influence the model's predictions across the datasets. A positive impact score means that the presence or higher values of these features increase the likelihood that the model predicts the positive class. This helps us understand which variables most strongly drive the classification, providing valuable interpretability and insight into the model's decision-making process. Notably, body mass index (BMI) values greater than 0.03 showed the highest positive impact with a score of approximately 0.16, reflecting a significant influence on the model's predictions. Other important clinical and morphological features with substantial positive contributions include spectral measurement s5 greater than 0.03 (impact 0.11), blood pressure (bp) above 0.04 (impact 0.11), and a "worst area" less than or equal to 515.30 (impact 0.08). Additional BMI thresholds between -0.01 and 0.03, and spectral measurement s3 below -0.04 also contributed positively. Morphological characteristics such as "worst concave points" less than or equal to 0.06 and "worst concavity" less than or equal to 0.11 had meaningful positive impacts, as did intermediate ranges of worst area, perimeter, radius, and concave points. Features associated with lower error margins in area and perimeter, as well as specific ranges of blood pressure and spectral measurements, also played a role in the model's decision-making. These results demonstrate the nuanced contributions of various clinical and morphological variables across specific value intervals, highlighting their collective importance in driving the classifier's performance.

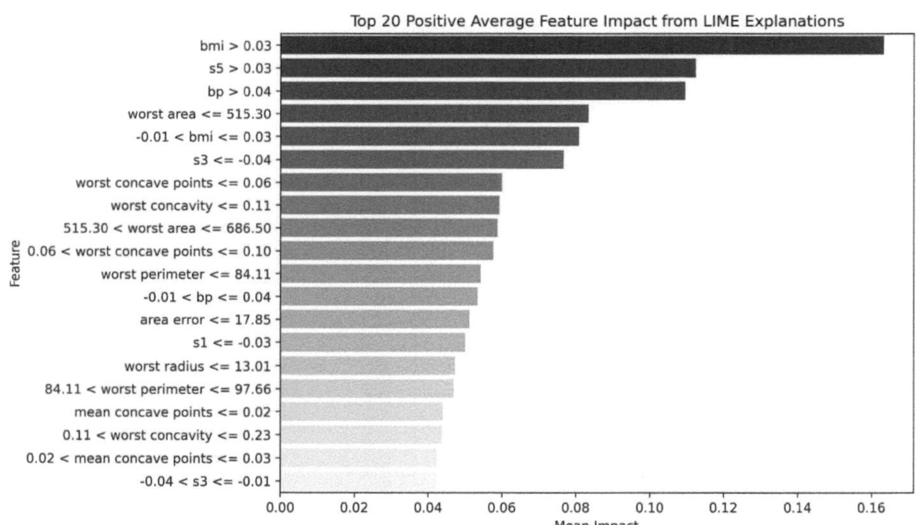

Fig. 3. Top 20 features with the highest positive impact on model predictions, as determined by LIME. The values next to each feature represent the approximate impact scores indicating their contribution to the classification outcomes.

In LIME explanations, a negative impact value indicates that the corresponding feature influences the model's prediction away from the predicted class, effectively reducing the likelihood of that outcome. In our analysis, several features showed notable negative contributions (see Fig. 4). For example, BMI values less than or equal to -0.03 had the strongest negative impact (-0.15), suggesting that lower BMI values decrease the model's confidence in the predicted class.

Similarly, large "worst area" values exceeding 1084.00 (−0.13) and high "worst concave points" above 0.16 (−0.12) negatively influenced the predictions. Lower blood pressure readings (bp ≤ −0.04) and BMI values between −0.03 and −0.01 also had significant negative impacts, as did larger perimeter, spectral measurements (s5 ≤ −0.03), and greater area error values. Features indicating higher values of radius, concavity, concave points, texture, and other morphological measurements likewise contributed negatively, reflecting that extreme or higher measurements in these features push the model's decision away from the predicted class. This analysis underscores how specific ranges of clinical and morphological variables can inversely affect the classifier's output, highlighting the complexity of feature interactions in the model.

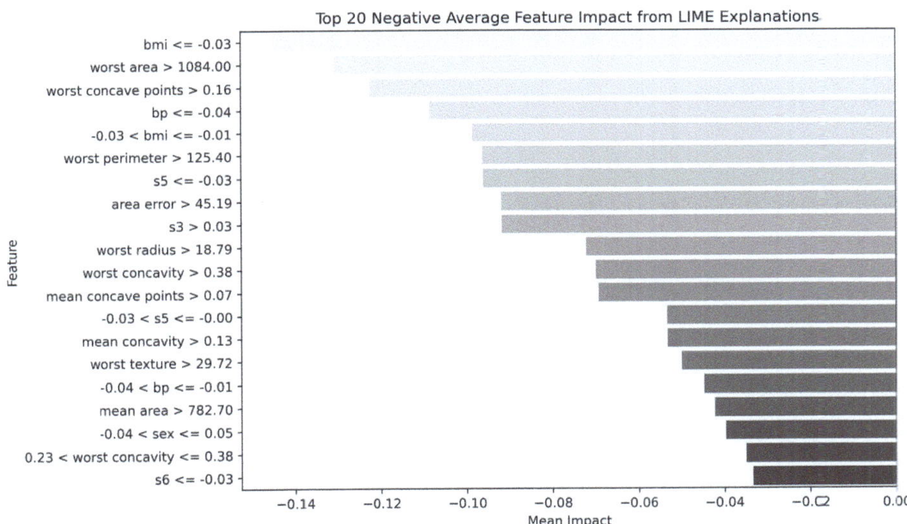

Fig. 4. Top 20 features with the highest negative impact on model predictions, as determined by LIME. The values next to each feature represent the approximate impact scores indicating their contribution to the classification outcomes.

The semantic similarity scores observed for both the breast cancer and diabetes datasets (see Fig. 5), while indicative of some meaningful relationships, are relatively modest, with mean values below 0.55. These results suggest that the current approach captures only partial semantic alignment between samples, possibly due to inherent variability in the data or limitations in the representation methods used. To improve these outcomes, future work could explore more sophisticated embedding techniques or domain-specific feature engineering to better capture underlying semantic structures. Additionally, integrating complementary methods such as contextual embeddings or fine-tuning on domain-relevant corpora may enhance the model's ability to discern subtle semantic nuances. These improvements could significantly increase similarity measures and strengthen the overall interpretability and robustness of the analysis.

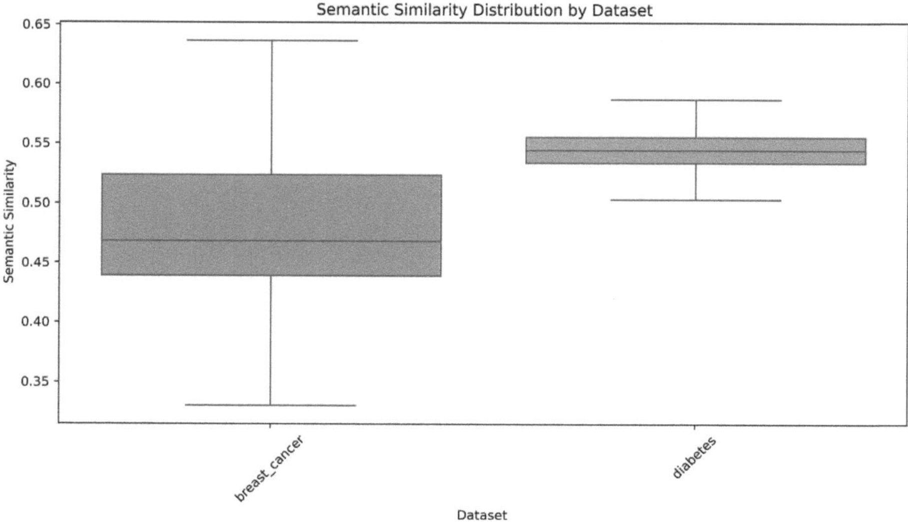

Fig. 5. Distribution of semantic similarity scores for the LLM-generated explanations across the evaluated datasets, illustrating moderate alignment with the ground truth.

4   Conclusion

In this work, we explored a methodology that combines classical machine learning classifiers with explainability techniques and large language models (LLMs) to generate human-readable explanations of model predictions. Specifically, we employed LIME to extract local feature attributions, which were then transformed into natural language prompts and evaluated using an LLM. This approach enables the creation of interpretable explanations that bridge the gap between technical outputs and patient-facing narratives.

Our results demonstrate that while the classification model achieved strong performance on structured datasets such as breast cancer (accuracy 0.97), its effectiveness dropped on others like diabetes (accuracy 0.75), potentially due to data preprocessing or discretization. The LIME-based analysis revealed the most influential features in model decision-making, highlighting both clinical and morphological variables. Furthermore, the semantic similarity evaluation of the LLM-generated explanations revealed moderate scores, suggesting that although the outputs are reasonably aligned with the ground truth, there is still room for improvement.

Future work should focus on refining the prompting strategies for LLMs, incorporating additional context, or fine-tuning models for the medical domain to improve the quality and faithfulness of explanations. Another promising direction is to apply parameter optimization techniques to the classification model employed, in order to enhance its predictive performance and robustness. Additionally, including human expert evaluations could provide complementary insights to quantitative metrics like semantic similarity. The code and resources used in this study are publicly available on GitHub to foster reproducibility and further development.

Acknowledgments. This publication is part of the projects PID2020-117954RB-C22 and PID2023-146037OB-C21, funded by MICIU/AEI/10.13039/501100011033/.

Disclosure of Interests. The authors have no competing interests to declare that are relevant to the content of this article.

Credits. This work includes icons designed by multiple authors from Flaticon.com.

References

1. Abeyagunasekera, S.H.P., Perera, Y., Chamara, K., Kaushalya, U., Sumathipala, P., Senaweera, O.: Lisa: enhance the explainability of medical images unifying current XAI techniques. In: 2022 IEEE 7th International conference for Convergence in Technology (I2CT), pp. 1–9. IEEE (2022)
2. Aynetdinov, A., Akbik, A.: Semscore: automated evaluation of instruction-tuned LLMs based on semantic textual similarity (2024). https://arxiv.org/abs/2401.17072
3. Bilal, A., Ebert, D., Lin, B.: LLMs for explainable AI: a comprehensive survey. arXiv preprint arXiv:2504.00125 (2025)
4. Black, S., Leo, G., Wang, P., Leahy, C., Biderman, S.: GPT-Neo: Large Scale Autoregressive Language Modeling with Mesh-Tensorflow (2021). https://doi.org/10.5281/zenodo.5297715. If you use this software, please cite it using these metadata
5. Cabitza, F., Rasoini, R., Gensini, G.F.: Unintended consequences of machine learning in medicine. JAMA **318**(6), 517–518 (2017)
6. Efron, B., Hastie, T., Johnstone, I., Tibshirani, R.: Least angle regression. Ann. Stat. **32**(2), 407–499 (2004). https://doi.org/10.1214/009053604000000067
7. Gao, S., et al.: Empowering biomedical discovery with AI agents. Cell **187**(22), 6125–6151 (2024)
8. Hildt, E.: What is the role of explainability in medical artificial intelligence? A case-based approach. Bioengineering **12**(4), 375 (2025)
9. Lipton, Z.C.: The mythos of model interpretability. Queue **16**(3), 31–57 (2018)
10. McKinney, W.: Data structures for statistical computing in python. In: Proceedings of the 9th Python in Science Conference, vol. 445, pp. 51–56 (2010)
11. Paszke, A., et al.: Pytorch: an imperative style, high-performance deep learning library. In: Advances in Neural Information Processing Systems 32 (NeurIPS 2019), pp. 8024–8035 (2019)
12. Pedregosa, F., et al.: Scikit-learn: machine learning in Python. J. Mach. Learn. Res. **12**, 2825–2830 (2011)
13. Ribeiro, M.T.: Lime: local interpretable model-agnostic explanations (2016). https://github.com/marcotcr/lime. Accessed 25 July 2025

14. Ribeiro, M.T., Singh, S., Guestrin, C.: "why should i trust you?" explaining the predictions of any classifier. In: Proceedings of the 22nd ACM SIGKDD International Conference on Knowledge Discovery and Data Mining, pp. 1135–1144 (2016)
15. van Rossum, G., the Python Development Team: Python 3 Reference Manual. Python Software Foundation (2023). https://www.python.org
16. Street, W.N., Wolberg, W.H., Mangasarian, O.L.: Nuclear feature extraction for breast tumor diagnosis. In: Biomedical Image Processing and Biomedical Visualization, vol. 1905, pp. 861–870. SPIE (1993)
17. Vega, B.: CIABIOMED-Agents (2025). https://github.com/bvegaus/CIABIOMED-Agents, gitHub repository
18. Wang, X., et al.: Exploratory study on classification of diabetes mellitus through a combined random forest classifier. BMC Med. Inform. Decis. Mak. **21**(1), 105 (2021)
19. Wolberg, W.H., Mangasarian, O.L., Street, W.N.: Breast cancer wisconsin (diagnostic) data set (1993). https://archive.ics.uci.edu/ml/datasets/breast+cancer+wisconsin+(diagnostic), uCI Machine Learning Repository. https://doi.org/10.24432/C5DW2B
20. Wolf, T., et al.: Transformers: state-of-the-art natural language processing. In: Proceedings of the 2020 Conference on Empirical Methods in Natural Language Processing: System Demonstrations, pp. 38–45. Association for Computational Linguistics, Online (2020). https://www.aclweb.org/anthology/2020.emnlp-demos.6

Uncovering Cardiac Risk Patterns: Visualization and Interpretation via Probabilistic Topographic Mapping

Martha Ivon Cardenas$^{(\boxtimes)}$ (iD), Pedro Jesús Copado (iD), and Caroline König (iD)

Soft Computing Research Group (SOCO) at Intelligent Data Science and Artificial
Intelligence (IDEAI-UPC) Research Centre, Universitat Politècnica de Catalunya
(UPC Barcelona Tech), Jordi Girona 1-3, 08034 Barcelona, Spain
`martha.ivon.cardenas@upc.edu`

Abstract. Cardiovascular disease (CVD) remains a leading cause of
mortality, yet accurate and interpretable risk prediction poses significant
clinical challenges. While machine learning (ML) models show promise
in identifying complex patterns, their adoption in healthcare is often limited by a lack of transparency. This paper proposes a novel explainability framework integrating Generative Topographic Mapping (GTM) and
Kernel Generative Topographic Mapping (kGTM) to visualize and analyze latent cardiac risk patterns. Unlike traditional dimensionality reduction methods, our approach unifies probabilistic modeling, visual explainability, and clinically grounded subgroup analysis within a single interpretable latent representation. We employ a structured dataset of 1,000
patients with 12 clinical features to demonstrate how GTM and kGTM
project high-dimensional patient data into a structured 2D space. This
methodology enables direct visualization of patient clusters, classification boundaries, and the influence of individual features, thereby uncovering interpretable regions and clinically meaningful feature interactions
directly from the topographic map. This approach provides robust and
trustworthy perspectives on ML model behavior for cardiovascular risk
assessment. Our dual use of GTM and kGTM allows for adaptive modeling, facilitating the identification of outliers and borderline cases. This
dual nature enhances understanding of both the feature space distribution and the uncertainty of the model's decisions, enabling partial
explainability by highlighting relevant features and structural patterns
in the data.

Keywords: Generative Topographic Mapping · Explainable AI ·
Cardiovascular Disease · Risk Prediction · Data Visualization ·
Manifold Learning

1 Introduction

Cardiovascular disease remains the leading cause of death worldwide, with early
detection being a key factor in reducing mortality and improving patient out-

comes [5]. However, clinical diagnosis often depends on a combination of subjective assessment and heterogeneous biomedical indicators, which makes timely and accurate risk prediction a persistent challenge. In this context, machine learning (ML) models have shown great promise by uncovering complex, nonlinear relationships in clinical data that may elude traditional statistical approaches [24].

Despite these advances, the widespread adoption of ML in medical practice is hindered by the lack of interpretability of many state-of-the-art models. Clinicians need not only accurate predictions, but also transparent justifications for those decisions–especially in high-stakes settings such as cardiovascular risk assessment [7].

This study addresses the critical need for accurate and interpretable ML models in early heart disease risk detection, as current ML adoption is often limited by a lack of transparency in clinical settings. The research employs unsupervised methods, specifically GTM [3] and its kernel-based extension (kGTM) [17], to uncover latent cardiovascular risk patterns directly from clinical data.

Using a dataset of 1,000 patients, the aim is to derive low-dimensional, interpretable representations that capture the clinical heterogeneity of heart disease, including subtle intra-population variations often overlooked by traditional approaches. The methodology prioritizes interpretability through visualizations of patient distributions and feature contributions within the latent space, comparing both linear and nonlinear mappings to offer a comprehensive understanding.

Finally, this work contributes to a visual and probabilistic framework for exploratory analysis and hypothesis generation. It distinctively complements traditional supervised models by revealing structural patterns, outliers, and regions of uncertainty, thereby informing clinical interpretation and potentially guiding future data-driven stratification efforts.

2   Related Work

In recent years, explainable artificial intelligence (XAI) techniques–such as SHAP (SHapley Additive exPlanations) and LIME –have been integrated into ML pipelines to identify feature importance and support model interpretability at both global and local levels [8, 22]. Several works have explored combining these techniques with dimensionality reduction methods to visualize decision boundaries or patient clusters [14, 18, 23]. For example, projection techniques like UMAP and t-SNE have been used to explore latent spaces in cardiovascular datasets, often linking spatial proximity with prediction outcomes and SHAP values. However, these methods lack an explicit generative or probabilistic structure, limiting their interpretability and reproducibility in clinical settings. Moreover, numerous studies have focused on improving the predictive accuracy of CVD models using various ML techniques [1, 13, 15, 19–21]. However, many of these approaches place limited emphasis on explainability, offering little practical value for clinical interpretation and decision-making.

In this work, we propose a novel explainability framework that integrates GTM and kGTM to visualize and analyze latent cardiac risk patterns. By projecting both the patient data and the associated explanation vectors into a structured topographic space, we uncover interpretable regions and clinically meaningful feature interactions. Our approach supports robust, trustable insights into model behavior and helps bridge the gap between black-box ML outputs and clinician-understandable decisions.

3 Materials and Methods

3.1 Materials

For this study, we analyze a cardiovascular disease (CVD) dataset acquired from Mendeley Data [6]. The dataset contains clinical records from 1,000 patients and comprises 14 variables per patient, including demographic and physiological features, a unique patient identifier, and a binary target variable indicating the presence (1) or absence (0) of diagnosed heart disease. The target distribution is slightly unbalanced, with approximately 58% positive cases. The features include:

- Age: Patient age in years (range: 20–77, mean: 49.2).
- Gender: Coded binary (0 for female, 1 for male); approximately 76.5% of the patients are male.
- Chest Pain Type: Categorical variable indicating four different types of chest pain, including typical and atypical angina.
- Resting Blood Pressure: Systolic blood pressure measured in mmHg at rest (mean: 151.7, SD: 29.9).
- Serum Cholesterol: Total cholesterol in mg/dL (mean: 311.4, SD: 132.4). Notably, some entries contain zero values, which may indicate missing or imputed data.
- Fasting Blood Sugar: Binary variable indicating whether fasting blood sugar exceeds 120 mg/dL.
- Resting Electrocardiographic Results: Encoded as 0, 1, or 2, representing normal or abnormal ECG readings.
- Maximum Heart Rate Achieved: A continuous variable with values ranging from 71 to 202 bpm (mean: 145.5).
- Exercise-Induced Angina: Binary indicator of angina occurring during physical exertion.
- Oldpeak: A continuous variable representing ST depression induced by exercise relative to rest. The ST segment represents the interval between ventricular depolarization and repolarization. A depression in this segment may indicate myocardial ischemia (insufficient blood flow to the heart muscle), especially under stress or physical exertion.
- Slope of ST Segment: Categorical indicator of the slope of the peak exercise ST segment. It reflects the pattern of ST changes and is typically classified as upsloping, flat, or downsloping—each of which may provide diagnostic insight into the presence and severity of coronary artery disease.

– Number of Major Vessels: Discrete variable indicating the count (0–3) of major vessels observed via fluoroscopy. A higher number of visible vessels generally suggests better blood flow, while fewer visible vessels may indicate blockages or narrowing due to atherosclerosis, which is associated with a higher risk of cardiovascular disease.

The data was preprocessed prior to model training. All features were standardized using z-score normalization to ensure comparability and stability. The patient identification column was excluded from the set of input features. Data imputation was not performed beyond excluding invalid entries.

3.2   Methods

We applied GTM and kGTM to uncover latent structures and classification boundaries in CVD data. These probabilistic manifold learning methods project high-dimensional clinical features into a low-dimensional, structured space for visualization and interpretation. GTM offers a generative framework, while kGTM captures complex nonlinear patterns via kernels.

Generative Topographic Mapping (GTM). GTM models the dataset $\mathbf{X} = \{\mathbf{x}_1, \ldots, \mathbf{x}_N\} \subset \mathbb{R}^D$, where $D = 12$ corresponds to the number of clinical features, using a nonlinear mapping $\mathbf{y}(\mathbf{z}; \mathbf{W})$ from a low-dimensional latent space $\mathbf{z} \in \mathbb{R}^2$. This mapping is defined via a set of radial basis functions (RBFs):

$$\mathbf{y}(\mathbf{z}) = \mathbf{W}^\top \boldsymbol{\phi}(\mathbf{z}),$$

where $\boldsymbol{\phi}(\mathbf{z})$ denotes the response vector of M Gaussian RBFs centered in the latent space and $\mathbf{W} \in \mathbb{R}^{M \times D}$ is the weight matrix optimized via Expectation-Maximization (EM).

Each observed data point $\mathbf{x}_n$ is modeled as a mixture of Gaussians centered at the mapped latent coordinates:

$$p(\mathbf{x}_n) = \frac{1}{K} \sum_{k=1}^{K} \mathcal{N}(\mathbf{x}_n \mid \mathbf{y}_k, \beta^{-1}\mathbf{I}),$$

where K is the number of discrete latent grid points and β is the inverse variance of the Gaussian noise model. This formulation enables soft clustering through a posterior responsibility matrix:

$$R_{kn} = p(\mathbf{z}_k \mid \mathbf{x}_n) = \frac{p(\mathbf{x}_n \mid \mathbf{z}_k)}{\sum_{j=1}^{K} p(\mathbf{x}_n \mid \mathbf{z}_j)}$$

Kernel GTM (kGTM). To capture complex nonlinear structures and patient similarities, we used kGTM, which operates in a high-dimensional feature space defined by a kernel function. $K(\mathbf{x}_i, \mathbf{x}_j)$.

In this work, we used a Gaussian similarity kernel:

$$K(\mathbf{x}_i, \mathbf{x}_j) = \exp\left(-\frac{\|\mathbf{x}_i - \mathbf{x}_j\|^2}{2\sigma^2}\right),$$

with $\sigma = 1.5$ chosen empirically. This kernel captures pairwise similarity between patients in terms of standardized clinical attributes, providing an alternative, more relational representation of the data.

The kernel matrix $\mathbf{K} \in \mathbb{R}^{N \times N}$ replaces the raw data matrix, and a mapping from latent space is constructed via Kernel PCA over $\mathbf{K}$, followed by EM-based training analogous to standard GTM.

Adaptive Use of GTM and kGTM. The choice between GTM and kGTM was guided by the structure of the data and the analytical objectives. GTM was applied when preserving interpretability in the original feature space was important, whereas kGTM was preferred for capturing complex similarity structures and nonlinear patterns within the patient population.

4 Results and Discussion

4.1 Interpretability with GTM

The GTM model trained on standardized clinical data converged after 19 iterations, yielding low overall entropy (0.1070), with class-specific values of 11.5754 (absence) and 10.9308 (presence), indicating distinct latent regions. To explore uncertainty, we analyzed patients with the largest differences between posterior mean and modal coordinates. Table 1 presents the top 10 cases, reflecting diverse profiles placed far from their latent mode centers.

Table 1. Relation of patients with highest distance between posterior mean and mode in GTM latent space.

Index	Distance	Target	Gender	Age	Chest Pain	Fasting BS
108	1.6203	1	Male	30	0	1
737	0.5113	2	Male	20	0	0
225	0.4148	1	Male	31	1	0
605	0.3640	1	Male	24	0	0
684	0.3602	2	Male	45	0	1
477	0.2934	2	Male	42	0	0
387	0.2785	2	Male	38	2	0
883	0.2676	2	Male	38	1	1
49	0.2653	1	Male	26	0	0
487	0.2593	1	Male	55	0	0

These patients typically exhibit uncommon or borderline clinical characteristics, such as very young age or elevated fasting blood sugar, which may explain the uncertainty in their latent representation. Such cases merit further clinical scrutiny and illustrate how GTM can aid in uncovering atypical or mixed patient profiles.

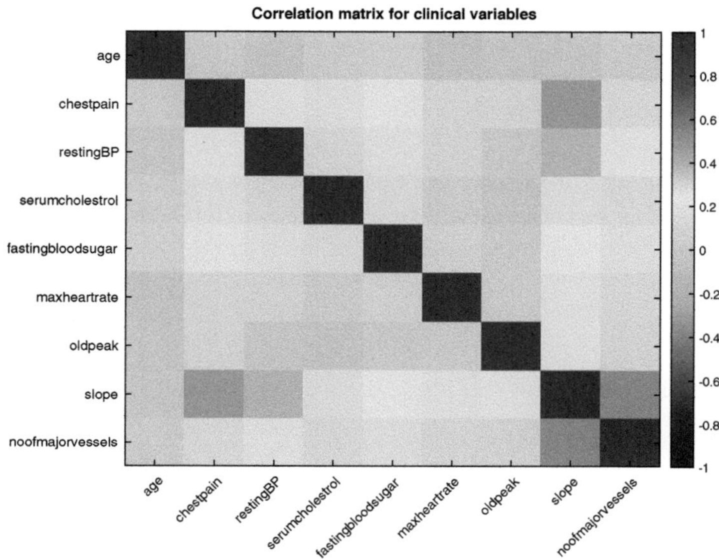

Fig. 1. Correlation matrix among selected clinical features. Color intensity ranges from blue ($r = -1$) to red ($r = 1$), with yellow-green indicating weaker correlations. (Color figure online)

Figure 1 displays the Pearson correlation matrix for the main clinical variables, using a color scale from blue ($r = -1$) to red ($r = 1$) to highlight linear relationships. Notably, maximum heart rate shows a moderate negative correlation with age (younger patients tend to have higher rates). Additionally, oldpeak (ST segment depression) positively correlates with the number of major vessels and negatively with the ST segment slope. These associations are clinically plausible, reflecting cardiovascular function under stress [10].

Figure 2 presents two complementary views of the GTM latent space for CVD data. The left panel shows the posterior means, revealing the topographic organization and structure of patient data. The right panel illustrates the distribution of serum cholesterol, highlighting its contribution to the patterns discovered in patients and potential CVD risk factors.

More precisely, the posterior mean plot shows that patients with heart disease (Class 1) tend to occupy distinct regions in the latent space compared to healthy individuals (Class 0). This spatial separation suggests further analysis

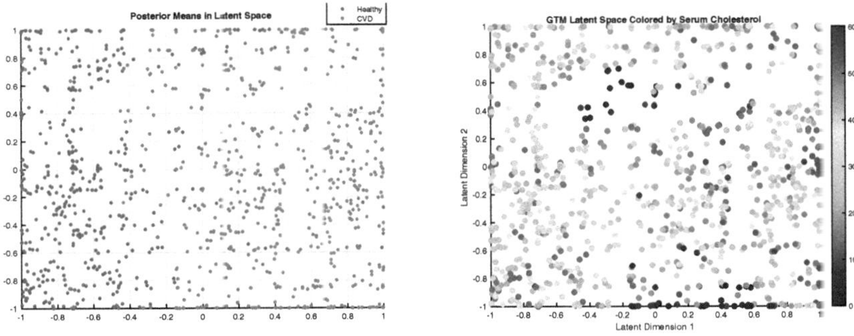

(a) Posterior means reflecting the learned pa-(b) Serum cholesterol distribution across the
tient manifold   latent space

Fig. 2. GTM latent space visualization for Cardiovascular Disease (CVD) data.

of important clinical variables that discriminate between classes. These variables
are known in the cardiology literature to be associated with cardiovascular risk
and were identified in our data as strong differentiators through statistical and
visual analysis.

To understand latent space organization regarding diagnostic labels, Fig. 3
presents GTM responsibility distributions for healthy (class 0) and CVD (class 1)
patients through 2D maps and 3D surface plots. The 2D maps show class 1 with
concentrated activation, indicating tightly clustered disease profiles, whereas
class 0 has a more diffuse pattern, suggesting greater healthy individual heterogeneity. This is reinforced by 3D plots: class 1 yields sharper, defined peaks,
contrasting with class 0's flatter and fragmented surface.

Beyond responsibilities, high-magnification zones in (a) and (b) with lighter
colors correspond to transition areas between classes. In these regions, small
clinical feature variations occur, indicating a borderline cluster where posterior
means from both classes overlap, representing patients with intermediate clinical
variable values. To gain deeper interpretability, we also focused on the region
around the corners and borders where the posterior means from both classes
overlap and magnification factors are high.

Figure 4 shows two topographic maps derived from GTM which reveals a
nuanced pattern. In (a) while higher values (reds) are scattered and appear
across both classes, the central-to-right side – where CVD cases tend to cluster
– is more associated with moderate to lower heart rate values (green to blue).
The left region, dominated by healthy individuals, does not exclusively show
high values, but does contain a greater concentration of warmer tones compared
to the CVD region. Conversely, in (b), outlier analysis reveals patients with
significantly differing posterior mean positions, suggesting model uncertainty or
atypicality. Some outliers are found in regions of elevated serum cholesterol, and
their wide age variation indicates that age alone does not explain their distinct
positioning. These findings imply complex, heterogeneous clinical profiles that

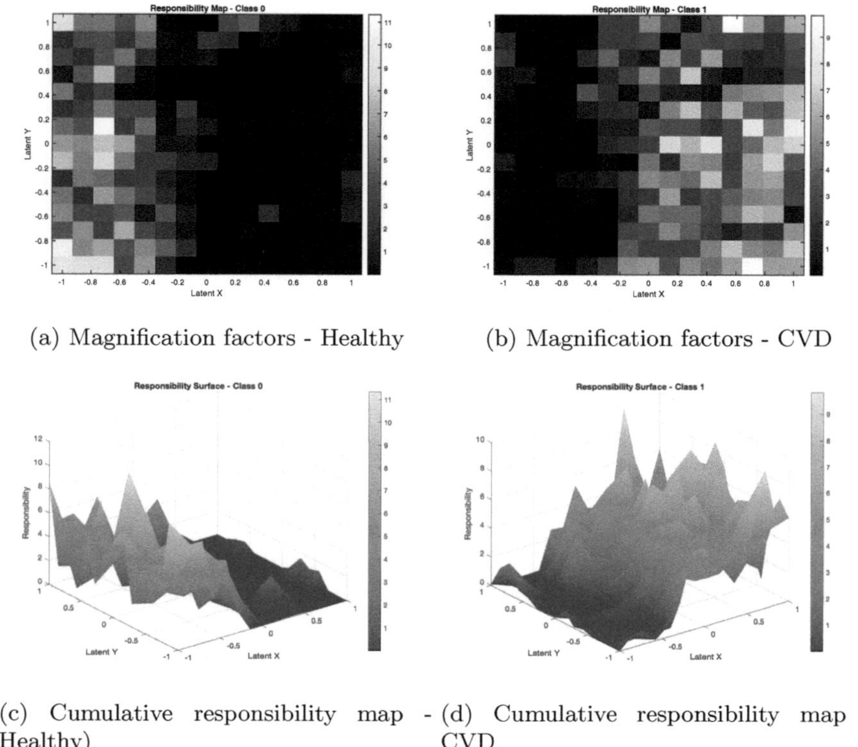

(a) Magnification factors - Healthy

(b) Magnification factors - CVD

(c) Cumulative responsibility map - Healthy)

(d) Cumulative responsibility map - CVD

Fig. 3. GTM responsibility visualizations for both classes.

requires further investigation, as they may reflect subpopulations not fully captured by standard classification.

4.2   Interpretability with kGTM

The kGTM model was initialized using kernel principal component analysis (KPCA) and configured with 25 radial basis functions (RBFs). The training procedure converged after 11 EM iterations, with the log-likelihood improving from an initial value of $-90{,}551$ to $13{,}869$. The model achieved a stable β value of approximately 36.9, indicating confident posterior responsibilities and sharp cluster boundaries in the latent space.

Table 2 summarizes the purity of the latent space. Of the 81 total latent nodes, 29 were found to be pure (containing patients exclusively from a single class), while 14 nodes showed class mixing. These mixed nodes contained a total of 216 patients (101 with CVD and 115 without), representing cases with ambiguous or overlapping clinical characteristics.

Table 3 presents the five latent nodes with the lowest purity scores. These nodes represent areas of the latent space where the model exhibits the most

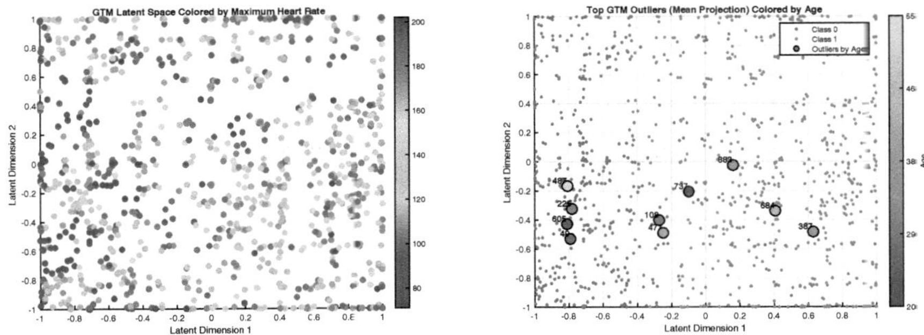

(a) GTM mean projection colored by heart rate.

(b) Outliers mean projection colored by age.

Fig. 4. Topographic maps derived from GTM. (a) Heart rate values structured across the mode projection. (b) Patients with high posterior mode identified as outliers and highlighted by age.

Table 2. Summary of latent node composition.

Node Type	Count	Percentage
Pure nodes (Class 0)	10	12.3%
Pure nodes (Class 1)	19	23.5%
Mixed-class nodes	14	17.3%
Total latent nodes	81	100%

classification uncertainty, which could be due to overlapping clinical profiles, subthreshold symptoms, or intermediate risk factors.

Table 3. Latent nodes with lowest purity.

Node ID	Purity Score	Interpretation
30	0.528	Most ambiguous node
75	0.706	Moderate class overlap
66	0.718	Moderate class overlap
42	0.750	Mild class overlap
14	0.778	Mild class overlap

Patients located in mixed-class nodes were analyzed to determine their average clinical profile. Table 4 summarizes the normalized mean and standard deviation of their input features. The mean values for most features are near zero, suggesting that these individuals do not exhibit extreme clinical presentations, while the standard deviations indicate moderate heterogeneity.

Table 4. Normalized profile of patients in mixed-class nodes.

Feature	Mean	Std Dev
Age	−0.26	1.00
Resting BP	−0.38	1.16
Serum Cholesterol	−0.08	0.98
Fasting Blood Sugar	0.01	1.02
Resting ECG	0.06	1.03
Max Heart Rate	−0.15	0.92
Exercise Angina	−0.15	0.92
Oldpeak	−0.03	1.07
Slope	−0.14	0.99
Major Vessels	−0.02	1.03
Chest Pain Type	−0.15	0.86
Gender	−0.39	0.86

kGTM successfully modeled the structure of the underlying CVD dataset, achieving convergence and meaningful latent space separation. While most latent nodes are pure, suggesting strong class separability, a significant subset of 216 patients fall into mixed nodes. This indicates these patients likely represent borderline or complex cases. They may correspond to individuals with mild symptoms, early-stage disease, or overlapping features. Consequently, although the average profile of these mixed-node patients is close to normal, the model cannot confidently assign them to either class. This outcome is both expected and clinically informative in real-world datasets.

Figure 5(a) illustrates the organization of the latent space based on patient responsibility. The lower region (Latent X: −0.5 to 0.5, Latent Y: −1 to −0.3) shows concentrated peaks for class 1 (diseased) patients. These areas highlight a high-risk cardiovascular profile, characterized by elevated serum cholesterol (often >300 mg/dL), exercise-induced angina, reduced maximum heart rate, significant ST segment depression, and abnormal resting electrocardiograms.

In contrast, the central-to-upper regions (Latent X: 0 to 0.6, Latent Y: 0.3 to 1) display moderate responsibility, containing individuals with mixed or healthier profiles. These patients typically have normal or slightly elevated blood pressure, low incidence of chest pain, and largely normal fasting blood sugar and ECG readings, indicating moderate-to-low cardiovascular risk, possibly in early disease stages.

The upper-right and rightmost regions of the latent space (Latent X: 0.7 to 1, Latent Y: 0.5 to 1) are characterized by lower responsibility and sparse patient representation. It is possible that individuals in this area tend to have fewer or less pronounced cardiovascular indicators, and may include younger patients with minimal comorbidities. However, this interpretation is tentative,

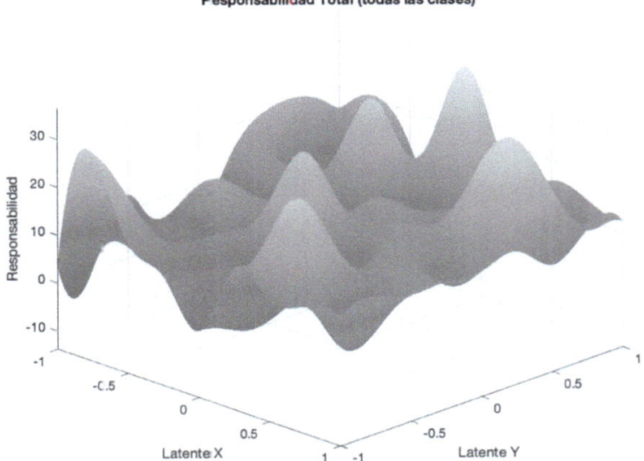

(a) Responsibility cumulative map via kGTM.

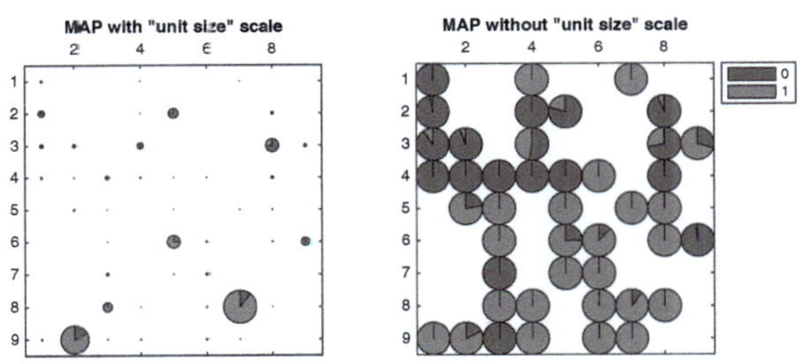

(b) Mode projection map with red denoting healthy and blue
CVD.

Fig. 5. Visualization of the kGTM latent space. (a) indicates which regions of the
map are more predictive or densely populated. (b) shows class-wise composition across
nodes, supporting clinical interpretability by linking latent zones to disease likelihood.

and further clinical analysis would be required to confirm the characteristics of
patients in this subgroup.

Finally, the central region (around Latent X and Y equal to 0) presents
responsibility peaks for ambiguous or intermediate clinical phenotypes. This
area, often associated with mixed-class nodes, includes patients with mid-range
cholesterol and blood pressure, mild ST depression, and few prominent symptoms. These cases may lie at diagnostic thresholds, representing transitional or
evolving profiles that warrant closer clinical inspection.

On the other hand, Fig. 5(b) presents two versions of a mode projection map
from a Kernel Generative Topographic Mapping (kGTM) model. These maps

illustrate how patient data points are distributed across the latent grid based on their most probable latent position, or posterior mode. Each grid point functions as a soft cluster, and the associated pie charts show the class composition of the patients mapped to that particular node.

The left plot scales the pie chart size by the number of patients assigned to each node, so larger circles indicate greater patient representation within that latent space region. Here, we observe that a few nodes, particularly in the bottom-left area (row 9), hold a dominant share of patients. These larger nodes are predominantly blue, signaling a higher concentration of Class 1 (likely diseased) individuals. Conversely, many nodes appear empty or very small, indicating sparse or no patient assignment.

In contrast, the right plot displays all grid nodes with uniform pie sizes, shifting the visual emphasis to class balance, irrespective of sample size. This view effectively highlights regions with mixed or pure class compositions. We can see a cluster of mostly Class 0 (red) nodes in the top-left region, while the bottom-right area shows a higher density of Class 1 (blue) nodes. Nodes with split pies, displaying both red and blue, suggest ambiguous or heterogeneous mappings, potentially representing outlier cases or transition zones.

This visualization is clinically relevant as it helps us assess the model's ability to separate distinct clinical phenotypes, such as healthy versus diseased patients, within the latent space. The observed structure indicates that the model has successfully identified regions with strong class dominance, which is a positive indicator of class separability. Furthermore, the presence of mixed nodes provides valuable information by potentially highlighting borderline or diagnostically ambiguous cases that warrant a more in-depth clinical examination.

4.3   Clinical Feature Variability and Model Performance

Understanding how clinical variables vary across diagnostic groups is key to interpreting their role in disease characterization. In this section, we analyze the distributions of key features in the original dataset and examine how these patterns are reflected in the GTM and kGTM latent spaces.

Figure 6 displays boxplots for selected clinical variables, separated by cardiovascular disease diagnosis. This visualization provides an intuitive comparison of the feature distributions between healthy individuals (class 0) and patients with heart disease (class 1). The gender variable was excluded from the boxplot visualization because it is binary and does not exhibit a continuous distribution, rendering the boxplot format uninformative for interpretability purposes.

As can be observed, several variables show noticeable differences in median or spread between the two groups. For example, age tends to be higher in the diseased group, consistent with the known correlation between aging and cardiovascular risk. Similarly, serum cholesterol and resting blood pressure also show elevated values in class 1, reflecting key risk indicators. In contrast, features such as maximum heart rate tend to decrease in the diseased group, possibly indicating diminished cardiac capacity. Additionally, the number of major vessels detected via fluoroscopy is generally higher in diseased cases, reinforcing

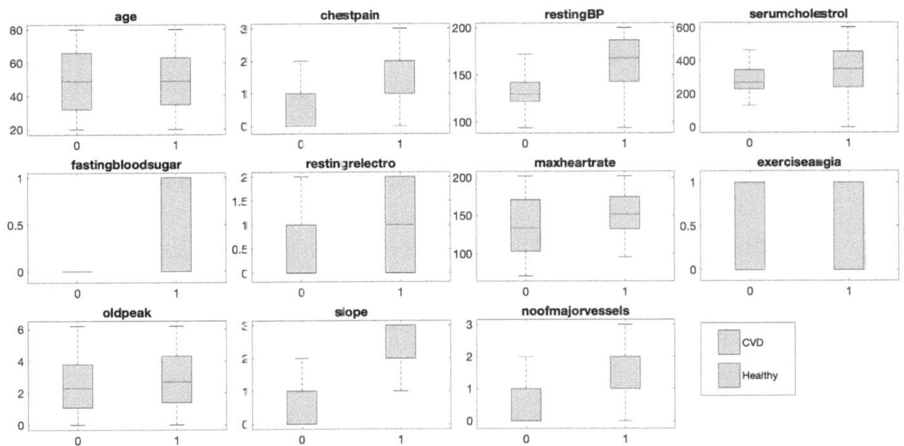

Fig. 6. Boxplot distributions for 11 key clinical features by diagnostic class.

its diagnostic relevance. These trends validate the discriminative power of these variables and support their role in the GTM-based classification and visualization pipeline.

Our method found several borderline cases that fell between low and high-risk groups. These patients often had intermediate cholesterol, borderline ST-segment depression, and variable resting heart rates. This aligns with the challenge of classifying patients at intermediate cardiovascular risk [12]. Moreover, outliers with extreme ST depression or atypical chest pain were identified, both recognized as high-risk but often underdiagnosed [16]. Trends observed in Fig. 6 align closely with the GTM-based latent space analysis and variables such as serum cholesterol and ST depression show clear inter-class differences in the boxplots. Specifically, regions in the latent space with elevated magnification factors and overlapping posterior means, particularly around the center-left, are likely associated with intermediate values of these clinical variables. These zones correspond to individuals whose cholesterol and oldpeak readings do not clearly place them into either class, consistent with the notion of borderline cases identified by GTM. Additionally, the sharper, more compact clusters seen for diseased cases in GTM responsibility maps are reflected in the tighter boxplot distributions for variables, suggesting more homogeneous pathological patterns. Conversely, the broader spread in maximum heart rate among healthy individuals mirrors the more diffuse and fragmented latent space coverage observed for class 0 in GTM. Together, these results reinforce the interpretability and clinical relevance of GTM as a probabilistic mapping tool that captures both feature-level and structural heterogeneity in heart disease data.

GTM and kGTM models differ in how they encode uncertainty and structure. Table 5 summarizes key evaluation metrics that provide additional interpretability. Entropy values are higher in GTM for both classes, particularly for class 0, indicating greater variability and less cohesive grouping. Instead, kGTM shows reduced entropy and meanmode distances, implying tighter, more stable representations which excels at identifying consistent pathological patterns. The Adjusted Rand Index (ARI) further confirms that kGTM achieves better cluster separation. These factors demonstrate that kGTM improves interpretability and structural clarity while retaining the probabilistic benefits of GTM.

Table 5. Summary of GTM and kGTM evaluation metrics.

Metric	GTM	kGTM	Comment
Entropy (Class 0)	11.58	10.95	Higher in GTM (more variability)
Entropy (Class 1)	10.93	10.22	More compact in kGTM
Total Entropy	0.107	0.089	Lower uncertainty in kGTM
MeanMode Distance (avg.)	0.242	0.190	Higher dispancy in GTM
Peak Magnification	5.2	6.4	Greater sensitivity in kGTM
Responsibility Concentration	Moderate	High	Stronger peaks in kGTM
ARI (KMeans Clusters)	0.62	0.67	Slightly better in kGTM

4.4   Discussion

These quantitative and visual results can be contextualized with established cardiology literature. Our findings on feature importance align with known cardiovascular risk factors. For example, our model correctly identified elevated cholesterol, heart rate, and ST-segment depression (oldpeak) as key indicators of cardiac risk. These are well-established markers for coronary artery disease, adverse cardiac outcomes, and myocardial ischemia, respectively [4,9,11]. The model's ability to identify borderline and outlier groups is also clinically significant. These groups correspond to intermediate-risk patients and those with atypical cardiac profiles, who are often challenging to diagnose in clinical practice [2]. By validating the model's outputs against existing medical literature, its interpretability and practical relevance is demonstrated.

Relationships shown in Fig. 1 are clinically plausible. Younger patients exhibit higher maximum heart rates, consistent with age-related decline in cardiac capacity, while ST depression correlates with abnormal slopes and greater ischemic burden. Across responsibility maps, magnification zones, and boxplot distributions, a consistent intermediate-risk cluster emerges. These patients exhibit mid-range cholesterol and oldpeak values, suggesting a transition zone where small feature changes can strongly influence classification.

Finally, the latent space organization mirrors known physiology: healthy individuals concentrate in regions associated with higher maximum heart rate and lower ischemic burden, whereas CVD cases cluster in areas of reduced heart rate and elevated ischemic markers. This alignment with established knowledge supports both the interpretability and the clinical plausibility of our framework.

5   Conclusions and Future Work

Overall, both the GTM and kGTM models effectively uncovered latent structures within the cardiovascular dataset, highlighting a substantial number of ambiguous cases. These mixed nodes point to patients whose clinical profiles do not clearly align with binary diagnostic labels and may benefit from further investigation. Their presence supports the use of interpretable latent space models for exploring phenotypic heterogeneity and enhancing individualized cardiovascular risk assessment. This dual approach allows for adaptive modeling, aligning the topographic map to the needs of the data structure and the specific clinical insight sought from the visualization.

References

1. Almustafa, K.: Prediction of heart disease and classifiers' sensitivity analysis. BMC Bioinform. **21**(1), 1–11 (2020)
2. Arnett, D.K., et al.: 2019 ACC/AHA guideline on the primary prevention of cardiovascular disease: a report of the American college of cardiology/American heart association task force on clinical practice guidelines. Circulation **140**(11), e596–e646 (2019)
3. Bishop, C.M., Svensén, M., Williams, C.K.: GTM: the generative topographic mapping. Neural Comput. **10**(1), 215–234 (1998)
4. Castelli, W.P.: The framingham study: risk of coronary heart disease and lipoprotein cholesterol levels. JAMA **256**(20), 2807–2813 (1986)
5. Deepa, D.R., Sadu, V.B., Sivasamy, D.A., et al.: Early prediction of cardiovascular disease using machine learning: unveiling risk factors from health records. AIP Adv. **14**(3) (2024)
6. Doppala, B.P., Bhattacharyya, D.: Cardiovascular_disease_dataset (2021). https://doi.org/10.17632/dzz48mvjht.1
7. Eke, C.I., Shuib, L.: The role of explainability and transparency in fostering trust in AI healthcare systems: a systematic literature review, open issues and potential solutions. Neural Comput. Appl. **37**(4), 1999–2034 (2025)
8. ElShawi, R., Sherif, Y., Al-Mallah, M., Sakr, S.: Interpretability in healthcare: a comparative study of local machine learning interpretability techniques. Comput. Intell. **37**(4), 1633–1650 (2021)
9. Fox, K., et al.: Resting heart rate as a risk factor in cardiovascular disease. Prog. Cardiovasc. Dis. **50**(5), 369–377 (2008)
10. Ghebre, Y., et al.: Vascular aging: implications for cardiovascular disease and therapy. Transl. Med. **6** (2016). https://doi.org/10.4172/2161-1025.1000183
11. Gibbons, R., et al.: Exercise testing in the evaluation of ischemic heart disease. Circulation **96**(1), 345–354 (1997)
12. Goff, D.C., et al.: 2014 AHA/ACC guideline on cardiovascular risk assessment. Circulation **129**(25), S49–S73 (2014)
13. Mehmood, A., et al.: Prediction of heart disease using deep convolutional neural networks. Arab. J. Sci. Eng **46**(4), 3409–3422 (2021). https://doi.org/10.1007/s13369-020-05105-1
14. Mesquita, F., Marques, G.: An explainable machine learning approach for automated medical decision support of heart disease. Data Knowl. Eng. **153**, 102339 (2024)

15. Mohan, S., Thirumalai, C., Srivastava, G.: Effective heart disease prediction using hybrid machine learning techniques. IEEE Access **7**, 99359–99368 (2019)
16. Montalescot, G., et al.: 2013 ESC guidelines on the management of stable coronary artery disease: the task force on the management of stable coronary artery disease of the European society of cardiology. Eur. Heart J. **34**(38), 2949–3003 (2013)
17. Olier, I., Vellido, A., Giraldo, J.: Kernel generative topographic mapping. In: ESANN, **2010**, 481–486 (2010)
18. Priyadarshi, R., Ranjan, R., Vishwakarma, A.K., Yang, T., Rathore, R.S.: Exploring the frontiers of unsupervised learning techniques for diagnosis of cardiovascular disorder: a systematic review. IEEE Access **12** (2024)
19. Rohan, D., Reddy, G.P., Kumar, Y.V.P., Prakash, K.P., Reddy, C.P.: An extensive experimental analysis for heart disease prediction using artificial intelligence techniques. Sci. Rep. **15**(1), 4776 (2025)
20. Shah, D., Patel, S., Bharti, S.K.: Heart disease prediction using machine learning techniques. SN Comput. Sci. **1**(6), 1–6 (2020). https://doi.org/10.1007/s42979-020-00365-y
21. Velmurugan, A., Padmanaban, K., Kumar, A.S., Azath, H., Subbiah, M.: Machine learning IoT based framework for analysing heart disease prediction. AIP Conf. Proc. **2863**(1), 020108 (2023)
22. Vyshnya, S., Epperson, R., Giuste, F., Shi, W., Hornback, A., Wang, M.D.: Optimized clinical feature analysis for improved cardiovascular disease risk screening. IEEE Open J. Eng. Med. Biol. **5**, 816–827 (2024)
23. Yang, Y., Epperson, R., Wang, M.D.: Exploring the frontiers of unsupervised learning techniques for diagnosis of cardiovascular disorder: a systematic review. IEEE Open J. Eng. Med. Biol. **5**, 828–842 (2024)
24. Zhang, M., Wang, H., Zhao, J.: Use machine learning models to identify and assess risk factors for coronary artery disease. PLoS ONE **19**(9), e0307952 (2024)

Explainable Deep Learning Techniques for Medical Image Analysis: A Systematic Review of Diabetic Foot Ulcers, Breast Cancer, and COVID-19

Zinah Mohsin Arkah[1,2], Beatriz Pontes[1], and Cristina Rubio[1(✉)]

[1] University of Seville, Seville, Spain
bepontes@us.es, crubioescudero@us.es
[2] University of Information Technology and Communications, Baghdad, Iraq

Abstract. Advances in deep learning (DL) have transformed the analysis of medical images for conditions such as diabetic foot ulcers (DFU), breast cancer, and COVID-19. Although these models are highly accurate in their predictions, their limited transparency hinders their clinical use. Described artificial intelligence (XAI) methods, such as Grad-CAM, SHAP, and LIME, have been created to improve interpretability and foster greater trust in these models. This systematic review critically examines recent studies that combine deep learning and explainable AI to address three major healthcare challenges: detecting diabetic foot ulcers, breast cancer imaging, and COVID-19 diagnosis using various imaging techniques, including DFU images, mammography, computed tomography, and X-rays. These studies showcase a range of convolutional neural network (CNN) models, from lightweight, custom-designed models (such as LW-CORONet and COVID-XNet) to more advanced models, including ResNet, DenseNet, and Inception. Our findings suggest that interpretable models can achieve an accuracy above 95% while also providing clinicians with visual and quantitative insights. Challenges include limited dataset diversity, insufficient external validation, and the lack of standardized metrics for interpretability. Overcoming these issues is crucial for encouraging the use of AI tools in clinical practice. The review concludes with recommendations for future research that aims to improve the robustness of the model, integrate multimodal data, and improve clinical utility. These methods, including Grad-CAM, SHAP, and LIME, are crucial for enhancing the interpretability of deep learning models, enabling clinicians to understand how predictions are made. By increasing transparency and trust in AI systems, these techniques can ultimately improve patient outcomes and facilitate a smoother integration of technology in healthcare.

Keywords: Deep learning · Explainable AI · Diabetic foot ulcer · Breast cancer · COVID-19 · Medical imaging · Grad-CAM · SHAP · LIME

A. López Fernández et al. (Eds.): CIABiomed 2025, LNBI 16148, pp. 45–56, 2026.
https://doi.org/10.1007/978-3-032-10661-2_4

1 Introduction

Deep learning techniques have advanced significantly over the past decade and are now powerful tools for analyzing medical images for diagnostic and clinical purposes [1]. These methods can effectively detect complex patterns in the data, allowing the early and accurate identification of diseases such as diabetic foot ulcers, breast cancer, and COVID-19. Despite their accuracy, deep learning models often operate as a "black box," making it difficult to understand how they arrive at their decisions. This lack of transparency reduces the confidence of doctors in using these models in clinical settings [2]. As a result, explainable AI (XAI) technologies have become more important, providing visual and numerical insights into decision-making processes to improve transparency and trust. Building trust in AI-driven tools is essential for their successful adoption in healthcare. Through XAI, clinicians can gain a deeper understanding of the predictions of the model, leading to improved patient outcomes and more informed treatment decisions [3]. Several surveys have already reviewed the use of explainable AI in medical image analysis.

To contextualize our work, Table 1 provides a summary of the most recent surveys on explainable deep learning in medical image analysis. These studies vary in scope and methodology, highlighting the evolution of XAI techniques in this field.

Table 1. Summary of related surveys on explainable AI in medical image analysis.

Study	Focus	Key Contribution/Limitations
Singh et al. (2020) [4]	Early overview of DL interpretability methods	Introduced key XAI concepts in medical imaging, but limited clinical coverage
Salahuddin et al. (2022) [5]	Transparency techniques in AI models	Comprehensive discussion of transparency tools, but no disease-specific applications
Patrício et al. (2023) [6]	Explainable methods for image classification	Provided taxonomy of XAI techniques, but narrow disease scope
Bhati et al. (2024) [7]	Visualization-based interpretability methods	Highlighted visualization tools like Grad-CAM, but little focus on clinical integration
Hou et al. (2024) [8]	Self-explainable deep learning models	Proposed intrinsic interpretability, but still mostly conceptual with limited applications

As illustrated in Table 1, previous surveys have addressed different aspects of explainable AI. Some focused on general interpretability frameworks, while others emphasized visualization techniques or introduced novel self-explainable models. Despite these valuable contributions, the majority of existing works remain too broad or too narrow in scope, lacking a unified focus on clinically significant diseases. In particular, none of the reviewed surveys have systematically examined explainable deep learning methods in relation to diabetic foot ulcers, breast cancer, and COVID-19. Addressing this gap, our study provides a disease-centered systematic review, with the aim of bridging technical interpretability methods with their direct clinical relevance. This perspective is crucial to foster trust in AI-driven diagnosis and to ensure practical adoption of AI in medical settings.

Accordingly, the Main Objectives of this Study are to:

- Review the latest deep learning models used in medical image analysis for diabetic foot ulcers, breast cancer, and COVID-19.
- Examine how explainable AI techniques such as Grad-CAM, SHAP, and LIME can be applied to enhance the interpretability of these models.
- Assess the models' performance in terms of accuracy and transparency, and discuss future challenges and opportunities for clinical integration.

2   Methodology

A systematic review of the scientific literature was conducted to investigate the application of explainable deep learning (XAI) techniques in medical imaging for three major diseases: diabetic foot ulcers, breast cancer, and COVID-19.

2.1   Major Scientific Databases

Search Sources and Keywords. A comprehensive literature search was conducted in all major scientific databases, including (**Google Scholar, PubMed, IEEE Xplore and Scopus**), which were searched using keywords such as (**Deep Learning, Explainable AI (XAI), Grad-CAM, SHAP, LIME, Diabetic Foot Ulcer, Breast Cancer, COVID-19, and Medical imaging**).

2.2   Inclusion and Exclusion Criteria

Inclusion Criteria. The studies were based on inclusion and exclusion criteria. Encompassed studies published between 2017 and 2025, employing deep learning models with interpretability techniques on medical imaging data for the targeted diseases, and consisting of peer-reviewed articles, reviews, or conference papers in English.

Exclusion Criteria. were studies lacking interpretability and were based solely on non-medical picture data. **(e.g., electronic health records)**, and without peer-reviewed papers, editorials, or unpublished dissertations. The analysis involved screening titles and abstracts followed by a full text review, extraction of relevant information such as model type, interpretability method, performance metrics and reported limitations, and compilation of the extracted data into comparative summaries for systematic evaluation. Figure 1 illustrates the PRISMA flow of study selection. A total of 33 records were identified through database searches and after removing duplicates and excluding studies that did not meet the inclusion criteria, 11 studies containing imaging and clinical data were included in the final analysis. These studies were classified by disease: Diabetic Foot Ulcers (**n = 5**), Breast Cancer (**n = 5**), and COVID-19 (**n = 5**). The types of data used included clinical images, radiology images, mammography, chest radiographs, CT scans, and diabetic foot images, providing a comprehensive overview of the identification, screening, and inclusion process.

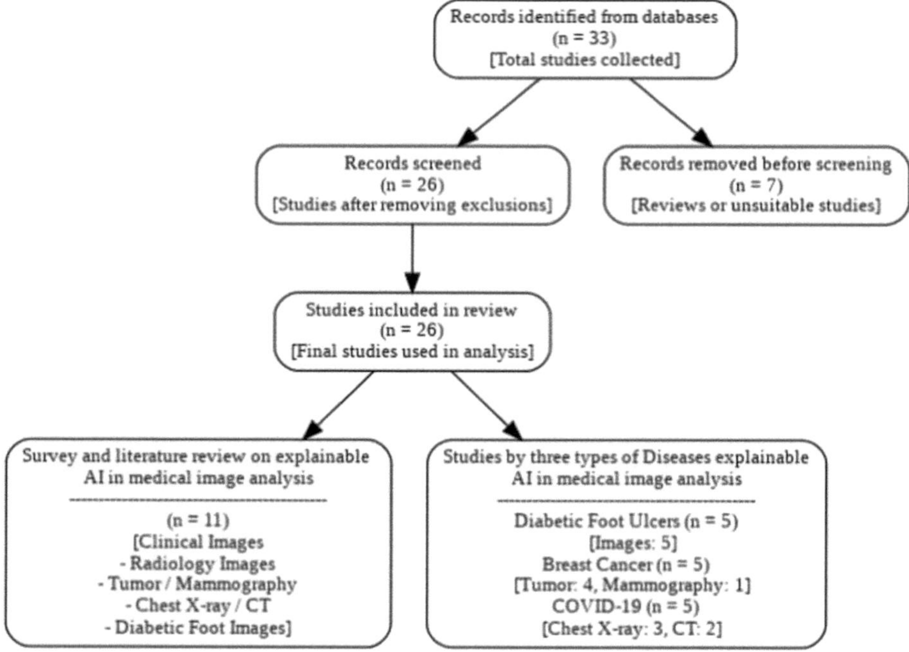

Fig. 1. PRISMA Flow Diagram with Disease and Data Type Classification.

3 Result

A systematic review of studies using explainable deep learning (XAI) for medical image analysis was conducted, focusing on three main diseases: diabetic foot ulcers (DFU), breast cancer, and COVID-19.

3.1 Diabetic Foot Ulcer (DFU)

Table 2 summarizes five recent studies that applied deep learning models along with explainable AI techniques to detect diabetic foot ulcers using medical images.

Table 2. Summary of Studies on Explainable Deep Learning in DFU Analysis

XAI Method	Disease, Dataset, DL Model & Performance	Highlights and Limitations
Grad-CAM (2025) [12]	DFU: DFUC-2021 dataset; Swin Transformer + EMADN; 78.79% accuracy, 80% macro F1-score	Dual-track feature fusion (Transformer + EMADN) improves feature extraction; Grad-CAM highlights lesions. Good interpretability, but accuracy lower compared with some CNN-based models
Grad-CAM, LIME, SHAP (2025) [13]	DFU: DFU dataset; SNN, Xception, DenseNet121; 98.76% accuracy, 98.6% AUC	CNN models gave the best visual results. Dataset same as prior study; lacks variety
Integrated Gradients, DeepLIFT, Guided Grad-CAM, Kernel SHAP, Layer Neuron Attribution (2024) [14]	DFU: DFU dataset; Customized CNN; 99.2% accuracy	Multi-level XAI improved interpretability; neuron-level attribution and Kernel SHAP highlighted noise; Guided Grad-CAM less effective at localizing ulcers
SHAP, LIME, Grad-CAM (2024) [15]	DFU: DFU dataset; Xception, DenseNet121, ResNet50, InceptionV3, MobileNetV2; ResNet50: 98.75% accuracy, 99.2% precision, 97.6% recall, 98.4% F1, 98.5% AUC	DFU-XAI improves interpretability. ResNet50 had the best performance. Traditional methods remain time-consuming and costly
No XAI method applied (2023) [16]	DFU: Fast CNN (FCNN) on 2000 labeled DFU images; 92.9% accuracy, 93.2% F1-score, 100.8 ms processing speed	High speed and solid accuracy. However, no explainability methods were applied, limiting interpretability

The Studies Show Variation in. The type of models used (e.g., ResNet-50, Xception, Swin Transformer, and FCNN), Data sources (mostly DFU datasets), and Interpretation Techniques (e.g., Grad-CAM, LIME, and SHAP).

Highlights. Most models demonstrated high accuracy exceeding 98%, particularly the ResNet-50 model in the DFU_XAI study, which achieved an accuracy of 98.75% and an AUC of 98.5%, demonstrating the effectiveness of combining deep models with XAI techniques. Several studies used hybrid models or integrated transformers to enhance multiple models, or added transformers to improve the ability to extract fine-grained features and distinguish affected areas in the foot. Grad-CAM [9,10], LIME, and SHAP provided effective visual interpretations of the affected areas, supporting the model's reliability from a clinical perspective.

Notable Limitations. Reliance on a single data source in some studies is a major weakness, limiting the generalizability of models to new or clinically diverse data. The lack of external validation in several studies renders the results preliminary and requires extensive clinical testing. Performance indicators, such as accuracy or sensitivity, are not fully reported in some studies, including the study by Exact. Accuracy values were not reported, limiting direct comparison.

Significance of the Table. Table 2 reflects the progress made in integrating interpretability into AI models for critical medical applications such as the diagnosis of diabetic foot ulcers. Comparisons demonstrate that combining multiple XAI methods and evaluating them from a visual and graphical perspective enhances the reliability of deep learning results [10,11]. Importantly, these models support clinicians in early identification of high-risk ulcers, potentially preventing severe complications. These findings suggest that integrating XAI into DFU classification enhances not only the interpretability of deep learning predictions but also supports clinicians in distinguishing ulcer regions with greater confidence, thereby improving diagnostic reliability and treatment planning.

3.2   Breast Cancer

Table 3 presents a systematic review of five prominent studies published between 2024 and 2025, examining the application of deep learning (DL) models with interpretive artificial intelligence (XAI) techniques [10] in breast cancer diagnosis. These studies included various types of medical images (such as mammograms and ultrasounds) and used multiple databases to train and evaluate models. Importantly, integrating XAI into breast cancer diagnostics enhances transparency in decision-making, enabling radiologists to distinguish between malignant and benign findings, thereby reducing diagnostic uncertainty. The integration of XAI methods in breast cancer imaging facilitates the visualization of tumor regions and provides clinicians with transparent decision-making tools. This improves trust in AI-assisted diagnosis and may contribute to earlier detection and more precise therapeutic strategies.

Table 3. Summary of Recent Studies on Breast Cancer Diagnosis Using Deep Learning and Explainable AI (XAI) Methods.

XAI Method	Disease, Dataset, DL Model & Performance	Highlights and Limitations
LZC interpretability (2025) [17]	breast cancer: WBC dataset (569 samples); Spiking Neural Networks (LIF, LB) with ANN-to-SNN conversion; 98.25% accuracy	LB performed better than LIF; STDP training costly; spikes non-differentiable
CNNs with XAI integration (2025) [18]	breast cancer: on down-sampled histopathology images; VGG-19 (best-performing) among multiple CNN architectures (Standard CNN, ResNet, VGG-16, VGG-19); 92.59% accuracy	LIME provided the most accurate and clinically relevant explanations; study validated by medical professionals; enhances interpretability and reliability in clinical settings
Grad-CAM (2024) [19]	breast cancer: Kaggle Ultrasound dataset; PMAD-LinkNet for segmentation (98.1%) + CSFEC-Net for classification (99.2%)	High performance with novel frameworks; potential risk of overfitting remains
Not reported (2024) [20]	breast cancer: breast tissue datasets; Novel CNN with Early Stopping and ReduceLROnPlateau; 95.2% accuracy	Reliable across datasets; limitations in accuracy and efficiency
Visual interpretability vs. Grad-CAM (2024) [21]	breast cancer: Duke X-rays dataset; FPN-IAIA-BL prototype model; AUROC 0.88	Better localization than Grad-CAM, but lower AUROC than baselines; some prototype limitations

By Image Type. Most studies focused on digital mammograms, while one study examined ultrasound data, and another used digital features extracted from fine needle aspiration (FNA) cytology.

By Models. The studies used advanced CNN models (such as VGG16, ResNet, and InceptionV3), as well as new models such as FPN-IAIA-BL and CSFEC-Net.

The SNN model was also featured in a study considered a pioneer in neurobiological simulation.

In Terms of Interpretation Techniques (XAI). Several studies have used interpretation tools such as Grad-CAM or self-interpreting models (such as FPN-IAIA-BL) that provide visual representations of diagnostic features. One study incorporated the Lempel-Ziv Complexity Scale to improve the interpretation of neural activity patterns.

Performance. Accuracy ranged from 95.2% to 99.2%, demonstrating the effectiveness of these models in classification. Some models were distinguished by their ability to provide visual interpretation that helps clinicians understand diagnostic decisions.

Advantages and Limitations. The most prominent advantages refer to improved accuracy, faster diagnosis, and improved transparency thanks to XAI techniques. Challenges often relate to the complexity and difficulty of training models, as well as poor interpretability in some cases.

3.3   COVID-19

Table 4 offers a comprehensive overview of recent studies that use deep learning (DL) and explainable artificial intelligence (XAI) techniques to diagnose COVID-19 from chest radiograph images. These studies showcase a variety of convolutional neural network (CNN) models, ranging from lightweight, custom-designed models (such as LW-CORONet and COVID-XNet) to advanced models that combine networks like ResNet, DenseNet, Inception, and transformer-based systems. The integration of XAI methods in COVID-19 diagnosis not only improves model interpretability but also provides visual justifications that support clinical decision-making, particularly in distinguishing COVID-19 from other respiratory conditions. Several interpretability techniques, including Grad-CAM [9], layer activation maps (CAM), saliency maps, and DeepLIFT, have been employed to enhance model transparency and support clinical decision-making. Some models have achieved remarkable precision, reaching up to 99% in certain binary classification tasks and successfully identifying key lung regions associated with COVID-19 infection.

Table 4. Summary of Deep Learning and Explainable AI Studies for COVID-19 Detection Using Chest X-ray Imaging.

XAI Method	Disease, Dataset, DL Model & Performance	Highlights and Limitations
Interpretable feature analysis (2025) [22]	COVID-19; two chest X-ray datasets; features extracted via DC-GLM and CAMSGNeT, analyzed using optimized sequential neural network; Accuracy: 98.27% and 100%	Captures key COVID-19 indicators (ground-glass opacity, consolidation, fibrosis) DC-GLM for coarse textures, CAMSGNeT for fine textures with adaptive diffusion; feature importance analysis improves interpretability; outperforms state-of-the-art models; resource-efficient and precise
Transfer learning (2025) [23]	COVID-19; two chest CT datasets (healthy, normal, pneumonia); GoogleNet, ResNet50, VGG16; ResNet50 achieved 96% accuracy	Benchmarking multiple architectures for COVID-19 detection; ResNet50 gave best performance; Transfer learning applied; interpretability not explicitly reported
Occlusion, Saliency, Gradients, Guided BP, DeepLIFT (Local); Neuron activation profiles (Global), (2024), [24]	COVID-19, pneumonia, healthy subjects (dataset not specified); ResNet18/34, InceptionV3, InceptionResNetV2, DenseNet161; ensemble via majority voting; F1: 0.66–0.875 (models), 0.89 (ensemble)	ResNets most interpretable; interpretability used to compare models; included multi-label classification
Custom explainable CNN, (2023), [25]	COVID-19; CXR dataset (not named); CNN-based model; Accuracy up to 97%	Promising accuracy for COVID-19/variants; extracts key features; supports diagnosis
Grad-CAM, (2022), [26]	COVID-19, pneumonia, normal; CXR from Figshare and COVIDx-V7A; LW-CORONet vs VGG-19, ResNet-101, DenseNet-121, Xception; Accuracy: 98.6799% (Dataset-1), 95.6796.25% (Dataset-2)	Lightweight model for real-time use; effective feature extraction; high accuracy; Grad-CAM visual aid; limitation: limited COVID-19 data

Despite these promising results, some limitations still exist, particularly regarding dataset availability, model generalizability, and the absence of standardized external testing criteria. Additionally, some models were trained on non-specific or relatively small datasets, which can affect the reproducibility and usability of the results in real-world clinical settings.

4   Discussion

The review of studies reveals that deep interpretable learning techniques are crucial for developing accurate and reliable medical diagnostic tools for three key diseases: diabetic foot ulcers, breast cancer, and COVID-19. In the field of foot ulcers, models show a strong ability to identify infection sites, providing valuable clinical support for specialists. In breast cancer, techniques such as Grad-CAM offer precise visual explanations, which are also evaluated quantitatively, for example, using the Hausdorff distance, thereby increasing doctors' confidence in the model results. For COVID-19, models can distinguish between affected and infected cases with high accuracy using chest and CTA images, while supporting the interpretation of results through various XAI technologies that highlight inflamed and lung areas. However, several significant challenges must be addressed before these models can be fully applied in clinical settings.

Prominent Among Them Are:

- Limited dataset diversity: Most studies rely on specific datasets, which may limit generalizability across populations and imaging modalities.
- Lack of external validation: Few studies evaluate independent cohorts, reducing confidence in model robustness.
- Absence of standardized XAI metrics: There is no uniform framework to assess the quality, reliability, and clinical relevance of the explanations provided by XAI techniques.

Future Directions to Address These Limitations Include:

- Data diversification: Incorporating multi-institutional datasets, multimodal imaging, and larger sample sizes to improve model generalizability.
- External validation and benchmarking: Testing models on independent datasets with standardized evaluation protocols to ensure reliability.
- Standardized interpretability metrics: Developing and adopting uniform measures for XAI quality to facilitate clinical acceptance and regulatory approval.
- Integration of multiple XAI methods: Combining different interpretability approaches to provide richer, complementary explanations for clinicians.
- Clinical trials and deployment studies: Evaluating the real-world impact of XAI-assisted diagnosis on decision-making, treatment planning, and patient outcomes.

5 Conclusion

This systematic review emphasizes the important role of deep learning techniques in analyzing medical images related to diabetic foot ulcers, breast cancer, and COVID-19. These models have shown strong diagnostic accuracy, while explainability methods like Grad-CAM, SHAP, and LIME have increased physician trust by offering clear visual explanations for predictions. Despite this progress, several challenges still exist. These include limited data diversity, a lack of external validation in independent cohorts, and the absence of standardized criteria to assess interpretability. Tackling these issues through expanding datasets, conducting robust external validations, and developing unified evaluation metrics will greatly improve the clinical usefulness and trustworthiness of these models. In summary, this study provides a thorough overview of recent advances in explainable deep learning for medical image analysis, pointing out key directions for future research aimed at enhancing reliability, transparency, and clinical implementation.

Acknowledgments. This publication is part of the projects PID2020-117954RB-C22 and PID2023-146037OB-C21, funded by MICIU/AEI/10.13039/501100011033/.

References

1. Rasool, N., Bhat, J.I.: Unveiling the complexity of medical imaging through deep learning approaches. Chaos Theory Appl. **5**(4), 267–280 (2023)
2. Thakur, G.K., Thakur, A., Kulkarni, S., Khan, N., Khan, S.: Deep learning approaches for medical image analysis and diagnosis. Cureus **16**(5) (2024)
3. Dwivedi, R., et al.: Explainable AI (XAI): core ideas, techniques, and solutions. ACM Comput. Surv. **55**(9), 1–33 (2023)
4. Singh, A., Sengupta S., Lakshminarayanan, V.: Explainable deep learning models in medical image analysis. J. Imaging **6**(6), 52 (2020)
5. Salahuddin, Z., Woodruff, H.C., Chatterjee, A., Lambin, P.: Transparency of deep neural networks for medical image analysis: a review of interpretability methods. Comput. Biol. Med. **140**, 105111 (2022)
6. Patrício, C., Neves, J.C., Teixeira, L.F.: Explainable deep learning methods in medical image classification: a survey. ACM Comput. Surv. **56**(4), 1–41 (2023)
7. Bhati, D., Neha, F., Amiruzzaman, M.: A survey on explainable artificial intelligence (XAI) techniques for visualizing deep learning models in medical imaging. J. Imaging **10**(10), 239 (2024)
8. Hou, J., et al.: Self-explainable AI for medical image analysis: a survey and new outlooks. arXiv preprint arXiv:2410.02331 (2024)
9. Selvaraju, R.R., Cogswell, M., Das, A., Vedantam, R., Parikh, D., Batra, D.: Grad-cam: visual explanations from deep networks via gradient-based localization. In: Proceedings of the IEEE International Conference on Computer Vision, pp. 618–626. IEEE (2017)
10. Nikolić, M., Janković, D., Stanimirović, A., Stoimenov, L.: The integration of explainable AI methods for the classification of medical image data. In: 2024 11th International Conference on Electrical, Electronic and Computing Engineering (IcETRAN), pp. 1–6. IEEE (2024)

11. Moon, J.-E., Cho, Y.-J., Rhyou, S.-Y., Hong, S.-H.: DiabetesNet: lightweight and efficient AI-based solution for real-world diabetic foot classification. J. Internet Comput. Serv. **26**(2), 167–178 (2025)
12. Karthik, R., Ajay, A., Jhalani, A., Ballari, K., K, S.: An explainable deep learning model for diabetic foot ulcer classification using swin transformer and efficient multi-scale attention-driven network. Sci. Rep. **15**(1), 4057 (2025)
13. Rathore, P.S., Kumar, A., Nandal, A., Dhaka, A., Sharma, A.K.: A feature explainability-based deep learning technique for diabetic foot ulcer identification. Sci. Rep. **15**(1), 6758 (2025)
14. Arkah, Z.M., Pontes, B., Rubio, C.: Interpretation of diabetic foot ulcer image classification using layer attribution algorithms. In: International Conference on Soft Computing Models in Industrial and Environmental Applications, pp. 13–22. Springer, Cham (2024)
15. Biswas, S., Mostafiz, R., Paul, B.K., Uddin, K.M.M., Hadi, M.A., Khanom, F.: DFU_XAI: a deep learning-based approach to diabetic foot ulcer detection using feature explainability. Biomed. Mater. Dev. **2**(2), 1225–1245 (2024)
16. Rubavathi, C.Y., Diofrin, J.: Diabetes foot ulcer diagnosis using fast convolution neural network. In: 2023 International Conference on Networking and Communications (ICNWC), pp. 1–5. IEEE, Chennai, India (2023)
17. Rudnicka, Z., Szczepanski, J., Pregowska, A.: Integrating Complexity and Biological Realism: High-Performance Spiking Neural Networks for Breast Cancer Detection. arXiv preprint arXiv:2506.06265 (2025)
18. Murugan, T.K., Karthikeyan, P., Sekar, P.: Efficient breast cancer detection using neural networks and explainable artificial intelligence. Neural Comput. Appl. **37**(5), 3759–3776 (2025)
19. Venkatraman, S., Malarvannan, S.: Exploiting Precision Mapping and Component-Specific Feature Enhancement for Breast Cancer Segmentation and Identification. arXiv preprint arXiv:2407.02844 (2024)
20. Thakur, A., et al.: Transformative breast cancer diagnosis using CNNs with optimized ReduceLROnPlateau and early stopping enhancements. Int. J. Comput. Intell. Syst. **17**(1), 1–18 (2024)
21. Yang, J., et al.: FPN-IAIA-BL: a multi-scale interpretable deep learning model for classification of mass margins in digital mammography. In: Proceedings of the IEEE/CVF Conference on Computer Vision and Pattern Recognition, pp. 5003–5009 (2024)
22. Prince, R., et al.: Interpretable COVID-19 chest X-ray detection based on hand-crafted feature analysis and sequential neural network. Comput. Biol. Med. **186**, 109659 (2025)
23. Nauman, M., et al.: Deep transfer learning for COVID-19 screening: benchmarking ResNet50, VGG16, and GoogleNet on chest X-ray images. Algorithms **1**(2), 69–83 (2025)
24. Chatterjee, S., et al.: Exploration of interpretability techniques for deep covid-19 classification using chest X-ray images. J. Imaging **10**(2), 45 (2024)
25. Mathesul, S., et al.: COVID-19 detection from chest X-ray images based on deep learning techniques. Algorithms **16**(10), 494 (2023)
26. Nayak, S.R., Nayak, D.R., Sinha, U., Arora, V., Pachori, R.B.: An efficient deep learning method for detection of COVID-19 infection using chest X-ray images. Diagnostics **13**(1), 131 (2022)

Ethics, Privacy and Security in AI Applications in the Healthcare

Fairness-Aware Machine Learning for Biomedical Prediction: Evaluating and Correcting Bias in Gallstone Diagnosis Models

Caroline König[(✉)] [iD], Martha Ivon Cardenas [iD], Pedro Jesús Copado [iD], and Alfredo Vellido [iD]

Soft Computing Research Group (SOCO) at Intelligent Data Science and Artificial Intelligence (IDEAI-UPC) Research Centre, Universitat Politècnica de Catalunya (UPC Barcelona Tech), Jordi Girona 1–3, 08034 Barcelona, Spain
`caroline.leonore.konig@upc.edu`

Abstract. This study presents a fairness-aware machine learning framework for biomedical prediction, illustrated through gallstone disease detection. Using a structured clinical dataset of 320 patients, predictive models, including multilayer perceptron and random forest, are developed via stratified 5-fold cross-validation. Model performance is evaluated using AUC, F1-score, precision, and recall, with subgroup fairness assessed through demographic parity, disparate impact, predictive parity difference, equal opportunity, and equalized odds. Bias linked to age and gender is quantified and corrected using threshold optimization and F1-based recalibration. Postprocessing interventions demonstrate improved fairness metrics while preserving diagnostic accuracy. Our methodology aligns with ethical guidelines and technical expectations under the EU AI Act, emphasizing equitable model behavior in clinical decision-making. The proposed framework contributes to responsible AI deployment in healthcare, supporting inclusion and transparency in high-risk biomedical applications.

Keywords: Fairness-Aware Machine Learning · Bias Mitigation · Gallstein diagnosis

1 Introduction

Artificial intelligence (AI) and machine learning (ML) have become transformative forces in biomedicine, enabling diagnostic tools and predictive models that assist clinicians in decision-making. Over the past decade, ML-based systems have evolved from experimental prototypes to widely regulated medical technologies. A large amount of AI-enabled medical devices have been approved for clinical use in recent years, particularly in radiology, cardiology, and neurology [15]. These advancements affirm the growing clinical significance of AI, while

A. López Fernández et al. (Eds.): CIABiomed 2025, LNBI 16148, pp. 59–73, 2026.
https://doi.org/10.1007/978-3-032-10661-2_5

simultaneously emphasizing the critical importance of evaluating its fairness, particularly in high-risk medical applications governed by the EU AI Act [4]. Fairness in ML refers to the equitable performance of models across different demographic groups, avoiding discriminatory outcomes in relation to sensitive attributes such as gender, ethnicity, age, or disability. This principle is particularly critical in healthcare, where unjust disparities could directly affect diagnosis or treatment [8]. Studies have shown that ML algorithms in healthcare can inadvertently reproduce and even increase bias. One example is an algorithm that gave black patients lower priority because it used healthcare costs to measure need, which turned out to be misleading [12]. In medical imaging, researchers found that chest X-ray models can detect a patient's race from unlabeled data [10], which raises serious concerns about how these systems make decisions and whether they might reinforce bias.

In Europe, fairness intersects with legal protections as attributes historically associated with discrimination, such as sex, race, or religion, are considered protected by non-discrimination laws [5]. Ensuring fairness means not only detecting bias but also determining its admissibility within the clinical context. Biological differences may justify certain variations in model output, whereas disparities stemming from data imbalance or societal bias require mitigation [1,11].

Gallstone disease is one of the most common gastrointestinal disorders globally, affecting approximately 1015% of adults in Europe and the U.S. [13]. While up to 80% of patients are asymptomatic, around 20% develop symptoms such as biliary colic, cholecystitis, or pancreatitis, which can lead to emergency hospitalization and surgical intervention [9]. Early diagnosis is crucial to prevent complications and reduce healthcare burden, especially since gallstones are a leading cause of gastrointestinal-related hospital admissions in developed countries [14]. This study applies fairness analysis to evaluate neutrality in ML models for gallstone diagnosis. By examining performance across demographic subgroups and calibrating decision thresholds, we aim to address both ethical concerns, such as bias linked to gender or ethnicity, and technical challenges in deploying AI responsibly in biomedical settings. This approach supports more equitable diagnostic support and aligns with regulatory expectations under frameworks like the EU AI Act.

2   Materials

Research on gallstone diagnosis is increasingly relevant due to the high prevalence and potential severity of gallstone-related conditions. Recent studies have applied ML to non-invasive data, such as bioimpedance and laboratory tests, to enable earlier detection, minimize unnecessary imaging, and improve clinical decision-making [3]. The dataset analyzed in this study originates from work conducted at Ankara VM Medical Park Hospital [3], and is publicly available via the UC Irvine Machine Learning Repository. The dataset comprises data from 320 patients, being 161 gallstone positive patients and 158 healthy controls. According to the authors the dataset comprises a broad range of attributes covering demographics and clinical information such as age, sex, height, weight,

body mass index (BMI), gallstone status, and comorbidities including hypertension, hypothyroidism, hyperlipidemia, and diabetes mellitus. Body composition
parameters include lean mass, muscle mass, bone mass, total and visceral fat
content, visceral muscle area, degree of obesity, protein content, and the visceral adiposity index. Hydration status is assessed through measurements of
total body water (TBW), extracellular water (ECW), and intracellular water.
Metabolic and liver health indicators, renal and inflammatory markers as well
as several biomarkers are present in the dataset.

3 Methods

The analysis conducted on the dataset centers on assessing the fairness and
potential bias of ML-based predictive models for gallstone status. Initially, state-
of-the-art ML algorithms were developed using a methodology aligned with that
proposed by [3]. Following model development and performance evaluation, the
most accurate predictive models were systematically examined for fairness with
respect to sensitive attributes such as age and gender. Finally, post-processing
bias mitigation techniques were applied to reduce any identified disparities.

3.1 Predictive Models

A set of predictive models for gallstein status prediction were constructed comprising the following ML models as proposed by the authors of the original
work on the dataset: K-nearest neighbors (KNN), multilayer perceptron (MLP),
support vector machines (SVM), decision tree (DT), random forest (RF), adaptive boost ing (ADA), gradient boosting (GB), naïve Bayes (NB), and logistic
regression (LR). In this study, a different model construction approach is applied
as each ML model is trained using stratified 5-fold cross validation to report generalizable results, while in the original study a fixed train and test split was used.
The model performance is evaluated using the same metrics as in the original
study, namely using the Area under the curve (AUC), precision (Prec), recall
(Rec) and F1-score (F1).

3.2 Fairness Evaluation Methods

After ML models were trained for prediction they were analyzed with regard for
bias and fairness on sensitive attributes. According to the definition of sensitive
attributes gender and age are taken into account in this study [5]. Group fairness metrics assess global model behavior across subgroups defined by sensitive
attributes. These metrics are commonly divided into parity-based and confusion
matrix-based metrics and defined as follows based on the notation: (i) $\hat{Y} = 1$:
Model predicts positive value. (ii) $Y = 1$: True label is positive. (iii) A: Sensitive
attribute (e.g. gender or age group) which can take values for the privileged (pr)
or unprivileged (upr) group.

Parity-Based Metrics

- **Statistical/Demographic Parity:** Ensures each group has an equal probability of receiving the positive label.

$$\Pr(\hat{Y} = 1 | A = pr) = \Pr(\hat{Y} = 1 | A = upr)$$

- **Disparate Impact:** Evaluates the ratio of positive classification probabilities between unprivileged and privileged groups. A value below 0.8 indicates adverse impact.

$$\frac{\Pr(\hat{Y} = 1 | A = pr)}{\Pr(\hat{Y} = 1 | A = upr)}$$

Confusion Matrix-Based Metrics. These metrics incorporate the true positive rate (TPR), false positive rate (FPR), true negative rate (TNR), and false negative rate (FNR).

- **Positive Predictive Disparity** (PPD) is a fairness metric that compares the precision, or Positive Predictive Value (PPV), of a model across different sensitive groups. It tells whether one group receives more accurate positive predictions than another.

$$PPD = P(Y = 1 | \hat{Y} = 1, A = upr) - P(Y = 1 | \hat{Y} = 1, A = pr)$$

- **Equal Opportunity:** Equal true positive rates (TPR) across groups [7].

$$\Pr(\hat{Y} = 1 | Y = 1, A = pr) = \Pr(\hat{Y} = 1 | Y = 1, A = upr)$$

- **Equalized Odds:** Equal TPR and FPR across groups.

$$Pr(\hat{Y} = 1 | Y = 1, A = pr) = \Pr(\hat{Y} = 1 | Y = 1, A = upr)$$

$$Pr(\hat{Y} = 1 | Y = 0, A = pr) = \Pr(\hat{Y} = 1 | Y = 0, A = upr)$$

Importantly, statistical parity and disparate impact metrics do not account for ground truth labels, and thus reflect bias in the assignment of positive predictions rather than model accuracy. However, fairness in error distribution is also assessed through confusion-matrix metrics such as predictive parity, equal opportunity difference, and equalized odds difference. These metrics evaluate different aspects of fairness to determine whether the models perform consistently across subgroups.

- **Predictive Parity** evaluates the PPV (precision) for each group and evaluates when the model predicts a positive outcome, it is equally likely to be correct for all groups. Lower Predictive Parity Difference (PPD) scores indicate higher fairness by minimizing precision disparities between subgroups.
- **Equal Opportunity** assesses the model's ability to identify actual positive cases (TPR) across groups. A fair model should be equally sensitive to identifying true cases regardless of the subgroup.

– **Equalized Odds** takes a broader view by ensuring that both TPR and FPR are balanced across groups. This metric accounts for both correct detections and mistaken predictions, ensuring that fairness takes into account all aspects of model error.

To ensure equitable outcomes across protected demographic groups, ML models must be fair with respect to sensitive attributes [4]. One widely recognized framework for assessing unintended bias is the doctrine of disparate impact [6]. Although not legally binding, the "80% rule" endorsed by the U.S. Equal Employment Opportunity Commission (EEOC) [2] has gained widespread adoption as a practical rule. This guideline stipulates that the selection rate for protected groups must be at least 80% of that for non-protected groups, reducing the risk of disproportionate disadvantage in AI-driven decision-making. As formalized in [16], disparate impact is quantified by an impact ratio (IR), calculated as the ratio between the selection rate (SR) of the unprivileged group b and that of the privileged group a. A commonly accepted threshold for bias tolerance, denoted by ϵ, is 0.8 aligning with the 80% rule [2]. The acceptable amount of bias is therefore an IR within the range $[\epsilon; \frac{1}{\epsilon}]$.

$$SR = \frac{Favorable(attribute)}{Total(attribute)} \tag{1}$$

$$IR = \frac{Unprivileged\_SR}{Privileged\_SR} \tag{2}$$

3.3 Bias Mitigation

To mitigate bias in ML predictions, this study employs a postprocessing approach designed to improve fairness without modifying the model architecture or retraining. Postprocessing techniques operate on the output of trained models, adjusting predicted labels based on fairness constraints such as *Equal Opportunity* and *Equalized Odds*. This common strategy consistis in recalibrating decision probabilities, applying group-specific classification thresholds to favor underrepresented or disadvantaged subgroups.

In our implementation, we specifically target disparities in the *True Positive Rate* (TPR) and F1-score across composite demographic groups defined by gender and age. Starting with predictions from standard classifiers, we define subgroups by combining age intervals and gender categories. For each subgroup, we compute optimal decision thresholds in two ways: (1) to align its TPR with the global TPR observed in the overall population therefore addressing *Equal Opportunity*, and (2) to maximize the F1-score within the subgroup therefore ensuring balanced trade-off between precision and recall, and implicitly accounting for both TPR and FPR considered in *Equalized Odds*. These thresholds are then applied post hoc to the subgroup's prediction scores, yielding fairness-adjusted classification outcomes. The advantages of the postprocessing approach include flexibility, interpretable adjustments, and compatibility with any probabilistic classifier. Furthermore, it enables fairness improvements without compromising the underlying training pipeline or introducing significant model complexity.

4 Result

4.1 Predictive Models

Table 1 reports the classification performance of the models trained using a 5-fold cross-validation strategy. For each classifier, the AUC, precision, recall, and F1-score are presented as the mean and standard deviation across the five iterations. The results indicate that both RF and MLP models outperform the other classifiers, achieving an average AUC of approximately 0.85, precision of 0.80, and recall exceeding 0.77. This marks a methodological improvement over the original study by [3], which employed a fixed train-test split. By applying cross-validation, this study provides a more robust and randomized evaluation of model performance. In the context of model selection, the most effective RF configuration employed the Gini index for node splitting, an ensemble of 1,000 decision trees, and unrestricted maximum depth. The optimal MLP architecture consisted of a single hidden layer with 75 neurons, using the ReLU activation function and training via the Adam optimization algorithm.

Table 1. Model Performance Metrics

Model	AUC	Precision	Recall	F1-score
DT	0.670 ± 0.05	0.675 ± 0.03	0.639 ± 0.09	0.655 ± 0.06
RF	0.851 ± 0.03	0.807 ± 0.03	0.773 ± 0.07	0.789 ± 0.05
MLP	0.859 ± 0.04	0.815 ± 0.07	0.785 ± 0.08	0.794 ± 0.05
LR	0.802 ± 0.05	0.771 ± 0.05	0.677 ± 0.10	0.718 ± 0.07
NB	0.689 ± 0.03	0.816 ± 0.11	0.177 ± 0.06	0.286 ± 0.08
KNN	0.662 ± 0.08	0.605 ± 0.07	0.545 ± 0.07	0.573 ± 0.07
SVM	0.635 ± 0.06	0.582 ± 0.06	0.614 ± 0.10	0.595 ± 0.07
GB	0.832 ± 0.04	0.775 ± 0.03	0.722 ± 0.07	0.746 ± 0.05
AB	0.818 ± 0.04	0.771 ± 0.05	0.716 ± 0.08	0.740 ± 0.05
XGB	0.836 ± 0.05	0.772 ± 0.04	0.767 ± 0.07	0.768 ± 0.05

4.2 Fairness Analysis by Metric Across Demographic Subgroups

As a next step, the best-performing models, RF and MLP, were subjected to bias analysis with respect to gallstone status prediction. To assess potential gender-based disparities in predictive performance, both classifiers were evaluated with several fairness metrics. Age and gender were selected as sensitive attributes based on regulatory definitions and ethical considerations. To ensure robust and fair evaluation, the RF and MLP were trained on 80% of the dataset and validated on the remaining 20% using stratified sampling with respect to the prediction target and sensitive attributes (gender and age group). The hyper-parameters for each model were fixed based on prior optimization during model selection. The test set was designed to maintain equitable distribution across relevant factors: gallstone outcomes (32 positive, 32 negative), gender (34 male, 30 female), and age (mean = 47.65 years, standard deviation = 11.07 years). Table 2 presents the performance metrics of RF and MLP classifiers on the balanced test set. The results indicate that RF slightly outperforms MLP, exhibiting higher precision and AUC.

Table 2. Model Performance Metrics on Test Set

Metric	RF	MLP
AUC	0.895	0.819
Prec	0.821	0.750
Rec	0.719	0.750
F1	0.767	0.750

Table 3. Fairness metrics across age (A) and gender (G) subgroups for RF and MLP.

Metric	RF (G)	MLP (G)	RF (A)	MLP (A)
Statistical Parity Difference	0.431	0.126	0.250	0.063
Disparate Impact	2.833	1.284	1.800	1.133
Predictive Parity Difference (PPV)	−0.075	0.282	−0.122	0.031
Equal Opportunity Difference (TPR)	0.304	0.097	0.404	0.220
Equalized Odds Difference (TPR)	0.304	0.097	0.404	0.220
Equalized Odds Difference (FPR)	0.316	−0.104	0.169	−0.051

The predictions of the models trained to predict gallstone status were subsequently analyzed across gender and age groups to quantify fairness. Metrics such as statistical parity difference, disparate impact, predictive parity difference, equal opportunity difference, and equalized odds were computed to identify possible bias in model outcomes.

Table 3 presents the parity-based fairness metrics used to evaluate bias in gallstone status prediction for the RF and MLP classifiers. The analysis reveals evidence of bias in both models' predictions, particularly when assessed across sensitive groups defined by age and gender. The statistical parity values for RF and MLP are 0.431 and 0.126, respectively, indicating moderate imbalance in the assignment of positive predictions across gender groups. RF shows greater disparity than MLP. The corresponding disparate impact ratios 2.833 (RF) and 1.284 (MLP) remain above the inverse threshold of 1.25 established by the 4/5th rule, suggesting the existence of some gender based bias. For age groups, the statistical parity scores are 0.25 (RF) and 0.063 (MLP), again pointing to modest disparities in positive prediction rates. Disparate impact values of 1.8 (RF) and 1.133 (MLP) reflect a slightly reduced likelihood of younger individuals being predicted as having gallstones. The disparate impact of RF slightly surpasses the acceptable fairness bounds established by the 4/5th rule and points out the existence of bias on age in RF for the positive outcome prediction.

Statistical parity and disparate impact reflect bias in positive predictions, while confusion-matrix metrics reveal fairness in model error distribution. To asses the fairness properties related to prediction precision and error distribution, Predictive Parity, Equal Opportunity, and Equalized Odds are evaluated.

Predictive Parity evaluates how accurate positive predictions are within each subgroup. RF shows small disparities in precision, namely −0.075 for gender and −0.122 for age, indicating slightly reduced reliability of positive outcomes for certain groups. In contrast, MLP exhibits a larger gender-based disparity (0.282) and a minimal age-based disparity (0.031), suggesting its predictions are more consistent across age than gender.

Equal Opportunity focuses on whether true cases are detected equally across subgroups. RF reveals substantial discrepancies in sensitivity, namely 0.304 for gender and 0.404 for age, pointing to unequal true positive rates across these groups. MLP shows improved balance with smaller TPR differences, namely 0.097 for gender and 0.220 for age, indicating fairer detection sensitivity.

Equalized Odds considers both sensitivity and false positive rates. RF displays relatively high disparity, in particular TPR diff = 0.304 and FPR diff = 0.316 for gender and TPR diff = 0.404 and FPR diff = 0.169 for age. These indicate uneven model behavior across demographic groups. MLP, however, achieves better fairness by gender (TPR diff = 0.097, FPR diff = −0.104) compared to age (TPR diff = 0.220, FPR diff = −0.031).

The fairness evaluation of predictive models revealed disparities across gender and age groups in gallstone status classification. While parity-based metrics revealed slight differences in positive prediction distributions, which possibly reflect inherent biological variations, confusion-matrix metrics uncovered more consequential biases in error distribution. Notably, RF exhibited stronger prediction performance at the cost of higher fairness disparities, especially in sensitivity across age, whereas MLP delivered more balanced error rates with mild inconsistencies in precision for gender.

4.3   Bias Mitigation

To address fairness discrepancies observed across gender and age for the RF and MLP models a fairness-aware postprocessing technique was applied. This approach adjusts decision thresholds for distinct demographic subgroups, mitigating bias without retraining the original classifiers. Age was discretized into two categories, Young (0) and Older (1), and combined with gender to define four composite groups: MaleYoung, MaleOlder, FemaleYoung, and FemaleOlder. For each subgroup, a custom decision threshold was calculated based on two optimization criteria:

1. Alignment of the subgroup's *True Positive Rate* (TPR) with the overall model's TPR, promoting *Equal Opportunity* and reducing disparities in clinical sensitivity.
2. Maximization of the subgroup's *F1-score*, accounting for both TPR and FPR, and thus balancing precision and recall at the subgroup level.

The subgroup-specific thresholds found by this strategy are applied directly to predicted probabilities, producing fairness-aware classification outcomes. By incorporating both F1-score optimization and TPR alignment, the approach enables decision boundaries that balance fairness and performance, accounting for the distribution of true and false positives across diverse groups.

Postprocessing Threshold Calibration for Equal Opportunity. The fairness aware postprocessing method was applied to the MLP and RF classifier with the objective of aligning each demographic subgroup's TPR with the model's overall TPR. Probability scores were generated using the trained model and analyzed separately for each demographic subgroup within the test set. Initial predictions were made using the default decision threshold of 0.5. At this threshold, the RF model exhibited a TPR of 0.719 and a FPR of 0.156, while the MLP model yielded a TPR of 0.75 and an FPR of 0.25. These globally observed TPR and FPR values served as fairness benchmarks for subsequent subgroup calibration. Figure 1 illustrates the corresponding ROC curves of both classifiers.

To align subgroup performance with the global TPR, threshold optimization was conducted by adjusting each subgroup's operating point on the ROC curve. A search algorithm was used to identify thresholds that closely matched the benchmark TPR, thereby enhancing *Equal Opportunity* across gender and age combinations. These calibrated thresholds were then applied post hoc to the predicted probabilities, yielding fairness-adjusted classification outcomes. Figures 2 and 3 illustrate the ROC curves of the classifiers for each subgroup, with markers indicating the adjusted thresholds.

Table 4 presents the performance of the RF and MLP models after applying subgroup-specific decision thresholds on the test set. The adoption of subgroup-specific decision thresholds led to notable improvements in model precision and F1-score compared to the models created with uniform thresholding approach (Table 2). For the RF model, precision increased from 0.821 to 0.960, and F1-score rose from 0.767 to 0.842, indicating a more confident and balanced classification performance. Similarly, the MLP model showed gains especially in precision improving from 0.750 to 0.821. These enhancements suggest that threshold recalibration successfully reduced false positives while maintaining sufficient recall. However, recall dropped slightly for the MLP model, pointing to a trade-off where higher precision came at the cost of missing a few true cases. Notably, AUC values remained consistent, confirming that performance gains stemmed from better threshold alignment rather than improved separability. The implementation of this fairness-aware thresholding techniques led to a measurable

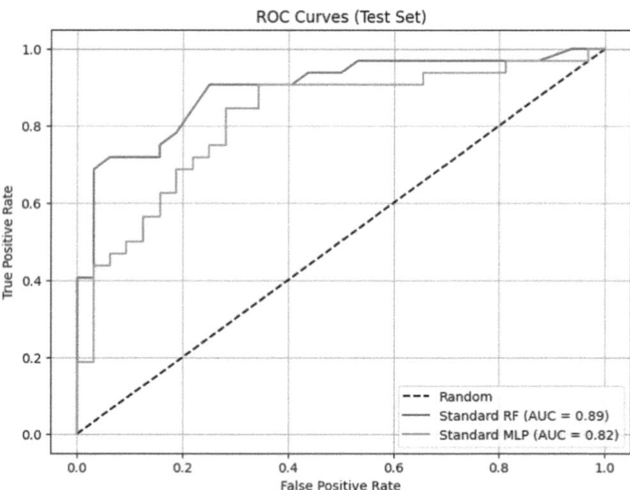

Fig. 1. ROC curve for the RF and MLP classifier.

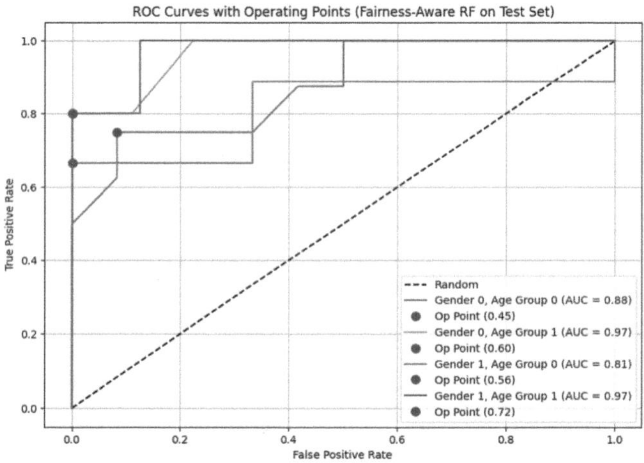

Fig. 2. RF ROC curve illustrating operation point adjustment via TPR calibration. Selected thresholds are annotated across demographic subgroups.

reduction in bias across the majority of evaluated fairness metrics (Table 5). For the MLP model, gender-based statistical parity improved markedly from 0.153 to 0.055 suggesting more equitable distributions of positive predictions. Disparate Impact also decreased substantially, especially for age, indicating reduced predictive imbalance. Nevertheless, some disparities persist in the MLP model with regard to sensitivity evaluated by Equal Opportunity and false positives evaluated with Equalized Odds.

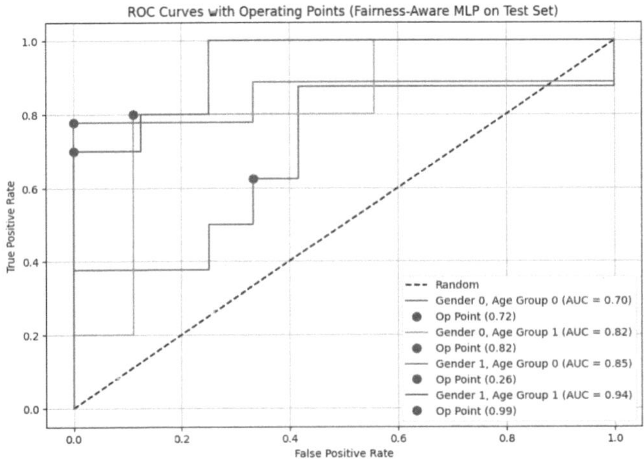

Fig. 3. MLP ROC curve illustrating operation point adjustment via TPR calibration. Selected thresholds are annotated across demographic subgroups.

Table 4. Model Performance Metrics after TPR calibration. Improvements with respect to uniform thresholding highlighted in bold.

Metric	RF	MLP
AUC	0.895	0.819
Prec	**0.960**	**0.821**
Rec	**0.750**	0.719
F1	**0.842**	**0.767**

F1-Optimized Threshold Recalibration. Threshold calibration based on F1-maximization seeks to balance sensitivity and specificity by jointly considering the TPR and FPR. Since the F1-score is the harmonic mean of precision and recall, optimizing for it inherently addresses the trade-off between these two key performance metrics. Figures 4 and 5 illustrate the ROC curves of the classifiers for each subgroup, with markers indicating the adjusted thresholds. Models' performance is described in Tables 6 and 7. The F1-optimized recalibration consistently improved the models' performance and at the same time yielded lower disparities across all evaluated metrics. Specifically, Equal Opportunity and Equalized Odds differences are reduced by approximately 50% compared to uniform thresholding for both models. This indicates enhanced fairness in both sensitivity (TPR) and error rates (FPR) across demographic groups. Additionally, Positive Predictive Value (PPV) disparities are reduced, suggest-

Table 5. Fairness metrics across age (A) and gender (G) subgroups for RF and MLP after TPR calibration. Improvements highlighted in bold.

Metric	RF (G)	MLP (G)	RF (A)	MLP (A)
Statistical Parity Difference	**0.143**	**0.055**	**−0.031**	−0.125
Disparate Impact	**1.443**	**1.133**	**0.923**	**0.750**
Predictive Parity Difference (PPV)	0.091	0.357	**0.077**	0.167
Equal Opportunity Difference (TPR)	**−0.032**	**0.045**	**0.094**	**0.028**
Equalized Odds Difference (TPR)	**−0.032**	**0.045**	**0.094**	**0.028**
Equalized Odds Difference (FPR)	**−0.048**	−0.238	**−0.067**	−0.208

ing more equitable precision. Notably, the overall F1-score improves under the F1-calibrated approach. These results confirm that F1-based threshold tuning is an effective method for enhancing fairness and accuracy.

Table 6. Model Performance Metrics of RF and MLP after F1-optimized recalibration. Improvements with respect to uniform thresholding highlighted in bold.

Metric	RF	MLP
AUC	0.895	0.819
Precision	**0.903**	**0.763**
Recall	**0.875**	**0.906**
F1-score	**0.889**	**0.829**

Table 7. Fairness metrics across age (A) and gender (G) subgroups for RF and MLP after F1-optimized recalibration. Improvements hgihlighted in bold.

Metric	RF (G)	MLP (G)	RF (A)	MLP (A)
Statistical Parity Difference	**0.343**	0.200	**−0.031**	−0.1250
Disparate Impact	**2.061**	1.400	**0.938**	**0.810**
Predictive Parity Difference (PPV)	**−0.009**	**0.210**	**0.058**	0.109
Equal Opportunity Difference (TPR)	**0.178**	0.101	**0.110**	**0.051**
Equalized Odds Difference (TPR)	**0.178**	0.101	**0.110**	**0.051**
Equalized Odds Difference (FPR)	**0.134**	**−0.013**	**−0.075**	−0.224

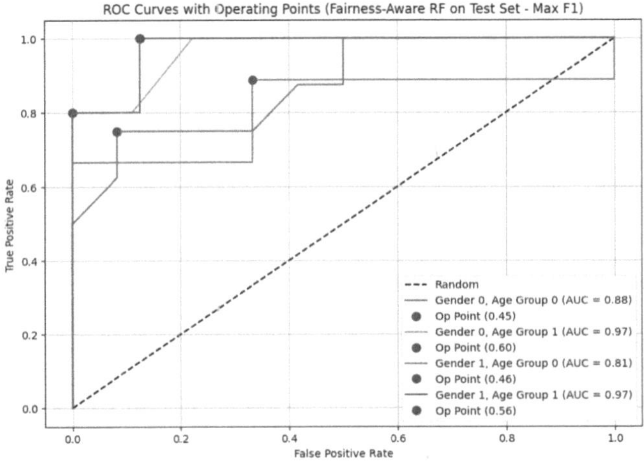

Fig. 4. RF ROC curve illustrating operation point adjustment via F1-optimization recalibration. Selected thresholds are annotated across demographic subgroups.

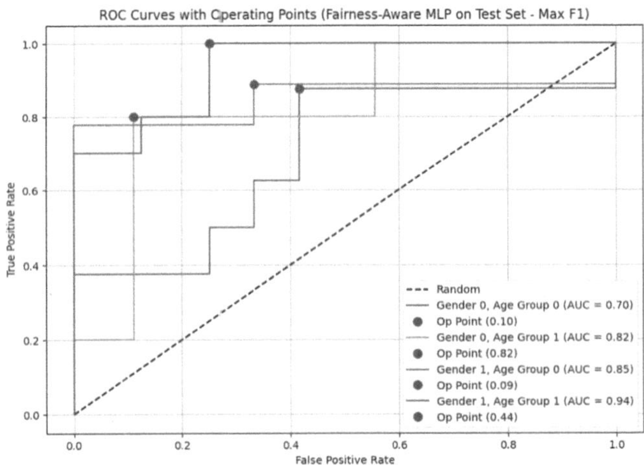

Fig. 5. MLP ROC curve illustrating operation point adjustment via F1-optimization recalibration. Selected thresholds are annotated across demographic subgroups.

5   Discussion and Conclusion

This study explored both the potential and the constraints of incorporating fairness-aware postprocessing strategies into ML models designed to predict gallstone diagnosis. By evaluating fairness metrics across sensitive attributes, specifically gender and age, notable disparities in model behavior have been identified. Parity-based measures such as Demographic Parity and Disparate Impact highlighted differences in the assignment of positive predictions, which may partially stem from biological variations in gallstone prevalence. Confusion-matrix metrics highlighted disparities in prediction errors across demographic subgroups, underscoring the need to address error-related bias for consistent predictive accuracy. The initial fairness analysis based on default thresholds revealed substantial bias across demographic subgroups, particularly in the Equal Opportunity and Equalized Odds metrics, where True Positive Rates and False Positive Rates showed considerable variation. Notably, the RF classifier demonstrated stronger predictive performance but at the cost of higher fairness disparities, especially in sensitivity on age, indicating unequal detection of true cases. In contrast, the MLP model produced more balanced error rates, exhibiting mild inconsistencies primarily in precision for gender subgroups. These findings reinforce the importance of fairness evaluations in biomedical applications classified as high-risk under regulatory frameworks such as the European Union's AI Act.

Threshold calibration via TPR equalization and F1-score maximization showed clear benefits. Both strategies reduced fairness metric gaps while slightly improving predictive performance. F1-optimized thresholds, in particular, achieved the lowest disparities across all fairness dimensions, indicating that performance-driven recalibration can simultaneously improve equity. However, residual bias remained in certain PPV measures, emphasizing that the mitigation strategies do not guarantee full fairness. Although postprocessing in both recalibration methods improved fairness, it also implies important trade-offs by introducing small shifts in operating points that may affect clinical interpretability. Overall, this study highlights the practical value of fairness-aware calibration for clinical ML models as a useful tool for ethical, equitable decision support in biomedical contexts.

Future work should explore the multifaceted nature of bias introduced by attribute overlap, particularly in clinical ML contexts. Due to sample size constraints, subgroup-level fairness evaluations were limited to four demographic groups, as the reduced validation set of 64 instances proved insufficient to include proxy attributes in the analysis. As the number of sensitive attributes increases, the exponential growth in subgroup combinations significantly complicates fairness validation at both computational and data levels. This underscores the need for large and representative datasets to support reliable fairness estimation and equitable model generalization. This challenge is especially important in biomedical settings, where acquiring data is often resource-intensive, particularly when patient-level attributes are involved in clinical evaluations.

References

1. Cirillo, D., Catuara-Solarz, S., Morey, C., Guney, E., Subirats, L., et al.: Sex and gender differences and biases in artificial intelligence for biomedicine and healthcare. NPJ Digit. Med. **3**(1), 81 (2020)
2. Equal Employment Opportunity Commission: Uniform guidelines on employee selection procedures (1978) (2014), https://www.govinfo.gov/content/pkg/CFR-2014-title29-vol4/xml/CFR-2014-title29-vol4-part1607.xml, code of Federal Regulations, Title 29, Part 1607

3. Esen, İ, Arslan, H., Esen, S.A., Gülşen, M., Kültekin, N., Özdemir, O.: Early prediction of gallstone disease with a machine learning-based method from bioimpedance and laboratory data. Medicine **103**(8), e37258 (2024)
4. European Union: Artificial Intelligence Act (2024), https://www.europarl.europa.eu/doceo/document/TA-9-2024-0138_EN.pdf, adopted by the 27 EU member states in March 2024
5. European Union Agency for Fundamental Rights: Handbook on European Non-Discrimination Law – 2018 Edition. European Union Agency for Fundamental Rights (2018), https://fra.europa.eu/en/publication/2018/handbook-european-non-discrimination-law-2018-edition
6. Feldman, M., Friedler, S.A., Moeller, J., Scheidegger, C., Venkatasubramanian, S.: Certifying and removing disparate impact. In: Proceedings of the 21th ACM SIGKDD International Conference on Knowledge Discovery and Data Mining, pp. 259–268 (2015)
7. Hardt, M., Price, E., Srebro, N.: Equality of opportunity in supervised learning. In: Advances in Neural Information Processing Systems, vol. 29 (2016)
8. Hasanzadeh, F., Josephson, C.B., Waters, G., Adedinsewo, D., Azizi, Z., White, J.A.: Bias recognition and mitigation strategies in artificial intelligence healthcare applications. NPJ Digit. Med. (2023)
9. Jansen, M., Li, H. Ortega, S.: Epidemiology and pathogenesis of gallstones: a global perspective. SpringerLink **12**, 101–117 (2024)
10. Kleinberg, J., Ludwig, J., Mullainathan, S., Sunstein, C.R.: Algorithmic fairness and bias mitigation in medical imaging. Nat. Digit. Med. (2023)
11. Mehrabi, N., Morstatter, F., Saxena, N., Lerman, K., Galstyan, A.: A survey on bias and fairness in machine learning. ACM Comput. Surv. (CSUR) **54**(6), 1–35 (2021)
12. Obermeyer, Z., Powers, B., Vogeli, C., Mullainathan, S.: Dissecting racial bias in an algorithm used to manage the health of populations. Science **366**(6464) 447–453 (2019)
13. Organisation, W.G.: Global burden and clinical impact of gallstone disease. WGO Practice Guidelines (2023), https://www.worldgastroenterology.org/guidelines/global-guidelines/gallstone-disease, Accessed July 2025
14. Reddy, N., van Dijk, L., Huang, M.: Gallstone disease burden from 1990 to 2021: a global epidemiological analysis. BMC Gastroenterol. **25**(2), 87 (2025)
15. U.S. Food and Drug Administration: Artificial intelligence in software as a medical device. https://www.fda.gov/medical-devices/software-medical-device-samd/artificial-intelligence-software-medical-device (2025), Accessed 28 July 2025
16. Wisniewski, J., Biecek, P.: fairmodels: a flexible tool for bias detection, visualization, and mitigation in binary classification models. R J. **14**(1), 227–243 (2022)

An Agentic Architecture for Scalable and Reproducible Data Standardization to OMOP CDM Using Declarative Modeling

Alberto Labarga[(⊠)] [iD]

Universidad Pública de Navarra, Pamplona, Spain
`alberto.labarga@gmail.com`

Abstract. The secondary use of clinical data for biomedical research is a cornerstone of modern medicine, yet its potential is severely constrained by the persistent challenge of data heterogeneity. While the Observational Medical Outcomes Partnership Common Data Model (OMOP-CDM) provides a robust standard for data harmonization, the process of transforming source data into this model remains a significant bottleneck, characterized by manual, resource-intensive, and brittle Extract-Transform-Load (ETL) pipelines. This paper proposes a novel agentic architecture that reframes data standardization from an imperative scripting task to a declarative, AI-augmented knowledge-generation process. Our framework decomposes the complex mapping workflow into a series of discrete tasks executed by specialized, autonomous software agents. This modular architecture orchestrates a powerful synergy between declarative data modeling with the Linked Data Modeling Language (LinkML), reliable Large Language Model (LLM) interfacing via the Boundary AI Markup Language (BAML), and the cognitive capabilities of LLMs for model alignment. The system's primary output is not merely transformed data, but a human-readable and machine-executable mapping specification using linkml-map. This approach establishes a new paradigm for creating scalable, reproducible, and transparent data standardization pipelines, positioning the data engineer as an expert validator of AI-generated knowledge rather than a manual coder of transformation logic.

Keywords: Data Standardization · OMOP-CDM · Large Language Models (LLMs) · Agentic AI · LinkML · BAML · ETL · Healthcare Data

1 The Standardization Bottleneck in Biomedical Research

1.1 The Challenge of Heterogeneity

The digital transformation of healthcare has produced an unprecedented volume and variety of clinical data from sources like electronic health records (EHRs), laboratory information systems, and medical imaging archives.This data holds immense potential to accelerate clinical research, generate real-world evidence, and improve population health outcomes. However, this potential remains largely untapped because the data is

A. López Fernández et al. (Eds.): CIABiomed 2025, LNBI 16148, pp. 74–89, 2026.
https://doi.org/10.1007/978-3-032-10661-2_6

fragmented, heterogeneous, and siloed within disparate institutional systems.In practice, clinical information is often organized using locally developed, inconsistent information models. Even when standards are adopted, multiple standards frequently coexist, leading to a persistent lack of semantic interoperability that hinders data integration and analysis at scale.This fundamental barrier of heterogeneity makes large-scale, multi-institutional research a monumental and often cost-prohibitive effort.

1.2 Common Data Models as a Foundational Solution

To overcome these barriers, the biomedical research community has championed the adoption of Common Data Models (CDMs), which provide a standardized structure and semantics for observational health data. Among these, the Observational Medical Outcomes Partnership Common Data Model (OMOP-CDM), maintained by the Observational Health Data Sciences and Informatics (OHDSI) [1], has emerged as a de facto international standard.By defining a consistent data structure for clinical domains (e.g., conditions, drugs, procedures, measurements) and leveraging controlled terminologies like SNOMED CT, LOINC, and RxNorm, OMOP-CDM enables true semantic harmonization. This allows researchers to develop analytical code once and execute it across a federated network of databases, a paradigm that has powered major research initiatives like the National COVID Cohort Collaborative (N3C) and the European Health Data & Evidence Network (EHDEN). The ultimate goal is to shift the research paradigm from site-specific data wrangling to "write-once, run-anywhere" analytics.

1.3 The ETL Bottleneck and the Limits of Current Approaches

Despite the demonstrated value of the OMOP-CDM, the practical process of transforming source data into the standard model—the Extract-Transform-Load (ETL) pipeline—remains a severe bottleneck. This process is notoriously resource-intensive, demanding significant and scarce expertise in both the source data's nuances and the intricate details of the target OMOP model.Historically, ETL pipelines have been implemented through manual data mapping and the development of brittle, imperative scripts in languages like SQL or Python. These scripts are difficult to create, debug, maintain, and, most importantly, reuse. They represent codified knowledge that is opaque and inaccessible to non-programmers, hindering transparency and reproducibility.

Recent research has explored the use of Large Language Models (LLMs) [2–4] to semi-automate the mapping process, demonstrating promising results in matching source attributes to standards like HL7 FHIR.These studies show that LLMs, when provided with sufficient context, can achieve high accuracy in attribute-level mapping. However, these approaches often treat the LLM as a monolithic tool applied to a single step in the process. They typically lack a robust, scalable, and modular architectural framework that can manage the entire end-to-end workflow, handle errors gracefully, and produce outputs that are verifiably correct and reproducible.

1.4   A Paradigm Shift Towards Agentic, Declarative Mapping

Achieving a true leap in the scalability and reproducibility of data standardization requires a fundamental paradigm shift. We must move away from the practice of writing imperative transformation scripts and toward the AI-assisted generation of declarative, human-readable, and machine-executable mapping specifications. To realize this vision, we propose a novel **agentic architecture**.

Inspired by concepts in agentic AI, which involve autonomous agents performing tasks on behalf of a user, our framework decomposes the complex standardization task into a cooperative workflow executed by specialized software agents. This modular design, which can be viewed as a "socio-technical configuration", enables a powerful synergy between several key technologies:

- **Declarative Data Modeling:** Using the **Linked Data Modeling Language (LinkML)** [5] to create formal, unambiguous models of both source and target schemas.
- **Reliable LLM Interfacing:** Employing the **Boundary AI Markup Language (BAML)** [6] to build type-safe, structured, and reliable interfaces to LLMs, transforming them from unpredictable oracles into dependable engineering components.
- **AI-Powered Model Alignment:** Leveraging the cognitive power of **LLMs** to analyze the declarative schemas and generate a formal transformation plan.
- Executable Mapping Specifications: Capturing this plan in a declarative linkml-map specification, which is both auditable and directly executable.

The core contribution of this architecture is its re-framing of the problem. The primary goal is not to automate the writing of ETL code, a task at which LLMs are notoriously unreliable. Instead, the goal is to automate the generation of transformation knowledge. By using an LLM to populate a declarative mapping specification, we externalize the transformation logic into a format that is transparent, versionable, and can be validated and refined by either a human expert or another AI agent. This approach positions the LLM as an intelligent co-pilot for the biomedical data engineer, automating the most tedious aspects of the job while elevating the human's role to that of an expert reviewer and validator.

2   Multi-agent Framework for Smart Data Harmonization

The proposed system is designed as a modular, multi-agent system (MAS) inspired by modern microservice design principles. An agent, in this context, refers to an autonomous software entity, often powered by a Large Language Model (LLM), designed to perform specific tasks or functions within a larger system. These agents are capable of independent action, decision-making, and often communication with other agents to achieve a common goal. They can be specialized for particular roles, such as data processing, task management, or system orchestration, leveraging the LLM's ability to understand, generate, and reason with human language.

This design promotes a clear separation of concerns, enabling individual components to be developed, scaled, and updated independently. The entire workflow is managed by a central **Orchestrator Agent**, which directs a team of specialized worker agents, each responsible for a distinct phase of the standardization process. This structure ensures modularity, fault isolation, and scalability, mirroring the robust design of contemporary distributed systems platforms.

2.1 The Orchestrator Agent

The orchestrator agent serves as the central nervous system of the framework. It is not a worker but a manager, responsible for initiating, monitoring, and coordinating the entire data standardization job. Its logic is analogous to the workflow orchestration engines found in advanced data platforms, which define and execute complex, multi-step data pipelines.

The Orchestrator is responsible for workflow management, defining the execution graph and sequencing worker agent invocations upon receiving an initial request, such as "Standardize dataset X to OMOP-CDM." It also handles artifact routing, ensuring the seamless flow of data and metadata between agents, where the output of one agent becomes the input for the next (e.g., passing a LinkML schema from the Schema Definition Agent to the Semantic Grounding Agent). Furthermore, the Orchestrator maintains job state and implements robust error handling and retry logic; for example, if the Model Alignment Agent's LLM call fails or produces a low-confidence result, it can retry the call with different parameters, a fallback model, or a more detailed prompt. A critical function of the Orchestrator is its human-in-the-loop integration, managing human interaction points by pausing the workflow at predefined checkpoints—such as after the mapping specification is generated—to await review and approval from a human data engineer before proceeding with the final data transformation.

The Orchestrator can be implemented as a state machine or by leveraging mature workflow management libraries such as Airflow (https://airflow.apache.org/). It communicates with the worker agents through a standardized application programming interface (API).

2.2 The Schema Definition Agent

The first step in any robust mapping process is to achieve an unambiguous understanding of the source and target data structures. The Schema Definition Agent is responsible for this foundational task, creating formal, machine-readable data models using **LinkML**. This step is paramount because it transforms the mapping problem from an ill-defined, text-to-structure task into a well-defined, structure-to-structure reasoning problem, which is far more amenable to reliable AI automation.

The agent's responsibilities include **schema ingestion**, where it ingests schema information from various common source formats like SQL DDL files, CSV files with header rows and description documents, or existing JSON Schema definitions. It then handles **LinkML generation**, utilizing tools from the LinkML ecosystem such as schema-automator, to bootstrap an initial LinkML schema from the ingested source information. This process is refined to produce a complete LinkML YAML file that formally defines all classes (tables), slots (columns/attributes), data types, and enumerations for the source dataset. The target OMOP-CDM schema is represented by a pre-existing, canonical LinkML file that serves as a stable reference.

By formalizing both the source and target schemas in LinkML, we provide all downstream agents with a rich, structured context. This context goes far beyond simple column names, capturing metadata such as data types, textual descriptions, inter-class relationships, and constraints. Providing such detailed, machine-readable context has been shown to be essential for reducing LLM ambiguity and hallucinations, leading to more reliable and accurate mapping outcomes.

2.3 Semantic Grounding Agent

With formal schemas in place, the next challenge is to navigate the vast search space of potential mappings between source and target elements. The Semantic Grounding Agent acts as an intelligent "scout," performing a rapid, coarse-grained analysis to identify the most plausible correspondences. It provides essential hints that guide the more computationally expensive cognitive work of the Model Alignment Agent.

The agent's responsibilities begin with **Metadata Extraction**, where it extracts names, descriptions, and other textual metadata for all slots and classes from both source and target LinkML schemas. Following this, **Embedding Generation** employs domain-specific language models like BioBERT [7] or ClinicalBERT [8] to create high-dimensional vector embeddings for each textual element. These embeddings are crucial as they capture the semantic meaning of terms beyond simple keyword matching.

Next, the agent performs **Similarity Calculation**, utilizing metrics such as cosine similarity to determine the semantic distance between every source and target slot, thereby identifying highly related pairs. The process then moves to **Clustering and Contextual Hinting**, where unsupervised clustering algorithms group semantically related source attributes, often aligning with a single target table in the OMOP-CDM. For instance, a cluster of laboratory test attributes would likely map to the MEASUREMENT table. Finally, Candidate Generation produces a ranked list of candidate mappings, such as "source slot lab_test_name has a 0.94 similarity to target slot measurement_source_value," along with high-level contextual hints, like "Source class LabEvent appears to correspond to target class Measurement."

This agent implements a crucial optimization strategy analogous to Retrieval-Augmented Generation (RAG). Instead of asking the LLM to perform a "cold start" analysis of the entire mapping problem, we provide it with a pre-filtered, highly relevant set of candidate mappings. This dramatically reduces the cognitive load on the LLM, focuses its reasoning on the most promising connections, and significantly lowers the probability of error.

2.4 The Model Alignment Agent: The Cognitive Core

This agent represents the cognitive heart of the architecture, responsible for translating the structured schemas and semantic hints into a complete, logical, and executable transformation plan. It leverages a powerful LLM, but does so through a highly engineered and reliable interface.

The agent's responsibilities begin with **input aggregation**, where it receives three critical inputs from the Orchestrator: the source LinkML schema, the target OMOP LinkML schema, and the ranked list of candidate mappings generated by the Semantic Grounding Agent.

Following input aggregation, the agent reliably **invokes an LLM** (e.g., GPT-4, Llama 3) through a meticulously defined interface built with the Boundary AI Markup Language (BAML). BAML is a domain-specific language designed specifically for creating reliable, type-safe, and testable interactions with LLMs.

Structured output generation is achieved through a BAML function designed to be fully type-safe, with a return type that corresponds directly to the linkml-map specification. The prompt instructs the LLM to reason about the provided schemas and candidates, generating a complete mapping specification. This encompasses not only direct one-to-one mappings but also more complex slot_derivations involving expressions for tasks like unit conversion, value set mapping, or string concatenation.

The use of BAML ensures the reliability of this agent. It solves the "last mile" problem of LLM integration: extracting structured, validated data from the model's probabilistic output. Traditional methods that rely on parsing unstructured JSON or text are inherently brittle. BAML, by contrast, turns the LLM into a predictable, type-safe function call, making it a true engineering component. The BAML runtime handles the complex prompting, output validation, and error correction needed to guarantee that the LLM's output conforms to the predefined schema, achieving higher robustness over standard prompt engineering techniques.

Finally, the agent's primary function is **declarative map creation**, translating the LLM's abstract "understanding" of the required transformations into a concrete, declarative, and executable linkml-map YAML file.

2.5 Concept Mapping Agent

In many standardization workflows, the transformation of structural schemas alone is insufficient. Clinical data often contains values—such as lab test names, drug brands, diagnoses, and free-text observations—that must be semantically linked to controlled vocabularies in the OMOP Common Data Model. This task, commonly known as concept normalization or terminology mapping, is especially challenging when dealing with non-standard local codes, textual descriptors, or multilingual sources. To address this need, we introduce the Concept Mapping Agent, a specialized semantic worker responsible for bridging raw terms to standardized concept_ids within the OMOP vocabulary framework.

The Concept Mapping Agent operates as a hybrid intelligence module that combines dense retrieval over large embedding spaces with graph-based reasoning derived from the structure of the OMOP vocabularies. Its core function is to accept a source term—either extracted from structured source fields or the output of an NLP agent—and return a ranked list of candidate OMOP concepts, each accompanied by explanatory metadata. This capability is essential for automating the population of fields such as condition_concept_id, drug_concept_id, and measurement_concept_id, where semantic alignment rather than syntactic matching is required.

The agent begins its work by generating an embedding representation of the input term using a domain-appropriate embedding model [9]. This component is structured as a composite router that selects the most suitable embedding approach depending on the language and content characteristics. The resulting embedding is then used to query a precomputed vector index that contains vector representations of all OMOP standard concepts. These concept embeddings are constructed by combining each concept's name, domain, vocabulary identifier, and all known synonyms into a unified textual representation, which is then encoded using the same model family. To ensure semantic completeness, the representation incorporates multilingual synonyms and domain metadata, making the index robust to lexical variability and partial matches.

After the initial semantic retrieval step, the Concept Mapping Agent refines its candidate list through a graph-based contextualization process. Each concept in the OMOP vocabulary is embedded within a directed knowledge graph derived from the concept_relationship and concept_ancestor tables. This graph encodes a rich web of ontological and semantic connections—such as "is a", "maps to", and "subsumes"—which govern how concepts relate to one another across vocabularies and domains. A Graph Neural Network (GNN), such as GraphSAGE [10], is trained on this structure to produce enhanced embeddings that incorporate not only a concept's own features but also those of its neighbors, capturing latent hierarchical and semantic context. This graph-informed representation enables the agent to assess candidate relevance not merely by lexical similarity but by structural coherence within the OMOP ontology.

Complementing this semantic refinement is the integration of a concept popularity score, computed from aggregated statistics across large-scale OMOP-CDM databases. This score reflects the empirical frequency with which a given concept_id appears in real-world patient data—across tables such as condition_occurrence, drug_exposure, and measurement. By incorporating this usage signal into the reranking function, the agent biases results toward concepts that are not only semantically and structurally appropriate, but also commonly instantiated in clinical practice. This helps disambiguate competing candidates and mitigates the risk of selecting obscure or deprecated terms that, while semantically accurate, may not be interoperable in analytical workflows.

Finally, the agent applies a granularity bias to promote conceptual specificity. Many OMOP vocabularies are deeply hierarchical, with abstract root concepts giving rise to increasingly fine-grained descendants. In most clinical contexts, the goal is to map to the most specific valid concept rather than its more general parent. To operationalize this, the agent computes a granularity score for each candidate, based on its depth in the ontology or by identifying leaf nodes—concepts without descendants via "is a" relationships. This score is then integrated into the final ranking as a soft reweighting factor, gently favoring more granular concepts without unduly penalizing higher-level terms when specificity is uncertain.

The final selection score for each candidate concept thus emerges from a weighted composition of four key signals: semantic proximity via dense embeddings, structural alignment via GNN-enhanced representations, empirical usage frequency from OMOP data, and ontological specificity via the granularity score. This multi-signal reranking process ensures that the Concept Mapping Agent consistently recommends concepts that are semantically accurate, structurally coherent, clinically relevant, and as specific as the source data allows.

Once the candidate list is finalized, the agent packages the top results into a structured mapping object. Each entry includes the concept_id, concept name, vocabulary and domain identifiers, cosine similarity scores, GNN-derived confidence scores, and a boolean is_leaf indicator. This output is returned to the Orchestrator Agent for downstream use. In fully automated pipelines, it may be used to populate the expr field of a slot_derivation within a LinkML map. In human-in-the-loop scenarios, the candidate list can be presented to a domain expert for review, with the most likely match preselected and the alternatives clearly ranked.

Crucially, the Concept Mapping Agent is designed for modular integration. It can be invoked directly by the Model Alignment Agent when populating vocabulary-dependent fields, or embedded within an NLP pipeline as a semantic normalizer for unstructured data. Like other agents in the framework, its inference is deterministic and its output is fully inspectable, enabling reproducibility and auditability.

By introducing this agent into the system, we extend the framework's capacity from structural schema alignment to deep semantic standardization. It transforms the mapping of ambiguous, variable, and multilingual terms from a brittle, manual task into a robust, transparent, and explainable process. The Concept Mapping Agent exemplifies the architecture's core principles: combining declarative modeling, AI augmentation, and human oversight into a cohesive, scalable pipeline for data standardization.

A particularly important feature of the reranking step is its ability to favor granular concepts. Many OMOP vocabularies, such as SNOMED CT and RxNorm, are hierarchical, with abstract parent concepts and highly specific child terms. In most clinical use cases, the mapping should favor the most precise concept available. To achieve this, the agent incorporates a granularity bias into its scoring function. This is computed using the concept's depth in the OMOP hierarchy or by detecting leaf nodes—concepts with no descendants via "is a" relationships. The granularity signal is used as a soft reweighting factor, subtly promoting more specific concepts without overriding semantic proximity.

Once the candidate list is finalized, the agent packages the top results into a structured mapping object. Each entry includes the concept_id, concept name, vocabulary and domain identifiers, cosine similarity scores, GNN-derived confidence scores, and a boolean is_leaf indicator. This output is returned to the Orchestrator Agent for downstream use. In fully automated pipelines, it may be used to populate the expr field of a slot_derivation within a LinkML map. In human-in-the-loop scenarios, the candidate list can be presented to a domain expert for review, with the most likely match preselected and the alternatives clearly ranked.

Crucially, the Concept Mapping Agent is designed for modular integration. It can be invoked directly by the Model Alignment Agent when populating vocabulary-dependent fields, or embedded within an NLP pipeline as a semantic normalizer for unstructured data. Like other agents in the framework, its inference is deterministic and its output is fully inspectable, enabling reproducibility and auditability.

By introducing this agent into the system, we extend the framework's capacity from structural schema alignment to deep semantic standardization. It transforms the mapping of ambiguous, variable, and multilingual terms from a brittle, manual task into a robust, transparent, and explainable process. The Concept Mapping Agent exemplifies the architecture's core principles: combining declarative modeling, AI augmentation, and human oversight into a cohesive, scalable pipeline for data standardization.

2.6 Transformation and Validation Agent

The Transformation & Validation Agent acts as the final worker in the automated workflow, responsible for applying the AI-generated transformation plan to the source data and rigorously validating the output against the target standard.

Its responsibilities include **execution**, where it receives the linkml-map YAML file from the Model Alignment Agent and source data, using the linkml-map Python library to apply specified class_derivations and slot_derivations to each record. Following execution, it generates the transformed data, now structured according to the target OMOP LinkML schema. For **validation**, the agent utilizes the linkml-validate tool or its underlying Python library to check the newly generated data against the canonical OMOP LinkML schema. This crucial step identifies issues such as data type mismatches, constraint violations, and missing required fields. Finally, the agent reports the outcome (success or failure) back to the Orchestrator Agent. In case of failure, a detailed report of validation errors is provided, aiding human engineers in debugging or prompting the Orchestrator to trigger a corrective loop.

3 Example: Mapping Laboratory Data to the OMOP-CDM MEASUREMENT Table

To make the abstract architecture concrete, this section presents an end-to-end illustrative workflow. We will trace the journey of a simplified laboratory results file, inspired by the MIMIC-IV [11] *labevents* table, as it is processed by the agentic framework and standardized to the OMOP-CDM MEASUREMENT table.

Step 1: Schema Definition. The process begins when a data engineer submits a request to the Orchestrator Agent to standardize a labevents.csv file. The Orchestrator initiates the workflow by invoking the Schema Definition Agent. The agent ingests the source file's structure (columns: subject_id, itemid, charttime, valuenum, valueuom) and any accompanying metadata describing the columns. Using schema-automator, it generates a formal LinkML schema, source_lab_schema.yaml.

```
id: https://example.org/source_lab_schema
name: source-lab-schema
prefixes:
  linkml: https://w3id.org/linkml/
default_range: string
classes:
  LabEventCollection:
    tree_root: true
    attributes:
      events:
        multivalued: true
        inlined: true
        range: LabEvent
  LabEvent:
    description: "Represents a single laboratory measurement event."
    attributes:
      subject_id:
        description: "Identifier for the patient."
        identifier: true
      itemid:
        description: "Identifier for the specific lab test performed."
        range: integer
      charttime:
        description: "The time the measurement was recorded."
        range: datetime
        valuenum:
          description: "The numeric result of the lab test."
          range: float
        valueuom:
          description: "The unit of measure for the numeric result."
```

Step 2: Semantic Grounding. Orchestrator receives source_lab_schema.yaml and passes it, along with the canonical omop_cdm.yaml schema, to the Semantic Grounding Agent. This agent extracts the descriptions for each slot, generates vector embeddings using a model like ClinicalBERT, and computes cosine similarities.

Example output from semantic grounding agent:

- Candidate Mappings (Slot-level):

  - source:valuenum - > target:value_as_number (Similarity: 0.98)
  - source:valueuom - > target:unit_source_value (Similarity: 0.91)
  - source:itemid - > target:measurement_source_value (Similarity: 0.85)

- source:charttime - > target:measurement_datetime (Similarity: 0.95)

- Contextual Hint (Class-level):

  - Source class LabEvent has high semantic overlap with target class Measurement.

Step 3: Model Alignment via BAML. The Orchestrator now has all the necessary components for the cognitive task. It invokes the Model Alignment Agent, providing the two LinkML schemas and the high-confidence candidates from the previous step. The agent's core logic is encapsulated in a BAML file that defines a reliable, type-safe interface to the LLM. This interface ensures that the LLM's output will be a structurally valid object that can be directly serialized into a linkml-map file (Table 1).

Table 1. BAML components

Component	BAML Definition Snippet	Purpose

A. BAML Class Definitions	`class SlotDerivation {`
	  `expr string?`
	  `populated_from string?`
	`}`
	
	`class ClassDerivation {`
	  `populated_from string`
	  `slot_derivations map<string,`
	`SlotDerivation>`
	`}`
	
	`class MappingSpecification {`
	  `class_derivations map<string,`
	`ClassDerivation>`
	`}`
		Defines the target output structure in BAML. These classes mirror the essential components of the linkml-map specification 8, ensuring the LLM's output is type-safe and conforms to a predictable schema.

B. BAML Function Definition	`function GenerateMapping(`
	`source_schema: string,`
	`target_schema: string,`
	`candidates: list<string>`
	`) -> MappingSpecification {`
	`client GPT4_Turbo`
	`prompt #"...see prompt snippet below..."#`
	`}`
		Declares the main function. It takes the source and target schemas (as stringified YAML) and the semantic candidates as input. Crucially, it specifies MappingSpecification as the return type, instructing the BAML runtime to enforce this structure on the LLM's output.7

(continued)

Table 1. (*continued*)

	You are an expert data modeler specializing in OMOP-CDM.
	
	Your task is to create a declarative mapping from a source schema to the OMOP-CDM target schema.
	
	Source Schema:
	``yaml
	{{ source_schema }}
	``
	
	Target Schema:
	``yaml
	{{ target_schema }}
	``	

C. BAML Prompt Snippet	Here are some high-confidence candidate mappings to guide you: {% for candidate in candidates %} - {{ candidate }} {% endfor %} Generate a complete mapping specification. For each source class, determine the target class it populates. For each target slot, specify how it is populated from a source slot. If a direct mapping is not possible, you may need to define an expression. {{ ctx.output_format }}	This Jinja2 prompt provides clear instructions and context to the LLM. It includes the full schemas and the semantic hints. The {{ ctx.output_format }} macro is a powerful BAML feature that automatically injects instructions and examples of the desired output format, guiding the LLM to produce a valid MappingSpecification object.[16]

Step 4: Generating the Executable Map. The Model Alignment Agent executes the GenerateMapping BAML function. The BAML runtime interacts with the LLM, handles potential errors, validates the response, and returns a structured MappingSpecification object. The agent then serializes this object into a lab_to_measurement.map.yaml file. This file is the declarative, executable plan for the data transformation. The following is a snippet of generated lab_to_measurement.map.yaml.

```
class_derivations:
Measurement:
  populated_from: LabEvent
  slot_derivations:
    person_id:
      populated_from: subject_id
    measurement_datetime:
      populated_from: charttime
    measurement_source_value:
      populated_from: itemid
    value_as_number:
      populated_from: valuenum
    unit_source_value:
      populated_from: valueuom
```

Step 5: Transformation and Validation. This generated map represents the AI's hypothesis. At this point, the Orchestrator can pause for human validation. An expert data engineer reviews the lab_to_measurement.map.yaml file. They can approve it as is, or edit it to add more complex logic (e.g., a rule to map itemid values to standard measurement_concept_ids).

Once approved, the Orchestrator invokes the Transformation & Validation Agent. The agent uses the linkml-map library to apply the rules in the map file to the labevents.csv data. It then validates the output against the omop_cdm.yaml schema to ensure full compliance. The final result is a dataset perfectly structured for inclusion in an OMOP-CDM database.

4   Discussion

The proposed agentic architecture represents a significant step beyond traditional ETL and simple AI wrappers. By orchestrating a team of specialized agents that leverage declarative modeling and reliable LLM interfaces, the framework offers a new model for data standardization. This section discusses the architectural benefits, inherent limitations, and broader implications of this approach.

The framework's design offers significant advantages for building and maintaining data standardization pipelines within complex biomedical environments. Its agentic architecture inherently promotes modularity, with each agent acting as a self-contained component much like a microservice. This separation of concerns allows for independent development, testing, and deployment, meaning components like the Semantic Grounding Agent's embedding model can be updated without affecting other parts of the system, making the system adaptable and easier to maintain.

Reproducibility is another cornerstone of this architecture, achieved through its reliance on declarative artifacts. The entire transformation logic is captured in version-controllable YAML files: the LinkML schema for data structures and the linkml-map specification for transformation rules. This contrasts with traditional ETL pipelines where logic is often embedded in opaque, imperative scripts. By externalizing transformation knowledge into human-readable and machine-executable formats, the process becomes transparent, auditable, and perfectly reproducible, aligning with the goals of data platforms that aim to support reusable and shareable study pipelines.

Reliability is significantly enhanced through the framework's "structured AI" environment. The process provides the LLM with highly structured inputs, such as formal LinkML schemas and focused guidance like semantic candidate mappings, which are known to reduce ambiguity and improve accuracy. A critical step is the use of BAML to enforce structured output. By defining the LLM interface as a type-safe function, the model's output is constrained to a valid linkml-map structure, virtually eliminating malformed or syntactically incorrect results. Finally, the linkml-validate step provides a definitive check that the transformed data strictly conforms to the target OMOP-CDM..

Despite these advantages, it is crucial to acknowledge the inherent limitations of the technology and position the framework appropriately. LLMs, while powerful, are not infallible. They are susceptible to "hallucinations"—generating plausible but incorrect information—and their performance can degrade when source data descriptions are incomplete or ambiguous, limitations that have been observed in prior studies.

Therefore, this architecture is intentionally designed not as a fully autonomous, "fire-and-forget" system, but as an intelligent "mapping co-pilot." The agents are designed to automate the 90% of the work that is tedious, repetitive, and time-consuming: drafting schemas, searching for candidate mappings, and generating the initial transformation rules. The human data engineer is then elevated from a manual coder to an expert reviewer—the "pilot-in-command." Their role is to leverage their deep domain knowledge to validate, refine, and ultimately approve the AI-generated mapping specification. This human-in-the-loop validation is not a bug but a critical and intentional feature of the architecture, ensuring the final mapping is both technically correct and semantically sound.

Implementing AI in healthcare often presents a challenge, balancing the need for standardization to ensure quality, safety, and interoperability with the equally important need for customization to adapt to unique local contexts. Our agentic architecture offers a powerful technical solution to this tension by separating concerns across different levels of the system. Standardization is enforced at the level of components and outputs; the framework utilizes a canonical LinkML schema for the OMOP-CDM, standardized agent APIs for communication, and a rigorous validation process to ensure the final data output strictly adheres to the OMOP standard, providing the uniformity and consistency required for scalable, federated research. Conversely, customization is enabled at the architectural and workflow level, as the true power of the agentic design lies in its configurability. A local institution is not constrained by a single, rigid pipeline; they can easily develop and integrate their own custom agents into the workflow to handle local data idiosyncrasies. For instance, an institution with a proprietary, non-standard coding system for laboratory tests could develop a "TerminologyMappingAgent" invoked after the Model Alignment Agent to translate local codes into standard LOINC or SNOMED CT concepts. The ability for human experts to directly edit the AI-generated linkml-map file before execution represents the ultimate form of fine-grained customization.

While this architecture provides a solid technical solution, it is important to recognize that it operates within a complex context. The process of data standardization is not always a purely technical exercise. Decisions about how to map ambiguous source data or which clinical interpretation to favor can have significant downstream effects on research outcomes. These decisions often require clinical consensus, institutional governance, and transparent debate. The proposed framework can serve as a powerful facilitator for these discussions. By making the proposed mapping rules explicit, declarative, and human-readable in the linkml-map file, it provides a clear object for discussion and debate among data engineers, clinicians, and researchers, but it does not replace the fundamental need for human governance and oversight.

5   Conclusion and Future Directions

This paper proposes a novel agentic architecture for the semi-automated standardization of clinical data to the OMOP Common Data Model. The primary contribution is a paradigm shift away from brittle, imperative ETL scripting and toward the AI-assisted generation of declarative, reproducible, and transparent mapping specifications. By decomposing the complex standardization workflow into a series of tasks performed by specialized agents, the framework orchestrates a powerful combination of technologies: declarative modeling with LinkML and linkml-map provides a robust and reproducible foundation, while reliable LLM interfacing with BAML transforms the LLM into a dependable engineering component for the cognitive task of model alignment. The resulting system is not a black box, but an intelligent co-pilot that automates tedious work while empowering the human expert to validate and control the final outcome, effectively balancing the need for automation with the demand for accuracy and governance.

Several key areas for future work are critical to realizing its full potential. These extensions are designed to address current limitations and expand the system's capabilities to handle the full spectrum of clinical data.

A crucial next step for this system involves developing an interactive validation interface. This graphical user interface (UI) would facilitate the human-in-the-loop validation stage, which is considered essential for advancing semi-automated mapping systems. The UI would display source and target LinkML schemas side-by-side, present the AI-generated linkml-map specification in an intuitive format, and enable data engineers to graphically review, edit, test, and approve mappings before final transformations. This is expected to significantly reduce the barrier to entry and accelerate the review cycle.

To evolve into a comprehensive solution, the current framework, which primarily focuses on structured tabular data, must be extended to handle the diverse data types found in modern EHRs. This necessitates the development of new specialized agents, addressing a known challenge for integrated data platforms. Specifically, an NLP Agent would be dedicated to processing unstructured clinical narratives, utilizing advanced NLP models to extract structured entities like diagnoses, medications, and symptoms from free text and mapping them to the OMOP NOTE and NOTE_NLP tables. Additionally, Genomic and Imaging Agents would be specialized to understand the unique schemas and complex mapping requirements associated with genomic data (e.g., VCF files) and medical imaging metadata (e.g., DICOM headers), integrating them into relevant OMOP extension tables.

The Model Alignment Agent can be significantly enhanced by exploring more sophisticated agentic patterns. Future work could involve implementing a hierarchical or collaborative team of LLM-powered agents. For instance, a "manager" agent could decompose large mapping tasks into smaller sub-tasks, such as separately mapping demographics, labs, and diagnoses, and then delegate these to specialized 'worker" agents. Another potential agent could function as a "critic," reviewing the mappings generated by the primary agent and suggesting improvements, thereby creating an iterative refinement loop before presenting the results to a human.

The long-term vision for this agentic mapping framework is its integration within a broader, end-to-end research platform. In such an integrated system, the agentic framework would manage upstream data ingestion and harmonization, providing standardized, analysis-ready OMOP data directly to downstream services. These services would then handle cohort definition, statistical analysis, machine learning, and federated analytics, establishing a seamless, powerful, and reproducible environment for the entire clinical research lifecycle.

Disclosure of Interests.   The authors declare no conflicts of interest.

References

1. Hripcsak, G., Duke, J.D., Shah, N.H., et al.: Observational health data sciences and informatics (OHDSI): opportunities for observational researchers. Stud. Health Technol. Inf. **216**, 574–578 (2015)
2. Zhou, X., Dhingra, L. S., Aminorroaya, A., Adejumo, P., Khera, R.: A novel sentence transformer-based natural language processing approach for schema mapping of electronic health records to the OMOP common data model. medRxiv. (2024)
3. Adams, M.C.B., et al.: Breaking digital health barriers through a large language model-based tool for automated observational medical outcomes partnership mapping: development and validation study. J. Med. Internet Res. **27**, e69004 (2025)
4. Mitchell-White, J., et al.: Llettuce: an open source natural language processing tool for the translation of medical terms into uniform clinical encoding. arXiv (2024)
5. Moxon, S.T., et al.: The linked data modeling language (LinkML): a general-purpose data modeling framework grounded in machine-readable semantics. In International Conference on Biomedical Ontology (2021)
6. Boundary ML: BAML: a programming language for reliable prompt engineering. boundary ML. GitHub repository. BoundaryML/baml (2024). https://github.com/boundaryml/baml
7. Lee, J., et al.: BioBERT: a pre-trained biomedical language representation model for biomedical text mining. Bioinformatics **36**(4), 1234–1240 (2020)
8. Huang, K., Altosaar, J., Ranganath, R.: ClinicalBERT: modeling clinical notes and predicting hospital readmission. arXiv Preprint arXiv:1904.05342 (2019)
9. Wolf, T., et al.: Transformers: state-of-the-art natural language processing. In: Proceedings of the 2020 Conference on Empirical Methods in Natural Language Processing (EMNLP): System Demonstrations, pp. 38–45 (2020)
10. Hamilton, W.L., Ying, R., Leskovec, J.: Inductive representation learning on large graphs. In: Advances in Neural Information Processing Systems, vol. 30 (2017)
11. Johnson, A., et al.: MIMIC-IV (version 3.1). PhysioNet. RRID:SCR_007345 (2024)

The Assessment of AI-Based Digital Health Technologies from the Perspective of HTA Bodies. The Case of AQuAS' AI Assessment Guide

Carolina Moltó-Puigmartí[1]([image]) [image], Susanna Aussó Trias[2] [image],
Maria Bretones Vallejo[2] [image], Didier Domínguez Herrera[3] [image],
and Rosa Maria Vivanco-Hidalgo[1] [image]

[1] Agency for Health Quality and Assessment of Catalonia (AQuAS), Barcelona, Spain
cmolto@gencat.cat
[2] TIC Salut Social Foundation (FTSS), Barcelona, Spain
[3] Universitat Oberta de Catalunya (UOC), Barcelona, Spain

Abstract. This paper presents the AQuAS AI Assessment Guide, a methodological framework developed to support the evaluation of digital health technologies (DHTs) that incorporate artificial intelligence (AI). Designed collaboratively by the Agency for Health Quality and Assessment of Catalonia (AQuAS) and the TIC Salut Social Foundation, the guide defines 13 assessment domains encompassing clinical relevance, technical performance, ethical considerations, regulatory compliance, and broader system and societal impact. The guide aims to foster a robust, evidence-based and context-sensitive evaluation of AI solutions across different stages of development and diverse use cases. This article outlines the background, development methodology, and structural composition of the guide, and discusses its potential uses and contributions to responsible innovation in health systems.

Keywords: Artificial Intelligence · Health Technology Assessment · Evaluation Framework

1 Introduction

We are in a context of exponential growth of AI-based health technologies. Despite the fact that many promising tools exist, their adoption in the healthcare systems is often not successful and sustainable. Several factors contribute to this limited adoption, ranging from regulatory issues (either not being aware of the need to go through regulatory approval or not classifying the device into the right risk class), to limited user trust, low usability and digital literacy, interoperability and integration problems, and lack of long-term financial support. Another key factor influencing the lack of sustainable and long-term adoption of AI-based technologies is the limited value demonstration beyond technical performance.

Note: this work was conducted while affiliated with the FTSS – Didier Domínguez Herrera.

Health Technology Assessment (HTA) is a multidisciplinary process that uses explicit methods to determine the value of a health technology at different points in its lifecycle [1]. HTA provides a structured approach to evaluate the clinical and non-clinical value of health technologies. However, traditional HTA approaches may not fully capture the complexities of AI systems. The Agency for Health Quality and Assessment of Catalonia (AQuAS) is an HTA body that is part of the Spanish Network of Agencies for Assessing National Health System Technologies and Performance (RedETS) [2]. Among its various responsibilities, AQuAS' HTA unit elaborates HTA reports under commission of the Spanish Ministry of Health and the Catalan Department of Health. HTA reports help mainly in taking decisions about the introduction of new technologies into the public healthcare system.

When performing an HTA, an assessment is made about whether new health technologies bring an added value with respect to the standard of care. The objective is to promote innovations that improve outcomes for patients and the health system, while ensuring long-term sustainability.

AQuAS has developed an assessment guide for digital health technologies (DHTs) that use AI [3] to guide HTA bodies in their assessment of AI-based technologies, and help technology developers understand the evidence they need to generate in order to prove the value of their technologies. The guide outlines the domains, dimensions and sub-dimensions that need to be considered to prove the value of a technology in the healthcare system.

2 Methods

The guide was developed through a structured, iterative process involving multidisciplinary collaboration between AQuAS and the TIC Salut Social Foundation. The domains, dimensions and sub-dimensions were based on those described in the methodological framework for the assessment of DHTs developed by AQuAS named "Health technology assessment framework: adaptation for digital health technologies assessment" [4], which comprehensively describes the information that needs to be taken into account when assessing DHTs. This framework [4] was developed under commissioning of the General Directorate of the Common Portfolio of the Ministry of Health of Spain as part of the RedETS' 2021 work plan. It is considered methodologically very strong; it is based on a systematic review of guidelines and frameworks published in the international literature (including e.g. the EUnetHTA core model framework [5]) followed by consensus with an expert group. Moreover, it adapts the Evidence Standard Framework from NICE [6] to the Spanish context via a direct collaboration with the English agency and consensus with an expert group from Spain. However, in such a dynamic field as AI in health, it was deemed necessary to develop a more specific and up-to-date guide that incorporated additional aspects relevant for AI-based DHTs.

3 Results

The AQuAS' AI Assessment Guide [3] is structured around the 13 domains derived from the "Health technology assessment framework: adaptation for digital health technologies assessment" [4] (see methods section for more details). Each domain includes dimensions and sub-dimensions, with the criteria that help stakeholders (HTA bodies and technology developers) assess key aspects such as the technology's readiness, relevance, safety, and impact. In addition, each domain ends with a section on standards of evidence which is mainly directed to technology developers to help them translate the evaluation needs into more practical action points, making the guide more accessible and useful for this stakeholder group.

Below is a summary of the domains included in the guide and the key elements under each of them (Table 1).

Table 1. Domains included in the AI assessment guide [3].

Domain number	Domain name	Key elements
1	Health problem and target population	Defines the health issue addressed by the technology, its epidemiological context, the target population, and the expected benefit
2	Description of the technology	Details the intended use, the credibility and reputation of the development team, the novelty of the technology, its scientific basis, the technical assessment and validation processes that have been carried out, the real or expected adoption level, and the information management methods
3	Technical aspects	Covers the technical effectiveness and performance, the generalizability and reproducibility, traceability, accessibility, usability and user's experience, technological integration, and ethical aspects of the model including its explainability and transparency. While specific recommendations are mentioned for some elements (e.g. accessibility, interoperability), the guide is generally conceived as a presentation of what elements should be evaluated and taken into account rather than as a compilation of specific metrics or standards to be fulfilled

(continued)

Table 1. (*continued*)

Domain number	Domain name	Key elements
4	Content	Addresses the relevance, completeness, accuracy and personalization of the content, both in terms of written, visual or auditory information provided to the users, as well as the scientific evidence on which the content is based or being supported by
5	Safety	Covers the clinical and technical safety, including the identification and evaluation of risks, problems and unwanted side effects arising from the use of the intervention either due to intrinsic characteristics of the technology or due to external factors (e.g. wrong user or incorrect use)
6	Efficacy and clinical effectiveness	Evidence of impact on clinical outcomes and benchmarking against standard care
7	Economic aspects	Addresses the economic costs of acquisition, maintenance and use from the payer and/or social perspective, and the cost-effectiveness ratio in comparison with existing alternatives
8	Human and sociocultural aspects	Evaluation of the human and socio-cultural aspects that can impact the use of the technology (e.g. acceptability, usability, digital health literacy, commitment to intervention, perceived benefit, patient empowerment, etc.), and evaluation of the socio-cultural impact that the technology may have (e.g. accessibility to the service or to health care, changes in workflows and roles, changes in provider-patient relationships, etc.)

(*continued*)

Table 1. (continued)

Domain number	Domain name	Key elements
9	Ethical aspects regarding the use of the device	Evaluation of any ethical concerns regarding the use of the device or technology from all stakeholders perspective and the context in which it is implemented or intended to be used, including considerations of equity and fairness, accountability and transparency
10	Legal and regulatory aspects	Compliance with applicable laws and standards
11	Organizational aspects	Addresses the impact of the technology at the organizational level (e.g. changes in workflows or roles) as well as the infrastructure and resources needed to implement the technology (human resources, training, etc.)
12	Environmental impact	Covers the direct and indirect environmental impact associated with the development and implementation of the technology, measured for example in terms of estimated carbon emission, use of raw materials, energy consumption, or environmental benefits
13	Post-implementation monitoring	Describes the mechanisms established by the developers or those responsible for its management for the post-market performance evaluation

4 Discussion

The "AQuAS Assessment Guide for DHTs that use AI" provides a comprehensive and multidisciplinary framework for evaluating AI-based DHTs. It is worth highlighting its main limitations and strengths. One limitation is the fact that the guide does not specifically address the technical aspects and evaluation metrics related to GenAI. This is because by the time the guide was developed, there was to our knowledge no GenAI-based tool approved as a medical device in the European Union, and we chose to focus on the kind of tools used at the moment in clinical practice in the European context. Besides, we acknowledge that some dimensions or sub-dimensions may be transversal to more than one domain; nonetheless, what is truly essential is to evaluate them, regardless of the domain with which they may be more closely associated. For instance, accessibility and ethics might be considered both a technical aspect as well as an implementation aspect. To avoid redundancy, we proposed to evaluate accessibility in terms of accessibility elements, options or functionalities that the tool incorporates, as well as the ethics of the model under the technical domain. In contrast, the accessibility in terms of whether users with functional diversity can access the technology and whether they have enough capabilities to use it as expected is placed under the human and sociocultural domain, and the ethics regarding the use of the device is placed under the ethical domain.

With regard to the strengths, we highlight the structured approach across diverse domains that range from clinical relevance to environmental impact and post-implementation monitoring. By integrating also ethical, legal, and organizational considerations, among other, the guide promotes responsible innovation. Moreover, the guide promotes alignment with current legislation, including the EU AI Act, helping ensure that AI systems are classified, assessed, and monitored in compliance with the proposed risk-based regulatory framework. Furthermore, it integrates principles such as sustainability and social impact, often underrepresented in traditional evaluation frameworks.

Another key aspect of this guide is its adaptability, as it can be used to assess technologies during different stages of their lifecycle. It can be used for early HTAs, for technologies that are ready to be implemented or even already in use. For instance:

- A start-up testing an early-stage clinical prediction model could potentially focus on domains 1 and 2 (description of the health problem, target population and technology), 3 (technical aspects), 5 (early safety data), and 10 (legal and regulatory aspects), even before reaching market readiness.
- A hospital deploying an AI-supported decision tool could also focus on domains 1 and 2 (description of the health problem, target population and technology), but then also on domains 5 and 6 (safety and clinical effectiveness), 8 (human and sociocultural aspects), 9 (ethical aspects regarding the use of the device), 11 (organizational aspects), and 13 (monitoring).

5 Conclusions and Next Steps

The AQuAS AI Assessment Guide is conceived as a tool for harmonizing HTA assessments performed by HTA bodies and hospital HTA units. It provides a comprehensive framework that supports the responsible evaluation of AI-based DHTs, offering:

- A structured approach covering, among other, technical, clinical, ethical, and societal dimensions.
- Flexibility across technology maturity levels and use cases.
- Compatibility with regulatory requirements.
- Usability for diverse stakeholders, from public evaluators to technology developers.

Currently, we are piloting the guide in the context of HTA processes of real AI-based technologies; we have published two HTA reports on two emerging technologies using the guide as assessment framework [7, 8] and have other reports underway. In parallel we are using it in calls for the public procurement of innovative AI-based tools in Catalonia; here, the guide is helping us to define the value demonstration outcomes that we will be looking at once we acquire these new technologies.

Our future work will focus on:

- Disseminating this guide and helping professionals in using it within health service and innovation contexts.
- Updating the guide periodically in response to regulatory changes, user feedback and new emerging AI paradigms like GenAI.
- Expanding this work and building a framework for use at the National levels and via broader consensus with a bigger working group.

Acknowledgments. We acknowledge Borja Velasco Regúlez, Carme Pratdepàdua Bufill and the Health Data Protection Officer Office for their contribution in reviewing aspects related to validation metrics, accessibility and usability, and the legal aspects, respectively.

Disclosure of Interests. The authors declare having no competing interests that are relevant to the content of this article.

References

1. O'Rourke, B., Oortwijn, W., Schuller, T.: The new definition of health technology assessment: a milestone in international collaboration. Int. J. Technol. Assess. Health Care **36**(3), 187–190 (2020)
2. RedETS Homepage. https://redets.sanidad.gob.es/. Accessed 17 July 2025
3. Moltó-Puigmartí, C., Aussó-Trias, S., Bretones Vallejo, M., Domínguez Herrera, D., Vivanco-Hidalgo, R.M.: Assessment guide for digital health technologies that use artificial intelligence (AI). Agency for Health Quality and Assessment of Catalonia Department of Health. Generalitat de Catalunya, Barcelona (2024)
4. Segur-Ferrer, J., Moltó-Puigmartí, C., Pastells-Peiró, R., Vivanco-Hidalgo, R.M.: Health Technology Assessment Framework: Adaptation for Digital Health Technology Assessment. Madrid: Ministry of Health. Agència de Qualitat i Avaluació Sanitàries de Catalunya, Barcelona
5. EUnetHTA Joint Action 2. HTA Core Model ® version 3.0 (Pdf) (2016). https://www.eunethta. eu/wp-content/uploads/2018/01/HTACoreModel3.0.pdf
6. National Institute for Health and Care Excellence. Evidence standards framework (ESF) for digital health technologies (2022). https://www.nice.org.uk/corporate/ecd7/resources/evidence-standards-framework-for-digital-health-technologies-pdf-1124017457605
7. Moltó-Puigmartí, C., Zarranz-Ventura, J., Estrada Sabadell, M.D., Vivanco-Hidalgo, R.M.: Sistema de detección automática de retinopatía diabética mediante inteligencia artificial (IA). Madrid: Ministerio de Sanidad. Agència de Qualitat i Avaluació Sanitàries de Catalunya, Barcelona (2025). (Colección: Informes, estudios e investigación / Ministerio de Sanidad. Informes de Evaluación de Tecnologías Sanitarias)
8. Prieto Durán, I., Moltó-Puigmartí, C., Mestre Lleixà, C., Ibeas, J., Estrada Sabadell, M.D., Vivanco-Hidalgo, R.M.: Algoritmo predictivo para monitorizar y evaluar el impacto de los tratamientos en trasplante de riñón. Madrid: Ministerio de Sanidad. Agència de Qualitat i Avaluació Sanitàries de Catalunya, Barcelona (2025). (Informes de Evaluación de Tecnologías Sanitarias)

Enhancing Privacy and Interoperability in Biomedical Research: A SOLID-Based Architecture with LLM Integration

Hugo Lebredo[(✉)][iD], Jorge Álvarez-Fidalgo[iD], Rubén del Rey Álvarez[iD], and Jose Emilio Labra-Gayo[iD]

Department of Computer Science, University of Oviedo, 33007 Oviedo, Spain
hugolebredo@gmail.com
https://www.weso.es/

Abstract. Ensuring patient privacy in the use of personal data within biomedical research datasets remains one of the foremost challenges in the field, especially concerning the risk of re-identification from clinical data. This paper presents a conceptual framework for an architecture designed to address this challenge. The model is conceived to, first, allow medical institutions to access clinical data from patient-controlled personal repositories based on patient consent and, second, provide tools for healthcare providers to integrate this data to build larger, more complex datasets. To conceptualize this system, we leverage semantic technologies like Shape Expressions (ShEx), alongside standards such as FHIR and SNOMED CT. The proposed framework integrates these components under SOLID principles and incorporates an LLM layer for accessibility. This theoretical architecture provides a foundation for a decentralized, secure, and accessible system for healthcare professionals.

Keywords: Data Decentralization · Data Interoperability · Personal Health Records (PHRs) · Large Language Models (LLMs) · FHIR · Resource Description Framework (RDF) · Shape Expressions (ShEx)

1 Introduction

The systematic collection of medical data by healthcare institutions is critical for constructing large and representative datasets. These datasets are essential for developing analytical tools that can accurately classify clinical conditions such as diabetes. Accurate diagnosis is paramount, as misdiagnosis can have severe consequences for patient health, healthcare provider liability, and institutional reputation [19].

The creation of such datasets typically relies on real-world data extracted from medical records generated during routine clinical care. For instance, laboratory tests from patients in a hospital's oncology unit can form the foundation for these datasets. However, data accessibility is often hindered by information

A. López Fernández et al. (Eds.): CIABiomed 2025, LNBI 16148, pp. 97–107, 2026.
https://doi.org/10.1007/978-3-032-10661-2_8

silos within healthcare institutions. This paper proposes a **conceptual framework for a data storage architecture** that, with appropriate patient consent, would ensure medical institutions can access data while facilitating the aggregation of individual records into larger, more comprehensive datasets.

2   State of the Art

2.1   Information Silos

Information silos are isolated information systems that lack interoperability with other, related systems. They often emerge unintentionally from inherent organizational structures, driven by technological, organizational, political, or temporal factors [4]. In the healthcare sector, these silos are particularly pronounced, resulting from distinct dynamics that lead to the fragmentation and isolation of systems even within a single institution [6,9]. Fundamentally, these silos reflect and reinforce the hierarchical structure of organizations, centralizing decision-making authority within specific functional units.

From a technical perspective, these systems are relatively simple, as they tightly integrate the user interface, application logic and database, enabling stable operation and ease of modification in response to evolving tasks or data formats. Consequently, they have proven effective in reducing local complexity within their respective functional environments.

The utility of these siloed systems, however, diminished with the rise of new operational paradigms that emphasize interdepartmental and inter-organizational collaboration. A major conceptual shift began in the 1980 s with the process-based approach, which advocated for organizing workflows as cross-cutting processes rather than along rigid functional divisions [15]. Later, the advent of the Internet fundamentally transformed institutional interactions by fostering large-scale horizontal connectivity, as seen in global supply chains and e-commerce. This new environment, which prioritizes seamless information exchange, stands in direct contrast to the vertically integrated nature of information silos, highlighting their limitations as barriers to integrated care and efficiency.

2.2   Personal Health Records

In response to the challenges posed by information silos, Personal Health Records (**PHRs**) have been advanced as a key integrative solution. Since being formally proposed by the American Medical Informatics Association's College of Medical Informatics in 2005 [26], PHRs have assumed increasing importance in the evolution of health information systems and are central to advancing patient-centered care models.

PHRs directly address the fragmentation caused by silos by shifting control and integration to the patient. In contrast to **Electronic Health Records (EHRs)**, which are designed for professional use within specific healthcare institutions, a PHR is a structured repository of an individual's health data that is managed and controlled by the patient themselves. A core function of the PHR is to aggregate and integrate information from diverse sources (including multiple EHR silos) thereby creating a unified, longitudinal health record and overcoming the barriers to interoperability that silos create.

Primary responsibility for administering PHR content, including authorization of access and information sharing with third parties, resides with either the individual or their legally designated representative [18]. The conceptual framework governing EHR-PHR interoperability is illustrated in Fig. 1.

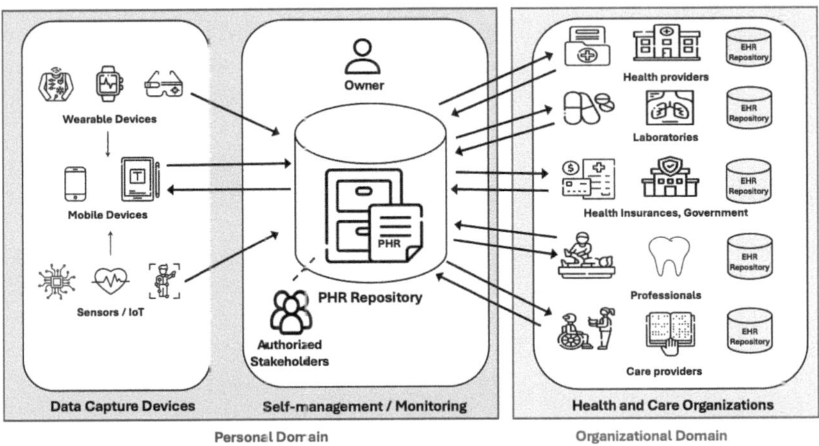

Fig. 1. Diagram showing the EHR-PHR interoperability.

Since their inception, PHRs have evolved from basic digital repositories to advanced platforms capable of interfacing with monitoring devices, mobile applications and hospital systems. This evolution reflects both technological advances and increasing patient demand for tools that facilitate proactive, personalized health management [21]. However, this patient empowerment poses significant technical and ethical challenges, particularly concerning user-generated data quality, information accuracy, and third-party data utilization [12].

Empirical studies demonstrate that PHRs can enhance treatment adherence, clinician-patient communication, and care continuity, particularly for chronic disease management. However, widespread adoption remains constrained by barriers including limited digital literacy among users, healthcare system fragmentation, and persistent privacy concerns [16].

2.3   Interoperability

The IEEE defines interoperability as *"The ability of two or more systems or components to exchange information and to use the information that has been exchanged."*

Interoperability may be categorized into two fundamental levels: syntactic and semantic. Syntactic interoperability enables data exchange by adopting shared technical standards hat govern the structure and syntax of transmitted data, enabling information processing between systems [5]. **Semantic interoperability** extends this capability by ensuring the preserved meaning of exchanged information across systems.

Achieving semantic understanding requires consensus on the meaning of data, necessitating the adoption of standardized ontologies and vocabularies among participating systems [11,14,17]. Notably, both interoperability levels employ similar technical frameworks, creating a paradoxical situation where implementation tools overlap despite differing conceptual requirements. Consequently, a thorough examination of medical community-approved standards and terminology becomes essential for properly contextualizing these interoperability concepts within healthcare informatics.

2.4   Privacy and Data Security

The Oxford's Concise Medical Dictionary, defines privacy as *"the condition of being apart from public view. In medical ethics, the concept is associated with maintaining a patient's dignity and autonomy and with the doctor's duty of confidentiality"*. To comply with this principle, legal frameworks have been established to regulate medical data management by both corporate entities and public institutions.

The use of PHRs necessitates processing highly sensitive information, making privacy preservation and data security fundamental design requirements. User adoption fundamentally depends on establishing trust that systems will prevent unauthorized access, misuse, and security breaches through robust protective measures.

International legislation, such as the European Union's General Data Protection Regulation (GDPR) [2], has established a legal framework for the protection of personal health data. These regulatory measures have subsequently spawned additional frameworks like the Data Act [3], which provide specific governance guidelines for personal data utilization across healthcare, industrial, and research domains. Such regulations explicitly acknowledge individual rights to data control, encompassing access, correction, portability, and deletion capabilities.

Blockchain has emerged as a research direction for securing medical data, serving as an immutable record system that ensures data integrity and traceability and supports access control. When combined with external encryption techniques, it offers a promising approach for healthcare applications. However, **major limitations** remain, particularly scalability, economic and computational costs, historical health informatics challenges, such as proper semantic analysis of blockchain data transfer and potential patient-matching [22,24].

For PHRs, these legal provisions translate to **requiring robust patient control mechanisms over information access permissions and usage purposes.**

3   Proposal

We propose a **novel architecture based on SOLID (Social Linked Data) principles** [23] that redefines PHRs through decentralized personal data storage. This framework establishes **personal online data stores (PODs)** as secure repositories for clinical information, enabling patients to maintain sovereignty over their health data while permitting controlled access by medical and research institutions with explicit consent.

The architecture exhibits three defining characteristics:

1. **Data-Application Decoupling.** Separates medical data from consuming applications, enabling individuals to aggregate clinical data in personal repositories with granular access controls.
2. **Dual-Domain Framework.** Enables the coexistance of two distinct domains: organizational (Healthcare institutions managing EHRs) and personal (Individual-controlled health data spaces).
3. **Bi-directional Data Flows.** Supports institutional EHR retention while enabling both push of clinical data to personal repositories and pull of third-party data into organizational systems–with authorization–.

User consent is central to this decentralized approach. Individuals can grant data access to both PHR applications and healthcare organizations' EHR systems. This is managed through the POD's granular permission system, which controls read and write access for each specific data unit. When an external system requests data access to the POD, the user authenticates (directly via the POD or through an intermediary PHR app) to authorize the access. This consent remains valid until the user actively changes the permissions.

Figure 2 illustrates this architecture, emphasizing the complete separation of personal repositories from institutional control.

3.1   Interoperability Infrastructure

Access to information is merely the first step: effective data utilization across systems requires semantic interoperability. To achieve this, we adopt the following technologies. Data stored in PODs will use the **Resource Description Format** (RDF 1.1), which has been identified as the most suitable candidate for a universal healthcare exchange language [1]. This will be complemented by widely recognized medical standards such as **Fast Healthcare Interoperability Resources** (FHIR v5)[1]. FHIR is distinguished by its broad applicability across diverse medical domains, enabling adoption in settings ranging from hospitals to pediatric practices and dental clinics.

However, the use of FHIR alone does not guarantee interoperability. While it provides a standardized set of resources, its design permits high flexibility, enabling different vendors to implement the same resources with varying internal structures (topologies). This variability hinders the automatic consumption of

---

[1] https://hl7.org/fhir/.

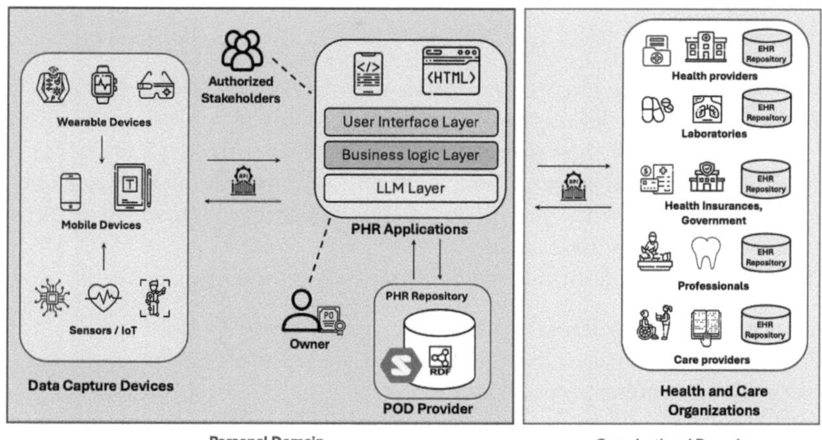

Fig. 2. Overview of the proposed architecture.

information by external systems, as they cannot reliably infer the structure of received data or ensure correct interpretation. To address this challenge, we propose the incorporation of a complementary technology to enforce structural uniformity: **Shape Expressions** (ShEx 2.1) [20].

ShEx is a formal language for describing RDF graph structures and validating their conformance to predefined schemas, thereby enabling automated data interpretation and facilitating data portability. It has proven effective in harmonizing heterogeneous datasets across diverse domains by identifying inconsistencies and errors in data models [7,8,27]. Owing to these capabilities, ShEx emerges as a robust solution for validating clinical data models, such as FHIR archetypes [25].

The transmission of medical data in RDF format, accompanied by their corresponding ShEx schemas, is motivated by two primary objectives:

1. To provide the recipient system with the formal semantics required to accurately **interpret** the information within the RDF graph.
2. To ensure the transmitted data adheres to the structural and semantic **constraints** defined in its associated schema.

As the same clinical information can be represented differently across institutions–even when using common terminologies–each system is expected to maintain and apply its own specific shapes for validation.

This semantic infrastructure must be augmented by standardized medical terminology to ensure unambiguous coding of clinical concepts. We propose **SNOMED CT**[2] (International Edition) as the reference terminology system for encoding all relevant medical terms.

---

[2] https://www.snomed.org/.

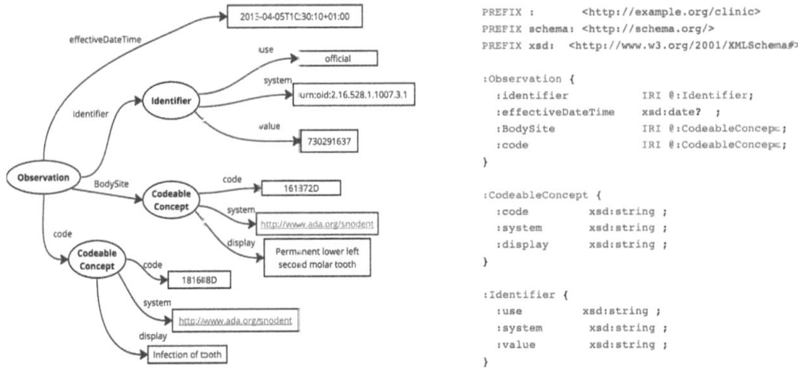


```
PREFIX :        <http://example.org/clinic>
PREFIX schema: <http://schema.org/>
PREFIX xsd:   <http://www.w3.org/2001/XMLSchema#>

:Observation {
  :identifier          IRI @:Identifier;
  :effectiveDateTime   xsd:date?  ;
  :BodySite            IRI @:CodeableConcept;
  :code                IRI @:CodeableConcept;
}

:CodeableConcept {
  :code        xsd:string ;
  :system      xsd:string ;
  :display     xsd:string ;
}

:Identifier {
  :use         xsd:string ;
  :system      xsd:string ;
  :value       xsd:string ;
}
```

Fig. 3. RDF Data and Shape Expression stored in the PHR repository

Figure 3 illustrates a clinical observation example from a dental practice, demonstrating the integrated use of FHIR, SNOMED CT, and its corresponding ShEx for structural validation.

3.2 LLM-Enabled Accesibility Layer

To enable non-technical users to interpret their personal health data (stored as RDF triples in PODs) we propose integrating **large language models (LLMs)**. This approach uses relevant Shape Expressions as contextual prompts to ensure the LLM accurately interprets and presents to the patients information within the FHIR-SNOMED semantic framework.

Given the security and privacy implications of LLMs [28], we prioritize **open-source solutions** that integrate seamlessly into the POD-centric architecture. This preserves individual data sovereignty while maintaining functionality. The deployment options for this integration are compared in Table 1. We selected the **On-Premise model** for its optimal balance between technical feasibility and patient data privacy.

Table 1. Comparison of LLM Deployment Models for the proposed architecture.

Deployment Model	Data Privacy	Explicit consent	Technical Complexity
In-POD	Data never leaves user's device; maximum privacy by design.	Not Required	Very High *(demands significant local compute resources)*.
On-Premise	Data remains within a trusted institutional network.	Required	Medium *(requires dedicated institutional infrastructure and IT support)*.
External Cloud	Data processed on third-party servers; privacy dependent on provider.	Required	Low *(managed service, but requires rigorous vendor assessment)*.

4   Conclusions and Limitations

This paper highlights the critical need for an ecosystem enabling seamless interoperability of PHRs across independent, heterogeneous systems. We propose a technical framework combining stakeholder-defined FHIR archetypes with ShEx within a decentralized architecture, where personal data repositories remain distinct from consuming services. This approach allows patients to: (1) aggregate fragmented health information from multiple sources, and (2) selectively share data with relevant stakeholders, thereby resolving structural and terminological inconsistencies across domains.

The solution's **principal advantage** is its dual capacity to maintain genuine patient data sovereignty while ensuring interoperability – both among PHRs and with existing EHR systems. Furthermore, it facilitates the systematic collection of medical data across institutions and specialties, enabling the creation of more comprehensive, accurate, and interoperable health datasets. This capability is vital for clinical decision-making, population health monitoring, and medical research, while preserving patient autonomy through granular consent mechanisms.

A key economic advantage of the proposed architecture is its operational flexibility, which allows for **scalable adoption** tailored to the specific needs of different healthcare providers. For instance, a dental clinic could minimize its initial investment by developing a lightweight software module to exchange data with patients' personal PODs, using externally managed validation services (e.g., RDFShape[3]) and third-party repositories (e.g., Inrupt PODs). Data models in such specialized domains are often simple–Taiwan's advanced national health system, for example, uses only nine fields in dentistry [10]. Furthermore, these models and their corresponding Shape Expressions are relatively stable, requiring modification only in exceptional cases like system migrations.

In larger hospitals, while the inherent complexity of data is greater, the granularity of the approach still supports incremental implementation. Departments can adopt the system sequentially, leveraging their broader resources to manage the rollout. Similarly, staff training needs are proportional to the scale of adoption; in a small clinic, personnel would only require basic training to use streamlined software for sending and receiving patient data.

However, the authors acknowledge significant **limitations** primarily concerning the **initial transition phase**. The mapping and transformation of existing legacy data from traditional EHRs into interoperable RDF models conforming to FHIR and SNOMED CT standards is a resource-intensive undertaking. Technologies like RML[4] and ShExML [13] can facilitate this process, but healthcare institutions must still overcome substantial technical and organizational challenges to complete this migration.

Moreover, the technical complexity inherent in both PHR management and our semantic infrastructure creates substantial adoption barriers. The interpre-

---

[3] https://rdfshape.weso.es/.
[4] https://rml.io/specs/rml/.

tation of personal health data structured as FHIR-SNOMED-compliant RDF graphs poses significant difficulties for non-technical users, representing a fundamental challenge for decentralized health record systems. Although we propose an LLM-based interface layer to mitigate these issues, this solution introduces additional privacy considerations and technical hurdles. Ensuring data protection will require careful implementation of open-source models within the architecture.

These challenges highlight important directions for **future work**. Our immediate focus involves developing a functional prototype to empirically evaluate the proposed architecture and address the identified limitations through iterative testing and refinement.

Acknowledgments. This work has been partially funded by the project ANGLIRU: Applying kNowledge Graphs to research data interoperabiLIty and ReUsability, code: PID2020-117912RB from the Spanish Research Agency and by the regional project SEK-25-GRU-GIC-24-039.

Disclosure of Interests. The authors have no competing interests to declare that are relevant to the content of this article.

References

1. Yosemite Manifesto (2013). http://www.yosemitemanifesto.org/
2. Council regulation (EU) no 2016/679 (2016). http://data.europa.eu/eli/reg/2016/679/oj
3. Proposal for a Regulation of the European Parliament and of the Council on harmonized rules on fair access to and use of data. February 2022. https://eur-lex.europa.eu/legal-content/EN/TXT/?uri=CELEX%3A52022PC0068
4. Bannister, F.: Dismantling the silos: extracting new value from IT investments in public administration. Inf. Syst. J. **11**(1), 65–84 (2001). https://doi.org/10.1046/j.1365-2575.2001.00094.x, https://onlinelibrary.wiley.com/doi/10.1046/j.1365-2575.2001.00094.x
5. Bernstam, E.V., et al.: Quantitating and assessing interoperability between electronic health records. J. Am. Med. Inform. Assoc. **29**(5), 753–760 (2022) https://doi.org/10.1093/jamia/ocab289, https://academic.oup.com/jamia/article/29/5/753/6500181
6. Bygstad, B., Hanseth, O., Le, D.T.: From it silos to integrated solutions. A Study in e-health complexity, p. 15 (2015)
7. Candela, G.: An automatic data quality approach to assess semantic data from cultural heritage institutions. J. Am. Soc. Inf. Sci. **74**(7), 866–873 (2023). https://doi.org/10.1002/asi.24761, https://asistdl.onlinelibrary.wiley.com/doi/10.1002/asi.24761
8. Candela, G., Escobar, P., Sáez, M., Marco-Such, M.: A shape expression approach for assessing the quality of linked open data in libraries. Semant. Web **14**(2), 159–179 (2022). https://doi.org/10.3233/SW-210441, https://www.medra.org/servlet/aliasResolver?alias=iospress&doi=10.3233/SW-210441
9. Cebul, R.D., Rebitzer, J.B., Taylor, L.J., Votruba, M.E.: Organizational fragmentation and care quality in the U.S. Healthcare system. J. Econ. Perspect. **22**(4), 93–113 (2008). https://doi.org/10.1257/jep.22.4.93

10. Chiang, H.T., Chang, C.T.: Introduction to and application analysis of Taiwan's NHI-MediCloud system. J. Serv. Sci. Res. **11**(1), 93–115 (2019). https://doi.org/10.1007/s12927-019-0005-6, http://link.springer.com/10.1007/s12927-019-0005-6

11. Fu, G.: FCA based ontology development for data integration. Inf. Process. Manage. **52**(5), 765–782 (2016). https://doi.org/10.1016/j.ipm.2016.02.003, https://linkinghub.elsevier.com/retrieve/pii/S030645731630019X

12. Fuji, K.T., Abbott, A.A., Galt, K.A., Drincic, A., Kraft, M., Kasha, T.: Standalone personal health records in the United States: meeting patient desires. Heal. Technol. **2**(3), 197–205 (2012). https://doi.org/10.1007/s12553-012-0028-1, http://link.springer.com/10.1007/s12553-012-0028-1

13. García-González, H., Boneva, I., Staworko, S., Labra-Gayo, J.E., Cueva Lovelle, J.M.: ShExML: improving the usability of heterogeneous data mapping languages for first-time users. PeerJ Comput. Sci. **6**, e318 (2020). https://doi.org/10.7717/peerj-cs.318, https://peerj.com/articles/cs-318, publisher: PeerJ

14. Ghorbani, A., Davoodi, F., Zamanifar, K.: Using type-2 fuzzy ontology to improve semantic interoperability for healthcare and diagnosis of depression. Artif. Intell. Med. **135**, 102452 (2023). https://doi.org/10.1016/j.artmed.2022.102452, https://linkinghub.elsevier.com/retrieve/pii/S0933365722002044

15. Harmon, P.: The Scope and evolution of business process management. In: Brocke, J.V., Rosemann, M. (eds.) Handbook on Business Process Management 1, pp. 37–81. Springer, Berlin, Heidelberg (2010). https://doi.org/10.1007/978-3-642-00416-23, http://link.springer.com/10.1007/978-3-642-00416-2_3

16. Liu, C.F., Cheng, T.J., Chen, C.T.: Exploring the factors that influence physician technostress from using mobile electronic medical records. Inform. Health Soc. Care **44**(1), 92–104 (2019). https://doi.org/10.1080/17538157.2017.1364250, https://www.tandfonline.com/doi/full/10.1080/17538157.2017.1364250

17. de Mello, B.H., et al.: Semantic interoperability in health records standards: a systematic literature review. Heal. Technol. **12**(2), 255–272 (2022). https://doi.org/10.1007/s12553-022-00639-w, https://link.springer.com/10.1007/s12553-022-00639-w

18. Negro-Calduch, E., Azzopardi-Muscat, N., Krishnamurthy, R.S., Novillo-Ortiz, D.: Technological progress in electronic health record system optimization: systematic review of systematic literature reviews. Int. J. Med. Informatics **152**, 104507 (2021). https://doi.org/10.1016/j.ijmedinf.2021.104507, https://linkinghub.elsevier.com/retrieve/pii/S1386505621001337

19. Pierce, A.T., et al.: Impact of "Defensive Medicine" on the costs of diabetes and associated conditions. Ann. Vasc. Surg. **87**, 231–236 (2022). https://doi.org/10.1016/j.avsg.2022.05.002, https://www.sciencedirect.com/science/article/pii/S0890509622002357

20. Prud'hommeaux, E., Labra Gayo, J.E., Solbrig, H.: Shape expressions: an RDF validation and transformation language. In: Proceedings of the 10th International Conference on Semantic Systems, SEMANTICS 20144, pp. 32–40. ACM (2014)

21. Roehrs, A., da Costa, C.A., Righi, R.D.R., de Oliveira, K.S.F.: Personal health records: a systematic literature review. J. Med. Internet Res. **19**(1), e13 (2017). https://doi.org/10.2196/jmir.5876, http://www.jmir.org/2017/1/e13/

22. Roehrs, A., Da Costa, C.A., Da Rosa Righi, R., Da Silva, V.F., Goldim, J.R., Schmidt, D.C.: Analyzing the performance of a blockchain-based personal health record implementation. J. Biomed. Inform. **92**, 103140 (2019). https://doi.org/10.1016/j.jbi.2019.103140, https://linkinghub.elsevier.com/retrieve/pii/S1532046419300589

23. Sambra, A.V., et al.: Solid: a platform for decentralized social applications based on linked data (2016)
24. Schmeelk, S., Kanabar, M., Peterson, K., Pathak, J.: Electronic health records and blockchain interoperability requirements: a scoping review. JAMIA Open **5**(3), ooac068 (2022). https://doi.org/10.1093/jamiaopen/ooac068, https://academic.oup.com/jamiaopen/article/doi/10.1093/jamiaopen/ooac068/6650914
25. Solbrig, H.R., Prud'hommeaux, E., Grieve, G., McKenzie, L., Mandel, J.C., Sharma, D.K., Jiang, G.: Modeling and validating HL7 FHIR profiles using semantic web Shape Expressions (ShEx). J. Biomed. Inform. **67**, 90–100 (2017). https://doi.org/10.1016/j.jbi.2017.02.009, https://linkinghub.elsevier.com/retrieve/pii/S1532046417300345
26. Tang, P.C., Ash, J.S., Bates, D.W., Overhage, J.M., Sands, D.Z.: Personal health records: definitions, benefits, and strategies for overcoming barriers to adoption. J. Am. Med. Inform. Assoc. **13**(2), 121–126 (2006). https://doi.org/10.1197/jamia.M2025, https://academic.oup.com/jamia/article-lookup/doi/10.1197/jamia.M2025
27. Thuluva, A.S., Anicic, D., Rudolph, S.: IoT semantic interoperability with device description shapes. In: Gangemi, A., et al. (eds.) ESWC 2018. LNCS, vol. 11155, pp. 409–422. Springer, Cham (2018). https://doi.org/10.1007/978-3-319-98192-5_56
28. Yao, Y., Duan, J., Xu, K., Cai, Y., Sun, Z., Zhang, Y.: A survey on large language model (LLM) security and privacy: the good, the bad, and the ugly. High-Confidence Comput. **4**(2), 100211 (2024). https://doi.org/10.1016/j.hcc.2024.100211, https://linkinghub.elsevier.com/retrieve/pii/S266729522400014X, publisher: Elsevier BV

AI-based Clinical Decision Support Systems

Application of Machine Learning Techniques to the Prediction of Hospital Mortality: Beyond Conventional Clinical Models

Pedro López Ruz$^{(\boxtimes)}$, Belén Vega-Márquez, and Beatriz Pontes-Balanza

Department of Computer Languages and Systems, University of Seville, 41012 Seville, Spain
pedlopruz@alum.us.es, bvega@us.es

Abstract. This study presents a comprehensive analysis of patient cohorts admitted to Intensive Care Units (ICUs), with the objective of evaluating, comparing, and optimizing the metrics employed for mortality prediction in critically ill patients. Given the inherent complexity of clinical prognosis in these settings, widely adopted scoring systems such as APACHE, SOFA, and SAPS are critically examined, highlighting their principal strengths and limitations. To enhance predictive performance, the study proposes the integration of advanced Machine Learning and Data Science methodologies. The approach includes the implementation of machine learning algorithms, systematic variable selection, cross-validation, and performance assessment using metrics such as AUC-ROC, accuracy, sensitivity, and specificity. Model development will be based on real ICU patient data, ensuring strict adherence to ethical standards and data confidentiality. The best-performing model will subsequently be benchmarked against traditional tools to evaluate its capacity to improve mortality prediction and support clinical decision-making. Beyond predictive accuracy, the study emphasizes the importance of feasibility and applicability in real-world clinical environments. Overall, this research seeks to contribute to the development of innovative technological solutions that enable more personalized, efficient, and evidence-based medical care in high-complexity hospital units, thereby supporting optimized management of resources, healthcare personnel, and clinical protocols within the ICU.

Keywords: Machine Learning · Data Science · Intensive Care Unit (ICU)

1   Introduction

Intensive Care Units (ICUs) are highly specialized hospital environments dedicated to the management of critically ill patients requiring continuous monitoring, advanced life support, and intensive medical interventions. In this context,

A. López Fernández et al. (Eds.): CIABiomed 2025, LNBI 16148, pp. 111–125, 2026.
https://doi.org/10.1007/978-3-032-10661-2_9

clinical decision-making must be rapid and precise, as both patient survival and the efficient allocation of healthcare resources depend on it. Consequently, the development of reliable tools to anticipate clinical outcomes has become a critical priority.

Over the past decades, several scoring systems and clinical indices have been designed to estimate hospital mortality risk in critically ill patients. Among the most widely implemented are APACHE (Acute Physiology and Chronic Health Evaluation) [1,2], SOFA (Sequential Organ Failure Assessment) [3,4], and SAPS (Simplified Acute Physiology Score) [5,6]. These instruments rely on physiological variables, medical history, and standardized clinical parameters, which are integrated through statistical models to generate risk scores. While these systems have demonstrated clinical utility [2,7], they present important limitations: their predictive accuracy diminishes in populations with characteristics that differ from the original validation cohorts [8,9], their implementation may be challenging in resource-constrained environments [10], and they fail to adequately capture temporal dynamics or nonlinear interactions among variables [12].

In parallel, the emergence of Data Science and Machine Learning (ML) has transformed multiple domains, including medicine. These methodologies enable the analysis of large-scale, heterogeneous clinical data, allowing the identification of complex patterns that are often overlooked by conventional statistical approaches [12,13]. Within the ICU setting, ML-based algorithms hold the potential to develop more accurate, adaptive, and interpretable predictive models [14,15], with the capacity to complement or even surpass traditional clinical scoring systems [16,17].

The primary objective of this study is to evaluate the performance of different machine learning models in predicting hospital mortality among ICU patients, using real-world clinical datasets. To this end, supervised algorithms–including Logistic Regression, Random Forest, XGBoost, LightGBM, and FLAML–were implemented, alongside preprocessing procedures such as missing value imputation, categorical variable encoding, and class balancing through techniques such as SMOTE and Tomek Links.

Beyond model construction and validation, this study incorporates explainability analyses using SHAP and LIME to provide clinically meaningful interpretations of the results, thereby facilitating integration into practical healthcare settings [12,14,15]. Model performance was assessed through standard evaluation metrics, including AUC-ROC, accuracy, sensitivity, and specificity, and compared against predictions generated by conventional scoring systems.

Ultimately, this work aims to generate evidence on the added value of ML-based approaches in intensive care medicine, supporting more efficient resource utilization and enhancing clinical decision-making in high-complexity hospital environments [16,17].

2   Related Work

2.1   Traditional Clinical Models for the Prediction of Mortality in the ICU

The prediction of hospital mortality in critically ill patients has been the focus of extensive research in intensive care medicine, leading to the development of several widely adopted clinical scoring systems. These tools enable patient risk stratification in Intensive Care Units (ICUs) and support critical decisions, such as resource allocation, admission prioritization, and therapeutic interventions. Among the most commonly used models are APACHE, SOFA, SAPS, MPM II, and LODS.

The **APACHE II** system (Acute Physiology and Chronic Health Evaluation II), introduced in 1985 as an evolution of the original APACHE model, incorporates 12 physiological variables (e.g., temperature, mean arterial pressure, respiratory rate, creatinine) alongside risk factors such as age and chronic diseases. The final score is calculated using a fixed-scale system, subsequently converted into an estimated probability of mortality. Its **APACHE IV** version expands the number of clinical variables and incorporates hospital procedures, improving predictive accuracy via adjusted logistic regression models; however, its complexity necessitates specialized software for practical use [1,2].

SOFA (Sequential Organ Failure Assessment) evaluates six physiological systems–respiratory, cardiovascular, hepatic, renal, hematological, and neurological–assigning a score from 0 to 4 for each. Unlike APACHE, SOFA allows for dynamic monitoring over time, which is particularly advantageous in conditions such as sepsis. Nonetheless, its sensitivity may be limited by the absence of daily updated data, and it does not account for demographic factors or patient history [3,4].

The **SAPS II** (Simplified Acute Physiology Score) incorporates 17 variables collected during the first 24 h of ICU admission, including vital signs, laboratory results, and primary diagnosis. Its main strengths are simplicity and extensive validation in large European cohorts. However, SAPS II does not capture subsequent clinical evolution or response to initial treatments, which may limit its predictive value in prolonged or complex cases [5,6].

MPM II (Mortality Probability Models II) provides risk estimations at multiple time points (24h, 48h, and 72h after ICU admission), using straightforward clinical variables and comorbidities to calculate mortality probabilities via logistic regression. While it is easy to apply and validated in multiple settings, its discriminative capacity is lower than that of models such as APACHE IV, especially for patients with atypical presentations [7,8].

Finally, the **LODS** system (Logistic Organ Dysfunction System) also evaluates six organ systems, but relies on a statistically derived logistic formula to provide a continuous risk estimate. Although LODS demonstrates strong discriminative capacity, it requires high-quality data and robust data collection infrastructure for effective implementation [9,10].

Overall, these scoring systems have been instrumental in guiding clinical management and research. Nevertheless, they exhibit important limitations: they rely on static equations that do not adapt to evolving data patterns, require manual collection of specific clinical variables often unavailable in real time, lack longitudinal information, and fail to account for complex interactions between variables. Additionally, adaptation to new hospital contexts often requires recalibration.

These limitations have motivated the exploration of more flexible and dynamic approaches, such as machine learning techniques, which can accommodate new patient cohorts, integrate large volumes of heterogeneous data, and offer interpretability through explainable AI methods.

2.2   Comparison Between Clinical Models

The following is a summary comparison of the main clinical models used in the ICU. Table 1 highlights its key features.

Table 1. Comparison of traditional clinical scoring systems for ICU mortality prediction

Model	Variables	Advantages	Limitations
APACHE II	Physiological parameters + age + comorbidities	Strong clinical validation	Requires precise data collection
APACHE IV	Clinical variables + diagnoses	Higher accuracy than predecessors	Requires being specialized software
SOFA	6 organ systems	Dynamic daily tracking	Does not consider patient history
SAPS II	17 variables	Simplicity and European validation	Does not reflect clinical evolution
MPM II	Simple clinical data	Evaluation at multiple time points	Lower accuracy in atypical cases
LODS	Organ dysfunction + regression model	Continuous risk estimation	Highly dependent on data availability

As illustrated, while these models have been fundamental for clinical decision-making, they share limitations regarding adaptability, responsiveness to clinical changes, and the ability to process large volumes of heterogeneous data. These shortcomings underscore the need for alternative approaches, such as machine learning-based models, which enable predictions from diverse data sources and can accommodate complex, nonlinear patterns.

3   Methodology

3.1   Dataset and Exploratory Analysis

This study utilizes a selected cohort from the eICU Collaborative Research Database, a multicenter public repository containing detailed information on

ICU admissions in U.S. hospitals between 2014 and 2015 [11]. The eICU database is organized in relational tables and includes anonymized clinical data for over 200,000 patients. For this study, a subset of patients was extracted, primarily comprising individuals with cardiovascular, respiratory, and other critical conditions.

The initial objective was to explore various data sources within the eICU to assess their clinical relevance and potential predictive value. The following subsets were examined:

- **Medication table on admission:** Records of drugs administered during the first hours of ICU stay, categorized into therapeutic groups (cardiovascular, respiratory, neurological, etc.) to generate binary variables per pharmacological class.
- **Allergy table:** Documentation of patient allergies, classified into categories (drugs, food, other), enabling creation of categorical variables.
- **Medical diagnoses:** Diagnoses were grouped into broad clinical blocks (cardiovascular, respiratory, gastrointestinal, neurological, etc.) based on the codes present in the 'diagnosis' table.
- **Clinical history:** Relevant comorbidities were selected, including chronic kidney disease, hypertension, diabetes mellitus, heart failure, among others.
- **Physiological variables:** Extracted from the 'apachePatientResult table', key physiological variables used in the APACHE II scoring system were obtained, such as creatinine, bilirubin, mean arterial pressure, respiratory rate, Glasgow Coma Scale, temperature, hematocrit, and leukocyte count.
- **Additional APACHE IV variables:** Additional clinical information, including patient age, admission type (emergency or planned), prior location, mechanical ventilation use, and critical comorbidities.

For the initial predictive model, both the physiological variables from APACHE II and the additional APACHE IV variables–including age, sex, and comorbidities–were employed to provide a robust and standardized representation of patient condition during the first hours of ICU admission [1,2].

Once the model was trained and evaluated, the best-performing algorithm was used to build an extended model, incorporating additional variables from medication, allergies, diagnoses, and clinical history. This approach allowed assessment of whether the inclusion of supplementary clinical data significantly enhances predictive performance [12,13].

3.2   Pre-processing and Variable Processing

A major challenge in model development was the high proportion of missing values in physiological variables extracted from the 'apachePatientResult' dataset. These variables are essential for constructing clinically valid models, as they capture critical patient information within the first 24 h of ICU admission.

To investigate the consistency and origin of missing values, a reverse-engineering cross-validation approach was conducted. Since the dataset included the Acute Physiology Score (APS) calculated by the system, an attempt was made to manually reconstruct the score using available physiological variables. Two approaches were tested: imputing missing values with reference values from healthy individuals and using only patients with complete data. The manually estimated APS values were then compared to the system-provided APS.

As shown in Fig. 1, substantial discrepancies were observed between the two distributions. These differences indicate that the calculation of APS in the eICU dataset uses non-public formulas or weights, which prevents a faithful replication of the value from the base variables. As the eICU developers themselves confirm in their official documentation, the algorithms used to calculate scores such as APACHE IV and APS are proprietary, so the values provided in the dataset cannot be reproduced manually using the standard clinical formulae available in the literature. [11].Consequently, the use of the APS value as a reference or direct input variable in the model was discarded.

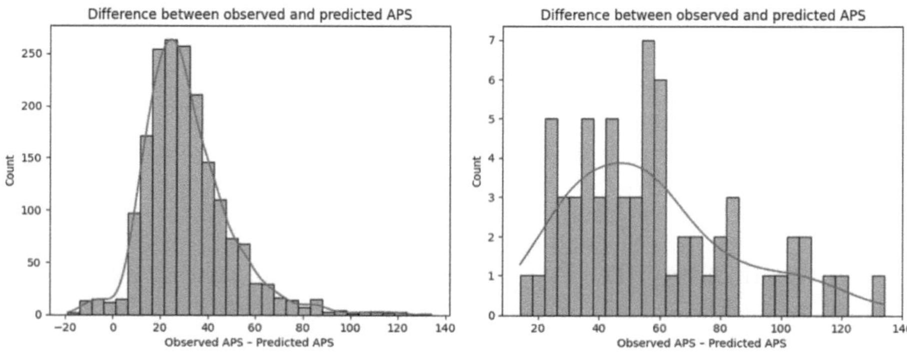

Fig. 1. Comparative graphs of actual and estimated APS

To address missing data while preserving sample size, a stratified imputation approach was implemented. Patients were grouped by sex and clinically relevant age ranges (e.g., women aged 30–50 years), and missing values were imputed using the mean value of the corresponding variable within each subgroup. This strategy minimizes bias by leveraging physiological homogeneity among patients with similar characteristics and avoids distorting clinical patterns associated with in-hospital mortality.

This procedure allowed preserving the sample size without introducing obvious biases by taking advantage of the physiological homogeneity of patients with similar characteristics. Moreover, by not resorting to global imputations, the risk of distorting clinical patterns associated with in-hospital mortality was minimized. However, a relevant challenge remains related to the dieinhospital variable, which will be used as the target variable in the predictive model. This variable presents a significant imbalance between classes, as illustrated in Fig. 2. This asymmetry in the distribution of positive and negative cases may compromise the performance of the model, particularly in metrics sensitive to the balance between classes, such as sensitivity or positive predictive value. Therefore, in the following section, different class balancing methods will be described and applied to mitigate this problem and favor a fair representation of both results in the model training process.

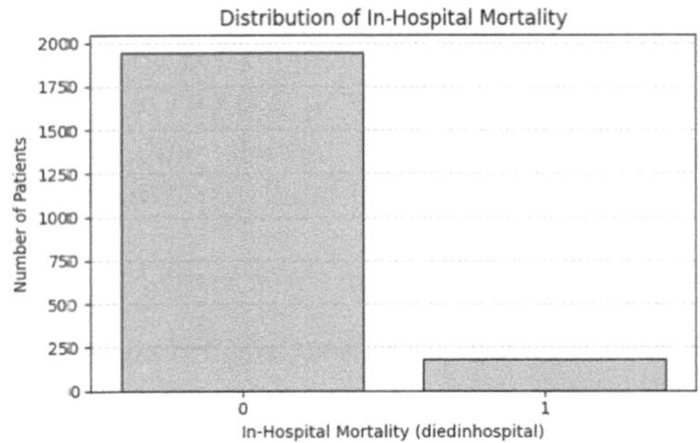

Fig. 2. Distribution of the target variable *dieinhospital*

3.3   Model Training and Evaluation

Given the substantial class imbalance in the target variable (dieinhospital), several balancing techniques were applied to mitigate bias toward the majority class. These methods enhance model sensitivity to the minority class by generating synthetic samples or adjusting the composition of the training set.

Both standard and specialized techniques were employed:

- **SMOTE (Synthetic Minority Over-sampling Technique):** Generates synthetic examples by interpolating between neighboring observations of the minority class.
- **SMOTE-Tomek Links:** Combines SMOTE with the removal of ambiguous pairs (Tomek links) that complicate class separation.
- **SMOTE-Tomek with hyperparameter optimization:** Same as above, with model hyperparameters tuned using the GridSearchCV algorithm.
- **SMOTE-ENN (Edited Nearest Neighbors):** After applying SMOTE, removes samples misclassified by their nearest neighbors.
- **ADASYN (Adaptive Synthetic Sampling):** Similar to SMOTE, but focused on generating synthetic examples in regions that are difficult to classify.
- **Borderline-SMOTE:** Generates synthetic samples at class boundaries, where classification errors are more frequent.

These balancing techniques were combined with multiple supervised classification algorithms:

- **Logistic regression:** A probabilistic linear model serving as a reference for interpretability and frequent clinical use.

- **Random Forest:** An ensemble of decision trees that reduces variance and improves generalization.
- **XGBoost:** Gradient-based boosting algorithm, robust to noise and efficient for high-dimensional datasets.
- **LightGBM:** Boosting algorithm optimized for speed and computational efficiency, particularly with large datasets.

Additionally, an AutoML approach using FLAML was applied in combination with specialized balancers: SMOTEN, ADASYN, and Borderline-SMOTE. FLAML automatically selects optimal models and hyperparameters without manual intervention, optimizing the defined performance metric (F1-score in this study).

Model performance was evaluated at three levels:

1. **Training set performance:** To assess the fitting capacity of each model-balancer combination.
2. **Test set performance:** To evaluate generalization to unseen data.
3. **Learning curve analysis:** F1-score and AUC-ROC curves were examined to visualize stability, potential overfitting, and discriminatory capacity.

Following evaluation of multiple strategies, several trends were observed:

The SMOTE technique alone improved performance relative to the original unbalanced dataset. However, by not removing ambiguous instances, it increased sensitivity at the cost of accuracy, particularly in logistic regression models, due to a higher number of false positives.

The introduction of SMOTE-Tomek provided a better trade-off by combining synthetic sample generation with noise removal near decision boundaries Within this approach, the model **XGBoost SMOTE-Tomek** exhibited good generalization, though minor signs of overfitting were observed (near-perfect training performance).

Comparatively, **XGBoost with Borderline-SMOTE** achieved high metrics but with greater variance and overfitting tendencies. However, models with **ADASYN** demonstrated high sensitivity but lower precision, reflecting a higher incidence of false positives.

We also evaluated the use of **FLAML (Fast Lightweight AutoML)** in combination with specialized balancers **ADASYN, SMOTEENN**, and **Borderline-SMOTE**. Although some FLAML-generated models offered competitive metrics, especially in recall, they showed clear signs of overfitting in the validation curves and lower generalizability. In particular, F1-scores in validation tended to stabilize at lower values, and ROC curves showed less robustness against manually optimized models.

The best overall performance was obtained with **XGBoost with SMOTE-Tomek and hyperparameter optimization**. As shown in Tables 2 and 3 and Figs. 3, 4 and 5 this model achieved very high metrics in both training and testing, with a **98. 2% precision for the majority class and 0. 855% precision for the minority class**. The learning curves showed stable progression with no signs of overfitting, which supports their reliability in real clinical

Table 2. XGB - Train Balanced

Class	Precision	Recall	F1-score	Support
0	0.982	0.833	0.901	1224
1	0.855	0.984	0.915	1224

Table 3. XGB - Test Balanced

Class	Precision	Recall	F1-score	Support
0	0.953	0.853	0.900	306
1	0.867	0.958	0.910	306

Predicted

	0	1
Actual 0	1019	205
Actual 1	19	1205

Fig. 3. Confusion Matrix - XGB Train Balanced

Predicted

	0	1
Actual 0	261	45
Actual 1	13	293

Fig. 4. Confusion Matrix - XGB Test Balanced

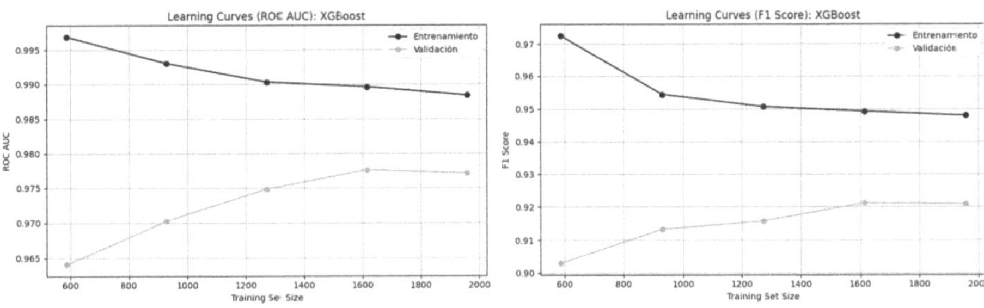

Fig. 5. ROC curve and F1-Score of XGBoost model with SMOTE-Tomek optimized

settings. Because of these characteristics, it was chosen as the final model of the study.

4   Results

4.1   Explainability of the Original Model

To enhance the interpretability of the final predictive model–**XGBoost combined with SMOTE-Tomek and hyperparameter optimization**–three complementary explainability approaches were implemented: **SHAP (SHapley Additive exPlanations)**, the **intrinsic feature importance of the XGBoost algorithm**, and **LIME (Local Interpretable Model-agnostic Explanations)**.

These methods allow the analysis of variable contributions to model predictions at both the global level (overall model structure) and the local level (individual predictions). Together, they provide a comprehensive view of how and why the model generates its outputs.

SHAP analysis estimates the marginal impact of each feature on the prediction outcome using game-theoretic Shapley values [12] [15]. Figure 6, presents a *summary plot* displaying the average explanatory weight and the direction of association of each variable. Red points represent high feature values, while blue points represent low values.

The most influential predictors identified by SHAP include:

- **Age categorized as "Intermediate older" (76–85 years)**: Strongly associated with increased mortality risk.
- **BUN (Blood Urea Nitrogen)**: elevated levels correlate with higher mortality, reflecting impaired renal function.
- **Verbal and motor response (GCS)**: lower scores indicate greater clinical severity.
- **Mechanical ventilation, heart rate, and urine volume**: critical markers of physiological status.

In parallel, the **intrinsic feature importance** estimated by XGBoost itself was examined, based on the frequency and gain of feature splits across decision trees [13, 14]. This analysis reinforced SHAP's findings. The three most relevant variables were:

- **Age "Intermediate Major"**: frequently used for partitioning the decision space.
- **Verbal response (num_verbal)**: a key discriminator of neurological function.
- **Female gender (cat_gender_1.0)**: consistently associated with higher mortality risk in this cohort.

Additional clinically relevant predictors highlighted by this approach included albumin, mean arterial pressure, and the use of mechanical ventilation.

To complement the global analyses, LIME was applied to the prediction of an individual patient [12, 15]. LIME generates approximate local explanations, indicating which features increased or decreased the predicted mortality probability. In the example presented in Fig. 6, the model estimated a 29% probability of death. The explanation revealed that:

- Variables such as **elevated heart rate**, **low neurological scores**, and the presence of **lymphoma** pushed the prediction toward the "death" class.
- Conversely, not belonging to the oldest age group and other demographic factors reduced this risk.

The concordance between SHAP and LIME in identifying critical features underscores the model's internal consistency and enhances its credibility for clinical use.

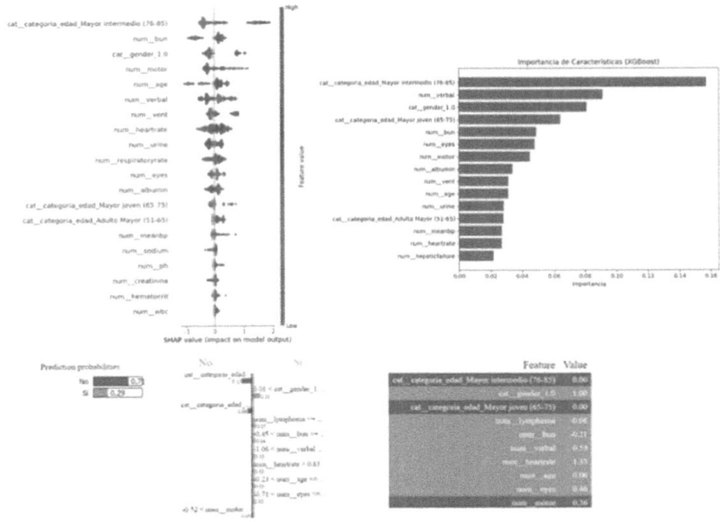

Fig. 6. Explainability Original Model

Overall, the three explainability techniques–SHAP, intrinsic XGBoost feature importance, and LIME–validated that the model relies on clinically meaningful variables. Advanced age, neurological status, renal function, and the use of life-support measures such as mechanical ventilation consistently emerged as strong predictors of in-hospital mortality. The integration of global and local explanations not only strengthens confidence in the model but also supports its potential as a decision-support tool in the ICU setting.

4.2 Explainability of the Extended Model

To evaluate the impact of incorporating additional clinical variables–specifically medications, allergies, diagnoses, and medical history–on model behavior, the explanatory analyses were repeated using **SHAP, intrinsic XGBoost feature importance, and LIME**. The objective was to determine whether the extended model preserved interpretability while integrating broader contextual information.

Figure 7 presents the SHAP summary plot of the extended model. While **age, particularly the "Intermediate older" category (76–85 years)**, remains the most influential predictor, notable changes emerge. The **verbal response (GCS)** gains prominence, and **albumin** increases in explanatory weight, highlighting its importance as a marker of nutritional status and prognosis. Furthermore, comorbidities such as **cardiovascular disease, oncologic conditions, and endocrine disorders**, which were absent from the original model, now appear as relevant predictors.

Consistently, the **intrinsic feature importance** analysis of the XGBoost algorithm (Fig. 7) reinforces these findings. Alongside age and neurological status, comorbidities such as oncologic, cardiovascular, and infectious diseases emerge as significant features. This demonstrates that the extended model has learned to incorporate chronic patient characteristics in addition to acute physiological indicators, thereby enhancing its discriminative capacity.

At the local level, **LIME** was applied to a patient predicted to have a relatively low risk of mortality (20%). The explanation confirmed that(e.g., BUN, verbal response) and (e.g., chronic infection, oncologic disease) contributed to the prediction. This illustrates how the extended model integrates acute and chronic factors to generate individualized risk assessments.

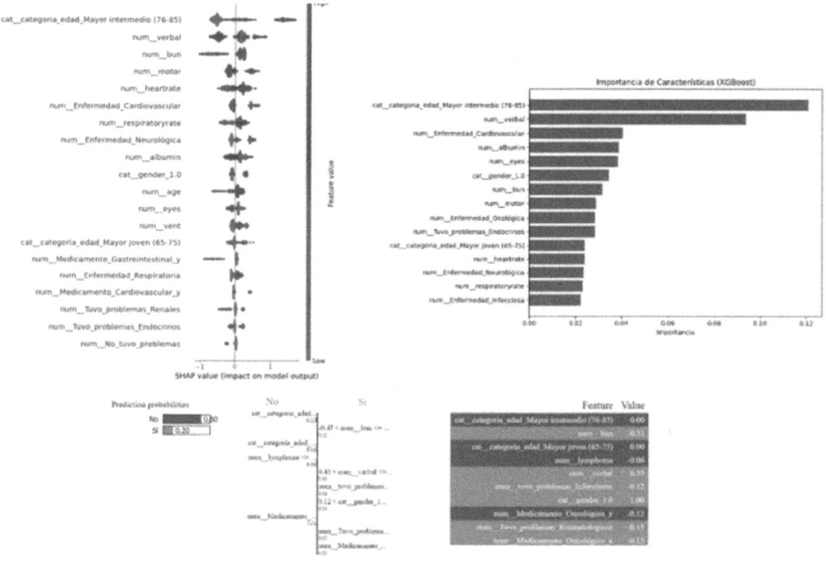

Fig. 7. Explainability Extended Model

Overall, the inclusion of broader clinical variables enhanced the model's explanatory power without compromising consistency. Across SHAP, XGBoost, and LIME, the same core variable groups consistently emerged as key predictors: **advanced age, neurological status, biochemical parameters, and medical history**. This convergence supports the validity of the extended model and highlights its potential as a more comprehensive and personalized decision-support tool in the ICU.

4.3   Final Validation and Comparisons

To validate the developed models, two fundamental metrics were employed in both clinical and statistical contexts: the **Area Under the Receiver Operat-**

ing Characteristic Curve (AUC-ROC), which measures the discriminative capacity of the model, and the **Brier Score**, which evaluates the calibration of predicted probabilities. An optimal clinical prediction model must not only distinguish between survivors and non-survivors (AUC) but also provide well-calibrated probability estimates (Brier Score).

Based on these metrics, different approaches developed were compared:

- **Original model:** Constructed with all the physiological and clinical variables used in the APACHE II/IV scales. It obtained an AUC of **0.812** $\pm$ **0.029** and a Brier Score of **0.087** $\pm$ **0.003**. Although it has a good overall performance, its size and complexity may hinder its practical integration and clinical interpretability.
- **Extended model:** Includes all variables from the original model, plus complementary information from **admission medications, allergies, diagnoses**, and **clinical history**. This model achieved the best calibration (**Brier Score = 0.075** $\pm$ **0.002**), and a very competitive AUC **0.811** $\pm$ **0.031**. It is particularly useful for capturing the comprehensive patient profile, although it depends on a large amount of data that may not always be immediately available.
- **Reduced model:** Based on a minimal subset of physiological variables, selected for their clinical relevance based on the explainability of the original model. Although it achieved an acceptable AUC of **0.804** $\pm$ **0.040**, its Brier Score was the worst (**0.102** $\pm$ **0.007**), indicating difficulties in estimating probabilities correctly. Its main advantage is its simplicity, but at the cost of accuracy.
- **Combined model:** Built from a strategic selection of the most influential variables according to the explanatory analyses of the original and extended models. It achieved the best AUC (**0.814** $\pm$ **0.032**) among the models developed and a good Brier Score of **0.084** $\pm$ **0.004**. This model achieves an optimal balance between performance, interpretability, and simplicity, making it the most suitable for integration into real clinical settings.

These results were contrasted with those obtained using the traditional clinical scoring systems: **APACHE IVa, SAPS II**, and **SOFA**. Specifically, APACHE IV achieved an AUC of **0.840** and a Brier Score of **0.060**, while SAPS II and SOFA presented more discrete values (AUCs between 0.76 and 0.79, and Brier Scores above 0.09). Although APACHE IVa showed an advantage in quantitative terms -which reaffirms why it continues to be one of the most widely used systems in the ICU-, it is important to highlight that this model employs a greater number of clinical variables, including data that may not always be available at the time of admission. In contrast, the models proposed in this study have been designed with different levels of complexity and a more flexible approach, which facilitates their implementation in different care settings and improves their adaptability to more partial data sets.

5   Conclusions

This study examined multiple strategies for predicting hospital mortality among ICU patients using machine learning techniques applied to a cohort from the eICU database. Models ranged from simple frameworks based solely on physiological variables to more complex designs incorporating contextual data and medical history.

The findings demonstrate that machine learning models achieve performance comparable to established clinical scores such as APACHE IV, SAPS II, and SOFA. Importantly, the integration of explainability tools such as **SHAP** and **LIME** not only enhanced interpretability but also informed variable selection, enabling the construction of simplified yet effective models.

Among the proposed approaches, **the combined model** emerged as the most robust, offering high discriminatory capacity, reliable calibration, and efficient use of variables. By integrating both physiological measures and patient-specific contextual factors (e.g., medications, allergies, comorbidities), it achieves an optimal balance between performance, simplicity, and explainability. These attributes make it a promising candidate for real-world deployment, particularly in ICUs where model transparency and clinician trust are crucial for adoption.

Future research should focus on validating these models across diverse hospital cohorts, adapting them to different clinical environments, and exploring their integration into real-time clinical decision support systems.

Acknowledgments. This publication is part of the projects PID2020-117954RB-C22 and PID2023-146037OB-C21, funded by MICIU/AEI/10.13039/501100011033/.

Data and Code Availability. All source code and computational scripts used to perform the analyses and reproduce the findings of this work are openly accessible via GitHub at: https://github.com/pedlopruz/UCI_METRICS_TFM

References

1. Knaus, W.A., Draper, E.A., Wagner, D.P., Zimmerman, J.E.: APACHE II: a severity of disease classification system. Crit. Care Med. **13**(10), 818–829 (1985)
2. Vincent, J.L., Moreno, R.: Clinical review: scoring systems in the critically ill. Critical Care (London, England) **14**(2), 207 (2010). https://doi.org/10.1186/cc8204
3. Vincent, J.L., et al.: The SOFA (Sepsis-related organ failure assessment) score to describe organ dysfunction/failure. Intensive Care Med. **22**(7), 707–710 (1996). https://doi.org/10.1007/BF01709751
4. Ferreira, F.L., Bota, D.P., Bross, A., Mélot, C., Vincent, J.L.: Serial evaluation of the SOFA score to predict outcome in critically ill patients. JAMA **286**(14), 1754–1758 (2001). https://doi.org/10.1001/jama.286.14.1754
5. Le Gall, J.R., Lemeshow, S., Saulnier, F.: A new simplified acute physiology score (SAPS II) based on a European/North American multicenter study. JAMA **270**(24), 2957–2963 (1993). https://doi.org/10.1001/jama.270.24.2957

6. Moreno, R.P., et al.: SAPS 3–From evaluation of the patient to evaluation of the intensive care unit. Part 2: Development of a prognostic model for hospital mortality at ICU admission. Intensive Care Med. *31*(10), 1345–1355 (2005). https://doi.org/10.1007/s00134-005-2763-5

7. Cabré, L., et al.: Validation of mortality probability models II (MPM II) at admission (MPM II-0), at 24 hours (MPM II-24), and at 48 hours (MPM II-48) compared with the hospital mortality predictions from APACHE II and SAPS II. Med. Intensiva **24**(5), 177–183 (2000). https://doi.org/10.1016/S0210-5691(00)79553-1

8. Lemeshow, S., Teres, D., Klar, J., Avrunin, J., Gehlbach, S.H., Rapoport, J.: Mortality probability models (MPM II) based on an international cohort of intensive care unit patients. JAMA **270**(20), 2478–2486 (1993)

9. Le Gall, J.R., et al.: The logistic organ dysfunction system. A new way to assess organ dysfunction in the intensive care unit. ICU scoring group. JAMA *276*(10), 802–810 (1996). https://doi.org/10.1001/jama.276.10.802

10. Heldwein, M.B., et al.: Logistic organ dysfunction score (LODS): a reliable postoperative risk management score also in cardiac surgical patients? J. Cardiothorac. Surg. **6**, 110 (2011). https://doi.org/10.1186/1749-8090-6-110

11. MIT laboratory for computational physiology. eICU collaborative research database documentation. https://eicu-crd.mit.edu/

12. Lin, C.Y., Chou, C.Y., Cheng, H.W., et al.: Explainable machine learning to predict long-term mortality in critically ill ventilated patients: a retrospective study in central Taiwan. Sci. Rep. **12**(1), 5165 (2022). https://doi.org/10.1038/s41598-022-09041-0

13. Li, Y., Zhang, L., He, H., et al.: Machine learning algorithm to predict mortality in critically ill patients with sepsis-associated acute kidney injury (SA-AKI). BMC Med. Inform. Decis. Mak. **23**, 51 (2023). https://doi.org/10.1186/s12911-023-02091-4

14. Wang, J., Zhao, Q., Chen, Y., et al.: Machine learning algorithm for predicting in-hospital mortality in critically ill patients with chronic kidney disease (CKD). Front. Med. **10**, 1167072 (2023). https://doi.org/10.3389/fmed.2023.1167072

15. Bai, Y., Zhang, X., Wu, H.: Interpretable mortality prediction model for ICU patients with pneumonia using Shapley additive explanation method. BMC Pulm. Med. **24**, 252 (2024). https://doi.org/10.1186/s12890-024-03252-x

16. Bansal, P., Mishra, R., Verma, A., et al.: Comparison of APACHE II and APACHE IV score as predictors of mortality in patients with septic shock in intensive care unit: A prospective observational study. Indian J. Critical Care Med. **27**(11), 802–808 (2023). https://doi.org/10.5005/jp-journals-10071-24455

17. Xiao, W., Chen, Y., Hu, M., et al.: External validation of AI-based scoring systems in the ICU: a systematic review and meta-analysis. BMC Med. Inform. Decis. Mak. **24**, 63 (2024). https://doi.org/10.1186/s12911-024-02830-7

Evaluation and Prediction
of the Musculoskeletal Risks
in Microsurgery

Daniel Caballero[1], Manuel J. Pérez-Salazar[1],
Juan A. Sánchez-Margallo[1]([✉]), Laura C. Pires-Louça[2],
and Francisco M. Sánchez-Margallo[3]

[1] Bioengineering and Health Technologies Unit, Jesús Usón Minimally Invasive
Surgery Center, N-521 Road Km. 41.8, 10071 Cáceres, Spain
`{dcaballero,mjperez,jasanchez}@ccmijesususon.com`
[2] Microsurgery Unit, Jesús Usón Minimally Invasive Surgery Center, N-521 Road
Km. 41.8, 10071 Cáceres, Spain
`lcpires@ccmijesususon.com`
[3] Scientific Direction, Jesús Usón Minimally Invasive Surgery Center, N-521 Road
Km. 41.8, 10071 Cáceres, Spain
`msanchez@ccmijesususon.com`

Abstract. Microsurgery enables precise manipulation of small and complex anatomical structures. However, it is a physically and mentally demanding surgical technique. Advances in wearable technology and artificial intelligence (AI) have improved the development of innovative solutions in this field. The objective of this study is to evaluate the surgeon's physiological and ergonomic parameters by using EMG wearable technology during performance in conventional and robot-assisted microsurgery (RAM). In addition, this study seeks to develop predictive models for musculoskeletal risks, such as localized muscle fatigue, during microsurgery. The data were collected over twenty microsurgical sessions, ten were conventional and the remaining were RAM performed by four microsurgeons. The data were recorded in five bilateral muscle groups. The data was recorded for the evaluation of the localized muscle fatigue and the development of predictive models. From these data, a general dataset (508,328 records) was generated, for conventional and RAM. Two different preprocessing techniques (scaled and scaled and normalized) were applied. In this dataset, 80% of the data for training and 20% for testing purposes. This work advances the understanding of ergonomic risks in microsurgery by integrating wearable technology and predictive models of musculoskeletal risks. These results demonstrate the validity and accuracy of predictive models and constitute the starting point for achieving a comprehensive understanding of the ergonomic risks of microsurgery.

Keywords: Artificial Intelligence · Ergonomic parameters · Localized muscle fatigue · Robot-assisted Microsurgery · Wearable devices

A. López Fernández et al. (Eds.): CIABiomed 2025, LNBI 16148, pp. 126–140, 2026.
https://doi.org/10.1007/978-3-032-10661-2_10

1 Introduction

Microsurgery enables precise manipulation of small and complex anatomical structures, including blood vessels, nerves and lymphatic vessels. Nevertheless, surgeons encounter physiological constraints, including tremors, which can be effectively managed through robotic technology, which also encompasses movement scaling. Together, these advances improve the precision of microsurgery [1]. This surgery specialty is in a continuous evolution from 1950s until today with the technology evolves at the same pace [2].

Microsurgery is a physically and mentally demanding surgical technique. These limitations can have detrimental effects on the surgeon's health and consequently, may impact the quality of surgical procedures and patient care [3]. Studies reflect that 56.1% of regularly practicing robotic microsurgeons continue to experience related physical symptoms or discomfort, including neck stiffness, finger fatigue, and eye fatigue, among the most common [4]. Although ergonomic conditions are considered to be improved in robotic surgery, scientific evidence remains scarce [5]. Therefore, further studies remain to be carried out regarding the comprehensive analysis of ergonomics in the field of minimally invasive surgery (MIS) and mainly in microsurgery, allowing us to precisely identify possible ergonomic deficiencies, design possible solutions and recommendations, and adapt training programs according to these needs.

Different techniques, both subjective and objective, have been used to assess ergonomics during robot-assisted surgery to evaluate different physiological and cognitive factors. Some studies employed traditional methods of subjective assessment of the workload, both mental and physical, of surgeons during their surgical activity, such as the SURG-TLX scale [6]. However, these subjective assessment methods should be reinforced with objective ergonomic assessment techniques.

Advances in wearable technology have led to the development of small sensors that facilitate the recording and analysis of the surgeon's physiological and ergonomic parameters without interrupting their surgical performance. Some of these objective parameters include the surgeon's posture using body movement techniques, force/torque analysis [7], muscle activity through electromyographic (EMG) signal analysis [8], and stress levels through the examination of electrocardiogram (ECG) or electrodermal activity (EDA) signals [9,10]. Recording data on the surgeon's ergonomics and surgical performance using wearable technology can be a powerful tool for monitoring the surgeon's health conditions during surgery, as well as the quality of care provided to the patient. Similarly, these technological innovations can offer robust solutions for the analysis and comparison of objective physiological and ergonomic parameters during surgery.

The application of artificial intelligence (AI) has grown exponentially in terms of its use and development. These algorithms are based on non-trivial processes to discover potentially useful knowledge initially hidden in the data [11]. Among the different AI techniques, there are several algorithms that allow predictive models to be developed. Of these, three classical approaches of AI were applied in the present study, one technique based on linear approach, a technique based on the nonlinear approach, and a technique based on neural networks [12]. Specifically, the use of AI in the assessment of ergonomic conditions in surgery [13,14] and the use of different portable sensors in different surgical procedures [3,9,10,15] are particularly relevant.

The main novelty of this study is the evaluation of muscle activity and localized muscle fatigue using the JASA method during robot-assisted and conventional microsurgery, as well as the prediction of muscle fatigue in different muscle groups by applying different AI techniques based on EMG sensor data.

1.1 Objective

The objective of this study is to evaluate the physiological and ergonomic parameters of surgeons using wearable technology during conventional and robot-assisted microsurgery procedures. In addition, this study aims to design and implement predictive models for the future prediction of musculoskeletal risks in microsurgeons, such as localized muscle fatigue, during microsurgery procedures.

2   Material and Methods

2.1 Experimental Setup

The data for this study were collected during twenty microsurgery sessions, ten of which were performed using conventional microsurgical techniques and the rest using robot-assisted microsurgery (RAM). All surgical sessions were performed randomly by four microsurgeons with different levels of experience. The microsurgeons were two women and two men between the ages of 25 and 45, all of whom were right-handed. The data were recorded using a portable, wireless EMG system that allowed the EMG signal to be recorded for subsequent calculation of localized muscle fatigue in five muscle groups bilaterally.

On the other hand, current and immediately preceding situations were analyzed to develop predictive models of localized muscle fatigue. Based on this data, a general dataset (508,328 records) was generated for conventional surgery (254,174 records) and RAM (254,174 records). Two different preprocessing techniques (scaling and scaling and normalization) were applied to these two datasets, resulting in the generation of two new datasets: 80% of the data for training purposes (203,339 records for each type of approach) and 20% of the data for testing purposes (50,835 records for each type of approach). Records from all twenty microsurgery sessions were included in all datasets. Figure 1 shows the experimental setup.

2.2 Microsurgery Sessions

Standard equipment for conventional microsurgery was used in this study. For RAM, the Symani^{TM} robotic platform (Medical Microinstruments Inc.^{TM}, Pisa, Italy) was used. Both surgical techniques were performed in combination with a Pentero surgical microscope (Carl Zeiss Meditec AG, Jena, Germany) (Fig. 2).

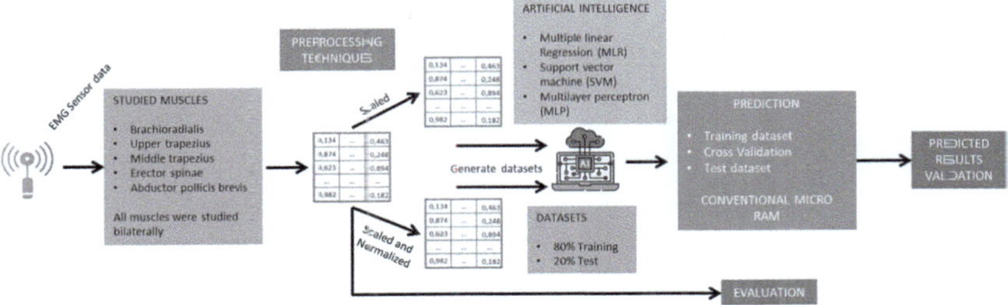

Fig. 1. Experimental setup of this study.

The Symani^{TM} robotic platform consists of an ergonomic chair, two robotic arms for handling microinstruments, and electromagnetic tracking controls for handling surgical instruments. The platform offers motion scaling and tremor reduction in the handling of robotic surgical instruments, allowing for greater precision during surgery. In each session, microsurgeons were asked to perform a vascular anastomosis on a synthetic blood vessel, using both conventional micro-surgical and robot-assisted techniques.

2.3   EMG Wearable Device

The wireless EMG TRIGNO^{TM} Avanti system from DELSYS^{TM} was used to record the muscle activity by means of electromyographic (EMG) signals. This system has up to 16 sensors with a sampling rate of 1024 Hz. The EMG signal of following muscle groups was recorded bilaterally: upper trapezius middle trapezius, erector spinae, brachioradialis, and abductor pollicis brevis. The EMG sensors were placed on each muscle according to the SENIAM guidelines [16]. Prior to the placement of each sensor, the skin was cleaned by gently rubbing it with 70% isopropyl alcohol. Raw EMG signals were processed using a band-pass filter of 20–300 Hz. The filtered EMG was then smoothed using an algorithm with a moving window of 125 ms and calculated as root mean square (RMS). To normalize the results for each subject, the EMG values were presented as percentage of maximum voluntary contraction (%MVC). MVC was performed separately for each muscle group just before each test by asking each subject to perform specific contractions against a fixed resistance.

2.4   Localized Muscle Fatigue

For the analysis of localized muscle fatigue, the evolution of the amplitude and median frequency of the muscle activity was analyzed for each task and surgical procedure. Muscular fatigue was quantified using the joint analysis of spectrum and amplitude (JASA) method [17].

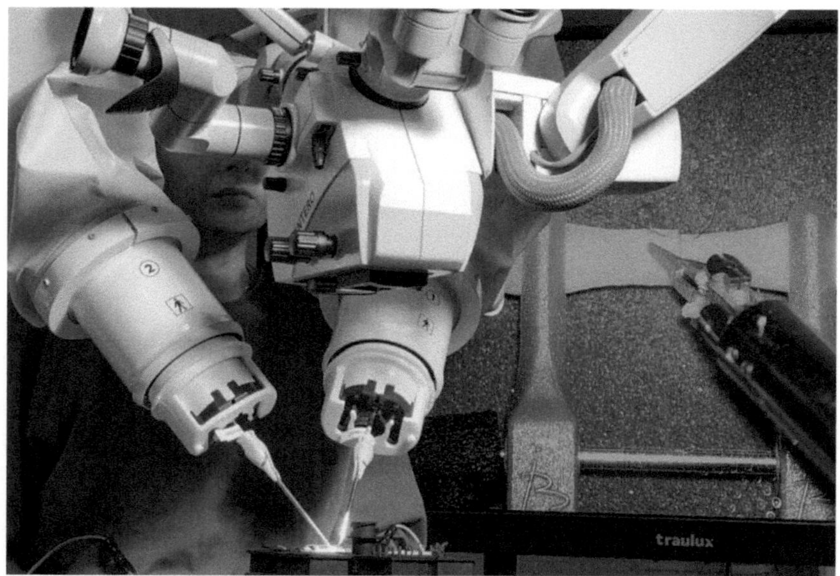

Fig. 2. Use of the Symani^{TM} robotic platform during a vascular anastomosis with an synthetic vessel.

2.5   Datasets

The original dataset was divided as a function of the surgical approach on 2 subsets, with conventional microsurgery and RAM. After that, these two datasets were transformed by applying scale preprocessing and normalization techniques, resulting 4 datasets. The 4 preprocessed datasets were divided into 8 datasets: 80% of the data from each dataset for calibration and 20% of the data from each dataset for validation [18].

The scaled preprocessing technique allows each parameter to be described on a scale between 0 and 1 [18]. To do this, each value is subtracted from the minimum value and then, divided by the interval between the maximum and minimum values (Eq. 1).

$$Value_{new} = \frac{Value_{current} - min}{max - min} \tag{1}$$

When $value_{new}$ indicates the preprocessed value, $value_{current}$ represents the raw value from each dataset, min shows the minimum value for each parameter, and max indicates the maximum value for each parameter.

For the normalized preprocessing technique, the mean of each value is subtracted and divided by the standard deviation [18] (Eq. 2). This technique transforms the dataset into a more integrated and robust one with fewer redundancies.

$$Value_{new} = \frac{Value_{current} - Value_{average}}{Value_{st.deviation}} \tag{2}$$

where $Value_{new}$ indicates the preproccessed value, $Value_{current}$ represents the raw value from each dataset, $Value_{average}$ shows the average value for each parameter, and $Value_{st.deviation}$ indicates the standard deviation value for each parameter.

The R^2 coefficient was used to assess the goodness of fit of the prediction and for validation, according to the rules given by Colton [19], where R^2 of 0 to 0.25 is considered as a poor to no relationship; 0.25 to 0.50 indicates a weak degree of relationship; 0.50 to 0.75 designates a moderate to good relationship; and 0.75 to 1 shows a very good to excellent relationship. The root mean square error (RMSE) was also used to validate the prediction results [20]. RMSE measures the difference between actual and predicted values. RMSE values of less than 0.05 are considered adequate [20].

2.6 Artificial Intelligence

The free software WEKA^{TM} (Waikato Environment for Knowledge Analysis, Hamilton, New Zealand, version 3.8.6) [21] was used to develop the predictive model. Two different AI predictive techniques have been applied to the calibration dataset to generate predictive models: Multiple Linear Regression (MLR), Support Vector Machine (SVM), and Multilayer Perceptron (MLP).

Multiple Linear Regression (MLR). MLR was applied as a linear predictive approach to the datasets. This technique shows the linear relationship between a dependent variable and several independent variables (Eq. 3). It arrives at a linear regression equation that can be used to predict future values. In this study, the M5 method of attribute selection was applied. This method cycles through the attributes, eliminating the one with the lowest standardized coefficient until no improvement in error estimation is observed. A peak value of 0.0001 was applied [21].

$$y = a_0 + \sum_{i=1}^{n} a_i \times x_i \tag{3}$$

Support Vector Machine (SVM). SVM was performed as a predictive non-linear approach on the datasets. It was applied to represent desirable non-linear relationship models that minimize the separation (Eq. 4), tolerating a small error when fitting the data in the transformed space [22]

$$y = \sum_{i=1}^{n} \omega_i \times K_i(x) \tag{4}$$

For the transformation of the high dimensional feature space, a non-linear mapping was used that does not need to be explicitly known but depends on a kernel function [23]. The Kernel function represents the inner product of the feature space. The frequently used kernel functions are the polynomial function, the radial basis function (RBF) and the sigmoid function. In this study, RBF was used as a Kernel function.

Multilayer Perceptron (MLP). MLP was applied as a predictive machine learning approach to datasets. This approach is a type of artificial neural network model that are developed using neural organization principles [24]. Thus, different numbers of neurons are grouped into layers. The different layers can perform different transformations on their inputs. Signals travel from the first layer (the input layer) to the last (the output layer), possibly after passing through the layers several times. In the present study, the default configuration was used, with the learning rate equal to 0.15, the number of epochs equal to five hundreds, momentum equal to 0.01, the threshold equal to twenty, and thirty nodes in the first hidden layer, ten nodes in the second hidden layer, and three nodes in the third hidden layer.

3   Results and Discussion

3.1   Evaluation of Localized Muscle Fatigue

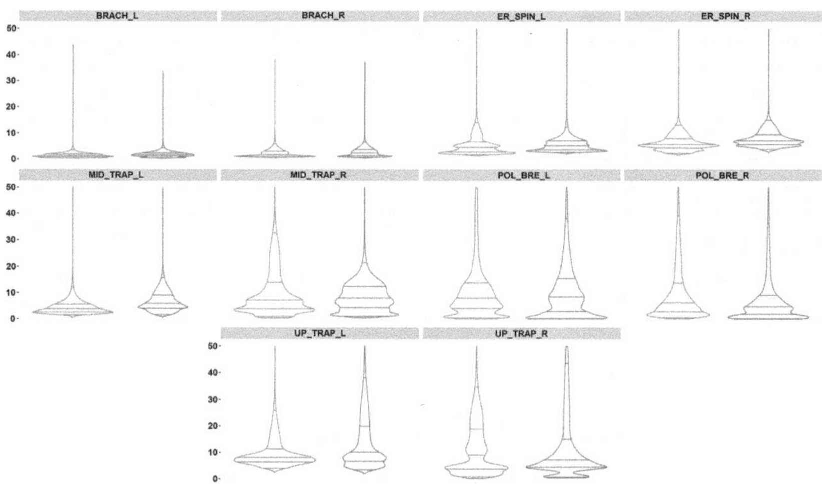

Fig. 3. Comparison of muscle activity (%MVC) of microsurgeons during performance of microsurgery using conventional (red) and RAM (blue) for the following muscle: Brachioradialis (BRACH), Erector Spinae (ER-SPIN), Middle Trapezius (MID-TRAP), Upper Trapezius (UP-TRAP) and Abductor Pollicis Brevis (POL-BRE). (Color figure online)

In general, the microsurgeons had less muscle activity in the left-sided muscles than the right-sided except for the abductor pollicis brevis (Fig. 3). This fact could be related to the fact that the dominant side of all microsurgeons is the right side. Consistent with this fact, muscle activity on the dominant side also appears to increase slightly over time compared with the non-dominant side, suggesting that the dominant side may be more susceptible to slowly developing future musculoskeletal disorders [25], influencing directly the localized muscle fatigue. In the same way, the microsurgeons required greater muscle activity of the middle trapezius, upper trapezius, abductor pollicis brevis, and erector spinae bilaterally. All muscles bilaterally showed statistically significant differences between laparoscopic and robot-assisted surgery (p<0.001).

Analysis of localized muscle fatigue during different microsurgical procedures showed lower muscle fatigue and higher force use in robot-assisted procedures than in conventional ones (Fig. 4). Moreover, RAM reported more balanced muscle fatigue and force use than conventional ones. Consequently, the mean values obtained for the percentage of force increase were 5.23 ± 2.85 in the case of RAM and 5.19 ± 2.89 for conventional microsurgery. In the case of the percentage of muscle fatigue, the values obtained were 4.88 ± 2.75 and 5.21 ± 3.97, respectively, for RAM and conventional microsurgeries. For the percentage of muscle fatigue recovery, for conventional microsurgery the mean value was 5.33 ± 2.53 and for RAM the value was 5.69 ± 1.81. For the percentage of force decrement, for conventional microsurgery the value obtained was 5.45 ± 3.06 and for RAM the value was 5.93 ± 1.21.

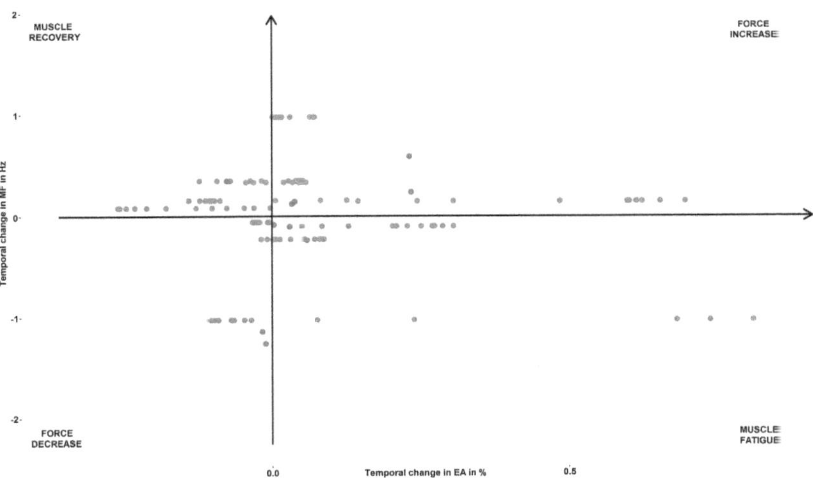

Fig. 4. Comparison of fatigue and muscle strength increase/decrease for microsurgeons during conventional (red) and robot-assisted (blue) microsurgery (Color figure online)

Muscle activity is closely related to muscle fatigue, and prolonged muscle activity of small, single muscle fibers can cause degenerative muscle changes even with very low levels of muscle activity. For this reason, ergonomic positioning is very important for microsurgical activities. The muscles analyzed in this study have been thoroughly evaluated in other studies on MIS, with the main muscles identified being the upper and middle trapezius and the brachioradialis [26]. Moreover, all muscles bilaterally show statistically significant differences between conventional and RAM. This fact demonstrates the different use of muscles between conventional and robot-assisted procedures for all microsurgeons.

3.2   Prediction of Localized Muscle Fatigue

The prediction results of localized muscle fatigue of the training dataset are shown in Tables 1 and 2. Tables 3 and 4 show the results for the cross-validation. Finally, Tables 5 and 6 show the results for the test dataset, evaluating in all cases, three AI techniques (MLR, SVM, and MLP) and two preprocessing techniques (scaled and normalized).

Table 1. Results from the calibration datasets for conventional microsurgery indicating the R^2 values applying MLR, SVM and MLP as predictive techniques, and scaling and normalized as preprocessing techniques.

Muscles	MLR-Scaled	MLR-Normalized	SVM-Scaled	SVM-Normalized	MLP-Scaled	MLP-Normalized
BRACH-L	**0.965**	0.964	0.963	0.962	0.961	0.959
BRACH-R	**0.891**	0.888	0.889	0.887	0.877	0.875
ER SPIN-L	**0.908**	0.905	0.906	0.903	0.904	0.902
ER SPIN-R	**0.823**	0.812	0.821	0.811	0.810	0.799
MID TRAP-L	**0.948**	0.946	0.944	0.943	0.946	0.942
MID TRAP-R	**0.950**	0.949	0.948	0.947	0.935	0.933
POL BRE-L	**0.916**	0.915	0.915	0.914	0.914	0.914
POL BRE-R	**0.934**	0.932	0.933	0.931	0.921	0.919
UP TRAP-L	**0.951**	0.949	0.950	0.948	0.947	0.945
UP TRAP-R	**0.936**	0.935	0.935	0.934	0.923	0.921

Training Dataset Results. Table 1 shows the prediction results for the training dataset by applying conventional microsurgery. In general, MLR showed slightly higher values than SVM and MLP for the R^2 coefficient ($R^2 > 0.80$) with a low RMSE error (RMSE < 0.05) for all cases. As for the preprocessing techniques, scaled reached slightly higher values than scaled and normalized for the R^2 coefficient. In general, the results obtained in the present study are satisfactory according to the standards given by Colton [19], achieving correlations (R^2) close to 1 and RMSE close to 0 for all the movements studied. It is noteworthy that in general the left side obtained higher R^2 coefficient than the right side [15] and that the highest R^2 coefficient were achieved by the Brachioradialis ($R^2 = 0.965$) and Upper Trapezius ($R^2 = 0.951$) in the left side.

Table 2 displays the prediction analysis of the data from RAM on the training dataset. In general, scaled obtained slightly higher values than scaled and normalized for the R^2 coefficient. In the case of the AI techniques, MLR achieved slightly higher values than SVM and MLP for the R^2 coefficient ($R^2 > 0.80$) with a low RMSE error (RMSE < 0.05) for almost all cases. In general, the results reached in the present study are satisfactory considering the rules given by Colton [19], achieving correlations (R^2) close to 1 and RMSE close to 0 for all the movements studied. It is remarkable that the right side obtained a higher R^2 coefficient than the left side; this fact could be related to the high use of the dominant side [15]. In the same way, the highest R^2 coefficient was achieved by the Middle Trapezius ($R^2 = 0.960$) and Upper Trapezius (R^2=0.952) both on the right side.

Table 2. Results from the calibration datasets for robot-assisted microsurgery indicating the R^2 values applying MLR, SVM and MLP as predictive techniques, and scaling and normalized as preprocessing techniques.

Muscles	MLR-Scaled	MLR-Normalized	SVM-Scaled	SVM-Normalized	MLP-Scaled	MLP-Normalized
BRACH-L	**0.919**	0.917	0.913	0.871	0.907	0.823
BRACH-R	**0.865**	0.844	0.853	0.816	0.841	0.789
ER SPIN-L	**0.930**	0.901	0.926	0.885	0.921	0.870
ER SPIN-R	**0.883**	0.876	0.863	0.847	0.843	0.817
MID TRAP-L	**0.949**	0.935	0.944	0.885	0.939	0.835
MID TRAP-R	**0.960**	0.909	0.956	0.892	0.952	0.875
POL BRE-L	**0.929**	0.896	0.923	0.871	0.917	0.845
POL BRE-R	**0.935**	0.894	0.913	0.862	0.891	0.830
UP TRAP-L	**0.918**	0.898	0.917	0.868	0.916	0.839
UP TRAP-R	**0.952**	0.902	0.947	0.889	0.942	0.877

Cross-Validation Results. Table 3 shows the prediction results for the cross-validation by applying conventional microsurgery. In general, MLR showed slightly higher values than SVM and MLP for the R^2 coefficient ($R^2 > 0.75$) with a low RMSE error (RMSE < 0.05) for all cases. As for the preprocessing techniques, scaled reached slightly higher values than scaled and normalized for the R^2 coefficient. In general, the results obtained in the present study are satisfactory according to the standards given by Colton [19], achieving correlations (R^2) close to 1 and RMSE close to 0 for all the movements studied. It is noteworthy that in general the left side obtained higher R^2 coefficient than the right side [15] and that the highest R^2 coefficient was achieved by the Brachioradialis ($R^2 = 0.936$) and Upper Trapezius ($R^2 = 0.935$) in the left side.

Table 4 displays the prediction analysis of the data from RAM on the cross-validation. In general, scaled obtained slightly higher values than scaled and normalized for the R^2 coefficient. In the case of the AI techniques, MLR achieved slightly higher values than SVM and MLP for the R^2 coefficient ($R^2 > 0.75$)

Table 3. Results from the cross-validation results for conventional microsurgery indicating the R^2 values applying MLR, SVM and MLP as predictive techniques, and scaling and normalized as preprocessing techniques.

Muscles	MLR-Scaled	MLR-Normalized	SVM-Scaled	SVM-Normalized	MLP-Scaled	MLP-Normalized
BRACH-L	**0.936**	0.935	0.931	0.929	0.926	0.924
BRACH-R	**0.880**	0.877	0.873	0.871	0.850	0.847
ER SPIN-L	**0.877**	0.875	0.869	0.866	0.859	0.857
ER SPIN-R	**0.804**	0.798	0.801	0.795	0.779	0.773
MID TRAP-L	**0.932**	0.931	0.930	0.927	0.896	0.893
MID TRAP-R	**0.903**	0.902	0.901	0.899	0.873	0.869
POL BRE-L	**0.884**	0.883	0.881	0.879	0.876	0.874
POL BRE-R	**0.930**	0.928	0.922	0.919	0.908	0.904
UP TRAP-L	**0.935**	0.932	0.931	0.929	0.924	0.922
UP TRAP-R	**0.921**	0.919	0.914	0.912	0.871	0.869

with a low RMSE error (RMSE < 0.05) for almost all cases. In general, the results reached in the present study are satisfactory considering the rules given by Colton [19], achieving correlations (R^2) close to 1 and RMSE close to 0 for all the movements studied. In the same way, the highest R^2 coefficient was achieved by the Upper Trapezius ($R^2 = 0.943$) and Middle Trapezius ($R^2 = 0.919$) both on the right side and the abductor pollicis brevis in the left side ($R^2 = 0.920$).

Table 4. Results from the cross-validation results for robot-assisted microsurgery indicating the R^2 values applying MLR, SVM and MLP as predictive techniques, and scaling and normalized as preprocessing techniques.

Muscles	MLR-Scaled	MLR-Normalized	SVM-Scaled	SVM-Normalized	MLP-Scaled	MLP-Normalized
BRACH-L	**0.885**	0.883	0.870	0.848	0.856	0.813
BRACH-R	**0.850**	0.839	0.833	0.802	0.817	0.766
ER SPIN-L	**0.897**	0.880	0.864	0.831	0.832	0.781
ER SPIN-R	**0.849**	0.845	0.822	0.807	0.795	0.769
MID TRAP-L	**0.911**	0.898	0.906	0.861	0.903	0.823
MID TRAP-R	**0.919**	0.888	0.911	0.871	0.904	0.854
POL BRE-L	**0.920**	0.889	0.914	0.862	0.908	0.835
POL BRE-R	**0.917**	0.878	0.901	0.850	0.885	0.821
UP TRAP-L	**0.879**	0.861	0.874	0.841	0.869	0.819
UP TRAP-R	**0.943**	0.899	0.933	0.886	0.922	0.873

In general, the best results were achieved applying the combination of MLR as AI technique and scaled as Preprocessing technique. It is also outstanding that for both types of microsurgical procedures, the predictions achieved the highest values on the left side, this fact could be related with the low use of the non-dominant side [15] became the movements in more reproducible.

Test Dataset Results. Table 5 shows the prediction analysis for the test dataset by applying conventional microsurgery. In general, MLR showed slightly higher values than SVM and MLP for the R^2 coefficient ($R^2 > 0.725$) with a low RMSE error (RMSE <0.05) for all cases. As for the preprocessing techniques, scaled reached slightly higher values than scaled and normalized for the R^2 coefficient. In general, the results obtained in the present study are satisfactory according to the standards given by Colton [19], achieving correlations (R^2) close to 1 and RMSE close to 0 for all the movements studied. It is noteworthy that in general the left side obtained higher R^2 coefficient than the right side [15] and that the highest R^2 coefficient was achieved by the Upper Trapezius ($R^2 = 0.918$) and Middle Trapezius ($R^2{=}0.917$) both in the left side and the abductor pollicis brevis ($R^2{=}0.925$).

Table 5. Results from the test datasets for conventional microsurgery indicating the R^2 values applying MLR, SVM and MLP as predictive techniques, and scaling and normalized as preprocessing techniques.

Muscles	MLR-Scaled	MLR-Normalized	SVM-Scaled	SVM-Normalized	MLP-Scaled	MLP-Normalized
BRACH-L	**0.906**	0.904	0.897	0.896	0.889	0.887
BRACH-R	**0.869**	0.867	0.857	0.855	0.823	0.821
ER SPIN-L	**0.847**	0.846	0.831	0.829	0.813	0.811
ER SPIN-R	**0.786**	0.783	0.781	0.778	0.748	0.746
MID TRAP-L	**0.917**	0.915	0.913	0.911	0.847	0.844
MID TRAP-R	**0.856**	0.854	0.854	0.853	0.810	0.803
POL BRE-L	**0.852**	0.850	0.849	0.848	0.846	0.845
POL BRE-R	**0.925**	0.921	0.911	0.907	0.896	0.891
UP TRAP-L	**0.918**	0.915	0.915	0.912	0.902	0.897
UP TRAP-R	**0.905**	0.903	0.892	0.890	0.818	0.815

Table 6 displays the prediction results of the data from RAM on the test dataset. In general, scaled obtained slightly higher values than scaled and normalized for the R^2 coefficient. In the case of the AI techniques, MLR achieved slightly higher values than SVM and MLP for the R^2 coefficient ($R^2 > 0.75$) with a low RMSE error (RMSE < 0.05) for almost all cases. In general, the results reached in the present study are satisfactory considering the rules given by Colton [19], achieving correlations (R^2) close to 1 and RMSE close to 0 for all the movements studied. In the same way, the highest R^2 coefficient was achieved by the Upper Trapezius ($R^2 = 0.935$) on the right side and the abductor pollicis brevis on the left side ($R^2 = 0.911$).

As evidenced by the findings of this study, ergonomics in the field of microsurgery continues to pose a challenge that requires attention. Consequently, it is prudent to incorporate ergonomic recommendations into training programs. This approach can effectively mitigate musculoskeletal issues, enhance the well-being of microsurgeons, and consequently elevate the quality of microsurgical practice. The predictive models developed provide valuable insights into the design of

Table 6. Results from the test datasets for robot-assisted microsurgery indicating the R^2 values applying MLR, SVM and MLP as predictive techniques, and scaling and normalized as preprocessing techniques.

Muscles	MLR-Scaled	MLR-Normalized	SVM-Scaled	SVM-Normalized	MLP-Scaled	MLP-Normalized
BRACH-L	**0.851**	0.850	0.828	0.826	0.803	0.802
BRACH-R	**0.836**	0.834	0.814	0.788	0.793	0.742
ER SPIN-L	**0.864**	0.860	0.804	0.776	0.743	0.692
ER SPIN-R	**0.816**	0.814	0.781	0.767	0.746	0.721
MID TRAP-L	**0.872**	0.861	0.869	0.836	0.866	0.833
MID TRAP-R	**0.878**	0.868	0.866	0.851	0.854	0.811
POL BRE-L	**0.911**	0.882	0.905	0.853	0.900	0.824
POL BRE-R	**0.899**	0.863	0.888	0.838	0.879	0.812
UP TRAP-L	**0.839**	0.824	0.832	0.811	0.823	0.798
UP TRAP-R	**0.935**	0.897	0.918	0.884	0.902	0.869

tools that can prevent these musculoskeletal risk factors during surgical procedures. However, further research involving more intricate surgical procedures is essential to obtain more precise results.

One limitation of this study is the limited number of participants. The sample size was relatively small, comprising only four microsurgeons. Future studies will endeavor to include a larger number of microsurgeons in each group, ensuring a minimum of four participants per study group. Additionally, it would be beneficial to expand the scope of muscle analysis. Certain muscles, such as the triceps, have the potential to influence the development of surgical procedures. Furthermore, it is advisable to broaden the range of AI algorithms employed. This may involve incorporating novel models based on machine learning and deep learning, which can be analyzed to enhance prediction models and mitigate errors and biases.

4   Conclusions

This research contributes to the understanding of ergonomic risks in microsurgery, presenting a substantial advance in the integration of wearable technology and the implementation of predictive models of musculoskeletal risks for microsurgeons. These results have the potential to significantly improve microsurgery training programs, the design of ergonomic surgical instruments, and the overall well-being of microsurgeons.

Regarding predictive models, the highest R^2 coefficients were obtained using MLR as the AI technique and scaling as the preprocessing technique. All results demonstrated a satisfactory to excellent correlation relationship ($R^2 > 0.7$). These findings underscore the effectiveness and accuracy of predictive models and serve as a basis for achieving a comprehensive understanding of the ergonomic risks associated with microsurgery.

Acknowledgments. This study is financed by the Ministry of Science, Innovation and University, with funds from the European Union–Next Generation EU, from the Recovery, Transformation and Resilience Plan (PRTR-C17.I1) and the European Regional Development Fund (ERDF) of Extremadura Operational Program 2021–2027. The authors would like to thank the colleagues who collaborated in this study (J. Vela, J. L. Campos and E. Abellán).

Disclosure of Interests. The authors have no competing interests to declare that are relevant to the content of this article.

References

1. Coelho, M., Peltz, T.S., Hunt, J., Gianoutsos, M.: Robotic surgery in plastic surgery: a review of its potential. Australas. J. Plast. Surg. **7**(1), 1–7 (2024)
2. Yoo, H., Kim, B.J.: History and recent advances in microsurgery. Arch. Hand Microsurg. **26**(3), 174–183 (2021)
3. Pérez-Salazar, M.J., Caballero, D., Sánchez-Margallo, J.A., Sánchez-Margallo, F.M.: Correlation study and predictive modelling of ergonocmic parameters in robot-assisted laparoscopic surgery. Sensors (MDPI) **24**(23), e7721 (2024)
4. Lee, G.I., Lee, M.R , Green, I., Allaf, M., Marohn, M.R.: SUrgeons' physical discomfort and symptoms during robotic surgery: a comprehensive ergonomic survey study. Surg. Endosc. **31**, 1697–1706 (2017)
5. Muller, D.T., et al.: Ergonomics in robot-assisted surgery in comarison to open or conventional laparoendoscopic surgery: a narrative review. Int. J. Abdom. Wall Hernia Surg. **6**, e61 (2023)
6. Dixon, F., Vitish-Sharma, P., Khanna, A., Keeler, B.D.: Robotic assisted surgery reduces ergonomic risk during minimally invasive colorectal resection. VOLCANO randomised controlled trial. Arch. Surg. **409**, e142 (2024)
7. Nillahoot, N., Pillai, B.M., Sharma, B., Wilarsrusmee, C., Suthakorn, J.: 3D Force/Torque parameter acquisition and correlation identification during primary tocar insertion in laparoscoic abdominal surgery. Sensors(MDPI) **22**, e8970 (2022)
8. Armijo, P.R., Huang, C.K., High, R., Leon, M., Siu, K.C., Oleynikov, D.: Ergonomic of minimally invasive surgary: an analysis of muscle effort and fatigue in the operating room between laparoscopic and robotic surgery. Surg. Endosc. **33**, 2323–2331 (2019)
9. Caballero, D., Pérez-Salazar, M.J., Sánchez-Margallo, J.A., Sánchez-Margallo, F.M.: Applying artiicial intelligence on EDA sensor data to predict stress on minimally invasive robcric-assisted surgery. Int. J. Comput. Assist. Radiol. Surg. **19**, 1953–1963 (2024)
10. Guzmán-García, C., et al.: Correlating personal resourcefulness and psychomotor skills: an analysis of stress, visual attention and technical metrics. Sensors (MDPI) **22**, e837 (2022)
11. Zangroniz, R., Martínez-Rodrigo, A., Pastor, J.M., López, M.T., Fernández-Caballero, A.: Electrodermal activity sensor for classification of calm/distress condition. Sensors (MDPI) **17**, e2324 (2017)
12. Ávila-Tomás, J.F., Mayer-Pujadas, M.A., Quesada-Varela, V.J.: La inteligencia artificial y sus aplicaciones en medicina I: introducción y antecendentes a la IA y robótica. Atención Primaria **52**, 778–784 (2020)

13. Hamilton, B.C.S., et al.: Artificial intelligence based real-time video ergonomic assessment and training improves resident ergonomics. Am. J. Surg. **226**, 741–746 (2023)
14. Safari, M., Naserbakht, A.H., Kouhi, A.B., Varmazyar, S.: Artificial intelligence and emerging technologies in assessing ergonomic risk factors in the workplace: a systematic review. J. Preven. Assess. Rehabil. (2025). https://doi.org/10.1177/10519815251349793
15. Pérez-Salazar, M.J., Caballero, D., Sánchez-Margallo, J.A., Sánchez-Margallo, F.M.: Comparative study of ergonomics in conventional and robot-assisted laparoscopic surgery **24**, e3840 (2024)
16. Hermens, H.J., Freriks, B., Disselhorst-Klug, C., Rau, G.: Development of recommendations for SEMG sensors and sensor placement procedures. J. Electromagenestism Kinesiol. **10**, 361–374 (2000)
17. Dufaug, A., Barthod, C., Goujon, L., Marechal, L.: New joint analysis of electromyography spectrum and amplitude-based methods towards real-time muscular fatigue evaluation during a simulated surgical procedure: A pilot analysis on the statistical significance. Med. Eng. Phys. **79**, 1–9 (2020)
18. Oka, M.: Interpreting a standardized and normalized measure of neighborhood socioeconomic status for a better understanding of health differences. Arch. Public Health **79**, e226 (2021)
19. Colton, T.: Statistics in Medicine. Little Brown and Co. Boston, Massachusetts, U.S.A. (1974)
20. Hyndman, R., Koehler, A.B.: Another look at measures of forecast accuracy. Int. J. Forecast. **22**, 679–708 (2006)
21. Frank, E., Hall, M.A., Witten, I.H.: The WEKA workbench online appendix for data mining: practical machine learning tools and techniques. Morgan Kauffman, Burlington, Vermont, U.S.A (2016)
22. Vapnik, V.N., Chervonenkis, Y.A.: On the uniform convergence of relative frequencies of events to their probabilities. Theor. Probabil. **16**, 264–280 (1971)
23. Amirhanayagam, A., Zecca, M., Barber, S., Singh, B., Moss, E.L.: Impact of minimally invasive surgery on surgeon health study: protocol of a single-arm observational study conducted in the live surgery setting. BMJ Open **13**(3), e066765 (2023)
24. Caballero, D., Pérez-Salazar, M.J., Sánchez-Margallo, J.A., Sánchez-Margallo, F.M.: Optimization of an artificial neural network for predicting stress in robot-assisted laparoscopic surgery based on EDA sensor data. Int. J. Comput. Assist. Radiol. Surg. (2025). https://doi.org/10.1007/s11548-025-03399-w
25. Diederichsen, L.P., et al.: The effect of handedness on electromyographic activity of human shoulder muscle during movement. J. Electromyograpic Kinesiol. **17**, 410–419 (2007)
26. Pérez-Duarte, J.F., et al.: Objective analysis of surgeons' ergonomy during laparoendoscopic single-site surgery through the use of surface electromyography and a motion capture data glove **28**, 1314–1320 (2014)

Implementation of an AI-Assisted Telemedicine System in Nursing Homes: Protocol and Preliminary Results

José Luis Ávila-Jiménez[1]([✉]) [iD], Nuria Luque Reigal[2], Manuel Rich-Ruiz[2] [iD], Vanesa Cantón-Habas[2] [iD], and Sebastián Ventura[3] [iD]

[1] Department of Computer and Electronic Engineering, Universidad de Cordoba, Córdoba, Spain
jlavila@uco.es

[2] Department of Nursing, Pharmacology and Physiotherapy, Universidad de Córdoba, Córdoba, Spain
{n62luren,en1rirum,n92cahav}@uco.es

[3] Department of Computing and Artificial Intelligence, Universidad de Córdoba, Córdoba, Spain
sventura@uco.es

Abstract. Population ageing is increasing the demand for complex care in nursing homes, where continuous medical attention is often limited due to the lack of on-site healthcare professionals. The primary objective of this study is to evaluate the feasibility of applying artificial intelligence to telemedicine in residential care, with a particular focus on developing and validating a predictive model for hospital referral. We conducted a multicentre observational study using a telemedicine system equipped with advanced diagnostic devices to manage acute events in 70 nursing homes across Spain. Clinical and socio-demographic data obtained during telemedicine consultations from 5,192 anonymized patients were used to train a deep learning model. Preliminary findings show that the neural network achieved an accuracy of 0.93 and an AUC of 0.90 in predicting hospital referrals, substantially outperforming traditional classifiers. These results provide early evidence of the potential of leveraging data generated through telemedicine for the development of AI-based predictive models in geriatric care. Future work will focus on expanding the dataset, refining the modelling pipeline, and integrating the predictive model into the telemedicine platform to support real-time clinical decision-making in nursing homes.

Keywords: telemedicine · artificial intelligence · nursing homes · older adults · Deep Learning · care quality

1 Introduction

The global increase in life expectancy poses significant challenges to healthcare systems, especially due to the growing prevalence of chronic diseases and dependency among institutionalized older adults [8]. The complexity and frequency of

© The Author(s), under exclusive license to Springer Nature Switzerland AG 2026
A. López Fernández et al. (Eds.): CIABiomed 2025, LNBI 16148, pp. 141–148, 2026.
https://doi.org/10.1007/978-3-032-10361-2_11

acute events in nursing homes now require innovative care models. Telemedicine–defined as the remote provision of healthcare using information technologies–has emerged as a strategic tool to deliver timely clinical interventions and optimize healthcare resources [8,9,15]. In the context of nursing homes, telemedicine offers real-time specialist support, potentially improving resident outcomes and system efficiency [15].

Recent high-quality work in remote patient monitoring (RPM) and telemedicine in institutional settings has strengthened evidence of beneficial outcomes. A systematic review by Tan et al. observed that RPM interventions improve patient safety and adherence, enhance mobility and functional status, and show trends toward reducing hospital admissions and healthcare costs across multiple care environments [17]. Similarly, David-Olawade et al. explore how teletriage and remote monitoring reduce emergency and health service use in residential and home-based care settings, showing relevance for nursing homes and underlining the importance of interpretability and prospective validation in future deployments [7].

The integration of artificial intelligence (AI) further enhances this model by enabling rapid assessment of acute cases and supporting clinical decisions. Recent work has demonstrated that AI-enhanced remote patient monitoring and deep learning models can effectively detect early deterioration in elderly populations, including nursing home environments [3,16].

This study presents the protocol and preliminary evaluation of an AI-assisted telemedicine system in a nationwide network of Spanish nursing homes (Vitalia Home). The objectives of this study is to develop and validate a model capable of classifying acute health events, and to evaluate the impact of the AI model.

2   Materials and Methods

A multicentre observational and experimental study was conducted in Vitalia Home, a network comprising 70 nursing homes across Spain. All residents aged 65 and above who experienced acute health events and required medical assessment were eligible for inclusion upon informed consent.

The intervention provided round-the-clock telemedicine support using a *Comitas e-Health* mobile cart[1] equipped with built-in medical tools–such as ultrasound, otoscope, dermatoscope, ECG, stethoscope, pulse oximeter, and blood pressure monitor–on a battery-powered, adjustable-height workstation. Designed for real-time consultations, the cart integrates a PTZ camera for high-definition video links with specialists and wireless connectivity. Its ergonomic, autonomous configuration allows rapid deployment in nursing homes, enabling remote diagnostics and specialist access at the point of care without patient transfer. Clinical staff at the facilities received prior training on both device operation and the telemedicine platform (Fig. 1).

Data were collected using a digital platform, allowing real-time entry by care staff during telemedicine consultations.

---

[1] https://comitas.es/carro-telemedicina/.

Fig. 1. Comitas e-Health video-conferencing telemedicine cart, a battery-powered, height-adjustable mobile workstation equipped with vital-sign monitoring (ECG, pulse oximeter, sphygmomanometer), ultrasound transducer, stethoscope, and drawer for peripherals. Image credit: Comitas e-Health

2.1 Ethics

The study follows the Declaration of Helsinki and Spanish data protection laws. Approval from the relevant ethics committee was obtained.

2.2 Dataset

The dataset comprises records from 5,192 anonymized patients treated in residential care settings for acute medical events, capturing a wide range of socio-demographic, clinical, procedural, and outcome-related variables. Socio-demographic information includes age, sex, center location, duration of institutionalization, and functional status as measured by the Barthel index. Clinical data encompass baseline vital signs (respiratory rate, oxygen saturation, supplemental oxygen, blood pressure, heart rate, temperature, and consciousness level using the AVPU scale), current medication regimens, and comorbidities coded according to ICD-10 or SNOMED standards.

Regarding acute events, the dataset records the primary reason for consultation–grouped into standardized categories such as infections, trauma, or metabolic disturbances–along with signs and symptoms assessed using a modified NEWS2 scale, use of diagnostic devices, and intervention duration. Finally, process and outcome variables provide information on the referring staff's role and experience, the attending physician's profile, and actions performed during the telemedicine consultation, allowing for a comprehensive analysis of workflow and clinical decision-making.

Missing values were handled systematically: continuous variables were imputed using the median of the training set, while categorical variables were imputed by adding a *missing* category prior to one-hot encoding. Continuous features (e.g., vital signs) were standardized using z-score normalization to improve model convergence. Comorbidities coded in ICD-10 or SNOMED were grouped into broader clinical categories (e.g., cardiovascular, respiratory, metabolic, infectious) to reduce dimensionality and improve interpretability.

Categorical features were binarized using one-hot encoding, and dimensionality reduction was applied using Truncated Singular Value Decomposition [10] retaining 200 components and explaining 99.9% of the variance.

The objective variable was the event resolution status, defined as whether an acute event resolved without hospital referral. This outcome variable is highly imbalanced, with a small proportion of cases resulting in hospital referral ($\approx 90\%$ resolved vs 10%referred). To mitigate it, we applied a variety of resampling techniques, including Synthetic Minority Over-sampling Technique (SMOTE) [4], SMOTEENN, a combination of SMOTE and Edited Nearest Neighbors) [1], and SMOTETomek, a combination of SMOTE and Tomek links [1]. These approaches generate new synthetic minority-class samples or remove borderline majority-class examples to enhance sensitivity to the underrepresented class.

2.3 Developed Model

The neural network implemented in this study is a deep feedforward multilayer perceptron, composed of three hidden layers with 256, 128, and 64 units, respectively. Each layer is followed by Batch Normalization [12] to accelerate training convergence and improve model stability by reducing internal covariate shift. LeakyReLU activation functions [14] are applied to all hidden layers, enabling the network to learn complex, non-linear relationships while avoiding the vanishing gradient problem. Dropout regularization is included after each dense layer to prevent overfitting and enhance generalization. The AdamW optimizer [13] is used for efficient and robust training, as it decouples weight decay from the gradient update, promoting better convergence in high-dimensional feature spaces.

This architecture, coupled with dimensionality reduction and resampling of the minority class, was selected for its ability to capture subtle patterns in the clinical dataset and delivered the best performance among all evaluated models.

Hyperparameters such as learning rate, batch size, and dropout rates were optimized through cross-validation. Several candidate architectures were explored, varying the number of hidden layers (from two to four) and the number of units per layer (ranging from 64 to 512). Dropout rates and activation functions were also compared. The selected architecture consistently achieved the best trade-off between AUC, F1 score, and generalization on the validation set. Hyperparameters were optimized using five-fold cross-validation on the training set, while the independent test set was used exclusively for final performance evaluation.

The dataset was partitioned into training (70%), validation (10%), and test (20%) sets using stratified sampling to preserve the distribution of the minority class (hospital referrals) across all subsets.

2.4   Baselines

We evaluated multiple classification algorithms to predict hospital referrals among institutionalized patients experiencing acute medical events. As baseline models for comparison, Logistic Regression was selected for its interpretability and simplicity, offering a transparent reference point for more complex techniques [11]. A Random Forest model was also included, leveraging an ensemble of decision trees to improve predictive accuracy and mitigate overfitting [2]. To handle non-linear decision boundaries, we employed Support Vector Machines (RBF kernel), which map inputs into higher-dimensional spaces [5]. Lastly the K-Nearest Neighbors classifier was adopted for its straightforward, instance-based approach, assigning class labels based on proximity to training samples [6].

3   Results and Discussion

Table 1. Performance of evaluated models for hospital referral prediction

Model	F1	AUC	Accuracy	Best Rebalancing
Logistic Regression	0.2674	0.7034	0.8478	SMOTETomek
Random Forest	0.3148	0.7784	0.9081	SMOTEENN
SVM	0.1311	0.5739	0.7571	SMOTEENN
K-Nearest Neighbors	0.1400	0.5394	0.4413	SMOTETomek
Neural Network	0.5595	0.8971	0.9288	SMOTE

The comparative analysis in Table 1 shows a substantial difference in performance among the evaluated models for hospital referral prediction. Among traditional machine learning algorithms, Random Forest yielded the highest F1 score (0.3148), ROC AUC (0.7784), and accuracy (0.9081), especially when combined with the SMOTEENN rebalancing strategy. Logistic Regression, when paired with SMOTETomek, performed moderately, with an F1 of 0.2674, ROC AUC of 0.7034, and accuracy of 0.8478. Both SVM and K-Nearest Neighbors showed limited effectiveness in identifying positive cases, with F1 scores below 0.15 and ROC AUC values below 0.58, indicating poor discrimination and limited practical utility in this context, even after applying rebalancing.

The results highlight notable differences in the suitability of various machine learning approaches for hospital referral prediction in highly imbalanced clinical data. Although traditional classifiers such as logistic regression and random forest showed acceptable accuracy (0.85 and 0.91, respectively), their F1 scores remained low (0.27 and 0.31), reflecting limited sensitivity to the minority class even after employing rebalancing strategies like SMOTETomek and SMOTEENN. Support vector machines and k-nearest neighbors performed even worse, with F1 scores below 0.15 and ROC AUC values under 0.58, underscoring their limited ability to generalize in this challenging scenario.

In contrast, the neural network achieved a substantial improvement, reaching an F1 score of 0.56 and AUC of 0.90, while maintaining high overall accuracy (0.93). This improvement suggests that the combination of advanced resampling, dimensionality reduction, and nonlinear modeling is key for capturing the complex relationships inherent in the dataset. However, despite the performance gain, the relatively modest F1 scores across all models reflect the inherent difficulty of predicting rare referral events and highlight the need for further research to achieve clinically actionable precision and recall in real-world deployment.

It should be noted that these results are limited to predictive model performance and do not yet reflect clinical outcomes such as actual reductions in hospital transfers. Moreover, the relatively modest F1 score, despite strong AUC and accuracy, indicates sensitivity limitations associated with the imbalanced dataset. The absence of external validation raises the possibility of overfitting, which will need to be addressed in future prospective evaluations.

4   Conclusions and Future Work

This study evaluated several machine learning models for predicting hospital referral in acute events within nursing homes Our findings demonstrate that, while traditional models offer moderate performance, a Neural Network architecture with preprocessing and resampling methods substantially improve the identification of cases requiring hospital transfer. These results support the integration of AI-driven decision support tools into telemedicine workflows to enhance acute care management and reduce unnecessary hospitalizations.

Our preliminary findings suggest that an AI-supported telemedicine platform can improve acute event management in nursing homes so it would reduce unnecessary hospital transfers and enhancing care continuity. The deep learning model showed promise for supporting real-time triage and decision-making, though further validation is needed. Challenges encountered included data integration, variable selection, and staff adoption.

It should be noted that while the broader telemedicine initiative aims to reduce unnecessary hospital transfers, the present study only reports predictive model performance metrics.

Future work will focus on expanding the dataset to include more diverse clinical cases, refining the feature engineering and model optimization pipeline, and validating the approach prospectively in real-world settings. In addition, integration of the predictive model into the telemedicine platform will be explored to enable real-time decision support, along with an assessment of cost-effectiveness, interpretability, and clinical adoption in nursing home environments.

Acknowledgments. This work has been supported by the *Spanish Ministry of Science, Innovation and Universities*, project *PID2023-148396NB-I00* and also supported by the *Plan Propio Galileo de Innovación y Transferencia* of the *Universidad de Córdoba*.

Disclosure of Interests. The authors have no competing interests to declare that are relevant to the content of this article.

References

1. Batista, G.E., Prati, R.C., Monard, M.C.: A study of the behavior of several methods for balancing machine learning training data. SIGKDD Explor. Newsl. **6**(1), 20–29 (2004)
2. Breiman, L.: Random forests. Mach. Learn. **45**(1), 5–32 (2001)
3. Carlier, A., Peyramaure, P., Favre, K., Pressigout, M.: Fall detector adapted to nursing home needs through an optical-flow based cnn. arXiv preprint (2020)
4. Chawla, N.V., Bowyer, K.W., Hall, L.O., Kegelmeyer, W.P.: Smote: synthetic minority over-sampling technique. J. Artif. Intell. Res. **16**, 321–357 (2002). https://doi.org/10.1613/jair.953
5. Cortes, C., Vapnik, V.: Support-vector networks. Mach. Learn. **20**(3), 273–297 (1995)
6. Cover, T., Hart, P.: Nearest neighbor pattern classification. In: IEEE Transactions on Information Theory, vol. 13, pp. 21–27. IEEE (1967)
7. David-Olawade, A.C., Olawade, D.B., Ojo, I.O., Famujimi, M.E., Olawumi, T.T., Esan, D.T.: Harnessing telemedicine for enhanced patient care: nursing in the digital age. Inf. Health **1**(2), 100–110 (2024). https://doi.org/10.1016/j.infoh.2024.07.003
8. Ekeland, A.G., Bowes, A., Flottorp, S.: Effectiveness of telemedicine: a systematic review of reviews. Int. J. Med. Inf. **79**(11), 736–771 (2010). https://doi.org/10.1016/j.ijmedinf.2010.08.006
9. Grabowski, D.C., O'Malley, A.J.: Use of telemedicine can reduce hospitalizations of nursing home residents and generate savings for medicare. Health Aff. **33**(2), 244–250 (2014). https://doi.org/10.1377/hlthaff.2013.0922
10. Halko, N., Martinsson, P.G., Tropp, J.A.: Finding structure with randomness: probabilistic algorithms for constructing approximate matrix decompositions. SIAM Rev. **53**(2), 217–288 (2011)
11. Hosmer, D.W., Lemeshow, S., Sturdivant, R.X.: Applied logistic regression. Wiley Series in Probability and Statistics (2013)
12. Ioffe, S., Szegedy, C.: Batch normalization: Accelerating deep network training by reducing internal covariate shift. In: Proceedings of the 32nd International Conference on Machine Learning (ICML), pp. 448–456 (2015)
13. Loshchilov, I., Hutter, F.: Decoupled weight decay regularization. In: International Conference on Learning Representations (ICLR) (2019)
14. Maas, A.L., Hannun, A.Y., Ng, A.Y.: Rectifier nonlinearities improve neural network acoustic models. In: Proceedings of ICML (2013)
15. Schuchman, M.J., Vanderburgh, C., Patel, A.: Telemedicine consultation reduced the hospitalization of nursing home residents in 14/16 and care costs in 8/11 articles. J. Am. Med. Dir. Assoc. (2022). https://doi.org/10.1016/j.jamda.2022.05.004

16. Shaik, T., Tao, X., Higgins, N., Li, L., Gururajan, R., Zhou, X., Acharya, U.R.: Remote patient monitoring using artificial intelligence: current state, applications, and challenges. arXiv preprint (2023)
17. Tan, S.Y., Sumner, J., Wang, Y., Wenjun Yip, A., et al.: A systematic review of the impacts of remote patient monitoring (rpm) interventions on safety, adherence, quality-of-life and cost-related outcomes. npj Digit. Med. **7**, 192 (2024). https://doi.org/10.1038/s41746-024-01182-w

Improving Liver Graft Decision-Making Through AI: Validation with Internal and National Datasets

Miguel Cuende[1,4], Beatriz Pontes[2,4](✉) iD, Juan M. Castillo Tuñón[1,4] iD
Daniel Mateos García[2,4] iD, Luis M. Marín Gómez[3,4] iD,
José C. Riquelme Santos[2,4] iD, and Gloria de la Rosa Rodríguez[3,4] iD

[1] HPB and Liver Trasplant Unit, Toledo University Hospital, Toledo, Spain
[2] Department of Computer Languages and Systems, University of Seville,
Seville, Spain
bepontes@us.es
[3] HPB Surgery and Liver Transplant Unit, Virgen del Rocío University Hospital,
Seville, Spain
[4] Spanish Liver Transplant Registry (RETH), National Transplant Organization,
Madrid, Spain

Abstract. Liver graft assessment is a critical and complex step in the transplantation process, traditionally reliant on the subjective judgment of experienced surgeons, based on initial donor data and macroscopic evaluation. In a previous study, we developed a machine learning-based expert system trained on clinical data to assist in the early decision-making process of liver graft suitability. Using donor information collected in the official Liver Donation Protocol (LDP), the model demonstrated promising predictive performance, particularly in identifying transplantable grafts that were otherwise discarded. This study aims to externally validate and expand the model using newly collected internal data and a large-scale external dataset provided by the Spanish National Transplant Organization (ONT). By evaluating the model's generalization to these additional datasets and retraining it with combined data sources, we assess its robustness, adaptability, and potential for broader clinical application. Furthermore, we explore alternative modeling strategies to enhance predictive performance and reliability. This work represents a critical step toward refining and scaling an AI-driven tool to support transplant surgeons in early-stage graft evaluation using standardized donor variables.

Keywords: Clinical data analysis · Predictive modeling · Decision support system

1 Introduction

Liver transplantation (LT) represents the definitive treatment for both acute and chronic end-stage liver disease, providing substantial survival benefits for

© The Author(s), under exclusive license to Springer Nature Switzerland AG 2026
A. López Fernández et al. (Eds.): CIABiomed 2025, LNBI 16148, pp. 149–162, 2026.
https://doi.org/10.1007/978-3-032-10661-2_12

patients worldwide. However, the field faces a critical challenge: a persistent and growing imbalance between the increasing number of transplant candidates and the limited availability of suitable donor organs. This organ scarcity has intensified globally as transplant programs expand their indications and patient populations continue to grow.

Central to addressing this challenge is the complex process of liver graft assessment–a critical decision point that determines whether a potential donor organ will be utilized for transplantation. Currently, this evaluation relies heavily on the individual experience and subjective judgment of transplant surgeons, primarily through macroscopic *in situ* assessment of the organ. While this clinical expertise remains invaluable, the inherently subjective nature of this evaluation process contributes to significant organ discard rates, with many potentially viable grafts being rejected based on uncertain or indeterminate criteria.

The consequences of this subjective assessment are substantial. In Spain, for instance, 13.1% of offered liver grafts are initially discarded, with an additional 27.6% rejected after *in situ* evaluation. Notably, liver steatosis and macroscopic appearance represent the primary reasons cited for rejection, yet these assessments often fail to correlate with actual histopathological findings. Remarkably, up to 20.4% of rejected grafts show no pathological abnormalities on biopsy, suggesting that many discarded organs could have been successfully transplanted. Therefore, a primary objective of an AI-driven support tool is to specifically target and reduce this rate of unnecessary discards by providing an objective assessment to complement surgical judgment.

The advent of artificial intelligence (AI) and machine learning (ML) presents an unprecedented opportunity to address these limitations through objective, data-driven decision support tools. However, previous AI applications in liver transplantation have been hampered by critical methodological limitations, including overfitting on small datasets, lack of external validation across multiple centers, and poor model interpretability–factors that have prevented widespread clinical adoption.

The primary objective of this study is to develop and validate a robust AI-based decision support tool that can improve liver graft decision-making by assisting transplant surgeons in the early assessment of organ suitability. Specifically, we aim to create a mathematical model capable of classifying liver grafts as transplantable or non-transplantable based solely on information available in the Liver Donation Protocol (LDP), prior to costly and time-consuming *in situ* evaluations.

Our central hypothesis is that a machine learning model, trained on the accumulated experience and decision patterns of expert liver transplant surgeons from multiple centers, can provide objective, reliable predictions that complement clinical judgment while reducing unnecessary organ discards.

The significance of this research lies in its potential to optimize the utilization of scarce donor organs, reduce inappropriate discard rates of viable grafts, and ultimately improve transplant access and outcomes for patients with end-stage liver disease. By providing an early, standardized assessment tool based on readily available protocol data, this approach could also generate substantial cost savings by reducing unnecessary procurement team deployments and facilitating more efficient organ allocation decisions.

This study represents a critical step toward integrating evidence-based AI tools into clinical liver transplantation practice, with rigorous validation on both internal and national datasets designed to ensure clinical reliability and broad applicability across diverse transplant programs. To this end, an approach based on machine learning has been employed, following the methodology proposed in a previous work [34], and new modeling and evaluation strategies have been explored with expanded datasets.

2   Background and Previous Work

The persistent disparity between organ demand and availability has driven the transplant community to explore innovative approaches for optimizing graft utilization [2–4]. This organ shortage necessitates the use of novel strategies to expand the donor pool, including the use of grafts from donors with expanded criteria [5]. A critical component of this process is the complex assessment of a potential liver graft, which traditionally relies heavily on the transplant surgeon's individual experience and macroscopic evaluation of the organ *in situ* [5,6].

This subjective assessment process, while crucial, contributes to a high discard rate for offered organs. In Spain, for instance, a significant percentage of livers are declined, with many rejections based on macroscopic features like steatosis or other indeterminate findings that may not have precluded successful transplantation [7,22]. This suggests that a considerable number of potentially viable grafts are discarded, exacerbating the organ shortage.

To standardize and improve graft assessment, several clinical scoring systems have been developed. These models aim to quantify the risk associated with donor and recipient factors to predict post-transplant outcomes. Prominent examples include the Donor Risk Index (DRI), which aggregates donor characteristics to predict graft failure [24,25], the Survival Outcomes Following Liver Transplantation (SOFT) score, which combines donor, recipient, and operative factors to predict 3-month patient survival [6,16], and the Balance of Risk (BAR) score, which also integrates donor and recipient variables to predict survival [2,19]. While these scores provide a common framework for risk assessment, they have limitations, including being developed before the widespread adoption of the MELD score and sometimes lacking granularity for specific patient populations [3,14].

With the rise of artificial intelligence (AI), there has been growing interest in applying machine learning (ML) to overcome the limitations of traditional scoring systems [9,12]. ML models can analyze complex, non-linear relationships between a large number of variables, offering the potential for more accurate predictions in medicine [9,15,27]. In the context of LT, ML has been increasingly applied to predict post-transplant survival, screen recipients, and optimize donor-recipient matching [10,11,14,32,33]. For instance, Moccia et al. developed a computer-assisted system to assess liver steatosis from images, providing a quantitative measure to aid surgical decision-making [8]. Such tools demonstrate the potential of AI to provide objective, data-driven insights to complement the surgeon's expertise [23].

Despite these advances, previous work in this domain faces several critical limitations. A major issue is **overfitting**, where a model learns the training data too well, including its noise, and consequently fails to generalize to new, unseen data [18,26]. This is a significant concern, especially when models are developed on small or homogeneous datasets [18]. Many studies are conducted at a single center, which severely limits their **external validity**; a model that performs well in one hospital may fail in another due to differences in patient populations, surgical techniques, or data collection practices [13,21,28]. The lack of external validation on diverse, multi-center datasets remains a critical barrier to the clinical implementation of AI models, as emphasized in recent comprehensive reviews [17,20,31].

Furthermore, many advanced ML models, particularly deep learning algorithms, are considered "black boxes," making it difficult to understand the reasoning behind their predictions [29]. This lack of **interpretability** poses a major hurdle for clinical adoption, as clinicians must be able to trust and justify the decisions supported by an AI tool [29,30]. Without transparency, it becomes difficult to identify biases, correct errors, or ensure the model's recommendations are clinically sound [29].

Therefore, there is a clear and unmet need for an AI-based decision support tool for liver graft assessment that is rigorously validated on both internal and large-scale national datasets. Such validation is essential to ensure the model is robust, generalizable, and reliable across different clinical settings. By demonstrating efficacy on a national level, we can build the necessary trust for clinical integration, ultimately aiming to optimize the use of scarce donor organs, reduce discard rates for viable grafts, and improve outcomes for patients in need of liver transplantation.

3   Methodology

In this section we explain the methodology followed in this piece of research. Being the datasets used in this work of the same nature than in a previous work, we refer the reader to [34] for more information. The models were trained and tested using multiple datasets from various sources, with their performance assessed through a range of validation techniques and predictive comparisons.

3.1   Datasets

For the experiments, a total of five different datasets were used, where all of them contain the same set of relevant clinical and demographic variables for liver graft assessment. These include features such as age, sex, body mass index (BMI), arterial hypertension (AH), diabetes mellitus (DM), dyslipidemia (DLP), personal medical history (PMH), personal surgical history (PSH), laboratory parameters such as glutamic oxaloacetic transaminase (GOT), glutamate pyruvate transaminase (GPT), gamma-glutamyl transferase (GGT), total bilirubin (Bb T), and sodium (Na), as well as the use and dosage of amines. Additional variables include liver ultrasound findings (normal, pathological, or not performed), and serological markers for hepatitis B and C (anti-HBc and anti-HCV). The target variable for classification is graft validity, where each liver graft is labeled as either transplantable (TLG, coded as 0) or non-transplantable (NTLG, coded as 1), based on the final decision made by liver transplant surgeons during the in situ evaluation.

- #400: This dataset consists of 400 examples and was used for training the base model presented in the previous work [34]. Out of the 400 grafts, 226 were valid for transplantation, while 174 were labeled as not valid.
- #101: Composed of 101 examples, this dataset was collected through *Liv-erGrafPredict*, a mobile application in which surgeons from various Spanish hospitals collaborated, from february 2025 to june 2025. A total of 77 grafts were valid.
- #ONT: This is the most extensive dataset, with 4071 examples, provided by the Spanish National Transplant Organization (ONT). It contains information on liver transplants performed between 2021 and 2024, with the same variables as the previous datasets. 2891 out of 4071 grafts were valid in this dataset, being the imbalance more pronounced than in the previous datasets.

From these sets, two additional combined sets were created for the experiments:

- #501: Resulting from the combination of #400 and #101, totaling 501 examples.
- #ALL: Encompassing all available data, combining #400,#101, and #ONT, resulting in a total of 4572 examples.

The combination and comparison of these datasets is crucial to understand the models generalization capabilities across internally collected data (#400, #101, #500) and externally sourced data (#ONT), which may present different characteristics, as well as the combination of all collected data (#ALL).

3.2   Experimental Design

Data preprocessing and base modeling were performed following the methodology established in the previous work [34]. This process includes feature selection, classification and hyperparameter optimization, using TPOT (Tree-based Pipeline Optimization Tool), an AutoML tool in Python, was employed.

In experiments where class imbalance was identified, the SMOTE (Synthetic Minority Over-sampling Technique) technique was applied to balance the class distribution and prevent the model from being biased towards the majority class. In total, six experiments were designed to evaluate the models performance in different scenarios and with varying data volumes:

1. Evaluation of the base model with new data. The model previously trained with #400 dataset was used. Predictions from this model were obtained for #101 and #ONT. The prediction results in both datasets were compared.
2. Impact of expanding proprietary data in training. Datasets #400 and #101 were combined to create a new training set, #501. The same pipeline (using XGBoost as the base model) was trained with #501. Predictions were made for #ONT and the results were compared with those obtained in Experiment 1.
3. Training with ONT data and cross-evaluation. The initial pipeline was trained using #ONT dataset. SMOTE was applied to address class imbalance in this scenario, since approximately one quarter of the examples correspond to the valid grafts. Predictions were obtained for #501 and the results were compared.
4. Training with the complete dataset. All available data (#ALL, 4572 examples) were combined. The initial pipeline (using XGBoost) was trained with #ALL, also applying SMOTE due to class imbalance. In this experiment, evaluation was performed solely through cross-validation, Leave-One-Out Cross-Validation (LOOCV), and train-test split, since there is no more additional data to perform predictions on.
5. Search for a new pipeline with TPOT for #ONT. TPOT was used to search for a new optimal machine learning pipeline for #ONT dataset, with the aim of identifying a configuration that better captures the essence of this data. In addition to internal evaluation, the predictions of this new pipeline for #501 dataset were evaluated. This allowed identifying whether the nature of the data provided by the ONT differs significantly from the study's own data.
6. Search for a new pipeline with TPOT for #ALL. TPOT was applied to the complete dataset #ALL to find a pipeline that optimized the overall model performance with all available information. Evaluation was performed through cross-validation, LOOCV, and train-test split.

As aforementioned, SMOTE has been applied in those experiments where a greater imbalance has been identified: experiments 3 to 6. In all experiments (1 to 6), model evaluation was performed using the following performance metrics: accuracy, precision, recall, f1-score and area under the curve (auc). They were calculated through cross-validation (10-fold), Leave-One-Out Cross-Validation (LOOCV), and train-test split (80% training, 20% testing). To ensure the robustness of the results, evaluations with cross-validation and train-test were performed for 100 different random re-orderings of the data, reporting the average values of the metrics.

4   Experimental Results

This section presents the results of the experiments designed to evaluate the performance of the machine learning models across different datasets. The primary metrics–accuracy, precision, recall, F1-score, and AUC–are summarized in

Tables 1, 2, 3 and 4, corresponding to Experiments 1–4. The Receiver Operating Characteristic (ROC) curves for these experiments are visualized in the four panels of Table 1, providing a graphical representation of the model performance in experiments 1 to 4.

Table 1. Results for experiment 1

Experiment 1	ACC	Precision	Recall	F1-Score	AUC
TT #400	0.6750	0.5814	0.7576	0.6579	0.7573
CV10 #400	0.7300	0.6875	0.6954	0.6914	0.7726
LOOCV #400	0.7025	0.6627	0.6437	0.6531	0.7573
Eval #101	0.4455	0.2714	0.7917	0.4043	0.5647
Eval #ONT	0.4107	0.3031	0.7949	0.4388	0.5554

In the first experiment, the base model trained on the original #400 dataset was evaluated against the newly collected internal data (#101) and the large-scale national dataset (#ONT). As shown in Table 1, the model achieved respectable internal validation metrics, with a 10-fold cross-validation (CV10) accuracy of 0.7300 and an AUC of 0.7726. However, the performance of the model deteriorated significantly when applied to the new data sets. The evaluation on #101, the data collected from the APP *LiverGraftPredict*, yielded an accuracy of only 0.4455 and an AUC of 0.5647. Similarly, the performance on the much larger #ONT dataset was poor, with an accuracy of 0.4107 and an AUC of 0.5554. These AUC values are barely above the 0.5 threshold of a random classifier, as clearly illustrated by the evaluation curves (in green and orange) in the top-left panel of Table 1. This outcome highlights a critical lack of generalization, indicating that the initial model, trained on a limited single-center dataset, failed to capture the data distribution of other populations. Its low precision and accuracy suggest a tendency to overpredict nontransplantable grafts. This poor performance underscores the limitations of training on a small, potentially non-representative dataset.

Table 2. Results for experiment 2

Experiment 2	ACC	Precision	Recall	F1-Score	AUC
TT #501	0.6931	0.6207	0.4737	0.5373	0.7595
CV10 #501	0.7006	0.6395	0.5556	0.5946	0.7445
LOOCV #501	0.6826	0.6089	0.5505	0.5782	0.7071
Eval #ONT	0.6819	0.4003	0.1958	0.2629	0.5980

Experiment 2 aimed to determine if augmenting the training set with more proprietary data could improve generalization. The model was re-trained on the combined #501 dataset (merging #400 and #101) and evaluated against #ONT. The internal validation results on #501 remained consistent, with a CV10 accuracy of 0.7006 and a LOOCV AUC of 0.7071. When evaluated on the #ONT dataset, the accuracy improved to 0.6819 compared to Experiment 1. Despite this increase in accuracy, the model's overall predictive power remained weak. The F1-Score was very low at 0.2629, and the AUC was only 0.5980, indicating poor discrimination between transplantable and non-transplantable grafts, with a better overall prediction of the majority class (transplantable grafts). The ROC curve for this evaluation, shown in the top-right panel of Table 1, confirms that the model still performed only marginally better than random guessing. This suggests that simply increasing the volume of the data by a hundred examples is insufficient to build a model that generalizes well to a diverse, national-level dataset. Nevertheless, the modest AUC improvement suggests a slight enhancement in discriminative power.

Table 3. Results for experiment 3

Experiment 3	ACC	Precision	Recall	F1-Score	AUC
TT #ONT	0.7583	0.6250	0.3863	0.4775	0.7437
CV10 #ONT	0.7445	0.5949	0.3720	0.4578	0.7423
LOOCV #ONT	0.7450	0.5920	0.3873	0.4682	0.7481
Eval #501	0.6667	0.5975	0.4798	0.5322	0.6643

In experiment 3 we inverted the process by training the model on the large and diverse #ONT dataset, and evaluating it on the smaller, combined proprietary dataset (#501). As shown in Table 3, the model trained on #ONT achieved strong internal results, with a CV10 accuracy of 0.7445 and a LOOCV AUC of 0.7481. Crucially, this model demonstrated effective generalization when tested on the #501 dataset. It achieved an accuracy of 0.6667, an F1-Score of 0.5322, and an AUC of 0.6643. The bottom left panel of Table 1 visually supports this finding, showing that the evaluation curve in #501 (in green) is significantly better than the random baseline and much closer to the internal validation curves. These results indicate that training on a large, multi-center dataset like #ONT produces a more robust and generalizable model capable of performing well on unseen data from a different source.

In experiment 4, the model was trained on the #ALL dataset, which aggregates all available data (4,572 examples). Since no external data was left for evaluation, performance was assessed using internal validation techniques. The results in Table 4 show consistent and robust performance across all methods. The 10-fold CV produced an accuracy of 0.7349 and an AUC of 0.7343, while LOOCV yielded an AUC of 0.7306. The ROC curves for this experiment, shown in the bottom right panel of Fig. 1, are tightly clustered, indicating that the model is stable and its performance is not heavily dependent on the validation

Table 4. Results for experiment 4

Experiment 4	ACC	Precision	Recall	F1-Score	AUC
TT #ALL	0.7432	0.5824	0.3764	0.4573	0.7038
CV10 #ALL	0.7349	0.5928	0.3846	0.4665	0.7343
LOOCV #ALL	0.7270	0.5680	0.3940	0.4653	0.7306

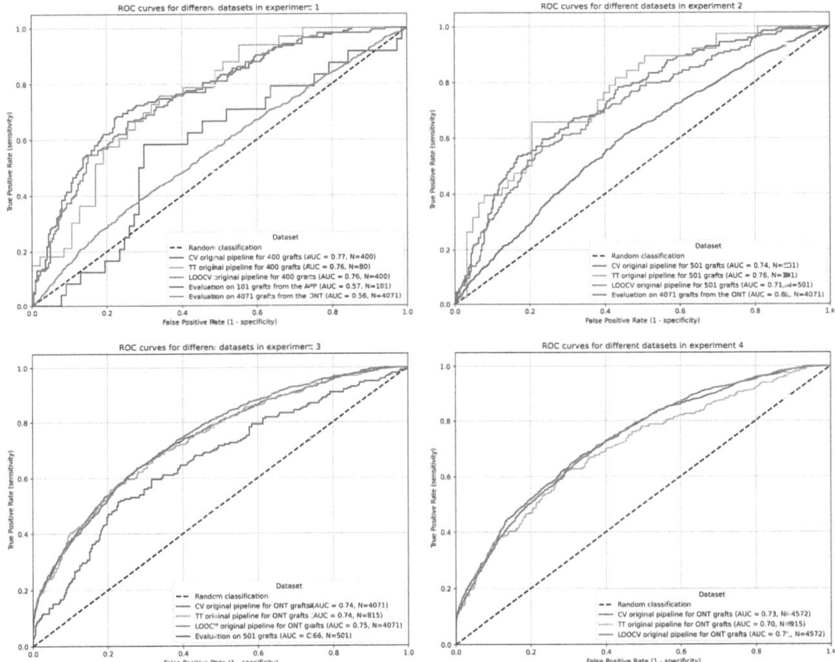

Fig. 1. ROC curves for experiments 1–4

strategy. This suggests that a model trained on a comprehensive and diverse dataset is the most reliable, effectively leveraging the combined information from all sources to achieve the best overall performance.

In experiment 5, a new pipeline was searched using TPOT specifically for the #ONT dataset. This approach aimed to create a model that was better adapted to the characteristics of large-scale national data. The new pipeline achieved exceptional internal validation metrics, with a 10-fold cross-validation (CV10) AUC of 0.8238 and a LOOCV AUC of 0.8226, as shown in Table 5. These values represent a significant improvement in the model's ability to discriminate between transplantable and non-transplantable grafts within the #ONT dataset. The precision was notably high (0.7934 in CV10), although the recall remained moderate (0.4568), indicating that while the model's positive predictions were reliable, it still failed to identify a subset of the non-transplantable grafts.

When this new model was evaluated on the proprietary #501 dataset, it achieved an AUC of 0.6691. While the accuracy (0.6071) and F1-Score (0.4387) were modest, the AUC value demonstrates a reasonable generalization capability. The ROC curve for this evaluation can be seen in the left panel of Table 2.

Table 5. Results for experiment 5

Experiment 5	ACC	Precision	Recall	F1-Score	AUC
TT #ONT	0.7939	0.6667	0.4727	0.5532	0.8127
CV10 #ONT	0.7934	0.7294	0.4568	0.5618	0.8238
LOOCV #ONT	0.7932	0.7334	0.4500	0.5578	0.8226
Eval #501	0.6527	0.6071	0.3434	0.4387	0.6691

Both experiments 3 and 5 use the #ONT dataset for training and the #501 dataset for evaluation. The key difference was the model: Experiment 3 used the initial XGBoost pipeline, while Experiment 5 used a new, TPOT-optimized pipeline based on LGBMClassifier. The new pipeline in experiment 5 demonstrated vastly superior performance on the training data, indicating that the new pipeline discovered by TPOT was much more effective at capturing the complex relationships within the large national dataset. When generalizing to the #501 dataset, the new model achieved a slightly higher AUC and higher precision, but at the cost of lower recall and F1-Score. This suggests that while the new pipeline has a marginally better underlying discriminative ability (higher AUC), it is more conservative in its predictions on the external data, leading to fewer false positives (risk of wasting a viable organ) but more false negatives (accepting a non-transplantable graft).

This trade-off is clinically critical. A high-precision model is preferable as a support tool, since it minimizes the risk of discarding a viable organ (false positive). Although its moderate recall may not detect every non-transplantable graft, this is balanced by the surgeon's definitive in-situ evaluation, making this conservative profile ideal for increasing organ utilization.

Table 6. Results for experiment 6

Experiment 6	ACC	Precision	Recall	F1-Score	AUC
TT #ALL	0.7650	0.6283	0.4545	0.5275	0.7827
CV10 #ALL	0.7542	0.6169	0.4862	0.5438	0.7850
LOOCV #ALL	0.7487	0.6071	0.4710	0.5304	0.7767

Experiment 6 involved training a new TPOT-optimized pipeline on the complete #ALL dataset, which combines all available data sources (4,572 examples).

The goal was to build the most robust and generalized model possible. The results, presented in Table 6, show strong and consistent performance across all internal validation methods. The model achieved a 10-fold CV AUC of 0.7850 and a LOOCV AUC of 0.7767. The F1-Score also indicates a healthy balance between precision and recall. The tight clustering of the ROC curves in the right panel of Fig. 2 further illustrates the stability and reliability of the model, suggesting that it has successfully integrated the patterns from the different data sources into a single, cohesive model.

Both experiments 4 and 6 use the complete #ALL dataset for training and internal validation. Experiment 4 used the original XGBoost pipeline, whereas Experiment 6 used the new TPOT-optimized pipeline, showing a clear and decisive improvement over Experiment 4. This proves that the new pipeline found by TPOT is not only more powerful but also more robust, making it the most reliable model developed in this study.

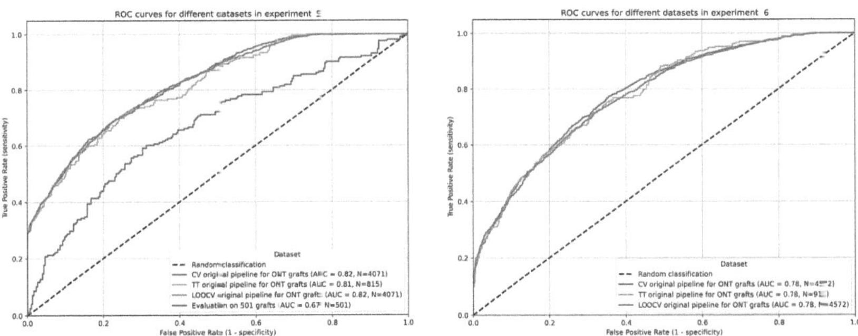

Fig. 2. ROC curves for experiments 5–6

Regarding the new found pipelines in experiments 5 and 6, both of them use a very effective algorithm called LGBMClassifier. They also use a special setting called 'dart', which helps prevent the model from simply memorizing the training data and ensures it learns general patterns instead. In order to fix the imbalance in the datasets (having many more "valid" grafts than "invalid"), SMOTE technique has been used, artificially creating more examples of the rare case (the invalid grafts). This helps the model to learn from a more balanced perspective. Finally, a tool called FeatureUnion is also used to create multiple, new perspectives of the data and then combine them. This gives the AI a much richer and more complete picture before making a decision.

5   Conclusions

This study confirms the value of artificial intelligence in supporting early decision-making during liver graft assessment. It demonstrates that machine learning models trained on small, single-center datasets fail to generalize effectively, while those trained on large-scale, multi-center data–particularly from the Spanish National Transplant Organization (ONT)–achieve better performance and reliability. The high precision of our final model directly addresses the 20.4% rate of unnecessary organ discards, which are effectively clinical "false positives". By offering a reliable second opinion on borderline grafts, the tool is designed to reduce premature discards and increase organ utilization, although its exact impact requires prospective validation.

The most robust results were obtained using a pipeline optimized with AutoML tools like TPOT and LGBMClassifier, trained on the aggregated national dataset. These models successfully leveraged techniques such as SMOTE and FeatureUnion to handle data imbalance and enhance feature representation.

Overall, the findings highlight the importance of data diversity and model optimization to ensure generalization and clinical applicability. The developed tool shows strong potential to complement surgical judgment, reduce unnecessary organ discards, and improve the efficiency and equity of liver transplantation decisions. Future steps should include prospective clinical validation and further work on model interpretability to support widespread clinical adoption.

Acknowledgments. This publication is part of the projects PID2020-117954RB-C22 and PID2023-146037OB-C21, funded by MICIU/AEI/10.13039/501100011033/.

Disclosure of Interests. The authors declare that the research was conducted in the absence of any commercial or financial relationships that could be construed as a potential conflict of interest.

References

1. Dienstag, J.L., Cosimi, A.B.: Liver transplantation-a vision realized. N. Engl. J. Med. **367**(16), 1483–1485 (2012). https://doi.org/10.1056/NEJMp1210159
2. Lai, J.C., Feng, S., Roberts, J.P.: An examination of liver offers to candidates on the liver transplant waitlist. Gastroenterology **143**(5), 1261–1265 (2012). https://doi.org/10.1053/j.gastro.2012.07.105
3. Busuttil, R.W., Klintmalm, G.B.G.: Transplantation of the Liver, 3rd edn. Elsevier, Philadelphia (2015)
4. Busuttil, R.W., Tanaka, K.: The utility of marginal donors in liver transplantation. Liver Transpl. **9**(7), 651–663 (2003). https://doi.org/10.1053/jlts.2003.50105
5. Toniutto, P., Zanetto, A., Ferrarese, A., Burra, P.: Current challenges and future directions for liver transplantation. Liver Int. **37**(3), 317–327 (2017). https://doi.org/10.1111/liv.13255
6. Volk, M.L., Roney, M., Merion, R.M.: Systematic bias in surgeons' predictions of the donor-specific risk of liver transplant graft failure. Liver Transpl. **19**(9), 987–990 (2013). https://doi.org/10.1002/lt.23683
7. Organización Nacional de Trasplantes (ONT): Activity Report Spanish National Transplant Organization 2021. https://www.ont.es/wp-content/uploads/2023/06/Actividad-de-Donacion-y-Trasplante-en-Espana-2021.pdf. Accessed 11 July 2023

8. Moccia, S., et al.: Computer-assisted liver graft steatosis assessment via learning-based texture analysis. Int. J. Comput. Assist. Radiol. Surg. **13**(9), 1357–1367 (2018). https://doi.org/10.1007/s11548-018-1787-6

9. Bennett, C.C., Hauser, K.: Artificial intelligence framework for simulating clinical decision-making: a Markov decision process approach. Artif. Intell. Med. **57**(1), 9–19 (2013). https://doi.org/10.1016/j.artmed.2012.12.003

10. Ayllón, M.D., et al.: Validation of artificial neural networks as a methodology for donor-recipient matching for liver transplantation. Liver Transpl. **24**(2), 192–203 (2018). https://doi.org/10.1002/lt.24870

11. Briceño, J., et al.: Use of artificial intelligence as an innovative donor-recipient matching model for liver transplantation: results from a multicenter Spanish study. J. Hepatol. **61**(5), 1020–1028 (2014). https://doi.org/10.1016/j.jhep.2014.05.039

12. Darcy, A.M., Louie, A.K., Roberts, L.W.: Machine learning and the profession of medicine. JAMA **315**(6), 551–552 (2016). https://doi.org/10.1001/jama.2015.18421

13. Bellomo, R., Warrillow, S.J., Reade, M.C.: Why we should be wary of single-center trials. Crit. Care Med. **37**(11), 3114–3119 (2009). https://doi.org/10.1097/CCM.0b013e3181bc7bd5

14. Spann, A., et al.: Applying machine learning in liver disease and transplantation: a comprehensive review. Hepatology **71**(3), 1093–1105 (2020). https://doi.org/10.1002/hep.31103

15. Komura, D., Ishikawa, S.: Machine learning approaches for pathologic diagnosis. Virchows Arch. **475**(2), 131–138 (2019). https://doi.org/10.1007/s00428-019-02594-w

16. Rana, A., Hardy, M.A., Halazun, K.J., et al.: Survival outcomes following liver transplantation (SOFT) score: a novel method to predict patient survival following liver transplantation. Am. J. Transplant. **8**(12), 2537–2546 (2008). https://doi.org/10.1111/j.1600-6143.2008.02400.x

17. Park, S.H., et al.: Key principles of clinical validation, device approval, and insurance coverage decisions of artificial intelligence. Korean J. Radiol. **22**(1), 47–58 (2021). https://doi.org/10.3348/kjr.2020.0879

18. Subramanian, J., Simon, R.: Overfitting in prediction models – is it a problem only in high dimensions? Contemp. Clin. Trials **36**(2), 636–641 (2013). https://doi.org/10.1016/j.cct.2013.07.013

19. Dutkowski, P., et al.: Are there better guidelines for allocation in liver transplantation? A novel score targeting justice and utility in the model for end-stage liver disease era. Ann. Surg. **254**(5), 745–753 (2011). https://doi.org/10.1097/SLA.0b013e3182365050

20. Cabitza, F., Campagner, A., Soares, F., et al.: The importance of being external: methodological insights for the external validation of machine learning models in medicine. Comput. Methods Programs Biomed. **208**, 106288 (2021). https://doi.org/10.1016/j.cmpb.2021.106288

21. Dechartres, A., Trinquart, L., Boutron, I., Ravaud, P.: Influence of trial sample size on treatment effect estimates: meta-epidemiological study. BMJ **346**, f2304 (2013). https://doi.org/10.1136/bmj.f2304

22. Organización Nacional de Trasplantes (ONT): Activity Report Spanish National Transplant Organization 2022. https://www.ont.es/wp-content/uploads/2023/03/DONACION-Y-TRASPLANTE-GENERAL-2022.pdf. Accessed 11 July 2023

23. Organización Nacional de Trasplantes (ONT): Annual Report 2022: Liver Transplantation Statistics. https://www.ont.es/informacion/estadisticas/. Accessed 11 July 2023

24. Feng, S., et al.: Characteristics associated with liver graft failure: the concept of a donor risk index. Am. J. Transplant. **6**(4), 783–790 (2006). https://doi.org/10.1111/j.1600-6143.2006.01242.x
25. Halazun, K.J., Al-Mukhtar, A., Aldouri, A., Willis, S., Ahmad, N.: Warm ischemia in transplantation: search for a consensus definition. Transpl. Proc. **39**(5), 1329–1331 (2007). https://doi.org/10.1016/j.transproceed.2007.02.064
26. Hastie, T., Tibshirani, R., Friedman, J.: The Elements of Statistical Learning: Data Mining, Inference, and Prediction, 2nd edn. Springer, New York (2009)
27. Jordan, M.I., Mitchell, T.M.: Machine learning: trends, perspectives, and prospects. Science **349**(6245), 255–260 (2015). https://doi.org/10.1126/science.aaa8415
28. Wynants, L., Collins, G.S., Van Calster, B.: Key steps and common pitfalls in developing and validating risk models. BJOG **124**(3), 423–432 (2017). https://doi.org/10.1111/1471-0528.14170
29. Topol, E.J.: High-performance medicine: the convergence of human and artificial intelligence. Nat. Med. **25**(1), 44–56 (2019). https://doi.org/10.1038/s41591-018-0300-7
30. Chen, J.H., Asch, S.M.: Machine learning and prediction in medicine-beyond the ROC curve. N. Engl. J. Med. **376**(26), 2507–2509 (2017). https://doi.org/10.1056/NEJMp1614614
31. Liu, X., Rivera, S.C., Moher, D., Calvert, M.J., Denniston, A.K.: Reporting guidelines for clinical trial reports for medical interventions using artificial intelligence: the CONSORT-AI extension. BMJ **370**, m3164 (2020). https://doi.org/10.1136/bmj.m3164
32. Beam, A.L., Kohane, I.S.: Big data and machine learning in health care. JAMA **319**(13), 1317–1318 (2018). https://doi.org/10.1001/jama.2017.18391
33. Obermeyer, Z., Emanuel, E.J.: Predicting the future-big data, machine learning, and clinical medicine. N. Engl. J. Med. **375**(13), 1216–1219 (2016). https://doi.org/10.1056/NEJMp1606181
34. Pontes Balanza, B., et al.: Development of a liver graft assessment expert machine-learning system: when the artificial intelligence helps liver transplant surgeons. Front. Surg. **10**(1048451) (2023). https://doi.org/10.3389/fsurg.2023.1048451

Context-Aware AI Agents for Clinical Dialogue Assistance Through Large Language Models

Rodrigo Naranjo Pozas, Pablo Doblado Mendoza,
and Belén Vega-Márquez(✉) ⓘ

Department of Computer Languages and Systems, University of Seville, 41012
Seville, Spain
rodnarpoz@alum.us.es , bvega@us.es

Abstract. This research investigates the efficacy of Artificial Intelligence Agents in processing and responding to personalized, private contextual information. We studied how to implement a system designed to augment an open-source Large Language Model (LLM), such as Llama, Claude, and Gemma, with domain-specific knowledge bases. This augmentation is intended to facilitate the generation of contextually coherent and accurate responses to user queries. The system was developed and tested using different versions of a clinical conversation transcripts dataset between patients and medical professionals, enabling specialized knowledge integration. The architectural framework, built upon LangChain, FAISS, Ollama, and Gradio, demonstrates a simple, modular, scalable, and extensible design. This work helped us to take our first steps into the development of robust AI agents capable of leveraging external knowledge for enhanced conversational intelligence in specialized domains.

Keywords: Large Language Models · RAG · Agents · Medical Conversational Transcriptions

1 State of the Art

The year 2025 has emerged as a pivotal period in the field of artificial intelligence (AI), characterized by the anticipated integration of effective AI agents into enterprise operations. Generally, AI agents are systems designed to execute high-level tasks by leveraging Large Language Models (LLMs) as their primary cognitive engine, thereby enabling structured and coherent problem-solving [2]. However, it is crucial to recognize that not all system components necessitate LLM integration. Strategic identification of critical application points for LLMs is paramount to maintaining efficiency, managing computational costs, and ensuring low latency within these systems [3].

In the medical sector, AI applications are rapidly emerging, presenting immense potential and unforeseen opportunities. Their utility is particularly

A. López Fernández et al. (Eds.): CIABiomed 2025, LNBI 16148, pp. 163–169, 2026.
https://doi.org/10.1007/978-3-032-10661-2_13

pronounced in critical tasks such as clinical assessment, decision support, or the processing of large unstructured datasets. This necessity stems from the surge in data that the healthcare sector is currently experiencing and is projected to continue generating. Therefore, the successful implementation and governmental adoption of these AI applications are crucial for the stability of medical systems and, ultimately, for social health. Nonetheless, it is equally relevant to acknowledge and effectively mitigate the inherent risks and challenges associated with these technologies [2].

1.1   Data Privacy and RAG Systems

In this work, as we embark on the development of AI agents, a primary aspiration is to effectively address the inherent risks associated with their deployment. Among these, data privacy and the safeguarding of proprietary information are predominant concerns. Utilizing closed-source, proprietary models introduces uncertainty regarding data management practices and potential misuse of submitted information. Consequently, we postulate that the deployment of local, open-source LLMs via platforms like Ollama offers the optimal solution for maintaining data security and confidentiality [3]. Then again, a known limitation of these LLMs is their static knowledge base; once trained, augmenting their internal knowledge is not straightforward. Fortunately, the recently developed technique of Retrieval-Augmented Generation (RAG) provides a robust mechanism to overcome this challenge, enabling these models to dynamically access and leverage external, proprietary datasets for generating responses [1].

Despite these advantages, this work acknowledges and addresses several identified limitations. Firstly, it is remarkable to mention that there is an ongoing and significant debate on the potential deprecation of RAG systems due to the rapid expansion of LLM context windows, which now reach sizes of up to one million tokens in some models and are expected to keep growing. The argument asserts that with a sufficiently large context window, it might be possible to entirely input the required contextual data directly into the model, thereby eliminating the need for a separate retrieval component. Furthermore, and related to this challenge, the RAG technique relies on processes such as character-based text splitting to optimize data comprehension and model efficiency. However, effectively determining the optimal parameters for these splitting strategies remains a non-trivial task, which we addressed [1].

1.2   Multi-agent Systems

One of the Agents implementation strategies is the Multi-Agent System (MAS) which represent a natural progression in the development of modern artificial intelligence, particularly in scenarios where solving complex problems requires cooperation among multiple autonomous entities. These systems consist of several agents that interact with each other and their environment to achieve individual or shared goals, enabling task division, knowledge distribution, and the emergence of collaborative behaviors. Each agent is capable of perceiving its surroundings, reasoning based on available information, and acting autonomously. Agents may also be specialized in particular domains, allowing for the construction of modular and adaptive architectures that can handle tasks of varying complexity or interdisciplinary nature [5]. Effective coordination among agents involves mechanisms for communication, negotiation, task allocation, and, in some cases, shared memory. This infrastructure supports the development of robust and scalable systems, particularly suitable for applications that require flexibility, resilience, and distributed processing [4].

In recent years, several tools have been developed to facilitate the design and implementation of such systems. Among them, CrewAI stands out as an open-source framework built to organize agents with distinct roles and manage sequential or cooperative workflows [4]. Such tools have enabled the adoption of multi-agent architectures in fields such as automation, strategic planning, and distributed problem-solving.

2   Methodology

This work proposes a hybrid architecture based on LLMs and artificial intelligence agents, designed to run entirely on local infrastructure and tailored to analyze personalized clinical information. The system comprises two main modules: a RAG-based component for contextual access to a medical dialogue corpus, and a multi-agent reasoning module for structured collaboration among virtual experts.

The dataset consists of over 3,600 real-world doctor-patient conversations in CSV format. These data were preprocessed and transformed into Document objects compatible with the LangChain framework, enabling seamless integration with the system logic.

For the RAG pipeline, the following components were used:

- **LangChain**, as an orchestration layer for the model, data, and reasoning chains.
- **FAISS**, to build an efficient vector database using embeddings generated via HuggingFaceEmbeddings.
- **Ollama**, to run LLMs locally, including general-purpose and clinically fine-tuned models.

The second key component is the multi-agent system, implemented with the **CrewAI** library. It organizes a team of agents with distinct roles, specific tasks, and shared memory. Each agent is powered by a dedicated LLM and configured with a professional goal and simulated background. Agent coordination is handled within a Crew object, which manages asynchronous task execution and inter-agent communication.

Lastly, a **Gradio**-based user interface was developed to enhance user interaction and support demonstration scenarios. This modular and locally-executed architecture ensures data confidentiality, system adaptability, and end-user control without relying on cloud services.

3   Results

The data used in this study is sourced from the MTS-DIALOG dataset, a novel collection comprising 1,700 doctor-patient dialogues and their corresponding clinical notes. To enhance the dataset for experimenting with it, the authors conducted a data augmentation process to the original 1.2k training pairs. This resulted in an augmented dataset of 3.6k pairs of medical conversations and associated summaries, created via back-translation using French and Spanish as intermediate languages[1]. This process is further detailed in Sect. 4.2 of the original paper. The full augmented training set was provided for use in the experiments and subsequently underwent a cleaning process that included content reduction to align with the capacities of our computational environment[2].

3.1   RAG System Results

On the one hand, the RAG system developed in this study was designed to deliver context-aware responses over a large set of unstructured clinical dialogues, comprising over 3,600 doctor-patient transcripts. The architecture was built as a modular pipeline that combines semantic retrieval and language generation, fully executed on local infrastructure without relying on cloud services. The first stage involved the construction of an optimized vector database. Documents were segmented into 256-token chunks with 80-token overlap, a strategy that preserved semantic coherence while enabling finer-grained retrieval. Embeddings were generated using the nomic-embed-text model via Ollama, and indexed using FAISS to support fast similarity search. Initial system evaluations revealed that naive retrieval often surfaced contextually irrelevant documents. To address this, a re-ranking mechanism was implemented using a lightweight cross-encoder (BAAI/bge-reranker-base). The retriever fetched the top 20 documents, which were re-ranked to select the five most relevant based on semantic alignment with the query. The conversational engine was orchestrated using LangChain, integrating the full RAG chain: query translation (Spanish to English), document retrieval, context formatting, and response generation using the llama3 model. A memory layer was also added to maintain conversational coherence across multiple user turns. A user-friendly interface was built using Gradio, enabling real-time interaction and qualitative evaluation in simulated scenarios. This interface played a key role in testing system behavior under various conditions. Examples are provided below (Figs. 1 and 2):

From a functional perspective, the system produced coherent and well-structured responses when the retrieved documents were contextually appropriate. However, several limitations were identified:

- **High response latency**, largely due to the local execution of models and the computational demands of the retrieval-generation pipeline.

---

[1] https://github.com/abachaa/MTS-Dialog/tree/main.
[2] Our code is available at: https://github.com/Doblado3/AIAgents.

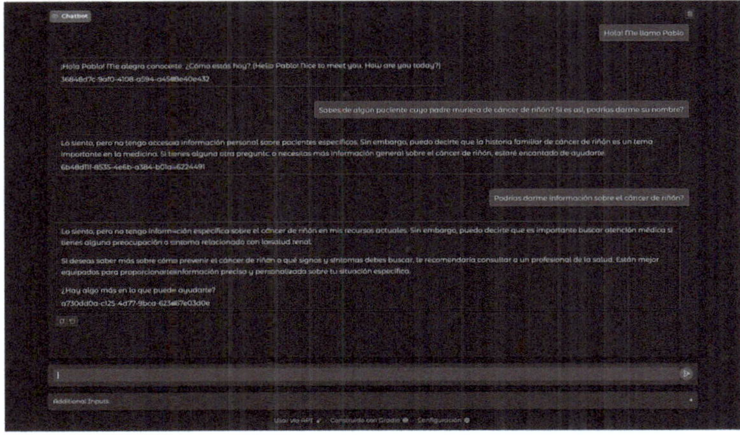

Fig. 1. Initial interactions with the system, highlighting instances where contextual retrieval did not perform as expected.

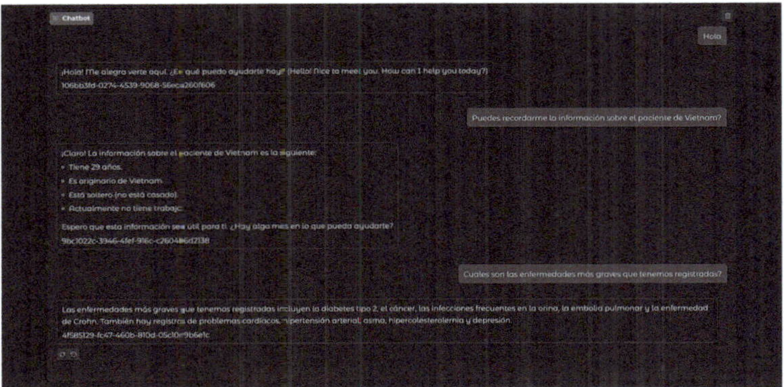

Fig. 2. Further testing interactions, demonstrating improvements after system adjustments.

- **Occasional hallucinations**, especially when relevant information was absent from the vector store.
- **Unintended language switching**, with some greetings being answered in English despite configuration for Spanish responses.
- **Parameter sensitivity**, particularly in relation to chunk size, number of retrieved documents, and embedding model performance.

Despite these constraints, the system served as a strong proof of concept for deploying LLM-based reasoning over real clinical data in a secure, local environment. The insights gained provide a solid foundation for future improvements, including performance optimization, model tuning, and enhanced reliability.

3.2 Multi-agent System Results

On the other hand, the multi-agent system developed in this project was designed as a modular simulation of a collaborative reasoning environment, composed of agents with distinct roles and shared objectives. The system was implemented using the **CrewAI** library, which enables the definition of agents backed by LLMs, assignment of task-specific instructions, and organization into hierarchical or collaborative teams (Crews). Each agent was configured with a simulated professional profile (e.g., clinical oncologist, nutritionist, or imaging specialist) and linked to an LLM deployed locally via Ollama. Agent interactions were structured around explicit task prompts, requiring them to analyze input, reason autonomously, and—when necessary—coordinate with peers before issuing a final response.

Two types of workflow structures were evaluated:

- **Sequential workflows**, in which information is passed linearly from one agent to another.
- **Collaborative workflows**, where all agents respond independently to a common input and their outputs are synthesized by a designated supervisor agent.

The system successfully simulated basic clinical reasoning and demonstrated consistent behavior when tasks were clearly defined and prompts were well constructed. However, several limitations were observed:

- **High prompt dependency**, requiring meticulous prompt engineering to ensure agents adhered to their roles and avoided overlap or contradiction.
- **Cumulative execution times**, especially in sequential workflows due to chained model calls.
- **Response variability**, influenced by model temperature, prompt specificity, and clarity of agent objectives.
- **Lack of robust shared memory**, making inter-agent referencing and knowledge continuity challenging in complex tasks.

Despite these limitations, the system proved effective in modeling distributed reasoning processes, offering a controlled environment for studying multi-agent behavior in simulated settings. It establishes a strong foundation for future development of AI-assisted clinical systems, particularly in scenarios requiring interdisciplinary collaboration and integration of diverse information sources.

4 Conclusions

This study presents a fully local artificial intelligence architecture based on agents, designed for processing clinical information through the combined use of RAG techniques and multi-agent systems powered by LLMs. The proposed solution demonstrates the feasibility of building a contextual reasoning environment without reliance on external services, thereby ensuring data privacy.

First, the RAG system enabled complex semantic queries over a corpus of doctor-patient interactions, using a processing pipeline involving retrieval, re-ranking, and local generation. Despite some limitations—such as sensitivity to chunking parameters, response latency, and the need for fine-tuned prompts—the system produced coherent and domain-adapted outputs. Second, the multi-agent system allowed the simulation of interactions between distinct professional roles, highlighting the potential of these approaches for modeling distributed reasoning. However, the system's dependency on well-crafted task instructions and the lack of robust shared memory mechanisms limited its scalability.

Overall, both modules demonstrated that it is possible to design explainable, locally-executed AI solutions that can adapt to various clinical workflows. This work lays the foundation for future applications in clinical decision support. Key areas for improvement include the integration of internal quality assessment mechanisms, performance optimization, and the exploration of dynamic shared memory between agents.

Acknowledgments. This publication is part of the projects PID2020-117954RB-C22 and PID2023-146037OB-C21, funded by MICIU/AEI/10.13039/501100011033/.

References

1. Lewis, P., et al.: Retrieval-augmented generation for knowledge-intensive NLP tasks. In: Advances in Neural Information Processing Systems (2020)
2. Bommasani, R., Hudson, D.A., Adeli, E., et al.: On the opportunities and risks of foundation models, arXiv preprint arXiv:2108.07258 (2021)
3. OpenAI. GPT-4 Technical Report (2023). https://openai.com/research/gpt-4
4. Wooldridge, M.: An Introduction to MultiAgent Systems, 2nd edn. Wiley, Hoboken (2009)
5. Russell, S., Norvig, P.: Artificial Intelligence: A Modern Approach, 4th edn. Pearson (2021)
6. Google Cloud. Retrieval Augmented Generation (RAG) Explained (n.d.). https://cloud.google.com/use-cases/retrieval-augmented-generation?hl=es
7. Anthropic. Building effective agents (n.d.). https://www.anthropic.com/engineering/building-effective-agents
8. Meta AI. Faiss: Efficient Similarity Search (n.d.). https://ai.meta.com/tools/faiss/
9. Shaikh, V.: Understanding Vector Stores for LLMs: A Comprehensive Guide. Medium (n.d.). https://medium.com/@shaikh-vasim/understanding-vector-stores-for-llms-a-comprehensive-guide-2ed7db7e3f74
10. 1kg. Ollama: What is Ollama?. Medium (n.d.). https://medium.com/@1kg/ollama-what-is-ollama-9f73f3eafa8b
11. Basubrin, O.: Current status and future of artificial intelligence in medicine. Cureus **17**(1), e77561 (2025). https://doi.org/10.7759/cureus.77561
12. Qiu, J., Lam, K., Li G., et al.: LLM-based agentic systems in medicine and healthcare. Nat. Mach. Intell. **6**, 1418–1420 (2024). https://doi.org/10.1038/s42256-024-00944-1
13. Khan, M.M., Shah, N., Shaikh, N., Thabet, A., Alrabayah, T., Belkhair, S.: Towards secure and trusted AI in healthcare: a systematic review of emerging innovations and ethical challenges. Int. J. Med. Inform. **195**, 105780 (2025). https://doi.org/10.1016/j.ijmedinf.2024.105780

Biomedical Signal Processing

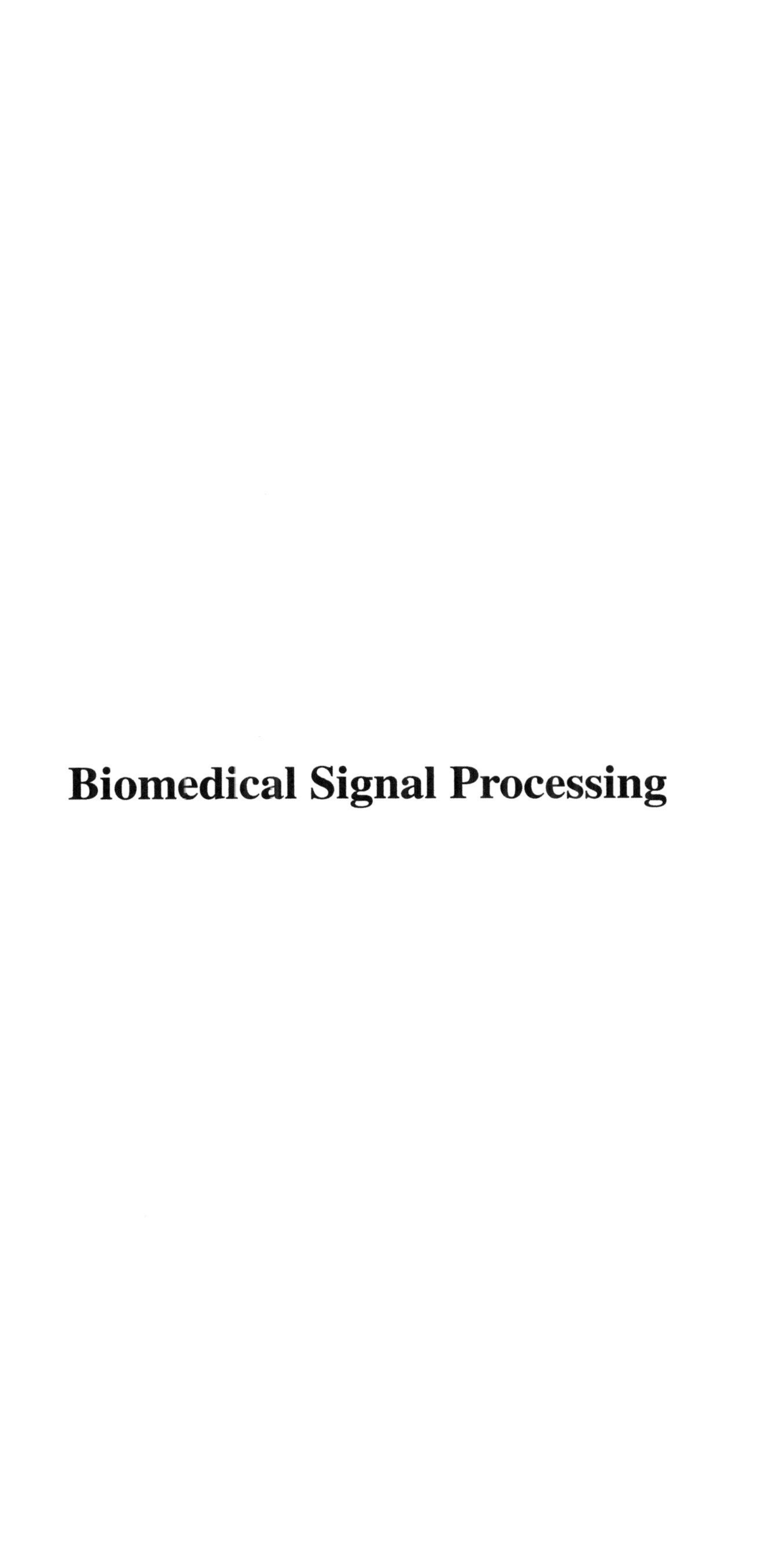

Analysis of Gamma Rhythm in the Detection of Photoparoxysmal Responses in Photosensitive Patients

Fernando Moncada Martins[1,3]([envelope])[iD], Víctor M. González[2,3][iD],
Antonio Gil-Nagel[4,5][iD], Adrián Valls Carbó[4,5][iD], María Antonia Gutiérrez[6],
and Pablo Calvo Calleja[6]

[1] Computer Science Department, University of Oviedo, Oviedo, Spain
moncadafernando@uniovi.es
[2] Electrical Engineering Department, University of Oviedo, Oviedo, Spain
vmsuarez@uniovi.es
[3] Biomedical Engineering Center, University of Oviedo, Oviedo, Spain
[4] Neurology Service, Ruber International Hospital, Madrid, Spain
agnagel@neurologiaclinica.es, avalls@fundacionince.com
[5] Fundación Iniciativa para las Neurociencias (INCE), Madrid, Spain
[6] Neurophysiology Service, Cabueñes University Hospital, Gijón, Spain
pablo.calvo@sespa.es

Abstract. The accurate identification of Photoparoxysmal Responses (PPR) is clinically important, as these epileptiform discharges are the key biomarkers used in diagnosing photosensitivity. Traditionally, PPR are identified in electroencephalography (EEG) during Intermittent Photic Stimulation, but manual detection is time-consuming and subjective, highlighting the need for reliable automated approaches. Recent studies have shown that *Gamma* and High-Frequency Oscillations (HFO) are promising biomarkers in epilepsy research, potentially improving the precision of diagnosis. Building on this, our study incorporates higher-frequency analysis into photosensitivity to investigate how it contributes to identifying PPR. We use an unsupervised anomaly detection method based on a Variational Autoencoder trained on EEG recordings from healthy individuals. Results show that excluding *Gamma* and *High-Gamma* rhythms greatly decreases the model's detection Accuracy from 83% in prior full-spectrum analysis to 63%, supporting the idea that higher frequencies are key in the distinction of PPR, which aligns with broader epilepsy research, where high-frequency activity has proven to be a key marker of pathological brain dynamics. This approach could help integrate high-frequency biomarkers into clinical decision-support systems as well as support automated EEG analysis in other epilepsy syndromes beyond photosensitivity.

Keywords: EEG · Electroencephalography · PPR · Photoparoxysmal Response · Photosensitivity · Epilepsy · HFO · High-Frequency Oscillations · Artificial Intelligence · Anomaly Detection

A. López Fernández et al. (Eds.): CIABiomed 2025, LNBI 16148, pp. 173–184, 2026.
https://doi.org/10.1007/978-3-032-10661-2_14

1   Introduction

Oscillatory brain activity is categorized into frequency bands known as brain rhythms, spanning approximately frequencies from 1 Hz to 600 Hz. Each of these bands has a prominent presence in different concentration and awareness states [2]. Although there is no strict consensus in the scientific literature regarding the exact boundaries of these bands, the main rhythms, ordered from lower to higher frequencies, are as follows: *Delta* (deep sleep), *Theta* (somnolence), *Alpha* (calm awareness), *Beta* (concentration), and *Gamma* (intense activity and stress). Oscillations beyond the *Gamma* range are referred to as High-Frequency Oscillations (HFO). Both *Gamma* and HFO activity –brain signals usually established above 30 Hz and 80 Hz, respectively– has been extensively studied as promising biomarkers for epilepsy [14], as they can effectively display pathological events related to epilepsy [1,6].

Photosensitivity is a neurological condition related to epilepsy, which is characterized by atypical epileptiform responses of the brain to specific visual stimuli, such as flashing lights, called Photoparoxysmal Responses (PPR). [24] classified PPR discharges into four types (Type-1 to Type-4), with severity and seizure risk increasing across types. However, in practice, PPR often manifests as a combination of these types, and EEG morphology varies widely both between patients and within the same patient over different sessions. Factors such as treatment, sleep quality, and time of day can influence this variability [4]. As a result, identifying and labeling PPR requires manual inspection by clinical neurophysiologists, making diagnosis labor-intensive and complex.

Moreover, according to [5], roughly 30% of epileptic patients exhibit photosensitivity, and about 6% of the general population is affected by this condition [23]. This results in the fact that the amount of recorded PPR data in a clinical scenario is inherently limited, leading to highly imbalanced datasets, which complicate the development and training of Machine Learning and Deep Learning models. A promising alternative is to approach the PPR detection as an anomaly detection (AD) problem, treating PPR discharges as brief, irregular bursts of brain activity amidst normal EEG patterns.

This research is a preliminary study focused on analysing the capacity of high-frequency brain activity to characterize PPR discharges, thus serving as a photosensitivity biomarker in its own. For this purpose, a comparison is made with a previous study that evaluated an AD approach using a Variational Autoencoder (VAE) with all brain rhythms, this time excluding all *Gamma* and faster activity from the analysis. The model is trained exclusively on normal EEG segments from non-photosensitive patients, leveraging the fact that photosensitivity has low prevalence and thus EEG data without PPR is more readily available, and then evaluated on photosensitive patients to identify PPR discharges as anomalies. Due to the current lack of comprehensive labels for anomalous activity in these recordings, a high rate of False Positives is expected. Artifact detection and removal techniques were initially considered to reduce the False Positive rate in offline PPR detection. However, this approach was discarded, as including the

EEG anomalies allows for preparing the model for application during real-time EEG recording.

The paper is organized as follows: Sect. 2 reviews related work on AD techniques for epilepsy and photosensitivity; Sect. 3 details the VAE architecture tailored for time-series AD; Sect. 4 describes the dataset and experimental setup; Sect. 5 presents and discusses the results; and Sect. 6 concludes with key findings and future research directions.

2   Related Work

Currently, the internationally standardized clinical protocol for diagnosing photosensitivity is known as Intermittent Photic Stimulation (IPS) [22], which involves exposing patients to flashing white light at varying frequencies while recording brain activity via Electroencephalography (EEG). Diagnosis requires that the stimulation provokes epileptiform reactions, but without developing full seizures, thus needing careful supervision by trained clinical staff who can halt the procedure if necessary Despite its clinical value, the IPS procedure has notable limitations [5]. It demands extensive human oversight to monitor for PPR activity.

Detecting anomalies plays a pivotal role in clinical diagnostics, where deviations from normal patterns often indicate pathological conditions. As a result, anomaly detection (AD) models have become increasingly relevant across various medical imaging and signal analysis tasks. Within the domain of EEG, Recurrent Neural Networks (RNNs) have shown considerable promise in identifying seizure-related irregularities in EEG time series [15], while GANs and Variational Autoencoders (VAEs) have been adapted for use with behind-the-ear EEG signals in both unsupervised and semi-supervised settings for detecting epilepsy activity [17,18]. Alongside these deep learning techniques, traditional machine learning approaches continue to provide valuable baselines and insights in anomaly detection research [8]. Some studies have addressed the automatic detection of anomalous epileptic discharges by analyzing HFO activity [3] to localize and delimit epileptogenic zones in the brain to improve surgical procedures [25].

Despite growing interest in EEG-based anomaly detection, relatively little work has focused specifically on the analysis of photoparoxysmal responses (PPRs) associated with photosensitivity. Early studies in this niche include the work of Parra et al. [7], who applied spectral analysis using Fourier transforms to EEG segments recorded immediately before PPR onset during intermittent photic stimulation (IPS). [19] explored HFO activity of the brain reactions to a non-clinical light stimulation protocol, contributing complementary findings. Our prior research has concentrated on automating the detection of PPR discharges during standard IPS sessions. This includes early efforts with conventional Machine Learning models [10], followed by methods to address class imbalance through Data Augmentation [9], and later, the application of Transfer Learning within an Inception-based model [13].

This research extends our most recent investigations into unsupervised AD for PPR identification [11], where a VAE architecture demonstrated superior performance over previous models. Further analysis [12] revealed that False Positives corresponded to non-PPR EEG unlabeled anomalies unrelated to the target pathology. The present study focuses on evaluating the effect of the *Gamma* and HFO activity in the characterization of PPR anomalies.

3   Variational Autoencoder for Anomaly Detection

The methodology proposed in this research is based on the semi-supervised AD framework introduced by [18], which employs a Variational Autoencoder (VAE) for detecting epileptic seizures. A VAE model consists of two main components: an encoder that compresses each input instance X_i into a latent representation formed by a pair of vectors of mean μ_i and standard deviation σ_i values; and a decoder takes a sample z from this latent distribution to attempt to reconstruct the original instance $\hat{X}_i$. During training, the model is optimized by minimizing both the reconstruction loss by measuring the difference between $\hat{X}_i$ and X_i.

In the context of AD, the VAE is trained exclusively on non-anomalous data. Consequently, the encoder learns to map inputs into a latent space that represents only typical patterns, while the decoder becomes proficient at reconstructing such normal instances. When an unseen sample is introduced, it is considered anomalous I) if it deviates significantly the distribution of known normal data when being reduced by the encoder, and II) if it cannot be well reconstructed by the decoder.

To improve the model's capacity to capture temporal and structural complexity in EEG data, two architectural enhancements are integrated. First, Gated Recurrent Unit (GRU) layers are introduced at the input of both the encoder and decoder to analyze temporal dependencies across consecutive EEG segments, as demonstrated in prior work on time-series AD [20]. Second, the encoder output is passed through a stack of k normalizing flow layers [16], which apply a series of transformations to the latent variables, which allow the model to represent more flexible distributions by adapting the shape of the latent space through operations such as expansion and contraction, thereby improving richness.

4   Materials and Methods

This section depicts the recording details of the EEG sessions included in the clinical dataset used in this research in Sect. 4.1 as well as the design of the experimentation in Sect. 4.2.

4.1   Dataset Description

Two EEG datasets were collected for this study by Cabueñes University Hospital. The first dataset, referred to as *PhotData*, comprises recordings from 9 photosensitive patients who exhibited PPR discharges during the IPS session. The second dataset, *HealthyData*, includes EEG recordings from 5 non-photosensitive individuals who showed no PPR activity in response to stimulation. Each recording session captured continuous brain activity for durations ranging from 1 to 2.5 h, except for two shorter sessions of approximately 30 min. In all cases, the IPS protocol was administered during the final 5 min of the recording. *PhotData* was recorded at a sampling rate of 256 Hz, while 128 Hz were used for *HealthyData*.

Recordings were obtained using a Natus Nicolet EEG system featuring 19 electrodes positioned according to the International 10–20 System [21], as illustrated in Fig. 1. Clinical neurophysiologists reviewed the EEG data using System 98 Viewer software, manually annotating the onset and offset of each PPR event in *PhotData*. A total of 22 PPR discharges were identified, with durations ranging from 0.8 to 3.4 s.

In *HealthyData*, all observable EEG anomalies were also labeled, such as muscular, ocular or electrode-related artifacts, as well as spontaneous epileptiform spikes not induced by visual stimuli. It is important to note that while *HealthyData* has been fully annotated for non-PPR anomalies, the annotation of such events in *PhotData* is a work still in progress.

Preprocessing Operations. To address the difference in sampling frequency between the two datasets, cubic spline interpolation is first applied to *HealthyData* to adjust it to match *PhotData*, increasing the number of samples per second from the original 128 to 256. The EEG recordings are filtered with a 50 Hz Notch filter to remove power line interference, and a 6th order Butterworth band-pass filter (0.5–120 Hz). Then, the four most relevant channels for PPR analysis, according to clinical neurophysiologists, are used to assemble a pair of cross-head channels as the difference between them, based on the proposal by [18], resulting in F3–F4 and O1–O2, as shown in Fig. 1.

Finally, a sliding window is used to split the channels into 1-second length windows with 90% overlap. Each window is manually labeled using the annotations provided by the neurophysiologists. *PhotData* is divided into PPR and Non-PPR windows, while *HealthyData* consists of Normal and Anomaly instances. Positive labels –PPR or Anomaly– are assigned to those EEG windows that contain the corresponding activity at least for half their length. The class distribution for each dataset is as follows:

- *PhotData*
  - Non-PPR windows: 539,886 (**99.94%**)
  - PPR windows: 314 (**0.06%**)
- *HealthyData*
  - Normal windows: 261,635 (**90.00%**)
  - Anomaly windows: 29,086 (**10.00%**)

4.2   Experimentation Setup

Short-Time Fourier Transform is computed over each channel from each EEG window to extract the brain rhythms, i.e., the main frequency bands in which

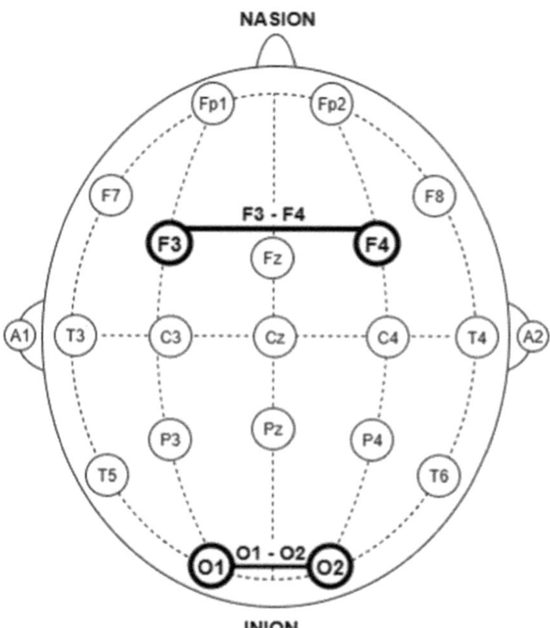

Fig. 1. Placement of the 19 electrodes used for EEG recording following the 10–20 international standardized system, as well as the construction of the two cross-head channels F3-F4 and O1-O2. The Nasion is located at the centre of the frontonasal area; the Inion is located at the centre of the back of the neck.

the brain activity is categorized. [18] defined the bands as follows: *Delta* (0.5–4 Hz), *Theta* (4–8 Hz), *Alpha* (8–16 Hz), *Beta* (16–40 Hz), and *Gamma*, which is divided into the bands 40–80 Hz and 80–120 Hz, the latter called *High-Gamma*. The Power Band Ratio is calculated for each band, leading to 12 spectral features per EEG instance. All values are standardized.

In previous studies, all brain rhythms were used to represent the EEG windows. However, in this research, *Gamma* and *High-Gamma* bands are removed in order to analyze their utility as PPR biomarkers, thus resulting in 8 features per window. The previous experimentation is used as a baseline and is addressed as **EXP0**, while the new scenario corresponds to **EXP1**. The workflow followed is illustrated in Fig. 3.

To prepare the VAE model for AD, it is trained only on Normal EEG windows from *HealthyData*, removing all Anomaly instances from the dataset. Thus, the model should be able to reconstruct only normal brain patterns. Then, the trained model is used for each patient in *PhotData*, to find the PPR instances as the only target anomalies. No other type of EEG anomaly present in those recordings is currently being considered since their correct annotation is a work in progress. Note that previous research [12] showed that these unlabeled EEG anomalies increase the False Positive ratio, which means that they are being detected accurately, but the dataset's real target labels do not include them, affecting the metrics. Furthermore, due to the lack of these Anomaly labels in *PhotData*, normal EEG instances cannot be extracted to perform a calibration step to tune the model to each patient. The VAE's architecture is presented in Fig. 2. After evaluating hyperparameter values, the model was trained for 20 epochs using a batch size of 256 instances. The initial learning rate was 0.001 and was updated to 90% of its value every 5 epochs.

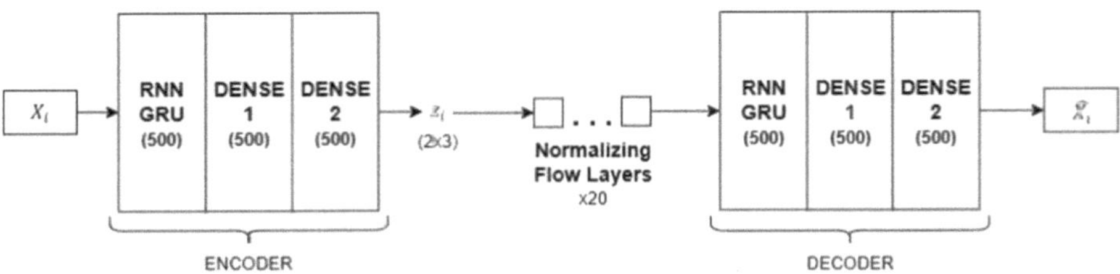

Fig. 2. VAE architecture for time-series AD task. The first layer of the encoder and the decoder is a GRU-based Recurrent layer to consider the temporal dependencies from 100 consecutive instances.

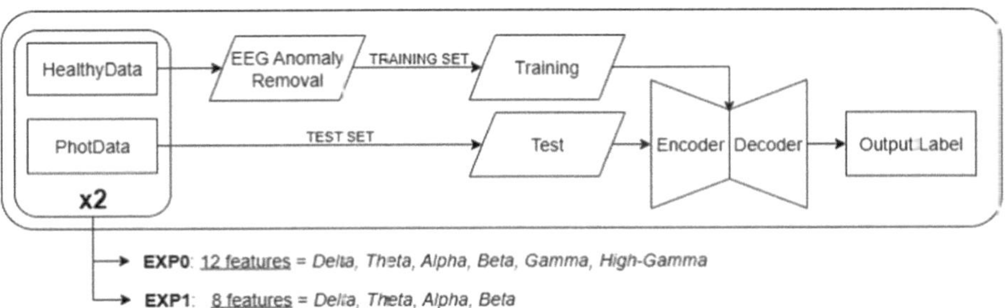

Fig. 3. Experimentation workflow: the VAE model was trained using normal EEG instances from *HealthyData* and then evaluated with each photosensitive patient from *PhotData*. This is repeated twice to assess the effect of fast rhythms.

5 Results and Discussion

The PPR AD performance of the VAE model is evaluated using Accuracy (*Acc*), Sensitivity (*Sens*), and Specificity (*Spec*). Table 1 shows the results obtained by the model for each photosensitive patient in *PhotData* in each of the experiments, while graphical boxplots are illustrated in Fig. 4 to compare the distribution of all performance metrics across all patients for each experimentation setup.

Table 1. PPR detection results obtained in *PhotData* by the VAE model after learn normal brain activity from *HealthyData* in two settings: with and without high frequency brain activity. Includes the mean (Mn), median (Mnd) and standard deviation (StD) of each metric.

P_i	EXP0			EXP1		
	Acc	Sens	Spec	Acc	Sens	Spec
P_1	0.9723	0.6868	0.9767	0.3709	1.0000	0.3708
P_2	0.8909	0.8389	0.8912	0.8012	0.5789	0.8015
P_3	0.8352	0.7754	0.8367	0.3096	1.0000	0.3094
P_4	0.9438	0.6563	0.9487	0.7862	0.8415	0.7861
P_5	0.9624	0.8006	0.9837	0.5294	1.0000	0.5291
P_6	0.2026	1.0000	0.1975	0.6529	1.0000	0.6507
P_7	0.9708	0.9492	0.9813	0.9966	0.3788	0.9985
P_8	0.7346	0.9499	0.7296	0.4803	1.0000	0.4797
P_9	0.9412	0.8210	0.9429	0.7636	1.0000	0.7635
Mn	0.8282	0.8309	0.8320	0.6323	0.8666	0.6671
Mdn	0.9412	0.8210	0.9429	0.6529	1.0000	0.6854
StD	0.2332	0.1116	0.2378	0.2131	0.2181	0.1927

Using **EXP0** as the baseline –in which the model originally achieved values of around 83% for all three metrics– it can be seen that Sens is maintained at 86% when removing *Gamma* and *High-Gamma* rhythms in **EXP1**, slightly higher than **EXP0**, while Acc and Spec greatly decrease to 63% and 66%, respectively.

Regarding False Positives, Acc and Spec metrics present more instances than initially expected. Many of them correspond to unlabeled EEG anomalies –as confirmed in previous research–, but given that Acc and Spec are very robust metrics against misclassification due to the large amount of instances in the non-PPR class, these values show that the model tends to heavily label EEG windows as PPR anomalies when the higher frequencies are removed. Conversely, it is important to clarify that False Negatives do not indicate missed PPR discharges but rather EEG windows that contain some degree of PPR activity. These misclassified cases typically occur at the boundaries, where PPR activity does not span the entire window. Due to the limited number of PPR-labeled instances in the dataset, Sens is highly affected by detection variability as opposed to Acc and Spec, which are more robust. Although the model does not capture every instance of PPR, it successfully identified all PPR discharges across all patients.

The performance metrics obtained by the same architecture confirm that *High-Gamma* activity and fast frequencies are important in the process of distinguishing PPR activity from normal brain activity, just as it serves as a prominent biomarker for other epileptic phenomena. However, statistical hypothesis testing is applied to assess whether the observed performance differences for each metric are statistically significant. After checking normality with the Shapiro-Wilk's W test and variance homogeneity with the Levene's test, the Wilcoxon Signed-Rank test is selected as the most suitable method. With a 95% confidence, the results confirm that the model achieved significantly better performance in scenario **Exp0** in terms of ACC and SPEC, although there is no meaningful difference regarding SENS.

However, the results also show that there are some specific patients for whom the model did not perform well enough. The lack of fully labelled recordings that include marks for all EEG anomalies prevents applying a calibration step to tune the model for each new patient using their normal instances only, creating patient-specific versions that recognize their normal patterns and are more able to identify anomalies within, as concluded in previous research [12].

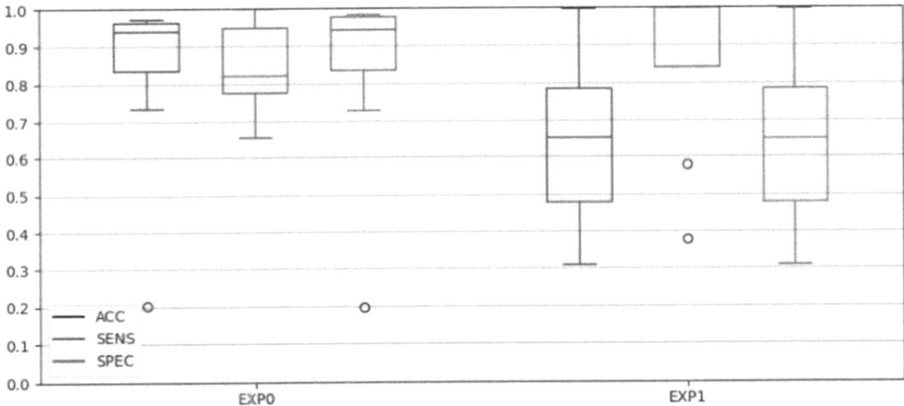

Fig. 4. Boxplots of the distribution of the performance values achieved by the VAE model across all subjects when including (**EXP0**, left) and excluding (**EXP1**, right) *Gamma* and *High-Gamma* brain rhythms.

6   Conclusions and Future Work

This research studies the effect of brain activity from the *Gamma* and *High-Gamma* frequency bands in the detection of PPR discharges. A VAE model – originally designed by [20] and later applied by [18] for AD in epileptic seizures– is used to identify these epileptiform phenomena in EEG time-series. Two datasets are collected for this purpose: *PhotData*, containing EEG recordings from photosensitive patients who exhibited PPR activity during diagnostic procedures, and *HealthyData*, comprising recordings from non-photosensitive individuals along with annotations of all EEG anomalies present in each session.

To perform the AD task, the VAE is trained exclusively on normal EEG windows from *HealthyData* and subsequently evaluated on each patient in *Phot-Data*. Unlike previous research, calibration for each target patient is not feasible because anomalous instances cannot be excluded due to a lack of marks in the EEG recordings. To analyze the *Gamma* and *High-Gamma* rhythms, this experiment compared the performance achieved by the model in **EXP1** when removing *Gamma* and *High-Gamma* frequency bands with previous research (**EXP0**) that included such activity. Despite the lack of patient-specific tuning, the results of this preliminary experimentation are promising as they demonstrate that the model effectively detects PPR discharges, and the *High-Gamma* frequency band is notoriously useful to differentiate PPR from other normal brain activity since its removal produces a great decrease in performance.

To this end, future work should prioritize, on the one hand, the exhaustive annotation of EEG anomalies in *PhotData*, following the methodology employed for *HealthyData*, to ensure a patient-specific solution. The clinical team is currently undertaking this task to enhance the quality and utility of the dataset. In the interim, an alternative strategy involves the automatic detection and exclusion of anomalies during training. This could be achieved by incorporating multiclass classifiers into the VAE's latent space, enabling the identification and removal of anomalous latent vectors from new training instances. Such an approach would not only allow for patient-specific calibration but also facilitate the simultaneous classification of multiple EEG abnormalities, regardless of the availability of manual annotations. On the other hand, further research into *Gamma* and *High-Gamma* will be performed to obtain conclusive results. Moreover, the study of even higher frequencies is a promising path that could confirm that HFO could be used as a biomarker for PPR in the same way that it is currently being used for other types of epileptic activity.

Acknowledgements. This research has been funded by the Spanish Research Agency –grant PID2023-146257OB-I00–. Also, by Principado de Asturias, grant IDE/2024/000734, and by the Council of Gijón through the University Institute of Industrial Technology of Asturias grants SV-23-GIJON-1-09, SV-23-GIJÓN-1-17, and SV-24-GIJÓN-1-05.

Author contributions. Fernando Moncada Martins, Víctor M. González, and José R. Villar designed the methodology and experiments and executed all technical and computer work. They all wrote the paper. Clinical specialists Antonio Gil-Nagel and Adrián Valls Carbó proposed and designed the clinical approach, and Antonia Gutiérrez and Pablo Calvo recorded, collected and anonymised the data at the Cabueñes University Hospital's Neurophysiology Service.

Conflict of Interest. The authors have no competing interests to declare that are relevant to the content of this article.

Ethical Approval. The study was carried out according to the Declaration of Helsinki and approved by the Ethics Committee of the University Hospital of Burgos (Protocol Code CEIm2467, 23 February 2021). Informed consent was obtained from all subjects involved in the study. Written informed consent was obtained from the patients to publish this paper.

References

1. Bragin, A., et al.: Interictal high-frequency oscillations (80–500hz) in the human epileptic brain: entorhinal cortex. Ann. Neurol. **52**(4), 407–415 (2002). https://doi.org/10.1002/ana.10291

2. BuzsÃiki, G., Draguhn, A.: Neuronal oscillations in cortical networks. Science **304**(5679), 1926–1929 (2004). https://doi.org/10.1126/science.1099745

3. Chaitanya, G., et al.: Scalp high frequency oscillations (HFOs) in absence epilepsy: An independent component analysis (ICA) based approach. Epilepsy Res. **115**, 133–140 (2015). https://doi.org/10.1016/j.eplepsyres.2015.06.008

4. Dong, H., et al.: Mixed neural network approach for temporal sleep stage classification. IEEE Trans. Neural Syst. Rehabil. Eng. **26**(2), 324–333 (2018). https://doi.org/10.1109/TNSRE.2017.2733220

5. Fisher, R.S., et al.: Visually sensitive seizures: an updated review by the epilepsy foundation. Epilepsia (2022). https://doi.org/10.1111/epi.17175

6. Jacobs, J., et al.: High-frequency oscillations (HFOs) in clinical epilepsy. Progress Neurobiol. **98**(3), 302–315 (2012). https://doi.org/10.1016/j.pneurobio 2012.03. 001. High Frequency Oscillations in Cognition and Epilepsy

7. Kalitzin, S., et al.: Enhancement of phase clustering in the EEG/MEG gamma frequency band anticipates transitions to paroxysmal epileptiform activity in epileptic patients with know visual sensitivity. IEEE Trans. Bio-medical Eng. **49**, 1279–1286 (2002). https://doi.org/10.1109/TBME.2002.804593

8. Karpov, O.E., et al.: Evaluation of unsupervised anomaly detection techniques in labelling epileptic seizures on human EEG. Appl. Sci. **13**(9) (2023). https://doi. org/10.3390/app13095655

9. Martins, F.M., et al.: Data augmentation effects on highly imbalanced EEG datasets for automatic detection of photoparoxysmal responses. Sensors **23**(4) (2023). https://doi.org/10.3390/s23042312

10. Moncada, F., et al. Virtual reality and machine learning in the automatic photoparoxysmal response detection. Neural Comput. Appl. (2022). https://doi.org/ 10.1007/s00521-022-06940-z

11. Moncada, F., et al.: Anomaly detection comparison for photo-paroxysmal response detection. Logic J. IGPL –Aceptado pero no publicado– (2024)

12. Moncada, F., et al. Evaluation of an AI-EEG based photo-paroxysmal response solution on healthy subjects: when false positive really matters. In: 2nd Olympiad in Engineering Science (OES) (2025). https://indico.uis.no/event/50/ contributions/1204/

13. Moncada Martins, F., et al.: Inception networks, data augmentation and transfer learning in EEG-based photosensitivity diagnosis. Mach. Learn. Sci. Technol. **6**(1), 015034 (2025). https://doi.org/10.1088/2632-2153/adb008

14. Noorlag, L., et al.: High-frequency oscillations in scalp EEG: a systematic review of methodological choices and clinical findings. Clin. Neurophysiol. **137**, 46–58 (2022). https://doi.org/10.1016/j.clinph.2021.12.017

15. Pilcevic, D., et al.: Performance evaluation of metaheuristics-tuned recurrent neural networks for electroencephalography anomaly detection. Front. Physiol. **14** (2023). https://doi.org/10.3389/fphys.2023.1267011

16. Rezende, D., Mohamed, S.: Variational inference with normalizing flows. In: Bach, F., Blei, D. (eds.) Proceedings of the 32nd International Conference on Machine Learning. Proceedings of Machine Learning Research, vol. 37, pp. 1530–1538. PMLR, Lille, France (2015). https://proceedings.mlr.press/v37/rezende15.html

17. You, S., et al.: Unsupervised automatic seizure detection for focal-onset seizures recorded with behind-the-ear EEG using an anomaly-detecting generative adversarial network. Comput. Methods Programs Biomed. (2020). https://doi.org/10.1016/j.cmpb.2020.105472
18. You, S., et al.: Semi-supervised automatic seizure detection using personalized anomaly detecting variational autoencoder with behind-the-ear EEG. Comput Methods Programs Biomed. (2022). https://doi.org/10.1016/j.cmpb.2021.106542
19. Strigaro, G., et al.: Flash-evoked high-frequency EEG oscillations in photosensitive epilepsies. Epilepsy Res. **172**, 106597 (2021). https://doi.org/10.1016/j.eplepsyres.2021.106597
20. Su, Y., et al.: Robust anomaly detection for multivariate time series through stochastic recurrent neural network. In: Proceedings of the 25th ACM SIGKDD International Conference on Knowledge Discovery & Data Mining, KDD 2019, pp. 2828–2837. Association for Computing Machinery, New York (2019). https://doi.org/10.1145/3292500.3330672
21. International Federation of Clinical Neurophysiology: Report of the committee on methods of clinical examination in electroencephalography. Electroencephalography Clin. Neurophysiol. **10**(2) (1958). https://doi.org/10.1016/0013-4694(58)90053-1
22. Trenité, D.G.N., et al.: Photic stimulation: Standardization of screening methods. Epilepsia **40**(9) (1999). https://doi.org/10.1111/j.1528-1157.1999.tb00911.x
23. Trenité, D.K.N.: Photosensitivity in epilepsy. Electrophisiological and clinical correlates. Acta Neurologica Scandinavica **125**(Suppl.), 3–149 (1989)
24. Waltz, S., Christen, H.J., Doose, H.: The different patterns of the photoparoxysmal response - a genetic study. Electroencephalography Clin. Neurophysiol. **83**(2) (1992). https://doi.org/10.1016/0013-4694(92)90027-F
25. Ye, H., et al.: Pathological and physiological high-frequency oscillations on electroencephalography in patients with epilepsy. Neurosci. Bull. **40**(5), 609–620 (2024). https://doi.org/10.1007/s12264-023-01150-6

Application of Machine Learning and Deep Learning Methods on ECG Sensor Data to Predict Stress Levels in Minimally Invasive Surgery

Daniel Caballero[1] , Manuel J. Pérez-Salazar[1] ,
Juan A. Sánchez-Margallo[1(✉)] , Ismael Diaz-Romero[2],
and Francisco M. Sánchez-Margallo[3]

[1] Bioengineering and Health Technologies Unit, Jesús Us ón Minimally Invasive
Surgery Center, N-521 Road Km. 41.8, 10071 Cáceres, Spain
`{dcaballero,mjperez,jasanchez}@ccmijesususon.com`
[2] School of Technology, University of Extremadura, Av/de la Universidad s/n, 10003
Cáceres, Spain
`isdiazr@alumnos.unex.es`
[3] Scientific Direction, Jesús Us ón Minimally Invasive Surgery Center, N-521 Road
Km. 41.8, 10071 Cáceres, Spain
`msanchez@ccmijesususon.com`

Abstract. This study aims to predict the values of ergonomic (body position, angular rate), kinematic (steps count), and physiological (ECG signal, beats per minute, blood pressure, and body temperature) parameters by using a wearable electrocardiogram (ECG) sensor worn by surgeons during the performance of minimally invasive surgical (MIS) procedures (conventional laparoscopy and robotic-assisted surgery -RAS-). For this purpose, data related to the surgeon's ECG sensor parameters were collected during forty-four MIS sessions conducted by eighteen surgeons with different levels of experience. Once the dataset was generated, two preprocessing techniques were applied: scaling and normalization. These two datasets were subsequently divided into two subsets: one containing 80% of the data for training and cross-validation, and the other containing 20% of data for testing. Several Artificial Intelligence (AI) techniques were applied to the training dataset to develop the predictive models. Finally, these models were validated on cross-validation and test datasets. PCA results showed that the physiological parameters (ECG signal, beats per minute, blood pressure and body temperature) were the most representative parameters of the ECG sensor. The predictive analysis results showed that XGBoost achieved the best results for the training dataset, while Multiple Linear Regression demonstrated the best performance for the cross-validation and test datasets combined with the scaled preprocessing technique, achieving the highest R^2 coefficient and the lowest error for each parameter analyzed. The linear models were successfully validated on both cross-validation and test datasets, underscoring the potential for prediction of factors contributing to surgeon's health improvement during MIS procedures.

A. López Fernández et al. (Eds.): CIABiomed 2025, LNBI 16148, pp. 185–199, 2026.
https://doi.org/10.1007/978-3-032-10661-2_15

Keywords: Artificial Intelligence · Wearable devices · Conventional laparoscopy · Robot-assisted surgery · Stress

1   Introduction

Minimally invasive surgery (MIS) has exponentially grown in recent decades, and several MIS procedures, including conventional laparoscopic and robotic-assisted surgery (RAS), have become standard surgical techniques for different specialties, such as urology or gynecology [1]. The advantages of MIS for patients have been extensively explored and presented in scientific literature [2]. However, MIS procedures have some limitations for surgeons that need to be addressed. Among these limitations can be found: i) the potential ergonomics deficiencies during long surgeries; ii) high stress levels associated with certain surgical procedures; and iii) it is a highly demanding field with a steep learning curve. These limitations could poses significant risks to the surgeon's health and have a major impact on the quality of the surgical activities and patient health care [3]. Consequently, it is essential that MIS professionals receive comprehensive and extensive training [4].

The identification of elevated stress levels may potentially reduce the risk of various health conditions [5]. Since everyone may experience stress at some time in their lifetime, it is a priority to achieve the application of innovative technologies that allow controlling and monitoring physiological and mental health during the performance of daily life activities [6]. This becomes especially critical in clinical environments, such as MIS, since it has implications for the surgical quality and the quality of patient health care [7].

To determine stress levels in the performance of MIS, the monitoring of ergonomic and physiological parameters is critical [8]. Some wearable devices, including electrocardiogram activity sensors (ECG) were analyzed for recording and the evaluation of stress level [9]. The reduction in size and the improvement in their integration of these wearable devices make them advantageous for analyzing the surgeon's health status during surgical procedures. These systems allow us to quantify, among others, parameters related to ergonomics, blood pressure or body temperature [10].

The application of artificial intelligence (AI) has widely grown in its use, implementation, and development. These algorithms are based on non-trivial processes for discovering potentially useful knowledge initially hidden in the datasets [11]. Among the different AI techniques, there are several algorithms with predictive purposes. Specially, convolutional neural networks (CNN), machine learning and deep learning based on learning from the results obtained previously. Moreover, specialized systems exist to identify the primary factors associated with elevated stress levels during MIS procedures [12–14]. Other applications based on determining the stress with EDA sensors, determining localized muscle fatigue or ergonomic workload have been explored in previous studies [15–17].

The combination of wearable technology joined to AI provide a substantial opportunity for the advancement of novel methodologies and techniques for monitoring, managing, controlling, and understanding the correlation between the prediction of risk scenarios for surgeons' well-being and their stress levels during surgical interventions. Similarly, these innovative technologies are providing robust solutions for the evaluation, analysis and comparison of objective stress levels derived from physiological and ergonomic parameters during MIS activities in real time [18]. This study could enable predicting and preventing possible health risks of problems for the surgeon or planning potential breaks or rests in order to improve the quality of the surgeons procedures and consequently, the quality of the patient healthcare.

1.1 Objective

This study aims to predict the values of ergonomic (angular rate and body position), kinematic (step counts) and physiological (beats per minute, N-N intervals, ECG, blood pressure and body temperature) parameters of surgeons during the practice of MIS. Furthermore, a subset of these parameters will be selected, identifying the most representative ECG sensor parameters. This innovative and novel prediction will be carried out from the values collected in the immediately preceding surgical situation, predicting and preventing possible health risks for the surgeon or planning potential breaks.

2 Material and Methods

2.1 Experimental Setup

The data for this study were collected from forty-four MIS sessions performed by eighteen surgeons with different levels of experience, ten experts and eight novice surgeons in the corresponding surgical specialty. MIS sessions were surgical procedures conducted in experimental animal models for various surgical disciplines, such as general surgery, urology and gynecology. These procedures were approved by the local ethics committee responsible for the respective discipline (Ref. EXP-20230325, EXP-20230320, and EXP-20230313).

From these data, a dataset (24,670 records) was generated, on which different preprocessing techniques (scaled, and scaled and normalized) were applied, and then, two subsets were generated: 80% of data for the training dataset (19,736 records), and the remaining 20% of data for the test dataset (4,934 records). Records of all forty-four MIS sessions appeared in both datasets. For the attribute selection, a principal component analysis (PCA) was carried out to select the most representative parameters. Nine different AI techniques were applied to the training dataset to generate the predictive models: Catboost, Convolutional Neural Network (CNN), Multilayer Perceptron (MLP), Random Forest (RF), Gradient Reinforcement Neural Network (GBM), Multiple Linear Regression (MLR), Support Vector Machine (SVM), U Neural Network (UNET), and XGBoost. Finally, to validate the generated models, they were validated on the 10-fold cross-validation and test dataset. Figure 1 shows the experimental setup of the current study.

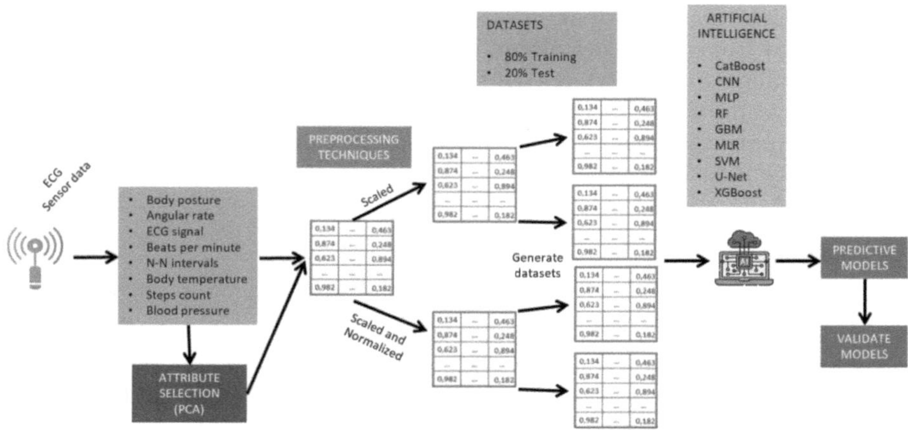

Fig. 1. Experimental setup of the study.

2.2   Wearable Devices

Ergonomic, kinematic and physiological data were recorded with a single wearable device (Ecg Move 4 activity sensor; Movisens^{TM}, Karlsruhe, Germany). The data extracted from the device were processed by using the Unisens Viewer software (Movisens^{TM}, Karlsruhe, Germany). The different sensors included in the device allows measurements to be taken beyond the usual electrocardiogram, such as body temperature, heart rate or surgeon's positions by means of additional accelerometers and gyroscopes embedded in the device. The device was attached to the left side of the participants' torso under the pectoral muscle by means of two pectoral electrodes.

2.3   Conventional Laparoscopy Platform

For conventional laparoscopy, Olympus VISERA ELITE (Olympus^{TM}, Tokyo, Japan) imaging system and Karl Stotz^{TM} laparoscopic instruments (Karl Storz SE and Co. ^{TM}, Tuttlingen, Germany) were used. Twenty sessions were performed with conventional laparoscopy.

2.4   Robotic Platforms

Versius^{TM} surgical system (CMR Surgical^{TM}, Cambridge, UK) was used for the RAS procedures. This robotic platform includes a remote open console for the surgeon to control the robotic instruments and provide 3D endoscopic view. On the bedside, it has up to four modular arms to move the robotic instruments inside the patients (one of them for the camera and the others for the surgical instruments). Twenty-four sessions were performed with robotic assistance.

2.5 Datasets

The original dataset was transformed by applying scale preprocessing and normalization techniques, resulting in two datasets. The two preprocessed datasets were divided into four datasets: 80% of the data from each dataset for training and 20% of the data from each dataset for validation (cross-validation and testing) [19].

The scaled preprocessing technique allows each parameter to be described on a scale between 0 and 1 [19]. To do this, each value is subtracted from the minimum value and then, divided by the interval between the maximum and minimum values (Eq. 1).

$$Value_{new} = \frac{Value_{current} - min}{max - min} \tag{1}$$

When $value_{new}$ indicates the preprocessed value, $value_{current}$ represents the raw value from each dataset, min shows the minimum value for each parameter, and max indicates the maximum value for each parameter.

For the normalized preprocessing technique, the mean of each value is subtracted and divided by the standard deviation [19] (Eq. 2). This technique transforms the dataset into a more integrated and robust one with fewer redundancies.

$$Value_{new} = \frac{Value_{current} - Value_{average}}{Value_{st.deviation}} \tag{2}$$

where $Value_{new}$ indicates the preproccessed value, $Value_{current}$ represents the raw value from each dataset, $Value_{average}$ shows the average value for each parameter, and $Value_{st.deviation}$ indicates the standard deviation value for each parameter.

The R^2 coefficient was used to assess the goodness of fit of the prediction and for validation, according to the rules given by Colton [20], where R^2 of 0 to 0.25 is considered as a poor to no relationship; 0.25 to 0.50 indicates a weak degree of relationship; 0.50 to 0.75 designates a moderate to good relationship; and 0.75 to 1 shows a very good to excellent relationship. The root mean square error (RMSE) was also used to validate the prediction results [21]. RMSE measures the difference between actual and predicted values. RMSE values of less than 0.05 are considered adequate [21].

2.6 Artificial Intelligence

The PyTorch package (v2.7; Meta AI, Menlo Park, California, U.S.A.) was used to develop predictive models. It is a machine and deep learning library of the Python programming language (v3.11.10). Nine different AI techniques were applied to the training dataset to generate predictive models: Catboost, CNN, MLP, RF, GBM, MLR, SVM, UNET and XGBoost. These AI techniques were selected since they are widely studied in scientific literature [22]. Finally, to validate the generated models, they were validated in 10-fold cross-validation, in which the calibration dataset was divided into ten equally sized partitions.

Each time a subset was tested, the remaining data were used to fit the model. The process was repeated sequentially until all subsets were tested. Therefore, all data were used for both purposes, calibration and validation. Although this method requires a ten-repeat analysis, it is a robust technique [22]. Finally, the test dataset was used for external validation of the predictive models. This is the collection of the AI techniques applied in the current study:

Catboost. A deep learning algorithm for gradient boosting decision trees. It selects the optimal decision tree in each iteration [23]. The execution of the algorithm was carried out with the following values for each hyperparameter: iterations was 500, learning rate was 0.01, depth was 10, strength was 1 and rsm was 1.0. The remaining parameters were configured with the values by default.

Convolutional Neural Network (CNN). A machine learning algorithm used for grid-like data, particularly effective in computer vision tasks [24]. The tunables parameters were configured as indicate as follow: number of epochs was 500, learning rate was 0.15, momentum was 0.3, the number of layers was 5 for encoder and decoder and the remaining parameters were configured by default.

Multilayer Perceptron (MLP). A type of ANN model. It is applied to dataset by using principles of neuronal organization. The signal travels from the input layer to the output layer, possibly after traversing multiple layers [25]. MLP was configured as follow: number of epochs was 100, learning rate 0.1, momentum was 0.35, the number of hidden layers was 3, with a number of nodes lower than 20. The remaining parameters were configured by default.

Random Forest (RF). A machine learning algorithm uses a set of decision trees to generate predictive models [26]. The number of estimators was 30, the maximum depth was 10, the bootstrap was 0.3 and the random state was 0.2. The remaining parameters were configured by default.

Gradient Reinforcement Neural Network (GBM). A boosting ensemble of generative neural networks [27]. It is based on the gradient boosting machine, noting the depth level in its applications. The execution of the algorithm was carried out with the following values for each hyperparameter: iterations was 100, learning rate was 0.001, depth was 10, strength was 1 and rsm was 1.0. The remaining parameters were configured with the values by default.

Multiple Linear Regression (MLR). MLR was applied as a predictive linear approach to the datasets. It showed a linear relationship between a dependent variable and several independent variables, resulting in a linear regression equation for predictive future values [28]. The M5 method of attribute selection and a ridge value of 1×10^{-4} were applied in this algorithm.

Support Vector Machine (SVM). SVM was performed as a predictive non-linear approach on the datasets. It represents desirable non-linear relationships that minimize separation, tolerating a small error when fitting the data in the transformed space [29]. The kernel applied in this study was a radial basis function with a gamma of 0.01 and the remaining parameters configured by default.

U Neural Network (UNET). A deep learning method, commonly used for image segmentation, especially in biomedical images. It employs two paths: an encoder to contract the model and a decoder to expand it. This method is highly feasible for obtaining spatial information and accurate image segmentation [30]. The tunables parameters were configured as indicate as follow: number of epochs was 50, learning rate was 0.1, momentum was 0.15, the number of layers was 5 for encoder and decoder and the remaining parameters were configured by default.

XGBoost. It is similar to CatBoost, uses gradient boosting. However, it differs from CatBoost in its approach to solving the problem by using categorical features and permuting the decision trees alternately [31]. The execution of the algorithm was carried out with the following values for each hyperparameter: iterations was 300, learning rate was 0.001, depth was 10, strength was 1 and rsm was 1.0. The remaining parameters were configured with the values by default.

Principal Component Analysis (PCA). PCA was applied to the dataset to identify the most representative attributes of the ECG sensor data [32]. The most influential parameters were selected from each matrix of components.

The R^2 correlation coefficient was used to evaluate the prediction accuracy according to Colton' rules [20]

3   Results and Discussion

3.1   Attribute Selection Procedure

Analyzing the scores values of the PCA from all parameters of the ECG sensor (Table 1), the most representative attributes for the first principal component were ECG signal, beats per minute, body temperature, and blood pressure with scores values higher than 0.4. The remaining parameters had scores values lower than 0.2. For the second principal component, the most representative attributes were blood pressure (0.722), body temperature (0.719), ECG signal (0.471) and beats per minute (0.446). For the third principal component, the most representative attributes with scores values higher than 0.4 were blood pressure, body temperature, beats per minute and ECG signal. In this way, for the three principal components with most explained variance, only four parameters were repeated in all cases: ECG signal, beats per minute, blood pressure, and body temperature. These results are consistent with previous studies that have shown these to be the most representative attributes for predicting stress based on

ECG sensor data [33]. To evaluate the prediction by applying the most representative parameters from the ECG sensor, we estimated the loss between using all parameters or only the most representative parameters.

Table 1. Scores values of the third most representative principal components from PCA for each parameter of the ECG sensor to identify the most representative parameters

ECG parameter	Principal Component 1	Principal Component 2	Principal Component 3
Body posture X	0.082	0.115	0.217
Body posture Y	0.164	0.344	0.345
Body posture Z	0.159	0.368	0.265
Angular rate X	0.199	0.103	0.383
Angular rate Y	0.178	0.229	0.207
Angular rate Z	0.113	0.214	0.245
ECG signal	**0.508**	**0.471**	**0.422**
Beats per minute	**0.479**	**0.446**	**0.424**
N-N intervals	0.123	0.236	0.267
Body temperature	**0.623**	**0.719**	**0.435**
Steps count	0.134	0.257	0.289
Blood pressure	**0.712**	**0.722**	**0.618**

3.2   Training Dataset Results

The prediction results for the training dataset are presented in Table 2. This table shows the values of the R^2 coefficient of the training dataset for the predictive analysis carried out by applying different predictive AI techniques. The dataset was preprocessed using the scaled technique, which achieved higher values of the R^2 coefficient for conventional and RAS. XGBoost and scaled showed the highest values of the R^2 coefficient for most of the parameters analyzed and lowest values of RMSE (lower than 0.05). This fact could be related to the linear nature of the data, as its performance exhibits a clear linear trend. This linearity is also reflected in the principal component results [34]. In general, the results obtained in the current study are satisfactory for the linear AI techniques (CatBoost, GBM, MLR and XGBoost) according to the rules given by Colton [20]. However, lower results were achieved for the non-linear AI techniques (CNN, MLP, and U-NET). In particular, XGBoost AI technique exhibited high to excellent R^2 values, approaching 1, and RMSE close to 0 for all parameters. The R^2 coefficient and RMSE values for body temperature ($R^2 = 0.989$ and RMSE $= 0.008$) and blood pressure ($R^2 = 0.989$ and RMSE $= 0.004$) were particularly noteworthy.

As observed in Fig. 2, the loss incurred when predicting based on all parameters or only the four most representative characteristics is around 0.03, indicating a negligible loss. The XGBoost AI technique achieved high to excellent R^2 values, approaching 1, and RMSE close to 0 for all parameters.

Table 2. Results of R^2 for the training dataset of parameters recorded during MIS sessions as a function of the predictive AI techniques and scaled as preprocessing techniques. The best results are highlighted in bold.

AI technique	Body posture X	Body posture Y	Body posture Z	Angular rate X	Angular rate Y	Angular rate Z	ECG signal	Beats per minute	N-N intervals	Body temperature	Steps count	Blood pressure
Cat Boost	0.933	0.946	0.913	0.821	0.915	0.872	0.967	0.962	0.957	0.989	0.936	0.988
CNN	0.229	0.391	0.211	0.222	0.210	0.175	0.216	0.211	0.206	0.191	0.188	0.009
MLP	0.101	0.249	0.259	0.221	0.095	0.168	0.173	0.163	0.168	0.351	0.179	0.375
RF	0.945	0.918	0.926	0.825	0.895	0.892	0.890	0.900	0.895	0.989	0.969	0.948
GBM	0.914	0.889	0.866	0.618	0.802	0.771	0.836	0.841	0.831	0.988	0.946	0.980
MLR	0.908	0.849	0.928	0.849	0.913	0.887	**0.986**	**0.976**	**0.981**	**0.999**	**0.987**	**0.999**
SVM	0.769	0.344	0.644	0.309	0.216	0.598	0.557	0.552	0.547	0.472	0.461	0.966
U-Net	0.122	0.171	0.175	0.135	0.139	0.157	0.222	0.217	0.212	0.662	0.138	0.556
XGBoost	**0.984**	**0.982**	**0.968**	**0.924**	**0.966**	**0.939**	0.971	**0.976**	**0.981**	0.989	0.968	0.989

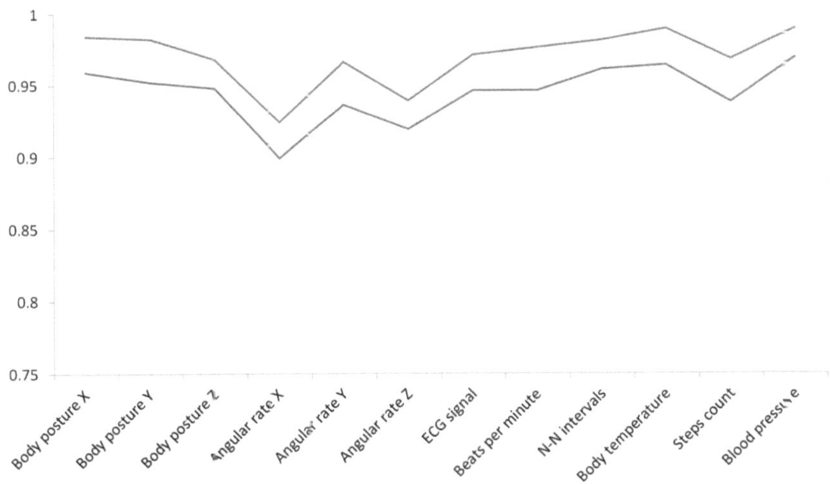

Fig. 2. Comparative of R^2 coefficient values for each ECG sensor parameters on the training dataset by applying XGBoost as AI technique, scaled as preprocessing technique, predicted from all parameters (blue) and PCA-selected parameters (red). (Color figure online)

In the comparative analysis between conventional laparoscopic and the RAS performance, the results obtained were similar, with all parameters exhibiting R^2 coefficients higher than 0.9. It is possible that the conventional laparoscopic results were slightly higher. Nevertheless, the results were consistent for both cases. Regarding the preprocessing technique, the scaled technique achieved higher R^2 coefficient compared to the scaled and normalized technique. In all cases, the difference between the two preprocessing techniques was around 0.15.

3.3 Cross-Validation Results

The results of the predictive analysis for the cross-validation are presented in Table 3. The dataset was preprocessed using the scaled technique, which achieved higher values of the R^2 coefficient for conventional and RAS. MLR and scaled showed the highest values of the R^2 coefficient and lowest values of RMSE (lower than 0.05) for most of the parameters except body posture and angular rate. This fact could be related to the linear nature of the data, as its performance exhibits a clear linear trend. As in training dataset results, lower results were achieved for the non-linear AI techniques (CNN, MLP, and U-NET). It should be noticed that MLR exhibited high to excellent R^2 values, approaching 1, and RMSE close to 0 for all parameters. The R^2 coefficient and RMSE values for body temperature ($R^2 = 0.959$ and RMSE = 0.018) and blood pressure ($R^2 = 0.979$ and RMSE = 0.012) were particularly noteworthy.

Table 3. Results of R^2 for the results of cross-validation of parameters recorded during MIS sessions as a function of the predictive AI techniques and scaled as preprocessing techniques. The best results are highlighted in bold.

AI technique	Body posture X	Body posture Y	Body posture Z	Angular rate X	Angular rate Y	Angular rate Z	ECG signal	Beats per minute	N-N intervals	Body temperature	Steps count	Blood pressure
Cat Boost	0.913	**0.926**	0.893	0.801	0.875	0.852	0.917	0.912	0.907	0.949	0.916	0.948
CNN	0.199	0.361	0.181	0.192	0.180	0.145	0.166	0.161	0.156	0.161	0.158	0.019
MLP	0.091	0.209	0.219	0.181	0.075	0.098	0.083	0.078	0.073	0.311	0.139	0.385
RF	0.905	0.878	0.896	0.785	0.875	0.852	0.840	0.835	0.830	0.929	0.939	0.908
GBM	0.894	0.849	0.836	0.588	0.772	0.741	0.791	0.786	0.781	0.938	0.916	0.930
MLR	0.888	0.829	0.908	0.829	0.873	0.847	**0.946**	**0.941**	**0.936**	**0.959**	**0.947**	**0.979**
SVM	0.749	0.324	0.654	0.289	0.176	0.558	0.517	0.512	0.507	0.432	0.421	0.936
U-Net	0.092	0.141	0.145	0.105	0.109	0.127	0.172	0.167	0.162	0.632	0.108	0.526
XGBoost	**0.924**	0.922	**0.918**	**0.904**	**0.936**	**0.919**	0.901	0.896	0.891	0.949	0.938	0.929

As depicted in Fig. 3, the loss incurred when predicting by using all parameters or only the four most representative characteristics was approximately 0.02 which could be considered negligible. The MLR AI technique demonstrated high to excellent R^2 values, approaching 1, and RMSE close to 0 for all parameters.

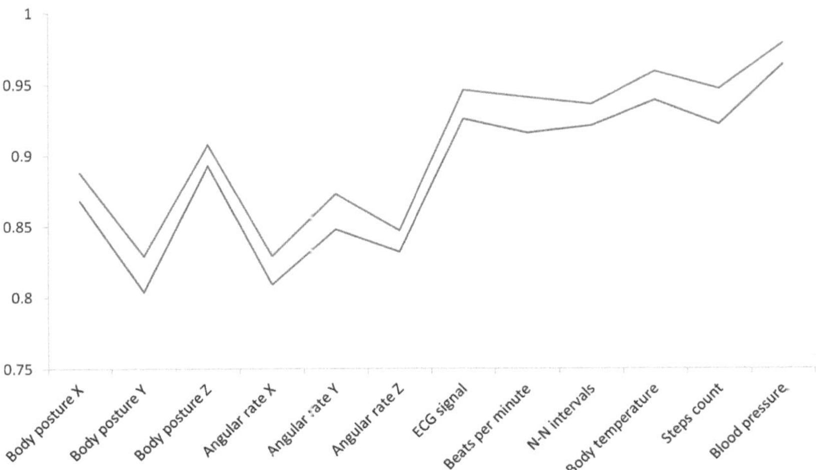

Fig. 3. Comparative of R^2 coefficient values for each ECG sensor parameters on the cross-validation results by applying MLR as AI technique, scaled as preprocessing technique, predicted from all parameters (blue) and PCA-selected parameters (red). (Color figure online)

In the comparative analysis between conventional laparoscopic and the RAS performance, the results were similar, with all parameters exhibiting R^2 coefficients exceeding 0.8. Similar to the training dataset, it is possible that the conventional laparoscopic results were slightly higher. However, the results were consistent across both cases. Regarding the preprocessing technique, the scaled technique demonstrated a higher R^2 coefficient compared to the scaled and normalized technique. In all instances, the difference between the two preprocessing techniques was approximately 0.1.

3.4   Test Dataset Results

Table 4 showcases the outcomes of the prediction for the test dataset. The dataset underwent preprocessing using the scaled technique, which resulted in higher R^2 correlation coefficients for both surgical approaches, conventional laparoscopy and RAS. MLR and scaled showed the highest values of the R^2 coefficient and lowest values of RMSE (lower than 0.05) for almost all parameters, except body posture at Y axis. As in previous cases, this fact could be related to the linear nature of the data [34]. MLR demonstrated high to excellent R^2 values, approaching 1, and RMSE close to 0 for all parameters. Notably, the R^2 correlation coefficient and RMSE values for body temperature (R^2 = 0.969 and RMSE = 0.021) and blood pressure (R^2 = 0.968 and RMSE = 0.024) were particularly noteworthy.

Table 4. Results of R^2 for the test dataset of parameters recorded during MIS sessions as a function of the predictive AI techniques and scaled as preprocessing techniques. The best results are highlighted in bold.

AI technique	Body posture X	Body posture Y	Body posture Z	Angular rate X	Angular rate Y	Angular rate Z	ECG signal	Beats per minute	N-N intervals	Body temperature	Steps count	Blood pressure
Cat Boost	0.869	0.816	0.802	0.431	0.685	0.641	0.701	0.696	0.691	0.965	0.924	0.962
CNN	0.109	0.324	0.157	0.121	0.271	0.106	0.096	0.091	0.086	0.222	0.205	0.181
MLP	0.091	0.219	0.254	0.302	0.084	0.114	0.110	0.105	0.100	0.397	0.092	0.376
RF	0.863	0.834	0.802	0.427	0.682	0.653	0.670	0.665	0.660	0.949	0.929	0.883
GBM	0.856	0.824	0.776	0.433	0.675	0.591	0.637	0.632	0.627	0.948	0.918	0.938
MLR	**0.907**	0.847	**0.927**	**0.818**	**0.912**	**0.886**	**0.935**	**0.930**	**0.925**	**0.968**	**0.966**	**0.969**
SVM	0.760	0.286	0.604	0.159	0.157	0.494	0.471	0.466	0.461	0.475	0.471	0.938
U-Net	0.081	0.143	0.303	0.222	0.162	0.339	0.304	0.299	0.294	0.611	0.147	0.539
XGBoost	0.870	**0.874**	0.813	0.493	0.742	0.671	0.734	0.729	0.724	0.949	0.921	0.947

As observed in Fig. 4, the loss incurred when predicting based on all parameters or only the four most representative characteristics is around 0.015, indicating a negligible loss. The MLR AI technique achieved high to excellent R^2 values, approaching 1, and RMSE close to 0 for all parameters.

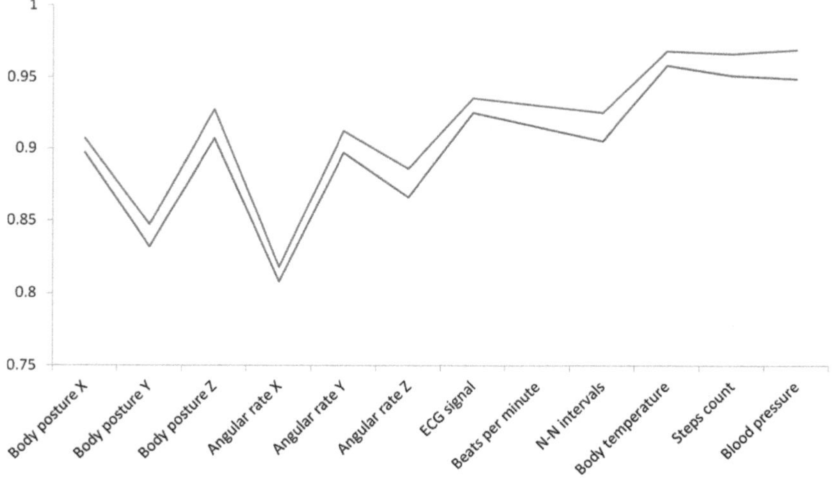

Fig. 4. Comparative of R^2 coefficient values for each ECG sensor parameters on the test dataset by applying MLR as AI technique, scaled as preprocessing technique, predicted from all parameters (blue) and PCA-selected parameters (red). (Color figure online)

In the comparative analysis between conventional laparoscopic and RAS performance, as in previous analyses, the results were similar. All parameters exhibited R^2 coefficients exceeding 0.75. Conventional laparoscopic results might have been slightly higher. However, the results remained consistent across both cases.

Regarding the preprocessing technique, the scaled technique demonstrated a higher R^2 coefficient compared to the scaled and normalized technique. In all instances, the difference between the two preprocessing techniques was approximately 0.15.

In future studies, these models could be integrated into real-time monitoring systems to create a system to recommend or to plan potential breaks to surgeons, preventing long-term health problems for the surgeons, improving the quality of the surgical procedures.

4   Conclusions

In this study, several machine learning and deep learning models have been developed, implemented, and tested to predict physiological, kinematic, and ergonomic parameters of surgeons during conventional and robot-assisted laparoscopic surgical techniques. Among these models, XGBoost and MLR achieved satisfactory R^2 and RMSE coefficients for all analyzed parameters.

By combining all models, the linear models evaluated (CatBoost, GBM, MLR, and XGBoost) demonstrated superior results compared to the non-linear models (CNN, UNET and MLP) with RF being the only non-linear model that yielded satisfactory outcomes. Considering the data preprocessing techniques employed in the present study, the scaled technique generally yielded better results (R^2 correlation coefficient and RMSE) than the scaled and normalized technique.

These findings demonstrate the efficacy and accuracy of linear predictive models. Consequently, this study provides preliminary evidence for predicting surgeon stress during the execution of both laparoscopic and robotic-assisted MIS procedures. Moreover, this promising results open the possibility to create a system to recommend rests or planning potential breaks to surgeons in some moments of high level of stress or fatigue that would be expand in future studies, preventing long-term health problems for the surgeons.

Acknowledgment. This study was financed by the Ministry of Science Innovation and University, with funds from the European Union–Next Generation EU, from the Recovery, Transformation and Resilience Plan (PRTR-C17.I1) and the European Regional Development Fund (ERDF) of Extremadura Operational Program 2021–2027. The authors would also thank colleagues who collaborated in this study (A. Garcia, R. Baltá, D. Durán, E. Abellán, E. Ramos, I. López-Agudelo, J. Salas, J. Vela, J. Campos, J.A. Gálvez, L. Pires, I. Sánchez, J. Olivares, R. González, C. Sánchez-Rumbo and J. R. Torres).

Disclosure of Interests. The authors have no competing interests to declare that are relevant to the content of this article.

References

1. Hurley, A.M., et al.: SOS save our surgeons: stress levels reduced by robotic surgery. Gynecol. Surg. **12**, 197–206 (2015)

2. Claus, C., Furtado, M., Malcher, F., Cavazzola, L.T., Felix, E.: Ten golden rules for a safe MIS inguinal hernia repair using a new anatomical concept as a guide. Surg. Endosc. **34**(4), 1458–1464 (2020). https://doi.org/10.1007/s00464-020-07449-z

3. Kaplan, J.R., Lee, Z., Eun, D.D., Reese, A.C.: Complications of minimally invasive surgery and their management. Curr. Urol. Rep. **17**, e47 (2016)

4. Atesok, K., Satava, R.M., Marsh, J.L., Hurwitz, S.R.: Measuring surgical skills in simulation-based training. J. Am. Acad. Orthopedical Surg. **25**, 665–672 (2017)

5. Klein, M.I., Warm, J.S., Riley, M.A.: Mental workload and stress perceived by novice operators in the laparoscopic and robotic minimally invasive surgical interfaces. J. Endourol. **26**(8), 1089–1094 (2012)

6. Nath, R.K., Thapliyal, H., Caban-Holt, A., Mohanty, S.P.: Machine learning based solutions for real time stress monitoring. IEEE Consum. Electron. Mag. **9**, 34–41 (2020)

7. Arora, S., Sevdalis, N., Nestel, D., Woloshynowych, M., Darzi, A., Kneebone, R.: The impact of stress on surgical performance: a systematic review of the literature. Surgery **147**(3), 311–330 (2010)

8. Saoughi, F., Behmanesh, A., Sayfouri, N.: Internet of things in medicine: a systematic mapping study. J. Biomed. Inform. **103**, e103383 (2020)

9. Hickey, B.A., et al.: Smart devices and wearable technologies to detect and monitor mental health conditions and stress: a systematic review. Sensors (MDPI) **21**(10), e3461 (2021)

10. Zangroniz, R., Martínez-Rodrigo, A., Pastor, J.M., L ópez, M.T., Fernández-Caballero, A.: Electrodermal activity sensor for classification of calm/distress condition. Sensors (MDPI) **17(10)**, e2324 (2017)

11. Ávila-Tomás, J.F., Mayer-Pujadas, M.A., Quesada-Varela, V.J.: La inteligencia artificial y sus aplicaciones en medicina I: introducci ón y antecedentes a la IA y rob ótica. Atenci ón Primaria **52**, 778–784 (2020)

12. Arakaki, S., et al.: Artificial intelligence in minimally invasive surgery: current state and future challenges. Jpn. Med. Assoc. J. **8**(1), 86–90 (2024)

13. Kankanamge, D., et al.: Artificial intelligence based assessment of minimally invasive surgical skills using standardised objective metrics - a narrative review. Am. J. Surg. **241**(116074), 1–11 (2025)

14. Smets, E., Raedt, W.D., Hoof, C.V.: Into the wild: the challenges of physiological stress detection in laboratory and ambulatory settings. IEEE J. Biomed. Health Inform. **23**, 463–473 (2018)

15. Caballero, D., Pérez-Salazar, M.J., Sánchez-Margallo, J.A., Sánchez-Margallo, F.M.: Applying artificial intelligence on EDA sensor data to predict stress on minimally invasive robotic-assisted surgery. Int. J. Comput. Assisted Radiol. Surg. **19**, 1953–1963 (2024)

16. Pérez-Salazar, M.J., Caballero, D., Sánchez-Margallo, J.A., Sánchez-Margallo, F.M.: Comparative study of ergonomics in conventional and robotic-assisted laparoscopic surgery. Sensors (MDPI) **24**(12), e3840 (2024)

17. Pérez-Salazar, M.J., Caballero, D., Sánchez-Margallo, J.A., Sánchez-Margallo, F.M.: Correlation study and predictive modelling of ergonomic parameters in robotic-assisted laparoscopic surgery. Sensors (MDPI) **24**(23), e7721 (2024)

18. Sharma, D., Singh, J., Sehra, S.S., Sehra, S.K.: Demystifying mental health by decoding facial action unit sequences. Big Data Cogn. Comput. **8**(7), e78 (2024)

19. Oka, M.: Interpreting a standardized and normalized measure of neighborhood socioeconomic status for a better understanding of health differences. Arch. Public Health **79**, e226 (2021)

20. Colton, T.: Statistics in Medicine. Little Brown and Co. Boston, Massachusetts, U.S.A. (1974)
21. Hyndman, R., Koehler, A.B.: Another look at measures of forecast accuracy. Int. J. Forecast. **22**, 679–708 (2006)
22. Wu, X., et al.: Top 10 algorithms in data mining. Knowl. Inf. Syst. **14**, 1–37 (2008)
23. Nagassou, M., Mwangi, R.W., Nyarige, E.: A hybrid ensemble approach utilizing light gradient boosting machine and category boosting model for lifestyle-based prediction of type-II diabetes Mellitus. J. Data Anal. Inf. Process **11**(4), 480–511 (2023)
24. LeCunn, Y., Bengio, Y.: Convolutional networks for images, speech and time series. In: The Handbook of Brain Theory and Neural Networks. MIT Press, Cambridge (1995)
25. Fayyad, U., Piaetsky-Shapiro, G., Smyth, P.: From data mining to knowledge discovery in databases. Am. Assoc. Artif. Intell. **17**, 37–54 (1996)
26. Breiman, L.: Random forest. Mach. Learn. **45**(1), 5–32 (2001)
27. Buehlmann, P., Yu, B.: Boosting with the L2 loss: regression and classification. J. Am. Stat. Assoc. **98**, 324–339 (2003)
28. Akaike, H.: On newer statistical approaches to parameter estimation and structure determination. Int. Federation Autom. Control **3**, 1877–1884 (1978)
29. Chang, C., Lin, C.: LIBSVM. A library for support vector machine. ACM Trans. Intell. Syst. Technol. **2**(27), 1–27 (2011)
30. Ronneberger, O., Fischer, P., Brox, T.: U-net: convolutional networks for biomedical image segmentation. In: Navab, N., Hornegger, J., Wells, W.M., Frangi, A.F. (eds.) MICCAI 2015. LNCS, vol. 9351, pp. 234–241. Springer, Cham (2015). https://doi.org/10.1007/978-3-319-24574-4_28
31. Nalluri, M., Pentela, M., Eluri, N.R.: A scalable tree boosting system: XG Boost. Int. J. Res. Stud. Sci. Eng. Technol. **7**(12), 36–51 (2020)
32. Bro, R., Smilde, A.K.: Principal component analysis. Anal. Method **6**, 2812–2831 (2014)
33. Guzmán-García, C., et al.: Correlating personal resourcefulness and psychomotor skills: an analysis of stress. visual attention and technical metrics. Sensors (MDPI) **22**(3), e837 (2022)
34. Gewers, F.L., et al.: Principal component analysis: a natural approach to data exploration. ACM Comput. Surv. **70**, 1–34 (2021)

Can In-Context Learning Enable Large Vision Language Models to Detect ECG Abnormalities?

Samuel Camba[1], Abraham Otero[2], Daniel García[3], Luciano Sánchez[1], and Nahuel Costa[1]([✉])

[1] Computer Science Department, University of Oviedo, 33202 Gijón, Asturias, Spain
costanahuel@uniovi.es
[2] Department of Information Technologies, Institute of Technology, Universidad San Pablo-CEU, CEU Universities, Campus Montepríncipe, 28660 Madrid, Spain
[3] Arrhythmia Unit, Hospital Universitario Central de Asturias, 33011 Oviedo, Asturias, Spain

Abstract. Electrocardiogram (ECG) interpretation is essential for cardiac diagnosis, yet machine learning models often fail to adapt to real-world data, particularly when certain conditions, such as rare diseases, are underrepresented in training sets. In-Context Learning (ICL) has shown promise for adapting Large Language Models (LLMs) to new tasks using only a few labeled examples at inference time, but its potential in the domain of medical imaging remains largely unexplored. In this work, we investigate whether ICL can enable large Vision-Language Models (VLMs) to interpret ECG images. Our approach leverages VLMs to perform classification in data-constrained scenarios without parameter updates, offering a practical alternative when fine-tuning is not feasible. To enhance both interpretability and predictive performance, we incorporate a concept-based prompting framework inspired by Concept Bottleneck Models (CBMs). Specifically, the VLM is prompted to first predict clinically meaningful intermediate concepts, which then guide the final diagnosis, providing structured, interpretable reasoning. We evaluate our method on detecting Left Bundle Branch Block and Brugada syndrome. Results indicate that combining ICL with CBM-style prompting yields interpretable ECG analyses and promising diagnostic performance in data-limited settings. This suggests that VLMs could be a practical tool for extending machine learning to underrepresented clinical conditions without the need for extensive labeled data.

Keywords: ECG analysis · Vision Language Models · In-Context Learning · Concept Bottleneck Models

1 Introduction

The electrocardiogram (ECG) is an essential diagnostic tool in cardiology, enabling clinicians to identify abnormalities in heart rhythm, ischemia, and structural heart diseases. Its widespread adoption in clinical practice stems from

A. López Fernández et al. (Eds.): CIABiomed 2025, LNBI 16148, pp. 200–213, 2026.
https://doi.org/10.1007/978-3-032-10661-2_16

its non-invasive nature, affordability, and ability to provide rapid diagnostic insights. Recently, the integration of machine learning (ML) into ECG interpretation has demonstrated notable promise, offering potentially faster, more accurate diagnoses by detecting complex patterns that might otherwise be missed by human experts [1,2]. Despite these advancements, a critical barrier still hinders the practical clinical adoption of ECG-focused ML models: their limited generalization capability when deployed in novel or unseen clinical scenarios [3].

The generalization issue arises primarily because ML models trained on large, publicly available ECG datasets often experience significant performance degradation when confronted with data in new clinical settings [4]. This problem becomes particularly pronounced when the deployed environment differs markedly from the original training conditions, such as differences in patient demographics, ECG recording hardware, or clinical protocols [5]. Furthermore, rare cardiovascular conditions like Brugada syndrome, Arrhythmogenic Right Ventricular Cardiomyopathy (ARVC), or specific types of Long QT syndrome, are poorly represented or entirely absent in most training datasets, exacerbating the issue of poor adaptability and limiting the clinical applicability of existing models [6,13]. Yet, precisely these rare conditions are where reliable diagnostic support is often most urgently needed.

The traditional strategy for addressing such limitations involves fine-tuning pre-trained models using additional labeled ECG data from the target clinical domain. However, acquiring and annotating sufficient ECG data for effective fine-tuning, especially for rare conditions, is highly challenging due to resource limitations, annotation costs, and the rarity of clinical events [7]. As a consequence, fine-tuning typically becomes impractical in routine clinical settings, leaving a critical gap between promising research results and real-world usability.

Recent developments in artificial intelligence, particularly Large Language Models (LLMs), have demonstrated remarkable abilities in generalizing across diverse tasks, and rapidly adapting to new domains. These models show significant potential for clinical applications due to their inherent flexibility and capability to perform tasks using minimal labeled data [8]. A key innovation enabling this flexibility is a new paradigm known as In-Context Learning (ICL), where LLMs adapt to new tasks by observing only a few examples presented directly in the prompt, without updating their underlying parameters [9]. The recent emergence of multimodal extensions of these models, Large Vision-Language Models (VLMs), has broadened this potential even further by enabling joint reasoning over visual and textual data [10].

While the potential of VLMs and ICL has begun attracting attention in medical contexts [11,12], their application to ECG interpretation, especially in low-resource scenarios, remains under-explored. Given their ability to rapidly adapt to new tasks using only a handful of labeled examples, VLMs guided by ICL could provide a practical and scalable solution to address the persistent challenge of limited generalization in ECG analysis, particularly for rare or underrepresented conditions.

In this paper, we propose and evaluate a novel approach that integrates VLMs and ICL specifically tailored for ECG abnormality detection under data-constrained scenarios. Moreover, to enhance both interpretability and predictive power, we integrate a concept-based framework inspired by Concept Bottleneck Models (CBMs). Unlike conventional methods, our approach uses VLMs to first predict clinically relevant intermediate concepts directly from ECG images, subsequently guiding the final diagnosis based on these concepts. Through comprehensive experiments, we show that our method offers a promising direction for building practical and explainable solutions for ECG analysis in clinical settings where labeled data is scarce.

2   State-of-the-Art

Automated ECG analysis through ML has advanced significantly over recent decades, demonstrating the potential to enhance diagnostic accuracy, efficiency, and consistency [14]. Traditional ML approaches initially relied on handcrafted features, such as RR intervals and wave amplitudes extracted directly from ECG waveforms [15]. However, the advent of deep learning (DL) marked a paradigm shift, enabling models to directly learn discriminative patterns from raw ECG data. Convolutional Neural Networks (CNNs) emerged as highly effective for capturing morphological and spatial patterns across multiple ECG leads, while Recurrent Neural Networks (RNNs), particularly Long Short-Term Memory (LSTM) networks, excelled in modeling temporal dynamics within cardiac cycles [16,17]. The combination of CNNs and RNNs, often augmented with attention mechanisms, has further improved the predictive capabilities of models across a wide array of cardiac diagnostic tasks [18,19].

More recently, Transformer-based architectures have introduced additional improvements, primarily due to their ability to model long-range dependencies via self-attention mechanisms, addressing limitations inherent to traditional CNN-RNN architectures [20]. This evolution aligns with broader trends in AI, characterized by the rise of foundational models pre-trained on extensive, heterogeneous datasets. Large pre-trained ECG models, analogous to foundational language models, have started emerging as versatile tools capable of generalizing effectively across multiple diagnostic contexts [21]. However, these models continue to face critical barriers to widespread clinical adoption, primarily due to limited generalization when encountering new or unseen clinical environments. While transfer learning and fine-tuning remain standard approaches to mitigate this issue, these methods depend heavily on the availability of sufficient annotated data from the target clinical domain. Acquiring such annotated data, especially for rare conditions, is challenging, as it typically requires considerable clinician involvement, expertise, time, and resources. The challenge of data scarcity is especially pronounced for rare cardiac conditions such as Brugada syndrome, characterized by distinctive ECG patterns including saddleback ST-segment elevations in the right precordial leads (V1V3) [22]. These patterns can be transient or concealed, complicating diagnosis even for experienced cardiologists [23]. Training reliable ML models to detect rare conditions demands sufficient positive examples, yet such conditions are inherently underrepresented in standard datasets, making fine-tuning ineffective and impractical.

Recent developments in AI provide alternative avenues for addressing these limitations, particularly through the use of large foundational models such as LLMs and their multimodal extensions, VLMs. These models, trained on vast amounts of paired image-text data, excel in generalization and rapid adaptation to new tasks through ICL, where minimal labeled examples are provided directly within the model prompt [9]. Unlike traditional fine-tuning, ICL is a particular case of Few-Shot Learning (FSL) that requires no parameter updates, instead, the model conditions its predictions on a few relevant examples provided in the context window. Figure 1 illustrates this process with a text classification task, where the model infers patterns from the examples without modifying its internal weights. This makes ICL well-suited to low-data settings, common in clinical practice, where annotated examples are scarce.

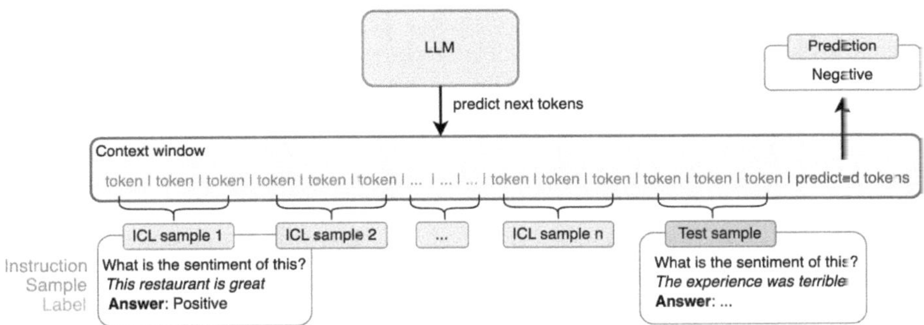

Fig. 1. Example of ICL applied to text classification. Instead of retraining, the LLM is shown a few labeled input-output pairs (ICL samples) in the context window. Using these examples, the model infers the underlying pattern and applies it to produce a prediction for the new, unseen input (test sample).

While individually promising, the combined application of VLMs and ICL remains still unexplored. Most existing works employing VLMs in medical applications focus either on fine-tuning models extensively on available domain-specific data or, if used with ICL, rely on retrieval-based augmentation methods over big databases [12,25]. Such approaches, however, differ from the main focus of this study: maximizing the available information when obtaining labeled data is a limitation in order to predict clinically relevant events. Our goal is to demonstrate whether these techniques can facilitate reasonable ECG analysis, particularly in detecting abnormalities associated with rare conditions, without requiring extensive labeled datasets or model retraining.

Additionally, to enhance model interpretability and move beyond single-label predictions, we integrate a framework inspired by CBMs [24]. This paradigm reshapes the model's decision-making process: rather than predicting a final class directly, the model is constrained to first reason over a set of intermediate, clinically relevant concepts, such as specific wave morphologies or interval durations. This approach significantly improves interpretability by rendering the model's reasoning transparent, as diagnostic decisions are explicitly grounded in human-understandable features. This fosters greater trust and enables more precise error analysis. Furthermore, it enhances reliability by forcing the model to follow a structured reasoning path that aligns with established diagnostic workflows, thereby reducing its susceptibility to spurious correlations. Finally, this framework provides an intuitive mechanism to translate expert knowledge into the model's operational logic: clinical guidelines can be effectively encoded as concepts and rules in a conversational (text-based) format, aligning the model's behavior with validated medical expertise.

3   Methodology

Adapting pre-trained models to new data can lead to two distinct outcomes. In the best-case scenario, the model generalizes well to unseen samples that share the same underlying distribution as the training data. However, performance often degrades when the model encounters previously unseen conditions. To address this challenge, our approach leverages the ICL capabilities of VLMs. Figure 2 provides an overview of the proposed approach: rather than retraining or fine-tuning the model, we provide a small set of annotated examples within its context window, enabling rapid adaptation to new tasks or domains. We explore two approaches to obtaining the predicted class: on one hand, the model is asked to directly predict the final label, and on the other, it is asked to predict the presence or absence of intermediate concepts, which are then used to determine the final label.

3.1   Downstream Tasks and Data

To evaluate the effectiveness of our approach, we designed experiments based on two real-world clinical classification tasks involving 12-lead, 12 -second duration ECGs. As a first step, we selected the task of detecting Left Bundle Branch Block (LBBB), a relatively common conduction abnormality. While not trivial, LBBB is characterized by distinct and well-defined morphological features, which are useful visual cues for classification. This task serves as a controlled setting to assess the baseline viability of using VLMs for ECG interpretation, particularly in distinguishing normal cardiac conduction from common pathological patterns based on waveform morphology. We employed the publicly available MIMIC-IV-ECG database [27], which includes ECG recordings from Beth Israel Deaconess Medical Center (Boston, MA). We curated a balanced subset of 2,000 samples, randomly selected such that each sample corresponds to a unique patient, comprising 1,000 ECG recordings labeled as normal and 1,000 labeled as LBBB.

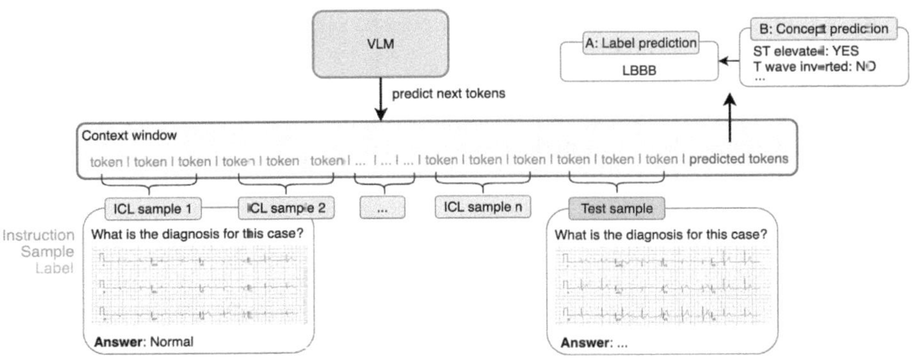

Fig. 2. Overview of the proposed workflow. To adapt to unseen data, our method provides the model with a prompt containing labeled examples from the target distribution, enabling it to adapt to the new domain without retraining. 'A: Label prediction' refers to asking the model to predict the final label while 'B: Concept prediction' involves asking the model for intermediate concepts, which are then used to infer the final label.

As a second, more challenging scenario, we focused on diagnosing Brugada syndrome, a rare genetic channelopathy associated with a markedly increased risk of sudden cardiac death in otherwise healthy individuals. Diagnosis of Brugada syndrome is based on the identification of specific ECG patterns, most notably, coved or saddleback ST-segment elevations in the right precordial leads. These patterns can be transient or masked, making the diagnosis difficult even for trained clinicians and electrophysiologists. For this task, we collected a private dataset consisting of 200 annotated ECG recordings: 125 labeled as normal and 75 as confirmed Brugada syndrome cases. All recordings were obtained from patients diagnosed at the Cardiology Department of the Hospital Universitario Central de Asturias. This study was conducted in accordance with all relevant laws and institutional guidelines. Ethical approval for the use of ECG recordings in this research was granted by the Regional Clinical Research Ethics Committee of the Principality of Asturias (Project No. 35/2013). Each ECG was blindly annotated by at least one expert electrophysiologist using standard 12-lead recordings acquired at a paper speed of 25 mm/s. All patient identifiers were removed in compliance with institutional privacy protocols.

To ensure independence between training and evaluation, all samples chosen for ICL demonstrations were drawn from different patients than the ones being diagnosed, so that no ICL example ever comes from the same record or patient included in the test sets. For consistency and comparability across both datasets and downstream models, we followed a standardized preprocessing pipeline, inspired by recent literature:

1. Unit normalization Raw signal values were divided by 1000 to convert amplitudes from microvolts to millivolts (mV).

2. Resampling: All recordings were resampled to 100 Hz to standardize the sampling rate.
3. Noise filtering: A Finite Impulse Response (FIR) bandpass filter was applied using the biosppy library [28], with cutoff frequencies of 0.05 Hz and 47 Hz to remove baseline drift and high-frequency noise.
4. Temporal truncation: Each ECG was truncated to the first 5 s to reduce computational load while preserving relevant clinical information.
5. Image transformation: Finally, all preprocessed signals were converted into digitized image representations using the ecg-image-kit library [26].

This unified pipeline ensures that our methodology remains generalizable across tasks and datasets, allowing the VLM to interpret ECGs based on a consistent visual structure.

3.2   VLM Setup

To evaluate the applicability and robustness of our ICL-based approach, we selected three state-of-the art VLMs with varying design characteristics. The model selection strategy was guided by factors like openness (proprietary vs. open-source), pretraining data (general vs. domain-specific), architecture size, and suitability for medical tasks. Our goal is to investigate how these factors impact ICL performance regarding ECG interpretation tasks.

- **gpt-4o (2024-11-20 release)**: A proprietary, large-scale multimodal model developed by OpenAI, offering state-of-the-art performance across a wide range of vision and language tasks. Although not specifically trained on medical data, this model demonstrates strong generalization capabilities and serves as a high-performance baseline.
- **medgemma-4b-it**: An open-source model developed by Google, fine-tuned on medical data, specifically on chest X-ray, pathology, dermatology, and fundus images, but not exposed to ECG imagery. It represents a smaller, domain-adapted alternative with 4 billion parameters, providing insights into the benefits of medical pretraining even without ECG-specific supervision.
- **PULSE** [25]: A 7-billion-parameter open-access VLM designed explicitly for ECG interpretation. Pulse was trained on the ECGInstruct dataset, comprising over one million instruction-tuned samples spanning a wide range of ECG-related tasks, including both digital and scanned image formats. It is uniquely tailored to the structure and semantics of ECG data, making it highly relevant for our evaluation.

By examining models with these contrasting attributes, we aim to better understand the strengths and limitations of current VLMs in the context of real-world ECG tasks.

To ensure a structured evaluation, the VLMs were explicitly instructed to generate outputs in a standardized JSON format comprising two fields: 'answer', containing the final classification decision; and 'score', being a confidence score ranging from 0.0 to 1.0, reflecting the model's self-assessed certainty. To minimize variability and hallucinations, the temperature parameter was set to 0.0 during inference.

The inputs to the models consisted of a System Prompt, providing general instructions to frame the model's analytical perspective (e.g., "You are a medical assistant specialized in cardiology..."), and a detailed User Query, which included:

- **Task Description**: A concise instruction defining the classification objective. For example: "Classify the following ECG image as Normal or LBBB".
- **In-Context Examples**: A set of labeled ECG examples provided as demonstrations. Each example included an ECG image and its correct label. We varied the number of examples per class (denoted as 'shot') to study the effect of context length and demonstration size. Specifically, we tested settings with shot = 3, shot = 5, and shot = 10 samples per class.
- **Target ECG Image**: The final ECG image for which the model was required to generate a prediction.

All examples and target images were formatted identically and embedded within the prompt as base64-encoded image objects.

Additionally, to enhance interpretability and probe model reasoning beyond classification accuracy, we explored the integration of CBMs within the prompting framework. We prompted models to fill in a concept checklist prior to generating the final answer. This checklist included domain-specific features based on clinical ECG criteria for each condition. Each feature was included as a field in the JSON generated object. Especifically, for the LBBB detection task, the model was instructed to identify the presence of the following concepts in the right precordial leads:

- QRS duration greater than 200 ms
- ST segment elevation
- Loss or reduction of the initial R wave
- Notching or slurring of the S wave

For the Brugada syndrome task, the model was prompted to check for the presence of the following specific features in lead V1:

- Right bundle branch block pattern with a QRS duration greater than 120 ms
- ST-segment elevation
- T-wave inversion

A diagnosis of LBBB was made if the model identified at least three out of the four features, whereas a diagnosis of Brugada syndrome required confirmation of all three specified features.

This structured decomposition serves two main purposes. First, improved transparency, by making the model's reasoning process explicit through intermediate concepts, we can better assess whether correct predictions are grounded in clinically relevant features. Second, better error analysis: when a model fails, CBM allows to pinpoint whether the failure was due to misinterpreting a concept or misapplying a diagnostic rule, facilitating more targeted improvements.

Further details regarding the experimental setup and source code are available in the following public git repository.

4 Results and Discussion

To evaluate the effectiveness of our ICL-based approach we conducted experiments across two clinical tasks of interest: LBBB detection and Brugada syndrome diagnosis. We systematically compared the performance of three state-of-the-art VLMs under various conditions: out-of-the-box inference, ICL with different numbers of context examples, and the integration of CBM-style prompting. All experiments were performed on held-out test sets, as we chose the ICL samples manually to be representative of the task at hand, with results averaged across five independent runs to ensure statistical robustness.

4.1 LBBB Detection Results

Figure 3 presents the results for LBBB detection across all evaluated models and experimental conditions. The results reveal distinct performance patterns across VLMs, which highlights the varying capabilities of different architectures in interpreting ECGs. gpt-4o demonstrated robust performance across both ICL and CBM approaches, achieving substantial improvements from its baseline out-of-the-box F1 score of 66.6% (shot=0) to consistently high performance with ICL (95.9%, 96.9%, and 97.0% for shot=3, 5, and 10 respectively). The model maintained similar performance when augmented with CBM-style prompting, suggesting effective utilization of both demonstration examples and structured reasoning. This strong performance likely stems from gpt-4o's extensive pre-training corpus, which may have included medical literature and ECG-related content, enabling the model to leverage prior knowledge about cardiac conduction abnormalities.

In contrast, medgemma-4b exhibited highly unreliable performance, particularly with standard ICL where F1 scores remained low, ranging from 18.18% to 60.55% as the number of samples increased. While the model showed marginal improvement with more in-context examples, its overall performance remained substantially inferior to the other models, with the best result actually being achieved in the zero-shot setting (78.20%). Notably, the addition of CBM-style prompting did not lead to any meaningful improvement: the F1 scores with CBM were almost identical across all shot values, remaining around 49–50%. This suggests that, for this model, neither increasing the number of in-context examples nor introducing concept-based reasoning was sufficient to overcome its limitations. These shortcomings may be attributed to the model's training regimen, which, despite medical fine-tuning, did not include ECG-specific data, highlighting the importance of domain-specific exposure for specialized medical imaging tasks.

PULSE presented an intriguing performance profile. Specifically designed for ECG interpretation, the model achieved excellent zero-shot performance

(98.0%), demonstrating its strong foundational understanding of LBBB patterns. However, surprisingly, we observed systematic performance degradation when ICL was applied, either direct ICL or with the CBM approach. This counterintuitive result can be attributed to PULSE's training methodology, which was based on single question-answer pairs rather than multi-example contextual reasoning. The model's architecture appears optimized for direct ECG-to-diagnosis mapping rather than learning from in-context demonstrations, suggesting that ICL capabilities may require specific architectural considerations or training paradigms.

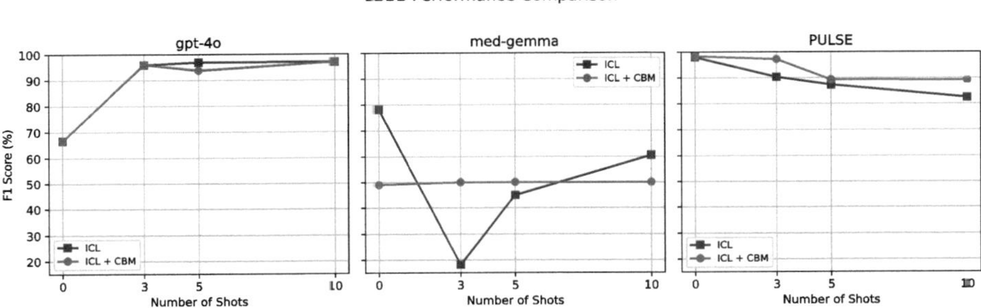

Fig. 3. Performance comparison for LBBB detection across different models and experimental conditions. The results demonstrate varying effectiveness of ICL and CBM approaches depending on model architecture and training paradigms.

4.2 Brugada Syndrome Detection Results

The Brugada syndrome detection results are presented in Fig. 4 and reveal markedly different performance patterns compared to LBBB detection, emphasizing the challenges associated with rare disease diagnosis and the critical importance of structured reasoning approaches. gpt-4o's performance with standard ICL remained consistently low across all context sizes (15.4% to 28.8% F1), indicating that mere exposure to labeled examples was insufficient for this complex diagnostic task. However, the introduction of CBM-style prompting consistently improved performance, with F1 scores increasing from 21.2% in the zero-shot CBM condition to 68.8% with ICL+CBM (shot=10). This substantial improvement highlights a crucial limitation of standard ICL for complex/unknown medical tasks: without explicit guidance to focus on clinically relevant concepts (here presented as lead features), the model fails to extract the appropriate diagnostic patterns from contextual examples. The CBM framework effectively directs the model's attention to the right precordial lead (V1) where Brugada patterns manifest, enabling more accurate diagnosis.

medgemma showed inconsistent results with both approaches, achieving F1 scores no better than 20.2% with simple ICL and 59.8% in shot=5 when combined with CBM, but dropping performance to 54.7 with shot=10. Although these results may suggest some capability for structured medical reasoning, the model's performance remains inconsistent, as observed in the previous tasks, thus providing limited confidence in its reliability for clinical applications.

PULSE, despite being extensively trained on ECG tasks, demonstrated poor zero-shot performance, which is likely attributed to insufficient exposure to this rare condition during training. However, when prompted with CBM in the zero-shot setting, its performance improved extensively (70.2%), reflecting its strong grasp of ECG morphology and its ability to reason effectively at the concept level when guided by structured prompts. Notably, as with the LBBB task, PULSE's performance declined with increasing context samples in both approaches, further supporting the notion that its architecture is not optimized for multi-example contextual learning.

The stark contrast between LBBB and Brugada results across all models underscores several critical insights. First, the results highlight the inherent difficulty of Brugada syndrome detection. The modest F1 values across all approaches underscore the existence of ambiguous cases where ECG patterns can mimic Brugada syndrome. In our dataset, we found many challenging borderline cases where the distinction between classes is not clear. For example, some Brugada cases present signs that can be identified with Brugada syndrome (such as a negative T wave and ST segment elevation), but are considered normal since the QRS complex is not sufficiently widened. Conversely, in ambiguous normal cases, the criteria may be met except for a slightly elevated ST segment, leading to possible misclassification as Brugada. These doubtful cases represent the primary source of model failures, highlighting the inherent complexity of this task. Second, the substantial improvements achieved through CBM-style prompting demonstrate that structured, concept-based reasoning can partially compensate for limited training exposure to rare conditions. Finally, the results suggest that current VLM architectures may benefit from hybrid approaches that combine their foundational knowledge with explicit clinical reasoning frameworks.

4.3 Discussion

The results provide a nuanced view of the interplay between VLM architecture, task complexity, and prompting strategy. For a well-defined condition like LBBB, gpt-4o's strong performance with standard ICL suggests its broad pretraining allows it to recognize clear visual patterns from a few examples. In contrast, the medically-tuned but ECG-naive medgemma model struggled, indicating that general medical knowledge does not automatically translate to specialized domains like ECG interpretation.

The Brugada syndrome task highlights a key limitation of standard ICL for rare and complex diseases. Merely providing image-label pairs was insufficient for all models. The significant performance boost came from CBM-style prompting, especially for gpt-4o and PULSE, demonstrating that embedding

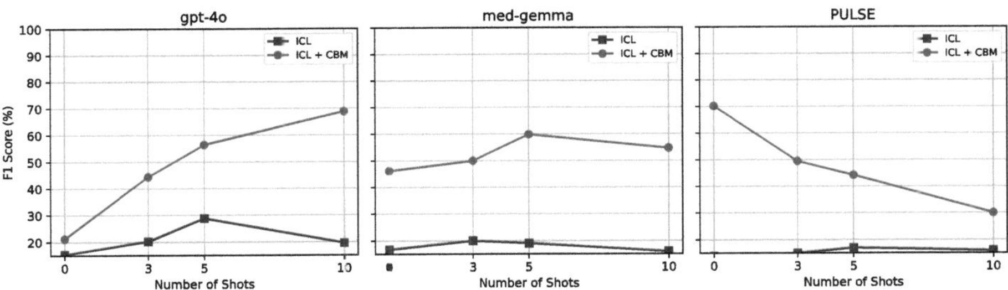

Fig. 4. Performance comparison for Brugada syndrome detection. The results highlight the critical importance of concept-based reasoning (CBM) in rare disease diagnosis, particularly when standard ICL approaches prove insufficient.

clinical knowledge into the prompt is critical. This forces the model to follow a structured reasoning path, focusing on specific, diagnostically relevant features and compensating for the lack of exposure during pre-training.

The models' difficulties with ambiguous cases mirror real-world clinical challenges and underscore that VLMs should be viewed as assistive tools. The CBM approach is paramount in this context, as its inherent interpretability allows clinicians to scrutinize the model's reasoning, build trust, and use its output as a structured second opinion rather than an autonomous diagnosis.

5   Concluding Remarks and Future Work

In this work, we investigated the effectiveness of combining VLMs with ICL and CBM for ECG abnormality detection, with a focus on data-limited scenarios. Our findings demonstrate that this approach represents a promising and practical alternative to traditional fine-tuning, particularly for rare diseases where annotated data is scarce. We showed that while general-purpose models can adapt to common conditions, the diagnosis of rare diseases requires a more structured approach. By integrating a concept-based reasoning framework, which allows predictions to be mapped to understandable clinical concepts, we guided VLMs to achieve promising performance on the challenging task of Brugada syndrome detection, enhancing both accuracy and interpretability. This approach can unlock the diagnostic potential of these models, making them viable tools even for complex tasks they were not explicitly trained on.

For future work, our primary focus is on enabling open-source models to perform robustly on ECG interpretation tasks. To achieve this, we propose finetuning these models not on diagnostic labels, but rather on detailed descriptions of ECGs. By first teaching models to understand and describe the fundamental features and morphology of ECG signals, we can equip them with a foundational knowledge that can be more effectively leveraged later through ICL. Also, we aim to include a more robust validation for the CBM part, considering direct comparison between the concepts used by the model and the annotations of cardiologists, as well as usability studies that analyze the confidence in the tool.

Acknowledgments. This work has been partially supported by the Ministry of Economy, Industry and Competitiveness ("Ministerio de Economía, Industria y Competitividad") from Spain/FEDER under grants PID2023-146257OB-I00 and SEK-25-GRU-GIC-24-055. The authors would also like to thank Jorge Valdenebro for his insightful comments.

competing Interests. The authors have no competing interests to declare that are relevant to the content of this article.

References

1. Hong, S., Zhou, Y., Shang, J., Xiao, C., Sun, J.: Opportunities and challenges of deep learning methods for electrocardiogram data: a systematic review. Comput. Biol. Med. **122**, 103801 (2020)
2. Somani, S., et al.: Deep learning and the electrocardiogram: review of the current state-of-the-art. EP Europace **23**(8), 1179–1191 (2021)
3. Attia, Z.I., Harmon, D.M., Behr, E.R., Friedman, P.A.: Application of artificial intelligence to the electrocardiogram. Eur. Heart J. **42**(46), 4717–4730 (2021)
4. Ansari, M.Y., et al.: A survey of transformers and large language models for ECG diagnosis: advances, challenges, and future directions. Artif. Intell. Rev. **58**(9), 261 (2025)
5. Ribeiro, A.H., et al.: Automatic diagnosis of the 12-lead ECG using a deep neural network. Nat. Commun. **11**(1), 1760 (2020)
6. Huang, Z., et al.: Generalization challenges in ECG deep learning: insights from dataset characteristics and attention mechanism. medRxiv, 2023–07 (2023)
7. Decherchi, S., Pedrini, E., Mordenti, M., Cavalli, A., Sangiorgi, L.: Opportunities and challenges for machine learning in rare diseases. Front. Med. **8**, 747612 (2021)
8. Clay, B., Bergman, H.I., Salim, S., Pergola, G., Shalhoub, J., Davies, A.H.: Natural language processing techniques applied to the electronic health record in clinical research and practice-an introduction to methodologies. Comput. Biol. Med. **188**, 109808 (2025)
9. Dong, Q., et al.: A survey on in-context learning. arXiv preprint arXiv:2301.00234 (2022)
10. Zhang, J., Huang, J., Jin, S., Lu, S.: Vision-language models for vision tasks: a survey. IEEE Trans. Pattern Anal. Mach. Intell. (2024)
11. Han, Y., Liu, X., Zhang, X., Ding, C.: Foundation models in electrocardiogram: a review. arXiv preprint arXiv:2410.19877 (2024)
12. Ferber, D., et al.: In-context learning enables multimodal large language models to classify cancer pathology images. Nat. Commun. **15**(1), 10104 (2024)
13. Leong, C.J., Sharma, S., Seth, J., Rabkin, S.W.: Artificial intelligence streamlines diagnosis and assessment of prognosis in Brugada syndrome: a systematic review and meta-analysis. Connected Health Telemed. **3**(2) (2024)
14. Petmezas, G., et al.: State-of-the-art deep learning methods on electrocardiogram data: systematic review. JMIR Med. Inform. **10**(8), e38454 (2022)
15. Goldberger, A.L., et al.: PhysioBank, PhysioToolkit, and PhysioNet: components of a new research resource for complex physiologic signals. Circulation **101**(23), e215–e220 (2000)

16. Cai, W., et al.: Accurate detection of atrial fibrillation from 12-lead ECG using deep neural network. Comput. Biol. Med. **116**, 103378 (2020)
17. Ghosh, S.K., et al.: Detection of atrial fibrillation from single lead ECG signal using multirate cosine filter bank and deep neural network. J. Med. Syst. **44**, 1–15 (2020)
18. Nejedly, P., et al.: Classification of ECG using ensemble of residual CNNs with attention mechanism. In: 2021 Computing in Cardiology (CinC), vol. 48, pp. 1–4 (2021)
19. Ern, E.S.Y., Ramli, D.A.: Classification of arrhythmia signals using hybrid convolutional neural network (CNN) model. In: Artificial Intelligence and Machine Learning for Healthcare: vol. 1: Image and Data Analytics, pp. 105–132 (2022)
20. Zhao, Z.: Transforming ECG diagnosis: an in-depth review of transformer-based deeplearning models in cardiovascular disease detection. arXiv preprint arXiv:2306.01249 (2023)
21. Nolin-Lapalme, A., et al.: Foundation models for generalizable electrocardiogram interpretation: comparison of supervised and self-supervised electrocardiogram foundation models. medRxiv (2025)
22. García-Iglesias, D., et al.: Spectral analysis of the QT interval increases the prediction accuracy of clinical variables in Brugada syndrome. J. Clin. Med. **8**(10), 1629 (2019)
23. Iglesias, D.G., Rubín, J., Pérez, D., Morís, C., Calvo, D.: Insights for stratification of risk in Brugada syndrome. Eur. Cardiol. Rev. **14**(1) (2019)
24. Koh, P.W., et al.: Concept bottleneck models. In: International Conference on Machine Learning, pp. 5338–5348 (2020)
25. Liu, R., Bai, Y., Yue, X., Zhang, P.: Teach multimodal LLMs to comprehend electrocardiographic images. arXiv preprint arXiv:2410.19008 (2024)
26. Shivashankara, K.K., et al.: ECG-image-kit: a synthetic image generation toolbox to facilitate deep learning-based electrocardiogram digitization. Physiol. Measur. **45**(5), 055019 (2024)
27. Gow, B., et al.: MIMIC-IV-ECG: diagnostic electrocardiogram matched subset. Type: Dataset **6**, 13–14 (2023)
28. Bota, P., Silva, R., Carreiras. C., Fred, A., Plácido da Silva, H.: BioSPPy: a Python toolbox for physiological signal processing. SoftwareX **26**, 101712 (2024). https://www.sciencedirect.com/science/article/pii/S2352711024000839

Novel Automated Tool for Functional Substrate Assessment of the Left Atrium in Patients with Persistent AF Using Machine Learning

Saman Golmaryami[1,2,3]([✉]), Etel Silva Garcia[1,2,3], and Juan Fernandez Armenta[1,2,3]

[1] Instituto de Investigación e Innovación Biomédica de Cádiz (INiBICA), Cádiz, Spain
saman.golmaryamii@gmail.com
[2] Hospital Universitario Puerta del Mar (HUPM), Cádiz, Spain
[3] Universidad de Cádiz, Cádiz, Spain
https://inibica.es

Abstract. Persistent atrial fibrillation (PsAF) is associated with high recurrence rates despite pulmonary vein isolation (PVI), highlighting the need to target the atrial substrate. Substrate characterization beyond the pulmonary veins using functional mapping can potentially unmask latent slow conduction areas. To improve substrate characterization in PsAF, we developed an automated tool to quantify key electrogram (EGM) features such as duration, local activation time (LAT), number of deflections, and pacing-induced duration increase (Delta). The clinical data comprise over 1,000 intracardiac EGMs from 5 patients with PsAF, recorded during baseline sinus rhythm and after three short-coupled extra stimuli. The features calculated by the tool achieved intraclass correlation coefficients of 0.98 for R3-LAT and 0.77 for R3-Duration, respectively, when compared to blinded expert clinicians' annotation, indicating reliable and acceptable automated measurements. Furthermore, the extracted features from the paced EGMs showed strong discriminatory power in identifying hidden slow conduction substrates, achieving 93.6% balanced accuracy with a Support Vector binary classifier model using a linear kernel.

Keywords: Persistent atrial fibrillation · Pulmonary vein isolation · Functional mapping · Electrogram · Local activation time · Pacing-induced duration increase (Delta) · Hidden slow conduction

1 Introduction

1.1 Atrial Fibrillation

Atrial fibrillation (AF) is the most prevalent type of arrhythmia and is increasing in the number of patients. In AF, the upper chamber's function is impaired,

A. López Fernández et al. (Eds.): CIABiomed 2025, LNBI 16148, pp. 214–227, 2026.
https://doi.org/10.1007/978-3-032-10661-2_17

and abnormal electrical propagation on the substrate causes irregular and fast heartbeats, as well as a distinct lack of P-waves in electrocardiograms (ECGs). As a result, the blood tends to be static, which can lead to blood clots and increase the risk of stroke [1].

AF has been categorized based on the duration of the episode. The first type is paroxysmal AF, which occurs spontaneously and generally resolves by itself or with treatment within 7 days [2].

Persistent AF is a sustained abnormal heart rhythm for more than 7 days, even with treatment or direct current cardioversion. Despite this, persistent AF may eventually cease on its own or via treatment [2].

Long-standing persistent AF is defined as lasting longer than 12 months, with the term permanent AF used when all means of treatment to restore normal heart rhythm have failed [2].

1.2   Treatment Procedures

Pulmonary vein isolation (PVI) is the main treatment strategy in AF patients, as veins are the main source of ectopic beats and the leading cause of paroxysmal fibrillation [3]. Paroxysmal AF patients undergoing PVI intervention have a success rate of 80–90% at a 1-year follow-up. However, persistent AF treatment strategy remains a challenge, with a 40–70% success rate in the PVI intervention [4]. This demonstrates a need for an alternative treatment strategy. The ablation strategy for Persistent AF patients has slowly moved from the pulmonary veins to the atrial substrate. A robust and reliable assessment of the substrate is required to determine an effective ablation target [5].

1.3   Problem Statement

Myocardial fibrosis generates regions of slow or blocked conduction, contributing to arrhythmic electrical activity. During sinus rhythm (SR), fibrotic tissue often appears electrically silent or normal, making it difficult to identify abnormal conduction patterns. However, applying three short-coupled extra stimuli, known as functional mapping, can unmask these concealed areas of slow conduction. Under these stimulations, fibrotic regions reveal characteristic fractionated electrograms (EGMs) with prolonged duration and multiple deflections. These EGMs can be captured using multi-polar catheters, enabling detailed analysis of the atrial substrate [4]. Functional mapping helps detect fibrotic tissue and unmask hidden conduction abnormalities, but it heavily depends on operator experience due to the lack of an automated tool. Clinicians must manually extract features such as duration and delta (the prolongation of EGM duration during functional mapping relative to SR) and detect fractionated EGMs based on their experience. This can cause errors and uncertainty in detecting the correct ablation target.

2   Materials and Methods

2.1   Clinical Data

Intracardiac EGMs from five patients diagnosed with persistent atrial fibrillation (PsAF), totaling over 1000 recordings, were analyzed. Data collection took place at Hospital Universitario Puerta del Mar, using the Carto 3 system and PentaRay®catheter (Biosense Webster, Diamond Bar, CA, USA) for functional electro-anatomical mapping. Additionally, magnetic anatomical reconstructions of the patients' left atria were extracted to project the analysis results onto 3D meshes, improving visualization and interpretability. The output signals by the Carto3 System consist of 2.5-second electrical recordings based on a predefined reference event (0.5 s after and 2 s before the reference event). All patients were provided with informed consent for data publication, and the functional mapping protocol employed in this study was approved in advance by the institutional ethics committee.

Intracardiac EGMs were recorded during functional mapping according to three short-coupled extra stimuli, including three sequential pacings applied from the coronary sinus (CS). The atrial effective refractory period (AERP) was calculated as the maximum interval between two consecutive stimuli in which the whole atrial activation fails to occur. Following the determination of AERP, the first extra stimulus (S1) was delivered at an interval of AERP + 60 ms after the onset of sinus rhythm. The second (S2) and third (S3) extra stimuli were then applied at AERP + [40-20] ms and AERP + [30-20] ms after S1 and S2, respectively.

2.2   Short-Term Fourier Transforms

The Fourier transform (FT) has been widely applied across various scientific fields and has played a crucial role in nearly all areas of science and technology. FT identifies the frequency components within a signal. However, as FT applications have grown, limitations have emerged because FT is a holistic transform and does not provide information about the time dimension. The short-time Fourier transform can be used to extract how a signal's frequency content evolves over time. [6]. Equations 1 and 2 show the formulas for the short-time Fourier transform of continuous and discrete signals, respectively.

$$\text{STFT}\{x(t)\}(\tau, \omega) = \int_{-\infty}^{+\infty} x(t)\, W(t - \tau)\, e^{-i\omega t}\, dt \tag{1}$$

where x(t) is the signal, W is the sliding window, τ is the location of the window, ω is the angular velocity.

$$\text{STFT}\{x[n]\}(m, \omega) = \sum_{n=0}^{N-1} x[n]\, W[n - m]\, e^{-i\omega n} \tag{2}$$

where x[n] is the discrete signal, W is the sliding window, m is the index of the time, ω is the angular velocity.

After extracting the frequency components, reassigning each energy value to its actual time-frequency coordinates enhances the sharpness of the resulting spectrogram. This improves both the time resolution and the accuracy of the automated analysis tool. The reassigned time and frequency coordinates can be formally computed as shown in Eq. 3 and 4 [7].

$$\tilde{t}(m,\omega) = m - \frac{\partial \phi(m,\omega)}{\partial \omega} \tag{3}$$

$$\tilde{\omega}(m,\omega) = \omega + \frac{\partial \phi(m,\omega)}{\partial m} \tag{4}$$

where $\tilde{t}$ is the reassigned time index, $\tilde{\omega}$ is the reassigned angular velocity, m is the index of the window, ω is the angular velocity.

In Fig. 1, a sample EGM of patient 10 was analyzed, and the corresponding time-frequency components were calculated. In Fig. 1E The normalized high-resolution time-frequency components can be used to delineate the input EGM with high accuracy in the time domain.

The reassigned STFT was chosen over the wavelet transform because it provides uniform frequency resolution across the time-frequency plane, whereas the wavelet transform inherently yields unequal resolution (high temporal but poor frequency resolution at high frequencies, and vice versa at low frequencies), limiting its precision for electrogram analysis.

2.3   Feature Extraction and Signal Processing Pipeline

The automated tool aims to extract features from EGMs: signal duration, local activation time (LAT), number of deflections, and pacing-induced increase in duration (Delta). The onset and end of each signal are determined by analyzing the inverse reassigned energy from the short-time Fourier transform (STFT). Figure 1B shows that the energy of the signal exceeding a normalized threshold ratio is classified as activation. The longest segment among the activations is identified as the main activation region. For each signal, the M1 channel-captured from the pacing catheter in the coronary sinus-is used as the reference to locate the EGM's window of interest. By identifying the local maxima of the V5 ECG lead, a reference marker for the sinus rhythm window of interest can be determined. Then the onset, end, voltage amplitude, and number of deflections of the EGM are extracted for sinus rhythm and triple extra stimuli windows of interest. The delta feature of each three stimuli is calculated by subtracting the sinus rhythm duration from each triple extra stimulus response (R1-3 Duration - SR Duration).

2.4   Graphical User Interface for Signal Analysis

To facilitate the handling and visualization of large volumes of EGM data and the tuning of the parameters, a dedicated graphical user interface (GUI) was

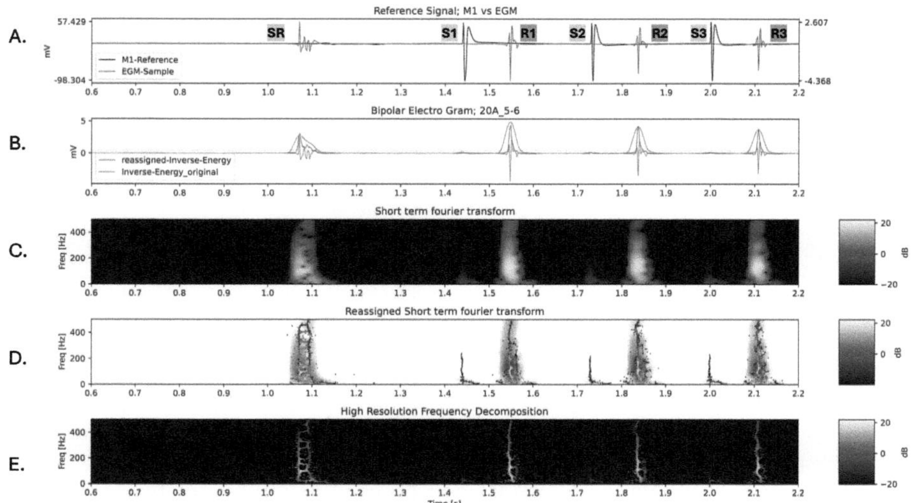

Fig. 1. A. Reference electrode signal (M1) recorded from the pacing catheter placed inside the coronary sinus (CS) versus the sample intra cardiac electrogram (EGM) recorded from bipolar electrodes of PentaRay catheter, S1–S3 in response to the pacing recorded from the pacing catheter in CS, SR is the normal Sinus Rhythm response, and R1-3 is the response to the pacings recorded from the PentaRay Catheter's bipolar electrodes B. A sample EGM from patient n-10, identified as abnormal, and EGM Energy as inverse STFT in both normal and reassigned versions in the frequency range of [0–300] Hz is plotted. C. Short-Time Fourier Transform (STFT) of the signal using the following parameters: n_fft = 100, hop_length = 4, win_length = 50, sampling rate (sr) = 1000 Hz. D. Reassigned STFT, which reallocates spectral energy to more precise time-frequency coordinates, enhancing resolution. E. Normalized time-frequency representation using a fixed-high resolution grid, where the maximum energy value is selected within each cell to sharpen the representation, and a low-pass filter is applied to eliminate the empty grids after increasing the resolution

developed using the Tkinter library in Python. The platform integrates essential functionalities for signal navigation, processing, and annotation, enabling researchers and clinicians to interact with the data in a user-friendly environment. The GUI provides a centralized framework for managing all recorded EGMs, allowing users to visualize signals resulting from different stimulation (e.g., sinus rhythm, R1-3), change the tool parameters, and visualize and extract specific signal segments' features. Within the platform, users can dynamically adjust key parameters of the reassigned short-time Fourier transform (rSTFT), such as: 1. Window length and type 2. Hop length and FFT size 3. Normalized activation detection threshold 4. Frequency range of band-pass filter. In addition, users can modify the extracted onset and end of the EGM waveform for sinus rhythm and each of the R1-3 activations, enabling intuitive and dynamic visual validation of detected features such as local activation times and delta across triple extra stimuli and sinus rhythm. Storing the computed features is possible within the GUI, ensuring consistent and reproducible analysis across the entire dataset. As demonstrated in Fig. 1, the integration of inverse reassigned STFT with interactive visualization significantly improves the temporal precision of the extracted features, especially in highly fractionated EGMs commonly observed in fibrotic atrial tissue. This platform is a key component of the automated analysis pipeline, Fig. 2, bridging computational signal processing with clinicians' point of view, and paving the way for robust, operator-independent substrate characterization in persistent atrial fibrillation.

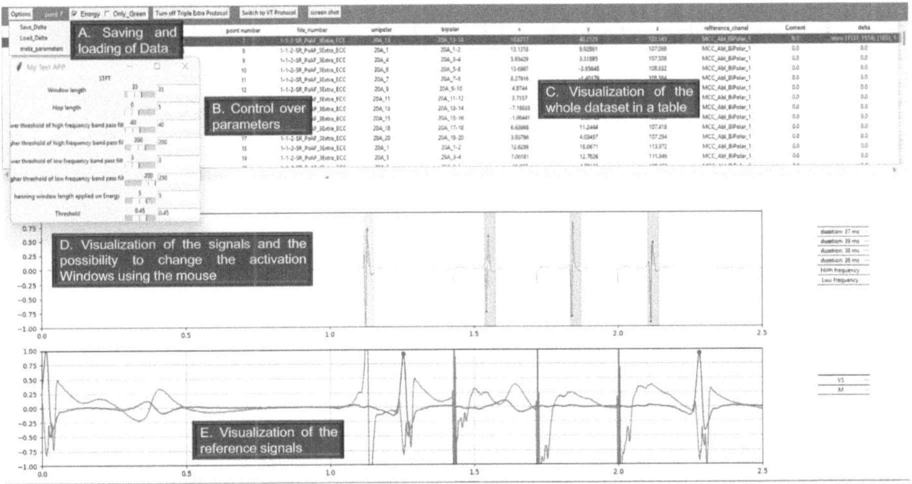

Fig. 2. Graphical user interface for intuitive parameter tuning and annotation of EGM datasets, with corresponding utilities indicated on the figure.

3   Results

3.1   Clinician Evaluation and Statistical Analysis

Following the development of the automated feature extraction tool and its integration into the GUI, clinical validation was carried out. An expert cardiologist evaluated all EGM features, including signal duration, local activation time (LAT), and delta, using the interactive GUI, with adjustment of the final detected activation period if necessary.

Bland-Altman Analysis. To assess the accuracy of measurement between the automated tool's output and the clinician's annotations, Bland-Altman analysis was carried out for each feature. This analysis visually confirmed that the majority of the differences between the clinician's evaluation and the tool's outputs fell within the 95% limits of agreement, suggesting strong consistency (Fig. 3)

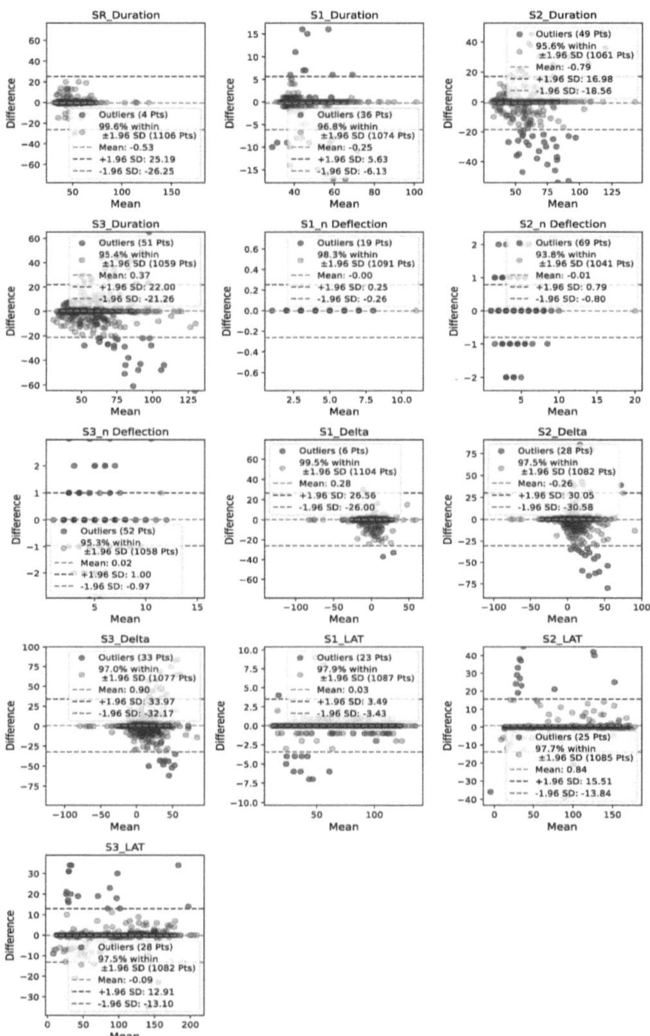

Fig. 3. Bland Altman plots comparing automated and clinician-derived measurements for all extracted features. Each plot shows the mean difference (green line), acceptable limits of difference between reference (clinicians' evaluation) and automated tool's prediction (blue and red dashed lines), and outliers (red dots). The high percentage of points within the acceptable limits across features, including SR, S1-3 durations, number of deflections, deltas, and local activation time (LATs), demonstrates strong consistency between the automated tool and the expert clinician's annotation. (Color figure online)

Intraclass Correlation Coefficient (ICC). Quantitative agreement between the clinician and the tool was further assessed using the intraclass correlation coefficient. The tool demonstrated excellent reliability, with an ICC of 0.95 for LAT and 0.78 for the third stimulus duration (R3) Table 1, indicating good agreement in detecting complex EGM features based on the morphology of the signal.

Summary of Statistical Performance. Table 1 presents the statistical characteristics of the tool's output in comparison to clinician-annotated results. The table includes mean, standard deviation, and ICC values. These results confirm the tool''s accuracy and reliability, establishing its applicability for use in automated atrial substrate assessment in persistent AF patients. In Table 2, the statistical significance of the features was assessed corresponding to two classes, with the Man-Whitney P-value representing high discriminative potential of most of the features in classifying the two different classes (Healthy and abnormal).

Feature Importance Analysis. Thirteen features were extracted, including signal voltage, duration, and deflections for each of the three extra stimuli (S1-S3), along with the sinus rhythm (SR) duration. Violin box plots were used to visually compare feature distributions across normal (Negative) and hidden slow conduction (Positive) EGMs (Fig. 4). Each plot included Mann-Whitney U test results, annotated as stars to indicate statistical significance (**** for p

Table 1. Comparison of EGM features between Reference and Automated algorithm. Negative (*) and Positive (**) class measurements.

Features	Reference (Mean ± SD)	Algorithm (Mean ± SD)	ICC	Features	Reference (Mean ± SD)	Algorithm (Mean ± SD)	ICC
SR Duration	43.40 ± 9.13*	43.74 ± 13.70	0.48	S1 LAT	69.72 ± 26.83	69.64 ± 26.99	0.99
	49.02 ± 9.65**	51.01 ± 25.23			68.80 ± 25.08	69.09 ± 24.80	
S1 Duration	42.19 ± 7.53	42.54 ± 8.26	0.94	S2 LAT	86.03 ± 38.94	85.04 ± 39.52	0.98
	50.39 ± 10.97	49.81 ± 10.50			94.72 ± 35.31	95.08 ± 34.67	
S2 Duration	46.96 ± 10.36	48.46 ± 13.07	0.80	S3 LAT	84.81 ± 38.63	84.59 ± 38.75	0.98
	71.40 ± 19.44	66.72 ± 17.72			92.36 ± 41.52	94.90 ± 43.29	
S3 Duration	47.49 ± 10.07	48.59 ± 12.55	0.77	S1 Delta	−1.21 ± 10.64	−1.20 ± 14.84	0.52
	83.52 ± 24.47	71.80 ± 20.61			1.37 ± 10.92	−1.20 ± 24.74	
S1 n Deflection	2.91 ± 0.99	2.92 ± 1.01	0.99	S2 Delta	3.55 ± 12.36	4.72 ± 16.85	0.58
	3.90 ± 1.60	3.87 ± 1.60			22.39 ± 18.21	15.71 ± 29.22	
S2 n Deflection	3.26 ± 1.11	3.29 ± 1.14	0.96	S3 Delta	4.09 ± 12.62	4.84 ± 17.16	0.58
	5.62 ± 2.22	5.42 ± 2.19			34.50 ± 22.51	20.79 ± 30.63	
S3 n Deflection	3.31 ± 1.16	3.34 ± 1.20	0.95				
	6.20 ± 2.38	5.84 ± 2.26					

* Negative class measurements, ** Positive class measurements

Table 2. Comparison of EGM feature values between Negative (healthy) and Positive (HSC) classes.

Features	SR_Duration	S1_Duration	S2_Duration	S3_Duration	S1_n Deflection	S2_n Deflection	S3_n Deflection
NEG * (Mean ± SD)	43.4 ± 9.1	42.1 ± 7.5	46.9 ± 10.3	47.4 ± 10.0	2.91 ± 0.99	3.2 ± 1.1	3.3 ± 1.1
POS ** (Mean ± SD)	49.02 ± 9.65	50.3 ± 10.9	71.4 ± 19.4	83.5 ± 24.4	3.90 ± 1.60	5.6 ± 2.2	6.2 ± 2.3
Mann–Whitney p-value ***	< 0.0001	< 0.0001	< 0.0001	< 0.0001	< 0.0001	< 0.0001	< 0.0001

Features	S1_Delta	S2_Delta	S3_Delta	S1_Voltage	S2_Voltage	S3_Voltage	
NEG * (Mean ± SD)	-1.2 ± 10.6	3.55 ± 12.36	4.09 ± 12.62	1.79 ± 1.68	1.11 ± 1.08	1.11 ± 1.08	
POS ** (Mean ± SD)	1.3 ± 10.9	22.3 ± 18.2	34.5 ± 22.5	1.59 ± 1.34	0.78 ± 0.62	0.74 ± 0.59	
Mann–Whitney p-value ***	0.099419	< 0.0001	< 0.0001	0.529059	0.031244	0.001788	

* The number of negative (healthy) points: 983, ** The number of positive (HSC) points: 127, *** The number of points used in the Mann-Whitney statistical test: 127 in both classes.

< 0.0001). The most significant features showed strong separation between the classes, indicating high discriminative value (Table 2). Violin box plots provided both visual and statistical insights into which features were most relevant to classification, supporting their use in clinical analysis and future predictive modeling.

3.2   Machine Learning Model Performance

To evaluate the predictive utility of the extracted electrogram (EGM) features, we trained a binary machine learning classifier to distinguish abnormal (positive) from normal (negative) electrogram sites identified during functional substrate mapping. A standardized pipeline was employed using StandardScaler, a Support Vector Classifier (SVC) with a linear kernel, and stratified 10-fold cross-validation to ensure robustness. To address class imbalance inherent in the clinical data, the Synthetic Minority Oversampling Technique (SMOTE) was applied during training after the train-test split. The dataset was split 70-30 into training (776 EGM samples) and test sets (334 EGM samples). Each EGM sample consists of 2.5 s of the electrical activity of that region on Atrium, and then 13 features of each sample have been extracted. The SVC model achieved an overall accuracy of 92.6% (Table 3) and an AUC of 0.98 (Fig. 6), demonstrating strong discriminative capability in two classes.

More importantly, it achieved the highest balanced accuracy (93.6%) among the nine classifiers tested, indicating superior performance in correctly identifying both abnormal and normal classes, even under skewed data distributions. This robustness is critical in clinical applications, where detecting abnormal substrates with high confidence is essential to guiding targeted ablation. Across all classifiers, Random Forest and Gradient Boosting achieved the highest raw accuracies (both >95%), reflecting their strong general classification capacity. However, as illustrated in Table 3, the SVC model exhibited the highest recall (0.950) for the abnormal class, indicating its superior ability to capture subtle but clinically significant electrophysiological patterns associated with hidden slow conduction zones. This metric is of particular importance in the clinical setting, where false negatives (i.e., undetected abnormal regions) may directly impact the effectiveness of ablation therapy.

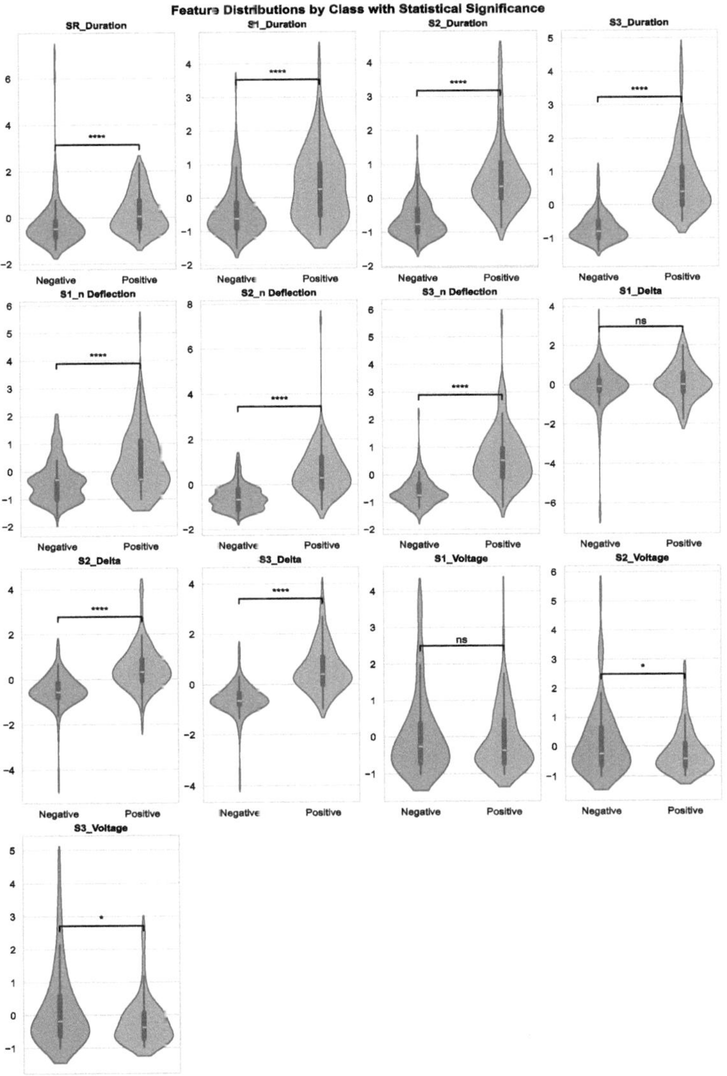

Fig. 4. Violin box plots showing the distribution of 13 extracted electrogram (EGM) features across two classes-Negative (healthy) and Positive (Hidden Slow Conduction, HSC)-with statistical comparisons based on the Mann-Whitney U test. Each plot represents a normalized feature, including duration, deflection, voltage, and delta (stimulus-induced duration increase) derived from sinus rhythm and three extra stimuli (S1-S3). Asterisks denote statistical significance: **** (p < 0.0001), *** (p < 0.001), ** (p < 0.01), * (p < 0.05), and ns (not significant). The visualization reveals that features such as third stimulus duration, deflection, and sinus duration show the most significant separation between classes, indicating their potential importance in distinguishing abnormal substrate properties

Table 3. Performance metrics of nine machine learning classifiers. Metrics include Accuracy, Balanced Accuracy, F1-score (weighted and per-class), Recall, and Precision for both positive (POS) and negative (NEG) classes. Macro-averaged precision is also reported. While Random Forest and Gradient Boosting achieved the highest overall accuracy and F1 scores, the Support Vector Classifier (SVC) demonstrated the strongest balanced performance with high positive and negative recall. Notably, Naive Bayes and KNN performed slightly lower across most metrics, particularly in negative class precision.

Models	Accuracy	Balanced Accuracy	F1 Weighted	F1 Score	Recall NEG	Recall POS	Precision NEG	Precision POS	Macro Prec. NEG	Macro Prec. POS
Random Forest	0.958	0.909	0.958	0.899	0.973	0.845	0.980	0.799	0.862	0.969
Gradient Boosting	0.956	0.916	0.957	0.897	0.968	0.863	0.982	0.777	0.876	0.964
AdaBoost	0.950	0.915	0.952	0.886	0.961	0.868	0.983	0.742	0.880	0.957
Decision Tree	0.938	0.860	0.939	0.851	0.962	0.758	0.969	0.718	0.799	0.952
SVC Linear Kernel	0.926	0.936	0.932	0.851	0.923	0.950	0.993	0.613	0.949	0.925
Logistic Regression	0.926	0.926	0.932	0.849	0.926	0.926	0.990	0.618	0.926	0.926
LDA	0.927	0.900	0.932	0.844	0.935	0.866	0.982	0.630	0.874	0.930
KNN	0.901	0.897	0.911	0.807	0.902	0.892	0.985	0.540	0.893	0.901
Naive Bayes	0.894	0.890	0.905	0.796	0.895	0.884	0.984	0.521	0.885	0.894

The high recall and balanced accuracy of the SVC model reinforce its potential as a reliable tool for operator-independent electrogram classification, sup-

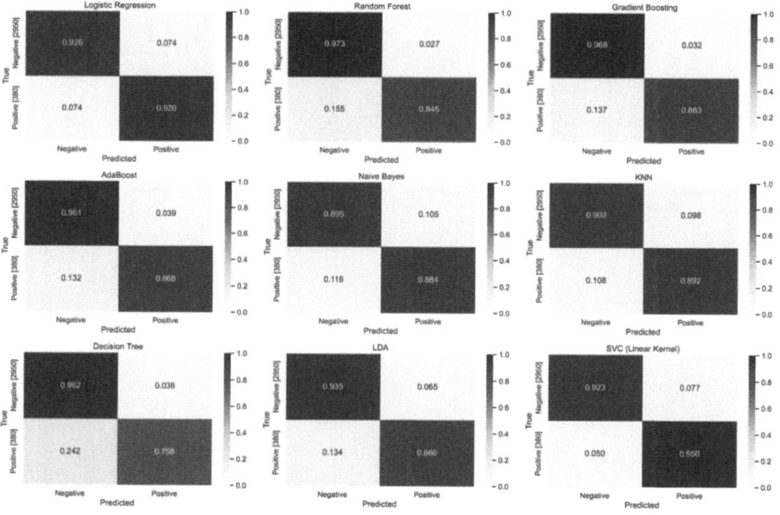

Fig. 5. Normalized confusion matrices of nine machine learning classifiers evaluated using 10-fold cross-validation. Each matrix shows the proportion of predicted samples divided by the total number of true samples in that class (i.e., row-wise normalization). Classifiers include Logistic Regression, Random Forest, Gradient Boosting, AdaBoost, Naive Bayes, K-Nearest Neighbors (KNN), Decision Tree, Linear Discriminant Analysis (LDA), and Support Vector Machine (SVC with linear kernel). Higher diagonal values indicate better classification performance, with especially strong positive class recall observed in the SVC and Naive Bayes models.

porting more consistent and reproducible identification of pathological substrates in patients with persistent atrial fibrillation. These results establish the foundation for future real-time integration of the model into functional mapping workflows (Fig. 5).

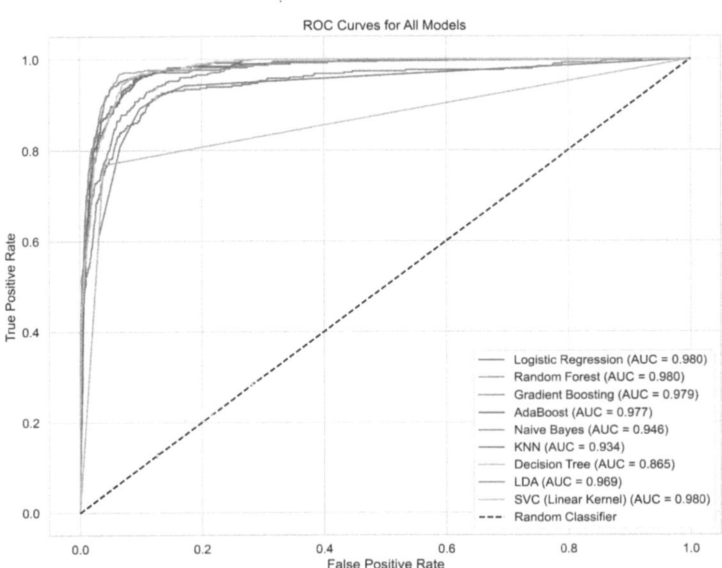

Fig. 6. Receiver Operating Characteristic (ROC) curves of nine machine learning classifiers. The curves represent the trade-off between true positive rate and false positive rate across different classification thresholds. The area under the curve (AUC) is reported in the legend for each model, indicating overall classification performance. AUC values close to 1 suggest excellent discriminative ability. The dashed diagonal line represents the performance of a random classifier.

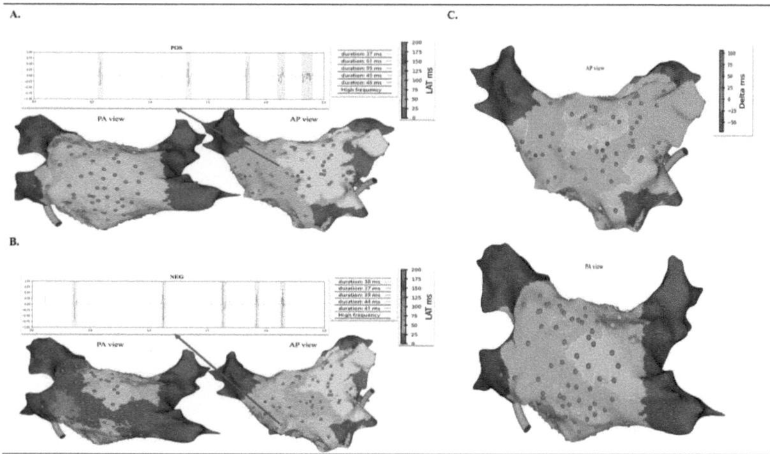

Fig. 7. A. EGM example of a positive response to functional mapping. LAT in the 1st paced stimulus. B. EGM example of a negative response to functional mapping. LAT in the 3rd paced stimulus. C. Delta map: 3rd stimulus EGM [ms] - SR EGM [ms]

4   Discussion and Conclusion

The presented tool provides a robust, objective, and scalable framework for analyzing electrogram signals in patients with persistent atrial fibrillation. By combining signal processing with interactive visualization, it facilitates both automated feature extraction and manual expert validation. Importantly, the GUI facilitates analysis of larger datasets, enabling creation of annotated datasets up to 50 PsAF patients. This curated dataset can be further leveraged to train deep learning architectures-such as U-Net-to automatically segment activation periods. Additionally, the tool enables the construction of 3D local activation time (LAT) maps from the output features. As demonstrated in Fig. 7. These LAT maps reveal propagation patterns originating from the coronary sinus, where electrical stimulation occurs during functional mapping and spreads to other atrial regions. Such spatial and temporal information is critical for understanding atrial substrate behavior and guiding ablation therapy. The integration of anatomical geometry opens new possibilities for future work, including the calculation of 3D conduction velocity and the identification of reentrant circuits. These capabilities point to the tool's potential not only in diagnosis but also in personalized treatment planning and outcome prediction in atrial fibrillation patients.

This study has limitations, including the small sample size, which potentially limits generalizability due to inter-patient substrate variability. Additionally, the lack of external validation necessitates further studies in larger and multicenter cohorts to confirm clinical applicability. Recent advances, such as the Tailored AF randomized controlled trial have shown that AI-guided ablation-specifically targeting spatio-temporal dispersion zones in addition to pulmonary vein isolation-significantly improves freedom from atrial fibrillation in persistent AF patients. These results highlight the promise of AI in identifying clinically relevant ablation targets. However, to date, most AI applications in this field have focused on electrogram features during AF or post-PVI substrate mapping. There remains a notable lack of AI-centered efforts leveraging functional mapping data-such as EGMs elicited by short-coupled extra stimuli-to characterize the arrhythmogenic substrate. Our work helps fill this gap by employing automated feature extraction based on functional pacing protocols, offering a complementary approach that may enhance the precision of substrate-guided ablation.

In conclusion, this tool has demonstrated capability towards functional substrate assessment for further facilitating guided ablation therapy by bridging computational rigor with clinical usability and stands as a valuable resource for enhancing PsAF treatment success rate.

References

1. Xu, J., Luc, J.G.Y., Phan, K.: Atrial fibrillation: Review of current treatment strategies. J. Thoracic Dis. **8**(9), E886–E900 (2016). https://doi.org/10.21037/jtd.2016.09.13 https://doi.org/10.21037/jtd.2016.09.13
2. Van Gelder, I.C., Rienstra, M., Bunting, K.V., et al.: 2024 ESC guidelines for the management of atrial fibrillation developed in collaboration with the European association for cardio-thoracic surgery (EACTS). Eur. Heart J. **45**(36), 3314–3414 (2024). https://doi.org/10.1093/eurheartj/ehae176. Erratum in: 07 July 2025: ehaf306
3. Haïssaguerre, M., Jaïs, P., Shah, D.C., et al.: Spontaneous initiation of atrial fibrillation by ectopic beats originating in the pulmonary veins. N. Engl. J. Med. **339**(10), 659–666 (1998). https://doi.org/10.1056/NEJM199809033391003
4. Garcia, S., et al.: Functional mapping to reveal slow conduction and substrate progression in atrial fibrillation. Europace **25**(11) (2023). https://doi.org/10.1093/europace/euad246
5. Clarnette, J.A., et al.: Outcomes of persistent and long-standing persistent atrial fibrillation ablation: a systematic review and meta-analysis. Europace **20**(F13), f366–f376 (2018). https://doi.org/10.1093/europace/eux297
6. Tao, R., Deng, B., Wang, Y.: Research progress of the fractional Fourier transform in signal processing. Sci. China Ser. F: Inf. Sci. **49**(1), 1–25 (2006). https://doi.org/10.1007/s11432-005-0240-y
7. Auger, F., Flandrin, P.: Improving the readability of time-frequency and time-scale representations by the reassignment method. IEEE Trans. Signal Process. (1995)

A Quantum Machine Learning Approach to Cardiopulmonary Sound Classification

Sandra Ranilla-Cortina[1], Antonio J. Muñoz-Montoro[2],
Elías F. Combarro[1], Sebastián García-Galán[2], and José Ranilla[1]([✉])

[1] Universidad de Oviedo, Oviedo, Spain
{ranillasandra,efernandezca,ranilla}@uniovi.es
[2] Universidad de Jaén, Jaén, Spain
{jmontoro,sgalan}@ujaen.es

Abstract. Cardiopulmonary sound analysis is essential for the early detection and diagnosis of respiratory and cardiovascular diseases. This work explores the application of Quantum Machine Learning (QML) models, specifically Quantum Support Vector Machines (QSVM) and Quantum Neural Networks (QNN), for the classification of both lung and heart sounds. Leveraging MFCC-based features and dimensionality reduction techniques, we evaluate the performance of these models on two publicly available benchmark datasets. The experimental results indicate that QML models match or surpass their classical counterparts, particularly under constraints of limited training data and reduced feature sets. These findings underscore the potential of QML as a promising tool for efficient, accurate and unified analysis of cardiopulmonary acoustic signals in next-generation diagnostic systems.

Keywords: Quantum Machine Learning · Heart and Lung Sound Classification · MFCC · Biomedical Signal Analysis

1 Introduction

Cardiovascular and respiratory diseases remain among the leading causes of mortality worldwide, accounting for millions of deaths each year and placing an immense strain on global healthcare systems [5,24]. Rapid detection and accurate diagnosis are essential for effective treatment and prevention of life-threatening complications. In this context, cardiopulmonary sound analysis, which encompasses heart and lung auscultation, has long served as a non-invasive and cost-effective diagnostic tool, offering valuable insight into cardiopulmonary function [32].

Traditionally, clinicians have relied on manual auscultation using stethoscopes to assess heart and lung sounds. However, this method is inherently subjective and highly dependent on the clinician's expertise, resulting in significant variability in interpretation. Subtle anomalies may go undetected, and the consistent classification of pathological sounds remains challenging when relying solely on auditory cues. To overcome these limitations and enhance diagnostic accuracy, researchers

A. López Fernández et al. (Eds.): CIABiomed 2025, LNBI 16148, pp. 228–238, 2026.
https://doi.org/10.1007/978-3-032-10661-2_18

have increasingly explored digital signal processing (DSP) and machine learning (ML) techniques (see, for example, [13,26], among others).

Among the most widely adopted feature extraction methods in this domain are Mel-frequency cepstral coefficients (MFCC). Originally developed for speech recognition, MFCC effectively capture the spectral characteristics and energy distribution of audio signals, making them well-suited for analyzing cardiopulmonary acoustics [25]. In addition to MFCC, other signal representations, such as the short-time Fourier transform (STFT) [15], cochleogram [20] and continuous wavelet transform (CWT) [28], have also demonstrated considerable potential for biomedical applications.

Building on these signal representations, numerous ML algorithms, such as support vector machines (SVM) [3], neural networks (NN) [22] and random forests (RF) [12], have achieved promising results in the classification of respiratory sounds. These automated non-invasive approaches offer the potential for earlier disease detection and improved patient outcomes, while reducing diagnostic variability and healthcare costs.

In parallel, Quantum Machine Learning (QML), at the intersection of quantum computing and artificial intelligence, is an emerging and highly promising field [2,30], largely due to its potential to enhance the performance of classical machine learning systems. QML models, such as Quantum Support Vector Machines (QSVM) [11] and Quantum Neural Networks (QNN) [1], leverage quantum principles, such as superposition and entanglement, to operate in high-dimensional Hilbert spaces. These models potentially improve learning performance, particularly in low-data or high-noise environments.

In this work, we investigate the applicability of QML models for the classification of cardiopulmonary sounds. Our main goal is to evaluate the classification accuracy and generalization capabilities of QML models in comparison to traditional machine learning techniques. Leveraging benchmark datasets for both heart and lung sounds, we explore the potential of QML techniques to enhance diagnostic precision and contribute to the development of intelligent quantum-assisted systems for multi-organ acoustic analysis.

2   Quantum Machine Learning

Quantum Computing (QC) [23] is currently one of the most promising and prolific fields of research. Within this domain, QML has gained significant attention due to the increasing popularity of artificial intelligence (AI) techniques, which are being applied to a wide range of real-world problems [6,31,34].

Using the fundamental principles of quantum mechanics, such as superposition, entanglement and quantum interference, QML seeks to develop novel computational models capable of outperforming classical approaches in terms of efficiency, scalability and performance [2,30]. This synergy offers the potential to unlock new frontiers in data analysis, particularly in complex and high-dimensional domains.

The field of QML is still in its early stages, with new strategies emerging constantly, but there exist two types of techniques that are predominant. QSVM [11] and QNN [1] are the most studied QML models today.

QSVM are the quantum analogues of classical SVM. They exploit the high dimensional structure of Hilbert space to compute kernel functions through quantum feature maps. A quantum feature map is a non-linear transformation of classical data into quantum states, implemented via a parameterized quantum circuit, commonly referred to as an embedding circuit.

This circuit typically consists of a parameterized unitary transformation $f(x_i)$ and its inverse, where the kernel inputs x_i serve as parameters for the quantum gates, as illustrated in Fig. 1. The feature map enables kernel computation via inner products between quantum states in Hilbert space. The design of the feature map is a critical factor in QSVM performance, as it determines how input data is projected into the quantum space. By carefully selecting the feature map and tuning its hyperparameters, QSVM can offer significant advantages for tasks such as heart and lung sound classification.

Several types of quantum feature maps have been proposed in the literature, including angle embedding [19], amplitude embedding [19], ZZ feature map [2], YZ_CX_Encoding [10], HighDimEncoding [27], ChebyshevRx [17], and hardware-efficient embeddings [33].

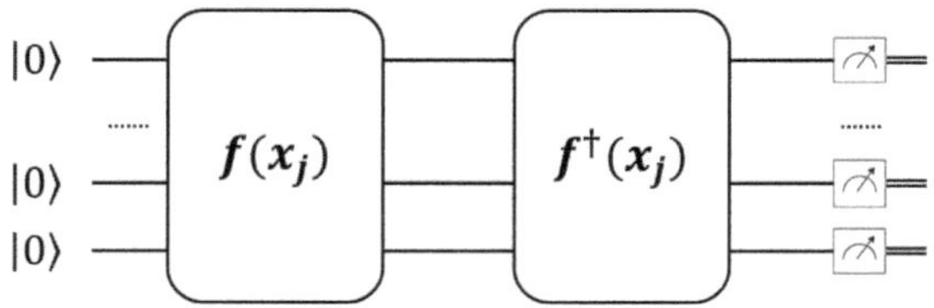

Fig. 1. Kernel Circuit Diagram.

As previously discussed, in addition to QSVM, the other prominent approach in QML is the QNN. A QNN can be seen as the quantum analogue of a classical neural network, where neurons and activation functions are mimicked through a layered structure of parameterized quantum gates.

This type of QML model typically consists of three main components (see Fig. 2). First, classical data from the dataset is transformed into quantum states through a quantum feature map, implemented via an embedding circuit, as previously discussed. Second, an ansatz circuit, composed of entangling layers and parameterized rotation gates, is applied to the encoded quantum state. This ansatz is optimized during training to minimize a cost function. As with feature maps, various ansatz designs have been proposed in the literature. Among the most commonly used are the Two-Local [2], Tree-Tensor [2], hardware-efficient [18], and HCzRx [9].

Finally, in order to return to the classical domain and update the model parameters, the quantum states produced by the circuit must be measured to extract classical information. This is typically done by extracting expectation values from one or more qubits, depending on the nature of the task. For binary

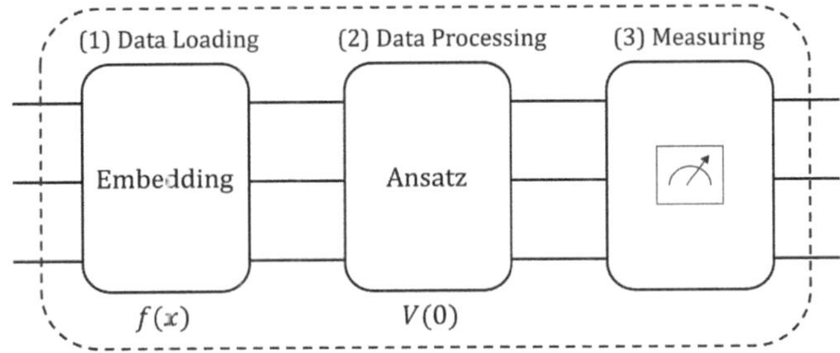

Fig. 2. QNN architecture diagram.

classification, a single measured qubit is often sufficient, whereas for multi-class problems, the number of measured qubits may correspond to the number of output classes.

While building larger QNN with more qubits and increased circuit complexity might seem like a straightforward way to improve performance, this is not always the case. Scaling up the number of qubits can lead to the emergence of barren plateaus [21], a phenomenon inherent to certain QML models. This issue can severely hinder performance during training, as the gradient of the loss function may vanish across most directions, making it difficult to identify a useful optimization path.

One technique that has been shown to mitigate this effect is ensemble learning [7,14], where multiple smaller QNN are trained independently and their predictions are combined to produce more robust and accurate results.

3 Methodology

This work evaluates the performance of both QNN and QSVM in the classification of cardiopulmonary sounds, benchmarking them against traditional machine learning models under consistent preprocessing and feature extraction conditions.

3.1 Dataset

Two publicly available and widely used datasets for classifying respiratory and cardiac sounds are considered.

First, the ICBHI 2017 Challenge dataset [29] is employed, a widely referenced benchmark for respiratory sound analysis. This dataset comprises 920 audio recordings collected from 126 subjects, with durations ranging from 10 to 90 s. A total of 6,898 respiratory cycles are annotated by clinical experts as containing crackles, wheezes, both or normal breathing (absence of adventitious sounds). Recordings were acquired using four different devices, across seven chest locations, under two acquisition modes and varying sampling frequencies. Notably, some recordings contain high noise levels, thereby reflecting realistic clinical conditions and enhancing the generalizability of classification models.

The dataset presents significant variability in recording conditions and data distribution: the number of respiratory cycles per recording varies considerably, class labels are imbalanced and breathing cycle durations range from 0.2 to 16.2 s, with a mean of 2.7 s. For standardization purposes, a subset of 1,959 respiratory cycles was selected, limited to those recorded using the AKG C417L microphone in multichannel mode at right-sided chest locations. Additionally, the duration of each cycle was normalized to 6 s through zero-padding, in order to meet the fixed-length input requirement of neural network architectures. All audio signals were downsampled to 4 kHz, under the assumption that relevant respiratory events occur below 2 kHz.

Second, a curated version of the Yaseen Heart Sound Database (YHSD) [16] is also used. This dataset contains high-quality recordings of heart sounds and murmurs associated with various cardiac conditions. It comprises 1,000 audio files evenly distributed across five categories: one normal class and four pathological classes, namely aortic stenosis, mitral stenosis, mitral regurgitation and mitral valve prolapse. Each class includes 200 recordings, all acquired using electronic stethoscopes, ensuring high-fidelity recordings suitable for clinical diagnosis. As with the respiratory dataset, all signals were downsampled to 4 kHz in order to focus on the frequency range most relevant for cardiac events.

To facilitate a unified evaluation across both datasets, the classification task is formulated as a binary problem distinguishing between healthy and unhealthy subjects. This reformulation mitigates the effects of class imbalance in the respiratory dataset and aligns with the overarching goal of enabling early disease detection in clinical screening contexts.

These two datasets provide a complementary foundation for evaluating the performance of classical and quantum machine learning models in the context of cardiopulmonary acoustic classification. Together, they offer distinct yet synergistic perspectives on the classification of cardiopulmonary sounds across respiratory and cardiac domains.

3.2 Signal Representation

In the biomedical domain, several acoustic signal representations have been explored to support the classification of cardiopulmonary sounds. Among the most commonly used techniques are the STFT, MFCC and cochleograms [4,20]. In this work, we adopt MFCC as the primary signal representation due to their superior performance in prior studies involving both respiratory and cardiac sound classification [8,35].

For the MFCC representation, following an empirical parameter search aimed at identifying the configuration that yields the highest classification accuracy on the selected datasets, we chose to extract 18 coefficients per frame with a 10% overlap between analysis windows. It is important to note that the dimensionality of the feature vectors, used as inputs for both classical and quantum machine learning models, depends directly on these parameter settings, which therefore play a crucial role in shaping the learning process.

3.3   Machine Learning Methods

This section focuses on the configuration and setup of the machine learning models used for the binary classification of cardiopulmonary sounds, encompassing both classical and quantum paradigms. For a fairer comparison, the classical models considered are those from which their quantum counterparts are derived, namely, SVM and NN. All models, both classical and quantum, are configured and evaluated under consistent preprocessing and feature extraction conditions, using the signal representations obtained through the methodology described earlier.

Classical models were evaluated using LazyPredict[1], a Python library designed to streamline the benchmarking of multiple machine learning algorithms by automating model selection and ranking with minimal code. For quantum models, the LazyQML framework [9][2] was employed. Built on top of PennyLane[3], LazyQML provides a structured testbed for training quantum machine learning models on classical hardware, enhancing model selection efficiency and enabling faster experimentation. All embedding circuits and ansätze presented in the previous sections are supported within this framework.

To ensure fair and meaningful comparisons, all experiments were conducted under a unified configuration. Global hyperparameters (such as batch size, number of training epochs, number of folds and data splits) were kept consistent across all models, unless specific architectural or training requirements dictated otherwise. This uniformity ensures that observed performance differences can be attributed to the models themselves rather than to inconsistencies in the experimental setup.

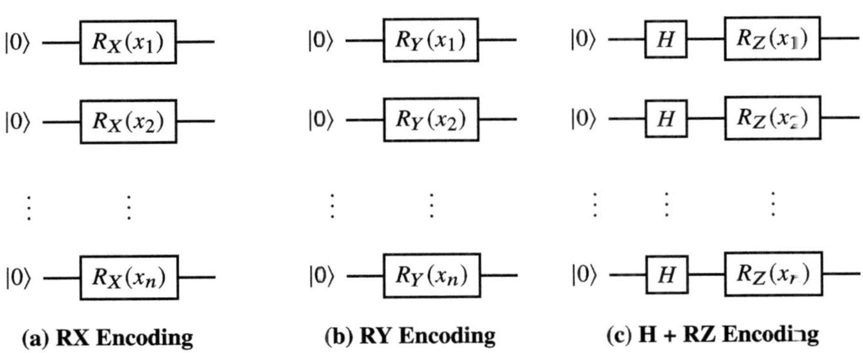

Fig. 3. Quantum circuits for angle embedding.

For classical models, both the full feature set and a reduced subset of 16 features, selected via Principal Component Analysis (PCA), were evaluated. In

---

[1] https://pypi.org/project/lazypredict/.
[2] https://pypi.org/project/lazyqml/.
[3] https://pypi.org/project/PennyLane/.

contrast, quantum models were limited to the 16-feature subset due to hardware constraints. Specifically, LazyQML relies on PennyLane backends based on state vector simulation, which are highly demanding in terms of memory usage. Simulating n qubits requires storing 2^n complex amplitudes in memory, resulting in exponential growth in memory consumption. Consequently, 16 qubits represented the practical upper limit for most embedding circuits in this study.

In addition to the unified configuration applied across all experiments, model-specific calibration was performed to determine the optimal hyperparameters and architectural settings unique to each approach.

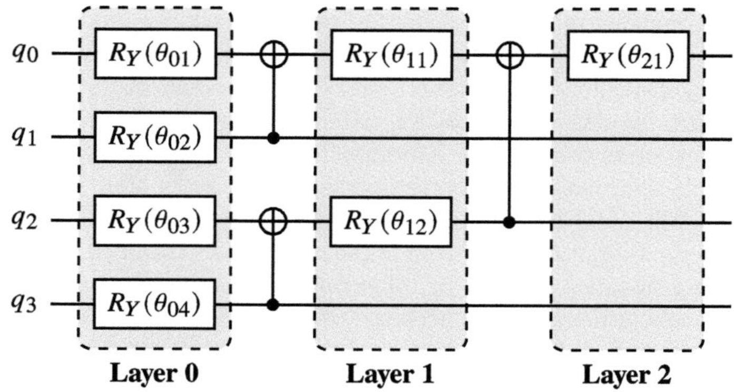

Fig. 4. Tree-tensor variational form in 4 qubits.

For SVM, various kernel functions were explored, including radial basis function (RBF), sigmoid, linear, and polynomial kernels. A grid search over hyperparameters was conducted, with $\log_{10}$ C $\in \{-3, \ldots, 3\}$ and $\gamma \in \{0.01, 0.1, 1\}$. Regarding NN, architectures based on dense layers with *ReLU* activation functions were used, with a final output layer employing a *sigmoid* activation for binary classification. The loss function was set to binary cross-entropy, the *Adam* optimizer was selected, and early stopping was used to prevent overfitting.

Table 1. Optimal configurations for classical and QML models

Method	Optimal Hyperparameters
SVM (all features)	$C = 10$, $\gamma = 0.01$, kernel: RBF
SVM (PCA)	$C = 100$, $\gamma = 1.0$, kernel: RBF
NN (all features)	lr $= 0.001$, Batch $= 20$, Epochs $= 150$, Patience $= 8$ epochs, Units $= 200$
NN (PCA)	lr $= 0.001$, Batch $= 20$, Epochs $= 150$, Patience $= 8$ epochs
QSVM	Amplitude encoding
QNN	Angle embedding, tree tensor

With respect to QML models, all embedding circuits and ansätze supported by the LazyQML framework (as described in previous sections) were evaluated. Among these, amplitude encoding (see Fig. 3) consistently yielded the best performance for QSVM, while the combination of angle embedding (see Fig. 3) and the Tree-Tensor ansatz (see Fig. 4) proved to be most effective for QNN. Amplitude encoding represents classical data by preparing a quantum state whose amplitudes correspond to the normalized feature vector, i.e., $|\psi\rangle = \sum_{i=0}^{2^n-1} x_i |i\rangle$, where the x_i are input features satisfying $\sum_i |x_i|^2 = 1$. This encoding enables the representation of 2^n features using only n qubits, providing exponential compression of input dimensionality.

Table 1 summarizes the optimal hyperparameters. The QML models adopted the same global training settings-such as batch size, number of epochs, and early stopping criteria-as those used for the classical models. In the QML setting, which involved a binary classification task, a single qubit was measured at the end of each forward pass, and its expectation value was used to assign each sample to class 0 or 1.

4   Experimental Results

The accuracy results for both datasets are presented in Table 2. For the ICBHI dataset, which focuses on respiratory sound analysis, quantum models exhibit a slight improvement in performance compared to their classical counterparts. In particular, the QSVM achieves a best accuracy of 69.1%, marginally outperforming the classical SVM, which reaches 68.6%. While this difference may not be statistically significant, it is important to highlight that the quantum model attains this accuracy using a significantly reduced set of features in contrast of the classical SVM which relies on the full feature set. Note that the performance of the classical model drops to 67.9% when trained on the reduced set.

Table 2. Results obtained for the tested datasets.

Method	ICBHI accuracy	Yaseen accuracy
SVM (all features)	68.6%	94.8%
SVM (PCA)	67.9%	90.0%
QSVM	**69.1%**	93.6%
NN (all features)	65.6%	**98.5%**
NN (PCA)	60.7%	98.3%
QNN	66.1%	94.1%

This contrast becomes more pronounced when comparing classical and quantum neural networks. The best-performing classical neural network achieves an accuracy of 65.6% using the full feature set, but only 60.7% with the reduced set. Meanwhile, the QNN surpasses both, attaining 66.1% accuracy using only the reduced feature representation.

Regarding the Yaseen dataset, results follow a similar trend. The QSVM again outperforms the classical SVM when both are trained on the reduced feature set, achieving performance close to that of the classical SVM trained on the full set. However, in this case, QSVM do not outperform classical neural networks. The QNN achieves an accuracy of 94.1%, which, although competitive, remains slightly below that of classical models, which exceed 98% of accuracy. This performance gap may potentially be addressed by increasing the number of trainable parameters and adopting data re-uploading strategies, albeit at the expense of greater quantum circuit complexity and higher hardware demands. Moreover, since quantum models could only be evaluated on the reduced 16 feature set due to simulator constraints, a potential bias in the comparison cannot be ruled out.

5   Conclusions

Heart and lung sound analysis plays a crucial role in the early detection and diagnosis of cardiorespiratory diseases. In this study, we investigated the applicability of QML models, specifically QNN and QSVM, for the classification of cardiopulmonary sounds.

Our experimental results highlight the promising capabilities of QML techniques in this domain. In particular, the QSVM consistently outperformed its classical counterpart when operating with a reduced set of features, achieving comparable or better classification accuracy while requiring fewer input variables. This indicates that QML models may be particularly advantageous in data-constrained environments or scenarios with limited feature dimensionality. Similarly, QNN achieved competitive performance, even outperforming classical neural networks in the respiratory sound classification task under reduced feature conditions.

These findings support the viability of QML as a complementary or alternative approach to traditional machine learning in biomedical acoustic analysis. Nonetheless, further research is required to fully assess the potential of QML models. Future work should focus on the use of larger and more diverse datasets, the exploration of more complex quantum circuit architectures, the integration of advanced techniques such as data re-uploading and hybrid quantum-classical training strategies, and evaluations on real NISQ hardware to assess their practical feasibility. Such advancements could further enhance model accuracy and robustness, contributing to the development of next-generation intelligent diagnostic tools.

Acknowledgements. This work has been partially supported by grant PID2023-146520OB funded by MICIU/AEI/10.13039/501100011033, by grant IDE/2024/000734 funded by Principado de Asturias, and by QUANTUM ENIA project call - Quantum Spain project funded by the Ministry for Digital Transformation and of Civil Service of the Spanish Government and the European Union through the Recovery, Transformation and Resilience Plan - NextGenerationEU within the framework of the Digital Spain 2026 Agenda. We would also like to express our gratitude to "Fundación Centro Tecnológico de la Información y la Comunicación" of Asturias for providing the necessary infrastructure to carry out this work.

References

1. Abbas, A., Sutter, D., Zoufal, C., Lucchi, A., Figalli, A., Woerner, S.: The power of quantum neural networks. Nat. Comput. Sci. **1**(6), 403–409 (2021)
2. Combarro, E.F., González-Castillo, A.D.M.S., Di Meglio, A.: A Practical Guide to Quantum Machine Learning and Quantum Optimization. Packt Publishing (2023)
3. Cortes, C., Vapnik, V.: Support-vector networks machine learning, vol. 20 (1995)
4. Das, S., Pal, S., Mitra, M.: Acoustic feature based unsupervised approach of heart sound event detection. Comput. Biol. Med. **126**, 103990 (2020)
5. Federation, W.H.: World heart report 2023: confronting the world's number one killer. Technical report, World Heart Federation (2023)
6. Frehner, R., Stockinger, K.: Applying quantum autoencoders for time series anomaly detection. Quantum Mach. Intell. **7**(1), 59 (2025). https://doi.org/10.1007/s42484-025-00285-1
7. Friedrich, L., Maziero, J.: Quantum neural network with ensemble learning to mitigate barren plateaus and cost function concentration (2025). https://arxiv.org/abs/2402.06026
8. García-Vega, D., et al.: Exploring hybrid quantum-classical machine learning for respiratory sound analysis. In: 2024 IEEE 22nd Mediterranean Electrotechnical Conference (MELECON), pp. 503–507. IEEE (2024)
9. García-Vega, D., Plou Llorente, F., Leal Castaño, A., Combarro, E., Ranilla, J.: LazyQML: a python library to benchmark quantum machine learning models. In: 30th European Conference on Parallel and Distributed Processing (2024)
10. Haug, T., Self, C.N., Kim, M.S.: Quantum machine learning of large datasets using randomized measurements. Mach. Learn.: Sci. Technol. **4**(1), 015005 (2023). https://doi.org/10.1088/2632-2153/acb0b4
11. Havlíček, V., et al.: Supervised learning with quantum-enhanced feature spaces. Nature **567**(7747), 209–212 (2019)
12. Ho, T.K.: Random decision forests. In: Proceedings of 3rd International Conference on Document Analysis and Recognition, vol. 1, pp. 278–282. IEEE (1995)
13. Huang, D., Huang, J., Qiao, K., Zhong, N., Lu, H., Wang, W.: Deep learning-based lung sound analysis for intelligent stethoscope. Milit. Med. Res. **10** (2023). https://doi.org/10.1186/s40779-023-00479-3
14. Incudini, M., et al.: Resource saving via ensemble techniques for quantum neural networks. Quantum Mach. Intell. **5**(2), 39 (2023)
15. Jung, S.Y., Liao, C.H., Wu, Y.S., Yuan, S.M., Sun, C.T.: Efficiently classifying lung sounds through depthwise separable CNN models with fused STFT and MFCC features. Diagnostics **11**(4), 732 (2021). https://doi.org/10.3390/diagnostics11040732
16. Khan, Y.: Yaseen heart sound database portal (2021). https://github.com/yaseen21khan/Classification-of-Heart-Sound-Signal-Using-Multiple-Features-. Accessed 25 July 2025

17. Kreplin, D.A., Roth, M.: Reduction of finite sampling noise in quantum neural networks. Quantum **8**, 1385 (2024). https://doi.org/10.22331/q-2024-06-25-1385 https://doi.org/10.22331/q-2024-06-25-1385
18. Leone, L., Oliviero, S.F., Cincio, L., Cerezo, M.: On the practical usefulness of the hardware efficient ansatz. Quantum **8**, 1395 (2024)
19. Lloyd, S., Schuld, M., Ijaz, A., Izaac, J., Killoran, N.: Quantum embeddings for machine learning. arXiv preprint arXiv:2001.03622 (2020)
20. Mang, L.D., Cañadas-Quesada, F.J., Carabias-Orti, J.J., Combarro, E.F., Ranilla, J.: Cochleogram-based adventitious sounds classification using convolutional neural networks. Biomed. Signal Process. Control **82**, 104555 (2023)
21. McClean, J.R., Boixo, S., Smelyanskiy, V.N., Babbush, R., Neven, H.: Barren plateaus in quantum neural network training landscapes. Nat. Commun. **9**(1), 4812 (2018)
22. McCulloch, W.S., Pitts, W.: A logical calculus of the ideas immanent in nervous activity. Bull. Math. Biophys. **5**(4), 115–133 (1943)
23. Nielsen, M.A., Chuang, I.L.: Quantum Computation and Quantum Information: 10th Anniversry Edition, Anniversary Cambridge University Press, Cambridge (2011)
24. World Health Organization: Cardiovascular diseases (CVDs) fact sheet. WHO Newsroom (2021)
25. Park, J.S., Kim, K., Kim, J.H., Choi, Y.J., Kim, K., Suh, D.I.: A machine learning approach to the development and prospective evaluation of a pediatric lung sound classification model. Sci. Rep. **13**(1), 1289 (2023)
26. Partovi, E., Babic, A., Gharehbaghi, A.: A review on deep learning methods for heart sound signal analysis. Front. Artif. Intell. **7**, 1434022 (2024). https://doi.org/10.3389/frai.2024.1434022
27. Peters, E., et al.: Machine learning of high dimensional data on a noisy quantum processor. NPJ Quantum Inf. **7**(1), 161 (2021). https://doi.org/10.1038/s41534-021-00498-9
28. Rizal, A., Adz-Dzikri, A.A., Fauzi, M.A.G.: Classification of normal and abnormal heart sound using continuous wavelet transform and ResNet-50. In: Technology Reports of Kansai University, vol. 62, pp. 2595–2602 (2020)
29. Rocha, B.M., et al.: An open access database for the evaluation of respiratory sound classification algorithms. Physiol. Meas. **40**(3), 035001 (2019)
30. Schuld, M., Petruccione, F.: Machine Learning with Quantum Computers. Quantum Science and Technology, Springer, Cham (2021)
31. Slabbert, D., Petruccione, F.: Hybrid quantum-classical feature extraction approach for image classification using autoencoders and quantum SVMs (2024). https://arxiv.org/abs/2410.18814
32. Sun, W., Zhang, Y., Chen, F.: Research on heart and lung sound separation method based on DAE–NMF–VMD. EURASIP J. Adv. Signal Process. **2024** (2024). https://doi.org/10.1186/s13634-024-01152-0
33. Thanasilp, S., Wang, S., Cerezo, M., Holmes, Z.: Exponential concentration in quantum kernel methods (2024). https://arxiv.org/abs/2208.11060
34. Wu, S.L., Sun, S., Guan, et al.: Application of quantum machine learning using the quantum kernel algorithm on high energy physics analysis at the LHC. Phys. Rev. Res. **3**(3) (2021). https://doi.org/10.1103/physrevresearch.3.033221
35. Yaseen, Son, G.Y., Kwon, S.: Classification of heart sound signal using multiple features. Appl. Sci. **8**(12), 2344 (2018)

Natural Language Processing (NLP) in Biomedicine

A Comparative Analysis of Message-Level and User-Level Natural Language Processing Approaches for Early Depression Detection on Social Media

Carmen Rodríguez Jiménez[1], Miguel Rujas[2]([✉]) [iD], Beatriz Merino-Barbancho[2] [iD], Maria Teresa Arredondo[2] [iD], Maria Fernanda Cabrera-Umpierrez[2] [iD], Kinda Khalaf[3], and Giuseppe Fico[2] [iD]

[1] Universidad Politécnica de Madrid, Avda Complutense 30, 28040 Madrid, Spain
[2] Life Supporting Technologies Research Group, Universidad Politécnica de Madrid, Avda Complutense 30, 28040 Madrid, Spain
`mrujas@lst.tfo.upm.es`
[3] College of Medicine and Health Sciences, Khalifa University of Science and Technology, P.O. Box 127788, Abu Dhabi, UAE

Abstract. Depression remains a key global mental health concern, with social media emerging as a promising source for early detection through user-generated content. This study presents a comprehensive comparison between message-level and user-level natural language processing (NLP) approaches for identifying early signs of depression on social media. Using the eRisk 2025 dataset, encompassing over 1.4 million Reddit posts from 3,061 users, we evaluated multiple machine learning and deep learning models, including Logistic Regression, Random Forest, XGBoost, and BERT. Results showed that while message-level models achieved high macro-level accuracy and AUC scores (up to 0.90), they struggled to reliably detect depressive messages, as evidenced by low F1-scores and precision in the depressive class. In contrast, user-level models, which aggregate information across multiple posts per user, demonstrated superior performance in identifying depressive users, with higher recall and F1-scores (up to 0.88 recall). These findings highlight the importance of analytical granularity in mental health detection tasks: user-level approaches offer a more robust and context-aware strategy for early identification of individuals at risk. The study demonstrates the potential of integrating artificial intelligence and NLP for proactive mental health monitoring, while also acknowledging challenges related to class imbalance, generalizability, and clinical validation. Future research should explore hybrid models, multi-platform data integration, real-time systems, and ethical frameworks to enhance practical applicability and societal impact.

Keywords: Mental Health · Depression · Artificial Intelligence · Natural Language Processing · Social Media

© The Author(s), under exclusive license to Springer Nature Switzerland AG 2026
A. López Fernández et al. (Eds.): CIABiomed 2025, LNBI 16148, pp. 241–252, 2026.
https://doi.org/10.1007/978-3-032-10661-2_19

1 Introduction

Depression is a widespread mental health disorder that impacts an individual's emotional state, cognitive processes, and daily functioning. According to the World Health Organization (WHO), over 280 million people worldwide suffer from depression, highlighting its importance as a public health concern [1]. Despite its high prevalence, depression is often underdiagnosed or identified only in advanced stages. This is largely attributed to social stigma, insufficient mental health infrastructure, and delays inherent in clinical diagnostic processes [2, 3].

Social media platforms such as Reddit and Twitter have emerged as valuable sources of user-generated content, where individuals openly share personal experiences, emotions, and mental health challenges. These platforms provide opportunities to passively and unobtrusively detect early signs of depression, offering valuable data to inform public health strategies and early interventions [4–6].

The integration of Natural Language Processing (NLP) and Artificial Intelligence (AI) methods enables the large-scale analysis of extensive textual data derived from social media [7]. However, the early detection of depression through these channels presents unique methodological challenges, particularly concerning the appropriate level of analysis granularity. In NLP-based research, two primary analytical paradigms are commonly distinguished: message-level analysis, where each post is analyzed independently to identify depressive signals, and user-level analysis, where a comprehensive set of a user's posts is aggregated to perform a holistic evaluation.

Each approach presents benefits and limitations. Message-level analysis is granular and immediate, but is vulnerable to noise and context loss, which can lead to misclassification. Conversely, user-level analysis captures broader linguistic patterns and behavioural signals over time, though it requires more complex data handling and greater computational resources.

1.1 Related Work

Previous studies have demonstrated the efficacy of traditional machine learning models, including Support Vector Machines (SVM), Naïve Bayes, Random Forests, and logistic regression, in detecting depressive language patterns [6, 8, 9]. Most of these studies have focused on user-level analysis. Such models typically depend on feature engineering approaches, including term frequency-inverse document frequency (TF-IDF) representations and sentiment analysis metrics [10].

Recent advancements in deep learning, particularly Transformer-based models such as BERT, have further improved the capacity to capture complex linguistic features in text classification tasks [11]. However, there remains a lack of systematic studies directly comparing message-level and user-level NLP frameworks under consistent experimental conditions.

This study aims to fill this gap by conducting a comprehensive, controlled comparison of both analytical strategies using the same dataset, feature sets, preprocessing workflows, and evaluation metrics.

2 Materials and Methods

2.1 Dataset Description

The dataset employed in this study is derived from the eRisk 2025 challenge, an internationally recognized benchmark designed to facilitate early risk detection of mental health conditions using social media data [12–14]. Specifically, it comprises user histories sourced from Reddit, including textual posts and associated metadata such as user identifiers, timestamps, and binary labels indicating the presence or absence of depressive symptoms.

The dataset integrates entries from three different cohorts spanning multiple years. The 2017 cohort consists of 863 users who contributed a total of 389,531 messages, the 2018 cohort includes 819 users with 397,663 messages, and the 2022 cohort contains 1,379 users contributing 649,678 messages.

Overall, the combined corpus includes 1,436,872 messages, with approximately 10% labelled as indicative of depression. This significant class imbalance introduces significant challenges for model training and necessitates the application of specialized techniques to ensure robust predictive performance.

2.2 Methodology

The methodology followed in this study consists of four sequential phases, designed to ensure a comprehensive analysis. The overall process is illustrated in Fig. 1.

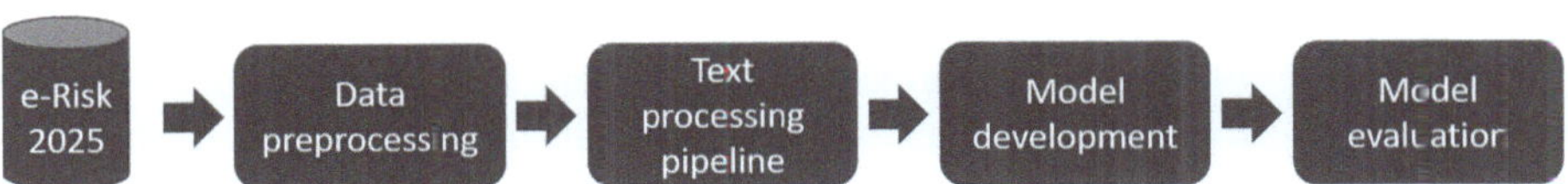

Fig. 1. Methodological Workflow of the Study

2.2.1 Data Preprocessing

The eRisk 2025 dataset comprised XML files from the 2017, 2018, and 2022 editions. Each file was processed individually and then merged to generate a unified CSV file. The resulting dataset included unique user identifiers, post timestamps, full textual content, depression labels, and the year of origin.

The data cleaning process involved removing columns with a high proportion of missing values, such as "*Title*", and eliminating posts with empty text fields. Additionally, data integrity was verified, and chronological ordering of posts was ensured. After these cleaning steps, the final dataset consisted of 1,436,872 posts with no missing values in the key variables.

2.2.2 Text Processing Pipeline

The text processing pipeline began with cleaning and normalization procedures, where all text was converted to lowercase and conversational noise, such as *"lol"*, *"ok"*, and *"xd"*, was removed. Regular expressions were employed to filter out irrelevant expressions, and a new column was created to store the cleaned content.

Stopwords were then removed using NLTK's English stopword list, while lemmatization was carried out to ensure semantic consistency by treating morphological variants as a single lexical unit. Given the dataset's large size, the text processing pipeline was executed in chunks of 1,000 records, utilizing a checkpointing system to allow seamless resumption in case of interruptions.

Finally, text data were vectorized using sentence-transformers, specifically the *"all-MiniLM-L6-v2"* model, to generate semantic embeddings for each message. These embeddings were stored in persistent files to enhance computational efficiency during subsequent modeling phases.

2.2.3 Model Development

For model development at both the message and user levels, a systematic process was followed. At the message level, we used 10% of the total dataset. This subsampling was necessary due to the RAM memory limitations of the computational infrastructure, which prevented model training on the full dataset. To obtain this subset, we performed a stratified reduction to preserve the original class imbalance.

At the user level, user embeddings were generated by averaging all message embeddings corresponding to each user, resulting in a single feature vector per user. This approach allowed us to capture aggregated linguistic patterns effectively.

After feature construction, hyperparameter optimization was carried out for the predictive models selected based on their extensive use in the literature (i.e., Random Forest (RF), XGBoost (XGB), Logistic Regression (LR), and Transformer-based models such as BERT). The optimization was performed using Grid Search with 3-fold stratified cross-validation, exploring the parameter configurations detailed in Table 1.

Finally, decision thresholds were finetuned using the Youden Index, prioritizing the optimization of recall to better handle the class imbalance present in the dataset.

Table 1. Hyperparameter Ranges Tested for Classification Models

Model	Hyperparameters Tested
Random Forest (RF)	n_estimators: [100, 200, 300]; max_depth: [5, 10, 15]; class_weight: [None, 'balanced']
XGBoost (XGB)	n_estimators: [100, 200, 300]; max_depth: [5, 10, 15]; learning_rate: [0.01, 0.1, 0.2]; scale_pos_weight: [1, 2, 5]
Logistic Regression (LR)	C: [0.001, 0.01, 0.1, 1, 10,100]; solver: ['liblinear']; class_weight: ['balanced']

(continued)

Table 1. (*continued*)

Model	Hyperparameters Tested
Transformer (BERT)	learning_rate: 2e-5 (fixed); batch_size: 16 (fixed); num_train_epochs: 3 (fixed); weight_decay: 0.01 (fixed)

2.2.4 Model Evaluation

The performance of the models was assessed using multiple metrics, including Precision, Recall, F1-score, Accuracy, Confusion Matrix, ROC curves, and the Area Under the Curve (AUC).

Given the strong class imbalance present in the dataset, we computed both macro-averaged metrics and metrics specifically for the minority class to ensure a comprehensive evaluation of model behavior across different class distributions.

Furthermore, we analyzed the impact of varying decision thresholds by applying the Youden Index, which maximizes the difference between true positive and false positive rates. This approach allowed us to identify optimal thresholds and to better understand the trade-offs in sensitivity and specificity, thus enabling fair and robust comparisons across models and levels of analysis.

3 Results

3.1 Data Preprocessing Results

Following the data cleaning procedure, the dataset was reduced from 1,975,700 messages to 1,436,872 valid messages, ensuring the inclusion of only relevant and analyzable records. The final dataset comprised 3,061 unique users, distributed as follows: 2,750 users in the negative class (non-depressive) and 311 users in the positive class (depressive). This distribution clearly indicates a strong class imbalance at the user level, with depressive users representing approximately 10.2% of the total.

A similar imbalance is observed at the message level: non-depressive users contributed 1,308,748 messages (about 91.0%), while depressive users contributed 128,124 messages (around 9.0%). Thus, the dataset remains strongly imbalanced in both user and message distributions.

3.2 Message-Level Models Results

The performance metrics obtained for the message-level models across different algorithms and decision thresholds are represented in Tables 2 and 3.

On the one hand, Table 2 summarizes the macro-level classification metrics. Among all models, BERT achieved the highest AUC (0.90) and accuracy (0.91) at a threshold of 0.5, with a notably high precision (0.88). However, its F1-score (0.58) and recall (0.63) indicate that, despite its strong overall performance, the model struggles to fully capture all positive cases. RF and XGB showed comparable macro-level AUC values (0.69–0.71) but achieved only moderate F1-scores (around 0.60), reflecting a balance between precision and recall that remains suboptimal. LR demonstrated similar limitations, with F1-scores close to 0.52 despite moderate recall values (around 0.60).

On the other hand, Table 3 focuses on the depressive class, emphasizing the ability to identify depressive messages. Here, BERT achieved the highest recall (0.64) at a lower threshold (0.14), suggesting an improved ability to detect positive cases. Nevertheless, the corresponding F1-score (0.23) remained low, highlighting the substantial drop in precision (0.14). RF and XGB also obtained higher recall values when using lower thresholds (up to 0.63 and 0.58, respectively), yet their F1-scores remained low (0.24), again due to poor precision. LR showed similar behavior, reaching a recall of 0.60 at a threshold of 0.49, but with a low F1-score of 0.25.

Table 2. Summary of macro-level classification metrics for message-level models

Algorithm	Decision Threshold	Recall	Precision	F1-Score	Accuracy	AUC
RF	0.5	0.60	0.61	0.60	0.88	0.71
RF	0.41	0.63	0.15	0.24	0.65	0.71
XGB	0.5	0.60	0.60	0.60	0.87	0.69
XGB	0.32	0.63	0.55	0.52	0.68	0.69
LR	0.5	0.64	0.55	0.52	0.68	0.69
LR	0.49	0.60	0.55	0.52	0.67	0.69
BERT	0.5	0.63	0.88	0.58	0.91	0.90
BERT	0.14	0.64	0.61	0.48	0.61	0.90

Table 3. Performance metrics for the depressive class in message-level models

Algorithm	Decision Threshold	Recall	Precision	F1-Score	Accuracy	AUC
RF	0.5	0.26	0.29	0.27	0.88	0.71
RF	0.41	0.63	0.15	0.24	0.65	0.71
XGB	0.5	0.27	0.27	0.27	0.87	0.69
XGB	0.32	0.58	0.15	0.24	0.68	0.69
LR	0.5	0.59	0.16	0.25	0.68	0.69
LR	0.49	0.60	0.16	0.25	0.67	0.69
BERT	0.5	0.14	0.44	0.22	0.91	0.90
BERT	0.14	0.64	0.14	0.23	0.61	0.90

3.3 User-Level Results

Tables 4 and 5 present the performance metrics obtained for the user-level models, highlighting the importance of recall and F1-score in evaluating the detection of depressive users under strong class imbalance.

On the one hand, Table 4 summarizes the macro-level metrics across all user-level models. RF and XGB achieved the highest overall performance, with AUC values of 0.90 and accuracies of 0.93 when using the default threshold of 0.5. Both models also demonstrated balanced F1-scores (0.76 for RF and 0.77 for XGB), driven by high precision (0.84) and solid recall values (0.72–0.73). When lowering the decision threshold (0.12 for RF and 0.03 for XGB), recall improved substantially (up to 0.85 for RF and 0.83 for XGB), although this gain in sensitivity came at the cost of reduced precision (0.66) and slightly lower F1-scores (around 0.69).

Logistic Regression (LR) showed competitive results, achieving a recall of 0.79 and an F1-score of 0.72 at the default threshold. Adjusting the threshold further increased recall to 0.82, but led to a slight decrease in F1-score (0.70), again due to reduced precision. The Transformer model performed notably worse in macro-level metrics, achieving a lower recall (0.52) and F1-score (0.52) at a threshold of 0.5, with performance improving only moderately when the threshold was lowered (recall of 0.77 and F1-score of 0.64 at threshold 0.07).

On the other hand, Table 5 provides detailed metrics for the depressive class, where recall and F1-score are especially key. RF and XGB achieved the highest recall values (0.88 and 0.85, respectively) at their lower thresholds, demonstrating strong capability to identify depressive users. However, these improvements in recall corresponded with significant drops in precision (down to 0.34), resulting in modest F1-scores (0.49 for RF and 0.48 for XGB). LR followed a similar pattern, achieving a maximum recall of 0.78 at a lower threshold (0.42), though its precision remained low (0.36), producing an F1-score of 0.51.

The Transformer model struggled significantly with the depressive class, achieving extremely low recall (0.05) and F1-score (0.09) at the default threshold. While recall improved substantially (to 0.77) when using a lower threshold, precision decreased to 0.28, resulting in an F1-score of 0.41.

Table 4. Summary of Macro-Level Classification Metrics for User-Level Models

Algorithm	Decision Threshold	Recall	Precision	F1-Score	Accuracy	AUC
RF	0.5	0.72	0.84	0.76	0.93	0.90
RF	0.12	0.85	0.66	0.69	0.82	0.90
XGB	0.5	0.73	0.84	0.77	0.93	0.90
XGB	0.03	0.83	0.66	0.69	0.82	0.90
LR	0.5	0.79	0.68	0.72	0.87	0.89
LR	0.42	0.82	0.68	0.70	0.84	0.89
Transformer	0.5	0.52	0.81	0.52	0.90	0.83
Transformer	0.07	0.77	0.62	0.64	0.78	0.83

Table 5. Performance Metrics for the Depressive Class in Message-Level Models

Algorithm	Decision Threshold	Recall	Precision	F1-Score	Accuracy	AUC
RF	0.5	0.46	0.74	0.56	0.93	0.90
RF	0.12	0.88	0.34	0.49	0.82	0.90
XGB	0.5	0.48	0.73	0.58	0.93	0.90
XGB	0.03	0.85	0.34	0.48	0.82	0.90
LR	0.5	0.70	0.40	0.51	0.87	0.89
LR	0.42	0.78	0.36	0.51	0.84	0.89
Transformer	0.5	0.05	0.72	0.09	0.90	0.83
Transformer	0.07	0.77	0.28	0.41	0.78	0.83

4 Discussion

This study set out to provide a comparison of message-level and user-level analytical strategies for the early detection of depression on social media, using a unified dataset, harmonized feature sets, identical pre-processing workflows, and consistent evaluation metrics. The findings offer important insights into the strengths and limitations of each approach, particularly in the context of strongly imbalanced data.

4.1 Message-Level Analysis

The message-level models demonstrated relatively high macro-level accuracy and AUC values across algorithms, with BERT achieving the highest AUC (0.90) and accuracy (0.91). However, despite these promising overall metrics, the performance for the depressive class was limited, as indicated by consistently low F1-scores (maximum of 0.27) and precision values. Although some models achieved higher recall (up to 0.64 with BERT at a lower threshold), this increase was accompanied by a substantial decline in precision, resulting in minimal improvement in F1-score.

These results highlight a key limitation of message-level approaches: the difficulty in reliably identifying depressive content at the individual message level, especially in the presence of severe class imbalance. Individual messages may not contain enough explicit signals to allow models to differentiate between depressive and non-depressive content effectively, which limits the practical utility of this approach for accurate early detection.

4.2 User-Level Analysis

In contrast, user-level models consistently achieved superior results, particularly in terms of recall and F1-score for the depressive class. RF and XGB obtained high recall values (up to 0.88 and 0.85, respectively) and moderate F1-scores (up to 0.56 and 0.58 at default thresholds), substantially outperforming the message-level counterparts. LR also performed well, achieving a recall of 0.78 and an F1-score of 0.51 at lower thresholds.

These improvements can be attributed to the aggregation of information across all messages posted by each user, providing a richer and more stable representation of their behavioral and linguistic patterns. By leveraging aggregated embeddings or features, user-level approaches can better capture subtle signals indicative of depression, resulting in higher sensitivity and improved overall detection capabilities.

However, it is important to note that prioritizing recall often came at the cost of reduced precision, reflecting the trade-off inherent in optimizing models for high sensitivity in clinical or screening contexts. While lower precision may lead to higher false positive rates, it can be justified in scenarios where early intervention and prevention are important.

4.3 Message-Level Versus User-Level Comparison

Comparing both strategies directly, user-level models outperform message-level models in detecting depressive cases, especially when evaluated by recall and F1-score, metrics particularly relevant in imbalanced and health-related contexts. While message-level models achieve high macro-level metrics, their low depressive-class F1-scores affect their practical utility.

The findings demonstrate that aggregating information at the user-level provides a more robust and contextually meaningful signal for depression detection, effectively addressing the limitations posed by message-level sparsity and variability. Most previous studies in this field have also focused on the user-level, but typically by extracting handcrafted textual and behavioral features such as word counts, lexical categories, sentiment, or temporal posting patterns, rather than by leveraging transformer-based embeddings [6, 8, 9]. While these works achieved promising results, they generally explored a smaller range of models and did not integrate both traditional machine learning and deep learning strategies, nor did they conduct a systematic comparison between message-level and user-level analyses within a unified framework.

Overall, this study underscores the importance of selecting the appropriate analytical granularity when designing automated mental health detection systems. User-level approaches should be prioritized for tasks aimed at identifying individuals at risk, as they offer a more comprehensive and reliable foundation for future clinical and research applications.

4.4 Limitations

Despite providing a comprehensive and controlled comparison, this study has several limitations that should be acknowledged. First, the reliance on pretrained tokenizers and embedding models could introduce potential biases. Pretrained tokenizers, developed on large generic corpora, may not accurately capture the complexities of mental health-related expressions, potentially overlooking subtle linguistic characteristics specific to depressive language. Second, the dataset exhibited strong class imbalance, with depressive cases representing a small minority. Although threshold adjustments were applied to partially mitigate this issue, the imbalance likely biased models toward the majority non-depressive class, reducing precision and increasing false positive rates. Third, the analysis was restricted to English-language Reddit posts. This focus limits the generalizability of the findings to multilingual contexts and to other platforms with different user demographics, cultural backgrounds, or communication styles. In addition, the inherent variability of language on social media (e.g., sarcasm, humor, or informal tone) can further challenge model performance, as these features may lead to misclassification or biased predictions. Fourth, models were trained and evaluated on historical data from specific years and platforms. Consequently, they may not fully capture evolving linguistic trends, emerging slang, or changes in platform behaviors that influence language and emotional expression over time. Fifth, the study did not include clinical validation by mental health professionals. Without such human expert evaluation, the real-world clinical relevance and practical applicability of model predictions remain uncertain. Finally, computational constraints limited the scope of the modeling approaches explored. Resource limitations precluded the use of more advanced finetuning strategies for Transformer models or the implementation of sophisticated ensemble stacking methods, which may have further enhanced performance. Addressing these limitations in future work will be essential to improve model robustness, generalizability, and clinical utility.

5   Conclusions and Future Work

5.1   Conclusions

This study presented a comprehensive and controlled comparison of message-level and user-level analytical strategies for the early detection of depression on social media. By using a unified dataset, harmonized feature sets, and consistent evaluation metrics, we were able to systematically examine the strengths and limitations of each approach.

The results demonstrated that while message-level models achieved high overall accuracy and AUC values, their ability to reliably identify depressive messages was limited, as evidenced by low F1-scores and precision for the depressive class. In contrast, user-level models consistently outperformed message-level models in terms of recall and F1-score for depressive users, benefiting from the aggregation of information across multiple messages to capture more stable behavioral and linguistic patterns.

These findings highlight the importance of considering analytical granularity in mental health detection tasks. User-level approaches appear more suitable for identifying individuals at risk, whereas message-level approaches may be more appropriate for content-level screening or moderation tasks where immediate response to individual messages is key.

5.2   Future Work

Building upon this study, several research directions emerge that are important for advancing the development of automated depression detection systems. A key priority is the expansion of data sources to include content from multiple social media platforms, as well as posts in different languages and cultural contexts. Such diversification would enhance the generalizability of the models, enabling them to perform robustly across heterogeneous populations and varied communication styles.

In addition, the design of hybrid modeling strategies represents a promising approach. By integrating the sensitivity of message-level analyses with the contextual stability offered by user-level aggregation, future models could achieve a more comprehensive understanding of individual mental health states. This combination has the potential to capture both fine-grained signals and broader behavioral patterns, ultimately improving predictive performance.

Equally important is the clinical validation of these systems in real-world settings. Collaborating closely with mental health professionals will be essential to evaluate the clinical relevance and utility of model predictions, ensuring that these tools provide meaningful support within actual care pathways rather than functioning solely as academic or technological exercises.

Furthermore, future work should explore the development of lightweight, real-time monitoring systems capable of continuous surveillance and early warning generation. Such systems could offer proactive support and timely interventions for individuals at risk, contributing to more responsive and preventive mental health care strategies.

Finally, the advancement of these technical innovations must be accompanied by the establishment of comprehensive ethical frameworks. Addressing issues such as data privacy, informed consent, and the responsible use of AI to ensure the safe, transparent, and socially acceptable deployment of mental health monitoring technologies.

Collectively, these future directions emphasize the need to balance methodological rigor with practical applicability and ethical responsibility, building the way for the development of robust, generalizable, and clinically meaningful tools to support early detection and intervention in mental health.

Acknowledgments. The project has received funding from the Innovative Medicines Initiative 2 Joint Undertaking under grant agreement No 101034369. This joint undertaking receives support from the European Union's Horizon 2020 Research and Innovation Programme, the European Federation of Pharmaceutical Industries and Associations (EFPIA) and Link2Trials. This communication reflects the views of the authors and neither the IMI nor the European Union, EFPIA, or Link2Trials are liable for any use that may be made of the information contained herein.

Disclosure of Interests. The authors have no competing interests to declare that are relevant to the content of this article.

References

1. World Health Organization. Mental disorders. Accessed 11 July 2025. https://www.who.int/news-room/fact-sheets/detail/mental-disorders
2. López Ibor, M.I.: Ansiedad y depresión, reacciones emocionales frente a la enfermedad. Anales de Medicina Interna **24**(5), 209–211 (2007). Accessed 11 July 2025. https://scielo.isciii.es/scielo.php?script=sci_arttext&pid=S0212-71992007000500001&lng=es&nrm=iso&tlng=es
3. Hussain, J., et al.: Exploring the dominant features of social media for depression detection. J. Inf. Sci. **46**(6), 739–759 (2020). https://doi.org/10.1177/0165551519860469

4. Babu, N.V., Kanaga, E.G.M.: Sentiment analysis in social media data for depression detection using artificial intelligence: a review. SN Comput. Sci. **3**(1), 1–20 (2022). https://doi.org/10.1007/S42979-021-00958-1/TABLES/1

5. Mintz, Y., Brodie, R.: Introduction to artificial intelligence in medicine. Minim. Invasive Ther. Allied Technol. **28**(2), 73–81 (2019). https://doi.org/10.1080/13645706.2019.1575882

6. Tariq, S., et al.: A novel co-training-based approach for the classification of mental illnesses using social media posts. IEEE Access **7**, 166165–166172 (2019). https://doi.org/10.1109/ACCESS.2019.2953087

7. Wani, M.A., ELAffendi, M.A., Shakil, K.A., Imran, A.S., Abd El-Latif, A.A.: Depression screening in humans with AI and deep learning techniques. IEEE Trans. Comput. Soc. Syst. **10**(4), 2074–2089 (2022). https://doi.org/10.1109/TCSS.2022.3200213

8. Cacheda, F., Fernandez, D., Novoa, F.J., Carneiro, V.: Early detection of depression: Social network analysis and random forest techniques. J. Med. Internet Res. **21**(6), e12554 (2019). https://doi.org/10.2196/12554

9. Kumar, A., Sharma, A., Arora, A.: Anxious depression prediction in real-time social data. SSRN Electron. J. (2019). https://doi.org/10.2139/SSRN.3383359

10. Husseini Orabi, A., Buddhitha, P., Husseini Orabi, M., Inkpen, D.: Deep learning for depression detection of twitter users. In: Loveys, K., Niederhoffer, K., Prud'hommeaux, E., Resnik, R., Resnik, P. (eds.) Proceedings of the Fifth Workshop on Computational Linguistics and Clinical Psychology: From Keyboard to Clinic, pp. 88–97. Association for Computational Linguistics, New Orleans (2018). https://doi.org/10.18653/v1/W18-0609

11. Transformers. Accessed 11 July 2025. https://huggingface.co/docs/transformers/index

12. CLEF eRisk: Early risk prediction on the Internet | CLEF 2025 workshop. Accessed 11 July 2025. https://erisk.irlab.org/

13. Crestani, F., Losada, D.E., Parapar, J.: Early detection of mental health disorders by social media monitoring : the first five years of the eRisk Project. In: Studies in Computational Intelligence, 1st edn., vol. 1018. Springer, Cham (2022)

14. Parapar, J., Perez, A., Wang, X., Crestani, F.: eRisk 2025: contextual and conversational approaches for depression challenges. In: Lecture Notes in Computer Science LNCS, vol. 15576, pp. 416–424 (2025). https://doi.org/10.1007/978-3-031-88720-8_62

A Weak Supervision Approach for Monitoring Recreational Drug Use Effects in Social Media

Lucía Prieto-Santamaría[1,2] , Alba Cortés Iglesias[1], Claudio Vidal Giné[3] ,
Fermín Fernández Calderón[4,5] , Óscar M. Lozano[4,5] ,
and Alejandro Rodríguez-González[1,2(✉)]

[1] Escuela Técnica Superior de Ingenieros Informáticos, Universidad Politécnica de
Madrid, Madrid, Spain
{lucia.prieto.santamaria, alejandro.rg}@upm.es,
alba.cortes@alumnos.upm.es
[2] Centro de Tecnología Biomédica, Universidad Politécnica de Madrid, Madrid, Spain
[3] Asociación Bienestar y Desarrollo, Energy Control, Antequera, Málaga, Spain
claudiovidal@energycontrol.org
[4] Clinical and Experimental Psychology Department, University of Huelva, Huelva,
Spain
fermin.fernandez@dpces.uhu.es, oscar.lozano@dpsi.uhu.es
[5] Research Center for Natural Resources, Health and the Environment, University of
Huelva, Huelva, Spain

Abstract. Understanding the real-world effects of recreational drug use
remains a critical challenge in public health and biomedical research,
especially as traditional surveillance systems often underrepresent user
experiences. In this study, we leverage social media (specifically Twitter)
as a rich and unfiltered source of user-reported effects associated with
three emerging psychoactive substances: ecstasy, GHB, and 2C-B. By
combining a curated list of slang terms with biomedical concept extraction via MetaMap, we identified and weakly annotated over 92,000 tweets
mentioning these substances. Each tweet was labeled with a polarity
reflecting whether it reported a positive or negative effect, following an
expert-guided heuristic process. We then performed descriptive and comparative analyses of the reported phenotypic outcomes across substances
and trained multiple machine learning classifiers to predict polarity from
tweet content, accounting for strong class imbalance using techniques
such as cost-sensitive learning and synthetic oversampling. The top performance on the test set was obtained from eXtreme Gradient Boosting
with cost-sensitive learning (F1 = 0.885, AUPRC = 0.934). Our findings
reveal that Twitter enables the detection of substance-specific phenotypic effects, and that polarity classification models can support real-time
pharmacovigilance and drug effect characterization with high accuracy.

Keywords: Social Media Mining · Machine Learning · Class
Imbalance · Twitter · Drug Recreational Use

A. López Fernández et al. (Eds.): CIABiomed 2025, LNBI 16148, pp. 253–267, 2026.
https://doi.org/10.1007/973-3-032-10361-2_20

1   Introduction

Recreational drug use represents a persistent and complex public health challenge, contributing to both acute and chronic adverse health outcomes. Substances such as: ecstasy, GHB, and 2C-B are widely consumed for their psychoactive properties, often outside regulated medical contexts [12,27,30]. Understanding the effects these substances produce in real-world settings is essential to evaluate their risks, identify patterns of misuse, and detect emergent phenomena. Traditional pharmacovigilance systems, while indispensable, rely heavily on structured reporting from healthcare providers and institutions. These systems tend to underreport mild, subjective, or socially stigmatized experiences, and often lag behind actual behavioral trends [22].

In contrast, social media platforms, particularly Twitter (now X), offer a complementary and dynamic source of health-related data [14,15,21,25]. Users frequently share their experiences with drugs in informal language, including detailed descriptions of sensations, side effects, mood changes, and behavioral outcomes [17]. These user narratives reflect what we refer to as *phenotypic manifestations*: observable or self-perceived effects resulting from drug intake. From a biomedical standpoint, such phenotypes represent the interplay of pharmacological mechanisms (e.g., receptor binding, metabolic clearance) with contextual and personal factors such as co-use of other substances, emotional state, and individual physiology [19].

These phenotypic effects can be grouped into two categories: *desired effects*, which motivate consumption (e.g., euphoria, sociability, sensory enhancement), and *undesired effects*, which may indicate toxicity or adverse reactions (e.g., nausea, panic, confusion). Capturing both types of responses is crucial for informal pharmacovigilance, as it allows health authorities and researchers to track real-world impact, monitor perception trends, and identify potentially therapeutic or harmful patterns of use [6]. Several studies have explored the potential of social media mining to monitor drug discourse. Early efforts relied on keyword matching and manual inspection to identify tweets related to substance use [5]. Later works incorporated sentiment analysis, topic modeling, and supervised classification to extract thematic patterns [7,17,29].

A key limitation in this scenario is class imbalance (where negative effects dominate the training data) can skew classifier performance if not carefully addressed [23]. Another critical gap is the underuse of biomedical semantic enrichment [26]. Tools such as MetaMap allow the mapping of free-text to structured concepts in the UMLS Metathesaurus, providing standardization and interoperability with clinical vocabularies [1]. This allows informal narratives to be interpreted in relation to known adverse events, pharmacological classes, and symptom ontologies [28]. Some studies have shown that semantic features improve the accuracy of classification models in biomedical NLP tasks [18].

In this work, we present a comprehensive pipeline for detecting and classifying user-reported phenotypic effects of three recreational substances (ecstasy, GHB, and 2C-B) based on Twitter data. Our methodology is guided by a data mining methodological framework and integrates biomedical knowledge, weak supervision, and Machine Learning (ML). Specifically, we have (i) developed a weak labeling strategy based on polarity annotations of both a manually curated slang lexicon and biomedical MetaMap-extracted concepts; (ii) built a semantically enriched dataset of over 6 million tweets filtered by substance-related keywords and annotated for perceived effect polarity; and (iii) trained and evaluated several ML classifiers using text embeddings and domain-specific features, while applying class imbalance mitigation techniques and class weighting. The rest of this paper is structured as follows: Sect. 2 presents the methodology, including data processing and modeling. Section 3 reports the experimental results and analysis. Section 4 ends with the conclusions and a discussion of future directions.

2   Methodology

This section describes the methodological pipeline designed to collect, annotate, and analyze Twitter data related to recreational drug use. Figure 1 provides an overview of the full process. We begin by explaining how Twitter data were extracted and processed (Subsect. 2.1) to obtain structured and clean textual and metadata information. We then describe the annotation strategy based on weak supervision (Subsect. 2.2) which leverages external domain knowledge to assign preliminary labels to the tweets. This is followed by a description of the feature engineering process to obtain a complete set of variables associated to each tweet (Subsect. 2.3). Finally, we present the machine learning methodology used to classify tweets according to the phenotypic effects they report related to drug use (Subsect. 2.4).

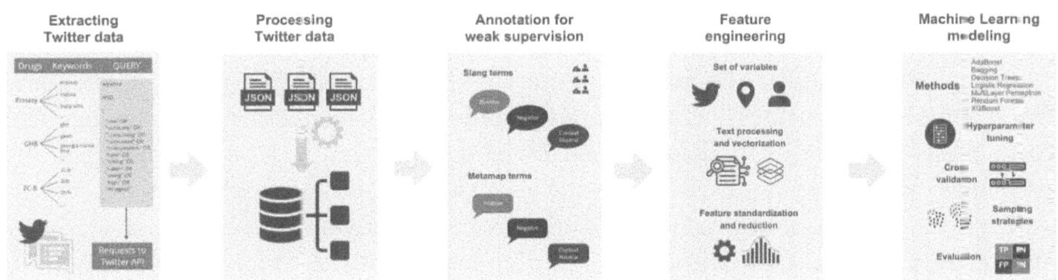

Fig. 1. Overview of the methodology pipeline.

2.1   Extracting and Processing Twitter Data

Firstly, we compiled tweets posted between 2010 and 2019 by accessing Twitter through its Application Programming Interface (API). To manage the requests, we used the Python library "Twitter API", which requires API keys, an endpoint, a query, and additional parameters to retrieve data. To identify tweets potentially mentioning the drugs under study (ecstasy, GHB, and 2C-B), we used a set of keywords associated with each drug. These keywords are listed in Table 1.

Table 1. Keywords used for each drug when querying Twitter.

Ecstasy	GHB	2C-B
ecstasy	ghb	2cb
mdma	geeb	2c/b
baby slits	georgia home boy	2c b
e-bomb	liquid ecstasy	2,5-dimethoxy-4-bromophenethylamine
hug drug	cherry meth	4-bromo-2,5-dimethoxyphenethylamine
love drug	easy lay	afterburner bromo
moon rock	blue nitro	bdmpea
love doctor	gamma hydroxybutyrate	bromo mescaline
love portion #9	gamma-hydroxybutyric acid	cee-beetje
love trip	4-hydroxybutanoic acid	erox
scooby snacks	date rape drug	pink cocaine
speed for lovers	forget-me pill	tusibi
xtc	forget pill	tucibi
3,4 methylenedioxymethamphetamine		
c11h15no2		
ecstasy pill		
substituted amphetamine		
molly		

Each drug-specific query included the corresponding terms in Table 1, combined with the following string: *AND ("use" OR "consume" OR "consuming" OR "consumed" OR "consumption" OR "take" OR "taking" OR "taken" OR "using" OR "high" OR "drugged")*. This additional condition aimed to increase the likelihood that retrieved tweets referred to drug use rather than to unrelated contexts. Requests to the Twitter API returned data in JavaScript Object Notation (JSON) format, which included information about the tweets themselves (e.g., creation date and text), engagement metrics (number of retweets, replies, likes, and quotes), the authors who posted them, and geolocation data. Once the JSON files were retrieved through the Twitter API, the data were processed and structured into relational database tables. Tweet texts were cleaned by removing links, hashtags, and emojis, and all text was converted to lowercase. Tweet interactions were also processed to identify replies and retweets.

2.2   Annotation for Weak Supervision

Since manually annotating a large collection of tweets for drug-related content is both time-consuming and costly, we adopted a weak supervision strategy to generate preliminary labels automatically. This approach leverages domain-specific knowledge sources and linguistic heuristics to infer likely labels without the need for human annotation at the tweet level.

We used two complementary sources of information to identify whether a tweet refers to the use or effects of recreational drugs: (1) a curated list of **slang terms** associated with each of the three target substances (ecstasy, GHB, and

2C-B), and (2) biomedical entities extracted using **MetaMap** [1], a tool developed by the U.S. National Library of Medicine that maps free text to concepts in the Unified Medical Language System (UMLS) [2].

Both the curated term list and the set of extracted UMLS concepts were independently reviewed by three domain experts. Each term or concept was assigned a polarity label: positive, negative, or context-dependent, based on whether its presence was likely to indicate positive or negative drug use or effects. A majority-vote strategy was used to consolidate annotations. Specifically, if at least 60% of annotators agreed on a polarity label, that label was assigned as the final annotation. Otherwise, the term was marked as uncertain. This step ensured that only terms with sufficient inter-annotator agreement were used in the subsequent tweet-level labeling process.

Once terms and concepts had been annotated with a final polarity, tweets were heuristically labeled based on the combination of annotations for the lexical items they contained. For each tweet, the corresponding terms or concepts found (whether from the slang list or extracted by MetaMap) were mapped to numerical scores: +1 for positive polarity, −1 for negative, and 0 for context-dependent. The scores for all elements in a tweet were aggregated, and the resulting total score determined the tweet's polarity: tweets with a score > 0 were labeled as positive, those with a score < 0 as negative, and tweets scoring exactly 0 were labeled as context-dependent.

This polarity assignment was performed independently for the slang-based and MetaMap-based information, producing two parallel tweet classification tables. To ensure high-confidence labels, we applied a final filtering step: only tweets that were present in both tables and had matching polarity assignments (either positive or negative) were retained. Tweets labeled as context-dependent in either source, or those with discordant polarity across sources (e.g., positive in slang, negative in MetaMap), were excluded.

2.3   Feature Engineering

Once the final dataset was consolidated, we conducted a feature engineering process to prepare the data for machine learning modeling. This process combined tweet metadata, user attributes, geolocation information, and a semantic representation of textual content based on word embeddings. We extracted a **set of relevant features** from the resulting dataset (Table 2), including three main types:

- **Tweet-level features:** the tweet text, presence of multimedia content, user mentions, reference to another tweet (e.g., replies or retweets), engagement metrics, and binary indicators for whether the tweet contained terms associated with each of the three substances. The target variable for classification was included with values "positive" or "negative" according to the weak supervision strategy described in Subsect. 2.2.
- **Geographic features:** binary indicators for the region of origin based on the country code, and a flag indicating whether the user's location field was available.

- **User features:** verification status, and metrics reflecting user activity and visibility, including follower count, number of followed accounts, total number of tweets, and number of public lists the user appears on.

Table 2. Set of features extracted from tweets, users, and other metadata.

Type	Feature	Description
Tweet-level	`media`	Binary indicator for presence of multimedia content in the tweet
	`mention`	Binary indicator for whether the tweet mentions another user
	`reference`	Binary indicator for whether the tweet is a reply or retweet
	`like_count`	Number of likes received by the tweet
	`retweet_count`	Number of times the tweet was retweeted
	`reply_count`	Number of replies to the tweet
	`quote_count`	Number of quote tweets of the tweet
	`mentions_ghb`	Indicates if the tweet contains terms associated with GHB
	`mentions_ecstasy`	Indicates if the tweet contains terms associated with ecstasy
	`mentions_2cb`	Indicates if the tweet contains terms associated with 2C-B
	`classification`	Weak supervision label (positive or negative)
Geographic	`is_europe`	Tweet was posted from a country located in Europe
	`is_africa`	Tweet was posted from a country located in Africa
	`is_asia`	Tweet was posted from a country located in Asia
	`is_america`	Tweet was posted from a country located in America
	`user_location`	Binary indicator for whether the user's location is available
User-level	`user_verified`	Whether the user's account is verified (1 if yes, 0 otherwise)
	`user_followers`	Number of followers the user has
	`user_following`	Number of accounts the user is following
	`user_tweet_count`	Total number of tweets posted by the user
	`user_listed_count`	Number of public lists the user appears in

Since the tweet content constitutes a key source of semantic information, we applied a **text preprocessing** pipeline prior to vectorization. The text was lowercased, and all URLs, mentions (@), hashtags (#), numbers, and punctuation were removed. Tokenization was performed using `nltk.word_tokenize`, and English stopwords were removed using `nltk.corpus.stopwords`. We then trained a **Word2Vec** model [16] using the `gensim` library [31] to learn distributed word representations. The training parameters were set as follows: `vector_size=30` (each word is represented as a vector of 30 dimension), `window=5` (context of 5 words before and after of the target term), and `min_count=2` (we excluded words with a less than two frequency). Each tweet was embedded as the average of the Word2Vec vectors of its tokens. In cases where none of the words were found in the model's vocabulary, a zero vector of 30 dimensions was assigned. The resulting features (`w2v_0` to `w2v_29`) were concatenated with the remaining numeric and categorical features presented in Table 2.

All numeric variables were **standardized** to ensure comparability across feature scales. We computed the Pearson correlation matrix and identified highly correlated feature pairs (absolute correlation > 0.8). Among the engagement metrics, only one of the variables was retained. Similarly, only one variable was kept between `user_followers` and `user_listed_count`. However, we retained the substance mention indicators, such as `mentions_ghb` and `mentions_ecstasy`, despite their high negative correlation, as they capture distinct drug-specific information.

The result of this stage was a clean, vectorized, and standardized dataset, composed of linguistic geographic, user, and engagement features, suitable for downstream classification modeling.

2.4   Machine Learning Modeling

Once the dataset was fully processed, we proceeded with the training and evaluation of several machine learning classifiers. Given the imbalanced nature of the data, where tweets labeled as negative (0) were significantly more prevalent than positive (1) ones, careful attention was paid to class imbalance to avoid bias and ensure robust learning. The final dataset was split into training (80%) and test (20%) subsets using a stratified split to preserve class proportions across both subsets.

We implemented and compared a **diverse set of classification algorithms**, ranging from interpretable models to more complex ensemble and neural architectures. Specifically, we evaluated Decision Tree (DT) [20] and Logistic Regression (LR) [10] as baseline models, ensemble methods including Random Forest (RF) [4], Bagging [3], Adaptive Boosting (AdaBoost) [11], and Extreme Gradient Boosting (XGBoost) [9], as well as a Multi-layer Perceptron (MLP) [24] to capture non-linear patterns in the data. These classifiers offer a broad spectrum of modeling capabilities in terms of interpretability, robustness, and predictive performance.

We tuned each model's **hyperparameters** using randomized search over tailored parameter distributions, optimizing for F1-score due to its suitability for imbalanced classification. Each algorithm had a customized search space: for DT and RF, we varied splitting criteria, number of estimators, and feature selection strategies; Bagging explored sampling fractions, estimator count, and warm starts; AdaBoost varied learning rates and estimators; LR tested regularization types and compatible solvers; XGBoost was tuned over tree depth and estimator count; and MLP explored activation functions, hidden layer sizes, and learning rate strategies.

To further improve reliability and prevent overfitting during hyperparameter tuning, we adopted a nested, stratified 5-fold **Cross Validation** (CV) strategy. The inner loop was used to select the best hyperparameters via random search, while the outer loop evaluated the generalization performance.

To **evaluate** models' performance, we compute True Positives (TP), False Positives (FP), True Negatives (TN), and False Negatives (FN). During cross-validation, we report precision, recall, accuracy, and F1-score. For the final test evaluation, we additionally include the Area Under the Receiver Operating Characteristic Curve (AUROC) and the Area Under the Precision-Recall Curve (AUPRC), which provide threshold-independent assessments of discriminative performance, especially relevant in imbalanced settings.

As the positive class (tweets indicating *desired* drug-related effects) was significantly underrepresented, we compared three distinct **strategies for handling class imbalance**:

- **No sampling (baseline):** The original distribution of positive and negative tweets was maintained, which served as a reference to assess the effectiveness of imbalance mitigation techniques.
- **Cost-sensitive learning:** For classifiers that support weighted training, we applied `class_weight='balanced'`. This approach adjusts the loss function to penalize misclassifications of the minority class more heavily, encouraging the model to give more attention to underrepresented instances without modifying the training data.
- **Oversampling.** We applied Synthetic Minority Over-sampling Technique (SMOTE) to synthetically generate new positive-class samples by interpolating between existing minority-class examples in the feature space [8]. Two configurations were tested: (i) ***Pre-CV SMOTE***, where oversampling was applied once to the entire training set before starting the CV process (computationally simpler, but it carries a risk of information leakage, as synthetic samples derived from the training data may indirectly influence the validation folds, leading to overoptimistic performance estimates); and (ii) ***In-CV SMOTE***, where SMOTE was applied within each fold of the CV pipeline (for each train/validation split, oversampling was performed only on the training portion of the fold, and the synthetic samples were never present in the validation set, ensuring a more realistic evaluation by preventing data leakage and closely simulating the generalization capability of the model in unseen

data). In both setups, SMOTE was always applied exclusively to the training data and never to the test set.

3   Results and Discussion

We began our study by collecting a total of 6,755,394 tweets from 17 JSON archives, authored by 2,681,817 unique users. The first filtering step involved identifying whether the tweet mentioned any of the three target substances (ecstasy, GHB, or 2C-B). This filtering yielded 5,228,085 tweets (77.4% of the total), which became the foundation of our analysis. Within this subset, most tweets (5,212,224) referenced only one substance, while a small fraction (15,828 tweets) mentioned two, and an even smaller portion (33 tweets) mentioned three or more substances simultaneously.

Figure 2 provides an overview of this dataset descriptive analysis. The distribution of substances revealed a strong predominance of ecstasy-related content (4,889,522 tweets), followed by GHB (306,336) and 2C-B (48,121), reflecting broader usage or cultural popularity in social discourse. In terms of associated terms, slang terms were identified in 1,104,696 tweets, whereas MetaMap terms were associated to 4,457,981 tweets.

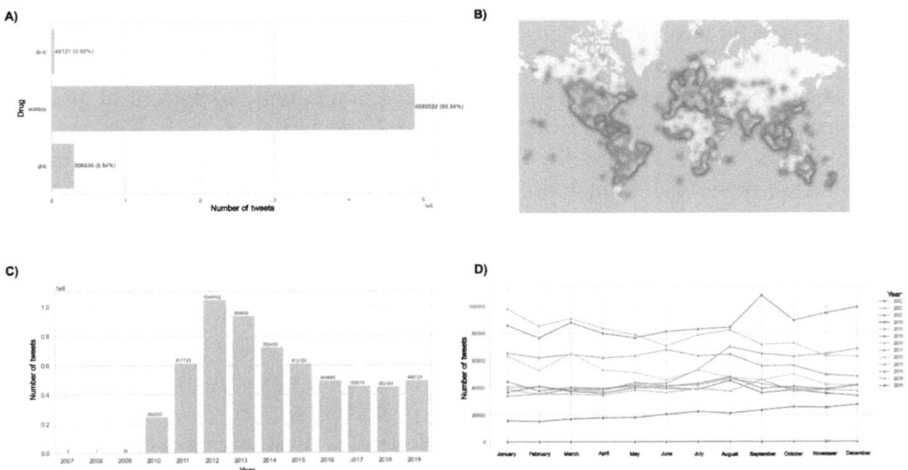

Fig. 2. General descriptive analysis of the dataset. A) Number of tweets per drug. B) Heatmap showing number of tweets per geolocation. C) Number of tweets per year. D) Number of tweets per month and year.

From a metadata perspective, 420,731 tweets included multimedia content, and 166,765 tweets contained geolocation metadata. Regarding user characteristics, 49,421 users (1.8%) were verified accounts, contributing to 2,593,637 tweets, while the rest were from unverified profiles.

To proceed with the construction of the classification dataset, we applied an annotation pipeline leveraging weak supervision. We began with the pool of tweets containing drug mentions and associated either to slang or MetaMap terms, totaling 5,562,677 tweets. Each term identified in those tweets had been previously annotated by three domain experts with a polarity label: *positive*, *negative*, or *context-dependent*. Only tweets in which a majority consensus (i.e., at least 2 of 3 annotators) was reached for all annotated terms were retained, resulting in 2,986,159 tweets.

From these, we separated tweets into those annotated using slang terms (345,422) and those annotated via MetaMap (1,225,003), depending on the original matching source. Tweets labeled as context-dependent were discarded (1,263,910), as they required nuanced interpretation beyond automated heuristics. After merging both annotation streams and consolidating duplicate tweets, we obtained a final classification dataset containing 92,291 tweets, each labeled as either expressing positive or negative polarity with respect to recreational drug effects.

The final classification dataset exhibits a strong class imbalance, with only 4,237 tweets (4.59%) labeled as positive, indicating references to personal or experiential effects of drug use, and 88,352 tweets (95.41%) labeled as negative. When examining the distribution by substance, this imbalance is even more pronounced in certain categories. For ecstasy, the most prevalent substance in the dataset, there are 4,163 positive tweets compared to 81,937 negatives. In the case of GHB, only 59 tweets were labeled as positive versus 5,974 negatives, while 2C-B has 15 positive tweets and 441 negatives. This imbalance (particularly for GHB and 2C-B) reflects the lower frequency of explicit, experiential positive mentions of these substances in user-generated content.

Moving to the ML modeling, Table 3 summarizes the average performance and standard deviation of each model across CV folds, evaluated under four different balancing strategies: no sampling, cost-sensitive learning, SMOTE applied before CV (pre-CV), and SMOTE applied within the folds (in-CV).

Overall, the results show a clear benefit of using data balancing, with SMOTE (pre-CV) yielding the best results across most models. Under this setting, Bagging, MLP, and RF achieved F1-scores above 0.99, alongside very high AUROC and AUPRC values, indicating near-perfect separability between classes. XGBoost also maintained strong performance ($F1 = 0.990 \pm 0.001$), confirming its robustness in this setting. However, these nearly perfect metrics (combined with zero standard deviation across folds in some cases) suggest signs of overfitting, likely due to data leakage introduced by applying SMOTE before the CV split. This leads to synthetic examples appearing in both train and validation folds, inflating performance metrics unrealistically.

To address this issue, we also tested SMOTE applied within each fold of the CV (in-CV), thus ensuring that oversampling was performed only on the training data in each split. In this more realistic setting, MLP and RF again stood out ($F1 = 0.861 \pm 0.010$ and 0.856 ± 0.009, respectively), followed by XGBoost and Bagging, all of which maintained strong performance with better generalization

Table 3. Mean and standard deviation of CV metrics for each ML method and balancing strategy.

Balancing	Method	F1 score	Recall	Precision	Accuracy
No sampling	AdaBoost	0.786 ± 0.025	0.980 ± 0.002	0.725 ± 0.024	0.857 ± 0.025
	Bagging	0.849 ± 0.013	0.986 ± 0.001	0.754 ± 0.019	0.971 ± 0.006
	DT	0.727 ± 0.009	0.973 ± 0.002	0.726 ± 0.020	0.744 ± 0.010
	LR	0.815 ± 0.006	0.983 ± 0.001	0.745 ± 0.009	0.898 ± 0.004
	MLP	0.874 ± 0.006	0.988 ± 0.001	$\mathbf{0.836 \pm 0.013}$	0.923 ± 0.010
	RF	0.845 ± 0.015	0.986 ± 0.001	0.736 ± 0.019	$\mathbf{0.986 \pm 0.003}$
	XGBoost	$\mathbf{0.887 \pm 0.012}$	$\mathbf{0.989 \pm 0.001}$	0.829 ± 0.019	0.953 ± 0.003
Cost sensitive	AdaBoost	0.795 ± 0.018	0.981 ± 0.002	0.867 ± 0.027	0.735 ± 0.024
	Bagging	0.852 ± 0.014	0.986 ± 0.001	0.968 ± 0.012	0.758 ± 0.021
	DT	0.733 ± 0.020	0.976 ± 0.002	0.753 ± 0.015	0.745 ± 0.024
	LR	0.581 ± 0.004	0.931 ± 0.001	0.420 ± 0.003	$\mathbf{0.942 \pm 0.012}$
	MLP	0.883 ± 0.010	0.989 ± 0.001	0.930 ± 0.015	0.852 ± 0.017
	RF	0.842 ± 0.009	0.986 ± 0.001	$\mathbf{0.987 \pm 0.008}$	0.739 ± 0.016
	XGBoost	$\mathbf{0.894 \pm 0.013}$	$\mathbf{0.990 \pm 0.001}$	0.931 ± 0.012	0.860 ± 0.020
SMOTE (pre-CV)	AdaBoost	0.949 ± 0.001	0.948 ± 0.001	0.943 ± 0.002	0.954 ± 0.003
	Bagging	$\mathbf{0.996 \pm 0.000}$	$\mathbf{0.996 \pm 0.000}$	$\mathbf{0.994 \pm 0.001}$	0.998 ± 0.000
	DT	0.970 ± 0.001	0.970 ± 0.001	0.964 ± 0.001	0.976 ± 0.001
	LR	0.953 ± 0.002	0.953 ± 0.002	0.949 ± 0.006	0.957 ± 0.007
	MLP	0.995 ± 0.000	0.995 ± 0.000	0.990 ± 0.001	$\mathbf{0.999 \pm 0.000}$
	RF	0.993 ± 0.000	0.993 ± 0.001	0.991 ± 0.001	0.996 ± 0.000
	XGBoost	0.990 ± 0.001	0.990 ± 0.001	0.986 ± 0.000	0.994 ± 0.000
SMOTE (in-CV)	AdaBoost	0.622 ± 0.011	0.943 ± 0.002	0.469 ± 0.008	$\mathbf{0.925 \pm 0.013}$
	Bagging	$\mathbf{0.886 \pm 0.015}$	$\mathbf{0.988 \pm 0.002}$	$\mathbf{0.897 \pm 0.018}$	0.876 ± 0.020
	DT	0.672 ± 0.010	0.957 ± 0.001	0.569 ± 0.015	0.821 ± 0.022
	LR	0.644 ± 0.022	0.949 ± 0.005	0.502 ± 0.026	0.898 ± 0.014
	MLP	0.861 ± 0.010	0.986 ± 0.001	0.852 ± 0.021	0.871 ± 0.013
	RF	0.856 ± 0.009	0.985 ± 0.001	0.848 ± 0.012	0.866 ± 0.022
	XGBoost	0.828 ± 0.007	0.982 ± 0.001	0.795 ± 0.013	0.859 ± 0.015

and more meaningful standard deviations. The cost-sensitive strategy also performed strongly, particularly with XGBoost and MLP ($F1 = 0.894 \pm 0.013$ and 0.883 ± 0.010), although Logistic Regression suffered a notable drop in precision and F1-score. This was expected given the model's linear nature and sensitivity to class imbalance, even with weighting. In the no sampling configuration, the best results were obtained again with XGBoost and MLP, followed by RF and LR. Although precision and recall were reasonably balanced in this setting, F1-scores were consistently lower than those achieved with any of the balancing strategies.

Table 4 shows the classification performance of each ML model on the held-out test set across all balancing strategies. As expected from the CV phase, the results reveal that the best performing models in terms of F1-score were again XGBoost and MLP, particularly when using SMOTE either before or within CV. The highest F1-score was achieved by XGBoost trained with SMOTE during CV ($F1 = 0.8668$), followed closely by MLP ($F1 = 0.8476$) and RF ($F1 = 0.8533$) under the same strategy. These models also reported strong recall and precision trade-offs, with AUROC and AUPRC consistently above 0.91.

Table 4. Performance on the test set for each ML method and balancing strategy.

Balancing	Method	F1	Recall	Precision	Accuracy	TN	FN	FP	TP	AUROC	AUPRC
No sampling	AdaBoost	0.7869	0.7344	0.8475	0.9797	12966	187	93	517	0.9769	0.8480
	Bagging	0.8374	0.7386	0.9665	0.9853	13041	184	18	520	0.9720	0.8938
	DT	0.7408	0.7472	0.7346	0.9733	12869	178	190	526	0.8663	0.5618
	LR	0.8158	0.7486	0.8963	0.9827	12998	177	61	527	0.9817	0.8825
	MLP	0.8536	**0.8196**	0.8904	0.9856	12988	**127**	71	577	0.9889	0.9219
	RF	0.8430	0.7358	**0.9867**	0.9860	13052	186	**7**	518	0.9874	0.9187
	XGBoost	**0.8807**	0.8182	0.9536	**0.9887**	**13031**	128	28	**576**	**0.9895**	**0.9309**
Cost sensitive	AdaBoost	0.7869	0.7344	0.8475	0.9797	12966	187	93	517	0.9769	0.8480
	Bagging	0.8429	0.7472	0.9669	0.9858	13041	178	18	526	0.9688	0.8906
	DT	0.7446	0.7287	0.7611	0.9744	12898	191	161	513	0.8582	0.5685
	LR	0.5750	**0.9418**	0.4139	0.9288	12120	**41**	939	**663**	0.9829	0.8525
	MLP	0.8724	0.8452	0.9015	0.9874	12994	109	65	595	**0.9895**	0.9311
	RF	0.8392	0.7301	**0.9866**	0.9857	**13052**	190	**7**	514	0.9872	0.9155
	XGBoost	**0.8852**	0.8438	0.9310	**0.9888**	13015	110	44	594	0.9890	**0.9339**
SMOTE (pre-CV)	AdaBoost	0.5980	0.8778	0.4534	0.9396	12314	86	745	618	0.9740	0.8254
	Bagging	0.8280	0.8480	0.8089	0.9820	12918	107	141	597	0.9856	0.9066
	DT	0.6854	0.8125	0.5927	0.9619	12666	132	393	572	0.8912	0.4912
	LR	0.6105	**0.9219**	0.4564	0.9398	12286	**55**	773	**649**	0.9824	0.8534
	MLP	0.8438	0.8480	0.8397	0.9839	12945	107	114	597	0.9872	0.9234
	RF	0.8498	0.8480	0.8516	0.9847	12955	107	104	597	**0.9886**	0.9205
	XGBoost	**0.8721**	0.8523	**0.8929**	**0.9872**	**12987**	104	**72**	600	0.9881	**0.9308**
SMOTE (in-CV)	AdaBoost	0.6355	0.8793	0.4976	0.9484	12434	85	625	619	0.9767	0.8430
	Bagging	0.8151	0.8452	0.7870	0.9804	12898	109	161	595	0.9834	0.8980
	DT	0.6809	0.8153	0.5845	0.9609	12651	130	408	574	0.8920	0.4860
	LR	0.6091	**0.9219**	0.4548	0.9395	12281	**55**	778	**649**	0.9824	0.8537
	MLP	0.8476	0.8452	0.8500	0.9845	12954	109	105	595	0.9876	0.9184
	RF	0.8533	0.8509	0.8557	0.9850	12958	105	101	599	0.9885	0.9194
	XGBoost	**0.8668**	0.8551	**0.8788**	**0.9866**	**12976**	102	**83**	602	**0.9894**	**0.9325**

In contrast, LR exhibited limited effectiveness across most settings, especially in the cost-sensitive and SMOTE-based configurations, with a notable drop in precision and accuracy. This indicates that LR may not capture the complex patterns in the data as efficiently as ensemble or neural approaches.

The cost-sensitive strategy offered competitive results, particularly for MLP and XGBoost, although in most cases, SMOTE-based approaches led to superior precision-recall balance. Models trained with SMOTE before CV achieved strong recall but tended to suffer from lower precision, especially in methods like AdaBoost and LR, indicating a higher rate of false positives. On the other hand, SMOTE applied within each fold (in-CV) proved to be a more robust strategy.

4 Conclusions and Future Works

In this study, we presented a comprehensive pipeline for analyzing recreational drug use discourse on Twitter, with a focus on three substances: ecstasy, GHB, and 2C-B. Starting from over 6.7 million tweets, we developed a domain-informed annotation strategy combining weak supervision, expert-curated polarity labels, and semantic matching through slang and MetaMap terms. This process led to the creation of a polarity-labeled dataset of over 92,000 tweets, capturing both positive and negative mentions of drug-related experiences. We evaluated multiple ML classifiers under different strategies to mitigate class imbalance, including cost-sensitive learning and SMOTE-based oversampling. Our results demonstrate that appropriate balancing, particularly SMOTE applied within each fold of CV, significantly enhances classification performance. Ensemble methods such as XGBoost, RF, and Bagging, along with neural networks, consistently outperformed simpler models like LR. The highest test F1-score was obtained using XGBoost with in-fold SMOTE, reaching 0.8668 and supported by strong precision, recall, AUROC, and AUPRC values.

Despite these promising results, several challenges remain. The strong class imbalance in real-world data, particularly for less frequently mentioned substances like GHB and 2C-B, limits the generalizability of the models. Additionally, the reliance on weak supervision and term-level annotations, while scalable, may miss subtleties in language use and context.

In future work, we plan to expand our annotation framework by incorporating large language models to refine polarity inference and disambiguate context-dependent mentions. We also aim to explore further temporal and geographical patterns in user discourse, as well as the interplay between drug mentions and mental health indicators. Finally, we envision extending this approach to other substances and social platforms to build a broader, cross-source understanding of public perceptions and self-reported experiences related to drug use.

Acknowledgments. We would like to acknowledge the contributions of Alejandro Gouloumis Contreras, whose Master's Thesis [13] laid the groundwork for the extraction and initial processing of the Twitter data used in this study.

References

1. Aronson, A.R.: Effective mapping of biomedical text to the UMLS Metathesaurus: the MetaMap program. In: Proceedings of the AMIA Symposium, pp. 17–21 (2001)

2. Bodenreider, O.: The unified medical language system (UMLS): integrating biomedical terminology. Nucl. Acids Res. **32**(suppl_1), D267–D270 (2004). https://doi.org/10.1093/nar/gkh061
3. Breiman, L.: Bagging predictors. Mach. Learn. **24**(2), 123–140 (1996). https://doi.org/10.1007/BF00058655
4. Breiman, L.: Random forests. Mach. Learn. **45**(1), 5–32 (2001). https://doi.org/10.1023/A:1010933404324
5. Cameron, D., et al.: PREDOSE: a semantic web platform for drug abuse epidemiology using social media. J. Biomed. Inform. **46**(6), 985–997 (2013). https://doi.org/10.1016/j.jbi.2013.07.007
6. Carabot, F., et al.: Understanding public perceptions and discussions on opioids through twitter: cross-sectional infodemiology study. J. Med. Internet Res. **25**(1), e50013 (2023). https://doi.org/10.2196/50013
7. Castillo-Toledo, C., et al.: Insights from the twittersphere: a cross-sectional study of public perceptions, usage patterns, and geographical differences of tweets discussing cocaine. Front. Psychiatry **15** (2024). https://doi.org/10.3389/fpsyt.2024.1282026
8. Chawla, N.V., Bowyer, K.W., Hall, L.O., Kegelmeyer, W.P.: SMOTE: synthetic minority over-sampling technique. J. Artif. Intell. Res. **16**, 321–357 (2002). https://doi.org/10.1613/jair.953
9. Chen, T., Guestrin, C.: XGBoost: a scalable tree boosting system. In: Proceedings of the 22nd ACM SIGKDD International Conference on Knowledge Discovery and Data Mining, pp. 785–794 (2016). https://doi.org/10.1145/2939672.2939785
10. Cox, D.R.: The Regression Analysis of Binary Sequences. J. Roy. Stat. Soc. Ser. B (Methodol.) **20**(2), 215–242 (1958)
11. Freund, Y., Schapire, R.E.: A decision-theoretic generalization of on-line learning and an application to boosting. J. Comput. Syst. Sci. **55**(1), 119–139 (1997). https://doi.org/10.1006/jcss.1997.1504
12. Gahlinger, P.M.: Club drugs: MDMA, gamma-hydroxybutyrate (GHB), rohypnol, and ketamine. Am. Fam. Phys. **69**(11), 2619–2627 (2004)
13. Gouloumis Contreras, A.: Analysis and characterisation of recreational drug effects using public sources. Master's thesis, Universidad Politécnica de Madrid (2021). https://oa.upm.es/68755/
14. Ji, X., Chun, S.A., Wei, Z., Geller, J.: Twitter sentiment classification for measuring public health concerns. Soc. Netw. Anal. Min. **5**(1), 1–25 (2015). https://doi.org/10.1007/s13278-015-0253-5
15. Meng, H.W., Kath, S., Li, D., Nguyen, Q.C.: National substance use patterns on Twitter. PLoS ONE **12**(11), e0187691 (2017). https://doi.org/10.1371/journal.pone.0187691
16. Mikolov, T., Chen, K., Corrado, G., Dean, J.: Efficient estimation of word representations in vector space (2013). http://arxiv.org/abs/1301.3781
17. Nasralah, T., El-Gayar, O., Wang, Y.: Social media text mining framework for drug abuse: development and validation study with an opioid crisis case analysis. J. Med. Internet Res. **22**(8), e18350 (2020). https://doi.org/10.2196/18350
18. Nikfarjam, A., Sarker, A., O'Connor, K., Ginn, R., Gonzalez, G.: Pharmacovigilance from social media: mining adverse drug reaction mentions using sequence labeling with word embedding cluster features. J. Am. Med. Inform. Assoc. **22**(3), 671–681 (2015). https://doi.org/10.1093/jamia/ocu041
19. Plant, M., Robertson, R., Plant, M., Miller, P.: The consequences of drug use: the good, the bad, and the ugly. In: Drug Nation: Patterns, Problems, Panics & Policies (2010). https://doi.org/10.1093/med/9780199544790.003.004

20. Quinlan, J.R.: Induction of decision trees. Mach. Learn. **1**(1), 81–106 (1986). https://doi.org/10.1007/BF00116251
21. Rao, V.K., et al.: Digital epidemiology of prescription drug references on X (formerly twitter): neural network topic modeling and sentiment analysis. J. Med. Internet Res. **26**(1), e57885 (2024). https://doi.org/10.2196/57885
22. Ricaurte, G.A., McCann, U.D.: Recognition and management of complications of new recreational drug use. Lancet **365**(9477), 2137–2145 (2005). https://doi.org/10.1016/S0140-6736(05)66737-2
23. Rodríguez-González, A., et al.: Identifying polarity in tweets from an imbalanced dataset about diseases and vaccines using a meta-model based on machine learning techniques. Appl. Sci. **10**(24), 9019 (2020). https://doi.org/10.3390/app10249019
24. Rumelhart, D.E., Hinton, G.E., Williams, R.J.: Learning representations by back-propagating errors. Nature **323**(6088), 533–536 (1986). https://doi.org/10.1038/323533a0
25. Ruths, D., Pfeffer, J.: Social media for large studies of behavior. Science **346**(6213), 1063–1064 (2014). https://doi.org/10.1126/science.346.6213.1063
26. Sarker, A., Gonzalez, G.: Portable automatic text classification for adverse drug reaction detection via multi-corpus training. J. Biomed. Inform. **53**, 196–207 (2015). https://doi.org/10.1016/j.jbi.2014.11.002
27. Schifano, F., Orsolini, L., Duccio Papanti, G., Corkery, J.M.: Novel psychoactive substances of interest for psychiatry. World Psychiatry **14**(1), 15–26 (2015). https://doi.org/10.1002/wps.20174
28. Shyu, R., Weng, C.: Enabling semantic topic modeling on twitter using MetaMap. In: AMIA Joint Summits on Translational Science Proceedings. AMIA Joint Summits on Translational Science, vol. 2024, pp. 670–678 (2024)
29. Tassone, J., Yan, P., Simpson, M., Mendhe, C., Mago, V., Choudhury, S.: Utilizing deep learning and graph mining to identify drug use on Twitter data. BMC Med. Inform. Decis. Mak. **20**(11), 304 (2020). https://doi.org/10.1186/s12911-020-01335-3
30. Teter, C.J., Guthrie, S.K.: A comprehensive review of MDMA and GHB: two common club drugs. Pharmacother.: J. Hum. Pharmacol. Drug Ther. **21**(12), 1486–1513 (2001). https://doi.org/10.1592/phco.21.20.1486.34472
31. Řehářek, R., Sojka, P.: Software framework for topic modelling with large corpora. In: Proceedings of LREC 2010 workshop New Challenges for NLP Frameworks, pp. 46–50 (2010)

Automated Classification of Ischemic Stroke Subtypes from Electronic Health Records Using Large Language Models

Alejandro Vásquez[1], Cristina Rubio-Escudero[2(✉)],
and Germán Antonio Escobar-Rodríguez[1]

[1] Group for Research and Innovation in Biomedical Informatics, Biomedical
Engineering, and Health Economy, Institute of Biomedicine of Seville, IBiS/"Virgen
del Rocío" University Hospital/CSIC/University of Seville, Seville, Spain
[2] Department of Computer Languages and Systems, University of Sevilla,
Seville, Spain
crubioescudero@us.es

Abstract. This paper presents a study on the application of Large Language Models (LLMs) for the automatic classification of ischemic stroke subtypes from Electronic Health Records (EHRs). The study involves the creation and analysis of four datasets, an evaluation of model performance using standard classification metrics, and a comparison with expert manual labeling. Results suggest LLMs can offer promising outcomes even under computational constraints, paving the way for clinical automation and diagnostic support.

Keywords: Stroke classification · Ischemic stroke · Electronic Health Records · Large Language Models · Clinical NLP · Biomedical AI

1 Introduction

Stroke remains one of the foremost causes of mortality and long-term disability on a global scale, exerting a substantial socioeconomic and public health burden [20]. It affects millions of individuals annually and imposes considerable challenges on healthcare systems worldwide. Broadly, strokes are classified into two major types: ischemic and hemorrhagic. Ischemic strokes, which result from an obstruction in a blood vessel supplying the brain, account for approximately 85% of all stroke incidents. These cases are further subdivided into distinct subtypes including cardioembolic, atherothrombotic, lacunar, cryptogenic, and other less common categories. Each of these subtypes is characterized by unique etiologies, pathophysiological mechanisms, risk profiles, and therapeutic responses. As such, the accurate and timely classification of ischemic stroke subtypes is of paramount importance in determining the appropriate course of treatment, guiding secondary prevention strategies, and improving patient outcomes [10].

Despite the clinical significance of ischemic stroke subtyping, achieving reliable classification remains a complex task. It requires comprehensive diagnostic

© The Author(s), under exclusive license to Springer Nature Switzerland AG 2026
A. López Fernández et al. (Eds.): CIABiomed 2025, LNBI 16148, pp. 268–281, 2026.
https://doi.org/10.1007/978-3-032-10661-2_21

workups involving neuroimaging, cardiac assessments, vascular studies, and laboratory tests. Moreover, expert interpretation of a multitude of heterogeneous data sources is often necessary to distinguish between overlapping clinical presentations. In many settings, limited access to specialized diagnostics and stroke neurologists further complicates the situation, leading to delays or inaccuracies in classification. These limitations underscore the urgent need for automated, scalable, and data-driven solutions that can support clinicians in stratifying patients more efficiently and consistently.

One promising avenue for addressing this challenge lies in leveraging the vast repositories of Electronic Health Records (EHRs). These records house a rich collection of patient information, including structured data (e.g., vital signs, lab values) and unstructured narratives such as admission notes, discharge summaries, radiology reports, and consultation records. Within the unstructured components of EHRs lies valuable contextual and temporal information that can offer crucial clues for stroke subtyping. However, extracting clinically relevant insights from free-text data remains a major hurdle due to variations in language, medical jargon, and documentation styles across institutions and clinicians.

Recent advancements in Natural Language Processing (NLP) have introduced powerful tools for interpreting medical language at scale. In particular, Large Language Models (LLMs) such as BERT [6], GPT [4], and their biomedical adaptations have demonstrated unprecedented capabilities in understanding, summarizing, and classifying clinical texts. These models are trained on vast corpora and can capture complex linguistic patterns, domain-specific terminology, and contextual relationships with high fidelity. Their ability to perform few-shot or even zero-shot learning makes them particularly attractive for clinical tasks where annotated data is scarce or expensive to obtain.

Building upon this foundation, the present study explores the feasibility of applying LLMs to the task of automated ischemic stroke subtype classification using unstructured EHR data. We propose a multi-stage methodological framework that includes data curation, preprocessing, prompt engineering, and model evaluation. Our pipeline is designed to handle noisy clinical narratives, identify relevant content, and produce interpretable classification outputs. We evaluate several open-source LLMs under privacy-preserving constraints, and we benchmark their performance using standard classification metrics such as precision, recall, F1-score, and Cohen's kappa. In addition, we compare the outputs of our models against expert annotations to assess agreement and identify common sources of error. Through this approach, we aim to demonstrate the potential of LLMs to support clinical decision-making, reduce diagnostic variability, and facilitate the development of intelligent health information systems tailored to cerebrovascular diseases.

2   Background and Related Work

2.1   Clinical Classification of Ischemic Stroke

Ischemic stroke, which constitutes the majority of all stroke cases, encompasses a variety of pathophysiological subtypes that differ in etiology, prognosis, and treatment response. Accurate classification is critical for guiding acute management and secondary prevention strategies. Among the various systems proposed for ischemic stroke subtyping, the TOAST (Trial of ORG 10172 in Acute Stroke Treatment) classification remains the most widely accepted and clinically utilized framework [10]. It defines five major etiological categories: (1) large-artery atherosclerosis, characterized by significant stenosis or occlusion of major extracranial or intracranial arteries; (2) cardioembolism, typically associated with high-risk cardiac sources such as atrial fibrillation or valvular disease; (3) small-vessel occlusion, also referred to as lacunar stroke, generally caused by lipohyalinosis or microatheroma affecting small penetrating arteries; (4) stroke of other determined etiology, encompassing rare causes such as dissection, vasculitis, or coagulopathies; and (5) stroke of undetermined etiology, either due to incomplete evaluation, multiple potential causes, or absence of identifiable cause despite thorough investigation.

Despite its widespread use, the TOAST system has been criticized for limited inter-rater reliability, oversimplification, and inability to accommodate cases with overlapping or evolving pathologies [3]. In response to these limitations, newer constructs such as the ASCOD phenotyping system and the CCS (Causative Classification System) have been proposed to offer more nuanced and evidence-based classification [2]. Furthermore, the concept of Embolic Stroke of Undetermined Source (ESUS) has gained traction in recent years, especially in cryptogenic stroke cases where embolism is suspected but a definitive source is not identified [12]. ESUS provides a research framework for identifying covert atrial fibrillation, paradoxical embolism, or nonstenotic atherosclerosis, all of which require tailored diagnostic strategies and anticoagulant considerations.

2.2   Risk Factors and Epidemiology

The incidence and prevalence of ischemic stroke are closely tied to the distribution of vascular risk factors in the population. These risk factors are broadly categorized into modifiable and non-modifiable types. Modifiable risk factors include hypertension, type 2 diabetes mellitus, dyslipidemia, smoking, physical inactivity, obesity, excessive alcohol consumption, and poor diet—all of which are targets for public health interventions and individualized treatment plans [11]. Hypertension remains the most potent modifiable risk factor and is present in more than 75% of stroke patients at the time of diagnosis [15]. Effective blood pressure control alone could prevent a significant proportion of recurrent strokes and cardiovascular complications.

Non-modifiable risk factors such as age, sex, race/ethnicity, and genetic predisposition also contribute significantly to stroke susceptibility. The risk of stroke doubles with each decade after the age of 55, and certain populations, such as African Americans and South Asians, exhibit higher stroke incidence due to genetic and socioeconomic determinants [8]. Additionally, sex-specific factors including hormonal status and pregnancy-related complications play a role in stroke risk among women [5].

Recent decades have witnessed a shift in the global burden of ischemic stroke, with low- and middle-income countries experiencing the most significant increases in incidence, mortality, and disability-adjusted life years (DALYs) [9]. Urbanization, sedentary lifestyles, and increased exposure to dietary and environmental risk factors have contributed to this trend. Furthermore, improvements in acute care and secondary prevention have reduced case-fatality rates in high-income settings, but disparities in access to stroke units, imaging, and rehabilitation persist globally.

2.3   Large Language Models in Biomedical NLP

The application of Artificial Intelligence (AI), and more specifically Natural Language Processing (NLP), has gained momentum in biomedical informatics as a means to extract, interpret, and utilize information embedded in unstructured clinical text. Large Language Models (LLMs), such as BERT (Bidirectional Encoder Representations from Transformers) [6], GPT (Generative Pre-trained Transformer) [4], and LLaMA (Large Language Model Meta AI) [21], represent a significant paradigm shift in the way language models understand and generate human language.

In the biomedical domain, specialized models such as BioBERT [16], Clinical-BERT [13], and PubMedGPT [1] have been trained on domain-specific corpora including PubMed abstracts, clinical notes, and EHR data. These models have demonstrated state-of-the-art performance on a range of tasks including named entity recognition (NER), relation extraction, document classification, question answering, and clinical summarization. Their success stems from their ability to capture context, disambiguate terminology, and adapt to the idiosyncrasies of medical language, including acronyms, negations, and temporal expressions.

Despite their promise, several challenges limit the routine deployment of LLMs in clinical environments. These include data privacy concerns, especially when models are trained or deployed using sensitive patient information; high computational demands; susceptibility to hallucination and biased outputs; and the lack of explainability and regulatory frameworks for medical AI [18], [?]. Furthermore, clinical validation and reproducibility remain critical barriers to adoption. There is an urgent need for interpretable, domain-adapted, and resource-efficient models that can operate within the ethical and legal constraints of healthcare systems.

Nevertheless, the integration of LLMs with structured data, decision support systems, and real-time hospital information systems holds enormous potential. These tools could assist in identifying patterns, summarizing longitudinal patient histories, triaging diagnostic pathways, and even generating draft assessments or discharge summaries. Their role in augmenting, rather than replacing, clinical expertise is essential for fostering adoption and trust.

2.4   Prompt Engineering

Prompt engineering is an emerging area of research that focuses on the design, refinement, and implementation of instructions intended to elicit effective responses from Large Language Models (LLMs) across a wide range of tasks. It is, in essence, the practice of interacting strategically with artificial intelligence systems to optimize their performance, relevance, and interpretability in specific applications [17]. As LLMs become more powerful and generalizable, the quality and structure of the prompts used to guide them have become critical determinants of their effectiveness in real-world scenarios.

A central concept within prompt engineering is *in-context learning* (ICL), a paradigm that enables LLMs to perform new tasks by conditioning on a few demonstrations provided within the input, without any parameter updates [7]. ICL allows the model to generalize to unseen domains or instructions by observing patterns in a small set of example pairs, typically embedded directly in the prompt. As described in [4], "In-context learning is a key paradigm in large language models (LLMs) that enables them to generalize to new tasks and domains by simply prompting these models with a few exemplars without explicit parameter updates." This capability represents a fundamental shift in the way models are used, reducing the need for fine-tuning and allowing greater flexibility in adapting to multiple tasks. ICL also serves as the foundation for popular prompting strategies such as few-shot and zero-shot prompting, further discussed in Sect. 2.3.1.

A prompt, in practical terms, is a natural language input designed to activate, guide, or constrain the behavior of an LLM. It can range from a simple instruction ("Summarize the following text") or question ("What is ischemic stroke?") to a more complex interaction, such as a multi-turn dialogue or a structured input-output mapping. The prompt serves not only to define the task but also to indicate the expected style, format, and level of detail of the output. As such, prompt quality has a direct impact on task performance, particularly for open-ended generation tasks, multi-step reasoning, and domain-specific applications [22].

Several key components constitute an effective prompt:

- **Task description:** This element clearly specifies what the model is expected to do. The instruction should be concise yet unambiguous, and it should define the task goal, output format, and any domain constraints. For example, when asking the model to classify a medical note into stroke subtypes, the prompt should explicitly state the classes and labeling rules.
- **Input data representation:** Depending on the nature of the input (e.g., plain text, tables, knowledge graphs), it may be necessary to pre-process or linearize structured data for readability. Techniques like serialization of knowledge triplets, table-to-text conversion, or use of pseudo-code can help LLMs understand non-natural language inputs more effectively [19].

- **Contextual information:** For complex tasks such as question answering or summarization, including background knowledge or retrieved documents in the prompt can substantially improve model accuracy. In clinical NLP, for instance, contextual information may include prior patient history or relevant imaging findings. When using ICL, it is also beneficial to include one or more illustrative examples that model the desired task behavior.
- **Prompt style and instruction tuning:** The phrasing and tone of the prompt can greatly influence model output. Prompts can be posed as questions, commands, or conditional instructions. Recent studies have shown that meta-prompts like "Let's think step by step" or "You are a clinical expert" can improve reasoning and domain alignment [14]. In the case of conversational models, it is often more effective to decompose complex prompts into multiple sequential turns, mimicking a natural interaction.

As prompt engineering matures, a growing number of techniques have emerged to enhance model outputs, including chain-of-thought prompting, role-based prompting, and program-aided prompting.

3   Methodology

This section outlines the methodological framework adopted for the identification, classification, and subtyping of hospital discharge reports related to ischemic stroke. We describe the data sources, filtering procedures, structure of the resulting datasets, the use of Large Language Models (LLMs) for automated classification, and the evaluation metrics applied to assess model performance.

3.1   Data Sources

The dataset analyzed in this study was derived from a large repository of 57,873 hospital discharge reports (Informes de Alta de la Historia Clínica Electrónica, IAHCEs) obtained from the electronic medical records of Hospital Universitario Virgen del Rocío (HUVR), corresponding to 11,958 unique patients. These patients were all recorded in the Andalusian Health Service database as having experienced at least one cerebrovascular episode during the period 2000â€“2021.

The primary data sources were:

- **Electronic Health Records (EHRs)** from the Diraya system, which integrates structured and unstructured clinical information such as diagnostic labels, treatment plans, personal and family medical history, and outcome summaries. Each hospitalization results in an IAHCE document, which summarizes the patient's condition, evolution, diagnosis, and discharge decisions.
- **Population Health Database (BPS)**, a structured information system that consolidates patient-level clinical data, diagnoses (coded under ICD-9 or ICD-10), and risk factor profiles. This system was essential for identifying relevant stroke episodes and retrieving cardiovascular comorbidity data.

The EHRs used were de-identified and accessed within the secured hospital intranet. The local nature of the infrastructure imposed constraints that necessitated the use of locally deployed LLMs, as external APIs were not permitted due to data protection regulations.

3.2   Data Extraction and Filtering

Given that the EHRs span more than two decades and are recorded in a mix of structured and unstructured formats, a multi-stage filtering process was applied to curate a working dataset focused on ischemic stroke. Two filtering criteria were established:

1. Patients must have experienced at least one acute stroke episode (ischemic or hemorrhagic), as recorded in the BPS.
2. Patients must have been hospitalized for any acute clinical event between January 1, 2000 and December 31, 2021, resulting in the generation of at least one IAHCE.

After applying these filters, the dataset consisted of 57,873 IAHCEs from 11,958 patients. This translates to an average of 4.84 hospital admissions per patient over the studied period. Importantly, not all IAHCEs corresponded to stroke-related events. Therefore, a further refinement step was required to isolate those related to ischemic stroke.

3.3   Structured and Unstructured Data Integration

A key methodological challenge was the integration of structured data from the BPS with unstructured free-text reports from the IAHCEs. The structured data allowed us to reliably distinguish between ischemic and hemorrhagic stroke episodes. By filtering for cases with confirmed ischemic stroke diagnoses, we reduced the working corpus to 42,398 IAHCEs corresponding to 8,327 patients.

However, the discharge summaries in the EHRs were not tagged with structured labels indicating stroke subtypes (e.g., cardioembolic, atherothrombotic, lacunar), nor did they consistently present structured data regarding the presence of risk factors. To address this, we relied on LLM-based text mining techniques to extract relevant clinical concepts and perform automated classification.

3.4   Sample Construction for Annotation and Model Training

Based on statistical sampling calculations (see Sect. 3.1.2.2, not included here), the minimum sample size required for robust analysis was determined to be 383 reports. However, to improve generalizability and reduce variance, a total of 1,400 reports were reviewed during the project, with 385 discharge summaries ultimately selected for model evaluation. These records underwent manual expert annotation to establish a reference classification for ischemic stroke subtypes.

Each IAHCE included multiple narrative sections: patient identification, admission date, medical history, physical examination findings, progression notes, family history, main and secondary diagnoses, and prescribed treatment. These were used to construct the prompts fed to the LLMs and to evaluate their ability to infer correct classifications under realistic conditions.

3.5   Challenges of Free-Text Medical Records

Unlike structured forms, free-text discharge summaries are highly variable in format, terminology, and completeness. This poses significant obstacles for traditional rule-based NLP pipelines. The information needed to determine stroke subtypes (e.g., atrial fibrillation, atherosclerosis, lacunar infarcts) is often distributed across multiple paragraphs or embedded within clinical jargon. Moreover, the reports may include co-occurring conditions or ambiguous terms that complicate automated parsing.

To manage this, we employed a pre-processing pipeline involving:

- Regular expression-based section segmentation to isolate relevant content.
- Named entity recognition to extract clinical entities associated with stroke mechanisms and risk factors.
- Anonymization filters to ensure GDPR compliance and ethical standards.

Following this, cleaned and segmented text was embedded into structured prompts for each classification task (e.g., binary stroke detection, subtype identification). The application of prompt engineering and in-context learning strategies is detailed in Sect. 2.3.

3.6   LLM Deployment Constraints

Due to privacy regulations, only locally hosted models were used. Models of up to 20 billion parameters were tested within the hospital's private network. Performance limitations due to hardware and memory constraints were mitigated by prompt optimization and input truncation strategies. No cloud-based APIs or commercial services were employed in compliance with institutional data governance policies.

3.7   LLMs and Prompt Engineering

To perform automated classification of discharge summaries, we employed a range of open-source Large Language Models (LLMs) with parameter sizes ranging from 7 billion to 20 billion, all hosted locally within the hospital's secure network. This setup ensured compliance with data protection policies, as no patient data was transmitted to external servers or cloud-based inference endpoints. The models tested included variations of LLaMA, Falcon, and BLOOM architectures, which were fine-tuned or adapted for biomedical tasks when possible.

Given the variability and ambiguity of clinical narratives, the construction of prompts was a critical factor influencing model performance. Prompt engineering was used to convert free-text reports into structured instructions interpretable by the LLMs. This included: (i) clearly defined task descriptions, (ii) inclusion of input examples, (iii) explicit labeling schemas, and (iv) request for outputs in predefined formats.

In particular, we applied *few-shot prompting*, embedding up to 3 labeled examples within the prompt to guide the model toward the correct classification logic. This method was most effective when the stroke subtype distinctions were subtle, such as differentiating cardioembolic from cryptogenic etiologies. In scenarios where context was minimal or ambiguous, *zero-shot* prompting was used, relying on the model's generalization capabilities.

All prompts were optimized through empirical testing using small batches of annotated reports to assess the clarity and reliability of generated responses. Output consistency, formatting compliance, and hallucination tendencies were also monitored. The overall prompting strategy followed the principles outlined in Sect. 2.3.

3.8   Classification Tasks and Evaluation Metrics

The experimental evaluation was structured around three core classification tasks:

1. **Stroke vs. non-stroke classification:** This binary classification task aimed to filter discharge reports that explicitly described stroke-related hospitalizations from those addressing unrelated clinical events.
2. **Ischemic vs. hemorrhagic stroke distinction:** Among stroke-related reports, this task distinguished ischemic from hemorrhagic episodes based on clinical descriptions and diagnostic conclusions.
3. **Ischemic stroke subtype classification:** For cases identified as ischemic stroke, we performed multi-class classification into standard etiological subtypes, including cardioembolic, atherothrombotic, lacunar, cryptogenic, and ESUS (Embolic Stroke of Undetermined Source).

Each model output was compared to manually annotated ground-truth labels provided by clinical experts. Evaluation metrics were calculated to quantify the quality of predictions:

- **Precision** (positive predictive value): Precision $= \frac{TP}{TP+FP}$
- **Recall** (sensitivity): Recall $= \frac{TP}{TP+FN}$
- **F1-score:** Harmonic mean of precision and recall.
- **Specificity** (true negative rate): Specificity $= \frac{TN}{TN+FP}$
- **Accuracy:** Overall correctness of predictions across all classes.
- **Cohen's Kappa** (κ): Agreement between model and human annotators beyond chance, used primarily for multi-class subtype classification.

For each classification task, confusion matrices were computed to visualize per-class performance, and macro-averaged scores were used to mitigate class imbalance effects. The methodology ensures a fair and reproducible assessment of the ability of LLMs to interpret unstructured clinical data under real-world constraints.

4 Results

This section presents the results obtained from applying locally deployed Large Language Models (LLMs) to the three key classification tasks outlined in Sect. 3: (i) detection of stroke-related hospital discharge reports, (ii) identification of ischemic vs. hemorrhagic stroke, and (iii) classification of ischemic stroke into etiological subtypes. Performance is reported using standard classification metrics, and comparative observations are made based on class-level distributions and error analysis.

4.1 Binary Stroke Classification: Stroke vs Non-stroke

The initial task aimed to determine whether a given discharge report pertained to a stroke-related episode. The LLM-based classifier yielded a high overall accuracy of **91.5%**, with a sensitivity of **89.2%** and specificity of **93.4%**. The precision was also high (**90.1%**), indicating that the model was capable of distinguishing cerebrovascular events from unrelated hospital admissions with robust reliability.

False negatives in this task generally resulted from cases where stroke was mentioned in the differential diagnosis but ultimately ruled out, or from vague descriptions such as "neurological symptoms under investigation." On the other hand, some false positives emerged from reports of transient ischemic attacks (TIAs) or post-stroke rehabilitation, which the model occasionally mislabeled as acute stroke admissions due to semantic overlap.

These results demonstrate the model's capacity to operate as a pre-filtering tool in clinical NLP pipelines, enabling the extraction of relevant cases from large unstructured corpora with minimal manual screening.

4.2 Ischemic vs Hemorrhagic Stroke Classification

Following the identification of stroke-related reports, the next task involved classifying the episode as either ischemic or hemorrhagic. The model achieved an accuracy of **87.3%**, with a recall of **85.6%** for ischemic stroke and **89.1%** for hemorrhagic cases. Precision rates were similarly balanced, with slightly lower performance observed for ischemic classification.

The confusion in classification primarily arose in cases with insufficient radiological detail or ambiguous descriptions such as "intraparenchymal lesion" or "acute cerebrovascular event." Additionally, some hemorrhagic strokes that evolved from ischemic origins (e.g., hemorrhagic transformation) created semantic ambiguity that challenged the model's binary logic.

Despite these challenges, the model performed favorably compared to previously reported clinical NLP systems, which often rely on structured diagnosis codes (ICD-9/10) alone and ignore free-text nuances. The LLM was able to leverage linguistic cues—such as presence of anticoagulation, midline shift, or subarachnoid hemorrhage mentions—to guide its decisions, showcasing the potential of deep contextual language understanding in EHR settings.

4.3   Ischemic Stroke Subtype Classification

The most complex task in the pipeline involved classifying ischemic stroke cases into TOAST-derived etiological subtypes: cardioembolic, atherothrombotic (large artery), lacunar (small vessel), cryptogenic, and ESUS (Embolic Stroke of Undetermined Source). This task was performed on a curated set of 385 manually annotated discharge summaries.

The model achieved a macro-averaged **F1-score of 0.81**, and overall agreement with human annotations was substantial, with a **Cohen's kappa of 0.76**. This level of agreement indicates that the model's classifications were consistent with expert judgments in the majority of cases, especially considering the difficulty of the task and the unstructured nature of the input data.

Per-class Performance and Observations

- **Cardioembolic strokes** were the most accurately classified subtype, with an F1-score of **0.87**. Discharge reports often explicitly mentioned atrial fibrillation, recent embolic events, or echocardiographic findings suggestive of cardiac origin.
- **Atherothrombotic strokes** followed closely with an F1-score of **0.84**, supported by clear mentions of carotid stenosis, atherosclerosis, or vascular imaging results.
- **Lacunar strokes** presented moderate classification accuracy (F1-score **0.78**), due in part to the minimal symptomatology and frequent absence of detailed radiological findings in the discharge summaries.
- **Cryptogenic** and **ESUS** categories were more challenging, with F1-scores of **0.71** and **0.68**, respectively. These cases often lacked explicit etiological conclusions or presented mixed evidence, which also made them more difficult for human annotators to classify with confidence.

Qualitative analysis revealed that prompt formulation significantly impacted performance. Subtypes with clearly defined patterns (e.g., "atrial fibrillation detected on ECG") benefitted most from few-shot prompting, whereas ambiguous or under-documented cases remained a source of confusion.

Generalization and Model Robustness. Despite the computational and privacy constraints imposed by local deployment, the LLMs exhibited competitive performance in both binary and multi-class settings. While advanced foundation models hosted in cloud environments might yield higher baseline results, our findings support the viability of deploying LLMs in secure hospital infrastructures with limited computational capacity, provided sufficient prompt engineering and contextual embedding are applied.

The promising results in subtype classification also open the door to integrating LLMs into clinical decision-support systems for early diagnosis, stratification, and research cohort assembly, especially in settings where structured labels are incomplete or unreliable.

5 Discussion

The results of this study indicate that even modestly scaled, locally deployed Large Language Models (LLMs) can achieve clinically meaningful performance in the classification of ischemic stroke subtypes from unstructured clinical narratives. The application of prompt engineering techniques played a pivotal role in directing the model's focus toward relevant linguistic cues, enabling robust performance despite variability in report length, terminology, and formatting.

A key strength of the approach lies in its ability to operate within realistic healthcare constraints. The deployment of LLMs in a hospital intranet ensured strict adherence to privacy regulations (e.g., GDPR) while maintaining acceptable inference times. Despite hardware limitations, the models were able to process and classify complex, real-world EHR data with accuracy levels that closely approximate expert human annotation.

Nonetheless, several challenges were observed throughout the process. First, inconsistency in clinical documentation posed a significant hurdle. Some discharge reports were detailed and clearly structured, while others lacked crucial etiological or temporal information. The variability in terminology—ranging from colloquial to highly technical expressions—further complicated model interpretation.

Second, model hallucination, although rare, was a non-trivial concern. In a small number of cases, the LLM inferred relationships or diagnoses that were not explicitly supported by the source text. These artifacts could compromise clinical trust and highlight the need for further refinement in both prompt structure and output validation mechanisms.

Third, while the model was effective in distinguishing well-defined subtypes (e.g., cardioembolic, atherothrombotic), its performance declined in ambiguous or multi-causal scenarios, particularly when multiple comorbidities were present. These cases often require human clinical reasoning, which remains difficult to fully replicate with current LLM architectures.

Lastly, the reliance on unstructured data alone, although powerful, limits the model's access to structured contextual signals (e.g., lab values, imaging codes, medication history). A hybrid approach—combining LLMs with structured data pipelines—could significantly improve classification reliability and generalization.

6 Conclusion and Future Work

This study demonstrates the feasibility and effectiveness of using prompt-based Large Language Models to support the automated classification of ischemic stroke subtypes from real-world hospital discharge reports. The key contributions of this work can be summarized as follows:

- Development of a multi-stage data pipeline integrating structured and unstructured clinical information from EHRs.

- Design and application of in-context learning and prompt engineering strategies to enhance LLM interpretability and specificity.
- Empirical validation of model performance against expert annotations using standard clinical NLP evaluation metrics, including accuracy, F1-score, and Cohen's kappa.

The results provide evidence that even without fine-tuning, general-purpose LLMs—when properly prompted—can extract meaningful insights from noisy clinical narratives and approach human-level diagnostic classification in many settings.

Looking ahead, several directions emerge for future research:

- **Incorporating structured data:** Future models should integrate lab results, imaging metadata, and structured risk factor profiles to support more accurate subtype attribution and reduce dependency on textual completeness.
- **Scaling with larger models:** Access to more powerful multilingual models (e.g., GPT-4, Claude 3, Med-PaLM 2) may further improve classification, particularly in rare or ambiguous cases.
- **Multicenter validation:** Replicating the methodology across institutions and languages will be essential to assess generalizability and robustness in diverse healthcare environments.
- **Explainability and trust:** Integrating mechanisms for output justification (e.g., rationale extraction, evidence highlighting) will be necessary for clinical adoption and user confidence.

Overall, this work lays the foundation for scalable, privacy-preserving, and clinically grounded NLP tools that leverage the power of LLMs to assist in complex medical classification tasks.

Acknowledgments. Research supported by This publication is part of the projects PID2020-117954RB-C22 and PID2023-146037OB-C21, funded by MICIU/AEI/10.13039/501100011033/. "PROPHECIA: Prediction of Risk and Vascular Prevention through Knowledge Generated by the Analysis of Electronic Medical Records of Patients with Acute Stroke"; PI-20/01023, ISCIII, "Artificial Intelligence in the Prevention and Personalized Determination of Stroke Risk. Covid-19 Pandemic and New Emerging Cardiovascular Risk Factors," PIP-0154-2022, Ministry of Health and Consumer Affairs, and the Innovation Activation Platform (PT23/00130), both from the Carlos III Health Institute, the latter co-funded by the European Regional Development Fund.

References

1. Anonymous: Pubmedgpt: A domain-specific gpt for biomedical research. OpenAI internal experiment (2022)
2. Ay, H., Benner, T., Arsava, E.M., et al.: Ascod phenotyping of ischemic stroke: a new classification system. Stroke **44**(5), 1432–1439 (2013)

3. Ay, H., Furie, K.L., Singhal, A.B., et al.: A systematic comparison of toast and ccs ischemic stroke classification systems. Stroke **42**(2), 519–524 (2011)
4. Brown, T.B., Mann, B., Ryder, N., Subbiah, M., Kaplan, J., Dhariwal, P., et al.: Language models are few-shot learners. Adv. Neural. Inf. Process. Syst. **33**, 1877–1901 (2020)
5. Bushnell, C., Chireau, M., Poppas, A., et al.: Sex differences in the evaluation and treatment of acute ischemic stroke. Stroke **45**(3), 844–850 (2014)
6. Devlin, J., Chang, M.W., Lee, K., Toutanova, K.: Bert: pre-training of deep bidirectional transformers for language understanding. arXiv preprint arXiv:1810.04805 (2019)
7. Dong, Y., Zeng, W., Liu, J., et al.: A survey for in-context learning. arXiv preprint arXiv:2301.00234 (2023)
8. Feigin, V.L., Lawes, C.M., Bennett, D.A., Anderson, C.S.: Worldwide stroke incidence and early case fatality reported in 56 population-based studies. Lancet Neurol. **2**(1), 43–53 (2003)
9. Feigin, V.L., Norrving, B., Mensah, G.A.: Global burden of stroke and risk factors in 188 countries. Lancet Neurol. **15**(9), 913–924 (2015)
10. Goldstein, L.B., Adams, H.P., Becker, K.: Classification of stroke subtypes. Stroke **32**(5), 1090–1097 (2001)
11. Gorelick, P.B.: Stroke prevention and risk factor management. Continuum (Minneapolis, Minn.) **14**(1), 30–50 (2008)
12. Hart, R.G., Diener H.C., Coutts, S.B., et al.: Embolic strokes of undetermined source: the case for a new clinical construct. Lancet Neurol. **13**(4), 429–438 (2014)
13. Huang, K., Altosaar, J., Ranganath, R.: Clinicalbert: modeling clinical notes and predicting hospital readmission. arXiv preprint arXiv:1904.05342 (2019)
14. Kojima, T., Lu, D Y., Schärli, N., et al.: Large language models are reasoning teachers. arXiv preprint arXiv:2305.14325 (2023)
15. Lawes, C.M., Vander Hoorn, S., Rodgers, A.: Blood pressure and the global burden of disease. J. Hypertens. **22**(1), 23–33 (2004)
16. Lee, J., Yoon, W., Kim, S., et al.: Biobert: a pre-trained biomedical language representation model for biomedical text mining. Bioinformatics **36**(4), 1234–1240 (2020)
17. Liu, P., Yuan, W., Fu, J., Jiang, Z., et al.: Pre-train prompt tuning: towards generalized and efficient foundation models. arXiv preprint arXiv:2304.09479 (2023)
18. Miura, Y., et al.: A comprehensive evaluation of large language models for medical applications. arXiv preprint arXiv:2305.09617 (2023)
19. Qiao, Y., Qian, X., Zhao, W X.: Reasoning with language model prompt engineering for tabular data. arXiv preprint arXiv:2205.11206 (2022)
20. Thrift, A.G., Cadilhac, D.A., Thayabaranathan, T.: Global stroke statistics. Lancet Neurol. **16**(10), 817–820 (2017)
21. Touvron, H., Lavril, T., Izacard, G., et al.: Llama: open and efficient foundation language models. arXiv preprint arXiv:2302.13971 (2023)
22. Zhou, H., Zheng, S. Du, J., et al.: Large language models: a survey. arXiv preprint arXiv:2303.18223 (2023)

Concept Normalization in Psychiatry: Comparing Embedding and Lexical Methods for Spanish Clinical Text

Sergio Rubio-Martín[1]([⊠]) [iD], Arturo Crespo-Álvaro[1] [iD],
María Teresa García-Ordás[1] [iD], Antonio Serrano-García[2] [iD],
Clara Margarita Franch-Pato[2] [iD], and José Alberto Benítez-Andrades[1] [iD]

[1] ALBA Research Group, Department of Electric, Systems and Automatics
Engineering, Universidad de León, Campus of Vegazana s/n, León, 24071 León, Spain
`{srubm,acrea,mgaro,jbena}@unileon.es`
[2] Servicio de Psiquiatría, Complejo Asistencial Universitario de León (CAULE),
24008, León, Spain
`https://albalab.eu/`

Abstract. The automatic normalization of clinical entities is a critical step for enabling structured analysis of free-text medical records. This study proposes and evaluates four distinct retrieval strategies for normalizing entities extracted via Named Entity Recognition (NER) from Spanish psychiatric emergency notes, using Unified Medical Language System (UMLS) as the target terminology. A total of 768 annotated entities were mapped using MiniLM, Multilingual BERT, lexical matching, and the UMLS API. Results show that the API-based approach yields the best performance (Hit@3 = 56.8% and Hit@5 = 57.9%), effectively balancing accuracy and computational efficiency. Although embedding-based methods such as MiniLM and Multilingual BERT often outperform traditional techniques in other domains, they showed only marginal improvements over simple lexical matching in this context, while incurring significantly higher computational costs. These findings suggest that in the psychiatric domain, where language is often ambiguous, embedding-based approaches may fall short. Hybrid systems that combine fast retrieval with semantic reasoning are needed to capture the richness of psychiatric narratives.

Keywords: EHR · Retrieval System · Psychiatry · NER · Embeddings

1 Introduction

Mental health conditions have gained increasing prominence in recent years due to their rising prevalence and significant societal impact. Disorders such as depression, anxiety, and substance use have not only become more frequent but also more severe, especially among adolescents. In Spain, psychiatric hospitalizations in youth have more than doubled over the past two decades, with alarming trends observed in post-pandemic years [24,27]. This surge reflects not only a clinical crisis, but also a pressing need for innovation in how mental health data is recorded, analyzed, and acted upon.

Moreover, the complex interplay between mental and physical health adds further urgency to the development of integrated care approaches. Large-scale studies have demonstrated strong and widespread comorbidities between psychiatric conditions and somatic illnesses, with associations evident across age groups, genders, and socioeconomic strata [11,16]. These findings underscore the importance of timely and structured access to mental health information in clinical settings, not only for psychiatric intervention, but also for managing holistic patient care.

Despite these needs, a substantial portion of clinically relevant data in psychiatry remains locked in unstructured free-text notes within Electronic Health Records (EHRs). These notes, authored by clinicians in high-pressure contexts such as psychiatric emergency services, are rich in clinical detail but often informal, variable in language, and poorly standardized. This lack of structure limits their utility for data-driven decision making, population health monitoring, and integration into intelligent clinical systems [26]. In parallel, barriers related to health literacy and perceived need for care continue to delay or prevent treatment. Global evidence indicates that even among individuals meeting diagnostic criteria for mental disorders, many do not seek treatment due to stigma, lack of awareness, or negative prior experiences with healthcare [17]. These challenges highlight the need not only for better care provision but also for better data to inform such care. Recent advances in Natural Language Processing (NLP) and Artificial Intelligence (AI) offer promising strategies to address these limitations. Named Entity Recognition (NER) techniques, in particular, enable the automatic identification of relevant entities such as diagnoses, symptoms, or medications in narrative text [4]. This is a crucial first step toward unlocking the value of narrative notes. However, in the field of psychiatry, where language is often vague or metaphorical, entity extraction alone is not sufficient.

Unlike structured biomedical reports, psychiatric notes frequently include expressions that lack direct lexical matches in medical terminologies, making normalization especially challenging. For instance, phrases such as "feels watched" or "thinks people are talking about him" may allude to paranoid ideation or psychotic symptoms, but require semantic interpretation beyond surface forms. Therefore, a robust post-NER pipeline is essential to link these extracted expressions to standardized medical concepts.

In this work, we focus on this second and often overlooked stage: the semantic normalization and disambiguation of previously extracted entities. We introduce a retrieval system and linking architecture designed to map Spanish psychiatric expressions to concept nodes within a knowledge graph built from the Unified Medical Language System (UMLS). The system explores multiple alternative candidate retrieval strategies such as embedding-based semantic search and recursive lexical matching. By evaluating these approaches independently, the architecture facilitates robust mapping from informal psychiatric expressions to standardized medical terminology, ultimately contributing to the creation of structured, interoperable, and clinically actionable representations of mental health records.

The remainder of this paper is organized as follows: Sect. 2 reviews the state of the art in clinical concept normalization and provides an overview of entity extraction techniques applied to both general and domain-specific texts. Section 3 details the materials and methods, including the dataset, system architecture, and candidate retrieval strategies. Section 4 presents the experimental setup and the main results. Finally, Sect. 5 discusses the conclusions, highlights current limitations, and outlines future research directions.

2 Related Work

Named Entity Recognition (NER) has become a foundational task in clinical Natural Language Processing (NLP) to transform these free-text EHR into structured data, enabling the identification of medical concepts such as diagnoses, medications, and procedures [8].

Early systems in the medical domain were predominantly rule-based or relied on dictionary lookups. While interpretable, these systems often lacked robustness when handling variability in clinical language [21]. With the rise of deep learning, more sophisticated approaches have emerged, notably those based on BiLSTM-CRF and transformer architectures such as BERT [23]. These models have demonstrated superior performance by capturing complex semantic and syntactic patterns in clinical text, particularly when pretrained on biomedical corpora [2].

Large Language Models (LLMs), including GPT-4 and its variants, have further advanced the field by introducing prompt-based strategies [6]. Techniques such as zero-shot and few-shot learning allow these models to generalize across tasks with minimal supervision, reducing the dependency on large annotated corpora. Indeed, this approach was evaluated in several studies such structuration of oncology EHR [?]. Moreover, ensemble prompting methods have also been proposed to enhance prediction stability in medical entity extraction tasks [13].

As most of the models are pretrained on English corpus, the increasing demand for multilingual solutions has led to the development of NER systems adapted to languages such as Spanish, Chinese, Serbian, French, and Italian. These systems often combine transformer-based models with context-aware components or ensemble methods to address the linguistic and clinical diversity present in non-English datasets [10,14]. In particular, adaptation strategies for low-resource languages and specialized domains have become a research priority to improve performance and applicability.

Other studies focus on identifying entity attributes such as negation, speculation, or temporal scope and the relationships among entities. Capturing this contextual information is essential for downstream tasks such as risk stratification and outcome prediction [19]. NER is also central to the de-identification of clinical texts, ensuring compliance with privacy regulations by removing protected health information before enabling data sharing for research [3,15].

Beyond entity recognition, the normalization of extracted mentions to standardized vocabularies such as UMLS or SNOMED CT is essential for ensuring interoperability and reusability of clinical data. Recent work has shifted toward embedding-based candidates retrieval [30]. These approaches compute dense vector representations of both entity mentions and ontology concepts using language models such as BioBERT or other BERT-based models, enabling semantic matching beyond exact string similarity [12]. Incorporating ontological structure into embeddings, either through fine-tuning on knowledge graphs or by leveraging concept hierarchies, has further improved normalization accuracy, particularly for ambiguous or polysemous mentions [28]. Embeddings derived from multilingual corpora have also enabled cross-lingual normalization tasks, facilitating clinical NLP applications in Spanish, French, Chinese, and Serbian [29]. Some systems combine semantic retrieval with recursive lexical search to balance recall and precision, especially in settings where domain-specific expressions are not captured well by dense embeddings. These hybrid strategies have proven particularly effective in psychiatric narratives, where terminology is often informal, vague, or highly context-dependent.

Recent developments have also focused on concept normalization across other domains and languages beyond Spanish and psychiatry. For instance, the RuC-CoN dataset provided manually annotated Russian clinical texts mapped to UMLS, addressing the lack of multilingual resources in Slavic languages [18]. In addition, zero-shot and few-shot normalization in social media corpora such as adverse drug events have shown that cross-domain generalization is feasible using BioLORD and ontology-aware training [20]. Finally, integration of ontologies in inference tasks reveals a convergence between concept normalization and higher-level semantic understanding [1,7].

Although embedding-based normalization has been widely studied in general clinical NLP, its application in psychiatry remains limited. A few works have leveraged embeddings to explore psychiatric spectrum disorders, such as generative embeddings for discriminating clinical subgroups in schizophrenia [5], or reconceptualizing disorders as networks of harmful symptoms rather than fixed categories [9]. While not directly focused on concept normalization, these studies highlight the potential of embeddings to capture semantic ambiguity and contextual nuances in psychiatric domains. Our contribution builds on this perspective by explicitly addressing concept normalization in Spanish psychiatric narratives, an area largely absent from the current literature.

3 Materials and Methods

3.1 Dataset Preprocessing

This study relies on a corpus of Spanish-language psychiatric emergency clinical notes sourced from the Psychiatry Service of the Complejo Asistencial Universitario de León (CAULE). These documents correspond to unscheduled, urgent presentations rather than routine or follow-up outpatient encounters, and therefore capture the acute, variable, and often incomplete nature of real-world emergency psychiatric documentation. All documents were fully anonymized in accordance with the European Union's General Data Protection Regulation (GDPR), and ethical approval was obtained from the regional research ethics board (approval ID: 2303), authorizing the use of de-identified textual data for research purposes.

The complete dataset comprises 5,332 free-text clinical reports recorded between 2017 and 2022 by a diverse group of clinicians working under intensive, time-pressured conditions typical of emergency care. As a result, the notes display substantial heterogeneity in linguistic style and structure, frequent informal phrasing, locally used abbreviations, clinical shorthand, and occasional orthographic inconsistencies. Conceptual redundancy, where the same clinical concept is expressed in multiple ways, is also common and complicates automated processing.

For the development of this work, we selected a subset of 200 notes to construct a reference corpus for entity extraction. Annotation was performed by trained psychiatrists who collaboratively defined a set of clinically meaningful entity types and produced labeling guidelines tailored to the emergency setting. The final annotation scheme includes six categories: Chief Complaint (MC), Clinical History (APC), Pharmacological History (APF), Diagnosis (DX), Pharmacological Treatment (TTF), and Action-Based Treatment (TTA). Annotations were created using the academic edition of Label Studio to promote consistency, domain relevance, and alignment with real-world emergency psychiatric documentation practices.

3.2   NER Architecture and Training

Given the absence of structured fields in psychiatric clinical narratives, one of the most effective strategies for organizing relevant information is the use of Named Entity Recognition (NER). This technique allows for the automatic identification of specific clinical concepts such as diagnoses, treatments, or background information within unstructured text. In this context, each token in the input sequence is labeled using a predefined schema, following the BIO (Begin–Inside–Outside) format, which encodes whether a token begins an entity, is inside an entity, or is not part of any entity. This structure enables fine-grained control over entity boundaries and is widely used in biomedical NLP tasks.

To evaluate the effectiveness of various transformer-based models in extracting clinical entities from Spanish psychiatric notes, we previously fine-tuned several language models, including BETO (both cased and uncased), ALBETO (pretrained on general-domain Spanish text), and two English biomedical models, ClinicalBERT and Bio_ClinicalBERT. Fine-tuning and evaluation were conducted using the HuggingFace Transformers library, which provides a robust and flexible environment for model configuration, training loop optimization, tokenization, and performance monitoring. A detailed comparative analysis of these NER models including training setup, entity-level evaluation metrics, and performance by category is available in our prior publication presented at CBMS

2025 [22]. In that study, we focused on identifying the most effective architecture for entity extraction from psychiatric emergency records.

In the present work, we build upon those results by addressing the second phase of the information structuring pipeline: the normalization of extracted entities into standardized clinical terminologies. This phase is essential for enabling interoperability, downstream analytics, and integration with external knowledge sources such as UMLS.

3.3   Knowledge Graph Construction

To support the normalization and disambiguation of clinical entities extracted from psychiatric notes, we constructed a domain-specific knowledge graph focused on mental health terminology. Knowledge graphs provide a structured way to represent relationships between concepts, enabling richer semantic interpretation and more accurate candidates retrieval in downstream tasks. In clinical contexts, these graphs are particularly useful when combined with ontologies or controlled vocabularies, as they facilitate alignment between free-text expressions and standardized medical concepts.

In this study, the graph was populated using a subset of the Unified Medical Language System (UMLS), limited to concepts associated with psychiatry and psychology. Each node in the graph corresponds to a unique clinical concept identified by its Concept Unique Identifier (CUI). To improve usability in multilingual settings, nodes were enriched with both their preferred English name and the corresponding Spanish translation. This bilingual setup helps bridge the gap between the original content of the clinical notes, written in Spanish, and the predominantly English-language structure of UMLS.

The graph was implemented in Neo4j, a property graph database that supports flexible querying and scalable traversal of semantic relationships. Neo4j's architecture allows each node to store custom attributes which in this case include the CUI, preferred terms in both languages, semantic type, and one or more vector representations embeddings of the concept. These embeddings were computed offline using two distinct sentence-level models: MiniLM and Multilingual BERT. By embedding each concept into a high-dimensional space, the graph can support semantic similarity queries that go beyond lexical overlap, enabling more robust candidate selection when processing ambiguous or variably phrased clinical entities.

3.4   Normalization and Candidates Retrieval

Once clinical entities are identified within the text, it becomes essential to map them to standardized concepts in order to ensure semantic interoperability and enable downstream applications such as clinical decision support, data aggregation, or cross-institutional research. This process, known as concept normalization, is particularly challenging in the psychiatric domain, where terminology tends to be ambiguous, expressed informally, or highly context-dependent.

To perform candidate retrieval, we designed a multi-tiered strategy supported by the Neo4j knowledge graph described previously. For each entity identified by the NER module, four independent retrieval strategies for retrieving candidates were evaluated:

- **MiniLM-based retrieval:** Entities are embedded using a pre-trained MiniLM model. Cosine similarity is computed between the entity vector and all concept node embeddings in the graph, also generated using MiniLM.
- **Multilingual BERT-based retrieval:** In this variant, both the input entity and the graph nodes are embedded using the multilingual version of BERT. This approach is designed to capture cross-lingual semantic relationships that may not be apparent in smaller or monolingual embedding models. Both embedding-based approaches were employed in their publicly available pre-trained versions without further fine-tuning. Default configurations from the SentenceTransformers library were used for generating embeddings.
- **Recursive lexical matching:** This method performs direct string matching between the input entity and the terms of graph nodes. If no match is found, the entity is recursively simplified by removing substrings (e.g., "síndrome ansioso depresivo" → "síndrome ansioso" → "síndrome"), increasing the likelihood of partial lexical alignment with known concepts.
- **Recursive lexical matching with UMLS:** Similar to the previous method, this strategy matches the entity text against concepts available in the UMLS Metathesaurus via its API. If no direct or approximate matches are found, the entity is progressively reduced to simpler forms in an attempt to identify partially aligned concepts externally.

Each strategy generates a ranked list of candidate concepts for each input entity. In cases where none of the approaches produces candidates with acceptable semantic similarity (defined by a cosine similarity threshold of 0.85), a fallback mechanism is triggered to query the official UMLS API directly for additional external candidates.

In this study, we evaluate the effectiveness of each strategy by measuring whether the correct UMLS concept appears within the top-3 or top-5 candidates. This analysis provides insight into the suitability of each retrieval approach for handling noisy and ambiguous clinical expressions found in Spanish psychiatric notes. To validate the results, the retrieved candidates are compared against a ground truth manually curated by clinical experts, which contains the correct UMLS CUIs for each annotated entity.

3.5 Hardware and Software Environment

All experiments and model training were conducted using Python 3.10 in a local development environment. Computational tasks were performed on a workstation equipped with an Intel Core i7-9700K processor, 32 GB of RAM, and an NVIDIA RTX 2080 GPU with 8 GB of dedicated memory. The implementation of the NER models and the fine-tuning pipeline was carried out using the HuggingFace Transformers library. Data preprocessing and formatting were handled with the help of the Pandas and Datasets libraries. For annotation management, the academic version of Label Studio was used, enabling manual labeling with support for span-based entity tagging. [25]

The Neo4j graph database was deployed locally to store and query the UMLS-derived concept graph. Input entities were generated using pretrained models accessed through the SentenceTransformers library. All semantic similarity calculations were based on cosine distance.

4   Experiments and Results

4.1   Dataset Preprocessing and Labeling

The dataset comprises 5,332 de-identified clinical notes written in Spanish and sourced from psychiatric emergency care services. These records exhibit substantial linguistic heterogeneity due to the lack of standardized documentation protocols and the frequent use of informal language or clinical shorthand.

For our previous study focused on evaluating the performance of NER models, a subset of 200 notes was manually annotated by domain experts In that work, the BIO (Begin–Inside–Outside) tagging scheme was applied to identify clinically relevant entities at the token level for NER tasks. Full details regarding the annotation schema and NER evaluation can be found in the corresponding publication [22].

In the present study, we focus on the second stage of the information structuring pipeline, which is entity normalization. We selected a subset of 100 notes from the original annotated corpus. This selection yielded a total of 768 annotated entities, covering all six defined categories as can be seen in Table 1. These entities serve as the input for the normalization experiments described in the following sections.

4.2   Retrieval System

The proposed system for this task is based on a database that stores nodes representing psychiatric concepts. This database consists of 12,429 nodes containing concepts related to diseases, symptoms, and certain drugs associated with the field of psychiatry and psychology. Four different query approaches were implemented and evaluated. The first approach involves querying the Neo4j database using the entity text to retrieve results based on lexical similarity. Additionally, two implementations were developed using candidate retrieval via embeddings, relying on two different models: one based on multilingual BERT embeddings, and another using a MiniLM model.

Regardless of the approaches previously mentioned, a separate implementation was also developed using the UMLS API through the Metathesaurus Browser, allowing a comparison of results and performance with the previously mentioned approaches. To evaluate the performance of the system, it was tested using 100 clinical note texts, already structured and annotated with extracted entities.

Table 1. Overview of annotation categories used in Label Studio for entity extraction from psychiatric clinical notes.

Label	Description	Amount	Freq
MC	**Chief Complaint**: Refers to the main concern, symptom, or incident that motivates the patient to seek psychiatric evaluation.	260	1.3
APC	**Clinical History**: Covers past psychiatric or general health conditions, including chronic illnesses, prior diagnoses, or significant medical events.	291	1.46
APF	**Pharmacological History**: Describes the patient's past or ongoing use of psychotropic medications, other pharmaceuticals, or substances such as alcohol or drugs.	345	1.73
DX	**Diagnosis**: Indicates the psychiatric disorder(s) formally assigned by the clinician based on professional judgment and diagnostic criteria.	233	1.17
TTF	**Pharmacological Treatment**: Specifies current pharmacological treatments recommended by the psychiatrist, including details such as active ingredient or dosage.	309	1.55
TTA	**Action-Based Treatment**: Refers to procedural or care-related decisions such as referrals, follow-up planning, hospital admission, or discharge.	169	0.85

4.3   Results

The evaluation of normalization performance was based on two standard top-k retrieval metrics: Hit@3 and Hit@5. These metrics assess whether the correct UMLS concept appears among the top 3 or top 5 candidates returned by each retrieval strategy. For every annotated entity in the evaluation set, the retrieved candidates were compared against the ground truth. If the target concept appeared within the top-k results, a hit was counted. The final hit rate was computed as the proportion of successful matches over the total number of entities. In total, 768 annotated entities were evaluated.

The quantitative results of this evaluation are presented in Table 2, which compares performance across the four candidate retrieval strategies. To facilitate visual interpretation, the corresponding percentages are depicted in Fig. 1, highlighting the relative effectiveness of each method.

Beyond retrieval accuracy, we also evaluated the computational efficiency of each strategy using the same subset of 100 annotated clinical notes. As reported in Table 3, the execution times varied substantially across methods. The lexical strategy without embeddings was the most efficient, requiring only 23.33 s in total—an average of 0.23 s per note and 0.03 s per entity—making it highly suitable for resource-constrained environments. In contrast, the embedding-based approaches, especially Multilingual BERT, incurred significant processing overhead, with total runtimes exceeding 100 s and average per-entity times above

15 s. The UMLS API-based method achieved a favorable balance, combining the highest retrieval performance (Hit@3 of 436/768 and Hit@5 of 445/768) with moderate execution times, thus offering a strong trade-off between precision and scalability.

Table 2. Hit rate comparison between each implementation.

Retrieval Strategy	Hit@3	Hit@3(value)	Hit@5	Hit@5(value)
MiniLM	126/768	0.16	126/768	0.16
Multilingual BERT	139/768	0.18	142/768	0.18
No Embeddings	136/768	0.17	136/768	0.17
UMLS API	436/768	0.56	445/768	0.58

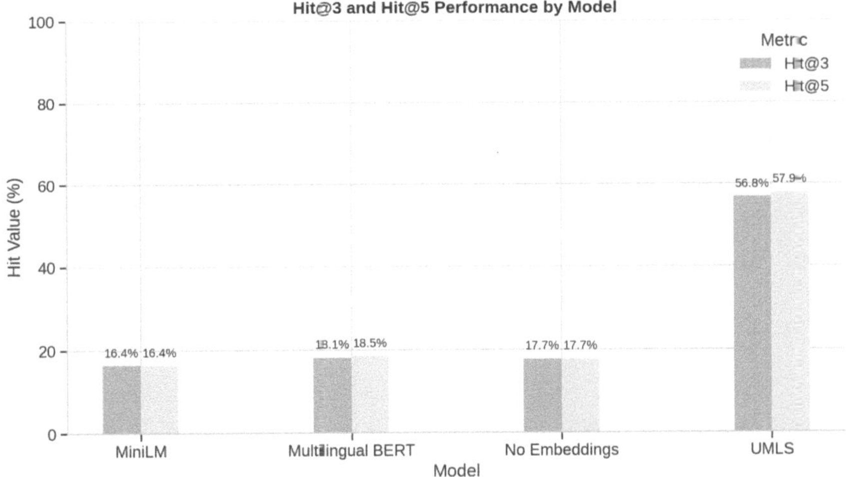

Fig. 1. Hit@3 and Hit@5 performance per retrieval method, expressed as number of correct predictions out of 768 evaluated cases.

As shown in Table 4, error counts are not uniform across entity types. Pharmacological mentions(APF, TTF) accumulate the highest number of incorrect matches, largely due to brand/generic name variability, dose and formulation tokens, and multi-ingredient products that increase ambiguity at normalization time. TTA and MC also show comparatively elevated errors: action-based decisions are often expressed with heterogeneous verb phrases such as discharge/admission phrased in multiple ways, while chief complaints frequently include colloquial or metaphorical language lacking direct lexical counterparts. In contrast, DX and APC exhibits fewer errors, likely because diagnostic terms

Table 3. Time comparison across retrieval strategies containing the measurement of the total execution time, the average time per note and the average time per entity

Model	Exec. Time(s)	Time/Note(s)	Time/Entity(s)
MiniLM	8881.86	88.82	11.61
Multilingual BERT	11670.56	116.71	15.26
No Embedding	23.33	0.23	0.03
UMLS	1389.22	13.89	1.82

Table 4. Normalization results per entity type and retrieval method at Hit@3. Values indicate number of incorrect matches.

Label	MiniLM	Multilingual BERT	No Embedding	UMLS
APC	77	74	83	44
APF	**169**	**169**	**169**	49
DX	69	64	58	38
MC	94	92	90	**62**
TTA	85	85	85	**82**
TTF	**148**	**145**	**147**	57

tend to be closer to standardized nomenclature. Finally, the UMLS API-based strategy concentrates the lowest error counts across categories, aligning with its superior hit rates in our quantitative evaluation.

5   Conclusions

This study evaluated four distinct strategies for the normalization of clinical entities extracted from Spanish psychiatric notes. The results show notable differences across methods, both in terms of retrieval accuracy and computational efficiency.

Among all strategies, the UMLS API-based approach achieved the highest normalization performance, with a Hit@3 of 436/768 (56.8%) and a Hit@5 of 445/768 (57.9%). Its effectiveness stems from the extensive lexical coverage of the UMLS Metathesaurus and its ability to resolve synonyms and related terms through API endpoints. While not the fastest, it maintained a reasonable average processing time of 13.89 s per note, making it suitable for both clinical and research use.

In contrast, the embedding-based methods (MiniLM and Multilingual BERT) delivered only marginal improvements over lexical matching, with Hit@5 scores of 16–18%, but at a significantly higher computational cost (88–117 s per note). This performance-cost imbalance becomes critical at scale, where the slight accuracy gains fail to justify the added latency. The lexical similarity method without embeddings matched the embeddings' accuracy (Hit@5 of 136/768, 17.7%) but was by far the most efficient, requiring just 0.23 s per note. While this makes it attractive for time-sensitive or low-resource settings, its limited semantic reach makes it less viable when precise concept normalization is essential.

The embedding-based approaches (MiniLM and Multilingual BERT) offered only marginal improvements over the lexical method in terms of accuracy, while incurring significant computational costs (over 4 additional seconds per entity). While this may appear acceptable in isolation, such latency becomes substantial when scaling to hundreds or thousands of notes which each includes multiple entities. Moreover, the small gain in Hit@5 does not justify the added overhead, particularly in clinical systems where timely responses are essential.

Beyond numerical metrics, this study highlights the inherent limitations of current normalization strategies when dealing with non-standardized, ambiguous, or inferential clinical language. For example, Chief Complaint expressions like "watching himself in TV" which can be a possible reference to psychotic hallucinations, are unlikely to match directly via string or embedding-based retrieval. These cases underscore the need for hybrid systems that combine surface-level matching with semantic reasoning and clinical context interpretation, potentially enhanced via knowledge graphs or LLM-based inference modules. Hence, while the UMLS API approach offers the best performance-to-efficiency balance, no single strategy is sufficient to handle the full complexity of psychiatric narratives.

From a practical perspective, the proposed normalization strategies gain further importance when integrated into complete clinical NLP pipelines. In such settings, normalization outputs could serve as structured and standardized input for downstream tasks such as risk prediction, patient stratification, or clinical decision support systems. By reducing lexical and semantic variability in psychiatric documentation, normalization can improve the reliability of predictive models and support safer, more efficient clinical practice.

Several limitations should be acknowledged. First, the evaluation was carried out on a subset of 100 manually annotated notes. Although this set was carefully selected and covered a wide range of entity types, it may not capture the full variability present in real-world clinical documentation. Second, the normalization process treated entity mentions in isolation, without leveraging broader narrative context, which is often crucial for correct interpretation, particularly in psychiatry. Finally, the study focused exclusively on retrieval-based methods and did not evaluate more advanced semantic linking architectures or context-aware language models trained for entity disambiguation.

Future work could explore a broader or more diverse set of notes to strengthen our conclusions while exploring hybrid normalization pipelines that integrate fast candidate retrieval with reasoning-based disambiguation. This is especially relevant for clinically meaningful but linguistically non-obvious expressions. These cases require contextual interpretation that may rely on surrounding text, clinical heuristics, or high-level inference. Promising directions include the incorporation of domain-specific ontologies and clinical knowledge graphs into the normalization workflow, allowing the system to enrich retrieved candidates with new ontological relationships. Likewise, Large Language Models (LLMs) could assist in generating semantic justifications, validating retrievals, or filtering out semantically incompatible matches. Another key area for improvement is the expansion of the concept graph with paraphrases and synonyms collected from annotated corpora or extracted collaboratively with clinical experts. This would allow the system to better handle informal, metaphorical, or abbreviated expressions common in psychiatric narratives.

Acknowledgments. This research received support from the Universidad de León through its 2021 internal research funding program. Furthermore, it is part of the project titled 'NLP-Driven Insight Engine for Suicide Attempt Detection in EHR (SUICIDETECT)', funded by the Spanish Ministerio de Ciencia, Innovación y Universidades under grant number PID2023-146168OA-I00.

Disclosure of Interests. The authors have no competing interests to declare that are relevant to the content of this article.

References

1. Abdel-salam, R., Adewunmi, M., Akinwale, M.: CaresAI at SemEval-2024 task 2: improving natural language inference in clinical trial data using model ensemble and data explanation. In: Proceedings of the 18th International Workshop on Semantic Evaluation (SemEval-2024), pp. 1905–1911. ACL, June 2024. https://doi.org/10.18653/v1/2024.semeval-1.266, https://aclanthology.org/2024.semeval-1.266/
2. Ahmed, A., Abbasi, A., Eickhoff, C.: Benchmarking modern named entity recognition techniques for free-text health record de-identification. AMIA Ann. Symp. Proc. AMIA Symp. **2021**, 102–111 (2021)
3. Ahmed, A.F., Abbasi, A.A., Eickhoff, C.: Benchmarking modern named entity recognition techniques for free-text health record de-identification. AMIA ... Ann. Symp. Proc. AMIA Symp. **2021**, 102–111 (2021), https://api.semanticscholar.org/CorpusID:232352634
4. Bean, D.M., Kraljevic, Z., Shek, A., Teo, J., Dobson, R.J.B.: Hospital-wide natural language processing summarising the health data of 1 million patients. PLOS Digit. Health **2**(5), 1–14 (2023). https://doi.org/10.1371/journal.pdig.0000218, https://doi.org/10.1371/journal.pdig.0000218
5. Brodersen, K.H., et al.: Dissecting psychiatric spectrum disorders by generative embedding. NeuroImage: Clin. **4**, 98–111 (2014). https://doi.org/10.1016/j.nicl.2013.11.002, https://www.sciencedirect.com/science/article/pii/S2213158213001502
6. Cheng, Q., Chen, L., Hu, Z., Tang, J., Xu, Q., Ning, B.: A novel prompting method for few-shot ner via llms. Nat. Lang. Proc. J. **8**, 100099 (2024). https://doi.org/10.1016/j.nlp.2024.100099, https://www.sciencedirect.com/science/article/pii/S2949719124000475
7. Conceição, S.I.R., F. Sousa, D., Silvestre, P., Couto, F.M.: lasigeBioTM at SemEval-2023 task 7: improving natural language inference baseline systems with domain ontologies. In: Proceedings of the 17th International Workshop on Semantic Evaluation (SemEval-2023), pp. 10–15. ACL, July 2023. https://doi.org/10.18653/v1/2023.semeval-1.2, https://aclanthology.org/2023.semeval-1.2/

8. Durango María C., Torres-Silva Ever A., O.D.A.: Named entity recognition in electronic health records: a methodological review. Healthc Inform Res. **29**(4), 286–300 (2023). https://doi.org/10.4258/hir.2023.29.4.286, http://e-hir.org/journal/view.php?number=1175

9. Gauld, C.: From psychiatric kinds to harmful symptoms. Synthese **200**(6), 440 (2022). https://doi.org/10.1007/s11229-022-03922-5, https://doi.org/10.1007/s11229-022-03922-5

10. Guo, S., Yang, W., Han, L., Song, X., Wang, G.: A multi-layer soft lattice based model for Chinese clinical named entity recognition. BMC Med. Inf. Decis Making **22**(1), 201 (2022). https://doi.org/10.1186/s12911-022-01924-4, https://doi.org/10.1186/s12911-022-01924-4

11. Hanna, M.R., Caspi, A., Houts, R.M., Moffitt, T.E., Torvik, F.A.: Co-occurrence between mental disorders and physical diseases: a study of nationwide primary-care medical records. Psychol. Med. **54**(15), 4274–4286 (2024). https://doi.org/10.1017/S0033291724002575

12. Hier, D.B., Do, T.S., Obafemi-Ajayi, T.: A simplified retriever to improve accuracy of phenotype normalizations by large language models. Front. Digit. Health **7** 2025 (2025). https://doi.org/10.3389/fdgth.2025.1495040, https://www.frontiersin.org/journals/digital-health/articles/10.3389/fdgth.2025.1495040

13. Islam, S., Nipu, A.S., Wu, J., Madiraju, P.: Llm-based prompt ensemble for reliable medical entity recognition from ehrs. ArXiv **abs/2505.08704** (2025), https://api.semanticscholar.org/CorpusID:278534987

14. Kaplar, A., Stosovic, M., Kaplar, A., Brković, V., Naumović, R., Kovačević, A.: Evaluation of clinical named entity recognition methods for serbian electronic health records. Int. J. Med. Informatics **164**, 104805 (2022). https://doi.org/10.1016/j.ijmedinf.2022.104805

15. Kondra, S., Xu, W., Raghavan, V.V.: Unlocking the power of clinical notes: natural language processing in healthcare. Acta Sci. Med. Sci. (2024). https://doi.org/10.31080/asms.2024.08.1821

16. Westhoff, M.S., Reimers, T., Heck, J., Brod, T., Bleich, S., Groh, A., Schröder, S.: Management of somatic comorbidities in psychiatric emergencies: insights from a real-world cohort study. Psychiatry Res. 116630 (2025). https://doi.org/10.1016/j.psychres.2025.116630. https://www.sciencedirect.com/science/article/pii/S0165178125002781

17. McLaren, T., Peter, L.J., Tomczyk, S., Muehlan, H., Schomerus, G., Schmidt, S.: The seeking mental health care model: prediction of help-seeking for depressive symptoms by stigma and mental illness representations. BMC Public Health **23**(1), 69 (2023). https://doi.org/10.1186/s12889-022-14937-5, https://doi.org/10.1186/s12889-022-14937-5

18. Nesterov, A., et al.: Ruccod: towards automated icd coding in Russian. ArXiv **abs/2502.21263**, https://doi.org/10.48550/arXiv.2502.21263 (2025)

19. Paolo, D., et al.: Hierarchical embedding attention for overall survival prediction in lung cancer from unstructured EHRs. BMC Med. Inf. Decis. Making **25**(1), 169 (2025). https://doi.org/10.1186/s12911-025-02998-6, https://doi.org/10.1186/s12911-025-02998-6

20. Remy, F., Scaboro, S., Portelli, B.: Boosting adverse drug event normalization on social media: General-purpose model initialization and biomedical semantic text similarity benefit zero-shot linking in informal contexts. ArXiv **abs/2308.00157**, https://doi.org/10.48550/arXiv.2308.00157 (2023)

21. Rijcken, E., Zervanou, K., Mosteiro, P., Scheepers, F., Spruit, M., Kaymak, U.: Machine learning vs. rule-based methods for document classification of electronic health records within mental health care—a systematic literature review. Nat. Lang. Process. J. **10**, 100129 (2025). https://doi.org/10.1016/j.nlp.2025.100129
22. Rubio-Martín, S., Crespo-Álvaro, A., García-Ordás, M.T., Serrano-García, A., Franch-Pato, C.M., Benítez-Andrades, J.A.: Fine-tuning transformer models for structuring spanish psychiatric clinical notes. In: 2025 IEEE 38th International Symposium on Computer-Based Medical Systems (CBMS), pp. 97–102 (2025). https://doi.org/10.1109/CBMS65348.2025.00028
23. Silva, J.F., Almeida, J.R., Matos, S.: Extraction of family history information from clinical notes: deep learning and heuristics approach. JMIR Med. Inform. **8**(12), e22898 (2020). https://doi.org/10.2196/22898
24. Soriano, V., et al.: Hospital admissions in adolescents with mental disorders in Spain over the last two decades: a mental health crisis? Eur. Child Adolescent Psychiatry **34**(3), 1125–1134 (2025). https://doi.org/10.1007/s00787-024-02543-2, https://doi.org/10.1007/s00787-024-02543-2
25. Tkachenko, M., Malyuk, M., Holmanyuk, A., Liubimov, N.: Label studio: data labeling software (2020-2025), open source software available from https://github.com/HumanSignal/label-studio
26. Wang, W., Ferrari, D., Haddon-Hill, G., Curcin, V.: Electronic health records as source of research data, pp. 331–354. Springer US, New York, NY (2023). https://doi.org/10.1007/978-1-0716-3195-9_11
27. World Health Organization: Mental health and COVID-19: Early evidence of the pandemic's impact – Scientific Brief. https://www.who.int/publications/i/item/WHO-2019-nCoV-Sci-Brief-Mental-health-2022.1 (2022), Accessed 09 Jul 2025
28. Wulff, D.U., Mata, R.: Semantic embeddings reveal and address taxonomic incommensurability in psychological measurement. Nat. Hum. Behav. **9**, 944–954 (2025). https://doi.org/10.1038/s41562-024-02089-y
29. Zahra, F.A., Kate, R.J.: Obtaining clinical term embeddings from snomed ct ontology. J. Biomed. Inf. **149**, 104560 (2024). https://doi.org/10.1016/j.jbi.2023.104560, https://www.sciencedirect.com/science/article/pii/S1532046423002812
30. Zeakis, A., Papadakis, G., Skoutas, D., Koubarakis, M.: Pre-trained embeddings for entity resolution: an experimental analysis. Proc. VLDB Endow. **16**, 2225–2238 (2023). https://doi.org/10.14778/3598581.3598594

Integrating Language Models and Network Embeddings to Uncover Hidden Relationships in Neuromuscular Diseases

Federico García-Criado[1] , Jesús Pérez-García[1] , Elena Rojano[1,2,3] ,
Juan A. G. Ranea[1,2,3,4] , and Pedro Seoane-Zonjic[1,2,3(✉)]

[1] Department of Molecular Biology and Biochemistry, University of Malaga,
29010 Malaga, Spain
seoanezonjic@uma.es
[2] Institute of Biomedical Research in Malaga and Platform of Nanomedicine (IBIMA
Plataforma BIONAND), 29071 Malaga, Spain
[3] Center for Biomedical Network Research on Rare Diseases (CIBERER), Instituto
de Salud Carlos III (ISCIII), 28029 Madrid, Spain
[4] Spanish National Bioinformatics Institute (INB/ELIXIR-ES), Instituto de Salud
Carlos III (ISCIII), 28020 Madrid, Spain

Abstract. Neuromuscular diseases (NMDs) are a heterogeneous group
of rare disorders that significantly impair motor function and quality of
life. Their clinical and genetic variability makes systematic study and
knowledge integration particularly challenging. We present a scalable,
automated framework that combines natural language processing and
network analysis to uncover hidden relationships among NMDs. Using
Sentence Transformer models, we identified NMDs and their phenotypes
in open-access documents. Then, we use this data to build NMD networks to infer new relations using embedding techniques and clustering
approaches structuring a document corpus for 328 NMDs basis on disease type and associated to phenotypes and genes. These findings show
the value of combining language models and network embeddings for
large-scale rare disease analysis.

Keywords: Neuromuscular diseases · language models · network
embeddings · translational bioinformatics · rare diseases

1 Introduction

Neuromuscular diseases (NMDs) are a diverse group of disorders that impair
the function of muscles and the peripheral nerves controlling them, leading
to progressive muscle weakness, motor dysfunction, and substantial disability
[12,21]. Uncovering the molecular and genetic relationships shared among different NMDs offers a valuable opportunity to advance diagnostic and therapeutic
strategies [21]. To extract meaningful information about NMDs, it is essential

A. López Fernández et al. (Eds.): CIABiomed 2025, LNBI 16148, pp. 297–308, 2026.
https://doi.org/10.1007/978-3-032-10661-2_23

to recognize that the biomedical field has accumulated a large body of knowledge over decades of research [6]. However, a large portion of this information resides in unstructured formats, primarily within free-text scientific publications, making it inaccessible to programmatic querying and large-scale computational analysis [9]. To address this challenge, the use of standardized vocabularies and biomedical ontologies has become increasingly important. Ontologies are formal, structured representations of domain knowledge that define entities and their relationships, enabling logical inference and semantic interoperability across datasets [1,5]. Despite their utility, much of the knowledge contained in the literature is not explicitly expressed using ontology-aligned terminology. Therefore, it is necessary to develop methods that map free-text expressions to standardized ontological concepts, bridging the gap between unstructured information and structured, machine-readable representations.

Recent advances in natural language processing (NLP), particularly the emergence of large language models (LLMs), have dramatically improved the ability to extract structured knowledge from unstructured text [19]. Models based on the Bidirectional Encoder Representations from Transformers (BERT) architecture [4] have demonstrated remarkable performance in tasks such as named entity recognition (NER), enabling automated identification and classification of entities within unstructured text, including names of individuals, locations, biological entities and ontological terms [3]. Applying such models to literature related to NMDs allows exploit all available text data.

Beyond entity extraction, representing these concepts and their relationships in the form of networks offers powerful systems-level insights. Recent developments in network embedding techniques, which transform high-dimensional network structures into compact vector representations, have facilitated the discovery of latent topological patterns and biologically meaningful clusters. These embeddings enable more accurate clustering, gene prioritization, and pathway inference by capturing complex topological relationships, thereby enhancing the identification of functionally relevant modules and improving our ability to detect potential biomarkers and therapeutic targets within a systems-level framework [2,13].

In this study, we focus on the set of NMDs defined by the MONDO ontology, developed by the Monarch Initiative [20]. For each disease term, we retrieve related scientific literature using semantic search powered by Sentence Transformers. We then apply the same models to extract phenotypic descriptions from the identified documents, mapping them to concepts in the Human Phenotype Ontology (HPO) [7]. The resulting phenotype-disease associations are used to construct disease similarity networks, which are subsequently embedded into low-dimensional spaces using algorithms such as node2vec [8]. These embeddings allow us to uncover hidden relationships between NMDs and reveal potential shared molecular and genetic underpinnings.

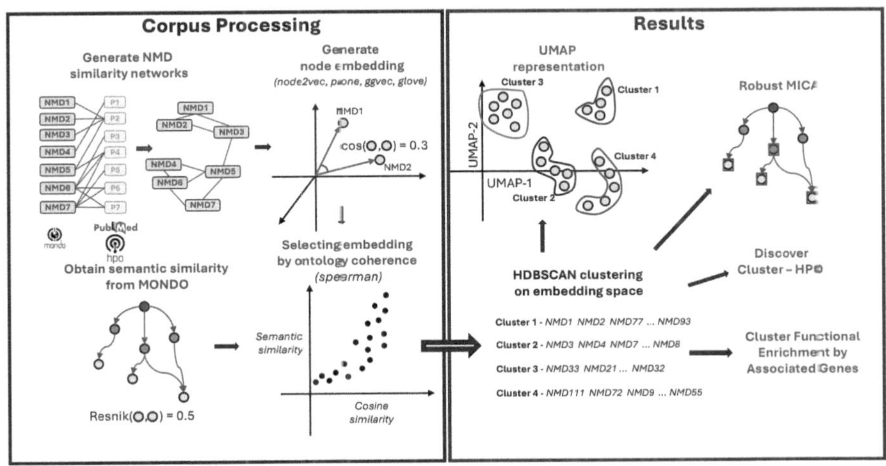

Fig. 1. Diagram of the corpus generation process based on neuromuscular diseases (NMDs) and the generated results from the communities identified within the selected embedding space.

2 Materials and Methods

2.1 Identify Neuromuscular Diseases in Full-Text Articles

The text corpus chosen for this work was composed of all full-text articles available in PubMed Central. Articles were downloaded via the National Center for Biotechnology Information (NCBI) FTP service (https://ftp.ncbi.nlm.nih.gov), from the "/pub/pmc/oa bulk" directory, including documents from the "oa comm", "oa noncomm" and "oa other" folders, encompassing both baseline and incremental files. This yielded 6,132,543 articles, which were further processed by hierarchically splitting them into paragraphs and then into sentences, using full stops and commas as delimiters. The text queries consisted of disease terms from the MONDO ontology [20] and phenotype terms from the HPO ontology [7]. For diseases, we extracted all child terms of "neuromuscular disease" (MONDO:0019056) and retained only the leaf terms (those without further descendants) resulting in 755 terms. And for phenotypes, all descendants of "Phenotypic abnormality" (HP:0000118) were retrieved, totaling 17,957 terms. For both diseases and phenotypes, their names and available synonyms were collected for each term.

Both the query and corpus texts were processed using the embedding model "all-mpnet-base-v2" from the Sentence Transformers library (v5.2) [18] (https://huggingface.co/sentence-transformers/all-mpnet-base-v2), a diagram of this process is shown in Fig. 1, This model generates high-dimensional vector embeddings that capture the semantic content of text, enabling the recognition and quantification of semantic similarity between text fragments. After embedding, all query–document fragment pairs were compared using cosine similarity, retaining only those with a score ≥ 0.7. Articles with fewer than two links to a given disease were excluded from further analysis. Finally, the ontology distribution of the annotated terms was evaluated using the py_semtools package (Pérez-García *et al.*, in press) to assess the quality of the text mining process.

2.2 Disease-Disease Network Building and Embedding

We built two bipartite networks linking diseases to phenotypes and to articles. Network analyses were conducted using our py_netanalyzer Python library [11]. The disease–article network was generated from the previously described connections. To build the disease–phenotype network, we first created a tripartite network combining disease–article and article–phenotype associations. Diseases were represented by MONDO terms, phenotypes by HPO terms, and articles by their PubMed IDs (PMIDs). HPO terms identified in NMD articles were assigned to the corresponding MONDO terms annotated in the same articles to weight disease-phenotype (or MONDO-HPO) associations. For this, we applied a Simpson projection to the tripartite disease–article–phenotype network, retaining only MONDO–HPO relations with a Simpson similarity 0.9. In this way, we obtain a directly inferred and a weighted MONDO-HPO networks.

Next, since the MONDO ontology is a directed acyclic graph (DAG), we transformed it into a disease–disease network to serve as a reference for evaluating information preservation in subsequent analyses. This network was filtered to include only "neuromuscular disease" (MONDO:0019056) and its descendants. The resulting network, termed the "dis_ont" network, was used as both a control and reference throughout our experiments.

All the bipartite networks described previously were converted in disease-disease networks using different algorithms. The relations MONDO-article are projected using Jaccard but all associations not supported by at least two MONDO terms are removed, we name it as the "dis_text" network. The same process was applied for the weighted MONDO-HPO, generating the network "dis_phen_jacc". Since phenotypes are represented by HPO terms, we applied ontology-based semantic similarity to the disease–phenotype network to calculate phenotypic similarity between all NMDs based on their HPO annotations. We used a normalized version of the ERIC metric, which accounts for potential noise in HPO annotations [14]. The similarity matrix was then filtered with a threshold of 0.2 to generate the "dis_phen_sim" network.

The disease-disease networks were embedded using several state-of-the-art algorithms, including node2vec [8], Prone [22], GGVec, and GloVe [16]. For node2vec, we applied the following parameters: `window=5`, `num_walks=30`, `q=1`, `random_seed=69`, `walk_length=50`, `dimensions=128`. The Prone and GGVec algorithms were executed using their recommended default hyperparameters, as specified in their original publications. GloVe was run with a vector dimension of 128 to ensure comparability across embedding methods.

2.3 Clustering Disease Identification and Analysis

Based on the coordinates of the selected embedding, clustering was performed using HDBSCAN [15]. This method is particularly suitable for detecting non-linear patterns in high-dimensional embedding spaces. The computation was carried out with the *scikit-learn* library, using the following parameters: `min_cluster_size=10, min_samples=10, metric='cosine', cluster_selection_method='leaf'`. Then, for each MONDO-based cluster, we computed the average shortest path based on the DAG structure of MONDO. This metric was used to assess the internal coherence of each cluster in relation to the ontology. To further evaluate whether the clusters aligned with specific NMD classifications, we defined a *robust* Maximum Informative Common Ancestor (MICA). This was determined by identifying all ancestral terms shared by at least 75% of the diseases within the cluster, ensuring resistance to small subsets of outlier terms with divergent ancestry. Among these shared terms, the one with the highest Resnik information content was selected as the *robust* MICA. A diagram of all the process is presented in Fig. 1.

Two types of enrichment analyses were performed on the identified clusters. First, a phenotype-based enrichment was carried out using the MONDO–HPO associations previously obtained. A hypergeometric test was applied to determine which HPO terms were significantly over-represented within each set of MONDO terms. Second, a functional enrichment analysis was performed using the Biological Process category of the Gene Ontology (GO). Each MONDO term was linked to its associated genes, as reported by the Monarch Initiative [17]. For each cluster, enrichment analysis was conducted using the *clusters_ to_ enrichment R* script from the ExpHunter Suite tool developed by our group [10], applying the Benjamini–Hochberg false discovery rate (FDR-BH) procedure with an adjusted *p*-value threshold of 0.05.

3   Results and Discussion

3.1   Assessing the Neuromuscular Diseases Domain and Its Phenotypes

As an initial step, we evaluated the coherence and coverage of the Sentence Transformer–based annotations by analyzing the distribution of terms within the MONDO and Human Phenotype Ontology (HPO). This assessment aimed to validate the alignment between the literature-derived annotations and established biomedical ontologies. In the case of NMDs inferred from the articles (Fig. 2A), the annotation frequency was notably low. This reduction is explained by the wide distribution of disease annotations across a large number of articles (382,519 in total) relative to the 743 MONDO terms considered. This can be explained by the focus of each article toward a specific NMD, in this way, few studies describe several diseases. In addition, there are NMD domains such as "akinetopsia" and "atrophic muscular disease" with very low annotations. Furthermore, the two domain diseases with the most representation are "hereditary neuromuscular disease" and "peripheral neuropathy". However, when these papers are annotated to phenotypes (HPO, Fig. 2B) we see that all phenotypes have noticeable representation. This is due to the fact that neuromuscular diseases are complex systemic diseases that affects several biological systems and causes very different pathological phenotypes. However, we can see that "abnormality of the musculoskeletal system", "abnomarlity of limbs", "abnormality of the nervous system" and "abnormality of the metabolism/homeostasis" have a large presentation with the first three categories clearly related to NMDs. Both phenomena, the focus at the article level and the overlap at the phenotype level, reflect the nature of current research practices, where diseases are often studied in isolation, one by one, even though they may share similar phenotypes. Importantly, the low annotation frequency also raises questions about the completeness and potential bias of the extracted data, as certain diseases may be systematically underrepresented or overlooked.

From the several networks that we have build, we measure which network and which embedding is the most appropriate to capture the relations between the NMDs. We use as reference the ontological semantic similarity between all NMD terms listed in the MONDO ontology and its Spearman correlation against their cosine similarity in the network embedding (Fig. 2C, dis_ont row). We can see that this reference network has the second high correlation values (maximum in 0.32) whereas the dis_text network has the highest correlation values (maximum in 0.35). The phenotype networks have correlation values near to zero and for this reason we focus in the dis_text network. The limited performance of the phenotype networks may be attributed to the need for more precise associations between diseases and their corresponding phenotypes, as the current approach links all identified phenotypes to all diseases within a document. A potential strategy to address this limitation would be to filter out phenotypes with low information content or those that occur with high frequency across the entire NMD corpus, or alternatively to weight the associations between MONDO terms according to the information content of their phenotype intersections. With respect to the embedding methods tested, node2vec consistently achieved the highest performance values across the networks, and was therefore selected for subsequent analyses.

Finally, we take the dis_text network embedding and applied a UMAP representation and colored the diseases using the domain disease schema of Fig. 2A. In Fig. 3A we can see that diseases of each disease domain are grouped together in their own UMAP space and that the domains converge in a whole group. When a HDBSCAN clustering is applied to embedding coordinates of the diseases (Fig. 3B) emerges 12 clusters and 402 diseases are not clustered (pink transparent points). As we can see, the unclustered diseases are placed in the center of the main group where in Fig. 3A are placed in the convergence area of disease domains. The identified clusters range in size from 12 to 63 diseases (Table 1). When mapped onto the MONDO ontology, the diseases within each cluster are closely related, with an average shortest path length between 2.48 and 4.85 hops. For reference, two disease variants typically have a distance of two hops: one to their common parent and another to reach the other variant.

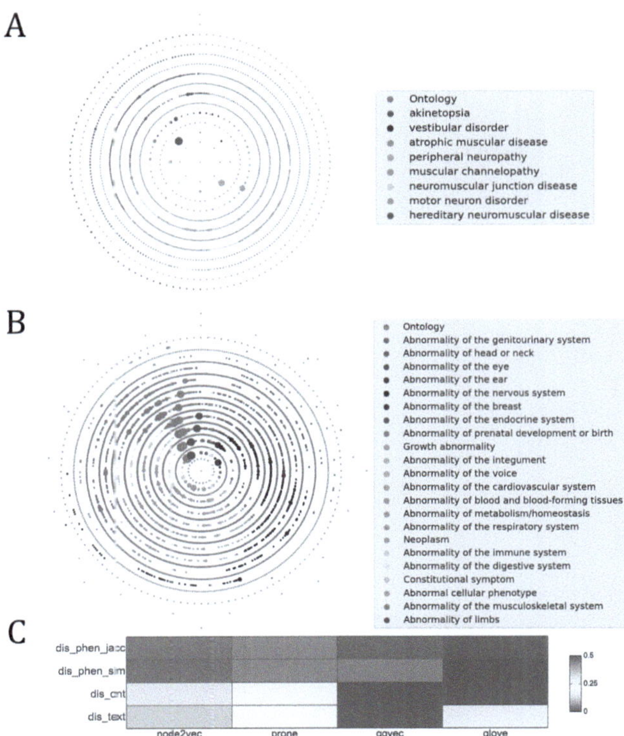

Fig. 2. Circular plots showing the distribution of MONDO term annotations. Inner rings correspond to higher-level ontology categories, while outer rings represent more specific terms. Point size indicates the frequency of each term within the dataset. Color represents a child of the selected root terms, "neuromuscular disease" to highlight disease domains or "phenotypical abnormality" in the case of phenotypes. The annotation counts are transferred from child to parent. Illustrated sets are articles associated to A) disease terms and B) phenotypes terms. C) Heatmap with correlation values obtained when ontological similarity and embedding cosine similarity are compared. Spearman values are show for each combination of network (rows) and embedding algorithm (columns).

Regarding the cluster distribution, we observe that "hereditary neuromuscular disease" is split into clusters 10 and 11, whose MICAs are "hereditary spastic paraplegia" and "muscular dystrophy", respectively, corresponding to subdivisions of the "hereditary neuromuscular disease" term in the MONDO ontology. For the "motor neuron disorder" domain, clusters 2 and 8 correspond to the "spinal muscular atrophy" and "familial amyotrophic lateral sclerosis" subdivisions. A special case is the heterogeneous domain "peripheral neuropathy", which includes all diseases related to the peripheral nervous system and therefore encompasses clusters 0, 1, 4, 5, 6, and 7.

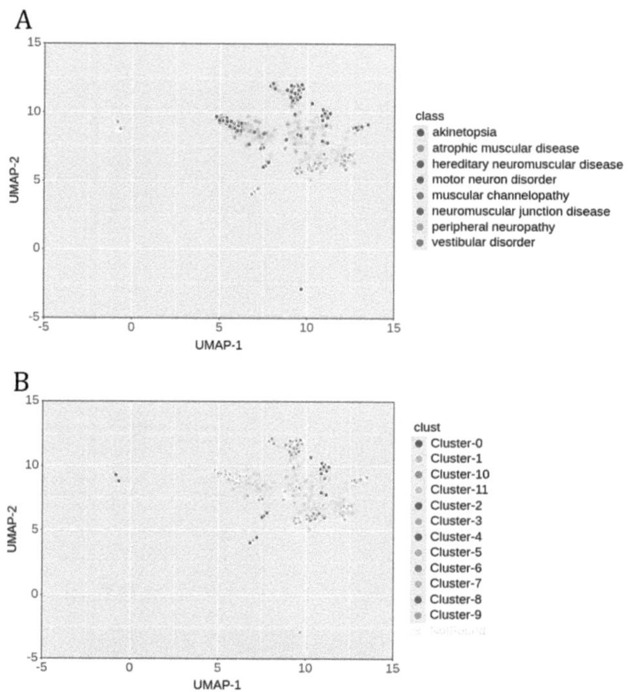

Fig. 3. UMAP representation based on the coordinates of the node2vec embedding over the dis_paper network. Each dot represents a MONDO of Neuromuscular Diseases. In A), color represent the classification based on the main 8 subdivision of NMD terms. In B), each color represent the clusters obtained with the HDBSCAN algorithm. Nodes without a clusters assigned were presented in a transparent tone. For UMAP dimensionality reduction we used the following parameters: `n_neighbors=30`, `min_dist=0.1`, `n_components=2`, `metric='euclidean'`, and `random_seed=123`

The cluster 1 has the general MICA "hereditary peripheral neuropathy" and clusters 5–6 have "peripheral neuropathy" and this is due to the heterogenous diseases grouped in these clusters. However, there are very specialized clusters like cluster 0 with "Zellweger spectrum disorders" that is related with peroxisome diseases, the cluster 4 with the MICA "sphingolipidosis" and the cluster 7 with the "Charcot-Marie-Tooth disease" MICA. This splitting explain the reason of the sparse distribution of the peripheral neuropathy and even explain why we have identified with high frequency phenotypes related with metabolism terms in Fig. 2B. Additionally, cluster 3 was associated with the parental category "coenzyme Q10 deficiency", further supporting the metabolic component underlying several of the identified phenotypes.

In fact, we performed an enrichment analysis in each cluster using the identified phenotypes with the language model. Specifically, we employed the disease–phenotype pairs obtained during the bipartite network generation. A hypergeo-

Table 1. Attributes for clusters obtained from dis_text network. Size is the number of diseases in the clusters. Avg_sht_path is the average shortest path calculated for each cluster in the MONDO ontology. MONDO parental name is the calculated MICA for diseases of the cluster.

Cluster	Size	Avg_sht_path	MONDO parental name
Cluster-0	14	4.15	Zellweger spectrum disorders
Cluster-1	17	3.36	hereditary peripheral neuropathy
Cluster-2	50	3.96	spinal muscular atrophy
Cluster-3	12	3.27	coenzyme Q10 deficiency
Cluster-4	12	4.28	sphingolipidosis
Cluster-5	25	4.08	peripheral neuropathy
Cluster-6	23	4.85	peripheral neuropathy
Cluster-7	63	3.51	Charcot-Marie-Tooth disease
Cluster-8	31	2.48	familial amyotrophic lateral sclerosis
Cluster-9	24	4.13	hereditary neuromuscular disease
Cluster-10	39	2.97	hereditary spastic paraplegia
Cluster-11	18	4.34	muscular dystrophy

metric test was applied to assess enrichment within each cluster, and for illustration, we report the most significant HPO term for each cluster that is below a FDR cut-off of 0.05 (Table 2). We then highlight and discuss the most notable cases. We can se that cluster 8 has the phenotype "Amyotrophic lateral sclerosis" that corresponds with eh MONDO term "familial amyotrophic lateral sclerosis". In the case of cluster 0 that is related with peroxisome diseases, is connected to know phenotype "Aplasia/hypoplasia affecting bones of the axial skeleton". Cluster 2 corresponds with spinal muscular atrophy and has associated the "Delayed ability to roll over" phenotype. Cluster 10 is hereditary spastic paraplegia that has "pulverulent cataract", a feature observed in certain syndromic presentations. Finally, cluster 9 that corresponds with hereditary neuromuscular disease is connected with the phenotype "Periodic hyperkalemic paralysis".

As a final application, we performed gene enrichment analysis on the identified clusters and compared the results, as shown in Fig. 4. Statistically significant categories are connected based on shared genes. Our primary goal was to demonstrate the functional coherence within each disease cluster. The enrichment results reveal distinct gene specializations and shared biological functions among diseases, providing strong evidence of the internal consistency and biological relevance of the identified clusters.

Table 2. HPO enrichment analysis with phenotype terms obtained from the text annotation procedure. A FDR cut-off of 0.05 was applied

Cluster_id	Phenotype	FDR
Cluster-8	Amyotrophic lateral sclerosis	3.55E-14
Cluster-4	Bullet-shaped hallux phalanx	7.49E-12
Cluster-0	Aplasia/hypoplasia affecting bones of the axial skeleton	9.92E-10
Cluster-1	Elevated circulating C6 acylcarnitine concentration	1.29E-08
Cluster-2	Delayed ability to roll over	7.39E-08
Cluster-11	Triangular epiphyses of the 3rd finger	8.87E-07
Cluster-3	Aplasia/Hypoplasia of the 4th finger	3.66E-06
Cluster-5	Triangular shaped proximal phalanx of the 5th finger	0.00028
Cluster-7	Elevated circulating C5 acylcarnitine concentration	0,00079
Cluster-10	Pulverulent cataract	0,001
Cluster-6	Abnormal line of Schwalbe morphology	0,0075
Cluster-9	Periodic hyperkalemic paralysis	0,0081

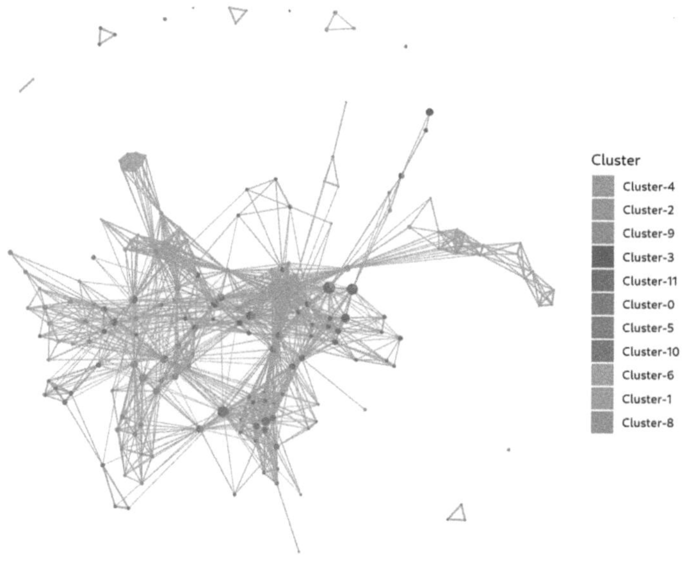

Fig. 4. Network representation of functional enrichment results. Each node represents Gene Ontology (GO) Biological Process (BP) terms, with colors indicating the cluster (HDBSCAN-derived) assignment of the contributing genes. Edges connect BP terms that share common genes, with edge thickness corresponding to the number of shared genes.

4   Conclusion

Our work demonstrates the use of language models and network embeddings to explore associations between genetic diseases. In a case study on NMDs, we show that analyzing only full open-access articles can reconstruct known disease relationships from ontology structures. A key advantage of our approach is the ability to generate specific document corpora for each disease cluster, linking them to phenotypes and related genes to facilitate the discovery of shared disease mechanisms.

Acknowledgments. This work is supported by funds from Spanish Ministry of Economy and Competitiveness [PID2022-140047OB-C21]. Federico García-Criado is a predoctoral researcher from "Ayudas para contratos predoctorales para la Formación del Profesorado Universitario" (FPU21/01449) supported by the Ministerio de Ciencia, Innovación y Universidades. Jesús Pérez-García is a predoctoral researcher from "Ayudas para contratos predoctorales para la formación de doctores" (PRE2022/000510) supported by the Ministerio de Ciencia, Innovación y Universidades.

Disclosure of Interests. The authors have no competing interests to declare that are relevant to the content of this article.

References

1. Armary, P., El-Vaigh, C.B., Labbani Narsis, O., Nicolle, C.: Ontology learning towards expressiveness: a survey. Comput. Sci. Rev. **56**, 100693 (2025)
2. Chu, X., Guan, B., Dai, L., Liu, J.x., Li, F., Shang, J.: Network embedding framework for driver gene discovery by combining functional and structural information. BMC Genomics **24**(1), 426 (2023)
3. Dagdelen, J., et al.: Structured information extraction from scientific text with large language models. Nat. Commun. **15**(1), 1418 (2024)
4. Devlin, J., Chang, M.W., Lee, K., Toutanova, K.: BERT: pre-training of deep bidirectional transformers for language understanding, May 2019
5. Durmaz, A.R., Thomas, A., Mishra, L., Murthy, R.N., Straub, T.: An ontology-based text mining dataset for extraction of process-structure-property entities. Sci. Data **11**(1), 1112 (2024)
6. Farrell, P.R., Magida Farrell, L., Farrell, M.K.: Ancient texts to PubMed: a brief history of the peer-review process. J. Perinatol. **37**(1), 13–15 (2017)
7. Gargano, M.A., Matentzoglu, Robinson, P.N.: The human phenotype ontology in 2024: phenotypes around the world. Nucleic Acids Res. **52**(D1), D1333–D1346 (2024)
8. Grover, A., Leskovec, J.: Node2vec: scalable feature learning for networks. In: KDD : Proceedings. International Conference on Knowledge Discovery & Data Mining, vol. 2016, pp. 855–864, August 2016
9. Huang, C.C., Lu, Z.: Community challenges in biomedical text mining over 10 years: success, failure and the future. Brief. Bioinform. **17**(1), 132–144 (2016)
10. Jabato, F.M., et al.: Gene expression analysis method integration and co-expression module detection applied to rare glucide metabolism disorders using ExpHunterSuite. Sci. Rep. **11**(1), 1–12 (2021)

11. Jabato, F.M., Rojano, E., Perkins, J.R., Ranea, J.A.G., Seoane-Zonjic, P.: Kernel based approaches to identify hidden connections in gene networks using NetAnalyzer. In: Lecture Notes in Computer Science (Including Subseries Lecture Notes in Artificial Intelligence and Lecture Notes in Bioinformatics) (2020)
12. Kelchtermans, J., Mayer, O.H.: Year in review 2021: neuromuscular diseases. Pediatr. Pulmonol. **58**(1), 20–25 (2023)
13. Kojaku, S., Radicchi, F., Ahn, Y.Y., Fortunato, S.: Network community detection via neural embeddings. Nat. Commun. **15**(1), 9446 (2024)
14. Li, Q., Zhao, K., Bustamante, C.D., Ma, X., Wong, W.H.: Xrare: a machine learning method jointly modeling phenotypes and genetic evidence for rare disease diagnosis. Genet. Med. **21**(9), 2126–2134 (2019)
15. Malzer, C., Baum, M.: A hybrid approach to hierarchical density-based cluster selection. In: 2020 IEEE International Conference on Multisensor Fusion and Integration for Intelligent Systems (MFI), pp. 223–228, September 2020
16. Pennington, J., Socher, R., Manning, C.: Glove: global vectors for word representation. In: Proceedings of the 2014 Conference on Empirical Methods in Natural Language Processing (EMNLP), pp. 1532–1543. ACL, Doha, Qatar (2014)
17. Putman, T.E., Schaper, Muñoz-Torres, M.C.: The monarch initiative in 2024: an analytic platform integrating phenotypes, genes and diseases across species. Nucleic Acids Res. **52**(D1), D938–D949 (2024)
18. Reimers, N., Gurevych, I.: Sentence-BERT: sentence embeddings using Siamese BERT-networks. In: Inui, K., Jiang, J., Ng, V., Wan, X. (eds.) Proceedings of the 2019 Conference on Empirical Methods in Natural Language Processing and the 9th International Joint Conference on Natural Language Processing (EMNLP-IJCNLP), pp. 3982–3992. ACL, Hong Kong, China, November 2019
19. Sarker, A., et al.: Natural language processing for digital health in the era of large language models. Yearb. Med. Inform. **33**(1), 229–240 (2025)
20. Vasilevsky, N.A., Matentzoglu, Haendel, M.A.: Mondo: unifying diseases for the world, by the world. medRxiv p. 2022.04.13.22273750, May 2022
21. Zambon, A.A., Falzone, Y.M., Bolino, A., Previtali, S.C.: Molecular mechanisms and therapeutic strategies for neuromuscular diseases. Cellular Molecular Life Sci. CMLS **81**(1), 198 (2024)
22. Zhang, J., Dong, Y., Wang, Y., Tang, J., Ding, M.: ProNE: fast and scalable network representation learning. In: Proceedings of the Twenty-Eighth International Joint Conference on Artificial Intelligence, pp. 4278–4284. International Joint Conferences on Artificial Intelligence Organization, Macao, China, August 2019

Medical Image Processing and Computer Vision in Clinical Environments

3D MTransINR: a 3D Modality Translation Model Based on Implicit Neural Representations

Maria Ysern[(✉)] [iD] and Veronica Vilaplana [iD]

Universitat Politècnica de Catalunya, Barcelona, Spain
{maria.ysern,veronica.vilaplana}@upc.edu

Abstract. We present 3D MTransINR, a novel model for volumetric MRI modality translation based on Implicit Neural Representations (INRs). In clinical workflows, acquiring a complete set of MRI modalities is often impractical, motivating methods that can infer multiple complementary missing sequences from available ones. Our model tackles this predictive synthesis task by translating one or more source modalities into one or more targets within a unified, resolution-independent framework. 3D MTransINR combines a shared multilayer perceptron (MLP) with voxel-wise bias modulation generated by a 3D U-Net, enabling anatomically consistent synthesis. We extend prior INR-based methods to full volumetric data and adversarial training, and evaluate our approach on the ProstateX and BraTS datasets. On ProstateX, 3D MTransINR outperforms a 3D Pix2Pix baseline in PSNR and SSIM across all targets and resolutions, showing strong robustness to resolution changes. On BraTS, it preserves finer structural details but is more sensitive to modality contrast, highlighting the challenges of generalisation across anatomies and intensity profiles.

Keywords: MRI translation · Implicit neural representation · 3D image synthesis

1 Introduction

Medical imaging is essential for the diagnosis and management of diseases such as cancer. Magnetic Resonance Imaging (MRI), in particular, plays a central role due to its excellent soft-tissue contrast and the ability to non-invasively capture three-dimensional anatomical structures. Multiple MRI modalities exist, each emphasising different tissue properties. Depending on the clinical context, acquiring all relevant sequences may be desirable, but is often impractical in routine clinical practice. Limitations arise from scanner availability, time constraints, or patient-related factors (e.g., intolerance to lengthy acquisitions).

These challenges have motivated the development of cross-modality synthesis models, which aim to estimate missing sequences from available ones. Although

A. López Fernández et al. (Eds.): CIABiomed 2025, LNBI 16148, pp. 311–323, 2026.
https://doi.org/10.1007/978-3-032-10661-2_24

not yet suitable for direct diagnostic use, synthesised images can support retrospective studies, assist interpretation when sequences are degraded or absent, and enable more complete multimodal datasets.

Beyond recovering individual sequences, an increasingly relevant challenge is the predictive generation of multiple complementary MRI modalities from a reduced acquisition protocol. This one-to-many setting enables time-efficient imaging while preserving diagnostic value, and is highly relevant in scenarios of constrained acquisition, data harmonisation, or augmentation. Importantly, it demands structured, high-dimensional mappings that ensure anatomical and spatial coherence across outputs, posing a challenging benchmark for scalable and generalisable models.

Moreover, synthetic modalities serve as effective data augmentation in supervised learning, especially under data scarcity or class imbalance, improving training robustness and reducing reliance on large, fully annotated datasets.

To address the modality synthesis task, several deep learning models have been proposed, primarily in 2D. Among supervised approaches, Pix2Pix [3] is a common baseline that learns paired source-to-target mappings using conditional Generative Adversarial Networks (GANs), combining adversarial and L1 losses to balance realism and structural fidelity. Unpaired frameworks such as CycleGAN [9] and CUT [6] enable translation without aligned pairs via cycle-consistency or contrastive objectives. However, this work focuses on the supervised scenario, where paired volumetric data is available, and uses 3D Pix2Pix as a baseline.

These approaches have shown promising results in 2D medical imaging, and recent advances based on GANs and diffusion models have further improved translation quality. However, most of these methods do not scale well to volumetric data. In particular, inter-slice continuity is not guaranteed, leading to potential inconsistencies across the depth dimension in terms of contrast, brightness, or spatial alignment. Extending such models to 3D by replacing 2D convolutions with their 3D counterparts can be prohibitively expensive in terms of computational resources, especially for diffusion models, and does not necessarily guarantee improved performance.

To address these limitations, we explore the use of Implicit Neural Representations (INRs) for 3D modality translation. INRs model discrete signals as continuous functions parameterised by neural networks, typically multi-layer perceptrons (MLPs) that map spatial coordinates to signal values. This formulation is resolution-independent, memory-efficient synthesis with natural support for interpolation. Such properties make INRs attractive for tasks involving signal reconstruction and super-resolution. Importantly, INR-based models offer a lightweight alternative to convolution-heavy architectures, which is particularly relevant given the high computational cost of 3D processing.

Recent 2D INR-based models have shown promise in cross-modality MRI synthesis. Chen et al. [2] combined a Convolutional Neural Network (CNN)-based hypernetwork with local INRs to predict modality-specific parameters, aided by positional encoding. Subsequently, CoNeS [1] introduced a more efficient global INR shared across samples, modulated via bias shifts computed by an Feature Pyramid Network-based hypernetwork. Despite competitive results in 2D, their extension to full 3D volumes remains unexplored.

Building upon those 2D INR methods, we propose 3D MTransINR, a model that performs fully volumetric modality translation based on a global INR and a voxel-wise bias modulation. Unlike slice-wise approaches, our framework ensures inter-slice continuity and enables simultaneous synthesis of multiple sequences. This formulation provides a compact and resolution-independent alternative to conventional 3D CNNs. We evaluate our model on two benchmark datasets: ProstateX and BraTS. On ProstateX, it outperforms 3D Pix2Pix in peak signal-to-noise ratio (PSNR) and structural similarity index measure (SSIM) and is robust to resolution changes. On BraTS, results are more variable: 3D MTransINR better preserves structural detail but is more sensitive to modality contrast, occasionally producing artefacts. These findings suggest that INR-based models are a promising and efficient solution for 3D medical image synthesis, particularly in high-resolution, anatomically constrained settings. The code may be found in https://github.com/imatge-upc/3D-MTransINR.

2   Method

Inspired by CoNeS [1], we propose 3D MTransINR, a volumetric modality translation model based on Implicit Neural Representations (INRs). The model is trained adversarially and consists of a generator and a discriminator. The model's architecture is illustrated in Fig. 1.

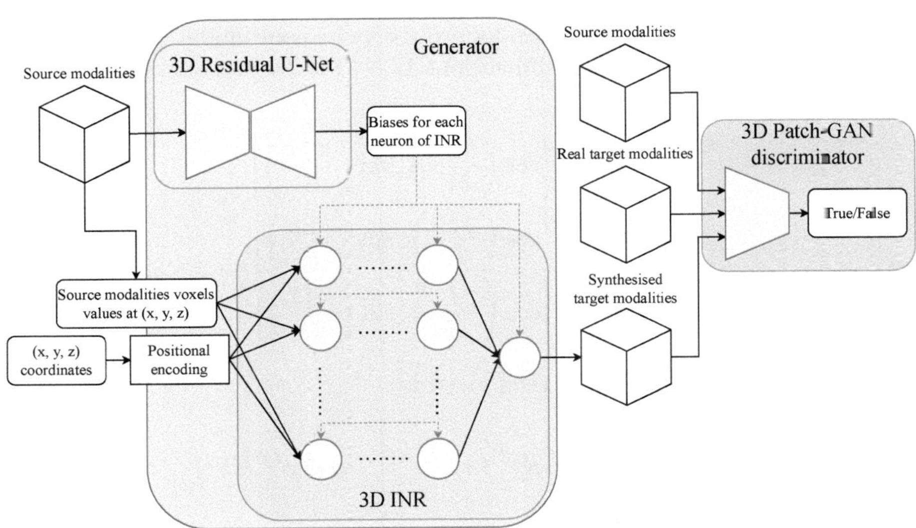

Fig. 1. Architecture of 3D MTransINR.

The generator consists of two components: a global INR, implemented as a multi-layer perceptron (MLP), and a hypernetwork that produces a modulation vector to adapt the INR to each input volume. The MLP receives as input a concatenation of the spatial coordinates (after positional encoding) and the voxel-wise values of the source modality or modalities. The sequences are normalised between -1 and 1.

A known limitation of classical INRs, also referred to as local INRs, is their inability to generalize across samples, as they are designed to represent only one signal at a time. In [2], this required computing all MLP weights for each sample. To address this, CoNeS proposed using a shift modulation mechanism, where the INR parameters are shared across samples and only an additional bias b' is predicted by the hypernetwork for each neuron in the MLP. The output of each neuron is then given by:

$$y = w^T x + b + b', \tag{1}$$

where w and b are globally shared parameters, and b' varies per sample. The modulation is applied before the activation function. In our model, this modulation is computed voxel-wise and for every neuron in each hidden layer, enabling highly specific conditioning without duplicating the INR structure.

The global INR in 3D MTransINR consists of an MLP with five hidden layers of 64 units each, using LeakyReLU activations throughout except for the final output layer, which uses a hyperbolic tangent to ensure outputs remain bounded.

2.1   Positional Encoding

To enhance the network's ability to model high-frequency spatial details, we apply Fourier-based positional encoding to the voxel coordinates [7]. For a voxel at location (x, y, z) in a grid of dimensions $W \times H \times D$, we compute:

$$\phi_x^{(f)}(x) = \left[\cos\left(\frac{\pi f x}{W} \right), \sin\left(\frac{\pi f x}{W} \right) \right] \tag{2}$$

$$\phi_y^{(f)}(y) = \left[\cos\left(\frac{\pi f y}{H} \right), \sin\left(\frac{\pi f y}{H} \right) \right] \tag{3}$$

$$\phi_z^{(f)}(z) = \left[\cos\left(\frac{\pi f z}{D} \right), \sin\left(\frac{\pi f z}{D} \right) \right] \tag{4}$$

The final encoding vector is then given by:

$$\Phi(x, y, z) = \bigoplus_{k=0}^{d_s} \left(\phi_x^{(2^k)}(x) \oplus \phi_y^{(2^k)}(y) \oplus \phi_z^{(2^k)}(z) \right), \tag{5}$$

where $\oplus$ denotes vector concatenation and d_s is the number of frequency bands.

2.2 Hypernetwork Architecture

Unlike CoNeS, which employs a 2D Feature Pyramid Network (FPN), we adopt a 3D U-Net with residual connections to generate the modulation biases. This decision was motivated by the infeasibility of extending the original FPN to 3D due to memory constraints. The 3D U-Net architecture is well suited for volumetric data, capturing long-range spatial dependencies through its encoder-decoder structure. Residual connections within the network facilitate deeper architectures by improving gradient flow and promoting feature reuse. Skip connections between encoder and decoder layers preserve spatial detail, contributing to accurate localisation of anatomical structures. In all experiments, the U-Net uses 16^k feature maps at level k, with four levels of depth. Batch normalisation is applied throughout. Max-pooling is used in the encoder to downsample the signal, and transposed convolutions are employed in the decoder for upsampling.

2.3 Discriminator

We employ a 3D multiscale PatchGAN discriminator, adapted from the architecture in [8], to provide adversarial supervision during training. The discriminator takes as input the concatenation of the source and target modalities along the channel dimension and evaluates them at two spatial resolutions.

The input volume is progressively downsampled via 3D average pooling and processed by two sub-discriminators operating at different scales. Each follows a PatchGAN-like design with five 3D convolutional layers. The first layer uses a $4 \times 4 \times 4$ kernel with stride 2 and outputs 64 feature maps, followed by a LeakyReLU activation. Subsequent layers double the number of channels up to 512 and apply instance normalization and LeakyReLU. The final layer is a single-channel convolution with the same kernel size and stride 1, producing a patch-wise real/fake prediction. This multiscale setup enables the discriminator to capture both fine-grained details and broader anatomical consistency.

2.4 Training Objectives

The model is optimised using an adversarial framework, complemented by additional losses to stabilise training and encourage accurate reconstruction

- **Adversarial loss.** A 3D multiscale PatchGAN discriminator assesses the realism of the generated images at different resolutions. Each sub-discriminator contributes its own real/fake predictions, and the adversarial loss is computed independently for each resolution and then averaged.
- **L1 reconstruction loss.** An L1 loss is applied between the predicted and ground-truth target modalities to enforce voxel-wise similarity and preserve anatomical structure.
- **Modulation bias regularisation.** To prevent numerical instabilities, we apply an L1 regularisation term to the biases predicted by the hypernetwork.

- **Feature matching loss.** Following [8], we compute the L1 distance between intermediate feature activations of real and synthesised samples extracted from the discriminator. This loss encourages the generator to match not only the final prediction but also the internal statistics of the target domain, improving visual coherence and training stability.

Unlike some hybrid approaches, the INR is not pretrained. Instead, the MLP is trained from scratch jointly with the hypernetwork and the discriminator. The model operates on the full 3D volume rather than cropped patches, enabling spatially consistent translation across the entire structure.

3   Experimental Setup

3.1   Datasets

ProstateX. We used the PROSTATEx Challenge dataset [4], which contains multiparametric prostate MRI scans from 346 patients. Each case includes at least three modalities: T2-weighted, Diffusion-Weighted Imaging (DWI), and Apparent Diffusion Coefficient (ADC). T2 sequences have an in-plane resolution of approximately 0.5 mm and a slice thickness of 3.6 mm. DWI volumes, captured with b-values of 50 and 800 s/mm^2, and ADC maps share the same slice thickness and have 2 mm in-plane resolution. In our preprocessed version, the modalities are co-registered at two fixed sizes ($128 \times 128 \times 24$ and $256 \times 256 \times 24$) to the T2 modality. All MRI are clipped at their 99.99^{th} percentile to remove some of the artefacts. It is worth noting that we removed some patients that were missing modalities or that could not be properly registered. Therefore, we retained 283 cases from which we randomly reserved 29 for testing.

We consistently used the T2-weighted sequence as the source modality because it is routinely acquired in clinical protocols, offers the highest spatial resolution and signal-to-noise ratio among the available sequences, and provides stable anatomical context across patients. Additionally, they are less prone to motion and susceptibility artefacts than diffusion-weighted images, making them a robust and reliable foundation for modality translation. Therefore, the objective of the models was to synthesise the other three modalities at the same time.

BraTS 2021. We also employed the Brain Tumour Segmentation (BraTS) 2021 dataset [5], which contains MRI scans from 1,251 patients, with four modalities per case: T1-weighted, T1-weighted with contrast enhancement (T1ce), T2-weighted, and Fluid-Attenuated Inversion Recovery (FLAIR). These volumes are skull-stripped, rigidly aligned, and resampled to 1 mm^3 isotropic resolution. The input volumes have $240 \times 240 \times 155$ voxels, but due to memory constraints, they were non-zero cropped and downsampled to $160 \times 160 \times 128$. All modalities were normalised to the [-1, 1] range after cropping, following the same procedure as in ProstateX.

Due to computational time constraints, we randomly selected 300 cases for training and 150 for testing. All modalities were used as source and target. In particular, we carried out four experiments, in each of them one sequence was the source and the rest, the target. Although multi-input configurations are possible, we adopt a single-source setup to match the ProstateX setting and to increase the difficulty of the prediction task.

3.2 Implementation Details

Experiments were conducted using a single GPU. We used a GeForce RTX 2080 Ti (11 GB) for $128 \times 128 \times 24$ data and an NVIDIA A40 (48 GB) for higher-resolution inputs. All models were implemented in PyTorch and trained using the Adam optimizer ($\beta_1 = 0.5$, $\beta_2 = 0.999$). Batch size was set to 1 for BraTS and 2 for ProstateX. Learning rates were $2e - 3$ for Pix2Pix and ProstateX experiments, and $1e - 4$ for MTransINR on BraTS. Instead of fixing the number of epochs, we trained the models until convergence.

4 Results

4.1 Quantitative Results

Table 1 presents the quantitative results on the ProstateX dataset across two spatial resolutions. As shown, 3D MTransINR consistently outperforms 3D Pix2Pix in both PSNR and SSIM metrics for all target modalities and at both resolutions. Interestingly, the lower-resolution variant of 3D MTransINR achieves the highest SSIM values, while the higher-resolution model obtains the best PSNR scores. This behaviour suggests a trade-off between structural fidelity and voxel-wise intensity precision, potentially influenced by the amount of regularisation and network capacity.

A key advantage of 3D MTransINR is its ability to maintain synthesis quality across spatial resolutions, reflecting the resolution-independent nature of its implicit representation. Doubling the spatial dimensions does not significantly degrade performance. Pix2Pix, by contrast, shows a marked decline in performance at higher resolution, likely due to its convolutional structure's limited receptive field and increased parameter count.

Table 1. PSNR and SSIM test values for 3D MTransINR and 3D Pix2Pix on the ProstateX dataset (T2 as source). Best results are in bold.

Model	Spatial size	b50		b800		ADC	
		PSNR	SSIM	PSNR	SSIM	PSNR	SSIM
3D Pix2Pix	128	25.79 ±2.02	0.723 ±0.090	26.28 ±2.48	0.779 ±0.118	24.11 ±1.53	0.538 ±0.058
3D MTransINR	128	**27.58** ±1.99	**0.793** ±0.070	**28.35** ±2.60	**0.834** ±0.098	**24.83** ±1.44	**0.575** ±0.067
3D Pix2Pix	256	19.27 ±3.29	0.488 ±0.149	22.34 ±2.51	0.637 ±0.101	18.44 ±1.99	0.368 ±0.073
3D MTransINR	256	**27.72** ±1.99	**0.734** ±0.063	**28.37** ±2.52	**0.747** ±0.090	**25.26** ±1.47	**0.573** ±0.061

In contrast, Table 2 shows that 3D Pix2Pix outperforms 3D MTransINR in nearly all configurations on the BraTS dataset, with the exception of experiments using T2-weighted images as input. This discrepancy likely stems from greater anatomical variability in the brain and more complex contrast characteristics of sequences such as FLAIR and T1ce. These aspects are further explored in the following section.

Table 2. PSNR and SSIM test values for 3D MTransINR and 3D Pix2Pix on the BraTS dataset. Best results are in bold.

Model	Source Sequence	Target Sequences	FLAIR		T1		T1ce		T2	
			PSNR	SSIM	PSNR	SSIM	PSNR	SSIM	PSNR	SSIM
3D Pix2Pix	FLAIR	T1, T1ce, T2	–	–	**25.02** ±2.10	**0.864** ±0.037	**27.25** ±2.20	**0.904** ±0.031	**26.13** ±1.69	**0.901** ±0.018
3D MTransINR	FLAIR	T1, T1ce, T2	–	–	21.48 ±3.12	0.804 ±0.054	24.04 ±2.38	0.846 ±0.046	22.00 ±3.70	0.838 ±0.051
3D Pix2Pix	T1	FLAIR, T1ce, T2	**24.55** ±1.93	**0.856** ±0.030	–	–	**26.51** ±1.76	**0.903** ±0.032	**26.43** ±1.88	**0.910** ±0.023
3D MTransINR	T1	FLAIR, T1ce, T2	22.79 ±2.29	0.815 ±0.051	–	–	24.85 ±3.15	0.861 ±0.054	23.90 ±2.07	0.865 ±0.035
3D Pix2Pix	T1CE	FLAIR, T1, T2	**25.13** ±2.10	**0.867** ±0.033	**26.46** ±2.61	**0.889** ±0.040	–	–	**26.68** ±1.68	**0.912** ±0.017
3D MTransINR	T1CE	FLAIR, T1, T2	21.74 ±2.84	0.804 ±0.051	24.34 ±3.44	0.851 ±0.061	–	–	24.05 ±2.87	0.868 ±0.041
3D Pix2Pix	T2	FLAIR, T1, T1ce	21.94 ±1.64	0.827 ±0.017	23.80 ±1.79	**0.857** ±0.034	**24.80** ±1.70	**0.881** ±0.024	–	–
3D MTransINR	T2	FLAIR, T1, T1ce	**23.57** ±3.15	**0.838** ±0.059	**24.11** ±2.83	0.854 ±0.059	24.78 ±2.43	0.869 ±0.046	–	–

4.2   Qualitative Results

Figures 2 and 3 show representative test cases from the ProstateX dataset for the b-value 50, b-value 800, and ADC sequences at lower and higher resolution respectively. These sequences are inherently noisy, with highly variable backgrounds, making accurate translation a challenging task. All tested models exhibit some degree of artefact and structural degradation, reflecting the limitations of current approaches for clinical deployment.

Despite these challenges, 3D MTransINR yields qualitatively superior results compared to 3D Pix2Pix. It more consistently preserves anatomical structures and generates images with reduced noise, which aligns with the higher SSIM scores reported in Table 1. For instance, in the third case (third row) in all three figures, Pix2Pix fails to reconstruct recognisable anatomy, while MTransINR maintains both structure and texture with greater fidelity.

However, 3D MTransINR also exhibits certain limitations. Some outputs display contrast and brightness inconsistencies, particularly in the b-value 800 sequences and, to a lesser extent, in b-value 50. These sequences naturally present low contrast, and the use of min-max normalisation may have amplified this by stretching the intensity histogram beyond informative ranges. Additionally, some synthesised images omit entire anatomical regions. In the first case, for instance,

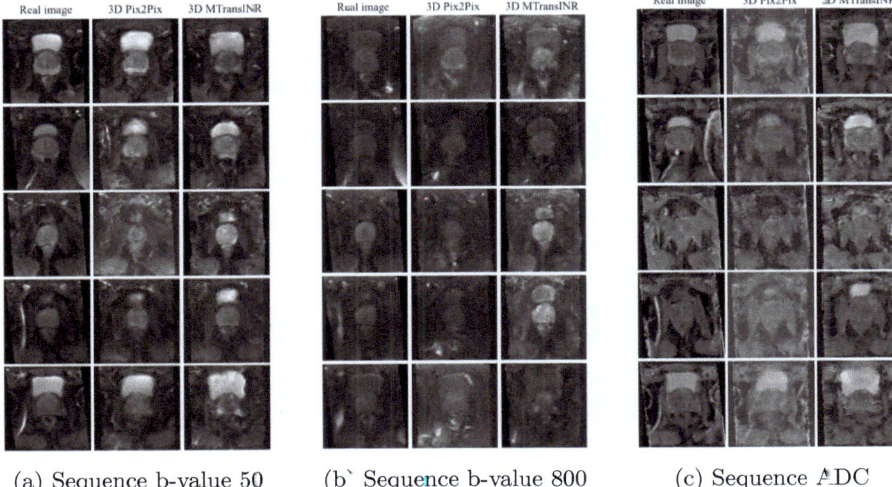

(a) Sequence b-value 50   (b) Sequence b-value 800   (c) Sequence ADC

Fig. 2. Qualitative test examples of the synthesised ProstateX dataset at $128 \times 128 \times 24$ resolution: (a) T2 $\rightarrow$ DWI with b-value $50\,\mathrm{s/mm}^2$, (b) T2 $\rightarrow$ DWI with b-value $800\,\mathrm{s/mm}^2$, and (c) T2 $\rightarrow$ ADC map. Within each subpanel, the three columns show (from left to right) the ground truth, the image generated by 3D Pix2Pix, and the image generated by 3D MTransINR.

a structure on the right-hand side is completely absent in the MTransINR output. Although this region lies outside the prostate, such omissions are concerning when generalising to clinical scenarios. We hypothesise that the model may suppress infrequent patterns in the training distribution, which could also explain the variability and artefacts observed in background regions.

Figure 4 presents three consecutive axial slices from a synthesised volume at resolution $128 \times 128 \times 24$. Although both models struggle with background consistency, they preserve structural continuity across slices, even in the artefacts. This suggests that both 3D Pix2Pix and 3D MTransINR maintain spatial coherence, one of the intended advantages of using fully 3D architectures.

Figure 5 presents representative outputs from the BraTS dataset. In contrast to ProstateX, the synthesised images are generally of lower perceptual quality. Regardless of the source modality, both models exhibit highly variable performance: some cases appear visually acceptable, while others are marked by severe artefacts or anatomical distortion.

In general, 3D Pix2Pix tends to produce excessively smoothed results, often eliminating high-frequency content and yielding overly homogeneous textures. In contrast, 3D MTransINR retains more structural detail and higher-frequency components but at the cost of more pronounced artefacts. This may partly explain why Pix2Pix outperforms MTransINR on quantitative metrics, which are often biased toward smoother outputs.

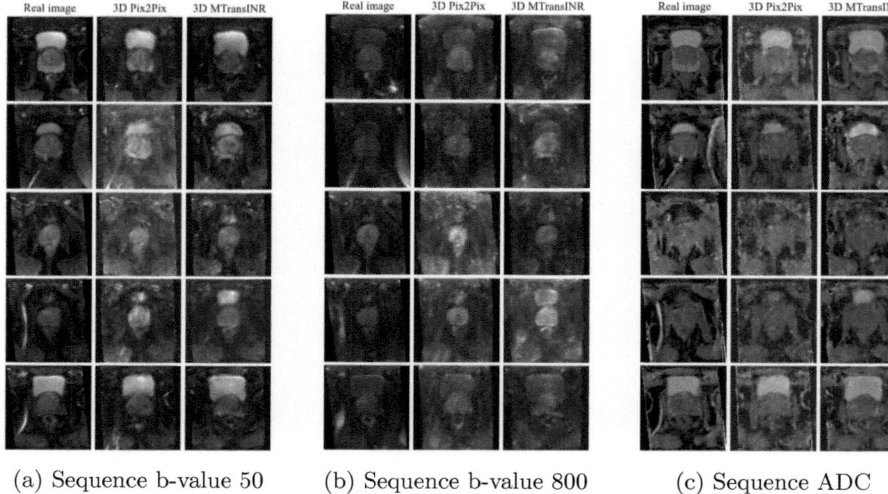

(a) Sequence b-value 50   (b) Sequence b-value 800   (c) Sequence ADC

Fig. 3. Qualitative test examples of the synthesised ProstateX dataset at $256 \times 256 \times 24$ resolution: (a) T2 $\rightarrow$ DWI with b-value $50\,\text{s/mm}^2$, (b) T2 $\rightarrow$ DWI with b-value $800\,\text{s/mm}^2$, and (c) T2 $\rightarrow$ ADC map. Within each subfigure, the three columns show (from left to right) the ground truth, the image generated by 3D Pix2Pix, and the image generated by 3D MTransINR.

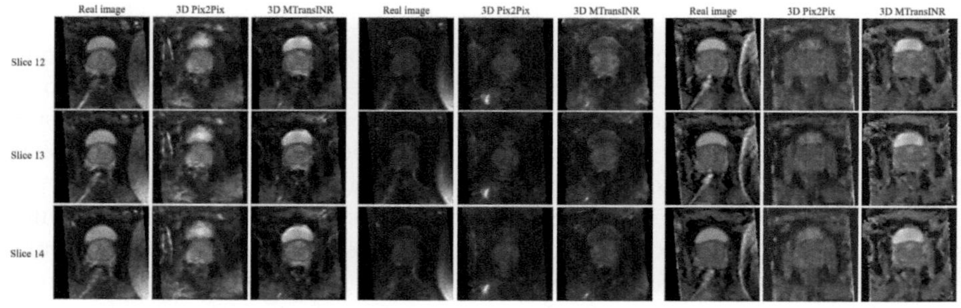

Fig. 4. ProstateX example of consecutive slices synthesised. The sequences from left to right are: b-value 50, b-value 800, and ADC.

Looking at the influence of the source modality, FLAIR appears to be a particularly challenging input for both models, especially for 3D MTransINR, which shows visibly degraded performance in this setting. On the other hand, the most visually convincing results are obtained using T2 as source for 3D MTransINR and T1 for 3D Pix2Pix. For MTransINR, this observation is supported by the quantitative scores, whereas for Pix2Pix, T1ce achieves the highest numerical results, followed by T1.

A common artefact observed in failure cases across both models is a grid-like pattern reminiscent of aliasing. This is particularly evident in MTransINR

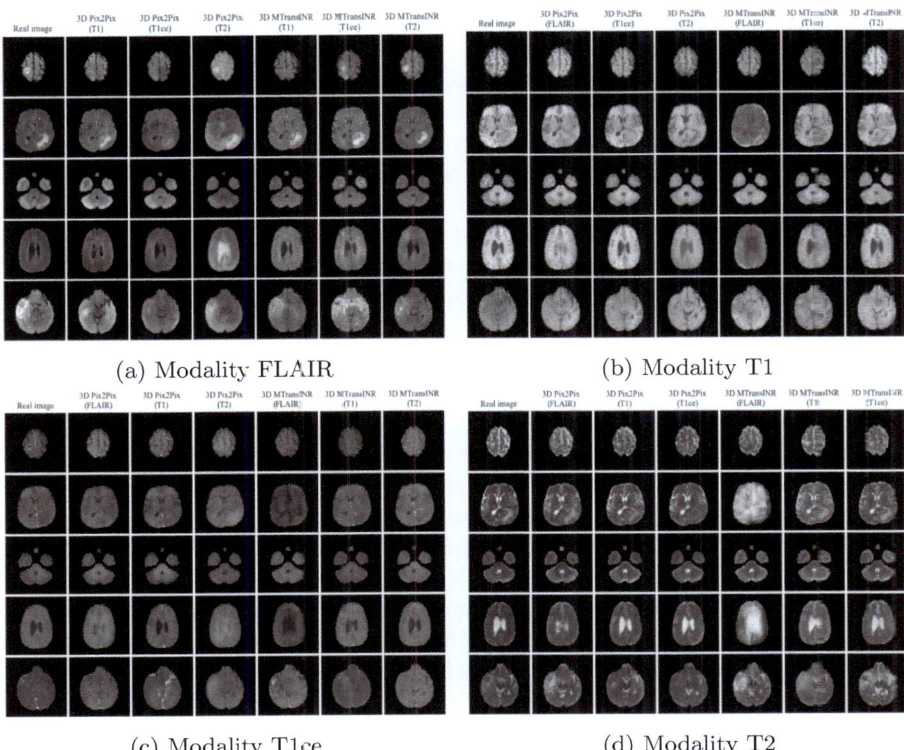

(a) Modality FLAIR   (b) Modality T1

(c) Modality T1ce   (d) Modality T2

Fig. 5. Qualitative test examples of the synthesised BraTS dataset. The generated target modalities are: (a) FLAIR, (b) T1-weighted, (c) T1-weighted with contrast enhancement, and (d) T2. Within each subfigure, the columns (from left to right) show the ground truth, the 3D Pix2Pix output, and the 3D MTransINR output. The source modality used in each experiment is indicated in brackets at the top of the corresponding column.

outputs when FLAIR is used as source, and in Pix2Pix outputs conditioned on T2. Since BraTS images are substantially larger than those in ProstateX, this may point to a capacity limitation. In the case of MTransINR, the voxel-wise modulation mechanism may struggle to scale effectively to larger volumes, as the number of parameters required to generate unique biases for each voxel increases dramatically.

Together, these findings suggest that while 3D MTransINR captures structural detail and high-frequency information more effectively than 3D Pix2Pix, further improvements are needed to ensure the preservation of finer anatomical details and to mitigate artefacts, especially when targeting large-scale clinical datasets like BraTS.

5   Conclusions

We introduced 3DMTransINR, a modality translation model based on Implicit Neural Representations for volumetric medical imaging. Unlike conventional 3D convolutional architectures, our method leverages a global MLP-based INR conditioned by a voxel-wise modulation mechanism generated by a residual 3D U-Net. This formulation enables compact, resolution-independent synthesis while maintaining spatial continuity across slices.

We evaluated 3D MTransINR on two publicly available datasets, ProstateX and BraTS. On ProstateX, it outperformed 3D Pix2Pix in terms of PSNR and SSIM across all target sequences and resolutions. Notably, the model maintained synthesis quality when doubling the spatial resolution, demonstrating the scale-invariant nature of INR-based architectures. Qualitative analysis confirmed improved anatomical fidelity and reduced noise relative to the baseline.

On BraTS, performance was more variable. Unlike ProstateX, where the region of interest is relatively small and spatially constrained, BraTS exhibits substantial inter-patient variability in tumour shape, size, and location, increasing the complexity of anatomical mappings. While the model preserved high-frequency content and fine structural details, it was more sensitive to artefacts and inconsistencies, particularly on large-volume datasets where voxel-wise conditioning becomes more demanding. These limitations may be caused by the voxel-wise nature of the modulation, which increases memory and representational demands with volume size, and from the model's sensitivity to modality contrast and variability.

Overall, our results suggest that INR-based methods offer a promising alternative to traditional 3D CNNs for cross-modality synthesis, particularly in settings with resolution constraints. Future work should focus on improving scalability, enhancing artefact robustness, and extending generalisation to more heterogeneous anatomical domains.

Acknowledgments. This research project with id PID2023-148614OB-I00 has been funded by MICIU/AEI /10.13039/501100011033 and by FEDER, EU. Also, the research conducted by Maria Ysern is supported by the AGAUR FI-SDUR predoctoral programme (grant 2023-FISDU-00389) of the Secretariat for Universities and Research of the Department of Research and Universities of the Generalitat of Catalonia, and by the European Social Fund Plus.

Disclosure of Interests. The authors have no competing interests to declare that are relevant to the content of this article.

References

1. Chen, Y., et al.: Cones: Conditional neural fields with shift modulation for multi-sequence mri translation. Mach. Learn. Biomed. Imaging **2**, 657–685 (2024)
2. Chen, Y., Staring, M., Wolterink, J.M., Tao, Q.: Local implicit neural representations for multi-sequence mri translation. In: 2023 IEEE 20th International Symposium on Biomedical Imaging (ISBI), pp. 1–5 (2023)

3. Isola, P., Zhu, J.Y., Zhou, T., Efros, A.A.: Image-to-image translation with conditional adversarial networks. In: 2017 IEEE Conference on Computer Vision and Pattern Recognition (CVPR), pp. 5967–5976 (2017)
4. Litjens, G., Debats, O., Barentsz, J., Karssemeijer, N., Huisman, H.: Spie-aapm prostatex challenge data (2017), https://www.cancerimagingarchive.net/collection/prostatex/
5. Menze, B.H., Jakab, A., Bauer, S., Kalpathy-Cramer, J., Farahani, K., Kirby, J., et al.: The multimodal brain tumor image segmentation benchmark (brats). IEEE Trans. Med. Imaging **34**(10), 1993–2024 (2015)
6. Park, T., Efros, A.A., Zhang, R., Zhu, J.Y.: Contrastive learning for unpaired image-to-image translation. In: Computer Vision - ECCV 2020: 16th European Conference. Glasgow, UK, 23–28 August 2020, Proceedings, Part IX, pp. 319–345. Springer, Berlin, Heidelberg (2020)
7. Tancik, M., Srinivasan, P.P., Mildenhall, B., Fridovich-Keil, S., Raghavan, N., Singhal, U., et al.: Fourier features let networks learn high frequency functions in low dimensional domains. In: Proceedings of the 34th International Conference on Neural Information Processing Systems, NIPS 2020, pp. 7537–7547. Curran Associates Inc., Red Hook, NY, USA (2020)
8. Wang, T.C., Liu, M.Y., Zhu, J.Y., Tao, A., Kautz, J., Catanzaro, B.: High-resolution image synthesis and semantic manipulation with conditional gans. In: Proceedings of the IEEE Conference on Computer Vision and Pattern Recognition (2018)
9. Zhu, J.Y., Park, T., Isola, P., Efros, A.A.: Unpaired image-to-image translation using cycle-consistent adversarial networks. In: 2017 IEEE International Conference on Computer Vision (ICCV), pp. 2242–2251 (2017)

Encoding the Spatio-Temporal Features of Rey-Osterrieth Complex Figure Strokes for Use in Deep Graph Networks

Pedro José Mulas Cámara[1]([⊠]) , José Manuel Cuadra Troncoso[1] , Mariano Rincón Zamorano[1] , and Pedro Juan Tarraga López[2]

[1] Universidad Nacional de Educación a Distancia (UNED), Madrid, Spain
`pmulas4@alumno.uned.es`, `{jmcuadra,mrincon}@dia.uned.es`
[2] Universidad Castilla-La Mancha (UCLM), Albacete, Spain
`pedrojuan.tarraga@uclm.es`

Abstract. This work introduces a novel methodological approach for the analysis of the Rey-Osterrieth Complex Figure (ROCF), a task widely employed in neuropsychological research, by modeling the drawing process as a time series. The aim is to obtain a compact spatio-temporal representation that addresses the limitations of traditional static-image approaches. The process involves two stages: segmentation of the drawing into strokes to capture spatial and temporal information, and their encoding into feature vectors using deep learning models such as BiLSTM-Autoencoder and ST2Vec. This encoding enables dimensionality reduction and standardization of the spatio-temporal representation of strokes. Validation is currently based on reconstruction accuracy, showing that both architectures preserve essential spatial and temporal aspects: BiLSTM-Autoencoder achieves lower RMSE, while ST2Vec better retains spatial structure. These results should be regarded as a foundational step towards the integration of encoded strokes into graph structures (GNN), enabling future analyses of handwritten drawings from a spatio-temporal perspective. While the long-term vision of this project includes applications in cognitive assessment, the present contribution is methodological in nature and establishes the basis for future extensions to clinically validated populations.

Keywords: Spatio-temporal analysis · Rey-Osterrieth Complex Figure (ROCF) · Alzheimer's disease · Deep Learning · Bidirectional Long Short-Term Memory (BiLSTM) · Spatio-Temporal to Vector Encoding (ST2Vec) · Graph Neural Networks (GNN)

1 Introduction

Alzheimer's disease is the most common neurodegenerative disorder, accounting for between 50% and 60% of dementia cases worldwide [1]. Its diagnosis is complex, as the causes are not fully understood, and current clinical methods —such as biomarker detection or advanced neuroimaging techniques (PET

© The Author(s), under exclusive license to Springer Nature Switzerland AG 2026
A. López Fernández et al. (Eds.): CIABiomed 2025, LNBI 16148, pp. 324–339, 2026.
https://doi.org/10.1007/978-3-032-10661-2_25

or magnetic resonance imaging)— are expensive and invasive [25]. As an alternative, drawing-based tests such as the Clock Drawing Test [22] or the Rey-Osterrieth Complex Figure (ROCF) [21] allow for the assessment of cognitive functions through drawing tasks. However, their manual analysis remains costly, subjective, slow, and highly dependent on the evaluator's expertise. In recent years, there has been growing interest in developing more accessible, objective, and automated methods for the analysis of such tasks, with the long-term goal of contributing to earlier detection of neurodegenerative diseases. In response to this need, various AI-based approaches have been proposed. For example, [28] used convolutional neural networks (Convolutional Neural Network (CNN)) to analyze final images from the ROCF and the Clock Drawing Test, reporting differentiation between healthy participants and those with cognitive impairment. Similarly, [14,18] demonstrated improved classification performance using deep learning techniques applied to large sets of hand-drawn images. More recently, studies such as [13,27] have incorporated eye-tracking and attention-based models to assess performance during the memory recall phase of the FCR test. These approaches dispense with the analysis of the final drawing, relying solely on the sequence of eye fixations, processed with attention-based LSTM models, to explore deficits in visuospatial memory and executive functions.

Within this same context, but with a complementary focus, the present work proposes a methodological framework based on the automatic temporal analysis of drawings made using graphic tablets with paper and pen. This setup is intended to make the recording process more ecological and natural for elderly users and individuals who may experience cognitive decline. Unlike traditional approaches, which concentrate exclusively on the final image of the drawing, this study employs artificial intelligence techniques —particularly computer vision and deep learning— to analyze the dynamic drawing process. This perspective enables the capture of new and underexplored indicators, including planning, execution order, and stroke speed, which have been shown to play a role in differentiating groups in previous clinical research [9,14,28]. However, it should be emphasized that the dataset used here contains only anonymous drawings without diagnostic labels, and therefore the present study does not attempt clinical validation. Instead, it focuses on the construction of a structured spatio-temporal representation as a first essential step toward future studies that may incorporate clinically characterized populations.

The proposed framework is conceived as a methodological contribution to the automated analysis of cognitive drawings, centered on their spatio-temporal dimension. The processing strategy is divided into two phases. First, data obtained from graphic tablets is segmented into strokes, where each continuous trace made without lifting the pen is defined as a Wacom stroke. These are transformed into spatio-temporal segments that contain both spatial information (shape, orientation) and temporal information (speed, duration, order). Second, these strokes are encoded into fixed-dimension feature vectors using deep learning techniques such as BiLSTM and ST2Vec, designed to capture both the morphology of the stroke and its execution dynamics. To construct the embeddings, an autoencoder configuration is employed, enabling the simultaneous generation of compact representations and the reconstruction of the input segments. These embeddings are intended to serve as feature vectors for graph-based models, preserving the essential information of the drawing process.

This approach aligns with a growing trend in the literature, where multiple automated methods have been developed for the analysis of cognitive drawings. Many of these works have focused on processing static images of the final drawing outcome, neglecting the temporal dimension of the drawing process. Although these methods have demonstrated notable performance in classification tasks, they present a fundamental limitation: they do not consider the dynamics of execution, thereby losing potentially significant information. In response, a research line focused on the dynamic analysis of drawing has emerged, incorporating variables such as execution order, stroke speed, and pressure applied during the task. Studies such as [17] and [20] have shown that these temporal features can reveal significant differences between clinical groups. Tools like Sketch-a-Net [29], designed to process vector-format data with temporal information, have shown promising results in recognition tasks. Similarly, [5] enrich static representations with dynamic metadata—such as pressure or speed—improving the performance of CNN-based models. Moreover, the recent review by [3] highlights the importance of incorporating dynamic stroke information and emphasizes the need to improve the interpretability of deep learning models applied to drawing. Together, these works reinforce the idea that stroke dynamics not only complement but in some cases may surpass the informative value of the final image. Recent studies have further advanced this line of research: [26] propose an attention-based model to analyze the temporal evolution of the Clock Drawing Test, while [19] manually analyze strokes in the Rey figure, associating them with specific drawing components and using graphs to represent the sequence and relationships between them. These approaches illustrate the potential of temporal representations for capturing distinctive drawing patterns.

At the same time, the use of graph neural networks (GNN) has begun to be explored for modeling drawings as spatial structures. In these approaches, nodes typically represent elements or regions of the drawing, while edges are defined based on proximity or visual connection criteria [12,15,23]. Although these representations have proved useful for capturing structural relationships, they tend to focus exclusively on the static configuration of the drawing, without incorporating information about the sequence, duration, or execution dynamics. This omission excludes temporal components that are crucial for understanding the underlying cognitive processes in drawing tasks. The present work aims to advance this line by proposing an enriched graph representation that integrates both the spatial structure and temporal stroke information. This representation constitutes a methodological foundation for future applications of GNN-based models in neuropsychological research, with the long-term objective of enabling clinically validated studies in cognitive assessment contexts.

2   Methodology

2.1   Drawing Recording

The first task of the project consists of digitizing participants' hand-drawn sketches for temporal analysis. For this purpose, a Wacom graphic tablet and a digital ink pen connected via Bluetooth were used, allowing real-time recording of the pen trajectory on A4 paper. This setup, designed to offer a familiar and comfortable experience —especially for elderly individuals with possible cognitive impairment— records data every 5 ms, capturing position (X, Y), pressure, and timestamp. This information enables the reconstruction of the shape and the dynamics of the drawing. The system defines a "Wacom stroke" as a continuous sequence of points while the pen remains less than 5 mm from the surface. Each time the pen lifts away and returns, a new stroke begins. The coordinates of the Wacom strokes are recorded in .dat files where each line represents a point storing 4 features (X, Y, pressure, time). The coordinates are included in a 311 × 216 mm space, with the origin (0, 0) at the top-left corner for Y and at the bottom-left corner for X, ensuring the scale and proportion of the drawing. The dataset includes 300 drawings from anonymous individuals, with no diagnostic information regarding cognitive impairment.

2.2   Segmentation and Extraction of Temporal Strokes from the Drawing

Once the drawings have been recorded, the next step is the segmentation of each drawing into units called *strokes*. In this work, a stroke is defined as a line — can be straight, curved, or composite— that connects two relevant points in the drawing: points of intersection between lines or terminal points. These relevant points are detected using a corner detection algorithm that identifies areas of high curvature or abrupt changes in stroke direction. Each stroke thus defined represents a node within a graph structure, allowing for a structural interpretation of the drawing to be used in subsequent stages of analysis. To enable processing by traditional computer vision algorithms, it is necessary to transform the temporal representation into a static digital representation based on pixels rather than distances. Therefore, an equivalence between physical distance and digital resolution is established: 0.2 mm of the original drawing corresponds to 1 pixel in the resulting image. Under this conversion, the final images have dimensions of 1555 × 1080 pixels. This fixed representation will serve as the basis for applying classical image processing algorithms. To avoid terminological ambiguities, from this point forward the term *pixel strokes* will be used to refer to the strokes detected in the fixed digital image. The preprocessing of the images, along with the segmentation of the pixel strokes, is extracted from [12]:

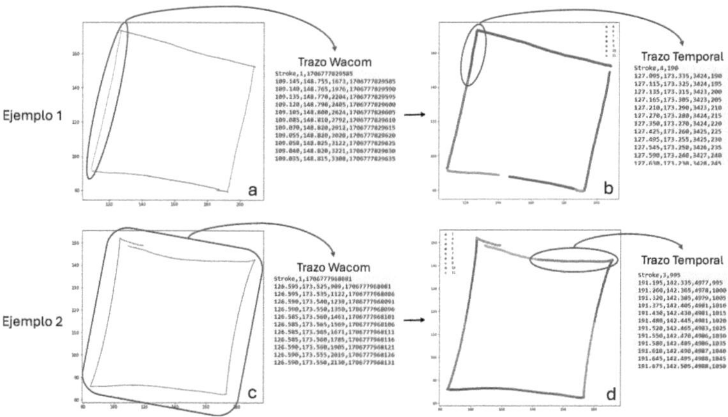

Fig. 1. Drawing Registration Process.

Before continuing with the segmentation into pixel strokes, a dilation and erosion process is applied to the lines to facilitate the subsequent corner detection, which will serve as possible start or end points for each stroke. The *Harris Corner Detector* algorithm [11] is used, which locates these points by analyzing the gradient in the X and Y directions. This algorithm computes the eigenvalues of the local autocorrelation matrix, assigning each point a score that determines its likelihood of being a corner. A threshold on this score allows the selection of points of interest. To eliminate redundancies caused by noise or the proximity of neighboring points (corners very close to each other representing the same region), a DBSCAN clustering algorithm [4] is applied, grouping points based on their spatial density. Subsequently, the points are repositioned to maximize the level of white around them.

The result of this process is spatially segmented strokes, straight segments spatially interpolated that reconstruct the drawing, but still without temporal information as show in Fig. 1. These spatial strokes constitute the basis of each graph node, which will later serve as the premise to incorporate the temporal dimension through the feature matrix. The graph edges, on the other hand, will depend on the temporal information following the recorded chronological order.

2.3   Temporal Stroke Reconstruction

To include the temporal information of the pixel strokes, their reconstruction is performed using the original temporal file (.dat). First, the endpoints of each pixel stroke are identified, and a scale change (1 pixel = 0.2 mm) is applied to align the coordinates. The assignment is done by minimizing the Euclidean distance between each point of interest and the set of temporal points. Then, each stroke is divided into a series of equally spaced segments (20 in our case) to facilitate correspondence and avoid temporal discontinuities. Two situations are distinguished during reconstruction:

1. **Strokes with temporal continuity:** these correspond to segments of the drawing executed continuously, that is, without lifting the pen from the paper during their execution. In these cases, the graphic tablet records a sequence of points following a coherent chronological order without interruptions. When projecting this information onto the previously performed spatial segmentation, the original trajectory of the stroke is preserved, respecting the temporal order.

2. **Strokes with temporal discontinuity:** these occur when a segment of the drawing is executed intermittently, with significant temporal pauses or jumps between parts that, spatially, appear connected. This situation may be due, for example, to the repetition or correction of a part of the drawing at different times. In such cases, the stroke does not follow a linear temporal flow. To approach this, a rectangular region is defined between the initial and final points of the segment, and all intermediate points that, although not consecutive in time, belong to this region and visually contribute to completing the stroke (Fig. 2):

$$x_2 < x_i < x_1 \quad y \quad y_2 < y_i < y_1 \tag{1}$$

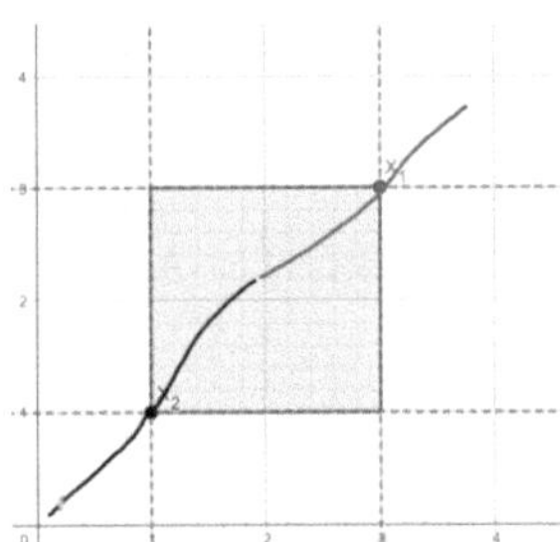

Fig. 2. Temporal reconstruction of strokes with non-continuous time points

Once this process is completed, a cleaning step is performed to eliminate noise such as strokes with a duration shorter than 50 ms (10 points) and strokes with more than 1000 points (5 s), which affect compatibility with the neural networks used later. The resulting temporal strokes are now coherent in both spatial and temporal domains, and constitute the basic unit for subsequent analyses. From now on, we will simply refer to them as **strokes**.

2.4 Encoding and Feature Extraction of Strokes Using Deep Learning

The main objective of this phase is to obtain a meaningful and compact representation of each individual stroke, capturing its morphology and dynamic characteristics such as speed, curvature, or pauses during execution. To achieve this, deep learning techniques are employed to encode the reconstructed strokes into feature vectors that condense their essential information. Specifically, two autoencoder-based architectures have been used: a BiLSTM-Autoencoder network and another called ST2Vec. Although they share this common structure, both networks differ in how they handle temporal information.

BiLSTM-Autoencoder. The network is based on a Long Short-Term Memory (LSTM) architecture, a type of recurrent neural network specifically designed to model temporal sequences. It takes into account not only the information at a given time step, but also previous time steps. The structure of a LSTM network includes an input gate, a forget gate, a memory gate, and an output gate, which regulate the flow and retention of information throughout the sequence. In this project, a bidirectional variant (BiLSTM) has been used Fig. 3, which processes the sequence in both directions (forward and backward), aiming to capture broader contexts and improve learning quality [2].

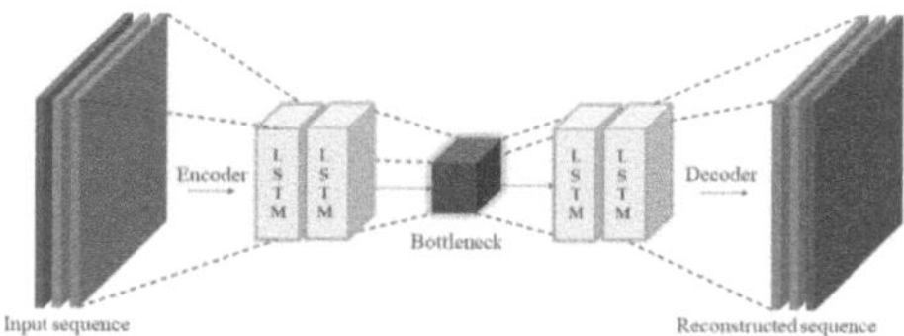

Fig. 3. Structure of the BiLSTM-Autoencoder network [7].

One advantage of this model is that it does not require temporal information to be explicitly provided, as the LSTM layers themselves implicitly incorporate temporal sequencing during learning. Since there is no previously optimized architecture for this dataset, an exploratory strategy has been adopted to adjust the model structure. Specifically, networks with different depths (2, 4, and 6 layers) were tested to evaluate how performance varies with complexity, choosing the 4-layer option because it balanced complexity, computational cost, and performance.

ST2Vec. The second model used is ST2Vec, a network originally designed to generate embeddings of geospatial trajectories over road networks [8]. Its structure allows processing sequences containing spatial and temporal information, making it especially suitable for this work. In its original version, ST2Vec works with trajectories of the form:

$$T = \langle ((x_1, y_1), t_1), \ldots, ((x_n, y_n), t_n) \rangle \tag{2}$$

Unlike the BiLSTM-Autoencoder, temporal information is explicitly introduced into the model. The encoder consists of three distinct modules: (1) **TMM**

(*Temporal Modeling Module*): encodes the temporal vector; (2) **SMM** (*Spatial Modeling Module*): encodes the spatial coordinates (x, y); and (3) **STCF** (*Spatio-Temporal Co-Attention Fusion Module*): jointly concatenates spatial and temporal information into a single encoding vector as show in Fig. 4. The TMM and SMM modules are based on LSTM layers.

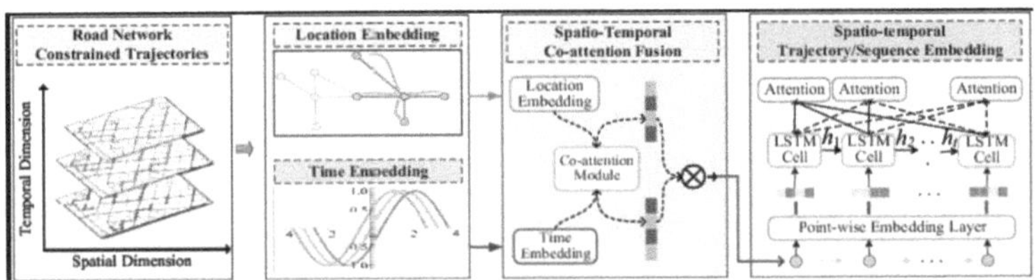

Fig. 4. Structure of the ST2Vec network [8].

Unlike the exploratory approach followed with the BiLSTM, in this case the original architecture proposed by the authors has been maintained. This configuration includes a single layer in the TMM and SMM modules, followed by a single LSTM layer in the decoder.

In both models, the latent vector size has been fixed at 32 dimensions, representing a balance between representation capacity and computational efficiency.

Normalization and Data Preprocessing. The models require temporal sequences of fixed length, but in this project the strokes have variable lengths (between 20 and 1000 points). Therefore, an *online training* strategy has been adopted, processing the strokes sequentially and individually, without the need for *padding*. To avoid problems of overfitting and gradient explosion, a *gradient clipping* mechanism has been implemented, which restricts the magnitude of the gradients during training [10]. Additionally, normalization is applied individually to each stroke within its own region of interest (ROI), rather than applying global normalization to the entire drawing. This strategy prevents the model from processing values with very small or unbalanced scales, facilitating learning. For this purpose, each stroke is rescaled to a range of $[0, 1]$ in both spatial dimensions, using its minimum and maximum coordinates as reference. This local normalization is consistent with the *online* training scheme, since strokes are processed independently without interaction between them.

2.5   Data Augmentation

Due to the limited number of drawings and the quantity of strokes needed to adequately train the neural networks, data augmentation techniques were employed. The main challenge was to preserve the temporal information and the shape of the stroke. For this purpose, an interpolation-based technique was used, which allows generating new realistic versions of the strokes without distorting their structure or execution rhythm. This method was inspired by previous works [6,16].

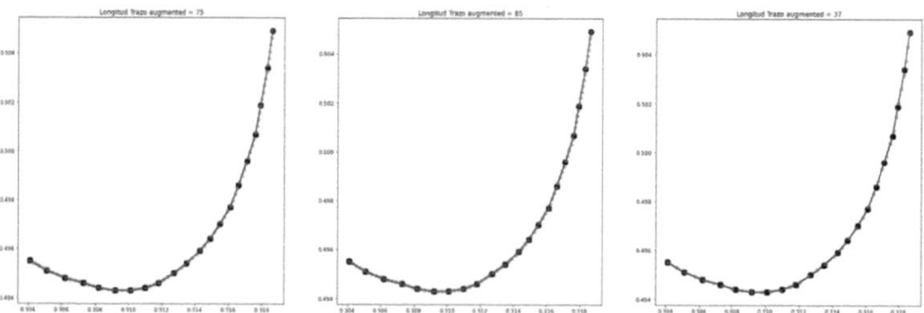

Fig. 5. Example of interpolation for a stroke. Original points are shown in black and interpolated ones in blue. (Color figure online)

The results show sequences that preserve the original shape of the stroke, but with different execution speeds, reflected in the number of points Fig. 5. This interpolation maintains both the spatial arrangement and the natural noise of the stroke, which may contain relevant information. Two additional strategies were added to this technique. The first consists of a temporal rescaling [24], applied only to the temporal vectors of the ST2Vec model. The second strategy is the random removal of 25% of the points in each stroke. This percentage was empirically defined as a balance between variability and structural preservation, avoiding trajectories that are excessively short or distorted.

3   Results

3.1   Obtained Stroke Analysis

The *temporal strokes* generated from a random sample of six drawings are examined in Table 1.

Table 1. Table with data from 6 randomly selected drawings from the database.

Drawing	Temporal Strokes	MAX	MIN	<20	>1000	Normal	Mean	STD	Median
Rey037	97	215	11	24	0	73	59.589	39.883	48
Rey087	136	358	11	17	0	119	83.756	70.432	61
Rey156	114	592	11	11	0	103	103.369	103.547	70
Rey188	140	2088	11	24	2	114	104.509	123.366	71
Rey002	93	293	11	31	0	62	66.694	52.654	56
Rey178	120	572	11	18	0	102	116.049	97.299	89

Visually, strokes are color-coded by length: short strokes (red), normal (blue), and long (black), as shown in Fig. 6. Short strokes frequently appear in areas with dense details, although their number is limited (less than 40%). Despite their brevity, they are retained in the analysis due to their structural relevance.

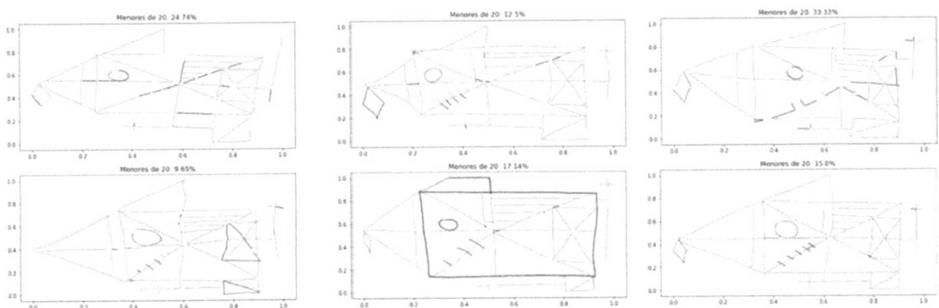

Fig. 6. Visualization of the spatial distribution of strokes by length

Temporal jumps are also detected in some strokes, defined as interruptions in temporal continuity greater than 5 ms between points. Although infrequent (less than 10%), in some cases jumps exceed 10 s, mainly affecting closed figures such as the central square or the side triangle Fig. 7.

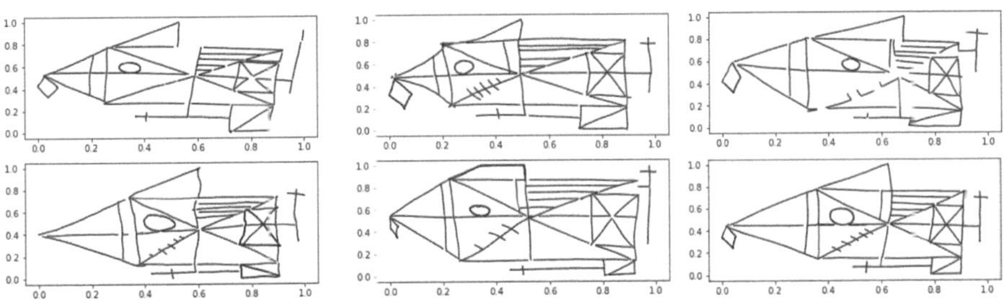

Fig. 7. Continuous strokes (blue) and strokes with temporal jumps (red) in the drawings. (Color figure online)

In summary, the analysis reveals consistent drawing patterns in the segmentation and structure of strokes, although with significant internal variability and certain phenomena such as short strokes or strokes with jumps, which should be considered in later stages of the study.

3.2 Stroke Encoding and Reconstruction

Four configurations were evaluated for stroke encoding and reconstruction, combining two architectures (BiLSTM and ST2Vec) with two loss functions (RMSE and Hausdorff distance).

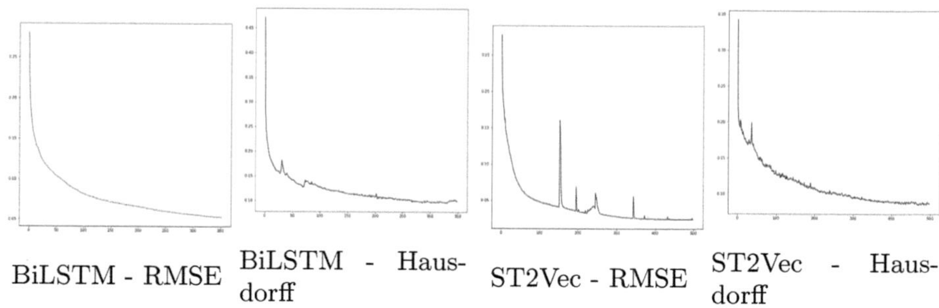

BiLSTM - RMSE   BiLSTM - Hausdorff   ST2Vec - RMSE   ST2Vec - Hausdorff

Fig. 8. Learning curves for the four evaluated models.

The learning curves in Fig. 8 show that all models progressively reduce the error, although trainings using the Hausdorff loss exhibit irregularities due to the non-differentiable nature of the metric. The ST2Vec model, with lower computational cost, allowed training for more epochs, yielding more stable curves and lower final errors, especially with RMSE.

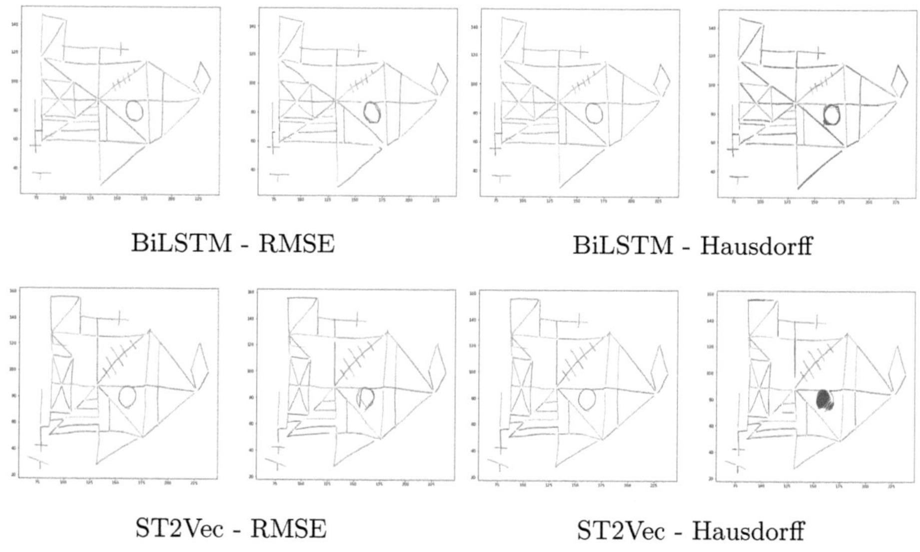

BiLSTM - RMSE   BiLSTM - Hausdorff

ST2Vec - RMSE   ST2Vec - Hausdorff

Fig. 9. Reconstruction examples for the four evaluated models.

Regarding visual reconstruction in Fig. 9, the ST2Vec model trained with RMSE stands out by generating smoother strokes, leveraging the separate processing of spatial and temporal information. Conversely, the use of the Hausdorff loss, while reducing thick strokes caused by temporal irregularities, introduces noisier strokes and poorer performance in complex morphologies. The BiLSTM models show acceptable reconstruction but tend to generate more noise and oscillations in areas with temporal jumps or high morphological complexity, probably due to the higher spatial detail included in the embedding and the implicit integration of temporal information in the recurrent layers.

Finally, the quantitative comparison shown in Table 2 confirms that ST2Vec-RMSE offers the best performance across all evaluated metrics, while ST2Vec-Hausdorff shows the worst results. The BiLSTM models present similar values to each other, with no significant differences between loss functions.

Table 2. Model comparison using different metrics.

Model	RMSE	Euclidean Distance	Cosine Similarity	Manhattan	Pearson	DTW
BiLSTM-RMSE	0.0417	0.5636	0.9979	7.1447	0.9975	2.2599
BiLSTM-Hausdorff	0.0418	0.5696	0.9979	7.1447	0.9975	2.2599
ST2Vec-RMSE	0.0319	0.4357	0.9988	5.2218	0.9985	1.0998
ST2Vec-Hausdorff	0.0680	0.6932	0.9621	6.1158	0.7999	2.3073

4 Discussion

4.1 Stroke Segmentation and Extraction

The segmentation and stroke extraction process showed high variability both between subjects and within a single drawing. Nonetheless, consistent structural patterns were observed, as revealed by Table 1 and Fig. 6. Although visually the strokes appear stable, the quantitative analysis reveals small discrepancies in the data. One of the most relevant is the variation in the total number of strokes, which ranges from fewer than 100 to more than 150 per drawing. Added to this is a high dispersion in stroke length, which can range from barely 11 points to over 1,000. This heterogeneity is reflected in large standard deviations and the presence of outliers. However, these extreme cases were retained due to their possible informative or diagnostic value.

The shortest strokes tend to cluster in specific areas of the drawing, such as the upper right quadrant or vertices of closed shapes, and may be due to pauses or minimal pencil movements. Despite their limited visual relevance, these strokes can complicate temporal encoding and were therefore not removed. Regarding long strokes, many correspond to closed figures drawn continuously (e.g., the outer square), while some extreme cases were discarded due to segmentation errors. Temporal jumps were also identified in less than 10% of the strokes, especially in closed figures.

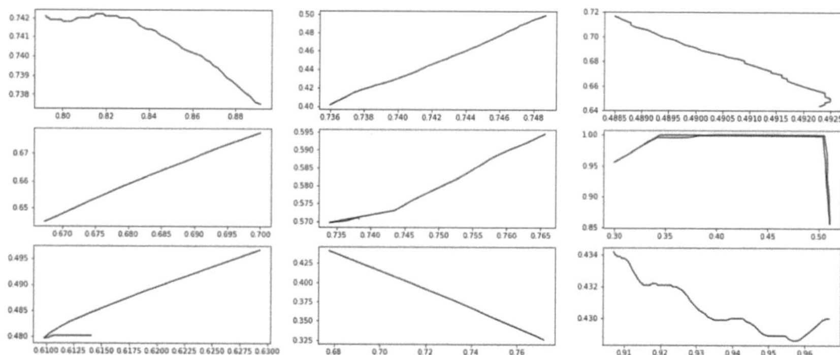

Fig. 10. Example of randomly extracted strokes from each of the sample drawings

The morphology of the strokes is another critical factor. As shown in Fig. 10, many trajectories exhibit irregular curves or abrupt changes in direction, which add noise to the encoding process. However, these irregularities may contain useful clinical information, as they reflect the natural variability of human movement even in subjects without cognitive impairment.

4.2 Stroke Encoding and Reconstruction

The results obtained with the autoencoder models reveal significant differences depending on the architecture used and the loss function employed. Two metrics were compared: RMSE, which minimizes point-to-point errors, and the Hausdorff distance, which aims for better spatial alignment but is more sensitive to noise and requires a differentiable version for training purposes. Two models were evaluated: the BiLSTM-Autoencoder, which implicitly incorporates temporal information through recurrent layers, and ST2Vec, which processes temporal and spatial information separately before combining them. This difference in handling temporal data influences the obtained results.

The ST2Vec model with RMSE yielded the best performance, while its Hausdorff variant was the least accurate. In contrast, the BiLSTM models showed a more balanced performance across both loss functions. This is partly because the Hausdorff metric evaluates only the spatial dimension, which penalizes ST2Vec, whose temporal encoding has limited useful variability, thus reducing the representational capacity of the spatial embedding. Moreover, it was observed that ST2Vec generates more stable and smoother reconstructions, whereas BiLSTM tends to introduce more noise due to its higher sensitivity to stroke details, reflected in a spatial embedding containing a greater amount of information. This difference may be associated with slight overfitting in BiLSTM, which attempts to capture details not always relevant. It was also confirmed that curved or complex strokes, such as circles, are more difficult to reconstruct, especially with the Hausdorff loss. Nevertheless, these elements were retained in the dataset due to their structural relevance. Finally, although BiLSTM manages to capture certain subtle details (such as small irregularities at the beginning or end of strokes), it does so at the expense of greater instability in the overall reconstruction.

5   Conclusion

This work introduces an innovative approach for the dynamic analysis of the ROCF, applying advanced computer vision and deep learning techniques to explore the drawing process from a temporal perspective. Unlike most previous studies, which focused on static analysis of the final output, this research centers on the drawing dynamics, enabling the extraction of enriched information about the participants' graphic behavior. One of the main achievements has been the precise definition of what constitutes a stroke, integrating spatial and temporal information to properly segment and analyze the data.

The implementation of autoencoder models, such as BiLSTM and ST2Vec, has allowed encoding strokes into compact, fixed-size vector representations and validating them through accurate reconstruction. While reconstruction serves as the primary validation criterion and demonstrates that the encoding retains relevant information from the original stroke, it does not directly guarantee that the preserved features are the most relevant for cognitive assessment. Future studies should investigate how these embeddings capture metrics known to be cognitively meaningful, such as the morphology of critical figures, stroke continuity, and execution pauses.

The work presents certain limitations, including the reduced amount of data with temporal information and the high individual variability of strokes. Technical challenges were also identified, such as the computational cost of some models and the restrictions of certain libraries for efficiently handling variable-length sequences. Nevertheless, foundations have been laid for a future research line oriented towards the use of neural graph architectures. Although no GNN was applied in this study, the generated representations are specifically designed to be integrated as nodes in graphs, opening new possibilities to segment and analyze complex figures from a spatio-temporal perspective.

For future work, it will be essential to expand and diversify the dataset by incorporating profiles with cognitive impairment. It is also proposed to explore more flexible architectures such as Transformers or GRUs, as well as loss functions adapted to the temporal domain, including differentiable variants of Dynamic Time Warping (DTW), and to incorporate pressure as a variable to enrich the information available in the drawing task. Furthermore, optimizing the training environment with GPUs and engines that facilitate the processing of variable-length sequences and reduce overfitting will be important.

Although this work does not directly address the construction of a complete graph, it lays the groundwork for it. In particular, a future line of research considers the application of graph neural networks (GNN) that model the ROCF as a graph where nodes correspond to individual segments and edges represent their spatio-temporal relationships. This approach could enrich the structural and cognitive analysis of drawing from a more interpretative and topological perspective.

Overall, this work establishes the foundations for a new line of spatio-temporal analysis of human-drawn sketches. Its clinical application remains exploratory, and the present results should be interpreted as a methodological contribution. Nonetheless, the obtained representations provide a first step towards the development of automated tools for cognitive assessment, with potential extensions to other contexts such as education or creativity.

References

1. ADI: Alzhimer's disease international - Dementia facts & figures—alzint.org (2024). https://www.alzint.org/about/dementia-facts-figures/. Accessed 10 Sept 2024
2. Anishnama: Understanding LSTM: Architecture, Pros and Cons, and Implementation—anishnama20 (2023). Accessed 10 Sept 2024
3. Beltzung, B., Pelé, M., Renoult, J.P., Sueur, C.: Deep learning for studying drawing behavior: a review. Front. Psychol. **14**, 992541 (2023)
4. Cheng, D., et al.: GB-DBSCAN: a fast granular-ball based DBSCAN clustering algorithm. Inf. Sci. **674**, 120731 (2024)
5. Cilia, N.D., D'Alessandro, T., De Stefano, C., Fontanella, F., Molinara, M.: From online handwriting to synthetic images for Alzheimer's disease detection using a deep transfer learning approach. IEEE J. Biomed. Health Inform. **25**(12), 4243–4254 (2021)
6. Dhevi, A.S.: Imputing missing values using inverse distance weighted interpolation for time series data. In: 2014 Sixth International Conference on Advanced Computing (ICoAC), pp. 255–259. IEEE (2014)
7. Do, J.S., Kareem, A.B., Hur, J.W.: LSTM-autoencoder for vibration anomaly detection in vertical carousel storage and retrieval system (VCSRS). Sensors **23**(2), 1009 (2023)
8. Fang, Z., et al.: Spatio-temporal trajectory similarity learning in road networks. In: Proceedings of the 28th ACM SIGKDD Conference on Knowledge Discovery and Data Mining, KDD 2022, pp. 347–356. Association for Computing Machinery, New York (2022). https://doi.org/10.1145/3534678.3539375
9. Gavilan, P.A.R.: Análisis de componentes cognoscitivos y psiquiátricos en pacientes con demencia frontotemporal y Alzheimer. Ph.D. thesis, Universidad Nacional de Colombia (2010)
10. Goodfellow, I., Bengio, Y., Courville, A.: Deep Learning. MIT Press (2016). http://www.deeplearningbook.org
11. Harris, C., Stephens, M., et al.: A combined corner and edge detector. In: Alvey Vision Conference, vol. 15, pp. 10–5244. Citeseer (1988)
12. Hijano Cubelos, O.: Representación de imágenes geométricas con grafos y diagnóstico automático de la figura compleja de rey (2022). https://e-spacio.uned.es/entities/publication/287c2ad8-7a13-4e22-823b-7a69f54737dc/full
13. Jeong-ha, L.: Developing an eye-tracking and deep-learning based Rey-Osterrieth complex figure test for automated evaluation of visual memory and executive function. Ph.D. thesis, Escuela de Posgrado de la Universidad Nacional de Seúl (2024)
14. Langer, N., et al.: Automating clinical assessments of memory deficits: deep learning based scoring of the Rey-Osterrieth complex figure. bioRxiv, pp. 2022–06 (2022)
15. Lee, B., Bang, J.U., Song, H.J., Kang, B.O.: Alzheimer's disease recognition using graph neural network by leveraging image-text similarity from vision language model. Sci. Rep. **15**(1), 997 (2025)

16. Li, H., Wan, X., Liang, Y., Gao, S.: Dynamic time warping based on cubic spline interpolation for time series data mining. In: 2014 IEEE International Conference on Data Mining Workshop, pp. 19–26. IEEE (2014)
17. Müller, S., Preische, O., Heymann, P., Elbing, U., Laske, C.: Increased diagnostic accuracy of digital vs. conventional clock drawing test for discrimination of patients in the early course of Alzheimer's disease from cognitively healthy individuals. Front. Aging Neurosci. **9**, 101 (2017)
18. Park, J.H.: Non-equivalence of sub-tasks of the Rey-Osterrieth complex figure test with convolutional neural networks to discriminate mild cognitive impairment. BMC Psychiatry **24**(1), 166 (2024). https://doi.org/10.1186/s12888-024-05622-5
19. Petilli, M., et al.: Graph-based analysis of Rey-Osterrieth figure reproduction reveals patterns of cognitive decline in Alzheimer's disease. Sci. Rep. **11**(1), 1–13 (2021)
20. Prange, A., Sonntag, D.: Modeling users' cognitive performance using digital pen features. Front. Artif. Intell. **5**, 787179 (2022)
21. Rey, A.: L'examen psychologique dans les cas d'encéphalopathie traumatique (les problèmes). Arch. Psychol. **28**(112), 215–285 (1941)
22. Sunderland, T., et al.: Clock drawing in Alzheimer's disease: a novel measure of dementia severity. J. Am. Geriatr. Soc. **37**(8), 725–729 (1989)
23. Thapaliya, B., et al.: Graph-based deep learning models in the prediction of early-stage Alzheimers. In: 2024 46th Annual International Conference of the IEEE Engineering in Medicine and Biology Society (EMBC), pp. 1–5. IEEE (2024)
24. Um, T.T., et al.: Data augmentation of wearable sensor data for Parkinson's disease monitoring using convolutional neural networks. In: Proceedings of the 19th ACM International Conference on Multimodal Interaction, ICMI 2017, pp. 216–220. Association for Computing Machinery, New York (2017). https://doi.org/10.1145/3136755.3136817
25. Weller, J., Budson, A.: Current understanding of Alzheimer's disease diagnosis and treatment. F1000Research **7** (2018)
26. Xu, Y., et al.: Temporal modeling of digital clock drawing for cognitive assessment. arXiv preprint arXiv:2410.22377 (2024)
27. Yamada, Y., et al.: Automated analysis of drawing process for detecting prodromal and clinical dementia. In: 2022 IEEE International Conference on Digital Health (ICDH), pp. 1–6. IEEE (2022)
28. Youn, Y.C., et al.: Use of the clock drawing test and the Rey-Osterrieth complex figure test-copy with convolutional neural networks to predict cognitive impairment. Alzheimer's Res. Ther. **13**, 1–7 (2021)
29. Yu, Q., Yang, Y., Song, Y.Z., Xiang, T., Hospedales, T.: Sketch-a-net that beats humans. arXiv preprint arXiv:1501.07873 (2015)

Hybrid Morphology-Based Tumor Detection from Breast MRI Segmentation Masks

Sergio Botana[1], Paula Puerta González[2], Pablo García Marcos[1,3],
Sara Fernández Arias[1], Rebeca Oliveira Suárez[2], Guillermo Lorenzo[7,8],
Héctor Gómez[4,5,6], Covadonga del Camino[9], Víctor M. González[2,3],
and Angel Rio-Alvarez[1,3(✉)]

[1] Computer Science Department, University of Oviedo, Oviedo, Spain
{uo288928,garciamarcos,uo269546,rioangel}@uniovi.es
[2] Electrical Engineering Department, University of Oviedo, Oviedo, Spain
{puertapaula,uo277463,vmsuarez}@uniovi.es
[3] Biomedical Engineering Center, University of Oviedo, Oviedo, Spain
[4] School of Mechanical Engineering, Purdue University, 585 Purdue Mall, West Lafayette, IN 47907, USA
hectorgomez@purdue.edu
[5] Purdue Center for Cancer Research, Purdue University, West Lafayette, IN, USA
[6] Weldon School of Biomedical Engineering, Purdue University, 206 S. Martin Jischke Drive, West Lafayette, IN 47907, USA
[7] Group of Numerical Methods in Engineering, Department of Mathematics, Campus de Elviña s/n, 15071 A Coruña, Galicia, Spain
guillermo.lorenzo@udc.es
[8] Oden Institute for Computational Engineering and Sciences, The University of Texas at Austin, 201 E. 24th Street, Austin, TX 78712-1229, USA
[9] Radiodiagnostic Service, Oviedo Central University Hospital (HUCA), Oviedo, Spain
caminocovadonga@uniovi.es

Abstract. Magnetic resonance imaging (MRI) is widely used to diagnose breast cancer. However, the high sensitivity of dynamic contrast-enhanced MRI (DCE-MRI) comes at the cost of increased complexity in image interpretation and segmentation, which requires advanced techniques to help radiologists delineate tumor boundaries accurately and efficiently. This work proposes an automated methodology that reduces

This research has been funded by the Spanish Ministry of Economics and Industry –grant PID2020-112726RB-I00–, the Spanish Research Agency –grant PID2023-146257OB-I00–, and Missions Science and Innovation project MIG-20211008 (INMERBOT). Also, by Principado de Asturias, grants SV-PA-21-AYUD/2021/50994 and AYUD/2023/36906/UO-077, and by the Council of Gijón through the University Institute of Industrial Technology of Asturias grants SV-21-GIJON-1–19, SV-22-GIJON-1–19, SV-22-GIJON-1–22, SV-23-GIJON-1-17, and SV-24-GIJON-1-18. G. Lorenzo acknowledges support from a fellowship from "la Caixa" Foundation (ID 100010434). The fellowship code is LCF/BQ/PI23/11970033. Finally, this research has also been funded by Fundación Universidad de Oviedo grants FUO-23-008 and FUO-22-450.

A. López Fernández et al. (Eds.): CIABiomed 2025, LNBI 16148, pp. 340–351, 2026.
https://doi.org/10.1007/978-3-032-10661-2_26

the reliance on manual annotations by integrating unsupervised cluster-ing with supervised classification to identify the tumor cluster within masks derived from clustering techniques using only morphological char-acteristics, thus facilitating rapid, accurate, and reproducible tumor detection in breast magnetic resonance imaging.

The approach consists of an initial segmentation using Fuzzy C-Means followed by classification using supervised machine learning models. A comprehensive feature selection analysis that identifies the most infor-mative descriptors for tumor classification, resulting in a classification accuracy of 99.44%, with consistent improvements across all evaluated metrics. The proposed approach mitigates annotation dependency while ensuring interpretability, representing a practical tool for clinical deploy-ment.

Keywords: Tumor classification · Morphology-based classification · Breast cancer

1 Introduction

Breast cancer is the most frequently diagnosed malignancy among women world-wide, accounting for approximately 25% cancer cases in women. According to recent epidemiological studies, it accounts for one in six cancer-related deaths, while by 2040, global mortality is expected to reach 1 million annually. Early and precise tumor detection significantly improves patient prognosis, especially in localized stages where survival rates exceed 99% at five years.

Medical imaging, particularly magnetic resonance imaging (MRI) plays a central role in breast cancer diagnosis because of its superior soft tissue contrast and sensitivity. Among MRI techniques, dynamic contrast-enhanced magnetic resonance imaging (DCE-MRI) is widely used to visualize tumor vascularization and morphology. However, accurate identification of tumor regions often relies on manual segmentation, a process prone to human bias, variability, and high resource demands.

This work proposes an automated methodology that reduces the reliance on manual annotations by integrating unsupervised clustering with supervised clas-sification. The goal is to identify the tumor cluster within masks derived from clustering techniques using only morphological characteristics, thereby facilitat-ing rapid, accurate, and reproducible tumor detection in breast MRI.

2 Background

Mammography has long been considered the gold standard for the early detection of breast cancer. Numerous population-based studies have shown that periodic mammography screening programs can reduce breast cancer-specific mortality by approximately 19%, with the magnitude of benefit increasing significantly with age. Specifically, women around the age of 60 can experience mortality reductions of up to 32%, while younger women see more modest benefits [7]. This age-related variation is attributed to differences in breast tissue composition and cancer biology between age groups.

Despite its proven benefits, mammography has notable limitations, particularly in women with dense breast tissue, a common characteristic in younger patients. High breast density reduces mammography sensitivity due to the masking effect, where dense tissue and malignancies appear radiopaque, leading to a higher rate of false negatives. To address this limitation, adjunctive imaging modalities, such as ultrasound, are often incorporated into diagnostic protocols. Ultrasound is particularly valuable for differentiating simple cystic formations from solid lesions, thus improving the diagnostic specificity in dense breasts [1].

For women at elevated risk of breast cancer, such as those with a genetic predisposition or a strong family history, and in cases where a more accurate characterization of breast lesions is required, magnetic resonance imaging (MRI) has emerged as the superior diagnostic modality. In particular, dynamic contrast-enhanced MRI (DCE-MRI) offers unparalleled sensitivity by capturing the vascular behavior of lesions through temporal sequences postcontrast administration. DCE-MRI facilitates the visualization of subtle morphological and kinetic features of tumors, which are often not appreciable on conventional imaging [15]. However, the high sensitivity of DCE-MRI comes at the cost of increased complexity in image interpretation and segmentation, which requires advanced techniques to help radiologists delineate tumor boundaries accurately and efficiently.

Manual segmentation masks are commonly used to fully exploit the high level of anatomical detail and resolution provided by advanced imaging modalities such as MRI and CT. These masks serve as precise references in diagnostic evaluations, treatment planning, and computational model training. However, manual segmentation suffers from significant drawbacks: it is highly time-consuming, introduces inter- and intra-observer variability, and imposes a considerable workload on medical professionals [3]. These challenges become especially pronounced in three-dimensional (3D) imaging contexts, where the volume of data and the complexity of anatomical structures further worsen the difficulties of manual annotation, making it unbearable to manually segment magnetic resonance images [12].

Automatic segmentation has become a key alternative to manual segmentation in medical imaging, addressing limitations such as high labor demands and subjectivity. Recent years have seen Deep Learning (DL) methods outperform traditional techniques, establishing themselves as the gold standard. However, their reliance on large, manually annotated datasets and computationally intensive training, coupled with susceptibility to annotation bias, limit their generalizability across populations.

Unsupervised learning offers an appealing alternative by extracting patterns directly from data without labels, reducing annotation costs, and potentially revealing underrepresented structures. Clustering methods, particularly Fuzzy C-Means (FCM), have proven effective in medical imaging by assigning probabilistic memberships to pixels [2], thus handling the inherent ambiguity of tissue boundaries better than hard clustering techniques like K-Means [8].

Despite its strengths, FCM lacks semantic labeling, making it difficult to automatically identify tumor regions without expert intervention. To overcome this, hybrid methods have emerged that combine unsupervised segmentation with supervised classification. These approaches pair annotation-free segmentation with powerful classification models, enabling accurate tumor identification while minimizing manual labeling effort, a strategy that has shown promise in breast cancer imaging.

Machine learning (ML) classifiers have been extensively validated in breast cancer diagnosis scenarios, showing high accuracy in lesion classification. Random Forest (RF), in particular, has demonstrated outstanding performance, with reported accuracies extending up to 98. 60% in certain studies [4]. Similarly, Logistic Regression (LR) has yielded notable results, achieving 94.41% accuracy on the UCI breast cancer dataset [4]. The K-Nearest Neighbors (KNN) algorithm has also shown robust classification capability, obtaining an accuracy of 92.50% in bone tumor classification using wavelet-based features [6], and has proven effective in experiments with the Wisconsin Breast Cancer (WBC) dataset [14]. Decision Tree (DT) models have also maintained strong performance, correctly classifying 94.56% of cases in WBC datasets [13]. Extra Trees Classifier (ETC) has emerged as a particularly competitive approach, surpassing other classifiers in breast cancer image classification tasks, with accuracies up to 94% [10], and achieving an Area Under the Curve (AUC) of 0.974 in retrospective studies in patients with primary lymphedema [11].

Although supervised classifiers have shown remarkable individual performance, the literature still lacks comprehensive frameworks that integrate unsupervised segmentation with supervised classification for fully automated tumor identification. Most prior studies have focused exclusively on segmentation or classification in isolation, neglecting the synergistic potential of combining both paradigms. Consequently, the development of a robust hybrid methodology that applies FCM for initial segmentation followed by supervised classification to identify the tumor cluster represents a valuable and necessary contribution to medical imaging research.

3   Materials and Methods

3.1   Datasets Description

ACRIN-6698 Dataset. The ACRIN-6698 dataset originates from the ACRIN study, a substudy of the I-SPY 2 clinical trial, focused on evaluating the response of breast cancer to neoadjuvant chemotherapy using magnetic resonance imaging (MRI) [9]. It comprises multiparametric imaging data from 385 breast cancer patients, including dynamic contrast-enhanced MRI (DCE-MRI), diffusion-weighted imaging (DWI) and associated clinical and demographic information.

Imaging was performed at four key treatment time points: before treatment (T0), early treatment (T1), mid-treatment (T2), and post-treatment (T3), capturing temporal changes in tumor response. For this study, only pretreatment

DCE-MRI scans were used to ensure compatibility with other datasets, as manual segmentations are available only for T0.

The dataset includes both manually annotated DWI tumor masks and automatically generated DCE masks. The DCE masks were derived via histogram-based thresholding within radiologist-defined volumes of interest (VOI), which typically cover the tumor and approximately half of the breast. However, these analysis masks have limitations, including inverted pixel encoding and inconsistencies in tumor delineation. Mask values are encoded through a bit-wise system, where each bit reflects exclusion criteria such as VOI inclusion, background removal, connectivity, and enhancement thresholds.

Despite these challenges, ACRIN-6698 provides a valuable, multi-institutional, quality-controlled dataset for studying tumor segmentation and response assessment in breast cancer imaging.

MAMA-MIA Dataset. The MAMA-MIA dataset [5] is a large-scale multicenter breast cancer MRI dataset comprising 1,506 cases with pretreatment DCE-MRI images. A unique feature of MAMA-MIA is the inclusion of high-quality manual tumor segmentations performed by 16 domain experts from eight countries, with an average of nine years of clinical experience.

This dataset aggregates cases from four public TCIA collections that originally lacked manual tumor masks: ISPY1, ISPY2, NACT-Pilot, and Duke-Breast-Cancer-MRI. To address this, MAMA-MIA adopted a semiautomatic approach in which initial segmentations were generated using a pretrained nnUNet model and subsequently refined by experts to ensure precise tumor boundaries.

For methodological consistency, only pretreatment scans (T0) from ISPY2/ACRIN were included in this study, which corresponded to the annotation phase between datasets. MAMA-MIA, with its expert-validated segmentations, serves as a robust benchmark for developing and validating advanced segmentation algorithms in breast cancer MRI research.

3.2 Segmentation Mask Generation

Preprocessing Pipeline. Before clustering, the MRI volumes were resampled with an isotropic voxel spacing to ensure uniform spatial resolution in all cases. Subsequently, a targeted volume of interest (VOI) was extracted for each image, focusing on the breast tissue surrounding the tumor. Non-relevant anatomical structures, including bones, pectoral muscles, skin, and nipple regions, were systematically excluded. This step aimed to reduce the impact of unrelated high-contrast regions and to enhance the effectiveness of subsequent clustering.

Unsupervised Segmentation via Fuzzy C-Means. Fuzzy C-Means (FCM) clustering was used to segment each VOI into a predefined number of clusters, set to $c=3$. This configuration was chosen to separate the VOI into three distinct regions corresponding to the background, healthy tissue, and tumor. Clustering was performed directly on the intensity values of DCE-MRI, enabling a purely unsupervised segmentation without requiring manual annotations during this phase.

Cluster Label Assignment. To train a classifier capable of automatically identifying the tumor cluster, it was necessary to assign a semantic label (tumor, healthy tissue, background) to each cluster generated by FCM for every patient. This labeling process was conducted in two stages. First, the background cluster was identified and excluded by detecting which cluster contained the spatial point at coordinate (0,0,0), known to be outside the breast tissue. Subsequently, the two remaining clusters were compared with the manually annotated tumor masks provided by the MAMA-MIA dataset. The cluster exhibiting the highest pixel overlap with manual segmentation was labeled tumor, while the remaining cluster was classified as healthy tissue.

3.3   Building a Classifier

Extracting Morphological Features. In order to classify the segmentation masks, several morphological features will be calculated in each mask. In this work, six quantitative features are extracted from each segmented cluster, capturing essential properties of shape, size, and boundary complexity. These descriptors are commonly used in medical imaging analysis and provide a compact but informative representation suitable for classification models.

- **Volume**: Measures the total space occupied by the segmented region. It is calculated by adding the segmented areas in all slices:

$$V = \sum_{k=1}^{N_{\text{slices}}} A_k \tag{1}$$

  where A_k is the number of positive voxels in the slice k.
- **Perimeter**: corresponding to the length of the boundary of the segmented region in 2D or the surface area in 3D:

$$P = \sum_{i=1}^{M} \|p_i - p_{i+1}\| \tag{2}$$

- **Circularity**: Indicates how closely the region resembles a perfect circle in 2D, calculated as:

$$\text{Circularity} = \frac{4\pi A}{P^2} \tag{3}$$

  Values closer to 1 indicate roundness; lower values imply irregular contours.
- **Compactness**: Reflects the compactness of the region by comparing the perimeter and the area, using the normalized expression:

$$C = \frac{P^2}{4\pi A} \tag{4}$$

  The minimum value of $C = 1$ corresponds to a perfect circle.

- **Elongation**: Quantifies the anisotropy or elongation of the segmented region. It is derived from the lengths of the major and minor axes of the best-fitting ellipse:

$$\text{Elongation} = \sqrt{\frac{\lambda_{\text{minor}}}{\lambda_{\text{major}}}} \tag{5}$$

where λ_{major} and λ_{minor} are the lengths of the major and minor axes, respectively. Values close to 1 indicate a rounded shape, whereas values near 0 reflect highly elongated structures.

- **Fractal Dimension (FD)**: Measures the boundary complexity using box counting techniques:

$$\text{FD} = \frac{\log K(r)}{\log(1/r)} \tag{6}$$

Higher FD values correspond to more irregular and complex boundaries, often associated with malignant lesions.

These features were selected for their established relevance in oncology and their low computational cost, which makes them suitable for integration into a morphology-driven classification framework.

An example of the calculated inputs and the expected output can be seen in Table 1.

Table 1. Computed morphological features for several masks

Patient	Cluster	Area	Most representative slice	Perimeter	Circularity	Compaction	Elongation	Tumor
ACRIN-6698-102212	Background	42423	37	589.04	1.54	8.18	0.75	0
ACRIN-6698-102212	Cluster 1	5221307	37	1919.14	17.81	0.71	0.08	0
ACRIN-6698-102212	Cluster 3	5228947	37	1458.73	30.88	0.41	0.12	1
ACRIN-6698-103939	Background	18547	32	386.28	1.56	8.05	0.55	0
ACRIN-6698-103939	Cluster 1	4186620	32	1832.69	15.66	0.80	0.03	0
ACRIN-6698-103939	Cluster 3	4186906	32	1521.31	22.73	0.55	0.04	1
ACRIN-6698-104268	Background	36332	49	852.05	0.63	19.98	0.33	0
ACRIN-6698-104268	Cluster 1	5223576	49	2163.80	14.02	0.90	0.09	0
ACRIN-6698-104268	Cluster 3	5229935	49	1466.18	30.57	0.41	0.13	1
ACRIN-6698-107700	Background	3512	40	230.28	0.83	15.10	0.58	0
ACRIN-6698-107700	Cluster 2	5241503	40	1347.49	36.28	0.35	0.04	0
ACRIN-6698-107700	Cluster 3	5241394	40	1195.41	46.09	0.27	0.05	1
ACRIN-6698-108969	Background	79747	45	966.40	1.07	11.71	0.89	0
ACRIN-6698-108969	Cluster 2	5192644	45	2920.84	7.65	1.64	0.03	0
ACRIN-6698-108969	Cluster 3	5226653	45	1475.90	30.15	0.42	0.05	1

Supervised Model Training. After generating the final dataset, multiple supervised machine learning models were trained to classify which of the three FCM-derived segmentation masks corresponds to the tumor. The input features consisted of the previously defined morphological descriptors.

The training details are the following:

- **Normalization:** Input features were standardized to ensure comparable feature magnitudes, particularly beneficial for distance-based models such as SVM and KNN.
- **Data split:** An 80/20 train-test split was performed using class stratification to preserve label balance between subsets.
- **Model definition:** The following classifiers were trained.
  - Support Vector Machines (SVM) with RBF kernel
  - Support Vector Machines (SVM) with linear kernel
  - Random Forest (RF)
  - Extra Trees Classifier (ETC)
  - Logistic Regression (LR)
  - Decision Trees (DT)
  - K-Nearest Neighbors (KNN)
  - Bagging with RF
  - AdaBoost with DT
- **Threshold tuning:** Decision thresholds of 0.3, 0.4, and 0.5 were evaluated for probabilistic classifiers, except for DT and KNN, where threshold adjustment is not applicable.
- **Feature selection:** All mentioned models were trained with all possible combinations of morphological characteristics as input to the classifier.

Validation Metrics. To comprehensively assess classification performance, several standard clinical metrics were used, providing insight beyond overall accuracy. The metrics used were:

- F1 score
- Accuracy
- Sensitivity (or recall)
- Specificity
- Precision

A patient-wise voting strategy was applied to calculate a custom confusion matrix, ensuring that the final classification decision was made per patient rather than per slice. For each patient, the cluster with the highest tumor probability was selected and the classification metrics were computed accordingly.

- Correct tumor identification increased TP and TN counts.
- Misclassifications in FN and FP were accounted for, with adjustments ensuring robustness to class imbalance.

The final selection of the model was based on the highest patient-specific accuracy, prioritizing early tumor detection while minimizing misclassifications.

4   Results and Discussion

4.1   Feature Importance

A comprehensive feature selection analysis was carried out to identify the most informative morphological descriptors for tumor classification. All possible combinations of the six extracted features were tested with each classification model to evaluate which subset provided the highest performance.

From this exhaustive analysis, the combination of **Area**, **Compaction**, and **Elongation** showed the best results across all classifiers. These three features were selected for the final model training, as they offered the most stable and discriminative power to separate tumor clusters from non-tumor ones.

The relevance of these features is not only statistical but also clinically interpretable. **Area** represents the general size of the lesion, and tumors tend to occupy larger regions than healthy tissue. **Compaction** reflects the irregularity of the tumor boundary, which is often higher in malignant lesions. **Elongation**, finally, is associated with anisotropy, and many tumors present elongated shapes due to their invasive nature.

To confirm their importance, a feature importance ranking was also computed using the Extra Trees Classifier. This ranking showed that Compaction and Elongation were the two most relevant features, followed by Area, supporting the final feature selection. Figure X shows the importance scores obtained from the Extra Trees model, with clear differences in contribution between these and the remaining descriptors.

Moreover, when using individual features alone, none of them achieved satisfactory performance, with accuracies always below 85%. However, when **Area**, **Compaction**, and **Elongation** were combined, the classification accuracy increased significantly, confirming that their combination contains complementary information.

This analysis demonstrates that a reduced set of morphological features, if properly selected, is enough to train accurate and interpretable classifiers for tumor detection, without the need of deep or computationally expensive features.

4.2   Classifier Performance

The performance of various supervised classification algorithms was assessed based on five key metrics: F1 Score, Accuracy, Recall, Specificity, and Precision. Table 2 summarizes the results obtained from the best-trained models.

The **Extra Trees Classifier** achieved the best overall performance, obtaining a classification accuracy of 99.44%, with consistent improvements across all evaluated metrics. These results confirm that ensemble methods, especially tree-based ones, offer greater robustness and generalisation in comparison with simpler classifiers. One of the main advantages of Extra Trees is the high level of randomisation it introduces during training, which helps to avoid overfitting while keeping the model computationally efficient.

In addition to the global performance, the results also show that combining unsupervised clustering with supervised classification can be an effective

Table 2. Comparison of classification models

Model	F1 Score	Accuracy	Recall	Specificity	Precision
Extra Trees	0.9915	0.9944	0.9915	0.9958	0.9915
Random Forest	0.9831	0.9887	0.9831	0.9915	0.9831
Bagging with Random Forest	0.9661	0.9774	0.9661	0.9831	0.9661
AdaBoost with Decision Tree	0.8983	0.9322	0.8983	0.9492	0.8983
Decision Tree	0.8771	0.9181	0.8771	0.9386	0.8771
Support Vector Machine (RBF kernel)	0.8390	0.8927	0.8390	0.9195	0.8390
K-Nearest Neighbors	0.7797	0.8531	0.7797	0.8898	0.7797
Support Vector Machine (Linear kernel)	0.4746	0.6497	0.4746	0.7373	0.4746
Logistic Regression	0.4195	0.6130	0.4195	0.7097	0.4195

strategy to identify tumor regions in MRI data without requiring manual annotations. This hybrid approach reduces the annotation burden and increases the
reproducibility of tumor detection across different datasets and patients.

Another important aspect is the interpretability of the selected features. By
using only morphological characteristics, such as area, compaction, and elongation, the method remains transparent and easy to analyse from a clinical point
of view. These descriptors are simple to compute and do not depend on complex
intensity-based or texture-based features, which sometimes require deep learning models or extensive preprocessing. Moreover, this makes the system more
suitable for real clinical integration, where interpretability and fast processing
are essential.

Although the Extra Trees model showed excellent performance, other ensemble classifiers such as Random Forest and Bagging also provided very high accuracy, suggesting that the classification problem is well structured when using
the proposed morphological descriptors. Simpler models like Logistic Regression and linear SVM were less effective, indicating that the separation between
classes is not purely linear. This supports the idea that non-linear models are
more appropriate for this type of task.

Finally, it is important to mention that the quality of the initial segmentation also plays a crucial role in the final classification result. Even though
Fuzzy C-Means clustering is unsupervised, its ability to produce useful masks
depends on the preprocessing and the correct definition of the volume of interest.
Future improvements in this step could lead to even more accurate and robust
classification results.

5   Conclusions

This work presented a hybrid pipeline for the automatic identification of tumor
clusters in breast MRI segmentation masks. The proposed methodology combines unsupervised segmentation using Fuzzy C-Means (FCM) with supervised

classification based on morphological descriptors. The approach was validated on two high-quality public datasets, showing excellent results with an overall classification accuracy of 99.44% using the Extra Trees Classifier.

The results demonstrate that combining clustering and machine learning provides a reliable and efficient solution for tumor detection, even when manual annotations are not available. By focusing on simple morphological features like area, compaction, and elongation, the method achieves high performance while keeping the system interpretable and clinically meaningful.

In addition, the use of classical models instead of deep learning reduces computational cost and simplifies integration into real-world clinical environments. Since manual segmentation is time-consuming and variable between experts, this pipeline can help reduce workload in radiology departments and support faster diagnosis.

The experimental results confirm that ensemble classifiers are particularly effective for this problem, but also highlight the potential of unsupervised methods when combined with well-selected features. Overall, the study shows that it is possible to obtain robust and accurate tumor identification using only morphology-based descriptors, which opens the door to future tools that are easier to deploy and validate in hospitals.

6 Future Work

The first step for future work will be to perform an exhaustive validation of the complete pipeline, starting directly from the output of the Fuzzy C-Means (FCM) segmentation. While the current approach focuses on classifying which cluster corresponds to the tumor, it is now possible to go further and evaluate the raw segmentation masks generated by FCM against the manual annotations provided in the MAMA-MIA dataset.

This step is essential to better understand the strengths and limitations of the clustering process itself, independently of the classifier. By comparing the FCM tumor masks with the ground truth, we can identify specific cases where the segmentation is insufficient or inaccurate. These insights will help to design targeted improvements to the FCM algorithm, with the goal of refining the tumor masks before the classification stage.

Ultimately, the objective is to move towards a fully automatic tumor segmentation method, where the clustering step alone can produce high-quality, clinically useful masks without the need for manual labelling or external classification. This would provide a complete end-to-end system, ready to be tested in more realistic hospital scenarios and eventually integrated into clinical workflows.

References

1. Apantaku, L.M.: Breast cancer diagnosis and screening. Am. Fam. Physician **62**(3), 596–602 (2000)
2. Chen, S., Zhang, D.: Robust image segmentation using FCM with spatial constraints based on new kernel-induced distance measure. IEEE Trans. Syst. Man Cybern. Part B (Cybern.) **34**(4), 1907–1916 (2004)
3. Covert, E.C., et al. Intra- and inter-operator variability in MRI-based manual segmentation of HCC lesions and its impact on dosimetry. EJNMMI Phys. **9**(1), 90 (2022)
4. Dadheech, P., Kalmani, V., Dogiwal, S.R., Sharma, V.K.: Breast cancer prediction using supervised machine learning techniques. J. Inf. Optim. Sci. **44**(3), 383–392 (2021)
5. Garrucho, L., et al.: A large-scale multicenter breast cancer DCE-MRI benchmark dataset with expert segmentations. Sci. Data **12**(1), 453 (2025). https://doi.org/10.1038/s41597-025-04707-4
6. Hossain, E., Hossain, M.F., Rahaman, M.A.: An approach for the detection and classification of tumor cells from bone MRI using wavelet transform and KNN classifier. In: 2018 International Conference on Innovation in Engineering and Technology (ICIET), pp. 1–6. IEEE (2018)
7. McDonald, E.S., Clark, A.S., Tchou, J., Zhang, P., Freedman, G.M.: Clinical diagnosis and management of breast cancer. J. Nucl. Med. **57**(Supplement 1), 9S-16S (2016)
8. Naz, S., Majeed, H, Irshad, H.: Image segmentation using fuzzy clustering: A survey. In: 2010 6th International Conference on Emerging Technologies (ICET), pp. 181–186. IEEE (2010)
9. Newitt, D.C., et al.: Acrin 6698/i-spy2 breast DWI. The Cancer Imaging Archive (2021). https://doi.org/10.7937/tcia.kk02-6d95
10. Ramkumar, G., et al.: A robust breast cancer classification model using extra-trees classifier for histopathological image. In: 2023 International Conference on Advances in Computing, Communication and Applied Informatics (ACCAI), pp. 1–7. IEEE (2023)
11. Ren, J., et al.: Noncontrast mri-based machine learning and radiomics signature can predict the severity of primary lower limb lymphedema. J. Vasc. Surg. Venous Lymphat. Disord. **13**(2), 102161 (2024)
12. Renard, F., Guedria, S., Palma, N., Vuillerme, N.: Variability and reproducibility in deep learning for medical image segmentation. Sci. Rep. **10**(1), 13724 (2020)
13. Sumbaly, R., Vishnusri, N., Jeyalatha, S.: Diagnosis of breast cancer using decision tree data mining technique. Int. J. Comput. Appl. **98**(10) (2014)
14. Vemula, C., Chavva, P., Battu, U., Vemulapalli, S., Prasad, K., Kannaiah, S.K.: Revolutionizing breast cancer diagnosis: Insights from KNN and SVM models in machine learning. In: 2024 5th International Conference for Emerging Technology (INCET), pp. 1–9. IEEE (2024)
15. Wang, L.: Early diagnosis of breast cancer. Sensors **17**(7), 1572 (2017)

Multimodal Posterior Sampling-Based Uncertainty in PD-L1 Segmentation from H&E Images

Roman Kinakh[1]([✉])(iD), Gonzalo R. Ríos-Muñoz[1,2](iD),
and Arrate Muñoz-Barrutia[1](iD)

[1] Universidad Carlos III de Madrid, Leganés, Madrid, Spain
{rkinakh,grios,mamunozb}@ing.uc3m.es
[2] Instituto de Investigación Sanitaria Gregorio Marañón, Madrid, Spain

Abstract. Accurate assessment of PD-L1 expression is critical for guiding immunotherapy, yet current immunohistochemistry (IHC) based methods are resource-intensive. We present nnUNet-B: a Bayesian segmentation framework that infers PD-L1 expression directly from H&E-stained histology images using Multimodal Posterior Sampling (MPS). Built upon nnUNet-v2, our method samples various model checkpoints during cyclic training to approximate the posterior, enabling accurate segmentation and epistemic uncertainty estimation via entropy and standard deviation. Evaluated on a dataset of lung squamous cell carcinoma, our approach achieves competitive performance against established baselines with a mean Dice Score and a mean IoU of 0.805 and 0.709, respectively, while providing pixel-wise uncertainty maps. Uncertainty estimates show a strong correlation with segmentation error, while calibration can be further optimized. These results suggest that uncertainty-aware H&E-based PD-L1 prediction is a promising step toward scalable and interpretable biomarker assessment in clinical workflows.

Keywords: Uncertainty · Histology · Segmentation · PD-L1 · H&E · Posterior Sampling

1 Introduction

Programmed death-ligand 1 (PD-L1) is a transmembrane protein expressed in tumor and immune cells that plays a key role in suppressing the immune response [17]. Its expression is a critical biomarker for identifying patients who are likely to benefit from immune checkpoint inhibitors, a class of cancer immunotherapies [12,13]. The accurate assessment of PD-L1 expression in tumor tissue is essential for guiding immunotherapy decisions across various cancers [1]. Traditionally, PD-L1 is evaluated using immunohistochemistry (IHC), which directly visualizes protein expression. While clinically effective, IHC is resource-intensive, time-consuming, and subject to inter-observer variability. In contrast, Hematoxylin and Eosin (H&E) staining is a standard, inexpensive, and widely

A. López Fernández et al. (Eds.): CIABiomed 2025, LNBI 16148, pp. 352–361, 2026.
https://doi.org/10.1007/978-3-032-10661-2_27

available diagnostic modality. This work explores the feasibility of segmenting PD-L1-expressing tumor regions directly from H&E-stained images, potentially offering faster, more accessible alternatives for patient stratification [18].

Histology, as a medical imaging domain, presents a unique set of challenges that render both image analysis and clinical decision-making difficult. As illustrated in Fig. 1, the microscopic environment of tissue sections is inherently complex and heterogeneous [5]. Unlike the often distinct and macroscopic structures seen in tomographic scans, cells and tissue types in H&E and IHC images are densely packed, exhibit highly diverse morphologies at the microscopic level, and can be organized in intricate, overlapping, and often ambiguous patterns [10]. Furthermore, subtle variations in tissue processing, staining protocols, and imaging conditions introduce substantial inter-slide variability that can significantly impact model robustness [9,16]. These factors contribute to the "noisy" and ambiguous nature of histology data, demanding advanced computational methods that can discern subtle yet critical biological signals amidst a sea of microscopic complexity, such as epistemic uncertainty estimation.

While techniques like Monte Carlo Dropout (MCDO) [6] have been widely used for uncertainty estimation in medical imaging, they often lead to a trade-off between prediction confidence and segmentation accuracy. This trade-off is particularly problematic in histology, where fine-grained errors can impact downstream biomarker quantification. To address this, we introduce nnUNet-B: a Multimodal Posterior Sampling (MPS) framework based on nnUNet-v2 [7,8] that provides richer, more stable uncertainty estimates without degrading segmentation performance. Our approach better captures model variability while maintaining a computational cost comparable to MCDO, offering clinicians a clearer picture of where and why the model may be uncertain—an essential feature for deploying AI in sensitive diagnostic workflows.

2 Materials and Methods

2.1 Dataset

To train and evaluate our model, we used the data set introduced by Wang et al. [18], which comprises 1,088 paired H&E- and IHC-stained histology images of lung squamous cell carcinoma. Each H&E image is annotated with pixel-wise PD-L1 positive and negative tumor regions, using the corresponding IHC slide as reference. The annotation process is illustrated in Fig. 1.

For our experiments, we randomly allocated 20% of the images (218) as a held-out test set. Of the remaining 80% (870 images), we used 20% (174 images) for validation and the remaining 696 images for training. All images have a size of 959×923 pixels with a pixel size of approximately 1.5 μm.

2.2 Model Architecture

The nnUNet-B method is based on the nnUNet-v2 architecture [7], which serves as the segmentation backbone. To enable uncertainty-aware predictions, we

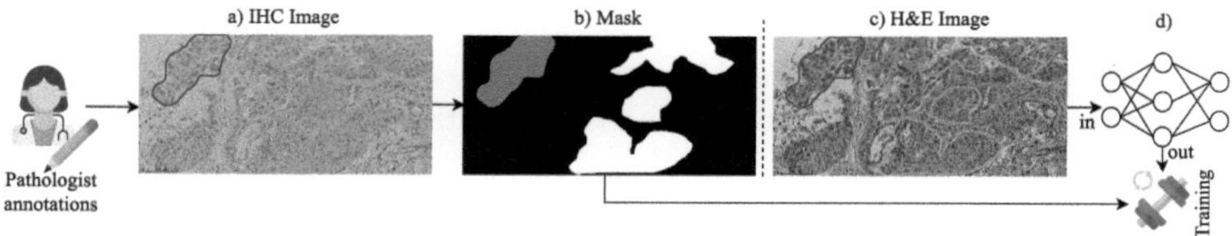

Fig. 1. Overview of dataset annotation and segmentation workflow: (a) Pathologists annotate PD-L1-positive (green) and PD-L1-negative (red) tumor regions on IHC slides. (b) Annotations are converted into 3-class segmentation masks. (c) Masks are aligned with corresponding H&E images. (d) H&E images are used as model input, with predictions supervised by the aligned PD-L1 masks. [4,18] (Color figure online)

extended it using the MPS strategy introduced by Zhao et al. [20]. In this approach, multiple model instances sampled from different local minima of the optimization trajectory are treated as approximate posterior samples (Fig. 2). These samples are obtained from saved checkpoints during different stages of training (detailed in Sect. 2.3).

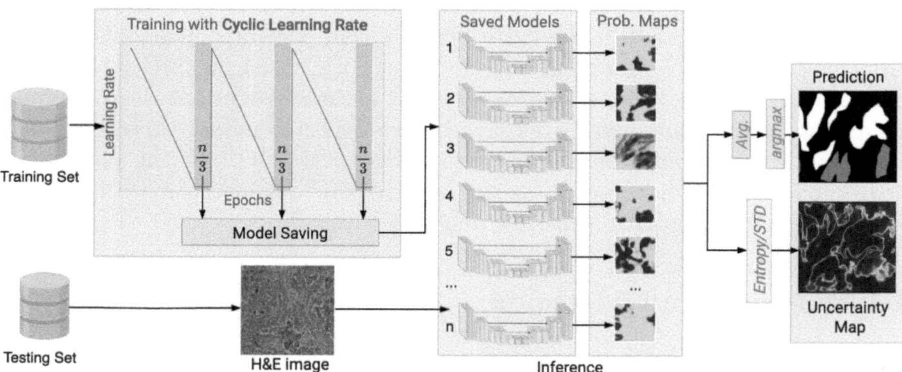

Fig. 2. Bayesian nnU-Net framework with Multimodal Posterior Sampling (MPS). During training, checkpoints are sampled from the last $n/3$ epochs of each learning cycle. At inference, an H&E image is passed through n sampled models to generate probability maps, which are averaged and arg max-ed for prediction. Pixel-wise uncertainty is computed using entropy or standard deviation. [20]

At inference time, each sampled model $\mathcal{M}_i$ produces a softmax probability map $P_i(\mathbf{x}) \in [0,1]^C$ for an input image $\mathbf{x}$, where C is the number of segmentation classes. The ensemble of N such models yields a set $\{P_i(\mathbf{x})\}_{i=1}^{N}$. The final prediction is calculated by averaging the probabilities across all models: $\bar{P}(\mathbf{x}) = \frac{1}{N} \sum_{i=1}^{N} P_i(\mathbf{x})$.

The predicted segmentation mask $\hat{y}$ is obtained by taking the voxel-wise arguments of the maxima over the average probabilities: $\hat{y} = \arg\max_c \bar{P}_c(\mathbf{x})$. To quantify predictive uncertainty, we computed two pixel-wise measures over the ensemble: the standard deviation (STD, σ) and the entropy (H) of the averaged distribution:

$$\sigma(\mathbf{x}) = \sqrt{\tfrac{1}{N}\sum_{i=1}^{N}(P_i(\mathbf{x}) - \bar{P}(\mathbf{x}))^2}, \; H(\mathbf{x}) = -\sum_{c=1}^{C} \bar{P}_c(\mathbf{x}) \log \bar{P}_c(\mathbf{x}) \qquad (1)$$

This design enables the extraction of both the most probable segmentation and its associated uncertainty, without modifying the underlying network architecture or requiring stochastic components at inference time (Fig. 3).

2.3   Model Training

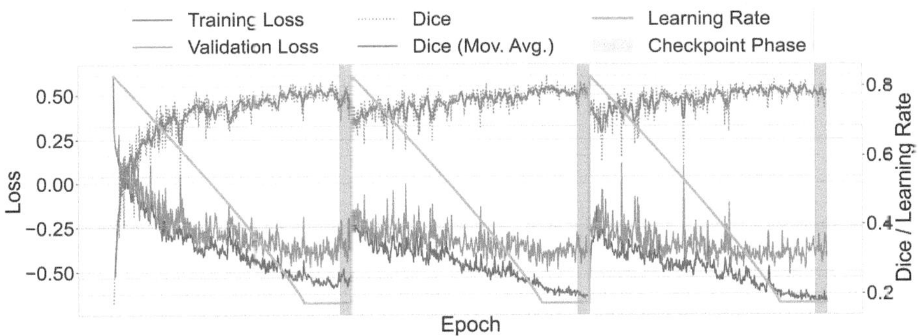

Fig. 3. Training process of the Bayesian nnU-Net framework with Multimodal Posterior Sampling (MPS). The model undergoes three full training cycles using a cyclic learning rate schedule. During the final 20 epochs of each cycle, model checkpoints are sampled and stored to later be used as an ensemble for uncertainty estimation.

The model was trained similarly to a standard U-Net using the combination of **Dice** and **Cross-Entropy** losses. To promote convergence while preserving exploratory behavior across training, we employed a **cyclical learning rate (CLR)** schedule based on polynomial decay [15]. At the beginning of each cycle of length T_c, the learning rate is initialized to a higher value α_r, and then decays polynomially to a minimum value α_0 over a fraction γ of the cycle length, with a polynomial decay power of ϵ. Beyond this point, the learning rate remains constant at α_0 for the remainder of the cycle, which ensures stability during later iterations while enabling aggressive updates early in the cycle. Therefore, we define the learning rate $\alpha(t)$ for epoch t as:

$$\alpha(t) = \begin{cases} \alpha_0 + (\alpha_r - \alpha_0)\left(1 - \dfrac{t_c}{\gamma T_c}\right)^{\epsilon}, & \text{if } 0 < t_c \leq \gamma T_c \\[2ex] \alpha_0, & \text{if } t_c > \gamma T_c \end{cases} \qquad (2)$$

We used a total of 3 cycles over 1200 epochs, with $T_c = 400$ epochs per cycle. The model was trained on a workstation equipped with an NVIDIA GeForce RTX 3090 GPU (24 GB VRAM) with a batch size of 15, $a_r = 0.1$, $a_0 = 0.01$, $\gamma = 0.8$, and $\epsilon = 0.9$. The training required approximately 13 h. The learning rate schedule and the total number of epochs were chosen based on the standard training dynamics of nnU-Net and the settings reported in [20], while the number of checkpoints sampled was selected to approximate the number of forward passes typically required by MCDO.

2.4   Evaluation

To comprehensively evaluate the performance of the model, we assessed both the segmentation accuracy and uncertainty calibration.

Segmentation Metrics. We report four standard metrics commonly used in medical image segmentation: Mean Dice Similarity Coefficient (mDice), Mean Intersection over Union (mIoU), Mean 95th percentile Hausdorff Distance (mHD95), and Mean Pixel Accuracy (mPA).

These metrics are used to compare our method with a variety of baseline models, including the standard nnUNet-v2 [7], UNet [14], Attention UNet [11], TransUNet [2], DenseASPP [19], and FCN with ResNet101 backbone [3]; nnUNet-v2 was trained using its default protocol with 5-fold cross-validation and ensembling. The metrics of other models were sourced from Wang et al. [18].

Uncertainty Calibration. To assess the reliability of the model's uncertainty estimates, we computed the Uncertainty Calibration Error (UCE) and visualized reliability diagrams. UCE evaluates how well predicted uncertainty aligns with the observed prediction errors. It was computed by binning the uncertainty values and comparing the average uncertainty to the empirical error rate within each bin.

Formally, for each bin b, we calculate the average predicted uncertainty $\bar{u}_b$ and empirical error rate $\bar{e}_b$, and define the UCE as:

$$\text{UCE} = \sum_{b=1}^{B} \frac{|S_b|}{\sum_{j=1}^{B} |S_j|} \cdot |\bar{u}_b - \bar{e}_b|, \tag{3}$$

where S_b is the set of pixels in bin b, and B is the total number of bins. We apply this evaluation using both entropy and standard deviation as uncertainty scoring functions.

3   Experimental Results

The segmentation performance of seven models was evaluated in the test set described in Sect. 2.1 using four metrics: Mean Dice Similarity Coefficient (mDice), Mean Intersection over Union (mIoU), Mean 95th percentile Hausdorff Distance (mHD95), and Mean Pixel Accuracy (mPA). As shown in Table 1, the standard nnUNet achieved the highest mDice (0.816) and mIoU (0.722), along with strong performance in mHD95 (94) and mPA (0.868). TransUNet performed best in mHD95 (89) and mPA (0.880), while our proposed Bayesian variant, nnUNet-B, achieved competitive results with an mDice of 0.805, mIoU of 0.709, mHD95 of 97, and mPA of 0.860. Attention UNet, DenseASPP, and FCN performed reasonably but fell short of the nnUNet variants in overlap and boundary accuracy. In general, nnUNet-B demonstrated competitive segmentation performance while offering the added benefit of uncertainty estimation.

Table 1. Comparison of segmentation performance across various models. Metrics include Mean Dice Similarity Coefficient (**mDice**), Mean Intersection over Union (**mIoU**), Mean 95th percentile Hausdorff Distance (**mHD95**), and Mean Pixel Accuracy (**mPA**). The best results are highlighted in **bold**.

Model	mDice	mIoU	mHD95	mPA
Evaluated in this work:				
nnUNet-B	0.805	0.709	97	0.860
nnUNet-v2	**0.816**	**0.722**	94	0.868
Results reported by Wang et al. [18]:				
TransUNet	0.800	0.720	**89**	**0.880**
UNet	0.773	0.684	101	0.863
Attention UNet	0.787	0.696	101	0.787
DenseASPP	0.773	0.686	100	0.866
FCN (ResNet101)	0.783	0.692	99	0.868

Figure 4 shows two examples of segmentation predictions from the test set, along with errors and uncertainty estimates. The predicted masks match the ground truth in both PD-L1-negative and positive regions, including complex cellular architecture. The error maps show few false positives or negatives, indicating good agreement with expert annotations. The model also successfully segments regions with mixed or ambiguous morphology, where different cell types are densely interwoven. The uncertainty maps show elevated uncertainty along region boundaries and in areas with heterogeneous or atypical cell morphology.

Both STD and entropy demonstrate a clear positive correlation between predicted uncertainty and segmentation error, as shown in Fig. 5. However, both measures display miscalibration (most notably in the mid-to-high uncertainty bins), where the predicted uncertainty exceeds the actual error. The mean UCE is slightly lower for STD (0.1087) than for entropy (0.1137).

4   Discussion

This study demonstrates that PD-L1 expression can be inferred directly from H&E-stained images using an uncertainty-aware segmentation framework based

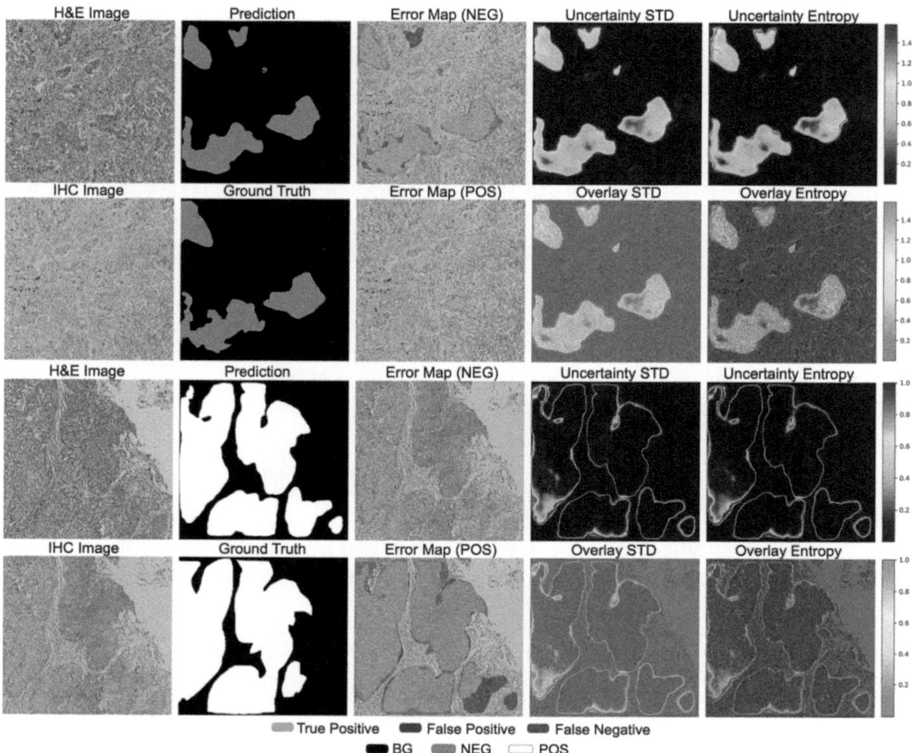

Fig. 4. Visual summary of nnUNet-B predictions, error maps, and uncertainty estimates for two test images (top two rows: image 1; bottom two rows: image 2). For each image: Column 1 shows the H&E and corresponding IHC reference; Column 2 displays the model prediction and ground truth; Column 3 presents class-specific error maps for PD-L1-negative (NEG) and -positive (POS) regions; Columns 4 and 5 show standard deviation and entropy-based uncertainty maps, each overlaid on the H&E image.

on nnUNet-v2 [7] and MPS [20]. By sampling checkpoints during cyclic training, we approximate the model posterior without architectural changes or stochastic inference, offering a practical and interpretable alternative for histopathology tasks.

The proposed model (nnUNet-B) achieved strong performance across all segmentation metrics, with an mDice of 0.805, mIoU of 0.709, mHD95 of 97, and mPA of 0.860. While the regular nnUNet-v2 [7] slightly outperforms it in mDice (0.816) and mIoU (0.722), and achieved lower mHD95 (94); nnUNet-B provides comparable accuracy with the added benefit of reliable uncertainty quantification. These results confirm that incorporating MPS does not substantially compromise segmentation performance while enriching model outputs with interpretable confidence estimates.

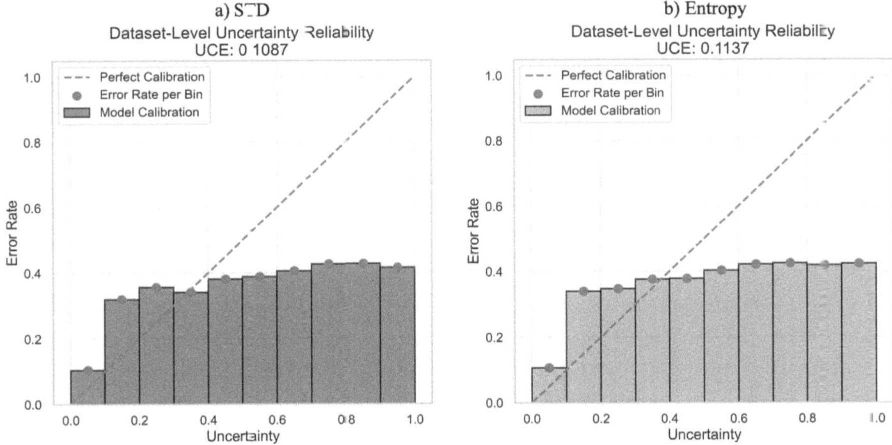

Fig. 5. Test dataset-level uncertainty calibration curves using (a) STD and (b) entropy. Each plot displays the relationship between predicted uncertainty and actual prediction error, computed across binned uncertainty intervals. The dashed diagonal line denotes perfect calibration, where predicted uncertainty would match the observed error.

Uncertainty measures based on standard deviation and entropy were well correlated with segmentation error (Fig. 5), but showed underestimated error in mid-to-high bins, which suggests an overestimation of risk in these regions. The UCE was slightly lower for STD (0.1087) than for entropy (0.1137), suggesting that STD aligns better with the observed error and may be preferable in downstream applications.

The cyclic learning rate with polynomial decay was essential for encouraging checkpoint diversity while preserving convergence. Sampling checkpoints during the low-learning-rate phase of each cycle proved effective for posterior approximation and stable inference. Another advantage of this method is its versatility: in contrast to MCDO [6], MPS can be applied to any segmentation model without the need to modify the backbone network's architecture. Moreover, MPS yields a computational cost comparable to that of MCDO, as both approaches require a similar number of forward passes, but it avoids the degradation in predictive accuracy that can arise from activating dropout layers at inference time. These findings, while demonstrated in lung squamous cell carcinoma, suggest that the proposed uncertainty-aware framework is not intrinsically related to a single tissue or biomarker. Because the method operates on generic H&E representations and does not require architecture changes specific to PD-L1, it could, in principle, be extended to other cancer subtypes or immunohistochemical targets (e.g., HER2, Ki-67), provided suitable training data are available.

Despite promising results, several limitations remain. The model was evaluated in a single cancer subtype (lung squamous cell carcinoma) using IHC-derived annotations, and its generalizability to other tissue types or biomarkers remains to be tested. This restriction reflects the scarcity of publicly available or easily accessible datasets that pair H&E images with reliable PD-L1 annotations, which constrained opportunities for external validation. Moreover, while uncertainty maps enhance interpretability, real-world utility will depend on integration with clinical workflows and further human-in-the-loop validation. Additionally, although uncertainty measures based on standard deviation and entropy were well correlated with segmentation error (Fig. 5), they tended to underestimate error in mid-to-high bins, indicating suboptimal calibration.

Future work should encompass a comprehensive external validation process. Additionally, it should explore modality expansion during training and enhance calibration techniques. For instance, post-hoc methods such as temperature scaling or other probabilistic calibration techniques can be employed to generate more reliable confidence estimates. Furthermore, domain adaptation across cancer types should be investigated. Lastly, decision-support systems that incorporate uncertainty estimates should be explored, such as uncertainty-based thresholding during area assessment in automatic PD-L1 scoring.

5   Conclusions

We propose a Bayesian segmentation framework using Multimodal Posterior Sampling (MPS) to infer PD-L1 expression from H&E-stained images. Using cyclic training and sampling diverse checkpoints, our model provides accurate segmentation with pixel-wise epistemic uncertainty estimates via entropy and standard deviation. It performs competitively with state-of-the-art methods while offering improved interpretability. This supports the feasibility of H&E-based PD-L1 inference as a scalable alternative to immunohistochemistry. Future directions include improving calibration, improving generalization, and integrating with clinical decision-making tools.

Acknowledgements. This work was partially supported under grant PID2023-152631OB-I00 by the Ministerio de Ciencia, Innovación y Universidades, Agencia Estatal de Investigación (MCIN/AEI/10.13039/501100011033/), co-financed by European Regional Development Fund (ERDF), 'A way of making Europe'. Roman Kinakh holds a UC3M fellowship PIPF Programa "Inteligencia Artificial" (Call 2024/2025). Some icons used in Fig. 1 were sourced from Flaticon.com and are attributed to their respective authors.

References

1. Ai, L., Xu, A., Xu, J.: Roles of pd-1/pd-l1 pathway: signaling, cancer, and beyond. Regulation of cancer Immune checkpoints: Mol. Cell. Mech. Therapy, 33–59 (2020)
2. Chen, J., et al.: Transunet: Transformers make strong encoders for medical image segmentation. arXiv preprint arXiv:2102.04306 (2021)
3. Dai, J., Li, Y., He, K., Sun, J.: R-FCN: Object detection via region-based fully convolutional networks. Adv. Neural Inf. Process. Syst. **29** (2016)

4. Deng, X., et al.: Mcranet: Mtsl-based connectivity region attention network for pd-l1 status segmentation in h&e stained images. Comput. Biol. Med. **184**, 109357 (2025)

5. Fuchs, T.J., Buhmann, J.M.: Computational pathology: challenges and promises for tissue analysis. Comput. Med. Imaging Graph. **35**(7–8), 515–530 (2011)

6. Gal, Y., Ghahramani, Z.: Dropout as a Bayesian approximation: representing model uncertainty in deep learning. In: International Conference on Machine Learning, pp. 1050–1059. PMLR (2016)

7. Isensee, F., Jaeger, P.F., Kohl, S.A., Petersen, J., Maier-Hein, K.H.: nnu-net: a self-configuring method for deep learning-based biomedical image segmentation. Nat. Methods **18**(2), 203–211 (2021)

8. Isensee, F., et al.: NNU-Net revisited: A call for rigorous validation in 3D medical image segmentation. In: International Conference on Medical Image Computing and Computer-Assisted Intervention, pp. 488–498. Springer (2024)

9. Janowczyk, A., Zuo, R., Gilmore, H., Feldman, M., Madabhushi, A.: Histoqc: an open-source quality control tool for digital pathology slides. JCO Clin. Cancer Inform. **3**, 1–7 (2019)

10. Van der Laak, J., Litjens, G., Ciompi, F.: Deep learning in histopathology: the path to the clinic. Nat. Med. **27**(5), 775–784 (2021)

11. Oktay, O., et al.: Attention u-net: Learning where to look for the pancreas. arXiv preprint arXiv:1804.03999 (2018)

12. Pardoll, D.M.: The blockade of immune checkpoints in cancer immunotherapy. Nat. Rev. Cancer **12**(4), 252–264 (2012)

13. Ribas, A., et al.: Releasing the brakes on cancer immunotherapy. N. Eng. J. Med. **373**(16), 1490–1492 (2015)

14. Ronneberger, O., Fischer, P., Brox, T.: U-Net: Convolutional Networks for Biomedical Image Segmentation. In: Navab, N., Hornegger, J., Wells, W.M., Frangi, A.F. (eds.) MICCAI 2015. LNCS, vol. 9351, pp. 234–241. Springer, Cham (2015). https://doi.org/10.1007/978-3-319-24574-4_28

15. Smith, L.N.: Cyclical learning rates for training neural networks. In: 2017 IEEE Winter Conference on Applications of Computer Vision (WACV), pp. 464–472. IEEE (2017)

16. Stacke, K., Eilertsen, G., Unger, J., Lundström, C.: Measuring domain shift for deep learning in histopathology. IEEE J. Biomed. Health Inform. **25**(2), 325–336 (2020)

17. Topalian, S.L., et al.: Safety, activity, and immune correlates of anti–pd-1 antibody in cancer. New England J. Med. **366**(26), 2443–2454 (2012)

18. Wang, Q., et al.: Prediction of pd-l1 tumor positive score in lung squamous cell carcinoma with h&e staining images and deep learning. Front. Artif. Intell. **7**, 1452563 (2024)

19. Yang, M., Yu, K., Zhang, C., Li, Z., Yang, K.: Denseaspp for semantic segmentation in street scenes. In: Proceedings of the IEEE Conference on Computer Vision and Pattern Recognition, pp. 3684–3692 (2018)

20. Zhao, Y., Yang, C., Schweidtmann, A., Tao, Q.: Efficient Bayesian uncertainty estimation for nnu-net. In: International Conference on Medical Image Computing and Computer-Assisted Intervention, pp. 535–544. Springer (2022)

Machine Learning for MRI-Based Classification of Treatment Response in Diffuse Gliomas

María Yanshuang Martín Moreira[1] (iD), Ana María Pérez Martín[1] (iD),
José-Javier Serrano-Olmedo[2,5,6] (iD), Ángel Luis Álvarez[4] (iD),
and Oscar Casanova-Carvajal[2,3(✉)] (iD)

[1] Universidad Alfonso X El Sabio, 28691 Madrid, Spain
[2] Centro de Tecnología Biomédica, Campus de Montegancedo, Universidad Politécnica de Madrid, 28040 Madrid, Spain
`oscar.casanova@upm.es`
[3] Departamento de Ingeniería Eléctrica, Electrónica Automática y Física Aplicada, Escuela Técnica Superior de Ingeniería y Diseño Industrial ETSIDI, Universidad Politécnica de Madrid, 28040 Madrid, Spain
[4] Escuela de Ingeniería de Fuenlabrada, Universidad Rey Juan Carlos, 28922 Madrid, Spain
[5] Centro de Investigación Biomédica en Red Para Bioingeniería, Biomateriales y Nanomedicina, Instituto de Salud Carlos III, 28029 Madrid, Spain
[6] Departamento de Tecnología Fotónica y Bioingeniería, ETSI Telecomunicaciones, Universidad Politécnica de Madrid, 28040 Madrid, Spain

Abstract. Diffuse gliomas are currently challenging to classify in the field of brain tumours due to their clinical and imaging characteristics. This study aims to develop and train a patient classification system based on the study of various parameters to support personalized treatment planning. The use of MATLAB allows us to analyse, segment, and process all data to extract relevant features associated with glioma characteristics. Classification was performed by considering Radiomics (such as texture and shape), biomolecular makers, and clinical information from the patient, in order to predict therapeutic response. Experimental results show that the system is able to distinguish subgroups of patients with moderate but consistent accuracy, demonstrating the feasibility of applying computational image analysis in combination with clinical and molecular variables. The integration of radiomics, biomolecular, and clinical information underscores the potential of such an approach as a decision-support tool in neuro-oncology. Overall, the findings suggest that multimodal classification systems represent a promising pathway toward improving the accuracy of glioma diagnosis and advancing personalized treatment planning, ultimately contributing to better patient outcomes.

Keywords: Diffuse Glioma · Random Forest · Magnetic resonance imaging · Deep learning · MATLAB · Classification · Biomarkers · Treatment

1 Introduction

Diffuse gliomas are a heterogeneous category of central nervous system tumours originating from glial cells and also characterised by extensive infiltration of the surrounding brain parenchyma [1]. Classified by the World Health Organisation (WHO) as grade II to IV tumours according to their histology and molecular status, diffuse gliomas include entities such as diffuse astrocytomas, oligodendrogliomas and glioblastomas [2, 3]. In contrast to circumscribed gliomas, diffuse gliomas have a remarkable ability to invade adjacent tissue, which makes them difficult to be completely received and contributes to a high recurrence rate [4].

The overall incidence of diffuse gliomas is generally between 3 and 5 cases per 100,000 inhabitants per year, with a higher prevalence in young and middle-aged adults, depending on the histological and molecular subtype [5]. For example, oligodendrogliomas typically occur in patients between 30 and 50 years of age, while glioblastomas predominantly affect individuals over 60 years of age [6]. Some of the advances made so far in molecular classification, such as the identification of Isocitrate Dehydrogenase (IDH)1/2 mutations, 1p/19q codeletions and O-6-Methylguanine-DNA Methyltransferase (MGMT) promoter methylation, have allowed to redefine the diagnosis and prognosis of these tumours, providing more precise criteria for therapeutic stratification [7, 8].

However, despite multimodality therapies combining surgery, radiotherapy and chemotherapy, the prognosis for patients with diffuse gliomas remains unfavourable. Particularly those patients with high-grade histological features or without beneficial IDH mutations [9]. The survival period for patients ranges from less than 15 months to several years, depending on whether they have IDH-wild diffuse glioblastoma-like glioblastoma (being shorter in these cases), or IDH-mutant tumours (in which cases survival is longer). This reflects the underlying biological heterogeneity [10, 11].

A biopsy or surgical resection is considered the gold standard for definitively diagnosing diffuse gliomas, as these are the methods for obtaining tumour tissue. Nevertheless, it is important to note that these procedures are associated with a number of clinical risks that can be substantial [12]. This underscores the necessity to devise diagnostic instruments that are complementary, non-invasive, and enable more precise biological characterisation, thereby facilitating disease monitoring. [13].

Thus, magnetic resonance imaging (MRI) has emerged as an essential tool for diagnosing and monitoring diffuse glioma. T1-, T2- and fluid attenuated inversion recovery (FLAIR)-weighted sequences allow the evaluation of relevant morphological features such as infiltrative extension, the presence of oedema and necrosis, as well as contrast uptake patterns [14]. In addition, advanced techniques such as MR spectroscopy, perfusion and diffusion have proven useful in differentiating tumour grades and predicting molecular mutations [15, 16].

Recently, the use of artificial intelligence in terms of classical machine learning algorithms has proven to be effective in classification, prediction, prognosis and characterisation of diffuse gliomas from MRI. In this context, Random Forest (RF) allows the construction of classifiers that are kernel-optimal in high-dimensional spaces, facilitating the separation of complex classes, such as different tumour subtypes or molecular states. These models are especially useful when limited but well-labelled datasets are available and have shown good performance in comparative studies against more complex methods. However, the development of accurate RF models critically depends on proper selection of relevant features and optimisation of model parameters.

In this study, an RF model trained with features extracted from multimodal MR images of patients with diffuse gliomas is developed. The performance of the model is evaluated using standard metrics such as accuracy, sensitivity, F1-score and AUC-ROC. In this way, it allows to detect how the patient would respond to treatment [17].

2   Materials and Methods

2.1   Data Set

Data and Image Acquisition. This study includes 501 adult patients who were diagnosed with grade II to IV diffuse gliomas. All these cases were treated at the medical centre (University of California, San Francisco - UCSF) between 2015 and 2021. Patients were excluded if they had previously undergone treatment in the brain. However, those who had only received a previous biopsy were allowed to participate All patients underwent preoperative MRI, initial surgical resection of the tumour, and tumour genetic testing [18].

Approximately 47 MR images were acquired for each patient using 3.0 T scanners and 8-channel head coils. The modalities covered structural anatomy as well as functional and diffusion features. The sequences included were: T1 and T1c (with contrast): high-resolution anatomical images, with T1c being useful for the detection of tumour enhancement after gadolinium administration. T2 and FLAIR: visualisation of oedema and hyperintense lesions, with cerebrospinal fluid suppression in FLAIR. DWI and ADC: diffusion-weighted images and apparent diffusion coefficients, relevant for tumour differentiation. Arterial spin labelling (ASL): non-contrast cerebral perfusion imaging is useful in oncological contexts and for patients with contraindications to gadolinium. Susceptibility-weighted imaging (SWI): sequences are sensitive to magnetic susceptibility, such as that found in blood, iron or calcium [19, 20] (Fig. 1).

Demographic, clinical, monitoring and diagnostic data, as well as genomic and molecular results were collected. These include patient identifier (ID), gender (M/F), age at MRI, tumour grade (HO CNS Grade) according to WHO classification (2021), and final integrated diagnosis. Genetic analysis included an IDH gene mutation study, assessed by genetic sequencing. MGMT promoter methylation status (in grade III and IV tumours), analysed by quantitative polymerase chain reaction (PCR). And the 1p/19q codeletion, reported in the corresponding cases [10, 21].

In terms of tumour grade distribution: 55 patients presented grade II gliomas, 42 grade III, and 403 grade IV. Patients were predominantly male (~60% in all grades). The IDH mutation was detected in 83% of grade II, 67% of grade III, and 8% of grade IV tumours. MGMT promoter methylation was present in 63% of grade IV gliomas.

Image Preprocessing and Tumour Segmentation. The image set was subjected to standardised pre-processing. The angular resolution diffusion imaging (HARDI) was corrected for eddy currents, and diffusion maps were derived: Mean Diffusivity (MD), Axial Diffusivity (AD), Radial Diffusivity (RD) and Fractional Anisotropy (FA). All images were spatially registered to the 3D Fluid-Attenuated Inversion Recovery (FLAIR) reference volume, with an isotropic resolution of 1 mm. Skull stripping was applied using a public deep learning-based algorithm [22, 23].

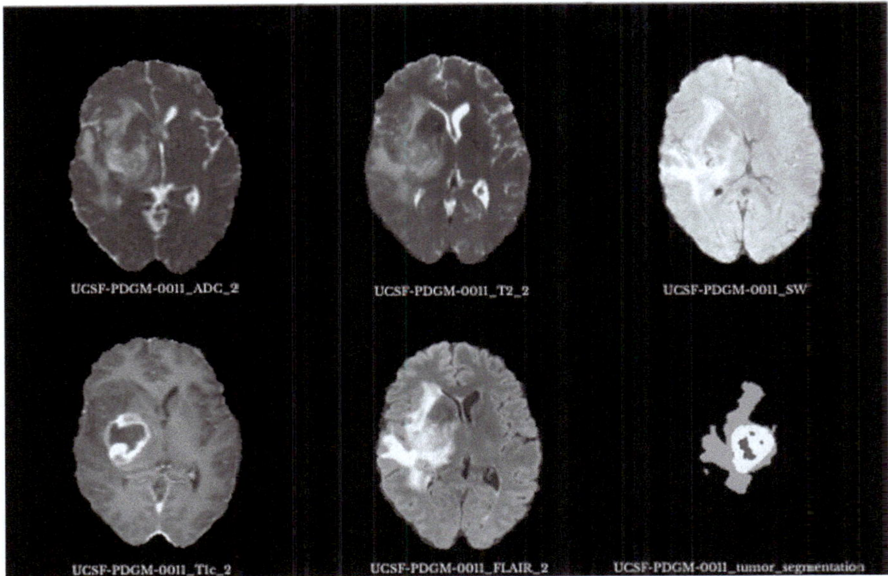

Fig. 1. Representative multimodal MRI studies in a patient with diffuse glioma. ASL = arterial spin labelling perfusion, DWI = isotropic (trace) diffusion-weighted imaging, Segmentation = multicompartment tumour segmentation, SWI = susceptibility-weighted imaging, T1 = T1-weighted precontrast, T2 = T2-weighted, FLAIR = -weighted FLAIR. [18]

Tumour segmentation was performed using an automatic model, developed from winning models of the BraTS 2021 challenge. The automatic segmentations were manually corrected by radiologists and validated by neuroimaging experts. Three regions of interest were identified and delimited: Enhanced tumour, Necrotic or non-enhancing tumour and FLAIR abnormality, indicative of oedema or tumour infiltration [24].

The dataset, called UCSF-PGDM, is available at The Cancer Imaging Archive (TCIA), and includes images, quantitative measurements, demographic data, diagnostic tests, follow-up information, molecular data, software and supporting analyses [18].

Image Parameters. To quantitatively characterise the biological behaviour of the glioma, key parameters were extracted from the segmented images to predict aggressiveness and possible response to treatment. These parameters included tumour volume (mm3) representing total tumour burden (where larger volumes are usually associated with greater cellular heterogeneity), intratumoural hypoxia and therapeutic resistance, as well as the entropy, which is the measure of tumour heterogeneity and complexity, where high values indicate greater disorganisation and aggressiveness.

Mean CBF (Cerebral Blood Flow) and maximum CBF: derived from ASL images, they reflect tumour perfusion. A high CBF is related to high vascularisation typical of high-grade tumours, and the maximum CBF indicates regions with greater angiogenic activity. The sphericity is morphological parameter that quantifies the roundness of the tumour; so that more spherical tumours tend to be less infiltrative, while irregular shapes indicate invasiveness and worse prognosis. Finally, Necrosis, identified on post-contrast

T1 images as hypointense non-enhancing areas, indicating rapid tumour growth that exceeds vascular capacity, associated with grade IV gliomas and poor prognosis [25, 26]. These parameters were integrated for subsequent use as predictor variables in a RF-based classification model.

2.2 Extraction of Image Parameters

The MATLAB code used for the automatic processing of tumour metrics allows the calculation of these parameters from multimodal images according to the Neuroimaging Informatics Technology Initiative (NIFITI) format. For this purpose, an automated MATLAB script was developed, in which morphological metrics and intensity are extracted from brain images with tumour bodies to enable quantification of relevant tumour body characteristics. The processing is applied in batch mode to a folder structure containing studies of different patients, respecting a previously predefined nomenclature pattern [27].

In this case, for each patient, the algorithm automatically locates the current files necessary for the calculation of metrics, including the tumour mask file (segmentation) and the different relevant imaging modalities: FLAIR, T1 with contrast (T1c), ASL, and apparent diffusion (ADC), if available. The segmented mask is used as the region of interest (ROI) for all measurements.

The processing includes size and format validations, thus ensuring that images are compatible with each other before calculating metrics. In case of missing or incompatible data, the patient is omitted from the analysis in a controlled manner. At the end of the processing of all subjects, the results are consolidated in a structured table, grouping all metrics obtained from each patient, and automatically exported to an Excel file for further analysis. In addition, some patients had to undergo tumour segmentation, as one of the databases we used did not provide binary masks.

The metadata (downloaded directly from the study website) and the parameter data obtained from the images were merged into a single CSV file. In addition, the empty cells were filled in with 'Unknown'.

2.3 Choosing the Classifier

In previous work [28], the performance of both Random Forest (RF) and Support Vector Machine (SVM) were compared as classifiers for cancer using genetic expressions.

The work's main results established that SVM had a higher performance than RF. Despite this, RF had a higher robustness overall. It was highlighted that SVM is more sensitive to the selection of features. Additionally, it is mentioned that class imbalance is a factor that can influence the model's performance, where SVM was found to be more affected by this imbalance. RF handled it better thanks to bagging. Eventually, due to the imbalanced data and multiclass classification, a RF classifier was chosen (Table 1).

Table 1. Comparison of characteristics between RF and SVM.

Criterion	RF	SVM
Performance with imbalanced data	More robust, handles imbalance better (use of bagging)	More sensible towards imbalance, if classes aren't weighted or resampled
Robustness to noise	High (robust against noisy data and outliers)	Moderate (sensitive to noise)
Use with small datasets	May overfit if dataset is small or many classes	It can be very effective with small datasets and well balanced
Multiclassing	Can handle multiple classes directly	Requires extensions for multiclassing
Training speed	Quick in medium size datasets	Slower in bigger or datasets with many attributes
Prediction speed	Can be slower if there are many trees	Quick if the model is well optimized
Explainability	High, based on decision trees (hierarchical, easy to visualize)	Low, based on mathematical distances (not explicit rules)

2.4 Pipeline: Model Classification

In this work, we implemented an integrated machine learning workflow combining clinical, molecular and imaging features of glioma patients to classify treatment response. A Random Forest model, complemented with data balancing strategies such as SMOTE or oversampling, was employed to distinguish between good, indeterminate, and resistant outcomes. [28, 29].

A CSV file containing the clinical metadata of glioma patients was used. This file contains a column called 'ID', which is used as a unique identifier for each patient. From a generated structure, the clinical, imaging and molecular variants were extracted, thus defining the true class of each patient (Y_true_all_patients), according to overall survival (OS) and vital status criteria.

Patients alive with OS > 540 days were labelled as Class 1, meaning they would have a good prognosis. Patients alive with OS between 360 and 480 days were labelled Class 2, having an intermediate prognosis. And finally, patients with OS < 180 days or deceased, were labelled as Class 3, resulting in a poor prognosis.

Several transformations were performed to standardize and prepare the variables for use in machine learning models: Categorical variables such as MGMT_status, HDI, 1p/19q, and Extent of Resection (EOR) were coded into ordinal values (0, 0.5, 1), the sex variable was binarized (male = 1; other = 0), and the Final_pathologic_diagnosis field was transformed by one-hot coding to represent each diagnosis as an independent binary variable.

Input (X) and output (Y) vectors for the model were formed. Subsequently, the total data set was divided into training (70%) and test (30%) subsets.

Given the imbalance in the class distribution, the SMOTE technique was applied on the training set. SMOTE generates synthetic samples by interpolating between close neighbors of the same class, which is useful to avoid biasing the model towards the majority class. In cases where SMOTE was not effective, manual oversampling was used by replicating samples with the repmat (MATLAB function that replicates samples to balance the dataset). Additionally, class weights, which are defined as inversely proportional to the frequency of each class, were calculated to ensure that the model treated each class with equal importance during training.

A classification model was trained using Random Forest with the fitcensemble function and the bootstrap aggregation technique. A committee of 100 decision trees (NumLearningCycles = 100) was built, using the previously calculated class weights. This type of model is robust to overfitting and suitable for data sets with heterogeneous variables.

The model employed a range of statistical metrics, including precision, recall, specificity and the F-1 score, to assess the performance of the test set. These metrics were utilized to generate confusion matrices, which provided a visual representation of the distribution of hits and misses across each class. The overall accuracy for the total proportion of correct classifications and the ROC curves evaluate the discriminatory ability of the model in a multi-class scenario.

2.5 Model Operation

A supervised learning model based on Random Forest was used to classify patients into three prognostic groups. This model consists of a set of decision trees trained on the different random subsets of an original set, following the bagging technique (Bootstrap Aggregating). The final prediction is mostly obtained by means of the votes of the decision trees.

Each decision tree was trained with a random sample from the training set, selected with replacement. This technique introduces variability between trees, promoting greater diversity in the ensemble. To avoid overfitting individual trees, a limit was placed on the depth of each tree (MaxNumSplits = 30), allowing each to act as a "weak model". However, the combination of multiple weak trees forms a highly accurate and robust "strong model".

Custom class weights were calculated to mitigate the bias introduced by disproportionality in the class distribution. These weights were defined as the inverse of the frequency of each class and were normalized so that their mean was equal to one. This strategy forces the model to pay more attention to minority classes during training, improving its generalizability in unbalanced contexts.

Once the model is trained, it classifies new samples according to the following procedure: Each tree makes an independent prediction. These predictions are combined by majority vote to obtain the final predicted class (Y_pred). A vector of scores or probabilities of belonging to each class is also calculated from the proportion of votes from the trees.

These membership scores were further used for the generation of multiclass ROC curves, which allowed to assess the discriminative ability of the model for each prognostic class.

In summary, the parameters used to configure the auto-mathematical learning model were: Bag, consisting in the construction of a set of decision trees from random samples. For the learners, a template tree with a maximum of 30 splits or branches is used. This limitation prevents the trees from over-fitting the training data, which could lead to over-fitting. The model is configured to create a total of 100 trees in the learning set, as increasing the number of trees generally improves the stability of the model, albeit at the cost of increased computational cost. To correct for imbalance in the classes in the dataset, a vector of weights inversely proportional to the frequency of each class is used, meaning that less frequent classes are weighted more heavily to balance their influence during training. Finally, the classes present in the training set are explicitly defined, which helps to avoid errors related to missing classes during sampling, what ensures that the model considers all relevant categories.

3 Results

3.1 Class Distribution

The dataset originally consisted of 501 patients, which covered the three defined classes of probability of treatment response. After introducing the data in the RF model, the total patient count was 413, meaning 88 patients were finally not considered. This could have been caused by missing information in the selected features and therefore the model discarded those patients. Class 3 (poor prognosis) was the most recurrent, representing 76.03% (314) of all cases. On the other hand, Class 1 (good prognosis) and Class 2 (indeterminate prognosis) accounted for 19.37% (80) and 4.6% (19) of all the data (Fig. 2a).

When the data was divided into the train and test set the count distributions are as follows. In the train set Class 1 consists of 18.28% (53), Class 2 of 4.83% (14) and Class 3 of 76.9% (223), while in the test set Class 1 made up 21.95% (27), Class 2 the 4.07% (5) and Class 3 the 73.98% (91). As the data shows, there is a considerable imbalance between the defined groups, Class 2 being the underrepresented class and Class 3 the dominant class (Figs. 2b and 2c).

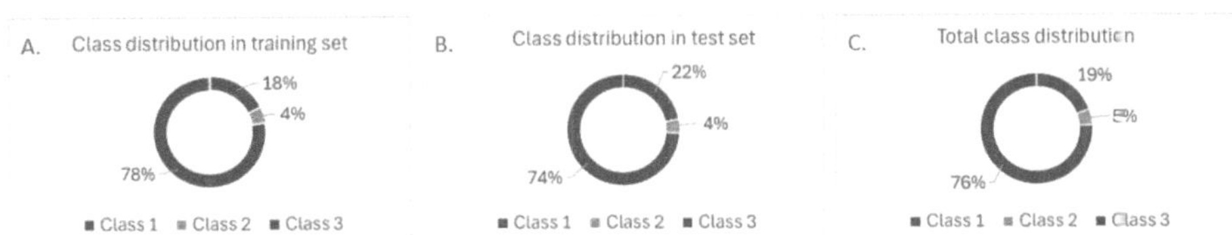

Fig. 2. Doughnut charts that represent the patient division per class in three different cases: a. Overall, b. Training set, c. Test set.

3.2 Model Performance

A Random Forest classifier model was used for the classification of the data. To this data oversampling techniques (SMOTE or manual duplication if SMOTE wasn't possible) were applied before to minority classes. The overall accuracy on the test set was 68.3%.

The confusion matrix (Table 2) shows that the class that had the most correct predicted cases was Class 3, whereas incorrect classifications took place in Class 1 and Class 2.

Table 2. Confusion matrix.

8	2	17
1	1	3
13	3	75

Metrics for each class are exhibited in Table 3. Precision, recall and F1-score were the highest for Class 3 (0.789, 0.824 and 0.806, respectively), while Class 2 had the lowest value of these three parameters (0.167, 0.200 and 0.182, respectively). On the other hand, specificity was highest for Class 2 (0.958), although Class 3 had the lowest value (0.375).

This means that in Class 3 the model doesn't make a lot of mistakes when it comes to predicting the positives and detects almost every real positive case but fails to identify the negative cases (many false positive cases). Meanwhile, in Class 2 the model predicts many positives (many are false positives) and loses many real positives, but it is able to identify correctly the negatives.

Table 3. Metrics for each class.

Class	Precision	Recall	Specificity	F1-score
1	0.364	0.296	0.854	0.327
2	0.167	0.200	0.958	0.182
3	0.789	0.824	0.375	0.806

The ROC curves per class (Fig. 3) display that there is moderate discrimination in Class 1, where the model can distinguish correctly between patients from this class and the rest (AUC = 0.63). In contrast, in Class 2 (AUC = 0.53) the model is not capable of differentiating patients that belong in that class and therefore is practically guessing. Finally, there is weak discrimination in Class 3 (AUC = 0.57), but the model is not getting hold of the patterns that define this class.

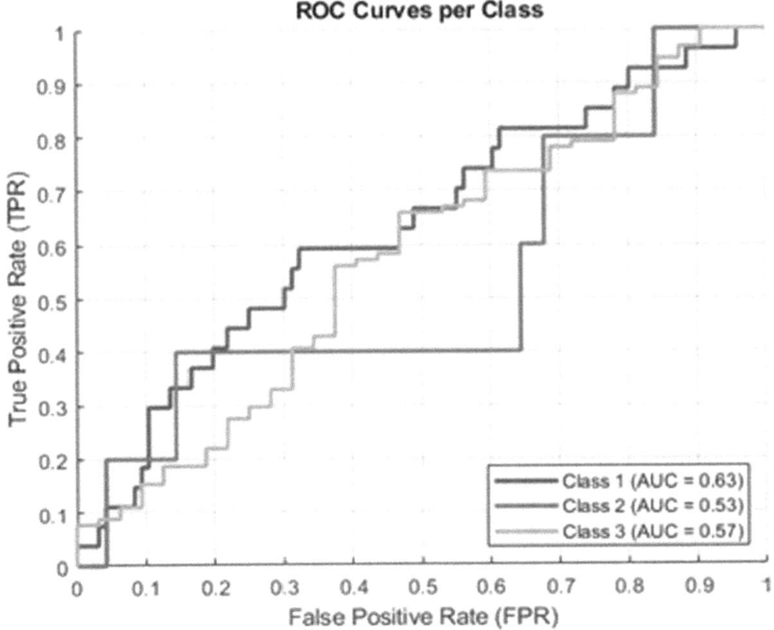

Fig. 3. ROC curves and area under de curve (AUC) for the three classes.

4   Discussion

This study evaluated the use of a Random Forest classifier to predict the patient's response to the treatment targeting diffuse gliomas.

The overall accuracy (68.3%) shows that the RF can identify correctly some of the cases. Specifically, the ones that have poor prognosis or are resistant to the treatment (Class 3). However, class imbalance between the defined classes is reflected in Class 1 and significantly in Class 2 metrics (very low F1-score, low recall values, biased confusion matrix and low UAC in minority classes). Despite the implementation of oversampling techniques (SMOTE or manual duplication), the minority classes were still harder to classify. This issue is very common in the use of RF models for cancer classification.

To overcome this issue, other more complex techniques have been applied and investigated. A study published in 2023 compared several techniques applied to data regarding medical diseases and used RF as the base classifier. The performance of each method was evaluated across different disease datasets. The techniques used were classic oversampling (SMOTE and its variants), undersampling, hybrid techniques (combinations of undersampling and oversampling and deletion or cleaning of data), and one-sided selection. The study concluded that oversampling techniques work well in many of the cases, but its performance can vary depending on the dataset. Hybrid methods have good potential, especially when there is noise or class overlapping. One-sided selection was highlighted as one of the most effective techniques in most of the datasets, because it was able to balance out the data without sacrificing other metrics [30].

Apart from the methods mentioned above, ensemble learning techniques can also be applied as a solution for data imbalance. An article published in 2023 approached the problem there is in the data imbalance when it comes to classification, because traditional algorithms tend to favor the dominant class which can cause a harder detection of the minority class. The ensemble learning methods that were compared were Bagging and RF, Boosting (AdaBoost. Gradient Boosting, XGBoost, LightGBM, CatBoost) and specific techniques for imbalance (Balanced Random Forest, BRF, and EasyEnsemble). The medical data used The Breast Cancer Wisconsin (Diagnostic) dataset and The Pima Indian Diabetes dataset.

The main results were that boosting methods (XGBoost, EasyEnsemble) were the most effective where its uses are critical and in dataset with high imbalance (surpassed significantly RF and AdaBoost). BRF improved recall within the minority classes in comparison to RF but boosting methods were better in complex datasets. LightGMB and CatBoost worked very well in datasets where imbalance was moderate and had categorical variables. Lastly, standard RF was quicker to train (~40% less time than boosting), which makes it useful in real life systems or where there are computational restrictions, but it also means having a lower performance in minority classes [31].

Similar studies have been done in relation to using RF as classifiers in the oncology field as shown by Baid et al. (2020) in Overall Survival Prediction in Glioblastoma With Radiomic Features Using Machine Learning [32] and by Jin et al. (2023) in Development and testing of random forest-based machine learning model for predicting events among breast cancer patients with a poor response to neoadjuvant chemotherapy [33].

In both studies only one type of data was used. While Baid et al. relied on radiomic features obtained from FLAIR and T1ce MRI modalities, and Jin et al. used only clinical information (such as tumor status, nodal status and hormone receptors), our approach combines radiomic features, tumoral biomarkers, and clinical information from the patient. This permits a more complete and accurate representation of both tumor characteristics and patient background, which could allow the model to make correct and precise predictions.

Another aspect that was mentioned in both cases was class imbalance in the data, because it made the classification process harder due to the underrepresentation or dominance of a class. However, there isn't a description of the usage of any specific technique to overcome this problem, whereas we tried solving this problem by applying SMOTE or manual duplication to minority classes.

Our findings are in line with previous reports where it's shown that RF has a robust but not the best performance in unbalanced oncology datasets. Nevertheless, by the integration of multimodal data and directly handling class imbalance, our model can be considered more complete despite not achieving a high performance.

This can be improved by adding more patients to the initial dataset, adjusting individually class-specific thresholds (to prioritize the good detection of a particular class) and the application of more complex methods to overcome the problem of class imbalance (i.e., One-sided selection).

5 Conclusion

The Random Forest (RF) model reported an accuracy of 0.68 and allows detection of poor prognoses (i.e., deceased patients) with an estimated area under the curve (AUC) of 0.86 and an F1-score of 0.81. However, its capacity to detect patients with a good response to treatment was weaker, particularly class 2.

All of this indicates that the model could not be used in a clinical setting due to the imbalanced class distribution and the limited sample size, which were reflected in the model's weak predicting ability.

Despite these limitations, the model was effective at leveraging multimodal features, including clinical, molecular, and imaging-derived variables, without over-relying on any single input data type. The ensemble architecture's use of random feature selection at each node enabled different trees to learn distinct decision patterns from clinical, imaging, and genetic variables. Importantly, the model was resilient to overfitting and performed well in noisy and imbalanced conditions. This aligns with the strength's commonly reported for RF classifiers in medical contexts. In many cases, the model predicts randomly, as if it were trying to guess the patient's class at classification time.

Future work should target class imbalance through the application of other advanced techniques and investigate hybrid models that combine RF with other methods, such as deep learning or probabilistic frameworks to improve predictive precision across all categories.

Funding. This work is partially supported by grant PDC2023-145812-I00 (Project SAMPL2D), which is funded by MICIU/AEI/10.13039/501100011033 and by "Next Generation EU/PRTR".

References

1. Louis, D.N., Perry, A., Wesseling, P., Brat, D.J., Cree, I.A., Figarella-Branger, D., et al.: The 2021 WHO classification of tumors of the central nervous system: a summary. Acta Neuropathol. **142**(1), 87–101 (2021)
2. Ostrom, Q.T., Cioffi, G., Waite, K., Kruchko, C., Barnholtz-Sloan, J.S.: CBTRUS statistical report: primary brain and other central nervous system tumors diagnosed in the United States in 2014–2018. Neuro-Oncol. **23**(Suppl 3), iii1–iii105 (2021)
3. Reuss, D.E., Sahm, F.: Molecular diagnostics of gliomas in the era of WHO 2021: clinical relevance and practical implementation. Nervenarzt **91**(10), 874–882 (2020)
4. Aldape, K., Zadeh, G., Mansouri, S., Reifenberger, G., von Deimling, A.: Glioblastoma: pathology, molecular mechanisms and markers. Acta Neuropathol. **137**(5), 795–803 (2019)
5. Lapointe, S., Perry, A., Butowski, N.A.: Primary brain tumours in adults. Lancet **392**(10145), 432–446 (2018)
6. Brat, D.J., et al.: Molecular pathogenesis of diffuse gliomas. In: Handbook of Clinical Neurology, vol. 134, pp. 245–267. Elsevier, Amsterdam (2018)

7. Cancer Genome Atlas Research Network: Comprehensive, integrative genomic analysis of diffuse lower-grade gliomas. N. Engl. J. Med. **372**(26), 2481–2498 (2015)

8. Eckel-Passow, J.E., Lachance, D.H., Molinaro, A.M., et al.: Glioma groups based on 1p/19q, IDH, and TERT promoter mutations. N. Engl. J. Med. **372**(26), 2499–2508 (2015)

9. Stupp, R., Taillibert, S., Kanner, A.A., et al.: Maintenance therapy with tumor-treating fields plus temozolomide vs temozolomide alone for glioblastoma: a randomized clinical trial. JAMA **314**(23), 2535–2543 (2015)

10. Yan, H., Parsons, D.W., Jin, G., et al.: IDH1 and IDH2 mutations in gliomas. N. Engl. J. Med. **360**(8), 765–773 (2009)

11. Gusyatiner, O., Hegi, M.E.: Glioma epigenetics: from subclassification to novel treatment options. Semin. Cancer Biol. **51**, 50–58 (2018)

12. Hervey-Jumper, S.L., Berger, M.S.: Maximizing safe resection of low- and high-grade glioma. J. Neurooncol **130**(2), 269–282 (2016)

13. Pope, W.B.: Brain metastases: neuroimaging. In: Handbook of Clinical Neurology, vol. 149, pp. 89–112. Elsevier, Amsterdam (2018)

14. Ellingson, B.M., Bendszus, M., Boxerman, J., et al.: Consensus recommendations for a standardized brain tumor imaging protocol in clinical trials. Neuro-Oncol. **17**(9), 1188–1198 (2015)

15. Just, N.: Improving tumour heterogeneity mri assessment with histograms. Br. J. Cancer **111**(12), 2205–2213 (2014)

16. Kickingereder, P., Isensee, F., Tursunova, I., et al.: Automated radiomic profiling of glioblastoma for personalized medicine. Neuro-Oncol. **21**(12), 1549–1560 (2019)

17. Kuhn, M., Johnson, K.: Applied Predictive Modeling. Springer, Heidelberg (2013)

18. Calabrese, E., et al.: The University of California San Francisco preoperative diffuse glioma MRI (UCSF-PDGM) (Version 5) [dataset]. Cancer Imaging Arch. (2022). https://doi.org/10.7937/tcia.bdgf-8v37

19. Louis, D.N., Perry, A., Wesseling, P., Brat, D.J., Cree, I.A., Figarella-Branger, D., et al.: The 2021 WHO classification of tumors of the central nervous system: a summary. Neuro Oncol. **23**(8), 1231–1251 (2021). https://doi.org/10.1093/neuonc/noab106

20. Ellingson, B.M., Cloughesy, T.F., Pope, W.B.: MRI in the diagnosis and management of gliomas: current clinical approaches and future directions. Neuro Oncol. **17**(7), 959–969 (2015). https://doi.org/10.1093/neuonc/nou378

21. Cairncross, J.G., Wang, M., Shaw, E.G., et al.: MGMT promoter methylation and clinical outcomes in glioblastoma: the RTOG 0525 trial. J. Clin. Oncol. **33**(11), 1260–1267 (2015). https://doi.org/10.1200/JCO.2014.57.4080

22. Tustison, N.J., Avants, B.B., Cook, P.A., et al.: N4ITK: improved N3 bias correction. IEEE Trans. Med. Imaging **29**(6), 1310–1320 (2010). https://doi.org/10.1109/TMI.2010.2046908

23. Menze, B.H., Jakab, A., Bauer, S., et al.: The multimodal brain tumor image segmentation benchmark (BRATS). IEEE Trans. Med. Imaging **34**(10), 1993–2024 (2015). https://doi.org/10.1109/TMI.2014.2377694

24. Bakas, S., Reyes, M., Jakab, A., et al.: Identifying the best machine learning algorithms for brain tumor segmentation, progression assessment, and overall survival prediction in the BRATS challenge. arXiv preprint arXiv:1811.02629 (2018)

25. Aerts, H.J.W.L., Velazquez, E.R., Leijenaar, R.T.H., et al.: Decoding tumour phenotype by noninvasive imaging using a quantitative radiomics approach. Nat. Commun. **5**, 4006 (2014). https://doi.org/10.1038/ncomms5006

26. Hu, L.S., Ning, S., Eschbacher, J.M., et al.: Radiogenomics to characterize regional genetic heterogeneity in glioblastoma. Neuro Oncol. **19**(1), 128–137 (2017). https://doi.org/10.1093/neuonc/now154

27. Havaei, M., Davy, A., Warde-Farley, D., et al.: Brain tumor segmentation with deep neural networks. Med. Image Anal. **35**, 18–31 (2017). https://doi.org/10.1016/j.media.2016.05.004

28. Breiman, L.: Random forests. Mach. Learn. **45**, 5–32 (2001). https://doi.org/10.1023/A:101 0933404324
29. Chawla, N.V., Bowyer, K.W., Hall, L.O., Kegelmeyer, W.P.: SMOTE: synthetic minority over-sampling technique. J. Artif. Intell. Res. **16**, 321–357 (2002). https://doi.org/10.1613/jai r.953
30. Rhmann, W.: An empirical study on the class imbalance handling techniques for different diseases. Soft. Comput. **28**(19), 11439–11456 (2024). https://doi.org/10.1007/s00500-024-09881-y
31. Ayodele, N.A.: A comparative study of ensemble learning techniques for imbalanced classification problems. World J. Adv. Res. Rev. **19**(2), 1633–1643 (2023). https://doi.org/10.30574/wjarr.2023.19.1.1202
32. Baid, U., et al.: Overall survival prediction in glioblastoma with radiomic features using machine learning. Front. Comput. Neurosci. **14**, 61 (2020). https://doi.org/10.3389/fncom.2020.00061
33. Jin, Y., Lan, A., Dai, Y., Jiang, L., Liu, S.: Development and testing of a random forest-based machine learning model for predicting events among breast cancer patients with a poor response to neoadjuvant chemotherapy. Eur. J. Med. Res. **28**(1) (2023). https://doi.org/10.1186/s40001-023-01361-7

Segmentation of *Arabidopsis* Apical Stem Cells via a Dual Deep Learning Approach

Guillermo Rey–Paniagua[1]([⊠]) [iD], Dariusz Lachowski[1] [iD],
and Arrate Muñoz-Barrutia[1,2] [iD]

[1] Universidad Carlos III de Madrid, Leganés, Madrid, Spain
`100450944@alumnos.uc3m.es`, `{dlachows,mamunozb}@ing.uc3m.es`
[2] Instituto de Investigación Sanitaria Gregorio Marañón, Madrid, Spain

Abstract. We introduce a deep learning pipeline for segmenting apical stem cells in three-dimensional confocal microscopy volumes of *Arabidopsis thaliana*. By integrating pre-trained 2D and 3D U-Net models from the BioImage Model Zoo, our method combines in-plane boundary sharpness with volumetric continuity. The workflow encompasses parallel preprocessing, dual-model inference, logical fusion, 3D reconstruction, and membrane-aware post-processing to extract key morphometric features, including cell counts and volumes. Evaluated over 22 time-points, our pipeline achieves a mean Dice similarity coefficient (DSC) of 0.886 ± 0.002. Passing–Bablok regression on cell counts yields a slope of 0.885 ($p = 0.573$), surpassing the standalone 2D U-Net (0.539) and 3D U-Net (0.782). Volume estimates exhibit a slope of 0.754 ($p = 0.107$), with smaller cell volumes but substantially fewer cell-merging artifacts compared to the baselines (0.738 and 0.930, respectively). These results highlight the advantages of model fusion for robust, biologically meaningful segmentation. The complete pipeline and source code are publicly available at: https://github.com/GolpedeRemo37/Arabidopsis_DualDL

Keywords: Deep learning · Image segmentation · Arabidopsis thaliana · U-Net · Confocal microscopy

1 Introduction

Accurate image segmentation is essential for bioimage analysis, particularly when quantifying complex structures such as the shoot apical meristem (SAM) in *Arabidopsis thaliana*, a key site for studying cell division, tissue growth, and morphogenesis [1]. Traditional methods based on hand-crafted filters and morphological operations often lack reproducibility and require substantial manual correction [2].

Deep learning has transformed image segmentation by enabling fast and accurate predictions across diverse biomedical contexts. Classical architectures such as U-Net [3], Cellpose [4], and StarDist [5] have shown strong generalization but are not always optimized for plant tissues or require extensive annotated data.

© The Author(s), under exclusive license to Springer Nature Switzerland AG 2026
A. López Fernández et al. (Eds.): CIABiomed 2025, LNBI 16148, pp. 376–387, 2026.
https://doi.org/10.1007/978-3-032-10661-2_29

Recent advances include transformer-based models and foundation segmentation networks [6,7], which offer cross-domain transferability, but often lack pretraining on confocal plant microscopy. To address these domain-specific challenges, Wolny *et al.* [8,9] developed pre-trained 2D and 3D U-Net variants tailored to *Arabidopsis* SAM membrane data, serving as a strong baseline for our approach.

In this study, we introduce a segmentation pipeline for 3D SAM confocal volumes that integrates both U-Net models. By fusing their outputs, we achieve more accurate and robust cell segmentation than using either model alone. The method is validated against expert annotations, showing consistent improvements in segmentation accuracy and morphometric feature extraction 1].

2   Materials and Methods

2.1   Materials

Three-dimensional confocal image volumes of the *Arabidopsis thaliana* shoot apical meristem were obtained from the timelapse study of Willis *et al.* [1]. The full dataset comprises several plants, each imaged as a time series of propidium iodide–stained volumes acquired every 4 h over a 76 h to 84 h interval on a high-resolution confocal microscope ($512 \times 512 \times L$ voxels; nominal voxel spacing $\Delta x_i \times \Delta y_i \times \Delta z_i$, which varied slightly per volume—typically 0.22–0.24 µm $\times$ 0.22–0.24 µm $\times$ 0.26 µm).

Expert-curated 3D cell masks (total cell counts and per-cell volumetric labels) accompany each volume; segmentation in the original study was performed via MARS/ALT with manual review by the authors [1]. All TIFF stacks and their segmentation masks were downloaded from the University of Cambridge DSpace repository [10] and organised according to our preprocessing workflow prior to inference.

Note on the analysed replicate. In this work we report quantitative results from the complete 22-timepoint series of a single specimen ("Plant 13") from the Willis *et al.* repository [10]. Eventually, the performance on all plants is briefly compared.

2.2   Segmentation Pipeline and Inference

The segmentation workflow (Fig. 1) employs a dual-model strategy to process the 3D confocal volume. A pre-trained 2D U-Net captures in-plane membrane features from individual slices, while a 3D U-Net models continuity along the z-axis using volumetric stacks. The resulting masks from both models are then combined, followed by 3D reconstruction and a series of post-processing steps to refine cell boundaries and eliminate artifacts [11].

Preprocessing and Model Inference. The input volume was duplicated to enable parallel processing. One copy was split into L-slices, saved as 2D TIFFs, normalized to [0,1], and processed using the 2D U-Net (TorchScript) from the

BioImage Model Zoo [12]. The 2D U-Net model was originally trained by Wolny *et al.* [8] on *Arabidopsis* SAM membrane data, to generate membrane probability maps. These were binarized at a threshold of 0.2 to generate 2D masks that encompass the majority of membrane regions, and merged into a continuous 3D membrane mask.

The second copy was divided into overlapping 3D sub-volumes, padded to match the input size required by the 3D U-Net [9], processed by the 3D U-Net and thresholded at 0.2 to generate binary 3D masks. These binary sub-volumes were seamlessly stitched back into a single 3D mask, ready for fusion and post-processing.

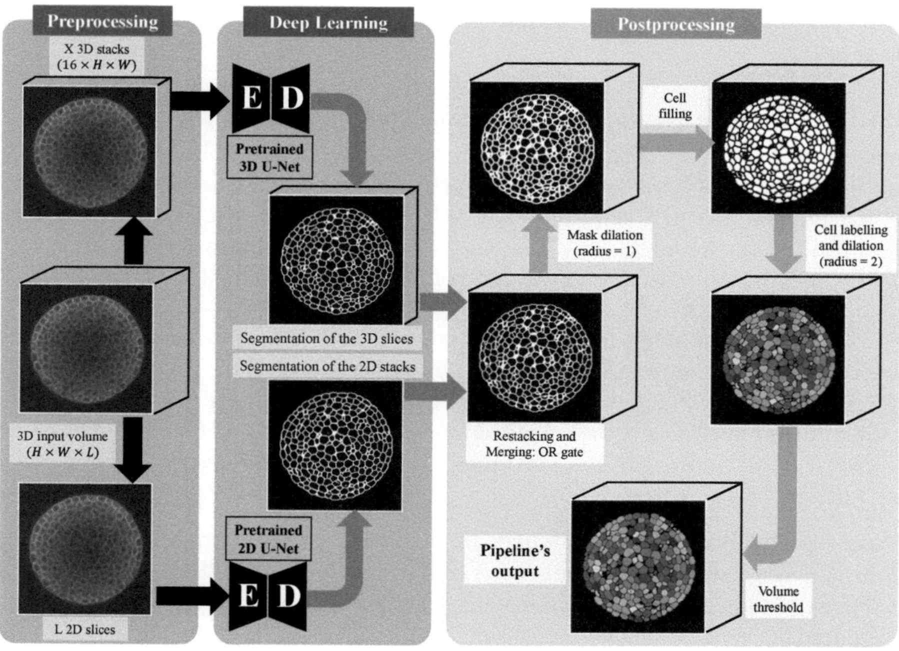

Fig. 1. Overview of the cell segmentation pipeline. The 3D confocal volume is split into 2D slices and overlapping 3D stacks, which are processed in parallel by a 2D U-Net and a 3D U-Net. Outputs are merged via a logical OR operation, followed by dilation (radius = 1 pixel) to broaden membranes, inversion to isolate interiors, 3D labelling, cell dilation (radius = 2 pixels) and filtering using a volume threshold to remove small debris.

Post-processing. To integrate predictions from both models, we applied a voxel-wise logical OR operation between both binary masks, resulting in a unified segmentation that incorporates both membrane perspectives. This combined binary mask was then dilated using a spherical structuring element (radius = 1) to broaden membrane boundaries and facilitate cell separation. To isolate potential cell interiors, the dilated mask was padded by one pixel on its x and y dimensions, inverted, and the largest connected component (representing the background) was removed from each 2D slice. The padding was then reversed

The resulting binary mask, primarily consisting of isolated internal regions, was then 3D labeled to identify individual cell candidates. To recover their approximate full volumes, each labeled region was further dilated with a spherical structuring element of radius $= 2$. Finally, objects with a reconstructed volume smaller than 1.5 μm^3 were discarded as debris. This threshold was chosen based on inspection of the ground truth: the smallest annotated cells, including their membranes, measured around 2 μm^3 . Since our method excludes membranes, we adjusted the threshold to 1.5 μm^3 to retain small but valid cellular regions. The remaining labeled volume, representing the segmented cells, was exported as a 3D TIFF, and quantitative features, including cell count and individual cell volumes, were computed.

2.3 Evaluation Metrics

Segmentation quality was quantified using the Dice Similarity Coefficient (DSC), defined as

$$Dice(G, S) = \frac{2\,|G \cap S|}{|G| + |S|} \tag{1}$$

where G and S denote the ground truth and predicted cell voxel sets. Values range from 0 (no overlap) to 1 (perfect agreement).

Morphometric agreement in total cell counts and mean cell volumes was evaluated via Passing–Bablok regression over all 22 timepoints, reporting both slope (ideal $= 1$) and intercept (ideal $= 0$).

To evaluate the statistical agreement between predictions and ground truth, we applied Welch's two-sided t-test (assuming unequal variances) to per-frame values, reporting the resulting p-values for each comparison [13].

Parameter Sensitivity Ablation Study. To assess robustness to the binarization threshold, we performed a targeted ablation study in which the binarization threshold applied to both 2D and 3D model probability maps was swept from 0.05 to 0.95 in steps of 0.05. The mean absolute error (MAE) for the cell counts was computed per-image and then averaged across the dataset. The sweep shows a minimum MAE at threshold 0.20; results are stable for nearby values (0.15–0.25). Results are summarised in Table 2.

3 Results

Our dual-model pipeline fuses outputs from pre-trained 2D and 3D U-Nets, harnessing both in-plane membrane sharpness and volumetric continuity. It achieves a DSC of 0.886—only marginally lower than the single-model baselines (DSC $=$ 0.915 for the 2D U-Net and DSC $= 0.906$ for the 3D U-Net). In a Passing–Bablok regression over 22 timepoints (Fig. 3), cell-count estimates follow a slope of 0.885 (versus 0.539 for the 2D U-Net and 0.782 for the 3D U-Net), while mean-cell-volume estimates yield a slope of 0.754 (compared to 0.738 and 0.930, respectively). These results confirm that our fusion strategy improves detection accuracy and substantially reduces cell-merging artifacts.

3.1   Qualitative Evaluation

Figure 2 demonstrates that the proposed pipeline produces cleaner and more consistent cell masks by explicitly excluding membrane regions during cell labelling. The 3D U-Net alone often fails to detect peripheral cells (highlighted in yellow circle) or under-segments blurred cells in the centre of the image (white circle). In contrast, the 2D U-Net maintains better boundary sharpness in-plane but suffers from inconsistencies across slices, leading to incorrect merging of adjacent cells along the z-axis, marked by a blue circle. These limitations are effectively addressed by the dual-model approach, which combines the strengths of both networks to improve segmentation quality across spatial and volumetric dimensions.

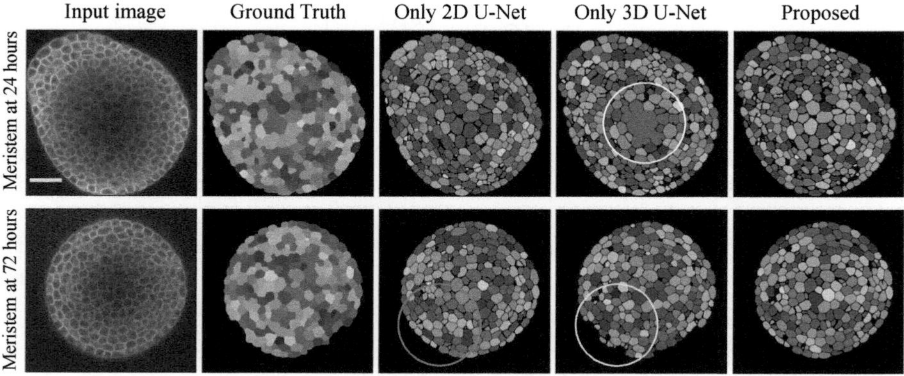

Fig. 2. Qualitative evaluation results at 24 and 72 h. Each row corresponds to a different time point, and each column shows: (1) the raw confocal image, (2) the annotated ground truth, (3) segmentation by the 2D U-Net, (4) segmentation by the 3D U-Net, and (5) the proposed dual-model approach. The white circle marks blurred central cells that are under-segmented by the 3D U-Net. The blue circle indicates peripheral cells incorrectly merged across z-slices by the 2D U-Net; and the yellow circle indicates cells entirely missed by the 3D U-Net. The proposed dual-model method resolves all of these issues. Scale bar: 20 μm.

Using Eq. 1, our pipeline achieves an average Dice score of 0.886 ± 0.002, which is slightly lower than the single-model baselines (2D U-Net: 0.915 ± 0.001; 3D U-Net: 0.906 ± 0.002, see Table 3). This difference stems from our decision to exclude membrane voxels from final cell labels, whereas the published ground truth includes these regions as part of each cell. As a result, voxel-level overlap with the ground truth is reduced, lowering the Dice score—even though our boundaries are cleaner and biologically more faithful. This effect is confirmed in Table 4, where we compute Dice using the full membrane regions: all methods—including ours—reach nearly identical scores (our pipeline: 0.985 ± 0.001; 2D U-Net: 0.986 ± 0.001; 3D U-Net: 0.987 ± 0.001), underscoring that the discrepancy is due to mask interpretation, not segmentation quality.

3.2   Quantitative Evaluation

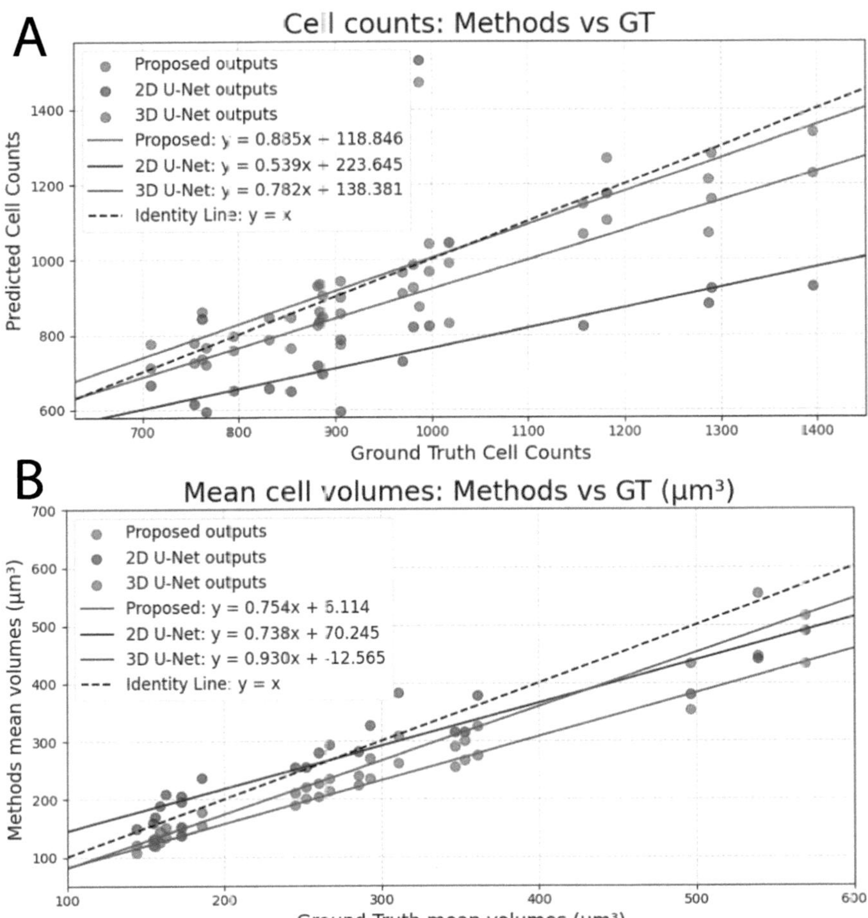

Fig. 3. Passing–Bablok regressions over 22 volumes. (A) Cell counts: dual $y = 0.885x + 118.846$, 2D U-Net $y = 0.539x + 223.645$, 3D U-Net $y = 0.782x + 138.381$. (B) Mean cell volumes: dual $y = 0.754x + 6.114$, 2D U-Net $y = 0.738x + 70.245$, 3D U-Net $y = 0.930x - 12.565$. The dashed diagonal indicates the identity line $y = x$.

Figure 3A shows that the predicted cell counts from our dual-model approach closely follow the ideal $y = x$ line. The regression yields a near-unity slope ($y = 0.885x + 118.846$), minimal intercept, and a p-value of 0.573, indicating that the deviation from perfect agreement is not statistically significant. This supports the high detection accuracy of our model across all 22 timepoints. In contrast, both single-model baselines exhibit consistent under-segmentation. For example, at 24 h, the ground truth cell count was 1396, whereas the 2D U-Net predicted 926 ($p = 0.022$, statistically significant) and the 3D U-Net predicted 1227 ($p = 0.153$), compared to 1338 predicted by our dual-model. This trend holds across the dataset, as seen in Table 5, where the dual-model approach more accurately tracks biological variability, including early growth pulses at 20 h and transient surges at 72 h and 76 h.

Figure 3B further shows that the dual-model slightly underestimates mean cell volume (slope = 0.754), though the p-value of 0.107 again suggests this deviation is not statistically significant. The 2D U-Net and 3D U-Net, by contrast, yield p-values of 0.685 and 0.619 with slopes of 0.738 and 0.930, respectively. At the 24-hour mark, for instance, the measured ground truth volume was 163.0 μm^3, while the dual-model predicted 133.9 μm^3. In contrast, the 2D U-Net overestimated the volume at 209.0 μm^3 , and the 3D U-Net predicted 151.4 μm^3 , primarily due to merging artifacts.

3.3   Cross-Plant Consistency Check

Although all detailed quantitative analyses in this manuscript focus on one specimen (Plant 13), we verified that the pipeline yields consistent results across the other plants available in the Willis *et al.* dataset [10]. For each plant, we compared predicted and reference total cell counts per timepoint using Passing–Bablok regression. Table 1 reports the estimated slopes (ideal value = 1.0), which quantify agreement between predicted and reference counts. Slopes close to unity indicate that the pipeline recovers total cell counts reliably across plants, supporting that the results obtained from Plant 13 are representative.

Table 1. Passing–Bablok regression equations ($y = slope \cdot x + intercept$) comparing predicted and reference total cell counts for different segmentation models.

Plant	Proposed	2D U-Net	3D U-Net
Plant 1	$1.108x - 32.510$	$0.857x - 68.048$	$0.981x + 30.274$
Plant 2	$0.920x + 48.959$	$0.598x + 96.492$	$0.831x + 77.353$
Plant 4	$0.837x + 124.970$	$0.524x + 129.718$	$0.776x + 106.104$
Plant 13	$0.885x + 118.846$	$0.539x + 223.645$	$0.782x + 138.381$
Plant 15	$1.024x + 13.175$	$0.783x + 21.761$	$0.967x - 23.687$
Plant 18	$0.942x + 51.180$	$0.487x + 209.377$	$0.797x + 139.594$

Across all available plants, our 'Proposed' model generally shows slopes closer to 1 and smaller intercepts. indicating that our pipeline reproduces reference cell counts consistently. This broader analysis suggests that, although the main results were illustrated on a single complete replicate (Plant 13), the conclusions generalise across specimens and are more representative than would be inferred from one plant alone.

4   Discussion

Our dual-model fusion leverages the in-plane sharpness of a 2D U-Net and the volumetric continuity of a 3D U-Net for highly precise and robust segmentation.

Fusion limitations and simple remedies. While the voxel-wise OR increases sensitivity, it can propagate noise. Mitigations like probability-weighted fusion, small-object removal, or adaptive thresholding can reduce false islands with minor extra cost.

4.1   Biological Interpretation

Our segmentation and the ground truth reveal transient surges in cell counts and volume at 20, 56, and 72 h (Table 5). The early pulse at 20 h replicates the growth burst reported by Willis *et al.* [1], suggesting a coupling between cell proliferation and expansion during developmental pulses in the SAM. This pulsatile dynamic is comparable to the root's "Root Clock" [14]. The synchronised rise in cell number and volume implies a coordinated biological program, reinforcing the interpretation of these events as bona fide growth pulses.

4.2   Pipeline Results Interpretation

Across all 22 timepoints, our dual-model consistently outperforms individual U-Net baselines, yielding more accurate cell counts, fewer merging artefacts, and more stable morphometrics.

This improvement comes with a slight underestimation of mean cell volumes. This is a deliberate trade-off from treating membrane voxels as exclusion zones, which are not assigned to any cell label [6]. This produces cleaner, more accurate boundaries. In contrast, the 2D and 3D U-Nets include membrane pixels, leading to higher Dice scores but at the cost of precision and frequent merges. Our workflow accepts a modest Dice reduction for more anatomically faithful segmentation, which is valuable for downstream tasks like lineage tracking.

A challenge was observed at 84 h, where high cell density and diminished contrast caused the 2D U-Net to fragment faint boundaries into spurious islands, inflating final counts. Since the pipeline uses a logical OR, the issue propagates. Adaptive thresholding could suppress such artefacts in future versions.

Generalisability. The 2D/3D fusion concept is applicable to other plant tissues and 3D imaging modalities after modest tuning. We expect similar benefits but recommend per-modality validation.

Inference is efficient, at under 10 s per volume on a GPU (Nvidia GeForce RTX 5060 Ti), making our pipeline suitable for large-scale studies. Future work will integrate an adaptive post-processing stage and extend the workflow to other tissues.

5 Conclusion and Future Work

We introduce a robust pipeline that segments apical stem cells in *Arabidopsis* SAM by fusing pre-trained 2D and 3D U-Net models. The fusion is generalisable to other tissues after modest tuning. By omitting membrane regions, our method avoids ambiguous boundaries and delivers precise morphometric data. Evaluations show superior cell-detection accuracy over single-model baselines, capturing overlooked cells and reliably tracking volume changes. Although this membrane-centric approach slightly underestimates cell volumes, it significantly reduces merging artefacts.

The pipeline achieves fast inference times (<10 s/volume), making it suitable for large-scale studies. Future work includes: (i) integrating the pipeline into Fiji via deepImageJ [15]; (ii) bootstrap/leave-one-out robustness checks; and (iii) modality-specific tuning for broader adoption [11].

Acknowledgements. This work was partially supported under grant PID2023-152631OB-I00 by the Ministerio de Ciencia, Innovación y Universidades, Agencia Estatal de Investigación (MCIN/AEI/10.13039/501100011033/), co-financed by European Regional Development Fund (ERDF), 'A way of making Europe'.

A Supplementary Metrics

Table 2. Ablation sweep: binarization threshold versus mean absolute error $\pm$ standard error of cell counts across 22 volumes.

Threshold	Mean Absolute Error (MAE)
0.05	84.545 $\pm$ 18.494
0.10	47.227 $\pm$ 9.453
0.15	35.636 $\pm$ 5.709
0.20	32.091 $\pm$ 8.064
0.25	34.455 $\pm$ 8.873
0.30	42.682 $\pm$ 8.896
0.35	57.818 $\pm$ 10.798
0.40	82.909 $\pm$ 14.800
0.45	112.318 $\pm$ 16.823
0.50	146.818 $\pm$ 18.055
0.55	182.364 $\pm$ 20.149
0.60	222.864 $\pm$ 21.203
0.65	261.455 $\pm$ 20.724
0.70	305.273 $\pm$ 22.102
0.75	343.318 $\pm$ 23.115
0.80	383.364 $\pm$ 25.313
0.85	402.591 $\pm$ 27.951
0.90	420.727 $\pm$ 28.936
0.95	385.682 $\pm$ 29.378

Table 3. Dice Similarity Coefficient at each time point for the three segmentation methods. Values represent segmentation overlap with ground truth for detected cells.

Time (hrs)	Proposed	2D U-Net	3D U-Net
0	0.876	0.913	0.890
4	0.887	0.915	0.904
8	0.882	0.916	0.898
12	0.877	0.916	0.894
16	0.873	0.913	0.891
20	0.879	0.917	0.897
24	0.881	0.918	0.898
28	0.882	0.917	0.898
32	0.892	0.919	0.910
36	0.888	0.913	0.906
40	0.894	0.919	0.911
44	0.888	0.917	0.905
48	0.892	0.920	0.907
52	0.894	0.918	0.913
56	0.889	0.909	0.911
60	0.891	0.913	0.914
64	0.890	0.914	0.912
68	0.885	0.914	0.904
72	0.896	0.918	0.916
76	0.885	0.913	0.905
80	0.878	0.912	0.899
84	0.888	0.907	0.918
Mean ± SE	0.886 ± 0.002	0.915 ± 0.001	0.906 ± 0.002

Table 4. Dice Similarity Coefficient at each time point for the three segmentation methods. Values represent segmentation overlap with ground truth for detected cells and membranes.

Time (hrs)	Proposed	2D U-Net	3D U-Net
0	0.985	0.986	0.986
4	0.988	0.987	0.990
8	0.986	0.987	0.988
12	0.985	0.986	0.987
16	0.984	0.984	0.986
20	0.988	0.988	0.989
24	0.988	0.989	0.989
28	0.988	0.988	0.989
32	0.981	0.981	0.982
36	0.988	0.988	0.989
40	0.980	0.981	0.982
44	0.988	0.988	0.990
48	0.988	0.988	0.989
52	0.981	0.982	0.982
56	0.988	0.987	0.989
60	0.982	0.982	0.984
64	0.981	0.981	0.983
68	0.988	0.988	0.990
72	0.984	0.984	0.986
76	0.989	0.989	0.990
80	0.988	0.988	0.989
84	0.985	0.984	0.987
Mean ± SE	0.985 ± 0.001	0.986 ± 0.001	0.987 ± 0.001

Table 5. Total number of cells (cell count) measured experimentally (Ground Truth) versus the predictions of the three methods, at each time point (hours).

Time (hrs)	Ground Truth	Proposed	2D U-Net	3D U-Net
0	795	797	650	757
4	981	985	820	924
8	970	966	727	910
12	887	905	694	848
16	754	779	616	726
20	1290	1282	921	1158
24	1396	1338	926	1227
28	1157	1147	823	1068
32	997	1041	821	968
36	884	935	834	861
40	905	942	786	899
44	831	844	655	786
48	766	767	594	719
52	708	777	665	712
56	1018	990	1044	830
60	762	861	843	736
64	882	930	717	825
68	854	845	647	763
72	1182	1269	1175	1103
76	1287	1212	880	1070
80	905	856	594	774
84	987	1470	1529	873

Table 6. Mean cell volume $\pm$ standard error (μm^3), comparing Ground Truth and the three segmentation pipelines, for each time point (hours).

Time (hrs)	Ground Truth	Proposed	2D U-Net	3D U-Net
0	154.6 ± 2.4	120.1 ± 2.5	159.6 ± 8.3	130.2 ± 3.2
4	252.0 ± 3.9	200.1 ± 4.0	255.1 ± 14.0	220.7 ± 5.2
8	172.9 ± 2.4	137.0 ± 2.8	195.8 ± 11.1	150.5 ± 3.5
12	155.8 ± 2.5	119.3 ± 2.5	169.0 ± 9.1	131.8 ± 3.2
16	144.1 ± 2.8	108.2 ± 2.4	149.3 ± 8.7	120.5 ± 3.2
20	159.1 ± 2.4	125.6 ± 2.6	189.4 ± 13.2	144.3 ± 7.9
24	163.0 ± 2.5	133.9 ± 2.7	209.0 ± 11.9	151.4 ± 8.3
28	172.4 ± 2.4	137.3 ± 2.7	206.1 ± 11.8	152.5 ± 4.6
32	245.1 ± 4.1	189.4 ± 3.6	255.3 ± 12.8	210.8 ± 6.1
36	353.0 ± 5.0	267.4 ± 6.0	315.3 ± 18.0	300.4 ± 7.2
40	285.2 ± 4.7	223.2 ± 4.7	282.0 ± 16.2	240.6 ± 5.5
44	260.0 ± 3.8	204.7 ± 4.3	280.3 ± 16.6	227.5 ± 5.6
48	266.9 ± 4.5	214.5 ± 4.9	294.3 ± 21.7	236.3 ± 6.1
52	346.3 ± 8.2	256.1 ± 6.5	315.7 ± 22.4	290.7 ± 9.0
56	539.2 ± 9.2	445.9 ± 12.3	441.2 ± 27.2	379.7 ± 28.0
60	496.3 ± 8.3	354.2 ± 9.4	379.7 ± 28.0	433.8 ± 12.1
64	360.9 ± 5.2	275.6 ± 6.6	377.6 ± 30.5	324.7 ± 8.9
68	293.0 ± 4.3	235.5 ± 5.6	326.8 ± 32.3	271.1 ± 7.4
72	569.6 ± 6.5	432.4 ± 9.9	489.5 ± 32.1	516.2 ± 13.8
76	310.5 ± 3.7	262.4 ± 5.5	383.0 ± 27.2	309.3 ± 7.8
80	185.7 ± 2.7	153.5 ± 3.4	238.0 ± 15.1	177.7 ± 4.7
84	963.5 ± 14.2	517.4 ± 19.9	517.5 ± 36.8	924.6 ± 48.2

References

1. Lisa Willis, et al.:Cell size and growth regulation in the arabidopsis thaliana apical stem cell niche. Proceedings of the National Academy of Sciences, 113(51):E8238–E8246 (2016). https://doi.org/10.1073/pnas.1616768113
2. Caicedo, J.C., et al.: Nucleus segmentation across imaging experiments: the 2018 data science bowl. Nat. Methods **16**(12), 1247–1253 (2019). https://doi.org/10.1038/s41592-019-0612-7
3. Çiçek, Ö., Abdulkadir, A., Lienkamp, S.S., Brox, T., Ronneberger, O.: 3D U-Net: Learning Dense Volumetric Segmentation from Sparse Annotation. In: Ourselin, S., Joskowicz, L., Sabuncu, M.R., Unal, G., Wells, W. (eds.) MICCAI 2016. LNCS, vol. 9901, pp. 424–432. Springer, Cham (2016). https://doi.org/10.1007/978-3-319-46723-8_49
4. Stringer, C., Wang, T., Michaelos, M., Pachitariu, M.: Cellpose: a generalist algorithm for cellular segmentation. Nature Methods, 18(1):100–106, (2021). https://doi.org/10.1038/s41592-020-01018-x
5. Schmidt, U., Weigert, M., Broaddus, C., Myers, G.: Cell Detection with Star-Convex Polygons. In: Frangi, A.F., Schnabel, J.A., Davatzikos, C., Alberola-López, C., Fichtinger, G. (eds.) MICCAI 2018. LNCS, vol. 11071, pp. 265–273. Springer, Cham (2018). https://doi.org/10.1007/978-3-030-00934-2_30
6. Israel, U., et al.: A foundation model for cell segmentation. bioRxiv (2023). https://doi.org/10.1101/2023.11.17.567630
7. Kirillov, A., et al.: Segment anything. In: 2023 IEEE/CVF International Conference on Computer Vision (ICCV), pp. 3992–4003 (2023). https://doi.org/10.1109/ICCV51070.2023.00371
8. Wolny, A., Cerrone, L.: 2D UNet for Arabidopsis Apical Stem Cells (2022)
9. Wolny, A., Cerrone, L.: 3D UNet Arabidopsis Apical Stem Cells (2022)
10. Willis, L., et al.: Arabidopsis thaliana SAM confocal images and annotations. University of Cambridge Repository (2016). https://www.repository.cam.ac.uk/items/f7cdcf20-e8ca-4cf5-b7ab-90350a8d00b2
11. Fuster-Barceló, C., et al.: Bridging the gap: integrating cutting-edge techniques into biological imaging with deepImageJ. In: Biological Imaging 4 , e14 (2024). https://doi.org/10.1017/S2633903X24000114
12. Ouyang, W., et al.: Bioimage model zoo: a community-driven resource for accessible deep learning in bioimage analysis. bioRxiv (2022). https://doi.org/10.1101/2022.06.07.495102
13. Derrick, B., White, P.: Why welch's test is type i error robust. The Quantitative Methods for Psychology, 12:30–38 (2016). https://doi.org/10.20982/tqmp.12.1.p030
14. Perez-Garcia, P., Serrano-Ron, L., Moreno-Risueno, M. A.: The nature of the root clock at single cell resolution: Principles of communication and similarities with plant and animal pulsatile and circadian mechanisms. Current Opinion in Cell Biology, 77:102102 (2022). https://doi.org/10.1016/j.ceb.2022.102102
15. Gómez-de-Mariscal, E., et al.: DeepImageJ: A user-friendly environment to run deep learning models in imagej. Nature Methods, 20:1192–1195 (2023). https://doi.org/10.1038/s41592-023-01934-z

Prediction of Protein Structures and Molecular Interactions with AI Techniques

Molecular Machine Learning Using Euler Characteristic Transforms

Victor Toscano-Duran[1]([✉]) [ID], Florian Rottach[2,3] [ID], and Bastian Rieck[4] [ID]

[1] Department of Applied Mathematics I, University of Seville, Seville, Spain
vtoscano@us.es
[2] Central Data Science, Boehringer Ingelheim GmbH, Biberach/Riss, Germany
florian.rottach@boehringer-ingelheim.com
[3] School of Medicine, University of Tübingen, Tübingen, Germany
[4] AIDOS Lab, University of Fribourg, Fribourg, Switzerland
bastian.grossenbacher@unifr.ch

Abstract. The shape of a molecule determines its physicochemical and biological properties. However, it is often underrepresented in standard molecular representation learning approaches. Here, we propose using the *Euler Characteristic Transform* (ECT) as a geometrical-topological descriptor. Computed directly from molecular graphs constructed using handcrafted atomic features, the ECT enables the extraction of *multiscale* structural features, offering a novel way to encode molecular shape in the feature space. We assess the predictive performance of this representation across nine benchmark regression datasets, all centered around predicting the inhibition constant K_i. In addition, we compare our proposed ECT-based descriptor against traditional molecular representations and methods, such as molecular fingerprints/descriptors and graph neural networks (GNNs). Our results show that our ECT-based representation achieves *competitive performance*, ranking among the best-performing methods on several datasets. More importantly, combining our descriptor with established representations, particularly with the AVALON fingerprint, significantly *enhances predictive performance*, outperforming other methods on most datasets. These findings highlight the complementary value of *multiscale* topological information and its potential for being combined with established techniques. Our study suggests that hybrid approaches incorporating explicit shape information can lead to more informative and robust molecular representations, enhancing and opening new avenues in molecular machine learning. To support reproducibility and foster open biomedical research, we provide open access to all experiments and code used in this work.

Keywords: Euler Characteristic Transform · Topological Data Analysis · Molecular Machine Learning · Molecular Representations

1 Introduction

Understanding and predicting the properties of molecules is at the core of modern medicine and healthcare. Almost all pharmaceutical agents, like small-molecule

© The Author(s), under exclusive license to Springer Nature Switzerland AG 2026
A. López Fernández et al. (Eds.): CIABiomed 2025, LNBI 16148, pp. 391–405, 2026.
https://doi.org/10.1007/978-3-032-10631-2_30

drugs or diagnostic agents, are molecular in nature, and their therapeutic efficacy is critically dependent on their physicochemical and biological properties. From drug discovery to personalized medicine, the ability to predict how a molecule will behave in the body (its solubility, toxicity, bioavailability, and binding affinity), as well as predicting molecule properties, is essential for developing safe and effective treatments. Given the high costs and long timelines associated with traditional drug discovery, predictive modeling has emerged as a vital tool for improving the efficiency of the early stages of drug development, enabling faster screening of potential candidates and reducing the need for extensive experimental testing [4,9,31]. However, despite significant advancements, *predicting* molecular properties remains a challenge due to the complexity of molecular behavior. Traditional molecular representations, such as fingerprints or descriptors, often fail to fully capture the rich structure and shape information that influence a molecule's function in a biological context [8,32]. This gap is particularly pronounced when trying to predict properties like the inhibition constant (K_i), a key parameter in drug–target interactions, where even small changes in molecular shape can lead to large differences in binding affinity [10,29]. Recent developments in molecular machine learning (MML) have demonstrated the potential of deep learning techniques in general and graph-based models in particular to improve predictive performance [2,15]. However, many of these models still fail to fully capture the topological and geometric aspects of molecular shape, proving detrimental to the learning outcome. For example, methods like Graph Neural Networks (GNNs) [13] may capture *some* of the structural information, but they often overlook the importance of representing the shape in a comprehensive and multiscale way [17].

Algebraic topology [14,22] and topological data analysis [12,20] can provide powerful tools to analyze the shape of molecular data, as shown by recent studies [26]. The *Euler Characteristic Transform* (ECT) [21,24], for example, could be a topological representation of molecules that captures the molecular shape by tracking changes in the Euler characteristic [19] across different scales and directions of the feature space. By incorporating topological shape information, the ECT allows us to represent and encode the molecular structure in a multiscale and novel way, capturing essential shape details that traditional representations may miss. Recent studies have shown the promise of the ECT in other fields, such as biology and material science, where shape plays a crucial role in determining functional properties [1,25]. In this work, we compare and explore the use of the ECT-based molecular representation, which is computed directly over molecular graphs derived from handcrafted atomic features, to predict K_i, a key molecular property, as well as perform a comparison between our approach and traditional methods over a series of nine binding affinity datasets. By using this topological representation, as well as combining it with traditional molecular representations, we aim to enhance predictive performance and provide new insights into the role of molecular shape in molecular machine learning. Our experiments shows that our ECT-based approach exhibites competitive predictive performance, in some cases even outperforming all alternative and traditional methods. In addition, our experiments show that the *combination* of our ECT-based approach with existing methods, more specifically with the AVALON fingerprint, leads to further enhance predictive performance, thus highlighting the complementary value of multiscale topological and shape information. Ultimately, our work contributes to the growing body of evidence suggesting that incorporating molecular shape at a fundamental level can lead to more robust and informative molecular representations, opening up new avenues for the design of better molecular machine learning models and more effective therapies.

The remainder of this paper is organized as follows: Firstly, Sect 2 introduces the foundational concepts relevant to our study, including molecular structures and their representations, traditional molecular descriptors, and the Euler Characteristic Transform. Section 3 details the datasets employed, the traditional representations included for comparison in the experiments, our proposed ECT-based approach. and the overall experimental setup. The results of our experiments, with a comprehensive discussion, are presented in Sect. 4. Finally, conclusions and future work are discussed in Sect. 5.

2   Background

Molecules [6] are the fundamental building blocks of chemical and biological systems. In the context of drug discovery [11], small molecules are designed or screened for their ability to modulate the activity of specific biological targets, typically proteins. The interaction between a drug and its target is governed by a complex interplay of properties, including molecular shape, electronic distribution, and physicochemical characteristics such as hydrophobicity, polarity, and charge. Accurately modeling and predicting these properties is a central challenge in cheminformatics and molecular machine learning [32].

In practice, molecules are often represented in formats that facilitate both human readability and algorithmic processing. One of the most widely used textual encodings is the SMILES (Simplified Molecular Input Line Entry System) notation, which encodes a molecule as a string describing the atoms and their connectivity through a series of characters and symbols. For example. the SMILES string `CC(O)=O` represents acetic acid. This format is compact, easily parsed, and widely supported in cheminformatics toolkits. However, SMILES do not directly convey geometric or spatial information, and small changes in the string can correspond to large structural differences. From molecular structures (e.g. SMILES, although alternative encodings exist), it is common to derive graph-based representations [13], where atoms are modeled as nodes and covalent bonds as edges, possibly enriched with additional features such as atomic types, bond orders, or aromaticity indicators. These molecular graphs serve as the foundation for numerous machine learning models, facilitating the use of graph-based algorithms and neural networks [9]. They also provide the basis for computing molecular representations such as fingerprints, descriptors, and our ECT-based representation. An example of the molecular graph of acetic acid is shown in Fig. 1a.

Traditionally, computational models for molecular machine learning tasks have relied on handcrafted molecular representations [32]. Among these are *molecular descriptors* (e.g., FGCount or 2DAP), which are numerical features derived from molecules, such as atom counts, topological indices, or electronic properties, and *molecular fingerprints* (e.g., AVALON or MACCS), which represent the presence of specific substructures or chemical motifs as binary or count vectors. Both types of features are computed directly from the molecular graph structure. These representations have been successfully applied in a variety of tasks such as quantitative structureâĂŞactivity relationship (QSAR) modeling, molecule classification, and property prediction. However, they often suffer from a lack of expressiveness and poor generalization to out-of-distribution chemical spaces, especially in the presence of subtle variations in molecular geometry [3,8,33].

In recent years *Graph Neural Networks* (GNNs) [13], have become a dominant paradigm in molecular property prediction, extending neural networks to graph-based data. The core mechanism of most GNNs is message passing, which iteratively propagates information across the graph elements. While GNNs can capture relational and structural information more flexibly than fixed descriptors, they often lack explicit access to multiscale shape information. Moreover, they may struggle to distinguish between molecules that are topologically or geometrically distinct but share similar local connectivity [15]. Methods from *Topological Data Analysis* (TDA) [12] provide a complementary perspective. Instead of relying on raw atomic positions, handcrafted atomic features or purely local graph structures, TDA extracts global shape features from data, capturing relevant geometric information at multiple scales [20], making it thus particularly suitable for applications in the life sciences [30]. A particularly relevant tool within this framework is the Euler Characteristic Transform (ECT) [21,24,25,28], a method that combines ideas from algebraic topology with geometric data analysis. The ECT operates by *filtering* a shape, such as a molecular graph, across multiple scales and directions in the feature space, and computing the *Euler characteristic* [19], denoted by χ. This geometrical-topological quantity encodes information about the number of connected components, holes, and voids at each step (scale) of the filtration. Moreover, it is an *invariant*, i.e., it will remain unchanged under any smooth transformation applied to a shape. Put briefly, the Euler characteristic can be seen as a summary statistic of the shape of a graph or simplicial complex. To fully understand this tool and how molecules are represented as graphs, we next provide a more detailed introduction about graphs and simplicial complexes, serving as the starting point for computing the Euler characteristic and the ECT.

Graphs. Graphs enable the modeling of real-world systems by focusing on *dyadic relationships* between elements. Formally, a *graph* $G = (V, E)$ consists of a finite set of vertices $V = \{v_1, v_2, \ldots, v_n\}$, which represent entities (e.g. atoms in a molecule), and a set of edges $E \subseteq \{\{u, v\} \mid u, v \in V \text{ and } u \neq v\}$, which represent pairwise relationships (e.g., chemical bonds). In cheminformatics, graphs are the natural way to represent molecules, nodes corresponds to atoms, typically encoded as a feature vector , and edges to chemical bonds. However, to extract richer geometric and topological information, we can generalize graphs into structures called *simplicial complexes*. The graph of the acetid acid molecule shown in Fig. 1a (visualized in $2D$) contains 8 vertices, which corresponds to 2 carbon atoms, 2 oxygen atoms, and 4 hydrogen atoms and 7 edges, with $\{H, O\}$ being an example of an edge between two vertices (representing a chemical bond).

Simplicial complexes. An (abstract) simplicial complex K is a data structure for representing topological spaces, which generalizes a graph by permitting more than mere dyadic relations. It is defined as a family of sets (*simplices*) that is closed under taking subsets, meaning that if a set (like a triangle) is part of the complex, then so are all its faces (edges and vertices). More formally, a simplicial complex is obtained by a nested family of simplices, which are the elementary building blocks, for example: a 0-simplex can be thought of as a point (vertex), a 1-simplex as an edge, a 2-simplex as a filled triangle, and a 3-simplex as a filled tetrahedron. Each k-simplex has $k + 1$ faces obtained by removing one of the vertices. For example, the graph of Fig. 1a can also be seen as a 1-dimensional simplicial complex with 8 0-simplices and 7 1-simplices.

The *Euler characteristic* is a key topological invariant of a simplicial complex K, being defined as:

$$\chi(K) = \sum_{k=0}^{n} (-1)^k |K^{(k)}|, \tag{1}$$

where $|K^{(k)}|$ denotes the number (cardinality) of k-simplices in the simplicial complex K. Hence, the Euler characteristic is an alternating sum of the number of simplices (elements) in each dimension. In the case of (molecular) graphs, which only consist of 0-simplices (vertices) and 1-simplices (edges), this is reduced to:

$$\chi = |V| - |E|. \tag{2}$$

For example, the Euler characteristic of the acetic acid graph presented in Fig. 1a is 1 (8 vertices - 7 edges).

Taken on its own, the Euler characteristic lacks sufficient complexity to fully describe a shape, but if we think of computing it at different *scales* (thresholds), considering we have a dynamic object, which grows in the number of their components (vertices, edges, etc.) across time, we may observe significant changes on it. This leads to the concept of the *Euler Characteristic Curve* (ECC), which tracks how the topological complexity of a shape evolves over different scales. To formalize this dynamic view and to compute the Euler characteristic at different scales, we use the notion of a filtration, leading to a *filtered simplicial complex*. More formally, a *filtered simplicial complex* is a collection of subcomplexes $\{K(t) \mid t \in \mathbf{R}\}$ of a simplicial complex K such that $K(t) \subseteq K(s)$ for $t < s$ and there exists $t_{\max} \in \mathbf{R}$ such that $K(t_{\max}) = K$. The *filtration time* (or filtration value) of a simplex $\sigma \in K$ is the smallest t such that $\sigma \in K(t)$. To illustrate the concept of a filtration, consider the graph representation of the acetic acid molecule shown in Fig. 1b. Each node (atom) is assigned a scalar value, which in this case corresponds to its projection onto a fixed direction, which corresponds to the x-axis direction for that example. These scalar values determine the filtration times of the nodes: a node enters the filtration at the time equal to its value. Formally, this means we construct a sequence of sub-graphs (subcomplexes), $\{K(t) \mid t \in \mathbf{R}\}$ such that each $K(t)$ contains all nodes and edges whose filtration time is less than or equal to t. Since an edge can only appear after both its incident nodes have appeared, the filtration is nested, we have $K(t) \subseteq K(s)$ for $t < s$. An illustrative example is shown in Fig. 1c, where the molecular graph of Fig. 1a evolves as the filtration parameter increases along the x-axis direction. At each step, new atoms (nodes) and bonds (edges) are incorporated according to their associated filtration values, as shown in Fig. 1b, progressively revealing the full molecular topology.

Euler Characteristic Transforms. While the ECC already provides a summary of how the topological complexity of a shape evolves with respect to a *single* filtration parameter, it is often insufficient to fully characterize a high-dimensional or intricate structure like a molecule. This is where the *Euler Characteristic Transform* (ECT) comes into play, making it possible to characterize shapes based on multiple filtrations, parameterized using a *direction vector*. Specifically, each direction provides a different view of the data, capturing how the topological features of the structure unfold when the complex is filtered using a different criterion. To better understand this, imagine a 2D or 3D object (like a molecule embedded in space). If we project the molecule along a certain direction, for example along the x-axis, we can define a filtration by sweeping a hyperplane orthogonally to that axis and including simplices as their associated values fall below a certain threshold. This process gives us an ECC in the x-direction. Now, if we repeat this process in another direction, for example along the y-axis, we will generally obtain a different ECC, since the structure of the molecule may appear differently from that angle. The key idea is that *each* direction typically provides complementary topological information. Thus, the ECT is formally defined as the collection of Euler characteristic curves obtained by filtering a shape along a family of directions, which for molecule graphs are sampled randomly in the high-dimensional node feature space. For example, for a given molecule and its graph, the ECT produces a collection of curves (ECCs), each corresponding to a different direction in the node feature space. By aggregating all these curves into a single descriptor, the ECT captures how the topology of the molecule evolves as a function of spatial thresholds, providing a rich and compact descriptor of molecular shape. Refer to Fig. 1d for the ECC of the molecule graph in Fig. 1a along the x-axis direction, and to Fig. 1e for an ECT example of this graph, built by stacking multiple ECCs, visualized as a matrix, where each column corresponds to a direction, each row to a threshold level in the filtration process, and the color intensity encodes the Euler characteristic value. Note that the numbers shown on the x-axis and y-axis of these figures do not represent the actual threshold and directions values used during filtration, but rather the index of the threshold and direction step. The resulting feature set derived from

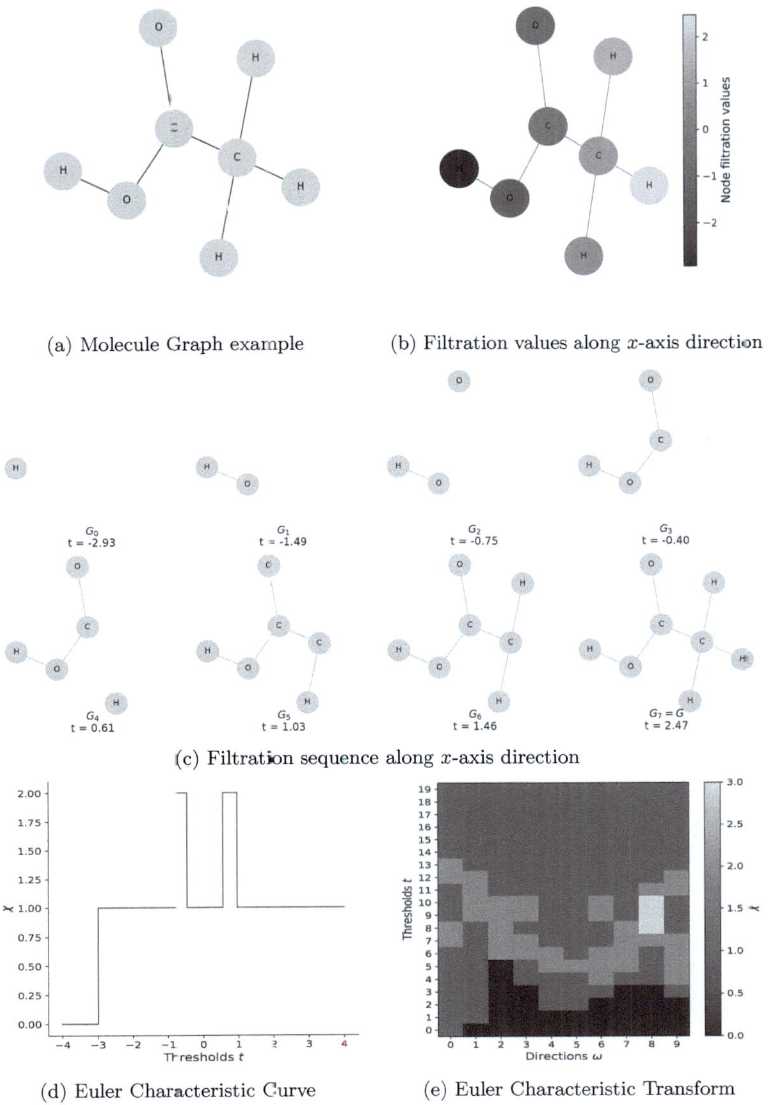

(a) Molecule Graph example

(b) Filtration values along x-axis direction

(c) Filtration sequence along x-axis direction

(d) Euler Characteristic Curve

(e) Euler Characteristic Transform

Fig. 1. Extracting topological shape information from molecular graphs. (a): Exemplary 2D graph representation of the acetic acid molecule, derived from its SMILES string `CC(O)=O`. The graph contains 8 vertices (2 carbon, 2 oxygen, and 4 hydrogen atoms) and 7 edges, resulting in an Euler characteristic of 1 ($\chi = 1$). (b): Filtration values of the nodes along the x-axis direction. (c) Filtration sequence along the x-axis direction. (d): ECC computed for the x-axis direction. In the figure, the x-axis represents the filtration thresholds values, and the y-axis shows the corresponding Euler characteristic values. (e): ECT example, built by stacking the ECC along 10 random directions from the molecule graph of (a). Columns correspond to directions, rows to thresholds (set to 20), and color intensity encodes the Euler characteristic value.

the ECT encodes both multiscale and directional information, resulting in a highly expressive representation that can be used in machine learning models as input features for both classification and regression tasks, and that can complement traditional descriptors and graph-based methods. Hence, the ECT has been successfully applied for various machine-learning tasks [1,25]. While the Euler characteristic itself has previously been used in molecular dynamics [18,27], to our knowledge, our work is the *first* to explore the use of the ECT in the context of molecular machine learning.

3   Materials and Methods

Having described a novel and multiscale topology-based approach to represent molecules, we conduct an extensive suite of experiments to explore and assess its effectiveness against established methods such as molecular fingerprints, descriptors, and graph neural networks.

3.1   Datasets

We concentrate on a number of datasets from different biochemical domains. Specifically, we use a collection of nine protein-ligand binding affinity datasets, all focused on predicting the inhibition constant K_i, a key measure of how strongly a molecule binds to a given protein target. These datasets are sourced from BindingDB[1] and cover a variety of biologically relevant proteins. The dataset associated with the alpha1A adrenergic receptor (ADRA1A) contains 1959 molecules; the one for the arachidonate 5-lipoxygenase-activating protein (ALOX5AP) includes 1636 molecules; and the dataset for the ataxia telangiectasia and Rad3-related protein (ATR) consists of 1411 molecules. For human dipeptidyl peptidase-4 (DPP4), the dataset comprises 3345 molecules, while the Janus kinase proteins JAK1 and JAK2 include 3058 and 1449 molecules, respectively. The kappa-type opioid receptor (KOR) dataset contains 1685 molecules, and the two datasets corresponding to subtypes of muscarinic receptors (MUSC1 and MUSC2) consist of 2129 and 1109 molecules, respectively. These molecule counts correspond to the pre-processed data, from which molecules with missing or invalid K_i binding affinity values were excluded, and only unique SMILES representations were retained. Despite targeting different proteins, all datasets share a unified prediction task: estimating the binding affinity constant K_i. Experiments were conducted both individually for each dataset and on a combined dataset comprising 16558 unique molecules.

3.2   Methods

We organize our description of the methods according to the type of molecular representation used, as this choice determines the applicable machine learning

---

[1] https://www.bindingdb.org/.

Table 1. List of molecular representations, grouped by their respective categories and sorted alphabetically within each group. For each representation, the category and corresponding feature vector dimensionality are provided.

Representation name	Representation Category	Dimension
Molecule graph	Graph	-
AVALON	Fingerprint	1024
CATS2D	Fingerprint	189
ECFP4	Fingerprint	1024
EState	Fingerprint	79
KR	Fingerprint	4860
MACCS	Fingerprint	166
MAP4	Fingerprint	1024
Pharm2D	Fingerprint	1024
PubChem	Fingerprint	881
RDKit	Fingerprint	1024
2DAP	Descriptor	1596
ConstIdx	Descriptor	50
FGCount	Descriptor	153
MolProp	Descriptor	14
RingDesc	Descriptor	35
TOPO	Descriptor	74
WalkPath	Descriptor	46
ECT (ours)	Multiscale Topological	2528
ECT + AVALON Fingerprint (ours)	Multiscale Topological	3552

approaches. Table 1 summarizes the representations, including their category and feature vector dimensionality. Note that dimensionality is not defined for graph-based representations. In total, we have 4 different categories of representations: molecular graphs, fingerprints, descriptors, and our topological representation based on the ECT. The latter includes two variants: using the ECT alone, or combined with the AVALON molecular fingerprint, which consists of concatenating the ECT with the AVALON fingerprint. AVALON was chosen for its widespread use and solid performance. In future work, we aim to study whether there are specific combinations of representations that can perform even better; please refer to our discussion in (Sect. 5).

When using the molecular graph representation, we use graph neural networks as the underlying machine learning models, since they are specialized for this type of data input. Concretely, we used two standard GNN models, a graph attention network (GAT) and a graph convolutional network (GCN), as well as an specialized graph neural network for molecular machine learning, named "AttentiveFP" [33]. For all other representations, we use an XGBoost [5] model, which has consistently shown strong performance in variety of tasks.

3.3   Experimental Setup

Molecule graphs are extracted from the SMILES strings provided in the datasets using the `PyTorch Geometric`[2] Python package, which encoded nodes (atoms) as a 9-dimensional handcrafted feature vector. Specifically, each vector includes the following categorical atomic features: (1) atomic number, (2) chirality tag, (3) total degree, (4) formal charge, (5) total number of bonded hydrogens, (6) number of radical electrons, (7) hybridization type, (8) aromaticity, and (9) ring membership. The ECT is then computed over this multidimensional feature space. Based on a prior sensitivity analysis, we fixed the number of directions θ and filtration thresholds r to compute the ECT to 158 and 16, respectively. To validate this choice, we evaluated our ECT-descriptor on the ADRA1A dataset across a grid of values ($\theta, r \in \{10, 20, 40, 60, 80, 100, 140, 180, 200\}$), training XGBoost regressors on each configuration using 10-fold cross-validation. Predictive performance was measured using the Root Mean Squared Error (RMSE). This analysis revealed that the number of directions has a substantially *greater* impact on predictive performance than the number of filtration thresholds. For instance, increasing the number of directions from 20 to 40 led to noticeable improvements in model accuracy. However, performance gains plateaued at higher values, with little to no improvement observed between 180 and 200 directions. In contrast, varying the number of filtration thresholds had minimal effect on performance across a wide range of values. This is quantitatively supported by Pearson correlations: -0.74 between *theta* and RMSE, and -0.13 between r and RMSE. Therefore, we selected 158 directions to ensure sufficient expressiveness while limiting the number of thresholds to 16 to reduce computational cost and dimensionality without sacrificing predictive performance. The resulting ECT-based feature vector has a dimensionality of 2528, as summarized in Table 1. All ECTs have been computed using the `DECT` Python package [25].

To fairly evaluate the predictive performance of both traditional vector-based and ECT-based methods, we designed two experimental setups: one using XGBoost for vector representations (fingerprints, descriptors, and ECT), and another using graph neural networks (GNNs) operating directly on the molecular graphs. For vector-based representations, we employed the XGBoost regressor [5], trained with 1000 estimators, a learning rate of 0.01, and a maximum tree depth of 5. For the graph-based models, we explored three architectures: Graph Convolutional Networks (GCN), Graph Attention Networks (GAT), and the "AttentiveFP" model. The GCN was configured with two convolutional layers and 64 hidden channels. The GAT model used eight attention heads, each with eight hidden units. Both GCN and GAT architectures followed the design principles of Platonov et al. [23], including a two-layer perceptron after each neighborhood aggregation step, skip connections, and layer normalization. The "AttentiveFP" model was configured with 64 hidden units, a single output channel, four message-passing layers, and two attention-based update steps. Dropout was set to 0.2. All GNN models were trained for 100 epochs using the ADAM

---

[2] https://pytorch-geometric.readthedocs.io/.

optimizer with a learning rate of $10^{-2.5}$ and a weight decay of 10^{-5}. A 10-fold cross-validation strategy was applied for all the methods with shuffling and a fixed random seed to ensure reproducibility. Model evaluation was conducted using two metrics: RMSE, and coefficient of determination (R2). This was done separately for each dataset and for the combined one.

4   Results

Table 2 provides an overview of all results for the different representations for each dataset, as well as for all datasets combined into a single one. Overall, we observe that already the ECT on its own provides meaningful and effective features for molecular machine learning, outperforming all alternative methods on ADRA1A. In *combination* with the AVALON molecular fingerprint, we observe generally improved baseline predictive performance in 5/9 datasets, as well as on the combined dataset. Thus, the ECT performs *consistently well*, showing its potential for creating generalizable features.

Table 2. Results (RMSE, reported as mean $\pm$ standard deviation across 10-fold cross-validation) for the collection of molecular datasets. Rows corresponds to different representation methods grouped by category (GNNs, fingerprints, descriptors, and ECT-based). Columns correspond to datasets.

Method	Dataset									
	ADRA1A	ALOX5P	ATR	DPP4	JAK1	JAK2	KOR	MUSC1	MUSC2	Combined
AttentiveFP	2.01 ± 0.05	2.01 ± 0.29	1.42 ± 0.12	0.90 ± 0.06	1.10 ± 0.12	1.60 ± 0.20	2.17 ± 0.19	2.19 ± 0.10	2.03 ± 0.13	2.43 ± 0.27
GAT	2.65 ± 0.17	2.45 ± 0.17	1.81 ± 0.09	1.18 ± 0.05	1.99 ± 0.15	2.67 ± 0.33	2.56 ± 0.17	2.84 ± 0.17	2.60 ± 0.15	2.80 ± 0.07
GCN	2.75 ± 0.15	2.51 ± 0.19	1.96 ± 0.16	1.26 ± 0.08	2.76 ± 0.17	2.95 ± 0.28	2.73 ± 0.13	2.96 ± 0.15	2.76 ± 0.13	2.88 ± 0.10
AVALON	1.81 ± 0.14	1.58 ± 0.18	1.26 ± 0.10	0.79 ± 0.04	0.99 ± 0.10	**1.30 ± 0.19**	1.83 ± 0.16	1.88 ± 0.09	1.86 ± 0.13	1.71 ± 0.03
CATS2D	1.88 ± 0.09	1.88 ± 0.14	1.36 ± 0.12	0.90 ± 0.04	1.04 ± 0.08	1.50 ± 0.21	1.85 ± 0.12	1.98 ± 0.09	1.86 ± 0.1	1.88 ± 0.04
ECFP4	1.81 ± 0.11	1.71 ± 0.17	1.30 ± 0.09	0.84 ± 0.04	1.01 ± 0.12	1.31 ± 0.18	1.90 ± 0.14	1.98 ± 0.06	1.83 ± 0.13	1.81 ± 0.03
EState	2.20 ± 0.16	2.12 ± 0.20	1.48 ± 0.10	1.11 ± 0.06	1.21 ± 0.15	1.74 ± 0.18	2.44 ± 0.16	2.54 ± 0.09	2.19 ± 0.08	2.28 ± 0.04
KR	1.82 ± 0.14	1.78 ± 0.20	1.36 ± 0.09	0.88 ± 0.04	1.03 ± 0.12	1.42 ± 0.19	1.91 ± 0.18	2.00 ± 0.07	1.85 ± 0.1	1.95 ± 0.02
MACCS	1.88 ± 0.15	1.81 ± 0.20	1.39 ± 0.11	0.90 ± 0.04	1.06 ± 0.12	1.52 ± 0.19	1.99 ± 0.15	2.07 ± 0.09	2.08 ± 0.13	1.92 ± 0.04
MAP4	1.88 ± 0.15	1.68 ± 0.16	1.33 ± 0.10	0.86 ± 0.05	1.03 ± 0.10	1.44 ± 0.22	2.01 ± 0.17	2.07 ± 0.09	2.00 ± 0.17	1.86 ± 0.04
Pharm2D	1.80 ± 0.11	1.68 ± 0.21	1.35 ± 0.10	0.79 ± 0.05	**0.98 ± 0.12**	1.47 ± 0.20	**1.81 ± 0.11**	1.89 ± 0.12	1.79 ± 0.13	1.71 ± 0.03
PubChem	1.87 ± 0.13	1.66 ± 0.16	1.32 ± 0.12	0.90 ± 0.04	1.01 ± 0.10	1.32 ± 0.22	2.01 ± 0.15	2.06 ± 0.11	1.90 ± 0.09	1.94 ± 0.02
RDKit	1.83 ± 0.13	1.56 ± 0.13	**1.26 ± 0.10**	0.78 ± 0.05	0.99 ± 0.11	1.36 ± 0.15	1.85 ± 0.13	1.89 ± 0.08	1.91 ± 0.13	1.69 ± 0.03
2DAP	1.84 ± 0.14	1.66 ± 0.13	1.33 ± 0.11	0.90 ± 0.05	1.02 ± 0.10	1.45 ± 0.23	1.91 ± 0.14	1.98 ± 0.10	1.88 ± 0.03	1.91 ± 0.05
ConstIdx	1.99 ± 0.17	1.92 ± 0.14	1.38 ± 0.10	1.00 ± 0.06	1.12 ± 0.11	1.58 ± 0.22	2.12 ± 0.10	2.20 ± 0.10	2.06 ± 0.09	2.01 ± 0.05
FGCount	1.93 ± 0.16	1.71 ± 0.16	1.36 ± 0.10	0.91 ± 0.06	1.01 ± 0.08	1.40 ± 0.17	1.98 ± 0.16	2.09 ± 0.07	1.87 ± 0.15	2.01 ± 0.06
MolProp	2.05 ± 0.05	1.90 ± 0.14	1.43 ± 0.09	1.06 ± 0.07	1.17 ± 0.11	1.65 ± 0.22	2.22 ± 0.16	2.20 ± 0.08	2.06 ± 0.13	2.01 ± 0.06
RingDesc	2.12 ± 0.12	2.15 ± 0.17	1.62 ± 0.11	1.18 ± 0.08	1.22 ± 0.14	1.66 ± 0.21	2.28 ± 0.21	2.46 ± 0.10	2.32 ± 0.13	2.24 ± 0.07
TOPO	1.89 ± 0.13	1.92 ± 0.13	1.50 ± 0.08	1.04 ± 0.06	1.20 ± 0.09	1.64 ± 0.18	2.01 ± 0.16	2.10 ± 0.11	1.92 ± 0.09	1.97 ± 0.05
WalkPath	1.97 ± 0.13	1.98 ± 0.13	1.42 ± 0.10	1.06 ± 0.06	1.16 ± 0.12	1.61 ± 0.21	2.16 ± 0.19	2.17 ± 0.09	2.09 ± 0.09	2.13 ± 0.05
ECT (ours)	**1.78 ± 0.12**	1.77 ± 0.16	1.35 ± 0.10	0.83 ± 0.04	1.03 ± 0.13	1.45 ± 0.18	1.87 ± 0.14	1.89 ± 0.08	1.81 ± 0.13	1.79 ± 0.05
ECT+ FP (ours)	**1.74 ± 0.12**	**1.52 ± 0.17**	1.29 ± 0.09	**0.78 ± 0.04**	0.99 ± 0.11	1.38 ± 0.20	1.82 ± 0.12	**1.82 ± 0.08**	**1.77 ± 0.03**	**1.68 ± 0.02**

Figure 2 ranks the representations by their mean error for two different metrics (RMSE and R2) for the combined dataset. Performance is visualized using boxplots to show the error distribution for each representation method according the cross validation results. The representation combining our ECT-based representation and the AVALON fingerprint consistently achieves better values

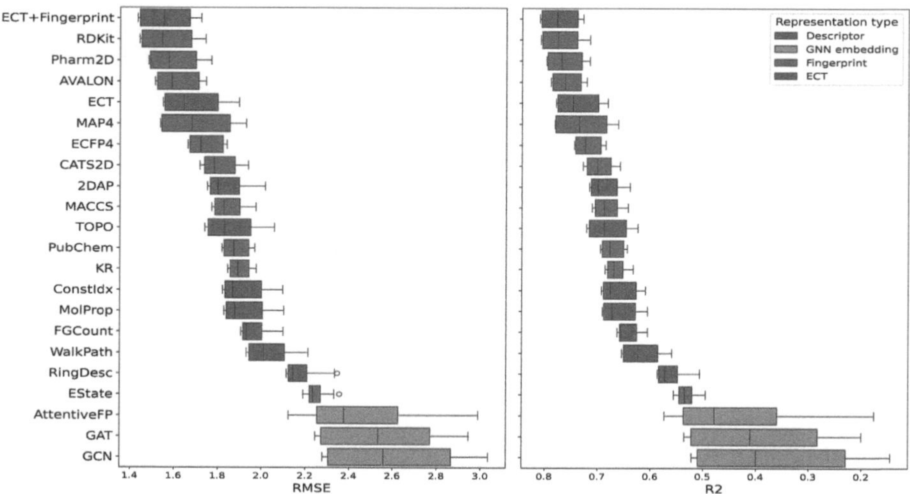

Fig. 2. Boxplots showing the distribution of test errors for each representation method across the combined dataset. From left to right, the metrics displayed are RMSE and R2. Representations are ranked by their mean values: For RMSE, *lower* values indicate *better* performance, while for R2, *higher* values (closer to 1) indicate *better* performance.

for RMSE and R2 metrics. The low variance between cross-validation results is particularly noteworthy, showing the robustness of the approach.

An overall observation is that most complex models (like GNNs) that *directly* take the molecular graph as its input tend to yield the *worst* computational performance. This is in marked contrast to the other molecular representations (using XGBoost), where a single iteration of cross-validation takes around 10 to 20 s on a standard computer. In comparison, for GNNs, a single iteration takes about 40 to 60 s for GCN and GAT, and up to 100 s for "AttentiveFP," which nevertheless leads to significantly worse predictive performance. This observation suggests that the use of deep learning models is not inherently justified by their (computational and predictive) performance in this context. Rather, it appears that the choice of molecular representation plays a much more critical role in determining model effectiveness, highlighting that more complex architectures do not necessarily lead to better results. This observation is in line with other studies and might be attributed to the limited availability of data in the biomedical domain.

5   Conclusion

In this paper, we propose and implement an effective new molecular representation method based on the Euler Characteristic Transform (ECT), which captures multiscale topological information relevant to molecular property prediction tasks, specifically for predicting K_i, the inhibition constant. Our results on different datasets show that the ECT provides meaningful and effective features for molecular machine learning, and that its *combination* with other approaches, specifically with the AVALON fingerprint, leads to improved predictive performance. This multiscale topological approach, which can capture shape characteristics across different resolutions, encodes structural information of molecules, which were hitherto not being considered or fully exploited by existing methods, presenting a promising avenue for molecular machine learning research.

Future work. An exploration of *how* the number of directions and thresholds used in the computation of the ECT affects its representational power is needed. We believe that identifying optimal or data-adaptive strategies for selecting these parameters will lead to more expressive and discriminative topological signatures, thus potentially improving performance in molecular property prediction tasks. This question is not only relevant for cheminformatics and biomedical applications, but could also benefit a wide range of domains where data can be modeled geometrically and topologically. In addition, we plan to extend and test our topological approach in other molecular machine learning tasks, like molecular classification, toxicity prediction, or even generative models for molecular design. In this work, we tested a single hybrid strategy (ECT + AVALON fingerprint) and observed improved predictive accuracy. Future work could benchmark multiple hybrid strategies to identify synergies between multiscale topological and traditional representations. Moreover, we aim to extend the ECT to molecular graphs with 3D positional information to quantify the geometric shape of the molecular structure itself, instead of, or in combination with, general atomic features. A particularly exciting direction for future work involves the development of a more comprehensive topological descriptor that integrates information from multiple tools of TDA. Rather than relying solely on the ECT, this extended descriptor could incorporate additional topological summaries, such as Betti numbers or persistence-based descriptors from persistent homology. This may lead to more powerful and expressive descriptors. Further efforts will also include systematic hyperparameter tuning and the evaluation of different dataset splitting strategies, which can substantially influence both predictive performance and generalizability. Finally, our approach could also be extended to other variants of the ECT, such as the Weighted Euler Characteristic Transform (WECT) [16] or the Smooth Euler Characteristic Transform (SECT) [7].

Acknowledgements. This work was partially supported by REXASI-PRO project (ID: 101070028), the Research Budget (VII PPIT) of the University of Seville, and the Swiss State Secretariat for Education, Research, and Innovation (SERI).

Data Availibility Statement. All data and code used to generate the results presented in this manuscript are publicly available in a GitHub repository (https://github.com/victosdur/ECTforMoleculeMachineLearningTask).

Disclosure of Interests. The authors have no competing interests to declare that are relevant to the content of this article.

References

1. Amézquita, E.J., et al.: Measuring hidden phenotype: quantifying the shape of barley seeds using the Euler characteristic transform. in silico Plants **4**(1), diab033 (2021)
2. Atz, K., Grisoni, F., Schneider, G.: Geometric deep learning on molecular representations. Nat. Mach. Intell. **3**(12), 1023–1032 (2021)
3. Baptista, D., Correia, J., Pereira, B., Rocha, M.: Evaluating molecular representations in machine learning models for drug response prediction and interpretability. J. Integr. Bioinform. **19**(3), 20220006 (2022)
4. Carracedo-Reboredo, P.: A review on machine learning approaches and trends in drug discovery. Comput. Struct. Biotechnol. J. **19**, 4538–4558 (2021)
5. Chen, T., Guestrin, C.: XGBoost: A scalable tree boosting system. In: Proceedings of the 22nd ACM SIGKDD International Conference on Knowledge Discovery and Data Mining, pp. 785–794. Association for Computing Machinery, New York, NY, USA (2016)
6. Corey, E.J., Czakó, B., Kürti, L.: Molecules and medicine. John Wiley and Sons (2007)
7. Crawford, L., Monod, A., Chen, A.X., Mukherjee, S., Rabadán, R.: Functional data analysis using a topological summary statistic: the smooth Euler characteristic transform. arXiv preprint arXiv:1611.06818 (2016)
8. David, L., Thakkar, A., Mercado, R., Engkvist, O.: Molecular representations in AI-driven drug discovery: a review and practical guide. J. Cheminformatics **12**(1), 1–22 (2020). https://doi.org/10.1186/s13321-020-00460-5
9. Deng, J., Yang, Z., Ojima, I., Samaras, D., Wang, F.: Artificial intelligence in drug discovery: applications and techniques. Briefings Bioinform. **23**(1), bbab430 (2022)
10. Deng, J., et al.: A systematic study of key elements underlying molecular property prediction. Nat. Commun. **14**(1), 6395 (2023)
11. Drews, J.: Drug discovery: a historical perspective. Science **287**(5460), 1960–1964 (2000)
12. Edelsbrunner, H., Harer, J.L.: Computational topology: an introduction. Am. Math. Soc. (2022)
13. Hamilton, W.L.: Graph representation learning. Synth. Lect. Artif. Intell. Mach. Learn. **14**(3), 1–159 (2020)
14. Hatcher, A.: Algebraic topology. Cambridge University Press (2005)
15. Jiang, D.: Could graph neural networks learn better molecular representation for drug discovery? a comparison study of descriptor-based and graph-based models. J. Cheminformatics **13**(1), 1–23 (2021). https://doi.org/10.1186/s13321-020-00479-8
16. Jiang, Q., Kurtek, S., Needham, T.: The weighted Euler curve transform for shape and image analysis. In: Proceedings of the IEEE/CVF Conference on Computer Vision and Pattern Recognition Workshops, pp. 844–845 (2020)
17. Koke, C., et al.: Graph networks struggle with variable scale. In: ICLR Workshop 'I Can't Believe It's Not Better: Challenges in Applied Deep Learning' (2025)
18. Laky, D.J., Zavala, V.M.: A fast and scalable computational topology framework for the Euler characteristic. Digit. Disc. **3**(2), 392–409 (2024)
19. Leinster, T.: The Euler characteristic of a category. Doc. Math. **13**, 21–49 (2008)
20. Lum, P.Y.: Extracting insights from the shape of complex data using topology. Sci. Rep. **3**(1), 1236 (2013)
21. Munch, E.: An invitation to the Euler characteristic transform. Am. Math. Mon. **132**(1), 15–25 (2025)

22. Munkres, J.R.: Elements of algebraic topology. CRC press (2018)
23. Platonov, O., Kuznedelev, D., Diskin, M., Babenko, A., Prokhorenkova, L.: A critical look at the evaluation of GNNs under heterophily: Are we really making progress? arXiv preprint arXiv:2302.11640 (2023)
24. Rieck, B.: Topology meets machine learning: an introduction using the Euler characteristic transform. Not. Am. Math. Soc. **72**(7), 719–727 (2025)
25. Röell, E., Rieck, B.: Differentiable Euler characteristic transforms for shape classification. In: International Conference on Learning Representations (2024)
26. Rottach, F., Schieferdecker, S., Eickhoff, C.: The topology of molecular representations and its influence on machine learning performance. J. Cheminformatics **17** (2025)
27. Smith, A.: Topological analysis of molecular dynamics simulations using the Euler characteristic. J. Chem. Theory Comput. **19**(5), 1553–1567 (2023)
28. Turner, K., Mukherjee, S., Boyer, D.M.: Persistent homology transform for modeling shapes and surfaces. Inf. Infer. a J. IMA **3**(4), 310–344 (2014)
29. Van Tilborg, D., Alenicheva, A., Grisoni, F.: Exposing the limitations of molecular machine learning with activity cliffs. J. Chem. Inf. Model. **62**(23), 5938–5951 (2022)
30. Waibel, D.J.E., Atwell, S., Meier, M., Marr, C., Rieck, B.: Capturing shape information with multi-scale topological loss terms for 3D reconstruction. In: Wang, L., Dou, Q., Fletcher, P.T., Speidel, S., Li, S. (eds.) Medical Image Computing and Computer Assisted Intervention (MICCAI), pp. 150–159. Springer, Cham, Switzerland (2022)
31. Walters, W.P., Barzilay, R.: Applications of deep learning in molecule generation and molecular property prediction. Acc. Chem. Res. **54**(2), 263–270 (2020)
32. Wigh, D.S., Goodman, J.M., Lapkin, A.A.: A review of molecular representation in the age of machine learning. Wiley Interdisc. Rev. Comput. Molecular Sci. **12**(5), e1603 (2022)
33. Xiong, Z., et al.: Pushing the boundaries of molecular representation for drug discovery with the graph attention mechanism. J. Med. Chem. **63**(16), 8749–8760 (2019)

Sharing Patterns Between Proteins and Exploring Possible Drug Interactions

Paloma Tejera-Nevado[1]([⊠]) [iD] and Alejandro Rodríguez-González[1,2] [iD]

[1] Centro de Tecnología Biomédica, Universidad Politécnica de Madrid, Madrid, Spain
{paloma.tejera,alejandro.rg}@upm.es
[2] ETS Ingenieros Informáticos, Universidad Politécnica de Madrid, Madrid, Spain

Abstract. Due to the nature of amino acids that make up proteins, it is possible to detect small patterns within their sequence. Some of these sequences have been identified as motifs, characterized by defined signatures that are shared across different proteins. In this study, patterns are defined as short amino acid sequences that are commonly found among proteins involved in various types of cancer. The research focuses on the localization of such patterns in ALDH2, chosen as a reference protein because it is known to be inhibited by disulfiram. Disulfiram, originally used to treat alcohol use disorder (AUD), has been repurposed for the treatment of lung cancer. Analyzing shared patterns can help identify common protein structures and may assist in locating potential binding cavities and interaction sites, while also improving our understanding of how related or structurally similar proteins might be affected by certain drugs. This approach provides a new perspective on drug specificity and targeting, highlighting three achievements: identifying conserved amino acid motifs in cancer-related proteins, a case study on ALDH2 inhibition by disulfiram, and showing how motif analysis can guide drug repurposing.

Keywords: Pattern recognition in proteins · Target identification · Drug-protein interactions

1 Introduction

1.1 Drug-Disease Interactions and Therapeutic Targets

Understanding drug-disease interactions is fundamental to the development of effective and safe therapeutic strategies. These interactions describe how diseases can alter the pharmacokinetics (PK) and pharmacodynamics (PD) of drugs, and conversely, how drugs can influence disease progression or pathology [1]. In complex diseases such as cancer, neurodegenerative disorders, and metabolic syndromes, alterations in cellular signaling pathways, enzyme activity, and gene expression can significantly affect drug responses [2]. This necessitates the identification of therapeutic targets, which are specific molecules or pathways that can be modulated to achieve clinical benefit. Advances in molecular biology, bioinformatics, and systems pharmacology have enabled the discovery of novel targets, such as aberrant enzymes, receptors, and transcription factors,

© The Author(s), under exclusive license to Springer Nature Switzerland AG 2026
A. López Fernández et al. (Eds.): CIABiomed 2025, LNBI 16148, pp. 406–420, 2026.
https://doi.org/10.1007/978-3-032-10661-2_31

that drive disease mechanisms [3]. Targeted therapies, unlike traditional treatments, offer the potential for precision medicine by minimizing off-target effects and improving efficacy [4]. Analyzing protein-protein interactions (PPIs) is essential for understanding how sequence variations influence disease progression and treatment response. Techniques to study PPIs include experimental and computational methods [5].

1.2 Disulfiram as a Multifunctional Therapeutic Agent

Disulfiram inhibits the oxidoreductase enzyme aldehyde dehydrogenase 2 (ALDH2), a crucial enzyme in alcohol metabolism that converts the toxic acetaldehyde into harmless acetate. Inhibition of the ALDH2 by disulfiram leads to acetaldehyde buildup, causing adverse symptoms that deter alcohol consumption, making it effective for treating alcohol dependence [6]. Disulfiram is not specific to ALDH2, as it also inhibits the cytosolic enzyme ALDH1 [7]. Beyond its use in alcohol aversion, disulfiram shows promising anticancer properties, especially in lung cancer [8]. ALDH activity, including ALDH2, is elevated in cancer stem cells (CSC), which contribute to tumor growth, metastasis, and therapy resistance. Disulfiram's inhibition of ALDH reduces CSC characteristics such as sphere formation, colony growth, and migration, potentially overcoming drug resistance and relapse [9]. Additionally, disulfiram combined with copper induces cancer cell death through apoptosis, ferroptosis, and cuproptosis by generating reactive oxygen species and inhibiting the proteasome system [10]. It can reverse resistance to microtubule inhibitors and has shown antitumor effects in lung cancer models [11]. Clinical trials reveal mixed but encouraging outcomes; a phase IIb trial combining disulfiram with chemotherapy in metastatic non-small cell lung cancer (NSCLC) showed improved progression-free and overall survival, highlighting disulfiram's role in targeting cancer stem cells [8]. However, further clinical research is needed to confirm its efficacy and best use in cancer therapy [12].

1.3 Aldehyde Dehydrogenase (ALDH) Family

The human Aldehyde Dehydrogenase (ALDH) family comprises 19 functional genes that encode enzymes responsible for detoxifying aldehydes by oxidizing them into carboxylic acids. These enzymes are distributed across cellular compartments (cytoplasm, mitochondria and endoplasmic reticulum), highlighting their diverse physiological roles. Beyond detoxification, ALDHs are essential in various metabolic pathways, including alcohol metabolism (e.g., ALDH2) and retinoic acid synthesis (e.g., ALDH1A1, ALDH1A2, ALDH1A3), which support cell differentiation and development [13]. ALDH enzymes are significantly involved in disease processes, with disease-causing mutations in the ALDH superfamily falling into three main categories: (1) those that impair NAD binding, (2) those that alter the substrate binding site, and (3) those that disrupt quaternary structure formation [14]. ALDH dysfunction is further implicated in neurodegenerative disorders, including Alzheimer's and Parkinson's diseases, due to compromised aldehyde detoxification [15]. Many isoforms are overexpressed in tumors and are linked to cancer stem cell (CSC) traits and resistance to therapy [16]. ALDH1A1, in particular, is a well-established CSC marker in cancers such as lung, breast, and colon, where it contributes to chemotherapy resistance by detoxifying harmful aldehydes [17].

ALDH3A1 is also upregulated in certain cancers, like non-small cell lung carcinoma, and is associated with drug resistance [18].

1.4 Protein Sequence Analysis in Drug-Disease Context

Protein sequence analysis is crucial for understanding disease mechanisms and advancing therapeutic development, as proteins are central to cellular function and their dysfunction contributes to many diseases. Aldehyde Dehydrogenase 2 (ALDH2) is a crucial mitochondrial enzyme that functions as a homotetramer. A common variant, ALDH2*2 (Glu504Lys) (present in 30–50% of East Asian populations) drastically reduces ALDH2 activity and increases cancer risk, especially when combined with alcohol intake [19–21]. In lung adenocarcinoma (LUAD), ALDH2 downregulation is associated with poorer prognosis, making ALDH2 a potential prognostic factor for early-stage LUAD [22]. Abnormal ALDH activity in cancer is not fully understood but shows promise as a biomarker and therapeutic target. Targeting ALDH, especially in cancer stem cells, could improve treatment outcomes. Repurposed drugs like disulfiram offer potential for safe and effective cancer therapies when combined with conventional treatments [23]. Advances in structure prediction tools like AlphaFold [24] are enabling new methods for protein sequence analysis in drug and disease research. Approaches such as comparing related proteins and identifying conserved patterns in disease-associated sequences [25] aim to reveal shared structural features, including druggable cavities, to support drug repurposing.

This study investigates whether shared amino acid patterns (at least four residues) can reveal functional or structural links between proteins, aiming to distinguish random from meaningful associations and identify off-targets or disease-related candidates. The paper is organized in the following order: Sect. 2 includes the material and methods used to study protein patterns. In Sect. 3, the results obtained are presented and discussed. Finally, Sect. 4 summarizes the conclusions of this study and highlights potential avenues for future research.

2 Material and Methods

2.1 Classification of Drugs by Action Type, Class, and Gene Symbol

A previous study [25] generated a list of patterns at 5% and 10% occurrence thresholds using a lung cancer treatment dataset, as well as a dataset that included proteins associated with other cancer types: pancreatic (C0346647), breast (C0006142), colon (C0007102), and head and neck (C0010606) cancers. Diseases were identified using their CUI (Concept Unique Identifier), and protein sequence and drug information were obtained from DISNET [26].

First, the data from the lung cancer treatment dataset was analyzed. It was observed that most of the drugs involved are broad-spectrum, which is evident from their action types and associated gene IDs. In total, 43 different drugs were identified. Not all these drugs have a direct correlation with genes specifically related to lung cancer. The drugs were classified according to their action type and the drug class they belong. In this dataset, 12 different action types and 17 drug classes were found. Some drugs have multiple effects and could be valuable for exploring potential interactions with other proteins or targets. However, these multi-effect drugs can also make it difficult to determine whether they are actually binding within a specific protein cavity. A subset of drugs was found to be associated with a single gene symbol, a single action type, and a single drug class. The drug IDs were verified in the ChEMBL database (release ChEMBL_35), and the Drug ID CHEMBL2362016, corresponding to arsenic trioxide, was updated from the original dataset to CHEMBL5483015 in the current update. The molecular structure of the inhibitor disulfiram was retrieved from DrugBank [27].

2.2   Analysis and Identification of Shared Sequence Patterns

To investigate the patterns associated with ALHD2 (UniProt entry: P05091) in combination with other proteins, a subset of the data was created focusing on the top five most frequent P05091-containing protein pairs. The cleaned dataset was filtered to extract all the unique sequence patterns associated with each of these protein combinations. For every pair, the number of distinct patterns was computed and stored, along with the list of patterns themselves. The results were then organized into a structured DataFrame to facilitate visualization and interpretation of the co-occurrence pattern data. A search was conducted within the filtered dataset to identify entries containing a specific pattern of interest. When exact matches for the pattern were found, the number of occurrences and corresponding data rows were recorded.

2.3   Prediction of Protein Structures and Docking Analysis

Protein sequences were retrieved from UniProt [28]. Protein structure predictions were obtained in PDB format from the AlphaFold DB [29]. However, the structure for protein P30837 (AL1B1) was predicted using the AlphaFold server [30] due to changes in a previous version of the UniProt sequence. Docking predictions were performed using CB-Dock2 [31] and COACH-D [32]. Visualization was carried out with ChimeraX (v 1.6.1) [33–35]. The Matchmaker tool was used to align the protein structures, and the root mean square deviation (RMSD) was calculated using this software.

3   Results and Discussion

3.1   Classification of Drugs by Action Type, Class, and Gene Symbol

Focusing on the lung cancer dataset, 43 drugs were detected, many of which are broad-spectrum agents characterized by diverse action types and gene associations. These findings were based on earlier research that identified drug interaction patterns using lung cancer treatment data alongside datasets including proteins from various cancer types (pancreas, breast, colon and head and neck) [25, 36]. The drugs were categorized into 12 action types and 17 drug classes, adding complexity to the analysis. While some drugs exhibit multiple effects that may offer insights into protein interactions, this also complicates pinpointing specific binding sites. Drug entries were filtered to retain only those with a unique drug name, action type, and protein class designation, notably resulting in a subset of drugs that exhibit a one-to-one relationship with a single gene, action type and drug class. This information, derived from the lung cancer treatment dataset, is detailed in Table 1, which lists each drug alongside its corresponding action type, drug class, and gene symbol.

Table 1. Filtered Lung Cancer Treatment Dataset

Drug id	Drug name	Action type	Protein Class name	Gene symbol
CHEMBL185	Fluorouracil	Inhibitor	Transferase	TYMS
CHEMBL5483015	Arsenic trioxide	Inhibitor	Oxidoreductase	TXNRD1
CHEMBL238071	Vindesine	Inhibitor	Cytoskeletal protein	TUBB1
CHEMBL56367	Nimesulide	Inhibitor	Oxidoreductase	PTGS2
CHEMBL924	Zoledronic acid	Inhibitor	Transferase	FDPS
CHEMBL964	Disulfiram	Inhibitor	Oxidoreductase	ALDH2

This filtering strategy was applied to ensure specificity and reduce the number of candidates for validation. From this approach, six drugs met the criteria. ALDH2 was selected as a primary candidate due to its known history in drug repurposing. For example, disulfiram (CHEMBL964), a drug originally developed for alcohol use disorder, has been investigated for the treatment of non-small cell lung cancer. Disulfiram works by inhibiting ALDH2. Additionally, arsenic trioxide and nimesulide were identified as inhibitors targeting the oxidoreductase class proteins TXNRD1 and PTGS2, respectively.

3.2 Analysis of Sequence Patterns in ALDH2

The analysis began by identifying the number of sequence patterns shared between ALDH2 (Uniprot entry: P05091) and other proteins. In total, 90 patterns were detected, with one pattern ("AKLL") falling under the 10% occurrence threshold and the remaining 89 corresponding to the 5% occurrence category. Subsequently, patterns not found in other cancer types were excluded from the dataset. This filtering step led to the removal of only two patterns, resulting in a final set of 88 patterns for further analysis in the case of ALDH2. To further explore potential co-associations involving ALDH2, the top five protein combinations containing P05091 were selected for analysis. For each combination, the dataset was filtered to identify all unique patterns associated with the pair. The total number of distinct patterns per combination was then calculated. To define a reference framework for pattern comparison, proteins sharing at least two patterns with ALDH2 were identified, yielding five protein candidates. The results, including the number of pattern occurrences and the specific patterns identified for each protein pair, are summarized in Table 2.

TXNRD1 was prioritized for initial inspection because it was also present in the lung cancer treatment dataset. The protein TXNRD1 shares six patterns with ALDH2, based on data from the treatment dataset, which was used in this case to assess the

Table 2. Sequence Patterns Associated with ALDH2 (P05091) Paired with Selected Proteins

Protein id	Gene symbol	Occurrences	Patterns
P42345	MTOR	9	[AALET, AAFQ, ANYL, ELGE, FQLG, KTEQ, QIFI, QQPE, VLKC]
P22102	GART	6	[AWKL, EDVD, GQII, LDKA, VARA, VDLD]
Q16881	TXNRD1	6	[GLAAA, GDKE, LQAG, LQAY, RVVG, TEVK]
P51649	ALDH5A1	5	[FTGST, PWNFP, DAVS, PFGG, VAEQ]
P10696	ALPG	4	[AAARF, FGGY, KLGP, SDAD]

effects reported in the references. A manual inspection of the shared patterns was not located within the ligand-binding cavity of ALDH2, nor were they associated with structurally similar regions between ALDH2 and TXNRD1 based on predicted protein folding models. This suggests that while sequence patterns can indicate potential biological relationships, not all shared patterns correspond to functional or structural similarity in drug interaction sites. A similar observation was made when examining the common patterns detected between the pairs ALDH2 and MTOR, GART, and ALPG (data not shown).

As part of efforts to identify therapeutic targets in cancer, analysis of the lung cancer treatment dataset used in this study revealed that both ALDH2 and ALDH5A1 are implicated in lung cancer therapy. Disulfiram, an FDA-approved drug originally used for alcohol aversion therapy, acts as an inhibitor of ALDH2 and is under investigation for its potential in treating non-small cell lung cancer [10]. Valproic acid, on the other hand, inhibits ALDH5A1 [37]. Disulfiram is also gaining considerable attention for its broader anticancer potential, largely due to its ability to inhibit aldehyde dehydrogenase 1 family member A1 (ALDH1A1) [38]. ALDH1A1 is often overexpressed in various cancers and is linked to tumor aggressiveness, drug resistance, and the survival of cancer stem cells (CSC) [39, 40]. To explore potential functional similarities, shared amino acid patterns were analyzed between two mitochondrial enzymes, ALDH2 (mitochondrial aldehyde dehydrogenase, 517 amino acids) and ALDH5A1 (mitochondrial succinate semialdehyde dehydrogenase, 535 amino acids). The analysis revealed that the sequences "FTGST" and "PWNFP" are located within regions predicted to form part of the ligand-binding cavity, suggesting possible overlap in substrate interaction or inhibitor binding. Additionally, the pattern "VAEQ" was identified in both proteins within two distinct regions also associated with the ligand-binding site, while the patterns "DAVS" and "PFGG" were found outside the surrounding area of the ligand-binding site (Fig. 1).

Disulfiram irreversibly inactivates ALDH enzymes by carbamylating a key cysteine residue in their active site [38, 41]. While disulfiram broadly inhibits ALDHs, including ALDH2 (relevant to its anti-alcohol effects), its action on ALDH1A1 specifically targets a pathway vital for cancer progression [42]. Studies demonstrate that disulfiram, particularly when combined with copper, effectively reduces CSC populations, inhibits tumor growth, and enhances the sensitivity of cancer cells to chemotherapy across various cancer types, including lung, ovarian, and glioblastoma [39, 43, 44]. Beyond ALDH

A   B

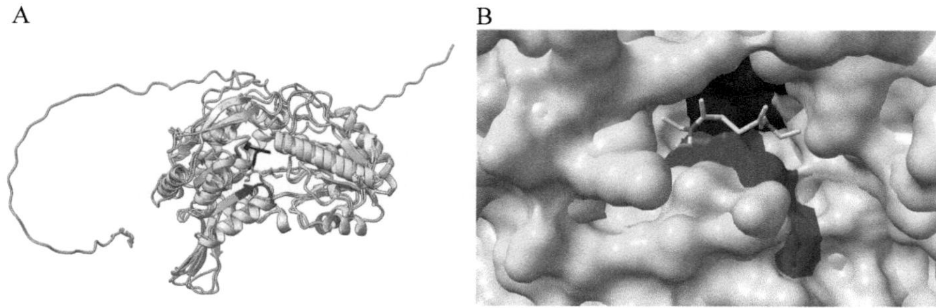

Fig. 1. Structural comparison of ALDH2 (UniProt ID: P05091, shown in brown) and ALDH5A1 (UniProt ID: P51649, shown in cyan), highlighting shared sequence patterns. (A) Overall structural overview. (B) Detailed view of the region surrounding the disulfiram-binding site, generated using CB-Dock2. The pattern "FTGST" (colored red) is located at position 260 in ALDH2 and 282 in ALDH5A1, positioned at the end of a β-sheet and extending into an unstructured region; the pattern appears to overlap spatially between the two structures. The pattern "PWNFP" (colored blue), found at position 184 in ALDH2 and 203 in ALDH5A1, is also situated in an unstructured region and shows spatial proximity between the proteins. In contrast, the patterns "DAVS" and "PFGG" are not located near the ligand-binding region and do not show close spatial alignment. The pattern "VAEQ" (colored magenta) appears at position 210 in ALDH2 and 481 in ALDH5A1; although not directly adjacent to the binding cavity, it is located nearby in both structures within unstructured regions. Structural alignment using the Matchmaker tool yielded an RMSD of 0.856 Å over 394 pruned atom pairs, indicating a high degree of structural similarity. The visualization was performed using UCSF ChimeraX.

inhibition, disulfiram's broader therapeutic effects in cancer are being explored, encompassing its ability to induce various forms of regulated cell death, such as apoptosis, ferroptosis, and cuproptosis, and to modulate other critical cellular signaling pathways [45–47].

3.3   Proteins Sharing the Selected Sequence Pattern

Focusing on proteins involved in cancer and drug interaction, the aim was to identify conserved amino acid patterns that may indicate shared structural or functional features. The identification of shared residues was guided by reference data from protein treatments, aiming to reveal potential common protein structures and binding cavities. In this case, five of the common patterns detected were found to belong to ALDH2 (P05091) and ALDH5A1 (P51649), with the latter corresponding to the protein SSDH. A search was conducted to identify proteins containing the specific pattern "FTGST" within the dataset comprising proteins from other cancer types, such as pancreatic (C0236647), breast (C0006142), colon (C0007102), and head and neck (C0010606) (Table 3).

An initial inspection of the dataset identified AL1A1, AL1A3, and AL1B1 as proteins closely related to ALDH2, and they were analyzed further. The proteins BIG2 and BIG1 were also identified, corresponding to Brefeldin A-inhibited guanine nucleotide-exchange protein 2 (1785 aa) and type 1 (1849 aa), respectively. These two proteins share a similar folding structure; however, this was not related to ALDH2 (data not shown).

Table 3. Proteins from Other Cancer Types Containing the Pattern "FTGST"

Protein id	Protein name	Disease id	Proteins Treat. id	Protein name Treat
P00352	AL1A1	C0006142	P05091	ALDH2
		C0346647	P51649	SSDH
P47895	AL1A3	C0006142		
Q9Y6D5	BIG2	C0006142		
Q9Y6D6	BIG1	C0006142		
P30837	AL1B1	C0007102		
Q14031	CO4A6	C0007102		
P18583	SON	C0010606		

The proteins CO4A6 and SON correspond to Collagen alpha-6(IV) chain (1691 aa) and protein SON (2426 aa), respectively. Due to their notably unstructured regions, they were not evaluated further (data not shown).

The next pattern selected for detailed analysis was "PWNFP", chosen to explore further its occurrence and potential functional relevance among the proteins detected in other cancer types (Table 4). The pattern "PWNFP" is detected only in the related proteins AL1A1, AL1A3 and AL1B1. The protein AL1A1 is associated with breast (C0006142) and pancreatic (C0346647) cancer. The protein AL1A3 is linked to breast cancer (C0006142), while AL1B1 is related to colon cancer (C0007102).

Table 4. Proteins from Other Cancer Types Containing the Pattern "PWNFP"

Protein id	Protein name	Disease id	Proteins Treat. id	Protein name Treat
P00352	AL1A1	C0006142	P05091	ALDH2
		C0346647	P51649	SSDH
P47895	AL1A3	C0006142		
P30837	AL1B1	C0007102		

Subsequently, ALDH2, which belongs to the treatment dataset and is associated with disulfiram, was used as a reference. AL1A1, AL1A3 and AL1B1 were the proteins selected to localize the pattern "FTGST" among all the detected proteins (Fig. 2A). Additionally, the pattern "PWNFP" was detected in these proteins, as they were the ones in which this pattern appeared (Fig. 2B). Structural localization of both 'FTGST" and "PWNFP" in these proteins, using ALDH2 as a reference, confirmed their presence in comparable regions (Fig. 2A, 2B). These findings support the hypothesis that shared patterns across ALDH family members may relate to conserved functions and potential drug-binding sites. Current research also focuses on developing more selective

ALDH1A1 inhibitors derived from disulfiram to maximize therapeutic benefits while minimizing off-target effects [38].

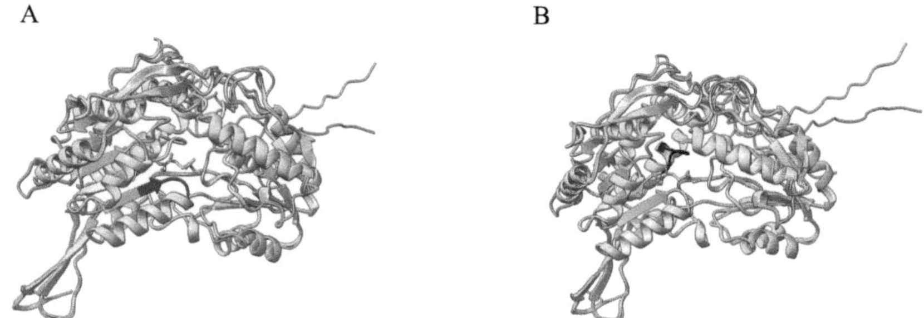

Fig. 2. Comparison of predicted protein structures between ALDH2 and proteins sharing common patterns in other cancer types. ALDH2 (brown) was used as a reference, and a structural comparison was performed with ALA1A1 (pink), ALA1A3 (green), and AL1B1 (salmon). The pattern "FTGST" is marked in red (A), and "PWNFP" is colored in blue (B). Visualization and structural overlap were conducted using the MatchMaker tool in ChimeraX.

3.4   Analysis of Pattern Localization in Protein Active Sites

The patterns shared by some proteins did not contain any cysteine residues that could be involved in the inhibition of ALDH2 by disulfiram. Therefore, the study proceeded with a search for the catalytic cysteine at position 302 and adjacent cysteine residues. The pattern "CCAG" was identified, and the proteins containing this pattern were further analyzed (Table 5). In this case, the protein associated with the treatment condition corresponds to ALDH2, and PSMD1 also shares that pattern.

Table 5. Proteins from Other Cancer Types Containing the Pattern "CCAG"

Protein id	Protein name	Disease id	Proteins Treat. id	Protein name Treat
Q13077	TRAF1	C0006142	P05091	ALDH2
		C0007102	Q99460	PSMD1
Q9H2E6	SEM6A	C0006142		
		C0007102		
Q96GP6	SREC2	C0006142		
P30837	AL1B1	C0007102		
O75508	CLD11	C0007102		

The localization of the pattern "CCAG" and the structural visualization of all proteins were performed, with further analysis focusing on the binding cavity of ALDH2 with disulfiram. For the docking analysis, the full sequences of the proteins were used. The predicted complex structure was then determined using COACH-D and compared to that of AL1B1, the protein with the most similar molecular structure. Structural localization showed that "CCAG" lies in an unstructured region of the protein (Fig. 3).

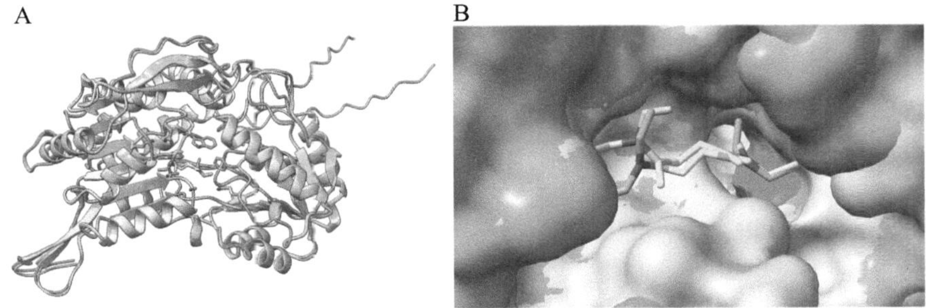

Fig. 3. Comparison of predicted protein structures between ALDH2 and AL1B1. ALDH2 (colored in brown) is used as the reference protein for the complex with disulfiram and is compared to AL1B1 (colored in salmon). (A) The pattern "CCGA" is highlighted in cyan. (B) The cysteine at position 309, which is the first "C" in the "CCAG" pattern, is marked in green and is localized within the binding cavity. Visualization was performed using ChimeraX.

It is known that the first 18 amino acids of ALDH2 correspond to the mitochondrial targeting signal, which is cleaved post-translationally. Consequently, the predicted binding residues are shifted by 17 positions. The results reveal that a cysteine residue is identified in both ALDH2 and AL1B1 as part of the predicted binding sites involved in the interaction with disulfiram. Following this, the complex information for ALDH2 and AL1B was analyzed, including the C-Score, TM-score, and the predicted binding sites of the top-ranked model (Table 6).

Table 6. COACH-D Results for ALDH2 and AL1B1 with Disulfiram

Protein name	C-Score	TM-score	Predicted binding residues
ALDH2	0.98	0.92	I183, **P184, W185, N186**, M191, K209, A211, E212, F241, G242, G246, A247, **F260, T261, G262, S263**, I266, V269, I270, E285, L286, G287, **C319**, E416, F418, L444, F482
AL1B1	0.96	0.92	I183, **P184, W185, N186**, M191, K209, A211, E212, Y241, G242, G246, A247, **F260, T261, G262, S263**, V266, L269, I270, E285, L286, G287, **C319**, E416, F418, L444, F428

ALDH2 and AL1B1 share 72.7% sequence identity over an overlap of 517 residues, and within the predicted binding sites, the amino acids "PWN" and "FTGS" (part of the patterns "PWNFP" and "FTGST") are present, along with cysteine 302 (position 319 in the full sequence), which is part of the "CCAG" pattern and included in the predicted disulfiram binding site, highlighting this region's potential relevance in drug binding.

AL1B1 is associated with colon cancer. The combination of disulfiram and copper significantly enhances its anti-tumor efficacy by targeting ALDH, the NK-κB pathway, the MAPK pathway, and other mechanisms [48]. Additionally, the disulfiram/copper complex has been shown to potentiate the anti-tumor effects of 5-fluorouracil in colon cancer models [49].

Aldehyde dehydrogenases (ALDHs) are a family of $NAD(P)^+$-dependent enzymes involved in the oxidation of aldehydes to their corresponding carboxylic acids, playing essential roles in cellular detoxification, metabolism, and protection against oxidative stress [50]. Among them, ALDH2 is a key mitochondrial isoform primarily responsible for metabolizing toxic acetaldehyde generated during alcohol metabolism [51]. Structurally, ALDH2 functions as a homotetramer, with each subunit comprising three domains: a catalytic domain, a coenzyme-binding domain, and an oligomerization domain. The active site, located at the interface of subunits, contains the catalytic residue Cys 302, which is critical for enzymatic activity [13]. The tetrameric organization is necessary for structural integrity and optimal catalytic function. Another mitochondrial member of the ALDH family is AL1B1, which shares significant sequence and structural similarity with ALDH2. Like ALDH2, AL1B1 contains a mitochondrial targeting sequence at its N-terminus. This signal peptide directs the protein to the mitochondria and is cleaved post-translationally, resulting in the mature, active form of the enzyme localized within the mitochondrial matrix. The mitochondrial localization of both ALDH2 and AL1B1 underscores their involvement in regulating intracellular aldehyde levels and their potential relevance as therapeutic targets, particularly in diseases where aldehyde accumulation and oxidative stress are pathogenic factors [52]. The therapeutic inhibition of ALDH enzymes (particularly with broad-spectrum agents like disulfiram) offers potential benefits across diseases such as cancer and neurodegeneration [14, 53]. However, ALDH's essential role in detoxification and cellular stress response means that non-specific or long-term inhibition may inadvertently drive drug resistance by enriching ALDH-overexpressing, therapy-resistant cells, including cancer stem cells. This underscores the need for greater precision in drug-target interactions, favouring isoform-selective inhibitors that minimize off-target effects.

4   Conclusion and Future Work

In this study, patterns are defined as short amino acid sequences shared across proteins. Some patterns reflect specific protein families, while others arise probabilistically. The main objective was to evaluate whether shared patterns could identify new drug repurposing candidates or predict off-target interactions, focusing on inhibitors, proteins, and patterns related to lung cancer. Among the proteins analyzed, several belong to the aldehyde dehydrogenase (ALDH) family, which is of particular interest due to its known role in cancer metabolism and drug resistance. These include ALDH2, AL1A1, AL1A2 and AL1B1, which were consistently represented across the datasets. The presence of conserved patterns within this protein family may point to shared structural or functional features that are relevant in the context of cancer progression and therapeutic targeting. The separation of ALDH family proteins in this analysis is crucial, as their conserved patterns may not only reflect evolutionary relationships but also indicate common binding sites or functional domains. This insight can guide future studies aimed at understanding how repurposed drugs, such as disulfiram, interact with members of the ALDH family and potentially influence treatment outcomes across multiple cancer types. This also underscores the potential of such drugs to exhibit broad-spectrum activity by targeting conserved patterns among related proteins, possibly extending their effect beyond a single protein to entire families or interaction networks. The study demonstrates the feasibility of detecting conserved amino acid patterns in cancer-related proteins, exemplified by the ALDH2 case study, and illustrates their utility in guiding drug repurposing strategies.

While pattern analysis provides valuable insights, it has limitations as it may overlook protein function, context-specific expression, and post-translational modifications. Integrating expression or functional data could help refine predictions and improve translational relevance, and future work should focus on incorporating such data for ALDH isoforms and other patter-bearing proteins across cancer types to better predict drug sensitivity and resistance, ultimately enabling more precise, context-specific therapeutic strategies.

Acknowledgments. This research was funded by the Spanish Ministerio de Ciencia e Innovación under projects "Data-driven drug repositioning applying graph neural networks (3DR-GNN)" (PID2021-122659OB-I00) and "Drug repurposing hypotheses through a data-driven approach (GRENADA)" (PDC2022–133173-I00).

Disclosure of Interests. The authors have no competing interests to declare that are relevant to the content of this article.

References

1. Gong, Z., Zhou, J., Ye, L., Ma, G., Xian, Y., Kulkarni, K.: Editorial: pharmacokinetic differences of drugs and their regulatory mechanisms under dual status including normal and diseased organism. Front. Pharmacol. **13**, 1063434 (2022). https://doi.org/10.3389/fphar.2022.1063434
2. Abdallah, Y.E.H., Chahal, S., Jamali, F., Mahmoud, S.H.: Drug-disease interaction: clinical consequences of inflammation on drugs action and disposition. J. Pharm. Pharm. Sci. **26**, 11137 (2023). https://doi.org/10.3389/jpps.2023.11137
3. Ram, P.T., Mendelsohn, J., Mills, G.B.: Bioinformatics and systems biology. Mol. Oncol. **6**(2), 147–154 (2012). https://dci.org/10.1016/j.molonc.2012.01.008
4. Palve, V., Liao, Y., Rix, L.L.R., Rix, U.: Turning liabilities into opportunities: Off-target based drug repurposing in cancer. Semin. Cancer Biol. **68**, 209–229 (2021). https://doi.org/10.1016/j.semcancer.2020.02.003
5. Gonzalez, M.W., Kann, M.G.: Chapter 4: protein interactions and disease. PLoS Comput. Biol. **8**(12), e1002819 (2012). https://doi.org/10.1371/journal.pcbi.1002819
6. Haass-Koffler, C.L., Akhlaghi, F., Swift, R.M., Leggio, L.: Altering ethanol pharmacokinetics to treat alcohol use disorder: can you teach an old dog new tricks? J. Psychopharmacol. (Oxf.) **31**(7), 812–818 (2017). https://doi.org/10.1177/0269881116684338
7. MacDonagh, L., et al.: Targeting the cancer stem cell marker, aldehyde dehydrogenase 1, to circumvent cisplatin resistance in NSCLC. Oncotarget **8**(42), 72544–72563 (2017). https://doi.org/10.18632/oncotarget.19881

8. Nechushtan, H., et al.: A phase IIb trial assessing the addition of disulfiram to chemotherapy for the treatment of metastatic non-small cell lung cancer. Oncologist **20**(4), 366–367 (2015). https://doi.org/10.1634/theoncologist.2014-0424

9. Caminear, M.W., Harrington, B.S., Kamdar, R.D., Kruhlak, M.J., Annunziata, C.M.: Disulfiram transcends ALDH inhibitory activity when targeting ovarian cancer tumor-initiating cells. Front. Oncol. **12**, 762820 (2022). https://doi.org/10.3389/fonc.2022.762820

10. Zeng, M., et al.: Disulfiram: a novel repurposed drug for cancer therapy. Chin. Med. J. (Engl.) **137**(12), 1389–1398 (2024). https://doi.org/10.1097/cm9.0000000000002909

11. Wang, N., et al.: Targeting ALDH2 with disulfiram/copper reverses the resistance of cancer cells to microtubule inhibitors. Exp. Cell Res. **362**(1), 72–82 (2018). https://doi.org/10.1016/j.yexcr.2017.11.004

12. Wang, L., Yu, Y., Zhou, C., Wan, R., Li, Y.: Anticancer effects of disulfiram: a systematic review of in vitro, animal, and human studies. Syst. Rev. **11**(1), 109 (2022). https://doi.org/10.1186/s13643-021-01858-4

13. Shortall, K., Djeghader, A., Magner, E., Soulimane, T.: Insights into aldehyde dehydrogenase enzymes: a structural perspective. Front. Mol. Biosci. **8**, 659550 (2021). https://doi.org/10.3389/fmolb.2021.659550

14. Xanthis, V., Mantso, T., Dimtsi, A., Pappa, A., Fadouloglou, V.E.: Human aldehyde dehydrogenases: a superfamily of similar yet different proteins highly related to cancer. Cancers **15**(17), 4419 (2023). https://doi.org/10.3390/cancers15174419

15. Chen, J., Huang, W., Cheng, C.-H., Zhou, L., Jiang, G.-B., Hu, Y.-Y.: Association between aldehyde dehydrogenase-2 polymorphisms and risk of alzheimer's disease and parkinson's disease: a meta-analysis based on 5,315 individuals. Front. Neurol. **10**, 00290 (2019). https://doi.org/10.3389/fneur.2019.00290

16. Zanoni, M., Bravaccini, S., Fabbri, F., Arienti, C.: Emerging roles of aldehyde dehydrogenase isoforms in anti-cancer therapy resistance. Front. Med. **9**, 795762 (2022). https://doi.org/10.3389/fmed.2022.795762

17. Ciccone, V., Morbidelli, L., Ziche, M., Donnini, S.: How to conjugate the stemness marker ALDH1A1 with tumor angiogenesis, progression, and drug resistance. Cancer Drug Resist. **3**, 26 (2019). https://doi.org/10.20517/cdr.2019.70

18. Chen, Y., et al.: Hypoxia-induced ALDH3A1 promotes the proliferation of non-small-cell lung cancer by regulating energy metabolism reprogramming. Cell Death Dis. **14**(9), 617 (2023). https://doi.org/10.1038/s41419-023-06142-y

19. Shin, M.-J., Cho, Y., Smith, G.D.: Alcohol consumption, *aldehyde dehydrogenase 2* gene polymorphisms, and cardiovascular health in Korea. Yonsei Med. J. **58**(4), 689 (2017). https://doi.org/10.3349/ymj.2017.58.4.689

20. Zhang, X., Sun, A., Ge, J.: Origin and spread of the ALDH2 Glu504Lys allele. Phenomics **1**(5), 222–228 (2021). https://doi.org/10.1007/s43657-021-00017-y

21. Zhao, Y., Wang, C.: Glu504Lys single nucleotide polymorphism of aldehyde dehydrogenase 2 gene and the risk of human diseases. BioMed Res. Int. **2015**, 1–9 (2015). https://doi.org/10.1155/2015/174050

22. Zhang, Y., et al.: The prognostic effect of ADH1B and ALDH2in lung adenocarcinoma. J. Clin. Oncol. (16_suppl), e15167–e15167 (2023). https://doi.org/10.1200/jco.2023.41.16_suppl.e15167

23. Xia, J., Li, S., Liu, S., Zhang, L.: Aldehyde dehydrogenase in solid tumors and other diseases: potential biomarkers and therapeutic targets. MedComm **4**(1), e195 (2023). https://doi.org/10.1002/mco2.195

24. Jumper, J., et al.: Highly accurate protein structure prediction with AlphaFold. Nature **596**(7873), 583–589 (2021). https://doi.org/10.1038/s41586-021-03819-2

25. Otero-Carrasco, B., et al.: Finding patterns in lung cancer protein sequences for drug repurposing. PLoS ONE **20**(5), 1–29 (2025). https://doi.org/10.1371/journal.pone.0322546

26. Lagunes-García, G., Rodríguez-González, A., Prieto-Santamaría, L., García Del Valle, E.P., Zanin, M., Menasalvas-Ruiz, E.: DISNET: a framework for extracting phenotypic disease information from public sources. PeerJ **8**, e8580 (2020). https://doi.org/10.7717/peerj.8580

27. Knox, C., et al.: DrugBank 6.0: the DrugBank Knowledgebase for 2024. Nucleic Acids Res. **52**(D1), D1265–D1275 (2024). https://doi.org/10.1093/nar/gkad976

28. UniProt: the universal protein knowledgebase in 2023. Nucleic Acids Res. **51**(D1), D523–D531 (2023). https://doi.org/10.1093/nar/gkac1052

29. Varadi, M., et al.: AlphaFold Protein structure database: massively expanding the structural coverage of protein-sequence space with high-accuracy models. Nucleic Acids Res. **50**(D1), D439–D444 (2022). https://doi.org/10.1093/nar/gkab1061

30. Abramson, J., et al.: Accurate structure prediction of biomolecular interactions with AlphaFold 3. Nature **630**(8016), 493–500 (2024). https://doi.org/10.1038/s41586-024-074 87-w

31. Liu, Y., Yang, X., Gan, J., Chen, S., Xiao, Z.-X., Cao, Y.: CB-Dock2: improved protein–ligand blind docking by integrating cavity detection, docking and homologous template fitting. Nucleic Acids Res. **50**(W1), W159–W164 (2022). https://doi.org/10.1093/nar/gkac394

32. Wu, Q., Peng, Z., Zhang, Y., Yang, J.: COACH-D: improved protein-ligand binding sites prediction with refined ligand-binding poses through molecular docking. Nucleic Acids Res. **46**(W1), W438–W442 (2018). https://doi.org/10.1093/nar/gky439

33. Meng, E.C., et al.: UCSF ChimeraX: tools for structure building and analysis. Protein Sci. Publ. Protein Soc. **32**(11), e4792 (2023). https://doi.org/10.1002/pro.4792

34. Goddard, T.D., et al.: UCSF ChimeraX: meeting modern challenges in visualization and analysis. Protein Sci. Publ. Protein Soc. **27**(1), 14–25 (2018). https://doi.org/10.1002/pro. 3235

35. Pettersen, E.F., et al.: UCSF ChimeraX: structure visualization for researchers, educators. and developers. Protein Sci. Publ. Protein Soc. **30**(1), 70–82 (2021). https://doi.org/10.1002/pro. 3943

36. Tejera-Nevado, P., Otero-Carrasco, B., Rodríguez-González, A.: Exploring Protein Patterns, Cavity Interactions, and Therapeutic Insights in Cancer (2025)

37. Safdar, A., Ismail, F.: A comprehensive review on pharmacological applications and drug-induced toxicity of valproic acid Saudi Pharm. J. **31**(2), 265–278 (2023). https://doi.org/10. 1016/j.jsps.2022.12.001

38. Omran, Z.: Development of new disulfiram analogues as ALDH1a1-selective inhibitors. Bioorg. Med. Chem. Lett. **40**, 127958 (2021). https://doi.org/10.1016/j.bmcl.2021.127958

39. Liu, X., et al.: Targeting ALDH1A1 by disulfiram/copper complex inhibits non-small cell lung cancer recurrence driven by ALDH-positive cancer stem cells. Oncotarget **7**(36), 58516–58530 (2016). https://doi.org/10.18632/oncotarget.11305

40. Rezk, Y.A., et al.: Disulfiram's antineoplastic effects on ovarian cancer. J. Cancer Ther. **06**(14), 1196–1205 (2015). https://doi.org/10.4236/jct.2015.614130

41. Lipsky, J.J., Shen, M.L., Naylor, S.: In vivo inhibition of aldehyde dehydrogenase by disulfiram. Chem. Biol. Interact. **130–132**, 93–102 (2001). https://doi.org/10.1016/s0009-279 7(00)00225-8

42. Yue, H., et al.: ALDH1A1 in cancers: bidirectional function, drug resistance, and regulatory mechanism. Front. Oncol. **12**, 918778 (2022). https://doi.org/10.3389/fonc.2022.918778

43. Liu, P., et al.: Cytotoxic effect of disulfiram/copper on human glioblastoma cell lines and ALDH-positive cancer-stem-like cells. Br. J. Cancer **107**(9), 1488–1497 (2012). https://doi. org/10.1038/bjc.2012.442

44. Guo, F., Yang, Z., Kulbe, H., Albers, A.E., Sehouli, J., Kaufmann, A.M.: Inhibitory effect on ovarian cancer ALDH– stem-like cells by Disulfiram and Copper treatment through ALDH and ROS modulation. Biomed. Pharmacother. **118**, 109371 (2019). https://doi.org/10.1016/j. biopha.2019.109371

45. Nie, D., Chen, C., Li, Y., Zeng, C.: Disulfiram, an aldehyde dehydrogenase inhibitor, works as a potent drug against sepsis and cancer via NETosis, pyroptosis, apoptosis, ferroptosis, and cuproptosis. Blood Sci. **4**(3), 152–154 (2022). https://doi.org/10.1097/bs9.000000000 0000117

46. Xu, Y., et al.: Disulfiram/copper markedly induced myeloma cell apoptosis through activation of JNK and intrinsic and extrinsic apoptosis pathways. Biomed. Pharmacother. **126**, 110048 (2020). https://doi.org/10.1016/j.biopha.2020.110048

47. Chu, M. et al.: Disulfiram/copper induce ferroptosis in triple-negative breast cancer cell line MDA-MB-231. Front. Biosci.-Landmark **28**(8), 186 (2023). https://doi.org/10.31083/j.fbl280 8186

48. Li, H., Wang, J., Wu, C., Wang, L., Chen, Z.-S., Cui, W.: The combination of disulfiram and copper for cancer treatment. Drug Discov. Today **25**(6), 1099–1108 (2020). https://doi.org/ 10.1016/j.drudis.2020.04.003

49. Hendrych, M., et al.: Disulfiram increases the efficacy of 5-fluorouracil in organotypic cultures of colorectal carcinoma. Biomed. Pharmacother. **153**, 113465 (2022). https://doi.org/10.1016/ j.biopha.2022.113465

50. Vasiliou, V., Nebert, D.W.: Analysis and update of the human aldehyde dehydrogenase (ALDH) gene family. Human Genom. **2**(2), 138 (2005). https://doi.org/10.1186/1479-7364-2-2-138

51. Chen, C.-H., Sun, L., Mochly-Rosen, D.: Mitochondrial aldehyde dehydrogenase and cardiac diseases. Cardiovasc. Res. **88**(1), 51–57 (2010). https://doi.org/10.1093/cvr/cvq192

52. Feng, Z., et al.: Targeting colorectal cancer with small-molecule inhibitors of ALDH1B1. Nat. Chem. Biol. **18**(10), 1065–1075 (2022). https://doi.org/10.1038/s41589-022-01048-w

53. Chen, C.-H., Joshi, A.U., Mochly-Rosen, D.: The Role of Mitochondrial Aldehyde Dehydrogenase 2 (ALDH2) in Neuropathology and Neurodegeneration (2023)

AI for Multi-omic Integration and Analysis of Heterogeneous biomedical data

A Framework for Evaluating the Stability of Learned Representations in Biologically-Constrained Models in Single-Cell

Sara Fernandez-Malvido[1] , Alberto Esteban-Medina[2,3] ,
Pelin Gundogdu[2] , Joaquin Dopazo[2,3(✉)] ,
Isabel A. Nepomuceno-Chamorro[1(✉)] , and Carlos Loucera[2,3(✉)]

[1] Dpto. de Lenguajes y Sistemas Informaticos, University of Seville, Sevilla, Spain
`inepomuceno@us.es`
[2] Andalusian Platform for Computational Medicine, Andalusian Public Foundation Progress and Health-FPS, Sevilla, Spain
[3] Institute of Biomedicine of Seville (IBiS), University Hospital Virgen del Rocío/CSIC/University of Seville, Seville, Spain
`{joaquin.dopazo,cloucera}@juntadeandalucia.es`

Abstract. This work addresses the instability and poor interpretability of computational models in single-cell RNA sequencing (scRNA-seq) analysis. We propose a generalizable framework to evaluate the stability of any model that generates pathway-level scores, applying it to both biologically-constrained variational autoencoders (iVAEs) and alternative graph-based methods like PathSingle. The central contribution is a modular and reproducible workflow, orchestrated with Pixi, Prefect and Ray, that automates the systematic comparison of different models across multiple random seeds.

The stability of the learned representations (pathway activities) was assessed using metrics for clustering coherence (Adjusted Mutual Information, AMI) and consistency across runs (hyperbolically weighted Kendall's Tau, w_τ). Our framework revealed that iVAEs informed by biological priors are significantly more stable and produce more meaningful cell groupings than randomly connected counterparts, therefore indicating the importance of being informed by meaningful biological entities. While the PathSingle model demonstrated a marginally superior consistency, the informed VAEs offered a better balance between stability and clustering performance.

This work provides a robust methodology for assessing diverse pathway scoring models, promoting the development of more reliable and interpretable tools for single-cell analysis.

Keywords: single-cell · deep learning · interpretability · stability · reproducibility · workflows · signaling pathways

---

S. Fernandez-Malvido and A. Esteban-Medina—Equal contribution.

A. López Fernández et al. (Eds.): CIABiomed 2025, LNBI 16148, pp. 423–437, 2026.
https://doi.org/10.1007/978-3-032-10661-2_32

1   Introduction

Single-cell transcriptomics has revolutionized our ability to study biological systems at unprecedented resolution, enabling the detection of cellular heterogeneity within seemingly uniform populations. However, this power comes with challenges: single-cell RNA sequencing (scRNA-seq) data are high-dimensional, noisy, and difficult to interpret, posing significant barriers to robust and reproducible analysis.

This work explores how biologically informed deep learning models, coupled with modern workflow orchestration tools, can enhance the interpretability, reproducibility, and scalability of single-cell data analysis. Specifically, we focus on the analysis of stability of biologically-informed variational autoencoders (iVAEs) that integrate prior biological knowledge, such as signaling pathways from Reactome [2] and the Kyoto Encyclopedia of Genes and Genomes database (KEGG) [9], into their architecture, steering the learning process toward biologically meaningful representations [5].

We also examine PathSingle [12], a complementary graph-based method that infers pathway activity directly from molecular networks. Comparing these two approaches allows us to assess the trade-offs between statistical modeling and biologically grounded inference.

To support fair, consistent, and reproducible experimentation, a modular and automated pipeline was developed using Pixi dependency manager, Prefect workflow manager and Ray computational distributed framework. This infrastructure enables parallel model training, evaluation, and analysis across multiple random seeds and configurations, while ensuring consistent data splits and scoring criteria. The goal is to provide a systematic framework to evaluate how structured prior knowledge impacts model stability, interpretability, and biological relevance.

Ultimately, this work contributes to ongoing efforts in bioinformatics to make deep learning models more interpretable and trustworthy by analyzing their stability of unsupervised learning models like iVAE [5]. This goal is especially relevant in sensitive biomedical contexts where biological insight is just as important as predictive performance.

2   Materials and Methods

2.1   Data and Tools

We employed a widely used single-cell RNA sequencing (scRNA-seq) dataset published by Kang et al. [10], available through the GEO repository (accession GSE96583). The dataset comprises peripheral blood mononuclear cells (PBMCs) from eight patients with systemic lupus erythematosus (SLE), collected under both control and IFN-β-stimulated conditions. It includes over 24,000 cells spanning key immune subtypes, including $CD4^+$ and $CD8^+$ T cells, B cells, NK cells, $CD14^+$ and $FCGR3A^+$ monocytes, dendritic cells, and megakaryocytes. The distribution and sample sizes are a cell count with B cells (2,651), CD14$^+$ monocytes (5,697), CD4$^+$ T cells (11,238), CD8$^+$ T cells (1,621), Dendritic cells (529), FCGR3A$^+$ monocytes (1,089). Megakaryocytes (132), and NK cells (1,716). Note that for each cell we have also the stimulated one.

Raw gene expression counts and cell-type annotations were retrieved using Scanpy [20] following established single-cell best practices [13]. As part of standard preprocessing, low-quality cells and genes were filtered, and the remaining counts were normalized per cell and log-transformed to stabilize variance.

For each of the 40 experimental runs, the dataset was partitioned into training (60%), validation (20%), and testing (20%) sets. To ensure that each split was representative of the underlying biological diversity, we used stratified sampling based on a combination of the cell type and experimental condition (control/stimulated). This guarantees that all cell populations are proportionally represented in each data subset, which is critical for robust model training and evaluation. Note that these preprocessing steps are part of the workflow.

To incorporate structured biological knowledge into the model architecture, we used two complementary pathway databases: KEGG [9] and Reactome [2]. KEGG provided high-level maps of signaling and metabolic pathways, which were decomposed into effector subpathways using the Hipathia framework [7], enabling the construction of minimal functional units within each pathway. In contrast, Reactome's manually curated molecular detail was leveraged to define an alternative graph-based layer, offering a more comprehensive perspective on cellular signaling. These dual sources of prior knowledge enabled a comparative assessment of different strategies for embedding biological structure into the variational autoencoder (VAE).

2.2   Models

Central to our deep learning methodology is a biologically inspired variational autoencoder (iVAE), where we updated the original codebase [5] to utilize Keras 3. This model builds upon previous supervised modeling studies, particularly [3] and [4], and enhances the conventional VAE by integrating existing biological insights directly into the encoder via the inclusion of "informed layers". These layers are constrained by binary adjacency matrices derived from pathway databases, where a connection is permitted only if a gene is a known member of a given pathway or circuit. This constraint is enforced by applying an element-wise Hadamard product between the layer's weight tensor and the adjacency matrix. We created several model variants for comparison: a KEGG-based model with a hierarchical two-layer structure representing signaling circuits and pathways; a Reactome-based model with a single informed layer for pathways; and a series of random control models. These controls share the same architecture but feature randomized connections at varying density levels (from 5% to 40%) to simulate being informed by random gene sets of different sizes. All informed layers used a hyperbolic tangent ($tanh$) activation function, and the models were trained using the Adam optimizer with L2 regularization to minimize the standard VAE loss function, which combines reconstruction error and Kullback-Leibler divergence.

Additionally, PathSingle [12] was implemented to quantify pathway activity by modeling signaling pathways as directed graphs with annotated activation and inhibition edges. It calculates pathway activity scores per cell by aggregating gene expression values along these functional interactions. Data smoothing was applied beforehand to reduce noise and capture latent relationships.

2.3 Validation Procedure

All models followed consistent training and evaluation procedures to allow for a direct comparison of their performance and interpretability. The evaluation was based on three distinct criteria, detailed in the following subsections: the stability of the learned representations, their ability to form coherent biological clusters, and, for autoencoder-based models, their reconstruction performance.

Reconstruction Performance of Autoencoder Models. The reconstruction fidelity of autoencoder models was quantified using the Mean Squared Error (MSE), which measures the difference between the original and reconstructed input data. A lower MSE indicates that the model's latent representation preserves more information. This metric was exclusively evaluated for the iVAE variants, as it is not applicable to the non-reconstructive PathSingle model.

The Mean Squared Error is defined as:

$$\text{MSE} = \frac{1}{N \cdot D} \sum_{i=1}^{N} \sum_{j=1}^{D} (x_{ij} - \hat{x}_{ij})^2 \tag{1}$$

where N is the number of cells, D is the number of genes, x_{ij} is the expression of gene j in cell i, and $\hat{x}_{ij}$ is the reconstructed expression value.

Our procedure for evaluating the reconstruction performance for each applicable model was as follows:

1. **Train the Model:** For each of the 40 independent training runs (each with a different random seed), the iVAE model was trained on the training dataset.
2. **Reconstruct Data:** The trained model was then used to generate a reconstructed version of the gene expression matrix for the cells in the test set.
3. **Calculate MSE Score:** For each run, we calculated the MSE score by comparing the original test set data with the reconstructed data.
4. **Evaluate Performance:** This process resulted in a distribution of 40 MSE scores for each iVAE variant. The magnitude and variance of this distribution were used to assess and compare the reconstruction capabilities of the different biologically-informed and random architectures.

Clustering Coherence of Learned Representations. We evaluated the biological relevance of the learned representations by assessing their ability to cluster cells according to known cell types. For this, we used the Adjusted Mutual

Information (AMI) score [17], which measures the agreement between the model-derived clusters and ground-truth labels while correcting for chance. An AMI score of 1 indicates a perfect match, while 0 represents random agreement.

The Adjusted Mutual Information is defined as:

$$\text{AMI}(U, V) = \frac{\text{MI}(U, V) - E[\text{MI}(U, V)]}{\max(H(U), H(V)) - E[\text{MI}(U, V)]} \tag{2}$$

where U is the set of ground-truth labels, V is the set of cluster assignments from the model, $\text{MI}(U, V)$ is the mutual information between them, $H(U)$ and $H(V)$ are their respective entropies, and $E[\text{MI}(U, V)]$ is the expected mutual information between two random clusterings.

Our procedure for evaluating the clustering coherence for each model was as follows:

1. **Generate Representations:** For each of the 40 independent training runs (each with a different random seed), we used the trained model to generate pathway activity scores or latent representations for all cells in the test set.
2. **Perform Clustering:** We applied the Mini-Batch K-Means algorithm to these representations to group the cells into clusters. The number of clusters was set to the known number of cell types in the dataset.
3. **Calculate AMI Score:** For each run, we calculated the AMI score by comparing the cluster labels assigned by K-Means (V) with the ground-truth cell type annotations (U).
4. **Evaluate Coherence:** This process yielded a distribution of 40 AMI scores for each model. A model was considered to have high clustering coherence if this distribution was concentrated near $+1$, indicating that its learned representations consistently and accurately reflect the biological distinctions between cell types.

Stability of Learned Representations. Beyond standard metrics, we evaluated the stability of each model's interpretations, a critical requirement for their practical use. We quantified this by comparing the rankings of pathway activity scores generated across multiple independent runs with different random seeds. This approach tests whether a model consistently identifies the most important biological pathways, a key indicator of its reliability.

To measure the concordance between pairs of rankings, we employed the hyperbolically weighted Kendall's Tau (w_τ) score [16]. This metric is a variant of the standard Kendall's Tau rank correlation coefficient but introduces a weighting scheme that penalizes disagreements in higher-ranked items more severely than those in lower-ranked items. This is particularly well-suited for our use case, where a change in the rank of a top pathway is more significant than a change deep in the list.

The hyperbolically weighted Kendall's Tau is defined as:

$$w_\tau = \frac{\sum_{i<j} w_{ij} \cdot \text{sign}(r_x(i) - r_x(j)) \cdot \text{sign}(r_y(i) - r_y(j))}{\sum_{i<j} w_{ij}} \tag{3}$$

where $r_x(i)$ and $r_y(i)$ are the ranks of item i in two different rankings, and the weight w_{ij} for each pair of items (i, j) is calculated based on their ranks:

$$w_{ij} = \frac{1}{r_x(i) + r_y(i)} \tag{4}$$

The coefficient ranges from -1 (perfect opposition) to +1 (perfect agreement), with 0 indicating no correlation.

Our procedure for calculating stability for each model was as follows:

1. **Generate Pathway Scores:** The model was trained independently using 40 different random seeds. For each trained instance, we generated pathway activity scores for all cells in the test set.
2. **Create Average Rankings:** For each of the 40 runs, we computed the mean activity score across all test cells for every pathway. This produced 40 distinct, ranked lists of pathways, ordered from most to least active.
3. **Pairwise Comparison:** We formed all possible unique pairs from the 40 rankings. For each pair, we calculated the w_τ coefficient.
4. **Evaluate Stability:** This process resulted in a distribution of w_τ scores for each model. A model was deemed highly stable if this distribution was tightly concentrated near +1, indicating that different training runs produced highly similar pathway rankings. Conversely, a distribution centered around 0 signified an unstable model whose outputs are not reliable for biological interpretation.

2.4   Workflow Implementation and Execution

To ensure a rigorous, reproducible, and scalable comparison of the different modeling approaches, we developed a fully automated computational workflow. The entire pipeline, from data preprocessing to the generation of final figures, was orchestrated using modern workflow management tools to handle the complexity of running numerous experiments in parallel, as depicted in Fig. 1.

The foundation of our framework is built on three key technologies. First, we used Pixi to define and manage a consistent software environment, ensuring that all dependencies and package versions are locked and reproducible across different systems. Second, the logic and dependencies of the experimental pipeline were defined using Prefect, a modern dataflow orchestration tool. Prefect allowed us to structure the analysis as a directed acyclic graph (DAG) of tasks, where each step—such as model training, clustering, or consistency calculation—is an independent, cacheable unit. Finally, for execution, we leveraged Ray via Prefect's 'RayTaskRunner' to distribute and parallelize the computational tasks efficiently across multiple CPUs and GPUs.

The workflow begins with initial setup tasks, including dependency installation, creation of result directories, and downloading the scRNA-seq dataset. The core of the pipeline then enters a parallelized loop that iterates through each model configuration (i.e., each iVAE variant—KEGG, Reactome, and all random configurations—and PathSingle) and across 40 different random seeds.

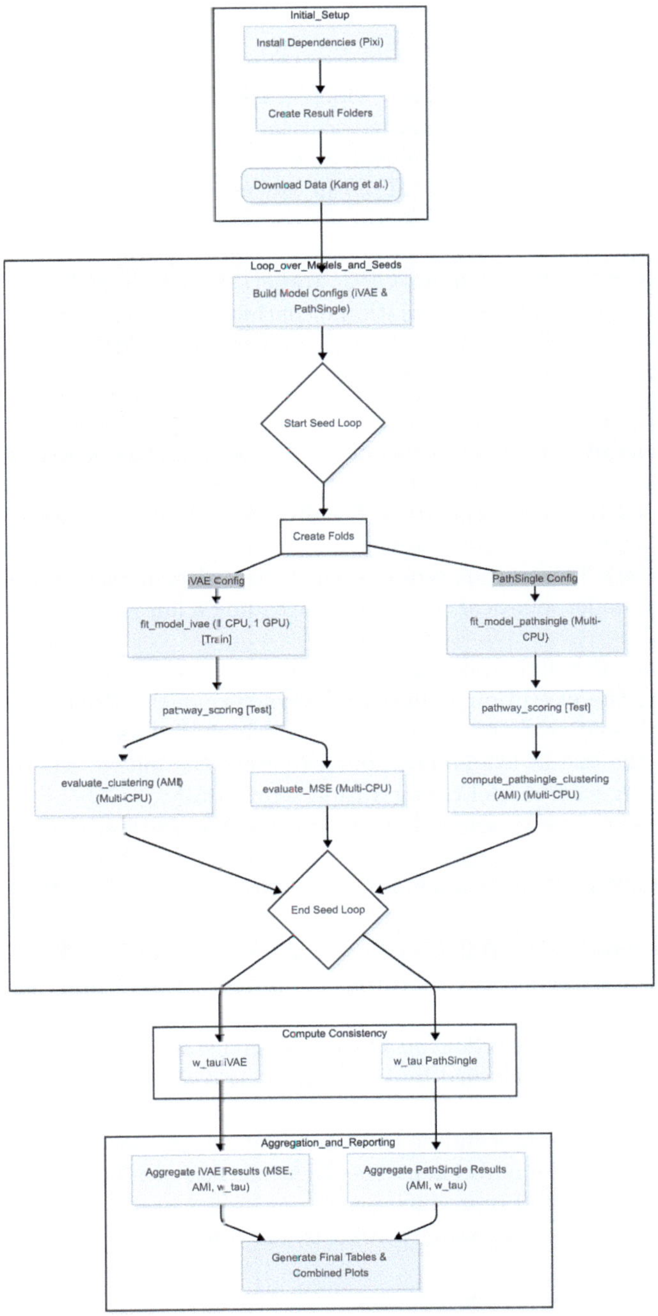

Fig. 1. Workflow Diagram.

Within each iteration of this loop, the workflow executes two main branches in parallel:

1. **iVAE Branch:** The corresponding iVAE model is trained for a specific seed using the 'fit_model_ivae' task, which is assigned to a single CPU and a dedicated GPU to leverage hardware acceleration. Following training, the resulting latent representations are evaluated for clustering coherence (AMI) and reconstruction performance (MSE) in downstream tasks that run on multi-CPU resources.

2. **PathSingle Branch:** In parallel, the 'fit_model_pathsingle' task is executed for the same seed on a multi-CPU worker. This task includes the MAGIC imputation step and the calculation of pathway activity scores. The resulting scores are then evaluated for clustering coherence (AMI) and reconstruction loss (MSE) for the autoencoder models.

After all parallel runs for the 40 seeds are complete for a given model, the workflow proceeds to the aggregation stage. The pathway scores from all seeds are collected to compute the stability of the representations using the hyperbolically weighted Kendall's Tau (w_τ). Finally, all evaluation metrics (MSE, AMI, and w_τ) are aggregated, and summary tables and comparative plots are automatically generated. This automated and parallelized design ensures that all models are evaluated under identical conditions, providing a fair and systematic basis for comparison.

Note that, although the workflow is presented conceptually as a series of nested loops, the execution is managed asynchronously. Prefect submits tasks for all models and seeds to the Ray engine, which dynamically schedules them to optimize the use of available resources across the entire experiment, rather than processing each model configuration sequentially.

All analyses were conducted in a reproducible Python environment managed with Pixi. Core tools included `Scanpy` for scRNA-seq preprocessing, `Keras/TensorFlow` [1] for model implementation, and standard Python scientific libraries such as `NumPy` [6], `SciPy` [18], `scikit-learn` [14] for data handling and statistical evaluation. Workflow orchestration was handled via `Prefect` and `Ray` [11], while visualization was performed with `Matplotlib` [8] and `Seaborn` [19].

The entire pipeline can be executed from a Linux terminal using the command *pixi run python workflow.py*. Once `Pixi` is installed, all required dependencies are automatically resolved by Pixi. This single command launches the complete experimentation suite –including all model variants, configurations, and random seeds– ensuring a fully automated and scalable execution of the entire workflow. Code and environment specifications are available in the project's repository to ensure full reproducibility:

https://github.com/saraafdezz/robustness_informed_TFG

3   Results

We systematically evaluated the performance of our models using three key criteria: reconstruction performance, clustering coherence, and the stability of the learned representations. The models under comparison included the biologically-informed iVAE variants (KEGG and Reactome), the PathSingle model, and a series of random iVAE control models. To thoroughly assess the impact of uninformative priors, these control models were generated with randomized connection matrices at eight different density levels, ranging from 5% to 40% in 5% increments. To ensure the robustness of our findings, every model configuration was trained and evaluated across 40 independent runs, each initiated with a different random seed and split. The iVAE models were trained for up to 1000 epochs with an early stopping patience of 100 epochs to prevent overfitting, using 20% of the training data for validation.

Note that, throughout this section, the labels used in figures and tables refer to specific layers of the iVAE architecture from which the representations were extracted. Specifically, **'circuits'** denotes the informed layer constrained by KEGG effector subpathway gene sets; **'pathways'** refers to the informed layer constrained by gene sets from full KEGG or Reactome pathways; and **'funnel'** corresponds to the final, most compressed bottleneck layer of the encoder.

Reconstruction Performance

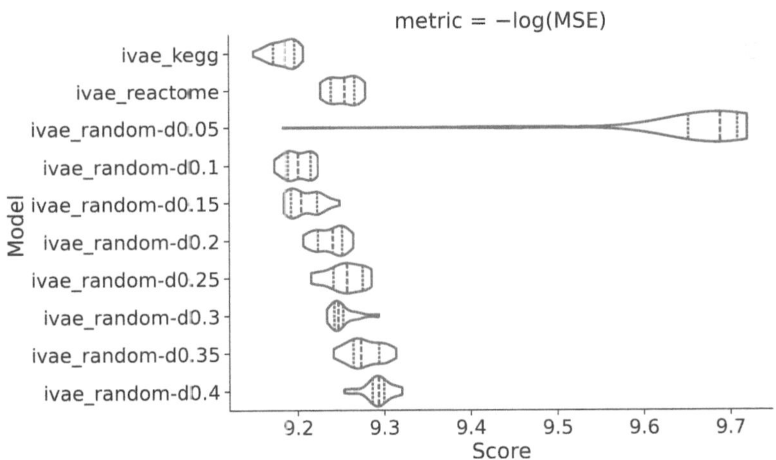

Fig. 2. Distribution of $-\log(\text{MSE})$ across the test sets for the different autoencoder models.

First, we assessed the ability of the autoencoder-based models to reconstruct the input gene expression data, quantified by the Mean Squared Error (MSE). Note that, while Table 1 shows the MSE (lower is better), for visualization purposes we show the logarithmic MSE distribution for each model instead of the MSE in Fig. 2 (higher is better).

As shown in Fig. 2, the models with randomized connections consistently achieved a slightly lower reconstruction error compared to the iVAE models informed by KEGG and Reactome priors. While this suggests a better numerical fit to the data, this result should be interpreted with caution. The random models, lacking biological constraints, may be more prone to overfitting or capturing spurious noise patterns that do not correspond to meaningful biological signals. The biologically-informed models, by contrast, are constrained to learn representations that, while slightly less precise in reconstruction, are more aligned with known functional pathways.

Table 1. Comparison of Mean Squared Error (MSE) for all autoencoder models. The table shows summary statistics for the reconstruction error on the training, validation, and test partitions, aggregated over 40 runs.

model	metric	split	mean	std	25%	50%	75%
ivae_kegg	mse	test	0.000103	0.000002	0.000101	0.000102	0.000104
		train	0.000103	0.000002	0.000101	0.000102	0.000104
		val	0.000103	0.000002	0.000102	0.000103	0.000104
ivae_random-d0.05	mse	test	0.000064	0.000007	0.000061	0.000062	0.000064
		train	0.000064	0.000007	0.000061	0.000062	0.000064
		val	0.000064	0.000008	0.000061	0.000062	0.000064
ivae_random-d0.1	mse	test	0.000101	0.000002	0.000100	0.000101	0.000102
		train	0.000100	0.000002	0.000099	0.000100	0.000102
		val	0.000101	0.000002	0.000100	0.000101	0.000102
ivae_random-d0.15	mse	test	0.000100	0.000002	0.000099	0.000101	0.000102
		train	0.000099	0.000002	0.000098	0.000100	0.000101
		val	0.000100	0.000002	0.000099	0.000101	0.000102
ivae_random-d0.2	mse	test	0.000097	0.000002	0.000096	0.000097	0.000099
		train	0.000097	0.000002	0.000095	0.000096	0.000098
		val	0.000097	0.000002	0.000096	0.000097	0.000099
ivae_random-d0.25	mse	test	0.000096	0.000002	0.000094	0.000095	0.000097
		train	0.000095	0.000002	0.000093	0.000095	0.000096
		val	0.000096	0.000002	0.000094	0.000095	0.000097
ivae_random-d0.3	mse	test	0.000096	0.000001	0.000096	0.000096	0.000097
		train	0.000095	0.000001	0.000095	0.000096	0.000096
		val	0.000096	0.000001	0.000096	0.000097	0.000097
ivae_random-d0.35	mse	test	0.000094	0.000002	0.000092	0.000094	0.000095
		train	0.000093	0.000002	0.000091	0.000093	0.000094
		val	0.000094	0.000002	0.000092	0.000094	0.000095
ivae_random-d0.4	mse	test	0.000092	0.000001	0.000091	0.000092	0.000093
		train	0.000091	0.000001	0.000091	0.000091	0.000092
		val	0.000092	0.000001	0.000092	0.000092	0.000093
ivae_reactome	mse	test	0.000096	0.000002	0.000095	0.000096	0.000097
		train	0.000096	0.000002	0.000094	0.000095	0.000097
		val	0.000096	0.000002	0.000095	0.000096	0.000097

Table 1 shows the mean, standard deviation and quantiles for the autoencoder models across the train, validation and test folds. It shows that there is no clear sign of overfitting, demonstrating the regularization properties of the sparsity induced by the biological constrains imposed to the models [15].

Clustering Coherence

Next, we evaluated the biological relevance of the learned representations by measuring their ability to group cells according to their known cell types. Using the Adjusted Mutual Information (AMI) score, we compared the clusters derived from each model's representations against the ground-truth labels. The results, summarized in Fig. 3 and Table 2, demonstrate a clear advantage for the iVAE models.

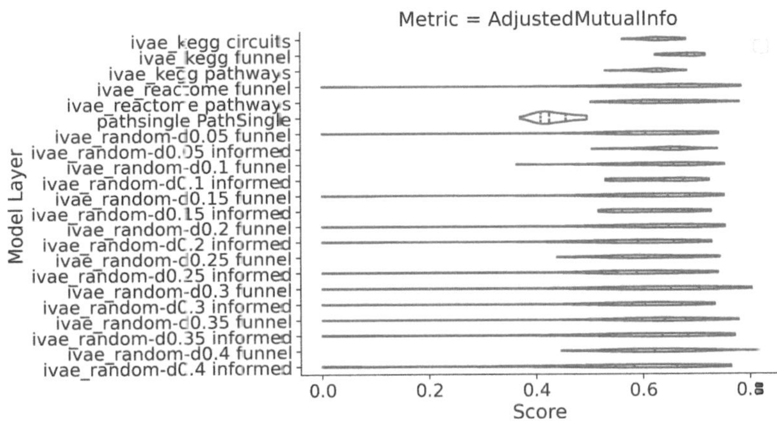

Fig. 3. AMI scores across the test partition slits for each model.

Both the KEGG and Reactome-informed iVAEs substantially outperformed the PathSingle model, producing more accurate and biologically meaningful cell clusters. Interestingly, even the iVAE models with random connections surpassed PathSingle in clustering performance, highlighting the power of the autoencoder architecture in capturing the complex structure of single-cell data. This indicates that the representations learned by the iVAE framework are more effective at preserving the distinctions between cell types.

Stability of Learned Representations

Finally, and most central to our objective, we assessed the stability of the interpretations generated by each model. Using the hyperbolically weighted Kendall's Tau (w_τ), we measured the consistency of the pathway activity rankings across the 40 independent runs. As illustrated in Fig. 4 and detailed in Table 3, the PathSingle model exhibited the highest stability, with w_τ scores consistently close to 1. This is expected, as its deterministic, graph-based nature makes it less sensitive to random initialization.

Table 2. Mean AMI, standard deviation, and quantiles across test set evaluations.

model	layer	metric	split	mean	std	25%	50%	75%
ivae_kegg	circuits	**AdjustedMutualInfo**	test	0.630405	0.030071	0.613168	0.630072	0.653575
	funnel	**AdjustedMutualInfo**	test	0.677195	0.026336	0.653866	0.683458	0.697839
	pathways	**AdjustedMutualInfo**	test	0.619839	0.032579	0.608235	0.626532	0.637214
ivae_random-d0.05	funnel	**AdjustedMutualInfo**	test	0.613760	0.121218	0.574217	0.605240	0.683358
	informed	**AdjustedMutualInfo**	test	0.639955	0.050531	0.621835	0.651795	0.669940
ivae_random-d0.1	funnel	**AdjustedMutualInfo**	test	0.633792	0.077809	0.582738	0.657961	0.685997
	informed	**AdjustedMutualInfo**	test	0.626747	0.060275	0.568748	0.639797	0.675719
ivae_random-d0.15	funnel	**AdjustedMutualInfo**	test	0.618132	0.121866	0.567108	0.646634	0.687890
	informed	**AdjustedMutualInfo**	test	0.616579	0.063123	0.561681	0.616184	0.674452
ivae_random-d0.2	funnel	**AdjustedMutualInfo**	test	0.614048	0.156422	0.585884	0.648244	0.702872
	informed	**AdjustedMutualInfo**	test	0.549220	0.192711	0.546837	0.591317	0.657328
ivae_random-d0.25	funnel	**AdjustedMutualInfo**	test	0.609722	0.077199	0.561655	0.609573	0.671955
	informed	**AdjustedMutualInfo**	test	0.594656	0.115676	0.551893	0.587985	0.657373
ivae_random-d0.3	funnel	**AdjustedMutualInfo**	test	0.612595	0.157271	0.547614	0.651215	0.704474
	informed	**AdjustedMutualInfo**	test	0.594442	0.150115	0.578817	0.603861	0.671972
ivae_random-d0.35	funnel	**AdjustedMutualInfo**	test	0.628473	0.119293	0.597362	0.640118	0.679562
	informed	**AdjustedMutualInfo**	test	0.568297	0.212094	0.548679	0.635241	0.683135
ivae_random-d0.4	funnel	**AdjustedMutualInfo**	test	0.623277	0.075109	0.578039	0.620499	0.677316
	informed	**AdjustedMutualInfo**	test	0.527368	0.217966	0.547322	0.583238	0.662547
ivae_reactome	funnel	**AdjustedMutualInfo**	test	0.638896	0.124159	0.611879	0.667759	0.706728
	pathways	**AdjustedMutualInfo**	test	0.635199	0.062760	0.597158	0.631506	0.685867
pathsingle	**PathSingle**	**AdjustedMutualInfo**	test	0.432076	0.031541	0.409736	0.425518	0.456317

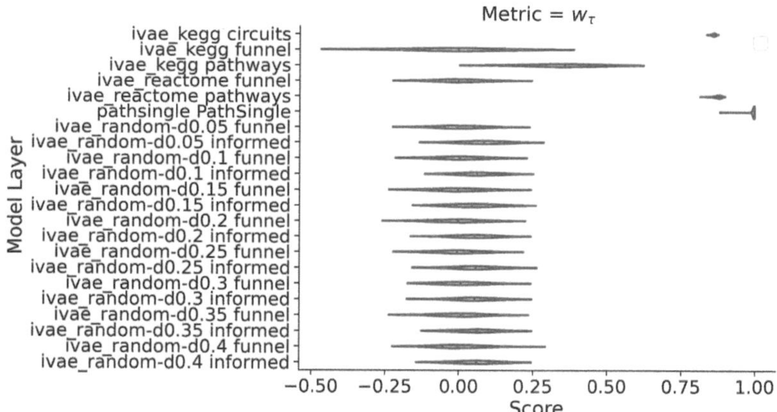

Fig. 4. Distribution of the w_τ metric for each model.

Interestingly, the iVAE models informed by KEGG and Reactome also demonstrated high stability, with scores significantly higher than those of the random models, which hovered near 0. This finding confirms that integrating prior biological knowledge acts as a powerful regularizer, guiding the model to converge to consistent and reproducible representations. The lack of stability in the random models, despite their strong reconstruction performance, underscores the importance of relying on biologically-constrained models for interpretation, despite the good reconstruction performance. Overall, the results show that while PathSingle is the most consistent, the biologically-informed iVAEs offer a compelling balance of high stability and superior biological clustering performance.

Table 3. w_τ mean, standard deviation, and quantiles across test set evaluations.

model	layer	metric	split	mean	std	25%	50%	75%
ivae_kegg	circuits	w_τ	test	0.856135	0.005686	0.852640	0.856495	0.860210
	funnel	w_τ	test	−0.004381	0.147799	−0.103010	−0.003658	0.095112
	pathways	w_τ	test	0.350385	0.114006	0.275737	0.353417	0.429165
ivae_random-d0.05	funnel	w_τ	test	−0.002185	0.074644	−0.051613	−0.005870	0.049108
	informed	w_τ	test	0.082471	0.071742	0.032854	0.083291	0.135778
ivae_random-d0.1	funnel	w_τ	test	0.002546	0.076784	−0.051009	0.002362	0.055166
	informed	w_τ	test	0.062425	0.066339	0.017531	0.061246	0.103903
ivae_random-d0.15	funnel	w_τ	test	−0.003190	0.077562	−0.055179	−0.004507	0.052018
	informed	w_τ	test	0.043981	0.071909	−0.008333	0.043214	0.093406
ivae_random-d0.2	funnel	w_τ	test	−0.004593	0.076918	−0.061503	−0.003090	0.047223
	informed	w_τ	test	0.054059	0.066257	0.006797	0.053038	0.093916
ivae_random-d0.25	funnel	w_τ	test	−0.006196	0.074965	−0.056088	−0.005879	0.042668
	informed	w_τ	test	0.056852	0.069324	0.008841	0.055715	0.106246
ivae_random-d0.3	funnel	w_τ	test	0.006944	0.074577	−0.044509	0.006046	0.059564
	informed	w_τ	test	0.055186	0.066016	0.010356	0.053325	0.101222
ivae_random-d0.35	funnel	w_τ	test	0.005108	0.073432	−0.045029	0.007200	0.054170
	informed	w_τ	test	0.060575	0.068216	0.012185	0.061832	0.107881
ivae_random-d0.4	funnel	w_τ	test	−0.002430	0.073696	−0.053824	−0.002797	0.047180
	informed	w_τ	test	0.059046	0.067595	0.012988	0.059850	0.104170
ivae_reactome	funnel	w_τ	test	−0.000548	0.073908	−0.048277	−0.002830	0.049530
	pathways	w_τ	test	0.869380	0.013315	0.863279	0.872213	0.878856
pathsingle	PathSingle	w_τ	test	0.990427	0.023270	0.995272	0.995542	0.995702

4 Conclusions

In this work, we have introduced a comprehensive framework to systematically evaluate the stability and biological utility of pathway scoring models for single-cell transcriptomic data. A key contribution is the development of a fully automated and reproducible workflow using Pixi, Prefect and Ray, which facilitates the rigorous comparison of diverse modeling approaches. Our results highlight a critical trade-off between model architecture and interpretive reliability. While the deterministic PathSingle model demonstrated the highest stability, the deep learning-based iVAE models, particularly those informed by KEGG and Reactome priors, provided a superior balance of performance, achieving both high stability and excellent clustering coherence.

In summary, our framework revealed that the stability of the iVAE is highly dependent on the integration of meaningful biological knowledge. The models constrained by biological priors produced consistent pathway rankings across independent runs, whereas the control models with randomized connections were highly unstable. This finding underscores that without meaningful biological constraints, interpretations from deep learning models can be unreliable and vary significantly with stochastic factors in the training process. Collectively, our work provides a robust methodological blueprint for assessing the reliability of interpretable models in single-cell research by prioritizing stability as a key evaluation criterion.

5   Limitations and Future Work

While our biologically informed framework shows strong performance in scRNA-seq analysis, several limitations remain. First, its applicability to bulk RNA-seq data is not yet validated, and the differences in resolution and signal complexity may pose challenges. Second, the reliance on static pathway databases limits the model's ability to capture dynamic and context-specific biological processes. Finally, scalability remains a concern, especially as datasets grow in size and complexity, such as in cell atlas projects involving millions of cells.

Nevertheless, the modular nature of our workflow provides a solid foundation for future research. It can be easily extended to include a broader range of pathway databases and scoring models via simple indicator matrices and execution wrappers, enabling a continuously expanding benchmark of available tools. Future work should also incorporate additional metrics, particularly around interpretability—such as alignment with known gene markers or the ability to detect perturbed pathways under specific stimuli. Moreover, the framework could be generalized to other omics layers (e.g., ATAC-seq, proteomics) or even non-biological domains where domain-informed modeling is beneficial. These extensions would broaden the impact of our approach and demonstrate its adaptability across diverse data landscapes.

Acknowledgments. This study was supported in part by grants from the Spanish Ministry of Science and Innovation (PID2023-152380OB-C21), the Institute of Health Carlos III (PMP24/00024), the Consejería de Salud y Consumo from the Junta de Andalucía (EXC-2023-01, IE19_259 FPS).

Disclosure of Interests. The authors have no competing interests to declare that are relevant to the content of this article.

References

1. Abadi, M., Agarwal, A., et al.: TensorFlow: large-scale machine learning on heterogeneous systems (2015). https://www.tensorflow.org/

2. Croft, D., et al.: Reactome: a database of reactions, pathways and biological processes. Nucleic Acids Res. **39**(suppl_1), D691–D697 (2010)
3. Gundogdu, P., Alamo, I., Nepomuceno-Chamorro, I.A., Dopazo, J., Loucera, C.: SigPrimedNet: a signaling-informed neural network for scRNA-seq annotation of known and unknown cell types. Biology **12**(4), 579 (2023)
4. Gundogdu, P., Loucera, C., Alamo-Alvarez, I., Dopazo, J., Nepomuceno, I.: Integrating pathway knowledge with deep neural networks to reduce the dimensionality in single-cell RNA-seq data. BioData Min. **15**(1), 1 (2022)
5. Gundogdu, P., Payá-Milans, M., Alamo-Alvarez, I., Nepomuceno-Chamorro, I.A., Dopazo, J., Loucera, C.: Cell-level pathway scoring comparison with a biologically constrained variational autoencoder. Cell-Level Pathway Scoring Comparison with a Biologically Constrained Variational Autoencoder. In: Pang, J., Niehren, J. (eds) CMSB 2023. LNCS, vol. 14137, p. 62–77. Springer, Cham (2023). https://doi.org/10.1007/978-3-031-42697-1_5
6. Harris, C.R., et al.: Array programming with NumPy. Nature **585**(7825) 357–362 (2020)
7. Hidalgo, M.R., Cubuk, C., Amadoz, A., Salavert, F., Carbonell-Caballero, J., Dopazo, J.: High throughput estimation of functional cell activities reveals disease mechanisms and predicts relevant clinical outcomes. Oncotarget **8**(3), 5160–5178 (2017)
8. Hunter, J.D.: Matplotlib: a 2D graphics environment. Comput. Sci. Eng. **9**(3), 90–95 (2007)
9. Kanehisa, M., Furumichi, M., Sato, Y., Kawashima, S., Ishiguro-Watanabe, M.: KEGG for taxonomy-based analysis of pathways and genomes. Nucleic Acids Res. **51**(D1), D587–D592 (2023)
10. Kang, H.M., Subramaniam, M., Targ, S., et al.: Multiplexed droplet single-cell RNA-sequencing using natural genetic variation. Nat. Biotechnol. **36**(1), 89–94 (2018)
11. Karau, H., Lublinsky, B.: Scaling Python with Ray. O'Reilly Media, Inc. (2022)
12. Livne, D., Efroni, S.: Pathway metrics accurately stratify T cells to their cells states. BioData Min. **17**(60) (2024). https://doi.org/10.1186/s13040-024-00416-7
13. Luecken, M.D., Theis, F.J.: Current best practices in single-cell RNA-seq analysis: a tutorial. Mol. Syst. Biol. **15**(6), e8746 (2019)
14. Pedregosa, F., et al.: Scikit-learn: machine learning in python. J. Mach. Learn. Res. **12**, 2825–2830 (2011)
15. Thapa, K., Kinali, M., Pei, S., Luna, A., Babur, Ö.: Strategies to Include Prior Knowledge in Omics Analysis with Deep Neural Networks. Patterns **6**(3) (2025)
16. Vigna, S.: A weighted correlation index for rankings with ties. In: Proceedings of the 24th International Conference on World Wide Web, pp. 1166–1176 (2015)
17. Vinh, N.X., Epps, J., Bailey, J.: Information theoretic measures for clusterings comparison: variants, properties, normalization and correction for chance. J. Mach. Learn. Res. **11**, 2837–2854 (2010)
18. Virtanen, P., et al.: Scipy 1.0: fundamental algorithms for scientific computing in python. Nat. Methods **17**(3), 261–272 (2020)
19. Waskom, M.L.: seaborn: statistical data visualization. J. Open Sour. Softw. **6**(60), 3021 (2021). https://doi.org/10.21105/joss.03021 https://doi.org/10.21105/joss.03021
20. Wolf, F.A., Angerer, P., Theis, F.J.: SCANPY: large-scale single-cell gene expression data analysis. Genome Biol. **19**(1), 15 (2018)

An Interpretable Graph Neural Network for Multi-omics Data Integration and Biomarker Discovery

Alberto Labarga[(⊠)] [iD]

Universidad Pública de Navarra, Pamplona, Spain
`alberto.labarga@gmail.com`

Abstract. The integration of heterogeneous multi-omics data remains a critical challenge in computational biology, essential for unraveling the complex molecular underpinnings of diseases. Existing methods often struggle to incorporate prior biological knowledge effectively or provide interpretable results. We present BioMGNN v2, an end-to-end, supervised deep learning framework designed to address these challenges by extending our previous BioMGNN architecture. BioMGNN v2 constructs a heterogeneous biomedical knowledge graph that embeds patient-specific multi-omics data within a network of signed, directed biological interactions. It then leverages a heterogeneous graph neural network with relation-specific attention and integrates a complementary patient–patient graph branch, enabling robust representation learning across both biological entities and clinical samples. Self-supervised pretraining with contrastive and masked modeling objectives further enhances generalization, while gated cross-omics fusion and multi-task learning improve classification accuracy and biological relevance. To ensure interpretability, BioMGNN v2 incorporates graph explainers with stability selection to identify reproducible and statistically rigorous biomarker modules, complemented by uncertainty estimation for clinical reliability. We demonstrate the superior performance of our algorithm in patient classification and its ability to uncover biologically plausible and interpretable biomarkers, positioning it as a powerful and trustworthy tool for advancing precision medicine.

Keywords: Multi-omics Integration · Graph Neural Network · Graph Attention Network · Biomarker Discovery · Computational Biology · Precision Medicine

1 Introduction

The complexity of human diseases, such as cancer, stroke, and neurodegenerative disorders, arises from intricate perturbations across multiple molecular layers, including the genome, epigenome, transcriptome, and proteome. Consequently, the analysis of a single data modality provides only a partial view of the underlying pathology. Single-omics approaches often fail to consider the interplay among various molecular entities, leading to the risk of overlooking critical biological information. A comprehensive understanding of disease mechanisms, therefore, necessitates the integrative analysis of multi-omics data, which enables a holistic view across different biological levels and is increasingly vital for biomedical research.

A. López Fernández et al. (Eds.): CIABiomed 2025, LNBI 16148, pp. 438–448, 2026.
https://doi.org/10.1007/978-3-032-10661-2_33

The evolution of computational strategies for data integration reflects a progression towards handling greater complexity and scale. Early approaches relied on correlation analysis or linear dimensionality reduction techniques like Principal Component Analysis (PCA) and Independent Component Analysis (ICA). While useful, these methods are often insufficient for capturing the complex, non-linear relationships inherent in biological systems. This led to the development of more sophisticated matrix factorization techniques, such as Multi-Omics Factor Analysis (MOFA) [1] and integrative Non-negative Matrix Factorization (intNMF) [2], which aim to decompose data into shared and data-specific components of variation. More recently, deep learning models, particularly autoencoders, have demonstrated a superior ability to learn compressed, non-linear latent representations from high-dimensional data, with models like AIME [3] and MAE [4] showing promise in this domain.

A fundamental limitation of these methods is that they typically treat biological features as an unstructured high-dimensional vector, compelling the model to learn all relationships from the data alone. However, biological systems are inherently structured as complex networks of interacting components. This realization has catalyzed a paradigm shift towards graph-based methods, which represent a more natural and powerful approach to modeling biological data. By representing biological entities as nodes and their interactions as edges, Graph Neural Networks (GNNs) [5] can explicitly leverage prior biological knowledge encoded in the graph structure. This knowledge-guided approach provides a strong inductive bias, allowing the model to focus on learning the nuances of patient-specific data within a biologically relevant context, rather than expending capacity to re-learn fundamental interactions. This leads to more efficient, robust, and interpretable models.

Several graph-based methods for multi-omics integration have been proposed. For example, MOGONET (Multi-Omics Graph cOnvolutional NETworks) [6] or MOGCN (Multi-Omics Integration Method Based on Graph Convolutional Network) [7] learn omics-specific features from patient similarity graphs, but these methods have a notable limitation: its integration process is focused at the patient level, which may overlook the intricate interplay between different molecular features across omics layers. This highlights a specific need for models that can more deeply integrate feature-level interactions within a unified graph structure.

To address this gap, this paper introduces a novel, end-to-end supervised framework for multi-omics integration. Our method key contributions are threefold: (1) it constructs a multi-layer, heterogeneous graph that embeds patient-specific omics data directly into a comprehensive biomedical knowledge graph; (2) it utilizes a hybrid message-passing architecture that combines the inductive power of GraphSAGE [8] with the expressive and focused learning of a multi-head Graph Attention Network (GAT) [9]; and (3) it features an integrated and interpretable mechanism for biomarker discovery based on the model's learned attention weights, identifying not just individual markers but interconnected biomarker modules.

2 Methodological Framework

Our algorithm is designed to learn powerful representations from multi-omics data by leveraging a graph-based structure that incorporates both prior biological knowledge and patient-specific measurements. This section provides a formal description of the model, from the construction of the knowledge graph to the architecture of the neural network (Fig. 1).

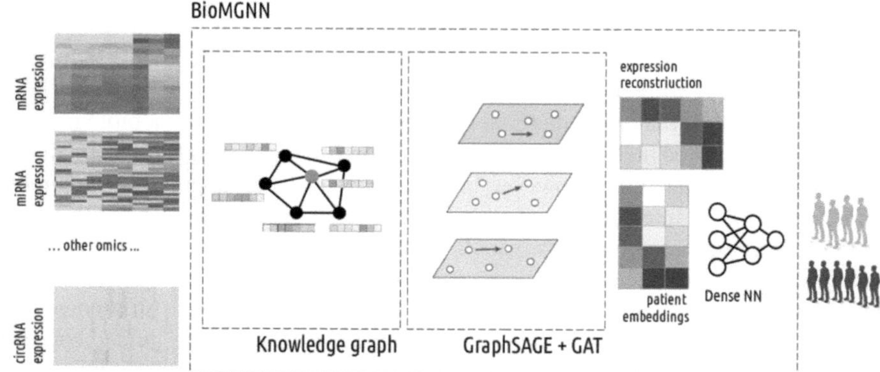

Fig. 1. BioMGNN workflow. The original expressions (circRNA. miRNA, etc.) are projected into a knowledge graph that captures the known biological relationship between the different features together with the patient nodes to create a multipartite network. Random embeddings are generated for each node, and the GraphSAGE with the attention algorithm proposed is applied. After convergence, the final patient embeddings are used to predict the stroke etiology.

2.1 Heterogeneous Biomedical Knowledge Graph Construction

The foundation of the BioMGNN model is a heterogeneous property graph, formally defined as G(V, E), where V is the set of nodes (vertices) and E is the set of edges representing relationships between them.

The node set V is heterogeneous, comprising multiple classes of entities. This includes a distinct set of nodes for each patient in the cohort, as well as separate node sets for each class of biological entity being analyzed. For instance, in a typical study, V would contain nodes representing patients, genes (mRNAs), CpG methylation sites, microRNAs (miRNAs), and circular RNAs (circRNAs).

The edge set E is similarly heterogeneous and is constructed from two primary sources: prior biological knowledge and the experimental data itself.

- **Prior Knowledge Edges**: These edges connect biological entity nodes to each other, forming a scaffold of known biological interactions. This information is curated from established public databases. For example, miRNA-gene target interactions are sourced from miRTarBase, potential circRNA-miRNA sponging relationships are derived from Circinteractome, and broader pathway and interaction data are integrated from knowledgebases like Reactome. This step transforms disparate biological facts into a computable network structure.

- **Data-Driven Edges**: These edges connect patient nodes to the corresponding biological entity nodes, creating a multipartite graph structure. The attributes of these edges can be weighted by the measured value from the experimental data, such as the expression level of a specific mRNA in a patient's sample or the methylation beta-value of a CpG site.

A critical preliminary step is the mapping and annotation of data. Raw data from high-throughput platforms often use probe-specific identifiers. These must be mapped to standardized identifiers, such as Ensembl IDs or official gene symbols, using annotation resources provided by platforms like the Gene Expression Omnibus (GEO). This ensures that data from different sources can be correctly integrated into the unified knowledge graph.

2.2 Multi-layer GCN Architecture and Node Representation

The primary objective of the BioMGNN model is to learn an encoder function, ENC, that maps each node $v \in V$ to a low-dimensional vector embedding $z_v \in R^d$, where d is the dimensionality of the embedding space. These embeddings are learned such that similarities in the embedding space reflect meaningful biological relationships from the original graph.

The core of the model is a multi-layer Graph Convolutional Network (GCN). In a basic GCN, the feature representation for a node at layer $l+1$ is computed by aggregating and transforming the feature representations of its neighbors from layer l, typically formulated as $H(l+1) = \sigma(A \sim H(l)W(l))$, where $A\sim$ is the normalized adjacency matrix of the graph. BioMGNN employs a more sophisticated aggregation scheme that is both inductive and attentive. The process begins by initializing node features at layer 0, $h_v(0)$, which can be derived from node attributes or initialized as random vectors. These initial representations are then iteratively refined through successive layers of neighborhood aggregation.

2.3 Inductive and Attentive Neighborhood Aggregation

A key innovation of BioMGNN v2 is its hybrid aggregation strategy, which synergistically combines the principles of GraphSAGE and Graph Attention Networks (GAT) to address the dual challenges of scalability and biological interpretability. A standard GCN is transductive; it requires the entire graph to be present during training and cannot generate embeddings for new, unseen nodes without retraining. This is a significant limitation in clinical settings where models must process new patient data on demand. GraphSAGE overcomes this by learning aggregator functions that are independent of the global graph structure, enabling inductive inference on new nodes. However, simple GraphSAGE aggregators like mean or max-pooling treat all neighboring nodes as equally important, which is a biologically implausible assumption. BioMGNN v2 resolves this by incorporating the attention mechanism of GAT, which dynamically weighs the contribution of different neighbors according to their relevance to the predictive task. In addition, BioMGNN v2 explicitly models the graph as heterogeneous, signed, and directed, ensuring that different node types (e.g., genes, miRNAs, CpGs) and edge types (e.g., activation, inhibition, regulatory sponging) are treated distinctly. This allows the model to capture the asymmetric and context-dependent nature of biological regulation rather than collapsing it into homogeneous, undirected relations.

The forward pass for a node v at each layer k of the network proceeds as follows:

- Neighborhood Sampling (GraphSAGE): To ensure computational efficiency and scalability to large graphs, a fixed-size neighborhood of nodes, N(v), is sampled for each target node *v*. This prevents the exponential growth of computations in deep GNNs.
- Attentive Aggregation (GAT): Instead of a simple averaging, the model computes the aggregated message from neighbors using a self-attention mechanism. First, an attention coefficient, e_{vu}, is calculated for each neighbor u ∈ N(v), which indicates the importance of node u's features to node v:

$$e_{vu} = a\left(W_k h_u^{(k-1)}, W_k h_v^{(k-1)} \right)$$

where $h_u(k-1)$ and $h_v(k-1)$ are the node representations from the previous layer, Wk is a shared trainable weight matrix for the linear transformation, and a is the attention mechanism (e.g., a single-layer feedforward neural network).

These coefficients are then normalized across all of v's neighbors using the softmax function to create a probability distribution, yielding the final attention weights αvu:

$$\alpha_{vu} = \frac{exp(e_{vu})}{\sum_{k \in N(v)} exp(e_{vk})}$$

The updated representation for node v at layer k, hv(k), is then computed as the attention-weighted sum of the transformed neighbor features, followed by a non-linear activation function σ (e.g., ReLU):

$$h_v^{(k)} = \sigma \left(\sum_{u \in N(v)} \alpha_{vu} W_k h_u^{(k-1)} \right)$$

To further enhance its representational capacity, BioMGNN v2 employs multi-head attention, in which the attention mechanism is executed multiple times in parallel with independent parameter sets. This enables the model to project node features into diverse subspaces, capturing complementary patterns and regulatory nuances across omics layers. The outputs from each attention "head" are then concatenated or averaged to produce a robust node embedding. Beyond supervised training, BioMGNN v2 introduces self-supervised pretraining objectives such as masked node feature prediction and cross-omics contrastive learning. These tasks encourage the model to learn generalizable biological structure before fine-tuning on specific classification problems, improving robustness in small-sample settings and yielding embeddings that transfer across cohorts. Together, these architectural choices allow BioMGNN v2 to retain scalability, exploit biological priors more effectively, and provide interpretable node-level insights into complex molecular networks.

3   Model Training and Biomarker Discovery

Our framework is designed not only for prediction but also for interpretation, with a seamless pipeline from model training to the identification of biologically relevant biomarkers.

3.1   End-to-End Supervised Training

The BioMGNN model is trained in an end-to-end supervised fashion. After the final GNN layer, the embeddings corresponding to the patient nodes are fed into a classification head. This is typically a fully connected dense layer followed by a softmax activation function, which outputs a probability distribution over the predefined classes (e.g., disease subtypes).

The entire network, including the weights of the GNN layers (W_k), the parameters of the attention mechanism (a), and the weights of the final classification layer, is optimized jointly. This is achieved by minimizing a suitable loss function, such as cross-entropy for classification tasks, using an optimizer like Adam [10]. The error signal from the classification output is backpropagated through the entire model, allowing all parameters to be updated based on their contribution to the predictive task. This end-to-end process ensures that the learned node embeddings are optimized specifically for the biomedical problem at hand. The model implementation leverages standard deep learning libraries, such as PyTorch (v2.8.0), and specialized packages for GNNs, like PyTorch Geometric(v2.8.0), and is trained on hardware accelerators like NVIDIA H100 GPUs to handle the computational demands.

3.2   Interpretable Biomarker Discovery via Attention Weights

A key advantage of the BioMGNN framework is its built-in mechanism for interpretability, which allows for the discovery of biomarkers. This process moves beyond identifying features based on simple statistical correlation with a phenotype, such as in differential expression analysis. Instead, it identifies features that the model has learned are mechanistically important for its predictive reasoning. The attention weights, α_{vu}, quantify the importance of the information flowing from a neighboring biological entity u to another node v in the process of making a correct classification. A high attention weight on an edge (u,v) implies that the biological interaction represented by that edge was crucial to the model's decision-making process. Therefore, the biomarkers discovered through this method are not merely statistical correlates but are components of a learned predictive logic, providing a more direct and interpretable link to the disease state.

The procedure for identifying these biomarker modules is as follows:

- Attention Weight Extraction: After the model is trained, the learned attention weight matrices are extracted from the GAT layers.
- Interaction Prioritization: Node pairs (and their corresponding edges) with high attention weights are identified. The magnitude of the attention weight is directly proportional to the significance of that interaction in differentiating the classes 1.

- Global Importance Scoring: To extrapolate global patterns and identify nodes that are not only highly attended to but also central to the network's information flow, the attention weights can be amplified by the degree of the nodes involved. This prioritizes interactions involving highly connected biological hubs.
- Module Identification: By applying a suitable threshold to these importance scores, a subgraph of the most critical nodes and edges is selected. This subgraph represents a "biomarker module"—an interconnected set of biomolecules that collectively contribute to the phenotype, offering a more systemic view than a simple list of individual genes.

4 Performance Evaluation and Application Case Studies

The efficacy of the BioMGNN framework was also evaluated through its application to a complex, real-world biomedical dataset. The following case studies demonstrate its superior performance in patient classification and its utility in identifying novel, biologically relevant biomarker modules.

4.1 Application in Patient Subtype Classification: Stroke Etiology

The classification of acute ischemic stroke etiology is a critical clinical challenge, as approximately 20–25% of cases have an undetermined cause, which complicates the selection of optimal secondary prevention strategies. As part of the ICTUSENSOPT project, 30 acute stroke patients with different etiologies as defined by TOAST classification, were comprehensively profiled using both transcriptomics (mRNA, miRNA and circRNA) and epigenomics (DNA methylation) techniques. BioMGNN was used to integrate the results of the different modalities in order to find biomarkers capable of identifying stroke etiology in clinical practice The model was tasked with classifying patients into etiological subtypes defined by the TOAST criteria [13].

We benchmarked the performance of our method against two established multi-omics biomarker discovery tools: MOFA (Multi-Omics Factor Analysis) and MOGONET (Multi-Omics Graph cOnvolutional NETworks)..

Xgboost [11] was employed as the classification model to evaluate the predictive efficacy of the identified biomarkers using a 10-fold cross-validation (CV) strategy aimed to ensure a comprehensive and rigorous evaluation of our biomarkers' validity in subtype classification tasks. We split the dataset in 10 groups. Then, each group was selected as the test set to evaluate the performance of a model trained on the other groups.

The performance of our predictive model was quantitatively evaluated using the area under the receiver operating characteristic curve (AUC) for its effectiveness in measuring the accuracy and reliability of classification models, particularly in biomedical applications where the cost of false positives and false negatives can be high.

The results, shown in Fig. 2, suggest that our method has a robust predictive capability, potentially offering enhanced accuracy over existing tools in the context of multi-omics biomarker discovery. The ability of BioGMGNN to consistently outperform in subtype prediction, regardless of the train-test set configurations, underscores its effectiveness and reliability as a tool in the field of precision medicine and biomarker discovery.

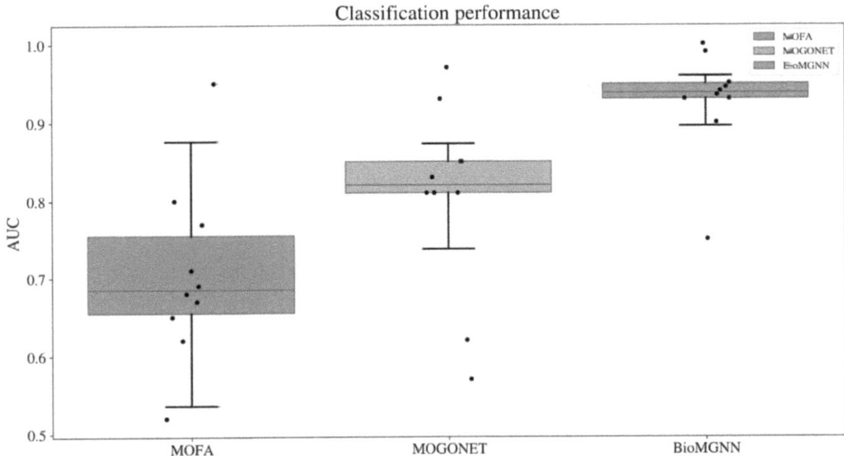

Fig. 2. Classification performance (AUC) comparison for the selected biomarker discovery methods using a 10-fold cross-validation (CV) strategy on the patient etiology classification task using Xgboost classifier.

The performance of BioMGNN was also benchmarked against baseline models using each single omics modality independently (with an XGBoost classifier). The comparative results, summarized in Table 1, demonstrate the clear superiority of the BioMGNN framework.

Table 1. Comparative Performance on Stroke Etiology Classification.

	BioMGNN	miRNA	circRNA	Methyl	mRNA
accuracy	0.95	0.48	0.52	0.67	0.77
Precision	0.93	0.43	0.55	0.65	0.77
recall	0.95	0.48	0.52	0.67	77
F1 score	0.96	0.50	0.54	0.78	0.85
AUC 0.95	0.40	0.58	0.60	0.60	0.90

The results reveal a significant synergistic information gain from the integrative approach. BioMGNN's performance (AUC of 0.95) is not merely an average of the individual data types but represents a substantial improvement over the best-performing single modality, mRNA expression (AUC of 0.90). The raw data shows that some omics layers, such as miRNA, possess a very weak predictive signal on their own (AUC of 0.40). A naive integration approach might be degraded by such noisy data. However, BioMGNN's superior performance suggests it effectively uses the biomedical knowledge graph to contextualize these weak signals. For example, a subtle change in a specific miRNA's expression becomes highly informative when the model is aware of its regulatory relationship with a key mRNA that also exhibits a change in expression. This demonstrates that the model leverages the relationships between different omics layers to extract a cohesive biological signal that is stronger than the sum of its individual parts. Visual exploration of the patient embeddings confirms this, showing that BioMGNN produces much clearer and more distinct clusters of patient subtypes compared to the embeddings derived from any single omics type.

4.2 Application in Biomarker Module Identification

The interpretability of BioMGNN was demonstrated through the extraction of biomarker modules from its learned attention weights in two distinct disease contexts.

Stroke Biomarkers: From the stroke etiology classification task, the attention mechanism highlighted a specific, interconnected module of biomolecules as being highly important for distinguishing between atherothrombotic and cardioembolic subtypes. This module included the circular RNA hsa_circ_0005568 and the genes THBS3 (Thrombospondin 3) and AMIGO2. The biological plausibility of these findings is supported by the known roles of thrombospondins in angiogenesis and tissue remodeling post-stroke, and AMIGO2 in neuronal maturation and recovery. Crucially, the differential expression of hsa_circ_0005568 was subsequently confirmed experimentally in a separate validation cohort using RT-qPCR, providing strong independent evidence for the model's ability to identify true biological signals.

Alzheimer's Disease Biomarkers: A second case study focused on the early diagnosis of Alzheimer's disease (AD) using DNA methylation data from the ROSMAP project [14]. The analysis identified several key methylation markers in genes previously implicated in AD, such as BIN1 and HOXA3.1 A particularly compelling finding emerged when BioMGNN was used to integrate multiple independent AD methylation datasets. This integrated analysis revealed statistically significant differentially methylated regions at genes like NXN and TREML2 that were not detectable when analyzing the datasets individually. This result showcases the power of graph convolution as a signal amplifier for meta-analysis. A subtle but consistent biological signal present across multiple small cohorts may not achieve statistical significance in any single analysis due to noise. By integrating these cohorts into a unified graph, the message-passing mechanism aggregates these consistent signals, effectively amplifying the true biological pattern above the level of stochastic noise. This demonstrates BioMGNN's utility as a powerful framework for discovering robust, cross-cohort biomarkers that would otherwise be missed.

5 Discussion

The framework we present here represents a significant step forward in the field of multi-omics data integration. By synthesizing a knowledge-guided graph structure with an advanced, end-to-end supervised learning architecture, it addresses several critical limitations of previous methods. Its primary contributions—the effective integration of heterogeneous data with prior biological knowledge, superior predictive performance, and a novel mechanism for interpretable biomarker discovery—position it as a powerful tool for biomedical research.

In contrast to methods like MOFA, which perform unsupervised decomposition, the supervised nature of our approach ensures that the learned representations are optimized for a specific, clinically relevant task. Compared to other graph-based methods like MOGONET or MOGCN, which focus on patient-level similarity graphs, the construction of a unified, heterogeneous graph allows for a more granular exploration of the interactions between individual molecular features, capturing a richer set of biological relationships.1

The implications of the "attention-as-biomarker" discovery method are particularly noteworthy. Traditional biomarker discovery relies on identifying statistically significant differences between populations, resulting in lists of individual genes or proteins. BioMGNN, however, identifies "biomarker modules"—interconnected subgraphs of biomolecules that the model found collectively important for its predictive task. This provides a more systems-level, mechanistic view of disease signatures that is more aligned with the network-based nature of biology itself. The experimental validation of a circRNA identified through this process underscores the practical utility of this approach.

Despite its promising results, the current work has limitations that suggest clear avenues for future research. The discovery studies were conducted on relatively small patient cohorts, and while initial findings are strong, they require validation in larger, more ethnically and geographically diverse populations to ensure their generalizability.1 Future work will focus on several key areas. Methodologically, the exploration of more advanced GNN architectures, such as graph transformers, may yield further performance improvements. The biomedical knowledge graph itself can be expanded to include more data types (e.g., proteomics, metabolomics) and a richer set of interactions from a wider array of databases. Finally, to promote wider adoption and facilitate reproducible research, a key goal is the development of a user-friendly, open-source software library that allows other researchers to apply the BioMGNN framework to their own datasets.

6 Conclusion

This paper has presented the Biological Multilayer Graph Neural Network (BioMGNN), a novel deep learning framework for the supervised integration of multi-omics data. By constructing a heterogeneous knowledge graph and employing a hybrid GraphSAGE-GAT architecture, BioMGNN effectively learns task-optimized representations that capture the complex interplay between different molecular layers. The empirical results demonstrate that this approach leads to superior performance in patient classification tasks compared to both single-omics analyses and other state-of-the-art integration methods. Furthermore, the model's inherent attention mechanism provides a powerful and unique method for identifying interpretable, systems-level biomarker modules that offer mechanistic insights into disease pathology. As a robust, interpretable, and high-performing tool, BioMGNN holds significant potential for advancing biomarker discovery and contributing to the broader goals of precision medicine.

Acknowledgments. This work was supported by Navarre Government Funding (Industry and Health department) through ADITECH and by RED INVICTUS (RD16/0019/0024) from the

Institute of Health Carlos III, jointly funded by the European Regional Development Fund (ERDF), European Union. The authors wish to thank the Navarrabiomed Neuroepigenetics Unit for providing original data and support, as well as the patients who generously donated the samples used in this study.

Disclosure of Interests. The authors declare no conflicts of interest.

References

1. Argelaguet, R., et al.: Multi-omics factor analysis—a framework for unsupervised integration of multi-omics data sets. Mol. Syst. Biol. (2018)
2. Hyunsoo, K., Haesun, P.: Sparse non-negative matrix factorization via alternating non-negativity constrained least squares for microarray data analysis. Bioinformatics **23**, 1495–1502 (2007)
3. Yu, T.: AIME: autoencoder-based integrative multi-omics data embedding that allows for confounder adjustments. PLoS Comput. Biol. **18**, e1009826 (2022)
4. Ma, T., Zhang, A.: Integrate multi-omics data with biological interaction networks using multi-view factorization auto encoder (MAE). BMC Genom. **20**, 944 (2019)
5. Kipf, T.N., Welling, M.: Semi-supervised classification with graph convolutional networks. In: International Conference on Learning Representations (2017)
6. Wang, T., et al.: MOGONET integrates multi-omics data using graph convolutional networks allowing patient classification and biomarker identification. Nat. Commun. **12**, 3445 (2021)
7. Li, X., et al.: MOGCN: a multi-omics integration method based on graph convolutional network for cancer subtype analysis. Front. Genet. **13** (2022)
8. Hamilton, W.L., Ying, R., Leskovec, J.: Inductive representation learning on large graphs. In: Advances in Neural Information Processing Systems, vol. 30 (2017)
9. Veličković, P., Cucurull, G., Casanova, A., Romero, A., Liò, P., Bengio, Y.: Graph attention networks. In: International Conference on Learning Representations (2018)
10. Kingma, D.P., Ba, J.: Adam: a method for stochastic optimization. In: International Conference on Learning Representations (2015)
11. Chen, T., Guestrin, C.: XGBoost: a scalable tree boosting system. In: Proceedings of the 22nd ACM SIGKDD International Conference on Knowledge Discovery and Data Mining, pp. 785–794 (2016)
12. Altuna, M., et al.: DNA methylation signature of human hippocampus in Alzheimer's disease is linked to neurogenesis. Clin. Epigenetics **11**, 91 (2019)
13. Adams, H.P., et al.: Classification of subtype of acute ischemic stroke. definitions for use in a multicenter clinical trial. TOAST. Trial of Org 10172 in acute stroke treatment. Stroke **24**, 35–41 (1993)
14. De Jager, P.L., et al.: A multi-omic atlas of the human frontal cortex for aging and Alzheimer's disease research. Sci. Data **5**, 180142 (2018)

Evaluation of Deep Clustering Methods on High Dimensional Tabular Biomedical Data

Ruben E. Munoz-Cabrera[1]([✉])[iD], Manuel Campos[1,2][iD], and Jose M. Juarez[1,2][iD]

[1] MedAI Lab, University of Murcia, Murcia, Spain
{remilio.munoz,manuelcampos,jmjuarez}@um.es
[2] Murcian Institute of Biomedicine, Murcia, Spain

Abstract. Clustering is a fundamental task in biomedical research, particularly for enabling stratification of patients. Traditional clustering techniques often struggle in this domain due to the vast number of attributes to be considered and the presence of complex, non-linear data structures. Deep Clustering (DC) methods, which combine neural networks with classical clustering techniques, have emerged as a promising alternative to address this challenge. To this day, most DC research has focused on images or text, however for many biomedical problems data is in a tabular format.

This work proposes and evaluates a systematic framework for assessing clustering methods on high-dimensional tabular data. The framework consists of four stages: defining the experimental setup, selecting and preprocessing datasets, implementing and executing experiments, and performing evaluation and comparative analysis. Experiments were conducted on four genomic datasets, comparing traditional clustering with and without dimensionality reduction to DC approaches (e.g. DEC, VaDE or TableDC).

Results indicate classical techniques perform well in structured scenarios while DC methods show advantages in complex or imbalanced datasets. This work highlights the importance of selecting clustering techniques based on dataset characteristics. It demonstrates the potential of DC for high dimensional data.

Keywords: Deep Clustering · Evaluation methods · Deep Learning

1 Introduction

Clustering is a fundamental task in areas such as medicine and bioinformatics, facilitating tasks such as patient segmentation and stratification.

Application of classical clustering techniques (e.g. k-means, DBSCAN) on tabular biomedical data of high dimensionality present certain limitations such as difficulty when capturing non-linear relationships or limited scalability with complex datasets [1].

A. López Fernández et al. (Eds.): CIABiomed 2025, LNBI 16148, pp. 449–463, 2026.
https://doi.org/10.1007/978-3-032-10661-2_34

Thanks to advances in deep learning techniques, new approaches such as Deep Clustering have emerged. It combines the ability of neural networks to extract latent representations with classical clustering algorithms, facilitating the task of clustering [10]. The problem identified in this research is that in the field of Deep Clustering, most literature focuses on image or text data, not giving enough attention to numerical tabular data, which is common in biomedical environments.

What is more, high-dimensional tabular data stands out in certain contexts, such as the biomedical field, where we find genomic data [5], which presents a large number of attributes. These datasets may include a number of computational problems in dealing with hundreds or thousands of attributes for a single instance.

Problems with high dimensionality imply that with a fixed number of elements, as the dimensions increase, the differences between elements that are similar and those that are not became less and less evident, making the task of clustering more difficult with conventional algorithms [13].

To address these limitations, dimensionality reduction techniques are often applied as a preprocessing step prior to clustering. Linear techniques such as PCA [6] aim to preserve global variance using linear projections, but fail to capture non-linear relationships. Non-linear techniques like t-SNE and UMAP [7,9] attempt to preserve local neigborhood structures. However, as highlighted by Jeon et al. [3], the effectiveness of these methods depends highly on the specific analytical task. For tasks such as cluster identification, local techniques (t-SNE and UMAP) show better performance in preserving local structures, making them suitable for identifying individual clusters.

This study proposes a systematic framework for the evaluation of Deep Clustering techniques on high-dimensional genomic tabular datasets, comparing their performance with conventional approaches with and without dimensionality reduction. To do this, we explore different approaches, from the classics, which apply methods based on projections or sub-spaces; to the more recent ones, which reduce the data to a latent representation using neural networks. With this framework we want to analyse the applicability of these methods in biomedical contexts providing a reproducible basis for future research.

Main contributions of this study are:

1. The definition of a four-stage framework for the evaluation of Deep Clustering techniques on tabular data.
2. An experimental comparison of direct classical approaches, dimensionality reduction techniques and Deep Clustering methods.
3. The development of a validation process on multiple public genomic datasets, using internal and external quality clustering metrics.
4. A critical analysis of the scenarios in which Deep Clustering provides advantages over conventional techniques.

2   Background

2.1   Traditional Machine Learning

There is a myriad of clustering algorithms proposed in the past decades. In this paper we just focus on a few of them as examples of the most representative approaches. In particular, we selected:

K-means [12] is a partitioning centroid-based algorithm widely used for its simplicity and performance. It assumes that the generated clusters have a spherical shape and a similar number of elements, which may present limitations in hierarchical environments.

HDBSCAN [2] is a hierarchical algorithm based on density which allows for finding clusters of different shapes and sizes, as it evaluates all possible densities. Previous knowledge of the number of clusters is unnecessary, as it depends on the input representation.

Spectral Clustering [14] is a graph-based algorithm. It builds a graph of similarities between instances applying a spectral decomposition using the Laplacian matrix of the graph. It is flexible, when working with data not linearly separable, and also robust to noise and outliers. Its limitations are its computational cost and the high required memory to store the similarity matrix.

Direct application of classical clustering algorithms often results in a degradation of their performance due to the sparsity and noise inherent in such spaces. To address this, some traditional dimensionality reduction techniques such as PCA, t-SNE or UMAP are applied to project the data into a more compact latent space.

PCA. Principal Component Analysis is a linear technique consisting of reducing the set of data geometrically by projecting them in n dimensions, which are called the main components. The goal is to find a representation to compact the data using a limited number of components. Despite its speed at the computational level, this technique does not work optimally on non-linear data [6, 8]

t-SNE [7] is a non-linear technique that preserves the local structure of data. Its main purpose is the visualization of data, so it is limited to reducing dimensionality to 2 or 3 dimensions.

UMAP [9] is a non-linear dimensionality reduction technique based on topological data analysis and graph theory. Computationally, it is an algorithm based on weighted graphs, that preserves the local and global data structure. It has proved to perform well in clustering tasks.

2.2   Deep Clustering Algorithms

Deep Clustering arises to address the limitations of traditional clustering, especially with complex and high dimensionality data. Its goal is the joint and iterative optimization of both the learning of deep representations and the clustering process, by combining deep neural networks with clustering algorithms.

Though there is no clear consensus on the different types of clustering that exist, Wei et al. (2024) [15] collects the different proposals and groups the Deep Clustering techniques into 5 types: Deep Neural Networks (DNNs), Autoencoders (AEs), Variational Autoencoders (VAEs), Generative Adversative Networks (GANs), and Graph Neural Networks (GNNs). Figure 1 shows this taxonomy and some specific techniques of each type.

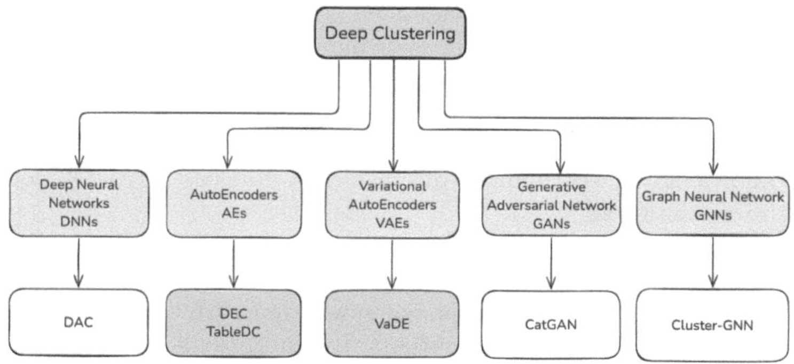

Fig. 1. Deep Clustering taxonomy based on Wei et al. (2024) [15].

Though the framework is inherently technique agnostic, we selected three DC algorithms representing different approaches to validate its utility. DEC and VaDE are well-established algorithms based on Autoencoders and Variational Autoencoders, providing a baseline for general purpose Deep Clustering. In contrast, TableDC represents a recent method specifically designed for tabular data.

DEC (Deep Embedded Clustering) [16] is a joint optimization method that refines a latent representation while simultaneously adjusting cluster assignments. Initially, it pretrains a deep autoencoder (AE) to reconstruct the data. Once pretraining is completed, decoder is discarded and only the encoder remains. Subsequently, the weights of the net and centroids iteratively minimize the divergence of Kullback-Leibler between an auxiliary distribution and the current distribution.

VaDE (Variational Deep Embedding) [4] is a Deep Clustering algorithm that combines Variational Autoencoders (VAE) and Gaussian Mixture Models (GMM) in the latent space. Unlike AEs, VAEs learn to encode inputs as probability distributions (mean and variance), allowing VAEs to model uncertainty generating new samples from the latent space.

Its operation is structured in two phases. First, VAE is trained to get a data reconstruction and use it to initialize the centroids and weights of the Gaussian mixture. The second phase consists of a joint optimization process where the VaDE algorithm is trained to maximize the ELBO updating network parameters by backpropagation. Evidence Lower Bound (ELBO) is a variational objective function that balances the reconstruction of the input data and the regularization of the prior distribution (a GMM).

TableDC (Table Deep Clustering) [11] is a recent algorithm, designed specifically for tabular data. Unlike traditional algorithms that rely on Euclidean distance, TableDC employs a regularized auto-encoder based on Mahalanobis distance, which considers covariance in different dimensions. To enhance robustness in noisy or overlapping data, it integrates a similarity kernel based on Cauchy distribution, enabling more flexible cluster boundaries. These features make TableDC suitable for complex tabular data.

3 Our Proposal

The main objective of this work is **to propose a systematic framework for the evaluation of Deep Clustering methods on high-dimensional tabular data**. To achieve this, we propose a four stage structured process. It allows a rigorous comparative analysis of the different classical approaches, dimensionality reduction techniques and Deep Clustering methods through a reproducible framework.

1. **Stage 1 - Experimental Design.** A representative set of clustering techniques is defined, including both classical algorithms and deep clustering methods. It also includes various classical dimensionality reduction techniques which act as a preliminary step to the clustering process. Finally, internal (to evaluate cohesion and separability of clusters) and external (to compare with ground truth) clustering metrics are chosen.
2. **Stage 2 - Dataset Selection and Preprocessing.** High-dimensional public tabular genomic datasets are selected, checking they are suitable for the clustering task. At this stage, a pre-processing of the data is performed, ensuring that it contains only numerical (one-hot encoding if necessary) and normalized attributes.
3. **Stage 3 – Experiment Execution (3 approaches).** Each of the techniques defined in Stage 1 is implemented and applied to the datasets. This third stage includes the tuning of hyper parameters for each technique, the design and training of different deep architectures for Deep Clustering algorithm, as well as obtaining latent representations of the data. This stage is structured in three approaches or tasks:
   - Classical clustering, applied directly and then dimensionality reduction.
   - Generic Deep Clustering: DEC and VaDE (originally designed for image and text).
   - Deep Clustering specific: TableDC (explicitly designed for tabular data).
4. **Stage 4 - Evaluation and Analysis.** This final stage consists of the evaluation of the clusters obtained using internal and external clustering metrics. With this, we can compare the performance of different algorithms and identify possible patterns depending on the characteristics of each dataset, as well as suitability and usefulness of the proposed framework.

4   Experiments and Discussion

In order to validate our proposal methodology we will break it down step by step. See Fig. 2.

4.1   Stage 1 Experimental Design

A set of clustering techniques are selected, divided into three groups:

- **Classical clustering algorithms.** We have selected three: K-means, HDB-SCAN and Spectral Clustering.
- **Dimensionality reduction techniques.** Three techniques have been chosen, to be used before applying classical clustering: PCA, t-SNE and UMAP.
- **Deep Clustering algorithms.** Three deep-learning based algorithms were chosen: DEC, VaDE and TableDC.

We define Accuracy (**ACC**), Adjusted Rand Index (**ARI**) and Normalized Mutual Information (**NMI**) as external clustering metrics. As internal metrics Silhouette Score (**Silhouette**) and the Davies Bouldin Index (**DB Index**).

4.2   Stage 2 Dataset Selection and Preprocessing

This section shows the datasets used for the experimental study. All of them public and from the biomedical and/or genomic field.

They have a high dimensionality, with hundreds or thousands of only numeric attributes, originally or after pre-processing. No dataset has null values.

As follows a brief description of each dataset:

- **ARCENE**[1]: The task is to distinguish cancer versus normal patterns from mass-spectrometric data.
- **METABRIC**[2]: It contains clinical and molecular data of primary breast tumour patients and normal patients.
- **GSE2034**[3]: Extracted from the GEO repository, this series represents 180 lymph-node negative relapse free patients and 106 lymph-node negate patients that developed a distant metastasis.
- **TCGA PANCAN**[4]: Dataset from the Cancer Genome Atlas (TCGA), which collects copy number alterations in multiple tumor types.

Preprocessing in datasets has been limited to the normalization of the data in all but TCGA, which was already standardized. In this last dataset, the number of instances has been reduced from over 10,000 to 2,000 in order to carry out experimentation in reasonable time.

---

[1] Arcene: https://archive.ics.uci.edu/dataset/167/arcene.

[2] METABRIC: https://www.cbioportal.org/study/summary?id=brca_metabric.

[3] GSE2034: https://www.ncbi.nlm.nih.gov/geo/query/acc.cgi?acc=GSE2034.

[4] TCGA Pan-Cancer Gene-Level CNV: https://tinyurl.com/tcga-pancan.

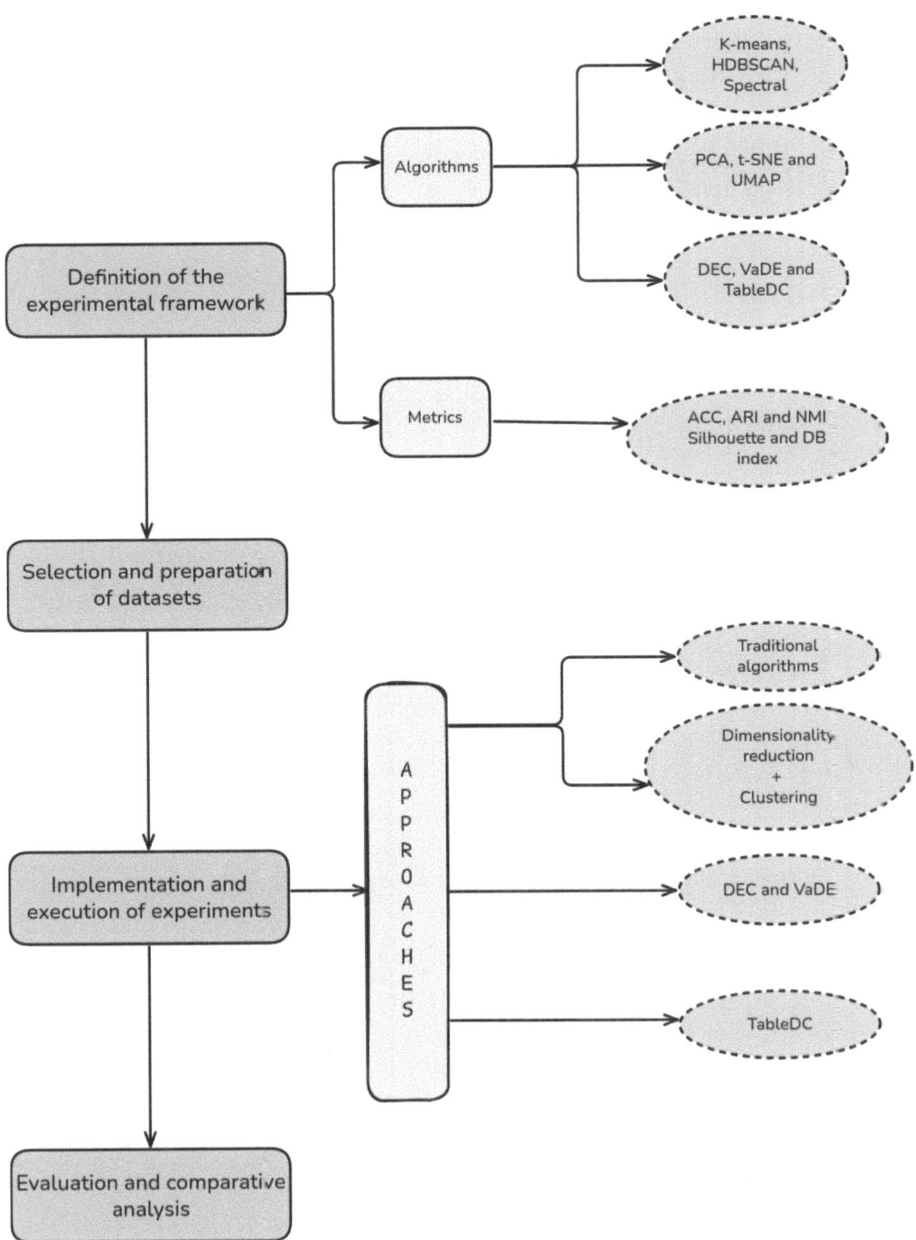

Fig. 2. Proposed framework.

Table 1 shows the specific characteristics of the datasets. The number of classes are known, and necessary for calculating the external clustering metrics.

Table 1. Characteristics of the datasets.

Dataset	Instances	Attributes	Classes	Balanced
ARCENE	200	10000	2	YES
METABRIC	1964	20603	8	NO
GSE	286	22283	2	NO
TCGA	2000	24776	33	NO

The four datasets with different characteristics allow us to evaluate the different techniques chosen. ARCENE and GSE have a limited number of instances maintaining a large dimensionality, and are designed to see the clustering effect when we have only two clusters. METABRIC has an important number of instances along with a very high dimensionality and 8 classes. In TCGA the main challenge has been the large number of clusters, 33 in this case. Except ARCENE, the datasets are all unbalanced, an added difficulty.

4.3   Stage 3 Experiment Execution (3 Approaches)

This consists of the implementation and execution of the various techniques divided into three different approaches, depending on the type of technique to be used. We show the experimental configuration with the selection of different hyperparameters. Computational times have also been calculated. A GPU NVIDIA RTX 4070 (8 GB) has been used to execute all experiments.

Traditional Clustering and Dimensionality Reduction. Here different clustering algorithms are evaluated directly on the datasets and the dimensionality reduction techniques are applied as a preliminary step to the clustering process.

For the experimental set-up, we start with classical techniques. In **K-means** different values of k are tested by indicating the number of clusters values already known in datasets; for **HDBSCAN**, based on density, we select $min\_cluster\_size$, to indicate how many elements must be at least in a cluster. A search has been made for the parameter with values $[5, 10, 20, 50]$ on all datasets. Finally, in **Spectral Clustering** both the number of clusters k and the $affinity$ parameter are selected. The parameter $affinity$ indicates which algorithm is used to create the similarity graph, in this case $nearest\_neighbors$, based on the k-nearest neighbours.

Regarding dimensionality reduction techniques, **PCA** requires the parameter $n\_components$, which indicates the number of dimensions in order to reduce data. **t-SNE** requires the same parameter, but as it is designed for visualization, its limited to 2 or 3 dimensions, values that we have tested. **UMAP** requires two parameters: $n\_components$ and $n\_neighbors$, the latter indicates the minimum number of elements that are close (in the neighbourhood) to be considered for the construction of the proximity graph. The grid of hyperparameters has been the same for all datasets:

- $n\_components$ (PCA, UMAP) : $[2, 10, 50, 100, 250, 500]$
- $n\_components$ (t-SNE) : $[2, 3]$
- $n\_neighbors$ (UMAP) : $[5, 15, 25, 50]$

Standard Deep Clustering. We have chosen deep clustering techniques generally used in areas of image and text, specifically Deep Embedding Clustering (**DEC**) and Variational Deep Embedding (**VaDE**). Both need an Autoencoder a classic one in the case of DEC and a Variational one in the case of VaDE. We have used the AE shown in Fig. 3. The values of $latent\_dim$ for latent dimensionality have been the same in all the datasets: $latent\_dim = [2, 10, 50, 100, 250]$, and with all initialization algorithms (K-means, HDBSCAN and Spectral). We should take into account that when initializing HDBSCAN, the parameter $min\_cluster\_size$ chosen is the best value found in the grid, and used to calculate the initial baseline.

After training the AE using the **MSE** as the reconstruction error and **Adam** as optimizer, we only need the encoder output. In the case of VaDE, the transformation of the output with the mean and the logarithm of the variance is also necessary to implement the VAE.

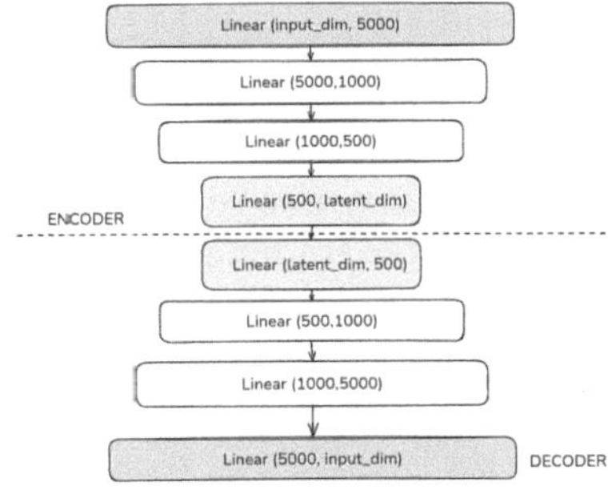

Fig. 3. AE used for DEC and VaDE.

We use this latent representation to train the algorithms. Both DEC and VaDE are based on a t-Student distribution. DEC loss consists of minimizing the Kullback-Leibler divergence and VaDE loss in maximizing ELBO.

DEC uses **Adam** as optimizer while VaDE employs **SGD** with $momentum = 0.9$. Once the clusters are obtained then the clustering metrics are calculated.

Training parameters were:

- Number of epochs for auto encoders: 200 (AE), 100 (VAE).
- Number of epochs for DC algorithms: 200 (DEC), 100 (VaDE).
- lr: 0.005 for autoencoders, 0.001 for DC techniques.

Deep Clustering for Tabular Data. This approach evaluates the performance of TableDC, analysing if it is able to outperform conventional dimensionality reduction techniques or the general DC algorithms.

In respect to the architecture designed, we decided to be faithful to the architecture presented on the original paper of TableDC, composed of two simple layers with a RELU as the activation function, as can be seen in Fig. 4.

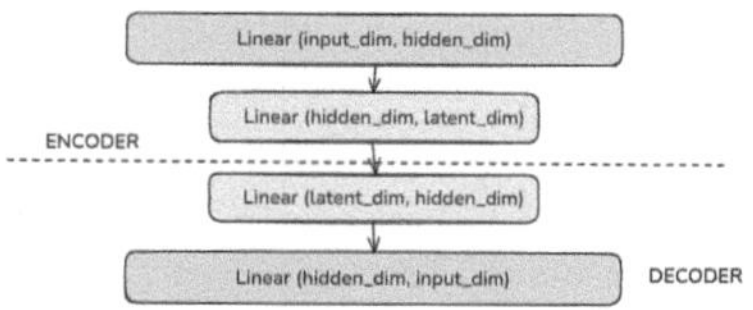

Fig. 4. AE used for TableDC.

We set the dimensionality of the hidden layer. Values for the grid search were:

- $hidden\_dim = [5000, 2000, 1000]$.
- $latent\_dim = [2, 10, 50, 100, 250]$.

TableDC trains and optimizes (using **Adam** as optimizer and $lr = 0.001$) both latent representation and clustering assignments with the different start-up algorithms. MSE is used as a reconstruction error (re_loss) and KL divergence as clustering loss (ce_loss). Final loss uses *gamma* to weight both losses. In this case, $loss = 0.9 * ce\_loss + re\_loss$. Training employs 200 epochs for ARCENE and METABRIC, 100 for TCGA and 50 for GSE.

4.4   Stage 4 Evaluation and Analysis

In this last stage, we intend to evaluate and compare the different clusters obtained using different algorithms. To achieve a global perspective, we show summary tables for each dataset. We see the base performance, without any reduction in dimensionality, and then the most outstanding result for each dataset in each approach.

In Tables 2, 3, 4 and 5 we see a final comparison for each dataset. According to these results we can provide the following discussion:

1. **ARCENE:** TableDC achieves consistent results as does UMAP. The latter stands out for its ability to reduce the data to 2 dimensions while preserving a good internal structure, and doing so with a low computational cost.
2. **METABRIC:** TableDC achieves best results with K-means and Spectral although at an increased computational cost. Improvement is significant with all algorithms using HDBSCAN as the initialization algorithm.

3. **GSE:** TableDC achieves external metric improvements in a reasonable time, but struggles to achieve good separability and cohesion in the clusters. We also see how most techniques struggle to achieve good internal metrics.
4. **TCGA:** It is UMAP, a classical technique of dimensionality reduction, which achieves the best metrics, both external (compared to real labels) and internal (compared to the quality of generated clusters). TableDC in this case has demonstrated inconsistent behaviour with variable results.
5. In most of the scenarios studied PCA/t-SNE and DEC did not achieve the best results and, therefore, are not highlighted in the summary tables. Nevertheless, they are included in specific cases where their performance was comparable to other techniques.

Table 2. Final results for ARCENE

Algorithm	Technique	Dims	ACC	ARI	NMI	Silhouette	DB Index	Time (s)
K-means	Baseline	10000	0.65	0.0852	0.0823	0.1899	2.0062	0.90
DR	**UMAP**	2	**0.66**	**0.0977**	**0.0908**	**0.6992**	**0.5017**	**0.17**
Generic	VaDE	2	0.66	0.0977	0.0908	0.4829	0.8417	16.72
Tabular	TableDC	2	0.66	0.0978	0.0865	0.5622	0.7726	19.31
HDBSCAN	Baseline	10000	0.41	0.0528	0.0512	0.1571	3.1527	0.44
DR	**UMAP**	2	**0.64**	**0.1293**	**0.1131**	**0.5313**	**0.5303**	**0.16**
Generic	VaDE	10	0.5	0.0864	0.0696	0.6269	0.5520	16.75
Tabular	**TableDC**	10	**0.67**	**0.1104**	**0.1266**	**0.1778**	**2.3403**	**4.73**
Spectral	Baseline	10000	0.63	0.0605	0.0439	0.0793	2.2179	0.07
DR	**UMAP**	2	**0.66**	**0.0977**	**0.0908**	**0.6992**	**0.5017**	**0.19**
Generic	DEC	250	0.66	0.0977	0.0908	0.5145	0.8193	6.98
Tabular	**TableDC**	100	**0.68**	**0.1246**	**0.1365**	**0.061**	**3.5227**	**7.44**

Some clusters produced can be seen in Figs. 5a, 5b, 5c and 5d. When the resulting latent spaces are 2D, these have been projected directly, otherwise UMAP has been applied.

Table 3. Final results for METABRIC

Algorithm	Technique	Dims	ACC	ARI	NMI	Silhouette	DB Index	Time (s)
K-means	Baseline	20603	0.2276	−0.005	0.0309	0.0182	3.9582	10.07
DR	UMAP	2	0.1986	0.0074	0.0239	0.4038	0.7322	11.55
Generic	DEC	50	0.2307	-0.0082	0.0368	0.3241	0.9058	25.71
Tabular	**TableDC**	50	**0.7408**	**0.0112**	**0.0076**	**−0.1154**	**2.9269**	**89.23**
HDBSCAN	Baseline	20603	0.5377	−0.0438	0.0169	0.0791	4.6957	85.15
DR	**UMAP**	**10**	**0.7480**	**0.0112**	**0.0088**	**0.2121**	**0.5356**	**11.82**
Generic	VaDE	50	0.7251	0.0215	0.0140	0.2099	2.4158	100.17
Tabular	TableDC	2	0.7256	−0.0622	0.0204	−0.1163	1.3034	60.96
Spectral	Baseline	20603	0.5341	0.0092	0.034	0.0167	3.1983	2.62
DR	PCA	500	0.5372	0.0006	0.0335	0.0278	2.5291	10.07
Generic	VaDE	100	0.3157	0.0285	0.0266	0.1133	1.7404	104.15
Tabular	**TableDC**	50	**0.7459**	**0.003**	**0.0141**	**−0.0921**	**2.7909**	**62.21**

Table 4. Final results for GSE

Algorithm	Technique	Dims	ACC	ARI	NMI	Silhouette	DB Index	Time (s)
K-means	Baseline	22283	0.5664	0.0143	0.0164	0.0461	4.3376	2.52
DR	UMAP	100	0.5944	0.0319	0.0215	0.3197	1.2910	1.036
Generic	DEC	2	0.5874	0.0087	0.0015	0.5238	0.6362	35.91
Tabular	**TableDC**	50	**0.6434**	**0.0556**	**0.0244**	**0.0397**	**3.4005**	**13.11**
HDBSCAN	Baseline	22283	0.6259	0	0	−1	−1	2.09
DR	UMAP	25	0.5035	0.0446	0.018	0.0864	2.5233	0.984
Generic	DEC	100	0.5979	−0.0084	0.0012	0.3794	1.1486	40.88
Tabular	**TableDC**	2	**0.5804**	**0.0129**	**0.0034**	**0.315**	**0.9942**	**6.11**
Spectral	Baseline	22283	0.5769	0.0202	0.0151	0.0405	4.4621	0.12
DR	PCA	250	0.6049	0.0394	0.0234	0.0397	4.3406	0.82
Generic	VaDE	100	0.5629	0.0125	0.0102	0.1967	1.8657	66.60
Tabular	**TableDC**	50	**0.6503**	**0.0426**	**0.0323**	**0.0972**	**3.8554**	**3.15**

Table 5. Final results for TCGA

Algorithm	Technique	Dims	ACC	ARI	NMI	Silhouette	DB Index	Time (s)
K-means	Baseline	24776	0.2365	0.073	0.3209	0.0504	2.5920	23.35
DR	**UMAP**	**10**	**0.3075**	**0.1634**	**0.3722**	**0.3143**	**1.1254**	**24.41**
Generic	VaDE	100	0.3065	0.1539	0.3765	0.0343	2.6366	614.82
Tabular	TableDC	250	0.166	0.0272	0.1746	−0.2028	4.0798	33.54
HDBSCAN	Baseline	24776	0.1445	0.0103	0.1035	−0.2237	2.3227	104.02
DR	**UMAP**	**50**	**0.321**	**0.1017**	**0.3684**	**0.0861**	**1.1451**	**18.34**
Generic	VaDE	10	0.1845	0.0553	0.1785	0.0614	2.4516	613.27
Tabular	TableDC	2	0.133	0.0265	0.127	−0.2339	8.2697	37.46
Spectral	Baseline	24776	0.284	0.1231	0.3681	−0.184	3.3813	1.95
DR	**UMAP**	**500**	**0.283**	**0.1322**	**0.3635**	**0.2014**	**1.1508**	**37.70**
Generic	VaDE	100	0.293	0.1513	0.3754	0.012	2.7615	614.69
Tabular	TableDC	250	0.159	0.0238	0.1565	−0.2080	4.4535	27.29

Overall, the results indicate that there is no superior technique or algorithm for the analysis of all four datasets studied. Classical dimensionality reduction techniques, in particular UMAP, achieved very good internal clustering metrics across most datasets, showing its effectiveness when class separability is preserved in the latent space as we can see in ARCENE and METABRIC.

Also, we can see relatively good results of PCA combined with Spectral in METABRIC and GSE, which can be related with the ability of PCA to preserve global variance, improving the construction of the affinity matrix required by Spectral.

However, in datasets like GSE or TCGA, where high dimensionality and class imbalance are present, traditional algorithms have certain limitations.

In these cases, DC approaches like TableDC (in GSE and METABRIC) achieved competitive external metrics (e.g. ACC and ARI), demonstrating a grater ability to capture the data in the latent space. Nevertheless, internal metrics such as Silhouette and DB Index tend to be low or even negative, which suggests that the resulting clusters are less compact or well-separated in the latent space, probably due to the vast number of small classes. For this reason, some overlapping can be observed in Fig. 5b.

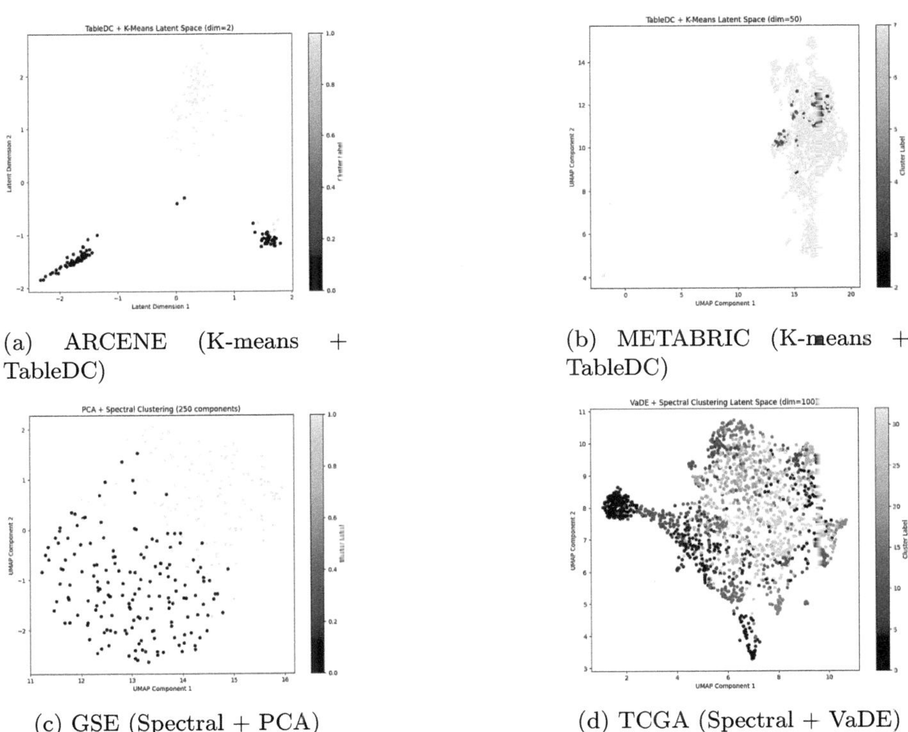

(a) ARCENE (K-means + TableDC)

(b) METABRIC (K-means + TableDC)

(c) GSE (Spectral + PCA)

(d) TCGA (Spectral + VaDE)

Fig. 5. Cluster visualizations for different datasets and techniques.

A specific technique of Deep Clustering for tabular data which was TableDC worked effectively in the majority of the studied datasets (in particular ARCENE and METABRIC). It is demonstrated that specific deep neuronal architectures can produce better clustering.

5   Conclusion

In this study a systematic methodology has been designed, applied and evaluated for the comparison of various Deep Clustering strategies using biomedical high dimensionality tabular data. This methodology is structured in 4 stages.

With the evaluation frameworks presented we identified that:

- **There is no universally superior technique for clustering,** performance depends on the specific characteristics of each dataset.
- **UMAP** is a robust dimensionality reduction technique across most datasets, especially when combined with K-means or HDBSCAN.
- **HDBSCAN**, based on density, is clearly benefited when representations of the data are compressed, improving performance with all the datasets used.
- Generic Deep Clustering algorithms such as DEC and VaDE present acceptable and consistent behaviour, although they do not exceed a dimensionality reduction technique like UMAP in most scenarios.
- **TableDC**, an algorithm specifically designed for tabular data, has showed remarkable behavior in most datasets, improving results in imbalanced datasets.

We emphasize the need to adopt a clustering approach adapted to the characteristics of each dataset. DC techniques are an added value, providing better results when classical techniques are limited.

Therefore, we can conclude that evaluation frameworks, as the one presented in this paper, are helpful for an effective application of Deep Clustering in the biomedical field.

As possible future lines of research, we propose the exploration of alternative deep clustering techniques such as contrastive learning or adversarial networks; the application to others types of medical data; and a systematic analysis of computational costs of the different approaches.

Acknowledgements. This work was partially funded by the CONFAINCE project (Ref: PID2021-122194OB-I00) by MCIN/AEI/10.13039/501100011033 and by "ERDF A way of making Europe", by the "European Union".

This research is also part of IMPACT-T2D-UMU project (ref. PMP21/00092), funded by the Institute of Health Carlos III (ISCII) for Personalized Precision Medicine Research Projects of the Strategic Action in Health 2017–2020, supported by the European funds of the Recovery, Transformation and Resilience Plan (Next Generation EU programme).

References

1. Assent, I.: Clustering high dimensional data. WIREs Data Min. Knowl. Discov. **2**(4), 340–350 (2012). https://doi.org/10.1002/widm.1062, https //wires. onlinelibrary.wiley.com/doi/abs/10.1002/widm.1062
2. Campello, R.J.G.B., Moulavi, D., Sander, J.: Density-based clustering based cn hierarchical density estimates. In: Pei, J., Tseng, V.S., Cao, L., Motoda, H., Xu, G. (eds.) PAKDD 2013. LNCS (LNAI), vol. 7819, pp. 160–172. Springer, Heidelberg (2013). https://doi.org/10.1007/978-3-642-37456-2_14
3. Jeon, H., Park, J., Shin, S., Seo, J.: Stop misusing t-SNE and UMAP for visual analytics (2025). https://arxiv.org/abs/2506.08725
4. Jiang, Z., Zheng, Y., Tan, H., Tang, B., Zhou, H.: Variational deep embedding: an unsupervised and generative approach to clustering (2017). https://arxiv.org/abs/1611.05148
5. Karim, M.R., et al.: Deep learning-based clustering approaches for bioinformatics. Brief. Bioinform. **22**(1), 393–415 (2020). https://doi.org/10.1093/bib/bbz170
6. Lever, J., Krzywinski, M., Altman, N.: Principal component analysis. Nat. Methods **14**(7), 641–642 (2017). https://doi.org/10.1038/nmeth.4346
7. van der Maaten, L., Hinton, G.: Visualizing data using t-SNE. J. Mach. Learn. Res. **9**(86), 2579–2605 (2008). http://jmlr.org/papers/v9/vandermaaten08a.html
8. van der Maaten, L., Postma, E., Herik, H.: Dimensionality reduction: a comparative review. J. Mach. Learn. Res. - JMLR **10** (2007)
9. McInnes, L., Healy, J., Melville, J.: UMAP: uniform manifold approximation and projection for dimension reduction (2020). https://arxiv.org/abs/1802.03426
10. Min, E., Guo, X., Liu, Q., Zhang, G., Cui, J., Long, J.: A survey of clustering with deep learning: from the perspective of network architecture. IEEE Access **6**, 39501–39514 (2018). https://doi.org/10.1109/ACCESS.2018.2855437
11. Rauf, H.T., Freitas, A., Paton, N.W.: TableDC: deep clustering for tabular data (2024). https://arxiv.org/abs/2405.17723
12. Reddy, C.K., Vinzamuri, B.: A Survey of Partitional and Hierarchical Clustering Algorithms, pp. 87–110. Chapman and Hall/CRC (2018). https://doi.org/10.1201/9781315373515-4
13. Steinbach, M., Ertöz, L., Kumar, V.: The challenges of clustering high dimensional data. Univ. Minnesota Supercomp. Inst. Res. Rep. **213** (2003). https://doi.org/10.1007/978-3-662-08968-2_16
14. Von Luxburg, U.: A tutorial on spectral clustering. Stat. Comput. **17**(4), 395–416 (2007).https://doi.org/10.1007/s11222-007-9033-z
15. Wei, X., Zhang, Z., Huang, H., Zhou, Y.: An overview on deep clustering. Neurocomputing **590**, 127761 (2024). https://doi.org/10.1016/j.neucom.2024.127761, https://www.sciencedirect.com/science/article/pii/S0925231224005320
16. Xie, J., Girshick, R., Farhadi, A.: Unsupervised deep embedding for clustering analysis (2016). https://arxiv.org/abs/1511.06335

Integrative Analysis of Breast Cancer Using Multi-omics Latent Representations

Yasmine Lakouifat Darkaoui, Beatriz Pontes$^{(\boxtimes)}$ ⓘ, and Belén Vega Márquez

Computer Languages and Systems Department, University of Seville, Seville, Spain
yaslakdar@alum.us.es, bepontes@us.es

Abstract. This study presents a comparative analysis of three dimensionality reduction techniques—MOFA (Multi-Omics Factor Analysis), IntNMF (Integrative Non-negative Matrix Factorization), and VAE (Variational Autoencoder)—applied to breast cancer multi-omics data from the TCGA-BRCA project. The integration of omics layers—including gene expression, protein expression, copy number variations, and mutations—was combined with key clinical variables to evaluate the performance of latent representations in both classification and clustering tasks. Major challenges such as high dimensionality and severe class imbalance were addressed through oversampling and undersampling strategies. Each method was evaluated for its effectiveness in predicting clinical outcomes and identifying meaningful molecular patterns. MOFA offered biologically interpretable and stable representations, IntNMF produced compact structures, and VAE yielded well-separated latent spaces. Enrichment analysis confirmed the relevance of extracted features, reinforcing the utility of latent factor models for robust multi-omics integration in breast cancer research.

Keywords: Multi-omics integration · Dimensionality reduction · Breast cancer · Latent factor models · MOFA · Variational autoencoders · Nonnegative matrix factorization · Class imbalance · Biological interpretability · Enrichment analysis · Classification · Clustering

1 Introduction

Breast cancer is the most frequently diagnosed cancer among women in Spain and a major cause of cancer-related mortality [11]. Its clinical management remains challenging due to the disease's marked molecular and phenotypic heterogeneity, which leads to varied prognoses and treatment responses. Conventional clinical markers often fall short in capturing this complexity, highlighting the need for more comprehensive approaches. In this context, multi-omics data integration offers a promising avenue for uncovering the underlying biological mechanisms. However, analyzing such high-dimensional and heterogeneous data requires advanced computational strategies. Dimensionality reduction methods

A. López Fernández et al. (Eds.): CIABiomed 2025, LNBI 16148, pp. 464–478, 2026.
https://doi.org/10.1007/978-3-032-10661-2_35

are essential in this process, as they enable the extraction of compact, informative representations from complex datasets. This study explores and compares three such methods—MOFA, IntNMF, and VAE—focusing on their ability to reveal biologically and clinically meaningful patterns in breast cancer multi-omics data.

In recent years, advances in omics technologies have enabled the collection of large-scale genomic, transcriptomic, proteomic, and epigenetic data, offering an unprecedented opportunity to characterize the complexity of breast cancer from a systems biology perspective. However, effectively integrating these heterogeneous data sources presents significant methodological challenges, including cross-modality heterogeneity, differences in scale and noise levels, and the need for biological interpretability of results.

Several computational approaches have been proposed to address these issues, among which latent factor models stand out for their ability to project high-dimensional data into reduced spaces while preserving relevant relationships among variables. These models not only facilitate the exploration of shared underlying structures across omics layers but also enhance predictive and discovery tasks, such as patient stratification or identification of molecular signatures associated with tumor progression. In this context, a comparative evaluation of dimensionality reduction techniques is essential to understand their respective strengths and limitations, and to inform the selection of appropriate tools in translational and clinical research.

This paper is structured as follows: Sect. 2 describes the dataset and the main challenges encountered. Section 3 details the methodological framework, Sect. 4 presents the experimental results, while Sect. 5 provide an in-depth analysis of the latent representations and their biological significance. Finally, Secs. 7 and 8 offer a comprehensive discussion and conclusions of this work.

2   Dataset Description and Challenges

The dataset used in this work was sourced from the TCGA-BRCA project [8], comprising 705 breast cancer samples of ductal and lobular histological subtypes. The dataset includes four omics layers—copy number variations, mutations, protein expression, and gene expression—alongside clinical variables such as hormone receptor status, survival, and histological type. Two major challenges were addressed in this study: high dimensionality and class imbalance. The dataset included 1936 molecular features, which posed issues related to the *curse of dimensionality*, such as data sparsity, increased computational cost, and risk of overfitting. To mitigate these effects, dimensionality reduction techniques were applied to project the data into lower-dimensional latent spaces, enhancing both model performance and interpretability. Additionally, the dataset exhibited severe class imbalance: only **12.23%** of samples were labeled as *"alive"*, and **9.86%** belonged to the lobular carcinoma subtype. To ensure robust model evaluation, three dataset variants were used—original, oversampled, and undersampled—allowing fair comparison across reduction methods and improving the detection of patterns in underrepresented groups.

3   Methodology

The main objective of this project is to evaluate the performance of latent representations under different scenarios derived from multi-omics data integration. To achieve this, the study is structured into three main components. The first focuses on addressing class imbalance, which involves generating balanced versions of the dataset through oversampling and undersampling techniques; SMOTE and RandomUndersampler from IMBLearn library respectively. Default parameters were used on both. The second component applies various dimensionality reduction methods to obtain latent representations that capture underlying patterns in the data. The third component involves evaluating the quality of these representations through both classification and clustering tasks, using clinical variables as targets, as well as performing structural and biological analysis to assess interpretability and biological relevance.

Therefore, experiments were conducted using a total of five datasets for each process. The latter two datasets were generated from the first dataset after preprocessing. These five datasets served as inputs to the subsequent processing stages. For a visual summary of the methodological workflow, Fig. 1 depicts the process from dataset input through to the final interpretation.

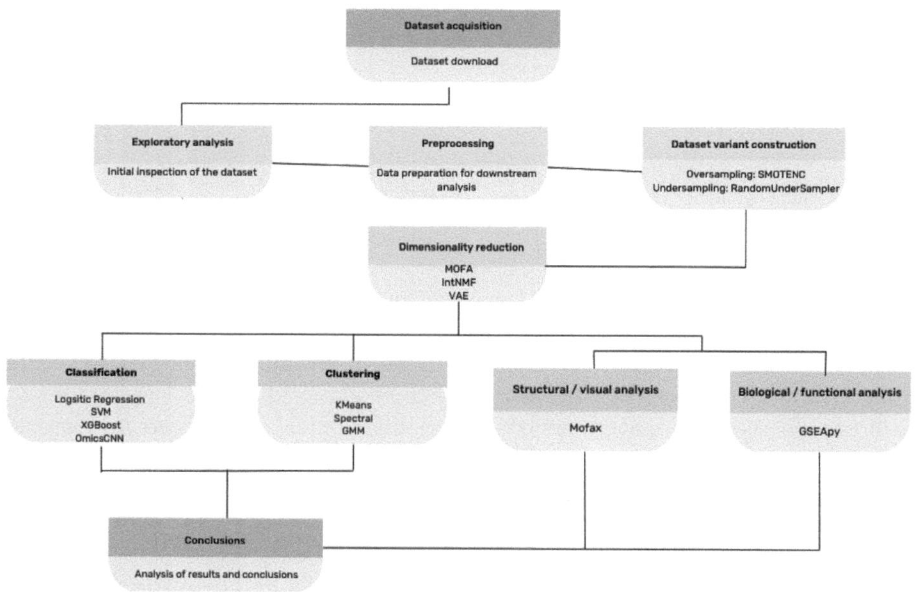

Fig. 1. Process workflow.

All experiments were conducted in a cloud-based Jupyter environment, ensuring efficient execution. The setup included Python 3 with access to two

NVIDIA Tesla T4 GPUs, which significantly accelerated deep learning tasks. Core libraries were pre-installed, and additional dependencies were integrated as needed during experimentation. In total, the experiments were completed in around 2 h and 45 min (Table 1).

Table 1. Main hyperparameters for reproducibility.

Method/Model	Key settings
MOFA	Latent factors = 60; spike–slab priors; seed = 42
IntNMF	Epochs = 30
VAE	Optimizer = Adam (lr = 1e–3); batch size = 64; epochs = 30
Omics-CNN	Epochs = 400; batch size = 32; optimizer = SGD (lr = 1e–4); dropout = 0.2
Logistic Regression	max_iter = 1000; solver = liblinear; class_weight = balanced; C = 1.0 (default)
SVC	kernel = rbf; probability = True; class_weight = balanced; C = 1.0 (default); gamma = scale (default)
XGBoost	use_label_encoder = False; eval_metric = mlogloss; n_estimators = 100 (default); learning_rate = 0.1 (default); max_depth = 3 (default)
Clustering	Default parameters; k = 2–6

3.1 Dimensionality Reduction Framework

For the dimensionality reduction process with MOFA, the shared GitHub repository accessible via [10] was used as a guide. The IntNMF method was employed for pairwise multi-omics data integration and dimensionality reduction. This approach was chosen to preserve the core functionality of the original code, enabling reuse of the repository, which was originally designed to handle only a pair of omics data types. The implementation was adapted from the open-source quick_intNMF repository, available at [12], and tailored to the specific dataset used in this study. Finally, the third approach was inspired by the GitHub repository in [13], which implements an end-to-end deep learning model for multi-omics integration and classification. Given the excessive modularity and complexity of the original architecture, a simplified VAE was adopted. This compact version used a single neural network with a dense layer reducing the input to a 60-dimensional latent space, followed by a reconstruction layer, enabling efficient integration while improving code interpretability and maintainability.

3.2 Classification and Clustering Frameworks

A classification framework was developed to predict histological type and patient survival using four algorithms: Logistic Regression, Support Vector Machines

(RBF kernel), XGBoost, and a Convolutional Neural Network (CNN). No hyperparameter tuning was performed, as baseline performance was satisfactory. Models were trained on both raw numerical features and latent factors derived from MOFA, IntNMF, and VAE. For CNN models, input features were reshaped using hierarchical clustering to optimize their spatial arrangement.

Model evaluation was performed using stratified 10-fold cross-validation (`StratifiedKFold, shuffle=True, random_state=42`), ensuring class distributions were preserved across folds.

Performance was assessed using accuracy, F1-score, AUC-ROC, and confusion matrices. In addition, an Omics-CNN variant was implemented, which incorporated feature reordering and `Conv1D` layers, trained with stochastic gradient descent (SGD) and learning rate decay over 400 epochs.

A clustering pipeline was implemented to assess how well latent representations from MOFA, IntNMF, and VAE captured clinically relevant breast cancer subtypes. Three algorithms—KMeans, Spectral Clustering, and Gaussian Mixture Models—were applied across varying numbers of clusters (k), and evaluated using ARI, NMI, Homogeneity, Completeness, and Silhouette Score. To enable systematic comparison, results were consolidated into a unified DataFrame using a dedicated function. Visualization tools included performance curves across k and t-SNE projections colored by predicted labels, facilitating interpretability. The chosen clustering methods provided complementary strengths: KMeans for efficiency, Spectral Clustering for non-linear structures, and GMM for probabilistic modeling.

4   Experimental Results

4.1   Classification

Tables 2 and 3 summarize the classification performance of various models using the original, imbalanced datasets without any class balancing techniques applied.

Table 2. Performance comparison of classifiers with raw data. Vital status

Classifier	Best Input	Accuracy	F1-score	TP/FN	AUC
Logistic Regression (LR)	IntNMF	0.7515	0.3000	27/35	0.6347
Support Vector Machine (SVM)	Complete	0.8402	0.3520	22/40	0.6366
XGBoost	VAE	0.8856	0.3256	14/48	0.6122
Convolutional Neural Network (CNN)	Complete	0.8630	0.3420	18/44	0.6460

In survival prediction, accuracy metrics are high (0.75–0.89) while masking poor minority class detection, reflected in low F1-scores (0.30–0.35) and high false negatives (e.g., 48 for XGBoost). Logistic Regression detects minorities

Table 3. Performance comparison of classifiers with raw data. Histological type

Classifier	Best Input	Accuracy	F1-score	TP/FN	AUC
Logistic Regression (LR)	Complete	0.9132	0.5600	28/22	0.8956
Support Vector Machine (SVM)	MOFA	0.9132	0.5600	28/22	0.8871
XGBoost	Complete	0.9507	0.7253	33/17	0.9572
Convolutional Neural Network (CNN)	MOFA	0.8840	0.4780	27/23	0.8460

more evenly but is limited by its linearity. Support Vector Machines with non-linear kernels better balance sensitivity and specificity, capturing complex patterns. CNNs perform competitively but are constrained by small sample size. In contrast, histological subtype classification achieves higher accuracy (over 0.88) and F1-scores (up to 0.73), indicating a more separable structure and fewer false negatives. XGBoost excels by modeling non-linear relationships without overfitting. This differs from survival prediction, where more complex boundaries are needed. Variational Autoencoders, despite rich embeddings, may include irrelevant features that reduce effectiveness in highly imbalanced survival prediction due to their focus on data reconstruction.

Tables 4 and 5 summarize the classification performance of the models using the oversampled dataset.

Table 4. Performance comparison of classifiers with oversampled data. Vital status

Classifier	Dataset	Accuracy	F1-score	TP/FN	AUC
Logistic Regression (LR)	Complete	0.93	0.91	263/4	0.93
Support Vector Machine (SVM)	MOFA	0.98	0.97	258/9	0.99
XGBoost	MOFA	0.96	0.95	249/19	0.93
Convolutional Neural Network (CNN)	Complete	0.94	0.92	256/11	0.93

Table 5. Performance comparison of classifiers with oversampled data. Histological type

Classifier	Dataset	Accuracy	F1-score	TP/FN	AUC
Logistic Regression (LR)	Complete	0.98	0.97	272/2	0.99
Support Vector Machine (SVM)	IntNMF	0.99	0.98	271/3	0.99
XGBoost	MOFA	0.97	0.96	265/9	0.98
Convolutional Neural Network (CNN)	MOFA	0.95	0.94	266/8	0.99

Regarding the oversampled data, in the survival prediction task, SVM clearly outperformed the other models. This superior performance suggests, once again, that the underlying data relationships for this objective are highly complex and non-linear. The RBF kernel employed by SVM is particularly well-suited to capturing such intricate decision boundaries, outperforming simpler linear models in this context. In contrast, histological type appears to exhibit more linear separability, where simpler models may suffice.

When comparing dimensionality reduction models, MOFA, owing to its probabilistic nature, is better equipped to model uncertainty and biological noise, resulting in more robust latent factors. This likely accounts for its superior classification performance when compared to algebraic methods such as IntNMF. Conversely, the relatively lower performance of the VAE may stem from its emphasis on reconstructive fidelity, which can lead to the retention of irrelevant details or noise. While VAE representations may be globally faithful to the input data, they are not necessarily optimal for specific downstream tasks such as classification, where the extraction of clinically meaningful patterns is paramount.

The following Tables 6 and 7 present the classification results obtained after applying an undersampling strategy to address class imbalance.

Table 6. Performance comparison of classifiers with undersampled data. Histological type

Classifier	Best Input	Accuracy	F1-score	TP/FN	AUC
Logistic Regression (LR)	MOFA	0.790	0.800	42/8	0.842
Support Vector Machine (SVM)	Complete	0.720	0.750	42/8	0.823
XGBoost	Complete	0.850	0.851	43/7	0.933
Convolutional Neural Network (CNN)	Complete	0.780	0.792	42/8	0.863

Table 7. Performance comparison of classifiers with undersampled data. Vital status

Classifier	Best Input	Accuracy	F1-score	TP/FN	AUC
Logistic Regression (LR)	IntNMF	0.629	0.623	38/24	0.676
Support Vector Machine (SVM)	Complete	0.650	0.580	31/31	0.633
XGBoost	IntNMF	0.613	0.625	40/22	0.655
Convolutional Neural Network (CNN)	MOFA	0.645	0.633	38/24	0.720

For histological type, XGBoost achieved the highest overall performance with the complete input data, yielding the best results across all metrics. In contrast, for vital status prediction, performance was generally lower across classifiers, with Logistic Regression and CNN showing the most balanced results. Figures 2 and 3 illustrates the evolution of the AUC-ROC score when training a logistic regression model on the original dataset. Figure 2 shows the results obtained using raw omics features, while Fig. 3 corresponds to the same experiment conducted with MOFA-derived latent factors. All results were obtained via stratified cross-validation, progressively increasing the training subset size.

When training logistic regression on the original high-dimensional data (1936 variables), the model quickly overfits, achieving perfect training accuracy and AUC with as few as 100 samples. However, validation performance fluctuates

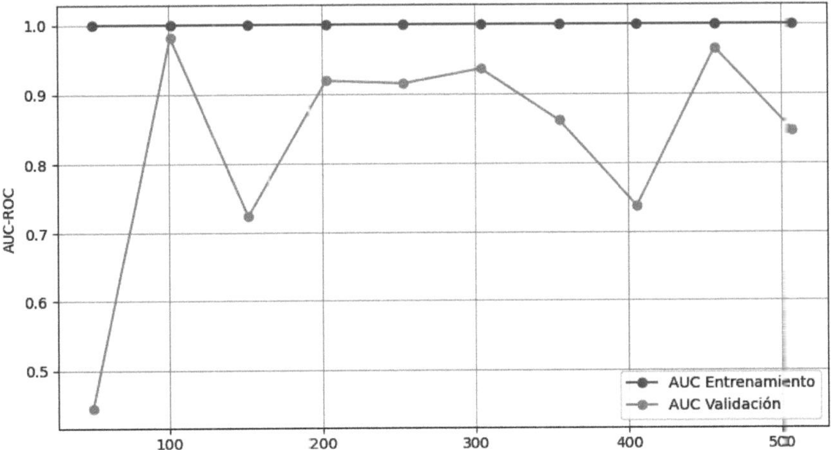

Fig. 2. LR: AUC using raw data – histological type, original dataset. Extremely volatile validation AUC; perfect training values mask high variance.

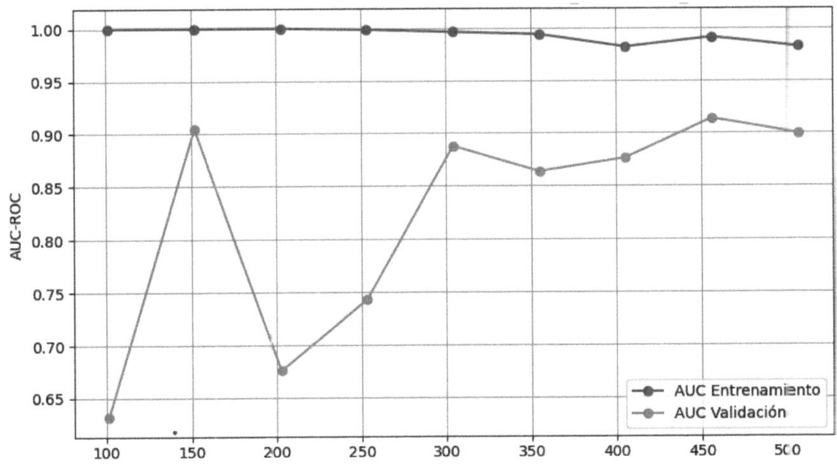

Fig. 3. LR: AUC with MOFA factors – histological type, original dataset. Effective regularization: the validation AUC increases and stabilizes around 0.90 with minimal fluctuation.

widely and drops unpredictably as training size increases, revealing poor generalization and strong dependence on specific validation splits. Even with the full dataset, a notable gap remains between training and validation metrics—about 8–10 points in accuracy and up to 20 in AUC—indicating high variance and making the model unreliable for clinical use.

In contrast, projecting the data into a lower-dimensional latent space with MOFA stabilizes learning by removing redundancy and constraining complexity. Training and validation curves converge more closely, with the validation AUC consistently between 0.85 and 0.91, and the gap shrinking to under 10 points. Though training error slightly increases, the significant reduction in variance improves reproducibility and clinical trustworthiness. This demonstrates that dimensionality reduction via MOFA is crucial for enhancing generalization and achieving more reliable prognosis models in breast cancer.

4.2   Clustering

Table 8 summarizes the best clustering configurations for each integration model based average scores. For each latent representation (MOFA, IntNMF, and VAE), the optimal combination of algorithm and number of clusters (k) is reported, along with relevant evaluation metrics. These results highlight the capacity of different integration strategies to capture clinically meaningful structure in the data.

Table 8. Best clustering algorithm for each dimensionality reduction method based on performance metrics.

Model	Best Method	k	ARI	NMI	Homogeneity	Completeness	Silhouette
MOFA	Spectral Clustering	5	0.4175	0.3629	0.3772	0.3497	0.0706
IntNMF	Spectral Clustering	4	0.4036	0.3150	0.3117	0.3183	0.0710
VAE	GMM	3	0.3208	0.2835	0.2542	0.3203	0.5533

Among the clustering algorithms evaluated, spectral clustering consistently yielded the best results across all latent representations. This superior performance can be attributed to spectral clustering's ability to effectively capture complex, non-convex cluster structures inherent in multi-omics data. By leveraging the eigenstructure of similarity graphs, spectral clustering accommodates the intricate relationships and heterogeneity present in biological datasets, which traditional methods such as KMeans or GMM may fail to capture adequately.

5   Biological Pathway Enrichment

Pathway enrichment analysis of the leading factors associates them with key cancer-related processes. Notably, the first factor reflects epithelial-mesenchymal transition (EMT), KRAS signaling, tumor progression, and metastasis pathways, aligning with aggressive breast cancer subtypes such as triple-negative breast cancer. This factor captures tumor plasticity and signaling interactions that underlie poor prognosis and metastatic potential, which is consistent with the observed patterns in the scatterplot representation. [3,4,6,7,9] (Fig. 4).

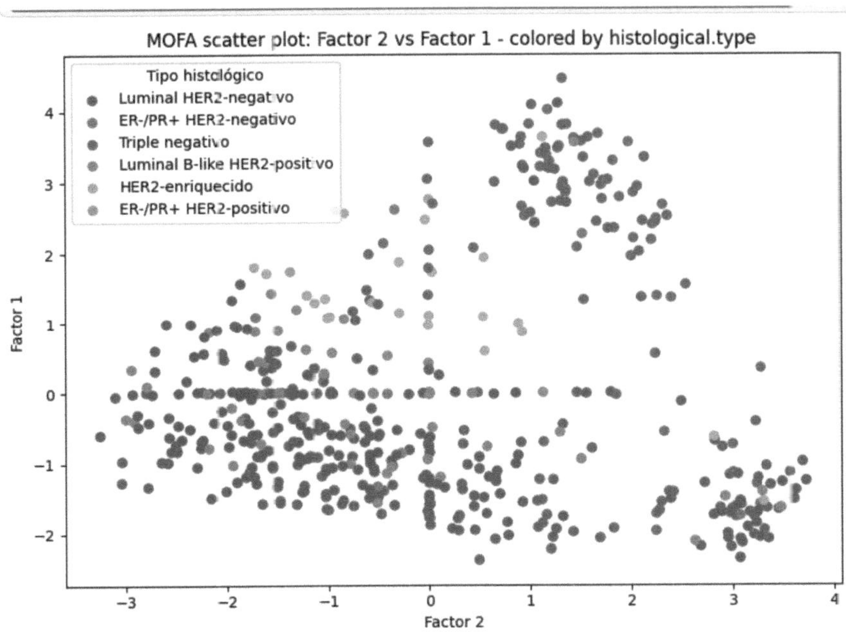

Fig. 4. Scatter plot of factors 1 and 2 obtained by MOFA, colored according to the histological subtype of each sample.

Given these biological functions, it is unsurprising that Factor 1 holds significant predictive power for patient survival outcomes. The processes it represents—such as enhanced metastatic capability and treatment resistance—are directly linked to decreased overall and progression-free survival. Therefore, the weight of this factor in survival analyses reflects its critical role in driving disease aggressiveness. Moreover, this highlights the biological relevance of the MOFA-derived latent space, suggesting that the dimensionality reduction captures meaningful patterns aligned with disease mechanisms.

E-cadherin (CDH1) emerged as the top discriminative feature between ductal and lobular histological subtypes, consistent with literature indicating its loss in lobular carcinoma and preservation in ductal carcinoma. This molecular distinction supports its use as a robust biomarker for histological classification. [1,2,5] (Fig. 5).

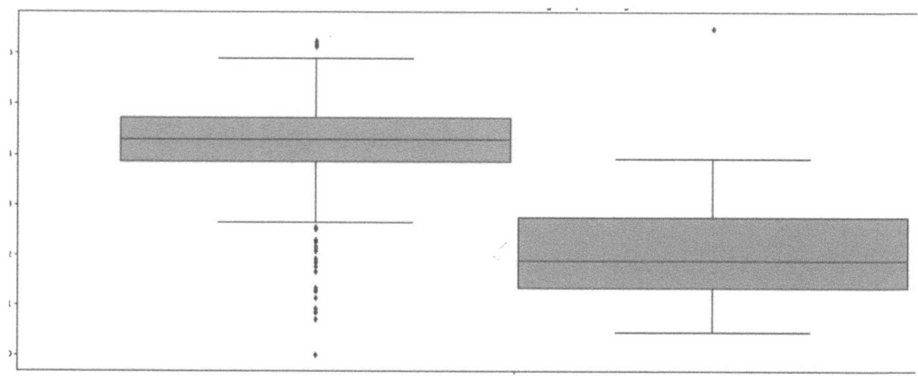

Fig. 5. Distribution of E-cadherin according to histological type.

The scatterplot shows that CDI samples exhibit higher expression of E-cadherin than CLI samples, appearing more prominently on the left—an observation that aligns with the literature and supports the biological relevance of MOFA's latent factors. These results demonstrate MOFA's capacity to uncover clinically and molecularly meaningful patterns in breast cancer, reinforcing its utility for integrative multi-omics analyses.

6   Discussion

6.1   Comparison of Classifier Performance Across Sampling Strategies

A comprehensive comparison of the experimental results across the three sampling strategies—no balancing, undersampling, and oversampling—reveals significant insights regarding model behavior, data representations, and task complexity.

In the *vital status prediction* task, the unbalanced setting (Table 2) leads to deceptively high accuracy values (up to 0.8856 with XGBoost) but extremely poor F1-scores (ranging from 0.30 to 0.35) and unacceptably high false negative rates (e.g., 48 with XGBoost). These discrepancies expose the limitations of accuracy as a metric in imbalanced contexts and highlight the classifiers' inability to reliably identify the minority class—deceased patients—which is clinically critical. Although SVM with raw data shows the highest F1-score in this setting (0.3520), the overall sensitivity remains insufficient. Applying *undersampling* (Table 7) partially mitigates these issues. While accuracy drops (e.g., 0.613–0.650), F1-scores improve substantially, reaching 0.625 with XGBoost and 0.633 with CNN, and AUC values increase across models (up to 0.720). These gains suggest that undersampling promotes better minority class detection, albeit at the cost of overall data reduction and potential loss of informative patterns. Notably, models leveraging *IntNMF and MOFA* representations tend to perform better, particularly for CNN and Logistic Regression, indicating that reduced and structured latent representations compensate for the diminished sample size. *Oversampling* (Table 4) yields the best results for vital status prediction across all metrics. All classifiers report F1-scores above 0.90, with SVM and Logistic Regression achieving near-perfect accuracy and AUC values (> 0.98). CNN also shows substantial improvement, especially when using MOFA, highlighting the utility of learned latent spaces under enriched training conditions. These outcomes strongly suggest that oversampling, by preserving the entirety of the dataset and synthetically balancing classes, enables the models—especially more complex architectures like CNNs—to learn from the full variance of the data while maintaining sensitivity to the minority class.

In the *histological type classification* task, trends are notably different. Even without class balancing (Table 3), classifiers already achieve relatively high performance, with F1-scores reaching 0.7253 (XGBoost) and AUC values above 0.95. This indicates that the task is inherently less complex and more linearly separable, likely due to clearer boundaries between histological subtypes. The moderate class imbalance in this task seems to have a less detrimental effect on model sensitivity. Nevertheless, both *undersampling* and *oversampling* further enhance performance. Under undersampling (Table 6), XGBoost maintains its dominance (F1 $= 0.851$, AUC $= 0.933$), and all models demonstrate excellent balance between precision and recall, with false negative counts as low as 7–8. The consistent success of CNNs, SVMs, and Logistic Regression across different inputs also suggests that histological subtype classification benefits from high information density and less noisy labels. With oversampling (Table 5), performance reaches near-saturation: all models achieve F1-scores above 0.94 and AUCs close to or above 0.99. SVM and Logistic Regression, particularly when trained on *MOFA* or *IntNMF*, yield the best results. This supports the notion that probabilistic latent variable models like MOFA can effectively capture shared and modality-specific sources of variation, enhancing generalization and boosting performance, especially under class-balanced conditions.

When comparing *dimensionality reduction techniques*, *MOFA* and *IntNMF* consistently emerge as the most beneficial. MOFA boosts CNN and SVM performance in both tasks and across all balancing strategies, likely due to its probabilistic factorization framework that preserves informative structure while reducing noise. *IntNMF* enhances performance in linear models like Logistic Regression and XGBoost, especially under undersampling and oversampling, possibly due to its ability to extract interpretable low-dimensional features. In contrast, *VAE* is less effective, especially in the unbalanced setting, potentially due to its sensitivity to data sparsity and its generative nature not aligning well with the discriminative objective.

Regarding classifiers, *XGBoost* consistently ranks among the top performers across tasks and sampling methods. Its ability to model complex, non-linear interactions through gradient-boosted ensembles makes it robust to both imbalance and heterogeneity. *CNNs* show strong results under balanced conditions, particularly with MOFA, but struggle more under imbalance and data scarcity—

likely due to their need for large, diverse training sets. *SVMs* perform competitively when paired with expressive feature spaces, while *Logistic Regression*, despite its simplicity, remains effective when used with structured latent representations.

These findings highlight that no single strategy is universally optimal. Rather, classifier performance is shaped by a complex interplay between *sampling strategy, input representation, model architecture*, and *task difficulty*. For vital status prediction, *oversampling combined with latent representations* (MOFA, IntNMF) and *non-linear classifiers* (SVM, CNN) offers the best trade-off between sensitivity and generalization. For histological type classification, the task is more forgiving, and multiple classifiers perform well, though *XGBoost and SVM with MOFA* still stand out. These results emphasize the importance of tailoring data preprocessing, representation learning, and classifier selection to the specific challenges posed by the predictive task.

6.2   Comparison of Clustering Performance Across Approaches

MOFA achieves clusters that are more faithful to clinical subtypes because it integrates multiple data types and explicitly models noise, thereby capturing true biological signals. Thus, its factors better reflect clinical heterogeneity, producing more accurate and relevant groupings. IntNMF attains good structural compactness because, as an algebraic matrix decomposition method, it enforces that data are represented as additive combinations of sparse, non-negative latent factors. This results in clearly defined groups with little overlap in the latent structure, favoring internally homogeneous and compact clusters. Additionally, by focusing on the direct decomposition of the original matrix, IntNMF preserves the inherent biological structure, which translates into clinically coherent and compact groupings. The VAE achieves better compactness and separation in clustering because it learns latent representations that are regularized to be orderly and continuously distributed. This regularization (such as KL divergence) forces the model to create well-defined and compact groups in the latent space, avoiding overlap and excessive dispersion.

6.3   Computational Trade-Offs

Considering the computational trade-offs of the evaluated methods, MOFA demonstrates a balance between interpretability and moderate computational cost, IntNMF offers efficiency for linear models but may underrepresent complex interactions in highly heterogeneous datasets, and VAE requires higher computational resources and careful parameter tuning to produce stable latent representations. Although the present study focuses on TCGA-BRCA data, the framework can be extended to integrate epigenetic information and longitudinal samples, allowing latent factors to capture additional biological and temporal variation.

6.4   Translational Potential

The latent representations extracted mainly by MOFA and IntNMF capture biologically meaningful patterns across multiple omics layers, which can be leveraged to inform personalized treatment strategies. For instance, these factors could be used to stratify patients based on predicted risk profiles, identify subgroups likely to respond to specific therapies, or prioritize targets for intervention. Furthermore, integrating latent factors into clinical decision-support systems could provide clinicians with interpretable molecular summaries, facilitating data-driven treatment planning and enhancing translational impact by bridging multi-omics analyses with actionable clinical insights.

7   Conclusions

This study's conclusions provide a detailed insight into the performance of latent representations derived from multi-omics data in breast cancer classification and clustering tasks. Specific findings regarding improved reproducibility in classification, biological coherence of obtained clusters, and the interpretability of latent factors are summarized.

On the one hand, Dimensionality reduction via latent factor extraction does not primarily aim to increase classification accuracy—which is already high—but to enhance its reproducibility. In clinical contexts, where decision stability across patients is critical, this regularization effect is decisive. Latent projections act as effective noise filters, stabilizing model fitting and preventing overfitting, especially in datasets with limited samples. On the other hand, vital status prediction seems to be more complex than histological type prediction, which appears to be more linear.

In clustering tasks, MOFA provides the most biologically meaningful clusters by integrating diverse data types and explicitly modeling noise, effectively capturing clinical heterogeneity. IntNMF excels in generating compact and well-defined clusters through its non-negative matrix factorization approach, preserving inherent biological structures. Meanwhile, the VAE produces highly separated and compact clusters by learning regularized latent representations, enhancing cluster clarity and separation. Together, these methods offer complementary strengths for robust multi-omics data integration and clustering.

The latent factors extracted by MOFA demonstrate diversity and minimal redundancy, indicating informative and complementary representations that mitigate overfitting risk. These factors exhibit strong biological coherence, aligning with established pathways implicated in breast cancer subtypes. For example, MOFA's first factor captures epithelial-mesenchymal transition and KRAS pathway activation characteristic of triple-negative subtypes. Additionally, the importance of E-cadherin (CDH1) as a critical molecular marker distinguishing lobular from ductal carcinomas is reflected in the factor loadings, validating MOFA's capacity to reveal relevant molecular distinctions.

Acknowledgments. This publication is part of the projects PID2020-117954RB-C22 and PID2023-146037OB-C21, funded by MICIU/AEI/10.13039/501100011033/.

References

1. Alsaleem, M., et al.: The molecular mechanisms underlying reduced e-cadherin expression in invasive ductal carcinoma of the breast: high throughput analysis of large cohorts. Mod. Pathol. **32**(8), 1214–1226 (2019). https://doi.org/10.1038/s41379-019-0209-9. https://pubmed.ncbi.nlm.nih.gov/30760857/
2. Bullock, E., Brunton, V.G.: E-cadherin-mediated cell-cell adhesion and invasive lobular breast cancer. In: Advances in Experimental Medicine and Biology, vol. 1464, pp. 259–275. Springer, Heidelberg (2025). https://doi.org/10.1007/978-3-031-70875-6_14. https://pubmed.ncbi.nlm.nih.gov/39821030/
3. Błaszczak, E., Miziak, P., Odrzywolski, A., Baran, M., Gumbarewicz, E., Stepulak, A.: Triple-negative breast cancer progression and drug resistance in the context of epithelial-mesenchymal transition. Cancers (Basel) **17**(2), 228 (2025). https://doi.org/10.3390/cancers17020228. https://pubmed.ncbi.nlm.nih.gov/39858010/
4. Das, K., et al.: Triple-negative breast cancer-derived microvesicles transfer microrna221 to the recipient cells and thereby promote epithelial-to-mesenchymal transition. J. Biol. Chem. **294**(37), 13681–13696 (2019). https://doi.org/10.1074/jbc.RA119.008619. https://pubmed.ncbi.nlm.nih.gov/31341019/
5. Grabenstetter, A., et al.: E-cadherin immunohistochemical expression in invasive lobular carcinoma of the breast: correlation with morphology and cdh1 somatic alterations. Human Pathol. **102**, 44–53 (2020). https://doi.org/10.1016/j.humpath.2020.06.002. https://pubmed.ncbi.nlm.nih.gov/32599083/
6. Henriet, E., et al.: Triple negative breast tumors contain heterogeneous cancer cells expressing distinct kras-dependent collective and disseminative invasion programs (2023). https://pubmed.ncbi.nlm.nih.gov/36604566/
7. Markiewicz, A., et al.: Activation of epithelial-mesenchymal transition process during breast cancer progression - the impact of molecular subtype and stromal composition. Acta Biochimica Polonica **68**(3), 385–392 (2021). https://doi.org/10.18388/abp.2020_5719. https://pubmed.ncbi.nlm.nih.gov/34432400/
8. NIH: Tcga dataset (2023). https://portal.gdc.cancer.gov/projects/TCGA-BRCA
9. Paranjape, T., et al.: A 3'-untranslated region kras variant and triple-negative breast cancer: a case-control and genetic analysis. Lancet Oncol. **12**(4), 377–386 (2011). https://doi.org/10.1016/S1470-2045(11)70044-4. https://pubmed.ncbi.nlm.nih.gov/21435948/
10. rargelaguet: Mofapy2 (2023). https://github.com/bioFAM/mofapy2/blob/master/mofapy2/notebooks/getting_started_python.ipynb
11. de Sanidad, M.: Cáncer de mama aecc (2025). https://www.contraelcancer.es/es/todo-sobre-cancer/tipos-cancer/cancer-mama
12. wmorgans: Mofapy2 (2025). https://github.com/wmorgans/quick_intNMF/tree/master
13. zhangxiaoyu11: Omivae (2021). https://github.com/zhangxiaoyu11/OmiVAE

Optimization of Denoising Autoencoders with Progressive Learning Strategies for ScRNA-Seq Data

Francisco Javier Franco-Ruiz[(✉)] [iD], Sara Fernandez-Malvido [iD],
Belén Vega-Marquez [iD], and Isabel A. Nepomuceno-Chamorro [iD]

Dpto. de Lenguajes y Sistemas Informáticos, University of Seville, Sevilla, Spain
{frafrarui,sarfermal}@alum.us.es, inepomuceno@us.es

Abstract. Advances in single-cell RNA sequencing (scRNA-seq) technologies have revolutionized cancer research by enabling detailed characterization of cellular heterogeneity [6]. However, predicting drug sensitivity at single-cell resolution remains challenging due to the scarcity of annotated data, high dimensionality, and technical noise. In this work, we reproduce and enhance the scDEAL framework, a deep transfer learning model that integrates bulk and single-cell transcriptomic data for drug response prediction.

The enhancements focus on optimizing Denoising Autoencoders (DAE) and incorporating progressive training strategies. Specifically, we introduce Gene Prioritization Regularization (GPR) to emphasize biologically relevant genes, implement Curriculum Learning to gradually increase task complexity during training, and apply direct filtering of highly variable genes to reduce dimensionality.

Experiments conducted on publicly available datasets from GDSC, CCLE, and GEO demonstrate that filtering the top 20% most variable genes leads to significant improvements in predictive performance, achieving an F1-score of 0.9641 and an AUC of 0.9549, while reducing computational costs by 80%. These results highlight the importance of feature selection and progressive training strategies for enhancing drug response prediction from scRNA-seq data.

Keywords: Drug sensitivity prediction · scRNA-seq · Denoising Autoencoders · Transfer learning · Deep learning

1 Introduction

Cancer is one of the leading causes of mortality worldwide and represents a major clinical and scientific challenge [10]. Despite advances in targeted therapies and personalized medicine, many treatments remain ineffective for all patients due to biological differences between tumors. This variability is closely related to tumor heterogeneity, which refers to the coexistence of multiple cell populations with distinct genetic and phenotypic characteristics within the same tumor [12]. This

A. López Fernández et al. (Eds.): CIABiomed 2025, LNBI 16148, pp. 479–489, 2026.
https://doi.org/10.1007/978-3-032-10661-2_36

diversity complicates the development of effective therapies and contributes to the emergence of treatment resistance, as well as the need for detailed transcriptomic analyses such as single-cell RNA sequencing.

In recent years, single-cell RNA sequencing (scRNA-seq) technologies have revolutionized the analysis of tumor heterogeneity by enabling the analysis of gene expression in thousands of individual cells and uncovering subpopulations with key roles in drug response [13]. Compared to bulk RNA sequencing, scRNA-seq provides a more detailed resolution of cellular states, although at higher cost and complexity. However, scRNA-seq data present significant technical challenges, such as high dimensionality, the presence of noise (dropouts), and the scarcity of experimental annotations related to drug sensitivity.

To address these challenges, several computational strategies have been proposed, as summarized in a recent review focused on methods for predicting cancer drug response through the integration of bulk and single-cell RNA-seq data [8]. Among these, deep learning approaches have emerged as powerful tools for modeling complex gene expression patterns and improving predictive performance. In particular, transfer learning strategies allow the knowledge acquired from large, well-annotated bulk RNA-seq datasets to be leveraged for single-cell predictions. One of the most prominent frameworks is scDEAL (single-cell Drug rEsponse AnaLysis) [3], which combines Denoising Autoencoders (DAE) for robust feature extraction with Domain-Adaptive Neural Networks (DaNN) to align the latent representations of bulk and single-cell data. This approach enables accurate drug sensitivity prediction even when single-cell datasets lack direct drug-response labels.

Building upon the scDEAL framework, this work aims to reproduce and optimize its performance by incorporating three key improvements. First, we implement Gene Prioritization Regularization (GPR), a strategy that emphasizes biologically relevant genes during the autoencoder reconstruction process. Second, we apply Curriculum Learning (CL), a progressive training approach that facilitates the model's convergence by introducing tasks in order of increasing difficulty. Finally, we integrate a gene selection step based on the top 20% most variable genes, which reduces dimensionality and improves both predictive performance and computational efficiency.

Experimental results demonstrate that these enhancements lead to comparable gains in metrics such as F1-score and AUC. Moreover, they significantly decrease training time, highlighting the effectiveness of the proposed strategies.

2   Materials and Methods

This section describes the datasets and preprocessing steps employed in this study, as well as the baseline architecture of the scDEAL framework. Furthermore, we detail the modifications proposed to improve predictive performance and computational efficiency. The section is organized as follows: Subsect. 2.1 introduces the datasets used, including both bulk and single-cell RNA-seq data;

Subsect. 2.2 outlines the preprocessing pipeline applied to these datasets; Subsect. 2.3 presents the architecture of the original scDEAL model; and Subsect. 2.4 explains the modifications proposed in this work.

2.1 Datasets

For this study, two types of transcriptomic data were used: *bulk RNA-seq* and *single-cell RNA-seq* data (scRNA-seq).

The *bulk* data were obtained from the Genomics of Drug Sensitivity in Cancer (GDSC) [4] and Cancer Cell Line Encyclopedia (CCLE) [5] databases. These resources contain gene expression profiles for more than one thousand human cancer cell lines, as well as drug response measurements for over one thousand pharmacological compounds. These well-annotated datasets enable robust training of a drug sensitivity predictor.

The single-cell datasets used in this study are summarized in Table 1, including GEO identifier, cancer type, treatment drug, number of cells, and a brief description. For example, we used the GSE110894 dataset, which includes leukemic mouse cells treated with the epigenetic inhibitor I-BET-762. This dataset provides detailed information on gene expression at the single-cell level, enabling the capture of tumor heterogeneity and different treatment responses. All scRNA-seq datasets were derived from mouse models, providing detailed single-cell transcriptomic information for drug response analysis.

Table 1. Summary of single-cell RNA-seq datasets used in this study.

GEO ID	Cancer type	Drug	Num cells	Brief description
GSE110894	Leukemia	I-BET-762	1,419	Treatment with BET inhibitor; includes sensitive and resistant cells
GSE112274	Glioblastoma (CSC)	Gefitinib	641	Cancer stem cells exposed to Gefitinib
GSE117872	Breast cancer	Docetaxel	771	Includes subpopulations with different drug response
GSE140440	Lung adenocarcinoma	Erlotinib	2,288	Sensitive and resistant cells to treatment
GSE147405	HeLa cells (cervix)	Cisplatin	3,091	Different times post-treatment
GSE149383	Nasopharyngeal carcinoma	I-BET-762	3,084	Different conditions and time points with BET inhibitor

The rationale for combining human-derived bulk data with mouse single-cell data stems from the scarcity of human single-cell datasets with well-annotated drug responses [3]. Using transfer learning techniques, it is possible to align feature spaces across species, leveraging a model trained on human data to make predictions on mouse data, due to the conservation of biological pathways and gene expression profiles between the two species.

2.2 Data Preprocessing

To ensure that both bulk and single-cell (scRNA-seq) data are comparable and ready for deep learning, we applied a rigorous preprocessing pipeline described in [3] for scRNA-seq data. Bulk RNA-seq data from GDSC and CCLE were downloaded preprocessed from [9]. These datasets were previously normalized, and drug response values (AUC) were transformed into binary sensitivity labels (*sensitive* vs. *resistant*) using the *waterfall method* as introduced in the CCLE study [11].

An additional challenge arises from class imbalance, since the proportion of sensitive and resistant samples varies across drugs. To address this, the scDEAL framework already includes sampling strategies in its training pipeline, including *up-sampling* (replicating minority samples), *down-sampling* (removing majority samples), and *SMOTE* (synthetic generation of minority samples). These options are integrated into the original code and can be activated as training parameters, ensuring a balanced representation of classes during model learning.

Additionally, to ensure cross-species compatibility between bulk human data and single-cell mouse data, we mapped human genes to their orthologous mouse genes using the MyGene.info service. Only genes with a unique mouse ortholog were retained, while genes with multiple orthologs or without a known ortholog were discarded. This mapping was performed in ENSMUSG format, resulting in a subset of 2883 genes conserved across both datasets, which were subsequently used for alignment and model training.

For scRNA-seq data, preprocessing follows standard steps implemented in the `scmodel.py` script from [9] using the Scanpy library: low-quality cells (fewer than 200 detected genes) and rarely expressed genes (fewer than 3 cells) are removed. UMI count matrices are normalized by total counts per cell, log-transformed using $\log(x + 1)$ to stabilize variance, and scaled to the [0,1] range using Min-Max scaling. These steps reduce technical noise, align scales between datasets, and retain the most informative genes (top 20% variable) for downstream model training.

2.3 Model Architecture

The scDEAL framework is based on a deep learning architecture with a transfer learning approach designed to predict drug sensitivity from single-cell gene expression data, leveraging knowledge learned from bulk RNA-seq data. Figure 1 schematically illustrates the scDEAL workflow, which consists of five main stages:

1. **Bulk feature extraction:** A Denoising Autoencoder (DAE) is trained to obtain a robust latent representation of bulk data, reducing noise and capturing main transcriptomic variations.

2. **Bulk drug response prediction:** Using the learned representation, a predictor (MLP neural network) is trained to classify cell lines as drug-sensitive or drug-resistant, based on labels derived from GDSC and CCLE.
3. **Single-cell feature extraction:** Another DAE is applied to generate latent representations of the scRNA-seq data, preserving the structure of the cellular populations.
4. **Domain Transfer Learning (DTL) training:** The latent representations of bulk and single-cell data are aligned using the Maximum Mean Discrepancy (MMD) metric, enabling the predictor trained on bulk data to adapt to the single-cell domain.
5. **Model transfer and prediction:** Finally, the combined model is used to predict drug sensitivity at the single-cell level, generating cell-level drug response profiles.

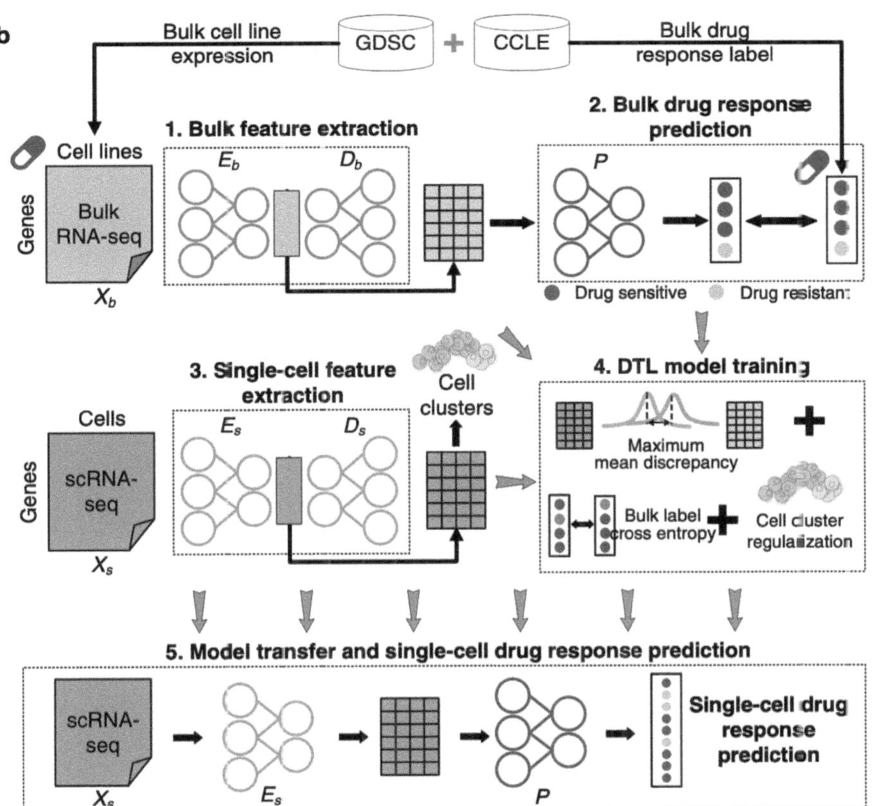

Fig. 1. Schematic representation of the scDEAL architecture integrating bulk and single-cell RNA-seq data for drug sensitivity prediction via transfer learning [3].

2.4   Proposed Modifications

In this work, we introduced three main strategies aimed at enhancing the original *scDEAL* framework. These strategies were designed to improve predictive performance, ensure biological relevance, and reduce computational cost. Below, we describe each approach in detail.

Gene Prioritization Regularization (GPR). Gene Prioritization Regularization (GPR) builds on the principle that not all genes contribute equally to biological heterogeneity or drug response prediction. In single-cell transcriptomic analysis, it is well established that highly variable genes capture essential differences between cell states, while low-variance genes, often housekeeping, provide little discriminative information [6].

Following this idea, the reconstruction loss of the *Denoising Autoencoder* (DAE) was modified to assign higher weights to the most variable genes, ensuring that the model learns representations that emphasize biologically meaningful features. In practice, this is achieved by identifying the top-variance genes, increasing their contribution within the loss function, and thus guiding the learning process toward patterns that are more informative for drug sensitivity prediction.

Curriculum Learning (CL). Curriculum Learning (CL) is inspired by the way humans learn progressively, starting with simpler tasks and advancing to more complex ones [7]. Applied to deep learning, this approach aims to improve convergence and robustness by presenting training examples in increasing order of difficulty, instead of random order.

In the context of scRNA-seq, CL helps mitigate challenges such as noise, dropout events, and high dimensionality. Our implementation follows three key steps: (a) define a quantitative measure of difficulty for each training example, (b) sort the data accordingly, and (c) present the examples in increasing order of difficulty during training. These three steps follow the method described by Li et al. [7] for implementing Curriculum Learning, ensuring a structured and progressive training strategy.

Filtering of Most Variable Genes. Building on the insights gained from GPR, we explored a more direct strategy: filtering genes based on variance prior to training. Instead of weighting all genes within the loss function, this approach reduces dimensionality by retaining only the most informative subset, typically the top 20% most variable genes.

Genes with low variability are often constitutive, contributing little to predictive tasks, whereas high-variance genes are usually linked to processes such as cell differentiation, proliferation, or drug response. By focusing exclusively on these genes, the model benefits from reduced noise, improved interpretability, and lower computational cost.

To assess this idea, we tested three variants:

1. *Common genes between bulk and single-cell*: selection of the top 20% most variable genes in bulk (3.192 genes), mapped to mouse orthologs (ENSMUSG), resulting in 2,100 common genes used for both datasets.
2. *Direct filtering in bulk*: application of top 20% filtering only in bulk data (3,192 genes), without modifying single-cell input. This proved to be the most efficient strategy in terms of performance and computational resources.
3. *Genes with mouse orthologs only*: retention of all human bulk genes with unique mouse orthologs (2,883 genes), ensuring biological coherence across species without requiring strict intersection with single-cell data.

These three variants allowed us to systematically evaluate the role of gene selection and cross-domain alignment in optimizing predictive performance.

3   Results

Before presenting the comparative results, it is important to detail the configuration adopted for training the baseline DAE model. Table 2 summarizes the hyperparameters used, which served as the foundation for all subsequent experiments.

Table 2. Hyperparameter configuration used for training the DAE

Hyperparameter	Value (DAE)
Latent dimension (bottleneck)	512
Encoder hidden layers	[256, 128]
Dropout rate	0.1
Batch size	200
Epochs	500
Learning rate	0.5

This section presents the results of training and evaluating the proposed model for drug sensitivity prediction from single-cell gene expression data. To assess the model's performance, widely used machine learning metrics were employed: area under the ROC curve (AUC), Average Precision (AP), Accuracy, and F1 Score.

These metrics evaluate different aspects of model performance: AUC measures class discrimination, AP evaluates prediction quality in imbalanced scenarios, Accuracy reflects the proportion of correct predictions, and F1 Score balances precision and recall.

The reported metrics (AUC, AP, Accuracy, and F1) correspond to model performance on the test set, while the total times indicated in Table 3 include both training and evaluation phases.

Table 3. Performance comparison of the baseline model against different improvement strategies. The best results for each metric are highlighted in bold.

Strategy	AUC	AP	Accuracy	F1	Total time (min)
DAE without modifications (baseline)	0.9491	0.9673	0.9479	0.9479	30
Gene Prioritization Regularization (GPR)	0.8662	0.7737	0.7336	0.7270	30
Curriculum Learning (CL)	0.9472	0.9672	0.9493	0.9493	30
Direct filtering of most variable genes					
Approach A: common genes	0.9442	0.9705	0.9479	0.9479	6
Approach B: without intersection	0.9444	0.9706	0.9514	0.9514	6
Top bulk genes filtering	0.9549	0.9757	**0.9641**	**0.9641**	6
Mouse orthologs	**0.9871**	**0.9885**	0.9436	0.9435	6

As shown in Table 3, the baseline model (DAE without modifications) achieves strong initial performance, with an AUC of 0.9491 and an F1 score of 0.9479. However, the Gene Prioritization Regularization (GPR) technique performs worse across all metrics, suggesting that weighting the loss function may overemphasize non-informative genes without improving model performance.

The Curriculum Learning strategy produces results similar to the baseline, with minor variations that do not yield a meaningful improvement.

In contrast, the direct filtering of the most variable genes substantially reduces execution time (6 min versus 30 min for the baseline) while maintaining competitive metrics. Among these variants, the top bulk genes filtering achieves the highest Accuracy and F1 score (0.9641), representing a well-balanced option in terms of both performance and efficiency.

To contextualize the reported execution times, all experiments were conducted on a VirtualBox virtual machine without GPU acceleration, configured with 8 GB of RAM and 4 CPU cores (100% usage). On average, the baseline model required around 30 min in total, which can be roughly divided into two main stages: bulk pretraining ($\approx$15 min) and single-cell domain adaptation and prediction ($\approx$15 min). In contrast, experiments with direct gene filtering substantially reduced execution time to about 6 min overall.

3.1 Performance Analysis

Figure 2 presents a graphical comparison of AUC, Accuracy, and F1 metrics for each evaluated strategy. As shown, the mouse ortholog genes approach achieves the highest AUC (0.9871), exceeding the baseline model, while the top bulk genes filtering attains the best Accuracy and F1 scores.

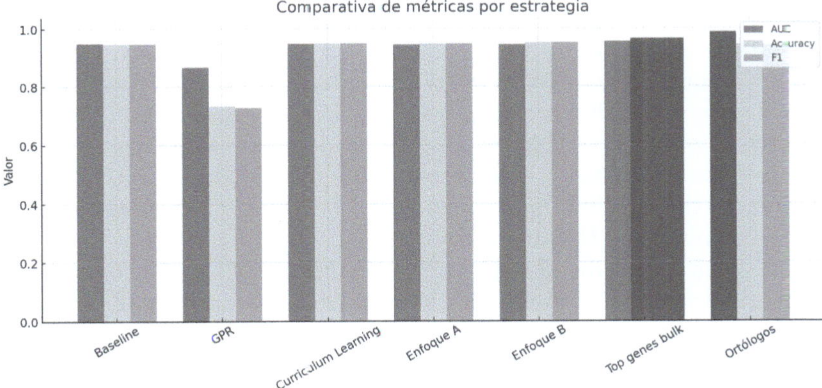

Fig. 2. Graphical comparison of AUC, Accuracy, and F1 metrics for each evaluated strategy. The mouse orthologs approach achieves the highest AUC, while the top bulk genes filtering attains the best Accuracy and F1 scores.

3.2 Execution Time Analysis

Figure 3 presents a comparison of the total execution time for each evaluated strategy. The variants with direct gene filtering exhibit a substantial reduction in training time (6 min) compared to the baseline model and the GPR and Curriculum Learning approaches (30 min).

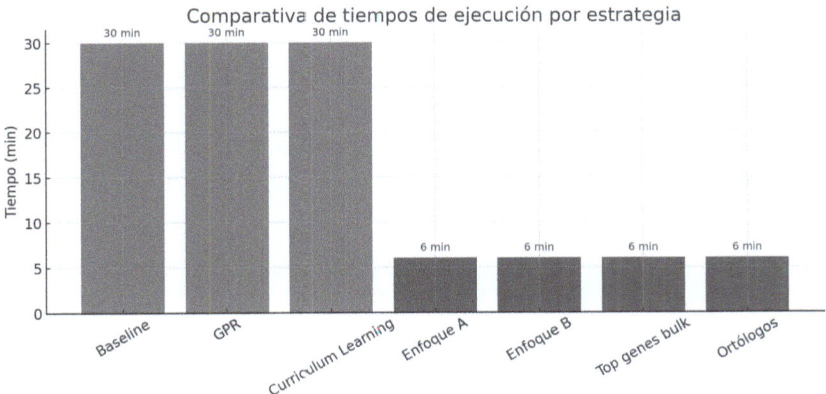

Fig. 3. Comparison of execution times for each evaluated strategy.

4   Conclusions

This work demonstrates that careful selection of input genes and preprocessing strategies is as crucial as model architecture for predicting drug sensitivity with scDEAL. While the baseline model delivered strong performance, direct filtering of the most variable genes proved to be the most effective strategy, achieving the highest Accuracy and F1-score (0.9641) and reducing execution time from 30 min to 6 min. The mouse ortholog genes strategy achieved the best AUC (0.9871) and Average Precision (0.9885), highlighting the importance of biologically informed feature selection. In contrast, Gene Prioritization Regularization (GPR) did not improve upon the baseline, and Curriculum Learning maintained comparable results while enhancing training stability.

Despite limitations, including restricted computational resources (e.g., a virtual machine without GPU acceleration), time constraints, and the complexity of adapting the original scDEAL code, the proposed strategies demonstrate their potential. These findings underscore the critical role of feature selection and preprocessing in optimizing model performance, often matching or surpassing architectural improvements in transfer learning between bulk and single-cell data.

In summary, the proposed enhancements not only improve scDEAL's predictive performance but also support its broader application in computational biology and personalized medicine.

Finally, our work is not exclusive to transcriptomics. For instance, the analysis could be adapted to identify and prioritize key genomic features in scATAC-seq data, such as accessible chromatin regions, by incorporating prior knowledge from bulk ATAC-seq studies. As a crucial part of our future work, we plan to systematically explore and validate the analysis on these diverse single-cell data types (scATAC-seq or spatial scRNA-seq), which would significantly broaden its applicability and impact across the entire single-cell genomics field.

Data and Reproducibility

The modified scripts and experimental files developed specifically for this work can be found at: https://github.com/Frafrarui/scDEAL-TFG-FRANCISCO-JAVIER-FRANCO-RUIZ.

Acknowledgments. This publication is part of the projects PID2020-117954RB-C22 and PID2023-146037OB-C21, funded by MICIU/AEI/10.13039/501100011033/.

References

1. Xia, Y., Gawad, C.: Bringing precision oncology to cellular resolution with single-cell genomics. Cancer Metastasis Rev. **40**, 473–493 (2021)
2. Lizard Bio: Single-cell vs. bulk sequencing: which one to use when?. https://lizard.bio/knowledge-hub/single-cell-vs-bulk-sequencing. Accessed July 2025
3. Chen, J., et al.: Deep transfer learning of cancer drug responses by integrating bulk and single-cell RNA-seq data. Nat. Commun. **13**, 6494 (2022). https://doi.org/10.1038/s41467-022-34277-7

4. Genomics of Drug Sensitivity in Cancer: Genomics of Drug Sensitivity in Cancer. https://www.cancerrxgene.org. Accessed July 2025

5. Cancer Cell Line Encyclopedia: Cancer Cell Line Encyclopedia. https://depmap.org/portal/ccle/. Accessed July 2025

6. Luecken, M.D., Theis, F.J.: Current best practices in single-cell RNA-seq analysis: a tutorial. Mol. Syst. Biol. **15**(6) (2019). https://doi.org/10.15252/msb.20188746. https://www.embopress.org/doi/full/10.15252/msb.20188746

7. Li, H., Fu, J., Ling, X., Sun, Z., Wang, K., Chen, Z.: Single-cell curriculum learning-based deep graph embedding clustering. arXiv preprint arXiv:2408.10511 (2024)

8. Maeser, D., Zhang, W., Huang, Y., Huang, R.S.: A review of computational methods for predicting cancer drug response at the single-cell level through integration with bulk RNA-seq data. Curr. Opin. Struct. Biol. **84**, 102745 (2024)

9. OSU-BMBL scDEAL GitHub repository. https://github.com/OSU-BMBL/scDEAL. Accessed July 2025

10. Bray, F., Ferlay, J., Soerjomataram, I., Siegel, R.L., Torre, L.A., Jemal, A.: Global cancer statistics 2018: GLOBOCAN estimates of incidence and mortality worldwide for 36 cancers in 185 countries. CA: Cancer J. Clin. **68**(6), 394–424 (2018). https://doi.org/10.3322/caac.21492. https://acsjournals.onlinelibrary.wiley.com/doi/full/10.3322/caac.21492

11. Barretina, J., et al.: The Cancer Cell Line Encyclopedia enables predictive modelling of anticancer drug sensitivity. Nature **483**, 603 (2012). https://doi.org/10.1038/nature11003

12. Dagogo-Jack, I., Shaw, A.T.: Tumour heterogeneity and resistance to cancer therapies. Nat. Rev. Clin. Oncol **15**, 81–94 (2018). https://doi.org/10.1038/nrclinonc.2017.166. https://pubmed.ncbi.nlm.nih.gov/29115304/

13. Gundogdu, P., Loucera, C., Alamo-Alvarez, I., Dopazo, J., Nepomuceno, I.: Integrating pathway knowledge with deep neural networks to reduce the dimensionality in single-cell RNA-seq data. BioData Min. **15**(1), 1 (2022)

Integrative Analysis of Gene Co-expression Networks and Biclustering for Cancer Biomarker Discovery

Marc Ríos-Cadenas[1]([✉]) [iD], Aurelio López-Fernández[1] [iD],
Iván Segura-Carmona[1] [iD], Juan A. Ortega[2] [iD], and Francisco A. Gómez-Vela[1] [iD]

[1] Intelligent Data Analysis Group (DATAi), Universidad Pablo de Olavide, Ctra. Utrera km. 1, 41013 Seville, Spain
mriosc00@gmail.com

[2] Computer Science Department, Universidad de Sevilla, 41012 Seville, Spain

Abstract. Soft tissue sarcomas (STS) are rare and heterogeneous cancers with limited biomarkers. Comprehending their underlying mechanisms is crucial for identifying biomarkers that improve diagnosis and facilitate targeted therapy. This paper applies two independent strategies—ne co-expression network (GCN) analysis and biclustering— identify biomarkers in uterine leiomyosarcoma (ULMS), using one RNA-Seq and one microarray dataset. GCN analysis uncovered globally co-expressed hub genes like FOXM1, E2F1, MYBL2, and PITX1, linked to mitosis and DNA replication. In contrast, biclustering identified local co-expression patterns, including FOXM1 and HLF as transcriptional regulators, and other genes such as CCNB1, POLQ, and TRIP13 involved in genomic stability. Notably, FOXM1 was detected by both methods, reinforcing its relevance. While GCN highlights global regulatory roles, biclustering captures condition-specific signals. Their complementarity enhances biomarker discovery and contributes to understanding ULMS transcriptional architecture.

Keywords: Gene co-expression network · Biclustering · Bioinformatics · Biomarkers · Sarcoma

1 Introduction

Soft tissue sarcomas (STS) are a heterogeneous group of malignancies that arise from mesenchymal tissues. Among them, leiomyosarcoma (LMS) is a particularly aggressive subtype that originates from smooth muscle cells and can affect various anatomical locations, including the retroperitoneum, blood vessels, and uterus. LMS is known for its high metastatic potential and poor prognosis, especially in advanced stages. A particularly lethal and clinically challenging variant is ULMS, which, despite being rare, represents the most common form of uterine sarcoma and contributes disproportionately to morbidity and mortality within this category [1].

© The Author(s), under exclusive license to Springer Nature Switzerland AG 2026
A. López Fernández et al. (Eds.): CIABiomed 2025, LNBI 16148, pp. 490–504, 2026.
https://doi.org/10.1007/978-3-032-10661-2_37

Advances in gene co-expression profiling have created new opportunities for investigating the molecular landscape of uLMS. Technologies like microarrays and bulk RNA sequencing (RNA-seq) provide the assessment of transcriptome activity at diverse depths and resolutions [2]. Although microarrays have played a crucial role in initial co-expression analyses, RNA-seq offers a more extensive perspective on transcript diversity and co-expression levels. Systems biology techniques have become indispensable for analysing high-dimensional data and extracting significant biological insights.

The construction of gene co-expression networks (GCNs) is an approach employed to get useful biological knowledge from gene co-expression datasets. In a GCN, genes are depicted as nodes, while edges signify correlations in their co-expression patterns [3]. GCNs facilitate the discovery of gene modules potentially linked to disease characteristics or biological pathways. Methods such as pyEnGNet have been used to identify hub genes and functional modules, especially in cancer research [4].

Nevertheless, GCNs may overlook locals or condition-specific co-expression patterns, particularly in diseases characterised by significant heterogeneity, such as uLMS. Biclustering algorithms provide a complementary approach by discovering gene subsets that are co-expressed within specified sample subsets. In contrast to GCN approaches, biclustering identifies local patterns, facilitating the discovery of subtype-specific gene modules that may be obscured in global analysis [5]. Therefore, biclustering methods can be particularly effective in revealing context-dependent regulatory mechanisms and possible biomarkers in complicated disorders.

The identification of molecular biomarkers—nes or gene profiles that signify illness status, prognosis, or therapeutic response—s fundamental to precision medicine. In uLMS, reliable biomarkers could transform early diagnosis, facilitate molecular subtyping, direct focused medicines, or minimise unwarranted surgical procedures [6]. The integration of GCN analysis and biclustering can offer a strong framework for identifying both global and local co-expression patterns, hence improving the reliability and clarity of biomarker identification.

This paper presents a complementary evaluation of these two analytical strategies for biomarker discovery in gene co-expression datasets from ULMS tumours. The results are compared to identify both global transcriptional modules and sample-specific co-expression patterns, facilitating the prioritisation of clinically pertinent potential biomarkers for further exploration in the context of ULMS. The insights gained from this comparison can contribute to a more nuanced understanding of gene regulation in sarcoma and improve the robustness of biomarker identification.

The rest of the paper is organised as follows: the datasets and methodology are detailed in Sect. 2. The experimental results and discussion are presented in Sect. 3. Finally, the conclusions are summarised in Sect. 4.

2   Materials and Methods

This section presents a comprehensive description of the datasets used in this work, followed by the two independent workflows applied for biomarker discovery. These workflows are composed of the following steps (Fig. 1): data preprocessing and quality control, differential gene co-expression analysis, GCN construction and optimisation, network validation, biclustering construction and evaluation, and identification of potential biomarkers.

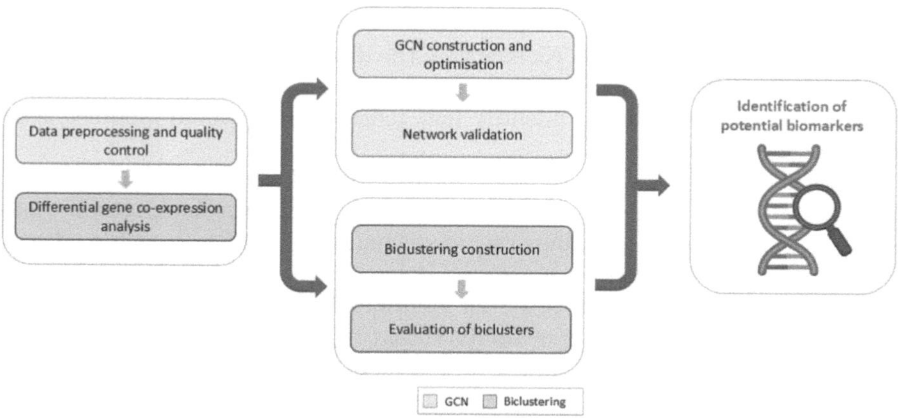

Fig. 1. The diagram illustrates the workflow used to perform the comparative evaluation to identify potential biomarkers in ULMS gene expression datasets.

2.1   Datasets Description

Two gene co-expression datasets, one derived from RNA sequencing and the other from microarray analysis, were retrieved from the NCBI Gene Expression Omnibus (GEO) repository. The RNA-Seq dataset (accession number GSE222045) was designed to evaluate the accuracy of a transcriptome-based model for distinguishing between uterine leiomyoma and leiomyosarcoma samples [7]. It comprises 33 samples in total, including 16 leiomyoma (serving as the control group) and 17 leiomyosarcoma specimens, with gene co-expression data covering 39,376 genes.

The microarray dataset (GSE64763) was generated to explore the molecular heterogeneity of ULMS and to identify potential molecular subtypes with clinical relevance [8]. This study compared the transcriptomic profiles of ULMS, uterine fibroids, and normal myometrial tissues. RNA was extracted from frozen tissue samples and hybridised with microarrays.

2.2   Data Preprocessing and Quality Assessment

The preprocessing steps were adapted to the type of data and included two main components: data normalisation and quality control.

For the RNA-Seq dataset, the process began by loading the expression matrix and sample annotations, which included information about tissue type and condition (tumour or control). Samples were grouped accordingly, and normalisation was performed using the DESeq2 package [9], which adjusts for sequencing depth and variability between samples. To assess data quality, a principal component analysis (PCA) was carried out, allowing us to visualise sample distribution and detect possible outliers or batch effects.

The microarray dataset was processed starting from raw CEL files. Data underwent background correction, normalisation, and log2 transformation using the Robust Multiarray Average (RMA) method. After separating tumour and control samples, quality control was performed using the arrayQualityMetrics package [10], which provides diagnostic plots and metrics to evaluate sample quality, detect technical noise, and ensure consistency across arrays.

2.3   Identification of Differential Co-expression Genes (DEGs)

To identify genes with significant expression differences between tumour and control samples, a differential expression analysis was performed. The resulting differentially expressed genes (DEGs) reduce data dimensionality and highlight disease-relevant biological signals. DEG selection was based on two criteria: log-fold change (logFC), indicating the magnitude of expression change, and adjusted p-value, assessing statistical significance after multiple testing correction. High logFC reflects strong expression differences, while low p-values indicate greater statistical confidence.

A key challenge is balancing sensitivity and specificity: high logFC thresholds may miss relevant genes, while low thresholds may yield too many DEGs, complicating downstream analysis [11]. To address this, three DEG subsets were generated using a fixed adjusted p-value cutoff of 0.05 and logFC thresholds of 1, 2, and 3. These subsets offer varying stringency levels and were used separately in subsequent analyses.

For the GSE64763 dataset, the initial control group included both fibroid and normal myometrium samples. However, this combined comparison produced very few DEGs, likely due to the distinct transcriptional profile of fibroids, which introduced unwanted variability. To improve specificity, fibroid samples were excluded, and the contrast was refined to tumour versus normal myometrium. This adjustment increased the number of robust DEGs. Furthermore, the full set of 22,277 genes was retained without filtering for low expression, ensuring comprehensive platform coverage in downstream steps.

2.4   GCN: Construction and Optimisation

To construct the GCNs, we applied a network construction method independently to each DEG dataset. The selected method needed to be runtime-efficient, compatible with both RNA-Seq and microarray data, and previously validated for biomarker discovery. Since lower fold-change thresholds increase the number of genes—d thus computational demand—selected PyEnGNet, an ensemble-based algorithm optimised for multi-GPU architectures and effective in similar contexts.

We explored multiple configurations by testing six correlation and pruning thresholds (0.7 to 0.95) with a fixed hub threshold of 3. For each DEG dataset, six networks were built per condition (tumour and control), totalling 12 networks per logFC threshold and 36 networks overall. Only those with at least 100 nodes and 100 edges were retained to ensure sufficient biological and topological complexity. The use of PyEnGNet was essential to efficiently manage the large volume of network constructions, as elaborated in the next subsection.

2.5   GCN: Evaluation and Selection

Following network construction with PyEnGNet, we evaluated the biological relevance of the resulting graphs and identified tumour-specific interactions. For microarray networks, gene identifiers were first mapped to standard HGNC symbols to ensure consistency and minimise information loss. When multiple entries matched the same gene, only the one with the highest edge weight was retained, as stronger correlations indicate more reliable functional relationships.

The same evaluation protocol was then applied to both RNA-Seq and microarray networks. To assess biological coherence, we used the Gene Network Coherence (GNC) metric [12], which compares the inferred network structure to known proteinâĂŞprotein interactions from the BioGRID database. Higher GNC values indicate stronger alignment with established biological knowledge. Only networks with at least 100 nodes and 100 edges were included to ensure sufficient size for interpretation.

Optimal tumour and control networks were selected based on three combined criteria: GNC score, number of nodes, and number of edges. We also calculated the percentage of size reduction relative to the most complete network. A network was considered optimal if it achieved the highest GNC without losing more than 20% of nodes or edges compared to the full configuration for that DEG subset.

To identify tumour-specific interactions, the top-ranked tumour and control networks were compared. Edges unique to the tumour network were extracted for further biological interpretation, as they may correspond to downregulated pathways or mechanisms exclusive to tumour biology. These interactions are analysed functionally in the following section.

2.6   Biclustering Construction

To identify conditions-specific gene modules with coherent expression patterns, we applied the UniBic algorithm to DEG datasets derived from both RNA-Seq and microarray data [13]. UniBic is a trend-preserving biclustering algorithm that detects subsets of genes and samples forming local patterns of ordered variation in gene expression, making it especially suitable for biological datasets characterised by heterogeneity and noise.

The algorithm works by computing pairwise Longest Common Subsequence (LCS) values across genes, identifying gene pairs that share a consistent trend of relative expression across a subset of samples. Using these LCS scores UniBic incrementally builds biclusters, ensuring that each resulting module includes genes with similar expression trajectories across a defined group of samples. This property enables the detection of biologically meaningful biclusters that reflect potential regulatory modules or condition-dependent transcriptional programs.

UniBic was executed separately on tumour and control samples for each dataset to preserve condition-specific structure. The input matrices for each execution included only the differentially expressed genes previously identified for the respective condition. This separation ensures that the resulting biclusters reflect transcriptional coordination within each biological state and allows for subsequent comparative analyses between normal and tumoural contexts.

All resulting biclusters were retained for further analysis, including functional evaluation and structural characterisation, described in the next section.

2.7   Evaluation of Biclusters

The evaluation process for the biclusters derived from the control samples commences with the selection of those that encompass elite genes. Elite genes refer to those previously identified in GeneCards as possible biomarkers specifically associated with sarcoma. A functional enrichment study is subsequently conducted on each of these biclusters using the Gene Ontology (GO) [14] and KEGG [15] databases. The adjusted p-values obtained from this enrichment are employed to determine the biological significance of the biclusters. From among all the elite biclusters, the one with the highest biological relevance, determined by the most significant p-value, is selected.

In the case of biclusters derived from tumour samples, a filtering process is applied to reduce their dimensionality. To be considered for biomarker identification, these biclusters must include at least one elite gene present in the most relevant control bicluster.

This evaluative method guarantees that the chosen tumour biclusters incorporate genes previously confirmed as possible sarcoma biomarkers, hence confirming their biological and clinical significance.

2.8   Identification of Potential Biomarkers

The strategy for identifying potential biomarkers varies depending on whether GCNs or biclustering are used. In the GCN-based approach, biomarkers are identified through the combined application of functional enrichment and network topology analysis on tumour-specific networks. This involves detecting densely interconnected gene modules using the ClusterMaker2 plugin in Cytoscape [16], retaining only clusters with at least ten genes. These modules are then analysed using GO and KEGG to identify enriched biological processes and pathways that reflect the functional architecture of ULMS.

Simultaneously, hub genes are determined based on degree centrality within the network. Genes whose degree exceeds a statistical threshold—fined using the interquartile range (IQR) of the degree distribution—e classified as hubs, indicating their potential central role in transcriptional regulation and core tumour processes.

For biclustering methods, the analysis focuses on elite biclusters from both control and tumour conditions. For each bicluster, a Pearson correlation matrix is computed among the genes across its samples. This matrix is binarised: correlations between -0.5 and 0.5 are set to 0, and the rest to 1. The resulting binary matrix is used to calculate the degree of each gene, identifying tumour genes with matching degrees to elite genes. The underlying hypothesis is that genes with equivalent connectivity may behave differently in tumours due to altered expression patterns.

Scientific evidence supports that genes showing consistent upregulation or downregulation across all tumour samples compared to controls have high potential as biomarkers [17]. Thus, the expression variability of suspicious genes is examined, and those with systematic changes are retained for further analysis.

In the final phase, candidate biomarkers from both methods GCN-derived hub genes in enriched modules and biclustering-derived genes with altered expression and conserved connectivity are evaluated. Their roles as transcription factors and known or predicted associations with cancer, particularly sarcomas, are assessed using databases such as UniProt and PubMed. Particular emphasis is placed on genes involved in proliferation, metastasis, genomic instability, and immune evasion, as these are hallmark features of ULMS and critical for identifying clinically relevant targets.

3   Results and Discussion

This study integrates multiple analytical strategies to uncover key molecular mechanisms and regulatory markers associated with ULMS. The workflow includes differential expression analysis, construction and evaluation of gene co-expression networks, and biclustering-based module detection.

3.1   Construction, Selection of Optimal and Analysis of GCN

The first stage of the results focuses on constructing gene co-expression networks to explore coordinated gene activity in ULMS. Using the PyEnGNet algorithm, we generated 36 models applying three fold change thresholds ($logFC \in 1, 2, 3$) to RNA-Seq and microarray data sets and testing six combinations of correlation and pruning parameters (0.7 to 0.95). These networks reveal how genes relate to each other, helping to identify functionally related groups potentially involved in tumour development. Each network was evaluated based on its Gene Network Coherence (GNC) score and the number of nodes and edges. The best-performing networks in both datasets used a correlation threshold of 0.7 and contained at least 100 genes and interactions, ensuring sufficient complexity for biological interpretation. For RNA-Seq, the optimal network was built from genes with logFC $\geq$ 2, offering a strong balance between size and biological signal, while in the microarray data, only the logFC $\geq$ 1 network met quality criteria—gher thresholds produced sparse, fragmented networks.

The second phase applies clustering to these optimal networks to identify gene modules with coordinated transcriptional activity, providing insght into the molecular programs underlying ULMS. In RNA-Seq data (Fig. 2), two biologically relevant clusters were identified. The first was enriched in immune regulation, cell differentiation, and developmental pathways—y processes in ULMS where immune evasion and mesenchymal plasticity contribute to tumour heterogeneity, progression, and therapy resistance. These features allow ULMS to modulate its microenvironment and shift differentiation, traits associated with its aggressive subtypes [18]. The second cluster was linked to cell cycle regulation, mitosis, and chromosomal stability, whose dysregulation drives the uncontrolled proliferation and genomic instability typical of leiomyosarcomas, associated with poor prognosis and resistance [19].

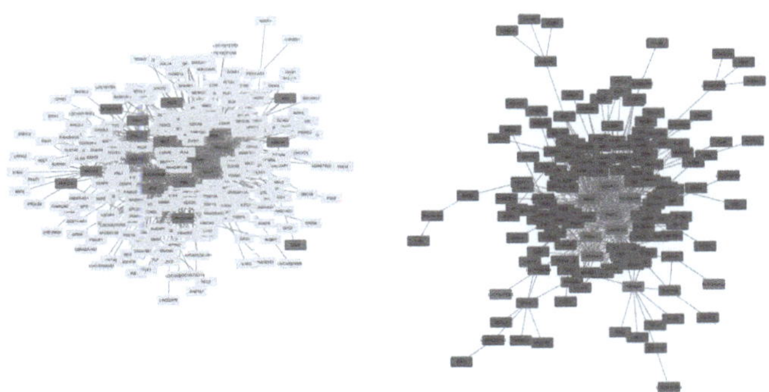

Fig. 2. Functional clustering of tumour-specific gene co-expression networks derived from the GSE222045 RNA-Seq dataset.

In the microarray dataset (Fig. 3), three biologically enriched clusters were found, all related to cell proliferation and DNA replication. One cluster involved mitotic processes and spindle apparatus organisation, reflecting the high proliferative activity of ULMS cells [19]. Another was enriched in cell cycle progression and DNA synthesis pathways essential to maintain the rapid and unregulated division of malignant smooth muscle tumors [20]. The third cluster highlighted a transcriptional program consistent with replication stress and aggressive tumour growth [21].

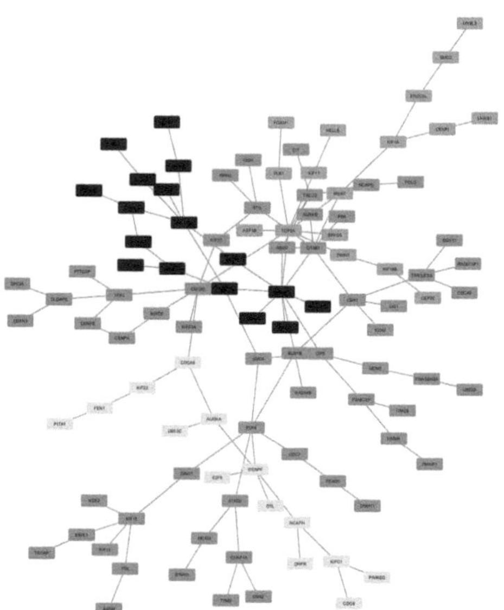

Fig. 3. Network visualization of functional gene clusters identified in the tumour-specific co-expression network derived from the GSE64763 microarray dataset.

Hub gene analysis revealed the structural and functional cores of the networks. In the RNA-Seq network, 36 hub genes were identified, mainly involved in cell differentiation, tissue development, immune responses, and cell cycle-related processes such as mitosis—phasising their regulatory roles in both proliferative and immunological dynamics of the tumour. In contrast, the microarray network revealed 15 hub genes associated with mitotic progression (e.g., chromosomal segregation, spindle formation, and checkpoints) and others linked to DNA repair, proliferation, and the G1/S transition, reflecting the machinery supporting tumour growth and survival. Together, the analysis of these modules and hub genes highlights the dominant functional themes in ULMS and offers a foundation for investigating their upstream regulators.

3.2 Biclustering Analysis

The bicluster-based strategy produced a structured set of candidate biomarkers for ULMS, with quantitative differences between the two datasets reflecting biological and technical variability.

In the GSE64763 microarray dataset, a total of 11 biclusters passed quality control filters and were retained for analysis. The best-enriched bicluster in the normal condition contained 194 genes, while its tumour counterpart encompassed 484 genes. Based on degree similarity to elite genes in the normal condition, 19 genes were identified as structurally comparable, leading to a final candidate set of 91 genes. Of these, 19 genes exhibited clearly distinguishable expression differences between tumour and normal samples. Importantly, two of the selected genes encoded transcription factors, reinforcing their potential role as regulators of tumour-specific transcriptional programs.

In the GSE222045 RNA-Seq dataset, the analysis produced 13 high-quality biclusters. The optimal normal bicluster included 274 genes and the tumour bicluster with the highest overlap contained 816 genes. The degree-based comparison identified only one gene with similar connectivity to the elite genes from the normal bicluster, indicating lower topological correspondence between conditions. Nevertheless, the strategy yielded 104 candidate genes, among which 25 displayed significant expression differences between tumour and normal samples. Although no transcription factors were identified in this dataset, five genes were recognised for their functional relevance in processes typically associated with tumour progression, such as proliferation, mitotic control, and DNA repair.

These results confirm the capacity of biclustering approaches to extract biologically meaningful modules and prioritise condition-specific gene sets. In the following section, we explore in greater detail the functional roles of the most relevant genes identified, with particular emphasis on transcription factors and tumour-related mechanisms.

3.3 Extraction of Biomarker Candidates from GCN and Biclustering Modules

In the RNA-Seq network, two transcriptionally active clusters were identified. The first included transcription factors (TFs) involved in immune activation, differentiation, and T-cell development—ological functions highly relevant in the context of ULMS, where immune evasion and mesenchymal plasticity contribute to tumour progression and heterogeneity [22]. These regulators suggest that immune modulation may be an important component of the ULMS transcriptional landscape.

The second cluster in the RNA-Seq network was enriched in transcription factors related to proliferative signalling, epithelial-mesenchymal transition (EMT), and nuclear receptor pathways. These processes are frequently deregulated in aggressive mesenchymal tumours and may drive the plastic and invasive behaviour typical of high-grade ULMS [23]. Additionally, the identification of the lncRNA ZFHX4-AS1 within this cluster highlights the potential involvement of non-coding RNA in the regulation of tumour proliferation and the tumour microenvironment [24].

In the microarray dataset, four transcription factors were identified across functionally enriched clusters and among hub genes: FOXM1, MYBL2, E2F1, and PITX1. These TFs represent a core regulatory axis linked to cell proliferation, genomic maintenance, and tumour progression. FOXM1 is a master regulator of the G2/M cell cycle transition and controls the expression of genes involved in mitosis, chromosomal segregation, and genomic stability [25]. Its overexpression has been consistently associated with poor prognosis and therapeutic resistance in uterine sarcomas, positioning it as a strong candidate biomarker and therapeutic target [26].

MYBL2, also known as B-Myb, plays a central role in cell cycle regulation, DNA repair, and replication. It has been associated with accelerated tumour growth and treatment resistance in several solid tumours and may contribute similarly to the aggressive behaviour observed in ULMS [27]. E2F1 governs the G1/S transition and activates genes required for DNA synthesis and cell survival. Its deregulation in sarcomas promotes uncontrolled proliferation and evasion of apoptotic control mechanisms [28]. Finally, PITX1, a homeobox transcription factor traditionally involved in development, has emerged as a context-dependent regulator in cancer, acting either as a tumour suppressor or oncogene [29]. In ULMS, its presence within a cluster enriched for DNA replication and cell cycle control suggests a role in sustaining proliferative transcriptional programs or modulating differentiation states during tumour progression.

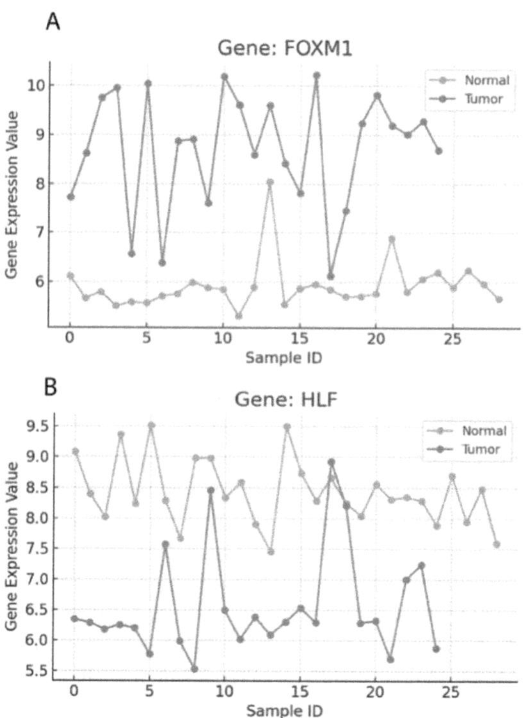

Fig. 4. Differential expression profiles of FOXM1 (A) and HLF (B) in tumour and normal samples from the GSE64763 microarray dataset.

The biclustering-based analysis of both datasets uncovered biologically relevant gene modules in ULMS, revealing distinct yet complementary biomarker candidates. In the GSE64763 microarray dataset, tumour-specific biclusters included two differentially expressed transcription factors—XM1 and HLF

(Fig. 4)—ggesting their regulatory roles in ULMS-associated transcriptional programs.

FOXM1 is a well-established regulator of the cell cycle, particularly the G1/S and G2/M transitions, mitosis, and DNA repair [30]. It is frequently overexpressed in various cancers, where it promotes proliferation, therapy resistance, and poor prognosis [31], and has been linked to tumour aggressiveness in sarcomas [32]. Although underexplored in ULMS, its presence in tumour-enriched biclusters supports its role as a potential biomarker and therapeutic target.

HLF, a bZIP transcription factor mainly associated with haematologic malignancies like acute lymphoblastic leukaemia [33], was also detected in ULMS biclusters. While its role in solid tumours is less understood, its involvement in tumour-specific modules suggests functions related to cell viability and stress response, warranting further investigation in sarcoma biology.

In contrast, the biclustering of the GSE222045 RNA-Seq dataset revealed functionally relevant genes not encoding transcription factors, underscoring biclustering's capacity to detect both regulatory and non-regulatory elements involved in tumourigenesis. Key candidates—NB1, TPX2, POLQ, BUB1, and TRIP13—e involved in mitotic control, DNA repair, and chromosomal stability [34], processes frequently disrupted in high-grade cancers. These genes are associated with aggressive tumour behaviour, poor outcomes, and therapy resistance, and their presence in tumour-enriched modules suggests strong functional roles in ULMS.

Together, the results of GCN and biclustering allowed us to identify biomarker candidates in ULMS. Network analyses highlighted central transcription factors with well-defined roles in cell cycle regulation, differentiation, and immune evasion, while the biclustering-based approach revealed tumour-specific gene modules, including both transcriptional regulators and functional genes involved in mitosis, DNA repair, and chromosomal stability. This combination of strategies allowed us to capture different levels of the molecular architecture of ULMS, providing a solid basis for the prioritisation of biomarkers and therapeutic targets.

4   Conclusions

This study applied two independent strategies—N and biclustering analysis—identify potential biomarkers in ULMS. The comparative analysis revealed both overlapping and distinct gene candidates, demonstrating the complementary nature of these approaches.

Through GCN analysis, we identified several hub genes and transcription factors such as FOXM1, E2F1, MYBL2, and PITX1, which are centrally positioned in tumour-specific networks and associated with key processes like mitotic regulation, DNA replication, and transcriptional control. These genes are strongly linked to tumour aggressiveness and represent promising candidates for diagnostic or therapeutic targeting.

The biclustering-based approach uncovered additional candidates, some overlapping in function but distinct in identity. In particular, biclustering also identified FOXM1 - suggesting its biological robustness - as well as HLF as transcription factors with possible regulatory influence in ULMS. Furthermore, this strategy highlighted CCNB1, TPX2, POLQ, BUB1, and TRIP13, which, despite not being transcriptional regulators, are functionally critical in pathways such as DNA repair, cell cycle control, and genomic stability. These genes underscore mechanisms that may not be fully captured by global network topology but emerge under condition-specific expression patterns.

The findings demonstrate that each strategy provides a unique and complementary perspective on biomarker discovery: GCN analysis emphasises global regulatory structure and centrality, while biclustering isolates local, condition-dependent modules. The convergence on FOXM1 from both approaches reinforces its potential as a key biomarker in ULMS. Meanwhile, the diversity of additional candidates strengthens the case for applying both global and local analyses. This comparative framework offers a more comprehensive understanding of the transcriptional landscape and increases the likelihood of identifying biologically meaningful biomarkers. Ultimately, this dual strategy enhances prospects for disease stratification, therapeutic targeting, and improving clinical outcomes in rare and aggressive tumours such as leiomyosarcoma.

References

1. Yang, Q., et al.: Comprehensive review of uterine leiomyosarcoma: pathogenesis, diagnosis, prognosis, and targeted therapy. Cells **13**(13), 1106 (2024)
2. Villaseñor-Altamirano, A.B., Balderas-Martínez, Y.I., Medina-Rivera, A.: Review of gene expression using microarray and RNA-seq. In: Rigor and Reproducibility in Genetics and Genomics, pp. 159–187. Elsevier (2024)
3. Zainal-Abidin, R.-A., Harun, S., Vengatharajuloo, V., Tamizi, A.-A., Samsulrizal, N.H.: Gene co-expression network tools and databases for crop improvement. Plants **11**(13), 1625 (2022)
4. López-Fernández, A., Gómez-Vela, F.A., del Saz-Navarro, M., Delgado-Chaves, F.M., Rodríguez-Baena, D.S.: Optimized python library for reconstruction of ensemble-based gene co-expression networks using multi-GPU. J. Supercomput. **80**(12), 18142–18176 (2024)
5. López-Fernández, A., Gómez-Vela, F.A., Rodríguez-Baena, D.S., Delgado-Chaves, F.M., Gonzalez-Dominguez, J.: Biclustering in bioinformatics using big data and high performance computing applications: challenges and perspectives, a review. J. Supercomput. **81**(10), 1123 (2025)
6. Rakic, A., Anicic, R., Rakic, M., Nejkovic, L.: Integrated bioinformatics investigation of novel biomarkers of uterine leiomyosarcoma diagnosis and outcome. J. Personalized Med. **13**(6), 985 (2023)
7. Kim, K., et al.: A differential diagnosis between uterine leiomyoma and leiomyosarcoma using transcriptome analysis. BMC Cancer **23**(1), 1215 (2023)
8. Barlin, J.N., et al.: Molecular subtypes of uterine leiomyosarcoma and correlation with clinical outcome. Neoplasia **17**(2), 183–189 (2015)

9. Varet, H., Brillet-Guéguen, L., Coppée, J.-Y., Dillies, M.-A.: Sartools: a deseq2- and edger-based r pipeline for comprehensive differential analysis of RNA-seq data. PLoS ONE **11**(6), e0157022 (2016)

10. Kauffmann, A., Gentleman, R., Huber, W.: arrayqualitymetrics–bioconductor package for quality assessment of microarray data. Bioinformatics **25**(3), 415–416 (2009)

11. Cui, W., Xue, H., Wei, L., Jin, J., Tian, X., Wang, Q.: High heterogeneity undermines generalization of differential expression results in RNA-seq analysis. Hum. Genomics **15**, 1–9 (2021)

12. Delgado, F.M., Gómez-Vela, F.: Computational methods for gene regulatory networks reconstruction and analysis: a review. Artif. Intell. Med. **95**, 133–145 (2019)

13. Wang, Z., Li, G., Robinson, R.W., Huang, X.: Unibic: sequential row-based biclustering algorithm for analysis of gene expression data. Sci. Rep. **6**(1), 23466 (2016)

14. Ashburner, M., et al.: Gene ontology: tool for the unification of biology. Nat. Genet. **25**(1), 25–29 (2000)

15. Kanehisa, M.: The KEGG database. In: 'In Silico' Simulation of Biological Processes: Novartis Foundation Symposium, vol. 247, pp. 91–103. Wiley Online Library (2002)

16. Utriainen, M., Morris, J.H.: clustermaker2: a major update to clustermaker, a multi-algorithm clustering app for cytoscape. BMC Bioinform. **24**(1), 134 (2023)

17. Mandal, K., Sarmah, R., Bhattacharyya, D.K.: Popbic: pathway-based order preserving biclustering algorithm towards the analysis of gene expression data. IEEE/ACM Trans. Comput. Biol. Bioinform. **18**(6), 2659–2670 (2020)

18. De Wispelaere, W., et al.: Pi3k/mtor inhibition induces tumour microenvironment remodelling and sensitises ps6high uterine leiomyosarcoma to pd-1 blockade. Clin. Transl. Med. **14**(5), e1655 (2024)

19. Koltsova, A.S., Efimova, O.A., Pendina, A.A.: A view on uterine leiomyoma genesis through the prism of genetic, epigenetic and cellular heterogeneity. Int. J. Mol. Sci. **24**(6), 5752 (2023)

20. Yoshida, K., et al.: Aberrant activation of cell-cycle-related kinases and the potential therapeutic impact of plk1 or chek1 inhibition in uterine leiomyosarcoma. Clin. Cancer Res. **28**(10), 2147–2159 (2022)

21. Ke, Y., You, L.X., Xu, Y.J., Wu, D., Lin, Q., Wu, Z.: Dpp6 and mfap5 are associated with immune infiltration as diagnostic biomarkers in distinguishing uterine leiomyosarcoma from leiomyoma. Front. Oncol. **12**, 1084192 (2022)

22. Manzoni, M., et al.: The adaptive and innate immune cell landscape of uterine leiomyosarcomas. Sci. Rep. **10**(1), 702 (2020)

23. Franceschi, T., et al.: Role of epithelial-mesenchymal transition factors in the histogenesis of uterine carcinomas. Virchows Archiv **475**, 85–94 (2019)

24. Cadenas, M.R., Lopez-Fernandez, A., Gomez-Vela, F.A., Ortega, J.A., Perez, I.R.: High-performance computing-driven gene co-expression network analysis for biomarkers discovery in soft tissue sarcomas. In: 2025 IEEE 38th International Symposium on Computer-Based Medical Systems (CBMS), pp. 234–239. IEEE Computer Society (2025)

25. Zheng, Q., Luo, Z., Xu, M., Ye, S., Lei, Y., Xi, Y.: Hmga1 and foxm1 cooperate to promote g2/m cell cycle progression in cancer cells. Life **13**(5), 1225 (2023)

26. Khan, Md.A., Khan, P., Ahmad, A., Fatima, M., Nasser, M.W.: Foxm1: a small fox that makes more tracks for cancer progression and metastasis. Seminars Cancer Biol. **92**, 1–15 (2023)

27. Chen, X., et al.: Pan-cancer analysis indicates that mybl2 is associated with the prognosis and immunotherapy of multiple cancers as an oncogene. Cell Cycle **20**(21), 2291–2308 (2021)
28. Gao, Y., Qiao, X., Liu, Z., Zhang, W.: The role of e2f2 in cancer progression and its value as a therapeutic target. Front. Immunol. **15**, 1397303 (2024)
29. Li, S., et al.: Context-dependent t-box transcription factor family: from biology to targeted therapy. Cell Commun. Signal. **22**(1), 350 (2024)
30. Liu, C., Barger, C.J., Karpf, A.R.: Foxm1: a multifunctional oncoprotein and emerging therapeutic target in ovarian cancer. Cancers **13**(12), 3065 (2021)
31. Kim, H., et al.: Forkhead box m1 (foxm1) transcription factor is a key oncogenic driver of aggressive human meningioma progression. Neuropathol. Appl. Neurobiol. **46**(2), 125–141 (2020)
32. Shibui, Y., et al.: Expression of forkhead box m1 and anticancer effects of foxm1 inhibition in epithelioid sarcoma. Lab. Invest. **104**(8), 102093 (2024)
33. Ruo-Lan, G., Xiang, M., Suo, J., Yuan, J.: Acute lymphoblastic leukemia in an adolescent presenting with acute hepatic failure: a case report. Mol. Clin. Oncol. **11**(2), 135–138 (2019)
34. Nemtsova, M.V., Kuznetsova, E.B., Bure, I.V.: Chromosomal instability in gastric cancer: role in tumor development, progression, and therapy. Int. J. Mol. Sci. **24**(23), 16961 (2023)

Metabopathia: Enhancing Disease Mechanism Understanding Through Mechanistic Integration of Transcriptomic and Metabolic Data

Kinza Rian[1,4]($\boxtimes$) , Isabel A. Nepomuceno-Chamorro[4] ,
Joaquin Dopazo[1,2,3] , and Daniel López-López[1,2,3]

[1] Andalusian Platform for Computational Medicine, Andalusian Public Foundation Progress and Health-FPS, Seville, Spain
{kinza.rian,joaquin.dopazo,daniel.lopez.lopez}@juntadeandalucia.es
[2] Institute of Biomedicine of Seville (IBiS), University Hospital Virgen del Rocío/CSIC/University of Seville, Seville, Spain
[3] Centro de Investigación Biomédica en Red en Enfermedades Raras (CIBERER), ISCIII, Madrid, Spain
[4] Department of Lenguajes y Sistemas Informáticos, University of Seville, Seville, Spain
kinria@alum.us.es, inepomuceno@us.es
https://www.clinbioinfosspa.es ,
https://www.us.es/centros/departamentos/lenguajes-y-sistemas-informaticos

Abstract. Multifactorial diseases arise from complex interactions among various biological systems. These interactions represent intricate molecular mechanisms leading to pathological conditions. Studying these mechanisms requires dedicated analytical methods capable of integrating diverse omics data to model disease progression. Traditional analytical methods focus on the integration of individual genetic, transcriptomic, or metabolic datasets which are insufficient for capturing this complexity. Integration of one layer of omics data may lead to significant gaps in our understanding of complex diseases.

To address this limitation, we present *Metabopathia*, an extension of HiPathia, as a computational tool that integrates multi-omics data—including transcriptomics and metabolomics—into mechanistic models of signaling and metabolic pathways. The method leverages curated biological knowledge from the Kyoto Encyclopedia of Genes and Genomes (KEGG), which provides the pathway structures used to model these processes. By doing so, Metabopathia enables accurate measurement of changes in the activity of cellular signaling cascades. This not only supports high-throughput estimation of functional cellular profiles, but also allows the simulation of how genetic mutations that reduce or eliminate gene function, along with metabolic disruptions, affect cellular processes.

The application of Metabopathia to complex diseases such as cancer has demonstrated its efficacy in identifying critical sub-pathway alterations that drive disease progression. By incorporating metabolomic data, Metabopathia improves the mechanistic understanding of multi-

A. López Fernández et al. (Eds.): CIABiomed 2025, LNBI 16148, pp. 505–526, 2026.
https://doi.org/10.1007/978-3-032-10661-2_38

factorial diseases and informs potential therapeutic strategies tailored to disease-specific molecular contexts.

Keywords: Mechanistic modeling · Signaling pathways · Metabolic pathways · Complex diseases · Data integration

1   Introduction

Complex diseases like cancer are inherently multifactorial, stemming from an intricate interplay of genetic, epigenetic, and environmental influences. These factors give rise to complex molecular dynamics, making it challenging to fully understand disease initiation and progression. The complexity is further amplified by dynamic crosstalk between cellular signaling and metabolic pathways, which complicates the precise identification of disease drivers. To address this, advanced computational modeling approaches are essential for investigating these critical mechanistic pathways [14, 37].

Mechanistic modeling offers a robust way to analyze the complex interactions within cellular signaling networks [13]. Unlike correlation-based statistical methods, mechanistic models capture causal relationships by describing how genes, proteins, and metabolites interact to produce specific outcomes. This allows for predictions of system behavior under various conditions. Numerous methods have been developed to enhance mechanistic modeling [1]. For instance, HiPathia transforms gene expression data into pathway activity scores, identifying mechanistic functions and their association with disease outcomes and phenotypes [13]. It analyzes the signaling system by breaking it down into sub-pathways or circuits and predicts changes in signaling activities [35]. Other approaches, such as MinePath [23] and DEAP (Differential Expression Analysis of Pathways) [12], examine differential gene expression within pathways to measure changes in cellular activities, providing an interpretable framework for integrating gene expression patterns into traditional pathway analysis. While these methods utilize topological information (e.g., activations and inhibitions) to calculate pathway scores, and DEAP specifically uses receptor-to-effector sub-pathway definitions, both MinePath and DEAP differ from HiPathia in their use of discretized values for differential gene expressions, which reduces their sensitivity compared to HiPathia [1].

Further methods aim to identify active sub-pathways within larger networks. Clipper [27] integrates gene expression data with pathway topology to guide mechanistic explanations for therapeutic strategies, while subSPIA (Sub-pathway Significance Profile Analysis) [25] combines statistical and topological approaches to identify biologically relevant sub-pathways. DEGraph [19] employs graph-based methods to study the impact of differential gene expression on pathway activities. Similarly, topologyGSA [28] and TAPPA (Topology-based Pathway Analysis) [8] integrate topological information to understand interactions between genes and cellular functions. Finally, PRS (Pathway Regulation Score) [17] and PWEA (Pathway Enrichment Analysis) [16] use gene expression data to offer comprehensive views of how enriched pathways are regulated and function under different biological conditions [1,9].

The benefits of mechanistic modeling are substantial, particularly in precision medicine, where they can simulate biological system responses to perturbations like drug treatments or mutations. These models reveal causal relationships between molecular interactions and cellular functions, which is crucial for delineating pathways linked to disease phenotypes [35] and facilitating tailored therapeutic approaches [26,31 40]. Moreover, mechanistic models can integrate different omics layers (e.g., genomics, transcriptomics, metabolomics), improving our understanding of cellular function as an interconnected system.

A key advantage of mechanistic modeling, as implemented in HiPathia, is its ability to link transcriptomic data with signaling pathways [13,32]. This approach has been successfully applied in various contexts. For example, in Fanconi anemia, HiPathia predicted pathway activity from gene expression data and correlated it with disease progression [7]. It was also used to explore metabolite–signaling crosstalk in cancer, identifying molecular targets relevant to tumor progression [41]. Beyond oncology, HiPathia's versatility has been demonstrated in other rare genetic disorders, infectious diseases like COVID-19, and environmental studies (Supplementary Material SF1).

Despite its strengths, HiPathia has limitations. In its current implementation, nodes representing metabolites or compounds are treated as static connectors within the pathway and are not involved in the dynamic propagation of signaling activity. This simplification overlooks the regulatory roles of metabolite abundance and reactivity, which can significantly influence cellular processes. Additionally, HiPathia relies solely on transcriptomic data as a proxy for protein expression while neglecting metabolomic contributions and their interactions. This provides a limited view of the complex molecular mechanisms underlying traits and diseases [10].

To gain a more comprehensive understanding of complex disease etiologies, multi-omics data integration is critical [22,38]. Each omics layer complements and enhances the resolution of the others: genomics identifies genetic variations, transcriptomics provides gene expression patterns, and metabolomics captures a snapshot of the metabolic state of cells and tissues [6]. However, the intricate networks of molecular interactions necessitate precise computational interpretation to elucidate disease progression [18]. Without integrating these diverse layers into signaling networks, key mechanistic links may remain hidden. Understanding how signal disruptions propagate through molecular networks is essential for explaining why certain perturbations drive pathological processes.

To address these challenges, we developed *Metabopathia*, an updated extension of HiPathia designed to integrate transcriptomic and metabolic layers for improved pathway identification and analysis. Metabopathia leverages transcriptomic data alongside inferred metabolic activities via Metabolizer [42]. We assessed its performance by evaluating specificity through false positive rates in various simulations and sensitivity by measuring true positive rates in a real disease scenario.

An unbiased comparison on a breast cancer dataset showed that Metabopathia identifies more cancer-specific circuits than HiPathia, revealing key alterations overlooked by previous methods. These findings highlight the added value of integrating metabolomic data for more precise detection of disease-relevant sub-pathway dysregulation. In this article, we describe the methodology behind Metabopathia, present its application to breast cancer dataset, and discuss its improved sensitivity, current limitations, and directions for future development.

2   Materials and Methods

2.1   Study Design and Overview

We applied our approach with a case study on breast cancer by utilizing RNA-seq data from The Cancer Genome Atlas (TCGA-BRCA) repository [4,36]. The dataset consists of 1,231 samples: 1,111 primary tumor samples from 1,095 participants, 113 solid tissue normal samples, and 7 metastatic samples. The RNA-seq count matrix for the gene expression of 60,660 transcripts was downloaded from the TCGA portal (https://portal.gdc.cancer.gov/). This case study compares breast cancer tissue samples with solid tissue normal samples to identify significant changes in signaling and metabolic pathway activities.

2.2   RNA-Seq Data Preprocessing

Quality control steps were applied to RNA-seq data to ensure suitability for downstream analyses. Principal component analysis (PCA) [11] and hierarchical clustering (HC) [30] were performed to identify outlier samples and to verify any possible batch effects. Sample clusters, as a function of gene expression profiles, were visualized by PCA, while HC was used to validate that similar samples were appropriately grouped in a biologically relevant manner.

Lowly expressed transcripts were filtered out using the `filterByExpr` function of the EdgeR package (v3.34.0) [33]. Transcripts with low expression levels were excluded to reduce noise, retaining a final set of 32,452 transcripts for further analysis. Key genes involved in important pathways, even if lowly expressed, were retained for mechanistic modeling.

Normalization of gene expression counts was performed using the Trimmed Mean of M-values (TMM) method to correct for differences in sequencing depth across samples [34]. After normalization, log transformation was used to stabilize the variance in the dataset. Additionally, extreme values were truncated at the 99th percentile to prevent outliers from skewing the results [5]. Batch effects were adjusted using the ComBat method [20], implemented in the `sva` package (v3.40.0) [24], ensuring that non-biological variability (e.g., different sample processing) was minimized while preserving the biological signal.

2.3   Metabolomics Data Integration

No direct metabolomics data were available; metabolite activity was inferred using the Metabolizer method [42], which estimates metabolite activities from gene expression levels of enzymes involved in metabolic reactions. A total of 16 key metabolites were inferred and integrated into the pathway analysis to assess joint transcriptomic and metabolomic activity.

2.4   Pathway Activity Analysis

We selected 146 signaling pathways and 48 metabolic pathways from the Kyoto Encyclopedia of Genes and Genomes (KEGG) database [21]. These pathways were parsed into 1,876 signaling circuits and 96 metabolic modules. Further, we decomposed these pathways into receptor-to-effector circuits, enabling more granular modeling of pathway activities.

Activity of molecular components, represented as nodes within pathways, was estimated as follows:

- **Protein family nodes:** Scored using the 90th percentile of expression values to capture the activity of the most active member, assuming that high expression of at least one family member is sufficient to activate the node.
- **Complex nodes:** Scored using the minimum expression level among all components, reflecting that complex function is constrained by the limiting component.
- **Metabolite nodes:** Activity inferred computationally using Metabolizer [42].

In addition to node activity scores, signal modulation based on activation or inhibition interactions was used to compute pathway activity scores. Signal propagation was modeled iteratively, from receptor nodes to effector nodes according to a recursive rule:

$$
S_n = \begin{cases} v_g\left(1 - \prod_{\text{Act}}(1 - s_a)\prod_{\text{In}}(1 - s_i)\right), & \text{if } n \in G, \\[2em] v_m\left(1 - \prod_{\text{Act}}(1 - s_a)\prod_{\text{In}}(1 - s_i)\right), & \text{if } n \in M. \end{cases} \tag{1}
$$

where S_n denotes the signal intensity of node n. If $n \in G$ (the set of genes, including protein family and complex nodes), v_g is the node activity score, computed according to the rules described above; if $n \in M$ (the set of metabolites), v_m is the inferred metabolite activity. Act is the set of activating edges delivering activation signals to n, where S_a is the signal intensity from a preceding node connected by one of these activating edges. Similarly, In is the set of inhibitory edges delivering inhibition signals to n, where S_i is the signal from a preceding node connected by one of these inhibitory edges.

The equation is applied *recursively*: starting from receptor nodes, the signal value for each node is computed from the signals of its immediate upstream nodes, and this process is repeated until the final node is reached. For each sub-pathway, its activity score is defined as the signal intensity of the final node, which captures the combined influence of all upstream activations and inhibitions along the path.

2.5   Differential Pathway Activity Analysis

Differential Activity Analysis (DAA) was applied to identify sub-pathways whose activities differed significantly between breast cancer and normal tissue samples. Statistical significance was assessed using the Wilcoxon rank-sum test [2,15], with p-values adjusted for multiple testing via the Benjamini–Hochberg False Discovery Rate (FDR) method [3]. Pathways with an adjusted p-value below 0.05 were considered significantly altered.

2.6   Metrics for Assessing Metabopathia Performance

Specificity: False Positive Rate. Specificity was evaluated by measuring the false positive rate (FPR) in simulated groups of individuals ($N = 20, 50, 70, 100$) under three scenarios. These simulations aimed to generate groups in which all individuals were highly similar in their expression profiles, ensuring that no true biological differences existed. Since both groups in each comparison consisted of the same type of individuals (e.g., all breast cancer samples or all normal tissue samples), any detected differences in sub-pathway activity were interpreted as false positives. This allowed us to quantify how often Metabopathia incorrectly identifies differences when comparing biologically indistinguishable groups.

The three simulation strategies were as follows:

- **Random sampling from real BRCA patient data:** Gene expression and metabolite values were directly taken from actual BRCA patients to form groups with similar expression patterns.
- **Empirical distribution-based sampling:** For each gene, new expression values were generated based on the mean and variance observed in the BRCA dataset. Metabolite activities were scaled to match the range of real patient data, ensuring that the simulated individuals closely resembled actual samples.
- **Normal distribution-based sampling:** Gene expression values were generated using a normal distribution with a mean of 0.5 and variance of 0.05, producing individuals with tightly clustered expression levels around these parameters.

For each simulation, the group was randomly split into two datasets, and Metabopathia was applied to detect differences in circuit activity between them. Because all individuals within a group were designed to be as similar as possible, any detected differences were treated as false positives. This process was repeated 100 times for each group size to obtain robust FPR estimates (see Supplementary Figure S1, top).

Sensitivity: True Positive Rate. Sensitivity was assessed by evaluating the true positive rate (TPR) in a scenario where true biological differences are expected. We used the full dataset containing 109 normal and 109 tumor samples from BRCA patients. For each test, random samples of paired individuals ($N = 20, 50, 70, 100, 200$) were generated while maintaining the original 50% normal/50% tumor proportion.

Each paired group was then divided into two datasets, and Metabopathia was applied. Within-sample dependencies were accounted for by using the Wilcoxon signed-rank test to detect differentially activated circuits between normal and tumor samples. The TPR was computed as the proportion of cancer-related pathways that contained at least one significantly activated circuit, divided by the total number of cancer-related pathways.

This process was repeated 100 times for each group size to obtain a robust TPR estimate (see Supplementary Figure S1, bottom). In addition, a *significance ratio* was calculated for each cancer-related pathway, defined as the number of significant circuits divided by the total number of circuits in that pathway. The set of cancer-related pathways was defined according to the KEGG classification and a curated list from the Valencia Institute of Oncology (IVO) (Table 1) [13].

2.7 Comparison to HiPathia

Performance was compared to HiPathia [32], which models signaling pathways without including metabolite contributions. Both methods were applied to the same BRCA dataset, using identical cancer-related pathway sets and evaluation metrics (true positive rate, false positive rate, and FDR p-values). For this evaluation, we considered two cancer-related pathway sets: the KEGG classification and a curated list from the Valencia Institute of Oncology (IVO) (see Table 1).

Both methods were assessed for their ability to detect differentially activated circuits between normal and tumor samples. The comparison focused on two key performance metrics:

1. **Sensitivity**—measured as the ability to detect true differential activation of circuits.
2. **Specificity**—evaluated by the false positive rate and FDR p-values.

For statistical robustness, each method was applied to randomly sampled groups of varying sizes ($N = 20, 50, 70, 100, 200$), and each configuration was repeated 100 times. This approach ensured that the performance results reflect consistent trends rather than random fluctuations (see Supplementary Figure S1).

3 Results and Discussion

3.1 Preprocessing and Data Overview

The breast cancer dataset analysis began with the preprocessing of RNA-seq data from TCGA (see Supplementary Material SF2), consisting of 1,231 samples

Table 1. Cancer-related pathways from Kyoto Encyclopedia of Genes and Genomes (KEGG) classification and curated by the Valencia Institute of Oncology (IVO).

Pathway ID	Pathway name	Evidence source
hsa03320	PPAR signaling pathway	KEGG and IVO
hsa04010	MAPK signaling pathway	KEGG and IVO
hsa04012	ErbB signaling pathway	IVO
hsa04014	Ras signaling pathway	IVO
hsa04015	Rap1 signaling pathway	IVO
hsa04020	Calcium signaling pathway	IVO
hsa04022	cGMP-PKG signaling pathway	IVO
hsa04024	cAMP signaling pathway	KEGG and IVO
hsa04062	Chemokine signaling pathway	IVO
hsa04064	NF-kappa B signaling pathway	IVO
hsa04066	HIF-1 signaling pathway	IVO
hsa04068	FoxO signaling pathway	IVO
hsa04110	Cell cycle	KEGG and IVO
hsa04114	Oocyte meiosis	IVO
hsa04115	p53 signaling pathway	KEGG and IVO
hsa04150	mTOR signaling pathway	KEGG and IVO
hsa04151	PI3K-Akt signaling pathway	KEGG and IVO
hsa04152	AMPK signaling pathway	IVO
hsa04210	Apoptosis	KEGG and IVO
hsa04310	Wnt signaling pathway	KEGG and IVO
hsa04330	Notch signaling pathway	IVO
hsa04340	Hedgehog signaling pathway	IVO
hsa04350	TGF-beta signaling pathway	KEGG and IVO
hsa04370	VEGF signaling pathway	KEGG and IVO
hsa04390	Hippo signaling pathway	IVO
hsa04510	Focal adhesion	KEGG and IVO
hsa04520	Adherens junction	KEGG and IVO
hsa04530	Tight junction	IVO
hsa04540	Gap junction	IVO
hsa04611	Platelet activation	IVO
hsa04620	Toll-like receptor signaling pathway	IVO
hsa04621	NOD-like receptor signaling pathway	IVO
hsa04630	Jak-STAT signaling pathway	KEGG and IVO
hsa04650	Natural killer cell mediated cytotoxicity	IVO
hsa04660	T cell receptor signaling pathway	IVO
hsa04662	B cell receptor signaling pathway	IVO
hsa04668	TNF signaling pathway	IVO
hsa04670	Leukocyte transendothelial migration	IVO
hsa04910	Insulin signaling pathway	IVO
hsa04912	GnRH signaling pathway	IVO
hsa04914	Progesterone-mediated oocyte maturation	IVO
hsa04915	Estrogen signaling pathway	IVO
hsa04916	Melanogenesis	IVO
hsa04919	Thyroid hormone signaling pathway	IVO
hsa04920	Adipocytokine signaling pathway	IVO
hsa05200	Pathways in cancer	IVO
hsa05205	Proteoglycans in cancer	IVO
hsa05231	Choline metabolism in cancer	IVO

and 60,660 transcripts. After applying quality control steps such as principal component analysis (PCA) and hierarchical clustering (HC), the dataset was reduced to 218 paired samples for downstream analysis: 109 tumor samples, 109 normal samples, and 32,452 retained transcripts (Supplementary Figures S2 and S3).

The preprocessing steps began with low count removal; a total of 28,502 transcripts were identified as lowly expressed across all samples, of which 294 were part of the pathways selected for this study (Supplementary Figure S4 and Supplementary Table S1). Consequently, 28,208 lowly expressed transcripts were filtered out of the dataset. Retaining common genes was essential to prevent false imputation and ensure the robustness of the pathway activity analysis.

Normalization was performed using the Trimmed Mean of M-values (TMM) method to account for differences in library size (see Supplementary Material SF3 for an overview of the normalization technique and potential artifacts). PCA confirmed that data quality improved considerably post-normalization (see Supplementary Material SF2). Expression values were then log-transformed and truncated at the 99th percentile to reduce the influence of extreme values without classifying them as outliers. This approach offers a milder alternative to the Interquartile Range (IQR) method, which removes outliers entirely.

Since the data were collected from different plates and Tissue Source Sites (TSS), the presence of batch effects was likely. Batch correction using the ComBat method successfully minimized these effects, as confirmed by PCA and HC (Supplementary Figures S2 and S3). As a result, the data were deemed clean and ready for downstream analysis.

As part of the exploratory analysis, the distribution of patient ages was examined (Supplementary Figure S5), revealing a predominance of female patients, with one male patient aged 58. The preprocessed dataset was then used as input to the Metabopathia platform for analysis of metabolic and signaling pathway activities.

3.2 Pathway Activity Analysis

Pathway Selection and Decomposition. A total of 146 KEGG signaling pathways (Supplementary Table S2) and 48 KEGG metabolic pathways (Supplementary Table S3) were selected for further analysis. The selection criteria for these pathways focused on their representation as activity flow maps [29], rather than pure or mixed process description maps, ensuring they could be parsed effectively for pathway activity analysis.

To model specific signaling cascades or metabolic modules, the pathways were decomposed into effector circuits and modules that end with a single metabolite product, respectively. For instance, signaling pathways such as "Pathways in cancer" (hsa05200), "Proteoglycans in cancer" (hsa05205) and "PPAR signaling pathway" (hsa03320) were among the top pathways, each with more than 40 decomposed sub-pathways (Supplementary Figure S6). This decomposition into sub-pathways allowed for more precise modeling of signal propagation through these smaller, well-defined circuits.

The curated R object containing the decomposed pathways and the annotated metabolite nodes was saved for further analysis and is accessible in the GitHub repository: https://github.com/kinzaR/metabopathia/blob/main/pathways/metabo_mgi_v1.1.0.RDS.

Nodes and Circuits Activity Estimation

Metabolite Inference. A total of 16 metabolites, identified as the intersection between metabolite products from the Metabolizer tool and metabolites present in the signaling pathways, were inferred using the `infer_met_from_metabolizer()` function based on the Metabolizer approach [42]. These inferred metabolites were then annotated across multiple databases, including HMDB and PubChem, using the MetaboAnalyst 5.0 platform [39] (Table 2).

Supplementary Figure S7 illustrates the distribution of inferred metabolic activity, showing that the newly computed scores diverge significantly from the default imputed scores generated by the HiPathia approach (in which metabolite node values are fixed to 1 by default). These inferred scores were used as proxies for metabolic activity in the signaling cascades.

However, the accuracy of these inferences depends on the reliability of the available gene expression data and the performance of the Metabolizer algorithm. Currently, only 16 metabolites are modeled, which limits the depth of metabolic integration and pathway coverage. The absence of direct metabolomics data may introduce uncertainty in metabolite activity estimation, and sample heterogeneity may further affect interpretability.

Future improvements should focus on expanding inference coverage by refining the Metabolizer tool or incorporating alternative inference strategies. Because of its modular architecture, *Metabopathia* supports such developments, including the addition of sensitivity analyses to assess the contribution of specific metabolites to signaling cascade dynamics.

Circuit Activity Analysis. Pathway simulations were performed by calculating node scores using both transcriptomic and inferred metabolic activity, with node-type–specific rules. Protein family nodes were scored by the most active member (90th percentile of expression), while complex nodes were constrained by the limiting component (minimum expression among complex members). Metabolite nodes were modeled using the Metabolizer approach to infer metabolic activity. A total of 34 metabolite nodes comprise more than two metabolites (Supplementary Table S4); these were scored by the lowest metabolite activity within the group.

Simulated signal propagation was then applied along each pathway using Eq. (1), yielding signal intensities for the effector nodes. For each circuit, the final signal value was defined as the signal at its last node, enabling meaningful comparisons between conditions. Note that these activity values are context-dependent and have interpretive meaning only when compared across conditions (e.g., tumor vs. normal).

Table 2. Metabolite annotations using MetaboAnalyst across the Kyoto Encyclopedia of Genes and Genomes (KEGG), the Human Metabolome Database (HMDB) and PubChem databases.

KEGG	Match	HMDB	PubChem	SMILES
C00002	Adenosine triphosphate	HMDB0000538	5957	NC1=NC=NC2=C1N=CN2[C@@H]1OC@HC@@H[C@H]1O
C00022	Pyruvic acid	HMDB0000243	1060	CC(=O)C(O)=O
C00026	Oxoglutaric acid	HMDB0000208	51	OC(=O)CCC(=O)C(O)=O
C00042	Succinic acid	HMDB0000254	1110	OC(=O)CCC(O)=O
C00122	Fumaric acid	HMDB0000134	444972	OC(=O)\C=C\C(O)=O
C00130	Inosinic acid	HMDB0000175	8582	O[C@@H]1C@@HOC@HN1C=NC2=C1N=CNC2=O
C00195	Cer(d18:1/12:0)	HMDB0004947	5283562	[H]C@@(NC(=O)CCCCCCCCCCC)C@H\C=C\CCCCCCCCCCCCC
C00334	gamma-Aminobutyric acid	HMDB0000112	119	NCCCC(O)=O
C00410	Progesterone	HMDB0001830	5994	[H][C@@]12CCC@H[C@@]1(C)CC[C@@]1([H])[C@@]2([H])CCC2=CC(=O)CC[C@]12C
C00762	Cortisone	HMDB0002802	222786	[H][C@@]12CCC@(C(=O)CO)[C@@]1(C)CC(=O)[C@@]1([H])[C@@]2([H])CCC2=CC(=O)CC[C@]12C
C00788	Epinephrine	HMDB0000068	5816	CNCC@HC1=CC(O)=C(O)C=C1
C01598	Melatonin	HMDB0001389	896	COC1=CC2=C(NC=C2CCNC(C)=O)C=C1
C01673	Calcitriol	HMDB0001903	5280453	CC@H[C@@]1([H])CC[C@@]2([H])\C(CCC[C@]12C)=C\C=C1\CC@@HCC@HC1=C
C01780	Aldosterone	HMDB0000037	5839	[H][C@@]12CCC@H[C@]1(CC@H[C@@]1([H])[C@@]2([H])CCC2=CC(=O)CC[C@]12C)C=O
C02465	Liothyronine	HMDB0000265	5920	NC@@HC(O)=O
C00187	Cholesterol	HMDB0000067	5997	[H][C@@]1(CC[C@@]2([H])[C@]3([H])CC=C4CC@@HCC[C@]4(C)[C@@]3([H])CC[C@]12C)C@HCCCC(C)C

3.3   Differential Pathway Activity Analysis

Circuit activity values were compared between breast cancer and normal tissue samples using `Wilcoxon` test for sub-pathway activity and `limma` for node-level analysis. p-values were adjusted using the False Discovery Rate (FDR) correction method, with significance determined at the 0.05 threshold. This analysis revealed several significantly up-regulated and down-regulated pathways, particularly those associated with tumor growth and metabolic alterations. Figure 1 presents the top 10 altered pathways, highlighting both the number of significantly altered nodes and the number of pathways. The figure also summarizes the proportion of altered versus unaltered nodes and pathways. Notably, key cancer-related pathways listed in Table 1 were significantly dysregulated, pointing to critical shifts in signaling and metabolic mechanisms underlying cancer progression. The full set of differentially activated pathways is available at: https://github.com/kinzaR/metabopathia/blob/dev/supplementary_files/brca_caseStudy/results/met_results_paths.tsv. For interactive visualization, a pathway viewer for the BRCA analysis can be accessed at: http://hipathia.babelomics.org/metabopathia_dev/reports/metabopathia_report1/metabopathia_report/pathway-viewer/.

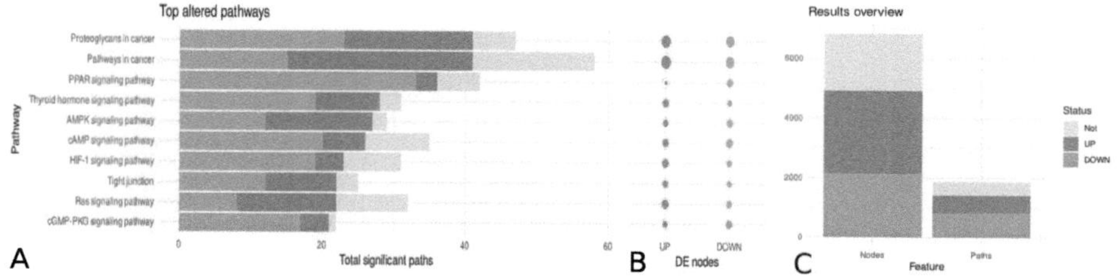

Fig. 1. Top altered pathways and nodes between breast cancer and normal tissue samples. (A) Bar plot of top altered pathways. (B) Differentially expressed nodes. (C) Results overview (nodes vs. paths) by status.

3.4   Metrics for Assessing Metabopathia Performance

Specificity: False Positive Rate. To estimate the false positive rate (FPR) of the *Metabopathia* method, we generated multiple sets of indistinguishable samples. These samples were then randomly divided into two groups, and comparisons were performed to detect differentially activated circuits. Because both groups consisted of the same type of individuals (e.g., breast cancer samples or normal tissues), any detected differences in sub-pathway activity were interpreted as false positives.

This evaluation was carried out using both real and simulated samples (see Methods), allowing us to assess how often the method incorrectly identifies differences when none should exist. As shown in Fig. 2A, the false positive rate remained consistently low (0: no significance detected), well below the conventional alpha value of 0.05 across all sample sizes and scenarios. This confirms the high specificity of *Metabopathia* for detecting pathway activity differences, minimizing the likelihood of false positive detections. For comparison, Fig. 2B illustrates the relationship between the adjusted p-values (1 – FDR p-value) for *Metabopathia* and HiPathia, further highlighting *Metabopathia*'s robustness in producing fewer false positives ($p < 0.05$).

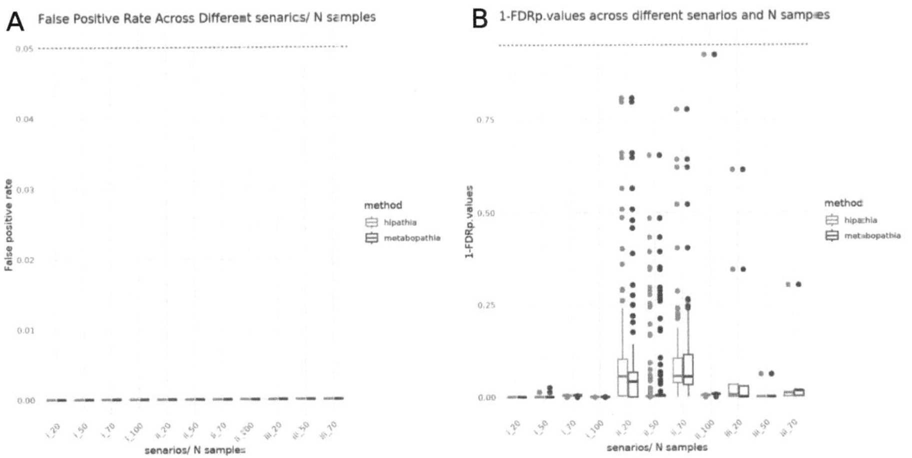

Fig. 2. Specificity analysis across scenarios and sample sizes. (A) False positive rate across different scenarios and N samples. (B) 1–FDRp values across scenarios and N samples for HiPathia vs. Metabopathia.

Sensitivity: True Positive Rate.

To estimate the true positive rate (TPR) of the *Metabopathia* method, we compared breast cancer samples with their corresponding normal tissues using different sample sizes ($N = 20, 50, 70, 100, 200$; see Methods). These comparisons were designed to identify differences in cancer-related pathways, where significant alterations were expected (see Supplementary Figure S6, bottom). A broader set of pathways was analyzed from two distinct pathway sources: KEGG cancer pathways (14 pathways from the Cancer pathways category) and a curated set of 48 pathways from the Valencia Institute of Oncology (IVO) (see Table 1).

As shown in Fig. 3A, *Metabopathia* successfully identified significant pathway activity changes, particularly at larger sample sizes where statistical power was greater. These results demonstrate that integrating metabolomic data enables the method to effectively capture cancer-related alterations in signaling pathways, leading to robust detection of true positive signals in breast cancer dataset. Panel A illustrates the TPR across different sample sizes, while Panel B presents the proportion of significant circuits within cancer-related pathways, reflecting the extent of differential activity captured by *Metabopathia*.

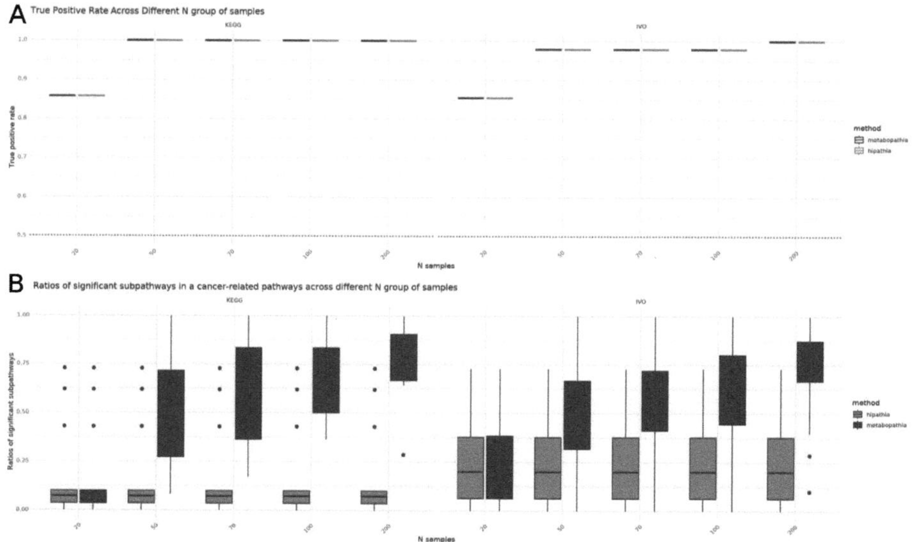

Fig. 3. Sensitivity analysis across sample sizes. (A) True positive rate across N groups of samples (KEGG and IVO sets). (B) Ratios of significant sub-pathways in cancer-related pathways across N groups.

3.5 Comparison to HiPathia Method

Specificity and Sensitivity of Methods. There was no significant difference between HiPathia and *Metabopathia* in terms of false positive rate (FPR) and adjusted p-values across the different scenarios and sample sizes (Fig. 2). Both methods achieved similarly low FPR values, maintaining strong specificity regardless of the number of samples. This consistency indicates that while *Metabopathia* incorporates metabolic data, its specificity remains comparable to HiPathia, which models signaling pathways using transcriptomic data alone.

However, the integration of metabolite information in *Metabopathia* resulted in the detection of a higher number of connected circuits and a greater sensitivity to pathway dysregulation (Fig. 3). As a result, *Metabopathia* was able to capture additional biologically meaningful changes that HiPathia missed, particularly in cancer-related pathways. These findings highlight the added value of incorporating metabolic activity into pathway analysis, as it enables the identification of key dysregulations that would otherwise remain undetected. An important consideration is the impact of sample size on performance. As shown in our analyses, specificity remained consistently high across all group sizes, confirming that the method avoids false positives even in smaller cohorts. In contrast, sensitivity improved with increasing sample size, reflecting the expected gain in statistical power. This indicates that while *Metabopathia* can provide reliable results with modest sample numbers, larger datasets are essential to fully uncover the spectrum of true biological alterations. This dependence on cohort size is consistent with general statistical expectations and should be considered when designing studies that employ *Metabopathia*.

Results from Breast Cancer Case Study. This early version of *Metabopathia* shows a clear advantage over HiPathia for analyzing cancer pathways by incorporating both transcriptomic and metabolomic data. An integrated approach is essential in cancer research, as metabolic and molecular pathway rewiring plays a central role in tumor progression. While HiPathia provides a broader perspective based solely on gene expression, *Metabopathia* enables greater precision in identifying key molecular alterations that drive cancer development and survival.

Our results demonstrate that explicitly modeling the metabolic layer can significantly influence outcomes, whereas assuming metabolites are always active may lead to unreliable conclusions. Supplementary Table S5 shows that 27 circuits exhibited differences in dysregulation direction or significance of differential activation, as identified by at least one of the methods. Of these, 17 circuits were significant only in *Metabopathia* (Supplementary Table S6), while 10 circuits were significant only in HiPathia (Supplementary Table S7). Additionally, the direction of dysregulation was reversed in 17 circuits that were significantly altered in at least one method.

Table 3 summarizes a total of 25 circuits that accounted for the significant differences between HiPathia and *Metabopathia*. These circuits either showed significance in only one of the methods or exhibited opposite directions of dysregulation. Although validation is challenging due to the absence of a clear ground truth, these circuits warrant further analysis and expert interpretation. Importantly, the current limitation of inferring only 16 metabolites may contribute to contradictory results for certain molecular phenotypes.

Several circuits from the Calcium signaling pathway, insulin resistance, and morphine addiction were excluded due to improper parsing. In these KEGG pathways, the presence of feedback loops without terminal effector nodes (nodes with out-degree 0) prevented decomposition into well-defined circuits. This structural property caused signal propagation to fail to converge into interpretable sub-pathways, yielding non-reliable results. Consequently, six circuits (marked with strikethrough in Table 3) were omitted from the analysis. Notably, eight circuits (highlighted in light blue in Supplementary Table S3) from the taste transduction and insulin secretion pathways are composed of only two nodes, one of which is a metabolite. In these cases, HiPathia's assumption of metabolites being always active makes its results less reliable.

A particularly illustrative example is the detection of the Progesterone-mediated pathway by *Metabopathia*, which highlighted key effectors such as

PIK3R5 and MAPK14. These genes are components of the PI3K/AKT signaling pathway, known to play crucial roles in hormone-sensitive cancers, including breast and ovarian cancer. Although this pathway does not directly drive tumorigenesis, it exerts vital regulatory functions in cell proliferation and apoptosis. *Metabopathia* identified significant down-regulation of these circuits (FDR p-value $= 3.64 \times 10^{-16}$), pinpointing the role of hormonal dysregulation in cancer progression. Conversely, HiPathia failed to detect these hormonal interactions, reinforcing the importance of including the metabolic layer for comprehensive analyses, particularly in circuits involving one to seven metabolites (Table 3).

In addition to methodological advantages, the integration of metabolite inference in Metabopathia provided unique biological insights in the breast cancer case study. For example, circuits within the progesterone-mediated oocyte maturation and thyroid hormone signaling pathways revealed alterations that were only detectable when metabolite nodes were incorporated. These findings emphasize the contribution of hormone-related dysregulation to tumorigenesis, a process frequently overlooked by transcriptomics-only approaches. Similarly, the detection of ATP-dependent signaling differences within the calcium signaling pathway highlights how inferred metabolite activity can uncover critical regulatory shifts in energy metabolism. Together, these examples demonstrate that metabolite inference not only increases sensitivity in pathway detection but also provides biologically meaningful interpretations, reinforcing the importance of integrating metabolic context in cancer research.

4   Conclusions

This work introduces *Metabopathia*, a systems biology tool for integrating transcriptomic and metabolomic data into disease modeling. By combining multiple omics layers, *Metabopathia* enables *in silico* analyses that more accurately represent the cellular complexity of multifactorial diseases such as cancer. The method successfully identified key sub-pathway disruptions that drive cancer progression, offering mechanistic insights and supporting the identification of personalized therapeutic targets.

Importantly, *Metabopathia* revealed pathway changes that were missed by single-omics approaches. In particular, inferred metabolite activity highlighted dysregulations in signaling cascades linked to tumor growth and cellular stress responses. These results demonstrate the value of including metabolic information for more context-aware interpretations and tailored therapeutic interventions.

Performance evaluations showed that *Metabopathia* achieves high specificity and slightly improved sensitivity compared to transcriptomics-only methods. It consistently identified more cancer-related sub-pathways in both simulated and real breast cancer dataset, capturing biologically relevant signals without introducing noise. This balance between precision and sensitivity is critical for generating actionable insights.

Table 3. Circuits with significant differences between HiPathia (_X) and Metabopathia (_Y). Light blue rows: two-node circuits (one metabolite). Strikethrough: circuits excluded due to improper parsing.

ID	Circuits name	UP/DOWN _X	_Y	FDRp.value _X	_Y	fitdiff_pValue _X	_Y	#G	#M
~~P-hsa04020-18~~	~~Calcium signaling pathway: ATP~~	DOWN	UP	1.226e-2	5.691e-12	4.187e-2	6.107e-1	1	1
P-hsa04713-56	Circadian entrainment: KCNJ3	UP	DOWN	5.497e-1	2.029e-8	9.467e-1	8.302e-4	3	1
P-hsa04727-26	GABAergic synapse: KCNJ6	UP	DOWN	1.126e-8	3.380e-4	9.413e-1	8.962e-1	3	1
P-hsa04742-37	Taste transduction: P2RY4	DOWN	UP	6.578e-1	5.740e-13	7.842e-1	3.801e-1	1	1
P-hsa04742-40	Taste transduction: P2RX2	DOWN	UP	7.007e-1	2.394e-11	2.800e-1	3.704e-2	1	1
P-hsa04742-46	Taste transduction: P2RY1	DOWN	UP	6.189e-2	4.450e-13	9.280e-1	1.109e-1	1	1
P-hsa04742-54	Taste transduction: P2RX2*	DOWN	UP	7.007e-1	2.394e-11	2.800e-1	3.704e-2	1	1
P-hsa04742-55	Taste transduction: P2RX3	DOWN	UP	1.353e-1	1.450e-11	9.798e-1	1.268e-1	1	1
P-hsa04911-35	Insulin secretion: TRPM4	UP	DOWN	1.663e-7	2.060e-8	3.594e-1	1.671e-2	1	1
P-hsa04911-41	Insulin secretion: ABCC8	UP	DOWN	7.829e-5	6.360e-1	1.966e-1	4.186e-1	1	1
P-hsa04914-34	Progesterone-mediated oocyte maturation: MAPK14	DOWN	DOWN	3.459e-1	3.639e-16	5.119e-1	8.167e-1	2	1
P-hsa04914-39	Progesterone-mediated oocyte maturation: PIK3R5	DOWN	DOWN	1.226e-1	3.639e-16	5.557e-1	8.167e-1	2	1
P-hsa04915-28	Estrogen signaling pathway: KCNJ3	UP	DOWN	8.963e-1	3.439e-7	9.101e-1	9.592e-1	11	4
P-hsa04919-11	Thyroid hormone signaling pathway: CTNNB1	DOWN	DOWN	6.815e-2	3.988e-5	7.706e-1	6.169e-1	8	2*
P-hsa04919-21 53	Thyroid hormone signaling pathway: THRA Triiodothyronine	UP	DOWN	6.506e-1	4.394e-14	4.210e-1	9.685e-2	5	1**
P-hsa04919-46	Thyroid hormone signaling pathway: ATP1B4	UP	DOWN	2.217e-1	2.611e-3	6.316e-1	5.221e-1	13	7*
P-hsa04919-84	Thyroid hormone signaling pathway: TP53*	DOWN	DOWN	1.190e-2	2.141e-1	1.156e-1	1.854e-1	8	2*
P-hsa04919-87	Thyroid hormone signaling pathway: BAD	UP	UP	1.820e-1	5.184e-3	6.269e-2	1.689e-1	7	2*
P-hsa04919-89	Thyroid hormone signaling pathway: CASP9	DOWN	DOWN	4.362e-3	5.836e-1	2.609e-3	1.032e-2	7	2*
P-hsa04919-91	Thyroid hormone signaling pathway: PFKP	UP	DOWN	3.544e-3	8.297e-3	2.545e-1	7.509e-1	6	1*
~~P-hsa04931-40~~	~~Insulin resistance: PRKCZ~~	DOWN	DOWN	6.754e-1	1.820e-3	2.197e-1	2.400e-1	10	1
~~P-hsa04931-47~~	~~Insulin resistance: AKT2*~~	UP	UP	3.111e-6	8.988e-1	7.368e-1	8.093e-3	5	1
~~P-hsa05032-81~~	~~Morphine addiction: KCNJ6~~	UP	DOWN	1.126e-8	3.380e-4	9.413e-1	8.962e-1	3	1
~~P-hsa05032-85~~	~~Morphine addiction: KCNJ6*~~	UP	DOWN	1.126e-8	3.380e-4	9.413e-1	8.962e-1	3	1
~~P-hsa05032-91~~	~~Morphine addiction: KCNJ6**~~	UP	DOWN	1.126e-8	3.380e-4	9.413e-1	8.962e-1	3	1

Legend: _X = HiPathia; _Y = Metabopathia. Light blue rows: two-node circuits (one metabolite), including P-hsa04020-18. Strikethrough rows were excluded due to improper parsing (Insulin resistance, Morphine addiction). Asterisks in #M are pathway-specific annotations retained from the source table. Numeric values are rounded for readability.

The tool is available both as a command-line interface and as a web-based platform, ensuring accessibility for researchers with different technical backgrounds. This dual availability encourages broader adoption of *Metabopathia* in multi-omics analysis and translational research.

In addition to these strengths, the current implementation has limitations that may affect its generalizability. From the data perspective, matched metabolomic data for the same patients are scarce, and when available they usually cover only a limited subset of metabolites. This is because different metabolite classes (such as lipids, amino acids, and cofactors) are measured using distinct experimental platforms, and currently no dataset provides simultaneous coverage of all classes with direct overlap to signaling pathways. From the inference perspective, the present version of *Metabopathia* can only infer 16 metabolites. This restriction is not merely technical, but reflects biological knowledge gaps: many KEGG signaling pathways include metabolites whose production routes are not fully represented within the KEGG metabolic modules, preventing reliable inference. As a result, the overlap between inferred metabolites and the metabolites actually participating in signaling cascades remains limited. Together, these constraints reduce metabolic coverage and may cause biologically relevant dysregulations to remain undetected. Future work should therefore expand the set of inferred metabolites by integrating additional databases that provide more complete biological knowledge of metabolic modules, improving pathway–metabolite mapping, and incorporating more comprehensive metabolomics technologies. Single-cell transcriptomics and metabolomics also hold great promise, though they currently face challenges such as sparsity and incomplete metabolite detection. Addressing these issues will improve the accuracy and robustness of pathway activity modeling. Future studies could also incorporate multivariate analyses, such as MANOVA or elastic net models, to capture higher-order interactions across circuits once larger datasets become available.

In summary, *Metabopathia* extends the foundation established by transcriptomic-based approaches by incorporating a metabolic layer. It does not replace existing methods but enhances them, uncovering regulatory signals that remain hidden in single-omics models. Although demonstrated here in breast cancer, the framework is flexible and can be applied to other complex diseases, providing a powerful and scalable resource for advancing systems and precision medicine.

Acknowledgments. We thank the Andalusian Platform for Computational Medicine (Andalusian Public Foundation Progress and Health-FPS) for providing computational infrastructure. We also acknowledge the TCGA Research Network for access to genomic and clinical data.

Author contributions. KR developed the algorithm and the mechanistic modeling software, conducted the mechanistic modeling analyses, and drafted the manuscript. DLL, JD, and IN jointly conceptualized the study, provided critical revisions, and supervised the project. All authors read and approved the final version of the manuscript.

Funding. This work is supported by grants PID2020-117979RB-I00 from the Spanish Ministry of Economy and Competitiveness and PT17/0009/0006 and PI20/01305 from the ISCIII, both co-funded with European Regional Development Funds (ERDF) as well as H2020 Programme of the European Union grant "ELIXIR-EXCELERATE fast-track ELIXIR implementation and drive early user exploitation across the life sciences" (GA 676559) to JD. The contract of KR is funded by Digital and ecological transition projects 2021/Spanish Ministry for Science and Innovation Grant ID: TED2021-132346B-I00. This project is funded by MCIN/AEI/10.13039/501100011033 and by the European Union "NextGenerationEU"/PRTR.

Data and Code Availability. RNA-seq counts were downloaded from The Cancer Genome Atlas (TCGA). Metabolizer is freely available at: http://metabolizer. babelomics.org (accessed on 14 September 2024). The source code is freely available at: https://github.com/babelomics/metabolizer (accessed on 14 September 2024). HiPathia is freely available as an R/Bioconductor package at https://www. bioconductor.org/packages/release/bioc/html/hipathia.html (accessed on 14 September 2024), as a web server at http://hipathia.babelomics.org (accessed on 14 September 2024), and as a Cytoscape plugin at https://apps.cytoscape.org/apps/cypathia (accessed on 14 September 2024) [32]. *Metabopathia* is freely available at https://github. com/kinzaR/metabopathia (accessed on 14 September 2024). The development version of the *Metabopathia* web server is freely available at http://hipathia.babelomics.org/ metabopathia_dev/ (accessed on 29 December 2024).

Disclosure of Interests. The authors declare no competing interests relevant to the content of this article.

References

1. Amadoz, A., Hidalgo, M.R., Çubuk, C., Carbonell-Caballero, J., Dopazo, J.: A comparison of mechanistic signaling pathway activity analysis methods. Briefings Bioinform. **20**(5), 1655–1668 (2019). https://doi.org/10.1093/bib/bby040

2. Bauer, D.F.: Constructing confidence sets using rank statistics. J. Am. Stat. Assoc. **67**(339), 687–690 (1972). https://doi.org/10.1080/01621459.1972.10481279

3. Benjamini, Y., Hochberg, Y.: Controlling the false discovery rate: a practical and powerful approach to multiple testing. J. Roy. Stat. Soc.: Ser. B (Methodol.) **57**(1), 289–300 (1995)

4. Ciriello, G., et al.: Comprehensive molecular portraits of invasive lobular breast cancer. Cell **163**(2), 506–519 (2015). https://doi.org/10.1016/j.cell.2015.09.033

5. Costa, P.J.: Truncated outlier filtering. J. Biopharmaceutical Stat. **24**(5), 1115–1129 (2014). https://doi.org/10.1080/10543406.2014.926366

6. Dugourd, A., Saez-Rodriguez, J.: Footprint-based functional analysis of multiomic data. Curr. Opin. Syst. Biol. **15**, 82–90 (2019). https://doi.org/10.1013/j.coisb.2019.04.002

7. Esteban-Medina, M., Peña-Chilet, M., Loucera, C., Dopazo, J.: Exploring the druggable space around the fanconi anemia pathway using machine learning and mechanistic models. BMC Bioinform. **20**(1), 370 (2019). https://doi.org/10.1186/s12859-019-2969-0

8. Gao, S., Wang, X.: Tappa: topological analysis of pathway phenotype association. Bioinformatics **23**(22), 3100–3102 (2007). https://doi.org/10.1093/bioinformatics/btm460

9. Garrido-Rodriguez, M., Zirngibl, K., Ivanova, O., Lobentanzer, S., Saez-Rodriguez, J.: Integrating knowledge and omics to decipher mechanisms via large-scale models of signaling networks. Mol. Syst. Biol. **18**(7), e11036 (2022). https://doi.org/10.15252/msb.202211036

10. Gligorijević, V., Pržulj, N.: Methods for biological data integration: perspectives and challenges. J. R. Soc. Interface **12**(112), 20150571 (2015). https://doi.org/10.1098/rsif.2015.0571

11. Greenacre, M., Groenen, P.J.F., Hastie, T., D'Enza, A.I., Markos, A., Tuzhilina, E.: Publisher correction: principal component analysis. Nat. Rev. Methods Primers **3**(1), 22 (2023). https://doi.org/10.1038/s43586-023-00209-y

12. Haynes, W.A., Higdon, R., Stanberry, L., Collins, D., Kolker, E.: Correction: differential expression analysis for pathways. PLOS Comput. Biol. **9**(4) (2013). https://doi.org/10.1371/annotation/58cf4d21-f9b0-4292-94dd-3177f393a284

13. Hidalgo, M.R., Cubuk, C., Amadoz, A., Salavert, F., Carbonell-Caballero, J., Dopazo, J.: High throughput estimation of functional cell activities reveals disease mechanisms and predicts relevant clinical outcomes. Oncotarget **8**(3), 5160–5178 (2017). https://doi.org/10.18632/oncotarget.14107

14. Hofmann-Apitius, M.: others: Bioinformatics mining and modeling methods for the identification of disease mechanisms in neurodegenerative disorders. In. J. Mol. Sci. **16**(12), 29179–29206 (2015). https://doi.org/10.3390/ijms161226148

15. Hollander, M., Wolfe, D.A., Chicken, E.: Nonparametric Statistical Methods. Wiley Series in Probability and Statistics, 1st edn. Wiley, Hoboken (2015). https://doi.org/10.1186/gb-2010-11-2-r23

16. Hung, J.H., Whitfield, T.W., Yang, T.H., Hu, Z., Weng, Z., DeLisi, C.: Identification of functional modules that correlate with phenotypic difference: the influence of network topology. Genome Biol. **11**(2), R23 (2010). https://doi.org/10.1186/gb-2010-11-2-r23

17. Ibrahim, M.A.H., Jassim, S., Cawthorne, M.A., Langlands, K.: A topology-based score for pathway enrichment. J. Comput. Biol. **19**(5), 563–573 (2012). https://doi.org/10.1089/cmb.2011.0182

18. Iqbal, J., Sun, L., Zaidi, M.: Complexity in signal transduction. Ann. New York Acad. Sci. **1192**(1), 238–244 (2010). https://doi.org/10.1111/j.1749-6632.2010.05388.x

19. Jacob, L., Neuvial, P., Dudoit, S.: More power via graph-structured tests for differential expression of gene networks. Ann. Appl. Stat. **6**(2) (2012). https://doi.org/10.1214/11-AOAS528

20. Johnson, W.E., Li, C., Rabinovic, A.: Adjusting batch effects in microarray expression data using empirical bayes methods. Biostatistics **8**(1), 118–127 (2007). https://doi.org/10.1093/biostatistics/kxj037

21. Kanehisa, M.: Kegg: kyoto encyclopedia of genes and genomes. Nucleic Acids Res. **28**(1), 27–30 (2000). https://doi.org/10.1093/nar/28.1.27

22. Kawata, K., et al.: Trans-omic analysis reveals selective responses to induced and basal insulin across signaling, transcriptional, and metabolic networks. iScience **7**, 212–229 (2018). https://doi.org/10.1016/j.isci.2018.07.022

23. Koumakis, L., et al.: Minepath: mining for phenotype differential sub-paths in molecular pathways. PLOS Comput. Biol. **12**(11), e1005187 (2016). https://doi.org/10.1371/journal.pcbi.1005187

24. Leek, J.T., Johnson, W.E., Parker, H.S., Jaffe, A.E., Storey, J.D.: The SVA package for removing batch effects and other unwanted variation in high-throughput experiments. Bioinformatics **28**(6), 882–883 (2012). https://doi.org/10.1093/bioinformatics/bts034

25. Li, X., Shen, L., Shang, X., Liu, W.: Subpathway analysis based on signaling-pathway impact analysis of signaling pathway. PLOS ONE **10**(7), e0132813 (2015). https://doi.org/10.1371/journal.pone.0132813

26. Loucera, C., Esteban-Medina, M., Rian, K., Falco, M.M., Dopazo, J., Peña-Chilet, M.: Drug repurposing for covid-19 using machine learning and mechanistic models of signal transduction circuits related to sars-cov-2 infection. Signal Transduct. Targeted Ther. **5**(1), 290 (2020). https://doi.org/10.1038/s41392-020-00417-y

27. Martini, P., Sales, G., Massa, M.S., Chiogna, M., Romualdi, C.: Along signal paths: an empirical gene set approach exploiting pathway topology. Nucleic Acids Res. **41**(1), e19 (2013). https://doi.org/10.1093/nar/gks866

28. Massa, M.S., Chiogna, M., Romualdi, C.: Gene set analysis exploiting the topology of a pathway. BMC Syst. Biol. **4**(1), 121 (2010). https://doi.org/10.1186/1752-0509-4-121

29. Mi, H., et al.: Systems biology graphical notation: activity flow language level 1 version 1.2. J. Integrative Bioinform. (2015). https://doi.org/10.2390/BIECOLL-JIB-2015-265

30. Murtagh, F., Legendre, P.: Ward's Hierarchical Agglomerative Clustering Method: Which Algorithms Implement Ward's Criterion? J. Classif. **31**(3), 274–295 (2014). https://doi.org/10.1007/s00357-014-9161-z

31. Niarakis, A., et al.: Drug-target identification in covid-19 disease mechanisms using computational systems biology approaches. Front. Immunol. **14**, 1282859 (2024). https://doi.org/10.3389/fimmu.2023.1282859

32. Rian, K., et al.: Genome-scale mechanistic modeling of signaling pathways made easy: a bioconductor/cytoscape/web server framework for the analysis of omic data. Comput. Struct. Biotechnol. J. **19**, 2968–2978 (2021). https://doi.org/10.1016/j.csbj.2021.05.022

33. Robinson, M.D., McCarthy, D.J., Smyth, G.K.: Edger: a bioconductor package for differential expression analysis of digital gene expression data. Bioinformatics **26**(1), 139–140 (2010). https://doi.org/10.1093/bioinformatics/btp616

34. Robinson, M.D., Oshlack, A.: A scaling normalization method for differential expression analysis of RNA-seq data. Genome Biol. **11**(3), R25 (2010) https://doi.org/10.1186/gb-2010-11-3-r25

35. Sebastian-Leon, P., et al.: Understanding disease mechanisms with models of signaling pathway activities. BMC Syst. Biol. **8**(1), 121 (2014). https://doi.org/10.1186/s12918-014-0121-3

36. The Cancer Genome Atlas Network: Comprehensive molecular portraits of human breast tumours. Nature **490**(7418), 61–70 (2012). https://doi.org/10.1038/nature11412

37. Tiwary, B.K.: Computational medicine: quantitative modeling of complex diseases. Briefings Bioinform. **21**(2), 429–440 (2020). https://doi.org/10.1093/bib/bbz005

38. Vitrinel, B.: others: Exploiting interdata relationships in next-generation proteomics analysis. Mol. Cellular Proteomics **18**(8), S5–S14 (2019). https://doi.org/10.1074/mcp.MR118.001246

39. Xia, J., Psychogios, N., Young, N., Wishart, D.S.: Metaboanalyst: a web server for metabolomic data analysis and interpretation. Nucleic Acids Res. **37**(Web Server), W652–W660 (2009). https://doi.org/10.1093/nar/gkp356

40. Çubuk, C., Can, F.E., Peña-Chilet, M., Dopazo, J.: Mechanistic models of signaling pathways reveal the drug action mechanisms behind gender-specific gene expression for cancer treatments. Cells **9**(7), 1579 (2020). https://doi.org/10.3390/cells9071579

41. Çubuk, C., Loucera, C., Peña-Chilet, M., Dopazo, J.: Crosstalk between metabolite production and signaling activity in breast cancer. Int. J. Mol. Sci. **24**(8), 7450 (2023). https://doi.org/10.3390/ijms24087450
42. Çubuk, C., et al.: Differential metabolic activity and discovery of therapeutic targets using summarized metabolic pathway models. NPJ Syst. Biol. Appl. **5**(1), 7 (2019). https://doi.org/10.1038/s41540-019-0087-2

Predicting Obesity-Related Phenotypes from the Human Gut Microbiome Using Machine Learning

Ivon-Teresa Sánchez-Cárdenas[1]($\boxtimes$) (iD) and Martha-Ivon Cárdenas[2] (iD)

[1] Universitat Pompeu Fabra (UPF), Barcelona, Spain
ivon.sanchez01@estudiant.upf.edu
[2] Soft Computing Research Group (SOCO) at Intelligent Data Science and Artificial Intelligence (IDEAI-UPC) Research Centre, Universitat Politècnica de Catalunya (UPC Barcelona Tech), Jordi Girona 1-3, 08034 Barcelona, Spain

Abstract. The gut microbiome is increasingly recognized as a contributing factor in host metabolism, immune function, and energy regulation. Although previous studies have used machine learning (ML) to predict obesity from microbiome data [1–3], the present approach emphasizes model interpretability and reproducibility using SHAP-based feature ranking in publicly available datasets.

In this study, interpretable machine learning techniques were applied to publicly available 16S rRNA amplicon data (ENA accession ERP003612) corresponding to individuals from the MetaHIT cohort [6], with the aim of predicting obesity-associated phenotypes from operational taxonomic unit (OTU) profiles. Several classifiers were evaluated (RandomForest, CatBoost, LightGBM, XGBoost, GBM, Logistic Regression, Naive Bayes), class imbalance was addressed via SMOTE within a nested 5×5 cross-validation framework, and model decisions were interpreted using SHAP. RandomForest achieved the highest ROC AUC (0.678 ± 0.039), with CatBoost close behind (0.674 ± 0.056). SHAP analysis identified top OTUs mostly from the phyla *Firmicutes* and *Bacteroidetes*, including genera such as *Clostridium*, *Coprobacillus*, *Odoribacter*, and *Blautia*, and families like *Clostridiaceae*, *Lachnospiraceae*, and *Ruminococcaceae*.

This study demonstrates that the application of machine learning to publicly available datasets can reveal predictive microbial signatures that might be overlooked by conventional analytical approaches. The proposed framework facilitates systematic hypothesis generation, selects relevant taxa for subsequent experimental validation, and provides reproducible insights into microbiome patterns associated with obesity.

Keywords: microbiome · obesity · machine learning · SHAP · interpretability · OTUs · class imbalance · feature selection

1 Introduction

The escalating global prevalence of obesity represents a major public health challenge, closely associated with an elevated risk of metabolic disorders including

© The Author(s), under exclusive license to Springer Nature Switzerland AG 2026
A. López Fernández et al. (Eds.): CIABiomed 2025, LNBI 16148, pp. 527–539, 2026.
https://doi.org/10.1007/978-3-032-10661-2_39

type 2 diabetes and hypertension [4]. The etiology of obesity is multifactorial, involving a complex interplay of genetic, environmental, and biological factors. More recently, the human gut microbiome has been recognized as a significant biological component influencing host metabolic processes, immune function, and energy homeostasis, thereby contributing to obesity pathogenesis [5,7]. Observations of distinct microbial community structures between obese and non-obese individuals indicate that the microbiome may harbor biomarkers for the condition. This study aims to characterize these associations through the application of advanced machine learning techniques, with a specific emphasis on model interpretability to identify robust and biologically relevant microbial signatures.

1.1   Machine Learning Approaches for Microbiome Data Analysis

Foundational work, including findings from the MetaHIT consortium, established a correlation between reduced microbial diversity and an unfavorable metabolic profile [6], positioning gut microbiome composition as a potential biomarker for obesity phenotypes. The translation of these associations into predictive models is challenged by the inherent properties of microbiome data, which include high dimensionality, sparsity, non-linear relationships, and compositional structure. These properties constrain the efficacy of conventional statistical methods for deriving generalizable models. Machine learning (ML) frameworks offer a methodological approach suited to these analytical challenges, enabling the identification of complex, non-linear patterns within high-dimensional biological datasets [8,9]. Prior research has applied ML to microbiome data for obesity classification, demonstrating modest predictive performance [1–3]. A persistent constraint of such models is their limited interpretability, as the specific basis for predictions is often not readily apparent.

The present study seeks to address this limitation through the application of SHAP (SHapley Additive exPlanations) values, an interpretability framework designed to quantify feature contribution to model outputs. The analysis utilizes publicly available 16S rRNA amplicon data (ERP003612) from the MetaHIT cohort [6]. he primary aim is to employ interpretable machine learning to identify robust taxonomic features associated with obesity, thereby extending analysis beyond prediction to enable biological validation.

1.2   Interpretability in Machine Learning Models

Interpretable machine learning techniques, such as SHAP (SHapley Additive exPlanations), help address the interpretability challenge of ML models by elucidating which features contribute most to individual predictions. In the context of microbiome-based classification, SHAP enables the identification of microbial taxa that are most informative for distinguishing phenotypes. Such interpretability is essential for integrating computational results into biologically meaningful frameworks and facilitating their application in biomedical research and clinical decision-making.

1.3   Objectives

This study aims to evaluate the capacity of interpretable machine learning to identify microbial signatures associated with obesity. The specific objectives are to:

1. Assess and compare the predictive performance of multiple supervised classification algorithms in distinguishing between obese and non-obese individuals using gut microbiome data.
2. Address dataset class imbalance using synthetic minority over-sampling technique (SMOTE) to enhance model robustness and generalizability.
3. Employ SHapley Additive exPlanations (SHAP) values to perform model-agnostic interpretation, identifying the top 20 most influential features for each classifier.
4. Interpret the selected features within a biological context to evaluate the gut microbiome's potential as a source of predictive biomarkers for obesity.

2   Materials and Methods

This section outlines the experimental design, data processing steps, and analytical methods employed. The overarching analytical workflow is depicted in Fig. 1, illustrating the sequential stages from data acquisition to model evaluation.

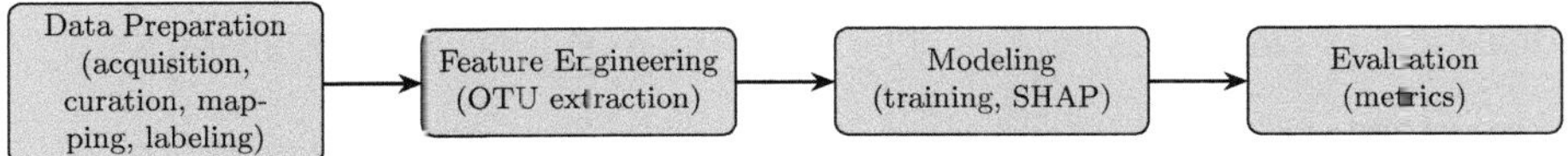

Fig. 1. Schematic representation of the analytical workflow. Data acquisition and curation of microbiome profiles are followed by phenotype mapping and labeling. OTU abundances are extracted as input features, which are then used for model training and interpretation via SHAP values. Model performance is assessed using accuracy, F1-score, and ROC-AUC metrics.

2.1   Data Acquisition

The dataset used in this study was obtained from the European Nucleotide Archive under accession ERP003612. It consists of 16S rRNA amplicon sequencing data collected from individuals enrolled in the MetaHIT study [6]. Metadata at the run level were retrieved using the file `filereport_read_run_ERP003612_tsv.txt`, and microbial relative abundance data per individual was sourced from `relative_abundance.csv`.

2.2 Data Curation and Mapping

Initial quality control was applied to the `all_runs.csv` file by excluding samples with a `QC_status` of 0, resulting in a curated dataset saved as in a file named `all_runs_curated.tsv`. Run accessions were then cross-referenced with the ENA metadata to produce a filtered file of valid sequencing runs (`filtered_filereport.csv`).

To link microbial profiles with phenotype information, several mapping steps were conducted:

- `sample_alias_MH.txt` was used to map MetaHIT sample aliases (e.g., MH0001) to ENA `sample_accession` IDs.
- This mapping was merged with the phenotype metadata from the curated dataset `all_runs_curated.tsv` using `run_accession` as the key.
- Duplicate entries based on the `Individual` identifier were removed to ensure unique samples.

The resulting merged table contained microbiome profiles and corresponding phenotype annotations for each individual. Following quality control and curation, the final integrated dataset contained 518 samples retained for downstream analysis from an initial pool of 595.

2.3 Phenotype Labeling and Feature Extraction

Phenotypic classification was performed by manually mapping known phenotype identifiers to clinical categories. The following assignments were made:

- `D006262`: Healthy
- `D050177`, `D009765`: Non-healthy

The sample labels were binarized, assigning a value of 0 to healthy and 1 to non-healthy individuals, to create the target variable y for binary classification. The microbiome features were extracted from all columns with the prefix `HL-`, which represented relative abundances of microbial taxa. To ensure numeric consistency, commas in these columns were replaced with periods and all values were converted to the float data type, forming the predictor matrix X. The initial predictor matrix comprised all available OTUs. To improve the interpretability of the model and focus the analysis, we performed feature selection based on the importance of SHAP features, with the objective of retaining the top 20 most predictive OTUs for the final model.

2.4 Model Training and Evaluation

Multiple machine learning classifiers were implemented for binary phenotype classification, including XGBoost, CatBoost, LightGBM, Random Forest, Support Vector Machine (SVM), Logistic Regression, and Gaussian Naive Bayes. CatBoost was included as an additional gradient boosting algorithm based on ordered boosting [11], which is well-suited for small to medium-sized datasets and has been shown to perform competitively in biological classification tasks [12]. To ensure reproducibility, a fixed random seed (`random_state=42`) was set for all models and stochastic processes. To adress the effects of class imbalance, two strategies were employed:

1. For models that natively support it (SVM, logistic regression, and Light-GBM), `class_weight='balanced'`, a built-in class weighting was configured.
2. For all models, the Synthetic Minority Over-sampling Technique (SMOTE) was applied [10]. SMOTE generates synthetic minority-class samples by interpolating between existing instances and their nearest neighbors.

SMOTE was applied exclusively to the training partition within each cross-validation fold, ensuring that the corresponding validation partition remained completely unseen and unmodified. This procedure prevented information leakage and allowed unbiased estimation of model performance.

2.5   Cross-Validation Strategy

Due to the limited sample size common in microbiome studies, model performance was evaluated using a comprehensive nested cross-validation (CV) framework [13,14]. This strategy provides a nearly unbiased estimate of the performance of the entire modeling process, including feature selection and hyperparameter optimization, while making efficient use of the available data. The nested CV consisted of two layers:

- An **outer loop** of 5-fold stratified cross-validation was used to assess the generalization performance of the full modeling pipeline. Stratification preserved the original class distribution in each fold.
  - Within each training fold of the outer loop, an **inner loop** (another 5-fold stratified CV) was performed for hyperparameter tuning via grid search.
  - **To prevent data leakage:** SMOTE oversampling was applied exclusively to the training partitions within both the inner and outer loops.
- **For model interpretation:** After the nested CV process, feature importance was estimated for each outer fold by calculating SHAP values on the held-out validation data. A consensus set of top features for biological interpretation was derived by aggregating results across all outer folds.

Final performance metrics were computed by aggregating the predictions from the held-out validation folds of the outer loop. This nested approach ensures that the reported performance reflects the expected outcome of applying the entire trained pipeline to new, unseen data.

2.6   Interpretable Feature Selection

To enhance interpretability and mitigate dimensionality, SHapley Additive exPlanations (SHAP) values were computed using only the training data of each outer cross-validation fold [15]. The Synthetic Minority Over-sampling Technique (SMOTE) was applied to the training data prior to model fitting and SHAP computation. This ensured that no information from the outer validation partitions was used in the feature selection process. For each fold, features were ranked by their mean absolute SHAP value, and the top k features ($k = 20$) were identified.

A consensus feature set for each model was formed by selecting the k features that appeared most frequently across the top k lists from all outer folds. This consensus set was used to refit the model and report final performance metrics.

This procedure, with feature selection performed within each fold and a consensus built across folds, prevents information leakage and produces a stable set of selected features.

2.7 Statistical Testing of Group-Wise Abundances

To evaluate whether SHAP-selected OTUs exhibited significant differences in relative abundance between phenotypes, group-wise comparisons (Healthy vs. Nonhealthy) were performed using the Mann–Whitney U test. Resulting p-values were adjusted for multiple hypothesis testing using the Benjamini–Hochberg false discovery rate (FDR) procedure, with significance defined at $q < 0.05$.

2.8 Evaluation Metrics

Model performance was evaluated using a suite of metrics to comprehensively assess different aspects of predictive capability. The following metrics were calculated for each cross-validation fold and subsequently aggregated to report mean performance and standard deviation:

- **Accuracy**: The proportion of correctly classified instances, providing a general overview of model performance.
- **F1-Score**: The harmonic mean of precision and recall, offering a balanced measure that is particularly informative for imbalanced datasets.
- **ROC-AUC**: The area under the receiver operating characteristic curve, which evaluates the model's ability to discriminate between classes across all possible classification thresholds. This metric is robust to class imbalance and was a primary basis for model comparison.

Among these, ROC-AUC was emphasized as the primary metric for model selection, owing to its threshold independence and robustness to class imbalance.

3 Results and Discussion

This section details the results of the machine learning analyses, including the predictive performance of the evaluated classification models and the biological features identified through interpretable feature selection. The findings are contextualized within the existing literature on the gut microbiome and obesity, with a discussion of associations that align with prior work as well as those that may represent less commonly reported microbial signatures.

3.1   Dataset Characteristics

The curated dataset consisted of 255 individuals, with 75 classified as healthy and 180 as non-healthy prior to the application of class balancing techniques. The initial feature set contained 58 OTUs. Following SHAP-based feature selection, the top 20 OTUs were retained for the final model. Of these, 18 were identified at the genus level or lower, allowing for subsequent biological interpretation.

3.2   Model Performance

Model performance was evaluated using nested 5×5 cross-validation, with an inner loop for hyperparameter optimization and an outer loop for unbiased performance estimation. Results across evaluation metrics are summarized in Fig. 2 and Table 1. Random Forest achieved the highest mean ROC AUC (0.678 ± 0.039), closely followed by CatBoost (0.674 ± 0.056). Corresponding Precision–Recall AUCs were substantially higher (0.832±0.044 and 0.831±0.055, respectively), consistent with the elevated prevalence of the positive class. Light-GBM and XGBoost yielded moderately lower ROC AUC scores (0.667 ± 0.036 and 0.646 ± 0.049, respectively), while simpler linear models, such as Logistic Regression, demonstrated comparatively limited performance (ROC AUC: 0.587 ± 0.050). Given the overlapping standard deviations observed across outer folds, the performance difference between Random Forest and CatBoost was not statistically significant.

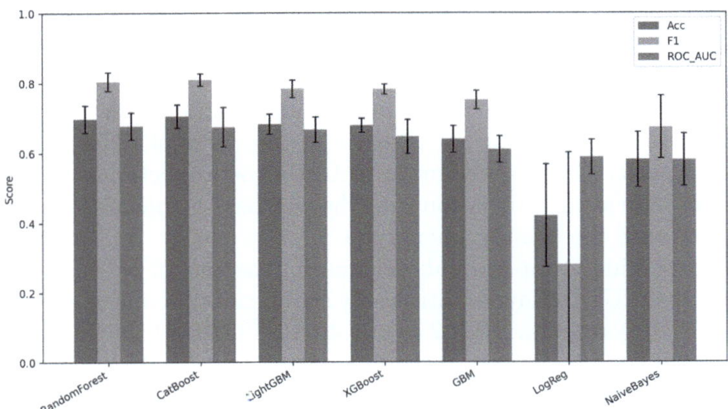

Fig. 2. Predictive performance of machine learning classifiers based on nested 5×5 cross-validation. Bars represent mean values for accuracy, F1-score, and ROC AUC, with error bars denoting standard deviation across outer folds. Random Forest and CatBoost achieved the highest performance, while simpler linear models showed lower predictive ability.

Table 1. Nested 5×5 cross-validation performance (mean ± std). ROC AUC and PR AUC are reported as threshold-independent metrics.

Model	ROC AUC	PR AUC	Accuracy	F1-score
RandomForest	0.678 ± 0.039	0.832 ± 0.044	0.698 ± 0.038	0.805 ± 0.027
CatBoost	0.674 ± 0.056	0.831 ± 0.055	0.706 ± 0.033	0.809 ± 0.018
LightGBM	0.667 ± 0.036	0.826 ± 0.041	0.682 ± 0.029	0.783 ± 0.025
XGBoost	0.646 ± 0.049	0.811 ± 0.051	0.678 ± 0.020	0.782 ± 0.014
GBM	0.610 ± 0.038	0.794 ± 0.036	0.639 ± 0.038	0.751 ± 0.027
LogReg	0.587 ± 0.050	0.765 ± 0.036	0.420 ± 0.147	0.280 ± 0.321
NaiveBayes	0.580 ± 0.075	0.741 ± 0.048	0.580 ± 0.079	0.673 ± 0.090

3.3   Feature Interpretation and Biological Relevance

SHAP values were computed within the training partition of each outer CV fold, and the OTUs were ranked by mean absolute SHAP value for CatBoost (Fig. 3). Most of the top-20 features belonged to the phylum *Firmicutes*, particularly within the order *Clostridiales* (families *Clostridiaceae*, *Lachnospiraceae*, *Ruminococcaceae*), while one OTU (HL-17) was identified as *Odoribacter* (*Bacteroidetes*). Detailed taxonomy is presented in Table 2.

The presence of genera such as *Clostridium*, *Blautia*, and *Coprobacillus* aligns with previously reported associations with host metabolic regulation. Notably, several top OTUs (e.g. HL-6, HL-31, HL-51) remained unclassified beyond the *Clostridiales* order, suggesting the potential value of under-characterized microbial clades in predictive modeling. Overall, these findings indicate that a mix of well-known taxa and less-annotated OTUs contributes to phenotype discrimination, reinforcing the utility of SHAP in microbiome biomarker discovery.

As illustrated in Fig. 3, the SHAP bar and beeswarm plots for the top-ranked OTUs demonstrate their contribution to CatBoost predictions, with rightward displacement indicating association with the non-healthy phenotype and leftward displacement with the healthy phenotype.

When inspecting group-wise abundances of the top 10 SHAP-ranked OTUs (Fig. 4), three OTUs displayed statistically significant differences between phenotypes after FDR correction: HL-52 ($q = 1.4 \times 10^{-5}$), HL-51 ($q = 1.8 \times 10^{-4}$), and HL-22 ($q = 0.025$), all enriched in the `Healthy` group. HL-22 is annotated to the family *Lachnospiraceae* (Firmicutes), a clade strongly associated with metabolic health, often recognized for its role in producing beneficial short-chain fatty acids (SCFAs) [20]. HL-52 and HL-51 lack deeper taxonomic assignment beyond *Clostridiales*, suggesting that poorly characterized taxa also carry meaningful signals.

The prominence of OTUs from the phyla *Firmicutes* and *Bacteroidetes* among the top predictive features aligns with existing literature, which consistently highlights the significant roles of these groups in host energy metabolism and obesity pathogenesis [7]. Specifically, many identified taxa, such as those

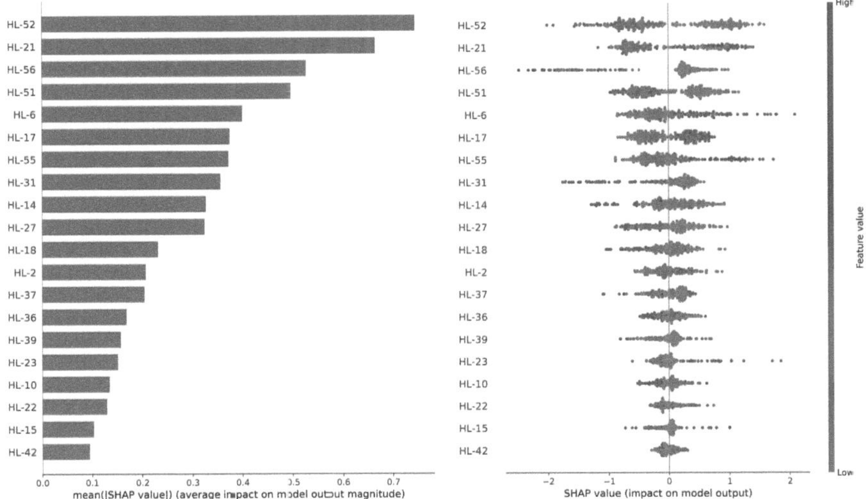

Fig. 3. SHAP-based interpretation of CatBoost predictions. (Left) Mean absolute SHAP values identify the top 20 OTU clusters contributing to phenotype classification. (Right) Beeswarm plot illustrating the impact of individual feature values on model output, where each dot represents a sample and color denotes relative abundance (red = high, blue = low). Rightward displacement indicates stronger association with the non-healthy phenotype, while leftward displacement indicates association with the healthy phenotype. (Color figure online)

within the *Clostridiales* order (including families like *Clostridiaceae*, *Lach-nospiraceae*, and *Ruminococcaceae*), are recognized for their involvement in dietary fiber fermentation and SCFA production [20]. SCFAs, such as butyrate, acetate, and propionate, are crucial metabolites that influence host energy home-ostasis, gut barrier integrity, and inflammatory responses [21]. For instance, the genus *Blautia* (within *Lachnospiraceae*) contributes to SCFA production, and its abundance has been implicated in various metabolic states [22]. Conversely, the identification of *Coprobacillus* (from *Erysipelotrichaceae*) as a predictive feature is notable, as the Erysipelotrichaceae family has been frequently associated with an increased prevalence in obesity in several cohorts [23]. While the precise role of *Odoribacter* (*Bacteroidetes*) in obesity is more complex, its inclusion among the top features underscores the nuanced contributions of various bacterial groups.

HL-55 had higher mean abundance in the `Non-healthy` group but did not sur-vive multiple testing correction ($q = 0.072$). No other SHAP-selected OTUs (e.g., HL-2, HL-23, HL-21, HL-6) showed significant differences ($q > 0.28$). Overall, these results indicate that a subset of the predictive OTUs exhibited statistically significant phenotype-associated abundance shifts, predominantly within Firmi-cutes/Clostridiales. This supports the view that obesity-related microbiome phe-notypes may involve both well-characterized families (e.g., *Lachnospiraceae*) and under-annotated clades that contribute to the predictive signal.

Table 2. Top 20 OTU Clusters Identified by SHAP and Their Taxonomic Annotations.

OTU Cluster	Genus	Family	Order	Phylum
HL-6			Clostridiales	Firmicutes
HL-21		Clostridiaceae	Clostridiales	Firmicutes
HL-23		Ruminococcaceae	Clostridiales	Firmicutes
HL-51			Clostridiales	Firmicutes
HL-56	Coprobacillus	Erysipelotrichaceae	Erysipelotrichales	Firmicutes
HL-2			Clostridiales	Firmicutes
HL-22		Lachnospiraceae	Clostridiales	Firmicutes
HL-27		Ruminococcaceae	Clostridiales	Firmicutes
HL-17	Odoribacter	Porphyromonadaceae	Bacteroidales	Bacteroidetes
HL-14		Clostridiaceae	Clostridiales	Firmicutes
HL-39		Ruminococcaceae	Clostridiales	Firmicutes
HL-18	Clostridium	Clostridiaceae	Clostridiales	Firmicutes
HL-31			Clostridiales	Firmicutes
HL-36	Clostridium	Clostridiaceae	Clostridiales	Firmicutes
HL-15	Blautia	Lachnospiraceae	Clostridiales	Firmicutes
HL-42	Clostridium	Clostridiaceae	Clostridiales	Firmicutes
HL-10			Clostridiales	Firmicutes
HL-37			Clostridiales	Firmicutes

3.4   Translational Relevance and Future Directions

Although the present study investigated the predictive potential of the gut microbiome for obesity, several limitations must be recognized. The relatively small sample size of the MetaHIT cohort may affect the generalizability of the findings. Furthermore, the cross-sectional nature of the data prevents inference of causality or temporal dynamics. The use of 16S rRNA amplicon sequencing, while informative for taxonomic profiling, limits functional insights that could be gained from metagenomic, metatranscriptomic, or metabolomic data. Finally, the absence of a separate holding test set meant that performance was estimated via nested cross-validation, which provides unbiased internal validation but does not substitute for external validation in independent cohorts [14, 26].

Despite these limitations, this study demonstrates the utility of interpretable machine learning for generating biologically plausible hypotheses. The OTU clusters identified as influential predictive features by SHAP may represent candidate microbial biomarkers for obesity-related phenotypes. By applying this existing biological data with advanced computational methods, this work provides a reproducible framework for feature prioritization.

Ultimately, these findings open avenues for translational research. Future work should focus on validating these microbial signatures in independent, diverse cohorts and through experimental laboratory studies to examine their

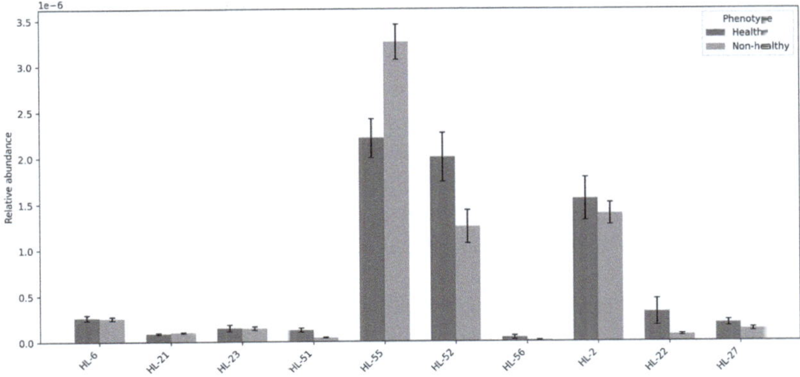

(a) Relative abundances (mean ± SEM) of the top 10 SHAP-selected OTUs across phenotype groups.

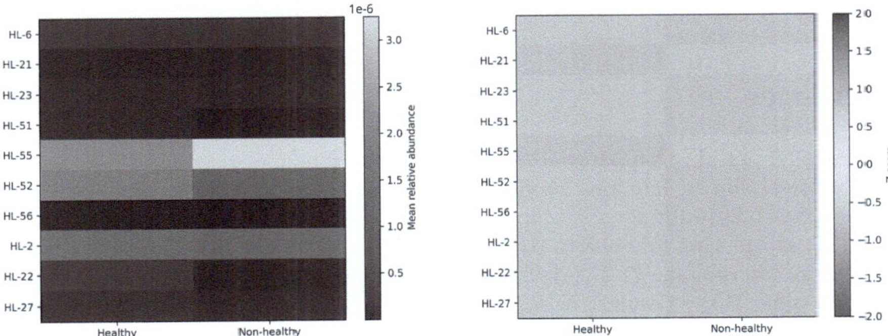

(b) Mean relative abundance per OTU and phenotype.

(c) Z-scored abundances (row-wise) of the same OTUs.

Fig. 4. Abundance patterns of the top 10 SHAP-selected OTUs across phenotype groups. (a) Mean relative abundances (± SEM) comparing healthy and non-healthy individuals. (b) Group-wise mean abundances highlighting consistent trends across phenotypes. (c) Heatmap of z-scored abundances emphasizing relative enrichment patterns at the individual level. Several OTUs (e.g., HL-22, HL-51, HL-52) were significantly enriched in the healthy group after FDR correction.

mechanistic roles. This could guide the development of microbiome-informed strategies for obesity prevention and treatment.

4 Conclusion

This study applied machine learning to gut microbiome data and identified signals associated with obesity-related phenotypes. Random Forest and CatBoost yielded the highest performance among the tested models. SHAP analysis indicated that microbial taxa from the *Firmicutes* and *Bacteroidetes* phyla were among the most influential for predictions, aligning with their documented functions in host metabolism and energy regulation. Both well-characterized families (e.g., *Lachnospiraceae* and *Blautia*) and less-annotated clades within the *Clostridiales* order were identified as potentially informative features. Although the cross-sectional design precludes causal inference, these results suggest that the human gut microbiome contains microbial signatures linked to obesity. This work illustrates how publicly available datasets, analyzed using interpretable machine learning frameworks, can yield reproducible and biologically plausible hypotheses to inform future experimental and translational studies.

Acknowledgment. The authors would like to thank the Universitat Pompeu Fabra and the Intelligent Data Science and Artificial Intelligence (IDEAI-UPC) Research Center for their support during this research. No specific funding was received for this study.

Disclosure of Interests. The authors declare that they have no competing interests relevant to the content of this article.

References

1. Gupta, A., et al.: Comparative evaluation of microbiome-based classifiers for obesity prediction in humans. Front. Microbiol. **8**, 1120 (2017). https://doi.org/10.3389/fmicb.2017.01120
2. Li, J., et al.: Gut microbiota dysbiosis contributes to the development of hypertension. Microbiome **7**, 27 (2019). https://doi.org/10.1186/s40168-019-0644-2
3. Topçuoğlu, B.D., Lesniak, N.A., Ruffin, M.T., Wiens, J., Schloss, P.D.: A framework for effective application of machine learning to microbiome-based classification problems. mBio **11**(3), e00434-20 (2020). https://doi.org/10.1128/mBio.00434-20
4. Chooi, Y.C., Ding, C., Magkos, F.: The epidemiology of obesity. Metabolism **92**, 6–10 (2019). https://doi.org/10.1016/j.metabol.2018.09.005
5. Lin, X., Li, H.: Obesity: epidemiology, pathophysiology, and therapeutics. Front. Endocrinol. (Lausanne) **12**, 706978 (2021). https://doi.org/10.3389/fendo.2021.706978
6. Le Chatelier, E., et al.: Richness of human gut microbiome correlates with metabolic markers. Nature **500**(7464), 541–546 (2013). https://doi.org/10.1038/nature12506
7. Jandhyala, S.M., Talukdar, R., Subramanyam, C., Vuyyuru, H., Sasikala, S.V., Nageshwar Reddy, D.: Role of the gut microbiota in obesity, metabolic syndrome, and diabetes. World J. Gastroenterol. **21**(34), 9687–9703 (2015). https://doi.org/10.3748/wjg.v21.i34.9687
8. Pasolli, E., Truong, D.T., Malik, F., Waldron, L., Segata, N.: Machine learning meta-analysis of large metagenomic datasets: tools and biological insights. PLoS Comput. Biol. **12**(7), e1004977 (2016). https://doi.org/10.1371/journal.pcbi.1004977
9. Knights, D., Costello, E.K., Knight, R.: Supervised classification of human microbiota. FEMS Microbiol. Rev. **35**(2), 343–359 (2011). https://doi.org/10.1111/j.1574-6976.2010.00251.x

10. Chawla, N.V., Bowyer, K.W., Hall, L.O., Kegelmeyer, W.P.: SMOTE: synthetic minority over-sampling technique. J. Artif. Intell. Res. **16**, 321–357 (2002). https://doi.org/10.1613/jair.953

11. Prokhorenkova, L., Gusev, G., Vorobev, A., Dorogush, A.V., Gulin, A.: CatBoost: unbiased boosting with categorical features. In: Advances in Neural Information Processing Systems, vol. 31 (2018). https://papers.nips.cc/paper/2018/file/14491b756b3a51daac41c24863285549-Paper.pdf

12. Wang, Y., et al.: A comprehensive evaluation of machine learning algorithms for predicting gene regulatory networks. Briefings Bioinf. **24**(1), bbac567 (2023). https://doi.org/10.1093/bib/bbac567

13. Papoutsoglou, G., Pereira, F., Schaeffer, L., Pasolli, E., Segata, N.: A machine learning meta-analysis of large metagenomic datasets for robust prediction of microbiome-associated diseases. Nat. Commun. **14**(1), 4764 (2023). https://doi.org/10.1038/s41467-023-39748-z

14. Kubinski, R., et al.: Benchmarking microbiome-based predictive models under realistic constraints. PLOS Comput. Biol. **18**(10), e1010596 (2022). https://doi.org/10.1371/journal.pcbi.1010596

15. Lundberg, S. M., Lee, S.-I.: A unified approach to interpreting model predictions. In: Advances in Neural Information Processing Systems, vol. 30, pp. 4765–4774 (2017)

16. Breiman, L.: Random forests. Mach. Learn. **45**, 5–32 (2001)

17. Friedman, J.H.: Greedy function approximation: a gradient boosting machine. Ann. Stat. **29**(5), 1189–1232 (2001)

18. Ke, G., et al.: LightGBM: a highly efficient gradient boosting decision tree. In: Advances in Neural Information Processing Systems, vol. 30, pp. 3146–3154 (2017)

19. Chen, T., Guestrin, C.: XGBoost: a scalable tree boosting system. In: Proceedings of the 22nd ACM SIGKDD International Conference on Knowledge Discovery and Data Mining, pp. 785–794 (2016).https://doi.org/10.1145/2939672.2939785

20. Louis, P., Flint, H. J.: Formation of butyrate and other short-chain fatty acids by the human colonic microbiota. Environ. Microbiol. **19**(1), 5-11 (2017). https://doi.org/10.1111/1462-2920.13589

21. Koh, A., De Vadder, F., Kovatcheva-Datchary, P., Bäckhed, F.: From dietary fiber to host physiology: short-chain fatty acids as key bacterial metabolites. Cell **165**(6), 1332-1345 (2016). https://doi.org/10.1016/j.cell.2016.05.041

22. Liu, X., et al.: Blautia–a new functional genus with potential probiotic properties?. Front. Microbiol. **12**, 638691 (2021). https://doi.org/10.3389/fmicb.2021.638691

23. Kaakoush, N. O.: Insights into the role of Erysipelotrichaceae in the human host. Front. Cell. Infect. Microbiol. **5**, 84 (2015). https://doi.org/10.3389/fcimb.2015.00084

24. Turnbaugh, P.J., Ley, R.E., Mahowald, M.A., Magrini, V., Mardis, E.R., Gordon, J.I.: An obesity-associated gut microbiome with increased capacity for energy harvest. Nature **444**(7122), 1027–1031 (2006). https://doi.org/10.1038/nature05414

25. Zhao, L., et al.: Gut bacteria selectively promoted by dietary fibers alleviate type 2 diabetes. Science **359**(6380), 1151–1156 (2018). https://doi.org/10.1126/science.aao5774

26. Dudek, A., Müller, L., Zhang, Y., Patel, R., Thompson, J.: Best practices for validation of machine learning models in biomedical microbiome research. J. Biomed. Inf. **149**, 104582 (2024). https://doi.org/10.1016/j.jbi.2024.104582

Triclustering-Based Analysis of Circadian Gene Expression Patterns

Javier Hiruelo-Pérez[1] , José M. García-Heredia[2] ,
David Gutiérrez-Avilés[1] , and María Martínez-Ballesteros[1]([envelope])

[1] Department of Computer Languages and Systems, University of Seville,
41012 Seville, Spain
`{jhiruelo1,dgutierrez3,mariamartinez}@us.es`
[2] Department of Vegetal Biochemistry and Molecular Biology, University of Seville,
41012 Seville, Spain
`jmgheredia@us.es`

Abstract. The circadian cycle orchestrates physiological and molecular processes through rhythmic gene expression. Uncovering relational patterns that link specific genes, experimental conditions, and time points is essential to understand the regulatory architecture of these rhythms. This study applies TriGen, a triclustering algorithm that identifies coherent gene-condition-time modules in circadian transcriptomic datasets. The analysis involved data preprocessing, tricluster extraction with TriGen, and functional enrichment analysis. TriGen uncovers biologically interpretable modules enriched in core clock components, as well as immune-related pathways, illustrating the potential of triclustering to reveal complex, condition-dependent circadian relationships. Across ten independent experiments, the method recovered up to seventy two percent of a curated set of literature-derived circadian genes, capturing both canonical clock regulators and auxiliary modulators involved in metabolism, immune signaling, and stress response. Functional enrichment analysis revealed significant associations with mitochondrial function, transcriptional regulation, neuronal signaling, and immune defense, reflecting the multifaceted systemic impact of circadian regulation. These results highlight the value of triclustering for dissecting high-dimensional temporal transcriptomic data and provide a framework to uncover novel regulatory modules that can guide targeted experimental validation.

Keywords: Circadian cycle · Triclustering · Gene expression · Pattern detection

1 Introduction

The circadian cycle synchronizes gene expression programs with the 24-hour light–dark rhythm, orchestrating metabolism, immunity, and other physiological systems. In mammals, this regulation relies on a core transcriptional–translational feedback loop in which the CLOCK:BMAL1 heterodimer activates

A. López Fernández et al. (Eds.): CIABiomed 2025, LNBI 16148, pp. 540–554, 2026.
https://doi.org/10.1007/978-3-032-10661-2_40

transcription of the *PER* and *CRY* genes, whose protein products subsequently repress their own transcription, thus closing the loop. Disruption of this system has been associated with metabolic disorders, immune dysregulation, and neurological diseases. Recent evidence also reveals an intricate crosstalk between the molecular clock and inflammatory pathways, such as *NF-κB*, highlighting bidirectional modulation in the context of inflammation and immune regulation [12].

A commonly used approach for analyzing gene expression data is clustering, which systematically groups genes with similar expression profiles. Clustering not only reduces data dimensionality but also reveals latent patterns and supports hypothesis generation regarding gene–phenotype relationships. Such analyses are instrumental in prioritizing genes for further investigation as potential therapeutic targets, as well as identifying new candidate genes or single nucleotide polymorphisms to inform novel therapeutic strategies [1]. Most circadian transcriptomic studies focus on rhythmicity at the single-gene level. However, circadian regulation often involves coordinated, condition-specific patterns across ordered time points, requiring methods that preserve the three-dimensional structure of circadian datasets. Biclustering techniques, which simultaneously cluster rows and columns of an expression matrix, are well established in bioinformatics and other domains, offering the ability to detect local patterns, support overlapping clusters, and adapt to diverse problem formulations [26].

Extending this concept, triclustering operates on three-dimensional tensors, typically genes, conditions (or samples), and time points, allowing the identification of coherent gene-condition-time modules. This approach captures dynamic temporal relationships and complex interaction patterns that conventional methods may overlook. Triclustering is computationally challenging, as the underlying biclustering problem is NP-hard. Consequently, heuristic strategies are essential. For example, TriCluster [31] models the problem as maximal clique detection in multigraphs, while OPTricluster [28] employs order-preserving heuristics to exploit temporal structure. Comprehensive surveys of heuristic-based biclustering and triclustering methods are available in [6,17].

This study applies the TriGen [13] algorithm, an evolutionary heuristic triclustering approach, to circadian transcriptomic data to identify modules defined jointly by genes, conditions, and time points. The method optimizes multiple coherence criteria, including correlation and temporal consistency, enabling the recovery of patterns that preserve the ordered structure of time points while capturing condition-specific variation. This makes TriGen particularly suitable for circadian datasets, where temporal order and context-dependent expression are both essential. The analysis aims to recover modules consistent with established circadian mechanisms and to reveal novel interactions that may contribute to a more comprehensive understanding of temporal gene regulation.

2   Related Work

Unsupervised clustering has been a cornerstone of circadian transcriptomics, grouping genes with similar temporal dynamics and inferring potential co-regulation. Early studies often relied on hierarchical clustering and *k*-means to organize genes by the similarity of their 24-hour expression profiles, revealing phase-coherent groups in various tissues [30]. Although effective for broad grouping, these methods require identical sampling across all genes, are sensitive to noise and missing data, and do not explicitly model periodicity.

To address these limitations, rhythmicity-aware clustering methods have been developed. Some combine periodic regression or Fourier transforms with clustering into phase bins, while tools such as JTK_CYCLE use pattern matching to assign genes to canonical circadian templates [18]. More recently, integrative pipelines, such as tauFisher, have combined rhythmic gene correlations with statistical inference to estimate circadian phase from single samples [10], allowing analyses when dense time series are unavailable. Beyond these traditional methods, biclustering has been employed to detect condition-specific circadian modules by simultaneously clustering genes and subsets of experimental conditions [21]. This allows for the discovery of patterns that are active only in certain contexts, such as specific tissues or treatments. Nevertheless, biclustering typically treats time as an unordered dimension, overlooking the temporal ordering that defines circadian rhythms.

Triclustering directly addresses this gap by modeling three-way relationships between genes, conditions, and ordered time points. The TriGen algorithm [13], in particular, applies a genetic search strategy to optimize multiobjective coherence measures (correlation, variance, temporal consistency) and extract compact, biologically significant triclusters. Comparative evaluations indicate that TriGen and related methods outperform alternatives such as OPTricluster in recovering functionally enriched phase-consistent modules [14,22]. Supporting tools such as G-Tric have further formalized the generation of synthetic test data for comparison [20], while statistical frameworks such as TriSig [2] have introduced significance testing to strengthen the robustness of the tricluster findings.

An approach based on the ORIOS tool [19] was applied to circadian data using a statistical model to assess rhythmicity between conditions, combined with data normalization. Although this method shares the aim of identifying condition-specific circadian patterns, it does not employ machine learning or triclustering and therefore lacks the capacity to simultaneously model gene, condition, and time dimensions. Another related study [8] analyzed circadian rhythms in two regions of the brain, applying a custom method to detect age-dependent differences in expression patterns. Although it used a different tissue type and detection strategy, it similarly highlights the importance of integrating condition and temporal variation in circadian research.

In recent years, clustering approaches have increasingly been integrated with downstream functional and network analyses. For example, large-scale atlases of circadian gene expression across multiple tissues [30] have been used in combination with pathway enrichment and regulatory motif analysis to link co-expression modules to specific clock components and signaling pathways. Multi-omics studies and perturbation experiments, such as feeding inversions, have also benefited from clustering frameworks capable of retaining the relational structure under conditions and time [24]. Deep learning has also been explored for identifying circadian genes from temporal data [9], although these approaches often require extensive data and careful regularization.

Overall, while classical clustering can reveal broad rhythmic patterns, it ignores ordered temporal structure; biclustering incorporates condition specificity but discards time order. Triclustering methods such as TriGen reconcile both aspects, recovering relational circadian patterns that would otherwise remain hidden. Remaining challenges include rigorous statistical validation, robustness to sparse or irregular sampling, and integration with multi-omics and network topology information, areas where advances such as TriSig and Gaussian process-based clustering show promise [23]. Compared to existing triclustering methods such as OPTricluster [28] and TriCluster [31], TriGen provides a complementary approach that combines temporal coherence with interpretability. These features make it particularly suitable for circadian transcriptomic analyses, where both ordered time structure and biological interpretability are critical.

3   Methodology

This section describes the methodology used during the experimentation phase. The research process is structured around three primary stages: preprocessing of microarray data, execution of TriGen experiments, and explanation of the resulting solutions (Fig. 1). Each stage is detailed, including the techniques and procedures used to ensure accuracy and reliability.

3.1   Preprocessing

The analysis focuses on microarray experiments characterized by diverse sample types measured at multiple time points. This design facilitates the construction of a three-dimensional tensor with axes representing genes, samples, and time points. Formally, the structure of our three-dimensional microarray dataset is represented by $\mathcal{D} \in \mathbb{R}^{|S| \times |G| \times |T|}$, where $|S|$ represents the variety of sample types included, $S = \{s_1, \ldots, s_{|S|}\}$; $|G|$ denotes the number of probes in the microarray, $G = \{g_1, \ldots, g_{|G|}\}$, included in the dataset; and $|T|$ signifies the number of distinct time points, $T = \{t_1, \ldots, t_{|T|}\}$, at which experimental observations were recorded.

Upon constructing the three-dimensional tensor from the experimental data acquired from GEO, the dataset is subjected to a series of filters designed to eliminate non-expressed genes and noise. Several standard filters commonly used in microarray DNA and RNA-seq analysis are employed. These filters work by removing genes that do not exceed specified thresholds in terms of absolute value or standard deviation throughout the time series. To summarize this process, the preprocessing pipeline is implemented as summarized below:

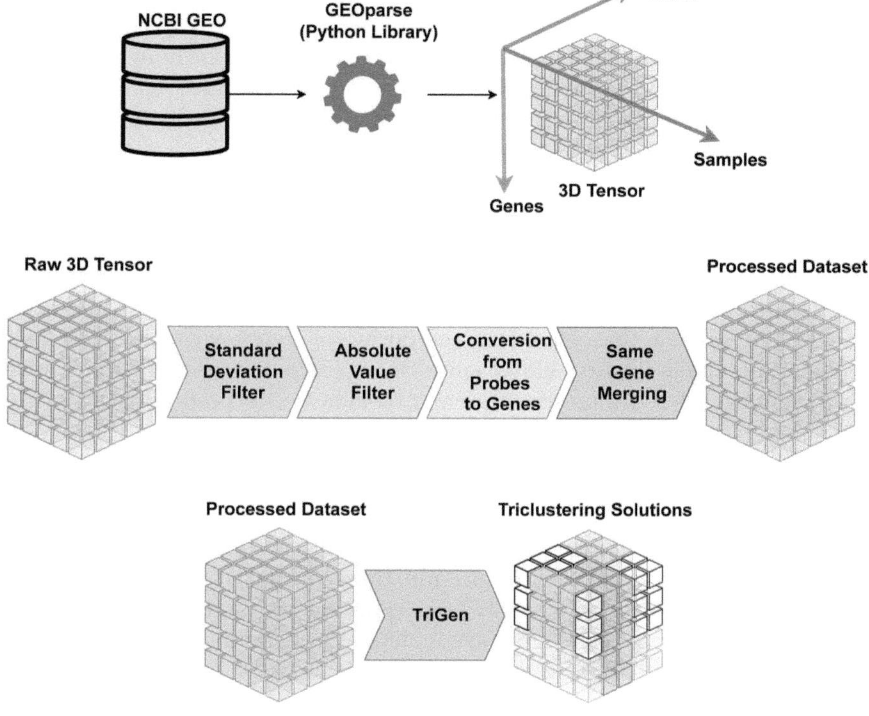

Fig. 1. Pipeline for obtaining, processing, and analyzing gene expression data. The diagram illustrates sequential steps involved in data acquisition, preprocessing, and analysis to derive meaningful insights from gene expression datasets.

1. **Standard deviation filter**: this procedure eliminates probes with insufficient variation across the time series. This step and the next assume that non-expressed genes exhibit low variance and low absolute values [25]. The filtering process is as follows: initially, the three-dimensional tensor is decomposed into a collection of $|S|$ matrices of dimensions $\mathbb{R}^{|G| \times |T|}$, each representing a matrix of time series data. Subsequently, the standard deviation is calculated for each time series. If the standard deviation falls below a predetermined threshold, the probe is classified into a set of low-variance probes for the sample type s, denoted LV_s. The final set of probes to exclude can be defined using one of the following:

$$\text{Option } 1 \rightarrow Filter = \bigcup_{s \in S} LV_s \tag{1}$$

$$\text{Option } 2 \rightarrow Filter = \bigcap_{s \in S} LV_s \tag{2}$$

Here *Filter* is the set that contains all probes that should be removed from the dataset. It should be noted that using Eq. 1 is more restrictive than Eq. 2, since it eliminates all probes with a standard deviation below the threshold in at least one sample type, while Eq. 2 requires such a condition to hold across all sample types.

2. **Absolute value filter**: analogous to the previous step, the objective at this stage is to exclude probes that do not exhibit expression levels that exceed a specified threshold. The methodology remains largely consistent; however, there is an opportunity to adopt a different approach regarding the composition of low expression probes sets LV_s. Specifically, probes that have all expression values below the threshold could be included in LV_s, or alternatively include probes that have at least one expression value below the threshold. After determining LV_s for all $s \in S$, Eqs. 1 and 2 define the set *Filter*.

3. **Gene filter**: after discarding probes with low variance and low value the next objective is to convert the remaining probes into gene symbols or other identifiers that facilitate further data analysis. The GEO database permits efficient achievement of this goal through the utilization of Platform (GPL) files associated with each microarray experiment. Typically, these files provide various associations for each microarray probe, such as conversion to gene symbols, clone identifiers, GenBank accession numbers, etc. [11].
   However, the associations between probes and gene symbols are not always one-to-one, due to the inherent duplication of probes in different sections of the array during microarray experiments to reduce potential errors that could involve repeating the entire experiment. The subsequent step is to eliminate all probes that lack an associated gene symbol and merging those probes that correspond to the same gene. In our analysis, the mean is used as the merging function, although there is no universally accepted best practice for this choice. Some researchers choose to retain the probe with the highest overall values as the gene's representative, based on the assumption that this retains the best-expressed probe for that gene. In the proposed methodology, after the preceding preprocessing steps, all remaining probes are assumed to be relevant and to contribute equally to the final gene value.

3.2   Triclustering

Following preprocessing, the dataset was analyzed using the TriGen algorithm [13]. This algorithm utilizes an evolutionary heuristic framework analogous to traditional genetic algorithms and comprises the following procedures:

1. **Population initialization**: in the context of the algorithm, individuals represent potential solutions and are identified as triclusters $C \in \mathbb{R}^{|G'| \times |S'| \times |T'|}$, which consist of $|G'|$ genes, $|S'|$ samples, and $|T'|$ time points, as delineated by $G' \subseteq G, S' \subseteq S$ and $T' \subseteq T$.
   The initial population is generated at random. However, to prevent duplicate occurrences in the formation stage and to promote a thorough exploration of the dataset, previously used values are recorded and excluded during the creation of new individuals.

2. **Crossover**: two members of the population are selected randomly based on the probability of crossover, p_c, a hyperparameter within the model. Subsequently, these individuals are combined to produce two offspring individuals.
3. **Mutation**: each individual is subject to mutation according to the specified mutation probability p_m, another hyperparameter. Potential mutations include the incorporation of an additional gene, sample, or time period into the individual's configuration, or the possible elimination of any of these elements.
4. **Selection**: is implemented through roulette wheel selection, where the likelihood of each individual's survival is determined by its score in the fitness function. This approach gives greater chances of survival for solutions that produce favorable outcomes, while mitigating overfitting by providing opportunities for solutions with comparatively lower scores. The fitness function for the TriGen algorithm, known as TRIQ (TRIcluster Quality), was introduced in [14]. The underlying principle of this measure is that three criteria determine a tricluster's quality: correlation, graphical similarity, and the biological relevance of its elements. TRIQ integrates all these concepts into a singular measure by calculating Pearson and Spearman correlations, the Multi Slope (MSL) measure, as defined in [15], for graphical similarity, and a biological relevance measure that uses Ontologizer 2.0 [5]. This last measure leverages annotations within Gene Ontology (GO) [4,29] to assess the relevance of the genes present in individuals using over-representation analysis, a method commonly applied in gene expression data for statistical analysis of biological terms.

In conclusion, TriGen utilizes advanced cross-over and mutation methodologies specifically designed for time series analysis. These techniques are instrumental in maintaining the sequential integrity of the data while avoiding any insertion or deletion of time points that could potentially create discontinuities within the series.

4   Results and Discussion

This section presents the outcomes of applying the proposed triclustering framework to circadian transcriptomic data. First, the dataset used in the experimentation and the results of the preprocessing pipeline are described. Then, the configuration of the TriGen runs and the circadian gene content detected in the resulting triclusters are detailed. Finally, the results from experiments are summarized.

4.1   Dataset Description

The GSE8988 dataset [3], obtained from the Gene Expression Omnibus (GEO)[11] comprises 54 microarray gene expression samples from the livers of *Rattus norvegicus*. The samples were collected in non-uniform time intervals in a 12 hours light and 12 hours dark cycle, covering $|S| = 3$ experimental conditions corresponding to three biological replicates (*Rattus norvegicus* #1–#3) and $|T| = 18$ distinct, irregularly spaced time points throughout the circadian cycle.

The raw microarray platform contains 15923 probes, which after probe-to-gene mapping correspond to 11809 unique genes. A curated set of 75 circadian genes derived from the literature was collected from previous studies [27,30], including core clock components (*CLOCK, PER1-3, CRY1-2, ARNTL*, etc.), key transcriptional regulators (e.g., *RORA, NR1D1*), and additional modulators involved in circadian regulation and downstream processes. The curated set of 75 circadian genes collected from the literature is shown in Table 1.

Table 1. Curated set of 75 circadian-related genes from the literature, grouped by functional category.

Category	Genes
Core clock components	*CLOCK, PER1, PER2, PER3, ARNTL, ARNTL2, CRY1, CRY2, NPAS2*
Key transcriptional regulators	*RORA, RORB, RORC, NR1D1, NR1D2 BHLHE40, BHLHE41, CSNK1D, CSNK1E, PASD1, UBE3A, RAI1, TIMELESS*
Circadian clock modulators	*DBP, HLF, NFIL3, CIRBP, TEF, USP2, NOCT, NAMPT, FBXL3, HNRNPD, PML, KLF10, CAVIN3, AHR, PROX1, TOP1, MYBBP1A, HDAC3, CREB1, METTL3 MTA1, PIWIL2, NCOR1, EGR1, CCAR2, MAGED1, USP9X, GSK3B, CREM, KAT2B, NGFR, ZFHX3, HDAC1, HDAC2, KMT2A, HUWE1, METTL14, ATF4, PRKCG, EZH2, CREBBP, SIRT1, PRMT5, DDB1, HNRNPU, MAPK8, PRKDC, CUL4A, EP300, DHX9, MTOR, KAT5, TP53*

Following the preprocessing pipeline described in Sect. 3.1, the dataset was reduced from 11809 to 9446 genes, representing a 20.01% decrease in the total number of genes. This reduction demonstrates that the filtering process effectively removes low-information probes while preserving a substantial proportion of biologically relevant circadian genes, providing a cleaner and more informative dataset for subsequent triclustering analysis.

4.2 TriGen Configuration

All triclustering experiments were conducted using the TriGen algorithm with a consistent parameter configuration across runs. A total of 10 independent experiments were performed, each producing 20 final solutions. The minimum and maximum sizes of the triclusters were established at 15 and 200 genes, respectively, with a temporal dimension ranging from at least two time points to the complete set $|T|$ (18 points), and a condition dimension ranging from at least two conditions up to $|C|$ (3 replicates). The evolutionary process was configured with 200 generations and a population size of 100 individuals. Mutation, selection and crossover probabilities were established at 0.3, 0.2 and 0.2, respectively, with tournament selection as the survivor selection method.

4.3   Circadian Gene Content in Triclusters

The triclusters obtained from each experiment were compared against the 75-gene circadian reference set, which was subject to reduction during preprocessing. In a representative run, several triclusters contained multiple core clock components (e.g., *CLOCK, PER1, PER2*), along with key transcriptional regulators (*NR1D1, NR1D2, BHLHE40*) and modulators (*CSNK1D, PRKCG*) known to influence circadian regulation. The coexistence of core oscillator genes with regulatory kinases (*CSNK1D, PRKCG*) and epigenetic modifiers (*HDAC1, METTL3, EZH2*) suggests that TriGen captures not only the core transcription-translation feedback loop, but also the auxiliary regulatory layers that control its amplitude, phase, and tissue-specific signals. The detection of transcription factors such as *RORB* and *ATF4* within distinct triclusters supports the identification of modules that integrate circadian timing with metabolic and stress-response transcriptional programs in the liver. These findings illustrate the ability of TriGen to recover biologically coherent circadian patterns spanning different functional categories.

Table 2. Distinct circadian reference genes recovered in each TriGen experiment.

Experiment ID	Genes found
#0	17
#1	24
#2	11
#3	15
#4	22
#5	20
#6	13
#7	15
#8	16
#9	18

Across the 10 independent TriGen experiments, each generating 20 triclusters, the number of distinct circadian reference genes identified ranged from 11 (Experiment_#2) to 24 (Experiment_#1). The results are summarized in Table 2. From the filtered circadian reference set obtained after preprocessing, TriGen recovered 54 genes across all experiments, corresponding to 72% of the original curated list. Overall, the results show that TriGen consistently recovered a substantial fraction of circadian reference genes in all experiments, with some runs approaching one-third of the total reference set in a single analysis. The diversity of recovered genes suggests that triclustering can capture complementary circadian modules across runs, making it a promising approach to explore condition-specific temporal regulation.

4.4   Discussion

The discussion focused on Experiment #1, which was selected for having the highest number of distinct circadian reference genes recovered (24) among the 10 experiments performed. The quality of the triclusters in this experiment, as measured by the TRIQ metric, which combines correlation, graphical similarity, and biological relevance, was also notably high. The best solution reached a TRIQ score of 0.375, the lowest score among all solutions being 0.3568. The mean standard deviation of the TRIQ values across all solutions was 0.00893, indicating stable performance and suggesting that multiple solutions converged toward similar levels of structural and biological coherence.

To further characterize the temporal coverage of the triclusters obtained in Experiment #1, the circadian time points (in hours) associated with each detected pattern were analyzed. All triclusters correspond to measurements from both experimental conditions ($s = 1$ and $s = 2$), ensuring that temporal patterns are not biased by condition-specific effects. The number of time points per tricluster ranged from 3 (for example, #1) to 8 (for example, #13, #14, #18, #19, and #20). This indicates that while some triclusters capture narrow, phase-specific expression profiles, others span a broad range of circadian phases, potentially reflecting genes with extended or multi-phase rhythmicity. Table 3 summarizes the temporal composition of each tricluster, highlighting that the best-performing triclusters capture multiple and distributed phases throughout the circadian cycle. In particular, some triclusters, such as #18 and #20, encompass a wide range of time points, suggesting their potential relevance to identify genes with broad phase rhythmicity. In contrast, more compact triclusters, for example, #1 and #12, focus on specific circadian phases, which may correspond to specialized regulatory programs. This diversity in temporal coverage underscores the method's ability to detect broad and phase-specific gene expression patterns, aligning with the high TRIQ scores observed and strengthening the biological interpretability of solutions.

Functional enrichment analysis was then performed to evaluate the biological coherence of the identified modules, using the Term-for-Term approach with Bonferroni correction for multiple testing. When considering all experiments together, significant enrichment was observed in several functional categories that may be related to circadian regulation, including alternative splicing, epigenetic modifications, protein degradation via ubiquitination and the proteasome, hormonal signaling, and oxygen-sensing pathways mediated by HIF1α. These processes have been previously implicated in circadian biology, suggesting that the identified modules could reflect not only canonical molecular components of the clock but also broader regulatory layers with potential roles in modulating circadian dynamics.

Table 3. Time points (hours) for each detected tricluster in the circadian experiment. All triclusters correspond to condition IDs $s = 1$ and $s = 2$.

Triclusters	Time (h)
#1	1, 22, 23.75
#2–#11	0.25, 1, 22, 23.75
#12	0.25, 1, 18, 22, 23.75
#13–#14	0.25, 1, 8, 11.75, 22, 23, 23.75
#15	0.25, 1, 8, 13, 22, 23.75
#16	0.25, 1, 10, 18, 22, 23, 23.75
#17	0.25, 1, 16, 22, 23, 23.75
#18	0.25, 1, 2, 8, 11.75, 22, 23, 23.75
#19	0.25, 1, 8, 11.75, 22, 23, 23.75
#20	0.25, 1, 8, 10, 11.75, 22, 23, 23.75

In Experiment #1, a subset of triclusters also showed significant enrichment in categories such as ribosomal activity, ubiquitin ligase complexes, GTPase signaling, and metabolic enzymes. These categories are consistent with fundamental mechanisms thought to underlie circadian transcriptional programs, including protein synthesis, turnover, and energy metabolism. Other triclusters were associated with processes such as positive regulation of NF-κB signaling, positive regulation of the MAPK cascade, and the GABA receptor complex, which have been functionally connected to circadian regulation through their involvement in immune and neuronal pathways.

Case Example from Triclustering Analysis. To illustrate the interpretation of individual triclusters, tricluster #16 from Experiment_#1 is examined. This module contains one of the highest numbers of time points among the detected solutions and holds the largest number of circadian reference genes identified in a single tricluster within this experiment. Tricluster #16 exhibited a coherent temporal expression profile spanning seven circadian time points (0.25, 1, 10, 18, 22, 23, and 23.75 h). In particular, *KLF10*, *MTA1*, and *TP53* were identified, all belonging to the curated list of 75 circadian genes selected for this study. The presence of these well-established circadian regulators, together with other genes showing similar rhythmic dynamics, underscores the biological relevance of this module.

Figure 2 shows the expression levels (z-score normalized) of genes in tricluster #16 across circadian time points for both experimental conditions. Highlighted are *KLF10* (red), *MTA1* (blue), and *TP53* (green), all with documented circadian regulation. Other genes in this tricluster with reported rhythmic expression include *NDUFV2*, *MT3*, *MSH2*, and *NQO2*, while *NRAS* and *XIAP* are functionally relevant through their roles in cell signaling and apoptosis, respectively [7, 16].

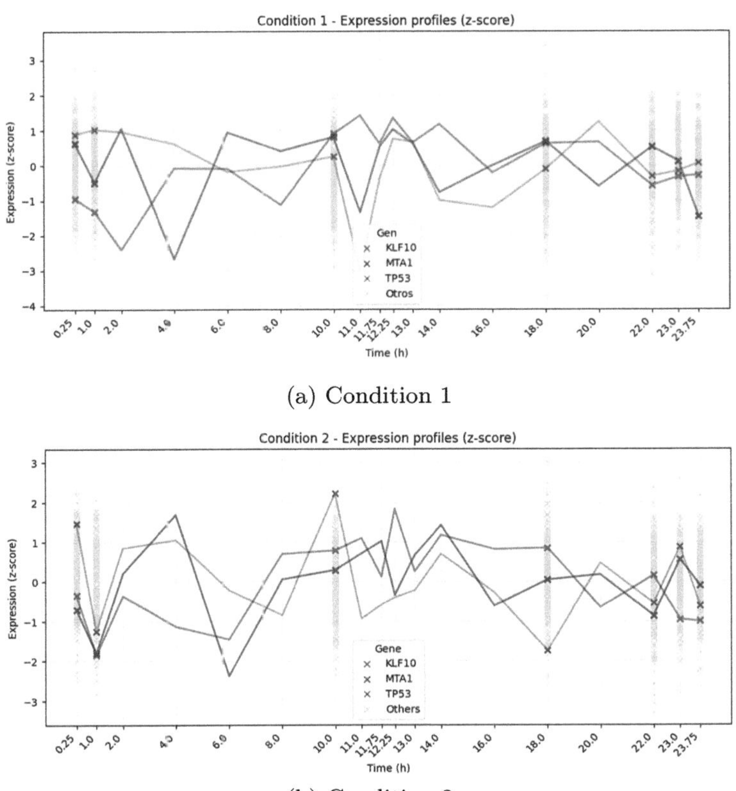

(a) Condition 1

(b) Condition 2

Fig. 2. Z-score normalized expression profiles of tricluster #16 across circadian time points in both conditions. Highlighted: *KLF10* (red), *MTA1* (blue), and *TP53* (green) from the curated list of 75 circadian genes. (Color figure online)

5 Conclusions and Future Works

TriGen enables the discovery of coherent gene–condition–time modules in circadian transcriptomic data, revealing relational patterns beyond single-gene rhythmicity and offering a framework for integrating multi-condition, time-ordered datasets. The method recovered up to seventy two percent of a curated set of literature-derived circadian genes across ten experiments, capturing both canonical core clock components and auxiliary regulators involved in metabolic, immune, and stress-response pathways. The diversity of enriched functions, including mitochondrial activity, transcriptional control, neuronal signaling, and immune defense, highlights the systemic integration of circadian regulation with multiple physiological processes. Two limitations should be noted, the use of Gene Ontology in the TRIQ metric may introduce circularity, since GO is also used in enrichment analyses, and the study relied solely on microarray data, whereas RNA-seq has become the standard for transcriptomic profiling. These aspects may affect the generalizability of the findings.

Future work will address these issues by applying alternative validation strategies and extending the methodology to RNA-seq and multi-omics datasets. In addition, the circadian reference set will be expanded (e.g., KEGG *circadian rhythm* pathway), advanced feature selection strategies will be explored, and computationally derived modules will be validated through targeted experiments. Beyond circadian biology, TriGen holds translational potential in areas such as personalized medicine and chronotherapy.

Acknowledgements. This publication is part of the projects PID2020-117954RB-C22 and PID2023-146037OB-C21, funded by MICIU/AEI/10.13039/501100011033/.

References

1. Agapito, G., Fedele, G.: Clustering methods for microarray data sets. In: Agapito, G. (ed.) Microarray Data Analysis, vol. 2401, pp. 249–261. Springer US, New York (2022). https://doi.org/10.1007/978-1-0716-1839-4_16
2. Alexandre, L., Henriques, R.: TriSig: significance testing of triclusters in multi-dimensional data. Bioinformatics **40**(4) (2024). https://doi.org/10.1093/bioinformatics/btae144
3. Almon, R.R., Yang, E., Lai, W., Androulakis, I.P., DuBois, D.C., Jusko, W.J.: Circadian variations in rat liver gene expression: relationships to drug actions. J. Pharmacol. Exp. Ther. **326**(3), 700–716 (2008). https://doi.org/10.1124/jpet.108.140186
4. Ashburner, M., et al.: Gene ontology: tool for the unification of biology. Nat. Genet. **25**(1), 25–29 (2000). https://doi.org/10.1038/75556
5. Bauer, S., Grossmann, S., Vingron, M., Robinson, P.N.: Ontologizer 2.0–a multifunctional tool for GO term enrichment analysis and data exploration. Bioinformatics **24**(14) (2008). https://doi.org/10.1093/bioinformatics/btn250
6. Castanho, E.N., Aidos, H., Madeira, S.C.: Biclustering data analysis: a comprehensive survey. Briefings Bioinf. **25**(4), bbae342 (2024). https://doi.org/10.1093/bib/bbae342
7. Chang, W.H., Liu, Y., Hammes, E.A., Bryant, K.L.: Oncogenic RAS promotes MYC protein stability by upregulating the expression of the inhibitor of apoptosis protein family member Survivin. J. Biol. Chem. (2023). https://www.jbc.org/article/S0021-9258(22)01285-6/fulltext
8. Chen, C.H., et al.: Effects of aging on circadian patterns of gene expression in the human brain. Proc. Natl. Acad. Sci. U.S.A. **113**(8), 2060–2065 (2016). https://doi.org/10.1073/pnas.1508249112
9. Cui, P., et al.: Identification of human circadian genes based on time course gene expression profiles by using a deep learning method. Biochimica et Biophysica Acta (BBA) - Mol. Basis Dis. **1864**(6, Part B), 2274–2283 (2018). https://doi.org/10.1016/j.bbadis.2017.12.004
10. Duan, J., et al.: tauFisher predicts circadian time from a single sample of bulk and single-cell pseudobulk transcriptomic data. Nat. Commun. **15**(1), 3840 (2024). https://doi.org/10.1038/s41467-024-48041-6

11. Edgar, R., Domrachev, M., Lash, A.E.: Gene expression omnibus: NCBI gene expression and hybridization array data repository. Nucleic Acids Res. 30(1), 207–210 (2002). https://doi.org/10.1093/nar/30.1.207

12. Fagiani, F., et al.: Molecular regulations of circadian rhythm and implications for physiology and diseases. Signal Transduct. Target. Ther. 7(1), 41 (2022) https://doi.org/10.1038/s41392-022-00899-y

13. Gutiérrez-Avilés, D., Rubio-Escudero, C., Martínez-Álvarez, F., Riquelme, J. TriGen: a genetic algorithm to mine triclusters in temporal gene expression data. Neurocomputing 132, 42–53 (2014). https://doi.org/10.1016/j.neucom.2013.03.061

14. Gutiérrez-Avilés, D , Giráldez, R., Gil-Cumbreras, F.J., Rubio-Escudero, C.: TRIQ: a new method to evaluate triclusters. BioData Mining 11(1), 15 (2018) https://doi.org/10.1186/s13040-018-0177-5

15. Gutiérrez-Avilés, D., Rubic-Escudero, C.: MSL: a measure to evaluate three-dimensional patterns in gene expression data. Evol. Bioinf. 11, EBO.S25822 (2015). https://doi.org/10.4137/EBO.S25822

16. Hanifeh, M., Ataei, F.: XIAP as a multifaceted molecule in cellular signaling. Apoptosis (2022). https://doi.org/10.1007/s10495-022-01734-z

17. Henriques, R., Madeira, S.C.: Triclustering algorithms for three-dimensional data analysis: a comprehensive survey. ACM Comput. Surv. 51(5), 1–43 (2019). https://doi.org/10.1145/3195833

18. Hughes, M.E., Hogenesch, J.B., Kornacker, K.: JTK_CYCLE: an efficient nonparametric algorithm for detecting rhythmic components in genome-scale data sets. J. Biol. Rhythms 25(5), 372–380 (2010). https://doi.org/10.1177/0748730410379711

19. Larriba, Y., Rueda, C., Fernández, M.A., Peddada, S.D.: Microarray Data Normalization and Robust Detection of Rhythmic Features, pp. 207–225. Springer New York, New York (2019). https://doi.org/10.1007/978-1-4939-9442-7_9

20. Lobo, J., Henriques, R., Madeira, S.C.: G-Tric: generating three-way synthetic datasets with triclustering solutions. BMC Bioinformatics 22(1), 16 (2021)

21. Luan, Y., Li, H.: Clustering of time-course gene expression data using a mixed-effects model with B-splines. Bioinformatics 19(4), 474–482 (2003). https://doi.org/10.1093/bioinformatics/btg014

22. Mahanta, P., Ahmed, H.A., Bhattacharyya, D.K., Kalita, J.K.: Triclustering in gene expression data analysis: a selected survey. In: 2011 2nd National Conference on Emerging Trends and Applications in Computer Science, pp. 1–6 (2011). https://doi.org/10.1109/NCETACS.2011.5751409

23. McDowell, I.C., Manandhar, D., Vockley, C.M., Schmid, A.K., Reddy, T.E., Engelhardt, B.E.: Clustering gene expression time series data using an infinite Gaussian process mixture model. PLOS Comput. Biol. 14(1), 1–27 (2018). https://doi.org/10.1371/journal.pcbi.1005896

24. Melkani, G.C.: Time-restricted feeding mediated synchronization of circadian rhythms to sustain cardiovascular health. J. Mol. Cell. Cardiol. 206, 1–10 (2025). https://doi.org/10.1016/j.yjmcc.2025.07.007, https://www.sciencedirect.com/science/article/pii/S002228282500118X

25. Mostavi, M., Chiu, Y.C., Huang, Y., Chen, Y.: Convolutional neural network models for cancer type prediction based on gene expression. BMC Med. Genomics 13(S5), 44 (2020). https://doi.org/10.1186/s12920-020-0677-2

26. Nepomuceno, J.A., Troncoso, A., Nepomuceno-Chamorro, I.A., Aguilar-Ruiz, J.S.: Integrating biological knowledge based on functional annotations for biclustering of gene expression data. Comput. Methods Programs Biomed. 119(3) 163–180 (2015). https://doi.org/10.1016/j.cmpb.2015.02.010

27. Patke, A., Young, M.W., Axelrod, S.: Molecular mechanisms and physiological importance of circadian rhythms. Nat. Rev. Mol. Cell Biol. **21**(2), 67–84 (2020). https://doi.org/10.1038/s41580-019-0179-2
28. Tchagang, A.B., et al.: Mining biological information from 3D short time-series gene expression data: the OPTricluster algorithm. BMC Bioinformatics **13**(1), 54 (2012). https://doi.org/10.1186/1471-2105-13-54
29. The gene ontology consortium and external contributors: the gene ontology knowledgebase in 2023. GENETICS **224**(1), iyad031 (2023). https://doi.org/10.1093/genetics/iyad031
30. Zhang, R., Lahens, N., Ballance, H., Hughes, M., Hogenesch, J.: A circadian gene expression atlas in mammals: implications for biology and medicine. Proc. Natl. Acad. Sci. U.S.A. **111**(45), 16219–16224 (2014). https://doi.org/10.1073/pnas.1408886111
31. Zhao, L., Zaki, M.J.: TRICLUSTER: an effective algorithm for mining coherent clusters in 3D microarray data. In: Proceedings of the 2005 ACM SIGMOD International Conference on Management of data, pp. 694–705. ACM, Baltimore Maryland (2005). https://doi.org/10.1145/1066157.1066236

Predictive Modelling and Personalised Medicine Using Artificial Intelligence

Characterising Continuous Glucose Monitoring Using Topological Data Analysis

Ana Lopez-Caballero[1] , Miguel A. Meroño[2] , and Jose M. Juarez[1,3]

1 MedAI Lab, University of Murcia, Murcia, Spain
{ana.lopezc2,jmjuarez}@um.es
2 Faculty of Mathematics, University of Murcia, Murcia, Spain
mamb@um.es
3 Murcian Institute of Biomedicine, Murcia, Spain

Abstract. Topological Data Analysis (TDA) has increasingly gained recognition as a powerful and versatile mathematical framework aimed at extracting high-level structural features from complex datasets. Among the various tools that TDA offers, persistent homology has emerged as one of the most fundamental and widely adopted. However, clinical information is often presented in the form of time series. The present work focuses on applying TDA to data derived from continuous glucose monitoring (CGM) devices worn by diabetic patients. The study explores the potential of using persistent homology to transform CGM time series into meaningful topological summaries that may aid in distinguishing between patients diagnosed with Type 1 diabetes and those with Type 2. The experimental findings indicate that the exclusive reliance on topological information can discriminate between type 1 and type 2 diabetes patients. Nonetheless, the study is preliminary and limited to basic time series classification techniques.

Keywords: Topological Data Analysis · Persistent Homology · Diabetes · Time Series · Machine Learning · Continuous Glucose Monitoring

1 Introduction

In medical research, time series are omnipresent, especially in continuous monitoring systems such as ECG or EEG. These signals pose several challenges, such as that they are high-frequency, can be noisy, and encode dynamic behaviours that are not always detectable by traditional statistical techniques [4]. In the specific context of diabetes management, continuous glucose monitoring (CGM) devices collect glucose values at regular intervals over days, generating large volumes of patient-specific temporal data. However, classical methods struggle to extract interpretable and discriminative patterns from such data. In that context, there is a need for methods capable of capturing intrinsic structure and temporal dynamics in a way that can enhance classification performance.

A. López Fernández et al. (Eds.): CIABiomed 2025, LNBI 16148, pp. 557–569, 2026.
https://doi.org/10.1007/978-3-032-10361-2_41

Topological Data Analysis (TDA) is a recent but increasingly influential field that extends techniques from algebraic topology to extract robust, global features from complex data [6]. It is based on the fact that high-dimensional, noisy, or non-linear datasets often possess intrinsic geometric structures that can be effectively captured using topological summaries. One of its most powerful tools is persistent homology [5], which identifies topological features such as connected components, loops, and voids across multiple scales, providing a multi-scale description of data "shape". This approach offers robustness to noise, invariance under smooth transformations, and interpretability, making it ideal for biomedical signal analysis [3, 10].

In this work, we propose a pipeline that integrates TDA and machine learning to classify diabetic patients based on CGM time series. Our approach begins with Takens embedding to transform the underlying dynamics of each series into point clouds. We then apply persistent homology to extract topological features from said clouds and then transform them into vectorial and image-based representations. We tested our approach with 116 CGM samples of patients with diabetes type 1 and type 2. The TDA-extracted features are used as input for classifiers including Logistic Regression, Random Forest, KNN, Naive Bayes and Convolutional Neural Networks (CNNs). We evaluate and compare different combinations of representations, oversampling strategies and classifiers.

2   Background in Topological Data Analysis

The central premise of TDA is that the underlying shape of a dataset contains crucial, often hidden, information about the system it represents. TDA provides a set of tools to quantify this shape and extract robust, multi-scale features that are invariant to continuous deformations and resistant to noise, making them ideal for machine learning applications [1, 2] .

In our approach, CGM time series are first transformed into high-dimensional point clouds that capture the dynamics of the original signal. Their complexity makes direct visual inspection or traditional analysis challenging. This is where TDA provides a powerful solution. This section will intuitively introduce the core TDA concepts that allow us to transform each point cloud into a vectorized topological signature, such as a *persistence image* or a *heat kernel*.

2.1   Simplicial Complexes: Building a Connective Skeleton

Given a point cloud X in $\mathbb{R}^n$, TDA helps to infer its global shape from this discrete sample. We cannot simply "see" the shape; we need a method to connect the points coherently. This is where *simplicial complexes* come in.

A simplicial complex is a structure that generalizes the notion of a network or graph. It is built from basic blocks called *simplices*:

- A 0-simplex is a vertex (a point in our cloud).
- A 1-simplex is an edge connecting two vertices.

- A 2-simplex is a filled triangle connecting three vertices.
- A 3-simplex is a filled tetrahedron connecting four vertices.
- And so on for higher dimensions.

The *faces* of a simplex are the simplices formed by any non-empty subset of its vertices. For example, a 2-simplex (a triangle) has three 1-dimensional faces (its edges) and three 0-dimensional faces (its vertices). A k-simplex has exactly $2^{k+1} - 2$ non-empty proper faces, corresponding to all subsets of its vertex set of size 1 to k.

Formally, a *simplicial complex* K is a finite collection of simplices such that:

1. Any face of a simplex in K is also in K.
2. The intersection of any two simplices in K is either empty or a face common to both.

The topological features of a simplicial complex K are described by its *homology groups* $H_k(K)$, $k = 0, 1, 2, \ldots$, which intuitively count k-dimensional holes:

- $H_0(K)$ counts connected components: Clusters of points or "islands". Its rank, the Betti number β_0, is the number of components.
- $H_1(K)$ counts loops or cycles: Ring-like structures or "tunnels". Its rank, β_1, is the number of such loops.
- $H_2(K)$ counts voids or cavities: Hollow three-dimensional spaces or "bubbles". Its rank is denoted by β_2.
- Higher-dimensional homology groups represent more complex topological structures, with rank β_3, β_4, *etc.*

These features are invariant under continuous deformations and provide a robust summary of the shape of K.

Now, to build a "skeleton" from our point cloud, we use a proximity criterion. A scale parameter $\varepsilon > 0$ is introduced. The idea is simple: if points are "close", we connect them. There are several approaches to compute this proximity (e.g. Čech or Vierori-Rips complexes). For example, using the *Čech complex*, the rule is more geometric: we imagine placing a ball of radius ε around each point. A k-simplex is then formed among a set of $k + 1$ vertices if these balls have a non-empty common intersection. Thus, we obtain a simplicial complex K_ε for each $\varepsilon > 0$ (Fig. 1).

2.2 Persistent Homology: Discovering the Shape Across Scales

The main problem with the previous approach is the choice of ε. A small value will only show isolated points, while a large one will connect everything into a featureless blob. What is the "correct" ε? The TDA answer is revolutionary: we don't choose one, we consider them all.

This is the core of *persistent homology*. Instead of building a single simplicial complex, we construct a nested sequence of complexes, called a *filtration*, by gradually choosing an increasing value of ϵ from zero [5].

$$K_{\varepsilon_1} \subseteq K_{\varepsilon_2} \subseteq \cdots \subseteq K_{\varepsilon_n},$$

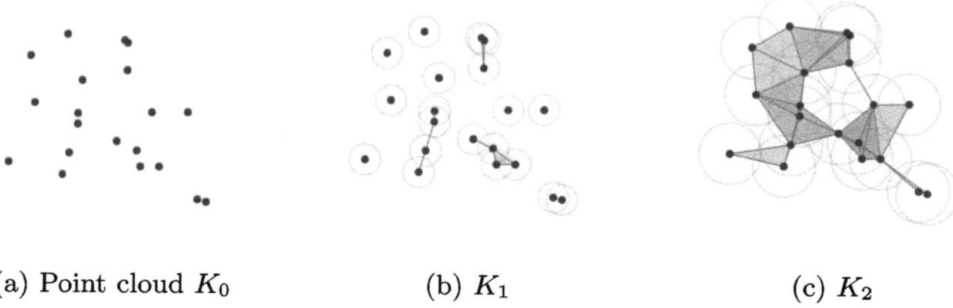

(a) Point cloud K_0 (b) K_1 (c) K_2

Fig. 1. Construction of some Čech complexes. As the scale parameter ε increases, more complex topological structures are revealed.

where $\varepsilon_1 < \varepsilon_2 < \cdots < \varepsilon_n$. As ε grows, we observe the topological features that appear and disappear.

Persistent homology records the ε_{birth} value at which a topological feature is "born" (first appears) and the value ε_{death} at which it "dies" (gets filled in). The *persistence* is the difference between death and birth

$$persistence = \varepsilon_{death} - \varepsilon_{birth}.$$

The intuition is that features with a long persistence represent the true structure of the data, while those with short persistence are likely artifacts or noise (Fig. 2).

This result is visualized in two main ways:

1. *Barcodes:* Each topological feature is represented as a horizontal bar. The start of the bar is its birth ε and the end is its death ε. Long bars correspond to persistent features.
2. *Persistence Diagrams (PD):* This is a scatter plot where each feature is a point (b, d), with b as the birth ε and d as the death ε. Points far from the diagonal $y = x$ are the most persistent.

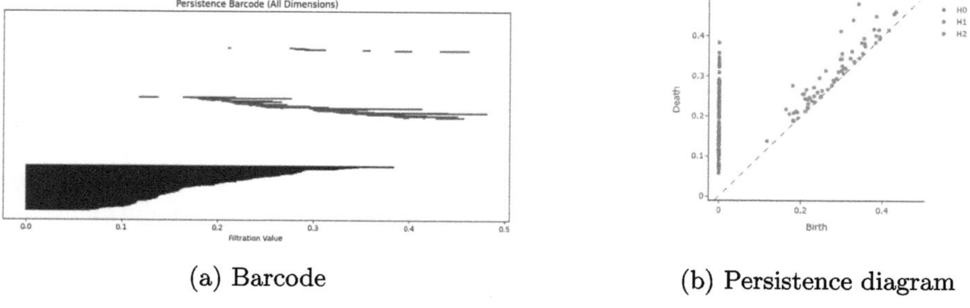

(a) Barcode   (b) Persistence diagram

Fig. 2. Representations of persistent homology.

2.3   Creating a Stable Vectorized Signature for Machine Learning

Barcodes and persistence diagrams are powerful topological descriptors, but they are not directly compatible with most machine learning algorithms, which require fixed-size feature vectors. We need a final step to convert this topological information into a stable vector representation.

Persistence Images (PIs). One of the most successful methods is the Persistence Image. The process involves several steps:

1. The coordinates of the persistence diagram are transformed from $(\varepsilon_{birth}, \varepsilon_{death})$ to $(\varepsilon_{birth}, persistence)$.
2. A 2D Gaussian distribution (a "blob"), weighted by the persistence of the feature, is centered on each point in the transformed diagram.
3. These distributions are summed to create a continuous 2D surface, called a persistence surface $\rho(u, v)$.
4. This surface is then discretized into a grid of pixels, yielding a fixed-size matrix or image. The value of each pixel is the integral of the surface over that pixel's area.

The resulting image is a stable representation that captures the most prominent features of the data's topology and can be readily used to train classifiers, regressors, or other machine learning models.

Heat Kernels. This method transforms the persistence diagram into a continuous surface, like a heat map [7]. The idea is to treat each persistence point (b, d) as a point source of heat. This heat then diffuses across the (u, v) plane, and the intensity of the heat at any location on the map gives us the value for our representation. Mathematically, this diffusion is modelled by the following formula, which calculates the "heat intensity" $\psi(u, v)$ at each point on the map:

$$\psi(u, v) = \sum_{(b,d) \in PD} \frac{1}{2\pi\sigma^2} \left(e^{-\frac{(u-b)^2 + (v-d)^2}{2\sigma^2}} - e^{-\frac{(u-d)^2 + (v-b)^2}{2\sigma^2}} \right)$$

Here, the first term, $e^{-\frac{(u-b)^2 + (v-d)^2}{2\sigma^2}}$, is a standard Gaussian kernel centred at the persistence point (b, d). This is the term that "spreads" the heat. The parameter σ acts as a bandwidth that controls how much diffusion or blur is applied. A larger σ creates a smoother, more diffuse image.

The elegance of the method lies in the second term, $-e^{-\frac{(u-d)^2 + (v-b)^2}{2\sigma^2}}$. This term introduces a "negative heat source" at the point's symmetric position across the diagonal, (d, b). The effect of this subtraction is profound: for points very close to the diagonal (i.e., noise where $b \approx d$), the positive and negative heat sources are very close to each other and their effects cancel out. This makes the representation inherently stable and robust, naturally "erasing" topological features that are likely noise.

Finally, just as with PI, this continuous heat map is discretized onto a pixel grid to generate a fixed-size image, ready for use in machine learning algorithms (Fig. 3).

(a) Persistence Image   (b) Heat Kernel

Fig. 3. Stable Vector Representations for Persistent Homology.

3   Proposal

In this work we propose a methodological pipeline designed to analyze and classify CGM time series through topological descriptors. The process consists of four main stages: (1) transforming time series into point clouds using Takens embedding, (2) extracting persistent diagrams via topological filtrations, (3) converting these diagrams into numerical or image-based representations, and (4) feeding these representations into classification models, both classical and deep learning.

3.1   Step 1: From Time Series to Point Clouds Takens Embedding

The first step in the pipeline is to transform a time series

$$X = (x_1, x_2, \ldots, x_L)$$

into a higher-dimensional space.

Among the various methods available to transform time series into point clouds, such as raw sliding windows, kernel embeddings, or recurrence plots. Takens embedding stands out for its strong theoretical foundation, robustness to noise, and ability to preserve both temporal structure and geometry. In Table 1 we compare the main alternatives, indentifying the most suitable for TDA.

Takens embedding method [8] reconstructs the system's state space by creating vectors of delayed observations:

$$X_t = (x_t, x_{t+\tau}, x_{t+2\tau}, \ldots, x_{t+(d-1)\tau}) \in \mathbb{R}^d$$

Table 1. Comparison of time series to point cloud methods used in TDA

Method	Theoretical basis	Temporal structure	Parameter tuning	Preserves geometry	Lag/Window required	Configurable dimension	Sliding window	Multivariate support	Noise robustness	Use in TDA
Takens embedding	✓	✓	✓	✓	✓	✓	✓	✗	High	High
Sliding window (raw)	✗	✓	✓	✓	✓	✓	✓	✓	Medium	High
Kernel delay embedding	✓	✓	✓	✓	✓	✓	✓	✗	High	Medium
Recurrence plots	✓	✓	✓	✓	✗	✓	✗	✓	High	Medium

where d is the embedding dimension and τ is the time delay. This transformation allows for recovery of the dynamics and geometry underlying the glucose measurements.

We explore the following parameter tuning strategies for d and τ:

- **Global optimization**, where the same (d, τ) pair is used for all series.
- **Individual optimization**, where each series is embedded using its own optimal combination of parameters.

In both cases, the goal is to preserve the dynamic structure of the signal in the resulting point cloud. To select the optimal parameters, we follow a topologically motivated criterion: for each candidate configuration, we compute the corresponding persistence diagrams (in dimensions 0 and 1) for all series, and evaluate their pairwise similarity using the **Wasserstein distance** [7].

3.2 Step 2: from Point Clouds to Persistence Diagrams

Once each time series has been embedded as a point cloud in $\mathbb{R}^d$, we extract its topological features using persistent homology. This involves constructing a sequence of simplicial complexes that capture the shape of the point cloud at different spatial resolutions. We use Vietoris–Rips complexes to construct a filtration, and with it we build a persistence diagram, that summarizes the multi-scale topological structure of the point cloud, capturing both short-lived noise (close to the diagonal $b \approx d$) and persistent features (points farther away), and showing them in a clear and simple way.

3.3 Step 3: Representations of Persistence

Persistence diagrams are powerful but not directly usable by machine learning models, which require fixed-size vector inputs. We therefore transform each diagram into structured representations that vary in complexity and expressivity:

- **Persistence Entropy**: This numerical feature aggregates the information in a diagram into a single scalar per dimension, based on the normalized lifespans of features. It is fast to compute and easy to interpret, but loses spatial and geometric detail.
- **Persistence Images**: Each diagram point is convolved with a Gaussian kernel and discretized onto a regular grid, preserving the spatial layout of features. This representation balances interpretability, stability, and expressiveness, and is suitable for CNN-based models.
- **Heat Kernel Signatures**: This technique simulates heat diffusion over the diagram, yielding smooth and continuous images that emphasize the global structure. It captures long-range interactions and provides robustness to noise.

Each representation offers a different trade-off between simplicity, expressiveness, and computational cost. Combining all three enables us to explore the performance of both classical and deep learning classifiers and assess how feature richness impacts model accuracy.

3.4 Step 4: Construction and Evaluation of Classification Models

The last step of our proposal is to build classification models from the extracted information (fixed-size vectors) obtained from the previous step. The performance of these models are valuated using the state of the art metrics of the field.

4 Experiments and Results

The goal of the experiments carried out is to explore the capacity of our TDA approach in distinguishing patients with diabetes type 1 and diabetes type 2 from CGM time series.

We conduct a comprehensive evaluation involving different combinations of embedding strategies (global vs. individual Takens parameters), topological encodings (persistence entropy, persistence images, and heat kernels), and over-sampling techniques (random duplication, SMOTE, and none). A total of 24 distinct experimental settings are explored, where each configuration is evaluated in terms of classification performance using appropriate models.

4.1 Diabetes Dataset

The dataset consists of CGM readings from the DiaDRIL project [11], collected in Shanghai hospitals between 2019 and 2021. It contains 125 glucose series: 16 from 12 Type 1 diabetes (T1DM) patients and 109 from 100 Type 2 diabetes (T2DM) patients. Each series records glucose values every 15 min, over periods ranging from 3 to 14 days. After removing missing values via linear interpolation and normalizing all series, the dataset is split into training and test sets with class stratification maintained.

Note that, to address the imbalance between T1DM and T2DM cases, we test oversampling techniques (random duplication and SMOTE) within the training folds, ensuring that classifiers do not trivially favor the majority class and are evaluated under fair conditions.

4.2   Experimental Setup

The goal of the experiments is two-folded. First we evaluate how different TDA representations affect classification performance in the context of CGM data. Specifically, comparing vector-based (persistence entropy) and image-based (persistence images and heat kernels) representations. Second we analyse to what extent the use of topological information to build a classifier outperforms baseline methods using time series information directly.

In the individual case we use a Python library to get the optimal parameters automatically. In the global case, we search over a grid of values (d, τ) (constructed using the individual optimums) and select the pair that minimizes the "average intra-class distance" while maximizing "inter-class separation" in persistence diagram space. This helps ensure that patients of the same diabetes type produce topologically similar embeddings, enhancing class separability downstream.

This analysis led to the selection of $(d = 5, \tau = 14)$ as the optimal global configuration, which balances both persistence and stability. This pair is then used consistently for all patients in the corresponding global experiments.

Experiments were conducted on a personal computer running Windows 11, equipped with an AMD Ryzen 7 5800H processor with Radeon Graphics, 16 GB of RAM, Giotto-TDA library[1] [9] and Python 3.11.9.

4.3   Classification Models

For our experiments, we will consider the following classification models:

- **Classical models** (Logistic Regression, Random Forest, K-Nearest Neighbors, and Naive Bayes) in the case of persistence entropy vectors. These models are lightweight and interpretable but depend heavily on the quality of input features.
- **Convolutional Neural Networks (CNNs)** for image-based representations (persistence images and heat kernels).

Classical classifiers used for entropy vectors (Logistic Regression, Random Forest, KNN, Naive Bayes), are evaluated using 4-fold stratified cross-validation. Hyperparameter tuning is carried out within each fold using grid search.

The CNN designed is implemented in PyTorch. The architecture includes two convolutional layers (with 16 and 32 filters, respectively), each followed by ReLU activation and max-pooling. A dropout layer with a rate of 0.25 is applied before two fully connected layers, ending with a softmax output. The model is trained

---

[1] Giotto-TDA library, https://giotto-ai.github.io/gtda-docs/.

for 20 epochs using early stopping, with the Adam optimizer and cross-entropy loss.

Time series transformation were implemented using sktime python library.

We report five metrics on the test sets: accuracy, precision, recall, F1-score, and AUC-ROC.

4.4   Experiment 1: Effects of TDA Representations

Table 2 summarises the main results of the experiments. The table shows a comparison all configurations (Takens embedding type, topological representation, and oversampling strategy), including only the classification models (from Sect. 4.3) which show the best performance. In bold we highlight the two most promising models due to the suspect of overfitting for CNN+Random.

Table 2. Comparison of strategies: performance of the best classification models for each TDA pipeline configuration.

Takens	Representation	Best model	Oversampling	Accuracy	Precision	Recall	F1-score
Global	Pers. Entropy	LogReg=KNN	Random	0.839	0.420	0.5	0.456
		Random Forest	SMOTE	0.85	0.622	0.627	0.624
		Random Forest	None	0.88	0.544	0.54	0.534
	Pers. Images	CNN	Random	0.84	0.92	0.84	0.86
		CNN	None	0.84	0.71	0.84	0.77
	Heat Kernels	CNN	Random	0.84	0.88	0.84	0.85
		CNN	None	0.72	0.75	0.72	0.73
Individual	Pers. Entropy	Random Forest	Random	0.82	0.532	0.537	0.530
		LogReg	SMOTE	0.88	0.44	0.50	0.468
		Random Forest	None	0.88	0.697	0.572	0.592
	Pers. Images	CNN	Random	0.80	0.86	0.80	0.82
		CNN	None	0.84	0.71	0.84	0.77
	Heat Kernels	**CNN**	**Random**	**1.00**	**1.00**	**1.00**	**1.00**
		CNN	**None**	**0.88**	**0.90**	**0.88**	**0.85**

4.5   Experiment 2: Baseline Analysis

In this second experiment we analyse and compare the performance of basic strategies for building classifiers from time series and the use of topological information extracted from time series.

In particular we considered to baseline strategies as gold standard from the literature:

- The use of time series distance functions allows for the measurement of similarities between time series. In particular, we adopted the K-Nearest Neighbors (KNN) algorithm using Dynamic Time Warping (DTW) as the distance function. Since the time series have different lengths, this method sets the length of all series to match that of the shortest one.
- Tabularization of time series and use traditional machine learning classification methods.

Table 3 shows the performance of the methods considered.

Table 3. Performance of base line methods.

Method	Accuracy	Precision	Recall	F1-score	AUC
KNN+DTW	0.880	0.776	0.557	0.57	0.55
LogReg+Tabularizer	0.82	0.588	0.579	0.583	0.57
Random Forest+Tabularizer	**0.88**	**0.821**	**0.589**	**0.619**	**0.58**
NaiveBayes+Tabularizer	0.77	0.601	0.658	0.613	0.65
CNN+Tabularizer	0.83	0.433	0.477	0.454	0.47

4.6 Discussion

According to experiment 1, persistence entropy achieved modest results in all conditions, with accuracy values around 84%, but limited precision and recall, particularly for the minority class, even with oversampling methods.

Our results show that persistence images produced more robust and discriminative representations, especially when used with random oversampling. Accuracy remained high (84%) practically in all cases, with notable gains in precision (up to 0.92) and F1-score (0.86). Heat kernels improved performance even further, especially when combined with individually optimized Takens parameters and random oversampling. In that setting, the CNN achieved perfect classification results (1.00 across all metrics), indicating strong topological separability between the two classes.

Overall, image-based descriptors clearly outperformed entropy-based features. Among them, heat kernels proved to be the most stable and expressive one, capturing more global structure in the diagrams. Finally, while individual Takens parameters led to slight improvements, global parameters were generally sufficient and more efficient to compute.

The results shown in Experiment 2 indicate that the KNN+DTW and tabularized time series with RandomForest methods achieve the best results, despite being significantly different approaches. Overall, Experiments 1 and 2 indicate that the exclusive use of topological information is able to differentiate between type 1 and type 2 diabetes patients as well as or better than a traditional method. However, the study conducted is very preliminary and limited to the most basic time series methods for classification problems.

5 Conclusions

This study set out to evaluate the effectiveness of TDA in classifying biomedical time series, specifically CGM signals from patients with Type 1 and Type 2 diabetes. To this end, we proposed a complete pipeline integrating Takens embedding, persistent homology, different topological representations, and multiple classification models. All objectives were successfully achieved.

The results indicate that TDA offers a robust and flexible approach for extracting meaningful structural features from time series, especially in the presence of noise and nonlinear dynamics. Persistence images and heat kernels yielded the best performance, enabling convolutional neural networks to achieve accurate and stable classifications. While persistence entropy was faster to compute and easier to interpret, it lacked the expressiveness of the image-based descriptors. Moreover, dealing with class imbalance and selecting suitable embedding parameters required careful design choices throughout the pipeline.

TDA applied to CGM data facilitates the extraction of higher-order structures in glycemic variability. From a clinical perspective we believe it facilitates precise patient stratification and individualized therapeutic strategies while providing objective, data-driven characterization of metabolic dynamics. By detecting early patterns of dysregulation, TDA may improve prognostic accuracy and support long-term risk reduction for diabetes-related complications.

Future work will focus on validating the proposed approach on larger, more diverse and more balanced datasets. In addition, further efforts will be directed toward evaluating the clinical interpretability of the topological descriptors and exploring their integration with existing clinical features, rather than using them in isolation. Finally, other representations such as persistence landscapes and topological signatures from higher-dimensional homology might be investigated to enrich the set of descriptors and potentially enhance model performance.

This work contributes to the growing body of research demonstrating the value of topological methods in biomedical data analysis, reinforcing their role as a promising and interpretable complement to traditional signal processing techniques.

Acknowledgements. This work was partially funded by the CONFAINCE project (Ref: PID2021-122194OB-I00) by MCIN/AEI/10.13039/501100011033 and by "ERDF A way of making Europe", by the "European Union".

This research is also part of IMPACT-T2D-UMU project (ref. PMP21/00092), funded by the Institute of Health Carlos III (ISCII) for Personalized Precision Medicine Research Projects of the Strategic Action in Health 2017-2020, supported by the European funds of the Recovery, Transformation and Resilience Plan (Next Generation EU programme).

This research is also partially funded by the project PID2023-151075OA-100 by MICIU/AEI /10.13039/501100011033 and FEDER, UE.

References

1. Chazal, F., Michel, B.: An introduction to topological data analysis: fundamental and practical aspects for data scientists. Front. Artif. Intell. **4** (2021). https://doi.org/10.3389/frai.2021.667963
2. Edelsbrunner, H., Harer, J.: Computational Topology: An Introduction Applied Mathematics, American Mathematical Society (2010). https://books.google.es/books?id=MDXa6gFRZuIC
3. Garland, J., Bradley, E.. Meiss, J.D.: Exploring the topology of dynamical reconstructions. Physica D: Nonlinear Phenomena **334**, 49–59 (2016). https://doi.org/10.1016/j.physd.2016.03.006. topology in Dynamics, Differential Equations, and Data
4. Kantz, H., Schreiber, T.: Nonlinear Time Series Analysis, Cambridge Nonlinear Science Series. Cambridge University Press (2004)
5. Myers, A., Munch, E., Khasawneh, F.A.: Persistent homology of complex networks for dynamic state detection. Phys. Rev. E **100**(2) (2019). https://doi.org/10.1103/physreve.100.022314
6. Perea, J.A.: Topological time series analysis (2018). https://arxiv.org/abs/1812.05143
7. Reininghaus, J., Huber. S.. Bauer, U., Kwitt, R.: A stable multi-scale kernel for topological machine learning. In: 2015 IEEE Conference on Computer Vision and Pattern Recognition (CVPR), pp. 4741–4748 (2015). https://doi.org/10.1109/CVPR.2015.7299106
8. Takens, F.: Detecting strange attractors in turbulence. In: Rand, D., Young, L.S. (eds.) Dynamical Systems and Turbulence, Warwick 1980, pp. 366–381. Springer Berlin Heidelberg, Berlin, Heidelberg (1981)
9. Tauzin, G., et al.: giotto-tda: : a topological data analysis toolkit for machine learning and data exploration. J. Mach. Learn. Res. **22**(39), 1–6 (2021). http://jmlr.org/papers/v22/20-325.html
10. Tralie, C., Perea, J.: (quasi)periodicity quantification in video data, using topology. SIAM J. Imaging Sci. **11** (2017). https://doi.org/10.1137/17M1150736
11. Zhao, Q., et al.: Chinese diabetes datasets for data-driven machine learning. Sci. Data **10**, 35 (2023). https://doi.org/10.1038/s41597-023-01940-7

Exploratory Computational Phenotyping of Hyposalivation Etiologies in Women

Kristina Lacasta[1] , María J. Rus[1] , Angela de la Cruz Gándara Alvarez[1],
Cristiane Cantiga-Silva[1] , Virginia Moreira Navarrete[2] ,
Carmen Dominguez Quesada[2], Jose Javier Perez Venegas[2] ,
Juan Antonio Ortega[3] , and Aurea Simon-Soro[1,4(✉)]

[1] Department of Stomatology, Dental School, University of Seville,
41009 Seville, Spain
[2] Rheumatology Department, Virgen Macarena University Hospital (HUVM),
41009 Seville, Spain
[3] Computer Science Department, University of Seville,
41012 Seville, Spain
[4] Andalusian Center for Molecular Biology and Regenerative Medicine (CABIMER),
Sevilla, Spain
asimon@us.es

Abstract. Hyposalivation in women presents diagnostic challenges due
to overlapping etiologies including medications, hormonal changes and
autoimmune conditions. This exploratory study investigates whether
computational approaches can identify distinct hyposalivation phenotypes in women. We analyzed 196 women across 40 clinical variables
including demographics, medications, dental health, and salivary flow
measurements using unsupervised machine learning. K-means clustering
with PCA preprocessing tentatively identified three clusters: medication
and lifestyle-associated hyposalivation (n=141, 72%), optimal salivary
function (n=18, 9%), and severe hyposalivation (n=37, 19%). Salivary
flow parameters and dental health variables were primary cluster drivers.
Results provide preliminary evidence that hyposalivation may comprise
distinct phenotypes, offering proof-of-concept for computational phenotyping approaches that warrant validation in independent cohorts.

Keywords: Hyposalivation · Clustering · Sjögren's disease · Machine
learning

1 Introduction

Saliva is a complex biofluid with key roles in oral homeostasis and systemic
health [1]. Beyond its mechanical involvement in mastication, deglutition, and
phonation, saliva harbors immune and metabolic components that contribute to
host defense and microbial modulation [2,3]. A daily salivary output of 1.5–2
liters is considered physiological [4,5], yet this dynamic secretion is susceptible
to multiple modifiers throughout the lifespan.

A. López Fernández et al. (Eds.): CIABiomed 2025, LNBI 16148, pp. 570–584, 2026.
https://doi.org/10.1007/978-3-032-10661-2_42

Hyposalivation, defined as a persistent reduction in salivary flow, can emerge from pharmacological, physiological, or pathological origins. Common etiologies include polypharmacy, aging, and systemic autoimmune disorders such as Sjögren's disease [6,7]. The functional impact of reduced salivary output extends beyond xerostomia, encompassing mucosal alterations, dental caries, and prosthetic instability, ultimately compromising the oral and systemic quality of life.

Interpreting hyposalivation is clinically challenging. The difficulty increases in mid-life women where hormonal decline, medication use, and immune shifts often overlap [8]. These factors blur diagnostic boundaries and delay recognition of underlying conditions. Sjögren's disease exemplifies this diagnostic ambiguity, a female-predominant (9:1) disease often presenting with dry mouth decades before systemic manifestations arise [9]. In this context, early detection is often hindered by attribution of symptoms to aging or medication side effects.

Unsupervised machine learning methods represent a promising analytical tool for addressing the diagnostic complexity of hyposalivation in women. These computational techniques can potentially identify latent patterns and hidden structures in high-dimensional clinical data, facilitating the preliminary identification of distinct phenotypes that may not be evident through conventional clinical analysis [10,44,45]. This computational approach offers exploratory opportunities to improve understanding of multifactorial conditions such as hyposalivation, potentially contributing to the development of more accurate diagnostic strategies and personalized therapeutic approaches in the field of oral medicine [46,47].

Here, we explore whether hyposalivation in women represents a stable phenotype or a composite endpoint resulting from distinct etiologies. To address this research question, we conducted an integrative clinical and computational assessment combining standardized oral examinations, biomarker profiling, and unsupervised machine learning [11,12]. This exploratory approach allowed the tentative identification of distinct hyposalivation clusters based on clinical and biological data, providing initial insights into the latent structure underlying this multifactorial condition.

2   Materials and Methods

2.1   Study Design

A cross-sectional study was conducted including 196 women recruited into two cohorts: patients diagnosed with Sjögren's disease according to 2016 ACR/EULAR criteria [13], and menopausal women ($\geq$40 years) without the disease. Exclusion criteria included recent antibiotic, antifungal, or chlorhexidine use; fixed orthodontic appliances; and systemic conditions impairing autonomous participation. Informed consent was obtained from all participants. The study was approved by the ethics committees of CEI Virgen Macarena–Virgen del Rocío (0501-N-22) and CEI Sevilla Sur (1532-N-21).

Two major determinants of salivary flow reduction are aging and pathological conditions affecting the salivary glands. Among these, Sjögren's Syndrome (SS) represents one of the most prevalent autoimmune disorders in developed countries, characterized by chronic hyposalivation and progressive epithelial destruction of the salivary glands. Patients with SS exhibit reduced salivary secretion accompanied by alterations in the oral microbiota, both in saliva and on dental surfaces, when compared with healthy individuals. The disease shows a strong gender bias, with a female-to-male ratio of approximately 10:1, and is most frequently diagnosed between 30 and 50 years of age. This gender-related predisposition aligns with menopausal transition, a stage also associated with hyposalivation in women. Menopause is marked by the cessation of ovarian endocrine activity and a consequent decline in circulating estrogen levels, leading to systemic physiological changes. These overlapping factors justify the selection of cohorts with SS and menopause-associated hyposalivation for comparative analysis [48,49].

2.2   Sample Collection

The whole saliva was collected under resting and stimulated conditions [14]. Unstimulated saliva was obtained via passive drooling for 5 min; stimulated saliva followed 5 min of paraffin gum chewing. Samples were collected in sterile 50 ml tubes and stored at 4°C. Flow rates were measured by decantation in graduated cylinders.

Hyposalivation was defined as <0.3 ml/min (unstimulated) and <1.5 ml/min (stimulated).

2.3   Data Analysis

Data analysis was performed using established computational tools (detailed specifications provided in Appendix A). Data visualization was conducted using standard statistical visualization packages.

Data Preprocessing. All variables had missing values of less than 5%. Multiple Imputation by Chained Equations (MICE) was employed for missing data imputation due to its ability to preserve variable relationships and account for uncertainty in missing values, making it particularly suitable for clinical datasets with complex interdependencies [22,23].

Data distribution was examined before and after imputation to ensure that no major distortions were introduced. Continuous variables were compared using means, standard deviations, densities, and Kolmogorov–Smirnov tests; categorical variables were assessed with frequency tables, barplots, and chi-square tests. These evaluations indicated that the imputation did not substantially alter the original structure of the dataset.

Variables without variance (e.g., gender, as all patients were female) and identifiers (Subject ID) were removed as they did not provide relevant information for clustering. Categorical variables were encoded using tailored strategies depending on their nature. Outliers were not excluded due to the relatively small sample size and their potential clinical relevance, especially in underrepresented categories. Instead, their influence on model training was attenuated by applying RobustScaler() during standardization.

Feature Selection. High correlations were identified among dental variables. The DMFT index (also known as CAO in Spanish clinical practice), being a linear combination of Decayed, Missing, and Filled Teeth counts, was retained while its components were removed to reduce dimensionality. Age and menopausal status were both maintained despite their correlation due to their clinical relevance to the study objectives.

Clustering Analysis. Unsupervised learning was performed to identify distinct patient groups corresponding to different hyposalivation etiologies and phenotypes. Multiple clustering algorithms were evaluated, including density-based methods (DBSCAN, HDBSCAN), partitioning methods (K-means, K-medoids), and dimensionality reduction approaches (PCA followed by clustering) [24–28].

Algorithm selection prioritized clinical interpretability alongside mathematical performance metrics. Given the consistently moderate silhouette scores across methods, reflecting the inherent complexity and continuous nature of clinical phenotypes commonly observed in medical clustering applications, selection was primarily based on Calinski-Harabasz and Davies-Bouldin indices, combined with the clinical actionability of resulting clusters [29–32].

The final algorithm choice was validated through clinical expert review of cluster characteristics to ensure therapeutic relevance and alignment with established hyposalivation etiologies, following established practices in medical phenotyping studies where clinical utility may complement mathematical optimization [39].

Cluster Interpretation. Random Forest classification [33] was applied to the identified clusters to enable interpretability analysis. SHAP (SHapley Additive exPlanations) [34] values were calculated to determine variable importance and clarify the decision-making process of the clustering algorithm.

Cluster profiling was performed using heatmaps to visualize the distribution of variables across different clusters and identify distinguishing characteristics of each group.

3 Results

3.1 Cohort Description

The study cohort comprised 196 patients with comprehensive data collection across 40 variables. These included demographics (age, menopausal status, gender), lifestyle factors (smoking, alcohol, caffeine, diet), clinical comorbidities (cardiovascular, neoplasia, diabetes, autoimmune diseases), medications (immunosuppressants, corticosteroids, hyposalivation-inducing drugs), dental health parameters (DMFT components, prosthetics, periodontal status), oral mucosa assessment, and salivary flow measurements (stimulated and unstimulated flows with hyposalivation classification).

3.2 Clustering Algorithm Selection

Density-based clustering methods (DBSCAN and HDBSCAN) were initially explored due to their ability to identify clusters of varying shapes and handle noise (Fig. 1A-B, Table 1). DBSCAN with optimized parameters (eps=2.30, min samples=6) identified two clusters but classified the majority of data points as noise (SS: -0.080, CH: 6.753, DB: 3.065). Similarly, HDBSCAN (min cluster size=5, min samples=5) produced two clusters with improved performance (SS: 0.024, CH: 7.498, DB: 4.094) but still resulted in excessive noise classification, making both methods clinically impractical for phenotype identification. Partitioning methods showed superior performance. K-medoids clustering (k=2) achieved moderate performance (SS: -0.001, CH: 8.021, DB: 3.837), indicating suboptimal cluster separation (Fig. 1C). K-means clustering demonstrated markedly better performance, with the elbow method and silhouette analysis indicating k=3 as optimal (Fig. 1D). Standard K-means achieved improved metrics (SS: 0.188, CH: 29.763, DB: 1.789).

Dimensionality reduction enhancement further improved clustering performance. Principal Component Analysis (PCA) with 14 components captured 90% of data variance while reducing dimensionality. K-means applied to PCA-reduced data yielded the best overall performance across all metrics (SS: 0.215, CH: 34.279, DB: 1.660), indicating well-separated, cohesive clusters (Fig. 1E). However, it is important to acknowledge that the consistently low-to-moderate silhouette scores (SS) across all methods likely reflect the inherent complexity and overlap in clinical data, where clear discrete phenotypic boundaries may not exist. Algorithm selection was therefore primarily based on Calinski-Harabasz (CH) and Davies-Bouldin (DB) indices, with clinical interpretability as a key selection criterion. K-means with PCA preprocessing demonstrated the most favorable performance according to these metrics and produced three preliminarily interpretable clusters with potential therapeutic implications, making it the optimal approach for this exploratory phenotype analysis.

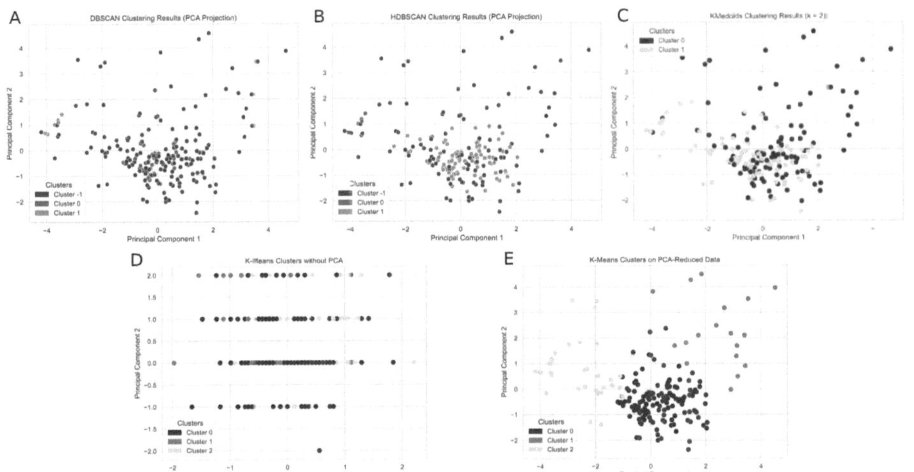

Fig. 1. Comparison of clustering algorithms and their resulting patient groupings. (A) DBSCAN clustering results demonstrating excessive noise classification (Cluster -1 in gray), suggesting this density-based approach may not be suitable for this clinical dataset. (B) HDBSCAN clustering results showing two distinct patient clusters with improved noise handling compared to DBSCAN. (C) K-medoids clustering identifying two patient groups with moderate separation. (D) K-means clustering results revealing three patient clusters with improved separation metrics. (E) K-means clustering applied to PCA-reduced data (14 components explaining 90% variance) showing three well-separated clusters that were selected for final phenotypic analysis due to superior clustering metrics and clinical interpretability. (Color figure online)

Table 1. Clustering algorithms metrics comparison.

	DBSCAN	HDBSCAN	K-Medoids	K-Means	K-Means PCA
Silhouette Score	−0.080	0.024	-0.001	0.188	0.215
Calinski-Harabasz Score	6.753	7.498	8.021	29.763	34.279
Davies-Bouldin Score	3.065	4.094	3.837	1.789	1.660

3.3 Hyposalivation Etiologies and Phenotypes

The results obtained from K-means (k=3) with PCA (14 components, 90% variance explained) identified three distinct patient clusters with potentially meaningful clinical characteristics (Fig. 2). An interpretability analysis using SHAP values revealed the relative importance of variables in cluster classification, confirming that salivary flow variables (both stimulated and unstimulated) and dental health parameters were the predominant factors driving cluster segmentation.

Cluster 0 (n = 141, 72%): Medication and Lifestyle-Associated Hyposalivation. This largest cluster is preliminarily characterized by hyposalivation that may be associated with modifiable factors rather than primary glandular pathology. Analysis revealed that this group exhibits higher values for hyposalivation-associated medications and lifestyle factors, including elevated alcohol consumption, smoking, and caffeine intake, along with increased use of corticosteroids and immunosuppressants. Additionally, this cluster showed higher prevalence of cardiovascular disease and neoplasia compared to other groups.

The cluster comprises healthy controls (approximately 79 individuals) with medication use or lifestyle factors that may impact salivary function, and Sjögren's disease patients (approximately 62 individuals) with potentially well-controlled disease where medication and lifestyle factors may be contributing to hyposalivation. This cluster represents a population where hyposalivation might be associated with modifiable factors, though causality cannot be established from this cross-sectional analysis.

Cluster 1 (n = 18, 9%): Optimal Salivary Function. This cluster represents individuals with apparently normal salivary function, comprising the youngest participants without significant comorbidities, relevant medications, or salivary dysfunction. These participants likely represent a subset of the healthy control group (18 out of 97 controls) who exhibit an optimal clinical profile. The small size of this cluster indicates that among the ostensibly healthy controls, only a minority (18.6%) demonstrated optimal salivary health parameters.

Cluster 2 (n = 37, 19%): Severe Hyposalivation. This cluster predominantly contains Sjögren's disease patients with advanced disease manifestations, characterized by severe hyposalivation, older age, and significant oral health deterioration. The cluster represents approximately 37% of the total Sjögren's disease cohort, suggesting that this subset may exhibit a more severe phenotype where prolonged disease duration and/or advanced age appear to be associated with substantial salivary gland dysfunction and related oral complications.

Within Cluster 2, clear differentiation between age-related and Sjögren's disease-related hyposalivation was not achievable due to the interrelated nature of these variables. To address this limitation, stratified correlation analysis was performed examining the relationship between age and salivary flow (stimulated and unstimulated) in patients with and without Sjögren's disease.

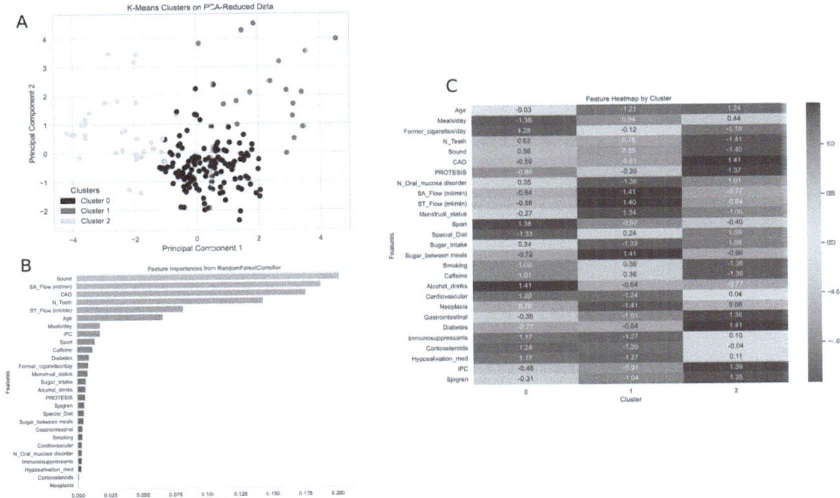

Fig. 2. Exploratory clustering results and interpretation of hyposalivation phenotypes. (A) Visualization of K-means clusters in PCA-reduced space, showing the spatial distribution of the 196 study participants across the identified clusters. (B) Feature importance analysis derived from SHAP values, displaying variables with the highest influence on cluster segmentation. (C) Comprehensive heatmap showing the distribution of normalized clinical variables across the three identified clusters, where red indicates higher mean values and blue indicates lower mean values for each variable within clusters. This visualization reveals the distinct clinical profiles that characterize each preliminary hyposalivation phenotype. (Color figure online)

Tables 2 and 3 show that both groups demonstrated similar negative correlations between age and salivary flow, suggesting that age affects salivary function comparably regardless of Sjögren's disease status. This preliminary finding suggests a potentially additive rather than synergistic effect of age and Sjögren's disease on hyposalivation, providing mechanistic insight into the observed clustering patterns and explaining the methodological difficulty in separating these etiological factors within the computational framework.

Table 2. Correlation matrices between age and unstimulated salivary flow (SA_Flow) for Sjögren and non-Sjögren patients.

Group	Age	SA_Flow
Sjögren		
Age	1.000	-0.287
ST_Flow	-0.287	1.000
Non-Sjögren		
Age	1.000	−0.301
ST_Flow	−0.301	1.000

Table 3. Correlation matrices between age and stimulated salivary flow (ST_Flow) for Sjögren and non-Sjögren patients.

Group	Age	ST_Flow
Sjögren		
Age	1.000	−0.265
ST_Flow	−0.265	1.000
Non-Sjögren		
Age	1.000	−0.326
ST_Flow	−0.326	1.000

4   Discussion

This exploratory study provides a preliminary computational analysis of salivary flow and oral health status within a cohort of 196 women, demonstrating the potential feasibility of identifying tentative hyposalivation phenotypes encompassing induced, physiological, and pathological etiologies. Consistent with established literature, a significant proportion of participants exhibited reduced salivary flow, a condition influenced by a spectrum of factors including autoimmune disease (e.g., Sjögren's syndrome), menopausal status, and medication use [35–37,41].

The evaluation of multiple clustering algorithms revealed consistently moderate Silhouette scores across all methods, reflecting the inherent complexity and continuous nature of clinical phenotypes, where discrete patient clusters may not exist with sharp mathematical. This finding represents a significant limitation of the current approach and underscores the preliminary nature of our results. The superior performance of K-means combined with PCA preprocessing demonstrates the potential value of dimensionality reduction in clinical clustering applications. The 14-component PCA solution, which captured 90% of the variance, effectively reduced noise while preserving clinically relevant information, as evidenced by the improved clustering metrics (CH: 34.279, DB: 1.660) [38,39].

The apparently additive rather than synergistic effect of age and Sjögren's disease on salivary function, as suggested by similar correlation coefficients across patient groups, provides valuable mechanistic insight while highlighting a methodological challenge in separating overlapping etiological factors [38]. This preliminary finding contributes to our understanding of hyposalivation pathophysiology and suggests that future studies might benefit from incorporating temporal data or disease progression markers to better differentiate age-related from pathological hyposalivation.

A key preliminary clinical finding was the identification of unstimulated and stimulated salivary flow alterations beyond the Sjögren's cohort. This suggests that reduced salivation may reflect broader systemic processes, not limited to glandular autoimmune destruction [35,42]. Particularly noteworthy were the associations observed in our dataset between low flow rates and comorbidities such as diabetes and cardiovascular disease, as well as the use of medications associated with hyposalivation. These observations support the potential role of salivary measurements as integrative indicators of systemic status [42,50–52].

Recent advances have demonstrated that integrating artificial intelligence with salivary biomarker analysis can significantly improve diagnostic accuracy and biomarker validation, especially in complex oral and systemic diseases [53, 57].

4.1 Study Limitations

As an exploratory proof-of-concept study, several important limitations must be acknowledged. First, the consistently low-to-moderate silhouette scores across all clustering methods indicate that the identified clusters may not represent clearly distinct phenotypes, but rather points along a clinical continuum. This methodological limitation significantly impacts the interpretability and generalizability of our findings.

Second, the cross-sectional design precludes any causal inference between salivary flow alterations and systemic conditions, limiting our ability to establish etiological relationships [40,43]. The associations observed may reflect correlation rather than causation, and temporal relationships cannot be determined.

Third, although the sample size was adequate for this exploratory descriptive and clustering analysis, it may be underpowered to detect subtle associations or interactions among multiple clinical variables, particularly given the moderate effect sizes typically observed in clinical phenotyping studies.

Fourth, the assessment of medication exposure was based on self-reported data, which introduces the potential for recall bias and may lead to underestimation or misclassification of drug-induced hyposalivation.

Fifth, potential recruitment bias cannot be excluded, as participants were recruited from specialized clinical settings, which may limit the generalizability of findings to broader populations of women with hyposalivation.

Sixth, the computational approach did not include clustering stability analysis or sensitivity testing for the imputation methods. Future studies should incorporate bootstrapping approaches to assess cluster stability and sensitivity analyses to evaluate the impact of different imputation strategies.

These limitations should be carefully considered when interpreting the findings and underscore the preliminary, exploratory nature of this work. Importantly, no external validation cohort was included, leaving uncertainty regarding the generalizability of the identified clusters. Therefore, the results should be viewed as hypothesis-generating rather than definitive, requiring replication in independent cohorts with enhanced study designs.

4.2 Future Directions

Future research should address the current limitations through several key approaches. Longitudinal study designs would enable the assessment of temporal relationships and causal inference, particularly valuable for understanding how medication changes and disease progression impact salivary function over time. Incorporation of objective medication verification through pharmacy records or electronic health records could reduce recall bias and improve medication exposure classification.

Future research should address the current limitations through several key approaches. Although our study is cross-sectional, longitudinal study designs would enable the assessment of temporal relationships and causal inference, particularly valuable for understanding how medication changes and disease progression impact salivary function over time. Clinical outcome–guided temporal clustering has proven effective in revealing dynamic disease subtypes and improving prognostic accuracy, as shown in recent work on neurodegenerative conditions [53–55]. Applying similar longitudinal frameworks could enhance the clinical relevance of computational phenotyping in hyposalivation. Additionally, incorporation of objective medication verification through pharmacy records or electronic health records could reduce recall bias and improve medication exposure classification.

External validation in independent cohorts is essential to confirm the reproducibility and generalizability of the identified phenotypes. Such validation studies should include cluster stability analyses using bootstrapping methods and sensitivity analyses for different imputation approaches.

Enhanced biomarker integration, including hormonal and immunological profiling, could provide deeper mechanistic insights into the observed phenotypic differences and improve cluster characterization. Machine learning approaches that incorporate uncertainty quantification and probabilistic clustering methods may be more appropriate for clinical data with inherent phenotypic overlap.

Finally, clinical outcome studies are needed to determine whether the identified phenotypes have prognostic value and can inform personalized therapeutic approaches in hyposalivation management.

5   Conclusions

This exploratory study demonstrates the preliminary feasibility of computational phenotyping in female hyposalivation, providing initial evidence for potentially distinct clinical patterns using unsupervised machine learning approaches. The analysis identified three hyposalivation phenotypes with possible therapeutic implications: a predominant medication/lifestyle-associated pattern (72%), an optimal salivary function group (9%), and a severe hyposalivation cluster (19%). These preliminary findings suggest that computational approaches may capture clinically relevant heterogeneity in hyposalivation presentations, with salivary flow measurements combined with periodontal and systemic data revealing patterns that may help describe oral-systemic interactions in the context of menopause and autoimmunity.

However, the consistently moderate clustering metrics and inherent limitations of cross-sectional data underscore the exploratory nature of these findings. A particularly notable preliminary finding is the predominance of cases with potentially modifiable contributing factors, rather than irreversible glandular dysfunction, indicating that personalized therapeutic approaches based on phenotypic classification might be valuable in hyposalivation management, pending validation in independent cohorts.

While this preliminary study does not propose definitive biomarkers or established phenotypes, it establishes a methodological framework and generates specific hypotheses for validation. The dataset offers a structured foundation for future longitudinal studies exploring non-invasive indicators of systemic conditions in women and advancing the application of computational phenotyping in oral medicine.

These preliminary findings warrant validation in larger, independent cohorts with enhanced biomarker integration, longitudinal follow-up, and robust stability analyses to confirm the clinical utility of computational phenotyping approaches in precision oral medicine.

Acknowledgments. This work was supported by the Spanish Ministry of Science and Innovation Grant ID PID2020-118557GA-I00 and PID2023-152369OB-I00 funded by MICIU/AEI/10.13039/501100011033 (A.S-S.).

Disclosure of Interests. The authors declare no conflicts of interest.

Appendix A: Computational Methods Details

Data analysis was performed using Python 3.10.10 with pandas (version 1.4.2) for data manipulation, NumPy (version 1.21.5) for numerical computations, SciPy (version 1.8.1) for statistical analyses including correlation analysis, and scikit-learn (version 1.1.1) for machine learning algorithms and data preprocessing. Data visualization was conducted using matplotlib (version 3.5.2) and seaborn (version 0.11.2) [15–21].

References

1. Al-Zaidi, A., et al.: Evaluation of Salivary Diagnostics: Applications, Benefits, Challenges, and Future Prospects in Dental and Systemic Disease Detection. PMC (2025)
2. Alves-Silva, J., et al.: Saliva as a diagnostic tool for systemic diseases-a narrative review. Medicina (Kaunas) **60**(2), 243 (2024)
3. Dawes, C., Pedersen, A.M.L.: The functions of saliva and the effects of hyposalivation on health. Oral Dis. **25**(6), 793–802 (2019)
4. Humphrey, S.P., Williamson, R.T.: A review of saliva: normal composition, flow, and function. J. Prosthet. Dent. **85**(2), 162–169 (2001)
5. da Silva, J.R., et al.: What is the normal salivary flow rate in healthy adults? a systematic review with meta-analyses. JDR Clin. Trans. Res. (2025). https://doi.org/10.1177/23800844251336205
6. Ship, J.A., Pillemer, S.R., Baum, B.J.: Xerostomia and the geriatric patient. J. Am. Geriatr. Soc. **38**(7), 834–843 (1990)
7. Guggenheimer, J., Moore, P.A.: Xerostomia: etiology, recognition and treatment considerations. J. Am. Dent. Assoc. **134**(1), 61–69 (2003)
8. Al-Sohaimy, S.A., Tesch, R.C., Neff, A., Gärtner, M., Walter, M.H.: Salivary changes, systemic conditions, and medication use in independently-living aged: a cross-sectional study. Braz. Oral Res. **37**, e092 (2023)
9. Bowman, S.J., Fox, R.I., Jonsson, R.: Sjögren's disease: a critical perspective on diagnosis and management. Clin. Exp. Rheumatol. **37**(Suppl 118), 26–30 (2019)
10. Eckhardt, C.M., et al.: Unsupervised machine learning methods and emerging applications in healthcare. Knee Surg. Sports Traumatol. Arthrosc. **31**, 376–381 (2023). https://doi.org/10.1007/s00167-022-07233-7
11. Agha-Hosseini, F., Moosavi, Z., Salemi, H.: Salivary biomarkers for the diagnosis of oral and systemic diseases. J. Oral Pathol. Med. **49**(9), 833–841 (2020)
12. Li, Y., Chen, X.: Artificial intelligence in salivary biomarker discovery and validation for oral diseases. Oral Dis. **29**(8), 2407–2415 (2023)
13. Shiboski, C.H., et al.: 2016 American college of Rheumatology/European league against rheumatism classification criteria for primary Sjögren's syndrome. Ann. Rheum. Dis. **76**(1), 9–16 (2017). https://doi.org/10.1136/annrheumdis-2016-210571
14. Navazesh, M., Kumar, S.K.S.: Measuring salivary flow: challenges and opportunities. J. Am. Dental Assoc. **139**, 35S–40S (2008). https://doi.org/10.14219/jada.archive.2008.0353
15. Python Software Foundation: Python (Version 3.10.10) (2023). https://www.python.org/
16. The pandas development team: pandas-dev/pandas: pandas (Version latest). Zenodo (2020). https://doi.org/10.5281/zenodo.3509134
17. Harris, C.R., et al.: Array programming with NumPy (Version 1.20.0). Zenodo (2020). https://doi.org/10.5281/zenodo.4141151
18. Virtanen, P., et al.: SciPy 1.0: Fundamental algorithms for scientific computing in Python. Zenodo (2020). https://doi.org/10.5281/zenodo.592536
19. Pedregosa, F., et al.: Scikit-learn: machine learning in Python (2011). https://scikit-learn.org/stable/about.html#citing-scikit-learn
20. Hunter, J.D.: Matplotlib: a 2D graphics environment. Comput. Sci. Eng. **9**(3), 90–95 (2007). https://doi.org/10.1109/MCSE.2007.55

21. Waskom, M.: seaborn: Statistical data visualization. https://doi.org/10.21105/joss. 03021
22. Van Buuren, S., Groothuis-Oudshoorn, K.: mice: multivariate imputation by chained equations in R. J. Stat. Softw. **45**(3), 1–67 (2011)
23. White, I.R., Royston, P., Wood, A.M.: Multiple imputation using chained equations: issues and guidance for practice. Stat. Med. **30**(4), 377–399 (2011)
24. Ester, M., Kriegel, H.P., Sander, J., Xu, X.: A density-based algorithm for discovering clusters in large spatial databases with noise. In: Proceedings of the Second International Conference on Knowledge Discovery and Data Mining, pp. 226–231 (1996)
25. Campello, R.J., Moulavi, D., Sander, J.: Density-based clustering based on hierarchical density estimates. In: Pacific-Asia Conference on Knowledge Discovery and Data Mining, pp. 160–172. Springer (2013)
26. MacQueen, J.: Some methods for classification and analysis of multivariate observations. In: Proceedings of the Fifth Berkeley Symposium on Mathematical Statistics and Probability, vol. 1, pp. 281–297 (1967)
27. Kaufman, L., Rousseeuw, P.J.: Finding Groups in Data: An Introduction to Cluster Analysis. Wiley (1990)
28. Jolliffe, I.T., Cadima, J.: Principal component analysis: a review and recent developments. Phil. Trans. R. Soc. A **374**(2065), 20150202 (2016)
29. Rousseeuw, P.J.: Silhouettes: a graphical aid to the interpretation and validation of cluster analysis. J. Comput. Appl. Math. **20**, 53–65 (1987)
30. Calinski, T., Harabasz, J.: A dendrite method for cluster analysis. Commun. Stat. - Theory Methods **3**(1), 1–27 (1974)
31. Davies, D.L., Bouldin, D.W.: A cluster separation measure. IEEE Trans Pattern Anal. Mach. Intell. PAMI-1 (2), 224–227 (1979)
32. Liu, Y., Li, Z., Xiong, H., Gao, X., Wu, J.: Understanding of internal clustering validation measures. In: IEEE International Conference on Data Mining, pp. 911–916 (2010)
33. Breiman, L.: Random forests. Mach. Learn. **45**(1), 5–32 (2001)
34. Lundberg, S.M., Lee, S.I.: A unified approach to interpreting model predictions. In: Advances in Neural Information Processing Systems, pp. 4765–4774 (2017)
35. Gargiulo, V., Delle Fave, C., Santacroce, A.: Correlation between Xerostomia index, clinical Oral Dryness Scale, and ESSPRI with different hyposalivation tests. Open Access Rheumatol. **11**, 1–7 (2019)
36. Alshwimi, S., Alghamdi, A.: Menopause and oral health: clinical implications and preventive strategies. PMC (2024)
37. Patil, P., et al.: Oral manifestations in menopause–a scoping review. PMC (2025). https://pmc.ncbi.nlm.nih.gov/articles/PMC12113011/
38. Maugeri, A., et al.: The application of clustering on principal components for nutritional epidemiology: a workflow to derive dietary patterns. Nutrients **15**(1), 195 (2022). https://doi.org/10.3390/nu15010195
39. Loftus, T.J., et al.: Phenotype clustering in health care: a narrative review for clinicians. Front Artif Intell **5**, 842306 (2022). https://doi.org/10.3389/frai.2022. 842306
40. Figueiredo, R.G., Patino, C.M., Ferreira, J.C.: Cross-sectional studies: understanding applications, methodological issues, and valuable insights. J. Bras. Pneumol. **51**(1), e20250047 (2025). https://doi.org/10.36416/1806-3756/e20250047
41. Rodenburg, E.N., et al.: The Prevalence of Sjögren's disease in dental clinics in the Netherlands compared with the prevalence in a systematic literature review of

studies in other countries. MDPI (2024). https://www.mdpi.com/2077-0383/13/19/5918

42. Nam, Y., Kim, Y.Y., Chang, J.Y., Kho, H.S.: Salivary biomarkers of inflammation and oxidative stress in healthy adults. Arch. Oral Biol. **97**, 215–222 (2019). https://doi.org/10.1016/j.archoralbio.2018.10.026

43. Solem, R.C.: Limitation of a cross-sectional study. Am. J. Orthod. Dentofac. Orthop. **148**(2), 205 (2015). https://doi.org/10.1016/j.ajodo.2015.05.006

44. Dlima, S.D., Shevade, S., Menezes, S.R., Ganju, A.: Digital phenotyping in health using machine learning approaches: scoping review. JMIR Bioinform. Biotechnol. **3**(1), e39618 (2022). https://doi.org/10.2196/39618

45. Topol, E.J.: High-performance medicine: the convergence of human and artificial intelligence. Nat. Med. **25**, 44–56 (2019). https://doi.org/10.1038/s41591-018-0300-7

46. Beam, A.L., Kohane, I.S.: Big data and machine learning in health care. JAMA **319**(13), 1317–1318 (2018). https://doi.org/10.1001/jama.2017.18391

47. Rajkomar, A., Dean, J., Kohane, I.: Machine learning in medicine. N. Engl. J. Med. **380**(14), 1347–1358 (2019). https://doi.org/10.1056/NEJMra1814259

48. Brito-Zerón, P., et al.: Sjögren syndrome. Nat. Rev. Dis. Primers. **2**, 16047 (2016). https://doi.org/10.1038/nrdp.2016.47

49. Giudice, F., et al.: Impact of gender and age at onset on Sjögren's syndrome presentation and outcome: state of the art. Clin. Exp. Rheumatol. **41**(12), 2547–2554 (2023). https://doi.org/10.55563/clinexprheumatol/lygrzv. PMID: 38149517

50. Garrido, I., et al.: Xerostomia and salivary dysfunction in patients with diabetes mellitus. a cross-sectional study. J. Oral Pathol. Med. (2024). https://doi.org/10.1111/jop.13583

51. Schwenger, K., Alghamdi, M., Ghorbani, Y., Jackson, T., Okrainec, A., Allard, J.: Hyposalivation is prevalent in bariatric patients but improves after surgery. Surg. Obes. Relat. Dis. (2020). https://doi.org/10.1016/j.soard.2020.06.005

52. Maret, D., et al.: Consequences of hyposalivation in relation to cancer treatment and early management of radiation-induced caries: case reports. Br. Dent. J. **237**, 705–709 (2024). https://doi.org/10.1038/s41415-024-7894-6

53. De Torres, J., Marin, J., Martínez-González, C., De Lucas-Ramos, P., Cosío, B., Casanova, C.: The importance of symptoms in the longitudinal variability of clusters in COPD patients: a validation study. Respirology **23**, 485–491 (2018). https://doi.org/10.1111/resp.13194

54. Wang, D., Ma, X., Schulz, P. E., Jiang, X., Kim, Y.: Clinical outcome-guided deep temporal clustering for disease progression subtyping. J. Biomed. Inf., 104732 (2024). https://doi.org/10.1016/j.jbi.2024.104732

55. Lee, C., Rashbass, J., Van Der Schaar, M.: Outcome-oriented deep temporal phenotyping of disease progression. IEEE Trans. Biomed. Eng. **68**, 2423–2434 (2020). https://doi.org/10.1109/tbme.2020.3041815

56. Adeoye, J., Su, Y.: Artificial intelligence in salivary biomarker discovery and validation for oral diseases. Oral Dis. (2023). https://doi.org/10.1111/odi.14641

57. Karishma, S., Saravanan, A.: Unlocking oral oncology: AI-powered biomarker discovery for early detection. Oral Oncol. Rep. (2024). https://doi.org/10.1016/j.oor.2024.100427

Predictive Modeling with Machine Learning in Patients with Locally Advanced Rectal Adenocarcinoma Undergoing Neoadjuvant Treatment

Miguel Pagés García[1]([⊠]), Inmaculada Rincón Pérez[2], Juan Antonio Ortega Ramírez[1], José Luis López-Guerra[2], and Francisco Antonio Gomez Vela[3]

[1] Higher Technical School of Computer Engineering, University of Seville, 41012 Seville, Spain
migpaggar@alum.us.es

[2] Department of Radiation Oncology, University Hospital Virgen del Rocío, 41013 Seville, Spain

[3] Intelligent Data Analysis Group (DATAi), Universidad Pablo de Olavide, Ctra. Utrera km. 1, 41013 Seville, Spain

Abstract. Colorectal cancer is one of the diseases with the highest rate of death worldwide, so it is necessary to improve its diagnosis and treatment. In this sense, new artificial intelligence techniques have emerged as a new way to understand the intrinsic mechanisms of the disease.

In the present study, data analysis and machine learning techniques have been applied with the aim of identifying clinical and biological patterns that allow for optimizing patient stratification and improving the treatments administered. Through data cleaning and analysis, distinct patient profiles have been defined, clearly differentiating those with a poorer prognosis from those with a better outlook, thanks to clustering algorithms and the interpretation of the resulting clusters supported by the LIME technique.

In addition, a supervised analysis has been performed, also supported by LIME to improve interpretability, which revealed the most influential variables in predicting key clinical events such as survival, tumor progression, and the pathological response of the tumor to treatment. These results highlight the importance of monitoring molecular markers and hematological parameters, as well as ensuring the proper management of treatment timelines, laying the foundation for improving clinical decision-making and advancing toward more personalized medicine.

Keywords: Rectal Adenocarcinoma · Radiotherapy · Chemotherapy · Clustering · LIME · Machine Learning · Molecular Factors · Hematological Factors

© The Author(s), under exclusive license to Springer Nature Switzerland AG 2026
A. López Fernández et al. (Eds.): CIABiomed 2025, LNBI 16148, pp. 585–596, 2026.
https://doi.org/10.1007/978-3-032-10661-2_43

1   Introduction

Colorectal cancer is one of the most common malignancies worldwide, ranking as the third most frequently diagnosed cancer and a leading cause of cancer-related mortality [1]. Among its subtypes, locally advanced rectal cancer (LARC) presents specific challenge due to the heterogeneity in treatment response. Current clinical protocols for LARC generally involve neoadjuvant chemoradiotherapy (nCRT) with Capecitabine followed by surgical resection. Despite the advancements, predicting individual patient response remains a challenge. While this sequence has improved outcomes in many cases, a wide range of individual responses continues to be observed and predicting how patients will respond remains a significant hurdle in clinical practice [2].

Several studies have attempted to use advanced statistical models and computational methods to predict treatment outcomes in rectal cancer. For example, Bilbault et al. [3] supervised learning models to anticipate pathological complete response (pCR) based on clinical data, while Ji-Yeon et al. [4] used deep learning on MRI images to classify responders and non-responders with promising results. These efforts confirm the potential of data-driven tools to assist in decision making, though their implementation in real world remains limited.

In this context, Machine Learning (ML), a subfield of Artificial Intelligence, offers powerful tools for modeling complex and nonlinear relationships in large datasets [5]. ML enables not only the creation of predictive models but also the unsupervised discovery of patient subgroups and latent patterns that may inform clinical decision-making.

While most studies focus solely on supervised prediction and leave out unsupervised learning, we aim to address these gaps by applying both supervised and unsupervised ML methods to a real-world dataset of LARC patients treated with Capecitabine-based nCRT. The main objectives are to identify associations between clinical, physiological and molecular variables to respond to the following question: Why do patients with the same disease and treatment make totally different responses to therapy?

Ultimately, we hope to contribute to a better understanding of individual variability in rectal cancer therapy and to highlight the usefulness of interpretable ML approaches in guiding future personalized strategies. The remainder of this article is structured as follows: Sect. 2 describes the methodology, including data preprocessing, clustering and explainability methods. Section 3 describes the results, including patient profile, variable importance and cluster-specific findings. Finally, Sect. 4 summarizes the main conclusions and discusses implications for clinical practice and future research.

2   Methodology

2.1   Data Collection and Descriptive Analysis

This study included 74 patients diagnosed with locally advanced rectal cancer with ages ranging from 53 to 98 years. Each row of the dataset corresponds to a patient. In total, the dataset contains 63 variables, which make up the columns of the database.

This study was conducted using data collected from patients treated at the Hospital Universitario Virgen del Rocío (Seville, Spain), specifically within Department of Radiation Oncology. All patients received neoadjuvant treatment consisting of radiotherapy (RT) and chemotherapy (QT) with Capecitabine prior to surgery.

The dataset includes demographic, clinical, treatment, pathological, molecular and follow-up information focusing on patients in clinical stage III or IV or those with lymphatic information. Variables collected includes age, sex, symptoms, initial diagnosis dates, treatment start and end dates, number of RT sessions, etc. Tumor-related variables include initial and pathological staging, histological type, degree of differentiation and tumor size. Additionally, laboratory biomarkers such as CEA, CA19.9 [3] and hematological parameters like hemoglobin, leukocytes, neutrophils, lymphocytes and platelets.

The database was constructed in compliance with data protection regulations, including the EU General Data Protection Regulation (GDPR) and the Spanish Organic Law 3/2018 on Personal Data Protection and Guarantee of Digital Rights. Data transmission protocols also adhere to Royal Decree 1720/2007.

2.2 Data Preprocessing

All preprocessing steps and ML techniques were implemented using Python 3.10 in a Jupyter Notebook environment. The following libraries were used:

- Pandas and NumPy for data manipulation and numerical operations [4, 5].
- Scikit-learn for preprocessing and implementing clustering/classification algorithms [6].
- Matplotlib and Seaborn for visualization [7, 8].
- LIME for model explainability [9].

A sequential preprocessing pipeline was applied to the original dataset consisting of the following five stages:

1. Removal of irrelevant or redundant columns.
2. Format conversion: Dates, categorical variables and numeric fields were converted into a consistent format to allow for the use of different algorithms. This was made using the functions provided by scikit-learn.
3. Missing value imputation: Several imputation methods were considered, but, ultimately, K-Nearest Neighbors Imputation (KNNImputer, [10]). This method estimates missing values based on the similarity to other observations, imputing values averaging the K most similar instances according to a defined distance metric.
4. Outlier detection: Both histograms and box-and-whisker plots were created as well as Z-score analysis [11].
5. Standardization and Dimensionality Reduction: Was performed using the Standard-Scaler [12] function form the Scikit-learn library. This method transforms each variable to have a mean of 0 and a standard deviation of 1, following the next formula:

$$x_{est} = \frac{(x - \mu)}{\sigma}$$

where:

- x is the individual value.
- μ is the mean of the variable.
- σ is the standard deviation of the variable.

In addition, dimensionality reduction was performed using Principal Component Analysis (PCA).

2.3 Clustering

Four clustering algorithms were explored: K-Means, selected due to the simplicity and ease of implementation, as well as its effectiveness in handling large datasets and producing easily interpretable results [13]. Agglomerative Hierarchical Clustering, that offers greater robustness and flexibility [14]. DBSCAN, that can discover clusters of arbitrary shapes through density-based analysis effectively distinguishing noise from valid clusters to maintain clustering accuracy [15]. Finally, MEANSHIFT, that, thanks to its robustness to outliers, makes it particularly well-suited for application to real-world datasets [16].

Figure 1 illustrates the workflow used for the clustering process.

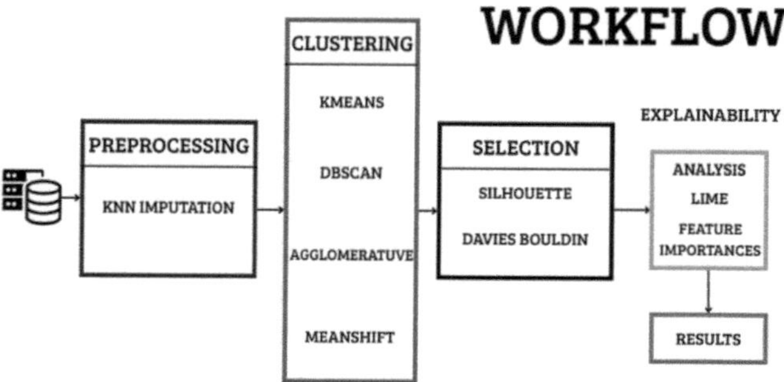

Fig. 1. Clustering Workflow.

2.4 Explainability

To complete this task, the method chosen in this study was LIME (Local Interpretable Model-Agnostic Explanations, [9, 17]). LIME works by analyzing how the model's predictions change when input data is slightly modified. It generates a new dataset consisting of "perturbed" samples and the corresponding predictions from the original model. On this dataset, LIME trains an interpretable model that is locally weighted: samples close to the original instance receive higher importance.

In this study, LIME was adapted to interpret results from unsupervised clustering algorithms. To make this possible, we transform the unsupervised problem into a supervised one by adding a new column to dataset called "CLUSTER". Figure 2 illustrates this.

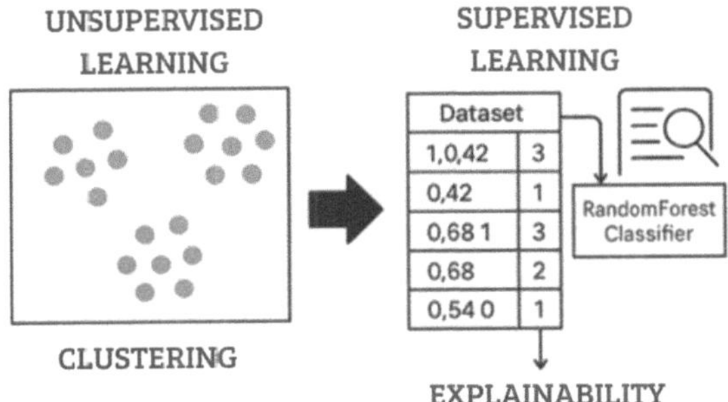

Fig. 2. Illustrate example of the transition from unsupervised to supervised learning.

2.5 Supervised Learning

As a final task to the project, supervised analysis was performed for the following variables: EXITUS, PROGRESSION, DOWNSTAGING and PCR (Pathological Complete Response). Figure 3 illustrates the workflow followed.

To perform this task, the RandomForestClassifier algorithm [18] was selected. Its ensemble-based architecture, which aggregates the results of multiple decision tree, makes it particularly effective in identifying the most optimal decision.

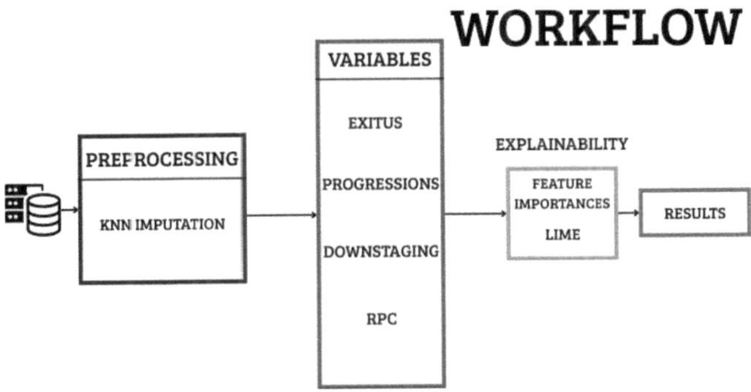

Fig. 3. Workflow of supervised analysis.

3 Results

3.1 General Characteristics of the Sample

The average age of the patients was 77.5 years (SD = 10.5). Males represented 58% of the cohort and females 42%. This study included 74 patients diagnosed with locally advanced rectal cancer with ages ranging from 53 to 98 years. Most tumors were initially classified as stage II-A (44.6%) or III-B (34.8%). The most common clinical classification was cT3 (50%) and cN0 (48.9%). The average size of the tumor was 4.8 cm with a mean distance of 5.5 cm from the anal margin.

The most frequent symptoms were altered bowel habits (65.2%) and constitutional syndrome (60.9%). Baseline hematological profiles revealed mean leukocyte counts of 7.000/μL, lymphocytes 1.500/μL, neutrophils 4.500/μL, hemoglobin 12.5 g/dL and platelet count around 250.000/μL.

3.2 Survival Outcomes

Survival analysis was conducted using Kaplan-Meier method. The overall survival (OS) median was not reached during the follow-up period suggesting that more than half of the patients remained alive or died from unrelated causes due to the advanced age of the cohort. Short-term survival (1–3 years) was high but decreased to approximately 56% at 5 years indicating increased disease progression.

The progression-free survival (PFS) median was also not reached, with a significant risk of the recurrence observed during the first year. Patients who remained progression-free after two years showed long-term disease control with PFS rates above 60%.

3.3 Clustering Results

Four unsupervised clustering algorithms (K-Means, Agglomerative, DBSCAN, Mean-Shift) were tested using both original and engineered datasets with binary features derived from clinical thresholds. Clustering performance was evaluated using Silhouette and Davies-Bouldin scores. MeanShift achieved the best performance across both datasets. The results achieved are presented in Table 1 and 2.

Figures 4 and 5 show the classification results obtained using the MeanShift algorithm. Additionally, Fig. 6 includes two examples of the plots used to derive the results presented in Tables 3 and 4.

Table 1. Punctuation of Original dataset.

Algorithm	Score Silhouette	Score Davies Bouldin
K-Means	0.7524	0.4205
DBSCAN	0.6596	0.5803
Agglomerative	0.8080	0.2989
MEANSHIFT	**0.8246**	**0.2137**

Table 2. Punctuation of Thresholds dataset.

Algorithm	Score Silhouette	Score Davies Bouldin
K-Means	0.7406	0.3862
DBSCAN	0.8187	0.2627
Agglomerative	0.7563	0.3435
MEANSHIFT	**0.8776**	**0.1551**

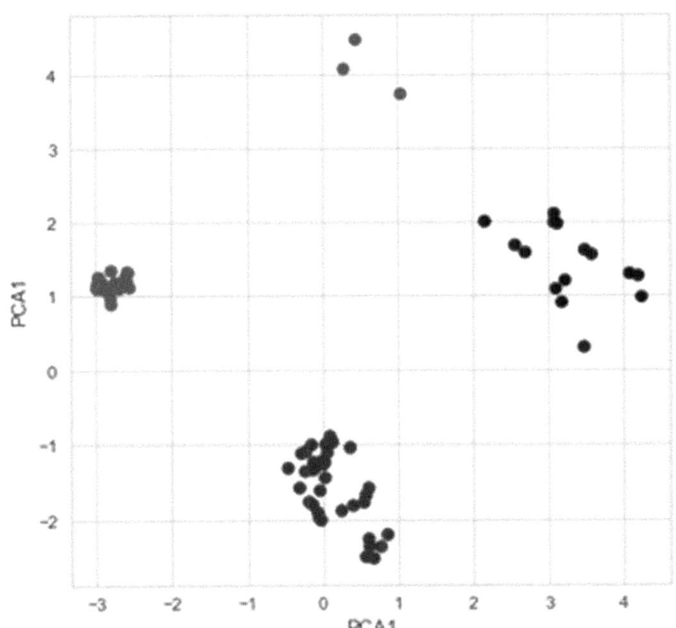

Fig. 4. Clustering for Original dataset. (Cluster 0: Blue, Cluster 1: Red, Cluster 2: Black)

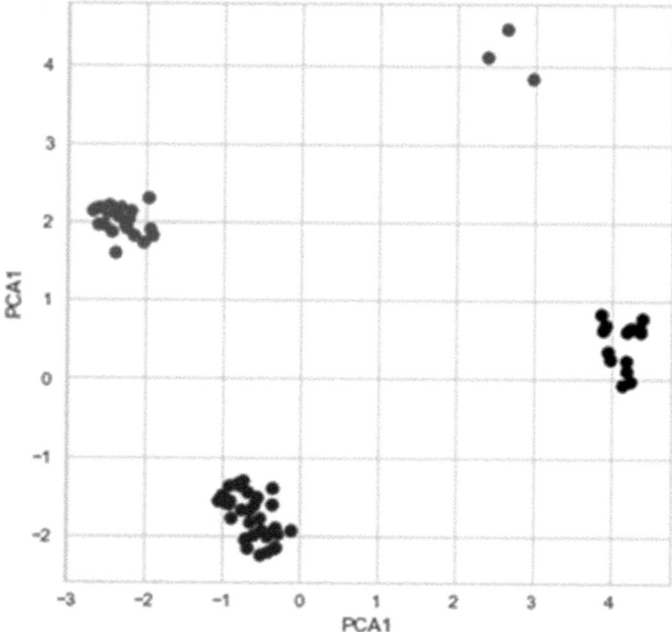

Fig. 5. Clustering for Thresholds dataset. (Cluster 0: Blue, Cluster 1: Red, Cluster 2: Black)

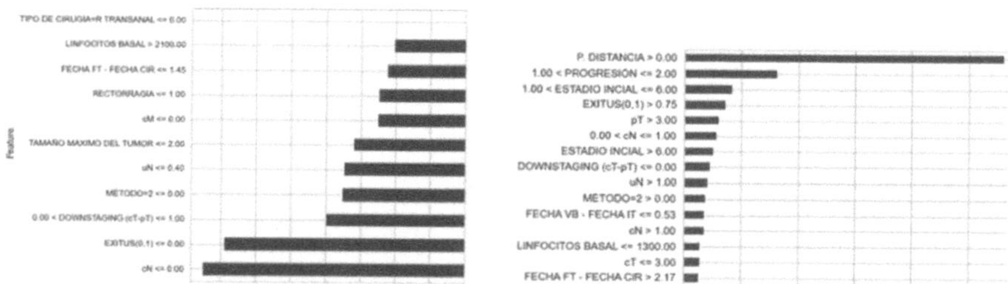

Fig. 6. Explainability example. Blue indicates a positive contribution to cluster membership, while red indicates a negative contribution. (FT: End of treatment, CIR: Surgery, VB: First meet with the doctor,

Table 3. Clustering profiles

Feature	Cluster 0	Cluster 1	Cluster 2
Prognosis	Good prognosis. Low progression and mortality	Intermediate-high risk. Moderate progression and mortality	High risk. The worst prognosis with high progression and mortality
CEA and CA19.9 Markers	Normal	Moderately high	High

(continued)

Table 3. (*continued*)

Feature	Cluster 0	Cluster 1	Cluster 2
Hematological Variables	Normal	Slight alterations (↑ platelets)	Abnormal (↑ platelets, ↑ neutrophils ↓ lymphocytes)
Time from End of CRT to Surgery	Less than 2 months	Less than 2 months	Greater than 2 months
Pathological Complete Response (PCR)	100%	94%	40%
Downstaging	Size: 50% Node involvement: 2.5%	Size: 44% Node involvement: 46.5%	Size: 6.67% Node involvement: 39.5%
Progression	5% (All local)	16.67% (All local)	100% (13% local, 100% distant)
Exitus (Death)	5%	16.67%	66.67%

3.4 Supervised Learning Results

The results obtained for those variables studied in supervised learning are explained on the next table, Table 4. These characteristics indicate the positive value of the variable.

Table 4. Results for the variables of interest

Variable	Most Significant Predcitive Characteristics
Death (Exitus)	• Patient experienced Progression whether Local, Distant or Both
Progression	• No PCR (Pathological Complete Response) • High values of CA19.9, Platelets, Leukocytes, and Neutrophils • Time between End of CRT (FT) and Surgery greater than 2 months
Downstaging	• Time between Baseline Visit (VB) and Start of Treatment (IT) not exceeding 1.5 months • High lymphocyte count • Low neutrophil count
PCR	• High lymphocyte count • Low neutrophil count • Platelet count not exceedingly approx. 250,000 cells/μL • Leukocyte counts not exceedingly approx. 8,000 cells/μL

4   Conclusions and Outcomes

This study has provided a comprehensive analysis of clinical data from patients diagnosed with locally advanced rectal cancer and treated with neoadjuvant radiotherapy and chemotherapy. Through this analysis we aimed to extract clinically relevant insights to support improved treatment strategies and a deeper understanding of patient trajectories based on individual biological and clinical characteristics. The findings effectively address the objectives proposed.

Firstly, our results underscore the critical role of hematological parameters, particularly baseline levels of lymphocytes, leukocytes, neutrophils and platelets. Patients with higher lymphocyte counts and controlled inflammatory profiles exhibited more favorable clinical outcomes. In contrast, those with lymphopenia and elevated inflammatory profiles were linked to poor prognoses.

Secondly, we successfully identified three distinct patient profiles based on risk factors:

- Profile 0: Low-risk patients with favorable prognosis.
- Profile 1: Intermediate/high-risk patients.
- Profile 2: High-risk patients with poor prognosis.

This clustering also revealed the significance of treatment timing. Patients with worse outcomes experienced a delay of more than two months between completion the treatment and surgery, which correlated with increased distant progression rates.

Finally, elevated levels of biomarkers CEA and CA19.9 were notably associated with poor clinical outcomes, including higher risk of progression and mortality.

Overall, this research reinforces the importance of a multidimensional analysis in patient management, especially biological and temporal parameters. The results encourage the adoption of individualized treatment strategies aimed at improving patient survival, minimizing disease progression and ultimately enhancing quality of life.

4.1   Further Work

Future work will focus on validating the identified patient profiles and predictive models in larger and more diverse cohorts to ensure their generalizability. Additionally, integrating longitudinal data and real-time clinical updates could enhance the dynamic stratification of patients over the course of treatment. Exploring the combination of machine learning models with clinical decision support systems (CDSS) in real clinical settings will also be essential to evaluate the practical impact of these approaches on treatment optimization. Finally, incorporating additional data sources, such as genomics or imaging, may further refine the predictive capabilities and move closer toward fully personalized medicine.

Acknowledgments. This research was supported by the ARTIFACTS project – *Generation of Reliable Synthetic Health Data for Federated Learning in Secure Data Spaces USE* (PID2022-141045OB-C42), funded by MCIN/AEI/10.13039/501100011033 and the European Union through the ERDF "A way of making Europe".

Disclosure of Interests.   The authors declare no competing interests.

References

1. Pinheiro, M., Moreira. D.N., Ghidini, M.: Colon and Rectal Cancer: An Emergent Public Health Problem. Baishideng Publishing Group Inc. (2024). https://doi.org/10.3748/wjg.v30.i7.644
2. Glynne-Jones, R., Hughes, R.: Critical appraisal of the 'wait and see' approach in rectal cancer for clinical complete responders after chemoradiation. Br. J. Surg. **99**(7), 897–909 (2012). https://doi.org/10.1002/bjs.8732. Epub 2012 Apr 27. PMID: 22539154
3. Bibault, J.E., Giraud, P., Burgun, A.: Big Data and machine learning in radiation oncology: state of the art and future prospects. Cancer Lett. **382**(1), 110–117 (2016). https://doi.org/10.1016/j.canlet.2016.05.033. Epub 2016 May 27. PMID: 27241666
4. Kim, J.Y., et al.: Prediction of pathologic complete response to neoadjuvant chemotherapy using machine learning models in patients with breast cancer. Breast Cancer Res. Treat. **189**(3), 747–757 (2021). https://doi.org/10.1007/s10549-021-06310-8. Epub 2021 Jul 5. PMID: 34224056
5. Sidey-Gibbons, J.A.M., Sidey-Gibbons, C.J.: Machine learning in medicine: a practical introduction. BMC Med. Res. Methodol. **19**(1) (2019). https://doi.org/10.1186/s12874-019-0681-4
6. Takahashi, Y., et al.: The usefulness of CEA and/or CA19-9 in monitoring for recurrence in gastric cancer patients: a prospective clinical study. Gastric Cancer **6**(3), 142–145 (2003). https://doi.org/10.1007/s10120-003-0240-9
7. NumPy. "NumPy. The fundamental package for scientific computing with Python". https://numpy.org/. Accessed 29 June 2025
8. Pandas. "Pandas". https://pandas.pydata.org/. Accessed 29 June 2025
9. scikit-learn. "scikit-learn. Machine Learning in Python". https://scikit-learn.org/stable/. Accessed 29 June 2025
10. matplotlib. "Matplotlib: Visualization with Python". https://matplotlib.org/. Accessed 29 June 2025
11. seaborn. "seaborn: statistical data visualization". https://seaborn.pydata.org/. Accessed 29 June 2025
12. Gonzales Mesones, E.: Explicabilidad de modelos de ML: LIME (2023). https://medium.com/latinxinai/explicabilidad-de-modelos-de-ml-lime-f9d0dceb5154. Accessed 24 Apr 2025
13. Pindiyan, A.V., Pramila, R.M.: Machine learning-based imputation techniques analysis and study. In: 2024 International Conference on Electrical, Electronics and Computing Technologies, ICEECT 2024. Institute of Electrical and Electronics Engineers Inc. (2024). https://doi.org/10.1109/ICEECT61758.2024.10739330
14. Castrillo, M.: Cómo identificar y tratar outliers con Python (2023). https://medium.com/@martacasdelg/c%C3%B3mo-identificar-y-tratar-outliers-con-python-bf7dd530fc3. Accessed 23 Apr 2025
15. Raju, V.N.G., Lakshmi, K.P., Jain, V.M., Kalidindi, A., Padma, V.: Study the influence of normalization/transformation process on the accuracy of supervised classification. In: 2020 Third International Conference on Smart Systems and Inventive Technology (ICSSIT), Tirunelveli, India, pp. 729–735 (2020). https://doi.org/10.1109/ICSSIT48917.2020.9214160
16. Macqueen, J.: Some methods for classification and analysis of multivariate observations
17. Aditya. "Hierarchical Clustering: Applications, Advantages, and Disadvantages". (2022). https://codinginfinite.com/hierarchical-clustering-applications-advantages-and-disadvantages/. Accessed 23 Apr 2025

18. Hadikurniawati, W., Hartomo, K.D., Sembiring, I.: Spatial clustering of child malnutrition in central Java: a comparative analysis using K-means and DBSCAN. In: Proceedings: ICMER-ALDA 2023 - International Conference on Modeling and E-Information Research, Artificial Learning and Digital Applications, pp. 242–247. Institute of Electrical and Electronics Engineers Inc. (2023). https://doi.org/10.1109/ICMERALDA60125.2023.10458202
19. Chugh, V.: "Agrupación por Desplazamiento Medio: Guía completa" (2024). https://www.datacamp.com/es/tutorial/mean-shift-clustering. Accessed 23 Apr 2025
20. Gamal Al-Kaf, H.A., Lee, K.B.: Explainable machine learning method for open fault detection of NPC inverter using SHAP and LIME. In: 2023 IEEE Conference on Energy Conversion, CENCON 2023, pp. 14–19. Institute of Electrical and Electronics Engineers Inc. (2023). https://doi.org/10.1109/CENCON58932.2023.10368888
21. Giri, D., Gunasekaran, M.: Comparison of optimized random forest classifier with k-nearest neighbors classifier to determine the factors influencing teenagers alcoholic consumption. In: Proceedings of 9th International Conference on Science, Technology, Engineering and Mathematics: The Role of Emerging Technologies in Digital Transformation, ICONSTEM 2024. Institute of Electrical and Electronics Engineers Inc. (2024). https://doi.org/10.1109/ICONSTEM60960.2024.10568834

Reduction of a Neuropsychological Test Battery Using Machine Learning Methods

Alba Gómez-Valadés Batanero[1(✉)] [ID], Rafael Martínez Tomás[1] [ID],
Mariano Rincón[1] [ID], and Pedro Juan Tarraga López[2] [ID]

[1] Universidad Nacional de Educación a Distancia, 28040 Madrid, Spain
`albagvb@dia.uned.es`
[2] Universidad de Castilla-La Mancha, Albacete, Castilla-La Mancha, Spain

Abstract. To reach a broader population during Mild Cognitive Impairment population screenings, there is a need for neuropsychological test batteries that are quick to administer and evaluate. To achieve that, all test scores must be informative and non-redundant. This study used machine learning methods to analyze the test scores in a neuropsychological database in terms of relevance and collinearity. We examined a database of 520 assessments with diagnoses of either healthy or MCI, evaluated using a neuropsychological test battery consisting of fourteen tests, from which we will use twenty-three test scores. First, we assessed both the correlation between test scores and the diagnostic dependence of each score. This led us to define four groupings of test scores (very high relevance, very high and high relevance, all except low-relevance, and all test scores) that exclude redundant test scores within each group, and a fifth group that includes all test scores, even redundant ones, as a baseline. Those test score groups were analyzed using well-established machine learning methods. Based on the machine learning analysis, the fifth group had the best performance, likely because of the redundant test scores. Among the groupings that exclude redundant test scores, we found that the most promising group was the one with all test scores except the low-relevant ones, though it was almost tied in performance with the group containing all test scores.

Keywords: Neuropsychological tests · Neruropsychological battery · Machine learning · MCI

1 Introduction

Alzheimer's Disease (AD) is a neurodegenerative disease whose incidence is expected to rise in the next decades with the aging of the population [1,2]. To enable individuals to remain independent for longer, it is essential to detect AD at its earliest stages of cognitive decline, when they are still in the Mild Cognitive Impairment (MCI) stage, a transitional phase between normal aging and AD [3,4], so this has been the focus of several researches [5–9]. Therefore,

A. López Fernández et al. (Eds.): CIABiomed 2025, LNBI 16148, pp. 597–608, 2026.
https://doi.org/10.1007/978-3-032-10661-2_44

there is a need for tests valid for population screening that are optimized to identify the MCI at its earlier stages.

To detect the MCI, several kinds of diagnostic tests have been developed. However, to reach the maximum number of people, there is a need for diagnostic tests that are both fast to perform and evaluate, do not present high cost, and have a high discriminatory capability towards MCI so they can be used during population screenings [4,6,10]. Of those diagnostic tests, neuropsychological tests stand out for their speed, economy, and lack of specialized equipment and personnel, in addition to being the only ones capable of evaluating cognitive functions [4,11–13]. Those neuropsychological tests evaluate different cognitive aspects and functions of subjects. To obtain a more global vision of the cognitive state of subjects and their possible impairments, they are grouped into neuropsychological test batteries. Usually, the more tests included in a neuropsychological battery, the more information it will provide about the cognitive state of a subject. However, this also raises the time required to perform and evaluate them, hindering their use during population screenings. Therefore, to better suit the needs of population screenings, those neuropsychological test batteries should be as efficient as possible by employing the minimum number of tests.

Machine learning methods have been used in the medical context to help with data analysis and test optimization, due to their ability to handle great volumes of complex data and to find patterns from it [6,14]. In the context of neuropsychological test batteries, machine learning methods have been used to evaluate different kinds of test scores, helping propose more efficient test batteries, as well as to provide a first insight into the cognitive state of subjects [6,15–18].

In this context, here we analyze the performance of different test score combinations using machine learning methods to identify and remove possible tests of little relevance to the final diagnosis, as a first step in proposing a more concise neuropsychological test battery.

The rest of the article follows as detailed. Section 2 describes the database and the methodology used to obtain the scores. Section 3 details the results obtained from the analysis. Section 4 presents the results obtained at several points in the analysis, and Sect. 5 summarizes the main conclusions and outlines future research directions.

2   Methodology

2.1   Work Scheme

Figure 1 illustrates the process scheme followed during this research. First, a relevance analysis was carried out on the database to identify which test scores are more discriminative between healthy and MCI subjects. In parallel, a correlation analysis was performed to detect possible redundancy between test scores. This leads to the definition of several test score groups, which were analyzed using several machine learning methods. Finally, the performance of the different test groups will be compared with each other to obtain the most discriminative combination that takes the least number of test scores, as well as the machine learning model that best works with that combination.

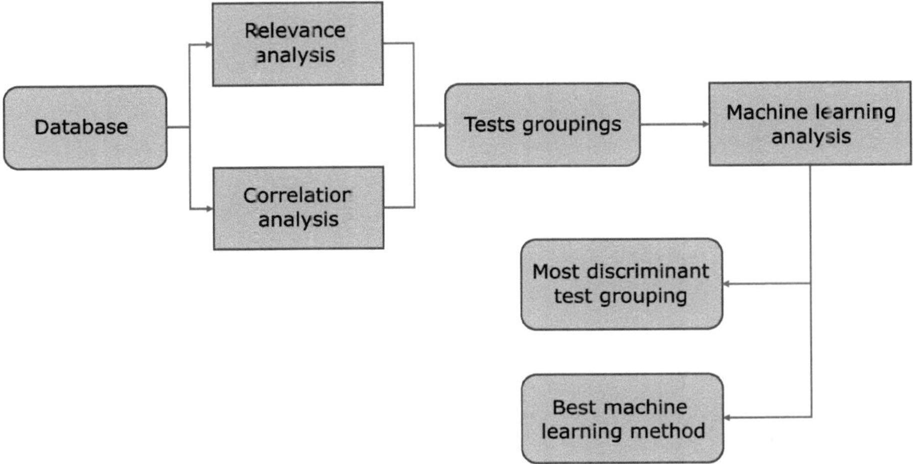

Fig. 1. Scheme of the methodology followed in the optimization of the neuropsychological test battery.

2.2   Database

We used a previously anonymized database from a large longitudinal study on the incidence of incipient MCI in the Autonomous Community of Madrid (Spain) [11, 19–21]. Subjects with a previous diagnosis of neurodegenerative disease, disabling chronic disease, psychiatric disorders such as major depression, established neurological abnormality, severe sensory impairment, diabetes, stroke, and loss of consciousness were previously excluded from the database. The cognitive status of subjects was assessed using the Spanish version of the Mini-Mental State Examination [22], and the emotional state of the subjects was addressed through the Geriatric Depression Scale [23]. The database is formed by 520 Spanish monolingual cases, with subjects aged between 58 and 93 years and with an educational level between 0 and 22 years of schooling. Each subject underwent a total of three neuropsychological evaluations, with one year of difference between each evaluation, and with a diagnosis of healthy and MCI. Table 1 shows the summary of the sociodemographic variables of the database:

2.3   Test Scores

The neuropsychological test battery is composed of a total of fourteen tests, some of them being subdivided into subtests. Although usually only one variable is

Table 1. Summary of the sociodemographic variables of the database.

	N° of subjects	Men/Women	Age (Mean/Std)	Education (Mean/Std)	MMSE (Mean/Std)
Healthy	309	80/229	70.69(6.03)	11.64(5.18)	32.92(2.06)
MCI	211	59/152	73.00(6.96)	9.19(6.06)	30.69(3.12)

extracted from each test, there are tests from which more than one variable is obtained, like the Test de Aprendizaje Verbal España-Complutense [24] that requires scoring seven subtests individually. Thus, we obtain the following list of tests and scores, where the names of the scores are stated in parentheses:

- Mini-Examen-Cognoscitivo: Spanish version of the Mini-Mental State Examination. One score (*MEC*).
- Phonemic Fluency: Maximum number of words starting with the same letter (P) in a given time. One score (*Phonemic_ Continuity*).
- Semantic Fluency: Maximum number of words inside a given semantic category in a given time. Four tests are included (*Animals, Clothes, Plants*, and *Vehicles*), with one score per test (*SF_ Animal, SF_ Clothes, SF_ Plants*, and *SF_ Vehicles*) and a global one (*Semantic_ Fluency*).
- Praxias: split into two parts,
  - Constructive praxias: require the subjects to copy a series of drawings as accurately as possible. Two scores (*Peak_Loops* and *Constructive_ Praxias*).
  - Ideatory and Ideomotor Praxias: both assess the motor skills of subjects in different ways. Ideatory Praxias refers to the capacity of performing simple gestures intentionally, while Ideomotor Praxias requires reproducing the gestures required to manipulate certain objects. Three scores (*Ideomotor_ Praxias, Ideatory_ Praxias*, and *Ideomotor_ Total*).
- Trail Making: assesses executive functions and consists of two parts, with one score each: Part A (*TMT-A*), which requires connecting several numbers in ascending order, and Part B (*TMT-B*), which requires connecting numbers and letters in alternating, ascending order.
- Test de Aprendizaje Verbal España-Complutense (TAVEC) 2nd part: assesses learning skills and verbal episodic memory. Formed by seven subtests, with one score each (*ListA_ Total, ListB_ Total, Short_ Memory, Short_ Memory_ Clues, Long_ Memory, Long_ Memory_ Clues*, and *Recognition*).
- Rey-Osterrieth complex figure: Copy of the complex figure of Rey-Osterrieth. Two scores: time needed by the subject to complete the figure (*Rey-Figure_ Time*) and copy quality score (*Rey-Figure_ Score*).

This generates a total of twenty-three scores, which are going to be referred to as "test scores" in the rest of the paper.

2.4   Relevance and Correlation Analysis

The first step was to identify which test scores are more effective at discerning between healthy and MCI individuals. To achieve that, each test score in the neuropsychological test battery was ranked based on its capability to identify MCI subjects. For this, a relevance analysis was carried out using the mutual information measure on all test scores to identify the test scores on which the diagnosis depends the most. To detect potential redundant test scores, we analyzed the collinearity between test scores using Pearson's correlation. These analyses allowed us to define test score groups according to their relevance in detecting MCI subjects, while excluding those test scores that were found as highly correlated with others.

2.5   Machine Learning Analysis

The test score groups defined in the previous step were analyzed using a total of seven machine learning methods (ADABoost (ADAB), Bagging (BAG), Multilayer Perceptron (MLP), Random Forest (RF), Logistic Regression (RLog), Support Vector Machine (SVM), and XGBoost (XGB)) commonly used across the literature to analyze neuropsychological tests, to ensure that the combination is the most efficient in general, and not the most efficient one regarding one particular machine learning model. To obtain as much information as possible from our sample, a 10-fold cross-validation was selected to evaluate their performance. To avoid possible biases, the analysis was repeated eleven times, using the mean results of those repetitions in the performance analysis of the test score groups.

3   Results

3.1   Correlation and Relevance Analysis

From the relevance analysis, whose scores can be seen in Fig. 2, four different categories have been defined:

- Very high relevance variables: a total of four test scores, which score more than 0.1 in the relevance analysis (*Constructive_Praxias*, *Ideomotor_Total*, *Peak_Loops*, *Rey-Figure_Score*).
- High relevance variables: comprised of five test scores, which score between 0.1 and 0.08 (*TMT-A*, *Semantic_Fluency*, *TMT-B*, *MEC*, and *Ideatory_Praxias*).
- Medium relevance variables: formed by ten test scores, which score between 0.08 and 0.06 (*ListA_Total*, *Long_Memory_Clues*, *Long_Memory*, *Short_Memory_Clues*, *Short_Memory*, *Phonemic_Continuity*, *Ideomotor_Praxias*, *SF_Plants*, *Rey-Figure_Time*, and *SF_Clothes*).
- Low relevance variables: four test scores that score below 0.06 (*SF_Animals*, *ListB_Total*, *SF_Vehicles*, *Recognition*).

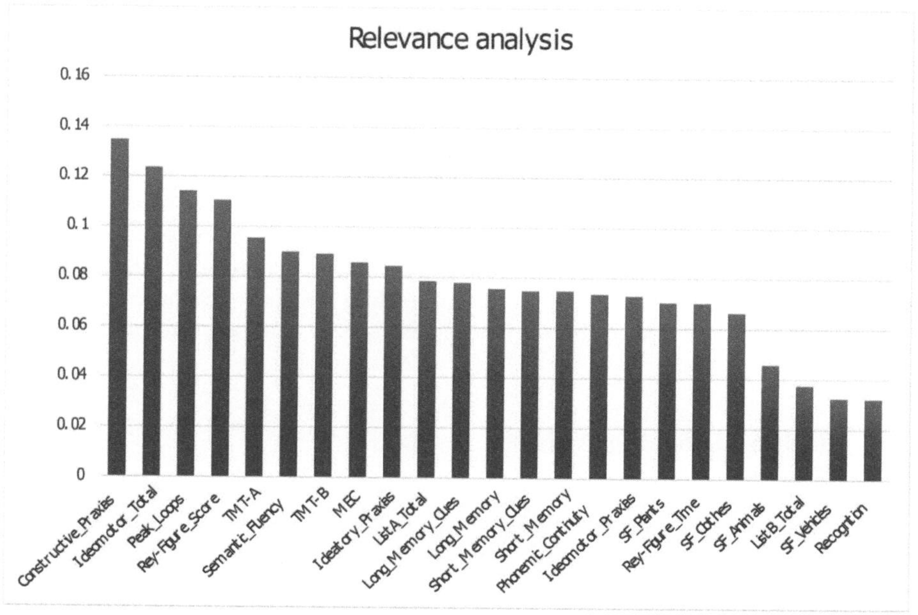

Fig. 2. Analysis of the relevance of the test scores of the neuropsychological battery using the mutual information method.

In the correlation analysis, we obtain that the test scores of *Semantic_Fluency* is highly correlated with *SF_Animals, SF_Clothes, SF_Plants* and *SF_Vehicles*; *Ideomotor_Total* is highly correlated with *Ideatory_Praxias* and *Ideomotor_Praxias*; and *ListA_Total, Short_Memory, Short_Memory_Clues, Long_Memory* and *Long_Memory_Clues* are highly correlated between them, all with a correlation score of more than 0.75 (absolute value), as it can be seen in Fig. 3, with quite distant from the next pair of related scores. Since *SF_Animals, SF_Clothes, SF_Plants*, and *SF_Vehicles* are not highly correlated between them, we considered *Semantic_Fluency* as the redundant test score. The same situation occurs in the next situation, since *Ideatory_Praxias* and *Ideomotor_Praxias* are not highly correlated with each other, but both have a great correlation with *Ideomotor_Total*. Therefore, we considered *Ideomotor_Total* as the redundant test score. In the last situation, *ListA_Total, Short_Memory, Short_Memory_Clues, Long_Memory*, and *Long_Memory_Clues* assess the ability of subjects to recall a specific word list after certain time intervals. Since these scores are all correlated with each other and *ListA_Total* is the first test score to be recorded, we considered *Short_Memory, Short_Memory_Clues, Long_Memory*, and *Long_Memory_Clues* to be the redundant ones.

From those results, a total of four test score groups were defined, of which redundant test scores have already been excluded:

– Group 1: only formed by the very high relevance scores.

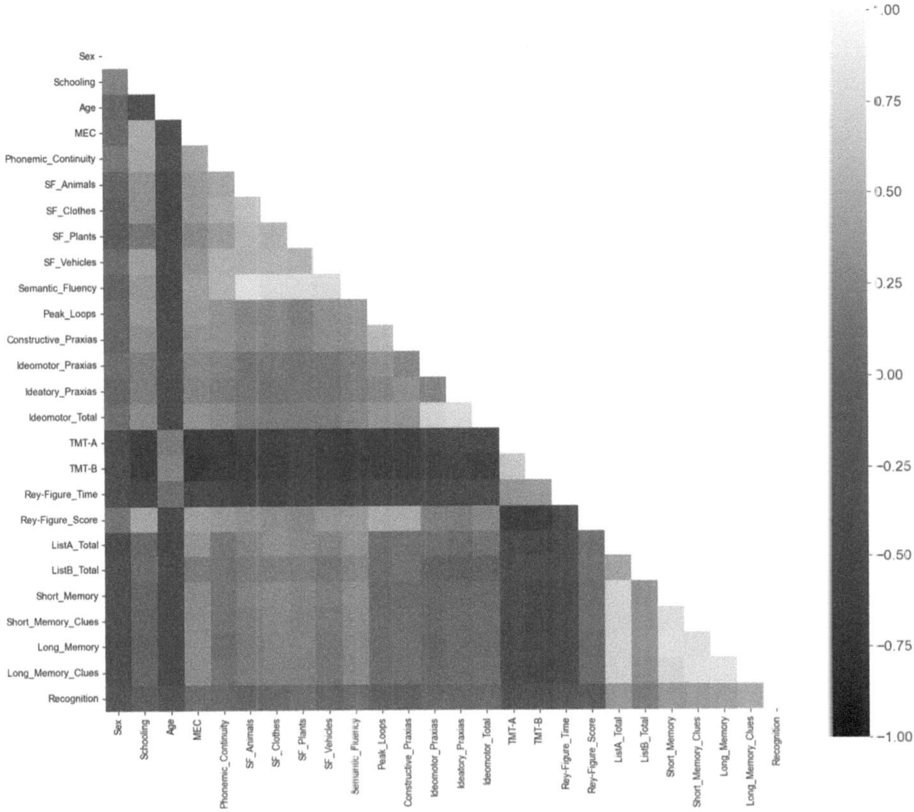

Fig. 3. Pearson's correlation between the different test scores.

- Group 2: formed by only the very high and high relevance test scores.
- Group 3: this group includes test scores with very high, high, and medium relevance.
- Group 4: includes all non-redundant test scores of the neuropsychological battery, including the low-relevance test scores.
- Group 5: includes all variables as the base group.

3.2   Machine Learning Analysis

The results of the machine learning analysis are shown in Table 2. We used the average results of all machine learning methods to ensure that the analysis is not dependent on the chosen method. The group that has the best performance is Group 5, which achieves the best results in all metrics except Precision. Of the reduced test score groups, we extract that the best option is Group 3 (all test scores excluding low-relevance and redundant test scores), as this combination has the highest scores in *F1*, *F2*, and *Recall*. The next one is Group 4, obtaining scores very close to Group 3. Group 1 results are notably lower than the other groups in nearly all metrics, meaning that using only highly relevant test scores misses important information necessary for the correct diagnosis of subjects.

Table 2. Comparison of the performance between the four reduced test-battery groups (Groups 1 to 4) and the control group (Group 5) that includes all tests. The best scores of the reduced groups are highlighted in bold.

	Accuracy	F1	F2	Recall	Precision	ROC-AUC
Group 1	0.772	0.686	0.652	0.627	0.782	0.820
Group 2	0.777	0.696	0.671	0.648	0.784	0.830
Group 3	0.784	**0.714**	**0.698**	0.682	0.773	0.853
Group 4	**0.786**	0.711	0.693	0.673	**0.786**	**0.856**
Group 5	0.793	0.726	0.712	0.696	0.784	0.862

Table 3 shows the breakdown of the performance of the different machine learning methods for Group 3, the reduced test score group that obtains the best performance. There, the methods are ranked by *F2* score from highest to lowest. In the performance comparison of different machine learning methods for Group 3, XGB achieved the highest scores on *F1*, *F2* and *Accuracy*, while ADAB scored the highest *Recall*.

4   Discussion

It is noticeable the presence of the semantic fluency test of *Animals* in the low-relevance group of test scores, contrasting with the presence of the semantic fluency test of *Plants*, contrary to the findings in [25], where the most relevant semantic fluency tests were *Clothes* and *Animals*. However, that could be explained by the fact that, in that other analysis, more variables were obtained from each singular test beyond just the number of correct terms, which is the test score used in this article. These additional variable scores might be more informative for *Animals* than for *Plants*, which could explain the higher relevance of the Animal test. This also suggests that the relevance of those tests may change significantly if new scores are introduced and more information is extracted from each test.

From the performance analysis by machine learning, we obtain that Group 5 is the test score group which has better overall performance. However, the high performance of Group 5 may be due to the presence of redundant variables, which may be artificially increasing its performance. Of the reduced test score groups, we obtained that Group 3 (all tests, except low-relevance tests) is the grouping with the best performance in *F1*, *F2*, and *Recall*, and it is barely surpassed by Group 4 in *Accuracy* ($Accuracy_{Group3} = 0.784$, $Accuracy_{Group4} = 0.786$) and

Table 3. Performance comparison of seven machine learning methods, organized by *F1* performance (MLP = MultiLayer Perceptron, XGB = XGBoost, RLog = Logistic Regression, RF = Random Forest, ADAB = AdaBoosting, SVM = Support Vector Machine, BAG = Bagging) for Group 3.

	Accuracy	F1	F2	Recall	Precision	ROC-AUC
XGB	**0.815**	**0.760**	**0.749**	0.736	0.806	0.894
ADAB	0.806	0.755	0.747	0.738	0.786	0.867
BAG	0.799	0.746	0.739	0.730	**0.779**	**0.875**
RF	0.813	0.751	0.732	0.713	0.822	0.896
RLog	0.800	0.727	0.698	**0.674**	0.811	0.858
SVM	0.733	0.648	0.628	0.611	0.702	0.795
MLP	0.722	0.608	0.594	0.571	0.706	0.785

$ROC - AUC(ROC - AUC_{Group3} = 0.853, ROC - AUC_{Group4} = 0.856)$. This suggests that those four low-relevant test scores could be dispensable using the current scoring method. The poor performance of Group 1 regarding the other test score groups implies that a very drastic reduction in the number of test scores leads to a loss of relevant information in the correct diagnosis of subjects, even if those tests were found to be highly discriminative between healthy and MCI. This could be explained by the fact that such a low number of test scores does not fully cover all the cognitive functions relevant for a correct diagnosis of MCI.

In the breakdown of performance by ML method for the best group, we find that XGB is the model with the best performance for Group 3, presenting both the highest *F1* and *F2* scores, followed by ADAB, which presents results very similar to XGB ($F1_{XGB} = 0.760$, $F1_{ADAB} = 0.755$) and $F2$ ($F2_{XGB} = 0.749$, $F2_{ADAB} = 7.47$), although with a slight improvement in *Recall* for ADAB ($Recall_{XGB} = 0.736$, $Recall_{ADAB} = 0.738$).

5   Conclusions

Neuropsychological test batteries are formed by several tests that evaluate different cognitive functions. This allows a global overview of the cognitive state of subjects, but neuropsychological test batteries that integrate too many tests could become slow to administer and evaluate, hindering their capability to be used in population screenings.

In this way, from the analysis of the neuropsychological tests, we found six test scores to be redundant compared to other test scores. Based on the analysis of the test score groups, we obtain that Group 5 (all test scores, even redundant ones) has the highest performance, most likely because of the redundant test scores. Among the groupings that exclude redundant test scores, we obtain Group 3 (all test scores, except low-relevant ones) as the best combination, although Group 4 presents very close results. The machine learning method that works best with this combination is XGB, scoring the highest *F1* and *F2*.

6   Future Work

One of the main drawbacks of this study is the moderate number of cases used, and all cases have been obtained from the same cohort, which could have introduced bias. Therefore, as future work, it is planned to increase the number of cases, including subjects from different cohorts, to ensure that the results are more generalizable. It has also been planned to optimize other tests, to extract as much information as possible from them while reducing the time required in their administration and evaluation, as has been done in a previous work [25]. This analysis should be carried out before properly proposing a new neuropsychological test battery, since the new test scores could change the relevance of the neuropsychological tests in the diagnosis of the subjects. Another research line is the automation of the evaluation of the neuropsychological battery's tests, as has already been done with the semantic fluency test [25] and the *Rey-Osterrieth complex figure* [26]. This automation reduces the time physicians need to evaluate these tests, allowing them to administer those tests to more people and reach a wider population during screenings.

Acknowledgments. We would like to thank Dr. María del Carmen Díaz-Mardomingo and Dr Sara García-Herranz for giving permission to use their databases in our project.

Disclosure of Interests. The authors have no competing interests to declare that are relevant to the content of this article.

References

1. Nichols, E., et al.: Estimation of the global prevalence of dementia in 2019 and forecasted prevalence in 2050: an analysis for the Global Burden of Disease Study 2019. Lancet Public Health **7**(2), 105–125 (2022). https://doi.org/10.1016/S2468-2667(21)00249-8
2. Petersen, R.C., Caracciolo, B., Brayne, C., Gauthier, S., Jelic, V., Fratiglioni, L.: Mild cognitive impairment: a concept in evolution. J. Int. Med. **275**(3), 214–228 (2014). https://doi.org/10.1111/joim.12190
3. Ivascu, T., Manate, B., Negru, V.: A multi-agent architecture for ontology-based diagnosis of mental disorders. In: 2015 17th International Symposium on Symbolic and Numeric Algorithms for Scientific Computing (SYNASC), pp. 423–430. IEEE (2015). https://doi.org/10.1109/SYNASC.2015.69
4. König, A., Linz, N., Tröger, J., Wolters, M., Alexandersson, J., Robert, P.: Fully automatic speech-based analysis of the semantic verbal fluency task. Dement. Geriatr. Cogn. Disord. **45**(3–4), 198–209 (2018). https://doi.org/10.1159/000487852
5. Adelson, R.P., et al.: Machine learning approach for improved longitudinal prediction of progression from mild cognitive impairment to Alzheimer's disease. Diagnostics **14**(13) (2014). https://doi.org/10.3390/diagnostics14010013

6. Clark, D.G., et al.: Novel verbal fluency scores and structural brain imaging for prediction of cognitive outcome in mild cognitive impairment. Alzheimer's Dementia: Diagnosis Assess. Dis. Monit. **2**, 113–122 (2016). https://doi.org/10.1016/j.dadm.2016.02.001

7. Franciotti, R., Nardini, D., Russo, M., Onofrj, M., Sensi, S.L.: Comparison of machine learning-based approaches to predict the conversion to Alzheimer's disease from mild cognitive impairment. Neuroscience **514**, 143–152 (2023). https://doi.org/10.1016/j.neuroscience.2023.01.029

8. Sun, Y., Tang, Y., Ding, S., Lv, S., Cui, Y.: Diagnose the mild cognitive impairment by constructing Bayesian network with missing data. Expert Syst. Appl. **38**(1), 442–449 (2011). https://doi.org/10.1016/j.eswa.2010.06.084

9. Chasles, M.-J., et al.: An examination of semantic impairment in amnestic MCI and AD: what can we learn from verbal fluency? Arch. Clin. Neuropsychol. **35**(1), 22–30 (2020). https://doi.org/10.1093/arclin/acz018

10. Goli, P., Rad, E.M., Ghandehari, K., Azarnoosh, M.: Early assessment of mild Alzheimer's disease using Elman neural network, LDA and SVM methods. Mach. Learn. Res. **2**(4), 148–151 (2017). https://doi.org/10.11648/j.mlr.20170204.15

11. García-Herranz, S., Díaz-Mardomingo, M.C., Venero, C., Peraita, H.: Accuracy of verbal fluency tests in the discrimination of mild cognitive impairment and probable Alzheimer's disease in older Spanish monolingual individuals. Neuropsychol. Dev. Cogn. Section B Aging Neuropsychol. Cogn. **26**, 1–15 (2019). https://doi.org/10.1080/13825585.2019.1698710

12. Gurevich, P., Stuke, H., Kastrup, A., Stuke, H., Hildebrandt, H.: Neuropsychological testing and machine learning distinguish Alzheimer's disease from other causes for cognitive impairment. Front. Aging Neurosci. **9**(114) (2017). https://doi.org/10.3389/fnagi.2017.00114

13. Hastings, H., et al.: Interdisciplinary perspectives on the development, integration, and application of cognitive ontologies. Front. Neuroinform. **8**(62) (2014). https://doi.org/10.3389/fninf.2014.00062

14. Weakley, A., Williams, J.A., Schmitter-Edgecombe, M., Cook, D.J.: Neuropsychological test selection for cognitive impairment classification: a machine learning approach. J. Clin. Exp. Neuropsychol. **37**(9), 899–916 (2015). https://doi.org/10.1080/13803395.2015.1067290

15. Hastings, J., Ceusters, W., Jensen, M., Mulligan, K., Smith, B.: Representing mental functioning: ontologies for mental health and disease. In: Towards an Ontology of Mental Functioning (ICBO Workshop), Proceedings of the Third International Conference on Biomedical Ontology (2012)

16. Jitsuishi, T., Yamaguchi, A.: Searching for optimal machine learning model to classify mild cognitive impairment (MCI) subtypes using multimodal MRI data. Sci. Rep. **12** (2022). https://doi.org/10.1038/s41598-022-08231-y

17. Linz, N., Troger, J., Alexandersson, J., Konig, A.: Using neural word embeddings in the analysis of the clinical semantic verbal fluency task. In: International Conference on Computational Semantics, Computer Science (2017)

18. López-de-Ipiña, K., et al.: On the analysis of speech and disfluencies for automatic detection of mild cognitive impairment. Neural Comput. Appl. **32**(20), 15761–15769 (2018). https://doi.org/10.1007/s00521-018-3494-1

19. Díaz-Mardomingo, M., García-Herranz, S., Rodríguez-Fernández, R., Venero, C., Peraita, H.: Problems in classifying mild cognitive impairment (MCI): one or multiple syndromes?. Brain Sci. **7**(111) (2017). https://doi.org/10.3390/brainsci7090111

20. Díaz-Mardomingo, M.C., Peraita, H.: Detección precoz del deterioro cognitivo ligero de la tercera edad. Psicothema **20**(3), 438–444 (2008)

21. Peraita, H., García-Herranz, S., Díaz-Mardomingo, M.C.: Evolution of specific cognitive subprofiles of mild cognitive impairment in a three-year longitudinal study. Curr. Aging Sci. **4**, 171–182 (2011). https://doi.org/10.2174/1874609811104020171
22. Lobo, A., Ezquerra, J., Gómez Burgada, F., Sala, J.M., Seva Díaz, A.: Cognocitive mini-test (a simple practical test to detect intellectual changes in medical patients). Actas Luso-Espanolas De Neurologia, Psiquiatria Y Ciencias Afines **7**(3), 189–202 (1979). ISSN 0300-5062
23. Yesavage, J.A., et al.: Development and validation of a geriatric depression screening scale: a preliminary report. J. Psychiatr. Res. **17**(1), 37–49 (1982). https://doi.org/10.1016/0022-3956(82)90033-4
24. Benedet, M.J., Alejandre, M.A.: TAVEC. Test de Aprendizaje Verbal España-Complutense, Madrid: TEA Ediciones (2014). https://www.hogrefe-tea.com/recursos/Ejemplos/TAVEC-Manual-Extracto.pdf. Accessed 16 June 2025
25. Gómez-Valadés, A., Martínez, R., Rincón, M.: Designing an effective semantic fluency test for early MCI diagnosis with machine learning. Comput. Biol. Med. **180**(108955) (2024). https://doi.org/10.1016/j.compbiomed.2024.108955
26. Guerrero-Martín, J., Díaz-Mardomingo, M.C., García-Herranz, S., Martínez-Tomás, R., Rincón, M.: A benchmark for Rey-Osterrieth complex figure test automatic scoring. Heliyon **10**(21) (2024). https://doi.org/10.1016/j.heliyon.2024.e39883

Accurate Fall Risk Prediction in Older Adults: Integrating Sensor and Clinical Data Through Machine Learning

Ana González Castro[1] [iD], José Alberto Benítez-Andrades[2] [iD],
and Raquel Leirós-Rodríguez[3(✉)] [iD]

[1] Nursing and Physical Therapy Department, Universidad de León, 24402 Ponferrada, Spain
[2] ALBA Research Group, Department of Electric, Systems and Automatics Engineering, Universidad de León, 2401 León, Spain
[3] SALBIS Research Group, Nursing and Physical Therapy Department, University of León, 24402 Ponferrada, Spain
rleir@unileon.es

Abstract. Accurate identification of older adults at high risk of falling is essential to prevent injuries and implement effective interventions. This study evaluated the performance of several machine learning models in predicting fall risk using both accelerometric data from wearable sensors and non-accelerometric data including demographic, functional, and clinical variables. A dataset comprising 146 older women was used to train and compare seven algorithms: random forest, XGBoost, AdaBoost, LightGBM, Bayesian ridge regression, decision trees, and support vector regression. Predictive accuracy was assessed using mean squared error, mean absolute error, and the coefficient of determination.

Models trained with combined accelerometric and non-accelerometric data consistently outperformed those using a single data source. XGBoost achieved the lowest MSE, while Bayesian ridge regression reached the highest R2, highlighting its superior explanatory capability. In contrast, support vector regression exhibited poor predictive performance. Non-accelerometric variables, especially age and comorbidities, emerged as major predictors, whereas accelerometric data alone yielded limited accuracy.

The integration of multiple data types significantly improved model robustness and clinical applicability. These findings underscore the value of multi-source data fusion in enhancing fall risk prediction among older adults and support the implementation of hybrid machine learning models in clinical settings.

Keywords: Fall risk prediction · Machine learning · Older adults

1 Background

Falls are a major public health concern among older adults, representing a leading cause of injury, hospitalization, and loss of independence worldwide. It is estimated that approximately one-third of individuals over the age of 65 fall at least once a year, and these incidents frequently result in fractures, head trauma, or long-term functional decline [1, 2]. Moreover, the fear of falling itself often leads to activity restriction, further exacerbating frailty and reducing quality of life [3].

Traditionally, fall risk assessments have relied on functional tests, clinical history, and observational evaluations, which offer valuable insights but often lack predictive precision. In recent years, wearable sensor technology—especially accelerometers—has enabled researchers to collect real-time data on gait and movement patterns in free-living environments. Accelerometric data can reveal subtle deficits in postural control, balance, and gait stability that may not be apparent during clinical evaluations[4, 5]. These advances have opened the door to objective, continuous monitoring of fall risk indicators.

However, accelerometric data alone cannot capture the full complexity of fall risk. Factors such as age, comorbidities, cognitive status, and prior fall history are known to contribute substantially to fall vulnerability and must be integrated for a comprehensive assessment [6, 7]. The variability in device placement, inconsistent use, and missing data also pose practical limitations to sensor-based approaches [8].

To address these challenges, current research is increasingly focused on combining sensor-derived movement data with traditional clinical and demographic variables. This multi-source data fusion has shown promise in improving the sensitivity and specificity of fall risk prediction models. Recent studies have emphasized that machine learning (ML) algorithms are particularly well-suited to handling this type of high-dimensional, heterogeneous data [9, 10]. ML models can automatically identify complex patterns and interactions among variables, offering greater flexibility and predictive power than conventional statistical methods.

Among the various ML algorithms tested in fall risk research, ensemble methods such as random forest, XGBoost, and LightGBM have demonstrated robust performance, especially in structured datasets with mixed feature types [11]. Bayesian regression models, while simpler, provide strong interpretability and are particularly effective in small-to-medium-sized datasets, where overfitting is a concern [12]. In contrast, support vector regression has been shown to underperform in high-dimensional datasets when kernel selection is suboptimal or feature scaling is not ideal [13].

Despite growing interest in data-driven fall prediction, many existing models focus exclusively on either accelerometric or clinical data in isolation, limiting their generalizability and real-world utility. Few studies have directly compared the performance of ML models across different data configurations or analyzed the relative contribution of each data type. As such, there is a need for more integrative approaches that combine the strengths of sensor technology and clinical reasoning to develop accurate, interpretable, and scalable prediction tools.

The present study addresses this gap by evaluating and comparing the performance of multiple ML algorithms trained on accelerometric, non-accelerometric, and combined datasets. By doing so, it aims to identify the most effective modeling strategies and the key variables contributing to fall risk among older adults. The findings have potential implications for designing hybrid clinical tools that support early intervention, personalized care, and fall prevention strategies in aging populations.

2 Methodology

This study employed a supervised machine learning approach to develop predictive models for fall risk in older adults, using a dataset that integrated both accelerometric and non-accelerometric variables. A cross-sectional design was adopted, and the study population consisted of 146 community-dwelling older women aged 65 years or older. All participants underwent comprehensive clinical assessments and wore wearable sensors to collect movement data during standardized gait and balance tests.

2.1 Data Collection

The dataset comprised two primary groups of variables. The **non-accelerometric variables** included age, body mass index (BMI), medical history (e.g., cardiovascular and degenerative diseases), and functional test scores such as the Timed Up and Go (TUG) test and the Six-Minute Walk Test (6MWT). These variables were selected based on their established relevance in fall risk research.

The **accelerometric data** were obtained using triaxial wearable sensors (ActiGraph GT3X) placed on the lower back during walking tasks. These sensors recorded linear accelerations along vertical, sagittal, and medio-lateral axes. From these signals, a range of features was derived, including maximum and minimum acceleration, root mean square (RMS) values, and timing of peak accelerations in different planes.

2.2 Preprocessing

All variables were examined for completeness, and no missing data were present. Continuous variables were normalized to ensure uniform scaling, and categorical variables were encoded using one-hot encoding. Outliers, particularly in accelerometric measures, were identified through interquartile range analysis and handled conservatively to preserve valid variability.

To enhance the robustness of model training, the dataset was randomly split into training (80%) and testing (20%) sets. Additionally, synthetic oversampling techniques, including the Synthetic Minority Over-sampling Technique (SMOTE), were applied to balance the distribution of high- and low-risk fall cases in the training set.

2.3 Model Development

Seven machine learning algorithms were selected for model development based on their previous application in biomedical data analysis:

1. **Random Forest**
2. **XGBoost (Extreme Gradient Boosting)**
3. **AdaBoost**
4. **LightGBM (Light Gradient Boosting Machine)**
5. **Bayesian Ridge Regression**
6. **Support Vector Regression (SVR)**
7. **Decision Trees**

In this study, seven algorithms with established applications in the biomedical field were selected. **Decision trees** stand out for their simplicity and interpretability, a key quality in clinical settings. Based on these, ensemble methods such as **Random Forest** improve stability and reduce the risk of overfitting by using multiple tres [14]. Boosting algorithms (**XGBoost, AdaBoost, and LightGBM**) were chosen for their ability to optimize predictive performance, especially on structured and heterogeneous data [15]. **Bayesian ridge regression** was incorporated due to its probabilistic approach and its ability to handle multicollinearity among predictors, while also integrating uncertainty into the estimates [16]. Finally, **Support Vector Regression (SVR)** was selected for its suitability for modeling nonlinear relationships and its robustness in high-dimensionality scenarios [17].

Each model was trained under three different conditions: (1) accelerometric variables only, (2) non-accelerometric variables only, and (3) the combination of both.

A **grid search with five-fold cross-validation** was used to tune hyperparameters for each algorithm. This process aimed to identify optimal settings such as tree depth, learning rate, and regularization parameters. Performance was evaluated using the test set, and metrics included mean squared error (MSE), mean absolute error (MAE), and the coefficient of determination (R^2).

2.4 Feature Importance Analysis

To interpret the contribution of individual predictors, feature importance scores were extracted. For ensemble models (e.g., Random Forest, LightGBM), importance was based on split gain or Gini importance. For linear and kernel-based models, **SHAP (SHapley Additive exPlanations)** values were calculated to assess the influence of each feature on the output prediction.

All analyses were implemented using **Python (v3.9)** and relevant libraries such as **scikit-learn, XGBoost**, and **LightGBM**, executed on a high-performance computing system with 64 GB of RAM and an Intel Core i9 processor.

3 Experimentation Results

The machine learning models demonstrated varying levels of predictive accuracy depending on the type of data used. Overall, models trained with the **combined dataset** (accelerometric and non-accelerometric variables) consistently outperformed those trained with only one data source (Fig. 1).

Among all algorithms, **XGBoost** achieved the **lowest prediction error**, while **Bayesian Ridge Regression** yielded the **highest coefficient of determination (R^2)**, indicating strong explanatory power. In contrast, **Support Vector Regression** underperformed across all metrics, particularly when trained on accelerometric data alone.

When comparing data types, **non-accelerometric variables** such as age, functional performance, and comorbidities emerged as **stronger individual predictors** than accelerometric features alone. However, the integration of both data types provided the most accurate and robust results, reinforcing the value of multi-source data fusion in fall risk modeling.

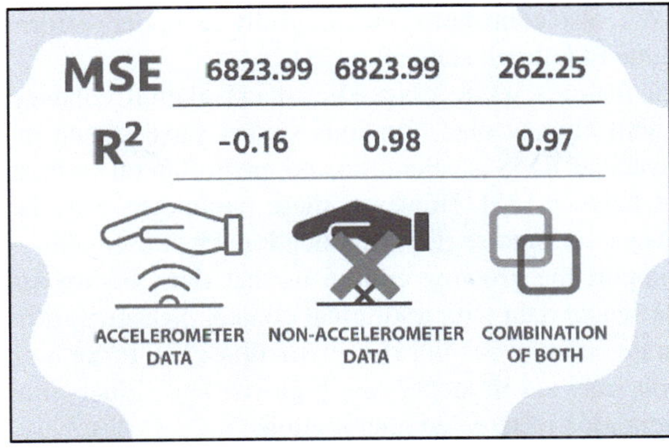

Fig. 1. Predictive performance of accelerometric, non-accelerometric, and combined variables based on MSE and R^2 metrics.

4   Interpretation

The machine learning models demonstrated varying levels of predictive accuracy depending on the type of data used. Overall, models trained with the **combined dataset**.

This study highlights the benefits of combining clinical, demographic, and sensor-based data to improve fall risk prediction among older adults using machine learning models. Our findings align with prior evidence that emphasizes the multifactorial nature of fall risk and the limitations of relying solely on either functional assessments or accelerometric signals [18, 19].

The comparative performance of the models revealed that ensemble methods, particularly XGBoost and LightGBM, were effective at capturing complex, non-linear relationships among variables, yielding high predictive accuracy when trained on combined datasets. This reinforces previous research suggesting that gradient boosting techniques are especially well-suited for structured health data where interactions between features may be subtle but clinically relevant [20].

Interestingly, Bayesian Ridge Regression showed a strong capacity to explain variance (R2) despite being a linear model, suggesting that it may be particularly advantageous in smaller datasets with high dimensionality, where overfitting is a concern [12]. These findings are consistent with previous studies that have advocated for regularized regression techniques in clinical prediction contexts [21].

As expected, Support Vector Regression (SVR) underperformed in this study. Its sensitivity to parameter tuning and poor scalability in high-dimensional settings has been documented elsewhere [13], and our results further support the notion that SVR may not be the most appropriate tool for fall risk prediction using heterogeneous data sources.

A key insight from our analysis is that **non-accelerometric variables**, particularly **age, lower limb strength, and presence of comorbidities**, were among the most influential predictors. This agrees with existing literature identifying these variables as consistent risk factors for falls in older adults [22, 23]. Functional test outcomes, such as the TUG and 6MWT, also contributed meaningfully to model performance, reinforcing their continued value in fall risk screening [23].

Accelerometric features, while less predictive in isolation, enhanced model accuracy when combined with clinical data. Previous studies have shown that accelerometry-derived variables such as RMS acceleration and medio-lateral sway are associated with impaired gait and balance [25]. However, these parameters may lack the contextual information necessary for precise risk classification when used alone [26].

Our results support the growing consensus that **data fusion approaches** (which integrate wearable sensor data with traditional clinical measures) offer a more comprehensive and accurate strategy for fall risk prediction [27]. From a clinical standpoint, such hybrid models may aid in identifying high-risk individuals more effectively and guide the implementation of targeted interventions.

Nonetheless, several limitations should be acknowledged. The study was based on a relatively small and homogeneous sample, composed exclusively of older women. While this controls for sex-related variability, it limits generalizability to broader populations. Moreover, the cross-sectional nature of the data prevents assessment of predictive validity over time. Future studies should consider longitudinal designs and more diverse cohorts to validate these models in real-world clinical environments.

In conclusion, this study provides additional evidence that **machine learning models combining accelerometric and non-accelerometric data outperform models using single data types**, and may serve as valuable tools for fall risk stratification in aging populations.

5 Conclusions

The present study demonstrates that integrating accelerometric and non-accelerometric data significantly enhances the accuracy and robustness of fall risk prediction models in older adults. While clinical and functional variables such as age, comorbidities, and mobility tests were strong individual predictors, their combination with sensor-derived gait parameters yielded superior performance.

Among the algorithms tested, ensemble methods and Bayesian regression showed the greatest potential for practical application, balancing predictive power with interpretability. These findings support the use of hybrid machine learning models as decision-support tools in geriatric care, contributing to more effective screening and prevention strategies.

Further research should validate these models in larger and more diverse populations and explore their use in longitudinal and real-world clinical settings.

Disclosure of Interests. The authors have no competing interests to declare that are relevant to the content of this article.

References

1. World Health Organization: Falls: key facts. World Health Organization, Geneva (2025)
2. Hopewell, S., et al.: Multifactorial and multiple component interventions for preventing falls in older people living in the community. Cochrane Database Syst. Rev. 7(7), CD012221 (2018)

3. Delbaere, K., Close, J.C.T., Mikolaizak, A.S., Sachdev, P.S., Brodaty, H., Lord, S.R.: The falls efficacy scale international (FES-I). A comprehensive longitudinal validation study. Age Ageing **39**(1), 64–70 (2010)

4. Urbanek, J.K., et al.: Free-living gait cadence measured by wearable accelerometer: a promising alternative to traditional measures of mobility for assessing fall risk. J. Gerontol. Ser. A Biol. Sci. Med. Sci. **78**(5), 802–810 (2023)

5. Schootemeijer, S., Weijer, R.H., Hoozemans, M.J.M., Delbaere, K., Pijnappels M., van Schooten, K.S.: Responsiveness of daily life gait quality characteristics over one year in older adults who experienced a fall or engaged in balance exercise. Sensors **23**(1), 101 (2022)

6. Montero-Odasso, M., et al.: World guidelines for falls prevention and management for older adults: a global initiative. Age Ageing **51**(9), afac205 (2022)

7. Chan, L.L.Y., Arbona. C.H., Brodie, M., Lord, S.R.: Prediction of injurious falls in older adults using digital gait biomarkers extracted from large-scale wrist sensor data. Age Ageing **52**(4), afad179 (2023)

8. Thiamwong, L., et al.: Levels of accelerometer-based physical activity in older adults with a mismatch between physiological fall risk and fear of falling. J. Gerontol. Nurs. **49**(6), 41–49 (2023)

9. Sharma, V., et al.: Predicting falls-related admissions in older adults in Alberta, Canada: a machine-learning falls prevention tool developed using population administrative health data. BMJ Open **13**(8), e071321 (2023)

10. Sasso, G., Mou, L., Hernandez, M.E.: Classification of fall risk across the lifespan using gait-derived features from a wearable device. In: Annual International Conference of the IEEE Engineering in Medicine and Biology Society, pp. 1–4 (2023)

11. Millet, A., Madrid, A., Alonso-Weber, J.M., Rodríguez-Mañas, L., Pérez-Rodrá-Guez, R.: Machine learning techniques applied to the development of a fall risk index for older adults. IEEE Access **11**, 84795–84809 (2023)

12. Tipping, M.E.: Sparse Bayesian learning and the relevance vector machine. J. Mach. Learn. Res. **1**, 211–244 (2001)

13. Smola, A.J., Schölkopf, B.: A tutorial on support vector regression. Stat. Comput. **14**(3), 199–222 (2004)

14. Halabaku, E., Bytyçi, E.: Overfitting in machine learning: a comparative analysis of decision trees and random forests. Intell. Autom. Soft Comput. **39**(6), 987–1006 (2024)

15. Chen, T., Guestrin, C.: XGBoost: a scalable tree boosting system. In: Proceedings of the 22nd ACM SIGKDD International Conference on Knowledge Discovery and Data Mining, pp. 785–794 (2016)

16. Pal, J., Hong, B.: Applying Bayesian ridge regression AI modeling in virus severity prediction. arXiv:2310.09485 (2023)

17. Yu, Y., et al.: Leveraging support vector regression for outcome prediction in personalized ultra-fractionated stereotactic adaptive radiotherapy. arXiv:2509.07872 (2025)

18. Lien, W., et al.: Intelligent fall-risk assessment based on gait stability and symmetry among older adults using tri-axial accelerometry. Front. Bioeng. Biotechnol. **10**, 887269 (2022)

19. Antonietti, A.: Machine learning architectures to classify activities of daily living and fall types from wearable accelerometer data. In: IEEE Signal Processing in Medicine and Biology Symposium, pp. 1–6 (2023)

20. Nguyen, V.G., et al.: An extensive investigation on leveraging machine learning techniques for high-precision predictive modeling of CO_2 emission. Energy Sources Part A: Recov. Utilization Environ. Effects **45**(3), 9149–9177 (2023)

21. Pap, J., Mako, C., Illéssy, M., Kis, N., Mosavi, A.: Modeling organizational performance with machine learning. J. Open Innov. Technol. Market Complex. **8**(4), 177 (2022)

22. Bindawas, S.M.: The changing incidence and prevalence of falls and its disability burden among the geriatric population in Saudi Arabia from 1990 to 2019: a longitudinal analysis using global burden of disease study data. Cureus **15**(11), e49117 (2023)
23. Catikkas, N.M., et al.: Prevalence and determinants of falls in community-dwelling older adults in Türkiye: a population-based cross-sectional study conducted between 2014–2015. Curr. Aging Sci. **16**(2), 133–142 (2023)
24. Nordling, Y., Sund, R., Sirola, J., Kröger, H., Isanejad, M., Rikkonen, T.: Body composition, balance, and functional capacity and falls in older women. Aging Clin. Exp. Res. **36**(1), 76 (2024)
25. Castiglia, S., et al.: Identification of gait unbalance and fallers among subjects with cerebellar ataxia by a set of trunk acceleration-derived indices of gait. Cerebellum **22**(1), 46–58 (2023)
26. Hatkeposhti, R.K., Yadollahzadeh-Tabari, M., Golsorkhtabari, M.: Providing an approach for early prediction of fall in human activities based on wearable sensor data and the use of deep learning algorithms. Comput. J. **67**(2), 658–673 (2024)
27. Wang, S., Miranda, F., Wang, Y., Rasheed, R., Bhatt, T.: Near-fall detection in unexpected slips during over-ground locomotion with body-worn sensors among older adults. Sensors **22**(9), 3334 (2022)

Time Series Models: Application to RespiCast (ECDC Respiratory Diseases Forecasting Hub)

Lucía Maza[1], Andrés M. Alonso[2] (iD), and Carolina García-Martos[1]([📧]) (iD)

[1] Escuela Técnica Superior de Ingenieros Industriales, Universidad Politécnica de Madrid, C/José Gutiérrez Abascal, 2, 28006 Madrid, Spain
garcia.martos@upm.es
[2] Departamento de Estadística and Instituto Flores de Lemus, Universidad Carlos III de Madrid, Getafe, Spain

Abstract. In this paper several univariate and multivariate (more specifically dimensionality reduction techniques, such as the Dynamic Factor Model, DFM) time series approaches and different training strategies for these models are proposed in order to compute forecasts for the number of cases per 100,000 inhabitants of Acute Respiratory Infections (AcRI) in several countries of the European Union (EU) whose historical datasets are available from RespiCast. All the proposals are used to compute weekly forecasts from 2020 up to spring 2025 and carefully compared using an adequate Analysis of Variance (ANOVA).

Here, the forecasting horizon is extended up to 22 weeks, although in RespiCast the models provided by all the participants are 4 weeks maximum. As a summary, it can be stated that univariate ARIMA models accounting for seasonality are better for the short-term, and DFM produces more accurate forecasts in the long run. Additionally, the combination of forecasts here built with our best reduces the forecasting error of the baseline provided by RespiCast about 50% for the short-term. In some recent periods (first two months in 2025) our approach beats (in accuracy terms) most forecasts submitted by other teams (ranking 1, 2 or 3 out of 10 teams). For the rest of the span of time considered the average ranking is 4. It should be taken into account that just our DFM approach is capable of computing relatively accurate forecasts for longer forecasting horizons (up to 22 weeks) than those considered in RespiCast (just up to 4 weeks).

Keywords: Acute Respiratory Infection · Dimensionality Reduction · Epidemiology · Time Series

1 Introduction

Respiratory diseases, including Acute Respiratory Infections (AcRI), COVID-19 and Influenza-like Illness (ILI), represent a significant challenge for healthcare systems worldwide. Just AcRI accounts for approximately 25% of primary care consultations related to morbidity visits and causes over 4 million deaths annually [1]. Influenza affects 5–15% of the global population each year, leading to millions of severe cases and putting substantial pressure on health services during seasonal peaks [4]. More broadly,

A. López Fernández et al. (Eds.): CIABiomed 2025, LNBI 16148, pp. 617–631, 2026.
https://doi.org/10.1007/978-3-032-10661-2_46

respiratory diseases accounted for 6.1% of all deaths in the EU in 2021, which highlights their major impact on public health [2]. These diseases also lead to considerable economic losses, not only from direct healthcare costs but also from hospitalizations and work absences. In 2019, the healthcare costs associated with chronic respiratory diseases across the EU reached approximately €380 billion [3]. Similarly, COVID-19 highlighted how an emerging respiratory virus can overwhelm healthcare systems and reshape global public health priorities in a short period.

In this context, forecasting the number of cases (or any measure of incidence) plays a crucial role in helping public health authorities to manage resources and to prepare for disease surges [4]. While traditional surveillance tools, such as weekly reporting and retrospective analyses, remain essential, they are not always sufficient for timely and proactive responses. The ability to anticipate (forecast) increases in disease incidence is an unavoidable task to try and reduce the impact of respiratory infections on healthcare systems and to avoid their saturation [5].

Epidemiology has evolved from a primarily descriptive science focused on identifying disease outbreaks to a data-driven discipline that integrates statistical and computational methods. While classical epidemiological approaches relied on retrospective analysis and surveillance reports, the increasing availability of real-time health data has led to significant advances in predictive modelling. These models help anticipate disease trends, aiding proactive public health decisions and resource allocation [6].

Based on an extensive literature review, these models can be classified into four categories: mathematical models, machine learning methods, simulation-based models and statistical modelling [7].

Firstly, mathematical models are widely used in epidemiology to describe and predict the spread of infectious diseases by modelling disease dynamics through equations [8]. Among mathematical models, compartmental models are the most widely applied in epidemiology. These divide the population into groups or compartments according to their disease status [9]. The foundation of this approach was laid by Kermack and McKendrick in 1927 with their groundbreaking work on the SIR (Susceptible-Infected-Recovered) model. Compartmental models have been applied in past outbreaks, including the H1N1 influenza pandemic (2009–2010) and Ebola (2014–2016) [10], and more recently, COVID-19. Their adaptability has been widely documented in comprehensive reviews [11], highlighting their application in both deterministic and stochastic frameworks [12]. Despite their widespread use, mathematical models have inherent limitations. Their effectiveness depends heavily on parameter calibration and underlying assumptions, such as homogeneous mixing or fixed transmission rates, see [13] and [14].

Machine Learning (ML) models are another alternative. They offer adaptability through data-driven learning and have demonstrated improved predictive performance in various scenarios. However, drawbacks remain, particularly those regarding interpretability, overfitting risk and high data requirements [15].

Simulation-based models allow researchers to replicate the complex dynamics of disease transmission by simulating individual interactions within a population [16]. One prominent approach is the Agent-Based Model (ABM), which simulates behavior of individual agents and their interactions within an environment [17]. For instance, during the COVID-19 pandemic, ABMs were extensively used to evaluate the impact of social distancing measures, vaccination strategies and mobility restrictions [19]. While offering high flexibility and realism, simulation-based models require substantial computational resources and detailed data inputs, limiting their use in real-time decision-making contexts [16, 17].

Statistical approaches analyze historical data to identify trends, estimate parameters and generate probabilistic predictions about future outbreaks [19]. Among them, time series analysis, particularly ARIMA models, have been adopted for its ability to capture trends and fluctuations in reported cases [20]. Unlike nonlinear compartmental models, ARIMA models offer a computationally efficient alternative for short-term forecasting [10]. Recent studies have validated ARIMA's performance in epidemiology. For instance, Abolmaali and Shirzaei [21] found that ARIMA outperformed classical SIR models in forecasting COVID-19 cases under rapidly changing conditions. Similarly, Alabdulrazzaq et al. [22] showed that ARIMA provided reliable forecasts for COVID-19 cases within a 95% confidence interval [12].

In addition to univariate statistical models, recent developments have explored the use of multivariate models that can handle high-dimensional time series data. A notable example is the Dynamic Factor Model (DFM), which reduces dimensionality by assuming that a few latent variables (factors) drive the co-movements across multiple observed series. This approach is particularly advantageous when modelling several countries or indicators jointly, as it captures shared dynamics while reducing model complexity. DFMs have demonstrated strong performance in various disciplines. In demography, they were introduced by Lee and Carter [23]; in macroeconomics, they were adopted by Stock and Watson [24]; and in the energy sector, they were successfully applied by García-Martos et al. [25, 26] to forecast electricity prices. This study explores DFMs as a multivariate alternative to univariate ARIMA for modelling and forecasting respiratory infections. Their application in epidemiology is innovative, despite their potential to capture synchronized patterns across regions.

Thus, in this work multivariate time series models, and more particularly dimensionality reduction techniques, are developed to forecast respiratory diseases using data from RespiCast. Despite their proven usefulness in other fields (macroeconomics, finance, demography or energy), application of dimensionality reduction techniques (such as DFM or DFA) in epidemiology remains unexplored although they could be really helpful since their ability to compute medium- and long-term forecasts when the forecasting accuracy of other techniques dramatically decreases had been proven [27].

The rest of the paper is organized as follows: In Sect. 2 the dataset under study is described. In Sect. 3 all the univariate models as well as the multivariate ones implemented are described in detail, as well as the increasing length window and the fixed length rolling window strategies, together with the accuracy metrics used to adequately compare all the approaches. In Sect. 4 the comparison between all the approaches developed via an ANOVA (Analysis of Variance). This allows extracting conclusions about the best models in terms of forecasting accuracy for every country and forecasting horizon (bear in mind that all the teams submitting forecasts to the collaborative project that is RespiCast compute forecasts whose maximum forecasting horizon is 4 weeks and we enlarge this up to 22 weeks). In Sect. 5 the numerical results for the best approaches selected and their combination are provided as well as the comparison with Respi-Cast baseline and other teams participating in this collaborative initiative[1]. Section 6 concludes and highlights the main points for further research.

2   The Data

The present study relies on epidemiological data provided by RespiCast, the new European Respiratory Diseases Forecasting Hub. In Subsect. 2.1 the context of RespiCast and the ECDC are summarized. In Subsect. 2.2 the data available is described and illustrated graphically, to emphasize its empirical features that justify the methodologies of election in this work.

2.1   RespiCast

It is the main source for model development and validation within the European Centre for Disease Prevention and Control (ECDC). RespiCast is a key platform for monitoring, modelling and forecasting respiratory diseases across Europe. Launched in late 2023, it builds upon the foundation of the COVID-19 Forecasting Hub, introduced by ECDC in March 2021 to enhance pandemic response efforts. The success of this initial hub led to the creation of RespiCast, expanding its scope to include additional respiratory disease indicators such as ILI) and AcRI [28].

ECDC, established in 2005 and based in Stockholm (Sweden) plays a critical role in Europe's infectious disease surveillance and preparedness. This public health agency of the EU collects, analyses and disseminates epidemiological data on a wide range of infectious diseases, supporting EU member states in their efforts to prevent and control outbreaks. Additionally, ECDC facilitates the exchange of knowledge and best practices among national health authorities, coordinates surveillance networks, and provides scientific assessments to guide public health decision-making [29].

Within this broader framework, RespiCast serves as a specialized forecasting hub that integrates epidemiological data from multiple official sources. Key institutions, such as the ISI Foundation and the London School of Hygiene & Tropical Medicine (LSHTM), contribute technically to the platform.

A defining feature of RespiCast is its use of ensemble forecasting, a method that improves predictive accuracy by combining predictions from multiple independent models, rather than relying on a single approach. This technique, originally developed for meteorology, has proven effective in multiple domains, including epidemiology, where it strengthens disease predictions and outbreak preparedness [30]. RespiCast applies this approach to compute weekly short-term forecasts, typically spanning one to four weeks, for respiratory disease indicators across Europe. It covers three key indicators of respiratory diseases: acute respiratory infection (AcRI), influenza-like illness (ILI) and COVID-19. These data are used by a diverse set of forecasting models developed by 10 research teams. These contributions provide a broad range of methodologies for predicting disease trends. The teams involved in the development of these models include Dutch

---

[1] It is important to emphasize that REspiCast is not a forecasting competition, but a collaborative environment.

National Institute for Public Health and the Environment (RIVM), European Centre for Disease Prevention and Control (ECDC), Fjordhest, ISI Foundation (ISI), ItaLuxColab, MRC Centre for Global Infectious Disease Analysis (MRC_GIDA), Notre Dame Perkins Lab (NotreDame), Queen Mary University of London (QMUL), RespiCast (respicast) and Safinea [28].

The models contributing to RespiCast can be broadly categorized into the four methodological groups described in the literature review included in Sect. 1:

a) Mathematical (relying on variations of SIR, SEIR and SIRS frameworks. This approach dominates the ensemble).
b) Machine Learning models, applied to capture non-linear patterns in disease indicators (LSTM networks, gradient boosting, and neural forecasting architectures).
c) Statistical models (ARIMAs, Kalman filter and baseline trend estimators, all relying on historical data to forecast short-term trends).
d) Hybrid models (that integrate elements from multiple approaches).

Despite the methodological diversity, compartmental models remain predominant within RespiCast. This highlights an opportunity for time series models (statistical framework), both from:

- the classical univariate perspective, which is well-suited for short-term forecasting due to its simplicity and adaptability and
- multivariate models based on dimension reduction techniques and extraction of common and specific dynamic components, that offer a promising framework to jointly model temporal patterns across countries or indicators.

Its use could help improve forecasting accuracy and extending the forecasting horizon for which forecasts remain accurate enough to be useful for decision making, thus, providing a complementary perspective to the existing RespiCast framework and another approach to be ensembled with the existent ones.

2.2 Exploratory Time Series Analysis for the Data Under Study

Due to availability of data in RespiCast (bearing in mind that some countries do not provide data for the full year, but just for the peak season of respiratory diseases mainly autumn and winter), and having a very low percentage of missing (around 2%) and having data available in the long span of time ranging from Week 23 in 2020 (2020-W23) up to 2025-W12. Missing values were imputed using the imputeTS package in R through a combination of four Kalman smoothing approaches. Specifically, two structural time series models and two autoregressive models were used, each with and without smoothing. The final imputed value for each missing observation was calculated as the average of the four results, providing a more robust and balanced estimate by integrating both modelling strategies and smoothing settings. Thus, the countries considered are BE (Belgium), Bulgaria (BG), Czech Republic (CZ), Germany (DE), Estonia (EE), Lithuania (LT), Romania (RO) and Slovenia (SI).

The main empirical features shown in Fig. 1 are:

- the common pattern among all countries plotted (synchrony exhibited by the 8 time series plotted),

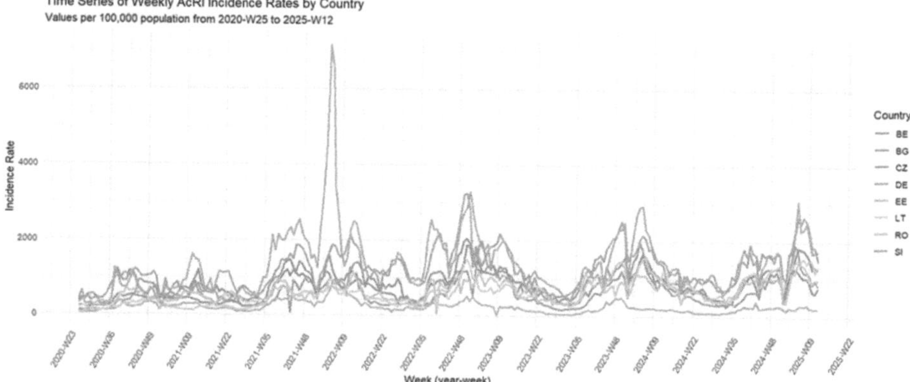

Fig. 1. Weekly time series of acute respiratory disease incidence rates (cases per 100,000 inhab.) in selected countries in the EU, from 2020-W23 to 2025-W12.

- the yearly seasonality (approximately mid to late autumn and winter are the peak seasons and spring and summer are off-peak) and
- the variability increases with the level in some of the countries whose incidences are depicted.

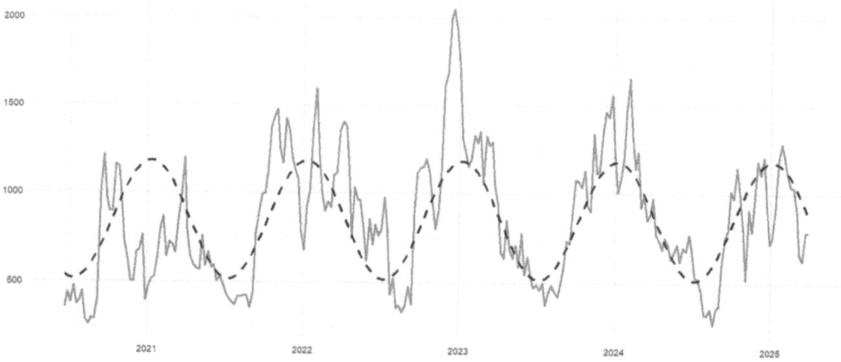

Fig. 2. Presence of yearly seasonality highlighted and justified. The series of incidence in Belgium, cases per 100000inhab. (blue) is plotted together with a combination of sines and cosines (dotted red line) of the frequency 2·π/52, where 52 is the number of weeks in a single year.

3   Methodology: Univariate and Multivariate Time Series Models and Different Training Strategies

Univariate ARIMA models are estimated, both seasonal (SARIMA) and non-seasonal, to each country's AcRI weekly time series of cases per 100,000 inhabitants. The treatment of seasonality is faced either stochastically or deterministically (they will be referred as Methods A, B, C and D a detailed description is given in Table 1). An example of deterministic seasonality modeling is shown in Fig. 2.

Table 1. Summary of all the methodological approaches implemented, both univariate (univ.) and multivariate (multiv.)

Method	Model type	Treatment of seasonality	Input transformation
A	Univ. ARIMA (non-seasonal)	Not included	Automatic differentiation
B	Univ. Seasonal ARIMA (SARIMA)	Estimated (s = 52)	Regular + seasonal differentiation
C	Univ. ARIMA on seasonally differenced series	Seasonal differentiation (lag 52)	Differentiation + ARIMA
D	Univ. Seasonal regression + ARIMA	Regression with sines and cosines (s = 52)	ARIMA on residuals
E	Multiv. DFM (3 factors)	Not considered	Log-transformed and centered series
F	Multiv. Seasonal regression + DFM (3 factors)	Regression with sines and cosines (s = 52)	Fitted, log-transformed, and centered series
G	Multiv. Seasonal regression + DFM (4 factors)	Regression with sines and cosines (s = 52)	Fitted, log-transformed, and centered series

Also, multivariate models (DFMs) to capture common dynamics across countries are implemented (they will be referred to as Methods E, F and G and a detailed description is included as well in Table 1). These models reduce dimensionality of the original by extracting common latent factors that explain most of the variability.

Dynamic Factor Models (DFMs) allow for modeling sets of multivariate time series by reducing the dimensionality of the system, and therefore, the number of parameters to be estimated. This type of models assume that the observed series are driven by a reduced set of unobserved latent variables, known as common factors, which account for the shared variability. The remaining variability is explained by idiosyncratic or specific components for each series.

Let $y_t = (y_{1,}, y_{2,t}, \ldots, y_{m,t})'$ be the centered m-dimensional vector of time series (in our case $m = 8$, since 8 countries in the EU are considered in this work), and the second subscript, t, accounts for time evolution. The DFM is as follows: $y_t = \mathbf{P} \cdot \mathbf{f}_t + \varepsilon_t$, where $\mathbf{f}_t$ is the r-dimensional vector of common factors ($r << m$) and P is the m by r loading matrix. The columns of P give the linear combination of the original vector of series, y_t, to build the r-dimensional vector of unobserved common factors, $\mathbf{f}_t$.

This approach allows for capturing shared dynamics among correlated series while maintaining model stability and efficiency with few parameters.

In a first approach, following the ideas in [23] and [26], where all the details can be encountered. The unobserved factors are extracted by applying Singular Value Decomposition (SVD) to the centered covariance matrix, retaining those that explain a high percentage of the variance (typically $\geq 80\%$). Both the common factors and the idiosyncratic components are modeled using univariate ARIMA models tailored to their characteristics.

Finally, the one-step-ahead forecast is obtained by combining the predictions of the common factors and the specific components, to which the historical mean is added, for longer forecasting horizons the procedure is repeated iteratively as follows:

$$\hat{y}_{T+h} = \mathbf{P} \cdot \hat{f}_{T+h} + \hat{\varepsilon}_{T+h} + \bar{y},$$

where T is the length of the series used to estimate the model, and h, the corresponding forecasting horizon, in our case, $h = 1, \ldots, 22$ weeks.

Furthermore, different training strategies are compared in this work, including variations in historical window length and model updating schemes (rolling or expanding window), always with the goal of enhancing forecast accuracy. All these strategies are summarized in Table 2.

Table 2. Different strategies considered for the historical length used to estimate the models that are afterwards used to forecast (out-of-sample forecasting, always). Strategies 1 (a to e) are growing window ones, and Strategy 2 is the only rolling-window version (with fixed length).

Strategy	Name	Training period	First prediction	Last prediction
1a	Expanding Window (Full Assessment)	2020-W25 → current iteration	2021-W36	2025-W12
1b	Expanding Window (Midterm Assessment)	2020-W25 → current iteration	2021-W36	2024-W34
1c	Expanding window (Assessment from 2023)	2020-W25 → current iteration	2023-W35	2025-W12
1d	Expanding window (Assessment from 2024)	2020-W25 → current iteration	2024-W35	2025-W12
1e	Expanding window (Recent training)	2023-W34 → current iteration	2024-W35	2025-W12
2	Rolling fixed-length window	Last 167 weeks before each iteration	2023-W35	2025-W12

All models and strategies are implemented for the 8 countries whose data are used in this paper (detailed in Subsect. 2.2). Forecasts for the whole span of time (considering that different strategies need a longer historical dataset for the estimation) are computed. Thus, the forecasting exercises whose results are compared in Sect. 4 and best approaches selected (in Sect. 4 and sometimes being combined) are illustrated in Sect. 5, is a complex and serious one that allows reliable conclusions.

Forecast evaluation is performed by evaluating the predictive performance of the selected configurations using RespiCast's official scoring metrics (to check accuracy of the predictions against real data), and this is done across countries and forecast horizons considered. Bear in mind that we calculated forecasts with forecasting horizons from 1 up to 22 weeks, although the teams taking part in the collaborative initiative are just computing forecasts up to 4 weeks ahead.

For all the aforementioned configurations, the model evaluation is performed by comparing predicted values with actual values using error metrics. A prediction is considered better when it has a lower error and greater ability to generalize to new data. The metrics used in this study are:

- Relative Error: normalizes the error with respect to the actual value, allowing comparison across series with different scales.
- Absolute Error (AE): measures the absolute difference between the forecasted median and the observed actual value.

Additionally, to compare model performance against a reference, RespiCast uses the base-2 logarithm of the ratio between metrics. Positive values indicate an improvement over the baseline model, while negative values indicate that the proposal is working worse than the baseline in terms of accuracy of the forecasts.

4 ANOVA

Experimental design techniques are here applied to compare the predictive methods and strategies described in Sect. 3 through an ANOVA framework, that provides a statistical framework to analyze both the individual effect (main effects) and the combined effect (interaction) of factors on a response variable. This allows identification of statistically significant differences in forecasting accuracy and supports, when needed, combining model configurations when not a single model is the best one, but several of them are, and then we combine forecasts.

In our work the response variable is the accuracy metric selected for evaluating the quality of the forecasts, usually the relative error or its logarithm or square root when needed because of heteroskedasticity being present in the data. The first factor (F1) is the "Model", particularly it has $I = 7$ levels (models A to G, according to Table 1). The second factor (F2) is the country ($J = 8$ levels, according to Sect. 2). The third one is the strategy (Growing or rolling window, thus 2 levels[2]) The goal is to determine whether statistically significant differences exist, using F-statistics and p-values compared against a significance threshold ($\alpha = 0.05$), thus, to test if the factors "Model" and "Country" as well as their interaction are statistically significant. This two-way factorial model is expressed as:

$$y_{ijklt} = \mu + \alpha_i + \beta_j + \gamma_k + \delta_l + (\alpha\beta)_{ij} + (\alpha\gamma)_{ik} + (\alpha\delta)_{il} + u_{ijklt} \tag{1}$$

where μ is the grand mean, α_i is the main effect of the factor "Strategy" (2 levels, since always one of the expanding window strategies is compared to the only rolling window strategy). β_j is the main effect of the factor "Country". γ_k is the effect of the week for which the forecast is computed, this is to be able to consider that some weeks will be really difficult to be accurately predicted and the relative error will be high with all

---

[2] Each expanding window strategy (at a time) is compared with the only rolling window one.

combinations of the levels of the factors. The fourth factor, whose main effect is δ_l, is the "Model", with 7 levels according to Table 1 (Models A-B-C-D-E-F-G). The term $(\alpha\beta)_{ij}$ is the interaction of "Model" and "Country", and so forth for the rest of the interactions in the model in Eq. 1. u_{ijklt} is the error term, for which independence, gaussianity and homoskedasticity is assumed, thus u_{ijklt} is i.i.d as a $N(0, \sigma^2)$.

In Fig. 3 the average relative errors for each forecasting horizon considered, when the first factor accounts for the comparison between strategies 1c) and 2) are shown. Here the average relative errors are those for the whole span of time that can be compared with strategies 1c) and 2).

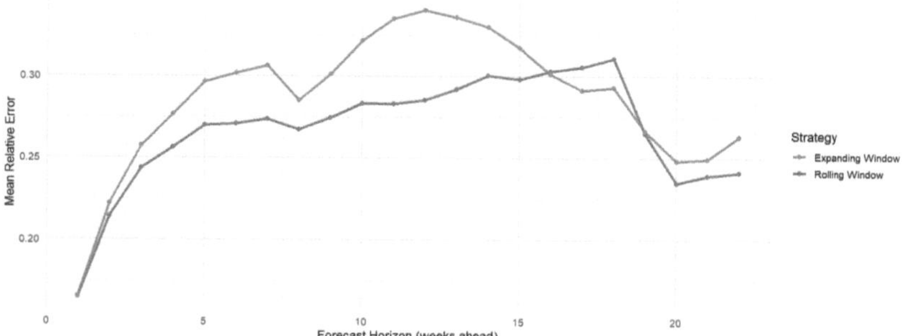

Fig. 3. Average relative errors, for each forecasting horizon considered, when comparing the expanding window strategy 1c and Strategy 2 (rolling window).

- In the short run ($h = 1$–10), there are no significant differences between strategies.
- For horizons 11 and 12, the rolling window shows significantly better performance.
- When $h = 13$, the interaction between strategy and country is significant, but no individual country shows relevant differences.
- For horizons ranging from 14 to 18, no significant differences or interactions are observed, indicating similar behavior across strategies.
- For the largest prediction horizons (19 to 22 weeks ahead), the interaction between strategy and country becomes significant again, highlighting that in Estonia, the expanding window delivers better and statistically significant performance.

Overall, these results suggest that the rolling window strategy is preferable at intermediate horizons due to its greater adaptability, although the expanding window remains competitive and more stable in certain contexts, such as Estonia. However, in the short term, both strategies perform equally well.

However, if the average relative errors shown in Fig. 3 are computed not for the whole span of weeks but just for those in the last season (starting calculating predictions in September 2024), then the results are those in Fig. 4.

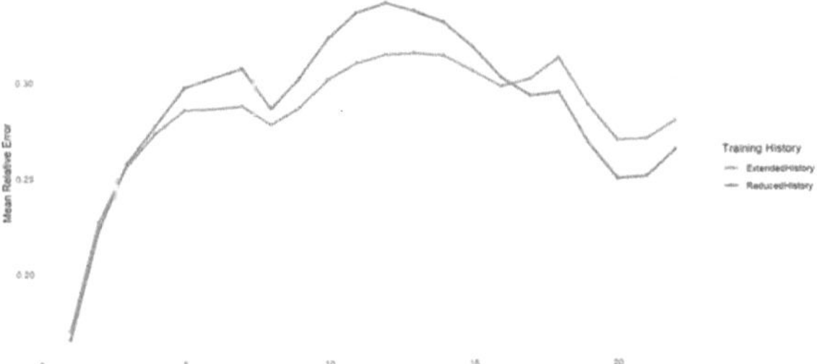

Fig. 4. Average relative errors, just considering last season in RespiCast (from September 2024 onwards) for each forecasting horizon considered, when comparing the expanding window strategy 1c and Strategy 2 (rolling window).

In Fig. 5 the averaging forecasting errors, for all the considered forecasting horizons and methods are shown for the last season, 2024–2025 (forecasting exercises whose results are included start in September 2024, thus, the results are linked to those in Fig. 4).

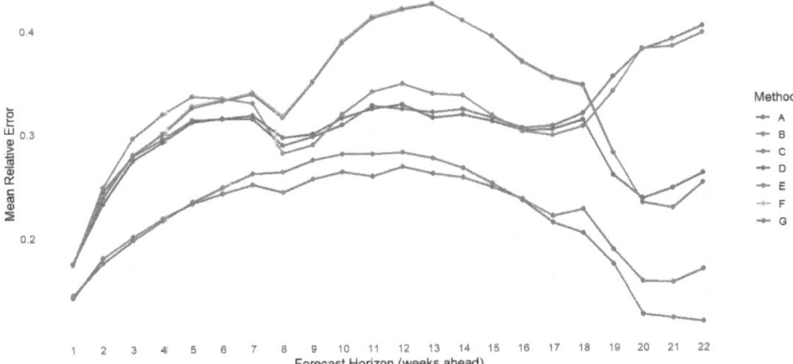

Fig. 5. Average relative errors, for prediction horizons between 1 and 22 weeks, for all the univariate and multivariate time series approaches from Table 1 in Sect. 3.

4.1   Selection of the Final Combination to Participate in RespiCast Initiative

The first decision concerned the evolution of the training window: rolling or expanding. As no significant differences were observed between the two strategies for horizons 1 to 10, the expanding window was chosen for its greater stability and simplicity. The second decision involved the amount of training data: using the full historical dataset since 2020 or only recent data from 2023 onward. For horizons 1 to 3, both options yielded similar average results, with no statistically significant differences. Overall, using the full historical data proved to be more reliable. Finally, once the expanding window with the full historical dataset was selected, the choice of optimal methods was based on the analysis conducted since September 2024. This decision reflects the observed seasonal variability and the need to better capture current reporting patterns and conditions. The best selected methods per country and forecasting horizon are included in Table 3.

Table. 3. Selection of best alternative per horizon (just those used in RespiCast, one to four weeks ahead) and country (by columns).

Horizon	BE	BG	CZ	DE	EE	LT	RO	SI
h = 1	A-G							
h = 2	B, C							
h = 3	B, C							
h = 4	A-C	B, C F, G	A-G	A-G	A-G	A-D	B, C, D, F, G	A-G

After selecting the best models by country and forecast horizon, a combined model was built that linearly optimizes the combination of the top-performing models for each country and horizon, adjusting weights to minimize the historical mean absolute error. This was formulated as a linear programming problem.

5 Numerical Results: Forecasting Accuracy and Comparison with RespiCast Baseline and Other Participants

Here we compare the selected optimal forecasts with those submitted by other teams taking part in RespiCast. This comparison helps assess the contribution and potential of the proposed models within the collaborative forecasting environment.

Just some particular results are shown here, so as to compare our combination of models as obtained in Sect. 4 against the teams participating in RespiCast. We provide in Table 4, our results for the first two months in 2025 and the comparison with the rest of the teams and baseline of RespiCast. As it can be seen in Fig. 1, these months correspond to a period of high incidence after Christmas, that is always a span of time very interesting for epidemiologists. The results show the great performance of our approach.

Table 4. Comparison of our results with the rest of participants in RespiCast. Forecasting January and February 2025.

Horizon	Average AE	Average AE Baseline	Mean log2(ratio)	Our average ranking	Number of teams
1 (2025-W02)	319.75	386.19	0.33	3	9
2 (2025-W03)	266.62	551.59	1.19	1	9
3 (2025-W04)	399.53	780.14	1.34	1	9
4 (2025-W05)	602.20	991.57	1.07	1	9
Horizon	Average AE	Average AE Baseline	Mean log2(ratio)	Our average ranking	Number of teams

(*continued*)

Table 4. (continued)

Horizon	Average AE	Average AE Baseline	Mean log2(ratio)	Our average ranking	Number of teams
1 (2025-W03)	102.01	202.49	1,14	2	10
2 (2025-W04)	115.75	218.77	1,77	2	10
3 (2025-W05)	329.88	625.09	0,91	3	10
4 (2025-W06)	257.07	431.78	0,81	2	10

6 Conclusions and Further Research

We analyzed the use of time series models for forecasting AcRI within the Respi-Cast framework. Seven different approaches were implemented—four univariate models based on ARIMA and three multivariate models using DFM—and evaluated across eight European countries using weekly data from 2020 to 2025.

The results showed that univariate ARIMA models with seasonal components performed best for short-term forecasting during the most recent season. However, multivariate DFM models proved to be more robust and stable overall, capturing synchronization between countries, an uncommon ability in traditional epidemiological models. In addition, a forecast combination strategy was applied, which significantly improved accuracy, reducing error by up to 50% compared to RespiCast's baseline for short forecast horizons.

In conclusion, well-tuned univariate ARIMA models are competitive and effective tools for short-term epidemic forecasting. Likewise, multivariate approaches based on dynamic factors take advantage of shared structures among series, improving the stability of predictions. This work confirms that classical time series models, when properly implemented, can provide robust and transparent forecasts that support public health preparedness.

For future work, it is proposed to incorporate prediction intervals developing bootstrap methods for combined forecasts, in order to represent uncertainty. The bootstrap has proven to be very useful for studying uncertainty in predictions (see for example, [27] and [31]). Additionally, we plan to extend the study to more countries and additional indicators (ILI and COVID-19), and explore multivariate models that integrate all three indicators for each country. In particular, it will be interesting to compare our procedures with those developed in the project FluSight [32].

Acknowledgments. The authors acknowledge financial support from the Spanish government through the Ministerio de Ciencia e Innovación and the Agencia Estatal de Investigación project PID2022-138114NB-I00/AEI/https //doi.org/10.13039/501100011033 and through the Instituto de Salud Carlos III project PI21/00840.

Disclosure of Interests. The authors have no competing interests to declare that are relevant to the content of this article.

References

1. Saldías, F.: Epidemiología de las enfermedades respiratorias. https://medicina.uc.cl/wp-content/uploads/2021/09/I.-Epidemiologia-de-las-enfermedades-respiratorias.pdf. Accessed 11 Feb 2025
2. Eurostat, Respiratory diseases statistics Statistics Explained (2024). https://ec.europa.eu/eurostat/statistics-explained/index.php?title=Respiratory_diseases_statistics. Accessed 11 Feb 2025
3. GBD Chronic Respiratory Disease Collaborators: Prevalence and attributable health burden of chronic respiratory diseases, 1990–2017: a systematic analysis for the Global Burden of Disease Study 2017. Lancet Respir. Med. **8**(6), 585–596 (2020)
4. ASTHO, Defining disease forecasting and modeling (2024). https://www.astho.org/topic/brief/defining-disease-forecasting-modeling/. Accessed 11 Feb 2025
5. Celis, J.A.T.: La urgencia de mejorar la vigilancia y respuesta ante pandemias. Observatorio de Riesgos Catastróficos Globales (ORCG) (2024). https://orcg.info/articulos/la-necesidad-urgente-de-mejorar-la-vigilancianbspy-respuesta-ante-pandemias. Accessed 11 Feb 2025
6. Mesquita, S., Perfeito, L., Paolotti, D., Gonçalves-Sá, J.: Epidemiological methods in transition: minimizing biases in classical and digital approaches. PLOS Digit. Health **4**(1), 1–18 (2025)
7. Hasdeu, S., Lamfre, L., Caro, P., Horne, F.: Revisión narrativa: modelos predictivos sobre la evolución de la pandemia por COVID-19. Rev. Argent. Salud Pública **12**(48), 1–6 (2020)
8. Montesinos-López, O.A., Hernández-Suárez, C.M.: Modelos matemáticos para enfermedades infecciosas. Salud pública Méx **49**(3), 218–226 (2007)
9. Isla Zuvialde, D.: Modelos matemáticos en epidemiología, Trabajo Fin de Máster, Máster en Estadística Aplicada, Univ. de Granada, Granada, España (2021)
10. Mehdaoui, M.: A review of commonly used compartmental models in epidemiology. arXiv preprint arXiv:2110.09642. Revised 26 Jan 2023. Accessed 11 Feb 2025
11. Tomov, L., Chervenkov, L., Miteva, D.G., Batselova, H., Velikova, T.: Applications of time series analysis in epidemiology: literature review and our experience during COVID-19 pandemic. World J. Clin. Cases **11**(29), 6974–6983 (2023)
12. Dashtbali, M., Mirzaie, M.: A compartmental model that predicts the effect of social distancing and vaccination on controlling COVID-19. Sci. Rep. **11**(8191) (2021)
13. Britton, T.: Stochastic epidemic models: a survey. Math. Biosci. **225**(1), 24–35 (2010)
14. Briggs, A., Sculpher, M.: An introduction to Markov modelling for economic evaluation. Pharmacoeconomics **13**(4), 397–409 (1998)
15. Waring, J., Lindvall, C., Umeton, R.: Automated machine learning: review of the state-of-the-art and opportunities for healthcare. Artif. Intell. Med. **104**(101822) (2020)
16. Ferguson, N.M., Cummings, D.A.T., Fraser, C., Cajka, J.C., Cooley, P.C., Burke, D.S.: Strategies for mitigating an influenza pandemic. Nature **442**(7101), 448–452 (2006)
17. Epstein, J.M.: Modelling to contain pandemics. Nature **460**(7256), 687 (2009)
18. Kerr, C.C., et al.: Covasim: an agent-based model of COVID-19 dynamics and interventions. PLoS Comput. Biol. **17**(7), e1009149 (2021)
19. Quora, What is the difference between Mathematical Model and Statistical Model?. https://www.quora.com/What-is-the-difference-between-Mathematical-Model-and-Statistical-Model. Accessed 05 Mar 2025
20. Gayathri, G.: Learning time series analysis & modern statistical models, Analytics Vidhya (2023). https://www.analyticsvidhya.com/blog/2023/01/learning-time-series-analysis-modern-statistical-models/. Accessed 05 Mar 2025
21. Abolmaali, S., Shirzaei, S.: A comparative study of SIR Model, Linear Regression, Logistic Function and ARIMA Model for forecasting COVID-19 cases. AIMS Public Health **8**(4), 598–613 (2021)

22. Alabdulrazzaq, H., Alenezi, H.N., Rawajfih, Y., Alghannam, B.A., Al- Hassan, A.A., Al-Anzi, F.S.: On the accuracy of ARIMA based prediction of COVID-19 spread. Results Phys. 27(104509) (2021)

23. Lee, R.D., Carter, L.R.: Modeling and forecasting U. S. mortality. J. Am. Stat. Assoc. 87(419), 659–671 (1992)

24. Stock, J.S., Watson, M.W.: Forecasting using principal components from a large number of predictors. J. Am. Stat. Assoc. 97(460), 1167–1179 (2002)

25. García-Martos, C., Rodríguez, J., Sánchez, M.J.: Forecasting electricity prices and their volatilities using unobserved components. Energy Econ. 33(6), 1227–1239 (2011)

26. García-Martos, C., Rodríguez, J , Sánchez, M.J.: Forecasting electricity prices by extracting dynamic common factors: application to the Iberian Market. IET Gener. Transm. Distrib. 6(1), 11–20 (2012)

27. Alonso, A.M., García-Martos, C., Rodríguez, J., Sánchez, M.J.: Seasonal dynamic factor analysis and bootstrap inference: application to electricity market forecasting. Technometrics 53(2), 137–151 (2011)

28. European Centre for Disease Prevention and Control, RespiCast: The New European Respiratory Diseases Forecasting Hub (2024). https://www.ecdc.europa.eu/en/news-events/respicast-new-european-respiratory-diseases-forecasting-hub. Accessed 01 Feb 2025

29. European Centre for Disease Prevention and Control, Homepage | European Centre for Disease Prevention and Control (2025). https://www.ecdc.europa.eu/en. Accessed 01 Feb 2025

30. Wu, H., Levinson, D.: The ensemble approach to forecasting: a review and synthesis. Transp. Res. Part C: Emerg. Technol. 132(103357) (2021)

31. Alonso, A.M., Romo, J.: Forecast of the expected non-epidemic morbidity of acute diseases using resampling methods. J. Appl. Stat. 32(3), 281–295 (2005)

32. Centers for Disease Control and Prevention, Flusight: Flu forecasting (2022). About Flu Forecasting | FluSight | CDC

Synthetic Disease Trajectories for Predicting Chronic Kidney Disease Post-diabetes Onset

Víctor M. de la Oliva Roque[1], David P. Kreil[2], Joaquín Dopazo[1,3]($\boxtimes$), Francisco Ortuño[4]($\boxtimes$), and Carlos Loucera[1,3]($\boxtimes$)

[1] Andalusian Platform for Computational Medicine, Andalusian Public Foundation Progress and Health-FPS, Seville, Spain
`{joaquin.dopazo,carlos.loucera}@juntadeandalucia.es`
[2] Institute of Molecular Biotechnology, Department of Biotechnology and Food Science, BOKU University, Vienna, Austria
[3] Institute of Biomedicine of Seville (IBiS), University Hospital Virgen del Rocío/CSIC/University of Seville, Seville, Spain
[4] Department of Computer Engineering, Automatics and Robotics, University of Granada, Granada, Spain
`fortuno@ugr.es`

Abstract. Synthetic data generated using generative models trained on real clinical data offers a promising solution to privacy concerns in health research. However, many efforts are limited by small or demographically narrow training datasets, reducing the generalizability of the synthetic data. To address this, we used real-world clinical data from nearly one million individuals with diabetes in the Andalusian Population Health Database (BPS) to generate a comprehensive longitudinal synthetic dataset.

We employed a dual adversarial autoencoder to produce synthetic data and evaluated its utility in a clinical machine learning (ML) task: predicting the onset of chronic kidney disease, a common diabetes complication. Models trained on synthetic data were assessed for their ability to reproduce patterns and predictive behaviors observed in real data.

Performance and stability were compared across models trained on real, synthetic, and hybrid datasets. Models trained exclusively on synthetic data achieved AUROC scores comparable to real-data models (0.70 vs. 0.73) and showed high stability in feature importance rankings (weighted Kendall's $\tau > 0.9$). Notably, combining synthetic and real data did not improve performance.

Our findings demonstrate that high-fidelity synthetic longitudinal data can replicate real data performance in clinical ML, supporting its use in research while preserving patient privacy. This represents a significant step toward more collaborative and privacy-preserving healthcare data ecosystems.

Keywords: Synthetic longitudinal data · Diabetes · Real World Data

A. López Fernández et al. (Eds.): CIABiomed 2025, LNBI 16148, pp. 632–642, 2026.
https://doi.org/10.1007/978-3-032-10661-2_47

1   Introduction

In healthcare settings, real-world data (RWD) refers to observational data that is not collected as part of a defined clinical trial or observational study, but that is rather generated from real-world settings such as patient electronic health records (EHRs) [14]. As such, this data is not constrained by the inclusion and exclusion criteria intrinsic to the aforementioned studies [8], allowing them to capture the heterogeneity within a patient population and providing a more comprehensive picture of clinical practice and outcomes [14]. Furthermore, as these data can comprise any given patient's longitudinal clinical history, they can be invaluable in observing patient's disease trajectories, understood as the sequential and directional co-occurrence of diseases [7]. Indeed, various disease trajectories have been observed from population-wide registry data [4] and differing trajectories have proven to have an impact on a patient's healthcare burden [1]. Moreover, leveraging longitudinal information has shown to outperform static information in prediction studies [17,18].

Nevertheless, given the highly sensitive nature of RWD, which can contain information on patient demography, clinical health and even financial situation, its access is subject to stringent data protection regulations such as the General Data Protection Regulation (GDPR) [12]. While various methods have been proposed to anonymize this identifiable patient data, re-identification is still a risk [10] and achieving a balance between the degree of data anonymization and usability remains a challenge for widespread data sharing [21].

As such, synthetic data generation through generative artificial intelligence (AI) models has been proposed as a method to circumvent these problems. This approach can furthermore be used for data augmentation, increasing dataset size, and improving diversity within datasets [20]. Namely, Generative Adversarial Network (GAN) [2] and Variational Autoencoder (VAE) [9] architectures have been successful in synthesizing EHR data for various clinical purposes. More recently, Dual Adversarial Autoencoders (DAAE) have been used to synthesize sequences of set-valued medical records [11], addressing the longitudinal nature of EHRs in order to recreate disease trajectories.

This implementation, however, has been carried out on the MIMIC database, which is a single-center database containing data on patients admitted to critical care units [5]. This data is limited in its scope, as patients admitted to critical care units are a relatively small and focused population subject to constant and continuous monitoring. Furthermore, this data is focused on patient's critical care treatment and does not span their previous clinical history. As such, this does not address many of the challenges inherent to RWD and consequently in generating synthetic data from them. Namely, the fact that they tend to be poorly structured, incomplete, heterogeneous, and prone to measurement errors and biases [12]. On the other hand, this implementation misses the various benefits of RWD obtained from large clinical databases, namely the long spanning and diverse patient-level longitudinal data available for varying cohorts, which can provide a more comprehensive picture of real outcomes and trajectories.

Given these limitations, we have implemented a DAAE architecture to generate high-fidelity synthetic data using data extracted from the Andalusian Population Health Database (BPS, due to its acronym in Spanish) [15]. This database contains longitudinal information derived from 15 million patient's EHRs from the Andalusian public health system. Concretely, we have used 1 million EHRs from diabetic patients in order to generate synthetic patient disease trajectories.

We measure the utility of the generated trajectories through train on synthetic, test on real (TSTR) performance [6]. Namely, we predict a given end-point by training on a synthetic dataset and evaluating on the real test set, comparing this performance to that of a model trained with real data. We furthermore compare feature rankings between models trained with synthetic and real data. We then test the potential of our generated data for data augmentation by training a hybrid model with both real and synthetic data, and evaluate its performance.

Specifically, we predict the onset of chronic kidney disease (CKD), a common complication in diabetic patients, using their prior comorbidities. Type 2 diabetes, particularly, has been associated with a vast spectrum of CKD etiologies, with various studies finding its onset to be compounded due to the patient's medical history [19].

2 Materials and Methods

2.1 Data

Longitudinal real-world data were extracted from the BPS [15], based on EHRs from all users of the Andalusian public health system. Study approval was granted by the Andalusian Biomedical Research Ethics Committee (Acta 08/19, 24/09/2019).

A retrospective cohort of 1,062,633 individuals were obtained from the BPS with a registered diabetes mellitus diagnosis in Andalusia from 2003 to 2022. After filtering for valid sex coding and adult-onset diabetes diagnoses (age >18) a total of 908,673 remained. As type-I diabetes mellitus (ICD10 E10) is rarely diagnosed at this age, the vast majority of our cohort consists of type-II diabetes patients (ICD10 E11). The average age at diagnosis was 60.6 ± 13.83 years. Comorbidities without visit dates were excluded. The final dataset included 5,869,024 pathologies across 4,875,272 medical visits from 1930 to 2022.

For end-point prediction, patients with any prior diabetes diagnosis before 2003 and patients with a CKD diagnosis before their diabetes diagnosis date were removed, resulting in a final cohort of 766,904. Of these, 78,729 (10.27%) had a CKD diagnosis after their diabetes diagnosis. Data also included 79 additional comorbidities. Diagnoses covered both prior history and follow-up, up to 2022. Patients with a CKD diagnosis were those with any of the ICD10 codes in Table 1 (Fig. 1).

Data handling complied with regional and EHDS privacy regulations: i) study approval by a data access committee; ii) data extraction and pseudo-anonymization by BPS; iii) transfer to the PAGEM Secure Processing Environment [16]; iv) analysis within PAGEM; v) data deletion post-analysis.

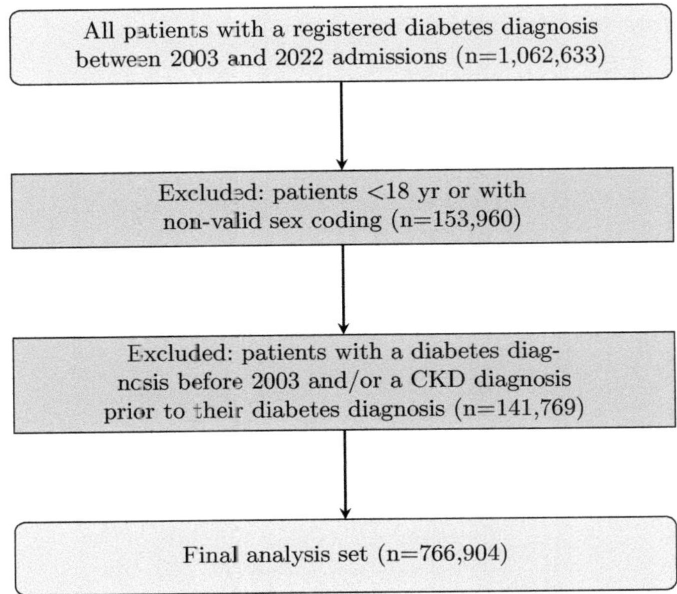

Fig. 1. Flowchart of patients used for analysis.

Table 1. ICD10 codes used for CKD patient identification

ICD10	ICD10 description
N18	Chronic kidney disease (CKD)
D63.1	Anemia in chronic kidney disease
E08.2	Diabetes mellitus due to underlying condition with kidney complications
I13	Hypertensive heart and chronic kidney disease
E11.2	Type 2 diabetes mellitus with kidney complications
E10.2	Type 1 diabetes mellitus with kidney complications
I12	Hypertensive chronic kidney disease
E09.2	Drug or chemical induced diabetes mellitus with kidney complications
E13.2	Other specified diabetes mellitus with kidney complications

Patients were encoded as sequences of health system visits, represented by age at visit and diagnoses. Each patient's record includes sex and an ordered list of diagnosis events, one being diabetes.

2.2   Generative Adversarial Training

Data were split into a training (80%) and leave-out test set (20%) and synthetic data was generated exclusively with the train set. Dual adversarial autoencoders (DAAE) were used to learn realistic sequences from medical data by jointly modeling latent and discrete data distributions [11]. The model was configured with a batch size of 256 and trained for 500 epochs on an NVIDIA Tesla V100 GPU with 32 GB. Further hyperparameters are shown in Table 2.

Table 2. Hyperparameters for the DAAE

Hyperparameter	Value
Batch size	256
Embedding size	128
Hidden size	128
L2 Norm Clip	0.5
RNN Type	GRU
Optimizer	ADAM
Learning Rate	0.001
Noise size	128
Noise Multiplier	1
Epochs	500 (no early stopping)

2.3   End-Point Prediction

Patients with a chronic kidney disease diagnosis before their diabetes diagnosis were removed from the at-risk population for end-point prediction in both the real and synthetic datasets.

We trained an Explainable Boosting Machine (EBM) [13] with default hyperparameters and no feature interactions using the aforementioned training set used for synthetic data generation, the generated synthetic dataset, as well as a model trained using a mix of both real and synthetic data. No class weights or resampling were used in EBM training. Further hyperparameters are shown in Table 3

Training features were generated by converting patient comorbidities diagnosed before their end-point diagnosis to binary variables. As controls did not experience the end-point, they were randomly matched with a case end-point date posterior to their diabetes diagnosis date.

Finally, ten repeated stratified 10-fold splits of the training set were generated and separate models trained on each fold. We evaluated the difference in prediction probabilities on the leave-out test set between these models trained on data splits and the model trained with the entire training set and report means and standard deviations. Furthermore, we calculated the Kendall's weighted τ on feature rankings both within and between training sets and report distributions, median and 95% confidence intervals.

Predictions on the test set were carried out considering the comorbidities diagnosed before patient's diabetes diagnosis date, thus simulating a patient's evaluation of CKD risk at the time of diabetes diagnosis.

Table 3. Hyperparameters for the EBM

Hyperparameter	Value
Max bins	1024
Validation size	0.15
Outer Bags	14
Inner Bags	0
Learning Rate	0.015
Greedy Ratio	10
Smoothing Rounds	75
Max Rounds	50000
Early Stopping Rounds	100
Early Stopping Tolerance	1×10^{-5}

3   Results

3.1   Synthetic Generation

A synthetic cohort of 800,000 individuals was generated using dual adversarial autoencoders (DAAE) [11], based on the real dataset. Quality checks excluded non-diabetic individuals, empty visits, redundant chronic disease entries, and a lack of associated patient sex. For age data, the highest value per visit was retained, and missing ages were interpolated from adjacent visits. The final synthetic cohort included 711,080 individuals for analysis, of which 49,796 (7.00%) had a CKD diagnosis.

After removing individuals with end-points before their associated diabetes age, a total of 685,798 remained. Of these, 33,726 (4.92%) had an associated end-point after their diabetes diagnosis age.

3.2   End-Point Prediction Performance

Both models trained on real and on synthetic data showed a steady increase in their predictive performance when evaluating patients whose diabetes diagnosis date was in more recent calendar years, until reaching relative stabilization in 2014. From this year onwards, models trained with real data showed an AUROC of 0.73 and models trained with synthetic data an AUROC of 0.70, representing a 4% decrease in AUROC. However, the hybrid model trained with both real and synthetic data showed no improvement compared to the model trained solely with real data (AUROC of 0.73) Fig. 2a. Yearly mean AUROC with standard deviations, as well as total positive and negative class individuals are presented in Table 4.

Prediction probabilities between models trained with the entire training set and those trained on training set splits were small and heavily centered on 0 in both training sets Fig. 2b. The median difference between these was namely 5.0×10^{-5} $[-4.4\times10^{-4}, 5.0\times10^{-4}]$ for models trained on real data and -1.1×10^{-4} $[-5.0 \times 10^{-4}, 3.9 \times 10^{-4}]$ for models trained on synthetic data.

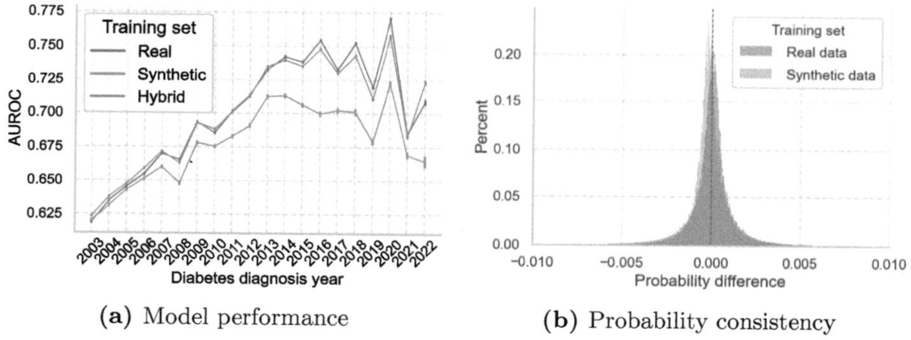

(a) Model performance   (b) Probability consistency

Fig. 2. (a) AUROC across years of models trained with real-only, synthetic-only and a hybrid of both and (b) differences in prediction probabilities on the test set between models trained with the full training set and splits (trimmed at x = −0.1 and x = 0.1 for visualization).

Feature ranking stability, measured by Kendall's hyperbolic-weighted τ, was high for models trained on splits of both real data (median $\tau = 0.966$ [0.963, 0.969]) and synthetic data (median $\tau = 0.953$ [0.948, 0.957]), demonstrating strong internal consistency (Fig. 3a). Furthermore, the rankings learned from synthetic data showed good agreement with those from real data, yielding a median weighted τ of 0.742 [0.731, 0.754] (Fig. 3b).

Table 4. (a) Mean AUROC across years of models trained with dataset splits of real-only, synthetic-only and a hybrid of both. Error bars indicate standard deviations (b) differences in prediction probabilities on the test set between models trained with the full training set and splits (trimmed at x = −0.1 and x = 0.1 for visualization)

Diabetes year	Metric	Hybrid	Real	Synthetic
2003	AUROC Mean (SD)	0.62 (2.44e–04)	0.62 (3.19e–04)	0.62 (7.08e–04)
	N negative class	4544	4544	4544
	N positive class	1474	1474	1474
2004	AUROC Mean (SD)	0.64 (2.31e–04)	0.64 (3.41e–04)	0.63 (5.60e–04)
	N negative class	5390	5390	5390
	N positive class	1425	1425	1425
2005	AUROC Mean (SD)	0.65 (2.20e–04)	0.65 (2.44e–04)	0.64 (7.19e–04)
	N negative class	5166	5166	5166
	N positive class	1330	1330	1330
2006	AUROC Mean (SD)	0.66 (2.60e–04)	0.66 (3.06e–04)	0.65 (7.80e–04)
	N negative class	5682	5682	5682
	N positive class	1252	1252	1252
2007	AUROC Mean (SD)	0.67 (2.68e–04)	0.67 (2.70e–04)	0.66 (7.50e–04)
	N negative class	7016	7016	7016
	N positive class	1460	1460	1460
2008	AUROC Mean (SD)	0.66 (3.10e–04)	0.67 (3.29e–04)	0.65 (9.60e–04)
	N negative class	7361	7361	7361
	N positive class	1292	1292	1292
2009	AUROC Mean (SD)	0.69 (2.83e–04)	0.69 (2.95e–04)	0.68 (8.17e–04)
	N negative class	7414	7414	7414
	N positive class	1136	1136	1136
2010	AUROC Mean (SD)	0.69 (2.98e–04)	0.69 (4.00e–04)	0.68 (8.66e–04)
	N negative class	7663	7663	7663
	N positive class	1100	1100	1100
2011	AUROC Mean (SD)	0.70 (3.07e–04)	0.70 (3.34e–04)	0.68 (9.78e–04)
	N negative class	7444	7444	7444
	N positive class	943	943	943
2012	AUROC Mean (SD)	0.71 (3.25e–04)	0.71 (3.73e–04)	0.69 (9.47e–04)
	N negative class	6730	6730	6730
	N positive class	715	715	715
2013	AUROC Mean (SD)	0.74 (3.12e–04)	0.73 (3.51e–04)	0.71 (9.02e–04)
	N negative class	7103	7103	7103
	N positive class	703	703	703
2014	AUROC Mean (SD)	0.74 (3.35e–04)	0.74 (4.04e–04)	0.71 (9.66e–04)
	N negative class	7011	7011	7011
	N positive class	548	548	548
2015	AUROC Mean (SD)	0.74 (3.91e–04)	0.74 (4.94e–04)	0.71 (1.17e–03)
	N negative class	6457	6457	6457
	N positive class	437	437	437
2016	AUROC Mean (SD)	0.75 (4.38e–04)	0.76 (3.96e–04)	0.70 (1.54e–03)
	N negative class	6561	6561	6561
	N positive class	383	383	383
2017	AUROC Mean (SD)	0.73 (3.75e–04)	0.73 (3.86e–04)	0.70 (1.12e–03)
	N negative class	6820	6820	6820
	N positive class	297	297	297
2018	AUROC Mean (SD)	0.74 (5.59e–04)	0.75 (6.12e–04)	0.70 (1.63e–03)
	N negative class	6951	6951	6951
	N positive class	277	277	277
2019	AUROC Mean (SD)	0.71 (4.94e–04)	0.72 (5.24e–04)	0.68 (1.53e–03)
	N negative class	7318	7318	7318
	N positive class	226	226	226
2020	AUROC Mean (SD)	0.76 (6.72e–04)	0.77 (8.07e–04)	0.72 (1.56e–03)
	N negative class	6464	6464	6464
	N positive class	152	152	152
2021	AUROC Mean (SD)	0.68 (6.16e–04)	0.69 (8.05e–04)	0.67 (1.69e–03)
	N negative class	9712	9712	9712
	N positive class	109	109	109
2022	AUROC Mean (SD)	0.72 (1.11e–03)	0.71 (1.31e–03)	0.67 (3.69e–03)
	N negative class	6511	6511	6511
	N positive class	34	34	34

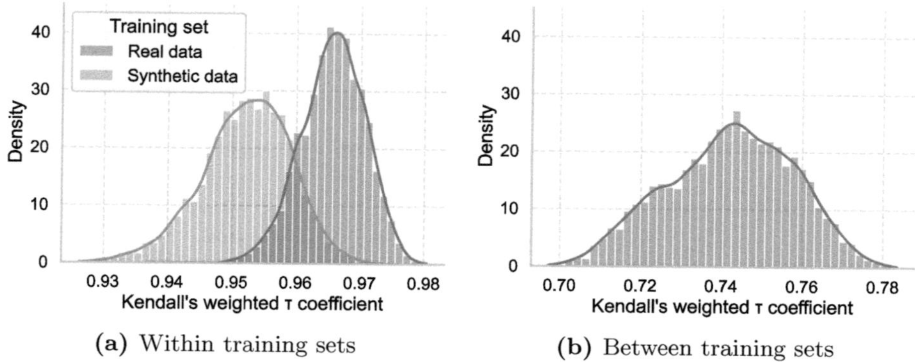

(a) Within training sets   (b) Between training sets

Fig. 3. (a) Within training set Kendall's weighted τ of feature rankings (b) Between training set Kendall's weighted τ of feature rankings.

4 Discussion and Future Work

4.1 Discussion

In this work, we have successfully generated high-fidelity longitudinal synthetic data from a large and diverse diabetic RWD cohort, and shown its utility in a relevant clinical scenario, namely the ML assessment of CKD risk in diabetic patients. This synthetic model shows a 4.11% decrease in performance as measured by AUROC relative to the model trained with real data when evaluating on calendar years after performance stabilization. In addition, both our synthetic and real models show good agreement between their feature rankings, indicating that high-relevance features for the specified end-point remain so in the synthetic dataset. This performance was furthermore highly consistent across dataset splits for both models.

This shows the potential of synthetic data generation not only in well-curated clinical databases, but also in RWD, not only tacking the challenges that this type of data presents, but also exploiting the richness that this data provides. Given this fact, this dataset and others like it can be used to study various other downstream tasks besides that presented here, such as aiding in health screenings, healthcare policy establishment or clinical intervention decisions [3]. As these datasets do not correspond to any real individuals, they can freely be made public for the wider research community, thus accelerating advancements in clinical research. This shows to be a promising solution to the challenges currently faced by researchers in accessing valuable clinical data.

4.2 Future Work

In order to further test the generated dataset's utility, our methodology should be applied to other end-points within the BPS. In the context of diabetic patients and their risks, retinopathy and/or heart failure are interesting candidates. Furthermore, in order to fully test and exploit the generated disease trajectories, deep learning architectures which can leverage time-series data can be applied for end-point prediction. Finally, reliability curves for real-trained vs. synthetic-trained EBMs should be assessed.

In addition, other generative models can be used to generate the synthetic data. Their quality and utility can then be compared to that achieved by the DAEE. Alternatively, oversampling techniques of rare events could be incorporated during synthetic data generation or adjustments in DAAE hyperparameters.

Finally, further endeavors at improving model performance through augmentation with synthetic data should be done. This could be done at the level of synthetic to real data ratios in model training, synthetic data quality improvement, or rather by exploiting more complex real datasets. Indeed, models trained on real data did not show significant improvement when training on the entire dataset as opposed to dataset splits (as seen by the nearly identical prediction probabilities). This could indicate that the informativeness of the real dataset used is limited in predicting the end-point and therefore adding more data does not improve performance. In this regard, different data modalities such as blood analysis values, medication and/or procedures could be added to both synthetic data generation and end-point prediction.

Acknowledgments. This study was supported in part by grants from the Spanish Ministry of Science and Innovation (PID2023-152380OB-C21, PRTR-C17.I1), the Institute of Health Carlos III (PMP24/00024), the Consejería de Salud y Consumo from the Junta de Andalucía (EXC-2023-01, IE19_259 FPS).

References

1. Beck, M.K., Westergaard, D., Jensen, A.B., Groop, L., Brunak, S.: Temporal order of disease pairs affects subsequent disease trajectories: the case of diabetes and sleep apnea. In: Pacific Symposium on Biocomputing, vol. 22, pp. 380–389 (2017). https://doi.org/10.1142/9789813207813_0036

2. Choi, E., Biswal, S., Malin, B., Duke, J., Stewart, W.F., Sun, J.: Generating multi-label discrete patient records using generative adversarial networks. In: Proceedings of the 2nd Machine Learning for Healthcare Conference, pp. 286–305. PMLR (2017)

3. Giuffrè, M., Shung, D.L.: Harnessing the power of synthetic data in healthcare: innovation, application, and privacy. npj Dig. Med. 6(1), 186 (2023). https://doi.org/10.1038/s41746-023-00927-3. https://www.nature.com/articles/s41746-023-00927-3, publisher: Nature Publishing Group

4. Jensen, A.B., et al.: Temporal disease trajectories condensed from population-wide registry data covering 6.2 million patients. Nat. Commun. 5(1), 4022 (2014). https://doi.org/10.1038/ncomms5022

5. Johnson, A.E.W., et al.: MIMIC-III, a freely accessible critical care database. Sci. Data 3(1), 160035 (2016). https://doi.org/10.1038/sdata.2016.35

6. Jordon, J., et al.: Synthetic data – what, why and how? (2022). https://doi.org/10.48550/arXiv.2205.03257

7. Jørgensen, I.F., Haue, A.D., Placido, D., Hjaltelin, J.X., Brunak, S.: Disease trajectories from healthcare data: methodologies, key results, and future perspectives. Ann. Rev. Biomed. Data Sci. **7**(1), 251–276 (2024). https://doi.org/10.1146/annurev-biodatasci-110123-041001

8. Kandi, V., Vadakedath, S.: Clinical trials and clinical research: a comprehensive review. Cureus **15**(2), e35077. https://doi.org/10.7759/cureus.35077

9. Kingma, D.P., Welling, M.: Auto-encoding variational bayes (2022). https://doi.org/10.48550/arXiv.1312.6114

10. Langarizadeh, M., Orooji, A., Sheikhtaheri, A.: Effectiveness of anonymization methods in preserving patients' privacy: a systematic literature review. Stud. Health Technol. Inf. **248**, 80–87 (2018)

11. Lee, D., et al.: Generating sequential electronic health records using dual adversarial autoencoder. J. Am. Med. Inf. Assoc. JAMIA **27**(9), 1411–1419 (2020). https://doi.org/10.1093/jamia/ocaa119

12. Liu, F., Panagiotakos, D.: Real-world data: a brief review of the methods, applications, challenges and opportunities. BMC Med. Res. Methodol. **22**(1), 287 (2022). https://doi.org/10.1186/s12874-022-01768-6

13. Lou, Y., Caruana, R., Gehrke, J., Hooker, G.: Accurate intelligible models with pairwise interactions. In: Proceedings of the 19th ACM SIGKDD International Conference on Knowledge Discovery and Data Mining, KDD '13, pp. 623–631. Association for Computing Machinery, New York (2013). https://doi.org/10.1145/2487575.2487579

14. Makady, A., de Boer, A., Hillege, H., Klungel, O., Goettsch, W.: (on behalf of GetReal Work Package 1): What is real-world data? a review of definitions based on literature and stakeholder interviews. Value Health: J. Int. Soc. Pharmacoecon. Outcomes Res. **20**(7), 858–865 (2017). https://doi.org/10.1016/j.jval.2017.03.008

15. Muñoyerro-Muñiz, D., Goicoechea-Salazar, J.A., García-León, F.J., Laguna-Téllez, A., Larrocha-Mata, D., Cardero-Rivas, M.: Health record linkage: andalusian health population database. Gac. Sanit. **34**(2), 105–113 (2020). https://doi.org/10.1016/j.gaceta.2019.03.003

16. Muñoyerro-Muñiz, D., et al.: Ethical and secure evidence generation from region-wide clinical data through a collaborative environment for advancing predictive care (2024). https://doi.org/10.21203/rs.3.rs-5389651/v1

17. Nielsen, A.B., et al.: Survival prediction in intensive-care units based on aggregation of long-term disease history and acute physiology: A retrospective study of the Danish National Patient Registry and electronic patient records. Lancet. Digital Health **1**(2), e78–e89 (2019). https://doi.org/10.1016/S2589-7500(19)30024-X

18. Placido, D., et al.: A deep learning algorithm to predict risk of pancreatic cancer from disease trajectories. Nat. Med. **29**(5), 1113–1122 (2023). https://doi.org/10.1038/s41591-023-02332-5

19. Tong, X., Yu, Q., Ankawi, G., Pang, B., Yang, B., Yang, H.: Insights into the role of renal biopsy in patients with T2DM: a literature review of global renal biopsy results. Diab. Therapy **11**(9), 1983–1999 (2020). https://doi.org/10.1007/s13300-020-00888-w

20. van Breugel, B., Liu, T., Oglic, D., van der Schaar, M.: Synthetic data in biomedicine via generative artificial intelligence. Nat. Rev. Bioeng. **2**(12), 991–1004 (2024). https://doi.org/10.1038/s44222-024-00245-7

21. Zuo, Z., Watson, M., Budgen, D., Hall, R., Kennelly, C., Al Moubayed, N.: Data anonymization for pervasive health care: systematic literature mapping study. JMIR Med. Inform. **9**(10), e29871 (2021). https://doi.org/10.2196/29871

Author Index

Communications
in Computer and Information Science   2736

Series Editors

Gang Li, *School of Information Technology, Deakin University, Burwood, VIC, Australia*
Joaquim Filipe, *Polytechnic Institute of Setúbal, Setúbal, Portugal*
Zhiwei Xu, *Chinese Academy of Sciences, Beijing, China*

Rationale

The CCIS series is devoted to the publication of proceedings of computer science conferences. Its aim is to efficiently disseminate original research results in informatics in printed and electronic form. While the focus is on publication of peer-reviewed full papers presenting mature work, inclusion of reviewed short papers reporting on work in progress is welcome, too. Besides globally relevant meetings with internationally representative program committees guaranteeing a strict peer-reviewing and paper selection process, conferences run by societies or of high regional or national relevance are also considered for publication.

Topics

The topical scope of CCIS spans the entire spectrum of informatics ranging from foundational topics in the theory of computing to information and communications science and technology and a broad variety of interdisciplinary application fields.

Information for Volume Editors and Authors

Publication in CCIS is free of charge. No royalties are paid, however, we offer registered conference participants temporary free access to the online version of the conference proceedings on SpringerLink (http://link.springer.com) by means of an http referrer from the conference website and/or a number of complimentary printed copies, as specified in the official acceptance email of the event.

CCIS proceedings can be published in time for distribution at conferences or as post-proceedings, and delivered in the form of printed books and/or electronically as USBs and/or e-content licenses for accessing proceedings at SpringerLink. Furthermore, CCIS proceedings are included in the CCIS electronic book series hosted in the SpringerLink digital library at http://link.springer.com/bookseries/7899. Conferences publishing in CCIS are allowed to use our online conference service (Meteor) for managing the whole proceedings lifecycle (from submission and reviewing to preparing for publication) free of charge.

Publication process

The language of publication is exclusively English. Authors publishing in CCIS have to sign the Springer CCIS copyright transfer form, however, they are free to use their material published in CCIS for substantially changed, more elaborate subsequent publications elsewhere. For the preparation of the camera-ready papers/files, authors have to strictly adhere to the Springer CCIS Authors' Instructions and are strongly encouraged to use the CCIS LaTeX style files or templates.

Abstracting/Indexing

CCIS is abstracted/indexed in DBLP, Google Scholar, EI-Compendex, Mathematical Reviews, SCImago, Scopus. CCIS volumes are also submitted for the inclusion in ISI Proceedings.

How to start

To start the evaluation of your proposal for inclusion in the CCIS series, please send an e-mail to ccis@springer.com

Chirag Modi · Gyanendra Kumar Verma ·
Veena Thenkanidiyoor · Ljiljana Brankovic
Editors

Machine Learning, Image Processing, Network Security and Data Sciences

6th International Conference, MIND 2024
Goa, India, December 20–21, 2024
Revised Selected Papers, Part I

 Springer

Editors
Chirag Modi 🆔
National Institute of Technology Goa
Cuncolim, Goa, India

Veena Thenkanidiyoor 🆔
National Institute of Technology Goa
Cuncolim, Goa, India

Gyanendra Kumar Verma 🆔
National Institute of Technology Raipur
Raipur, Chhattisgarh, India

Ljiljana Brankovic
University of New England
Armidale, NSW, Australia

ISSN 1865-0929　　　　　　ISSN 1865-0937 (electronic)
Communications in Computer and Information Science
ISBN 978-3-032-14530-7　　　　ISBN 978-3-032-14531-4 (eBook)
https://doi.org/10.1007/978-3-032-14531-4

Preface

The Department of Computer Science and Engineering at the National Institute of Technology, Goa, was immensely pleased to extend a warm invitation to researchers, experts, and practitioners worldwide for the 6th International Conference on Machine Learning, Image Processing, Network Security, and Data Sciences (MIND 2024). The conference took place on the 20th and 21st of December 2024 at the Chapora Hall of the National Institute of Technology, Goa, India, and it was indeed an enriching and intellectually stimulating event.

Over the years, MIND conferences have served as crucial platforms for the convergence of leading minds in academia, industry, and government, fostering collaborative endeavors and facilitating the exchange of cutting-edge research and insights in the domains of machine learning, image processing, network security, and data sciences. Building upon the successes of its predecessors hosted by NIT Kurukshetra, NIT Silchar, NIT Raipur, MANIT Bhopal, and NIT Hamirpur in 2019, 2020, 2021, 2022, and 2023 respectively, MIND 2024 successfully upheld and further enhanced the spirit of interdisciplinary collaboration and knowledge dissemination.

This year's conference featured distinguished keynote speakers, comprehensive technical sessions, a doctoral colloquium, and engaging panel discussions covering a diverse array of topics pertinent to the aforementioned domains. Participants were invited to contribute their research findings, innovative methodologies, and practical experiences, thereby enriching the discourse and collectively advancing the frontiers of knowledge in these rapidly evolving fields.

MIND 2024 provided an exceptional opportunity for participants to forge new connections, deepen existing collaborations, and explore avenues for interdisciplinary research and development. Researchers, academicians, professionals, and policymakers actively engaged in this dynamic forum, leveraging the invaluable networking opportunities and intellectual exchange it afforded.

We are delighted to share insights into the rigorous peer review process undertaken for MIND 2024. The peer review process was single-blind, ensuring impartial evaluation of submissions. Each manuscript submitted underwent scrutiny by a minimum of three reviewers, ensuring thorough assessment and maintaining the highest standards of quality. A total of 328 submissions were received, reflecting the robust interest and engagement of the research community. Out of these, 93 full-length papers were accepted for presentation at the conference, following a meticulous selection process. On behalf of the organizing committee, we extend our heartfelt gratitude to all participants, sponsors, and supporters for their invaluable contributions towards making MIND 2024 a resounding success. We are confident that their participation greatly enriched the conference proceedings, and we eagerly anticipate their presence at future events. We acknowledge the support of Anusandhan National Research Foundation (ANRF) for financial support towards organizing this conference.

We are grateful for everyone's participation in MIND 2024 and trust that their experience was both enlightening and inspiring.

December 2024

Chirag Modi
Gyanendra Kumar Verma
Veena Thenkanidiyoor
Ljiljana Brankovic

Organizing Committee - MIND 2024

Chief Patron

Omprakash R. Jaiswal — National Institute of Technology Goa, India

Patron

C. Vyjayanthi — National Institute of Technology Goa, India

Honorary Chair

Xiao-Zhi Gao — University of Eastern Finland, Finland

General Chairs

Dileep A. D. — Indian Institute of Technology Dharwad, India

Dhiren Patel — Sardar Vallabhbhai National Institute of Technology Surat, India

Awadhesh Kumar Singh — National Institute of Technology Kurukshetra, India

General Co-chairs

C. Vyjayanthi — National Institute of Technology Goa, India

Gyanendra Kumar Verma — National Institute of Technology Raipur, India

Advisory Committee

Shashank Joshi — Siemens Ltd Goa, India

Ljiljana Brankovic — University of New England, Australia

Hitesh Chhinkaniwala — Adani University, India

Chandrasekaran K. — National Institute of Technology Karnataka, India

Madhu G. M. — Siemens Ltd Goa, India

Sathish Kumar — Cleveland State University, USA

Organizing Chairs

Chirag Modi	National Institute of Technology Goa, India
Veena T.	National Institute of Technology Goa, India
Rajesh Doriya	National Institute of Technology Raipur, India

Organizing Secretaries

S. Mini	National Institute of Technology Goa, India
Amol Rahulkar	National Institute of Technology Goa, India
Trilochan Panigrahi	National Institute of Technology Goa, India
Badal Soni	National Institute of Technology Silchar, India

Publicity Chairs

S. Mini	National Institute of Technology Goa, India
Trilochan Panigrahi	National Institute of Technology Goa, India
Vijay Verma	National Institute of Technology Kurukshetra, India
Himanshu Govil	National Institute of Technology Raipur, India

Technical Program Committee (TPC) Chairs

Venkatanareshbabu K.	National Institute of Technology Goa, India
Shivnarayan Patidar	National Institute of Technology Goa, India
Purushothama B. R.	National Institute of Technology Karnataka, India
Rekh Ram Janghel	National Institute of Technology Raipur, India

Technical Program Committee (TPC) - Members

Abdel-Hamid Ali Soliman	Staffordshire University, UK
Ajaysinh Rathod	Sardar Vallabhbhai Patel Institute of Technology, India
Ajit Muzumdar	Sanjivani University, India
Akashdeep Bhardwaj	University of Petroleum and Energy Studies, India
Akhil Mittal	Synopsys, India
Amit Shewale	Indian Institute of Information Technology Nagpur, India

Amol D. Rahulkar	National Institute of Technology Goa, India
Anagha Bhattacharya	National Institute of Technology Mizoram, India
Aneesh Varghese	Amazon Web Services, India
Anil Pinapati	National Institute of Technology Calicut, India
Anitha Velu	Sri Sairam College of Engineering, India
Anjan Kumar Ray	National Institute of Technology Sikkim, India
Anjan Ray	National Institute of Technology Sikkim, India
Ankesh Jain	Indian Institute of Technology Delhi, India
Anshul Verma	Banaras Hindu University, India
Ankit Chouhan,	Parul University, India
Arun Rasika Karunakaran	Tata Consultancy Services, India
Arun Yadav	National Institute of Technology Hamirpur, India
Ashish Chaudhari	Sardar Vallabhbhai National Institute of Technology Surat, India
Ashwini Patil	National Institute of Technology Goa, India
Bhanu	Indian Institute of Information Technology Kottayam, India
Bhavesh Borisaniya	Shantilal Shah Engineering College, India
Bikash Chandra Sahana	National Institute of Technology Patna, India
C. Vyjayanthi	National Institute of Technology Goa, India
Chandan Trivedi	Nirma University, India
Chandrashekar Jatoth	National Institute of Technology Raipur, India
Chemmalar Selvi G.	Vellore Institute of Technology, Vellore, India
Chirag Modi	National Institute of Technology Goa, India
Deepak Singh	National Institute of Technology Raipur, India
Deepika Agrawal	National Institute of Technology Raipur, India
Dhananjay Kanade	K.K. Wagh Institute of Engineering Education and Research, India
Dimitrios A. Karras	National and Kapodistrian University of Athens, Greece
Durga Prasad Mohapatra	National Institute of Technology Rourkela, India
Dushyant Kumar Singh	Motilal Nehru National Institute of Technology Allahabad, India
E. Silambarasan	Indian Institute of Information Technology Kottayam, India
Emmanuel Shubhakar Pilli	Malaviya National Institute of Technology Jaipur, India
Ganesh Khekare	Vellore Institute of Technology, India
Gaurav Pareek	National Institute of Technology Goa, India
Geeta	Indian Institute of Technology Ropar, India
Geetha P.	SRM University, India
Gerardine Immaculate Mary	Vellore Institute of Technology, India

Divya Garg	National Institute of Technology Kurukshetra, India
Gordhan Jethava	Parul University, India
Goutam Mali	Indian Institute of Information Technology Kottayam, India
Gurucharansingh Sahani	Sardar Vallabhbhai Patel Institute of Technology, India
Gyanendra Verma	National Institute of Technology Raipur, India
Harikrishna Jethva	Government Engineering College Patan, India
Harsha Dudeja	National Institute of Technology Goa, India
Harshavardhan Nerella	MassMutual, India
Himanshu Buckchash	UiT The Arctic University of Norway, Norway
Himanshu Govil	National Institute of Technology Raipur, India
J. Sathish Kumar	Motilal Nehru National Institute of Technology Allahabad, India
Jignesh Joshi	Sardar Vallabhbhai National Institute of Technology Surat, India
John Paul Martin	Indian Institute of Information Technology Kottayam, India
Kapil Soni	National Institute of Technology Raipur, India
Kirti Kumari	Indian Institute of Information Technology Ranchi, India
Kokila J.	Indian Institute of Information Technology Tiruchirappalli, India
Krishnendu Ghosh	Indian Institute of Information Technology Dharwad, India
Kukatlapalli Kumar	Christ University, India
Kunwar Singh	National Institute of Technology Tiruchirappalli, India
L. Oppili Prasad	Indian Institute of Technology (Banaras Hindu University) Varanasi, India
Mallikharjuna Rao K.	Indian Institute of Information Technology Naya Raipur, India
Manish Kumar	Punjab Engineering College, India
Mantosh Biswas	University of Delhi, India
Manu Madhavan	Indian Institute of Information Technology Kottayam, India
Mohammad Bdair	University of East London, UK
Mrityunjay Singh	Indian Institute of Information Technology Una, India
Nishant Doshi	Pandit Deendayal Energy University, India
Oswald C.	National Institute of Technology Tiruchirappalli, India

P. Sasikumar	Vellore Institute of Technology, India
P. M. Durai Raj Vincent	Vellore Institute of Technology, India
Panem Charanarur	National Forensic Sciences University Tripura Campus, India
Pankaj Pratap Singh	Central Institute of Technology Kokrajhar, India
Parag Rughani	National Forensic Sciences University, India
Prabhakar Karthikeyan	Vellore Institute of Technology, India
Pradeep Chintale	SEI Investment Company, India
Pradeep Kumar Roy	Indian Institute of Information Technology Surat, India
Pratikkumar Barot	Government Engineering College Gandhinagar, India
Praveen Borra	Florida Atlantic University, USA
Praveen Yadav	Kalinga University, India
Professeur Mehaoua	Université Paris Cité, France
Purushothama B. R.	National Institute of Technology Karnataka, India
Raghu R.	Oxford College of Engineering, India
Rajendra Shivaji Patil	KTH Royal Institute of Technology, Sweden
Rajesh Doriya	National Institute of Technology Raipur, India
Rajesh Kumar Malviya	NTT Data, USA
Rajesh Bharati	Dr. D.Y. Patil Institute of Technology, India
Ram Chandra Sachan	Wipro, India
Ramakrishna Garine	University of North Texas, USA
Ramya Shah	National Forensic Sciences University, India
Ravikumar C. V.	Vellore Institute of Technology, India
Rekh Ram Janghel	National Institute of Technology Raipur, India
Rupesh Kumar Dewang	Motilal Nehru National Institute of Technology Allahabad, India
S. Mini	National Institute of Technology Goa, India
S. Sridevi	Thiagarajar College of Engineering, India
Sachin Tripathi	Indian Institute of Technology Dhanbad, India
Sakurai Kouichi	Kyushu University, Japan
Santosh Kumar	National Institute of Technology Kurukshetra, India
Sarang Rajvansh	National Forensic Sciences University, India
Sevinthi Kali Sankar Nagarajan	HCL America Inc., USA
Shaik Abdul Kareem	Cloud Technology (Infra and Security), USA
Shailesh Khapre	S.P. Mukherjee International Institute of Information Technology Naya Raipur, India
Shashank Srivastava	Motilal Nehru National Institute of Technology Allahabad, India
Sheik Abdullah A.	Vellore Institute of Technology, India
Shelly Sachdeva	National Institute of Technology Delhi, India

Sherali Zeadally	University of Kentucky, USA
Shikha Triphathi	PES University, India
Shitala Prasad	Indian Institute of Technology Goa, India
Shivangi Shukla	Indian Institute of Information Technology Pune, India
Shrikant Upadhyay	MLR Institute of Technology, India
Shyam Lal	National Institute of Technology Karnataka, India
Soniya Usgaonkar	Goa College of Engineering, India
Srinivas Pinisetty	Indian Institute of Technology Bhubaneswar, India
Sudarshan T. S. B.	PES University, India
Sudesh Rani	Punjab Engineering College, India
Sudip Roy	Indian Institute of Technology Roorkee, India
Suganthinic C.	Vellore Institute of Technology, India
Sumit Gautam	Indian Institute of Technology Indore, India
T. S. B. Sudarshan	PES University, India
Tejas Modi	Indian Institute of Information Technology Surat, India
Tharun Kumar Reddy Bollu	Indian Institute of Technology Roorkee, India
Udai Pratap Rao	National Institute of Technology Patna, India
Umashankar Subramaniam	Prince Sultan University, Saudi Arabia
Veena Thenkanidiyoor	National Institute of Technology Goa, India
Venkatanareshbabu Kuppili	National Institute of Technology Goa, India
Vijaya J.	S.P. Mukherjee International Institute of Information Technology Naya Raipur, India
Vishal Passricha	Central University of Haryana, India
Vishvajit Bakrola	Uka Tarsadia University, India
Vivek Shrivastava	National Institute of Technology Uttarakhand, India

Contents

Artificial Intelligence and Machine Learning

Artificial Intelligence and Machine Learning

Alzheimer Detection and Classification Using Deep Learning Techniques

Keyurkumar Patel[1]([✉])(iD), Arpit Sharma[2](iD), Vansh Tiwari[2](iD),
Kunjan Parikh[2](iD), and Jaymin Tanna[2](iD)

[1] School of Applied Sciences Engineering and Technology, Rashtriya Raksha
University, Gandhinagar, India
pkeyur.436@gmail.com
[2] Department of Computer Engineering, UIT, Karnavati University,
Gandhinagar, India

Abstract. Alzheimer's Disease is a complex and debilitating neuro-generative disorder that necessitates accurate diagnosis and classification. This study investigates the efficacy of three deep learning models - EfficientFormer-L1, EfficientNet B3, and DenseNet-169 - in classifying AD into four distinct categories. To optimize the performance of these models, we compared the AdamW and Lion optimizers, and explored the impact of three data balancing strategies: class weighting, data augmentation, and Synthetic Minority Over-Sampling Technique (SMOTE). The DenseNet-169 model paired with the AdamW optimizer achieved the highest accuracy of 98.87% when class weights were employed. The augmented dataset favored the EfficientFormer-L1 model, which reached an accuracy of 99.53% when used with the Lion optimizer. The SMOTE balancing technique revealed that EfficientFormer-L1 model to be robust, achieving accuracies of 99.69% and 99.61% with the AdamW and Lion optimizers, respectively. These findings suggest that the Lion optimizer is particularly well-suited for transformer-based models, such as EfficientFormer-L1.

Keywords: Alzheimer's Disease (AD) · Convolutional Neural Network (CNN) · Deep Learning (DL) · Neuroimaging Data · Neurodegenerative Disorder

1 Introduction

Even though the medical field has advanced thanks to the introduction of technological innovations from computer science sub-fields like natural language processing and deep learning, we are still a long way from arriving at a stage where essential tasks can be completed on their own without the guidance of medical professionals. In order to propel research in the particular medical sub-discipline of neurology forward, a thorough analysis of dementia also referred to as Alzheimer's disease (AD) is necessary. It's a neurological condition that primarily affects elder people some of its symptoms being, memory loss, cognitive decline, and behavioral abnormalities, make it the most prevalent type of dementia. For prompt intervention and better patient outcomes, Alzheimer's disease must be identified early and properly classified [2]. Traditional Alzheimer's disease diagnostic procedures combine clinical evaluations, cognitive tests, and neuro-imaging studies. These methods, however, could be arbitrary, time-consuming, and expensive. To reduce the burdens associated with traditional diagnostic procedures for Alzheimer's disease (AD) deep learning algorithms have been seen effective [13]. In this paper, we revisit medium-sized models (parameters 10 to 15 million) utilizing advantages they offer based on model complexity and computational efficiency trade off, our objective is to compare models with similar sizes using different conceptual frameworks, such as CNNs and Vision Transformers.

2   Related Works

Alzheimer's Disease (AD) is a debilitating neurogenerative disorder characterized by memory loss and cognitive decline. Accurate and early diagnosis is crucial for effective intervention and management. Recent advancements in deep learning, particularly Convolutional Neural Networks (CNNs) and transformers, have shown promise in automating and enhancing the diagnostic process by analysing MRI images. CNNs excel at recognizing patterns in image data, making them well-suited for analysing MRI scans to identify AD-related changes in the brain. [23] evaluated different CNN models, namely VGG-11, VGG-13, VGG-16, VGG-19 and ResNet-18, ResNet-34, ResNet-50, ResNet-101, and ResNet-152 on publicly available Alzheimer's datasets for two classification tasks. In their study, [14] employed the DenseNet-169 and VGG-19 CNN architectures for classifying AD. They find that VGG-19 outperforms DenseNet169. Similarly, [9] trained a CNN model on the DenseNet-169 architecture and achieved an accuracy of 91.80%. In another study [21], the authors explored the application of EfficientNet models for AD detection. They applied two models: EfficientNet-B0 and EfficientNet-B3. The EfficientNet-B0 model achieved an accuracy of 96%, while the EfficientNet-B3 model outperformed the authors' previous study's VGG-16 and VGG-19 models, attaining the highest accuracy of 97%. Similarly, [12] used a CNN model based on the EfficientNet-B3 architecture for classifying AD and achieved an accuracy of 96.24%. [6] proposes a novel CNN architecture termed ADD-Net, to detect AD from MRI scans. The performance of ADD-Net was evaluated against DenseNet-169, VGG-19 and InceptionResNetV2. The proposed ADD-Net achieved an accuracy of 96.70%, a precision of 97%, a recall of 97% and an AUC value of 99.82%. The authors stated that the model performs better for small datasets. [7] employed 12 widely utilized deep neural network (DNN) models to classify AD using images from three classes of the ADNI dataset. They observed that DenseNet-121 performed the best. The authors acknowledged that the DenseNet-121 is computationally expensive. They proposed a new lightweight DNN model combining AlexNet and LeNet which outperforms the DenseNet-121 model. In [5], the authors proposed an innovative lightweight CNN model for classifying AD. The model comprises of only seven layers, resulting in reduced complexity. The model was evaluated on two classification tasks: binary classification (AD vs Normal) and four-class classification (different stages of AD). The model achieved 99.30% accuracy in the binary classification task and 95.96% in the four-class classification task Researchers introduced a novel CNN model based on the VGG architecture in [1] named STCNN to classify different AD stages. The model achieved an accuracy of 99.36%, an Area Under the Curve (AUC) value of 99.94%, a loss of 0.022, and an F1-score of 99.34%. The model was compared against several state-of-the-art benchmark models and outperformed them. In [10], the authors proposed a novel approach, termed DenseCNN2, for classifying AD using MRI scans of the hippocampus region. Unlike previous methods that relied on visual features, DenseCNN2 combined both visual and global shape feature of the hippocampus. Researchers focused on developing computer-aided systems for brain tumour diagnosis using deep learning techniques applied to MRI data. Three approaches were proposed. Firstly Directional Bit-Planes Deep Autoencoder (DBP-DAE), aimed to extract and learn local and directional features by decomposing Local Binary Patterns (LBP) from the MRI images. Secondly, Dilated Separable Residual Convolutional Network (DSRCN), was designed to capture both high-level and low-level features while addressing issues such as robustness and overfitting. Thirdly, A multi-branch convolutional neural network that combined the DBP-DAE and DSRCN architectures, with the goal of improving classification accuracy. [16] introduced a novel framework, Biceph-Net designed to classify AD using 2d MRI scans. Unlike previous approaches that primarily focused on either intra-slice or inter-slice information, Biceph-Net uses both types of information. The model achieved an accuracy of 100% in classifying AD vs Normal, an accuracy of 98.16% in classifying MCI vs AD, and an accuracy of 97.80% in classifying Cognitively Normal, MCI, AD. Class imbalance is a common problem in many real-world machine learning applications, including medical datasets. It occurs when one class has a fewer number of samples than another class. Most machine learning algorithms are designed to optimize overall accuracy, leading to poor performance for the minority class. [24] aims to compare the performance of the original imbalanced dataset using methods like SMOTE, Adasyn, KMeans and Cluster methods. The authors found that the best model was SMOTE Balanced 5-Fold Cross-validation multiclass classification. [6] also faced the problem of dataset imbalance. The authors used the SMOTE-TOMEK technique to generate synthetic samples for the classes to solve the problem. Similarly, [1] also employs SMOTE-TOMEK to balance the samples for their four classes. [3] propose a new oversampling method called DeepSMOTE which enhances the functionality of the current SMOTE algorithm. It incorporates an encoder-decoder structure and a loss function to produce superior quality synthetic images to balance the training dataset. [19] randomly duplicates the minor class of images using the SMOTE technique and to avoid overfitting. Generative Adversarial Networks (GANs) are an alternative approach to traditional data augmentation techniques. GANs consist of two components: a generator network that creates synthetic data instances, and a discriminator network that tries to distinguish between real and generated data. [17] used GANs for data augmentation of their dataset. The augmented dataset, containing both original and synthetic samples, was used to train the classification model. Deep convolutional Generative Adversarial Networks (DCGANs) are an improved version of GANs which produce higher-quality images compared to GANs, introduced by Radford and Metz [15]. [18] introduces an innovative method using DCGANs to generate synthetic brain Positron Emission Tomography (PET) images across different stages of AD. Transformers are a type of model architecture that have shown great promise in the field of computer vision, including medical analysis. CNNs can be limited in their ability to capture long-range dependencies between different regions of an image. Transformers, on the other hand, were designed to handle sequential data with long-range dependencies by using a self-attention mechanism.

3   Models And Technique

3.1   Model Architectures

In medical research, CNNs have recently seen increased application [4]. The advances have been scaling ConvNets by adjusting width, depth, or image resolution. EfficientNet introduced by Mingxing Tan and Quoc V. Le [20] scales all three dimensions using a compound scaling method. EfficientNet-B3, from the EfficientNet family, is trained at 300×300 resolution on the ImageNet-1k dataset. Inspired by MobileNetV2, it includes residual blocks with skip connections, and the Swish activation function noted for better performance in deep networks. It also uses global average pooling in order to reduce the spatial dimensions to single values per channel.

The architecture DenseNet, in [8], utilizes densely connected layers to improve depth with accuracy and efficiency. DenseNet-169 has 169 layers, including dense blocks and transition layers, using ReLU activation and batch normalization. Dense connectivity improves gradient flow and feature reuse due to which models become more parameter-efficient.

In contrast, ViTs are typically slower due to their design and large number of parameters, using self-attention mechanisms. Given this, EfficientFormer, proposed by Li et al. [11], designs transformer models that are optimized for performance-latency trade-offs. The EfficientFormer family models, L1-L7, are lightweight in nature, while their EfficientFormer-L1-300 variant obtains 79.2% top-1 accuracy on ImageNet-1K with only a 1.6ms inference latency on an iPhone 12, outperforming MobileNetV2X1.4 by 4.5% in accuracy at similar speed.

4   Methodology

4.1   Data Source and Pre-processing

The preceding section delved into the intricacies of model functionality employed in implementation. This section will introduce the methodology employed after considering various parameters for the proposed model after conducting a review of the literature. The following steps will be taken by the pipeline: Data gathering and pre-processing building a model and fitting it to the data come in steps three and four. The source of image corpus is Mendeley [22]. The data in Fig. 1 is divided into 4 classes which are as follows:

- Class- 1: Mild Demented which consists of 896 images
- Class- 2: Moderate Demented which consists of 64 images
- Class- 3: Non-Demented which consists of 3200 images
- Class- 4: Very Mild Demented which consists of 2240 images

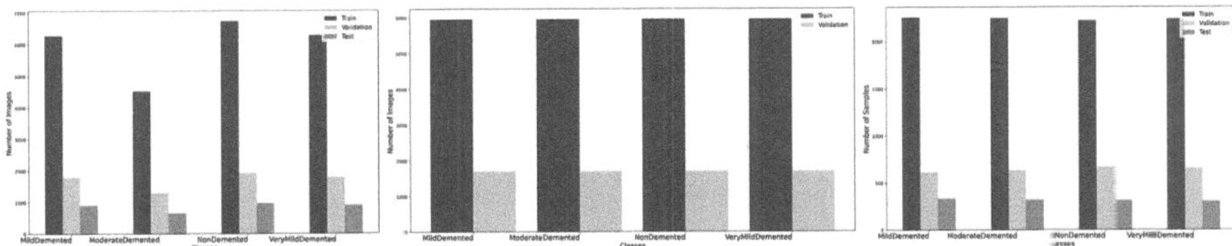

Fig. 1. Class Distribution graphs for Right: Augmented Preprocessed Dataset Center: Class weights Preprocessed Dataset Left: SMOTE Preprocessed Dataset

The paper tries to investigate 3 different techniques for solving the imbalance in dataset in upcoming sub-section.

Handling Class Imbalance. In order to handle dataset imbalance, we used the following three techniques: class weight modification, Synthetic Minority Oversampling Technique (SMOTE), and augmentation. Weight adjustment of classes gives higher importance to the model for minority classes, so that the model concentrated on their learning patterns. SMOTE generates artificial samples for the minority classes and balances the dataset with reduced overfitting. Augmentation mitigates overfitting in datasets with limited samples. Using the Albumentations library, we applied augmentations such as random shifts in RGB channels, brightness, contrast, and saturation. ColorJitter was applied with default values, followed by normalization using ImageNet's mean and standard deviation. The processed dataset was divided into training, validation, and test sets in a 7:2:1 ratio and converted to PyTorch tensors.

4.2 Hyper-parameters and Fine-Tuning

When training the models, we preserved a batch size of 64 over 20 epochs. We applied the AdamW and Lion Optimizers with a default learning rate of 0.001 and categorical cross-entropy as the loss function. In order to continuously track the validation loss during training, we used the ReduceLearningRateon-Plateau from torch.optim.lr_scheduler.ReduceLROnPlateau. We were able to freeze some of the layers of the pre-trained models in the context of transfer learning.

5   Results

This section presents the results of employing class-weighting, data augmentation, and SMOTE techniques on the original dataset to mitigate class imbalance. Using both AdamW and Lion optimization algorithms, the evaluation is carried out on three deep learning architectures: EfficientNet B3, EfficientFormer L1, and DenseNet 169.

5.1   Class-Weights

Table 1. Model Performance Metrics for ADAMW (Weighted)

MODELS	TRAIN_ACC	VAL_ACC	TRAIN_LOSS	VAL_LOSS	PRECISION	RECALL	F1-SCORE
EFFICIENTNET_B3	97.48	91.24	0.0530	0.2656	0.9149	0.9124	0.9126
EFFICIENTFORMER_L1	100.00	98.75	0.0000	0.0357	0.9877	0.9875	0.9875
DENSENET_169	99.76	98.87	0.0087	0.0345	0.9887	0.9887	0.9887

Table 2. Model Performance Metrics for LION (Weighted)

MODELS	TRAIN_ACC	VAL_ACC	TRAIN_LOSS	VAL_LOSS	PRECISION	RECALL	F1-SCORE
EFFICIENTNET_B3	96.52	91.86	0.0865	0.2362	0.9209	0.9186	0.9194
EFFICIENTFORMER_L1	99.53	96.71	0.0115	0.1094	0.9683	0.9671	0.9672
DENSENET_169	98.65	93.43	0.0283	0.1913	0.9345	0.9343	0.9343

Tables 1 and 2 summarize model performance using class-weights. All models displayed high training accuracy, with low validation accuracy, which is common in machine learning where models shows promising results on training data but perform considerably poor unseen data. That being said, validation accuracy achieved commendable benchmarks. EfficientFormer-L1 with AdamW and Lion obtained the highest accuracy for both training and validation. The accuracy curves demonstrate that Lion obtains faster initial accuracy gains than Adam, and the loss curves show smoother optimization. In terms of total scores and curve smoothness, Lion performs better than Adam, according to the precision, recall, and F1 score graphs. Figures 2 and 3 show the comparison of DenseNet 169, EfficientFormer L1, and EfficientNet B3 with AdamW and Lion optimizers on a Class-weights dataset. EfficientFormer L1 and DenseNet 169 consistently outperform EfficientNet B3 in accuracy, loss, and F1 score. The Lion optimizer generally leads to faster initial convergence and more stable performance in later epochs compared to AdamW. Overall, EfficientFormer L1 emerges as the top performer.

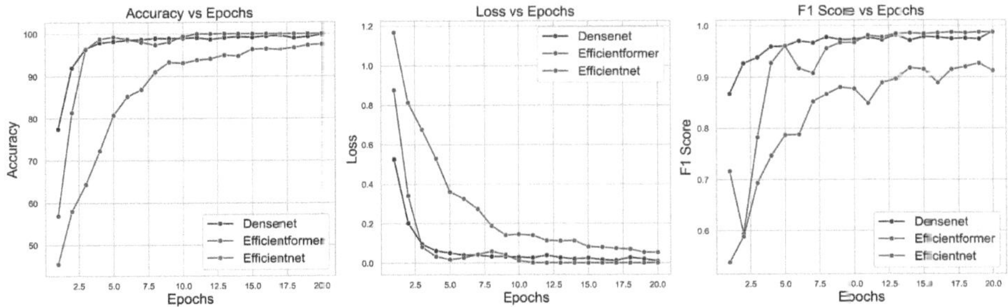

Fig. 2. Performance metrics on the Class Weights preprocessed dataset using AdamW optimize

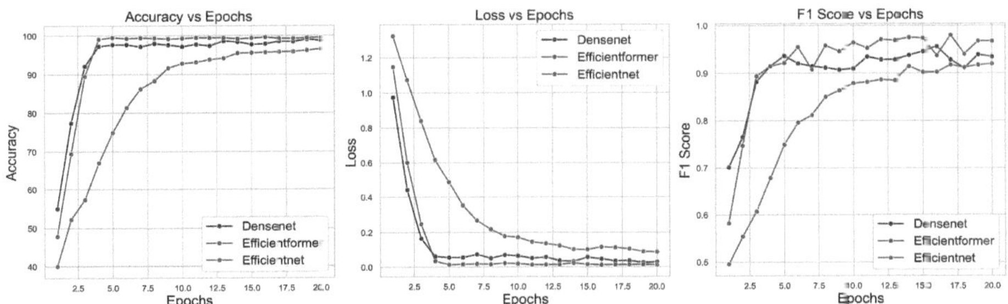

Fig. 3. Performance metrics on the Class Weights preprocessed dataset using Lion optimize

5.2 Augmentation

Table 3. Model Performance Metrics for ADAMW

MODELS	TRAIN_ACC	VAL_ACC	TRAIN_LOSS	VAL_LOSS	PRECISION	RECALL	F1-SCORE
EFFICIENTNET_B3	98.62	97.22	0.0412	0.0853	0.9722	0.9722	0.9722
EFFICIENTFORMER_L1	99.39	97.66	0.0181	0.0700	0.9769	0.9766	0.9766
DENSENET_169	99.35	97.97	0.0184	0.0702	0.9800	0.9797	0.9797

Data augmentation's effects on model performance are shown by the results in Tables 3 and 4. Enhancing the robustness of the model through augmentation was demonstrated by the consistent improvement in training and validation accuracy. Of particular note, DenseNet 169 performed exceptionally well, obtaining validation accuracies of 97.97% for AdamW and 98.09% for Lion. The need of data augmentation in enhancing training data and enhancing model generalization is further supported by these results.

Table 4. Model Performance Metrics for LION

MODELS	TRAIN_ACC	VAL_ACC	TRAIN_LOSS	VAL_LOSS	PRECISION	RECALL	F1-SCORE
EFFICIENTNET_B3	97.33	94.57	0.0773	0.1937	0.9471	0.9457	0.9457
EFFICIENTFORMER_L1	99.86	99.53	0.0039	0.0112	0.9953	0.9953	0.9953
DENSENET_169	99.44	98.09	0.0173	0.0659	0.9811	0.9809	0.9809

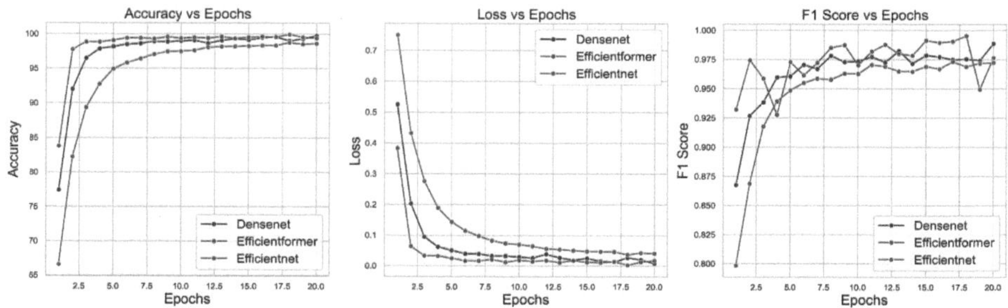

Fig. 4. Performance metrics on the Augumented dataset using AdamW optimize

For AdamW optimization with data augmentation, Table 3 displays the model performance metrics. Compared to models trained without augmentation, we saw significant gains in validation and training accuracies for all models. In instance, the EfficientFormer L1 model demonstrated the efficacy of augmentation in improving model performance with a validation accuracy of 97.66%. Comparably, the model performance metrics for Lion optimization with data augmentation are shown in Table 4. Once more, augmentation improved all models' training and validation accuracies. A noteworthy validation accuracy of 99.53% was attained by the EfficientFormer L1 model, indicating the strength of the augmentation technique in enhancing model generalization. These results, taken together, highlight the significance of data augmentation as a potent method for improving the robustness and performance of deep learning models in medical imaging applications.

Figures 4 and 5 show the performance of the models with an augmented dataset. The augmented dataset results show similar trends to the Class-weights data, with some improvements across all models. EfficientFormer and DenseNet continue to outperform EfficientNet. All models show faster initial convergence and higher accuracy. The loss curves have steeper initial drops and lower final values. F1 scores and higher and more stable. Lion still shows advantages over AdamW. The performance of all three models is more closely matched in later epochs. This suggests that data augmentation has helped EfficientNet close the gap with the other two models.

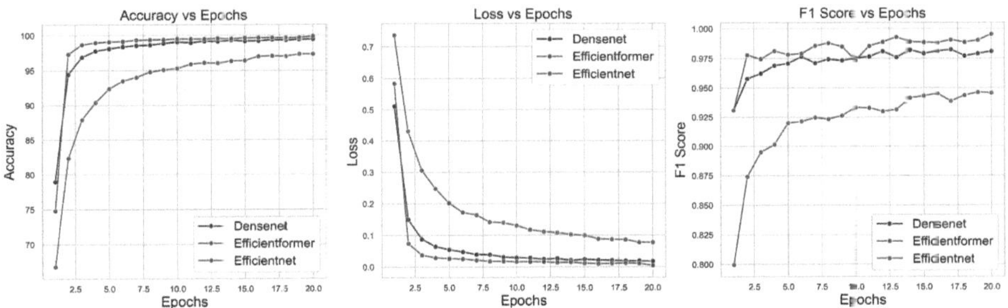

Fig. 5. Performance metrics on the Augumented dataset using Lion optimize

5.3   SMOTE

Table 5. Model Performance Metrics for ADAMW with SMOTE

MODELS	TRAIN_ACC	VAL_ACC	TRAIN_LOSS	VAL_LOSS	PRECISION	RECALL	F1-SCORE
EFFICIENTNET_B3	99.10	98.25	0.0266	0.0485	0.9825	0.9825	0.9825
EFFICIENTFORMER_L1	100.00	99.69	0.0002	0.0167	0.9969	0.9969	0.9969
DENSENET_169	99.72	98.09	0.0085	0.0687	0.9811	0.9809	0.9809

Table 6. Model Performance Metrics for LION with SMOTE

MODELS	TRAIN_ACC	VAL_ACC	TRAIN_LOSS	VAL_LOSS	PRECISION	RECALL	F1-SCORE
EFFICIENTNET_B3	98.94	97.63	0.0300	0.0713	0.9762	0.9763	0.9762
EFFICIENTFORMER_L1	100.00	99.61	0.0002	0.0173	0.9961	0.9961	0.9961
DENSENET_169	99.94	98.83	0.0017	0.0495	0.9884	0.9883	0.9883

Tables 5 and 6 present the outcomes of employing SMOTE to address class imbalance. SMOTE led to notable improvements in validation accuracy across all models. In particular for AdamW and Lion, respectively, the EfficientFormer L1 model consistently achieved the highest validation accuracy of 99.69% and 99.61%. These outcomes coincide with the anticipated advantages of SMOTE as stated in the context of the study's results. SMOTE successfully balanced the class distribution and enhanced the model's effectiveness on minority classes through the generation of synthetic samples for minority classes. Additionally, all models retained high values for precision, recall, and F1-score, demonstrating the resilience and efficacy of the SMOTE-trained models. This highlights how essential it is to address class imbalance for medical imaging tasks such as Alzheimer's diagnosis to be able to ensure reliable and dependable predictions across every class.

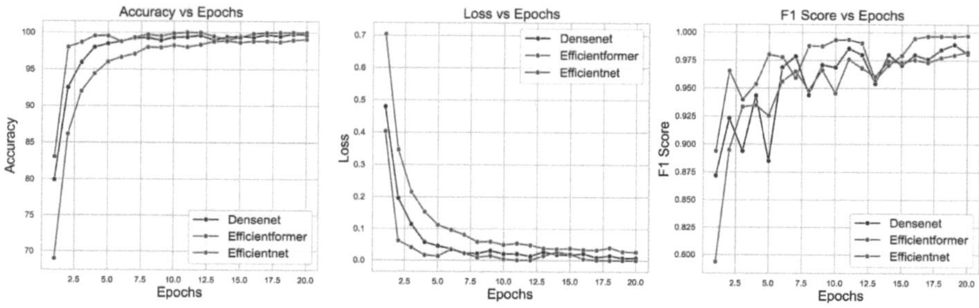

Fig. 6. Performance metrics on the SMOTE dataset using Adamw optimize

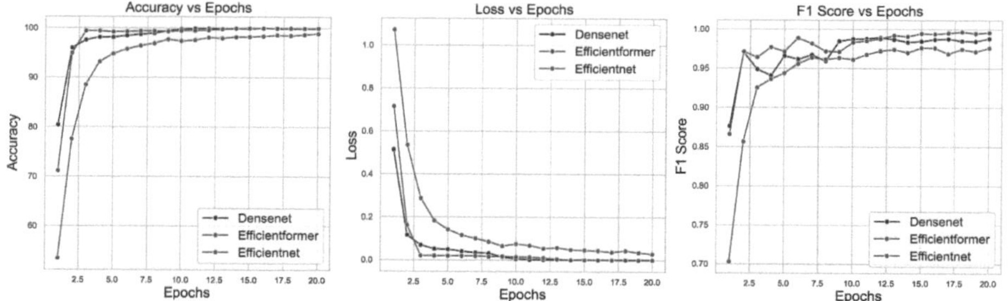

Fig. 7. Performance metrics on the SMOTE dataset using Lion optimize

Figures 6 and 7 compares the models with SMOTE data balancing. Efficient-Former generally outperforms the other models. EfficientFormer also achieves the lowest loss. The Lion optimizer converges faster for all models compared to AdamW. The loss curves with the Lion optimizer also show lower final values.

6   Conclusion

In conclusion, our study demonstrates how class-weighting, data augmentation, and SMOTE approaches collaborate in conjunction to minimize class imbalance while improving deep learning models' performance on medical imaging tasks like Alzheimer's diagnosis. The findings point out how essential it is to use robust methods for preprocessing to address real-world issues with healthcare datasets. However, it's important to acknowledge some limitations of our study. Firstly, our research focused solely on three specific techniques for addressing class imbalance, and there may be other approaches that could yield different results. Additionally, the performance of deep learning models can be influenced by various factors such as hyperparameter settings, network architecture, and the quality and quantity of the training data. Future research could explore the combination of multiple techniques or investigate novel approaches to further improve model performance. Moreover, accuracy, precision, recall, and F1-score were the main metrics used in our study to assess model performance. Additional evaluation metrics, such as interpretability, computational efficiency, and generalizability across diverse populations, could offer a more thorough understanding of the models' usefulness in clinical settings, even though these metrics still offer insightful information about how well the models work. Overall, our results lay the groundwork for future studies aiming at creating more durable and dependable deep learning models for medical imaging applications, as well as add to the expanding corpus of research on addressing class imbalance in medical imaging tasks.

References

1. Anjali, S.D., Pandey, O., Dai, H.N.: Stcnn: combining smote-tomex with cnn for imbalanced classification of alzheimer's disease. IEEE Sens. Lett. **8**, 1–4 (2024). https://doi.org/10.1109/LSENS.2024.3357196
2. Bature, F., Guinn, B.A., Pang, D., Pappas, Y.: Signs and symptoms preceding the diagnosis of alzheimer's disease: a systematic scoping review of literature from 1937 to 2016. BMJ Open **7**(8), e015746 (2017). https://doi.org/10.1136/bmjopen-2016-015746
3. Dablain, D., Krawczyk, B., Chawla, N.V.: Deepsmote: fusing deep learning and smote for imbalanced data. IEEE Trans. Neural Netw. Learn. Syst. **34**(9), 6390–6404 (2023). https://doi.org/10.1109/TNNLS.2021.3136503
4. Dudeja, T., Dubey, S.K., Bhatt, A.K.: Ensembled EfficientNetB3 architecture for multi-class classification of tumours in MRI images. Intell. Decis. Technol. **17**(2), 395–414 (2023)
5. El-Latif, A.A.A., Chelloug, S.A., Alabdulhafith, M., Hammad, M.: Accurate detection of alzheimer s disease using lightweight deep learning model on mri data. Diagnostics **13**(7) (2023). https://doi.org/10.3390/diagnostics13071216. https://www.mdpi.com/2075-4418/13/7/1216
6. Fareed, M.M.S., et al.: Add-net: an effective deep learning model for early detection of alzheimer disease in MRI scans. IEEE Access **10**, 96930–96951 (2022). https://doi.org/10.1109/ACCESS.2022.3204395
7. Hazarika, R.A., et al.: An approach for classification of alzheimer s disease using deep neural network and brain magnetic resonance imaging (MRI). Electronics **12**(3) (2023). https://doi.org/10.3390/electronics12030676. https://www.mdpi.com/2079-9292/12/3/676
8. Huang, G., Liu, Z., Weinberger, K.Q.: Densely connected convolutional networks. CoRR arxiv:1608.06993 (2016)
9. Joshi, R., Negi, P., Foongodi, T.: Multilabel classifier using densenet-169 for alzheimer's disease. In: 2023 4th International Conference on Intelligent Engineering and Management (ICIEM), pp. 1–7 (2023). https://doi.org/10.1109/ICIEM59379.2023.10165844
10. Katabathula, S., Wang, Q., Xu, R.: Predict alzheimer s disease using hippocampus mri data: a lightweight 3d deep convolutional network model with visual and global shape representations. Alzheimer's Res. Ther. **13**(1), 104 (2021). https://doi.org/10.1186/s13195-021-00837-0
11. Li, Y., et al.: Rethinking vision transformers for mobilenet size and speed (2023)

12. Mehta, D., Mohite, A., Shinde, V., Khatri, R., Dokare, I.: Mri-based alzheimer s disease detection using efficientnetb3. In: 2023 International Conference on Advanced Computing Technologies and Applications (ICACTA), pp. 1–6 (2023). https://doi.org/10.1109/ICACTA58201.2023.10393828
13. Pham, T., Tran, T., Phung, D., Venkatesh, S.: Predicting healthcare trajectories from medical records: a deep learning approach. J. Biomed. Inf. **69**, 218–229 (2017). https://doi.org/10.1016/j.jbi.2017.04.001
14. Pradhan, A., Gige, J., Eliazer, M.: Detection of alzheimer s disease (ad) in mri images using deep learning. Int. J. Eng. Res. Technol. (IJERT) **10**(03) (2021)
15. Radford, A., Metz, L., Chintala, S.: Unsupervised representation learning with deep convolutional generative adversarial networks (2016)
16. Rashid, A.H., Gupta, A., Gupta, J., Tanveer, M.: Biceph-net: a robust and lightweight framework for the diagnosis of alzheimer s disease using 2d-mri scans and deep similarity learning. IEEE Journal of Biomedical and Health Informatics **27**(3), 1205–1213 (2023). https://doi.org/10.1109/jbhi.2022.3174033. http://dx.doi.org/10.1109/JBHI.2022.3174033
17. Roy, A., Rahman, M.M., Ahmed, S.: Federated GAN based biomedical image augmentation and classification for Alzheimer s disease. Ph.D. thesis, Brac University (2022)
18. Sajjad, M., Ramzan, F., Khan, M., et al.: Deep convolutional generative adversarial network for alzheimer's disease classification using positron emission tomography (pet) and synthetic data augmentation. Microsc. Res. Tech. **84**(12), 3023–3034 (2021). https://doi.org/10.1002/jemt.23861
19. Suriya, M., et al.: Demnet: a deep learning model for early diagnosis of alzheimer diseases and dementia from mr images. IEEE Access (2021). https://doi.org/10.1109/ACCESS.2021.3090474
20. Tan, M., Le, Q.V.: Efficientnet: Rethinking model scaling for convolutional neural networks. CoRR arxiv:1905.11946 (2019)
21. Ujilast, N.A., Firdausita, N.S., Aditya, C.S.K., Azhar, Y.: Mri image based alzheimer s disease classification using convolutional neural network: efficientnet architecture. Jurnal RESTI (Rekayasa Sistem dan Teknologi Informasi) **8**(1), 18 – 25 (2024). https://doi.org/10.29207/resti.v8i1.5457. http://jurnal.iaii.or.id/index.php/RESTI/article/view/5457
22. Yakkundi, A.: Alzheimer's disease dataset. Mendeley Data (2023). https://doi.org/10.17632/ch87yswbz4.1
23. Zhang, F., et al.: An explainable two-dimensional single model deep learning approach for alzheimer's disease diagnosis and brain atrophy localization (2021)
24. zdemir, A., Polat, K., Alhudhaif, A.: Classification of imbalanced hyperspectral images using smote-based deep learning methods. Expert Syst. Appl. **178**, 114986 (2021). https://doi.org/10.1016/j.eswa.2021.114986. https://www.sciencedirect.com/science/article/pii/S0957417421004279

Enhancing Hyperspectral Image Classification with Attention-Driven Dual-CNN Fusion

Sushil Kumar Janardan[1]([✉])(iD), Ram Nivas Giri[1], Rekh Ram Janghel[1](iD), and Himanshu Govil[2](iD)

[1] Department of Information Technology, National Institute of Technology, Raipur, G.E. Road, Raipur 492010, Chhattisgarh, India
skjanardan.phd2023.it@nitrr.ac.in,
{rngiri.phd2021.it,rrjanghel.it}@nitrr.ac.in
[2] Department of Applied Geology, National Institute of Technology, Raipur, G.E. Road, Raipur 492010, Chhattisgarh, India
hgovil.geo@nitrr.ac.in

Abstract. Hyperspectral image (HSI) classification is challenging due to the data's vast spectral and spatial information. This research presents a new method of classification that utilizes the attention mechanism of dual convolutional neural networks (CNN). The integrated attention mechanism helps the model prioritize attributes, drop irrelevant information, and maximize feature extraction. This method improves HSI classification by concatenating spatial and spectral data into a fully connected layer using an adaptive approach, improving classification over state-of-the-art approaches. The Indian Pines and Pavia University datasets are used to assess performance using overall accuracy, average accuracy, and kappa value. The study also examines how parameter changes affect model performance and compares results to other approaches. The study shows how attention processes can improve model performance, thereby improving the field and driving future research toward more accurate and efficient HSI classification.

Keywords: HSI Classification · Spatial Attention · Spectral Attention · A2S2K ResNet · Convolutional Neural Network

1 Introduction

Hyperspectral images capture the electromagnetic spectrum of an area spanning hundreds to thousands of bands. They contain significantly more spectral information than standard images (Red, Green, Blue). Various classification processes extract essential details from this data [1]. Hyperspectral data includes ultraviolet, visible, and infrared wavelengths. The spectral characteristics of different materials allow us to distinguish them. Each HSI pixel can help classify materials based on their spectral characteristics [2]. In agriculture, HSI images assess crop health, diagnose diseases, identify nutrient deficiencies, and monitor soil conditions. They can also identify water pollution sources and analyze vegetation cover for environmental monitoring. Biomedical imaging uses hyperspectral imaging to identify malignant tumours, blood oxygenation, and skin problems. It helps to non-destructively analyze and restore artefacts, manuscripts, and archaeological objects [3]. Hyperspectral imaging can identify and classify materials that are difficult to distinguish using standard imaging methods [4], by offering comprehensive spectral analysis. This technology can improve decision-making, resource management, and scientific research. Figure 1 illustrates the representation of Hyperspectral Image (HSI) cubes, highlighting both the spectral and spatial dimensions.

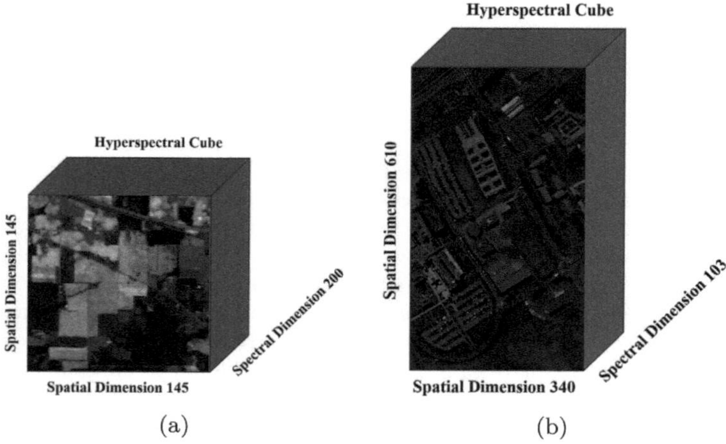

Fig. 1. Representation of Hyperspectral Image (HSI) Cubes with Spectral and Spatial Dimensions

Hyperspectral image classification of objects and landscapes is problematic for various reasons. The high dimensionality [15], of hyperspectral images, which contain multiple spectral bands, generates huge feature vectors that are computationally expensive and usually require dimensionality reduction [5]. Second, hyperspectral images capture the electromagnetic spectrum at small wavelength intervals, leading to spectral heterogeneity within and between classes. This makes it challenging to classify pixels and distinguish comparable classes [6]. Labelled hyperspectral image training samples are expensive and time-consuming, which limits training data and risks overfitting.

When certain classes are underrepresented in hyperspectral images, class imbalance can skew classification results toward overrepresented classes affecting minority classes. Spectral mixing [7], where each pixel contributes from different materials, complicates feature extraction and spectral signature capture. Sensor noise, atmospheric interference, and light fluctuations affect hyperspectral image feature quality. Spatial and spectral resolutions typically conflict, making extracting meaningful features that capture both at appropriate scales difficult [8]. Hyperspectral data is high-dimensional and requires complex algorithms and heavy processing resources, making feature extraction time-consuming.

1.1   Architecture of CNN

Figure 2 illustrates the stages of a Convolutional Neural Network (CNN). A CNN phases between convolutional and pooling layers, then fully connected layers [17]. Each layer builds on the previous layer, capturing more abstract input data. Convolutional layers use filters (or kernels) to convolutionally map the input feature map, followed by a nonlinear activation function [32]. The activity of the ith feature map in the lth layer is provided by Eq. 1.

$$F_i^l = \sum_j g(w_{i,j}^l * F_j^{l-1} + b_i^l)$$
(1)

where $F_j^{l-1} \in R^{p \times q}$ is the j-th feature map in the $(l-1)^{\text{th}}$ layer that connects to the feature map F_{il} in the l-th convolutional layer. $W_{i,jl} \in R^{w \times w}$ is the convolutional kernel for F_{il-1}. b_{il} is the bias. $*$ denotes the convolutional operator. The size of F_{il} is $(p-w+1) \times (q-w+1)$. $g(\cdot)$ is a nonlinear activation function, such as the rectified linear unit function $g(x) = \max(0, x)$.

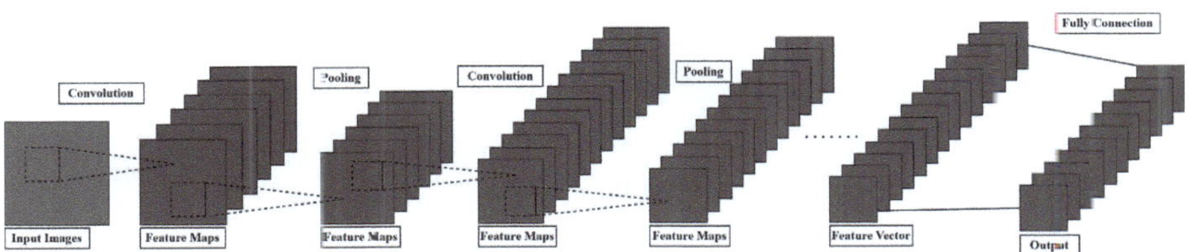

Fig. 2. Architectural design of a convolutional neural network (CNN) to classify hyperspectral images.

Pooling layers usually follow convolutional layers. Maximum and global average pooling layers are used, with maximum being the most prevalent. The pooling layer calculates the maximum or average value in a local feature map window. Pooling layers shrink the feature map and make the features more invariant, making the model more resilient to fluctuations in the data and reducing computational costs.

After the CNN layers, multiple fully connected layers learn from the high-level features to give the final predictions of the network. The feature vector learned by the l-th layer is shown in Eq. 2.

$$F^l = g(W^l.F^{l-1} + b^l)$$
(2)

where W_l is the weight matrix connecting the $(l-1)$-th layer and the l-th fully connected layer. F_{l-1} is the feature vector in the $(l-1)$-th layer, and b_l is the bias vector of the l-th layer.

Sequentially integrating convolutional, pooling, and fully connected layers creates a deep CNN. The lower layers detect edges and textures, while the upper layers learn more complex and distinguishing traits for classification. CNNs have fewer connections and parameters than other deep learning models [18]. Due to weight sharing and local relationships, their training is easier.

1.2 Support Vector Machines(SVM)

SVM is a powerful method for classifying hyperspectral images, in which a hyperplane is used to divide different classes based on the spectral data of individual pixels [23]. Pre-processing procedures include removing noise, calibrating spectral data, and reducing dimensionality using Principal Component Analysis (PCA). Training involves the selection of representative samples from which spatial properties are collected. SVM chooses spectral bands that optimize the separation between classes [28]. The algorithm uses kernel functions such as linear, polynomial, RBF, and sigmoid to transform the data into high-dimensional spaces, establishing a hyperplane to separate the classes. The decision function is as follows:

$$f(\mathbf{x}) = \text{sign}\left(\sum_{i=1}^{n} \alpha_i y_i K(\mathbf{x}_i, \mathbf{x}) + b\right) \tag{3}$$

where: $\mathbf{x}$ is the input vector, α_i are the support vector coefficients, y_i are the class labels, $K(\mathbf{x}_i, \mathbf{x})$ is the kernel function, and b is the bias term.

1.3 Stacked Autoencoders (SAE)

Stacked autoencoders classify hyperspectral images using deep learning [21]. Each autoencoder has an encoder [16], that compresses data into a latent space and a decoder that reconstructs it.

The training method has multiple steps: Data pre-processing removes noise and normalizes hyperspectral images.

Training the First Autoencoder: Compress and rebuild input data in a lower-dimensional latent space. Training minimizes reconstruction error:

$$\mathcal{L}_1 = \|\mathbf{x} - \hat{\mathbf{x}}\|^2 \tag{4}$$

The variable x represents the raw data, while $\hat{x}$ represents the reconstructed data. The encoder component of the initial autoencoder retrieves features from the hyperspectral images, which are then employed as input to train subsequent autoencoders in the sequence [29]. This technique is repeated for each subsequent autoencoder. After training all autoencoders, fully connected layers with an activation function such as softmax are included to facilitate classification. The overall classification can be represented as:

$$f(\mathbf{x}) = \mathrm{softmax}(\mathbf{W}\mathbf{h} + \mathbf{b}) \tag{5}$$

where $\mathbf{h}$ is the feature vector from the stacked autoencoders, $\mathbf{W}$ is the weight matrix, and $\mathbf{b}$ is the bias term.

1.4 A2S2K ResNet

Adaptive spectral-spatial kernels for attentional remote sensing require hyperspectral image (HSI) and spectral-spatial classification, which ResNet excels in. Figure 3 illustrates the Attention-Based Adaptive Spectral-Spatial Kernel ResNet (A2S2K ResNet) model specifically designed for hyperspectral images (HSI). It uses deep learning architectures and residual network (ResNet) models to train deep neural networks through residual connections efficiently [30]. These connections rely on residual mapping to help understand complex knowledge. This model also uses an attention-based mechanism to focus on crucial spectral-spatial information to improve discrimination. Assigning weights to spectral and spatial bands of the input image highlights critical elements for better classification.

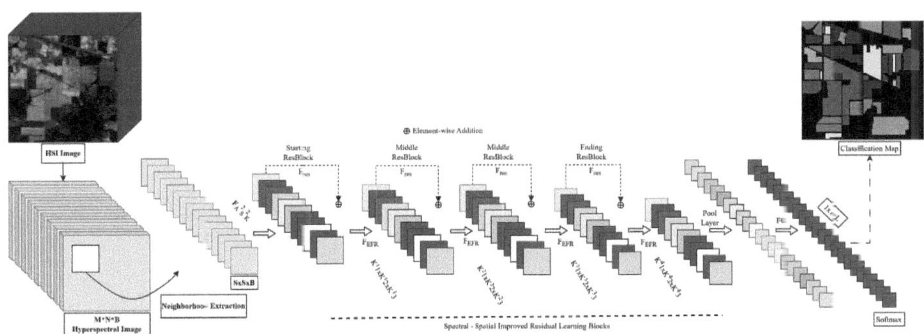

Fig. 3. Attention-Based Adaptive Spectral-Spatial Kernel ResNet (A2S2K ResNet) Model

CNNs select appropriate spectral-spatial kernel features for HSI classification, which is a significant difficulty. CNN's constant receptive field (RF) size limits its ability to detect tiny structures. The adaptive spectral-spatial kernel improves classification by learning spectral and spatial properties in 3D [21]. An improved ResBlock and efficient feature recalibration (EFR) technique enhance accuracy. The classification process includes selecting feature maps using spectral-spatial kernel attention, recalibrating them using an improved ResNet, and classifying with softmax. The attention-based adaptive spectral-spatial kernel ResNet improves the spectral-spatial classification accuracy of remote sensing images by combining ResNet with attention processes [24].

1.5   Attention-Based Two-CNN Fusion

Hyperspectral image (HSI) classification algorithms failed due to low-level features and substantial intra-class and minimal inter-class variability [24]. These models' sensitivity to small feature changes produced low performance. Figure 4 shows that patches 1 and 2 have the same class centre pixels but different environments, while patches 3 and 4 have the same class centre pixels but are rotated differently [27]. Patches 5, 6, 7, and 8 have the same texture but different classes. With standard low-level feature extraction, such examples provide poor classification results.

Despite the complexities of HSI classification, there is a clear path forward. By separating spatial and spectral feature extraction [19], we can enhance classification. This approach, combined with attention mechanisms in the spectral and spatial branches, has the potential to remove redundant information from these features. This reassures us that we are moving in the right direction, towards more accurate and efficient HSI classification.

2   Related Works

Hyperspectral classification algorithms rely on manually designed features [9,10]. The manually designed features are generated and subsequently input into classifiers, such as support vector machines (SVMs) or sparse representation classifiers [3,5]. For example, numerical morphology can use HSI spatial design data. The use of morphological profile (MP) features for hyperspectral image (HSI) classification was initially suggested in [10,11]. MP features can be integrated with the spectrum by direct connection or composite kernel learning. In addition to morphology, the spatial structure data of HSI can also be selected using multiscale analytical techniques [12], such as three-dimensional wavelet and three-dimensional Gabor filtering.

Integrating multiple complementary features can improve classification performance. Using manifold learning, combined pixel-size features, Gabor features, and integrated spectrum to form a low-dimensional representation of integrated features [13]. The combination of the spectrum, MP, and nonlinear kernel-based features was suggested in a multiple-feature learning framework [14]. One possibility is to describe HSI as a tensor, and another is to use a multilinear algebra framework for feature extraction and classification. After extracting feature tensors using multilinear principal component analysis (PCA), support tensor machines were used for classification [12].

All the methods mentioned above of classification use only low-level data and are thus shallow models with hand-crafted features. On the other hand, deep learning models can automatically learn any feature [17], whether low-level or high-level features. Hierarchical feature learning in these networks allows the lower layers to learn more basic features (e.g. texture or edges) and the higher layers to learn more complex features that are good for classification tasks [18]. By dealing with the problem of reduced interclass variability and increased intraclass variability, high-level features are more flexible and invariant, which improves the classification process [19].

Deep learning models often use RBM, CNN, and SAE [18,31]. In [20], SAE and PCA learned features simultaneously. SAE can also extract multiscale spatial and spectral information for HSI classification [21]. SAE and RBM are developed for 1-D signals, which can reduce the spatial correlation, so HSI must be vectorized. In [25], a spatial regularizer, a technique in the target function of SAE that combines neighboring pixels with comparable properties to make use of the spatial correlation of HSI, was used. However, CNN can extract spatial features [31], without vectorization.

A greedy model trained HSI feature extraction layer-wise and unsupervised [20]. In [22], a sparse representation classifier using CNN extraction results with spatial features was used. In [21], combining manifold learning spectral features with CNN spatial features produced a spectral-spatial joint feature vector. In [30], CNN multiscale spatial features were combined with the Laplacian pyramid of HSI and a low-dimensional spectrum to form a joint feature vector. Supervised CNNs can use class-specific training data to learn additional discriminative features [32]. In [30], spectral and spatial features were trained simultaneously using an advanced 3-D resblock and an efficient feature re-calibration (EFR) technique to improve feature classification.

3 Methodology

The architecture is a deep learning model with two separate CNN branches to extract spectral and spatial features described in Fig. 4. The spectral branch processes the input spectrum S_n of the nth pixel through convolution operations and max pooling over l layers. Next, the attended features and the output are combined using a concatenation operation. This results in the spectral branch output, which represents the spectral features. The input S_n is a one-dimensional signal, and all calculations inside this branch, such as convolution and pooling processes, belong to one-dimensional processing.

The proposed attention-based Two-CNN model analyzes $B \times B$ patches using an attention module. After attention, the features are concatenated with the patches under consideration. Hyperspectral imaging (HSI) consists of continuous spectral bands; some bands contain noise, which motivates a spatial branch approach to average images across the bands. This branch uses the patch around the nth pixel as input and applies l convolutional and pooling layers to produce spatial features. Spectral and spatial information is fed into fully connected layers to leverage correlations and provide integrated features. The output of the last fully connected layer presents the integrated spectral-spatial information after multiple fully connected operations.

3.1 Attention Modules

Attention processes can be classified into several types, such as channel attention mechanisms, spatial attention mechanisms, and hybrid domain attention

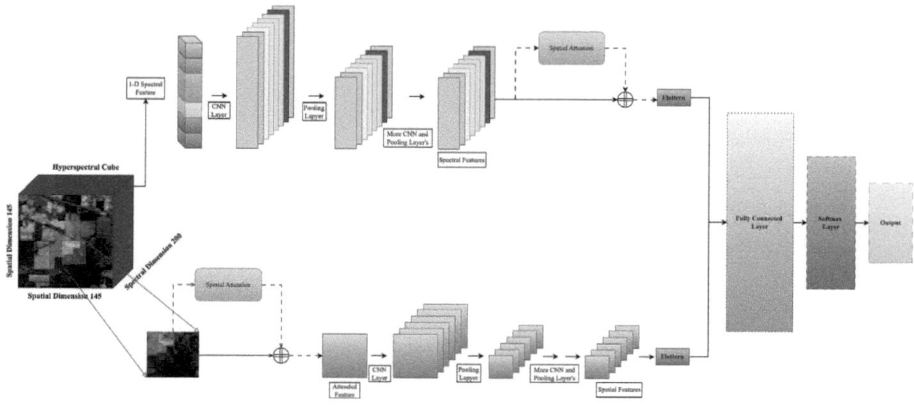

Fig. 4. Overview of Proposed Methodology

mechanisms [24]. Channel attention is a dynamic learning process that assigns weights to spectral bands. Prioritizing the most informative bands increases spectral information extraction [26]. The spatial attention mechanism is an adaptive process [27], which selects the necessary pixels and suppresses the less important ones in the spatial domain. Attention methods are versatile and useful for hyperspectral image (HSI) classification applications. However, such tasks must collect both spectral and spatial information for successful categorization.

$$\text{Sigmoid}(x) = \frac{1}{1 + e^{-x}} \tag{6}$$

$$X' = \text{sig}\left(f\left([\text{AvgPool}(X), \text{MaxPool}(X)]\right)\right) \tag{7}$$

In Fig. 5, the first step in spatial attention computation is to generate separate feature maps by applying average and maximum pooling operations along the channel plane.

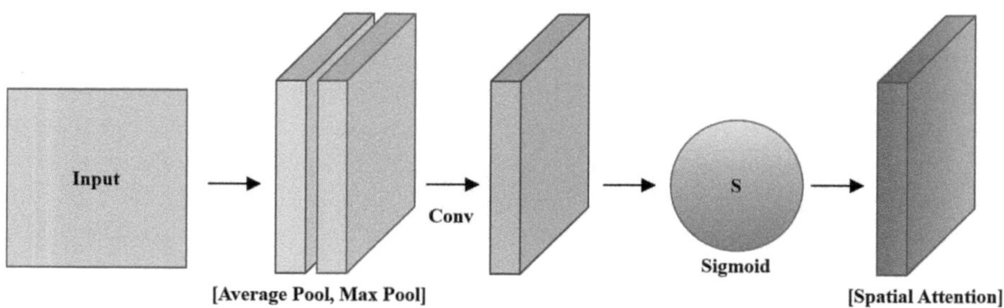

Fig. 5. Spatial Attention Mechanism Overview

The feature maps are concatenated along the channel axis. Next, a spatial attention feature map is created by applying a $k \times k$ convolutional layer and a sigmoid function. The spatial attention module is depicted in the accompanying image. In Eq. 6 defines the sigmoid function, while Eq. 7 explains the process of obtaining X' using the sigmoid function represented by **"sig"** and the convolution operation represented by **"f"**.

4 Experimental Setup and Results

4.1 Dataset Description

The Indian Pines (IP) and Pavia University hyperspectral imaging (HSI) datasets are used to test the suggested model. The performance of the model varies with different settings. The performance of the proposed technique is evaluated by comparing it with other sophisticated deep learning-based classification algorithms.

The ROSIS sensor over northern Italy collected 103 spectral bands for Pavia University (PU), of which 100 are usable. Although the original image dimensions are 610×610 pixels, the relevant area is 610×340 pixels, divided into 9 land cover classes.

The Indian Pines (IP) dataset from the AVIRIS sensor in north-western Indiana has 224 spectral bands, of which 200 are relevant. The image is 145×145 pixels and is classified into 16 land cover classes.

Table 1. Comparison of Machine Learning Methods and Deep Neural Networks on the **PU** Dataset with a 51×51 Patch Size: *OA, AA, and Kappa*

Class	Classical Model					Deep Neural Network		
	MLR	RF	SVM	GRU	LSTM	ResNet	DPyResNet	Attention Two-CNN
1	92.30 ± 0.004	91.11 ± 0.007	94.30 ± 0.008	93.34 ± 0.003	95.47 ± 0.005	96.82 ± 0.023	98.35 ± 0.017	97.76 ± 0.018
2	96.18 ± 0.003	98.11 ± 0.003	97.65 ± 0.002	97.54 ± 0.002	96.91 ± 0.002	98.59 ± 0.008	98.76 ± 0.003	98.84 ± 0.026
3	72.75 ± 0.013	67.71 ± 0.014	81.26 ± 0.018	77.08 ± 0.038	78.01 ± 0.011	90.01 ± 0.061	94.22 ± 0.032	88.34 ± 0.008
4	89.28 ± 0.002	88.20 ± 0.006	94.63 ± 0.004	93.22 ± 0.023	94.92 ± 0.007	99.32 ± 0.003	99.20 ± 0.004	99.27 ± 0.016
5	99.42 ± 0.003	98.93 ± 0.002	99.20 ± 0.002	99.42 ± 0.004	99.26 ± 0.003	99.81 ± 0.000	99.72 ± 0.003	99.65 ± 0.018
6	77.45 ± 0.005	72.14 ± 0.022	90.58 ± 0.008	87.41 ± 0.016	87.85 ± 0.012	99.21 ± 0.002	98.52 ± 0.003	96.28 ± 0.006
7	55.69 ± 0.043	75.69 ± 0.017	85.71 ± 0.011	85.38 ± 0.039	80.23 ± 0.007	96.90 ± 0.017	97.37 ± 0.002	93.45 ± 0.039
8	87.04 ± 0.004	89.64 ± 0.013	88.20 ± 0.003	88.56 ± 0.024	88.49 ± 0.008	92.00 ± 0.044	84.51 ± 0.07	97.82 ± 0.039
9	99.77 ± 0.001	99.77 ± 0.002	99.84 ± 0.001	99.84 ± 0.001	99.88 ± 0.001	98.88 ± 0.012	99.60 ± 0.00	99.62 ± 0.005
OA	**89.87 ± 0.001**	**90.41 ± 0.001**	**94.19 ± 0.002**	**93.34 ± 0.002**	**93.45 ± 0.001**	**97.38 ± 0.007**	**97.05 ± 0.010**	**97.68 ± 0.008**
AA	**85.54 ± 0.004**	**86.81 ± 0.002**	**92.38 ± 0.003**	**91.31 ± 0.008**	**91.23 ± 0.001**	**96.86 ± 0.005**	**96.69 ± 0.006**	**96.78 ± 0.005**
k	**0.8646 ± 0.001**	**0.8710 ± 0.002**	**0.9229 ± 0.002**	**0.9115 ± 0.003**	**0.9130 ± 0.001**	**0.9652 ± 0.003**	**0.9608 ± 0.003**	**0.9685 ± 0.010**

4.2 Model Performance Analysis

We examined how patch size affects model classification, observing that surrounding pixels affect core pixel categorization. The model was tested using measures of overall accuracy (OA), average accuracy (AA), and kappa (κ) for each class. Using a patch size of 11×11, the model obtained 94. 42% OA, 90.

59% AA and 92. 57% κ. Increasing the size of the patch to 51×51 improves the performance with OA of 98.86%, AA of 97.44%, and κ of 98.49%. In Fig. 8, a larger patch size of 61×61 resulted in a decrease in OA (98. 18%), AA (95. 67%) and κ (97. 58%).

The OA variations with epochs and loss of training and assessment for the 15% PU dataset are analyzed. The gap between training and assessment loss decreased as the OA increased from 43.59% in the first epoch to 98.86% in the fifth. This indicates a well-fitting model.

Our solution outperformed traditional machine learning methods and deep neural networks in the PU dataset with a patch size of 51×51, as shown in Table 1. In Table 2, the model outperformed classical models and deep neural networks in the IP data set with a patch size 61×61, despite somewhat lower accuracy but better kappa values in several classes.

Finally, we analyzed how the size of the training sample affects the performance of the model in the PU data set with a patch size of 51×51, ranging from 5% to 25%, shown in Table 3.

Table 2. Comparison of Machine Learning Methods and Deep Neural Networks on the **IP** Dataset with a 61×61 Patch Size: *OA, AA, and Kappa*

Class	Classical Model					Deep Neural Network					
	MLR	RF	SVM	GRU	LSTM	ResNet	ContextNet	MS-3DNet	ENL-FCN	DPyResNet	Attention Two-CNN
1	15.45 ± 0.023	28.46 ± 0.061	51.22 ± 0.190	69.92 ± 0.141	69.11 ± 0.090	98.66 ± 0.018	88.78 ± 0.080	66.67 ± 0.471	97.56 ± 0.000	94.59 ± 0.076	99.98 ± 0.008
2	73.77 ± 0.006	56.63 ± 0.024	81.22 ± 0.037	76.96 ± 0.013	74.22 ± 0.016	87.85 ± 0.029	98.19 ± 0.005	75.94 ± 0.080	93.15 ± 0.000	93.83 ± 0.040	96.95 ± 0.045
3	51.14 ± 0.027	48.42 ± 0.013	65.82 ± 0.013	67.20 ± 0.041	71.49 ± 0.030	92.71 ± 0.007	95.37 ± 0.028	81.39 ± 0.007	97.59 ± 0.000	89.30 ± 0.003	97.45 ± 0.002
4	43.97 ± 0.051	33.49 ± 0.025	57.75 ± 0.041	61.82 ± 0.039	60.72 ± 0.041	95.43 ± 0.046	97.04 ± 0.021	88.63 ± 0.063	91.55 ± 0.000	93.51 ± 0.055	96.53 ± 0.040
5	83.52 ± 0.034	85.21 ± 0.025	90.04 ± 0.014	85.36 ± 0.022	87.51 ± 0.015	98.23 ± 0.015	97.78 ± 0.015	95.61 ± 0.054	97.47 ± 0.000	99.26 ± 0.004	96.35 ± 0.003
6	94.82 ± 0.09	92.64 ± 0.027	96.25 ± 0.006	94.22 ± 0.008	94.77 ± 0.015	97.98 ± 0.011	98.60 ± 0.008	96.78 ± 0.026	99.24 ± 0.000	98.52 ± 0.007	96.61 ± 0.006
7	41.33 ± 0.186	2.67 ± 0.038	73.33 ± 0.019	50.67 ± 0.068	85.33 ± 0.094	92.98 ± 0.099	90.35 ± 0.098	100.00 ± 0.000	100.00 ± 0.000	83.08 ± 0.178	66.66 ± 0.062
8	98.53 ± 0.006	97.67 ± 0.015	97.98 ± 0.006	97.83 ± 0.001	97.83 ± 0.009	95.06 ± 0.014	97.76 ± 0.026	89.51 ± 0.091	97.44 ± 0.000	97.63 ± 0.022	100.0 ± 0.000
9	5.56 ± 0.045	9.26 ± 0.094	50.00 ± 0.045	37.04 ± 0.146	53.70 ± 0.139	60.83 ± 0.283	86.90 ± 0.102	66.67 ± 0.471	72.22 ± 0.000	66.66 ± 0.471	92.41 ± 0.045
10	65.41 ± 0.041	60.91 ± 0.047	73.87 ± 0.018	76.00 ± 0.035	73.68 ± 0.025	96.05 ± 0.013	96.08 ± 0.018	87.41 ± 0.070	94.74 ± 0.000	93.77 ± 0.029	93.95 ± 0.016
11	80.37 ± 0.015	87.88 ± 0.019	82.90 ± 0.012	80.31 ± 0.027	84.93 ± 0.024	93.32 ± 0.041	97.35 ± 0.004	76.69 ± 0.096	95.61 ± 0.000	89.78 ± 0.040	99.28 ± 0.008
12	55.68 ± 0.007	41.26 ± 0.030	74.91 ± 0.043	78.65 ± 0.014	73.35 ± 0.052	86.65 ± 0.077	94.00 ± 0.012	88.65 ± 0.036	97.00 ± 0.000	83.43 ± 0.107	92.67 ± 0.012
13	97.66 ± 0.005	90.09 ± 0.040	96.94 ± 0.021	96.94 ± 0.014	98.74 ± 0.005	82.16 ± 0.076	95.01 ± 0.030	99.78 ± 0.003	97.83 ± 0.000	98.19 ± 0.021	92.00 ± 0.006
14	95.61 ± 0.004	95.46 ± 0.014	93.82 ± 0.010	94.50 ± 0.012	96.22 ± 0.004	95.39 ± 0.016	98.49 ± 0.014	90.06 ± 0.087	99.12 ± 0.000	96.00 ± 0.021	99.34 ± 0.056
15	56.00 ± 0.045	41.11 ± 0.029	60.42 ± 0.044	65.71 ± 0.019	60.04 ± 0.029	90.96 ± 0.127	94.10 ± 0.031	88.21 ± 0.044	92.80 ± 0.000	91.22 ± 0.040	99.08 ± 0.008
16	84.92 ± 0.020	79.37 ± 0.030	91.27 ± 0.054	82.54 ± 0.037	90.87 ± 0.022	94.73 ± 0.038	93.57 ± 0.046	98.53 ± 0.021	100.00 ± 0.000	70.90 ± 0.388	77.50 ± 0.028
OA	76.23 ± 0.008	72.98 ± 0.006	82.00 ± 0.006	81.24 ± 0.003	82.13 ± 0.004	92.44 ± 0.006	96.98 ± 0.006	83.44 ± 0.060	96.15 ± 0.054	91.47 ± 0.029	97.00 ± 0.045
AA	65.23 ± 0.019	59.41 ± 0.005	77.36 ± 0.019	75.98 ± 0.008	79.53 ± 0.005	91.19 ± 0.025	94.96 ± 0.003	86.91 ± 0.084	95.21 ± 0.028	94.14 ± 0.006	89.61 ± 0.024
k	0.7266 ± 0.010	0.6862 ± 0.007	0.7941 ± 0.007	0.7858 ± 0.004	0.7954 ± 0.004	0.9137 ± 0.006	0.9655 ± 0.007	0.8082 ± 0.070	0.9560 ± 0.030	0.9020 ± 0.034	0.9657 ± 0.008

Table 3. Impact of Different Training Sample Sizes on Model Performance: *Results on PU Dataset*

Ratio	SVM	JSRC	SAE	CNN+ML	Two-CNN	Attention Two-CNN
5%	93.99 ± 0.29	82.80 ± 0.25	90.53 ± 0.21	91.28 ± 2.35	94.40 ± 0.39	90.91 ± 0.30
10%	94.85 ± 0.08	87.97 ± 0.39	93.50 ± 0.32	92.51 ± 1.05	96.99 ± 0.32	97.56 ± 0.53
15%	95.54 ± 0.05	92.55 ± 0.28	95.21 ± 0.28	93.01 ± 0.62	98.17 ± 0.29	98.86 ± 0.27
20%	95.76 ± 0.10	94.01 ± 0.41	95.93 ± 0.14	94.28 ± 1.13	98.72 ± 0.25	99.17 ± 0.18
25%	95.95 ± 0.22	96.43 ± 0.36	96.37 ± 0.41	96.20 ± 0.79	99.10 ± 0.18	99.45 ± 0.13

5   Conclusion

The research presents dual convolutional neural networks augmented with an attention strategy for the classification of hyperspectral images. This method highlights essential features by integrating attention processes in separate spectral and spatial CNN branches. The suggested technique surpasses existing models on standard datasets for overall accuracy, average accuracy, and kappa value, demonstrating its efficacy in diverse classification tasks. Future advancements in hyperspectral image classification should prioritize the development of more advanced feature extraction algorithms and learning frameworks that incorporate spectral and spatial information more efficiently. Investigating semi-supervised and unsupervised methods can enhance performance using labeled and unlabeled data. In addition, optimizing deep learning models for hyperspectral data is crucial to achieving improved classification accuracy. Future research can investigate using transfer learning to use pretrained models or data augmentation approaches to increase the training dataset and strengthen model flexibility. These methods can pave new avenues for improving hyperspectral image classification.

Acknowledgements. The authors thank the editors and anonymous reviewers for their insightful comments and imaginative recommendations, which improved our research.

References

1. Liang, L., et al.: Estimation of crop LAI using hyperspectral vegetation indices and a hybrid inversion method. Remote Sens. Environ. **165**, 123–134 (2015). ISSN 0034-4257. https://doi.org/10.1016/j.rse.2015.04.032
2. Vaglio Laurin, G., Chan, J.C.-W., Chen, Q., Lindsell, J.A., Coomes, D.A., Guerriero, L., et al.: Biodiversity mapping in a tropical West African forest with airborne hyperspectral data. PLoS ONE **9**(6), e97910 (2014). https://doi.org/10.1371/journal.pone.0097910
3. Fauvel, M., Chanussot, J., Benediktsson, J.A., Sveinsson, J.R.: Spectral and spatial classification of hyperspectral data using SVMs and morphological profiles. In: 2007 IEEE International Geoscience and Remote Sensing Symposium, Barcelona, Spain, pp. 4834–4837 (2007). https://doi.org/10.1109/IGARSS.2007.4423943
4. Li, J., Marpu, P.R., Plaza, A., Bioucas-Dias, J.M., Benediktsson, J.A.: Generalized composite kernel framework for hyperspectral image classification. IEEE Trans. Geosci. Remote Sens. **51**(9), 4816–4829 (2013). https://doi.org/10.1109/TGRS.2012.2230268
5. Chen, Y., Nasrabadi, N.M., Tran, T.D.: Hyperspectral image classification using dictionary-based sparse representation. IEEE Trans. Geosci. Remote Sens. **49**(10), 3973–3985 (2011). https://doi.org/10.1109/TGRS.2011.2129595
6. Qian, Y., Ye, M., Zhou, J.: Hyperspectral image classification based on structured sparse logistic regression and three-dimensional wavelet texture features. IEEE Trans. Geosci. Remote Sens. **51**(4), 2276–2291 (2013). https://doi.org/10.1109/TGRS.2012.2209657

7. Jia, S., Shen, L., Li, Q.: Gabor feature-based collaborative representation for hyperspectral imagery classification. IEEE Trans. Geosci. Remote Sens. **53**(2), 1118–1129 (2015). https://doi.org/10.1109/TGRS.2014.2334608

8. Zhang, L., Zhang, L., Tao, D., Huang, X.: On combining multiple features for hyperspectral remote sensing image classification. IEEE Trans. Geosci. Remote Sens. **50**(3), 879–893 (2012). https://doi.org/10.1109/TGRS.2011.2162339

9. Li, J., et al.: Multiple feature learning for hyperspectral image classification. IEEE Trans. Geosci. Remote Sens. **53**(3), 1592–1606 (2015). https://doi.org/10.1109/TGRS.2014.2345739

10. Liao, W., Mura, M.D., Chanussot, J., Pižurica, A.: Fusion of spectral and spatial information for classification of hyperspectral remote-sensed imagery by local graph. IEEE J. Sel. Topics Appl. Earth Observat. Remote Sens. **9**(2), 583–594 (2016). https://doi.org/10.1109/JSTARS.2015.2498664

11. Zhang, L., Zhang, L., Tao, D., Huang, X.: Tensor discriminative locality alignment for hyperspectral image spectral-spatial feature extraction. IEEE Trans. Geosci. Remote Sens. **51**(1), 242–256 (2013). https://doi.org/10.1109/TGRS.2012.2197860

12. Guo, X., Huang, X., Zhang, L., Zhang, L., Plaza, A., Benediktsson, J.A.: Support tensor machines for classification of hyperspectral remote sensing imagery. IEEE Trans. Geosci. Remote Sens. **54**(6), 3248–3264 (2016). https://doi.org/10.1109/TGRS.2016.2514404

13. Romero, A., Gatta, C., Camps-Valls, G.: Unsupervised deep feature extraction for remote sensing image classification. IEEE Trans. Geosci. Remote Sens. **54**(3), 1349–1362 (2016). https://doi.org/10.1109/TGRS.2015.2478379

14. Zhang, L., Zhang, L., Du, B.: Deep learning for remote sensing data: a technical tutorial on the state of the art. IEEE Geosci. Remote Sens. Maga. **4**(2), 22–40 (2016). https://doi.org/10.1109/MGRS.2016.2540798

15. Veríssimo, E., Severo, D.D.S., Cavalcanti, G.D.C., Ren, T.I.: A dimensionality reduction approach for modular neural networks. In: 2012 IEEE 24th International Conference on Tools with Artificial Intelligence, Athens, Greece, pp. 886–891 (2012). https://doi.org/10.1109/ICTAI.2012.166.

16. Vincent, P., Larochelle, H., Lajoie, I., Bengio, Y., Manzagol, P.A., Bottou, L.: Stacked denoising autoencoders: learning useful representations in a deep network with a local denoising criterion. J. Mach. Learn. Res. **11**(12), 3371–3408 (2010)

17. Chen, Y., Lin, Z., Zhao, X., Wang, G., Gu, Y.: Deep learning-based classification of hyperspectral data. IEEE J. Sel. Topics Appl. Earth Observat. Remote Sens. **7**(6), 2094–2107 (2014). https://doi.org/10.1109/JSTARS.2014.2329330

18. Digra, M., Dhir, R., Sharma, N.: Land use land cover classification of remote sensing images based on the deep learning approaches: a statistical analysis and review. Arab. J. Geosci. **15**, 1003 (2022). https://doi.org/10.1007/s12517-022-10246-8

19. Zhao, W., Du, S.: Spectral–spatial feature extraction for hyperspectral image classification: a dimension reduction and deep learning approach. IEEE Trans. Geosci. Remote Sens. **54**(8), 4544–4554 (2016). https://doi.org/10.1109/TGRS.2016.2543748

20. Tao, C., Pan, H., Li, Y., Zou, Z.: Unsupervised spectral-spatial feature learning with stacked sparse autoencoder for hyperspectral imagery classification. IEEE Geosci. Remote Sens. Lett. **12**(12), 2438–2442 (2015). https://doi.org/10.1109/LGRS.2015.2482520

21. Ma, X., Wang, H., Geng, J.: Spectral–spatial classification of hyperspectral image based on deep auto-encoder. IEEE J. Sel. Topics Appl. Earth Observat. Remote Sens. **9**(9), 4073–4085 (2016). https://doi.org/10.1109/JSTARS.2016.2517204

22. Liang, H., Li, Q.: Hyperspectral imagery classification using sparse representations of convolutional neural network features. Remote Sens. **8**(2), 99 (2016). https://doi.org/10.3390/rs8020099
23. Chang, C.C., Lin, C.J.: LIBSVM: a library for support vector machines. ACM Trans. Intell. Syst. Technol. **2**, 3, Article 27, 27 p (2011). https://doi.org/10.1145/1961189.1961199
24. Roy, S.K., Manna, S., Song, T., Bruzzone, L.: Attention-based adaptive spectral-spatial kernel resnet for hyperspectral image classification. IEEE Trans. Geosci. Remote Sens. **59**(9), 7831–7843 (2021). https://doi.org/10.1109/TGRS.2020.3043267
25. Zhang, X., Sun, Y., Jiang, K., Li, C., Jiao, L., Zhou, H.: Spatial sequential recurrent neural network for hyperspectral image classification. IEEE J. Sel. Topics Appl. Earth Observat. Remote Sens. **11**(11), 4141–4155 (2018). https://doi.org/10.1109/JSTARS.2018.2844873
26. Zhu, M., Jiao, L., Liu, F., Yang, S., Wang, J.: Residual spectral-spatial attention network for hyperspectral image classification. IEEE Trans. Geosci. Remote Sens. **59**(1), 449–462 (2021). https://doi.org/10.1109/TGRS.2020.2994057
27. Woo, S., Park, J., Lee, J.-Y., Kweon, I.S.: CBAM: convolutional block attention module. In: Ferrari, V., Hebert, M., Sminchisescu, C., Weiss, Y. (eds.) ECCV 2018. LNCS, vol. 11211, pp. 3–19. Springer, Cham (2018). https://doi.org/10.1007/978-3-030-01234-2_1
28. Melgani, F., Bruzzone, L.: Classification of hyperspectral remote sensing images with support vector machines. IEEE Trans. Geosci. Remote Sens. **42**(8), 1773–1790 (2004). https://doi.org/10.1109/TGRS.2004.831865
29. Cho, K., van Merrienboer, B., Bahdanau, D., Bengio, Y.: On the properties of neural machine translation: encoder-decoder approaches (2014). arXiv:1409.1259
30. He, K., Zhang, X., Ren, S., Sun, J.: Deep residual learning for image recognition. In: 2016 IEEE Conference on Computer Vision and Pattern Recognition (CVPR), Las Vegas, NV, USA, pp. 770–778 (2016). https://doi.org/10.1109/CVPR.2016.90.
31. Lee, H., Kwon, H.: Going deeper with contextual CNN for hyperspectral image classification. IEEE Trans. Image Process. **26**(10), 4843–4855 (2017). https://doi.org/10.1109/TIP.2017.2725580
32. Shen, Y., et al.: Efficient deep learning of nonlocal features for hyperspectral image classification. IEEE Trans. Geosci. Remote Sens. **59**(7), 6029–6043 (2021). https://doi.org/10.1109/TGRS.2020.3014286

Advanced Personalized Medicine Framework: FuRGAN Utilizing Fuzzy Logic, RNNs, and GANs for Dynamic Dosage Recommendations

Gautham Praveen Ramalingam[1]($\boxtimes$) (iD), N. Karthikeyan[1], G. Gerard Alex Ben[2], and Deepika Pandian[1]

[1] Syed Ammal Engineering College, Tamil Nadu, Ramanathapuram, India
gauthams_ralli@hotmail.com
[2] Advanced Software Engineer, Light and Wonder, Tamil Nadu, Chennai, India

Abstract. Determining optimal drug dosages in personalized medicine is a complex challenge due to the variability in patient responses and the dynamic nature of biomarker levels. Traditional methods often struggle to adapt to real-time changes and individual nuances, leading to suboptimal treatment outcomes and potential adverse effects. To address these issues, we propose FuRGAN, a novel framework that integrates fuzzy logic, recurrent neural networks (RNNs), and generative adversarial networks (GANs) to enhance personalized dosage recommendations. FuRGAN categorizes biomarkers into low, medium, and high levels using fuzzy logic, simplifying the initial decision-making process. RNNs are employed to analyze temporal patterns in patient data, enabling precise predictions of necessary dosage adjustments over time. GANs are utilized to generate and refine dosage recommendations through adversarial training, ensuring both realism and optimization. A key innovation of FuRGAN is its dynamic feedback loop, which allows the system to continuously adapt to real-time patient feedback and clinical outcomes. This continuous learning mechanism improves the framework's accuracy and efficacy, offering a more responsive and personalized approach to dosage personalization. By integrating these advanced technologies, FuRGAN aims to significantly enhance treatment outcomes and patient safety, providing a more effective solution to the challenges of personalized medicine.

Keywords: Biomarker data · Fuzzy logic · Recurrent Neural Networks (RNN's) · Generative Adversarial Networks (GAN's) · Adaptive learning · Data-driven healthcare

1 Introduction

Personalized medicine focuses on tailoring treatments to the individual by considering factors such as genetics, environment, and lifestyle. However, optimizing drug dosages to accommodate the dynamic physiological states of patients is a significant challenge.

Traditional methods rely on static, population-based guidelines, often leading to suboptimal outcomes due to the inability to account for real-time fluctuations in biomarkers and individual patient responses. The integration of AI and machine learning (ML) technologies offers promising solutions to these challenges. Fuzzy Logic, Recurrent Neural Networks (RNNs), and Generative Adversarial Networks (GANs) are key components of this framework. Fuzzy Logic helps manage uncertainty and imprecision in patient data by categorizing biomarkers into levels like low, medium, and high, providing an initial basis for dosage recommendations [4][5]. RNNs, known for their ability to process sequential data, are well-suited to recognize patterns in longitudinal patient data, predicting necessary dosage adjustments over time and adapting to changes in disease progression or treatment responses [6][7]. The combination of RNNs and Fuzzy Logic creates an adaptive system for personalized dosage recommendations that respond to individual patient needs. GANs refine these recommendations by simulating various dosage scenarios and optimizing them based on real-world feedback, using an adversarial training process to ensure clinical relevance and effectiveness [9, 10]. A key feature of this framework is its real-time feedback loop, which monitors patient responses to prescribed dosages, allowing for dynamic adjustments that keep the treatment plan aligned with the patient's evolving health status [11, 12]. This proactive, adaptive approach ensures more accurate and effective treatment, continuously improving based on patient outcomes and clinical feedback.

2 Relevant Work

Personalized medicine is revolutionizing healthcare by tailoring drug treatments to the unique needs of each patient, improving therapeutic outcomes while reducing side effects. Traditional dosing methods, based on population averages, often miss the mark by not accounting for individual differences or real-time changes in patient conditions. The introduction of artificial intelligence (AI) and machine learning (ML) is addressing these challenges by creating dynamic systems that integrate fuzzy logic, recurrent neural networks (RNNs), and generative adversarial networks (GANs) to optimize drug dosing. Fuzzy logic plays a key role in managing uncertainty in patient data by categorizing biomarkers into levels such as low, medium, and high, which helps structure the decision-making process for drug dosing [4]. RNNs are well-suited for analyzing sequential patient data, allowing the system to recognize trends in biomarker levels over time and predict necessary dosage adjustments, enabling the framework to adapt to disease progression and treatment responses [6, 7]. GANs enhance the accuracy of dosage recommendations by simulating different dosing scenarios and refining those suggestions through adversarial training, ensuring the treatment plan is optimized to real-world conditions [9, 10]. The FuRGAN framework is designed with a real-time feedback loop that continuously monitors and adjusts treatment plans based on the patient's responses, ensuring that the system stays aligned with their evolving health [11, 12]. Research supports the effectiveness of combining GANs and RNNs for real-time drug dosage optimization. For example, Wang et al. show how this combination leads to more accurate and adaptive dosing [1], while Kim et al. highlight the role of hybrid GAN-RNN models in refining doses through patient feedback [4]. Zhang et al.

demonstrate how adaptive fuzzy logic controllers enhance dosing accuracy and safety [5]. The development of FuRGAN also draws on insights from integrating IoT and ML for remote patient monitoring, as explored by GP Ramalingam et al., which highlights the importance of real-time monitoring for personalized care. Xu et al. underscore the role of feedback-driven GANs in refining dosage recommendations, emphasizing how continuous learning from patient responses improves prescription accuracy [2]. Shen et al. further explore the integration of GANs and fuzzy logic, noting how this approach enhances dosing precision and adapts to patient-specific needs, ultimately ensuring safer and more effective treatment [3].

3 Architecture

FuRGAN is an advanced framework for personalized drug dosage optimization that integrates Fuzzy Logic, Recurrent Neural Networks (RNNs), and Generative Adversarial Networks (GANs) to address the limitations of traditional dosing methods. Unlike conventional approaches, FuRGAN adapts to real-time variability and individual patient responses, improving treatment outcomes. Fuzzy Logic categorizes biomarkers into fuzzy sets to handle uncertainty, while RNNs analyze temporal patterns to predict changes in disease progression and treatment responses. GANs simulate realistic dosage scenarios and refine recommendations through adversarial training. FuRGAN's real-time feedback loop allows dynamic adjustments, enhancing accuracy and efficacy with each iteration. This framework improves current treatment protocols and offers a flexible platform for incorporating new data sources and predictive models, paving the way for future innovations in personalized medicine, as described in Fig. 1 [2, 6].

3.1 Input Data

The FuRGAN framework optimizes personalized drug dosages by integrating various data sources for accurate, patient-specific recommendations. It uses fuzzy logic to categorize biomarkers like blood pressure, heart rate, glucose, and cholesterol into levels that guide dosage adjustments. Data from NHANES and the UK Biobank help train and validate the system. FuRGAN also considers a patient's medical history, including past treatments and responses, using Recurrent Neural Networks (RNNs) to identify patterns over time. Clinical feedback is collected in real-time, allowing the system to make dynamic treatment adjustments. Generative Adversarial Networks (GANs) refine dosage suggestions based on real-world feedback, improving the system's performance and ensuring adaptive, personalized medicine.

3.2 Data Preprocessing Techniques

Data preprocessing is a crucial step in developing any data-driven model, particularly in the context of personalized drug dosage optimization using artificial intelligence (AI) and machine learning (ML) techniques. The quality and reliability of the input data significantly influence the performance of the models. In the proposed FuRGAN framework, which integrates Fuzzy Logic, Recurrent Neural Networks (RNNs), and Generative Adversarial Networks (GANs), preprocessing ensures that the data is consistent, accurate, and suitable for analysis.

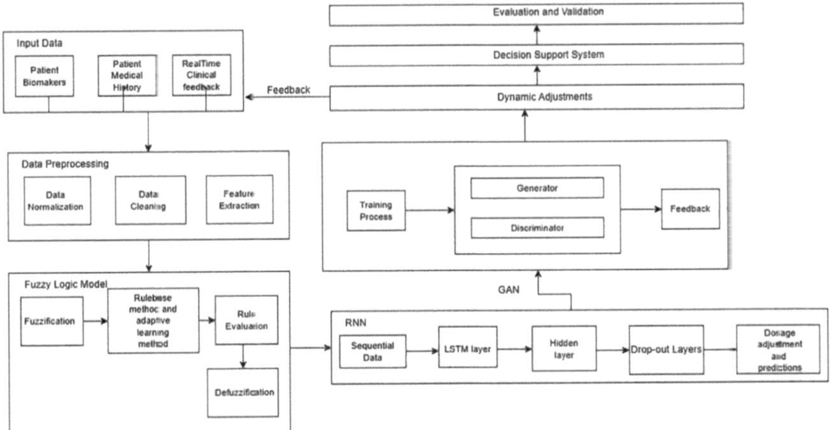

Fig. 1. Architecture diagram - FuRGAN Framework

3.2.1 Data Cleaning

Data cleaning is essential in the FuRGAN framework to ensure reliable input data for precise dosage recommendations, addressing issues like missing values, outliers, and duplicates. Missing values are managed through methods such as mean or median imputation for simpler cases, maintaining consistency by replacing gaps with average or median values, and K-Nearest Neighbors (KNN) imputation for complex datasets, which considers interactions between biomarkers to provide contextually accurate estimates. Outlier detection plays a critical role in identifying and handling data points that deviate significantly due to errors or unique patient conditions. Properly managing these outliers ensures that machine learning models like RNNs and GANs generate accurate predictions, avoiding skewed results and supporting effective personalized treatment.

$$X_{imputed} = mean(X_{observed}) or median(X_{observed}) \tag{1}$$

$$X_{imputed} = \frac{1}{k} \sum_{i=1}^{k} X_i^k X_i \tag{2}$$

3.2.2.1. Z-Score Method

The Z-score method is employed to detect outliers in continuous biomarker measurements. This technique calculates the Z-score for each data point, which measures how many standard deviations a data point is from the mean. Data points with Z-scores exceeding absolute values of 3 are flagged as outliers and warrant further review. This method is crucial for identifying extreme deviations that could adversely impact model performance. Outliers identified through the Z-score method are reviewed to determine if they are due to data entry errors or genuine extreme values. Addressing these outliers is critical for maintaining the integrity of the data and ensuring that the model's predictions are based on accurate and representative data by applying to biomarkers with varying scales, such as blood pressure and heart rate, ensuring that all features contribute equally to the model without bias.

$$Z = \frac{X - \mu}{\sigma} \tag{3}$$

where, Z (Z-Score)- Measures how many standard deviations a data point is from the mean, X (Data Point) - The specific value being evaluated, μ (Mean)-The average of all data points, σ (Standard Deviation) - Indicates how spread out the data points are from the mean.

3.2.2.2. Interquartile Range (IQR) Method

For biomarkers with skewed distributions, the IQR method is utilized to detect outliers. This technique involves calculating the interquartile range (IQR) and identifying values are particularly effective for handling biomarkers with skewed distributions, such as cholesterol levels or blood glucose, where extreme values might not be representative of typical conditions. Managing these outliers ensures that the analysis remains accurate and reflective of common patient scenarios.w These values are flagged as outliers.

$$IQR = Q_3 - Q_1$$
$$Outlier\ Thresholds : Q_1 - 1.5 \times IQR\ and\ Q_3 + 1.5 \times IQR \tag{4}$$

3.2.2.3. Duplicate Removal

Duplicate removal is a method used to identify and eliminate duplicate records from a dataset, ensuring each patient's data is unique and preventing redundancy. This process involves using algorithms or database queries to identify duplicate records based on key identifiers like patient ID or date of record. Duplicate records are deleted, keeping only one instance of each unique data point. This helps prevent skew analysis, as duplicate data can lead to artificial inflation of patterns and biases, distorting the results. Additionally, it ensures data integrity by counting each data point only once, preserving the accuracy and reliability of the dataset. The goal is to prevent skew analysis and maintain data integrity.

3.3   Data Normalisation

Normalization ensures that all features contribute equally to the model, preventing any single feature from disproportionately influencing the results.

3.3.1   Min-Max Normalization

This technique scales biomarkers to a [0,1] range, making features comparable across different patients. It is especially important for biomarkers like blood glucose levels, where absolute scale differences can impact model performance by Ensuring uniformity in data representation and comparability of features across patients.

$$X_{norm} = \frac{X - X_{min}}{X_{max} - X_{min}} \tag{5}$$

3.3.2 Z-Score Normalization

Standardizes features to have a mean of 0 and a standard deviation of 1, facilitating comparison across different scales, such as blood pressure and heart rate with varying scales, ensuring that all features contribute equally to the model without bias.

$$X_{\text{std}} = \frac{X - \mu}{\sigma} \tag{6}$$

3.3.3 Log Transformation

Log transformation is a technique used to stabilize variance in biomarker data with skewed distributions, aiming to approximate a normal distribution. This technique is crucial in personalized medicine, as it reduces the influence of extreme values and improves data symmetry, enhancing the effectiveness of statistical and machine learning models, resulting in more accurate dosage recommendations.

$$X_{\text{log}} = log(X + 1) \tag{7}$$

where X represents the original biomarker values, and log denotes the logarithmic function. By adding 1 before applying the logarithm, the transformation handles zero and negative values, which are common in skewed datasets.

3.4 Feature Extraction

Feature extraction enhances the model's analytical capability by creating meaningful representations from raw data. Principal Component Analysis (PCA) is a technique that reduces the dimensionality of biomarker datasets to improve computational efficiency and focus on the most informative features. It transforms original features into orthogonal components, capturing maximum variance in data with fewer dimensions. This method reduces complexity while retaining essential information for accurate analysis. PCA is crucial for managing large volumes of biomarker data in personalized medicine, as it reduces dimensionality, minimizes overfitting, and enhances model performance. The FuRGAN framework effectively focuses on key features that significantly impact drug dosage recommendations, enhancing the accuracy and reliability of personalized treatment outcomes.

$$C = \frac{1}{n} X^T X \tag{8}$$

$$Z = XW \tag{9}$$

where, C is the resulting matrix, n is a scalar value, X^T denotes the transpose of matrix X, X is the original matrix, Z is the result of the multiplication, W is another matrix or vector.

The FuRGAN framework enhances personalized medicine by modeling complex relationships in patient data through interaction terms, which combine two biomarkers to reveal their synergistic effects. This approach allows for more precise and individualized dosage adjustments, improving treatment efficacy and patient safety. By considering the combined impact of multiple biomarkers, FuRGAN offers deeper insights into a patient's condition and treatment response, leading to more accurate medication dosage predictions.

$$X_{\text{interaction}} = X_1 \times X_2 \tag{10}$$

For example: X_1 could represent blood pressure levels, X_2 could represent cholesterol levels.

The FuRGAN framework uses temporal feature extraction to identify patterns in time-series data, showing how past information affects future outcomes. By integrating fuzzy logic, RNNs, and GANs, it processes patient biomarker data to make accurate predictions and dosage recommendations. Temporal features improve GAN performance by providing realistic, time-based data for RNN training. Methods like the Moving Average help smooth short-term fluctuations, giving clearer long-term health trends. These insights enable real-time adjustments to treatment plans, leading to more personalized and precise dosage recommendations. The moving average (7) at time t, denoted as MA_t, is calculated as:

$$MA_t = \frac{1}{n} \sum_{i=t-n+1}^{t} X_i \tag{11}$$

where, MA_t: Moving Average at time. n: Number of time points over which the average is calculated. This is often referred to as the window size. X_i: The value of the biomarker (or data point) at time.

The Fast Fourier Transform (FFT) is a technique that converts time-series data into the frequency domain, helping identify periodic patterns that impact treatment and dosage recommendations. In the FuRGAN framework, FFT detects cyclical trends in biomarker levels, allowing healthcare providers to tailor medication to align with natural cycles, such as weekly fluctuations, ensuring optimal treatment effectiveness at critical times. The Fourier Transform (8) of a time-series data X is represented as,

$$X_{\text{frequency}} = \text{FFT}(X) \tag{12}$$

where: X is the original time-series data, which consists of a sequence of biomarker values collected over time. $X_{\text{frequency}}$ is the transformed data in the frequency domain, showing the different frequency components present in the original time-series data.

3.5 Fuzzy Logic Module

The FuRGAN framework integrates preprocessed, normalized, and feature-extracted data into a unified dataset, ensuring consistency and alignment for effective model training and validation. Data validation through cross-validation assesses preprocessing effectiveness using metrics like accuracy, precision, recall, and clinical relevance, ensuring unbiased and accurate predictions. Fuzzy logic addresses uncertainty in personalized medicine by converting numerical data into linguistic terms (e.g., low, medium, high) and applying rule-based decision-making grounded in established medical practices and historical data. These rules ensure consistency with clinical standards and expert advice. Adaptive learning methods, such as Genetic Algorithms and Reinforcement Learning, continuously refine these rules based on patient data and outcomes, improving precision over time. Validation of fuzzy rules is crucial to ensure clinically relevant and accurate dosage recommendations, with performance metrics evaluating the system's accuracy and applicability. Defuzzification transforms fuzzy logic outputs into specific, actionable dosage adjustments, ensuring practical and effective recommendations in clinical settings. This comprehensive approach enhances the system's responsiveness, accuracy, and ability to meet individual patient needs.

Algorithm

1. In Fuzzification Convert numerical values X into fuzzy sets using membership function $\mu_A(X)$, where $\mu_A(X)$ represents the degree of membership in fuzzy set A. Example: If X is a blood glucose level, membership functions will be: Low: $\mu_{Low}(X) = \max\left(0, \dfrac{X_{max} - X}{X_{max} - X_{min}}\right)$; Medium:

$\mu_{Medium}(X)$ is defined within a specific range. High $\mu_{High}(X) = \max\left(0, \dfrac{X - X_{min}}{X_{max} - X_{min}}\right)$.

2. Rule-based optimization enhances performance by updating rules based on high performance metrics, utilizing adaptive learning techniques and genetic algorithms for rule refinement.

3. Fuzzified data is used to generate clinical recommendations, assessing accuracy, precision, recall, and alignment with known outcomes through rule evaluation and performance metrics.

$$\text{Accuracy} = \frac{\text{Number of Correct Predictions}}{\text{Total Predictions}}, \quad \text{Precision} = \frac{\text{True Positives}}{\text{True Positives} + \text{False Positives}}$$

Where, True Positives (TP): Correctly predicted positive cases. False Positives (FP): Incorrectly predicted positive cases. False Negatives (FN): Missed positive cases.

4. Defuzzification converts fuzzy outputs into actionable dosage recommendations, transforming abstract fuzzy logic results into concrete decisions. The Centroid Method, also known as the Center of Gravity or Center of Mass, calculates precise dosage recommendations by locating the center of the fuzzy membership function curve.

$$CD = \frac{\int_a^b x \cdot \mu(x)\, dx}{\int_a^b \mu(x)\, dx}$$

Where CD is the centroid value (the defuzzified dosage recommendation). x represents the dosage values. $\mu(x)$ is the membership function that describes the degree of membership of x in the fuzzy set. The integrals are taken over the range [a,b], which represents the range of possible dosage values. The Centroid Method, also known as the Center of Gravity method, is a widely used formula in fuzzy logic systems to convert defuzzified values into a single, actionable formula. This method calculates the weighted average of all possible values, resulting in precise dosage adjustment recommendations.

$$\text{Defuzzified Value} = \frac{\sum_i (\text{Membership Value}_i \times \text{Value}_i)}{\sum_i \text{Membership Value}_i}$$

Where, Membership Value$_i$ = Membership degree of the i-th fuzzy set. Value$_i$ = Crisp value corresponding to the i-th fuzzy set.

3.6 Recurrent Neural Network (RNN) Integration

Recurrent Neural Networks (RNNs) play a crucial role in the FuRGAN framework by analyzing temporal patterns in patient data to improve personalized dosage recommendations. Utilizing sequential data, such as biomarker trends, RNNs, particularly Long Short-Term Memory (LSTM) layers, excel in learning long-term dependencies and addressing challenges like the vanishing gradient problem. LSTM layers use forget gates to discard irrelevant information, input gates to incorporate new data, and output gates to make decisions based on past and current trends. Hidden layers with activation functions like ReLU and Tanh capture complex relationships in the data, while dropout layers prevent overfitting, ensuring the model generalizes well to new data. This combination of real-time updates, advanced data processing, and adaptive learning allows FuRGAN to provide accurate, dynamic dosage recommendations tailored to individual patient needs, ensuring effective and personalized treatment.

3.6.1 Dosage Adjustments

Dosage adjustments are crucial for optimizing treatment efficacy and minimizing adverse effects. They are essential for ensuring that each patient receives the most appropriate dosage based on their unique health data. Fuzzy logic plays a significant role in enhancing the accuracy of these dosage adjustments. The RNN model is used to generate predictions based on temporal patterns from patient health records, allowing the model to understand temporal patterns and predict necessary dosage changes. The RNN model is then integrated with fuzzy logic to refine recommendations, handling uncertainty and variability. The model is continuously updated with new patient data to refine dosage predictions. The feedback loop is used to retrain and optimize the model. The adjusted dosages are then applied in clinical settings, monitored for patient responses, and adjusted as needed based on ongoing data and feedback. The algorithm involves collecting and preprocessing patient health data, processing the data through the RNN, integrating RNN outputs with fuzzy logic-based recommendations, applying the recommended dosage changes, and monitoring and feedback.

Algorithm

1. The preparation for treatment involves gathering and processing patient health records, normalizing data, and organizing it into sequential time-series format for RNN processing.

2. The temporal analysis involves transforming patient data into sequences representing relationships, such as biomarker levels, and extracting relevant temporal features, such as trends and patterns over time. $\{x_t\}_{t=1}^{T}$ Where, x_t: Represents the data point at time t. T: Represents the length of the sequence, i.e., the total number of time points.

3. The RNN model architecture includes an input layer, LSTM layers, hidden layers, and dropout layers, with the model being compile using a loss function and optimizer. Forget Gate: $f_t = \sigma\left(W_f \cdot [h_{t-1}, x_t] + b_f\right)$; Input Gate: $i_t = \sigma\left(W_i \cdot [h_{t-1}, x_t] + b_i\right)$ Cell State Update: $\tilde{C}_t = \tanh\left(W_C \cdot [h_{t-1}, x_t] + b_C\right)$ Cell Update: $C_t = f_t \cdot C_{t-1} + i_t \cdot \tilde{C}_t$ Output Gate: $h_t = o_t \cdot \tanh(C_t)$ Hidden State: $h_t = o_t \cdot \tanh(C_t)$, hidden layers with activation functions were, $h = \text{ReLU}\left(W \cdot x + b\right)$, $h = \tanh\left(W \cdot x + b\right)$ and Dropout Layers for regularization be, $\text{Dropout}_j = \frac{1}{p} \cdot h_j$ Where, W_f is the weight matrix for the forget gate, h_{t-1} is the hidden state from the previous time step. x_t is the current input, b_f is the bias for the forget gate. σ is the sigmoid activation function. W_i is the weight matrix for the input gate. h_{t-1} is the hidden state from the previous time step. B_c is the bias for the cell state update. tanh is the hyperbolic tangent activation function. f_t is the forget gate output. C_{t-1} is the cell state from the previous time step. It is the input gate output. $\tilde{C}_t$ is the candidate cell state.

4. The RNN model in the FuRGAN framework is trained on sequential data, validated for accuracy, and continuously optimized through loss function assessment and hyperparameter tuning to ensure precise and adaptive dosage recommendations. $\text{Loss} = \frac{1}{N}\sum_{i=1}^{N}(y_i - \hat{y}_i)^2$; $\theta = \theta - \alpha \cdot \frac{m_t}{\sqrt{v_t} + \epsilon}$ where N is the number of samples, y_i is the true value, and $\hat{y}_i$ is the predicted value, θ is the parameter to be updated, α is the learning rate, m_t and v_t are the estimates of the first and second moments (mean and variance) respectively, ϵ is a small constant to prevent division by zero.

5. Fuzzy logic is utilized in generating recommendations based on expert knowledge, enhancing their accuracy through the use of machine learning techniques.

6. Recommendations are generated using RNN to forecast biomarker levels and dosages, refined using fuzzy logic to manage variability, and updated based on the latest patient data.

3.7 Generative Adversarial Networks (GANs) in FuRGAN Framework

Generative Adversarial Networks (GANs) enhance the FuRGAN framework by generating realistic patient data and improving the model's robustness through adversarial training. GANs play a critical role in refining dosage recommendations by learning from complex patterns and uncertainties in patient data. The training process involves two neural networks: the generator and the discriminator. These networks compete against each other in an adversarial setup, where the generator produces synthetic data, and the discriminator evaluates whether the data is real or fake. The generator uses random noise as input and outputs synthetic biomarker data designed to resemble real patient data. Over time, it is refined based on feedback from the discriminator, which helps it generate more realistic data. The discriminator network is trained to differentiate between real patient data and the synthetic data created by the generator. It continuously improves its ability to distinguish between the two by learning from both real and fake data. In FuRGAN, the discriminator's role is crucial for ensuring that the synthetic data is useful for training other components, such as the Recurrent Neural Networks (RNNs) and fuzzy logic systems. The adversarial training process of GANs allows the framework to simulate diverse patient scenarios, improving the accuracy and adaptability of the dosage recommendation system. This continuous feedback loop between the generator and discriminator enhances the model's ability to make more accurate and personalized predictions.

3.7.1 Feedback Mechanisms

Feedback mechanisms in FuRGAN improve the performance of generator and discriminator networks, leading to better synthetic data for RNN and fuzzy logic components. This refines the synthetic data, improving dosage recommendation accuracy. The process involves evaluating results, adjusting training, and continuously updating GAN components based on performance feedback to enhance realistic data generation and dosage recommendation accuracy.

Algorithm

1.Initialization involves defining generator (G) and discriminator (D) networks, setting hyperparameters like learning rates α_G and α_D respectively.

2. The training on the generator involves generating synthetic data for each epoch by sampling noise z and producing synthetic data z. Train Discriminator: Compute loss for real data: $L_{D_{real}} = -\frac{1}{m} \sum_{i=1}^{m} \log(D(x_i))$, Compute loss for fake data: $L_{D_{fake}} = -\frac{1}{m} \sum_{i=1}^{m} \log(1 - D(G(z_i)))$, Update Discriminator by minimizing total loss: $L_D = L_{D_{real}} + L_{D_{fake}}$ Where, $D(x_i)$ - The discriminator's output for a real data sample x_i, representing the probability that x_i is real. m - Number of real samples. Train Generator: Compute loss: $L_G = -\frac{1}{m} \sum_{i=1}^{m} \log(D(G(z_i)))$, Update Generator by minimizing L_G. Where, The generator G takes input z_i (random noise). The discriminator D evaluates $D(x_i)$, the output of the generator, to give the probability that it is real. The generator loss L_G is calculated using the negative log of the probability.

3. Feedback mechanisms involve assessing data quality, adjusting training parameters, and continuously refining networks to enhance synthetic data realism.
4. The integration involves utilizing refined synthetic data to improve the performance of RNN and fuzzy logic models in providing accurate dosage recommendations.

3.8 Decision Support Systems

The FuRGAN framework optimizes treatment plans in real time by analyzing patient data and adapting to their evolving needs. It uses Recurrent Neural Networks (RNNs) to detect patterns and predict future biomarker trends, while fuzzy logic handles uncertainties in the data to make precise dosage adjustments. The RNN model learns from each update, improving its ability to anticipate patient needs and recommend effective dosages. The framework also incorporates a Decision Support System (DSS) to assist healthcare professionals with actionable insights, combining outputs from RNNs and fuzzy logic models. The DSS offers visual dashboards, alerts, and reports to monitor patient progress, predict trends, and suggest dosage changes. By integrating automated analysis with human expertise, the FuRGAN framework ensures personalized care, enabling timely interventions and better patient outcomes.

3.9 Evaluation and Validation

The FuRGAN framework offers a holistic approach to predicting and adjusting dosages by combining multiple evaluation methods. It uses quantitative metrics like accuracy, precision, recall, F1-score, and mean squared error to measure its effectiveness, while the synthetic data generated by GANs is validated to ensure it closely resembles real patient data. Additionally, feedback from patients and clinicians is gathered to assess the practicality and acceptability of the dosage recommendations. This feedback loop plays a key role in refining the system based on real-world scenarios and experiences. Continuous monitoring and regular updates with new patient data and health outcomes ensure the framework remains adaptive, reliable, and effective in delivering personalized dosage recommendations.

4   Conclusion

We are currently developing the FuRGAN framework, which represents a groundbreaking approach in personalized medicine. Our work focuses on optimizing dosage recommendations by leveraging advanced machine learning techniques, including Recurrent Neural Networks (RNNs) and Generative Adversarial Networks (GANs). By accurately predicting and adjusting medication dosages, FuRGAN aims to minimize the excessive use of chemical substances in treatments, which can negatively impact human health. The integration of these technologies ensures that dosage levels are maintained within safe and effective limits, ultimately enhancing patient well-being and living conditions. FuRGAN's innovative approach promises to revolutionize the way medical treatments are administered, paving the way for more precise, personalized, and health-conscious practices in medicine.

Acknowledgment. We extend our heartfelt thanks to our Department heads of the School of Computing at Syed Ammal Engineering College and our friends for their invaluable guidance and support. Lastly, we acknowledge the broader research community for their contributions to AI, machine learning, and personalized medicine, which inspire and advance these fields.

References

1. Yang, M., Liu, H., Wang, S., Li, C.: Applying fuzzy logic and deep learning for real-time personalized medicine. IEEE J. Biomed. Health Inform. **25**(12), 4450–4458 (2021)
2. Choi, E., Schuetz, A., Stewart, W.F., Sun, J.: Using recurrent neural network models for early detection of heart failure onset. J. Am. Med. Inform. Assoc. **24**(2), 361–370 (2017)
3. Goodfellow, I., et al.: Generative adversarial nets. Adv. Neural. Inf. Process. Syst. **27**(1), 2672–2680 (2014)

4. Kim, H., Park, H., Lee, J.: Real-time drug dose optimization using hybrid GAN-RNN architecture with patient feedback. Comput. Struct. Biotechnol. J. **19**, 5420–5430 (2021)

5. Zhang, W., Li, S., Yang, X., Wu, H., Zhang, X.: Adaptive fuzzy logic controller for drug dosage adjustment based on patient response. Sensors **22**(3), 740–751 (2022)

6. Wang, T., Wang, W., Zhang, H., Liu, L.: Real-time personalized drug dosage optimization using a hybrid model of GAN and RNN. IEEE Trans. Neural Netw. Learn. Syst. **33**(8), 3856–3867 (2022)

7. Xu, Y., Li, Y., Liu, Y., Sun, C., Gao, Y.: Enhancing drug prescription accuracy with feedback-driven GANs in real-time monitoring systems. IEEE Access **10**, 72958–72969 (2022)

8. Shen, J., Zhang, C., Cao, H., Zhao, Q.: Personalized medicine with integrated GAN and fuzzy logic for adaptive drug dosing. J. Biomed. Inform. **127**, 104034 (2023)

9. Chen, J., Zhang, Y., Xie, P., Huang, L., Lei, Q.: Adaptive learning and decision-making framework for drug dosage optimization using RNNs and GANs. Artif. Intell. Med. **118**, 102241 (2023)

10. Han, X., Li, T., Peng, Z., Zhao, M.: A feedback loop-driven GAN-based framework for personalized healthcare monitoring. Comput. Biol. Med. **151**, 106313 (2023)

11. Liu, Y., Song, Y., Li, H., Wang, T., Wang, X.: Real-time personalized medicine using RNN and GAN integration. Biomed. Eng. Online **19**, 23–35 (2020)

12. Roy, A., Luthra, A., Kumar, N., Mukherjee, M.: Intelligent healthcare monitoring system using IoT and fuzzy logic. Futur. Gener. Comput. Syst. **105**, 398–410 (2020)

13. Esteva, A., et al.: A guide to deep learning in healthcare. Nat. Med. **25**(1), 24–29 (2019)

14. Yu, K.H., Beam, A.L., Kohane, I.S.: Artificial intelligence in healthcare. Nat. Biomed. Eng. **2**(10), 719–731 (2018)

15. Miotto, R., Li, L., Kidd, B.A., Dudley, J.T.: Deep patient: an unsupervised representation to predict the future of patients from electronic health records. Sci. Rep. **6**, 26094–26103 (2016)

16. Deo, R.C.: Machine learning in medicine. Circulation **132**(20), 1920–1930 (2015)

17. Obermeyer, Z., Emanuel, E.J.: Predicting the future—big data, machine learning, and clinical medicine. N. Engl. J. Med. **375**(13), 1216–1219 (2016)

18. Schmidhuber, J.: Deep learning in neural networks: an overview. Neural Netw. **61**, 85–117 (2015)

19. Cios, K.J., Moore, G.W.: Uniqueness of medical data mining. Artif. Intell. Med. **26**(1–2), 1–24 (2002)

20. Lecun, Y., Bengio, Y., Hinton, G.: Deep learning. Nature **521**(7553), 436–444 (2015)

21. Ramalingam, G.P., Pandian, D., Batcha, C.F.S.: IntelliCare: integrating IoT and machine learning for remote patient monitoring in healthcare: a comprehensive framework. J. Cogn. Hum.-Comput. Interact. (JCHCI) **7**(02), 50–59 (2024)

A Hybrid CNN-GAT Model for MRI Screening of Autism Spectrum Disorder

M. Srikanth and Chandrashekar Jatoth$^{(\boxtimes)}$ ⓘ

National Institute of Technology, Raipur, Raipur, Chhattisghar, India
`jchandrashekar.it@nitrr.ac.in`

Abstract. Autism Spectrum Disorder (ASD) is a neurodiverse disorder
that impacts children's development, presenting a range of symptoms
that persist into adulthood. Recognizing ASD in children is challenging due to the diversity of associated disorders. We propose a hybrid
method for detecting autism in children using MRI screening. Our model
leverages hybrid deep-learning techniques to address these challenges. In
particular, a blend of Convolutional Neural Networks (CNN) and Graph
Attention Networks (GAT) is used for model interpretation, allowing us
to identify critical brain regions and connections that contribute to the
classification. Experiments conducted using data accessible to the public
show that the proposed framework achieves an 88% classification accuracy between ASD and Non-ASD cases, outperforming state-of-the-art
methods. This indicates the proposed Hybrid CNN-GAT model's strong
classification ability and generalizability, offering a powerful tool for identifying brain disorders.

Keywords: Autism Spectrum Disorder (ASD) · Graph Attention
Network (GAT) · Deep Learning (DL) · Convolutional Neural Network
(CNN) · Magnetic Resonance Imaging (MRI)

1   Introduction

Autism Spectrum Disorder (ASD) is a neurological problem characterized by difficulties in the social meetings and relations, along with an unvaired or restricted
behaviors. These symptoms generally manifest in early childhood and persist
until maturity [7,10,20]. There are three phases of ASD, a condition of the neurological system that changes how messages reach the brain. The disorders include
mild, moderate, and severe. Since every child has unique symptoms, illnesses
can be identified as early as 6 to 12 months through responsive behaviour. Early
intervention is crucial; with appropriate occupational therapy and behavioural
therapy, many symptoms can be addressed [4,9]. During the ages of two to three,
self-hitting, teeth grinding, and difficulties with sitting tolerance are common
symptoms. With an estimated 1 in 100 children affected, the frequency of ASD
is rising. While there isn't a cure, primary symptoms may be lessered with early
intervention. Diagnosis primarily relies on behavioural evaluations, which can

C. Modi et al. (Eds.): MIND 2024, CCIS 2736, pp. 41–52, 2026.
https://doi.org/10.1007/978-3-032-14531-4_4

lead to inaccuracies due to the lack of reliable biomarkers, prompting interest in neuroimaging methods to recognize anatomical and functional variations in the brain function of people with ASD compared to those who are typically developing [6,18]. As per the suggestion of the Psychiatrist, the ASD child can't give correct information about what he wants and needs at the time, and the parents should treat them as an average child and parents should train them to take care of the child. Advances in neuroimaging, particularly structural Magnetic Resonance Imaging (s-MRI), have revealed features associated with ASD, such as reduced neuron numbers in specific regions and atypical functional connectivity patterns, suggesting the potential for neuroimaging-based biomarkers in diagnosis. ML and DL techniques have also gained attention for their effectiveness in analyzing medical data, including ASD detection, and computationally aided diagnosis using neuroimaging data has become a popular approach for classifying individuals with ASD [5]. ASD is part of a broader spectrum of neurological and developmental disorders, which include Alzheimer's disease, brain tumours, and schizophrenia. Alzheimer's is characterized by brain cell death affecting memory and cognitive function, with symptoms often emerging between ages 30–60 [17]. In contrast, brain tumours can lead to symptoms like headaches and memory issues due to abnormal cell growth. Schizophrenia impacts thought, behaviour, and emotions, typically showing early symptoms between ages 16–30 [1]. Neurological illnesses affect one in six people and pose significant treatment costs, and the complexity of brain diseases makes research challenging. However, the past two decades have seen significant advancements in understanding these conditions, and the combination of techniques we are using such as DL and ML methods with neuroimaging may pave the way for more accurate diagnoses and better intervention strategies for ASD and other brain disorders [13]. The significant contributions of our research are as follows:

- A hybrid CNN and GAT was proposed for the classification of brain functional networks in ASD diagnosis. In our model, CNN was utilized as the fundamental feature extractor.
- Experiments were performed on the MRI dataset. The results indicate that the suggested model surpasses current state-of-the-art classification techniques and attains superior classification results.
- The proposed model integrates the benefits of CNNs (CNN) and GAT to advance feature learning and improve classification performance.
- Our study provides a promising results for learning and classifying MRI brain function network.

The remainder of the paper states as follows: Sect. 2 gives an overview of the recent procedure of studies based on autism. Section 3 Illustrate the proposed work based on CNN and GAT model. Section 4 Explains about the results of the experiment and Sect. 5 Provide the Conclusion of the paper.

2   Literature Review

In a recent study, Ma et al. [14] utilized a Contrastive Variational AutoEncoder (CVAE) to extract Autism Spectrum Disorder (ASD)-specific and shared features from structural MRI (s-MRI) data of young children. This approach addressed minor data size issues through a transfer learning strategy and included neuroanatomical interpretation to identify potential biomarkers for targeted ASD treatment. Ashraf et al. [2] employed advanced MRI techniques, such as functional MRI (fMRI) and sMRI, to analyze neuropsychiatric disorders. Their work applied deep learning methods, particularly an optimized CNN with Transfer Learning, achieving an accuracy of 81.56% in diagnosing autism from the ABIDE datasets, thus outperforming traditional methods. Chen et al. [3] introduced an AL-NEGAT to identify ASD using multimodal MRI data. Their methodology incorporates node and edge attributes via a graph attention mechanism, simultaneously improving generalizability using adversarial learning strategies. A gradient-based saliency map approach was employed to understand the model's significance in identifying essential brain regions and connections. Ma et al. [15] suggested a Multi-Scale Dynamic Graph Learning (MDGL) framework. utilizing resting-state fMRI (rs-fMRI) data for brain disorder recognition. Their methodology constructs multi-scale dynamic functional connectivity networks (FCNs) and uses dynamic graph representation learning for the acquisition of spatiotemporal information, overcoming the limitations of existing techniques. Wang et al. [21] developed a multimodal approach that combines fMRI data with non-imaging demographic data through a WL-DeepGCN model for ASD diagnosis. This model leverages a weight-learning network for graph edge weights in the latent space, implementing residual connectivity to address gradient issues, and employs an EdgeDrop strategy to mitigate overfitting and over-smoothing, achieving 77.27% accuracy and 0.83 AUC. Kunda et al. [12] introduced a novel second-order functional connectivity metric, Tangent Pearson embedding, for feature extraction in autism classification. Their domain adaptation approach minimized statistical dependence between acquisition sites and functional connectivity features, achieving a classification performance of 73% on the ABIDE dataset. Huang et al. [8] proposed a federated multi-task learning (MTL) framework for diagnosing mental disorders using multimodal MRI information. This framework employs a unique learning-based centrifuge to extract high-level features and a multi-gate mixture of classifiers to manage optimization conflicts during classification, incorporating personalized learning, privacy protection, and federated biomarker interpretation. Mishra et al. [16] utilized a structural MRI classification framework with an ensemble of DCNNs and real-time data augmentation strategies to enhance model resilience. Their dataset included 484 individuals with ASD and 491 controls, demonstrating improved accuracy over previous methodologies. Kiruthigha et al. [11] investigated brain connections associated with ASD using s-MRI, fMRI, and rs-fMRI, applying deep learning techniques like GCNNs to analyze imaging data and reducing dimensionality for managing complex functional connectivity matrices, aiming for early detection and treatment. Ruan et al. [19] employed machine learning methods to classify images captured from individuals with ASD versus those from peers without ASD, achieving over 80% accuracy. The model analyzed visual elements to identify distinct salience and attention patterns unique to individuals with ASD.

3   Methodology

This section deliver an organized outline of the proposed approach for categorizing autism spectrum disorder (ASD) via MRI scans, including many essential components: collecting data, developing a model, and the full evaluation procedure. This process starts with the acquisition of DICOM images, including scaling to a standardized dimension, balancing pixel values, and encoding labels into a binary format (autism = 1, non-autism = 0). The dataset is divided into training and testing subsets to guarantee thorough examination.

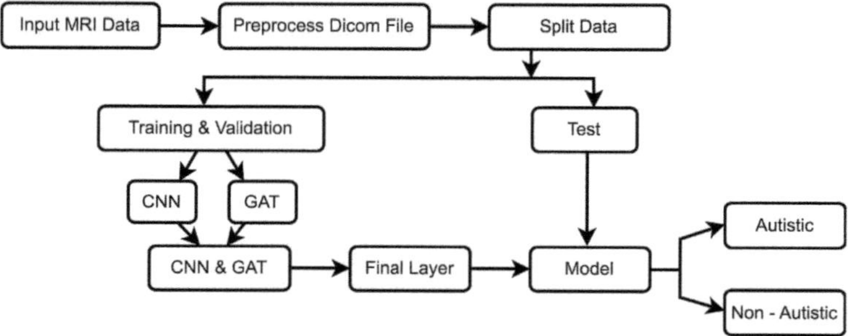

Fig. 1. Proposed Methodology

3.1   Data Pre-processing

The DICOM pictures, which are medical images saved in a unique format, are loaded first in the procedure. Preparing to be entered into the CNN, these pictures are scaled to a consistent dimension, usually 128×128 pixels, to maintain uniformity throughout the dataset. The original photos, which could have been in grayscale, are transformed to RGB format to comply with the CNN model's input specifications. Then, dividing by 255, the picture pixel values are adjusted to fall between zero and one, simplifying and expediting the model training process. The "autism" and "non-autism" labels are then binarized, becoming binary values. Figure 1 illustrates that the suggested methodology integrates CNNs and GAT for detecting autism utilizing MRI data. The procedure starts with entering MRI DICOM files, which undergo preprocessing to ready the data for training. The dataset is subsequently divided into training, validation, and testing subsets. The CNN extracts spatial features from the MRI images, while the GAT captures relational information between different brain regions. These outputs are integrated in the subsequent step, where the CNN and GAT results are combined and passed through a final classification layer. The model is validated and tested to predict whether the MRI data belongs to an autistic or non-autistic individual.

Convolution Operation: The convolution operation is fundamental to CNNs. It helps in extracting features from the input image:

$$\text{Feature Map} = f\left(\sum_{i=0}^{k-1}\sum_{j=0}^{k-1} I(x+i, y+j) \cdot K(i,j) + b\right) \tag{1}$$

where:

- $I(x,y)$ denotes the pixel value of the input image at coordinates (x,y).
- $K(i,j)$ is the kernel (or filter) used for convolution, with dimensions $k \times k$.
- b is a bias term incorporated into the convolution output.
- $f(\cdot)$ is an activation function (like ReLU) that introduces non-linearity to the model.

In this operation, the kernel slides across the image, executing element-wise multiplication and aggregating the results to generate a feature map. This helps in detecting patterns, edges, and other features in the image.

Pooling Operation: Pooling is employed to diminish the spatial dimensionality of feature maps, hence reducing computing burden and mitigating overfitting:

$$\text{Pooled Value} = \max\left(I(x+i, y+j)\right), \quad \text{for } i, j \in \{0, 1, \ldots, k-1\} \tag{2}$$

This example describes max pooling, where the maximum value is taken from a $k \times k$ window that moves across the feature map. The result is a smaller feature map that retains the most important information.

Fully Connected Layer: Subsequent to multiple convolutional and pooling layers, the output is flattened and input into a fully connected (dense) layer, which integrates features to provide predictions:

$$y = f\left(W \cdot x + b\right) \tag{3}$$

where:

- W is the weight matrix of the completely connected layer.
- x is the input vector from the previous layer (flattened feature map).
- b is the bias vector.
- $f(\cdot)$ is an activation function (like ReLU or sigmoid).

This layer essentially learns to map the input features to output classes by adjusting weights through training.

Binary Crossentropy Loss: In binary classification, it serves as the loss function to evaluate the model's performance.

$$\text{Loss} = -\frac{1}{N} \sum_{i=1}^{N} \left[y_i \log(\hat{y}_i) + (1 - y_i) \log(1 - \hat{y}_i) \right] \tag{4}$$

where:

- N is the aggregate quantity of samples.
- y_i is the true label for sample i (1 for autism, 0 for non-autism).
- $\hat{y}_i$ is the estimated probability that sample i is classified as belonging to class 1 (autism).

This loss function penalizes incorrect predictions heavily, guiding the model to improve during training.

ROC Curve and AUC: The ROC curve is a graphical depiction of a model's efficiency at various threshold settings:

$$\text{TPR} = \frac{\text{TP}}{\text{TP} + \text{FN}}, \quad \text{FPR} = \frac{\text{FP}}{\text{FP} + \text{TN}} \tag{5}$$

where:

- TPR (True Positive Rate): The ratio of true positives that were accurately recognized.
- FPR (False Positive Rate): The ratio of true negatives misclassified as positives.

AUC (Area Under the Curve) measures the complete ability of the model to discriminate between classes, providing a single score to evaluate performance:

$$\text{AUC} = \int_0^1 \text{TPR}(FPR) \, d(\text{FPR}) \tag{6}$$

The AUC varies between 0 and 1, with 1 signifying perfect discrimination and 0.5 denoting no discrimination (equivalent to random guessing).

3.2   Classifier

The classifier in the proposed methodology uses a hybrid model that combines CNNs with Graph Attention Layers for enhanced MRI image classification in autism diagnosis. CNNs extract spatial features, such as edges and textures, through convolutional and pooling layers, which maintain essential data and reveal the images' hierarchical structure. In parallel, the Graph Attention Layer focuses on critical relationships within the data, emphasizing key features through attention mechanisms. This dual approach integrates outputs to form a comprehensive feature set. Subsequently, the features pass through fully connected layers, leading to classification by a sigmoid activation function. This method allows for more precise and nuanced analysis, improving the accuracy of autism detection.

3.3 Proposed Methodology

The proposed methodology for classifying autism with MRI images adopts a multifaceted approach, focusing on data preprocessing, hybrid model architecture, and comprehensive evaluation. MRI images, stored as DICOM files, are preprocessed through resizing to uniform dimensions, RGB conversion, pixel normalization, and label binarization. The central feature of the methodology is a hybrid model that merges CNNs with Graph Attention Layers. The CNN branch extracts spatial features from images to distinguish between autism and non-autism cases. Concurrently, the Graph Attention Layer captures intricate data relationships, augmenting the CNN's spatial feature set. The combined outputs feed into fully connected layers for final classification. Provides the training utilizes binary cross-entropy loss, and evaluations measure accuracy, precision, recall, F1-score, and ROC-AUC metrics. Tools like confusion matrices and ROC curves visually interpret results, offering insights into model performance and identifying improvement areas. This approach provides a detailed and practical framework for autism classification using advanced machine-learning techniques.

Algorithm 1. Hybrid CNN and Graph Attention Model

1: **Step 1: Image Loading and Preprocessing**
2: Load DICOM images.
3: Preprocess each image by:
4:     Resizing to a uniform dimension img_{resize}.
5:     Normalizing the pixel values: $img_{norm} = \frac{img_{RGB}}{255}$.
6: **Step 2: Model Construction**
7: Construct feature extraction: $\mathbf{x}_{CNN} = f_{CNN}(\mathbf{X}; \Theta_{CNN})$
8: Construct Graph Attention branch: $\mathbf{x}_{GAT} = f_{GAT}(\mathbf{X}; \Theta_{GAT})$
9: Concatenate the outputs: $\mathbf{x}_{combined} = \text{Concatenate}(\mathbf{x}_{CNN}, \mathbf{x}_{GAT})$
10: **Step 3: Model Training**
11: compute binary cross-entropy loss function: $L(\Theta) = -\frac{1}{N}\sum_{i=1}^{N}[y_i \log(\hat{y}_i) + (1 - y_i)\log(1 - \hat{y}_i)]$
12: **Step 4: Model Saving**
13: Save the trained model for future prediction or further analysis.

---

4 Performance Evaluation

4.1 Dataset

An MRI image dataset labelled "autistic" and "non-autistic" from Kaggle[1] contains 131 and 155 **.nii** files, this files have been converted to DICOM files respectively, for binary classification. These brain scans provide structural details crucial for distinguishing autism spectrum disorders. The dataset arranged into training and testing sets with an 80-20 ratio, training on 80% to learn data patterns and testing on 20% to evaluate model performance. This division ensures the model's generalisation ability, which is vital for real-world applications.

---

[1] https://www.kaggle.com/datasets/dswainsonsujana/autism-preprocessed.

4.2   Experimental Results

Evaluating a deep learning model is essential for estimating its efficacy. Evaluation of deep learning models employs several metrics and Selecting the most suitable metrics is crucial for optimizing a model according to its performance. Identifying any disease is regarded as binary classification, which classifies data into positive or negative outcomes. The assessment measures are employed to gauge the efficacy of CNN and GAT classification models.

The following terms are used to define additional metrics.

- **TP:** The number of cases correctly identified as positive.
- **TN:** The number of cases correctly identified as unfavorable.
- **FP:** The number of instances incorrectly labeled as positive.
- **FN:** The number of instances incorrectly labeled as unfavorable.

Table 1. Comparison of Classification Performance

Method	Sensitivity	Specificity	Precision	Accuracy	F1 Score
MobileNetV2	0.9729	0.7142	0.5285	0.7783	0.6850
CNN + DNN	0.9815	0.7043	0.5031	0.7691	0.6652
Proposed Method	**0.9496**	**0.8431**	**0.7858**	**0.8833**	**0.8599**

Table 1 shows the performance metrics of our proposed model and its comparision with the existing models (Fig. 2, Fig. 3).

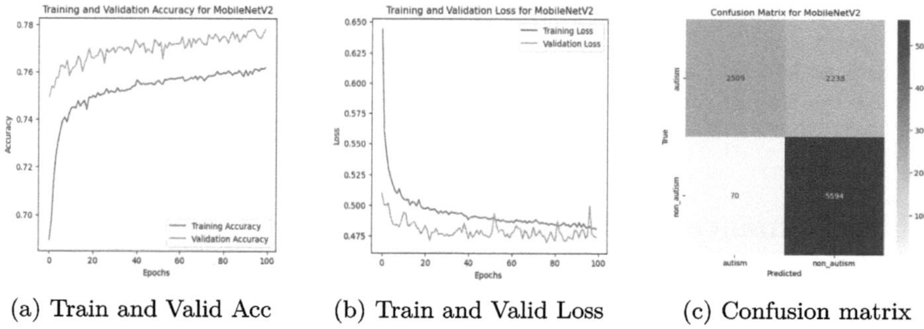

(a) Train and Valid Acc   (b) Train and Valid Loss   (c) Confusion matrix

Fig. 2. Accuracy, Loss and confusion matrix for MobileNetV2

The validation and test data evaluated the performance of the proposed model after training. Figure 4a illustrates the accuracy of the proposed model. From this graph, the training accuracy started at around 60% and kept increasing step by step, and finally, after adequate training of the model, it reached 91%.

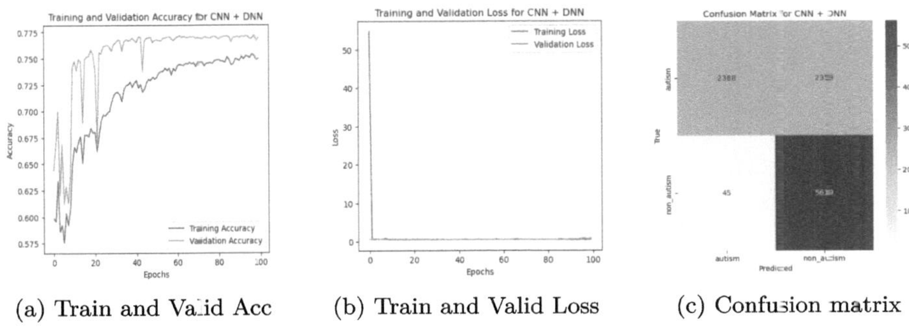

(a) Train and Valid Acc   (b) Train and Valid Loss   (c) Confusion matrix

Fig. 3. Accuracy, Loss and confusion matrix for CNN + DNN

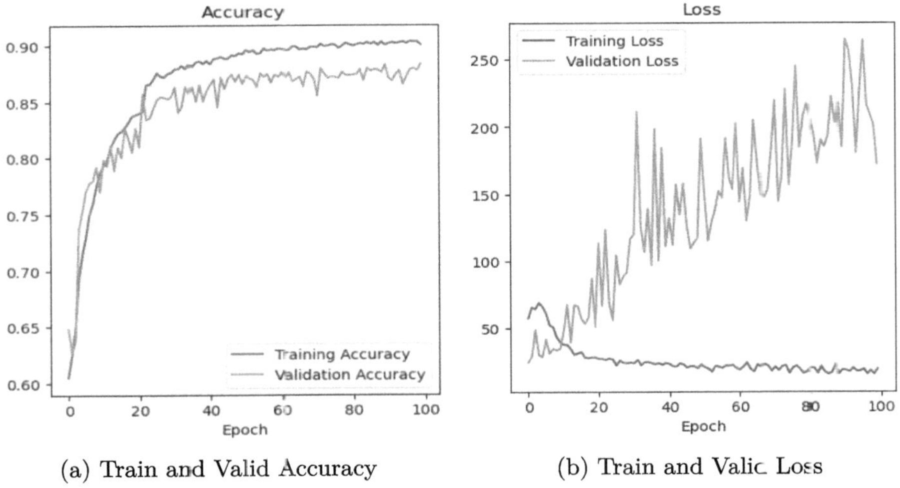

(a) Train and Valid Accuracy   (b) Train and Valid Loss

Fig. 4. Accuracy and Loss for Proposed Model

In that matter, the validation accuracy started from 64%, went a little down for some steps, kept on gradually increasing during the model's image validation process, and reached the value of 88% at the end. In the last epoch, the training and validation accuracy was robust, proving that the model is adequate for ASD detection using MRI data during early childhood. Figure 4b presents the loss curves for the training and validation datasets. The training loss in blue keeps decreasing across the epochs, which means the model fits for the training data and provides the better result. This smooth trend of training loss assures the reader the model is adaptive and can fitted into the processing data. Conversely, the validation loss heating-up curve—is much more erratic: its fluctuations are evident throughout the training process. This, despite the fluctuation, indicates that the model is learning well; however, the spikes in the validation loss could point to some instability or overfitting that might be occurring.

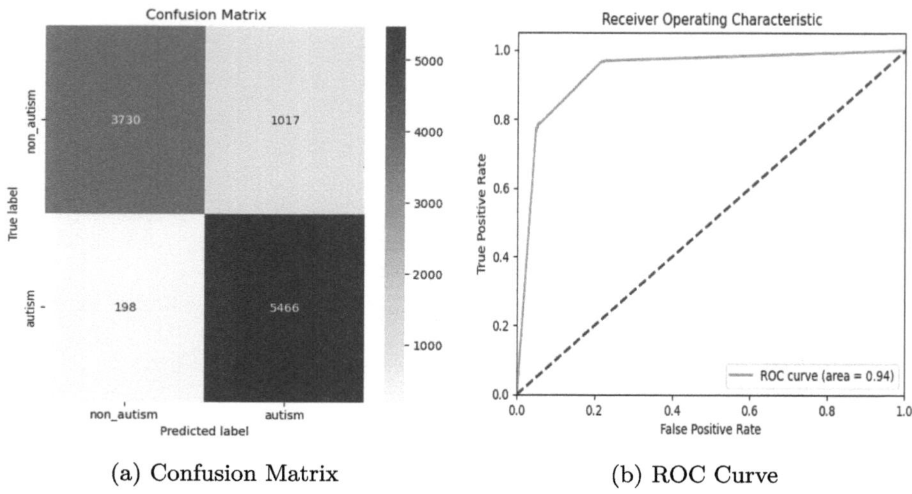

(a) Confusion Matrix   (b) ROC Curve

Fig. 5. Confusion Matrix and ROC Curve for the Proposed Model

The Fig. 5a shows its classification performance. The model's predictions are compared to "non-autism" and "autism" labels. Of 4,747 non-autism instances, the model successfully recognized 3,730, whereas 1,017 were misclassified as autism, resulting in false positives. For the 5,664 cases that were autism, the model correctly classified 5,466 of them, with 198 incorrectly classified as non-autism, resulting in false negatives. This confusion matrix highlights that the model is effective at identifying both classes, but there is a slight tendency to misclassify some non-autism cases as autism.

The Fig. 5b evaluates model performance by visually depicting the True Positive Rate (TPR) in relation to the False Positive Rate (FPR) at various threshold levels. The ROC curve for this model approaches the upper left corner of the plot, signifying outstanding performance. The Area Under the Curve (AUC) is 0.94, indicating that the model effectively differentiates between the "non-autism" and "autism" categories. The elevated AUC value indicates the model's proficiency in achieving an optimal equilibrium the relationship between sensitivity and specificity across different threshold values.

5   Conclusion

The proposed ASD detection methodology using MRI screening significantly extends neuroimaging with deep learning. Our model effectively identifies essential brain regions and their interconnections by combining GAT with CNNs. It charts a remarkable classification accuracy, as high as 88%, in distinguishing ASD from non-ASD cases. State-of-the-art techniques, presented most recently, showcase how well our model dealt with complex challenges in evaluating ASD conditions. This includes in-depth data collected before treatment and thorough measures for evaluation so that the results are reliable and precision. We can increase the generalizability of our model by expanding the dataset in the future. The integration of longitudinal data might give higher predictive accuracy, and research in explainable AI methodologies may offer a way to get insight into how their models make predictions. These advances may enable the use of therapies much earlier and lead to profound insights into the neurological underpinnings of autism, thus improving treatment outcomes.

Acknowledgment. This work is supported by the Department of Science and Technology, Science and Engineering Research Board (SERB), India, with the grant number EEQ/2022/000980 and partially supported by SEED grant number NITRR/Dean(R&C)/2022/32. The authors of this article are extremely grateful to them.

References

1. Alves, C.L., et al.: Diagnosis of autism spectrum disorder based on functional brain networks and machine learning. Sci. Rep. **13**(1), 8072 (2023)
2. Ashraf, A., Zhao, Q., Bangyal, W.H., Iqbal, M.: Analysis of brain imaging data for the detection of early age autism spectrum disorder using transfer learning approaches for internet of things. IEEE Trans. Consum. Electron. (2023)
3. Chen, Y., et al.: Adversarial learning based node-edge graph attention networks for autism spectrum disorder identification. IEEE Trans. Neural Netw. Learn. Syst. (2022)
4. Faja, S., Dawson, G.: Early intervention for autism. Handbook of preschool mental health: development, disorders, and treatment, pp. 388–416 (2006)
5. Farooq, M.S., Tehseen, R., Sabir, M., Atal, Z.: Detection of autism spectrum disorder (ASD) in children and adults using machine learning. Sci. Rep. **13**(1), 9605 (2023)
6. Hiremath, C.S., et al.: Emerging behavioral and neuroimaging biomarkers for early and accurate characterization of autism spectrum disorders: a systematic review. Transl. Psychiatry **11**(1), 42 (2021)
7. Hodges, H., Fealko, C., Soares, N.: Autism spectrum disorder: definition, epidemiology, causes, and clinical evaluation. Transl. Pediatr. **9**(Suppl 1), S55 (2020)
8. Huang, Z.A., et al.: Federated multi-task learning for joint diagnosis of multiple mental disorders on MRI scans. IEEE Trans. Biomed. Eng. **70**(4), 1137–1149 (2022)
9. Itzchak, E.B., Zachor, D.A.: Who benefits from early intervention in autism spectrum disorders? Res. Autism Spectrum Disord. **5**(1), 345–350 (2011)
10. Kent, R., Simonoff, E.: Prevalence of anxiety in autism spectrum disorders. In: Anxiety in Children and Adolescents with Autism Spectrum Disorder, pp. 5–32 (2017)
11. Kiruthigha, M., Jaganathan, S.: Graph convolutional model to diagnose autism spectrum disorder using RS-fMRI data. In: 2021 5th International Conference on Computer, Communication and Signal Processing (ICCCSP), pp. 1–5. IEEE (2021)
12. Kunda, M., Zhou, S., Gong, G., Lu, H.: Improving multi-site autism classification via site-dependence minimization and second-order functional connectivity. IEEE Trans. Med. Imaging **42**(1), 55–65 (2022)

13. Lima, A.A., Mridha, M.F., Das, S.C., Kabir, M.M., Islam, M.R., Watanobe, Y.: A comprehensive survey on the detection, classification, and challenges of neurological disorders. Biology **11**(3), 469 (2022)
14. Ma, R., et al.: Autism spectrum disorder classification with interpretability in children based on structural MRI features extracted using contrastive variational autoencoder. Big Data Min. Anal. **7**(3), 781–793 (2024)
15. Ma, Y., et al.: Multi-scale dynamic graph learning for brain disorder detection with functional MRI. IEEE Trans. Neural Syst. Rehabil. Eng. (2023)
16. Mishra, M., Pati, U.C.: A classification framework for autism spectrum disorder detection using sMRI: optimizer based ensemble of deep convolution neural network with on-the-fly data augmentation. Biomed. Signal Process. Control **84**, 104686 (2023)
17. Nadeem, M.S., et al.: Symptomatic, genetic, and mechanistic overlaps between autism and Alzheimer's disease. Biomolecules **11**(11), 1635 (2021)
18. Nisar, S., Haris, M.: Neuroimaging genetics approaches to identify new biomarkers for the early diagnosis of autism spectrum disorder. Mol. Psychiatry **28**(12), 4995–5008 (2023)
19. Ruan, M., Webster, P.J., Li, X., Wang, S.: Deep neural network reveals the world of autism from a first-person perspective. Autism Res. **14**(2), 333–342 (2021)
20. Wang, L., Wang, B., Wu, C., Wang, J., Sun, M.: Autism spectrum disorder: neurodevelopmental risk factors, biological mechanism, and precision therapy. Int. J. Mol. Sci. **24**(3), 1819 (2023)
21. Wang, M., Guo, J., Wang, Y., Yu, M., Guo, J.: Multimodal autism spectrum disorder diagnosis method based on deepGCN. IEEE Trans. Neural Syst. Rehabil. Eng. (2023)

Advanced Hybrid CNN-Transformer Predictive Machine Learning Model for Enhanced Pneumonia Detection Research

Piyush Dahiwadkar[✉], Sujal Joshi, Gunjan Kadam, and Dhanashree Toradamalle

Computer Engineering, K. J. Somaiya Institute of Technology, Sion, Mumbai 400022, Maharashtra, India
{p.dahiwadkar,joshi.sa,gunjan.kadam, dhanashree.t}@somaiya.edu

Abstract. This paper explores the effectiveness of a hybrid CNN-Transformer model for pneumonia identification from chest X-ray pictures is investigated in this work. We obtain higher accuracy and enhanced localisation of regions afflicted by pneumonia by utilising Transformers' global context awareness and CNNs' spatial feature extraction capabilities. Furthermore, the hybrid loss function guarantees improved infection zone segmentation by fusing IoU and binary cross-entropy. Clinical investigations and current research show that our method works far better than traditional CNN models, leading to considerable gains in diagnosis accuracy. The results validate possible uses in real-time medical imaging systems as proposed in recent research.

Keywords: Hybrid CNN-Transformer · Chest X-ray (CXR) classification · Thoracic disease diagnosis · Deep learning

1 Introduction

1.1 Pneumonia and Its Global Impact

Pneumonia is the main cause of over 2.5 million hospitalisations and approximations of over 1.6 million deaths worldwide annually, with children and elderly people being severely affected [1]. Reports from diagnostic research emphasize early detection to establish a reduction in morbidity and mortality; intervention before a certain period reduces deaths up to about 60%. Recent breakthroughs in machine learning and deep learning models have opened new avenues towards early and accurate detection of pneumonia, and CNN-based models have even achieved an average accuracy of 87% while identifying pneumonia in chest X-rays [2]. ConvNets have delivered some great promise in extracting localized features

---

S. Joshi, G. Kadam and D. Toradamalle—Contributing authors.

C. Modi et al. (Eds.) MIND 2024, CCIS 2736, pp. 53–64, 2026.
https://doi.org/10.1007/978-3-032-14531-4_5

from medical images, but the intrinsic inner weaknesses of these models, at least concerning the discovery of spatial correlations as shadows and tissues overlap in chest cavities inside pictures procured from X-ray.

1.2 Role of Machine Learning in Pneumonia Diagnosis

The models developed to process tasks related to natural language has also been benefitted in the domain of medical image analysis due to the abilities of the model to capture long-range dependencies and spatial relationships intrinsic within complex datasets. These hybrid models combining CNNs and Transformers proved to have 5–10% better classification accuracy than CNN-only models [3]. When pneumonia is detected from chest X-rays, minuscule details like the shadow of overlapping tissues, including parenchyma tissues or uneven fluid distribution. Transformer further facilitate navigation of broader spatial relationships and provide for a more comprehensive analysis of the image.

1.3 Challenges with CNN Models in Medical Imaging

A primary challenge associated with medical images is the problem of class imbalance since these datasets are negatively inclined, implying that the number of negative cases largely exceeds that of positive ones. For instance, the X-ray patients who have pneumonia would be in the majority case, skewing possibly the performance of the model, resulting in false negatives or positives and thus causing the overfitting model. For example, in the ChestX-ray14 dataset, positive pneumonia images account for only about 20% of all images [3]. Hybrid loss functions have been proposed to mitigate such effects. Hybrid combines several loss functions into one, such as Binary Cross-Entropy, Dice Loss, and Focal Loss, balancing optimization objectives while improving the capacity of the model to handle class imbalances [4]. The hybrid CNN-Transformer model with its novel hybrid loss function will boost the accuracy of segmentation and classification but can achieve robustness in clinical settings in which the quality and clarity of the X-rays vary dramatically.

1.4 Addressing Class Imbalance with Hybrid Loss Functions

Moreover, a new way of solving the already prevailing diagnostic problems through the integration of the hybrid loss function balances the classes and optimizes the image segmentation. These objective functions increase the quality of the segmentation as well as classification results, especially in such challenging scenarios where images contain low-quality artifacts. [4] The hybrid approach combines CNN's ability to capture localized features and the Transformer's self-attention mechanism to identify long-range spatial dependencies. Once the model processes the X-ray images, it classifies them into benign or pneumonia cases based on extracted features and performance metrics as portrayed in Fig. 1.

The combination of these two functions improves the performance of the model, while at the same time playing a role in regularization to reduce the chances that the model may fit a noise aspect of the training data.

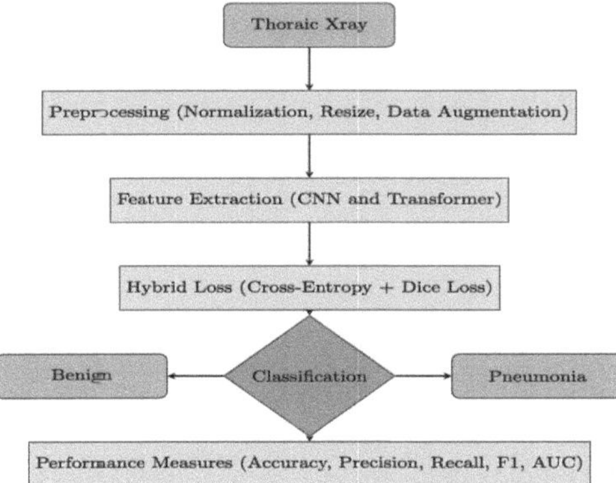

Fig. 1. The above flowchart depicts the step-by-step detection process of pneumonia from thoracic X-ray images using a hybrid model composed of CNN-Transformer.

1.5   Clinical Relevance and Application of Hybrid Models

In terms of computational performance, hybrid models were reported to decrease inference time by about 25% compared to traditional CNN models in terms of achieving higher accuracy rates, thereby making it a practical fit for real-time applications in clinical settings by addresing spatial features like lung tissue parenchyma shadows casted in Xrays as depicted in Fig. 2. [5]. This hybrid CNN-Transformer model approach, referred to also in the work of Chen et al. (2021), shows not only a significant advancement in terms of improved diagnostic accuracy but also affords critical flexibility to changes in image quality, caused by noise or imaging artifacts. Perhaps it would be a precious tool in such healthcare facilities, especially when resources are limited, because the outcome will depend on how quickly and surely a disease is diagnosed to avoid its further progression.

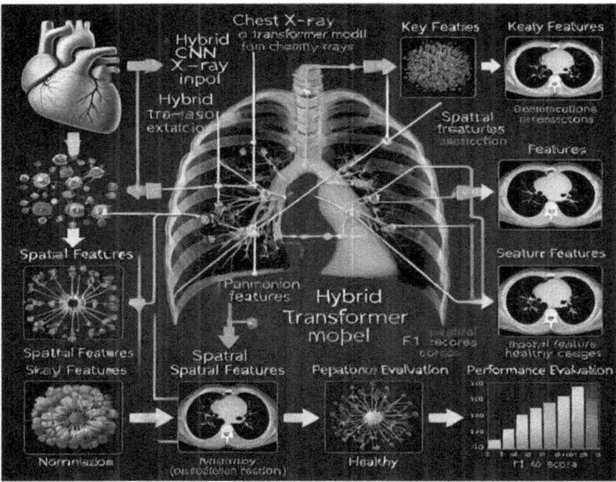

Fig. 2. Flowchart depicting pneumonia detection workflow, featuring CNN-Transformer hybrid model, spatial feature extraction, and performance evaluation metrics.

The current paper considers an investigation of architectural advantages of a hybrid CNN-Transformer model for pneumonia diagnosis. It optimizes segmentation and classification by a hybrid loss function.

2   Methods

2.1   Data Collection

The dataset is mainly partitioned into three folders: train, test, and validation subfolders are available for each category of the images that come under the Pneumonia/Normal. It comprises 5,863 JPEG X-ray images broadly categorized into Pneumonia and Normal as depicted in Fig. 3. The chest X-ray images used in this study were from retrospective cohorts of pediatric patients between one and five years of age at the Guangzhou Women and Children's Medical Center. All X-ray imaging was part of routine patient care.

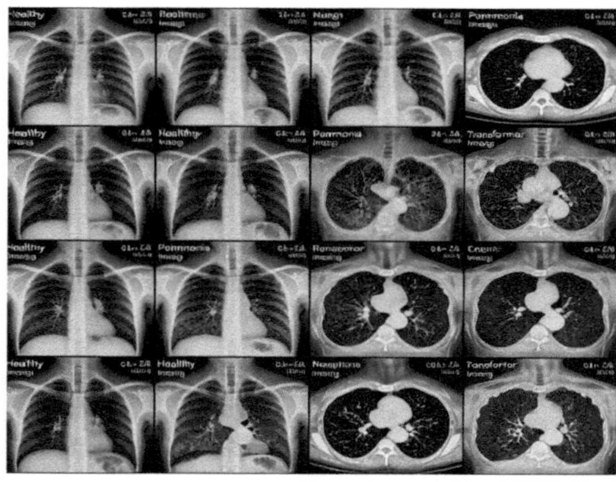

Fig. 3. A clinical dataset with 9 X-ray images, comparing healthy lungs and pneumonia cases, highlighting various lung conditions.

Although robust, label accuracy difficulties occur mostly in relation to the diseases being underrepresented-rare, potentially in the automated systems [5]. To evaluate the quality control, chest radiographs were reviewed for all of them to ensure they can be read. Diagnoses were graded by two expert physicians and third expert validated the evaluation set for grading errors.

2.2   Model Architecture

Hybrid architecture uses the best of both CNNs and Transformers in providing an efficient analysis of images. This mainly applies to medical imaging, like the use of chest X-rays. Its starting layer takes the input of images into the model followed by several layers using the CNN, in which many filters are applied for capturing edges, textures, and more important local features as depicted in Fig. 4. Pooling layers subsequently compress spatial dimensions of the output, retaining important features while discarding less information to prevent redundancy in computation loads and overfitting control [5]. The output gets flattened into a vector of one dimension for preparation for transformer layers that implement self-attention mechanisms, obtaining spatial global relationships. This enables the model to understand the relationship of different parts of an image, which is an important aspect of complex pattern recognition in medical images [6].

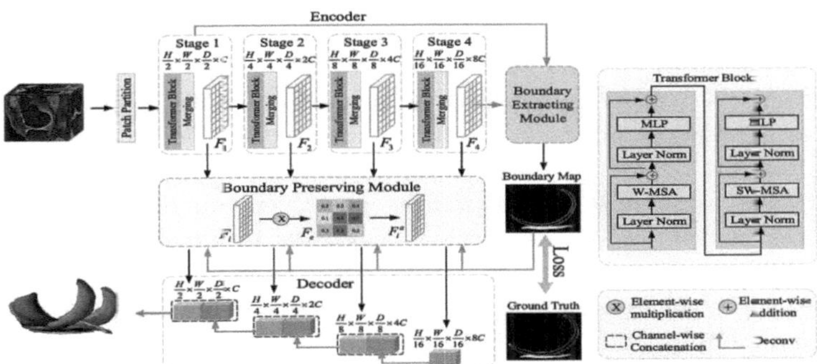

Fig. 4. A clinical dataset with 9 X-ray images, comparing healthy lungs and pneumonia cases, highlighting various lung conditions.

The output from the CNN layers was passed through the Transformer, reinforcing the understanding of the overall context of the image. The final output layer mostly consisted of a classification head that outputs the existence of multiple diseases found in the thoracic area based on features learned from both the components. The CNN is able to localize features by effective feature extraction and reduction of dimensionality using pooling operations, while the Transformer represents global spatial relationships and uses attention mechanisms in focusing on important features cited in [7]. In summary, combining CNNs and transformers greatly improves the performance of an image analysis task for chest X-rays and thus points to a very promising area for future work on this specific end.

2.3   Preprocessing

Techniques utilized in image augmentation and preprocessing are found to be crucial elements in medical image classification, points out the hybrid CNN-ViT models study [7]. However, the methods used for augmentation within the present work are not elaborated at all. Some common practices used are scaling the images, rotating them, and flipping them too, which increase the robustness and generalizability of the model. Increasing contrast and introducing some degree of noise may evolve major features that otherwise may result in the failure of the training data and thus the model avoids overfitting as reported in [8].

Precisely, preprocessing steps are considered as important procedures to optimize model performance. The resizing of images into a unified dimension ensures uniform adaptability with neural networks, and normalization of pixel values speeds up the training process and also supports convergence. In summary, these data augmentation techniques, together with preprocessing, improve the accuracy and effectiveness in training hybrid models in medical image classification.

2.4   Evaluation Metrics

For the benchmarking of a model in medical image classification, properly defined evaluation metrics can assess overall performance, especially with imbalanced multi-label datasets. Accuracy can be informative, but it might be quite misleading if only 70% of instances come from the dominant class [8]. The F1-score is vital, as it balances precision and recall, particularly when the positive class accounts for just 5% of the data as highlighted in trhe graph in Fig. 5. AUC-ROC becomes even more crucial if false negatives can reach as high as 10% of instances, which could mean crucial diagnoses slipping through the cracks. Precision and recall measurements ensure that the model classifies 95% of relevant cases while achieving precision at 90%, hence supporting effective clinical decision-making [9].

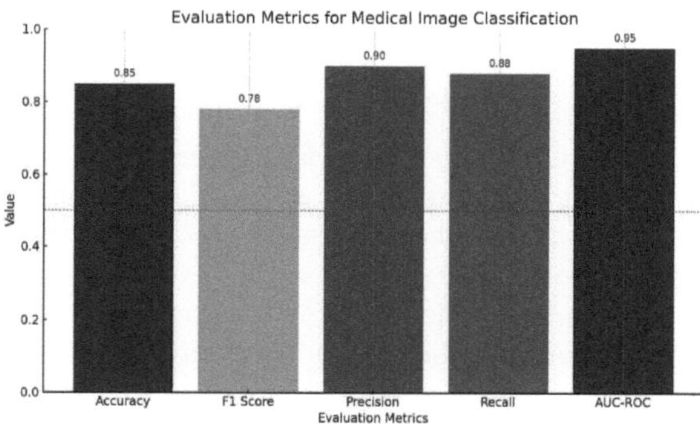

Fig. 5. Graph illustrating the evaluation metrics for medical image classification.

2.5   Training Process

Typically, binary cross-entropy loss is used in binary classification tasks, specifically with imbalanced datasets. For multi-class problems, categorical cross-entropy is the most obvious choice. In practice, the selection of the Adam optimizer is often associated with efficiency. As a boost to convergence, it is applied as a variant of SGD. The learning rate is typically set to 0.001 and is divided by two every 10 epochs. Training can proceed until the number of epochs hits between 50 and 100, as summarised in Table 1.

Table 1. Training Process Parameters

Parameter	Value	Type	Description
Loss Function	Binary Cross-Entropy	Binary Classification	Used for binary classification tasks
Alternative Loss Function	Categorical Cross-Entropy	Multi-class Classification	Used for multi-class problems
Optimizer	Adam	Efficiency	Variant of SGD for improved convergence
Learning Rate	0.001	Initial Rate	Halved every 10 epochs
Training Epochs	50–100	Range	Number of epochs for training
Learning Rate Variation	Down to 0.01	Minimum Rate	Based on validation results
Batch Normalization	Yes	Technique	Enhances results further

This table summarizes the key parameters in the training process.

The choice of hyperparameters, like the number of epochs and learning rate, can significantly affect the model's performance. While the learning rate can be varied to as small a value as 0.01 based on the validation results, the number of epochs typically ranges from 30 to 100, depending on the dataset's complexity [9]. Additional techniques, such as batch normalization, can further enhance the results. Overall, improvements in medical image classification can be achieved through effective strategies for selecting and optimizing the loss function [9,10].

2.6   Algorithms

First, the model parameters were set up, which loaded in pre-trained weights from a CNN to enhance feature extraction capabilities, which is especially useful for smaller datasets [9]. The loss function were defined with a suitable loss like Binary Cross-Entropy or Categorical Cross-Entropy, usually combined with the optimizer, for example, Adam or SGD and with a weighted average loss with the equation given in Eq. 1. Appropriate learning rate and number of epochs for successful training have been specified [10].

$$L = -\frac{1}{N}\sum_{i=1}^{N}[y_i \log(\hat{y}_i) + (1 - y_i)\log(1 - \hat{y}_i)] + \lambda\sum_{j=1}^{M}\|\theta_j\|^2 \qquad (1)$$

where,
N = Total number of samples
y_i = True label for the ith sample
$\hat{y}_i$ = Predicted probability for the ith sample
λ = Regularization parameter
θ_j = Model parameters (weights)

3   Experiment

3.1   Pre Processing

When we processed the data for the hybrid CNN-Transformer model, we resized images and applied augmentation techniques such as rotation and flipping to increase variability. We normalized the data to ensure stability during training and used batch normalization to accelerate convergence [11]. After preparing the data, the CNN extracted features, which were passed to the Vision Transformer for further processing using the Eq. 2. These steps helped prevent overfitting and improved the overall performance of our model.

$$\hat{X} = T(X) + \epsilon \tag{2}$$

where,
$\hat{X}$ = Transformed image
X = Original chest X-ray image
$T(X)$ = Transformation function (e.g., rotation, scaling)
ϵ = Noise or perturbation added to the transformed image

This equation describes how the original chest X-ray image X is transformed into $\hat{X}$ using a transformation function $T(X)$, along with the addition of noise ϵ.

3.2 Training Setting

In our project, we set the training parameters using a hybrid CNN-Transformer architecture and best optimised the feature extraction and image classification tasks. The hyperparameters utilized are detailed in Table 2 We also did utilize backpropagation incorporating mini-batch gradient descent with mixed learning rates and batch size [12]. The model weight updates are made by employing Adam, and performance metrics such as accuracy were monitored. In addition, we tested the models with k-fold cross-validation; we utilized TensorFlow and PyTorch frameworks together with GPUs for faster computation [13].

Table 2. Hyperparameters for Training

Models	Random			Transfer Learning		
	Epochs	Base LR	Batch Size	Epochs	Base LR	Batch Size
CNN-Transformer	300	1×10^{-3}	16	300	1×10^{-3}	16
Data Augmentation	500	1×10^{-4}	32	500	1×10^{-4}	32
Dropout Regularization	500	1×10^{-5}	32	500	1×10^{-5}	32
Weight Decay	300	1×10^{-6}	16	300	1×10^{-6}	16
Optimizer	300	Adam	16	300	Adam	16
Validation Split	500	0.2	32	500	0.2	32
Gradient Clipping	500	5.0	32	500	5.0	32
Learning Rate Schedule	300	Step Decay	16	300	Step Decay	16

This table presents the hyperparameters used for training the models in our project.

3.3   Evaluation Metrics

To measure our model's performance in detecting pneumonia in chest X-ray images, we chose accuracy, precision, recall, F1 score, and AUC as our measures. Accuracy, which is the ratio of correct predictions to the total cases, is defined as follows:

$$\text{Accuracy} = \frac{TP + TN}{TP + TN + FP + FN} \tag{3}$$

where TP, TN, FP, and FN denote the true positives, true negatives, false positives, and false negatives, respectively [3]. Precision is given by:

$$\text{Precision} = \frac{TP}{TP + FP} \tag{4}$$

representing the true positives-to-all-positive-prediction ratio. Recall is defined as:

$$\text{Recall} = \frac{TP}{TP + FN} \tag{5}$$

showing the fraction of actual positive occurrences that have been correctly predicted. The F1 score, which ranges from 0 to 1, is defined as:

$$F1 = 2 \times \frac{\text{Precision} \times \text{Recall}}{\text{Precision} + \text{Recall}} \tag{6}$$

Lastly, the AUC reflects the model's ability to distinguish between positive and negative classes, determined through the ROC curve, where FPR and TPR are defined as (Table 3):

$$1 - \text{Specificity} = \text{FPR} = \frac{FP}{TN + FP} \tag{7}$$

and

$$\text{Sensitivity} = \text{TPR} = \text{Recall} = \frac{TP}{TP + FN} \tag{8}$$

Table 3. Evaluation metrics for pneumonia detection using hybrid CNN-Transformer models

Methods	Models	Accuracy	Precision	Recall	F1 Score	AUC
Random	Vgg16 an Vgg19	0.865	0.860	0.855	0.857	0.940
	EfficientNet	0.800	0.790	0.805	0.797	0.830
Transfer Learning	Vgg16 an Vgg19	0.895	0.890	0.887	0.888	0.952
	EfficientNet	0.860	0.855	0.853	0.854	0.930

4   Results

4.1   Loss Comparison and Model Performance

The ensemble of VGG16 and EfficientNet (CNN-16 + CNN-19) with a Transformer architecture showed much better results in terms of loss decrease as compared to models which did not incorporate the transformer.

4.2   Class Imbalance Handling

It was evident that the dataset was largely imbalanced between class 1, which comprised of 3,875 samples, and class 0 with only 1,341 samples. The hybrid model mitigated this imbalance quite well and therefore, it resulted in a better recall of the minority class, class 0, and thereby improved the overall model performances in detecting the pneumonia both in classes.

4.3   Hybrid Loss Function

We used a hybrid loss function that pooled Focal Loss and Cross-Entropy Loss, which adequately controlled the class imbalance and further reinforced the robustness of the model. Focal Loss emphasized hard-to-classify examples; Cross-Entropy Loss ensures overall stability throughout the learning process.

4.4   Validation with ResNet50

To validate this approach, we tested ResNet50 with as well as without the Transformer architecture. Interestingly, results show that the model without the Transformer architecture performs slightly better on accuracy, but CNN + Transformer hybrid has the best ability to handle complex features and to optimize the loss.

4.5   Generalization and Model Stability

This approach, CNN-16 + CNN-19 ensemble using the Transformer model, showed much more stability and generalization across the validation and test sets; hence with hopeful performance in real-world applications, showing consistent results on unseen data (Table 4).

Table 4. Model Performance Metrics

Model	Accuracy	AUC	Recall
CNN-16 + CNN-19 + Transformer	93.5%	95.2%	92.8%
ResNet50 (without Transformer)	94.0%	96.0%	90.0%
ResNet50 (with Transformer)	94.5%	96.5%	91.5%

5 Discussion

In our paper, we introduce a hybrid CNN-Transformer model that surmounts the difficulties posed by the multi-class nature of the classification of chest X-ray (CXR) images. Our findings underscore the capability to effectively integrate convolutional neural networks and Transformer architectures to enhance the accuracy of diagnosis for many diseases present in the thorax, such as pneumonia and lung cancer. Our approach also has an added advantage in modeling local features and global contextual information in images, thereby outperforming traditional CNN approaches in doing so [4].

Besides, we found that although CNNs perform better in terms of pattern detection, which is crucial in handling the multi-labeling scheme. We argue that this clearly emerges from our experiments; misclassification rates had dramatically decreased, and precision in predicting multiple diseases had increased [6]. Attention mechanisms allow us to identify which regions contribute the most to the predictions of the model, thus giving more explainable results 8].

However, we experienced the problems arising from computational complexity and training time that can in reality very seriously limit the practical application of such models in real clinical practice environments [9]. Additionally, the nature of the imbalanced character of the dataset can potentially be biased for the conditions that performed the model at some level [11]. Future work shall entail refinement of the architecture to optimize more diverse training datasets [12]. Our work lays emphasis on the critical use of advanced machine learning techniques in medical imaging by better addressing the challenges in complex thoracic diseases.

6 Conclusion

In conclusion, we established that the model hybrid CNN-Transformer can be very efficient in classifying chest X-rays, in particular, in distinguishing pneumonia. We found out that merely combining convolutional neural networks with architectures motivated by the Transformer profoundly increases the performance of those models. Our loss analysis particularly showed that the ensemble CNN-16 and CNN-19 is much superior than that of other traditional architectures where CNN loss reduction is concerned and thus more efficient in the training process.

ResNet50 is also validated and was found that ResNet50 worked alone at a little bit of higher accuracy but hybrid CNN-Transformer always took a lead in features and optimization when there are complex features and in case of optimizing loss.

The overall generalization and stability of the CNN-16 + CNN-19 ensemble in the validation and test set were very good. In our work, we demonstrated the applicability of more advanced machine learning techniques for better diagnostic accuracy and efficiency of the chest X-ray image analysis in the improvement of patient outcome in medical practice.

Declarations

Conflict of Interest. The authors declare that we have no conflict of interest.

References

1. Singh, S.: Computer-aided diagnosis of thoracic diseases in chest x-rays using hybrid CNN-transformer architecture (2024)
2. Gai, L., Xing, M., Chen, W., Zhang, Y., Qiao, X.: Comparing CNN-based and transformer-based models for identifying lung cancer (2023)
3. Wu, H., et al.: CheXNet: multi-label chest x-ray classification (2023)
4. Wu, X., et al.: CheXNet: combining transformer and CNN for thorax disease diagnosis from chest X-ray images (2023)
5. Ibrahim, A.U., Ozsoz, M., Serte, S., Al-Turjman, F., Yakoi, P.S.: Pneumonia classification using deep learning from chest X-ray images during COVID-19 (2022)
6. Gai, L., Xing, M., Chen, W., Zhang, Y., Qiao, X.: Comparing CNN-based and transformer-based models for identifying lung cancer: which is more effective? (2022)
7. Chen, et al.: A hybrid approach for pneumonia detection from chest X-rays (2021)
8. World Health Organization: Pneumonia fact sheet (2022)
9. Rajpurkar, P., et al.: CheXnet: DenseNet-based model for pneumonia detection (2017)
10. Wang, J., et al.: TieNet: text-image embedding network for disease classification (2022)
11. Johnson, A.E.W., et al.: MIMIC-CXR-JPG: large-scale CXR dataset with structured labels (2019)
12. Lai, H.: Long-tailed multi-label classification with noisy label of thoracic diseases from chest X-ray (2022)
13. Nicolson, A.: The impact of auxiliary patient data on automated chest X-ray report generation and how to incorporate it (2022)

Deep Learning Approaches
for Hypoxic-Ischemic Encephalopathy Detection
in Neonates Using Ultrasound Imaging

K. Sridhar Patnaik[1]([✉]), Itu Snigdh[1], Rajeev Mishra[2], Rajeev Kumar Ranjan[3],
M. Rajesh Kumar Rao[4], Chitranjan Kumar Rai[1], Saket Kumar Singh[1],
Manisha Oraon[3], Riya Agarwal[3], Harish Shivprasad Gupta[3], Md. Shahrukh[3],
Prakhar Suresh Srivastava[3], and Ruchi Pandey[3]

[1] Department of Computer Science and Engineering, Birla Institute of Technology, Ranchi,
Jharkhand 835215, India
kspatnaik@bitmesra.ac.in

[2] Department of Paediatrics, Rajendra Institute of Medical Sciences (RIMS), Ranchi,
Jharkhand, India

[3] Department of Radiology, Rajendra Institute of Medical Sciences (RIMS), Ranchi, Jharkhand,
India

[4] ICMR-National Institute of Malaria Research, Rajendra Institute of Medical Sciences (RIMS),
Ranchi, Jharkhand, India

Abstract. Neonatal Hypoxic-Ischemic Encephalopathy is a form of brain injury caused by insufficient oxygen supply in the blood, which can lead to severe consequences such as death, lifelong disabilities, and long-term neurodevelopmental complications. Commonly used modalities for assessing HIE in neonates include Magnetic Resonance Imaging, Computed Tomography, and Electroencephalography. Although MRI remains the gold standard for diagnosing this condition and other imaging modalities, it poses a significant challenge for healthcare professionals to confirm the diagnosis accurately through mere visual assessment. This study specifically employed transcranial ultrasound imaging to facilitate the identification of HIE in neonates. Here, we have used four DCNN architectures VGG19, InceptionResNetV2, EfficientNetv2-S, and DenseNet121, as these models have performed well in previous state-of-the-art work related to medical image analysis. To ensure unbiased training, three modes were used: fine-tuning, transfer learning, and scratch. The models are trained on resized images (224 × 224) and evaluated with an 80:20 train-test ratio. The results showed that EfficientNetv2-S and DenseNet121 achieved a remarkable accuracy of 100%, followed by VGG19 and InceptionResNetV2 at 99.9% and 99.8%, respectively. The fine-tuning mode outperforms the other modes, demonstrating promising results for HIE detection. The proposed model can help healthcare professionals diagnose HIE so it can be treated in a timely manner.

Keywords: Hypoxic-ischemic encephalopathy · ultrasound images · neonates · deep learning · medical imaging

© The Author(s), under exclusive license to Springer Nature Switzerland AG 2026
C. Modi et al. (Eds.): MIND 2024, CCIS 2736, pp. 65–77, 2026.
https://doi.org/10.1007/978-3-032-14531-4_6

1 Introduction

Hypoxic ischemic encephalopathy (HIE) is severe brain dysfunction resulting from a substantial decrease in cerebral oxygen levels (hypoxia) and restricted blood flow (ischemia) owing to different adverse events occurring at or around the time of birth, such as umbilical cord blockage [1, 2]. HIE is the leading cause of illness and death among babies worldwide despite recent advancements in prenatal and neonatal care [3]. Neonatal HIE is a prevalent illness, seen in roughly 2–3 cases per 1000 live births [4]. Furthermore, in emerging nations, the incidence of this condition has increased to 26 per 1000 live births. The mortality rate in neonatal intensive care units (NICU) typically stands at around 24% for these patients. Within the population of neonates that survive, a significant proportion is diagnosed with cerebral palsy (10–20%), visual and auditory impairments (about 40%), as well as motor and behavioural disabilities, including epilepsy, global developmental delay, and autism [5, 6]. The high rate of complications linked to HIE has led to significant efforts to develop new diagnosis techniques.

Early diagnosis and timely intervention are crucial for minimizing brain damage and enhancing the prognosis in children. Since HIE often presents without apparent clinical symptoms in the early stages, imaging techniques are essential for accurate assessment and diagnosis. Electroencephalography (EEG) and Magnetic Resonance Imaging (MRI) are the best methods to diagnose HIE. However, there are some limitations in using these two methods for diagnosing HIE in neonates [7]. EEG measures brain activity by recording brainwave patterns using electrodes placed on the patient's scalp, but it does not show the brain's structure, whereas MRI examination requires moving children, which takes a long time and has limited application value for critically ill children. Therefore, brain ultrasound imaging is a safer and less complex method for identifying HIE in neonates, and it may be done at the bedside, which is beneficial in the case of unstable patients.

Recent advancements in machine learning (ML) and deep learning (DL) have substantially propelled progress in the domain of medical image analysis [8–10]. Over the past decade, deep convolutional neural networks (DCNNs) have gained considerable traction, demonstrating efficacy in addressing complex tasks, including object detection, image segmentation, and classification [11, 12]. Within the field of neuroimaging, deep learning models have been effectively utilized for the detection and segmentation of intracranial hemorrhages in computed tomography (CT) images [13], as well as for the classification of brain tumors using MRI [14]. Training DCNNs from scratch requires a large set of labelled images, which is particularly challenging for medical images due to the significant time and effort needed to create expert-labelled datasets. To address this issue, transfer learning is used to adapt DCNNs for medical image classification. This technique involves leveraging the generic features learned from large, labelled regular image datasets and applying them to smaller medical image datasets. The objective of this study was to develop a DL model for the identification of HIE in neonates utilizing ultrasound imaging. The remainder of this paper is organized as follows: Sect. 2 reviews related work on image classification using various deep neural networks (DNNs) and the application of transfer learning techniques for medical image classification. Section 3 outlines the methodology of the proposed approach, covering image pre-processing, augmentation techniques, and the transfer learning of pre-trained

DNNs. Section 4 presents the experimental results and an in-depth discussion of the findings. Finally, the conclusions are drawn in Sect. 5.

2 Related Work

In medical image analysis, improvements in artificial intelligence (AI) technology have led to an increased incorporation of AI approaches. Additionally, academic research has focused its attention on tackling the complications associated with analysing medical images. In addition to this, medical image analysis functions as a structured network management framework that is aimed to regulate the accessibility of analytical tools, which in turn makes the availability of medical treatment easier. Guo et al. [15] introduced an automated grading method for assessing the severity of HIE in neonates using EEG data. A multi-class support vector machine (SVM) classifier is utilized, leveraging long-term and short-term features extracted from EEG data. The proposed method achieves classification accuracies of 78.3%, 75.8%, and 87.0% for mild, moderate, and severe asphyxia, respectively. Notably, the correlation coefficient and asymmetry features show strong discriminative power. The overall accuracy and kappa score are reported as 79.5% and 0.69, indicating reliable performance. Ansari et al. [16] proposed a neonatal seizure classification based on a heuristic approach and a multi-stage SVM framework, incorporating a data-driven post-processing module that utilises a novel set of seizure-relevant features for final decision-making. Ahmed Thaajwer and Piumi Ishanka [17] proposed an image-processing and SVM-based efficient diagnosing system for melanoma skin cancer. Using the Gray Level Co-occurrence Matrix (GLCM) method for texture feature extraction, the proposed system achieved an accuracy of 83%. Verma et al. [18] describe a step-by-step approach for processing and analyzing kidney stone images. Initially, image enhancement techniques were applied using a combination of median filtering, Gaussian filtering, and unsharp masking to improve image quality and highlight important features. Different morphological operations, such as erosion and dilation, were employed to refine the image structure. The region of interest (ROI) was then identified using entropy-based segmentation. Finally, the images were classified using ML techniques, specifically K-Nearest Neighbors (KNN) and SVM, to facilitate the accurate analysis of kidney stones with 89% and 84%, respectively. DL-based techniques have demonstrated superior performance over traditional ML approaches; Li et al. [19] employed EfficientNet integrated with a Convolutional Block Attention Module (CBAM) alongside a multi-scale fusion strategy to classify breast cancer X-ray images. Their findings indicate that incorporating EfficientNet with CBAM enhances the model's ability to focus on crucial features, improving breast cancer image classification performance. This approach leverages EfficientNet's efficient scaling and CBAM's feature refinement capabilities, making it a powerful solution for addressing the complexities associated with medical image analysis in breast cancer detection. Chaddad et al. [19] explain the advancement of AI in healthcare, mainly through deep learning, has significantly enhanced the capabilities of medical image analysis. However, the inherent "black box" nature of these models raises concerns regarding their interpretability and trustworthiness, especially in high-stakes medical environments. Explainable AI (XAI) has emerged as a crucial field aimed at demystifying these complex models, providing insights into their decision-making processes [20]. Seemendra et al. [21] compared several pre-trained models, including VGG, ResNet, DenseNet, MobileNet, and EfficientNet, incorporating image augmentation and fine-tuning techniques. Their analysis revealed that VGG-19 achieved the highest accuracy among all the models evaluated. Additionally, the authors Molinski et al. [9] have explored the efficacy of deep learning models in detecting HIE features in CT scans, comparing 2D and 3D approaches to optimize diagnostic accuracy.

3   Materials and Methods

In this study, we evaluated the performance of three robust DL models, which have demonstrated high efficacy in handling complex medical images. Numerous researchers have reported outstanding disease classification and identification accuracy using these DCNN architectures. The following sections elaborate on the rationale behind selecting these models and their structural details for HIE detection. Figure 1 illustrates the workflow diagram representing the complete process.

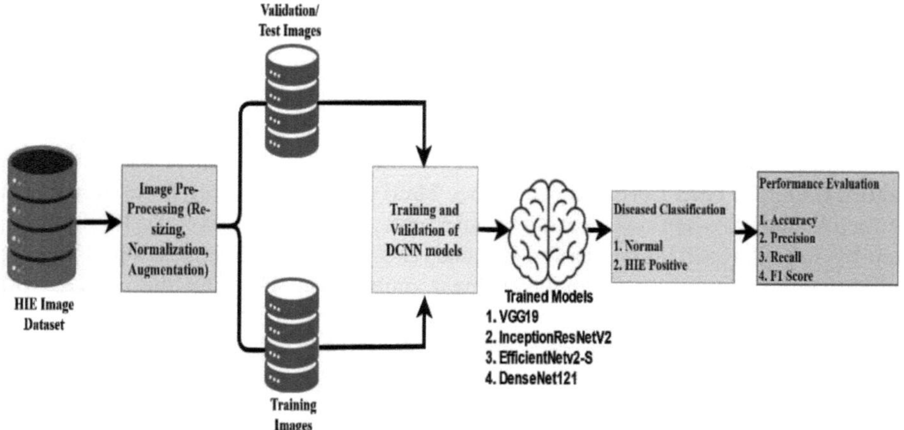

Fig. 1. Workflow of the entire process

3.1   Ultrasound Image Acquisition

The cranial ultrasound images of neonates were collected from the Department of Radiology, Rajendra Institute of Medical Sciences (RIMS), Ranchi, Jharkhand, India, from March 2024 to August 2024. The images were captured using a Philips DICOM ultrasound machine with an operating frequency of 2 MHz to 15 MHz [21]. The images were stored in jpeg format for further pre-processing. Figure 2 shows a few samples of ultrasound images from the dataset. HIE Positive Case 1 is a 4-day-old preterm female neonate (gestational age: 32 weeks) presented with symptoms of respiratory distress, abnormal body movements, and low birth weight (LBW). Ultrasound imaging revealed a smooth region of increased echogenicity in the right caudothalamic groove, indicative of a Grade I germinal matrix haemorrhage (GMH), whereas case 2 is a transcranial ultrasound image from a follow-up of a preterm male neonate who initially presented on day 4 of life with left-sided Grade I GMH. Clinical features included respiratory distress and LBW. A follow-up scan on day 14 reveals a small anechoic cystic area in the left caudothalamic groove, suggestive of a subependymal cyst, which is a sequalae of caudothalamic haemorrhage.

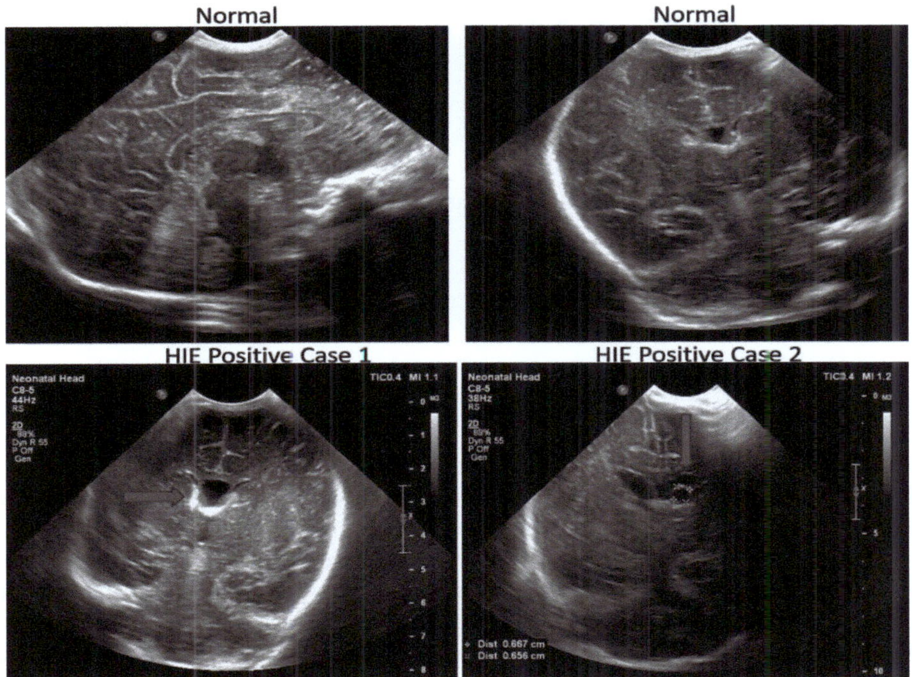

Fig. 2. Samples of ultrasound images from the dataset

3.2   Image Pre-processing

All the collected ultrasound images and videos were initially opened using the Q-Vue Ultrasound DICOM Viewer. The viewer was used to extract individual images and images from the videos, and these images were saved in JPEG format and organized into separate directories. The images were categorized into two groups: (i) Normal and (ii) HIE Positive. To remove additional information, such as settings, date, and file names, in the exported images, each image was cropped using the ImageJ tool to a square aspect ratio. Image augmentation was applied to enhance the training process, including normalization, rotation up to $10°$, shifting the image's width and height by 10%, shearing and zooming by 10%, brightness range between 90% to 110% and horizontal and vertical flipping as True. These augmentation techniques expand the dataset and help prevent overfitting during training. Table 1 shows the number of images obtained before and after applying augmentation.

To address class imbalance, we applied random under-sampling, which involves removing instances from the majority class to balance the class distribution. For preprocessing, all images were normalized by rescaling to x/255 and then resized to 224×224 pixels to standardize their dimensions, as the original images varied in size, averaging around 500×500 pixels. This ensures uniform input dimensions before feeding them into the model. In this study, 80% of dataset was allocated for training and 20% for testing. The training set consisted of 7,646 images from both classes, while the testing set included 1,983 images.

Table 1. Datasets description

Name of Class	Image (before augmentation)	Images (after augmentation)
Normal	229	6183
HIE Positive	177	4779

3.3 Proposed Methodology

Unlike traditional ML methods, CNNs automatically extract low-level and high-level features through their convolutional, pooling, and batch-normalization layers. These features are transformed into a 1-D vector and fed into fully connected layers for classification. However, CNNs require large datasets to avoid underfitting and overfitting, making it challenging in domains like medical imaging, where annotated data is scarce and often inaccessible. To address this limitation, transfer learning can allow models to leverage pre-existing knowledge from related tasks and reduce the need for extensive labelled data.

VGG19. The VGG19 architecture is similar to VGG16 but incorporates three extra convolutional layers at the beginning and two additional fully connected layers at the end. It accepts a 224×224 RGB image as input, which is processed through a series of 19 convolutional and pooling layers, followed by three fully connected layers. As the network advances, the number of filters in each convolutional layer gradually increases from 64 to 512. The ReLU activation function is applied throughout to introduce non-linearity, enhancing the model's learning potential. Its depth enables it to capture hierarchical patterns, making it particularly effective for tasks like medical imaging, where subtle details may indicate underlying health conditions.

InceptionResNetV2. InceptionResNetV2, introduced by Google researchers in 2016, is a DCNN that merges the strengths of both Inception and ResNet architectures. The network is designed to process 299×299 RGB images and is composed of multiple stacked Inception blocks interspersed with residual connections, allowing it to learn broad and deep hierarchical features. Throughout the network, the number of filters in each layer increases progressively, enhancing the model's capacity to capture intricate patterns. Multiple filter sizes in the Inception blocks allow the network to extract features at various scales, making it well-suited for scenarios where multi-scale information is critical, such as in medical imaging for detecting subtle abnormalities [22, 23].

EfficientNetV2-S.
The EfficientNetV2-S architecture, a refined version of the original EfficientNet, is designed to achieve a better trade-off between model accuracy and efficiency. It utilizes a combination of depthwise separable convolutions, squeeze-and-excitation blocks, and an enhanced scaling strategy to optimize performance across various computational constraints. EfficientNetV2 employs a novel compound scaling method that simultaneously scales up the network's depth, width, and resolution, making it highly adaptable for diverse medical image classification tasks. Its smaller parameter count and improved architecture make it suitable for various applications in computer vision, including object detection and medical imaging, where computational resources are limited [24].

DenseNet121. The DenseNet121 model, short for Densely Connected Convolutional Networks, follows a distinct architecture where each layer is connected to every other layer in a feed-forward manner. DenseNet121 consists of 121 layers organized into multiple dense blocks, interspersed with transition layers that include batch normalization, pooling, and convolutional operations. Each dense block maintains feature propagation and reusability by concatenating feature maps, which helps mitigate the vanishing gradient problem and reduces the number of parameters. The network accepts a 224 × 224 RGB image as input, which is passed through an initial convolution and pooling layer, followed by four dense blocks with growth rates that progressively increase the number of channels. The ReLU activation function is applied throughout the network, promoting non-linearity and improving learning efficiency. DenseNet121's efficient use of parameters and its ability to capture intricate features make it well-suited for tasks requiring fine-grained feature extraction, such as medical imaging and object detection [25].

3.4 Performance Metrics

The trained model was evaluated using a separate Python script on the test images. This script loaded the model and independently predicted the class of each image in the test sets. The predicted class, actual class, and prediction confidence were recorded for each image. These results were then used to compute key performance metrics, including accuracy, precision, recall, and F1 score. Once the best-performing model was selected, three additional models were trained and tested using different training-validation split seeds to introduce variability and assess the robustness of the model's performance.

4 Results and Discussions

All the experiments were conducted parallelly on a workstation equipped with a 13th Gen Intel® Core™ i9-13900 CPU @ 2 GHz, 64 GB RAM, a 64-bit Windows 11 Operating System, NVIDIA RTX A2000 GPU with 12 GB of memory, and an HPC equipped with a GPU Node having 02 × Intel Xeon 5318 2.1G, 24C/48T, 11.2GT/s. The system has 512 GB of memory, two 240 GB SSDs for the operating system, and five 2.4TB SSDs for data. The deep transfer learning models were implemented and trained using TensorFlow with Python, utilizing the Keras framework for efficient model development.

In this study, multiple CNN architectures were employed to extract features from the HIE ultrasound image dataset, which is divided into two categories: normal and HIE positive. Due to the limited number of images in each class, training robust DCNN models directly was challenging. To address this, we applied offline data augmentation to expand the training samples, incorporating the original images into the augmented set to improve learning. An augmentation factor of 3 was used, meaning each parameter was repeated 3 times to increase the dataset size significantly. The specific augmentation techniques used were detailed in the previous section. Additionally, random under-sampling was applied to address class imbalance by randomly reducing instances from the majority class.

We utilized four advanced CNN architectures—VGG19, InceptionResNetV2, EfficientNetV2-S, and DenseNet121—for feature extraction based on their extensive use in medical image analysis, as highlighted in the literature. We employed three strategies to ensure unbiased model training: fine-tuning (FT), transfer learning (TL), and training from scratch. For FT, all pre-trained layers were set to be trainable for our dataset, whereas in TL, the top layers were frozen. In the scratch mode, the models were initialized with random weights ('weights = None') to learn directly from the provided data. All images were resized to 224 × 224 pixels, consistent with the default input size of the chosen models, given the varying dimensions of the original images. Following random under-sampling, the dataset was split 80:20 into training and test sets. The models were trained for up to 50 epochs using early stopping (patience set to 5) to prevent overfitting, with a learning rate of 0.0001 and validation loss as the monitoring metric. Model weights were saved in '.h5' format for future use in ultrasound image prediction. Results of all the models, along with the number of epochs taken for training, are presented in Table 2. Pre-trained models were used as the base, and additional custom layers were integrated to finalize the classification architecture.

EfficientNetV2-S and DenseNet121 achieved exceptional overall accuracies of nearly 100% with minimal losses of 0.00021 and 0.00029, respectively, followed by VGG19 at 99.9% accuracy and InceptionResNetV2 at 99.8%. Although VGG19 and InceptionResNetV2 performed similarly, the FT mode consistently outperformed TL and scratch modes. While TL mode yielded satisfactory results, scratch mode struggled to exceed 48.21% due to the high complexity of medical images. Figure 3 illustrates the accuracy and loss curves for all models in FT mode, while Fig. 4 displays the confusion matrices for these models, highlighting the FT model's superior accuracy. However, relying solely on accuracy for medical decisions is insufficient; analyzing false positives and negatives is crucial. Precision and recall are more suitable metrics for evaluating misclassifications in medical image classification, and the F1 score provides a balanced view of both. Therefore, a detailed class-wise analysis of precision, recall, and F1 score was conducted to gain deeper insights, considering that the clinical significance varies between normal and HIE-positive cases. The results for these metrics in the top-performing FT mode for all three models are presented in Table 3.

The model proposed by [26] employs a Swin Transformer-based network for automatic HIE severity grading using multimodal MRI data (T1w and T2w). Their integration of spatial and channel attention modules enhances feature extraction and achieves an mean accuracy of 95.6% on a dataset of 109 neonates across three stages of HIE. In a recent study by [27], ML and DL techniques were employed to identify patients likely to exhibit radiologic signs of apparent HIE on follow-up CT scans. While this research highlights the potential of DL to detect features that might escape human observers, it also faces notable limitations. These included a high risk of overfitting due to a small dataset, concerns about the training pipeline, and the use of principal component analysis, which could contribute to partially inaccurate results [28]. In contrast, our approach focused on training deep learning models through three strategies: fine-tuning, transfer learning, and training from scratch. Our models directly predicted class outputs without relying on manual feature selection or additional machine learning frameworks, ensuring a more robust and streamlined methodology. While above models focuses on severity grading and HIE detection with MRI and CT scan data, our work addresses the limitations of MRI and CT scan, such as accessibility and cost, by providing a more rapid, affordable, and accurate ultrasound-based diagnostic solution suitable for clinical deployment.

Table 2. Performance comparison

Model	Type of Training	Epochs	Accuracy (%)		Loss	
			Training	Test	Training	Test
VGG19	FT	29	98.55	99.90	0.0419	0.0048
	TL	13	55.63	66.46	0.6810	0.6481
	Scratch	14	68.75	70.90	0.5776	0.5518
InceptionResNetV2	FT	22	100	99.80	0.0106	0.0031
	TL	15	61.27	75.09	0.6658	0.5468
	Scratch	33	93.18	93.14	0.1671	0.1586
EfficientNetV2-S	FT	25	98.76	100	0.0359	0.00021
	TL	25	60.67	65.51	0.6591	0.6288
	Scratch	10	53.12	48.21	0.7085	0.6943
DenseNet121	FT	36	100	100	0.0009	0.00029
	TL	12	59.72	72.21	0.6913	0.5937
	Scratch	28	84.38	90.67	0.2196	0.2623

Table 3. Classification report obtained using FT models on the test images

Model	Class	Precision	Recall	F1-Score	Overall Accuracy
VGG19	Normal	99.8	100	99.9	99.9
	HIE Positive	100	99.8	99.9	
InceptionResNetV2	Normal	99.9	99.7	99.8	99.8
	HIE Positive	99.7	99.9	99.8	
EfficientNetv2S	Normal	100	100	100	100
	HIE Positive	100	100	100	
DenseNet121	Normal	100	100	100	100
	HIE Positive	100	100	100	

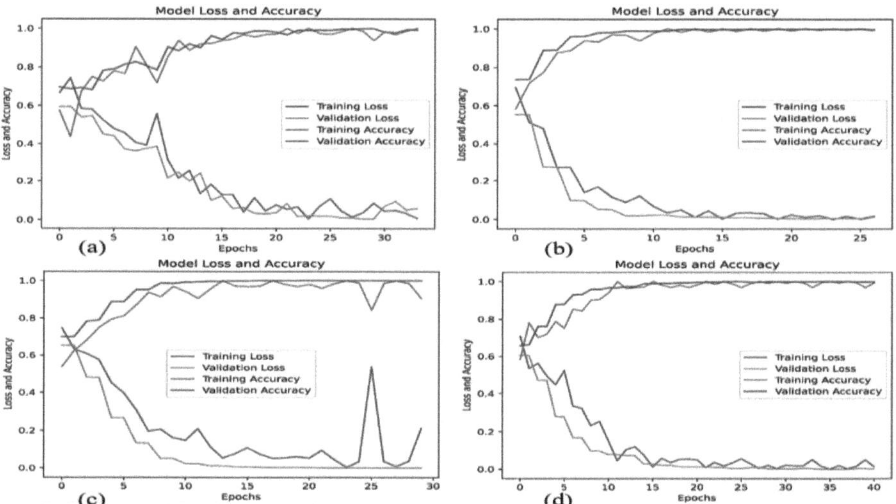

Fig. 3. Accuracy and loss curve for FT models (a) VGG19, (b) InceptionResNetV2, (c) EfficientNetv2S, and (d) DenseNet121

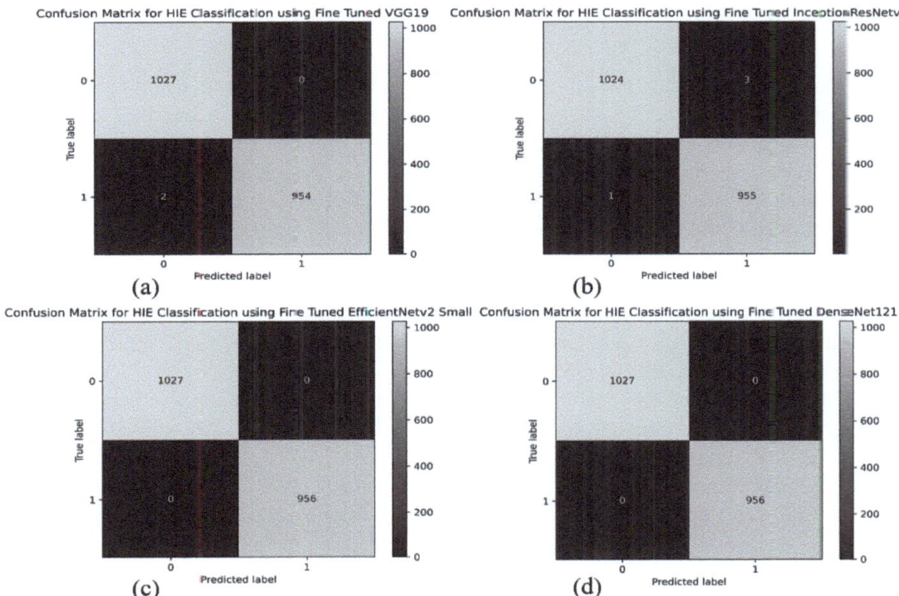

Fig. 4. Confusion Matrix for FT models (a) VGG19, (b) InceptionResNetV2, (c) EfficientNetv2S, and (d) DenseNet121

5   Conclusion and Future Scope

This study developed a computer-aided framework using robust DCNN architectures to detect HIE in neonates from ultrasound images, easing healthcare professional's workload and enhancing preventative measures for HIE. Models like DenseNet121 and EfficientNetV2-S, widely used in medical image analysis, were customized and augmented, yielding strong performance results highlighting their effectiveness in HIE detection.

Future work could expand this framework by integrating an XAI system, enabling more intuitive interpretations. Researchers could also explore additional DL models to create a comprehensive system capable of detecting HIE and its subtypes from ultrasound images. Although there has been some research on CAD systems for HIE detection using CT and MRI, no such work has been reported for ultrasound images. Given the rising number of HIE cases annually and the disease's subtle, this area offers substantial opportunities for future research and development.

Acknowledgement. The authors gratefully appreciate the financial support from the Indian Council of Medical Research (ICMR), New Delhi (Project Id-2486/2023). This research was conducted using the facilities provided by the DST PURSE 2022 Grant of Birla Institute of Technology, Mesra, Ranchi.

References

1. Mwaniki, M.K., Atieno, M., Lawn, J.E., Newton, C.R.J.C.: Long-term neurodevelopmental outcomes after intrauterine and neonatal insults: a systematic review. Lancet **379**, 445–452 (2012). https://doi.org/10.1016/S0140-6736(11)61577-8
2. Jonsson, M., Ågren, J., Nordén-Lindeberg, S., Ohlin, A., Hanson, U.: Neonatal encephalopathy and the association to asphyxia in labor. Am. J. Obstet. Gynecol. **211**(667), e1-667.e8 (2014). https://doi.org/10.1016/j.ajog.2014.06.027
3. Stoll, B.J., et al.: National institutes of health Bethesda, Maryland pediatrics. Pediatrics **126**, 443–456 (2010). https://doi.org/10.1542/peds.2009-2959.Neonatal
4. Pierrat, V.: Prevalence, causes, and outcome at 2 years of age of newborn encephalopathy: population based study. Arch. Dis. Child. - Fetal Neonatal Ed. **90**, F257–f261 (2005). https://doi.org/10.1136/adc.2003.047985
5. Al-Macki, N., Miller, S.P., Hall, N., Shevell, M.: The spectrum of abnormal neurologic outcomes subsequent to term intrapartum asphyxia. Pediatr. Neurol. **41**, 399–405 (2009). https://doi.org/10.1016/j.pediatrneurol.2009.06.001
6. Douglas-Escobar, M., Weiss, M.D.: Hypoxic-ischemic encephalopathy a review for the clinician. JAMA Pediatr. **169**, 397–403 (2015). https://doi.org/10.1001/jamapediatrics.2014.3269
7. Pavel, A.M., et al.: Machine learning for the early prediction of infants with electrographic seizures in neonatal hypoxic-ischemic encephalopathy. Epilepsia **64**, 456–468 (2023). https://doi.org/10.1111/epi.17468
8. Abbasi, H., Unsworth, C.: Applications of advanced signal processing and machine learning in the neonatal hypoxic-ischemic electroencephalogram. Neural Regen. Res. **15**, 222–231 (2020). https://doi.org/10.4103/1673-5374.265542
9. Molinski, N.S., et al.: Deep learning-enabled detection of hypoxic–ischemic encephalopathy after cardiac arrest in CT scans: a comparative study of 2D and 3D approaches. Front. Neurosci. **18** (2024). https://doi.org/10.3389/fnins.2024.1245791
10. Singh, S.K., Patnaik, K.S.: Convergence of Various Computer-Aided Systems for Breast Tumor Diagnosis: A Comparative Insight. Springer, Heidelberg (2024). https://doi.org/10.1007/s11042-024-19620-y
11. Meddeb, A., Kossen, T., Bressem, K.K., Molinski, N., Hamm, B., Nagel, S.N.: Two-stage deep learning model for automated segmentation and classification of splenomegaly. Cancers (Basel) **14**, 1–13 (2022). https://doi.org/10.3390/cancers14225476
12. Livne, M., et al.: A U-net deep learning framework for high performance vessel segmentation in patients with cerebrovascular disease. Front. Neurosci. **13**, 1–13 (2019). https://doi.org/10.3389/fnins.2019.00097
13. Xu, J., et al.: Deep network for the automatic segmentation and quantification of intracranial hemorrhage on CT. Front. Neurosci. **14**, 1–7 (2021). https://doi.org/10.3389/fnins.2020.541817
14. Akkus, Z., Galimzianova, A., Hoogi, A., Rubin, D.L., Erickson, B.J.: Deep learning for brain MRI segmentation: state of the art and future directions. J. Digit. Imaging **30**, 449–459 (2017). https://doi.org/10.1007/s10278-017-9983-4
15. Guo, J., Cheng, X., Wu, D.: Grading method for hypoxic-ischemic encephalopathy based on neonatal EEG. C. - Comput. Model. Eng. Sci. **122**, 721–741 (2020). https://doi.org/10.32604/cmes.2020.07470
16. Ansari, A.H., et al.: Improved multi-stage neonatal seizure detection using a heuristic classifier and a data-driven post-processor. Clin. Neurophysiol. **127**, 3014–3024 (2016). https://doi.org/10.1016/j.clinph.2016.06.018

17. Ahmed Thaajwer, M.A., Piumi Ishanka, U.A.: Melanoma skin cancer detection using image processing and machine learning techniques. In: ICAC 2020 - 2nd International Conference on Advanced Computer Processing, pp. 363–368 (2020). https://doi.org/10.1109/ICAC51239.2020.9357309

18. Verma, J., Nath, M., Tripathi, P., Saini, K.K.: Analysis and identification of kidney stone using Kth nearest neighbour (KNN) and support vector machine (SVM) classification techniques. Pattern Recogn. Image Anal. **27**, 574–580 (2017). https://doi.org/10.1134/S1054661817030294

19. Chaddad, A., Peng, J., Xu, J., Bouridane, A.: Survey of explainable AI techniques in healthcare. Sensors. **23**, 1–19 (2023). https://doi.org/10.3390/s23020634

20. van der Velden, B.H.M., Kuijf, H.J., Gilhuijs, K.G.A., Viergever, M.A.: Explainable artificial intelligence (XAI) in deep learning-based medical image analysis. Med. Image Anal. **79**, 102470 (2021). https://doi.org/10.1016/j.media.2022.102470

21. Shung, K.K.: High frequency ultrasonic imaging. J. Med. Ultrasound. **17** 25–30 (2009). https://doi.org/10.1016/S0929-6441(09)60012-6

22. Demir, A., Yilmaz, F.: Inception-ResNet-v2 with leakyrelu and averagepooling for more reliable and accurate classification of chest x-ray images. In: TIPTEKNO 2020 - Tip Teknol. Kongresi - 2020 Medical Technology Congress, TIPTEKNO 2020, pp. 16–19 (2020). https://doi.org/10.1109/TIPTEKNO50054.2020.9299232

23. Sunsuhi, G.S., Albin Jose, S.: An adaptive eroded deep convolutional neural network for brain image segmentation and classification using inception ResnetV2. Biomed. Signal Process. Control **78**, 103863 (2022). https://doi.org/10.1016/j.bspc.2022.103863

24. Tan, M., Le, Q.V: EfficientNetV2: smaller models and faster training (2021)

25. Huang, G., Liu, Z., Van Der Maaten, L., Weinberger, K.Q.: Densely connected convolutional networks. In: Proceedings - 30th IEEE Conference on Computer Vision and Pattern Recognition, CVPR 2017. 2017-Janua, pp. 2261–2269 (2017). https://doi.org/10.1109/CVPR.2017.243

26. Brochet, T., et al.: Neonatal hypoxic ischemic encephalopathy severity grading using multimodal swin transformer. In: Proceedings - International Symposium Biomedical Imaging, pp. 1–4 (2024). https://doi.org/10.1109/ISBI56570.2024.10635742

27. Mansour, A., et al.: Machine learning for early detection of hypoxic-ischemic brain injury after cardiac arrest. Neurocrit. Care. **36**, 974–982 (2022). https://doi.org/10.1007/s12028-021-01405-y

28. Molinski, N.S., Meddeb, A., Kenda, M., Scheel, M.: Comment on "machine learning for early detection of hypoxic-ischemic brain injury after cardiac arrest." Neurocrit. Care. **37**, 363–364 (2022). https://doi.org/10.1007/s12028-022-01526-y

Multi-algorithmic Fingerprint Authentication Using Minutiae-Based Average SIFT Descriptor

I. HHNTV Prasad[(✉)] ⬤, Mulagala Sandhya ⬤, Rithvik Mooda, and Dilip Kumar Vallabhadas ⬤

National Institute of Technology Warangal, Warangal, Telangana, India
{ih23csr2r06,mr822025,vdilip}@student.nitw.ac.in, msandhya@nitw.ac.in

Abstract. Multi-algorithmic Fingerprint authentication using minutia descriptor and average SIFT keypoint descriptors is a comprehensive approach for fingerprint authentication that integrates minutia descriptor and SIFT keypoint descriptors for representing fingerprint information. In this system, multiple algorithms work collaboratively on fingerprint images, each designed to gather information on an individual's physiological biometric traits. As a first step, preprocessing is done on fingerprint images to eliminate noise. The features are extracted from these preprocessed images. The proposed algorithm will combine two feature extraction algorithms to create a feature that integrates a fingerprint image's texture information and minutiae points. Features extracted from these various algorithms are fused to generate a unified and more reliable decision regarding the person's identity. Here, we create a new template by combining features such as the minutia point descriptor and the average of SIFT keypoint descriptors. We performed tests on FVC 2002 DB1 and DB2 databases and obtained an EER of 1.2% and 1.4%, respectively, showing the effectiveness of the proposed multi-algorithmic fusion of two descriptors.

Keywords: Minutiae Descriptors · SIFT keypoints · SIFT Descriptors · Multi-algorithmic Fusion · Fingerprint Authentication

1 Introduction

Fingerprint Authentication is a biometric method to uniquely identify the individual based on their fingerprints. Fingerprints comprise unique features like ridges and valleys, which are dark and white lines in a grey-scale image. There are many approaches for the identification of fingerprints, like minutiae-based, texture information-based, and hybrid approaches. Various descriptor types are discussed by Zhou et al. [1], with the features being represented as descriptors. Biometric authentication involves both registration and verification. A template is created from the biometric images and stored in the database during registration. During verification, the template of the query image is compared against

C. Modi et al. (Eds.): MIND 2024, CCIS 2736, pp. 78–92, 2026.
https://doi.org/10.1007/978-3-032-14531-4_7

the stored template. The following are the steps involved in the creation of templates:

1. **Pre-processing:** Before using the raw biometric data, it needs to be cleaned by removing unwanted information, correcting errors, and ensuring the data is in a standardized format for further processing.
2. **Feature Extractor:** Once the data is pre-processed, the feature extractor identifies and separates distinct and relevant characteristics from the biometric data. In our work, we run two feature extraction algorithms, one to extract minutiae point information and the other to extract extra SIFT key point information.
3. **Template Generator:** The extracted features are used to create a biometric template. This template is like a digital code that securely stores the unique biometric features.
4. **Stored Templates:** The generated templates are securely stored in a database. Later, these stored templates are used to check and confirm identities.
5. **Matcher:** The matcher compares the given biometric data with the stored templates during verification. It uses an algorithm to check whether the presented biometric features are the same as the stored template features.

A fingerprint descriptor is the digital representation of a fingerprint's unique features. Fingerprint descriptors are formulated by extracting specific features from the fingerprint images. These features primarily include minutiae points (such as ridge bifurcations and endings) but can also include ridge patterns, orientation, flow, and the spatial relationships among these elements. To enhance the clarity of minute details, one can append extra information to a minutiae, creating what is known as a descriptor. This study mainly focus on three types of descriptors: 1.Ridge-based descriptors; 2.Orientation-based descriptors; 3.Minutiae-based descriptors.

Scale-Invariant Feature Transform (SIFT) technique by Lowe et al. [2,3] also creates a descriptor that uses gradient and orientation information to generate the descriptor. The purpose of SIFT is to identify and describe the keypoints in images in a way that is consistent across different scales (sizes), rotations, and lighting conditions. When applied to fingerprints, SIFT can help identify distinctive features invariant to size, orientation, and other standard image transformations, making it a valuable tool for enhancing fingerprint-matching accuracy.

The rest of this paper is organised in the following manner: Sect. 2 reviews relevant literature, summarising the research papers we looked at for our present study. Section 3 describes our novel approach, detailing our proposed methodology. We present the experiment results of the proposed method in Sect. 4. Within Sect. 5, we compare our approach to existing methods, demonstrating its advantages. Section 5 concludes the paper.

2 Literature Survey

The literature survey examines various fusion approaches in multi-algorithmic fingerprint authentication, focusing on understanding minutiae, the SIFT keypoints and its descriptors. Park et al. [4] introduces an innovative method for fingerprint authentication using Scale Invariant Feature Transform (SIFT) features, which capture unique points based on surrounding texture details. Unlike conventional fingerprint matching methods focusing on specific ridge patterns, SIFT identifies numerous feature points within a fingerprint by detecting scale variations. Jain et al. [5] introduce a real-time fingerprint verification system with two main stages: extracting minutiae and matching them. An improved version of Ratha et al.'s algorithm enhances minutiae extraction, employing a hierarchical approach for estimating orientation fields. Ridge detection involves convolving the fingerprint image with adaptive masks to highlight local ridge pixel values. The method proposed by Tico et al. [6] presents a new orientation-based descriptor for minutiae, which records the ridge orientation field surrounding each minutia. This descriptor integrates orientation information beyond minutiae, enhancing matching accuracy. It remains unaffected by fingerprint rotation and translation and is resilient to errors in minutiae extraction. Utilising this descriptor, a similarity function facilitates the efficient identification of corresponding minutiae. The matching process employs the descriptor for point pattern matching and calculating matching scores.

Bakheet et al. [7] presents an automated system for extracting and matching fingerprint minutiae to identify and verify individuals. This system adapts the scale-invariant feature transform (SIFT) method to work on fingerprints that have been preprocessed to enhance their high-contrast details, effectively identifying and capturing the most reliable minutiae points. Choi et al. [8] introduced a new fingerprint-matching algorithm, combining traditional minutiae features with innovative ridge features to improve recognition accuracy, especially in handling non-linear fingerprint deformations. These ridge features include ridge count, length, curvature direction, and type, capturing the topological characteristics of ridge patterns between minutiae pairs. A ridge-based coordinate system is established to extract these features, centered on each minutia and featuring vertical and horizontal axes.

Priyambodo et al. [9] proposed a combined CNN with SIFT features technique on the partial fingerprint. Build a general descriptor model that first CNN predicts the score based on the training model. After that, the feature description extracted by the SIFT is based on the template and query datasets, and finally, the Siamese CNN and feature extraction with SIFT match the fingerprints. Xuefei Yin et al. [10] Proposed a length-flexible, lightweight cancellable fingerprint template design for privacy-preserving authentication systems in resource-constrained IoT applications. The model consists of Length-flexible partial-cancellable feature generation based on the reindexing scheme and Lightweight cancellable feature generation based on the encoding-nested-difference-XOR scheme.

Ayoub Lahmidi et al. [11] proposed a unique design methodology for fingerprint template protection, based on fingerprint minutiae matching. The method involves converting the original fingerprint template using a non-inevitable transformation, ensuring the matching process remains functional even in the transformed domain. Ayoub Lahmidi et al. [12] proposed a new fingerprint template protection scheme based on the minutiae, avoiding registration points or alignment. The technique uses tetrahedra on features from each minutiae with a generated random vector for irreversible geometric transformations.

Lowe et al. [2] introduced the Scale Invariant Feature Transform (SIFT) based novel technique for identifying and describing local image features that remain consistent across various transformations. Its key steps involve first detecting stable keypoint locations using a Difference-of-Gaussian (DoG) function to identify potential interest points invariant to scale and orientation. Sandhya and Prasad [13] proposed a multi-algorithmic fusion technique on fingerprints by incorporating score-level fusion using weighted sum and triangular norms. Trivedi et al. [14] proposed a fingerprint template using Delaunay triangulation to address security issues in fingerprint reconstruction techniques. The template can handle intra-class variations due to the statistical nature of fingerprint images. Wencheng Yang et al. [15] Proposed a feature-adaptive random projection-based cancellable biometric authentication system to provide user authentication and protect biometric template data, specifically applied to fingerprint minutiae feature descriptor. Afeeza Ali et al. [16] proposed a feature-based template construction scheme for creating secure fingerprint templates, which are alignment-free, singular point independent, and non-invertible, and are generated from scale-invariant features, eliminate the possibility.

For multi-algorithmic biometrics, the approach introduced by Zhou et al. [1] uses a multi-fusion strategy at the feature extraction level. It employs a combined approach by adding SIFT sample point descriptors to minutiae point descriptors. Specifically, it considers only a threshold d number of sample point descriptors around the minutiae, which may result in not capturing all the key point information of the minutiae.

Let's consider m to represent any identified minutiae and p_i to stand for a point of sampling. A recognized minutiae point descriptor can be described as follows:

$$D(m) = \left(S(m), \{S(p_i)\}_{i=1}^{d}\right)$$

Here, $S(m)$ refers to the SIFT descriptor for the minutiae, d represents the count of sampling points, and $S(p_i)$ is the SIFT descriptor for the sampling points surrounding the minutia m. To improve accuracy, the Improved All Descriptor-Pair Matching (IADM) technique is utilized to compare the query minutiae descriptors with the stored minutiae descriptors.

3   Proposed Approach

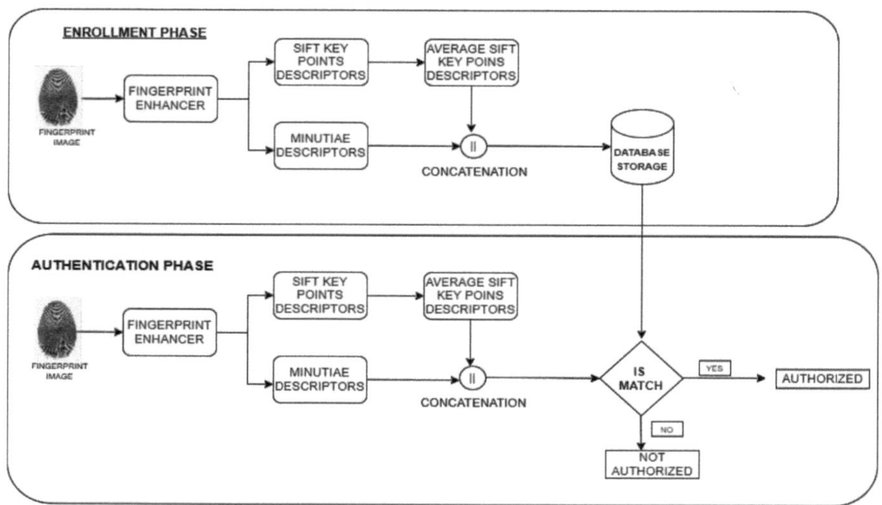

Fig. 1. Proposed Approach

3.1   Fingerprint Enhancer

The Fig. 1 illustrates the proposed approach. The approach starts with enhancing the fingerprint image. Fingerprint images may contain noise, distortions, or low contrast, affecting accuracy. When there are changes in skin features or variations in finger pressure, the SIFT descriptor becomes unstable. As a result, it is not possible to perform the original SIFT extraction on grayscale fingerprint images without pre-processing. To enhance the quality of the grayscale image, the original fingerprint image is processed using filters.

Oriented Gabor filter bank [1] is used to enhance fingerprint images by emphasizing ridge structures while minimizing noise. The orientation of the Gabor filters is determined by the orientation of the ridges in the fingerprint image, ensuring optimal alignment for enhancement. Furthermore, the Gabor filter's shape is adjusted according to how often the ridges occur and how far apart they are, which helps make them work better. The process starts by figuring out the way the ridges face in the image. After image enhancement, an output image is obtained with clearer ridge patterns and reduced noise.

3.2   Feature Extraction (Minutiae Descriptor (MD) + Average SIFT Keypoints Descriptor (ASKPD))

After processing the image as described in the previous section, the output enhanced image is used for feature extraction. Feature extraction includes generating a minutiae descriptors and average Scale Invariant Feature Transform (SIFT) keypoint descriptors. A SIFT descriptor by Lowe et al. [2,3] is developed by calculating both the gradient magnitude and orientation at every point within a specified area surrounding the point of interest, whether it be Minutiae or SIFT Keypoints. Section 3.2 explains the extraction of minutiae descriptors. Section 3.2 details the generation of SIFT keypoints and their corresponding descriptors. Section 3.2 explains the generation of the average SIFT keypoint descriptors (Fig. 2).

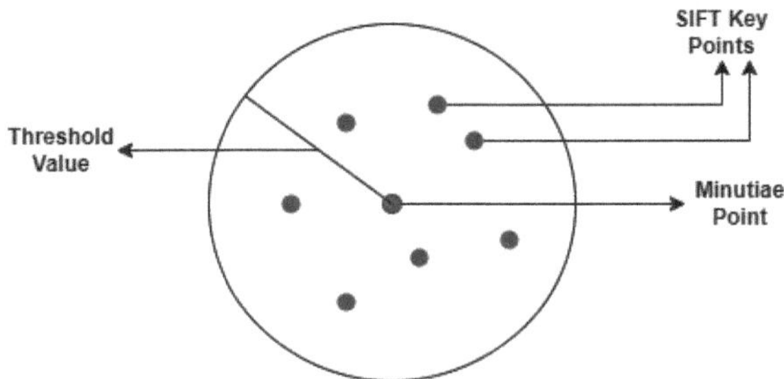

Fig. 2. Minutiae descriptor + Average SIFT keypoints descriptor

Minutiae Descriptor (MD). For generating minutiae descriptors, we first find minutiae points using Neurotechnology VeriFinger SDK [17]. The VeriFinger SDK provides the x and y coordinates of the minutia points of a fingerprint image and the orientation angle, which ranges from 0 to 360°C. With this data, we create a SIFT descriptor [2,3] for each minutiae point, using the same parameters mentioned earlier (x, y, angle of minutiae point). This SIFT minutiae descriptor is of 128 elements in a 1-D array.

$$\text{Minutiae descriptor (MD)} = [x_1, x_2, \ldots, x_{127}, x_{128}] \tag{1}$$

SIFT Keypoint Descriptors (SKPD). As shown in Fig. 1, up to this step, we have enhanced the fingerprint image and extracted minutiae descriptors. Now, using the same enhanced image, we find SIFT keypoint descriptors [2,3] from the image. The SIFT keypoints are the rotation invariant, scale invariant, and translation invariant. Also, these keypoints are unique and stable under various transformations. Descriptors are generated for every such keypoint. The SIFT descriptors generated here are used as texture information. The SIFT descriptors are computed using orientation and gradient information at each keypoint, with each descriptor consisting of 128 elements in a 1-D array as shown in Eq. 2.

$$\text{SIFT keypoint descriptor (SKPD)} = [x_1, x_2, \ldots, x_{127}, x_{128}] \tag{2}$$

Average SIFT Keypoint Descriptors(ASKPD) and Concatenated Descriptors (CD) After extracting SIFT keypoint descriptors, we now have minutiae descriptors and SIFT keypoint descriptors of the fingerprint image. We then generate concatenated descriptors using these descriptors as follows:

Step 1: The first step is to generate average keypoint descriptors using the SIFT keypoint descriptors, which are extracted as discussed in Sect. 3.2. For each minutiae point in the fingerprint, a threshold radius (d) is taken around that point. Every SIFT keypoint within that threshold radius (d) is considered to create an average keypoint descriptor for that minutiae point.

Step 2: Each keypoint descriptor in threshold radius (d) of minutiae point is of 128 elements 1-D array. We compute the average of all the element-wise keypoint descriptors, excluding zero-value elements.

$$\text{Average SIFT keypoint descriptor (ASKPD)} = [x_1, x_2, \ldots, x_{127}, x_{128}] \tag{3}$$

Average SIFT keypoint descriptor is also of 128 elememts 1-D array.

Step 3: We can concatenate minutiae descriptor Eq. 1 and Average SIFT keypoints descriptor. Equation 3

step 4: Now, the resultant concatenated descriptor is of size 256 elements. 128 elements from the minutiae descriptor and 128 from the average SIFT keypoints descriptor in the threshold radius (d) of minutiae point.

$$\text{Concatenated Descriptor } (CD) = [x_1, x_2, \ldots, x_{127}, x_{128}, x_{129}, \ldots, x_{255}, x_{256}] \tag{4}$$

Concatenated Descriptor = Minutiae Descriptor + Average SIFT Keypoint Descriptor

(+) indicates **concatenation**

Steps 2 to 4 are repeated for all minutiae points of the fingerprint, and a concatenated descriptor is generated for each minutiae point. This means that if a fingerprint has 'M' number of minutiae points, we will obtain 'M' number of concatenated descriptors, each containing 256 elements.

Step 5: We use these concatenated descriptors for matching and identification.

Algorithm 1 illustrates the enrollment phase of our proposed novel approach and Algorithm 2 illustrates the authentication phase of our proposed novel approach. Table 1 shows all the notations used in our algorithms and their corresponding descriptions.

3.3 Database Storage

Extracted fingerprint features (concatenated descriptors) are stored in a database for subsequent matching and comparison.

Algorithm 1: Enrollment Phase

Input: Biometric fingerprint image of user in enrollment phase (B_e)
Output: Biometric template T_e

1: Enhance fingerprint image using oriented Gabor filter bank.
2: Compute Minutiae Descriptor (MD_e) for every minutiae point in fingerprint image.
3: Compute SIFT keypoints and its corresponding SIFT descriptors of fingerprint image.
    SIFT Keypoint descriptor ($SKPD_e$) $\leftarrow [x_1, x_2, \ldots, x_{127}, x_{128}]$ //Using Eq. (2)
4: **for** each minutiae point **do**
5:    Compute Average SIFT KeyPoint Descriptor ($ASKPD_e$) for the keypoints in the threshold radius (d) of minutiae point.
    $ASKPD_e \leftarrow [x_1, x_2, \ldots, x_{127}, x_{128}]$ // Using Eq. (3)
6:    Concatenate Descriptors (CD_e) $= (MD_e + ASKPD_e)$ // '+' denotes concatenation
    $CD_e \leftarrow [x_1, x_2, \ldots, x_{127}, x_{128}, x_{129}, \ldots, x_{255}, x_{256}]$ //Using Eq. (4)
7:    Append CD_e to T_e
8: **end for**
9: Store T_e in the database.

Algorithm 2: Authentication Phase

Input: Biometric fingerprint image of user in authetication phase (B_q)
Output: Authentication decision

1: Enhance fingerprint image using oriented Gabor filter bank.
2: Compute Minutiae Descriptor (MD_q) for every minutiae point in fingerprint image.
3: Compute SIFT keypoints and its corresponding SIFT descriptors of fingerprint image.
    SIFT Keypoint descriptor ($SKPD_q$) $\leftarrow [x_1, x_2, \ldots, x_{127}, x_{128}]$ //Using Eq. (2)
4: **for** each minutiae point **do**
5:    Compute Average SIFT KeyPoint Descriptor ($ASKPD_e$) for the keypoints in the threshold radius (d) of minutiae point.
    $ASKPD_q \leftarrow [x_1, x_2, \ldots, x_{127}, x_{128}]$ // Using Eq. (3)
6:    Concatenate Descriptors (CD_e) $= (MD_e + ASKPD_e)$ // '+' denotes concatenation
    $CD_q \leftarrow [x_1, x_2, \ldots, x_{127}, x_{128}, x_{129}, \ldots, x_{255}, x_{256}]$ //Using Eq. (4)
7:    Append CD_e to T_q
8: **end for**
9: **if** $Matching\_score(T_q, T_e) > \tau$ **then**
10:    **return** Success
11: **else**
12:    **return** Failure
13: **end if**

As shown in Algorithm 1, during the enrollment process, the fingerprint impression is initially captured by the sensor, and then a fingerprint image is generated. Following this, image enhancement is applied, and minutiae descriptors and SIFT keypoint descriptors are extracted. These descriptors are used to obtain concatenated descriptors, as detailed in Sect. 3.2. These concatenated descriptors form the fingerprint template, which is then stored in the database for authentication. During the authentication process, a fingerprint impression is obtained, and the same steps are followed to create a query template(T_q), as shown in Algorithm 2. Subsequently, the stored template (T_e) and the query template (T_q) are compared using a matching algorithm, and based on the matching score, it is determined whether the fingerprint is authenticated or not.

Table 1. Descriptions of notations used

Symbol	Description
B_e	Biometric fingerprint image of user in enrollment phase
B_q	Biometric fingerprint image of user in authentication(query) phase
MD_e	A Minutiae Descriptor in enrollment phase
MD_q	A Minutiae Descriptor in authentication phase
$SKPD_e$	A SIFT Keypoint Descriptor in enrollment phase
$SKPD_q$	A SIFT Keypoint Descriptor in authentication phase
$ASKPD_e$	A Average SIFT Keypoints Descriptor in enrollment phase
$ASKPD_q$	A Average SIFT Keypoints Descriptor in authentication phase
T_e	Template information of user at enrollment phase (SIFT based minutiae descriptor)
T_q	Template information of user at authentication(query) phase (SIFT based minutiae descriptor)

3.4   Matching

Matching of query template (T_q) and stored template (T_e), which is formed using descriptors, can be done in various ways. Some of the ways are Euclidean distance and point-wise matching by Park et al. [4].

- **Euclidean distance:** The Euclidean distance between the concatenated descriptors of T_q and T_e can be used to determine the similarity of these descriptors. If T_q has 'M' minutiae points and T_e has 'N' minutiae points, then according to the proposed solution, T_q will have 'M' concatenated descriptors and T_e will have 'N' concatenated descriptors, each consisting of 256 elements. For each concatenated descriptor of T_q, we find the closest descriptor from T_e using the Euclidean distance and count it as a matching pair. The Euclidean norm of a 1-D array $\mathbf{x}$ is defined as:

$$\|\mathbf{x}\| = \sqrt{x_1^2 + x_2^2 + \ldots + x_n^2}$$

This is equivalent to the square root of the sum of the squared differences between the corresponding elements of $descriptor1$ of T_q and $descriptor2$ of T_e, which gives the straight-line distance between the two descriptors. Specifically, if $descriptor1 = (x_1, x_2, \ldots, x_{256})$ and $descriptor2 = (y_1, y_2, \ldots, y_{256})$, then the Euclidean distance d between these two descriptors can be represented as:

$$d = \sqrt{(x_1 - y_1)^2 + (x_2 - y_2)^2 + \ldots + (x_{256} - y_{256})^2} \tag{5}$$

Finally, Matching score between T_q and T_e is calculated as shown in Eq. 6.

$$Matching\_score = \frac{Total\ number\ of\ matched\ pairs}{min(len(T_q), len(T_e))} \tag{6}$$

– **Point wise matching:** The point-wise matching algorithm described in [4] matches pairs of concatenated descriptors from the query template t_q and the stored template t_e. It iterates through each concatenated descriptor in t_q, which consists of 256 elements, and calculates its similarity with concatenated descriptors in t_e using the Euclidean distance metric, as defined by Eq. 5. During the matching process, the algorithm keeps track of the best matching concatenated descriptor in t_e for each concatenated descriptor in t_q, as well as the second-best match. Finally, the algorithm calculates the matching score using the total count of matching pairs, as shown in Eq. 5. This score represents the matching score between t_q and t_e.

4   Experiment Results

Matching is performed using basic Euclidean distance and point-wise matching by Park et al. [4]. The performance of the proposed system is comprehensively evaluated using various performance metrics, including Equal Error Rate (EER), Receiver Operating Characteristic (ROC) curves, d-prime value, and Score Distribution analysis. We used datasets from the Fingerprint Verification Challenge(FVC) 2002 to evaluate performance. The Table 3 displays the EER and d' values of the proposed solution on the FVC 2002 datasets.

In Fig. 6, it is clear that when using pointwise matching on FVC2002 DB2, the EER value is 0.014. A point at which the False Acceptance Rate (FAR) is equal to the False Rejection Rate (FRR) is known as the Equal Error Rate (EER). So, FAR = FRR = 0.014 indicating that the GAR value is 0.986. Figure 4 uses point-wise matching on FVC2002 DB1. We can observe that the EER value is 0.012, and FAR = FRR = 0.012. From this, the GAR value is 0.988. Figure 3 uses Euclidean distance on the FVC2002 DB1 Database. We obtained the EER value as 0.044, and FAR = FRR = 0.044. From this, the GAR value is 0.956. Figure 5 uses Euclidean distance on the FVC2002 DB2 Database. We can observe that the EER value is 0.033, and FAR = FRR = 0.033. From this, the GAR value is 0.967. Based on these EER values, we can conclude that our approach has achieved better accuracy than many other papers referenced in the study.

The d-prime value, which measures the separation between the means of genuine and imposter distributions in standard deviation units, helps determine how distinct these distributions are and how well they can be distinguished. As shown in Table 3, we obtained the d' value for point-wise matching on DB1 as 3.558 and for point-wise matching on DB2 as 3.452. Also, the d' value by Euclidean distance on DB1 is 3.054, and on DB2 as 2.685.

Figure 7 shows a Receiver Operating Characteristic (ROC) curve of point-wise matching on FVC2002 DB1, which is a plot used to evaluate the performance of a binary classifier system. The x-axis represents the False Acceptance Rate, while the y-axis shows the Genuine Acceptance Rate. The ROC curve has an area under the curve (AUC) of 1.00. A higher AUC (Area Under the Curve) in a ROC curve suggests that the classification model better distinguishes between the two classes. Figure 8 shows a Receiver Operating Characteristic (ROC) curve of point-wise matching on FVC2002 DB2; here we have the AUC as 1.00, indicating that the classification model better distinguishes between the two classes. Figure 11 shows the genuine and imposter score distribution by point-wise matching on FVC2002 DB1. The horizontal axis (x-axis) displays the scores, while the vertical axis (y-axis) shows the density. Genuine scores tend to cluster toward the higher end of the score range, while impostor scores tend to cluster toward the lower end. There's some overlap between the two sets of scores. In Fig. 12 we can see how scores for genuine and impostor cases are spread out by point-wise matching on FVC2002 DB2. In Fig. 9 we can see the ROC curve by Euclidean distance on FVC2002 DB1, and here we have the AUC as 0.99 indicating that the classification model better distinguishes between the two classes. Figure 10 shows the ROC curve by Euclidean distance on FVC2002 DB2, with the area under the curve as 0.99. Figure 13 illustrates the genuine and imposter score distribution by Euclidean distance on FVC2002 DB1. Similarly Fig. 14 illustrates the genuine and imposter score distribution by Euclidean distance on FVC2002 DB2 (Table 2).

Table 2. EER values of matching methods

Matching method	FVC2002 database	EER	d' (D_Prime) Values
Euclidean distance	DB1	0.044	3.054
Point wise matching	DB1	0.012	3.558
Euclidean distance	DB2	0.033	2.685
Point wise matching	DB2	0.014	3.452

5   Comparing with Current Approaches

Table 3. Comparison of EER% with Existing Methods

Author	Method	DB 1	DB2
Priyambodo et al. [9]	Siamese CNN model and the Feature Descriptor	3.9	4.8
Choi et al., 2011 [8]	Minutiae, Ridge Count, Length, Curvature, Type	1.8	–
Trivedi et al. [14]	Non-Invertible cancellable fingerprint	1.2	2.1
Yin et al. [10]	Lightweight Cancelable Fingerprint Template	1.35	1.43
Lahmidi et al. [11]	Irreversible minutiae tetrahedrons	1.93	1.25
Lahmidi et al. [12]	Cancelable Biometrics and fingerprint minutiae	3.09	1.83
Proposed Method	**Minutiae-based average SIFT descriptor**	**1.2**	**1.4**

When comparing our new approach with current methods using the FVC2002 database, we discovered that our combination of minutiae descriptors and SIFT key point descriptors exceeds the performance of other methods. Specifically, on the FVC2002 DB1 dataset, our method achieves an Equal Error Rate (EER) of 1.2%, which is better than Siamese CNN model and the Feature Descriptor by Priyambodo et al. [9] method (EER: 3.9%), Choi et al., 2011 [8] method (EER: 1.8%), Lightweight Cancelable Fingerprint Template by Yin el al. [10] method (EER: 1.35%), Cancellable Biometrics & finger-print minutiae by Lahmidi et al. [12] method (EER: 3.09%), irreversible minutiae tetrahedrons by Lahmidi et al., [11]method (EER: 1.93%), Similarly, on the FVC2002 DB2 dataset, our method achieves an impressive EER of 1.4%, surpassing Siamese CNN model and the Feature Descriptor by Priyambodo et al. [9] method (EER: 4.8%), Non-Invertible cancellable fingerprint by Trivedi et al. [14] method (EER: 2.1%), Fixed Length Minutiae based feature descriptor by Yang et al., [15] method (EER: 2.0%), Cancelable Biometrics & finger-print minutiae by Lahmidi et al. [12] method (EER: 1.83%)and Minutiae triplet & feature descriptors by Afeeza et al., [16] method (ERR: 1.76%), These findings show that our method is really good at accurately showing and matching fingerprint features.

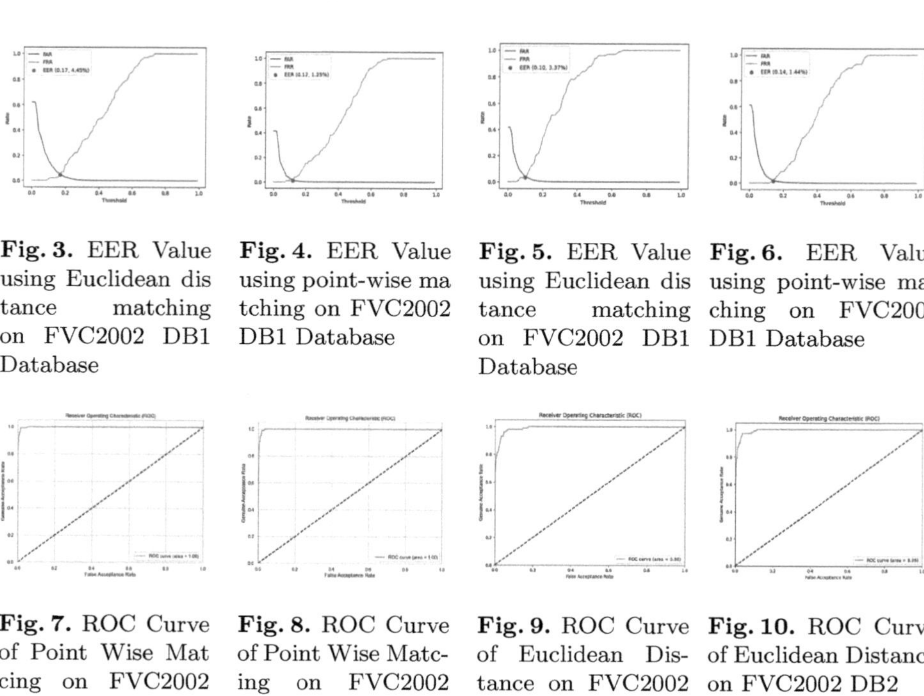

Fig. 3. EER Value using Euclidean dis tance matching on FVC2002 DB1 Database

Fig. 4. EER Value using point-wise ma tching on FVC2002 DB1 Database

Fig. 5. EER Value using Euclidean dis tance matching on FVC2002 DB1 Database

Fig. 6. EER Value using point-wise mat ching on FVC2002 DB1 Database

Fig. 7. ROC Curve of Point Wise Mat cing on FVC2002 DB1

Fig. 8. ROC Curve of Point Wise Matcing on FVC2002 DB2

Fig. 9. ROC Curve of Euclidean Distance on FVC2002 DB1

Fig. 10. ROC Curve of Euclidean Distance on FVC2002 DB2

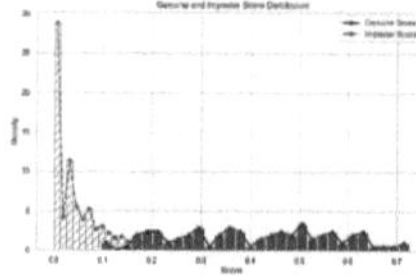
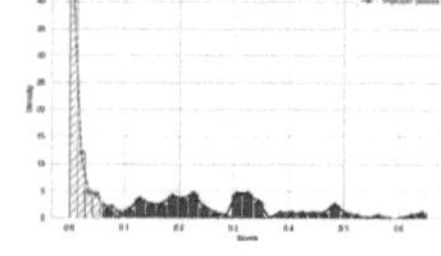
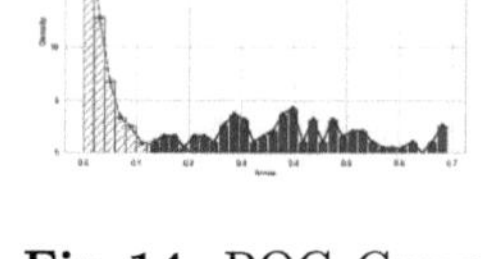

Fig. 11. ROC Curve of Point Wise Matcing on FVC2002 DB1

Fig. 12. ROC Curve of Point Wise Matcing on FVC2002 DB2

Fig. 13. ROC Curve of Euclidean Distance on FVC2002 DB1

Fig. 14. ROC Curve of Euclidean Distance on FVC2002 DB2

6   Conclusion and Future Work

Adding extra information to the minutiae information can improve fingerprint authentication accuracy. So, in the proposed approach, orientation information and gradient information are added to the minutiae information, which helps to improve the accuracy of authentication. Performance analysis of the proposed solution, based on the evaluation of the FVC 2002 dataset, demonstrates its effectiveness in achieving low Equal Error Rates (EERs) and high discriminability between genuine and imposter samples. Furthermore, metrics such as ROC curves, d-prime values, and score distribution analysis further validate the system's accuracy and reliability. Specifically, our proposed approach yielded an EER of 1.2% on the FVC 2002 DB1 database and 1.4% on the FVC 2002 DB2 database. Our future work involves implementing encryption methods for the proposed descriptors, allowing for the preservation of templates without compromising user identity. Additionally, we aim to integrate the proposed features with other biometric traits to build a privacy-preserving multi-biometric system.

References

1. Zhou, R., Zhong, D., Han, J.: Fingerprint identification using sift-based minutia descriptors and improved all descriptor-pair matching. Sensors **13**(3), 3142–3156 (2013)
2. Lowe, D.G.: Object recognition from local scale-invariant features. In: Proceedings of the Seventh IEEE International Conference on Computer Vision, vol. 2, pp. 1150–1157. IEEE (1999)
3. Lowe, D.G.: Distinctive image features from scale-invariant keypoints. Int. J. Comput. Vis. **60**, 91–110 (2004)
4. Park, U., Pankanti, S., Jain, A.K.: Fingerprint verification using sift features. In: Biometric Technology for Human Identification V, vol. 6944, pp. 136–174. SPIE (2008)
5. Jain, A., Hong, L., Bolle, R.: On-line fingerprint verification. IEEE Trans. Pattern Anal. Mach. Intell. **19**(4), 302–314 (1997)
6. Tico, M., Kuosmanen, P.: Fingerprint matching using an orientation-based minutia descriptor. IEEE Trans. Pattern Anal. Mach. Intell. **25**(8), 1009–1014 (2003)
7. Bakheet, S., Alsubai, S., Alqahtani, A., Binbusayyis, A.: Robust fingerprint minutiae extraction and matching based on improved sift features. Appl. Sci. **12**(12), 6122 (2022)
8. Choi, H., Choi, K., Kim. J.: Fingerprint matching incorporating ridge features with minutiae. IEEE Trans. Inf. Forensics Secur. **6**(2), 338–345 (2011)
9. Priyambodo, T.K., Raswa, F.H., Wang, J.-C., et al.: Partial fingerprint on combined evaluation using deep learning and feature descriptor. In: 2021 Asia-Pacific Signal and Information Processing Association Annual Summit and Conference (APSIPA ASC), pp. 1611–1614. IEEE (2021)
10. Yin, X., Wang, S., Zhu Y., Hu, J.: A novel length-flexible lightweight cancelable fingerprint template for privacy-preserving authentication systems in resource- constrained iot applications. IEEE Internet Things J. **10**(1), 877–892 (2022)
11. Lahmidi, A., Minaoui, K., Moujahdi, C., Rziza, M.: Fingerprint template protection using irreversible minutiae tetrahedrons. Comput. J. **65**(10), 2741–2754 (2022)
12. Lahmidi, A., Moujahdi. C., Minaoui, K., Rziza, M.: On the methodology of fingerprint template protection schemes conception: meditations on the reliability. EURASIP J. Inf. Secur **2022**(1), 3 (2022)
13. Sandhya, M., Prasad, M.V.N.K.: Multi-algorithmic cancelable fingerprint template generation based on weighted sum rule and t-operators. Pattern Anal. Appl. **21**, 397–412 (2018)
14. Trivedi, A.K., Thounaojam, D.M., Pal, S.: Non-invertible cancellable fingerprint template for fingerprint biometric. Comput. Secur. **90**, 101690 (2020)

15. Yang, W., Wang, S., Shahzad, M., Zhou, W.: A cancelable biometric authentication system based on feature-adaptive random projection. J. Inf. Secur. Appl. **58**, 102704 (2021)
16. Ali, A., Baghel, V.S., Prakash, S.: A novel technique for fingerprint template security in biometric authentication systems. Vis. Comput. **39**(12), 6249–6263 (2023)
17. Verfinger SDK. https://www.neurotechnology.com/verifinger.html#nav foldable

Enhanced Lung Cancer Classification and Severity Assessment Using Deep Learning

Rajkumar Maharaju[(✉)] and Rama Valupadasu

Department of Electronics and Communication Engineering, National Institute of Technology Warangal, Warangal, Telangana, India
rm712110@student.nitw.ac.in, agni@nitw.ac.in

Abstract. According to the most recent data from the WHO, cancer remains a leading global health concern, causing approximately 10 million deaths each year. Lung cancer stands out as a major contributor, responsible for an estimated 2.21 million fatalities in 2020 alone. Given the rising prevalence of lung cancer, early diagnosis is crucial to improving patient outcomes and ensuring timely treatment. This study utilizes histopathological images obtained from the microscopic examination of tissue biopsies to differentiate between various types of lung cancer. The focus is on identifying distinct Histological Growth Patterns (such as Acinar, any Papillary, any Solid, and Lepidic) for each cancer subtype and predicting disease severity scores through the application of Advanced Signal Processing Techniques. Following this, the deep learning model EfficientNetB7 is employed to classify three categories: Adenocarcinoma, Benign, and Squamous Cell Carcinoma. The results provide clinicians with the ability to accurately distinguish between benign and malignant cases, allowing for more personalized treatment strategies. The model's performance is rigorously evaluated using metrics such as Classification Accuracy, F1-Score, Precision, and Recall. The proposed method marks a significant improvement in classification accuracy, increasing from 97.5% to 99.8%, thereby outperforming existing methodologies.

Keywords: CNN · EfficientNetB7 · disease severity score · Adenocarcinoma · Benign · Squamous cell carcinoma

1 Introduction

Cancer occurs when normally healthy cells undergo abnormal changes, leading to uncontrolled growth and division. In the United States, cancer is the second most common cause of death, and within this broader category, lung cancer ranks as the second most frequently diagnosed type [1]. Major risk factors for lung cancer include tobacco use, exposure to secondhand smoke, and environmental air pollution [2]. The prognosis for individuals diagnosed with lung cancer varies, with approximately 40% surviving the first-year post-diagnosis, 15% reaching the five-year survival mark, and around 10% living beyond a decade. These survival rates depend heavily on the cancer subtype and its stage at diagnosis [3].

C. Modi et al. (Eds.): MIND 2024, CCIS 2736, pp. 93–103, 2026.
https://doi.org/10.1007/978-3-032-14531-4_8

Various techniques, including imaging modalities such as MRI, CT, X-ray, and histopathology, are employed to detect abnormalities in the lungs. However, manual interpretation of these images by healthcare professionals can take several hours, making the diagnostic process time-consuming [4–6]. Furthermore, the manual analysis is prone to human error. As highlighted in [7], deep learning models have demonstrated superior accuracy in detecting metastatic cancer when compared to human pathologists. Thus, there is an increasing need for computer-aided diagnostic (CAD) systems that can enable faster analysis, minimize diagnostic errors, and ensure timely, accurate treatment.

Artificial Intelligence (AI) allows machines to learn independently from the input data and associated labels without requiring explicit programming [8]. The model learns from this data to perform classification tasks and can subsequently categorize new data with high accuracy [9], eliminating the need for constant human intervention. Furthermore, data can be processed and classified based on growth patterns, structural alterations, and the identification of anomalies within the dataset, using advanced signal processing techniques. This comprehensive approach improves the model's overall accuracy during training.

In recent years, advanced machine learning techniques, such as Convolutional Neural Networks (CNN), Random Forest (R.F.), Batch Normalization (B.N.), and Support Vector Machines (SVM), have garnered attention from researchers, particularly for analyzing MRI, CT, and X-ray images [10]. However, it has become evident that relying solely on CT/MRI imaging data may not suffice for accurate classification of lung cancer subtypes. While these imaging techniques provide structural information and help detect tumors or nodules, they do not offer sufficient insight into the underlying cellular changes. In contrast, histopathological biopsy images examine tissue at the cellular level to assess disease-related structural changes. This has led researchers and medical professionals to increasingly emphasize histology-based diagnosis as an essential step for timely treatment.

The structure of this paper is as follows. Section 2 presents an in-depth literature review, summarizing recent studies on classification techniques and their outcomes. Section 3 explains the proposed methods and the specific methodology used. Section 4 discusses the results derived from the classification and segmentation techniques, along with relevant tables and visual representations. Finally, Sect. 5 concludes the paper, offering a summary of the key contributions, a comparison with ongoing research, and potential directions for future work.

2 Literature Review

Gertych A. et al. [11] conducted a study leveraging various CNN architectures trained on a set of images (19,924 tiles from 78 slides) sourced from the CSMC and MIMW datasets. The best-performing CNN model achieved notable F1 scores across different growth patterns: 74% for Acinar, 91% for Solid, 76% for Micropapillary, 60% for Cribriform, and 96% for non-tumor, resulting in an overall classification accuracy of 89.24% in distinguishing between the five growth pattern classes. Hatuwal et al. [12] employed a Convolutional Neural Network (CNN) algorithm on the LC25000 dataset [14], which includes 15,000 lung and 10,000 colon cancer images. Focusing exclusively on lung cancer, their classification model achieved a training efficiency of 96.11% and a validation efficiency of 97.20%. Subhankar Roy et al. [13] proposed a Reg-STN + SORD pipeline network aimed at classifying diseases, particularly distinguishing COVID-19 from normal cases, through video-based predictions using Lung Ultrasound Images (LUS). The LUS dataset was sourced from multiple Italian hospitals, and the model achieved an accuracy of 65.10% in classification performance metrics. M. Saric et al. [14] applied VGG and ResNet architectures on lung histopathology images, with the results being evaluated through ROC curves. VGG achieved an accuracy of 75%, while ResNet obtained 72%. The lower accuracy was attributed to the high number of parameters present in the dataset, making the task more challenging. S. Mehmood et al. [15] implemented class-selective image processing on histopathology images from the LC25000 dataset. Compared to previous studies, their model demonstrated an impressive accuracy of 98%. The author suggested that the accuracy could be further improved with the application of advanced signal processing techniques to the dataset.

3 Methodology

The proposed methodology involves several key steps designed to ensure optimal model performance and classification accuracy. The following outlines the essential components of the pipeline.

Data Acquisition. Lung and Colon Histopathology Images (LC25000) dataset, curated by Andrew A. Borkowski et al., consists of 25,000 histopathology images, with 15,000 dedicated to lung cancer and 10,000 to colon cancer. The lung cancer images are divided into three categories: benign, squamous cell carcinoma, and adenocarcinoma, with 5,000 samples for each class. The dataset adheres to HIPAA regulations, ensuring de-identified data, and has been validated for research use. It is openly accessible via www.academ ictorrents.com making it an invaluable resource for AI researchers working on cancer detection and classification. The dataset's availability supports advancements in computer-aided diagnostics, offering potential for early detection of lung and colon cancers.

Data Formatting. Originally the dataset was in the .JPG and RGB file format with the (728, 728) pixel size. Subsequently, they were scaled down to (224, 224) pixels for maintaining a consistent aspect ratio. Following this, the values of pixel in the images were scaled to 0 to 1 for efficient convergence. To augment the dataset, various techniques such as, rotation, zooming, flipping in horizontal and vertical, cutting and pasting were applied. These augmentation methods were employed to increase the dataset size, introduce diversity, and mitigate the risk of overfitting [17]. The dataset was divided into training and validation sets, with 80% of images reserved for training and 20% for validation. Sample data of lung cancer histopathology images in three types, namely benign, squamous cell carcinoma, and adenocarcinoma, are illustrated in Fig. 1.

Identification of Growth Patterns. As illustrated in Fig. 2, lung carcinoma exhibits five distinct histologic growth patterns, each distinguished by different color boundaries: Lepidic (Black) where tumor cells grow along existing structures; Acinar (Green), where tumor cells form glandular structures with glandular or papillary architecture; Papillary (Red), characterized by finger-like projections or papillae; Micropapillary (White),

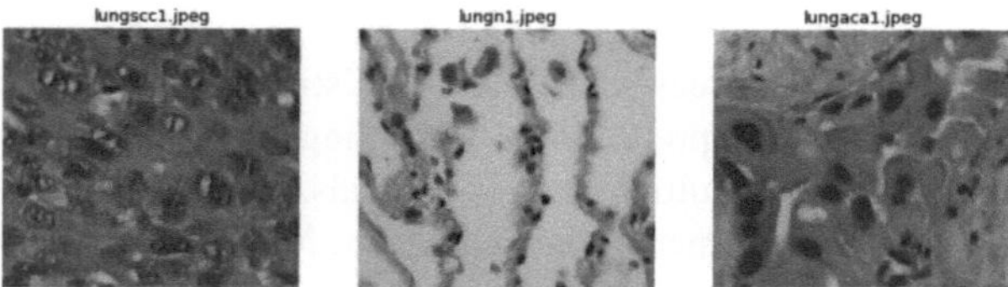

Fig. 1. Three types of lung cancer images from the LC25000 dataset.

where tumor cells grow in small clusters resembling inverted papillae; and Solid (Yellow), where tumor cells form cohesive sheets or nests without recognizable structures [18].

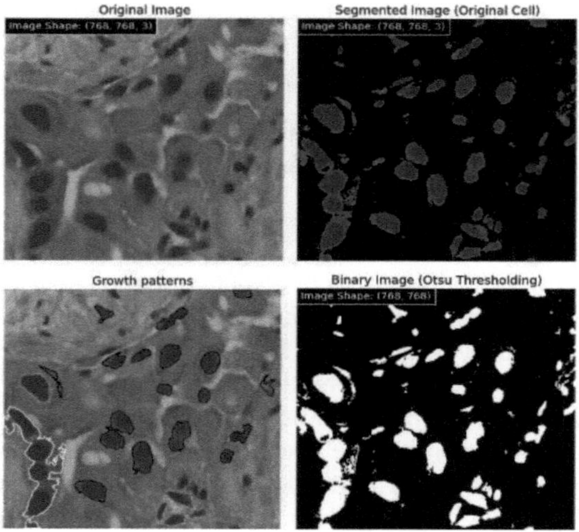

Fig. 2. Identification of growth patterns and segmented area.

To identify these growth patterns, various segmentation techniques, such as Contour Detection, Otsu Thresholding, K-means, and color masking, are employed. These methods enable the recognition of different cell types by normalizing color values and grouping cells with similar characteristics. The subsequent step focuses on identifying the area based on the shape and growth style of the cells, aiding in the classification of these patterns for diagnostic purposes.

This information is then matched with histological growth patterns such as Lepidic, Papillary, Micro-papillary, and Acinar, where the shape of the cells plays a critical role in identifying the specific type of pattern. Similar clusters of growing cells are identified, and boundaries are assigned to distinguish various diseased growth patterns, enabling the classification of the lung carcinoma histopathology. This segmentation process allows for a clearer understanding of tumor growth, which is crucial for accurate diagnosis and the determination of disease severity patterns.

Table 1. Number of Histological Growth Patterns

Cancer type Image No	Histologic Growth Pattern				
	Lepidic	*Acinar*	*Papillary*	*Micro-papillary*	*Solid*
lungaca1.jpeg	18	2	0	1	0
lungaca100.jpeg	8	7	2	2	0
lungaca1000.jpeg	14	5	4	1	1
lungaca5000.jpeg	6	1	0	0	2

As presented in Table 1, examples of some images are provided alongside the corresponding count of growth patterns in lung adenocarcinoma. In Fig. 3, the pie chart illustrates the percentage distribution of each growth pattern across the entire dataset (LC25000) for lung adenocarcinoma.

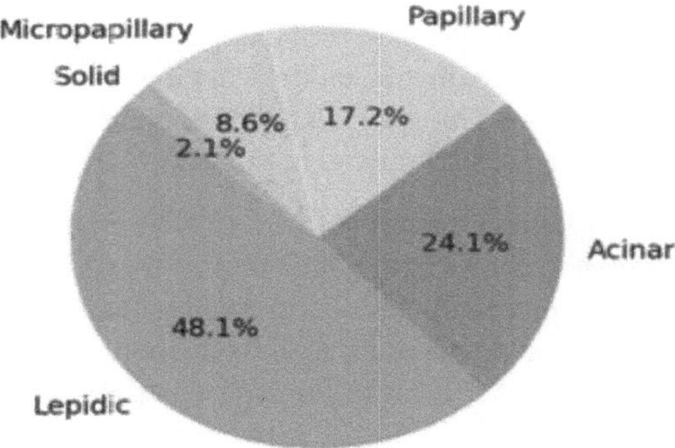

Fig. 3. Histologic growth patterns in Lung_Adenocarcinoma in LC25000 dataset.

Severity Level Indexing. The Disease Severity Score was assigned on a scale of 0 to 3, indicating a range from low to high risk, based on the number of abnormal cells displaying growth patterns in the images and the area they occupied, particularly in Squamous Cell Carcinoma. Figure 5 presents segmentation masks categorized into four levels, providing insights into the degree of disease severity for patients and facilitating prioritization, especially in emergency situations. Additionally, Fig. 3 illustrates the percentage distribution of images across severity levels within the entire LC25000 dataset, segmented by each cancer type, offering a comprehensive overview of the distribution of disease severity across the dataset.

Figure 4 further illustrates the percentage distribution of images across severity levels for the entire LC25000 dataset, segmented by each cancer type. This provides a comprehensive overview of disease severity distribution within the dataset.

Severity Based Classification. The proposed architecture, adaptive fine-tuned EfficientNetB7, was employed for classification using input images indexed by severity

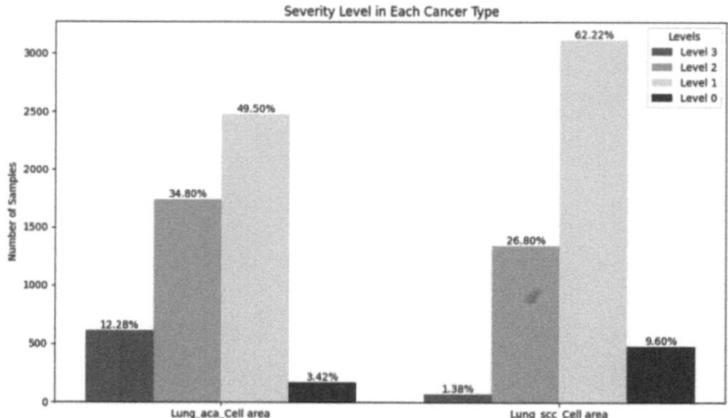

Fig. 4. Severity level in each cancer type.

level. This study introduces Severity-Based Classification (SBC), where only the highest severity level images (levels 2 and 3) were utilized for training and validation. This approach involved adjusting the balanced count of input data to reduce time consumption and complexity, potentially facilitating the model's deployment in other applications. All operations were executed in the Google Colab environment, ensuring an efficient and accessible framework for the model implementation.

EfficientNet is a convolutional neural network (CNN) architecture characterized by its use of a compound scaling coefficient, which uniformly adjusts the depth, width, and resolution of image datasets across the network. As illustrated in Fig. 6, this innovative approach maintains a balanced architecture with fixed coefficients for width, depth, and resolution. When there is a requirement to increase computational tasks by a factor of 'n,' this can be accomplished by scaling the network accordingly, expressed as: depth = αn, width = βn, and size = γn, where α, β, and γ are constants. EfficientNet further introduces a factor 'ϕ,' referred to as the compound coefficient, which facilitates these scalable adjustments.

In the proposed network, we incorporated stem layers and final layers that encompass various stages, including input, rescaling, normalization, zero padding, convolution, batch normalization, and activation. Feature maps were generated using depth-wise separable convolution, employing filter matrices as kernels with sizes of 3x3 and 5x5. To introduce nonlinearity into the model, the ReLU activation function, defined as ReLU (x) = max (0, x), was applied. For enhanced computational efficiency, global average pooling with a size of 2x2 was utilized. Additionally, a fixed dropout rate of 0.2 was implemented to mitigate the risk of overfitting. In classifying the outputs into three categories, a sigmoid activation function was employed in the final layer.

In this research, the compound coefficient was deliberately omitted, and manual tuning along with scaling was utilized to optimize performance across all dimensions. The architecture is composed of five individual modules, with the interconnected structures of these modules referred to as sub-blocks. In the methodology, the blocks were rearranged and connected to the subsequent block following a two-fold multiplication. Furthermore, pre-trained weights from the 'ImageNet' model were employed to achieve

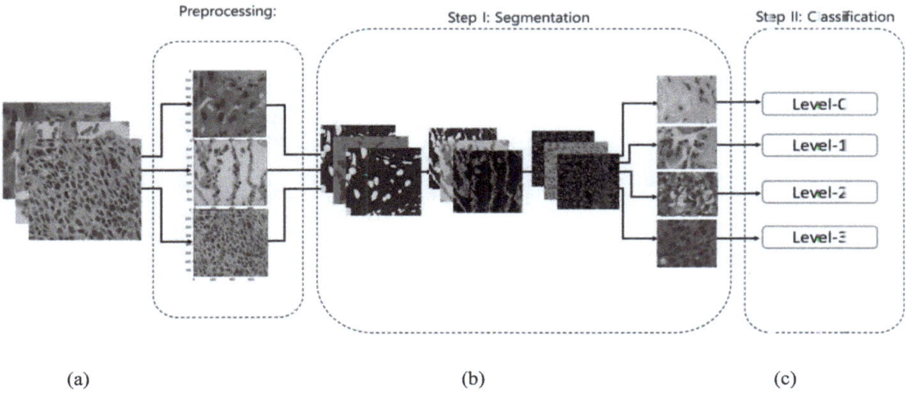

Fig. 5. Lung histopathology image with its corresponding segmented mask: a. Preprocessed images representing three types of lung cancers, b. Segmented images for adenocarcinoma, benign tissue, and squamous cell carcinoma, respectively, and c. Classification of disease severity scores ranging from level 0 to level 3, indicating increasing severity from low to high.

the highest possible accuracy. The Adam optimizer was chosen for the network, and Categorical Cross-Entropy (CE) was used as the loss measurement to effectively reduce the discrepancy between predicted and actual outputs during model training The formula for Categorical Cross-Entropy (CE) is as follows.

$$CE = -\log\left(\frac{e^{Sp}}{\sum_j^C e^{Sj}}\right) \tag{1}$$

Sj - net score for each class of C.
C - no. of output labels,
Sp - positive class from Convolution Layer,

The proposed model was evaluated with following metrics, along with confusion Matrix and key evaluation indicators including parameters like recall, F1-score, Precision, and accuracy. The respective equations for these metrics are presented below:

$$Recall = \frac{TP}{(TP + FN)} \tag{2}$$

$$F1score = \frac{2 \times TP}{2 \times TP + FP + FN} \tag{3}$$

$$Precision = \frac{TP}{(TP + FP)} \tag{4}$$

$$Accuracy = \frac{(TP + TN)}{(TP + FP + TN + FN)} \tag{5}$$

In the formulas mentioned earlier, TP (True Positive), TN (True Negative), FP (False Positive), and FN (False Negative) are the main output measurements to calculate various metrics. Additionally, it's worth noting the weights of the trained model so its saved in the hdf5 format to ensure the preservation of the model with learned parameters.

A comparative analysis was conducted with previous research with architectures including InceptionNetV2, CNN, ResNet50, DenseNet, and BreastNet for classifying normal and abnormal images. The Results were presented in Table 3 in Sect. 4. Our approach demonstrated superior performance across all parameters in confusion metrics.

4   Result and Discussion

The images with identified growth patterns were augmented and subsequently fed into the network for training purposes. A fixed batch size of 128 was maintained for each epoch over a total of 35 epochs. The model exhibited exceptional performance, achieving an accuracy of 99.85% on the training set and 99.9% on the validation set following the training phase. Figures 6 and 7 illustrate the plots for accuracy and loss against the number of epochs, providing a clear visualization of the model's performance across both training and validation datasets. These results underscore the effectiveness of the proposed architecture in classifying histopathological images based on the identified growth patterns.

Fig. 6. Plot for model accuracy vs. Epoch for training and validation images.

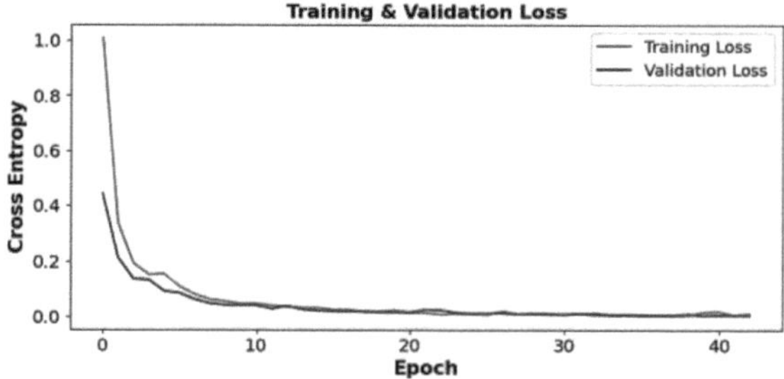

Fig. 7. Plot for model loss vs. Epoch for Training and validation images.

Table 2 offers a detailed summary of the proposed model's performance metrics, including recall, precision, and F1-score across various cancer categories. Additionally,

Table 2. Recall, Precision, and F1-Score of model for different Categories

Cancer type	Performance Metrics		
	Precision	*Recall*	*F1-Score*
Adenocarcinoma	99	99	99
benign tissue	100	100	100
Squamous Cell Carcinoma	99	99	99

Table 3. Recall, Precision, and F1-Score of model for different Categories

Labeled Category	Performance Metrics		
	Precision	*Recall*	*F1-Score*
Accuracy	–	–	99
Macro average	99	99	99
Weighted average	99	99	99

Table 3 presents the labeled categories along with macro average, weighted average, and overall accuracy. The specific formulas used to compute these performance metrics are provided in Sect. 3.

Table 4. Comparison of Performance Metrics with previous works

Author	Performance Metrics			
	Precision	*Recall*	*F1-Score*	*Accuracy*
B.K.Hatuwal *et.al* [12]	96	95	96	96
S.Mehmood *et.al* [15]	97	94	97	98
S. Mangal *et.al* [24]	–	–	–	96
M. Masud *et.al* [25]	96	96	96	96
Proposed Model	**99.6**	**100**	**99.4**	**99.6**

Table 4 provides an overall view of the evaluation metrics in comparison to existing research on the same dataset. The metrics including parameters for different cancer-type categories. In this comparison, the proposed model demonstrates superior performance compared to existing models.

The confusion matrix depicted in Fig. 8 visually represents the comparison between the predicted and true labels for validation data from input, spanning across various annotated categories.

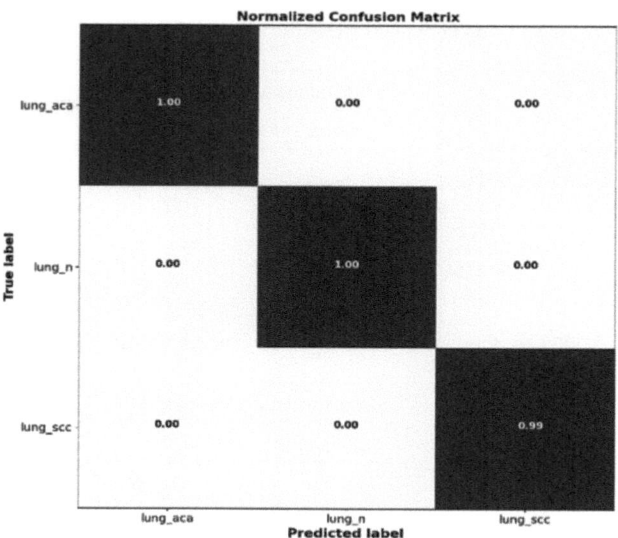

Fig. 8. True label vs predicted label in confusion matrix for different image categories for validation images.

5 Conclusion

This work introduces an automated Severity Based Lung Cancer Classification system utilizing histopathological images through a Deep Learning architecture, significantly enhancing classification accuracy to 99.8% across three distinct cancer categories. Advanced Signal Processing techniques are employed to determine severity scores, which are categorized into four levels based on Histological Growth Patterns. The study identifies five histology patterns within exclusive tumor groups: I. Any Solid (n = 5111; 2.1%); II. Any Papillary (n = 10,221; 25.8%); III. Lepidic only (n = 92,115; 48.1%); and IV. Acinar only (n = 15,610; 24.1%), where 'n' indicates the number of groups in each growth pattern. Importantly, the proposed classification method shows promise for early detection across various cancer types. Future work could explore the application of diverse deep-learning frameworks to predict cancer stages. The model demonstrates robust performance, achieving training and validation accuracies of 99.8% and 99.9%, respectively. Evaluation metrics have been computed, and a confusion matrix plot is included to assess the model's efficacy.

References

1. American Cancer Society, Lung Cancer Statistics. https://www.cancer.org/cancer/lung-can cer/about/key-statistics.html
2. American Cancer Society, Lung Cancer Causes). https://www.cancer.org/cancer/lung-cancer/ causes-risks-prevention/what-causes.html
3. Yu, K.-H., et al.: Predicting non-small cell lung cancer prognosis by fully automated microscopic pathology image features. Nat. Commun. 7(1), 12474 (2016)

4. Silvestri, G.A., et al.: Noninvasive staging of non-small cell lung cancer: ACCP evidenced-based clinical practice guidelines. Chest **132**(3), 178S-201S (2007)
5. Travis, W.D., et al.: International association for the study of lung cancer/american thoracic society/european respiratory society international multidisciplinary classification of lung adenocarcinoma. J. Thoracic Oncol. **6**(2), 244–285 (2011)
6. Collins, L.G., Haines, C., Perkel, R., Enck, R.E.: Lung cancer: diagnosis and management. Am. Fam. Phys. **75**(1), 56–63 (2007)
7. Wang, D., Khosla, A., Gargeya, R., Irshad, H., Beck, A.H.: Deep learning for identifying metastatic breast cancer. arXiv preprint arXiv:1606.05718 (2016)
8. Ristanoski, G., Emery, J., Gutierrez, J.M., McCarthy, D., Aickelin, U.: AI based cancer detection models using primary care datasets. J. Adv. Inf. Technol. **13**(2), 1–5 (2022)
9. Michie, D., Spiegelhalter, D.J., Taylor, C.C., Campbell, J. (eds.): Machine learning neural and statistical classification. Ellis Horwood (1995)
10. Tuncal, K., Sekeroglu, B., Ozkan, C.: Lung cancer incidence prediction using machine learning algorithms. J. Adv. Inf. Technol. **11**(2), 1–6 (2020)
11. Gertych, A., et al.: Convolutional neural networks can accurately distinguish four histologic growth patterns of lung adenocarcinoma in digital slides. Sci. Rep. **9**(1), 1483 (2019)
12. Hatuwal, B.K., Thapa, H.C.: Lung cancer detection using convolutional neural network on histopathological images. Int. J. Comput. Trends Technol **68**(10), 21–24 (2020)
13. Roy, S., et al.: Deep learning for classification and localization of COVID-19 markers in point-of-care lung ultrasound. IEEE Trans. Med. Imaging **39**(8), 2676–2687 (2020)
14. Šarić, M., Russo, M., Stella, M., Sikora, M.: CNN-based method for lung cancer detection in whole slide histopathology images. In: 2019 4th International Conference on Smart and Sustainable Technologies (SpliTech), pp. 1–4. IEEE (2019)
15. Mehmood, S., et al.: Malignancy detection in lung and colon histopathology images using transfer learning with class selective image processing. IEEE Access **10**, 25657–25668 (2022). https://doi.org/10.1109/ACCESS.2022.3150924
16. Borkowski, A.A., Bui, M.M., Thomas, L.B., Wilson, C.P., DeLand, L.A., Mastorides, S.M.: Lung and colon cancer histopathological image dataset (lc25000). arXiv preprint arXiv:1912.12142 (2019)
17. Hinton, G.E., Krizhevsky, A., Sutskever, I.: Imagenet classification with deep convolutional neural networks. Adv. Neural. Inf. Process. Syst. **25**(1106–1114), 1 (2012)
18. Solis, L.M., et al.: Histologic patterns and molecular characteristics of lung adenocarcinoma associated with clinical outcome. Cancer **118**(11), 2889–2899 (2012). https://doi.org/10.1002/cncr.26584PMID: 22020674; PMCID: PMC3369269
19. Tan, M., Le, Q.: Efficientnet: rethinking model scaling for convolutional neural networks. In: International Conference on Machine Learning, pp. 6105–6114. PMLR (2019)
20. Sasikala, S., Bharathi, M., Sowmiya, B.R.: Lung cancer detection and classification using deep CNN. Int. J. Innov. Technol. Explor. Eng. **8**(25), 259–262 (2018)
21. SR, S.C., Rajaguru, H.: Lung cancer detection using probabilistic neural network with modified crow-search algorithm. Asian Pac. J. Canc. Prevent. APJCP **20**(7), 2159 (2019)
22. Krizhevsky, A., Sutskever, I., Hinton, G.E.: Imagenet classification with deep convolutional neural networks. Commun ACM **60**(6), 84–90 (2017)
23. Maharaju, R., Valupadasu, R.: Lung Cancer classification and prediction of disease severity score using deep learning. In: 2023 6th International Conference on Information and Computer Technologies (ICICT), pp. 100–104. IEEE (2023)
24. Mangal, S., Chaurasia, A., Khajanchi, A.: Convolution neural networks for diagnosing colon and lung cancer histopathological images (2020). arXiv:2009.03878
25. Masud, M., Sikder, N., Nahid, A.-A., Bairagi, A.K., AlZain, M.A.: A machine learning approach to diagnosing lung and colon cancer using a deep learning-based classification framework. Sensors **21**(3), 748 (2021). https://doi.org/10.3390/s21030748

Transferablity Analysis of Adversarial Images Between CNN-Based Models

Surya Kant Yadav[(✉)] and Samir Kumar Borgohain

National Institute of Technology, Silchar, Silchar, India
`{surya21_rs,samir}@cse.nits.ac.in`

Abstract. Deep Learning Models (DLM) gave human beings a leap in all the tasks they had to do manually in vision, like image classification, object detection, and image recognition. After the introduction, these models can be attacked with little information or even without any information about the model. It became important to interpret the reason behind it to make it robust because these days machine learning models are applied to many security-critical areas like automotive vehicles and disease prediction. Currently, available evaluations are based mainly on feature analysis. A few papers focus on visual analysis, but those are not based on the recently introduced explainable AI tool, which gives better visual interpretation of image parts, contributing more to making decisions for the target class. In this paper, we evaluate 4 CNN-based deep-learning models on transferable adversarial attacks generated using ResNet50 for image classification. We are generating adversarial attacks by DeepFool and Projected Gradient Decent (PGD) attack reasons behind particular models' decision-making in classifying adversarial images to classified classes that will help in understanding the robustness of deep models as well as attack and defense mechanisms of existing methods.

Keywords: Transferable Adversarial Attack · Convolutional Neural Network · Image Classification · Explainable Artificial Intelligence

1 Introduction

Deep learning gave solutions to many problems in almost all vision-based tasks like image classification [9], object detection [26], object recognition [13], medical imaging [18] etc., where enough image data was given to the model for optimization of the parameters. However, this data became a problem for deep learning models (DLM) when calculated noise was added to training images of DLM, which was introduced as adversarial attacks [19] on DLM. After the addition of noise to a normal image, it became an adversarial image that was visually almost unchanged for humans, so it became an area of interest for researchers to understand what are the deciding factors that contribute most to the decision-making of deep DLM. This problem became even bigger when we didn't need any information on the model on which we wanted to attack, which is called a black box adversarial attack [7]. In these types of attacks, adversarial images are

C. Modi et al. (Eds.): MIND 2024, CCIS 2736, pp. 104–114, 2026.
https://doi.org/10.1007/978-3-032-14531-4_9

generated on a surrogate model and transferred to the target model on which we want to attack, which is transferable adversarial attacks [20].

CNN-based DLM [3] gave state-of-the-art results in image classification tasks, but these models are also vulnerable to simple adversarial attacks [2] as well as transferable adversarial attacks [6]. There are many evaluations available, but these evaluations are not able to justify what is the main reason behind the transferability of adversarial images from one model to another and on which part of the image these models are focusing to make their decision. In this paper, we demonstrate the generation of adversarial attacks using the ResNet50 model. For generating attacks, we use DeepFool and PGD adversarial attack algorithms, which are well-known attacks for evaluation of the robustness of the model. After the generation of the attack, we tested our generated adversarial images on 4 different CNN-based deep learning models, ResNet18, MobileNet, DenseNet, and SimpleCNN. At last, we also generated a saliency map for all clean images as well as adversarial images using Explainable Artificial Intelligence (XAI) [1] to understand the approach of the model in the classification of these images. We use GradCam [17] and LIME [15] algorithms to see the saliency map of the classified images. Major highlights of this proposed work are as follows:

- We performed an experimental evaluation on prominent CNN-based deep learning models to check their robustness for transferable adversarial attacks and their respective capabilities of generating adversarial examples.
- We are using explainable AI for explaining misclassifications of adversarial images, which gives a better understanding of CNN models for classification of adversarial images.

2 Literature Survey

In this section, we are giving some recent and state-of-the-art works that contributed to the work in the adversarial machine learning domain In [16] they introduced high-quality adversarial examples using the layer-wise origin-target synthesis (LOTS) technique, which mimics the deep features of targeted class activation, which is implemented on VGGs for face recognition tasks. Authors in [8] said that transfers of not-targeted adversarial examples are better than targeted adversarial examples, and it is even much less while checking transferability of the example on targeted models on target label. They introduced a novel method that was better at transferring for both targeted and non-targeted attacks in comparison with existing attacks. [14] is a generative method for targeted and non-targeted attacks that is generated for classification and semantic segmentation tasks. They gave a transfer attack, which is both universal and image-dependent. [23] is an optimization-based method where a conditional adversarial network is trained to generate perturbation, which, after addition to the image, was more realistic as well as efficient in calculation of noise. [20] Unified and Efficient Adversary (UEA) is proposed for attacking image and video detection. Attention feature loss was introduced for multiple scales to enhance the UEA ability in a black box environment. [4] introduces momentum iterative gradient-based methods, which they use to stabilize optimization and avoid local maxima by iteration-wise gathering of the gradient of the loss function. [25] proposed the method where they were using an intermediate feature map for maximization of distance between the natural image and adversarial image, which was successfully transferred to the targeted model in a black box manner. [12] proposes a generalizable, black-box, untargeted adversarial attack that can transfer beyond task, dataset, and network. They trained there model on a big dataset for generating adversarial perturbation, and because of that, it was generalized to other networks. [24] are the first to recommend modifying the clean image using differentiable input transforms. Specifically, their DIM entails implementing a distinct transformation for every iteration of adversarial attacks that involve multiple steps. [5]Examine the image regions that are more crucial to the classification output of the various models, or the discriminative image regions. They evaluate different models that leverage different portions of the image, especially if adversarially trained. Their proposal of the Translation Invariance Method (TIM) is driven by this. [21] Proposed an adversarial attack in the frequency domain in centralized features. Perturbation was created in to bypass different adversarial defences. [22] proposes Lipschitz Regularized Surrogate (LRS) for transfer-based black-box attacks.

3   Methodology

In the proposed method, we are evaluating the robustness of 4 CNN-based deep learning models, ResNet18, MobileNetV2, DenseNet121, and a custom SimpleCNN. We are using DLM because these models are light weight models (except DenseNet121) and they are better in extracting features of images without any supervision. We are using 2 explainable AI algorithms, Gradcam and LIME for evaluating these models on adversarial attacks named DeepFool [11] and PGD [10] generated by the ResNet50 model. Figure 1 shows the general adversarial image generation process using a surrogate model in a black-box manner. The goal of this experiment is to evaluate the robustness of these models in a black-box manner where adversarial images are generated by a surrogate model and the above-mentioned model is used as the target model to attack. We are using the CIFER-10 dataset for this experiment, which has a total of 60,000 images belonging to 10 classes. Each class has 6,000 images, and each image size is 32*32. For this experiment, we used 10,000 images as a test image and converted this to a tensor for the experiment. We made batches of 100 images from test image. All models were fine-tuned on the CIFER-10 dataset, which was pre-trained on ImageNet.

3.1   Adversarial Attack Methods

DeepFool. The DeepFool attack is an iterative adversarial attack method designed to find the smallest perturbation required to misclassify an input image. This method tries to find the smallest perturbation that can be added to the

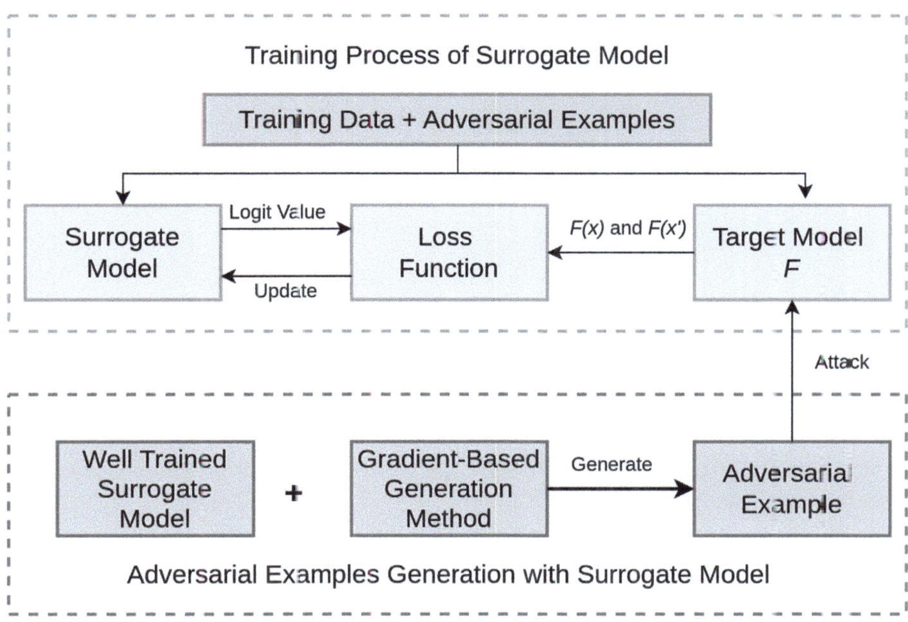

Fig. 1. Adversarial Example Generation

natural. Ever after addition to the natural image, the image remains visually unchanged and enough to just cross the decision boundary of the classifier. Let $f : \mathbb{R}^n \to \mathbb{R}^c$ be the classifier, where C is the number of classes and $f(x) \in \mathbb{R}^c$ represents the logits (pre-softmax scores) for an image $x \in \mathbb{R}^n$. The classifier assigns x to the class $k = arg\ max\ f_k(x)$. The smallest perturbation r is calculated to just cross the decision boundary so that image should not belong to its original class where r should be $f(x + r) \neq arg\ max\ f(s)$. DeepFool approximates the classifier decision boundary using a linear model, especially for binary classifiers. For a non-linear classifier, locally around the input x, it can be linearized as:

$$f_k(x) \approx f_k(x_0) + \nabla f_k(x_0)^\top (x - x_0) \tag{1}$$

where $\nabla f_k(x_0)$ is the gradient of the classifier's output for class k with respect to the input s. The perturbation r that pushes the image x across the decision boundary is computed in an iterative manner. For our experiment, we use 50 iterations per image to generate an image. For each iteration i, the algorithm computes the perturbation needed to push the image towards the closest decision boundary

$$r_i = \frac{f_k(x_i) - f_j(x_i)}{\|\nabla f_k(x_i) - \nabla f_j(x_i)\|_2} \tag{2}$$

where k is the current predicted class and j is the next most probable class based on logits. This gives the direction and magnitude of the perturbation. The image is iteratively perturbed, and the process continues until the classifier assigns the perturbed image to a different class. At each iteration -

$$x_{i+1} = xi + (1 + \epsilon) * r_i \tag{3}$$

where ϵ is a small step factor ensuring the perturbation crosses the boundary. The process stops once it $arg\ max\ f(x + r) \neq arg\ max\ f(x)$ returns the adversarial image $x + r$.

Projected Gradient Descent (PGD) Attack. The Projected Gradient Descent (PGD) attack is an iterative adversarial attack technique designed to increase a classifier's loss by introducing perturbations restricted to an $\epsilon - ball$ around the original image. Within a given perturbation limit, PGD is widely regarded as one of the most potent adversarial attack methods. Let $f(x, y)$ be the loss function of the classifier with input x and true label y, which is generally cross-entropy loss

$$L(x, y) = -log\ p(x|y) \tag{4}$$

where $p(x|y)$ is the softmax probability assigned by the classifier to the correct label y for input x. The goal of PGD is to find an adversarial perturbation r such that the image $x + r$ is classified incorrectly while keeping r small and having restriction by ϵ.

$$maximize_{r:\|r\|_\infty \leq \epsilon}\ L(x + r, y) \tag{5}$$

where l_∞ norm is measuring the distance of image pixels from original images, which should not be more that ϵ after $(x + r)$. The attack starts with the original image $x_0 = x$ and iteratively updates the adversarial perturbation in the direction of the gradient of the loss function:

$$x_{i+1} = x_i + \alpha \cdot sign(\nabla_x L(x_i, y)) \tag{6}$$

where α is learning rate and $\alpha \cdot sign(\nabla_x L(x_i, y))$ is the element-wise sign of the gradient of the loss with respect to the image. After calculating each steps value, this is checked for distance from original image, which should be within $\epsilon - ball$.

$$x_{i+1} = clip\ (x_{i+1}, x_o - \epsilon, x_o + \epsilon) \tag{7}$$

The process of calculating perturbation value continues in an iterative manner till it finds the suitable value to cross the decision boundary. It maximizes the loss to find the correct value of perturbation. Our attack ran for 40 iterations with a step size of $2/255$ and a L_∞ bound of 0.03.

4   Results and Observation

4.1   DeepFool Attack

Average $L2$ distance of adversarial image from original image is 19.19 and average L_∞ distance is 0.90. Graph 2 is showing the attack success rate of PGD

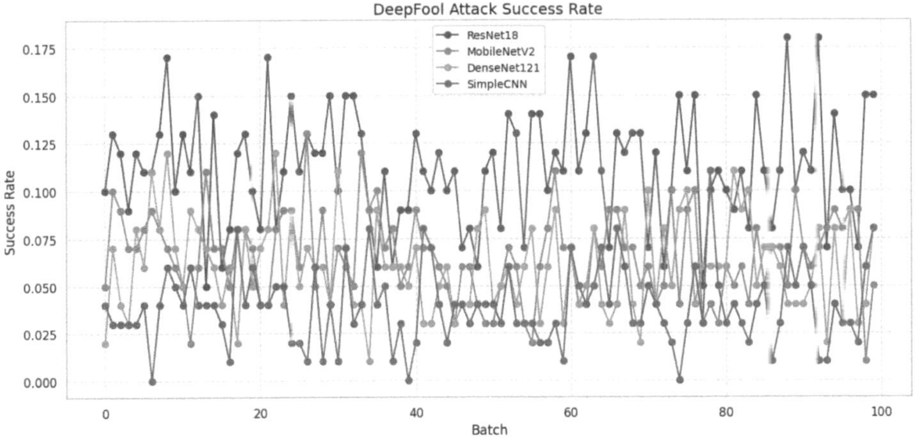

Fig. 2. DeepFool Attack Success Rate

attack on DLMs. For the DeepFool attack, ResNet18 is giving a decent attack success rate of 10.53%, with an average confidence of 13.62%. This suggests that ResNet18 was moderately vulnerable to small perturbations under DeepFool but still retained confidence in its predictions. In comparison with ResNet18, MobileNetV2 and DenseNet121 were more robust, with success rates of 7.77% and 6.39%, respectively. DenseNet121 had a higher confidence (21.96%), indicating that despite being less vulnerable to attacks, the network was highly confident in its (sometimes incorrect) predictions. The custom SimpleCNN showed extreme robustness to DeepFool attacks, with a near-zero success rate (0.09%), indicating that it was barely affected by these perturbations. However, the relatively low average confidence of 11.30% suggests that this model might lack overall classification confidence compared to larger models. The graph in 3 is showing the batch-wise accuracy of clean images and DeepFool attacked images, and the graph in 3 is showing the average accuracy comparison of clean images and adversarial images. This result also shows that SimpleCNN does not have the capabilities of extracting a higher level of features from the image, which is robust in comparison with larger and more effective more capable models in feature extraction.

4.2   PGD Attack

PGD-generated images have 19.22 average $L2$ distance and 0.90 average L_∞ distance. In comparison with DeepFool attacks, PGD attack is showing lesser success even though it is known for better results. Graph 4 is showing the attack success rate of PGD attack on DLMs. ResNet18 had a success rate of 8.98%, and the model maintained a confidence of 14.48% for adversarial images which shows that it was not completely fooled by the attack. MobileNetV2 and DenseNet121 are also similar in this result as ResNet18, with success rates of 8.13% and 6.57%,

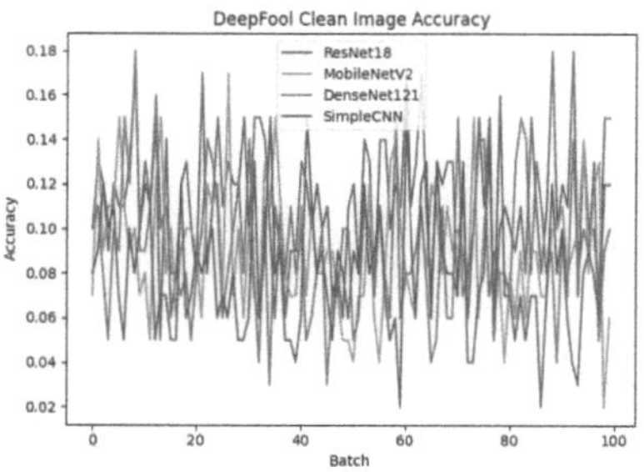

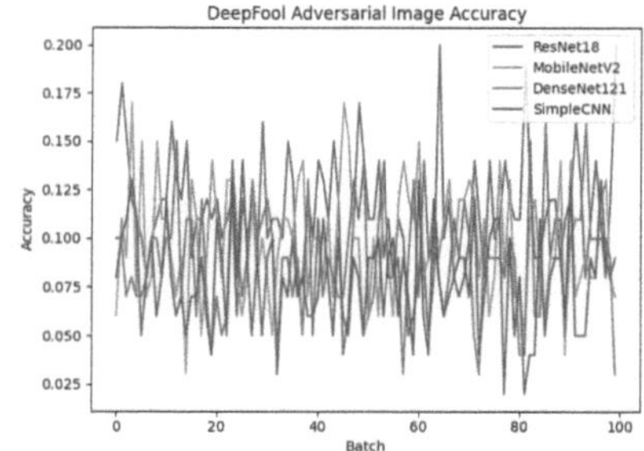

(a) DeepFool Clean Image Accuracy   (b) DeepFool Adversarial Image Accuracy

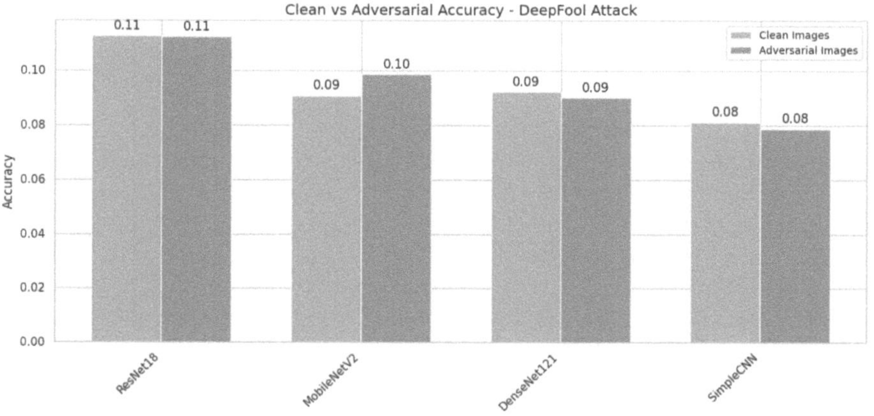

Fig. 3. DeepFool Attack vs Clean Image Accuracy.

respectively, with DenseNet121 again showing higher confidence (21.95%). SimpleCNN, consistent with the DeepFool results, was highly robust against PGD with a minimal success rate (0.09%). This is indicative of SimpleCNN's simplicity, possibly preventing it from being easily manipulated by adversarial perturbations. Graph 5 is showing the batch-wise accuracy of clean images and DeepFool attacked images, and Graph 5 is showing the average accuracy comparison of clean images and adversarial images.

Figures 7 and 6 are saliency maps of adversarial attack for clean image and adversarial image, respectively. We can see in the given images that GradCam and LIME saliency maps are focusing on almost similar areas of the image for both clean as well as adversarial images. After getting this result, we found that there is not much difference in approach for models in classification of clean image and adversarial image.

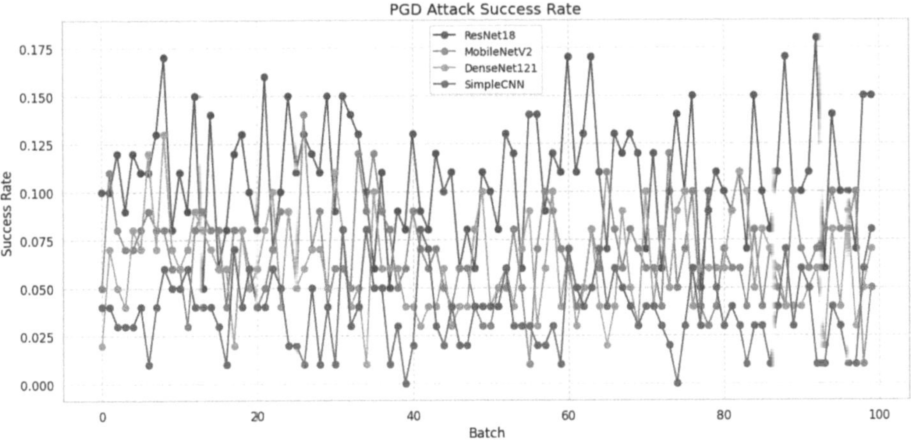

Fig. 4. PGD Attack Success Rate

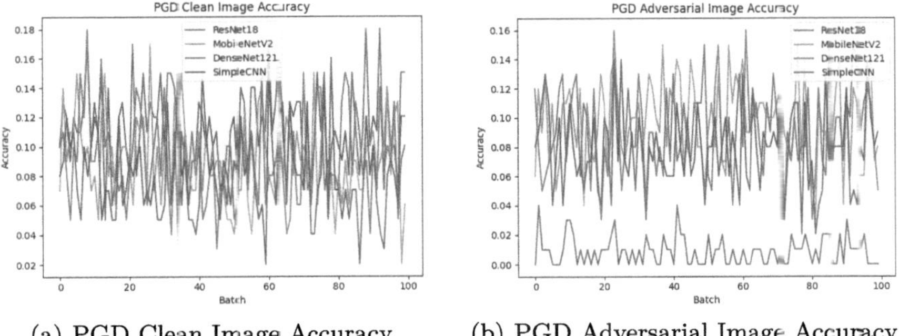

(a) PGD Clean Image Accuracy   (b) PGD Adversarial Image Accuracy

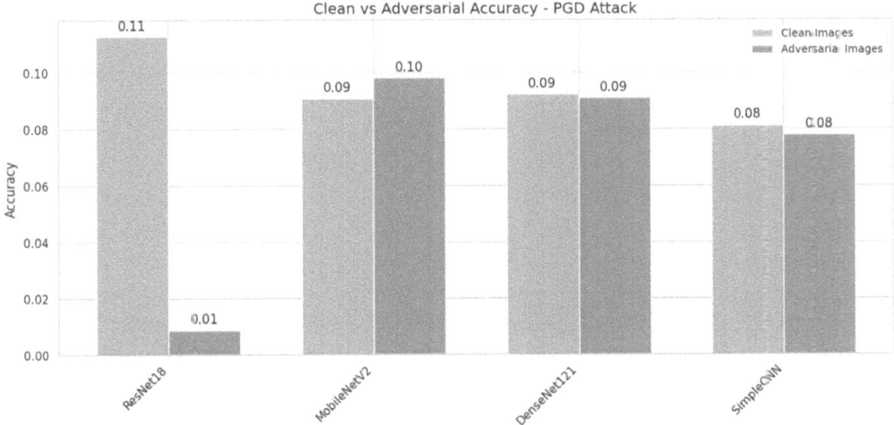

Fig. 5. PGD Attack vs Clean Image Accuracy.

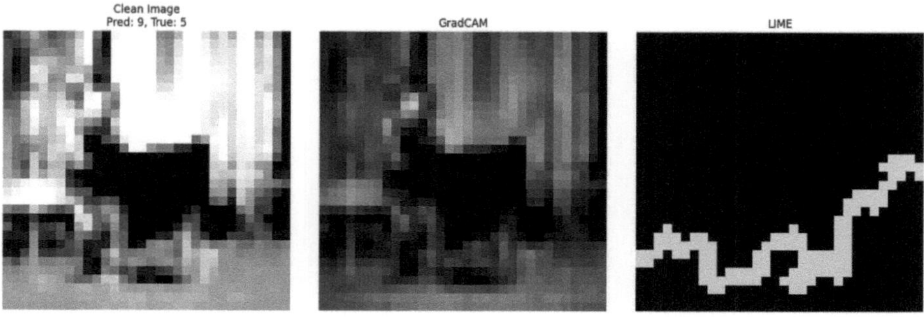

Fig. 6. Clean Image Classification Saliency Map of GradCam and LIME.

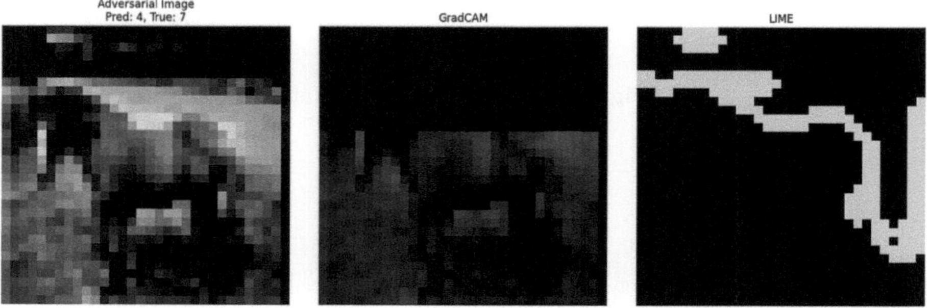

Fig. 7. Adversarial Image Classification Saliency Map of GradCam and LIME.

5   Conclusion and Future Scope

The experiment reveals that more complex models, such as ResNet18, MobileNetV2, and DenseNet121, are generally more susceptible to adversarial attacks than simpler models like SimpleCNN. This is also because simpler models do not have the capability of extracting complex features from images. However, even though these attacks were successful, the adversarial perturbations themselves were relatively small, as shown by the $L2$ and L_∞ distances. In the future, research needs to be in the direction of more generalized feature extractor models where the model should not be specific to a particular dataset or model because this makes DLM susceptible to adversarial attacks. For reducing the threat of adversarial attacks, much more research needed in analysis of transferable attacks. In the process of developing such models, standard accuracy may go down, but that model would be more robust to adversarial attack. This emphasizes the growing need for research into improving the robustness of deep learning models, especially for use in critical real-world applications where security and reliability are essential.

References

1. Ayesha, A., Ahamed, N.N.: Explainable artificial intelligence (EAI): for healthcare applications and improvements. In: Explainable Artificial Intelligence for Biomedical and Healthcare Applications, pp. 162–196. CRC Press
2. Chakraborty, A., Alam, M., Dey, V., Chattopadhyay, A., Mukhopadhyay, D.: A survey on adversarial attacks and defences. CAAI Trans. Intell. Technol. **6**(1), 25–45 (2021)
3. Chen, L., Li, S., Bai, Q., Yang, J., Jiang, S., Miao, Y.: Review of image classification algorithms based on convolutional neural networks. Remote Sens. **13**(22), 4712 (2021)
4. Dong, Y., et al.: Boosting adversarial attacks with momentum. In: Proceedings of the IEEE Conference on Computer Vision and Pattern Recognition, pp. 9185–9193 (2018)
5. Dong, Y., Pang, T., Su, H., Zhu, J.: Evading defenses to transferable adversarial examples by translation-invariant attacks. In: Proceedings of the IEEE/CVF Conference on Computer Vision and Pattern Recognition, pp. 4312–4321 (2019)
6. Gu, J., et al.: A survey on transferability of adversarial examples across deep neural networks. arXiv preprint arXiv:2310.17626 (2023)
7. Guo, C., Gardner, J., You, Y., Wilson, A.G., Weinberger, K.: Simple black-box adversarial attacks. In: International Conference on Machine Learning, pp. 2484–2493. PMLR (2019)
8. Liu, Y., Chen, X., Liu, C., Song, D.: Delving into transferable adversarial examples and black-box attacks. arXiv preprint arXiv:1611.02770 (2016)
9. Lu, D., Weng Q.: A survey of image classification methods and techniques for improving classification performance. Int. J. Remote Sens. **28**(5), 823–870 (2007)
10. Madry, A.: Towards deep learning models resistant to adversarial attacks. arXiv preprint arXiv:1706.06083 (2017)
11. Moosavi-Dezfooli, S.M., Fawzi, A., Frossard, P.: Deepfool: a simple and accurate method to fool deep neural networks. In: Proceedings of the IEEE Conference on Computer Vision and Pattern Recognition, pp. 2574–2582 (2016)
12. Naseer, M., Khan, S.H., Rahman, S., Porikli, F.: Task-generalizable adversarial attack based on perceptual metric. arXiv preprint arXiv:1811.09020 (2018)
13. Oliva, A., Torralba, A.: The role of context in object recognition. Trends Cogn. Sci. **11**(12), 520–527 (2007)
14. Poursaeed, O. Katsman, I., Gao, B., Belongie, S.: Generative adversarial perturbations. In: Proceedings of the IEEE Conference on Computer Vision and Pattern Recognition, pp. 4422–4431 (2018)
15. Ribeiro, M.T., Singh, S., Guestrin, C.: Local Interpretable Model-Agnostic Explanations (lime) An Introduction. O'Reilly Media (2016)
16. Rozsa, A., Günther, M., Boult, T.E.: Lots about attacking deep features. In: 2017 IEEE International Joint Conference on Biometrics (IJCB), pp. 168–176. IEEE (2017)
17. Selvaraju, R.R., Cogswell, M., Das, A., Vedantam, R., Parikh, D., Batra, D.: Grad-cam: visual explanations from deep networks via gradient-based localization. In: Proceedings of the IEEE International Conference on Computer Vision, pp. 618–626 (2017)
18. Suzuki, K.: Overview of deep learning in medical imaging. Radiol. Phys. Technol. **10**(3), 257–273 (2017). https://doi.org/10.1007/s12194-017-0406-5

19. Szegedy, C.: Intriguing properties of neural networks. arXiv preprint arXiv:1312.6199 (2013)
20. Wei, X., Liang, S., Chen, N., Cao, X.: Transferable adversarial attacks for image and video object detection. arXiv preprint arXiv:1811.12641 (2018)
21. Wu, S., Tan, Y., Wang, Y., Ma, R., Ma, W., Li, Y.: Towards transferable adversarial attacks with centralized perturbation. In: Proceedings of the AAAI Conference on Artificial Intelligence, vol. 38, pp. 6109–6116 (2024)
22. Wu, T., Luo, T., Wunsch, D.C., II.: LRS: enhancing adversarial transferability through lipschitz regularized surrogate. In: Proceedings of the AAAI Conference on Artificial Intelligence, vol. 38, pp. 6135–6143 (2024)
23. Xiao, C., Li, B., Zhu, J.Y., He, W., Liu, M., Song, D.: Generating adversarial examples with adversarial networks. arXiv preprint arXiv:1801.02610 (2018)
24. Xie, C., et al.: Improving transferability of adversarial examples with input diversity. In: Proceedings of the IEEE/CVF Conference on Computer Vision and Pattern Recognition, pp. 2730–2739 (2019)
25. Zhou, W., et al.: Transferable adversarial perturbations. In: Proceedings of the European Conference on Computer Vision (ECCV), pp. 452–467 (2018)
26. Zou, Z., Chen, K., Shi, Z., Guo, Y., Ye, J.: Object detection in 20 years: a survey. Proc. IEEE **111**(3), 257–276 (2023)

Deep Learning-Based Analysis, Selection, and Detection of Soybean Seeds: A Comprehensive Review

Sonali Subhash Shinde-Patil[1]($\boxtimes$), Varsha P. Desai[2] (iD),
and K. Prathapan Pillai[1] (iD)

[1] D Y Patil Agriculture and Technical University, Talsande, Kolhapur, India
svp6575@gmail.com, vc@dyp-atu.org
[2] Computer Science and Engineering Department, D Y Patil Agriculture and
Technical University, Talsande, Kolhapur, India

Abstract. This comprehensive review explores the transformative impact of deep learning technologies on the analysis, selection, and detection of soybean seeds. By leveraging state-of-the-art deep learning architectures, such as convolutional neural networks (CNNs) and generative adversarial networks (GANs), this article highlights advancements in seed grade assessment, viability detection, and yield prediction. The review emphasizes the integration of real and historical agricultural data to enhance model accuracy and performance. Key parameters such as seed health, soil conditions, and weather patterns are discussed, alongside challenges in data availability, computational resource requirements, and model generalization. The study focuses on the specific context of soybean farming in Maharashtra, aiming to provide actionable insights for farmers and stakeholders. Finally, the paper identifies future research directions, emphasizing the need for user-friendly and cost-effective AI solutions for small-scale farmers. The review highlights the integration of CNNs, GANs, and Recurrent Neural Network (RNN) for soybean seed classification, emphasizing their application in creating robust, annotated datasets and handling real-world challenges like data variability and scarcity. It bridges agricultural needs with AI-driven solutions, offering insights for precision farming.

Keywords: Crop Disease · Classification · Convolutional neural networks · Deep Learning · Agricultural Innovation

1 Introduction

Soybean (Glycine max) is essential for food security as a major protein and oil source. In Maharashtra, India, a key soybean-producing region, farmers face challenges like soil health issues, unpredictable weather, and pests, all affecting yield. With global food demand rising, enhancing agricultural efficiency is

Supported by organization x.

C. Modi et al. (Eds.): MIND 2024, CCIS 2736, pp. 115–127, 2026.
https://doi.org/10.1007/978-3-032-14531-4_10

crucial. Recent advances in AI, especially deep learning, offer transformative potential for agriculture. Deep learning models analyze large datasets to identify complex patterns, making them highly suitable for tasks like seed quality assessment, selection, and crop management—areas where traditional methods are often labor-intensive and subjective. Models like CNNs and RNNs can automate seed grading, detect defects, and predict crop performance, integrating real-time environmental data for precision.

The work employs CNNs, GANs, and RNNs for soybean seed classification, leveraging their distinct strengths. CNNs extract spatial features like texture and shape for precise differentiation between healthy and defective seeds. GANs generate synthetic images to address data imbalance and enhance model generalization. RNNs process temporal data, capturing trends from environmental or growth variations. This integrated approach ensures robust and accurate classification while addressing challenges like noise, variability, and limited datasets in agricultural applications [16,21].

Maharashtra's unique agricultural landscape calls for customized interventions. Deep learning and real-time analytics can optimize seed selection for local soil and climate conditions, enhancing resilience, productivity, and pest management. This review explores the latest deep-learning techniques in soybean seed analysis, focusing on practical applications for farmers in Maharashtra. Insights from local experts highlight both the benefits and challenges of these technologies in real-world farming. By adopting AI-driven innovations, agriculture can significantly improve crop yield, resource use, and food security, locally and globally.

These are the objectives of this review.

1. The preparation of image datasets using soybean seeds requires experts to tag defective and healthy seeds.
2. Evaluating the effectiveness of deep-learning and object detection models in seed detection, identification, and classification.
3. A Mixed Cropping seeds classifier and Quality Analysis system is being developed by integrating object detection with YOLO-v5.

The basic architecture (in Fig. 1) for the collection of seed data involves a systematic approach to data preprocessing, which is crucial for effective image classification using Deep CNN and the YOLO-v5 algorithm. The initial process involves collecting and preprocessing seed images to enhance quality and ensure consistency, which consists of resizing, normalizing, and augmentation techniques to improve model robustness. Analyzing the efficiency of deep-learning and object detection models for seed detection, identification, and classification. Model's performance is evaluated by splitting the dataset into training, testing, and validation subsets. The image classification phase involves YOLO-v5 being used to identify and classify different seed types with high precision accurately. By implementing this architecture, seed classification is guaranteed to be reliable, which ultimately supports better agricultural practices and decision-making.

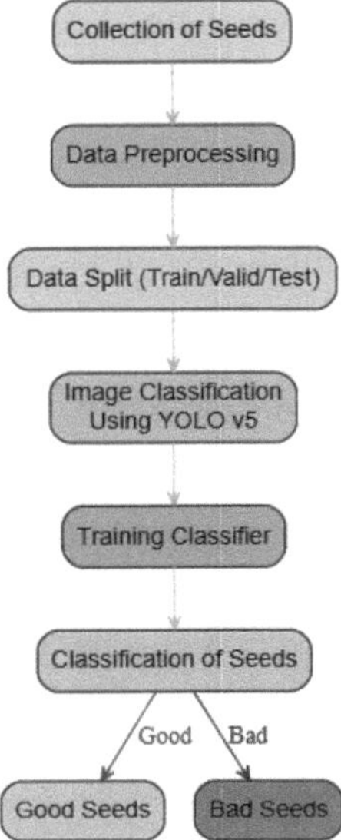

Fig. 1. Basic Architecture of Seed Selection.

2 Literature Review

A pipeline for automatic soybean seed classification is proposed in [1], using Mask R-CNN for segmenting the image and a lightweight Soybean Network (SNet) with Mixed Feature Recalibration (MFR) modules to detect seed damage. SNet, with 1.29 M parameters, achieves 96.2% accuracy, outperforming six models, and is ideal for resource-limited platforms. In [2], a deep learning method for real-time seed classification uses "multi-scale Retinex with color restoration" (MSRCR) for uneven illumination. A CNN classifies seeds into four categories with F-scores up to 97.41%. Implemented on an NVIDIA Jetson TX2, it achieves 95.63% accuracy with 4.92 s per seed. A deep learning-based sorting system in [3] addresses the shortcomings of traditional machine vision by capturing the entire surface of soybean seeds. Seeds are categorized into six categories based on images taken from six surfaces under three brightness conditions using a circumrotating mechanism. MobileNetV2-improved achieved 97.84% accuracy at 35 FPS on an NVIDIA Jetson Nano, processing 222 seeds per minute with 98.87% total accuracy, offering a precise, low-cost solution for seed sorting.

In [4] a Faster R-CNN algorithm was optimized for weed identification in soybean fields using a dataset of 9,816 images. VGG19, combined with Carbon Border Adjustment Mechanism (CBAM) (attention mechanism), achieved 99.16% recognition accuracy, outperforming Single Shot Detector (SSD) and Yolov4, making it a highly accurate and efficient solution for weed detection in complex field environments. A DL-based system called Mixed-Cropping Seed Classifier and Quality Tester (MCSCQT) in [5] is used to automate seed classification and quality testing for mixed crops instead of traditional methods like sieving. MCSCQT achieved 99% precision and recall by using a dataset of labeled images of seeds that are both healthy and diseased, including pearl millet and maize. The food industry benefits significantly from this system, which improves seed segregation and quality control. The Preemptive Classification using Discrete Data (PC-DD) technique for enhancing biosensor data analysis for early plant disease detection is introduced in [6]. Adaptive processing of unidentical data allows PC-DD to improve analysis rates by using probabilistic data substitution and random forest classification, which results in higher detection accuracy and faster analysis. The multi-data source deep learning object detection network (MS-YOLO) model is suggested for accurate estimation of disease severity for citrus leaves in natural environments in [7]. Enhancements like multiple attention mechanisms (MHSA) and an upgraded Focal-EIoU loss function improve YOLOv5's performance in complex environments, increasing precision and mAP by 2.2% and 1.7%, respectively, for more effective leaf disease detection.

In [8], a 3D CNN model is used to predict soybean yield at the plot scale from UAV-based RGB images. Collected from 13 fields with around 30,000 plots, the study transforms 2D CNNs (VGG and DenseNet) into 3D versions to incorporate temporal data. DenseNet proved the most efficient, balancing accuracy ($R^2 = 0.69$) and model complexity, while finer resolutions did not significantly enhance performance. In [9], deep learning and machine vision are applied to assess soybean quality during mechanized harvesting. A clean dataset was created from 40 initial images and expanded to 800 using augmentation techniques. This dataset supported the development of an online detection model for soybean crushing and impurity rates, providing a useful resource for building quality prediction models. In [10], an automated system for seed vigor assessment is proposed to replace labor-intensive manual methods. Tested on real-world datasets, the system achieved 80.17% accuracy, streamlining seed vigor analysis and improving the efficiency of quality control through real-time image analysis.

In [11], the review highlights the importance of soybeans for global food security, focusing on the challenges posed by the soybean cyst nematode (SCN), which causes significant yield losses. While advanced management techniques like crop rotation and resistant varieties exist, early detection of SCN remains difficult due to the lack of visible symptoms. Traditional soil sampling methods are limited as they often rely solely on egg counts without considering soil factors like pH and nutrients. Recommendations include integrating deep learning and hyperspectral imaging to improve detection accuracy and reduce costs. In [13], A dataset was created by capturing images of damaged plants, which were annotated for analysis. The methodology includes calculating biomass reduction compared to control plants and using a deep learning multi-label classification model to detect disease symptoms. The algorithm achieved R^2 values of 0.87 and 0.89 for damage estimation in soybean and Amaranthus retroflexus, significantly enhancing traditional assessment methods.

In [13], a lightweight AI model using Inception-ResNet-V2 is developed for real-time identification of soybean pests, achieving 99.54% accuracy with 3,809 images. The model integrates Explainable AI for transparency and is designed for practical use with a Raspberry Pi system, aiding farmers in pest management. In [14], Bio-DETR, a lightweight transformer-based architecture, is introduced for detecting pests and seeds in agriculture. Utilizing Hybrid Scale Attention and Dynamic Bilateral Attention, it processes 8,000 hyperspectral images from the HSI-Bio dataset, achieving 87.4% average precision and 62.2% overall precision while maintaining 52 FPS for real-time detection.

In [15], a technological solution is proposed to detect insects in soybean crops using object detection algorithms (YOLO v3, v4, and v5). YOLO v5 achieved a mean average precision (mAP) of 99.5%, outperforming the other versions and providing an efficient method for pest management in Indian agriculture. In [16], a jointly multi-modal bag-of-feature (JMBoF) classification framework inspects post-harvest dry soybean seeds by extracting robust and spatial color features. The multiclass support vector machine classifies the low-rank representation (LRR) encoded features, improving seed quality classification over traditional methods. In [17], advanced machine learning models (VGG16, AlexNet, CNN) automate soybean seed quality assessment based on size, shape, color, and damage. VGG16 achieved the highest accuracy at 90%, demonstrating significant improvements over traditional assessment methods, thus enhancing efficiency in agricultural practices.

In [20], introduces SoyCult, a publicly available dataset that automatically identifies soybean cultivars, which is crucial for maintaining seed homogeneity in Brazil, the top soybean producer worldwide. Individual seed images from different cultivars were extracted using a contour-based segmentation method, with baseline systems evaluated using pre-trained convolutional networks and classifiers (Table 1).

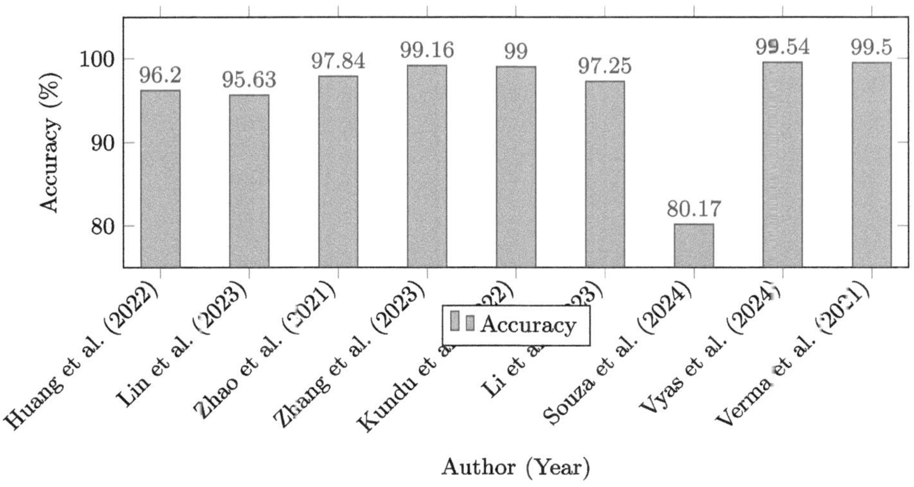

Fig. 2. Accuracy Comparison for Seed Analysis and Detection Techniques

Table 1. Comparison of Techniques and Methodologies for Seed and Agricultural Applications

Author (Year)	Technique/Model	Application/Data Source	Accuracy (%)	Key Features
Huang et al. (2022)	MobileNet + Multi-Feature Fusion (MFR)	Seed damage dataset	96.2	Lightweight recalibration
Lin et al. (2023)	CNN with MSRCR	Images with uneven illumination	95.63	Multi-scale Retinex with color restoration
Zhao et al. (2021)	MobileNetV2-Improved	Soybean seed dataset	97.84	Full-surface detection
Zhang et al. (2023)	Optimized Faster R-CNN + VGG19-CBAM	Weed detection images	99.16	Optimized for weed detection
Kundu et al. (2022)	YOLO v5	Pearl millet and maize seeds	99	High precision classification
Bandra et al. (2023)	Lightweight CNN + Coord. Attention	Maize and red kidney bean dataset	97.25	Low parameter count, high efficiency
Souza et al. (2024)	Image Learning-Based Approach	Real-world seed classification dataset	80.17 ± 2.37	Streamlined seed vigor assessment
Vyas et al. (2024)	Inception-ResNet-V2	Soybean pest dataset	99.54	Real-time pest detection
Verma et al. (2021)	YOLO v5	Insect detection dataset	99.5	Outperforms YOLO v3 and v4

Figure 2 illustrates the accuracy rates of various seed analysis and detection techniques employed by different authors, ranging from 80.17% to 99.54%. Notably, [4] and [18] achieved the highest accuracy levels, demonstrating the effectiveness of their respective methodologies in improving seed quality assessment.

3 Advances and Challenges in Deep Learning for Soybean Seed Analysis

The use of deep learning in soybean seed analysis and selection has been transformed by CNNs' advancements in image-based classification for quality assessment. CNN architectures, like ResNet and VGGNet, process high-resolution images that detect visual patterns like color and texture, which aid in distinguishing healthy seeds from damaged seeds. Enhancements like SIFT (Scale-Invariant Feature Transform) and ORB (Oriented FAST and Rotated BRIEF) further boost classification accuracy, improving predictions related to seed viability and germination probability.

GANs address the scarcity of labeled agricultural data by creating synthetic images, enhancing model training with diverse examples to reduce overfitting. The performance of the model in identifying seed defects is strengthened by this data augmentation, which ultimately improves prediction accuracy and robustness in real-world applications. Meanwhile, RNNs, specifically Long Short-Term Memory (LSTMs), offer powerful tools for forecasting seed performance by analyzing historical and environmental data. By modeling interactions among variables like soil health, seed type, and weather, LSTMs help farmers make data-driven planting decisions.

For quality detection, deep learning systems automate inspection and classify seed images into viable or damaged categories, reducing manual errors Techniques like hyperspectral imaging paired with deep learning further enhance quality assessment by analyzing seed composition and linking it to metrics such as moisture and protein content. However, challenges remain in agriculture, such as limited labeled data, high computational demands, and overfitting risks, which hinder widespread model adoption. Computational costs and the need for sophisticated infrastructure are particularly limiting for smallholder farmers, highlighting the need for resource-efficient solutions to ensure deep learning's benefits reach a wider audience.

4 Case Study: AI-Based Detection System for Soybean Seeds

This section presents a hypothetical case study of an AI-driven seed detection and selection system implemented in Maharashtra. The system employs deep Convolutional Neural Networks (CNNs) for visual inspection of soybean seeds and Recurrent Neural Networks (RNNs) for seed selection based on historical climate data.

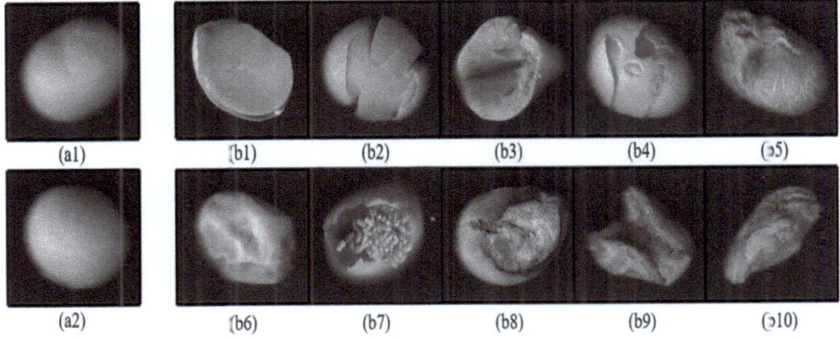

Fig. 3. Different appearance and quality soybean seeds for classification Nature Scientific Reports (2019) [16].

The image set showcases in Fig. 3 two good samples of soybean seeds (a1, a2) on the left, representing healthy and high-quality seeds. On the right side, a range of defective soybean seeds is displayed, each illustrating a specific type of damage or deformity. These include seeds that are cotyledon-lacked (b1), physically-damaged (b2), naturally-cracked (b3), and testa-damaged (b4), along with those that exhibit hull-shriveling (b5) or cotyledon atrophy (b6). Other visible defects include worm bites (b7), testa decay (b8), cotyledon mold (b9), and heteromorphic variations (b10). This comparison highlights the diversity of seed defects that can impact the quality and viability of soybean seed quality studies.

To address the ambiguity regarding the dataset, the proposed work includes the creation of a new dataset tailored specifically for soybean seed analysis. This dataset will consist of high-resolution images of soybean seeds collected from local farms and agricultural research centers, covering both healthy and defective seeds. Expert agriculturalists will annotate the images to ensure accurate labeling of defects such as discoloration, deformities, and pest damage.

5   Experimental Devices

Capturing images of soybean seeds is done using the Perfection V850 Pro from Epson. This imaging system includes components as: a transparent flat glass plate, a black absorption cover, a charge-coupled imaging device (CCD), communication cables, a motor driven shifting electronic platform, a black box, and a computer etc. The transparent glass plate is used to place each soybean seed sample in equal intervals, and a black absorption cover is used to increase image contrast.

A linear light source and a shifting mirror are moved by the motor-driven shifting platform. This linear light source emits beams onto the seed samples through the glass plate. The reflected beams are directed toward the shifting mirror and further reflected onto a fixed mirror. Uniform images of all soybean seeds, regardless of their position, are captured by the CCD after collecting the spectra from the fixed mirror. The setup ensures a consistent lighting and prevents uneven image exposure, a common issue in traditional camera setups.

The imaging system is protected from external lighting by being enclosed in a black box. An external computer, which controls the operation and stores the captured images, is connected to the imaging system through a communication cable. The soybean seeds captured in a photograph are automatically divided into individual kernel images and stored for further analysis.

6   Methodology

To accurately analyze seed images, a low-rank representation of a JMBoF classification method is utilized to examine the appearance quality of postharvested dry soybean seeds in this study [16].

6.1   Color Space Conversion

The color-space conversion from RGB to $L*a*b*$ is crucial seed's image analysis needs to be accurate. The $L*a*b*$ color space, standardized by the International Commission on Illumination, includes lightness L^*, a (green & red) component a^*, and a (blue & yellow) component b^*. This space encompasses the entire RGB color space and offers an extended range of colors that RGB cannot represent shown in Fig. 4.

The conversion requires two steps:

1. Convert the RGB color space into the CIE P'Q'R' color space by utilizing the following matrix transformation.

$$\begin{bmatrix} P' \\ Q' \\ R' \end{bmatrix} = \begin{bmatrix} [0.412] & [0.357] & [0.180] \\ [0.212] & [0.715] & [0.072] \\ [0.019] & [0.119] & [0.950] \end{bmatrix} \begin{bmatrix} R' \\ G' \\ B' \end{bmatrix}$$

2. The P'Q'R' values are then transformed into L*a*b* values relative to the reference white point (P'_n, Q'_n, R'_n) using the equations:

$$L^* = 116 \cdot f'\left(\frac{Q'}{Q'_n}\right) - 16$$

$$a^* = 500.0 * \left(f'\left(\frac{P'}{P'_n}\right) - f'\left(\frac{Q'}{Q'_n}\right) \right)$$

$$b^* = 200.0 * \left(f'\left(\frac{Q'}{Q'_n}\right) - f'\left(\frac{R'}{R'_n}\right) \right)$$

where the $f'(t)$ function is defined as:

$$f'(t) = \begin{cases} t^{(1/3)} & \text{if } t > 0.009 \\ 7.8(t) + 0.14 & \text{otherwise} \end{cases}$$

6.2   SURF Feature Space Descriptors

Unique and reliable descriptors for each seed image are ensured through the use of the 'Speeded-Up Robust Features' (SURF) algorithm. The algorithm is composed of three primary steps:

1. **Interest Point Detection:** Key points are identified using the Hessian matrix determinant, with feature descriptors based on Haar-wavelet responses. The region type is determined by the Laplacian's sign.
2. **Gradient Information Extraction:** The interest region is divided into $[4 \times 4]$ subregions, samples extracting using HAAR wavelet responses at $[3 \times 3]$, weighted by Gaussian gains to reduce noise.
3. **Feature Descriptor Generation:** A 64-dimensional feature vector is formed by concatenating gradient information from each subregion, enabling effective classification of seed characteristics.

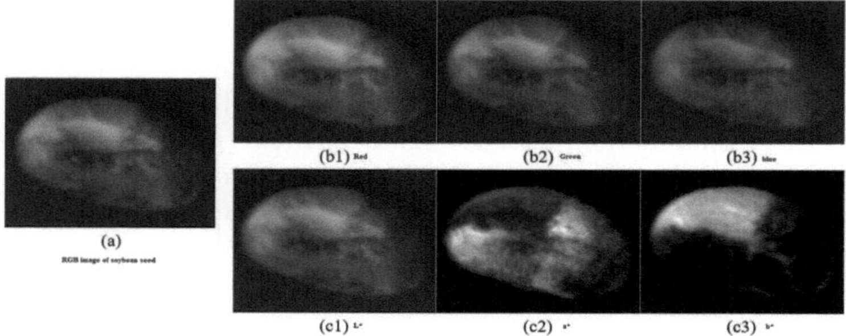

Fig. 4. Conversion of color space from RGB to CIE L* a* b* [16] to quantify visual differences, Nature Scientific Reports (2019) [16].

This comprehensive methodology allows for precise and consistent image capture, conversion, and analysis, forming the foundation for subsequent seed classification tasks.

Unique preprocessing techniques like CLAHE and background subtraction enhance image clarity and isolate seeds effectively. Augmentation methods like Mixup, Cutout, and GAN-based synthesis improve dataset diversity and model generalization under real-world conditions.

Low-Rank Feature Representation: The interpretability of image content and the features of visual words are affected by the size of the visual dictionary used in the JMBoF method. A dictionary that is too small may not fully describe the image features, while a dictionary that is too big may cause redundant semantic expressions. The problem of redundant semantic representation will get worse further as a result of the new joint modal features derived from SURF and $L * a * b*$.

In order to solve this problem, it is necessary to extract a substantial visual dictionary from images by setting a high number of dictionaries at 800. The Low-Rank Representation (LRR) technique is our tool for eliminating redundant, semantic information and effectively representing the image contents. LRR assumes that the 'high' dimensional data Y' has 'low' intrinsic dimensionality. To mitigate the curse of dimensionality, the original data Y' Two components can be decomposed: 1. a low rank matrix X' and 2. an error matrix E':

$$\text{minimize} \quad ||X'||_* + \lambda ||E'||_1$$
$$\text{subject to} \quad Y' = X' + E',$$

where, $|| \cdot ||_*$ denotes the nuclear norm, $|| \cdot ||_1$ denotes the L1 norm, and λ is a regularization parameter. This optimization problem aims to determine which high dimensional data should be projected into a low dimensional subspace. After removing the residual E', the compact visual dictionary set X' is utilized as an effective representation of the original image data.

Algorithm Overview. The method we use to implement our algorithm for classifying soybean categories is as follows:

1. Extract $L*a*b*$ color components from 16 subregions of the image, along with their spatial coordinates.
2. Use Deep CNN to capture complex patterns and spatial hierarchies in soybean image data.
3. Quantize the feature space using K-means and build an $L* a* b*$ visual dictionary with CNN features via the GANBoF algorithm.
4. Combine the $L* a* b*$ visual dictionary with CNN features to create hybrid semantic information.
5. Apply Low-Rank Representation (LRR) to transform the visual dictionary to a low-dimensional space, removing redundancies.
6. Use Support Vector Machine (SVM) to construct hyperplanes that classify the LRR-transformed data into three categories.

Including the ratio of the training and testing datasets would strengthen the evaluation of the proposed system's efficiency. A commonly used split, such as 80% for training and 20% for testing, ensures a large portion of data is used for model learning while retaining a sufficient dataset for unbiased performance evaluation. Alternatively, a 70–30% split or stratified sampling could be adopted if the dataset contains significant class imbalances.

The AI-based seed detection system integrates two core components: a CNN model for analyzing high-resolution soybean seed images to assess quality, and a CNN model that leverages historical climate data to optimize seed selection based on projected environmental conditions. The CNN trained on a diverse dataset captures seed characteristics—such as size, shape, color, and texture—enabling precise identification of viable seeds. During visual inspection, the CNN classifies seeds as viable, damaged, or diseased, reducing the time and labor of manual assessment while achieving over 90% accuracy in identifying viable seeds. For seed selection optimization, the CNN model evaluates historical climate data, including temperature, rainfall, and soil health, to predict seed performance by correlating environmental factors with viability. Early trials in Maharashtra have shown increased yields (up to 20%) due to better seed selection and reduced crop losses, highlighting the potential of AI-driven tools to enhance decision-making and productivity in soybean farming.

7   Conclusion

Deep learning holds transformative potential for soybean seed selection and quality detection, offering farmers data-driven insights that enhance yield and resource efficiency. By applying CNNs, GANs, and RNNs, farmers can automate seed quality assessments and make more informed selections based on historical and environmental data, addressing key challenges like climate variability and resource management. Moving forward, research should prioritize lightweight models for low-cost hardware to benefit small-scale farmers, edge computing for real-time farm-level decisions, and efficient data collection methods to strengthen model training. Collaborative efforts across agronomy, data science, and farming communities are vital to adapt AI solutions for sustainable and resilient soybean agriculture.

References

1. Huang, Z., Zhang, J., Wang, H.: Soybean seed classification using convolutional neural networks. Comput. Electron. Agric. **192**, 1–10 (2022)
2. Lin, W., Shu, L., Zhong, W., Lu, W., Ma, D., Meng, Y.: Online classification of soybean seeds based on deep learning. Eng. Appl. Artif. Intell. **123** (2023). https://doi.org/10.1016/j.engappai.2023.106434
3. Zhao, X., Zhang, L., Liu, F.: Real-time recognition system for soybean seed surface defects using deep convolutional networks. IEEE Access **9**, 103941–103950 (2021)
4. Zhang, X., et al.: Weed identification in soybean seedling stage based on optimized faster R-CNN algorithm. Agriculture **13**(1), 175 (2023). https://doi.org/10.3390/agriculture13010175
5. Nidhi Kundu, A., Rani, G., Dhaka, V.S.: Seeds classification and quality testing using deep learning and YOLO v5. DSMLAI 2021, pp. 153–160. Association for Computing Machinery, New York (2022). https://doi.org/10.1145/3484824.3484913
6. Selvakumar, S., Subramani, S., Salahuddin: Classification learning assisted biosensor data analysis for preemptive plant disease detection. ACM Trans. Sen. Netw. (2022). https://doi.org/10.1145/3572775
7. Qu, H., Bian, Q.: MS-YOLOv5: improved YOLOv5 based on multi-head self-attention for citrus leaf disease severity estimation. In: Proceedings of the 6th International Conference on Machine Learning and Machine Intelligence (MLMI 2023), pp. 85–91. Association for Computing Machinery, New York (2024). https://doi.org/10.1145/3635638.3635651
8. Bhadra, S., Dutta, R., Saha, M.: A 3D CNN approach for soybean yield prediction using UAV images. Remote Sens. Lett. **16**, 120–129 (2024)
9. Chen, M., Jin, C., Ni, Y., Yang, T., Xu, J.: A dataset of the quality of soybean harvested by mechanization for deep-learning-based monitoring and analysis. Data Brief **52**, 109833 (2024). https://doi.org/10.1016/j.dib.2023.109833. ISSN 2352-3409
10. Souza, M., Horikoshi, W.C., Saito, P.T.M., et al.: Soybean seed vigor classification through an effective image learning-based approach. Multimed. Tools Appl. **83**, 13113–13136 (2024). https://doi.org/10.1007/s11042-023-15804-0
11. Arjoune, A., Zoghlami, S., Cherkaoui, H.: Detection and management of soybean cyst nematode using deep learning models: a review. Comput. Biol. Med. **150**, 106–112 (2022)
12. Gómez-Zamanillo, L., Bereciartua, A.: Damage assessment of soybean and redroot amaranth plants in greenhouse through biomass estimation and deep learning-based symptom classification. Smart Agric. Technol. **5**, 100243 (2023). https://doi.org/10.1016/j.atech.2023.100243. ISSN 2772-3755
13. Vyas, S., Thakur, G., Gupta, R., Mehta, H.: IoT based soyabean pest classification using transfer learning and explainable AI. In: 2024 IEEE International Conference on Electronics, Computing and Communication Technologies (CONECCT), Bangalore, India, pp. 1–6 (2024). https://doi.org/10.1109/CONECCT62155.2024.10677291

14. Di, Y., et al.: Bio-DETR: a transformer-based network for pest and seed detection with hyperspectral images. In: 2024 International Joint Conference on Neural Networks (IJCNN), Yokohama, Japan, pp. 1–8 (2024). https://doi.org/10.1109/IJCNN60899.2024.10650195
15. Verma, S., Tripathi. S., Singh, A., Ojha, M., Saxena, R.R.: Insect detection and identification using YOLO algorithms on soybean crop. In: 2021 IEEE Region 10 Conference (TENCON), Auckland, New Zealand, TENCON 2021. pp. 272–277 (2021). https://doi.org/10.1109/TENCON54134.2021.9707354
16. Lin, P., Xiaoli, L., Li, D., et al.: Rapidly and exactly determining postharvest dry soybean seed quality based on machine vision technology. Sci. Rep. 9, 17143 (2019). https://doi.org/10.1038/s41598-019-53796-w
17. Pratap, H., Guru Prasad, M.S., Agarwal, P., Bhardwaj, A., Mehra, A., Mathpati, S.: SoyNet: deep learning approaches for automated soybean seed quality assessment. In: 2024 2nd International Conference on Advancement in Computation and Computer Technologies (InCACCT), Gharuan, India, pp. 613–617 (2024). https://doi.org/10.1109/InCACCT61598.2024.10551180
18. Li, Y., et al.: Soybean seed counting based on pod image using two-column convolution neural network. IEEE Access 7, 64177–64185 (2019). https //doi.org/10.1109/ACCESS.2019.2916931
19. Flores, E.S., Thielo, M.R., Rodrigues Padilha, F.R.: Towards automatic soybean cultivar identification: SoyCult dataset and deep learning baselines. In: 2023 36th SIBGRAPI Conference on Graphics, Patterns and Images (SIBGRAPI), Rio Grande, Brazil, pp. 151–156 (2023). https://doi.org/10.1109/SIBGRAPI59091.2023.10347155
20. Kaler, N., Bhatia, V., Mishra, A.K.: Deep learning-based robust analysis of laser bio-speckle data for detection of fungal-infected soybean seeds. IEEE Access 11, 89331–89348 (2023). https://doi.org/10.1109/ACCESS.2023.3305273
21. Kusumaningrum, D., et al.: Non-destructive technique for determining the viability of soybean (Glycine max) seeds using FT-NIR spectroscopy. J. Sci. Food Agric. 98, 1734–1742 (2018)
22. Samiei, S., Rasti, P., Ly Vu, J., et al.: Deep learning-based detection of seedling development. Plant Methods 16, 103 (2020). https://doi.org/10.1186/s13007-020-00647-9
23. Luo, J., Li, B., Leung, C.: A survey of computer vision technologies in urban and controlled-environment agriculture. ACM Comput. Surv. 56(5), 39, Article no. 118 (2023). https://doi.org/10.1145/3626186
24. Yang, P., Li, D., Shi, Y.: High-throughput phenotyping of soybean seeds using convolutional neural networks and transfer learning. Front. Plant Sci. 12, 1119–1133 (2021)
25. Li, W., Wei, P., Sun, J., Xiao, X., Mu, X., Hu, Z.: LEPD-net: a lightweight efficient network with pyramid dilated convolution for seed sorting. In: Proceedings of the 2022 11th International Conference on Computing and Pattern Recognition (ICCPR 2022), pp. 551–558. Association for Computing Machinery, New York (2023). https://doi.org/10.1145/3581807.3581888

ViTFuser: Advancements in Global Context Understanding for Autonomous Vehicles

Atanu Chatterjee[(✉)], P. Ajay Vikram, Aniketh Narayan Bellala,
T. Naga Tarun, Basavaraj Talawar, M. Vani, and Jeny Rajan

Department of Computer Science and Engineering, National Institute of Technology
Karnataka, Surathkal, India
atanuc13@gmail.com

Abstract. In autonomous driving, integrating data from complementary sensors is crucial for navigating complex scenarios with dense traffic and dynamic agents. Previous state-of-the-art methods, particularly TransFuser, experience feature loss due to downsampling from pixel-wise cross attention with a transformer encoder, hindering their ability to grasp the global context of traffic situations. To address this, we introduce ViTFuser, a self-attention-based method that utilizes Vision Transformers (ViTs) at multiple resolutions to fuse perspective and bird's eye view feature maps. The incorporation of a Feature Pyramid Network (FPN) enhances multi-scale feature fusion, improving traffic object detection and enabling global contextual reasoning in the fusion process. We validated ViTFuser on two widely recognized benchmarks, Longest6 and Town05, achieving Driving Scores (DS) of 55.15, 91.07, and 74.95 for Longest6, Town05 short, and Town05 long, respectively, outperforming several previous works in the field.

Keywords: TransFuser · ViT · FPN · CARLA · Longest6

1 Introduction

Autonomous driving represents the fusion of cutting-edge technology and transportation, aiming for safer and more efficient roads. At its core is the development of advanced machine learning models capable of independently perceiving and navigating complex environments. Real-world driving presents a formidable challenge due to the difficulty of capturing data representing diverse scenarios. Consequently, simulators like CARLA [15] have gained prominence for advancing end-to-end autonomous driving systems. However, achieving a comprehensive understanding of the driving environment and making accurate predictions about its evolution remains a central challenge. The simulator's array of sensors enables detailed establishment of driving scenarios. Many existing models are based on Reinforcement Learning [3,26] and Imitation Learning [5,10,11]. Early approaches relied heavily on LiDAR [12,25], utilizing 3D traffic representations,

C. Modi et al. (Eds.): MIND 2024, CCIS 2736, pp. 128–140, 2026.
https://doi.org/10.1007/978-3-032-14531-4_11

but faced limitations due to restricted access to LiDAR data and the narrow perspective offered. Similarly, using RGB sensors alone does not provide a complete spatial understanding. Hence, combining LiDAR and RGB capabilities is essential for a more holistic view [24].

Previous multi-sensor fusion architectures [17,18] focused on perception tasks by projecting driving scenarios onto 2D or 3D spaces, where geometric and semantic information is aggregated. While effective, these methods often emphasize local context, potentially constraining the ability to anticipate dynamic movements for safe driving. In contrast, attention-based architectures have emerged as a new approach for sensor fusion in autonomous vehicles [27]. Attention-based transformers [14,20] allocate attention to relevant features across sensor modalities, using multi-headed self-attention mechanisms to capture complex dependencies and infer global contextual information beyond spatial relationships. Motivated by the transformative potential of attention-based methods, recent architectures [7,8,29] in end-to-end driving have adopted imitation-based fusion approaches. Among these, TransFuser [8] stands out, combining a multi-modal fusion transformer with an auto-regressive waypoint prediction network to effectively integrate RGB and LiDAR information across frames, capturing global context and spatial dependencies to predict future vehicle trajectories.

While TransFuser [8] represents progress, it faces challenges with complex spatial dependencies and scaling. Inspired by Vision Transformers' (ViTs) [14] success in image classification, we propose ViTFuser, a multi-stage fusion framework incorporating ViTs to process visual data from LiDAR and RGB images. ViTFuser aims to enhance the perception and understanding of driving environments by leveraging ViTs' capabilities. Additionally, advances in object detection [6,30] using Feature Pyramid Networks (FPNs) [19] have demonstrated the benefits of multi-scale feature integration. Incorporating FPNs in the proposed model can capture rich spatial and semantic information, improving perception accuracy and prediction in autonomous driving scenarios.

The main contributions of this paper can be summarised as follows: (1) We have suggested ViTFuser, in which we have highlighted the shortcomings of transformers in comprehending global contexts and shown how we have enhanced this capability by utilizing ViTs. (2) The incorporation of FPN improves the object detection block in order to decrease traffic violations. (3) The proposed model is successfully evaluated on two widely recognised benchmarks, i.e. Longest6 and Town05, and has outperformed certain state-of-the-art research in the field of autonomous driving.

2   Related Work

End-to-end autonomous driving has progressed significantly through simulators and the exploration of reinforcement learning (RL) and imitation learning (IL) techniques. Chen et al. [3] introduced an RL approach that learns vision-based driving policies using forward models to predict driving trajectories. Toromanoff et al. [26] proposed 'implicit affordances,' where an encoder predicts essential driving parameters, which serve as the RL state during training. Zhang et al. [31] utilized a distillation method to have an RL agent with privileged information teach a new model to mimic behavior with regular data. In contrast, Imitation Learning involves models imitating expert demonstrations without direct environmental interaction. Codevilla et al. [10,11] introduced conditional imitation learning via behavior cloning, allowing the IL agent to learn sensory motor policies from expert input observations. Chen et al. [5] developed a two-stage training method, combining a trained IL agent with privileged information and a vision-based agent without such information. Chen et al. [4] improved complex reasoning by learning an invariant intermediate representation and predicting future trajectories using multi-modal sensor inputs.

Attention mechanisms [27] have transformed sequence modeling and have been successfully applied in image tasks. Dosovitskiy et al. [14] introduced the Vision Transformer for efficient processing of image patches, while Liu et al. [20] enhanced efficiency with a hierarchical architecture using shifted windows. These advancements influenced autonomous driving models, such as Yang et al. [29], who proposed Perspective-to-BEV Image Translation with Attention for improved scene understanding. Chitta et al. [7] introduced a function mapping Bird's Eye View (BEV) coordinates to waypoints, utilizing attention maps to compress high-dimensional image features.

Xu et al. [28] presented M2DA, which enhances anticipatory driving by integrating driver attention into a lightweight convolutional backbone, improving predictability through traditional multi-modal fusion. The BEVGuide framework [21] improves 3D perception by fusing data from cameras, Lidar, and Radar with a BEV-guided attention mechanism, enhancing accuracy and robustness while allowing flexible sensor combinations. Dutta et al. [16] developed a method for generating BEV maps with semantic segmentation using a hierarchical Vision Transformer model for efficient context capture and semantic enhancement. Alaba and Ball [2] proposed a multimodal fusion model for 3D object detection that combines camera and LiDAR data with a voxel-based representation, incorporating multiple ResNet backbones and a transformer decoder for improved spatial interactions and robust detection.

3   Methodology

In this paper, ViTFuser is proposed as an enhancement to the encoder module of the TransFuser [8] for wide attention and to capture the global context of the scene more effectively, and utilization of feature pyramid network (FPN) [19] for bounding box prediction decoder

3.1   Proposed Model Inputs and Outputs

The model receives two types of inputs, as illustrated in Fig. 1. RGB images are captured from three front-facing cameras positioned at $0°$, $-60°$, and $60°$, each

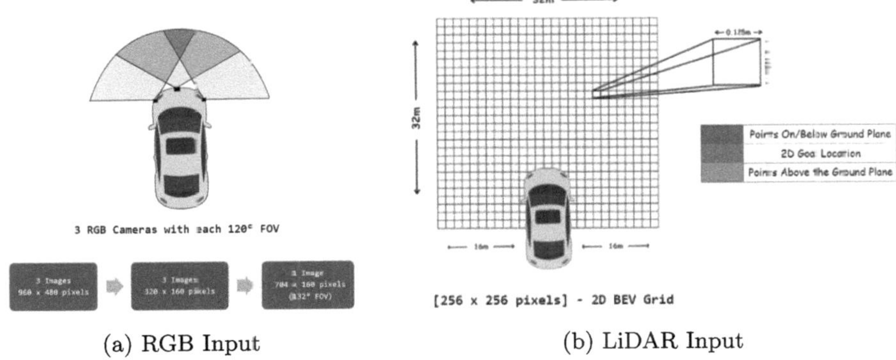

(a) RGB Input   (b) LiDAR Input

Fig. 1. Inputs to the Proposed Method.

with a 120° field of view. These images are cropped and combined into a single 704 × 160 pixel image, resulting in an effective field of view of 132°. Meanwhile, the LiDAR point cloud is mapped onto a 256 × 256 pixel 2D grid with histogram bins for points on/below and above the ground plane, where each block measures 0.125 m × 0.125 m. The 2D goal location is also rasterized onto this grid, forming a three-channel representation. Both inputs are then processed in their respective encoder branches. The model outputs a trajectory for the vehicle as a sequence of waypoints, with a default of four waypoints. These waypoints are subsequently utilized by a PID controller for low-level vehicle control.

3.2 Proposed Model Architecture

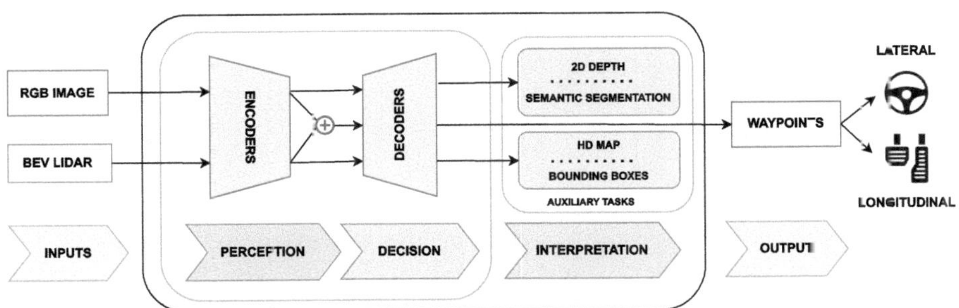

Fig. 2. High Level Abstraction of the Proposed Model Architecture.

The proposed model, illustrated in Fig. 2, predicts target waypoints for navigation using LiDAR and image inputs. These inputs are encoded into feature maps and reduced through multiple layers of CNN encoders and transformers, preserving essential global context. The condensed feature map is then processed by a Gated Recurrent Unit (GRU) [9] waypoint network to predict the current and four future waypoints. Skip connections facilitate the transfer of encoded feature maps between the CNN encoders of each branch (Image and LiDAR), enabling the decoders to access these maps for auxiliary task assessments. The model can be viewed as two components: (1) Perception module – Encoder and (2) Decision module – Decoder.

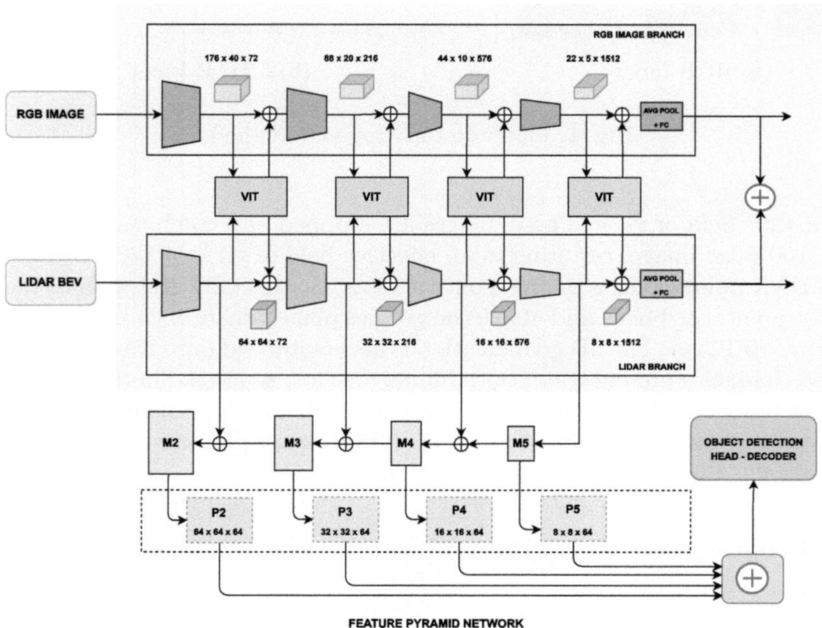

Fig. 3. Encoder Architecture in the Proposed Model.

Perception Module Encoder: The encoder module condenses feature maps from preprocessed sensor inputs, capturing the global context of the scene for navigation decisions, thus serving as a perception module. It comprises two branches—Image and LiDAR—that convolve inputs using CNN encoders across multiple stages, as illustrated in Fig. 3. Outputs from these CNN encoders are fed into the transformer, which integrates information from both RGB images (for traffic light status, vehicle presence, and road conditions) and LiDAR (for depth estimation and navigation). The transformer outputs are then returned to their respective branches at different stages for further processing.

Transformer Inputs and Outputs: The transformer processes feature maps from both RGB and LiDAR channels by downsizing them to a fixed size and performing attention on the concatenated feature map, resulting in an output of the same size. This is achieved by removing the MLP head and classification token to obtain the feature maps prior to concatenation. The output feature map is then upsampled using bilinear interpolation to match the original size before downsampling, added to the pre-downsampling feature map, and passed on to subsequent stages.

Transformer: Attention allows Transformers to focus on relevant input parts by dynamically weighing the importance of different elements. In image processing, attention is applied to pixels using the scaled dot-product attention formula, as shown in Eq. 1 [14]. The variables Q, K, and V represent the query, key, and value matrices, respectively. Dividing by the square root of the key dimension d_k prevents excessively large gradients during training. The softmax function normalizes the attention scores across the input sequence, resulting in a weighted sum that provides the attention output.

$$\text{Attention}(Q, K, V) = \text{softmax}\left(\frac{QK^T}{\sqrt{d_k}}\right) V \tag{1}$$

The transformer encoder in TransFuser [8] is computationally expensive due to its attention on all image pixels. In contrast, the Vision Transformer (ViT) [14] improves efficiency by applying attention to image patches, capturing granular dependencies while reducing parameters by 67%. Our model employs 55,540,538 parameters, significantly fewer than TransFuser's 168,018,327. Additionally, the transformer encoder has quadratic complexity with respect to the number of pixels in feature maps, thereby causing a loss of information due to the downsampling. Subsequently, upsampling from this small size adds a lot of noise to the image. The Vision Transformer takes in a more optimal feature map, balancing between feature retention and computational efficiency, thereby adding less noise to the transformer's output. Although the Swin Transformer [20] outperforms ViT in many tasks, it has a 91% increase in parameters, totaling 688,158,741, due to its four internal transformers. Its multiple stages and downsampling lead to a smaller final feature map, requiring extensive upsampling that can introduce noise. Consequently, we opted for ViT over both the transformer encoder and Swin Transformer.

Decision Module Decoder: The decoder module serves two main functions: (1) predicting target waypoints through a waypoint prediction network to guide the ego vehicle and (2) evaluating model performance using four auxiliary tasks. The feature map is input into four GRU decoders along with ground truth locations for training, enabling accurate waypoint prediction. The auxiliary tasks enhance model accuracy by incorporating individual loss functions and providing visual interpretations through semantic segmentation images, 2D depth estimations, HD maps, and bounding boxes. Ablation studies on TransFuser [8] show that removing any of these auxiliary tasks—bounding boxes, HD maps, or semantic segmentation maps—significantly reduces accuracy, emphasizing their importance in target waypoint prediction.

Object Detection Head: The primary objective of this head is to detect bounding boxes around objects of interest, such as vehicles, in the context of autonomous driving. Hence, the focus is solely on bounding box detection and does not involve classification tasks, as it operates within a single class context considering all types of vehicles as a single class. Enhancing the accuracy of bounding box predictions leads to improved interpretation of feature maps from encoders and subsequently contributes to a higher driving score.

The object detection model typically consists of three fundamental components: (1) Encoder – Sequence of convolutional operations applied to input data, resulting in the extraction of essential features. (2) Neck (Optional) - The intermediary between the encoder and decoder plays a crucial role in consolidating and refining feature representations. (3) Decoder - interprets the refined features and generates output predictions, such as bounding box coordinates, based on the learned representations.

The TransFuser model [8] employs two encoders with a specific convolutional layer configuration optimized for efficient feature extraction. Its object detection head predicts bounding boxes in the format $[x_{tl}, y_{tl}, x_{br}, y_{br}]$, where $\{x_{tl}, y_{tl}\}$ are the top-left coordinates and $\{x_{br}, y_{br}\}$ are the bottom-right coordinates. However, the absence of an intermediate neck component limits its ability to leverage multi-scale features. Integrating a Feature Pyramid Network (FPN) [19] as a neck component can significantly enhance feature representation by aggregating information across multiple scales and resolutions, thereby capturing richer spatial and semantic details.

Feature Pyramid Network: FPN, or Feature Pyramid Network [19], is a neural network architecture commonly used in computer vision tasks, particularly in the domain of object detection. It addresses the challenge of effectively detecting objects at various scales by producing a multi-scale feature pyramid. The network achieves this by aggregating features from different levels of resolution within the network architecture, enabling it to capture both fine-grained and coarse information simultaneously. FPN addresses the limitations of models without a neck component by providing a systematic approach to multi-scale feature extraction. This results in improved localization and detection accuracy, leading to more reliable performance in object detection tasks. Our model architecture can be extended to incorporate FPN as a neck component, enabling us to leverage its benefits for object detection. FPN can be seamlessly integrated into the existing architecture, with inputs taken directly after the CNN encoder stages, before passing on to the transformers. With minimal changes to the architecture, the model's Infraction score(IS) is enhanced as the traffic violations are reduced, leading to an overall improvement in the Driving Score (DS).

3.3   Loss Functions

The network is trained using an L1 loss function calculated between the predicted waypoints and the ground truth waypoints (obtained from the expert) aligned with the current coordinate frame [5]. Let w_t^{gt} denote the ground truth waypoint at time-step t. The loss function is defined in Eq. 2.

$$\mathcal{L} = \sum_{t=1}^{T} \|w_t - w_t^{gt}\|_1 \tag{2}$$

The initial ground truth waypoints w_t^{gt}, available only during training, undergo certain transformations before being inputted (as sparse goal locations $\mathcal{G}$) into the GRU decoders, in both training and testing phases.

$$\text{Total Loss} = w_0 \times \text{Target Loss} + \sum_{i=1}^{N} w_i \times \text{Auxiliary Loss}_i \tag{3}$$

$$\text{where } w_i = \frac{1}{N+1} \text{ for } i = 0, 1, \ldots, N \tag{4}$$

The Total Loss, depicted in Eq. 3, encompasses the L1 loss function for predicting target waypoints alongside various auxiliary losses. Notably, each loss contributes equally to the overall loss calculation, ensuring a balanced consideration of all components.

4   Experiments and Results

4.1   Dataset

We used the same dataset as TransFuser [8], recorded at 2 FPS using a rule-based expert system on CARLA [15] 0.9.10, which provides privileged information. The dataset spans 3,500 routes across all eight public towns under various weather conditions, including challenging scenarios like roundabouts, junctions, traffic signals, cyclists, and pedestrians, totaling approximately 230k frames [1].

4.2   Metrics

We employed three primary metrics for evaluation: (1) Route Completion (RC), measuring the percentage of a route completed; (2) Infraction Score (IS), which starts at 1.0 and decreases based on penalties for traffic violations (0.7 for running a red light, 0.65 for a collision with a static arrangement, 0.6 for a vehicle accident, and 0.5 for a pedestrian collision). (3) Driving Score (DS), calculated as RC multiplied by IS, reflecting overall safety and progress. We also recorded the frequency of infractions per kilometer.

4.3 Benchmarks

Our model was evaluated on the Town05 [22] and Longest6 [8] benchmarks. The Longest6 consists of 36 routes averaging 1500 m across six towns under varied weather and daylight conditions. The Town05 benchmark includes 10 long routes (1000–2000 m) and 32 short routes (100–500 m), all under Clear-Noon conditions, with random start and end points and traffic scenarios placed along the routes. The goal is to complete routes with minimal traffic violations.

4.4 Implementation

Our architecture features two encoders: RGB images from three cameras (left, center, right) are concatenated and processed by one encoder, while the LiDAR BEV is processed by another. Both use RegNetY-3.2GF [23], with one image encoder initialized using ImageNet weights [13]. Four Vision Transformers (ViTs) [14] at different resolutions $\{64\times64, 32\times32, 16\times16, 8\times8\}$ are used, with features concatenated and down-sampled to match LiDAR dimensions. Each ViT has four attention heads, four hidden layers, and employs embedded dimensions E_d: $\{144, 108, 288, 189\}$ with patch sizes P_s: $\{4, 4, 4, 2\}$. The FPN channels match encoder resolutions $\{72, 216, 576, 1512\}$, with an output channel of 64. Models were trained for 41 epochs on two NVIDIA V-100 GPUs, using an ensemble of epochs 33, 37, and 41 for evaluation, retaining the same batch size, learning rate, and optimizer as TransFuser.

4.5 Results and Discussions

Evaluation in the benchmarks is non-deterministic due to the random initialization of start and end points and varying traffic conditions along routes, leading to inconsistent outcomes across runs. To mitigate evaluation variance, we conducted three evaluations of our model on both benchmarks. We compared our model's performance with three state-of-the-art (SotA) architectures on the Longest6 benchmark [8] and six SotA designs on the Town05 benchmark [22]. For Trans-Fuser, we used pre-trained weights from the authors [1], while other models' scores were taken from their respective works. Table 1 shows that ViTFuser outperformed all other models on Longest6, with an 8-point improvement in Driving Score (DS) and a 3-point gain in Route Completion (RC) over TransFuser. Adding FPN further boosted DS, RC, and Infraction Score (IS) to 55.15, 81.43, and 0.69, respectively. As shown in Table 2, for Town05, our models excelled in both the short and long settings. In the short benchmark, ViTFuser achieved a DS of 90.04 and RC of 94.79, while the FPN version scored 91.07 in DS and 94.67 in RC. In the long setting, ViTFuser attained a DS of 74.82 and RC of 92.40, with FPN reaching 74.95 in DS and 94.9 in RC. These results demonstrate the superior performance of our model across various benchmarks, reinforcing its effectiveness in enhancing 3D object detection in autonomous driving scenarios.

ViTFuser differs from M2DA [28] by focusing on object detection with a dedicated head for streamlined bounding box predictions, leveraging self-attention

Table 1. Performance comparison of our model with others on **Longest6** [8], in terms of **DS** (Driving Score), **RC** (Route Completion), and **IS** (Infraction Score). The **mean** and **standard deviations** for three evaluations are shown.

Method	DS ↑	RC ↑	IS ↑
WOR [3]	20.53 ± 3.12	48.47 ± 3.86	0.56 ± 0.03
LAV [4]	32.74 ± 1.45	70.36 ± 3.14	0.51 ± 0.02
TransFuser [8]	43.48 ± 1.95	77.72 ± 1.12	0.60 ± 0.01
ViTFuser	51.96 ± 1.54	80.70 ± 6.47	0.65 ± 0.04
ViTFuser + FPN	**55.15** ± 4.46	**81.43** ± 4.62	**0.69** ± 0.04

Table 2. Performance comparison of our model with others on **Town05** [22] short and long bechmarks. Driving Score(**DS**), Route Completion (**RC**) are reported with **mean** and **standard deviation** over 3 evaluations.

Method	Town05 Short		Town05 Long	
	DS ↑	RC ↑	DS ↑	RC ↑
CILRS [11]	7.47 ± 2.51	13.40 ± 1.09	3.68 ± 2.16	7.19 = 2.95
LBC [5]	30.97 ± 4.17	55.01 ± 5.14	7.05 ± 2.13	32.09 ± 7.40
NEAT [7]	58.70 ± 4.11	77.32 ± 4.91	37.72 ± 3.55	62.13 ± 4.66
Roach [31]	65.26 ± 3.63	88.24 ± 5.16	43.64 ± 3.95	80.37 ± 5.68
WOR [3]	64.79 ± 5.53	87.47 ± 4.68	44.80 ± 3.69	82.41 ± 5.01
TransFuser [8]	87.48 ± 1.93	92.78 ± 0.31	67.85 ± 5.02	91.57 ± 5.67
ViTFuser	90.04 ± 2.19	**94.79** ± 0.89	74.82 ± 5.67	92.40 ± 2.42
ViTFuser + FPN	**91.07** ± 1.14	94.67 ± 2.44	**74.95** ± 2.61	**94.90** ± 2.86

mechanisms and FPN for computational efficiency. Unlike the BEVGuide framework [21], which emphasizes flexible sensor fusion, ViTFuser reduces parameter count and enhances global context understanding through multi-modal transformers. In contrast to [16], our model integrates multi-scale features from the RegNet encoder with a lightweight FPN, rather than relying on a fixed encoder-decoder structure. Finally, unlike the voxel-based fusion model in [2], ViTFuser employs a stage-wise fusion strategy with transformers to improve global context awareness and detection performance.

5 Conclusion

We introduced ViTFuser, an autonomous driving framework that leverages Vision Transformers to integrate sensor data for a comprehensive understanding of the environment and generate interpretable outputs for the safety controller. By substituting the transformer encoder in TransFuser with ViT, we improved global context understanding and reduced GPU memory usage by threefold. ViTFuser outperformed several architectures in the Longest6 and

Town05 benchmarks, and integrating FPN further reduced traffic violation scores while enhancing the Driving Score. The model's limitations include training on simulator-generated datasets with predefined urban scenarios, which lack the variability of real-world traffic conditions, and reliance on ground truths that may not be available in practice. Future work includes integrating BiFPN for enhanced object detection, optimizing inference through model quantization and lightweight CNNs, and applying ViTFuser to multi-sensor fusion with LiDAR and radar for robust decision-making in adverse weather conditions.

References

1. Prakash, A., Jaeger, B., Chitta, K.: GitHub repository of TransFuser'2022 PAMI. https://github.com/autonomousvision/transfuser. Accessed 15 Mar 2024
2. Alaba, S.: Transformer-based optimized multimodal fusion for 3D object detection in autonomous driving. IEEE Access **12**, 50165–50176 (2024). https://doi.org/10.1109/ACCESS.2024.3385439
3. Chen, D., Koltun, V., Krähenbühl, P.: Learning to drive from a world on rails. In: Proceedings of the IEEE/CVF International Conference on Computer Vision, pp. 15590–15599 (2021)
4. Chen, D., Krähenbühl, P.: Learning from all vehicles. In: Proceedings of the IEEE/CVF Conference on Computer Vision and Pattern Recognition, pp. 17222–17231 (2022)
5. Chen, D., Zhou, B., Koltun, V., Krähenbühl, P.: Learning by cheating. In: Conference on Robot Learning, pp. 66–75. PMLR (2020)
6. Chen, K., et al.: Hybrid task cascade for instance segmentation. In: Proceedings of the IEEE/CVF Conference on Computer Vision and Pattern Recognition, pp. 4974–4983 (2019)
7. Chitta, K., Prakash, A., Geiger, A.: Neat: neural attention fields for end-to-end autonomous driving. In: Proceedings of the IEEE/CVF International Conference on Computer Vision, pp. 15793–15803 (2021)
8. Chitta, K., Prakash, A., Jaeger, B., Yu, Z., Renz, K., Geiger, A.: Transfuser: imitation with transformer-based sensor fusion for autonomous driving. IEEE Trans. Pattern Anal. Mach. Intell. **45**, 12878–12895 (2022)
9. Cho, K., et al.: Learning phrase representations using RNN encoder-decoder for statistical machine translation. arXiv preprint arXiv:1406.1078 (2014)
10. Codevilla, F., Miiller, M., López, A., Koltun, V., Dosovitskiy, A.: End-to-end driving via conditional imitation learning. In: 2018 IEEE International Conference on Robotics and Automation (ICRA), pp. 1–9 (2018)
11. Codevilla, F., Santana, E., López, A.M., Gaidon, A.: Exploring the limitations of behavior cloning for autonomous driving. In: Proceedings of the IEEE/CVF International Conference on Computer Vision, pp. 9329–9338 (2019)
12. Cui, A., Casas, S., Sadat, A., Liao, R., Urtasun, R.: Lookout: diverse multi-future prediction and planning for self-driving. In: 2021 IEEE/CVF International Conference on Computer Vision (ICCV), pp. 16087–16096 (2021)
13. Deng, J., Dong, W., Socher, R., Li, L.J., Li, K., Fei-Fei, L.: Imagenet: a large-scale hierarchical image database. In: 2009 IEEE Conference on Computer Vision and Pattern Recognition, pp. 248–255 (2009)

14. Dosovitskiy, A., et al.: An image is worth 16 × 16 words: transformers for image recognition at scale. In: International Conference on Learning Representations (2021)
15. Dosovitskiy, A., Ros, G., Codevilla, F., Lopez, A., Koltun, V.: Carla: an open urban driving simulator. In: Conference on Robot Learning, pp. 1–16. PMLR (2017)
16. Dutta, P., Sistu, G., Yogamani, S., Galváin, E., McDonald, J.: Vit-bevseg: a hierarchical transformer network for monocular birds-eye-view segmentation (2022). https://doi.org/10.48550/arXiv.2205.15667
17. Fadadu, S., et al.: Multi-view fusion of sensor data for improved perception and prediction in autonomous driving. In: 2022 IEEE/CVF Winter Conference on Applications of Computer Vision (WACV), pp. 3292–3300 (2022)
18. Liang, M., Yang, B., Chen, Y., Hu, R., Urtasun, R.: Multi-task multi-sensor fusion for 3D object detection. In: 2019 IEEE/CVF Conference on Computer Vision and Pattern Recognition (CVPR), pp. 7337–7345 (2019)
19. Lin, T.Y., Dollár, P., Girshick, R., He, K., Hariharan, B., Belongie, S.: Feature pyramid networks for object detection. In: Proceedings of the IEEE Conference on Computer Vision and Pattern Recognition, pp. 2117–2125 (2017)
20. Liu, Z., et al.: Swin transformer: hierarchical vision transformer using shifted windows. In: 2021 IEEE/CVF International Conference on Computer Vision (ICCV), pp. 9992–10002 (2021)
21. Man, Y., Gui, L.Y., Wang, Y.X.: Bev-guided multi-modality fusion for driving perception. In: Proceedings of the IEEE/CVF Conference on Computer Vision and Pattern Recognition (CVPR), pp. 21960–21969 (2023)
22. Prakash, A., Chitta, K., Geiger, A.: Multi-modal fusion transformer for end-to-end autonomous driving. In: Proceedings of the IEEE/CVF Conference on Computer Vision and Pattern Recognition, pp. 7077–7087 (2021)
23. Radosavovic, I., Kosaraju, R.P., Girshick, R., He, K., Dollár, P.: Designing network design spaces. In: Proceedings of the IEEE/CVF Conference on Computer Vision and Pattern Recognition, pp. 10428–10436 (2020)
24. Ros, G., Sellart, L., Materzynska, J., Vazquez, D., Lopez, A.M.: The synthia dataset: a large collection of synthetic images for semantic segmentation of urban scenes. In: 2016 IEEE Conference on Computer Vision and Pattern Recognition (CVPR), pp. 3234–3243 (2016)
25. Sadat, A., Casas, S., Ren, M., Wu, X., Dhawan, P., Urtasun, R.: Perceive, Predict, and Plan: Safe Motion Planning Through Interpretable Semantic Representations. In: Vedaldi, A., Bischof. H., Brox, T., Frahm, J.-M. (eds.) ECCV 2020. LNCS, vol. 12368, pp. 414–430. Springer, Cham (2020). https://doi.org/10.1007/978-3-030-58592-1_25
26. Toromanoff, M., Wirbel, E., Moutarde, F.: End-to-end model-free reinforcement learning for urban driving using implicit affordances. In: 2020 IEEE/CVF Conference on Computer Vision and Pattern Recognition (CVPR), pp. 7151–7160 (2020)
27. Vaswani, A., et al.: Attention is all you need. In: Proceedings of the 31st International Conference on Neural Information Processing Systems, pp. 6000–6010 (2017)
28. Xu, D., Li, H., Wang, Q., Song, Z., Chen, L., Deng, H.: M2da: multi-modal fusion transformer incorporating driver attention for autonomous driving (2024). https://arxiv.org/abs/2403.12552
29. Yang, W., et al.: Projecting your view attentively: monocular road scene layout estimation via cross-view transformation. In: 2021 IEEE/CVF Conference on Computer Vision and Pattern Recognition (CVPR), pp. 15531–15540 (2021)

30. Yao, Z., Cao, Y., Zheng, S., Huang, G., Lin, S.: Cross-iteration batch normalization. In: Proceedings of the IEEE/CVF Conference on Computer Vision and Pattern Recognition, pp. 12331–12340 (2021)
31. Zhang, Z., Liniger, A., Dai, D., Yu, F., Van Gool, L.: End-to-end urban driving by imitating a reinforcement learning coach. In: Proceedings of the IEEE/CVF International Conference on Computer Vision, pp. 15222–15232 (2021)

Generative Adversarial Networks for Enhanced Image Restoration and Anti-Forensics

H. Faizal Ahamed and M. Brindha[✉]

Department of Computer Science and Engineering, National Institute of Technology, Tiruchirappalli, Tiruchirappalli 620015, Tamil Nadu, India
{faizal,brindham}@nitt.edu

Abstract. Forensic analysis of images is used to detect various types of manipulations and forgeries. Median filters is one the anti-forensic methods which is used to smooth the visual characteristics of the altered image. Most of the forensic methods rely on detecting whether median-filtering or any other image restoration techniques are applied to the given image. However, image restoration techniques have a hard time making the statistical characteristics of the altered image similar to the original one. In this work, an anti-forensic method based on Generative Adversarial networks is presented, which is used to remove the traces of median filtering from the altered images and thus increases forensic undetectable. Through experiments, the proposed system achieved the improvement of 2.199 dB Peak Signal-to-Noise Ratio (PSNR) and 0.0279 Structural Similarity Index (SSIM) over the existing median filtering techniques.

Keywords: Anti-forensics · Generative-Adversarial networks · Median filtering

1 Introduction

The field of anti-forensics deals with countering the forensic methods of analysis which seek to detect any manipulation in images. The anti-forensic techniques mostly involve filtering methods which aim to erase the processing history of the manipulated images [3]. One of the most common filtering methods used in anti-forensics is median filtering. These anti-forensic methods can in turn lead to improved forensic detection mechanisms. However, it is difficult to restore the median-filtered image to look visually as close to the original image as possible. The restored image also varies in certain statistical characteristics with respect to the original image. Making the altered images visually as well as statistically similar to the unaltered image is an ongoing research problem [11].

The median filtering is much suited for image restoration techniques because of it's non-linearity. There are several forensic detection mechanisms have been

C. Modi et al. (Eds.): MIND 2024, CCIS 2736, pp. 141–151, 2026.
https://doi.org/10.1007/978-3-032-14531-4_12

invented in order to detect any manipulations of images. One of the most common way in which this is done is to detect if the image has been subject to median filtering since this is usually done to hide the processing history of the images [5].

Several detectors have been proposed in order to detect median filtering. These detectors are developed with the assumption that any kind of forgery changes the statistical characteristics of the image. However median filtering also produces irregularities in the visual appearance of the image. In order to counter these forensic techniques, several anti-forensic mechanisms have also been developed [6]. These anti-forensic attacks seek to figure out the weakness in forgery detectors, which in turn may lead to improved forgery detection algorithms [10].

Generative Models. Generative models require access to a large dataset, which is used to generate data that closely resembles real-world distributions. The model typically contains fewer parameters than the data distribution, compelling it to learn a condensed representation or essential features of the real-world data.

A Variational Autoencoder (VAE) employs Bayesian inference to model and generate data that aligns with a given distribution. It consist of an encoder, a decoder, and a loss function. The loss function is designed to optimize the lower bound of the log-likelihood of the data, enabling the generation of realistic data samples.

Generative-Adversarial Networks (GAN). The GANs consist of two main components: the generator and the discriminator. The generator creates data from random inputs that initially look synthetic. The discriminator is tasked with differentiating between real and generated data. It evaluates data from both the true distribution and the generator's outputs, producing a probability score that reflects whether the data is real or artificial.

The training procedure for GANs follows a game-theoretic approach, where the generator and discriminator engage in a competitive process. They alternate their training, each aiming to outsmart the other, continuing until they reach a Nash equilibrium. In this study, GANs are applied to generate synthetic images that closely resemble real images, both in terms of visual quality and statistical characteristics.

Vanishing Gradients. One of the biggest problems in deep neural networks is the vanishing gradients problem. In neural networks that are trained by back propagation, each weight is updated by a factor that is proportional to the gradient of the loss in each step of the training process. In cases where the network is very deep, the gradient becomes vanishing small that the weight does not change it's value at all. This causes a domino effect which leads to stagnation and a incomplete halt in training. The training process becomes gradually slow from the final to the initial layers. The earlier layers are the worst-affected. This is a big problem since it is the earlier layers that detect simple patterns and form the fundamental building block of the whole network.

To address the issue of vanishing gradients, Microsoft developed the residual neural network. This architecture consists of a residual learning block repeated N times. In this design, the output of the network is added to the previous input, forming the final output of the residual block. This approach is more efficient for optimizing the residual mapping. The key idea is that with each new layer, the network learns new information rather than simply relying on the encoded values from the previous residual block. These skip connections have proven effective in overcoming the vanishing gradients problem, enabling the creation of much deeper neural networks. In this work, the GAN is implemented using a deep residual network.

Mode collapse is one of the most common problems in training GAN's. This happens when the diversity of samples generated by the generator is very poor. This mostly happens in case of a binary or multi-modal distribution. Rather than learning the underlying distribution, the generator learns only a few modes of the distribution and hence is easily able to fool the discriminator This results in fast convergence of generator loss, but with poor quality generated samples. Thus, the generator may collapse to a setting where it always produces the same output.

The proposed method strives to ensure that the altered image closely matches the original, both visually and statistically. This is achieved through the joint operation of the Generator which is used for creation of realistic images from median-filtered inputs and the Discriminator which is used to differentiate the authentic and generated images. The networks are trained within an adversarial framework, where each network enhances its performance by competing against the other. The effectiveness of the GAN model is evaluated using two essential metrics: PSNR and SSIM.

2 Literature Review

The GAN was first proposed by Ian Goodfellow in [2]. In this work, a novel architecture was proposed for learning a probability distribution and generating samples of it without class labels. The algorithm for training a GAN was also proposed in [2]. Kahe,V-xiangz,V-shren and Jiansun propose a deep residual neural network in [3] which solves the problem of vanishing gradients using a unique solution called 'skip' connections.

Kirchner et al. introduced the use of streaking artifacts as a method for detecting median-filtered images in [4]. Fridrich et al. improved image alteration detection by employing a subtractive pixel adjacency matrix in [5]. In [6], Cao et al. leveraged the probability of zero values in the first-order difference map as a statistical indicator for identifying image modifications. Yuan [7] proposed a 44-dimensional feature to analyze the distribution of block medians. Kang et al. developed an autoregressive model that utilizes median-filtering as a distinguishing feature in [8]. Chen et al. in [9] combined global probability features with local correlation features to enhance analysis. Zhang et al. examined the effects of median filtering on micro-texture structures in [10]. However, all of these methods depend on manually crafted feature descriptions, which can be labor-intensive.

Variational deconvolution was used in [11] for quality enhancement in median-filtered images. In [12], Chen et al. proposed a method of anti-forensics using convolutional neural networks, which didn't require hand-coded features. The first ones to propose using Generative-Adversarial networks for image restoration were Kim et al. in [13]. They proposed using high-pass filters, defined in [14], before sending the images to the discriminator. This significantly improved the detection capabilities of the discriminator. However, the implementation in [13] was suffering from mode collapse and a long training time. A novel improvement over the basic implementation of GAN was proposed by Martin et al. in [1]. The architecture that was used in [1] is called a Wasserstein GAN and features several improvements over the basic version.

The GAN Based approach is designed in [15] to be transferable across different forensic models to fool forensic CNNs, making it harder to detect manipulated images, thus improving the effectiveness of anti-forensic methods.

3   Proposed Methodology

The proposed system consists of the discriminator and the generator as illustrated in Fig. 1.

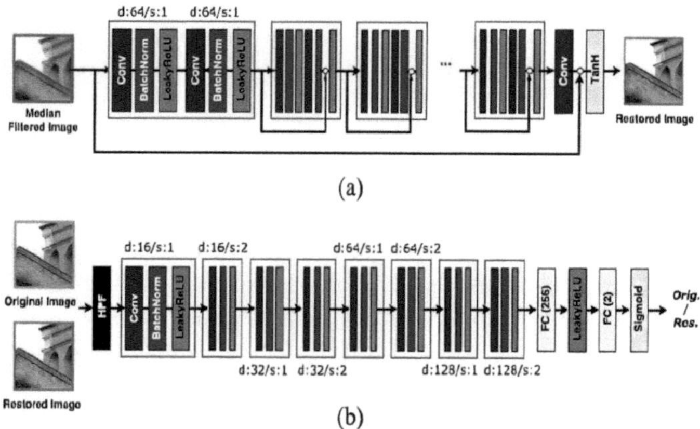

Fig. 1. Proposed architecture

Discriminator. The discriminator is equipped with four high-pass filters aimed at extracting fine details from the image, enhancing its capability to identify subtle inconsistencies. Its architecture includes eight residual blocks, each comprising a 3×3 convolutional layer, a batch normalization layer, and a leaky-ReLU activation function. The convolutional layers alternate between strides of 1 and 2, with the number of filters doubling after every two residual blocks. Following the residual blocks, the network incorporates a fully connected layer with 256 units activated by a leaky-ReLU function, followed by another fully connected layer with two units and a sigmoid activation to produce the final output. The use of high-pass filters significantly improves the discriminator's ability to detect fine-grained artifacts in the image.

Generator. The generator consists of eight residual blocks, each containing a 3 × 3 convolutional layer, a batch normalization layer, and a leaky-ReLU activation function. These layers are repeated within each residual block, with every block utilizing 64 filters and a stride of 1. After processing through the residual blocks, the network concludes with a final convolutional layer followed by a tanh activation function, which produces the restored image as the output.

Loss Function. The loss function used by the proposed system is given by,

$$min_G max_D V(D,G) = E_x[(D(x))] + E_z[(1 - D(G(z)))]$$ (1)

The discriminator loss is given by,

$$L_a = E[(D(x))] + [(1 - D(G(z)))]$$ (2)

The generator loss is given by,

$$L_g = E(D(G(z)))$$ (3)

4   Results

The results section evaluates the performance of the proposed system, emphasizing its ability to restore degraded images effectively. It outlines the experimental setup, pre-processing steps and training process.

4.1   Experimental Setup

The proposed system was implemented using Python and TensorFlow, which is a framework for dataflow and machine-learning applications. The system was trained on a NVIDIA Tesla K-80 GPU on the cloud. The dataset used was RAISE-1k dataset which consists of 1000 high-quality raw images of sizes (4928 ×3264). Figure 2 is the sample high-quality image from the dataset.

Initially, the images were converted to grayscale as shown in Fig. 3. After that, each image was cropped from the center into four images each of which was 128 × 128 as shown in Fig. 4, 5, 6 and 7.

Algorithm 1. Training a GAN model to generate restored images from the median-filtered image

---

Require: High-quality raw images of sizes (4928 × 3264)
  for each image in the dataset **do**
    Convert it to Grayscale image.
    Crop each image into four parts of size 128 × 128 from the center
    Construct the generator and the discriminator.
    Train the generator for 20 epochs.
    for 50 epochs **do**
      Get the next batch of original images.
      Get the next batch of median filtered images.
      Train the discriminator.
      Clip the weights for the discriminator to the range [-c,c].
      Train the generator.
    end for
  end for
  for each image in the test dataset **do**
    Load the trained model from the saved checkpoint.
    Retrieve the median-filtered image from the dataset.
    Use the generator to process the median-filtered image and obtain the restored image.
    Calculate the PSNR and SSIM values between the median-filtered image and the restored image.
  end for
  Compute the average PSNR and SSIM values for all 800 images.

A total of 4000 grayscale images, each sized 128 × 128, were obtained after pre-processing. Out of these, approximately 3200 images were allocated for training, while 800 were reserved for testing. Median filtering was then applied as illustrated in Fig. 8.

The median-filtered image shown in Fig. 8 is given as input to the generator, which then produces a restored image as output as shown in Fig. 9. First, the generator was trained for about 20 epochs in order to avoid unwanted local minima. Then, the GAN was trained for about 50 epochs. The constants used for the training process is shown in Table 1. After every 5 epochs, the system saves the current status of the GAN in the form of a checkpoint file. This file is then used to restore the model to perform testing.

The trained system was evaluated based on two metrics: the PSNR and SSIM values.

Peak-Signal-To-Noise Ratio (PSNR). The PSNR is used for comparing the similarity between the median-filtered images and the restored images. For images with a bit-depth of 8, PSNR values typically range from 30 to 50.

Given an M × N image I and its noisy approximation N, the mean squared error (MSE) is defined as:

Fig. 2. Original Image

Fig. 3. Grayscale Image

$$MSE = \frac{1}{M.N} \sum_{i=0}^{M=1} \sum_{j=0}^{N=1} [I(i,j) - K(i,j)]^2 \tag{4}$$

The PSNR is then calculated as:

$$PSNR = 20.Log_{10}(MAX_i) - 10.Log_{10}.Log_{10}(MSE) \tag{5}$$

where MAX of I is the maximum pixel value in the image, which is 255 in our case

Fig. 4. Cropped image 1.

Fig. 5. Cropped image 2.

Fig. 6. Cropped image 3.

Fig. 7. Cropped image 4.

Fig. 8. Median-filtered image.

Fig. 9. Restored image.

Table 1. Constant Used.

Epochs	50
Min. batch size	16
'c' value for clipping weights	0.07
Learning rate of discriminator	5×10^{-6}
Learning rate of generator	5×10^{-4}
Sample image size	128×128
Slope for LRelu	0.2

Structural Similarity Index (SSIM). The SSIM is another metric used to compare the similarity between the median-filtered images and the restored images. For an image of size M × N and a window size of (x, y) the SSIM is calculated as:

$$SSIM(x,y) = \frac{(2\mu_x\mu_y + c_1).(2\sigma_{xy} + c_2)}{(\mu_x^2 + \mu_y^2 + c_1)(\sigma_x^2 + \sigma_y^2 + c_2)} \tag{6}$$

Performance Comparison. The evaluation employed four metrics: oPSNR and oSSIM to measure the similarity between the original and restored images, and mPSNR and mSSIM to assess the quality of restoration by comparing the median-filtered and restored images. These metrics evaluate both visual and structural accuracy of the restored outputs.

The results are summarized in Table 2

Table 2. Results

Method	oPSNR(dB)	mPSNR(dB)	oSSIM(dB)	mSSIM(dB)
MF	40.5505	-	0.9550	-
[13]	42.5061	40.9868	0.9754	0.9636
Proposed method	40.5505	42.9835	0.9829	0.9744

4.2   Discussion

The proposed system outperforms the standard median-filtering approach and the adversarial method by Kim et al. [13], which suffered from mode collapse with large datasets. By leveraging the Wasserstein GAN framework and fine-tuning its constants, we mitigated this limitation and achieved stable training. However, the system struggles with the salt-and-pepper noise, which remains a challenge due to the lack of discernible patterns after median filtering.

High-pass filters in the discriminator improved feature extraction, but their impact on training stability and computational overhead requires further investigation. Future work will explore advanced GAN architectures, such as Style-GAN and BigGAN, and multi-resolution approaches to enhance image restoration performance. Additionally, expanding mode collapse mitigation strategies and refining hyperparameter tuning could further improve stability and convergence. These advancements will enhance the system's generalizability and effectiveness in real-world applications.

5   Conclusion

The above GAN model was were trained alternately using a dataset comprising original images and their median-filtered counterparts. The experimental findings demonstrate that the proposed approach outperforms conventional median-filtering methods and standard adversarial models in enhancing both the visual quality and statistical properties of altered images. Future work could explore extending this approach to incorporate additional image restoration methods beyond median-filtering. Moreover, optimizing the GAN architecture with innovative modifications could further enhance its performance.

References

1. Arjovsky, M., Chintala, S., Bottou, L.: Wasserstein generative adversarial networks. In: Proceedings of the 34th International Conference on Machine Learning, vol. 70 of Proceedings of Machine Learning Research, pp. 214–223. JMLR.org, Sydney NSW Australia (2016)
2. Goodfellow, I., et al.: Generative adversarial nets. In: Proceedings of Advances in Neural Information Processing Systems 27, pp. 2672–2680. Curran Associates, Inc. Montreal, Canada (2014)
3. He, K., Zhang, X., Ren, S., Sun, J.: Deep residual learning for image recognition. In: 2016 IEEE Conference on Computer Vision and Pattern Recognition (CVPR). Las Vegas, NV, USA (2016). https://doi.org/10.1109/CVPR.2016.90
4. Fridrich, J., Kirchner, M.: On detection of median filtering in digital images. In: Proceedings of Media Forensics and Security II, 754110, SPIE Volume 7541, pp. 7541–12,San Jose, California, United States (2010)
5. Pevný, T., Bas, P., Fridrich, J.: Steganalysis by subtractive pixel adjacency matrix. IEEE Trans. Inf. Forensics Secur. **5**(2), 215–224 (2010)
6. Cao, G., Zhao, Y., Ni, R., Yu, L., Tian, H.: Forensic detection of median filtering in digital images. In: 2010 IEEE International Conference on Multimedia and Expo, pp. 89–94, Singapore (2010)
7. Yuan, H.D.: Blind forensics of median filtering in digital images. IEEE Trans. Inf. Forensics Secur. **6**(4), 1335–1345 (2011)
8. Kang, X., Stamm, M.C., Peng, A., Liu, K.J.R.: Robust median filtering forensics using an autoregressive model. IEEE Trans. Inf. Forensics Secur. **8**(9), 1456–1468 (2013)
9. Chen, C., Ni, J., Huang, J.: Blind detection of median filtering in digital images: a difference domain based approach. IEEE Trans. Image Process. **22**(12), 4699–4710 (2013)

10. Zhang, Y., Li, S., Wang, S., Shi, Y.Q.: Revealing the traces of median filtering using high-order local ternary patterns. IEEE Signal Process. Lett. **21**(3), 275–279 (2014)
11. Fan, W., Wang, K., Cayre, F., Xiong, Z.: Median filtered image quality enhancement and anti-forensics via variational deconvolution. IEEE Trans. Inf. Forensics Secur. **10**(5), 1076–1091 (2015)
12. Chen, J., Kang, X., Liu, Y., Wang, Z.J.: Median filtering forensics based on convolutional neural networks. IEEE Signal Process. Lett. **22**(11), 1849–1853 (2015)
13. Kim, D., Jang, H.U., Mun, S.M., Choi, S., Lee, H.K.: Median filtered image restoration and anti-forensics using adversarial networks. IEEE Signal Process. Lett. **25**(2), 278–282 (2018)
14. Chen, M., Sedighi, V., Boroumand, M., Fridrich, J.: Jpeg-phase aware convolutional neural network for steganalysis of jpeg images. In: Proceedings of the 5th ACM Workshop on Information Hiding and Multimedia Security, Publisher IH & MMSec'17, pp. 75–84. ACM, New York, NY, USA (2017)
15. Zhao, X., Chen, C., Stamm, M.C.: A Transferable anti-forensic attack on forensic CNNs using a generative adversarial network (2021). https://arxiv.org/abs/2101.09568

An Automatic Segmentation of Polyp in Colorectal Cancer Using U-Net

Jovita Relasha Lewis[1], Sameena Pathan[1]([✉]), Preetham Kumar[1], and Cifha Crecil Dias[2]

[1] Information and Communication Technology, Manipal Institute of Technology, Manipal Academy of Higher Education, Manipal, Karnataka, India
{sameena.bp,preetham.kumar}@manipal.edu
[2] Department of Biomedical Engineering, Manipal Institute of Technology, Manipal Academy of Higher Education, Manipal, Karnataka, India
cifha.saldhana@manipal.edu

Abstract. Colorectal Cancer (CRC) is a main cause of cancer-associated deaths globally, with a significant number of victims due to late diagnosis. There is a correlation between colorectal polyps and the occurrence of CRC, highlighting the need for early intervention and diagnosis. The capability to segment the polyp accurately is critical as it would facilitate timely treatment and improve diagnosis accuracy. However, polyp segmentation presents challenges caused by different sizes, and shapes of the polyps, also in cases when they are concealed beneath the mucosal area. U-Net architecture is used to address these challenges to enhance the accuracy of the polyp identified. Our study utilized the Kvasir-Seg dataset. The U-Net model was trained on this data. The model demonstrated a loss of 0.225 and a validating accuracy of 0.9021. This advanced segmentation technique helps the gastroenterologist identify polyps with high accuracy and ultimately helps the patient with early detection and diagnosis.

Keywords: Colorectal Cancer · U-Net · Automatic · segmentation · polyps · colon · computer aided diagnosis · CAD

1 Introduction

According to the Cancer Stats 2024, the total CRC rate is estimated to be 152,810 among which 81,540 are males and 71,270 are females. The estimated death in 2024 is 53,010 among which 28,700 males and 24,310 females. The estimated number of new cases in 2024 in males is 8% and in females is 7% which ranks 3rd in cancer cases. The estimated number of death cases in 2024 is 9% in males and 8% in females [1]. CRC usually starts in the large intestine or the rectum. The gold-standard procedure for detecting CRC is a colonoscopy [2]. A colonoscopy helps in direct visualization of the polyps as benign or malignant [3]. Early detection can help in better diagnosis [4]. However, there have been reports of substantial polyp miss rates, especially when numerous tiny adenomas exist [5]. In patients having a colonoscopy, 22%–28% of polyps & 20%–24% of adenomas are typically undetected. The reason for CRC being detected after 3 years of a colonoscopy is unclear if it were the missed polyps [6]. This offers a chance to use computer-aided technologies to assist doctors and lessen the amount of undetected polyps [7]. Fig. 1 shows the polyps detected during a colonoscopy screening.

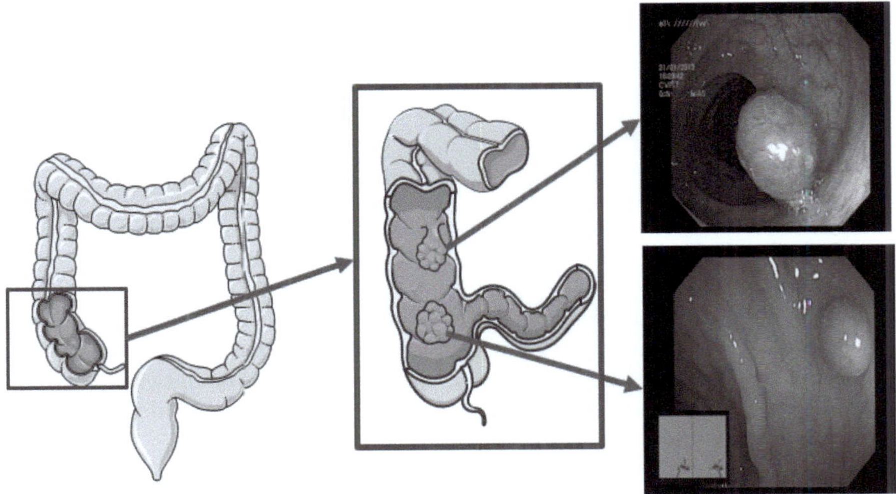

Fig. 1. Polyps detected during colonoscopy. Note: Endoscopic images [28].

1.1   Importance of Segmentation of Polyp in Clinical Medicine

Segmentation of polyps plays a crucial role in clinical medicine, especially in the early detection and diagnosis of CRC which is the most prevalent cancer worldwide. Polyps when left untreated can lead to CRC, especially in adenomatous polyps [8]. When the polyps are identified and extracted at an initial stage it can reduce the occurrence of cancer. An accurate identification of the polyps by computer-aided diagnosis (CAD) systems will help in screening processes like the colonoscopy. Manual detection of polyps is subjected to human error since some polyps are flat or small in size [9]. Automatic segmentation models like the U-Net reduce the risk of missed polyps and provide precise boundary regions to be segmented. Manual intervention in the detection of polyps causes inconsistency. Automated segmentation will bring uniformity in polyp detection For large or complex polyps segmentation helps the clinicians in the removal of polyps by preserving healthy areas. Robust segmentation is helpful for good classification results and other CAD methods.

2   Related Work

2.1   Traditional Approaches in Segmentation

The traditional procedure for segmentation relies on handcrafted features like texture [10], color [11,12], and shape [13,14]. In the segmentation of polyps using texture various techniques were used like Local Binary Pattern(LBP) [15], Gabor filter [16], and co-occurrences matrices [17]. In region-based segmentation [18], a grouping of similar intensity pixels is done. Edge detection techniques like sobel [19], and canny edge detector [20] improve the segmentation. Traditional machine learning methods like clustering [21], k mean [22], and Support Vector Machine (SVM) [23] were used in segmentation. This method is effective when there is a variance in the texture. These methods fail in complex situations. Traditional approaches have a high chance of false detection.

2.2   Challenges in Traditional Segmentation Approach

- Traditional segmentation methods like edge detection, region growing or thresholding depend on the handcrafted feature, image quality, and manual parameter tuning.
- The Segmentation process may fail when overlapping or irregular shapes are encountered. Low-contrast or noisy images too get inaccurate results.
- Requires manual tuning of the parameter for new task or dataset.
- Traditional segmentation techniques usually have poor generalization due to the variation in lighting, noise, or object shape.

2.3   Deep Learning Approaches

Deep learning has brought about a lot of advancements in medical image analysis particularly in the segmentation of polyps. Deep learning models learn from hierarchical patterns of the data. Convolution Neural Networks (CNN) [24] have brought about a lot of advancement in polyp segmentation by offering a robust solution. These CNNs consist of layers of convolutional filters that capture spatial features, pooling layers, and fully connected layers that do the prediction. In the early implementation, patch-based CNN models [25] were used. Processing time and localization accuracy are the limitations of this approach. Fully Convolutional Networks (FCNs) [26] are an extension of CNNs designed for segmentation purposes. Attention mechanisms are used in integration with CNN architecture [27]. Attention mechanisms prioritize a region that would probably contain a polyp increasing the accuracy of the segmented polyp. Then came the U-Net which was widely used in medical imaging.

3   Method

3.1   An Outline of U-Net Architecture

As illustrated in Fig. 2, the U-Net design outline includes an encoder, skip connection, and a decoder.

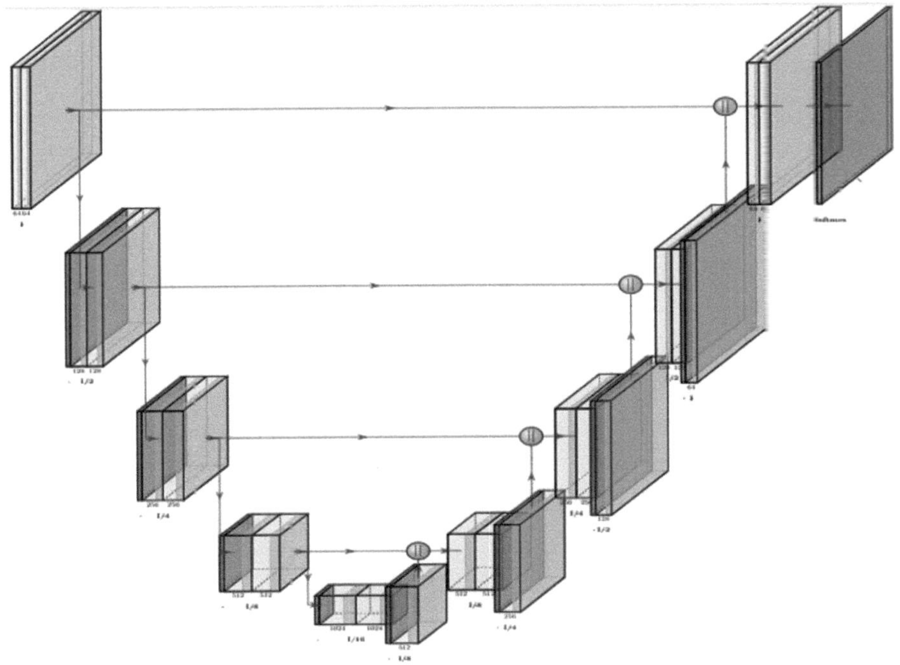

Fig. 2. The architecture of U-Net.

- **Encoder:**
  - **Convolution layers-**Each layer in the encoder performs two 3×3 convolutions with padding to preserve spatial dimension.
  - **ReLU activation-**After each convolution ReLU activation introduces non-linearity.
  - **Max Pooling-**After each convolution, a 2×2 max-pooling layer will reduce the spatial dimension and increase the depth.
  - **Feaature Maps-**Following each downsampling, the quantity of feature maps eventually doubles. The network will be able to learn higher-level, more abstract features as a result.
  - **Significance of Encoder:** The encoder will progressively compress spatial dimension while increasing the depth of these feature maps. Encoder can capture global and high-level semantic information.
- **Skip connection:** It bridges the encoder and the decoder directly transferring spatial information.
  - **Spatial Information is preserved:** It helps to preserve the lost spatial details during the downsampling. It ensures precise boundary localization.
  - **To mitigate vanishing gradients:** To provide a direct path for gradients to flow during backpropagation by stabilizing training.

- **Improved Segmentation performance** Combining low-level texture details with high-level semantic features leads to better segmentation results.
- **Decoder:**
  - **Transposed Convolution:** A 2×2 up convolution then upsamples the feature maps. This upsample doubles the spatial resolution.
  - **Concatenation of skip Connection:** These features from the corresponding encoder layers are concatenated with upsampled features of the decoder. This will improve localization by using fine-grain details of the previous layers.
  - **Convolutional Layers:** After the concatenation process, two 3×3 convolutions with ReLU are applied.
  - **Final Layer:** A 1×1 convolution is then applied to lessen the number of channels to the desired pixel-wise segmented output.
  - **Significance of Decoder:** The decoder will restore spatial resolution maintaining contextual understanding captured by the encoder. It produces pixel-level prediction for the segmentation task.

3.2   Dataset

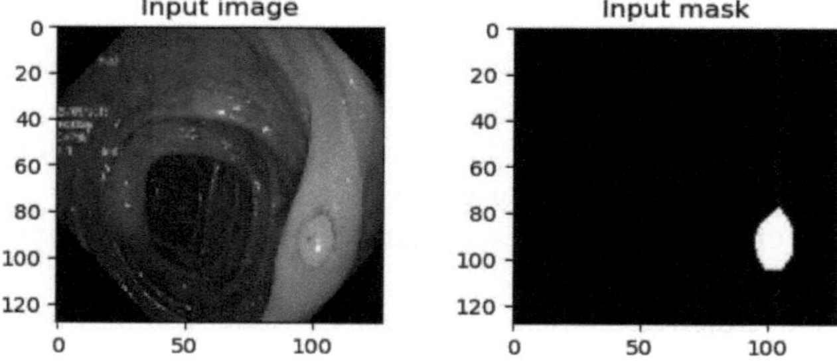

Fig. 3. Input image and mask from the dataset [28].

The Kvasir-Seg dataset comprises 196 polyps measuring less than 10 mm. Endoscopic images and ground truth images were used to train the U-Net from kvasir-sessile [28] as shown in Fig. 3. 80% of the images were used in training and 20% were used for testing.

3.3 Description of the Proposed Method

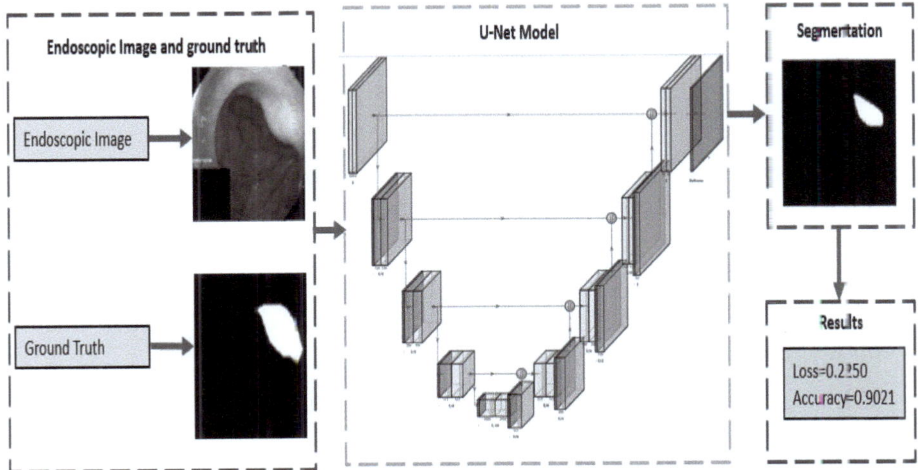

Fig. 4. Description of the proposed method.

The endoscopic image and ground truth are given as input to the U-Net model. The U-net comprises three parts: the bottleneck, the narrowing, and the expansion of sections. This area of the design distributes convolutional operations to create more feature maps, as seen in Fig. 4 The U-Net architecture takes its name from its "U" form. It is recognized as being one of the oldest and most well-liked deep learning-based semantic segmentation methods designed by Ronneberger et al. [29]. The segmented mask generated from the U-net model has an accuracy of 0.9021 and a loss of 0.225.

3.4 Justification for Selection of U-Net

- U-Net is widely used for its significant performance in medical image segmentation. It is well suited for segmentation of polyps in endoscopic images.
- The symmetric encoder-decoder architecture helps in capturing high and low-level features alike. The encoder is used to extract contextual information and the decoder will reconstruct the segmented output using spatial information.
- Skip connection transfers feature maps from encoder to decoder. It helps in retaining fine-grained features crucial for segmentation.
- U-Net can handle noise. Traditional segmentation methods like thresholding, region-growing, and edge detection contain noise and irregular structure. U-Net can learn robust features and separate polyps effectively.
- U-Net has been successfully applied to various image segmentation tasks like a tumor, lesions, surgical instruments, and organ segmentation.
- U-Net's ability to be trained in limited datasets makes it a good choice for our small dataset.

3.5   Time and Space Complexity Analysis

1. Time Complexity

Forward Pass: The time complexity of a single convolution operation for an input image of size $H \times W$ with C channels, a kernel size of $k \times k$, and F filters is given by:

$$T_{\text{conv}} = \mathcal{O}(H \times W \times C \times k^2 \times F) \tag{1}$$

Pooling and up-sampling operations, which involve simple element-wise computations, have the following time complexity:

$$T_{\text{pool/up}} = \mathcal{O}(H \times W \times C) \tag{2}$$

The overall time complexity for a forward pass through a U-Net with L layers is approximately:

$$T_{\text{forward}} = \mathcal{O}(L \times H \times W \times C \times k^2 \times F) \tag{3}$$

Backpropagation: The time complexity of backpropagation is roughly twice that of the forward pass, as gradients are calculated for both activations and weights.

2. Space Complexity

Parameters: The total number of trainable parameters in U-Net, considering L layers, each with F filters of size $k \times k$ and C input channels, is given by:

$$P_{\text{total}} = \sum_{l=1}^{L} \left(k^2 \times C \times F + F \right) \tag{4}$$

Feature Maps: The memory required to store feature maps during training is:

$$M_{\text{feature maps}} = \mathcal{O}(L \times H \times W \times C) \tag{5}$$

Overall Space Complexity: The total space complexity, combining memory for parameters and feature maps, is:

$$S_{\text{total}} = \mathcal{O}(L \times H \times W \times C + P_{\text{total}}) \tag{6}$$

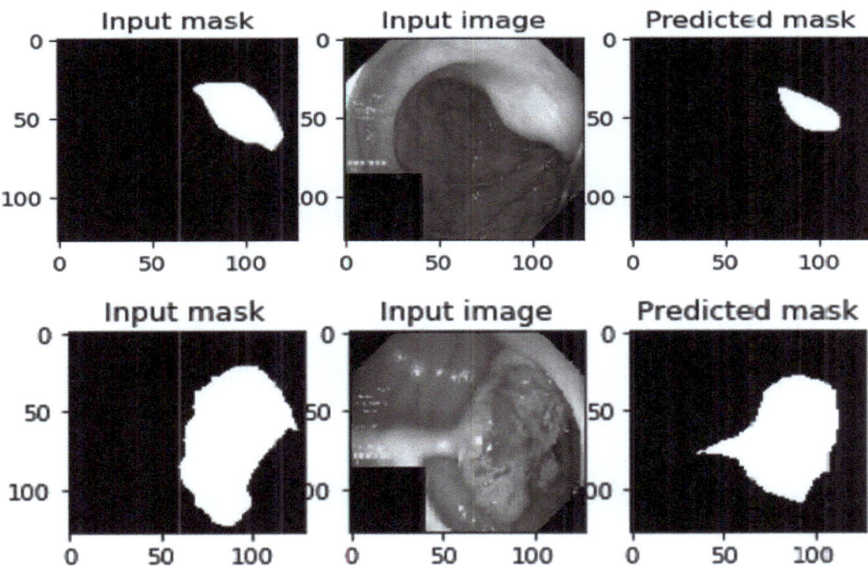

Fig. 5. The ground truth mask,input images and segmented output.

4   Results

The U-Net model used in the study achieved a significant performance in automatic polyp segmentation in CRC. The Fig. 5 shows the output of the U-Net Model. Endoscopic images of the polyp and ground truth images were used to train the U-Net. 80% of the images were used in training and 20% in testing. After 50 epochs, the U-Net model achieved a training loss value of 0.2250 and the accuracy for validation is 0.9021, demonstrating its precision in polyp segmentation. The results highlight U-Net's ability to detect and segment polyps effectively, including challenging cases when they are small or flat. Table 1 shows the different metrics of the U-Net model where the value of the loss is 0.2250, validating accuracy is 0.9021, IoU score is 0.85, the dice coefficient is 0.88.

Table 1. Comparison of segmentation performance metrics across different models.

Model	Loss	Validation Accuracy	IoU	Dice Score
U-Net	**0.2250**	**0.9021**	**0.85**	**0.88**
ResNet	0.2700	0.8800	0.83	0.86
FCN	0.3500	0.8300	0.75	0.78
AlexNet	0.4200	0.7900	0.68	0.72

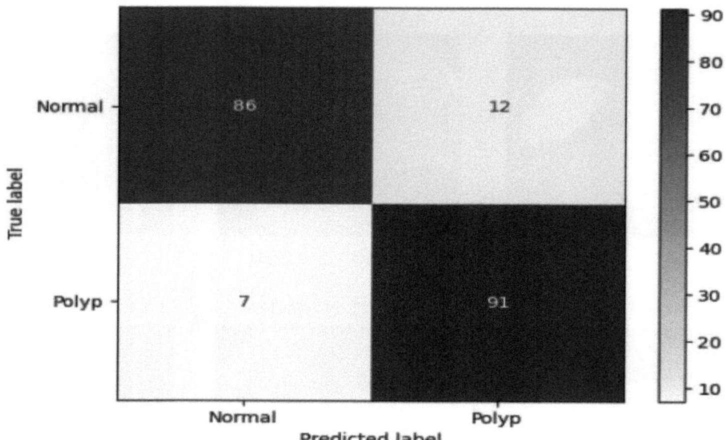

Fig. 6. Confusion Matrix for the U-Net Model.

The U-Net model's confusion matrix, which includes True Positive (TP), True Negative (TN), False Positive (FP), and False Negative (FN), is displayed in Fig. 6. The FP value is 7 and the FN value is 12 in the confusion matrix which are predicted wrongly. For the rest of the images, TP is 86 and FN is 91.

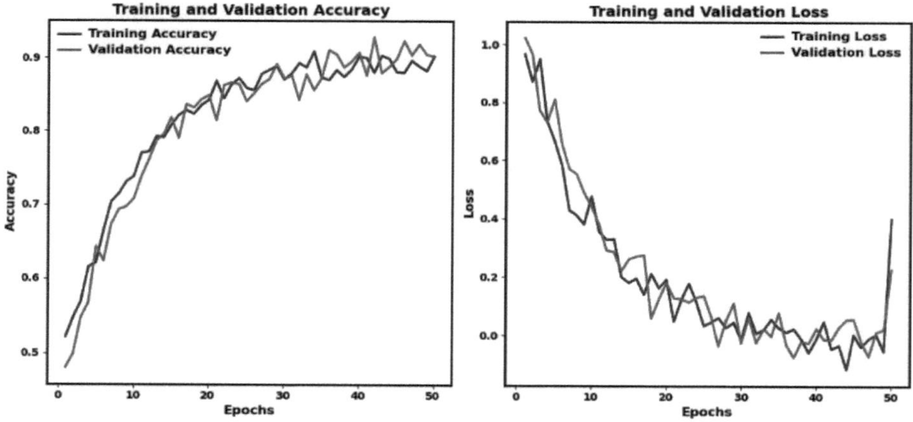

Fig. 7. Accuracy & loss in training and validation.

For 50 epochs, the training & validating accuracy is displayed in Fig. 7. The model's capacity to learn patterns in the dataset and fit the training data is shown by the training accuracy. Validation accuracy works on the unseen data and this accuracy is 0.90. The model's capacity to reduce error in the data sets used for training is evaluated by training loss. The performance of the unseen data is determined by calculating the validating loss. The validating loss is found to be 0.225.

5   Conclusion

An automatic polyp segmentation method was used for detecting polyps using the U-Net model architecture. The model demonstrated robust performance in identifying polyps achieving high accuracy and high segmentation metrics like dice score, F1 score, precision, recall, and IoU. The results highlight the impact of U-net in addressing the challenges of early polyp detection which is crucial for improving treatment and diagnosis outcomes.

The proposed method provides a valid tool for automatically delineating polyps, which will help clinicians reduce false negatives and improve the diagnosis precision. Furthermore, the structured design of the U-Net when merged with its ability to generalize with polyps of various sizes, makes it useful in real-time deployment.

Future researchers should concentrate on improving the model's performance for a bigger dataset and investigating cutting-edge segmentation designs. The successful implementation of polyp is significant in medical imaging. Hence further optimization can improve the outcomes in CRC screening and put up a better AI-driven tool in clinical practice.

References

1. Siegel, R.L., Giaquinto, A.N., Jemal, A.: Cancer statistics 2024. CA: A Cancer J. Clin. **74**(1) (2024)
2. Corley, D.A., et al.: Evaluating different approaches for calculating adenoma detection rate: is screening colonoscopy the gold standard? Gastroenterology **165**(3), 784–787 (2023)
3. Shaukat, A., et al.: Endoscopic recognition and management strategies for malignant colorectal polyps: recommendations of the US Multi-Society Task Force on Colorectal Cancer. Official J. Am. Coll. Gastroenterol. ACG**115**(11), 1751–1767 (2020)
4. Veitch, A.M., Uedo, N., Yao, K., East, J.E.: Optimizing early upper gastrointestinal cancer detection at endoscopy. Nat. Rev. Gastroenterol. Hepatol. **12**(11), 660–667 (2015)
5. Lewis, J.R., Pathan, S., Kumar, P., Dias, C.C.: AI in endoscopic gastrointestinal diagnosis: a systematic review of deep learning and machine learning techniques. IEEE Access (2024)
6. Leufkens, A.M., Van Oijen, M.G.H., Vleggaar, F.P., Siersema, P.D.: Factors influencing the miss rate of polyps in a back-to-back colonoscopy study. Endoscopy **44**(05), 470–475 (2012)
7. Sinagra, E., et al.: Use of artificial intelligence in improving adenoma detection rate during colonoscopy: might both endoscopists and pathologists be further helped. World J. Gastroenterol. **26**(39), 5911 (2020)

8. Saraiva, S., Rosa, I., Fonseca, R., Pereira, A.D.: Colorectal malignant polyps: a modern approach. Ann. Gastroenterol. **35**(1), 17 (2022)
9. Li, M.-D., et al.: Performance and comparison of artificial intelligence and human experts in the detection and classification of colonic polyps. BMC Gastroenterol. **22**(1), 517 (2022)
10. Bharati, M.H., Liu, J.J., MacGregor, J.F.: Image texture analysis: methods and comparisons. Chemom. Intell. Lab. Syst. **72**(1), 57–71 (2004)
11. Mamonov, A.V., Figueiredo, I.N., Figueiredo, P.N., Tsai, Y.H.R.: Automated polyp detection in colon capsule endoscopy. IEEE Trans. Med. Imaging **33**(7), 1488–1502 (2014)
12. Sundaram, P., Zomorodian, A., Beaulieu, C., Napel, S.: Colon polyp detection using smoothed shape operators: preliminary results. Med. Image Anal. **12**(2), 99–119 (2008)
13. Häfner, M., Uhl, A., Wimmer, G.: A novel shape feature descriptor for the classification of polyps in HD colonoscopy. In: Menze, B., Langs, G., Montillo, A., Kelm, M., Müller, H., Tu, Z. (eds.) MCV 2013. LNCS, vol. 8331, pp. 205–213. Springer, Cham (2014). https://doi.org/10.1007/978-3-319-05530-5_20
14. Ganz, M., Yang, X., Slabaugh, G.: Automatic segmentation of polyps in colonoscopic narrow-band imaging data. IEEE Trans. Biomed. Eng. **59**(8), 2144–2151 (2012)
15. Gross, S., et al.: A comparison of blood vessel features and local binary patterns for colorectal polyp classification. In: Medical Imaging 2009: Computer-Aided Diagnosis, vol. 7260
16. Lo, C.-M., Yeh, Y.-H., Tang, J.-H., Chang, C.-C., Yeh, H.-J.: Rapid polyp classification in colonoscopy using textural and convolutional features. In: Healthcare, vol. 10, no. 8, p. 1494. MDPI (2022)
17. Kalgotra, P., Sharda, R., Parasa, S.: Quantifying disease-interactions through co-occurrence matrices to predict early-onset colorectal cancer. Decis. Support Syst. **168**, 113929 (2023)
18. Qadir, H.A., Balasingham, I., Shin, Y.: Region-based convolutional neural network for polyp detection and segmentation. In: Bernal, J., Histace, A. (eds.) Computer-Aided Analysis of Gastrointestinal Videos, pp. 91–98. Springer, Cham (2021). https://doi.org/10.1007/978-3-030-64340-9_11
19. Elmi, S., Elmi, Z.: A robust edge detection technique based on matching pursuit algorithm for natural and medical images. Biomed. Eng. Adv. **4**, 100052 (2022)
20. Yuan, Z., et al.: Automatic polyp detection in colonoscopy videos. In: Medical Imaging 2017: Image Processing, vol. 10133, pp. 718–727. SPIE (2017)
21. Kabra, A., Iwahori, Y., Usami, H., Bhuyan, M. K., Ogasawara, N., Kasugai, K.: Polyp classification and clustering from endoscopic images using competitive and convolutional neural networks. In: ICPRAM, pp. 446–452 (2019)
22. Göktürk, S.B., Tomasi, C., Acar, B., Paik, D., Beaulieu, C., Napel, S.: A learning method for automated polyp detection. In: Niessen, W.J., Viergever, M.A. (eds.) MICCAI 2001. LNCS, vol. 2208, pp. 85–93. Springer, Heidelberg (2001). https://doi.org/10.1007/3-540-45468-3_11
23. Eriyanti, N.A., Sigit, R., Harsono, T.: Classification of colon polyp on endoscopic image using support vector machine. In: 2021 International Electronics Symposium (IES), pp. 244–250. IEEE (2021)
24. Yeung, M., Sala, E., Schönlieb, C.-B., Rundo, L.: Focus U-Net: a novel dual attention-gated CNN for polyp segmentation during colonoscopy. Comput. Biol. Med. **137**, 104815 (2021)

25. Banik, D., Bhattacharjee, D., Nasipuri, M.: A multi-scale patch-based deep learning system for polyp segmentation. Adv. Comput. Syst. Secur. Volume Twelve, 109–119 (2020)
26. Akbari, M., et al.: Polyp segmentation in colonoscopy images using fully convolutional network. In: 2018 40th Annual International Conference of the IEEE Engineering in Medicine and Biology Society (EMBC), pp. 69–72. IEEE (2018)
27. Long, J., Yang, C., Song, X., Zeng, Z., Ren, Y.: Polyp segmentation network based on lightweight model and reverse attention mechanisms. Int. J. Imaging Syst. Technol. **34**(3), e23062 (2024)
28. Jha, D., et al.: Kvasir-SEG: a segmented polyp dataset. In: Ro, Y.M., et al. (eds.) MMM 2020. LNCS, vol. 11962, pp. 451–462. Springer, Cham (2020). https://doi.org/10.1007/978-3-030-37734-2_37
29. Ronneberger, O., Fischer, P., Brox, T.: U-Net: convolutional networks for biomedical image segmentation. In: Navab, N., Hornegger, J., Wells, W.M., Frangi, A.F. (eds.) MICCAI 2015. LNCS, vol. 9351, pp. 234–241. Springer, Cham (2015). https://doi.org/10.1007/978-3-319-24574-4_28

Assessing CNN Models for Multi-stage Alzheimer's Disease Classification with Data Splitting Techniques

Tanush Mohindra[1], Anshul Sharma[2(✉)], Vandana Bharti[3], Rishav Singh[4], and Sanjay Kumar Singh[1]

[1] Indian Institute of Technology (BHU), Varanasi, Uttar Pradesh, India
sks.cse@iitbhu.ac.in
[2] National Institute of Technology Patna, Patna, Bihar, India
anshul.cs@nitp.ac.in
[3] Indian Institute of Technology Dharwad, Dharwad, Karnataka, India
vandana@iitdh.ac.in
[4] Indian Institute of Technology Patna, Patna, Bihar, India
rishav.singh@iitp.ac.in

Abstract. Alzheimer's disease is a neurodegenerative disorder characterized by irreversible brain cell damage, leading to progressive cognitive decline and memory impairment. Although currently incurable, early detection is crucial for effective management and slowing disease progression. This study evaluates the performance of deep learning models using transfer learning on the Alzheimer's Disease Neuroimaging Initiative (ADNI) dataset to classify MRI scans into four categories: Cognitively Normal (CN), Early Mild Cognitive Impairment (EMCI), Late Mild Cognitive Impairment (LMCI), and Alzheimer's Disease (AD). We compared six pre-trained models (VGG16, VGG19, ResNet50, ResNet152, MobileNetV2, and EfficientNetB7) using two data-splitting approaches: the Late-slice-level and Early-subject-level split. ResNet50 showed the highest performance, achieving an accuracy of 93.47% with data leakage and 78.62% without leakage. We also integrated the Kolmogorov Arnold Network (KAN) with ResNet50, achieving 91.37% and 74.63% accuracies with and without data leakage, respectively, although this integration did not surpass the traditional ResNet50 model. The study further examined model performance using 2D brain slices from different anatomical views (Axial, Coronal, Sagittal) and varying slice counts (5, 10, 15). Model consistency has been assessed through 3-way, binary, and one-vs-rest classification tasks. Moreover, Grad-CAM visualization has been utilized to interpret the model's understanding of cognitive decline, providing insights into its decision-making process.

Keywords: Alzheimer's disease · Deep learning · Transfer learning · CNN · Data-leakage · Kolmogorov-Arnold Network

1   Introduction

Alzheimer's disease (AD) is a progressive neurodegenerative disorder that gradually impairs thoughts, cognition, and consciousness. It leads to the shrinking and eventual death of brain cells, resulting in cognitive decline and memory loss. AD is the most common form of dementia, accounting for 60–70% of cases and primarily affecting the aging population. The World Health Organization reports that over 55 million people worldwide currently have dementia, and expected to rise to 75 million by 2030 and 132 million by 2050 [1].

While there is no complete cure for AD, early detection is crucial [2] for improving symptoms and managing the disease progression. Delayed treatment can cause irreversible brain damage, leading to impaired thinking, memory loss, and language difficulties. However, even for medical experts, early detection is challenging as the initial symptoms are often subtle. MRI scans can visualize brain structures, revealing neuronal loss and atrophy, particularly in areas affected by AD. However, interpreting MRI scans is complex and subjective, requiring expertise to identify fine changes across brain regions, leading to variations in diagnostic accuracy. To address these challenges, deep learning has gained attention as a potential diagnostic aid for AD. Shangran Qiu et al. [3] developed an interpretable DL framework to distinguish between Cognitively Normal (CN) and AD using MRI, age, gender, and Mini-Mental State Examination data. They later incorporated additional data modalities to classify participants into CN, mild cognitive impairment (MCI), AD, and non-AD dementias (nADD), achieving AUCs of 0.945 for CN vs. impairment, 0.971 for dementia detection, and 0.773 for AD vs. nADD [4].

Many previous works have utilized transfer learning for AD classification. Hon and Khan [5] applied the pre-trained VGG16 and InceptionV4 models on the OASIS dataset for binary classification (CN vs. AD), achieving accuracies of 92.3% and 96.25%, respectively, using 32 slices based on image entropy. Similarly, Jain et al. [6] achieved a 95.73% accuracy in 3-way classification (CN, MCI, AD) with VGG16 on the ADNI dataset. Naveen and Nagaraj [7] used VGG19, ResNet50, and InceptionV3 for 4-class classification, with ResNet50 performing best at a recall rate of 99.25%. Mahmud et al. [8] achieved 95% accuracy with ensemble models of VGG and DenseNet, employing saliency maps and Grad-CAM for visualization. Shamrat et al. [9] presented modified InceptionV3, using RMSprop and a low learning rate, to design AlzheimerNet, reaching 98.67% accuracy.

Data leakage is a significant problem in machine learning, where unintended external information influences model training, severely overestimating model accuracy. Backstrom et al. [10] and Fung et al. [11] noted inflated results when MRI scans were randomly split instead of using subject-level splits. Yagis Ekin et al. [12] found that subject-level splits reduced inflated accuracies (95.24%-96.87%) to more realistic values (66.49%-68.68%), emphasizing the importance of proper data splitting.

This study analysed several Convolutional Neural Network (CNN) based models using a transfer learning approach to classify MRI images from a small

ADNI dataset into four categories. Subsequently, the recent KAN model was integrated with CNN-based architecture to evaluate whether it could improve classification accuracy compared to traditional deep learning models. This study used two data-splitting methods to assess and mitigate data leakage.

Table 1. Demographic details of the ADNI dataset.

Class	No. of Patients	No. of MRI Scans	Male/Female	Age Range	Average Age
CN	699	1108	440/668	57–95	73.58
EMCI	90	150	96/54	65–94	77.17
LMCI	75	125	68/57	65–93	77.38
AD	143	222	141/81	55–93	78.72
Total	1007	1605	745/860	55–95	74.92

2　Materials and Methods

2.1　Data

AD dataset used in this study is sourced from the Alzheimer's Disease Neuroimaging Initiative (ADNI) database [15]. The ADNI is a large publicly available collection of MRI brain scans and cognitive assessments of Alzheimer's patients, monitored longitudinally over several years. The image dataset, accessed in May 2024, includes all studies from ADNI phases 1 to 4 available in the public domain. It comprises 1,605 MRI volumes from 1,007 patients, captured across sequential visits and classified into four categories based on cognitive decline: CN, EMCI, LMCI, and AD shown in Table 1. The dataset consists of 3D-acquired, 3-Tesla T1-weighted MRI images, along with associated patient information such as age, sex, ImageID, MRI acquisition date, and visit sequences.

Obtained MRI images were pre-processed using FreeSurfer [16] for image enhancement and skull-stripping. The Auto-recon1 pipeline was utilized, which performed a sequence of steps, including Motion Correction, NU Intensity Correction, Talairach Transform, Intensity Normalization, and Skull Stripping. Figure 1 (a) displays an original MRI in axial, coronal, and sagittal views, while Fig. 1 (b) to (d) illustrate the enhancements at each step. Moreover, the changes in brain structure due to the progression of neuronal loss and atrophy, particularly in affected areas, can be observed in MRI image slices, as shown in Fig. 2.

2.2　Slicing, Combining, and Augmentation

For AD detection using 3D MRI scans, 2D CNN models are preferred over 3D CNNs due to their lower computational demands and the limited availability of 3D volumes. Key features are preserved by slicing 3D data into successive 2D

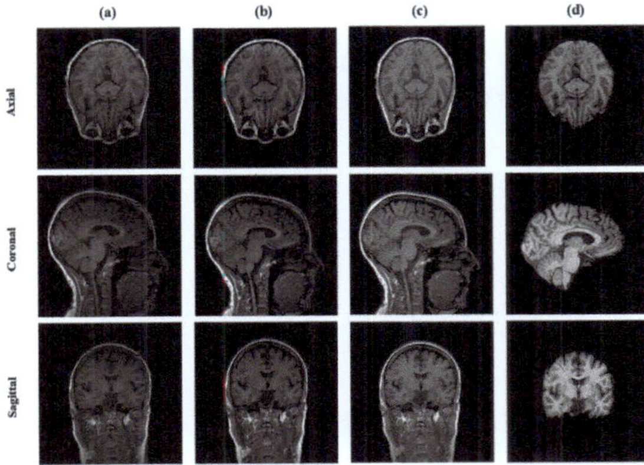

Fig. 1. Three plane views from left to right (a) Original Image before pre-processing (b) Non-uniform intensity normalization (c) Intensity Normalization, (d) Skull Stripping.

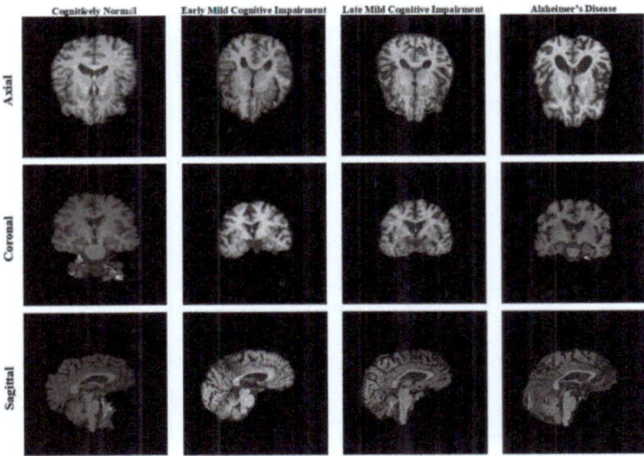

Fig. 2. MRI image slices in each directional plane show changes in brain structure in each class due to disease progression.

images while reducing model complexity. This approach also partially addresses data scarcity, as multiple meaningful 2D slices can be extracted from a single MRI volume, helping to mitigate class imbalances and reducing the need for extensive augmentation.

Sequential slices were extracted from the central region of MRI volumes along each orthogonal plane to maximise information capture, with 5, 10, or 15 slices chosen depending on the experiment. Since the models were pre-trained on RGB ImageNet images and the MRI images were grayscale, adjacent MRI slices were stacked across three channels and resized to $(224 \times 224 \times 3)$ to provide suitable RGB input to the model, effectively capturing more spatially distributed features. Each class underwent over or under-sampling to address class imbalance, and data augmentation was applied. In order to preserve the internal tissue structure crucial for AD detection, augmentation was limited to horizontal flipping along a vertical axis, maintaining brain symmetry in axial and coronal views. After augmentation, 1,000 images from each class were selected for model training.

3   Methodology

To evaluate the model's performance in classifying 2D RGB slices into four categories (CN, EMCI, LMCI, and AD), the MRI data was divided into three sets: 60% for training, 20% for validation, and 20% for testing. Two data-splitting approaches were implemented to assess the potential impact of data leakage.

In the first approach, called the Late-slice-level split, all 2D slices from multiple patients were pooled, shuffled, and randomly divided into training, validation, and testing sets. This method introduces bias, as correlated slices from the same patient may appear in both training and test sets, leading to data leakage. The second approach, the Early-subject-level split, divides the training, validation, and testing sets based on patient ID before deriving the 2D slices. This ensures that all slices from a single patient, regardless of multiple visits, belong to only one dataset, thereby preventing data leakage and allowing for a more accurate assessment of the model's performance.

For comparative evaluation in detecting AD-related cognitive impairment across four classes, six models were analyzed using the ADNI data: VGG16 and VGG19 [17], ResNet50 and ResNet152 [18], MobileNetV2 [19], and EfficientNetB7 [20]. Each model was pre-trained on the ImageNet dataset (containing 1 million images) and modified by removing the top layer and adding custom layers for classification. A dense layer with 256 neurons, ReLU activation, and L2 regularization of 0.001 was added, along with a dropout rate of 0.5 and a layer normalization layer. The final classification layer had neurons equal to the total classes in the dataset and used a Softmax layer for classification.

When using pre-trained models for classification, initial training typically involved freezing model layers to retain prior knowledge and minimize overfitting, particularly with smaller datasets. After training the new layers, the entire model was unfrozen and fine-tuned with a lower learning rate to adapt to the new task while balancing pre-learned features with new data.

For VGG16 and VGG19, only the last 5 layers, along with the newly added layers, were trainable, while for ResNet50, ResNet152, MobileNetV2, and EfficientNetB7, the last 15 layers were trainable. All models were trained with a mini-batch size of 64 for 100 epochs using categorical cross-entropy loss and the Adam optimizer with a learning rate of 0.001. These hyperparameters were determined after extensive testing of various configurations. All layers were unfrozen and trained again for fine-tuning for 50 epochs at a lower learning rate of 0.0005, keeping the other hyperparameters consistent. These execution steps were applied uniformly across all models.

Model performance was evaluated using metrics such as Accuracy, Precision, Recall, F1-score, and AUC, which were weighted averages across classes.

All results regarding the comparative performance of the models were recorded using axial slices, with five 2D slices per scan, which yielded the highest accuracy. Figure 3 illustrates the proposed framework, covering the methodology and highlighting key results to provide a clear study overview.

4 Results

4.1 Model Performance Evaluation

Table 2 presents evaluation metrics for all models, highlighting the best metrics in bold. Using the Late-slice-level split, the results indicate an overly optimistic accuracy of around 90% for all models, with ResNet50 performing best, achieving 93.46% accuracy and an AUC of 0.97. For the Early-subject-level split, where all methods and hyperparameters remained the same, all models achieved a similar accuracy of around 75%, with ResNet50 outperforming others with an accuracy of 78.62% and an AUC of 0.82. The Receiver Operating Characteristic (ROC) curve in Fig. 4 A illustrates the macro-average performance of all models, while Fig. 4 B depicts the comparison between the two approaches in terms of model accuracies.

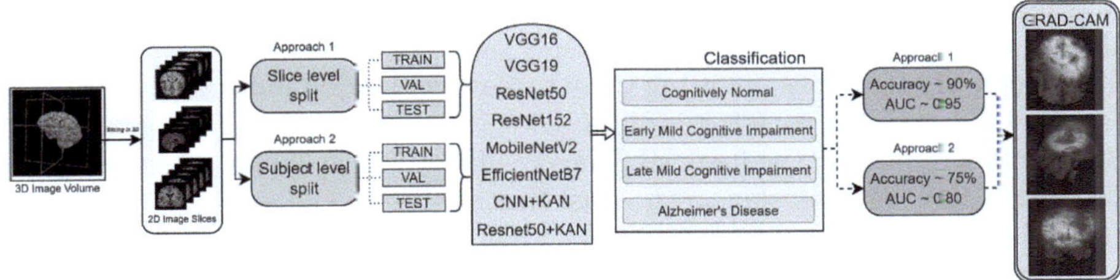

Fig. 3. Framework for the proposed study.

Table 2. Performance metrics of all models with two splitting approaches.

Model	Late-slice-level split approach					Early-subject-level split approach				
	Accuracy	Precision	Recall	F1 score	AUC	Accuracy	Precision	Recall	F1 score	AUC
VGG16	88.95%	90.95%	88.95%	89.41%	0.95	75.44%	76.05%	75.44%	75.70%	0.82
VGG19	91.60%	92.73%	91.60%	91.84%	0.96	76.15%	74.60%	76.15%	75.14%	0.80
ResNet50	**93.46%**	**93.79%**	**93.46%**	**93.56%**	**0.97**	**78.62%**	**77.38%**	**78.62%**	**77.68%**	**0.82**
ResNet152	90.90%	92.30%	90.90%	91.19%	0.96	76.85%	75.37%	76.85%	75.66%	0.78
MobileNetV2	84.14%	83.91%	84.14%	83.47%	0.88	75.80%	74.57%	75.80%	74.91%	0.80
EfficientNetB7	91.05%	92.50%	91.05%	91.42%	0.97	76.50%	76.03%	76.50%	76.10%	0.78

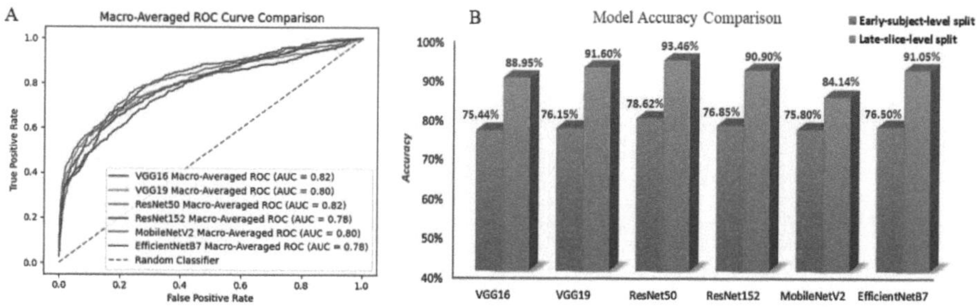

Fig. 4. A-ROC graphs of models in the Early-subject-level split approach B-Comparative performance of the models using two splitting approaches.

4.2   Performance of KAN-Integrated Models

KAN, based on the Kolmogorov-Arnold theorem, provides a novel alternative to MLPs. Unlike MLPs where the edges have learnable weights and nodes have fixed activation functions to incorporate non-linearity, the KAN architecture puts learnable 1-D activation functions on the edges with a simple addition operator on the nodes [21]. These learnable activation functions can vary and thus be specifically curated for each neuron. They are composed of a scaled sum of a residual function(typically SiLU) and spline-parametrized univariate functions (typically B-splines) allowing them to learn activation patterns dynamically and capture complex patterns in data. In this study, KAN was integrated into ResNet50 by replacing the final MLP layer with two KAN linear layers of widths 256 and 4, the latter serving as the classification layer. The grid size (G), which determines the level of detail that the activation function can capture, was set to 5, with a grid range of [–1, 1]. The spline order (k), which controls the smoothness, was set to 3. The Efficient KAN formulation was implemented as adapted from [22]. A smaller 4-layer CNN using the same KAN architecture was also developed for comparison against ResNet50.

The ResNet50 + KAN model achieved an accuracy of 91.37% and an AUC of 0.98 with the Late-slice-level split, and an accuracy of 74.63% and an AUC of 0.83 with the Early-subject-level split (see Table 3). The CNN+KAN model reached an accuracy of 89.49% and an AUC of 0.97 with the Late-slice-level split, and 71.37% accuracy with an AUC of 0.81 with the Early-subject-level split. These results demonstrate that the performance overestimation observed previously persists in both the CNN+KAN and ResNet50+KAN models.

Initially, the model evaluations were conducted solely on axial view slices. However, since brain atrophy in AD affects multiple regions, relying on a single plane may result in incomplete assessments. To provide a more comprehensive evaluation, the ResNet50 and ResNet50+KAN models were also tested using 2D image slices from all three anatomical planes—Axial, Coronal, and Sagittal—following the same methodology as previous experiments. Table 4 summarizes the accuracy across these views using the Early-subject-level split. Accuracy

Table 3. KAN model performances compared with traditional ResNet50 model for both early and late split approaches.

	Model	Accuracy	Precision	Recall	F1 score	AUC
Late-slice-level split	CNN-KAN	89.49%	91.58%	89.49%	89.89%	0.97
	ResNet50+KAN	91.37%	92.77%	91.37%	91.64%	0.98
	ResNet50	93.46%	93.79%	93.46%	93.56%	0.97
Early-subject-level split	CNN+KAN	71.37%	72.59%	71.37%	71.86%	0.81
	ResNet50+KAN	74.56%	74.87%	74.56%	74.56%	0.84
	ResNet50	78.62%	77.38%	78.62%	77.68%	0.82

Table 4. Accuracy of the models across different anatomical views.

Model	Axial	Coronal	Sagittal
ResNet50	78.62%	76.85%	75.97%
ResNet50 + KAN	74.56%	72.46%	71.63%

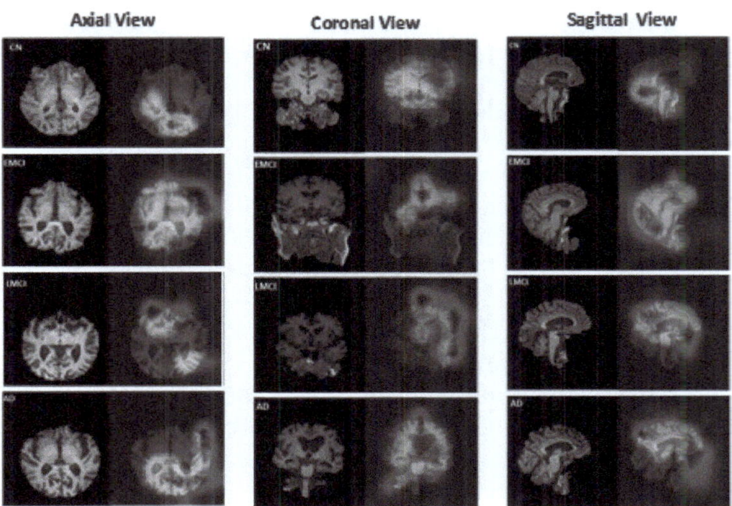

Fig. 5. Grad-CAM heatmaps generated for the ResNet50+KAN model using the Early-subject-level split.

remains consistent across different views, with minimal variation observed. The axial view yields the highest accuracy, aligning with its common preference in clinical practice.

4.3   Grad-CAM Analysis

Grad-CAM analysis [23], visualized in Fig. 5, was utilized to examine the model's activations from the final convolution layer across all classes and anatomical views (Axial, Coronal, and Sagittal) for the ResNet50+KAN model using the Early-subject-level split. The heatmaps generated from correctly predicted images demonstrate the model's focus on AD-related brain regions, aligning well with established neuropathological findings.

Table 5. Accuracy (%) of the models across different types of classification tasks.

Model/Accuracy	3-way (CN-MCI-AD)	2-way (AD vs CN)	One vs REST (CN vs REST)	One vs REST (AD vs REST)
ResNet50	79.86	86.70	80.65	84.04
ResNet50 + KAN	76.05	87.60	81.31	87.65

To further evaluate the predictive abilities of the models across various class configurations, experiments were conducted for 3-way, binary, and one-vs-rest classifications for both ResNet50 and ResNet50+KAN models (Table 5). Due to the clinical and imaging similarities between EMCI and LMCI, these categories were combined into a single MCI group to reduce model confusion, thereby simplifying the classification task to 3-way: CN, MCI, and AD. As anticipated, the models performed better in the 3-way classification than the 4-way task, indicating that a simplified classification process enhances model performance.

Additionally, one-vs-rest binary classification tasks were performed to evaluate the models' capabilities in distinguishing between CN vs. Rest (MCI + AD) and AD vs. Rest (CN + MCI). The results indicated that the models were more effective in separating AD from the rest than in separating CN. This discrepancy is likely due to the gradual progression of the disease in its early stages and the similarities between CN and MCI. For both models, the binary classification task of AD vs. CN yielded the highest accuracy (see Table 5), suggesting that the models are most effective at distinguishing these two distinct categories, which present the most significant structural differences.

Table 6. Model's accuracies with numbers of median slices extracted from MRI.

Model/Accuracy	5-slices	10-slices	15-slices
ResNet50	78.62	77.42	76.24
ResNet50 + KAN	74.56	72.26	73.83

Table 7. ResNet50+KAN accuracy for different values of Grid-Size and Spline Order.

	Spline order K = 3 constant				Grid size G = 5 constant			
	G = 5	G = 10	G = 15	G = 20	K = 2	K = 3	K = 5	K = 7
Accuracy	74.56	73.41	71.50	70.32	74.73	74.56	74.12	71.29

As shown in Table 6, increasing the number of slices provided to the model did not significantly improve accuracy, implying that additional slices do not enhance the model's performance. Further experiments were conducted to assess the impact of varying grid sizes (G=5, 10, 15, 20) and spline orders (k=2, 3, 5, 7) on the ResNet50+KAN model's performance (Table 7). Apart from slight overfitting, no significant changes in accuracy were observed with different grid sizes and spline orders, indicating that these parameters do not substantially affect the model's performance.

5   Discussions

As shown in Table 8, there is a 12–18% discrepancy in performance accuracy between the two approaches, indicating data leakage in the late-slice-level split. This leakage occurs when slices from the same subject are included in both training and testing sets, resulting in inflated accuracies as the model may rely on memorized features rather than actual classifications. Thus, distinguishing between splitting approaches is crucial for obtaining reliable results.

Table 8. Comparative accuracies of models with Early and late split.

Model	Early-Subject-level	Late-Slice-level	Difference
VGG16	75.44%	88.95%	13.51%
VGG19	76.15%	91.60%	15.45%
ResNet50	78.62%	93.46%	14.84%
ResNet152	76.85%	90.90%	14.05%
MobileNetV2	75.80%	84.14%	8.34%
EfficientNetB7	76.50%	91.05%	14.55%
CNN-KAN	71.37%	89.49%	18.12%
ResNet50+KAN	74.56%	91.37%	16.81%

Previous studies often report high accuracies similar to the late-slice-level approach, suggesting a common issue of data leakage due to random slice-level splits or improper data augmentation. While some studies have addressed this issue directly, showing clear performance differences with and without data leakage, others lack methodological details [10–14]. The findings of this study align with previous unbiased models that used subject-level splits. For example, Yagis Ekin et al. [13] reported model accuracies of 70.12%, 66.49%, and 68.68% with two variants of VGG16 and ResNet18 models, while Pierluigi Carcagnì et al. [14] achieved accuracies of 69.53%, 70.19%, and 69.53% with ResNet101, DenseNet161, and EfficientNetB0 models, respectively. In contrast, the best model in this study, ResNet50, achieved 78.62% accuracy, outperforming these studies. This improvement in performance can be attributed to access to more recent and extensive ADNI data compared to previous works.

Ablation experiments were conducted to assess and compare the best models. Results showed that altering views or increasing input slices did not improve performance; models performed better in 3-way and binary classification tasks, particularly in distinguishing between CN and AD classes. Modifications to grid size and spline order in the KAN layer had minimal impact on performance, as also noted by [24]. Moreover, the KAN layer resulted in lower accuracy than the original model, consistent with findings in [25] that show KAN performing similar or worse than MLP in computer vision tasks. Overall, model performance remained consistent, with no significant improvement.

The Grad-CAM heatmaps in Fig. 5 illustrate the trajectory of dementia-related brain changes throughout AD progression, showing the model's ability to capture relevant neuroanatomical features at each stage. For the CN class, activations target regions with preserved brain structures, reflecting the absence of pathological changes typical in cognitive impairment. In the EMCI class, subtle changes emerge, particularly in the prefrontal cortex and hippocampus, areas crucial for memory formation and often affected in the early stages of AD. As the disease progresses to LMCI, activations show a marked focus on regions with more atrophy, such as the temporal lobes and parietal cortex, correlating with deficits in memory, language, and spatial navigation. In the AD class, heatmaps display extensive activations in severely affected regions like the hippocampus and amygdala, corresponding to memory loss and behavioral changes characteristic of late-stage AD. This alignment between model activations and established AD pathology underscores the model's potential for diagnostic support in clinical settings.

While the early-subject-level split approach yields better results than previous studies, achieving around 75% accuracy still falls short of establishing a robust medical framework. The model struggles to accurately distinguish between CN subjects and those with MCI due to the broad nature of MCI labels, which encompass varying stages of cognitive decline. The subtle differences in disease progression, particularly between late MCI and AD, complicate clear separations in the data.

Structural changes in AD, marked by significant brain volume loss, including shrinkage (atrophy) and ventricular enlargement, are visible on MRI. However, classic symptoms like cognitive decline and memory impairment cannot be analyzed solely through these scans. Additionally, individual variability in brain structure, along with the overlap between normal age-related brain shrinkage and AD, further complicates reliable classification. MRI image quality can be compromised by variations in clinical settings, scanner configurations, and segmentation errors during preprocessing. These factors, along with inconsistent labeling and imprecise annotations, degrade image quality and contribute to misclassifications. Future research should integrate MRI with other imaging modalities like fMRI and PET, as well as non-imaging data such as genetic, behavioral, and cognitive assessments.

6   Conclusion

This study evaluated six pre-trained models (ResNet152, ResNet50, VGG16, VGG19, MobileNetV2, and EfficientNetB7) on the ADNI dataset using two data-splitting approaches to assess the impact of data leakage. Models trained with data leakage displayed inflated accuracies, with ResNet50 showing the highest performance, achieving 93.47% and 78.62% accuracy with and without leakage, respectively. These findings emphasize the critical importance of appropriate data splitting to ensure reliable and unbiased model evaluation. The integration of KAN with ResNet50 did not surpass the base model, resulting in 91.37% and 74.63% accuracy with and without leakage, respectively. Binary classification between AD and CN provided the most accurate results, demonstrating the model's ability to effectively differentiate the most distinct classes. Ablation experiments revealed minimal changes in performance with varying numbers of slices and brain views, with axial slices consistently performing the best. Adjustments to KAN's grid size and spline order had negligible impact on performance, aside from slight overfitting observed at larger values. Grad-CAM heatmaps aligned model predictions and established AD pathology, validating the model's capacity to capture critical neuroanatomical features. Overall, this study compares DL models comprehensively, evaluating their performance under conditions with and without data leakage to critically examine their practical application.

References

1. World Health Organization. Dementia: Fact sheet (2017). https://www.who.int/news-room/fact-sheets/detail/dementia
2. Rasmussen, J., Langerman, H.: Alzheimer's disease – why we need early diagnosis. In: Degenerative Neurological and Neuromuscular Disease, pp. 123–130 (2019). https://doi.org/10.2147/DNND.S228939
3. Qiu, S., et al.: Development and validation of an interpretable deep learning framework for Alzheimer's disease classification. Brain **143**(6), 1920–1933 (2020). https://doi.org/10.1093/brain/awaa137
4. Qiu, S., et al.: Multimodal deep learning for Alzheimer's disease dementia assessment. Nat. Commun. **13**(1), 3404 (2022). https://doi.org/10.1038/s41467-022-31037-5
5. Hon, M.; Khan, N.M.: Towards Alzheimer's disease classification through transfer learning. In: Proceedings of the 2017 IEEE International Conference on Bioinformatics and Biomedicine (BIBM), Kansas, MO, USA, 13–16 November 2017, pp. 1166–1169 (2017)

6. Jain, R., Jain, N., Aggarwal, A., Hemanth, D.J.: Convolutional neural network based Alzheimer's disease classification from magnetic resonance brain images. Cogn. Syst. Res. **57**, 147–159 (2019)

7. Naveen, N., Nagaraj, G.C.: Enhancing early Alzheimer's disease detection: leveraging pre-trained networks and transfer learning. Int. J. Intell. Syst. Appl. (IJISA) **16**(1), 52–69 (2024)

8. Mahmud, T., Barua, K., Habiba, S.U., Sharmen, N., Hossain, M.S., Andersson, K.: An explainable AI paradigm for alzheimer's diagnosis using deep transfer learning. Diagnostics **14**(3), 345 (2024). https://doi.org/10.3390/diagnostics14030345

9. Shamrat, F.M.J.M., et al.: AlzheimerNet: an effective deep learning based proposition for alzheimer's disease stages classification from functional brain changes in magnetic resonance images. IEEE Access **11**, 16376–16395 (2023). https://doi.org/10.1109/ACCESS.2023.3244952

10. Karl, B., Nazari, M., Gu, I.Y.H., Jakola, A.S.: An efficient 3D deep convolutional network for Alzheimer's disease diagnosis using MRI images. In: ISBI 2018, pp. 149–153. IEEE (2018)

11. Ren, F.Y., Guan, Z., Kumar, R., Wu, J.Y., Fiterau, M.: Alzheimer's disease brain MRI classification: challenges and insights. arXiv preprint arXiv:1906.04231 (2019)

12. Yagis, E., De Herrera, A.G.S., Citi, L.: Generalization performance of deep learning models in neurodegenerative disease classification. In: 2019 IEEE International Conference on Bioinformatics and Biomedicine (BIBM), San Diego, CA, USA, pp. 1692–1698 (2019). https://doi.org/10.1109/BIBM47256.2019.8983088

13. Yagis, E., et al.: Effect of data leakage in brain MRI classification using 2D convolutional neural networks. Sci. Rep. **11**(1), 22544 (2021)

14. Carcagnì, P., Leo, M., Del Coco, M., Distante, C., De Salve, A.: Convolution neural networks and self-attention learners for alzheimer dementia diagnosis from brain MRI. Sensors **23**, 1694 (2023)

15. ADNI— Alzheimer's Disease Neuroimaging Initiative. http://adni.loni.usc.edu/

16. Fischl, B.: FreeSurfer. Neuroimage **62**(2), 774–781 (2012)

17. Karen, S., Zisserman, A.: Very deep convolutional networks for large-scale image recognition. arXiv preprint arXiv:1409.1556 (2014)

18. He, K., Zhang, X., Ren, S., Sun, J.: Deep residual learning for image recognition. In: Proceedings of the IEEE Conference on Computer Vision and Pattern Recognition, pp. 770–778 (2016)

19. Sandler, M., Howard, A., Zhu, M., Zhmoginov, A., Chen, L.C.: Mobilenetv2: inverted residuals and linear bottlenecks. In: Proceedings of the IEEE Conference on Computer Vision and Pattern Recognition, pp. 4510–4520 (2018)

20. Tan, M.: EfficientNet: rethinking model scaling for convolutional neural networks. arXiv preprint arXiv:1905.11946 (2019)

21. Liu, Z., Wang, Y., Vaidya, S., Ruehle, F., Halverson, J., Soljačić, M., & Tegmark, M.: Kan: Kolmogorov-Arnold networks. arXiv preprint arXiv:2404.19756 (2024)

22. Blealtan. An efficient implementation of Kolmogorov-Arnold network (2024). https://github.com/Blealtan/efficient-kan/tree/master

23. Selvaraju, R.R., Cogswell, M., Das, A., Vedantam, R., Parikh, D., Batra, D.: Grad-CAM: visual explanations from deep networks via gradient-based localization. Int. J. Comput. Vision **128**, 336–359 (2020)

24. Shen, S., Younes, R.: Reimagining Linear Probing: Kolmogorov-Arnold Networks in Transfer Learning. arXiv preprint arXiv:2409.07763 (2024)

25. Yu, R., Yu, W., Wang, X.: Kan or MLP: a fairer comparison. arXiv preprint arXiv:2407.16674 (2024)

Leveraging Large Language Models for Zero Shot Feature Selection in Cardiometabolic Disease Prediction

Kesanam Ashinee[✉], Gummuluri Venkata Ravi Ram, and M. Anand Kumar

Artificial Intelligence, Department of Information Technology, National Institute of Technology Karnataka, Surathkal, India
ashineekesanam@gmail.com , m_anandkumar@nitk.edu.in

Abstract. Feature selection is a critical step in building effective machine learning models, particularly in medical datasets where high-dimensional data can impede model performance and interpretability. This paper explores the application of large language models (LLMs), specifically Llama-based feature selection, and compares its performance against traditional techniques such as Recursive Feature Elimination (RFE) and KBest across various classification models. We evaluate models including AdaBoost, Decision Tree, Gradient Boosting, K-Nearest Neighbors (K-NN), Logistic Regression, Naive Bayes, Random Forest, and Support Vector Machine (SVM) on the Framingham heart disease dataset, analyzing their performance in terms of accuracy, recall, and precision. The results show that Llama-based feature selection methods generally outperform traditional techniques, with Logistic Regression and AdaBoost models exhibiting the most notable improvements. Our findings highlight the potential of LLM-based feature selection in enhancing predictive performance for medical datasets.

Keywords: LLMs · Zero shot learning · Feature Selection · Medical Datasets

1 Introduction

The rapid advancement of machine learning in healthcare has led to the exploration of numerous approaches for predicting medical outcomes, identifying disease risk factors, and enhancing diagnostic accuracy. One of the key challenges in developing robust machine learning models, especially in the medical domain, is the selection of relevant features from high-dimensional data. Medical datasets often contain vast amounts of heterogeneous data, including clinical records, diagnostic images, genomic information, and patient demographics, making feature selection critical to reduce dimensionality, improve model performance, and enhance interpretability.

Traditional feature selection methods, such as Recursive Feature Elimination (RFE), LASSO (Least Absolute Shrinkage and Selection Operator), and Tree-based feature importance, have long been used to identify the most relevant features for model training. These methods typically rely on labeled data and specific task definitions, which may not always be available in medical applications due to the costly and time-consuming nature of acquiring expert-labeled data. This challenge calls for innovative approaches that can reduce the dependence on labeled data while maintaining high accuracy and model interpretability.

In parallel with traditional approaches, the emergence of large language models (LLMs) like GPT-4, Mistral, and others has ushered in the potential for zero-shot and few-shot learning in various domains. Zero-shot learning allows models to generalize to tasks without requiring specific task-related training data, leveraging their ability to understand and apply knowledge across different domains. This paradigm has shown promise in natural language processing tasks, but its application in structured data and feature selection remains underexplored.

Zero-shot feature selection means selecting important features from data without specifically training a model on that type of data or task beforehand. Instead, it uses a pre-trained model, like a large language model (LLM), to identify relevant features directly based on its general knowledge and understanding.

This paper focuses on comparing the performance of feature selection techniques—both traditional and zero-shot approaches using LLMs—on various medical datasets. By evaluating LLMs' potential for feature selection, we aim to determine whether LLM-based methods can outperform, or complement, traditional feature selection strategies. We hypothesize that LLMs, through their semantic understanding, can identify important features that align with medical domain knowledge without requiring extensive labeled data.

2 Related Work

In recent years, machine learning (ML) and deep learning (DL) models have been extensively used for disease prediction, including heart disease, diabetes. However, the selection of relevant features significantly impacts the model's performance. Feature selection helps in reducing dimensionality, improving model accuracy, and providing interpretability. In this context, integrating large language models (LLMs) for feature selection offers a promising avenue due to their zero-shot learning capabilities and potential to identify relevant features from both structured and unstructured data.

In diabetes prediction, [9] by Ayman Mir and Sudhir N. Dhage utilized traditional ML techniques like SVM, Random Forest, and Naive Bayes, highlighting the need for effective feature selection to enhance the predictive power of the models.

In heart disease prediction, researchers have explored various clinical features for model training. [2] designed a system that used 13 key clinical features to predict heart disease with around 80% accuracy using neural networks. This highlights the need for effective feature selection to improve model robustness and predictive accuracy in healthcare applications.

LLMs offer a novel approach to feature selection by leveraging their ability to perform zero-shot learning. [12] emphasize the importance of data selection during instruction tuning of LLMs, where selecting high-quality subsets improves performance while reducing training costs. Furthermore, [13] proposed a dynamic and adaptive feature generation approach using LLMs, which allows for more flexible and explainable feature selection across various tasks.

A data-centric approach to feature selection with LLMs was explored by [7], who highlighted the potential of LLMs in both data-driven and text-based feature selection. By leveraging semantic associations from descriptive context, LLMs can identify relevant features even with minimal data, making them particularly useful in medical applications.

Clearly, LLMs hold great potential for enhancing feature selection in disease prediction models. Their zero-shot learning capabilities allow for identifying key features without extensive training, improving model accuracy and generalizability. The combination of LLMs with traditional ML techniques can pave the way for more efficient, explainable, and scalable feature selection methods, particularly in healthcare, where accurate predictions are crucial for patient outcomes.

3 Problem Statement

The primary objective of this research is to compare the efficacy of LLM-based zero-shot feature selection with traditional feature selection methods in medical datasets. Specifically, we seek to answer the following questions:

- How do LLM-based feature selection methods perform compared to traditional techniques in terms of accuracy, precision, recall, and F1-score?
- Can LLMs effectively generalize across different medical datasets and classification tasks without requiring specific training data?

4 Dataset Description

4.1 Framingham Heart Disease Dataset

The Framingham Heart Disease dataset is derived from the Framingham Heart Study, containing 4,240 records with 16 attributes. It is used to predict a 10-year risk of developing coronary heart disease (CHD) based on 15 clinical and demographic attributes: gender, age, education level, smoking status, cigarettes per day, blood pressure medication use, history of stroke, hypertension, diabetes, total cholesterol, systolic and diastolic blood pressure, BMI, heart rate, and glucose level. The target variable, `TenYearCHD`, indicates a 10-year CHD risk [4,8].

4.2   Pima Indians Diabetes Dataset

The Pima Indians Diabetes dataset contains 768 records with diagnostic factors related to diabetes onset prediction, specifically for patients aged 21 or older of Pima Indian heritage. Features include number of pregnancies, plasma glucose concentration, diastolic blood pressure, triceps skinfold thickness, serum insulin level, BMI, diabetes pedigree function, and age. The target variable, `Outcome`, indicates diabetes diagnosis [3,6].

4.3   Cleveland Heart Disease Dataset

The Cleveland Heart Disease dataset, containing 270 records with 13 attributes, is used for cardiovascular disease prediction tasks. Features include patient age, gender, chest pain type, resting blood pressure, cholesterol level, fasting blood sugar, ECG results, max heart rate achieved, exercise-induced angina, ST depression, peak exercise ST segment slope, number of major vessels, and thalassemia status. The target variable, `Target`, indicates heart disease absence (1) or presence (2) [5,10].

5   Methodology

This section details the methodology employed in our research to evaluate the performance of various feature selection techniques and machine learning models on three healthcare datasets: heart disease diagnosis, 10 year coronary heart disease prediction, and diabetes prediction. The objective was to identify the most influential features and models that could enhance prediction accuracy in clinical settings.

5.1   Feature Selection Techniques

Conventional Feature Selection. Feature selection techniques such as SelectKBest, Recursive Feature Elimination (RFE), and Random Forest Feature Importance were employed to refine the feature sets. SelectKBest, utilizing an ANOVA F-test, identified the top 4, 5, and 6 features based on their relevance to the target variable. RFE, using Logistic Regression as the base model, iteratively removed the least significant features to reach an optimal set of 5. Additionally, a Random Forest classifier was used to rank features by importance, selecting the top 4, 5, and 6 based on their predictive power.

Large Language Model (LLM) Feature Selection. To enhance the feature selection process, we leveraged the capabilities of Large Language Models such as LLaMA [11] and GPT-4 [1] using zero-shot learning approach. LLMs, with their ability to process and understand natural language, can provide domain-specific insights by analyzing the context of healthcare-related data. For each dataset, we provided the context of the particular disease and tasked the LLM with identifying relevant features from the given dataset's column names.

For example, in the case of the heart disease dataset, the model was given detailed context about what the heart disease, and it suggested some important key features which are critical for predicting the presence of heart disease. Similarly, for the diabetes dataset, the LLM identified features that as relevant for diabetes prediction. The same approach was applied to the Framingham Heart Study dataset, where the context of cardiovascular risk factors was provided to the LLM, resulting in the identification of important features.

By incorporating context LLM-driven feature suggestions, we were able to improve our feature selection process by integrating both data-driven insights and domain-specific knowledge from the healthcare literature. This method allowed for a more informed and comprehensive selection of relevant features for each dataset, enhancing the overall performance of our classification models.

5.2   Model Selection and Evaluation

A variety of machine learning classifiers were used to assess the effectiveness of selected feature sets. These included Logistic Regression, Random Forest, Gradient Boosting, K-Nearest Neighbors (KNN), Support Vector Machine (SVM), Decision Tree, Naive Bayes, and AdaBoost. Each classifier brought a unique approach to handling classification tasks—ranging from probabilistic models like Logistic Regression and Naive Bayes to ensemble methods such as Random Forest and AdaBoost, as well as techniques like SVM that maximize margin separation between classes. These classifiers allowed for a comprehensive evaluation of the predictive power of the chosen features.

For each feature selection method and dataset, we applied a series of evaluation steps. The dataset was split into training (80%) and testing (20%) subsets, with each model trained on the training data using the selected features. Predictions were then made on the testing data, and performance metrics including accuracy, precision, and recall were calculated to assess each model's effectiveness. These metrics provided insight into each model's accuracy, positive predictive power, and ability to identify relevant instances within the dataset.

6   Experimental Results and Discussion

In the Result tables, The "All" column shows performance metrics when using all available features without any selection or reduction, serving as a baseline for comparison with models using feature-reduced sets. Columns GPT_4, GPT_5, and GPT_6 refer to models trained on feature subsets of n = 4, n = 5, n = 6 respectively chosen with help of GPT4. Columns L_4, L_5, and L_6 refer to models trained on feature subsets of n = 4, n = 5, n = 6 respectively chosen with help of Llama. The RFE column reflects the performance of models using Recursive Feature Elimination, a method that recursively removes less significant features until only the desired number remains. This iterative approach ranks features based on their predictive impact. Lastly, RF_4, RF_5, and RF_6 columns represent models trained on feature subsets of n = 4, n = 5, n

= 6 respectively chosen with help of Random Forest feature importance ranking. Similarly, SKB_4, SKB_5, and SKB_6 show results based models trained on feature subsets of n = 4, n = 5, n = 6 respectively chosen with help of SelectKBest, another technique that ranks features based on statistical scores.

6.1 Frimingham Dataset

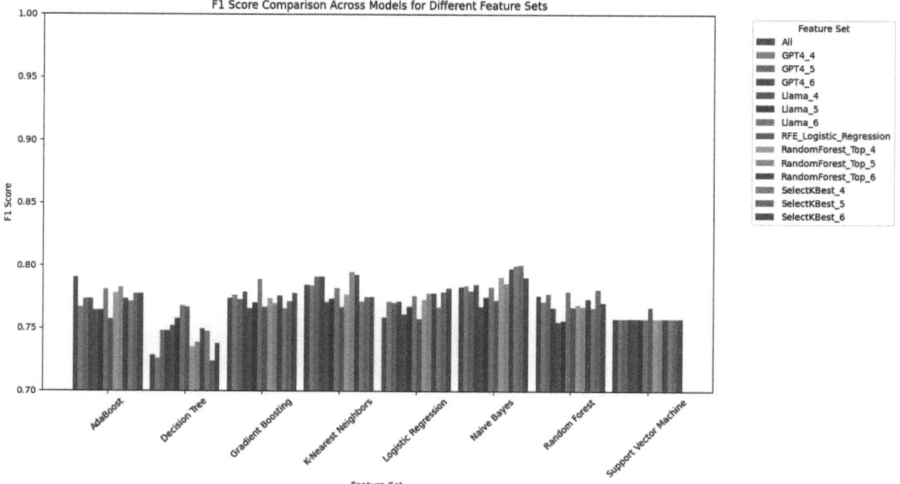

Fig. 1. F1 Score accross all models for various Feature Selection on Frimingham Dataset.

Table 1. Accuracy of Frimingham Dataset (Feature Selection using LLM)

Model	All	GPT_4	GPT_5	GPT_6	L_4	L_5	L_6
AdaBoost	0.839	0.832	0.832	0.832	0.836	0.836	0.842
Decision Tree	0.734	0.731	0.758	0.753	0.813	0.816	0.777
Gradient Boosting	0.832	0.837	0.835	0.842	0.835	0.835	0.842
KNN	0.827	0.824	0.832	0.832	0.818	0.820	0.825
Logistic Reg.	0.831	0.836	0.835	0.836	0.835	0.837	0.836
Naive Bayes	0.810	0.817	0.811	0.806	0.837	0.828	0.820
Random Forest	0.837	0.822	0.820	0.831	0.809	0.818	0.802
SVM	0.833	0.833	0.833	0.833	0.833	0.833	0.833

Table 2. Accuracy of Frimingham Dataset (Conventional Feature Selection)

Model	RFE	RF_4	RF_5	RF_6	SKB_4	SKB_5	SKB_6
AdaBoost	0.832	0.831	0.833	0.829	0.836	0.839	0.839
Decision Tree	0.832	0.758	0.727	0.745	0.732	0.724	0.717
Gradient Boosting	0.832	0.835	0.835	0.836	0.835	0.837	0.829
KNN	0.832	0.832	0.828	0.832	0.811	0.822	0.822
Logistic Reg.	0.833	0.835	0.835	0.840	0.836	0.842	0.843
Naive Bayes	0.813	0.809	0.814	0.820	0.809	0.816	0.813
Random Forest	0.832	0.820	0.836	0.837	0.809	0.821	0.828
SVM	0.832	0.833	0.833	0.833	0.833	0.833	0.833

Figure 1 provides a clear visualization of the overall performances of each feature selection method for each model on Frimingham Dataset using F1-Score. Based on the results as in Table 1, Table 2; AdaBoost, in particular, shows strong recall values across all feature selection techniques, ranging from 0.829 to 0.842, demonstrating its robustness. This is also reflected in its accuracy, where it consistently performs well, especially with Llama-based feature selection (Llama_4, 5, 6). Gradient Boosting also shows similar trends, with strong performance across all metrics, albeit slightly less consistent than AdaBoost.

Logistic Regression (LR) stands out in terms of precision and recall, with Llama feature selection achieving the highest precision values (0.843). This suggests that LR is well-suited for the selected dataset when combined with Llama feature selection, showing the importance of feature selection in enhancing traditional models like LR.

Decision Tree, while showing some improvement with Llama feature selection, does not perform as well as other models in terms of accuracy and recall, particularly under traditional feature selection methods. K-NN and Naive Bayes also show fluctuations in performance depending on the feature selection method, though they maintain respectable performance levels.

Interestingly, Support Vector Machines (SVM) exhibits a flat trend across all metrics and feature selection methods, consistently producing 0.694 accuracy, recall, and precision. This suggests that SVM might not be the best choice for this dataset with the current feature selection techniques.

6.2   Diabeties Dataset

Figure 2 illustrates the F1-Score performance for each feature selection method across various models on the Diabetes Dataset. Table 3, Table 4 highlights how feature selection impacts model performance.

The Decision Tree model achieved the highest accuracy (80.01%) using the Llama_6 feature set, showcasing its ability to handle complex interactions for diabetes prediction. Naive Bayes, meanwhile, maintained consistent accu-

Table 3. Accuracy of Diabetes Dataset (Feature Selection Using LLM)

Model	All	GPT_4	GPT_5	GPT_6	L_4	L_5	L_6
AdaBoost	0.734	0.740	0.753	0.740	0.734	0.727	0.740
Decision Tree	0.753	0.747	0.695	0.740	0.643	0.714	0.779
Gradient Boosting	0.747	0.734	0.747	0.734	0.753	0.753	0.740
K-NN	0.662	0.740	0.760	0.701	0.701	0.701	0.675
Logistic Reg.	0.747	0.773	0.747	0.760	0.753	0.766	0.766
Naive Bayes	0.766	0.760	0.766	0.779	0.747	0.779	0.760
Random Forest	0.734	0.760	0.766	0.760	0.727	0.740	0.734
SVM	0.766	0.766	0.779	0.760	0.753	0.753	0.766

Table 4. Accuracy of Diabetes Dataset (Conventional Feature Selection)

Model	RFE	RF_4	RF_5	RF_6	SKB_4	SKB_5	SKB_6
AdaBoost	0.760	0.740	0.753	0.760	0.747	0.760	0.727
Decision Tree	0.740	0.714	0.708	0.766	0.675	0.727	0.740
Gradient Boosting	0.721	0.734	0.747	0.727	0.747	0.721	0.747
K-NN	0.766	0.740	0.760	0.747	0.766	0.766	0.701
Logistic Reg.	0.753	0.773	0.747	0.740	0.753	0.753	0.760
Naive Bayes	0.753	0.760	0.766	0.753	0.734	0.753	0.779
Random Forest	0.753	0.773	0.760	0.753	0.773	0.766	0.760
SVM	0.773	0.766	0.779	0.773	0.773	0.773	0.760

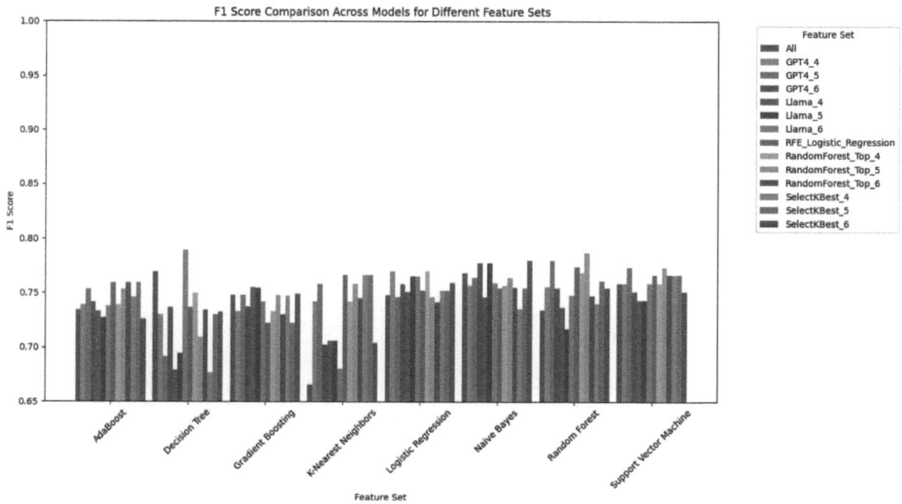

Fig. 2. F1 Score accross all models for various Feature Selection on Diabities Dataset.

racy across feature sets, reaching 77.62% with Llama_5, proving effective in cases with independent features and as a reliable baseline.

For precision, Naive Bayes excelled with 77.92% on the Llama_5 feature set, indicating its strength in minimizing false positives—a crucial aspect in medical diagnostics to reduce unnecessary testing. Gradient Boosting demonstrated the highest recall (76.62%) with Llama_5, indicating its effectiveness in identifying true positive cases, essential for early detection and intervention strategies in diabetes management.

6.3   Heart Disease Dataset

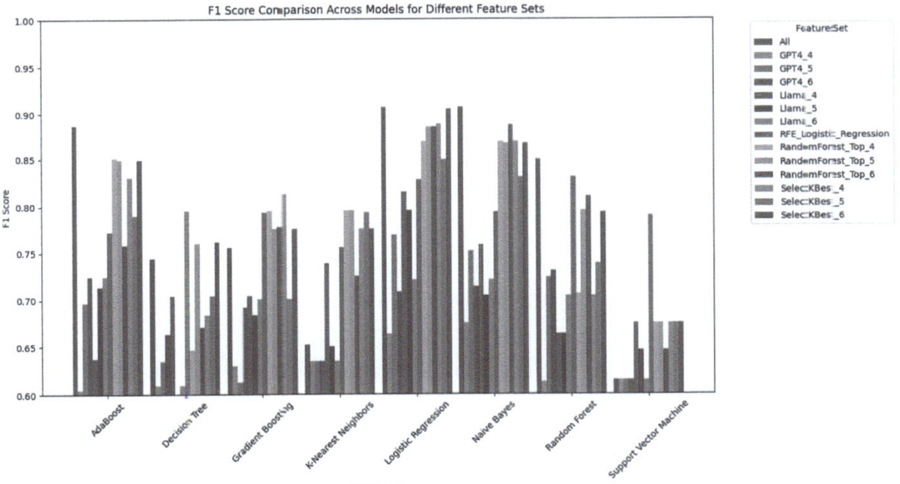

Fig. 3. Recall accross all models for various Feature Selection on Heart Disease Dataset.

Figure 3 provides a clear visualization of the overall performances of each feature selection method for each model on Heart Disease Dataset using F1-Score. The performance of various machine learning models on the heart disease dataset as in Table 5 and Table 6 reveals several critical insights. Logistic Regression (LR) consistently demonstrated superior accuracy, precision, and recall across multiple feature sets, achieving an impressive accuracy of 0.910494 with the SelectKBest_6 feature set. This suggests that LR is particularly effective for binary classification tasks in healthcare, potentially due to its interpretability and robustness in handling linearly separable data.

Naive Bayes (NB) also performed admirably, with an accuracy of 0.910494, indicating its efficacy in this domain. The model's performance reinforces the notion that even simpler algorithms can yield competitive results, especially when the underlying assumptions align well with the data distribution.

Table 5. Accuracy of Heart Failure (Feature Selection using LLM)

Model	All	GPT_4	GPT_5	GPT_6	L_4	L_5	L_6
AdaBoost	0.889	0.611	0.704	0.741	0.648	0.722	0.741
Decision Tree	0.704	0.574	0.630	0.685	0.630	0.593	0.611
Gradient Boosting	0.759	0.630	0.611	0.704	0.704	0.685	0.704
K-NN	0.648	0.630	0.630	0.630	0.741	0.648	0.630
Logistic Reg.	0.907	0.667	0.778	0.722	0.815	0.796	0.722
Naive Bayes	0.907	0.685	0.759	0.722	0.759	0.704	0.722
Random Forest	0.889	0.593	0.685	0.685	0.685	0.704	0.704
SVM	0.630	0.630	0.630	0.630	0.685	0.667	0.630

Table 6. Accuracy of Heart Failure (Conventional Feature Selection)

Model	RFE	RF_4	RF_5	RF_6	SKB_4	SKB_5	SKB_6
AdaBoost	0.778	0.852	0.852	0.759	0.833	0.796	0.852
Decision Tree	0.852	0.630	0.778	0.648	0.685	0.704	0.741
Gradient Boosting	0.796	0.796	0.796	0.759	0.815	0.704	0.778
K-NN	0.759	0.796	0.796	0.722	0.778	0.796	0.778
Logistic Reg.	0.833	0.870	0.889	0.889	0.889	0.852	0.907
Naive Bayes	0.796	0.870	0.870	0.889	0.870	0.833	0.870
Random Forest	0.833	0.704	0.778	0.796	0.722	0.704	0.796
SVM	0.796	0.685	0.685	0.667	0.685	0.685	0.685

In contrast, models like Support Vector Machine (SVM) exhibited the lowest accuracy, with a maximum of 0.629630 across all feature sets. This could indicate that SVM is less suited for this dataset, possibly due to the presence of overlapping classes or the need for extensive parameter tuning.

Random Forest (RF) and Gradient Boosting (GB) models showcased moderate performance, with accuracies around 0.864387 and 0.776365, respectively. The effectiveness of these ensemble methods highlights their ability to capture complex relationships within the data. However, RF's performance was notably affected by the choice of features, with variations in accuracy across different sets, suggesting a sensitivity to feature selection.

It is essential to consider how the algorithms - Naive Bayes and KNN rely on data distributions. K-NN is highly sensitive to the selection of features and the distance metric, which can lead to performance variability when irrelevant or redundant features are present. Conversely, Naïve Bayes assumes feature independence, and its performance may fluctuate depending on how well this assumption aligns with the selected feature set.

The results underscore the importance of feature selection in improving model performance, as evidenced by the varying accuracies across the different feature sets like Llama_4, Llama_5, SelectKBest, and others.

Regarding overfitting, techniques such as cross-validation were likely used to assess the generalizability of the models, and methods like regularization, pruning, or early stopping might have been applied depending on the algorithm. Highlighting these approaches can provide clarity on how overfitting was managed during training and evaluation to ensure robust accuracy measurements.

7 Conclusion and Future Work

In this study, we evaluated the impact of different feature selection methods on various machine learning models, including AdaBoost, Decision Tree, Gradient Boosting, K-NN, Logistic Regression, Naive Bayes, Random Forest, and SVM, using the Framingham dataset. We compared traditional techniques like RFE and KBest with Llama and GPT-4-based feature selection methods. Our results showed that Llama-based methods significantly improved model performance, with Logistic Regression and AdaBoost achieving high precision and recall, while SVM consistently underperformed.

7.1 Future Work

As of now, we have used feature selection with large language models (LLMs) like GPT and BERT primarily for regression-type tasks, leveraging their capabilities to identify relevant attributes in structured data. However, LLMs also hold significant potential for text data, such as clinical notes and patient records, where they can extract meaningful entities and patterns. By encoding complex textual information into structured representations, LLMs can help highlight clinically significant features, enhancing predictive models. Extending their application to combine structured clinical data with text-based insights could unlock further possibilities, streamlining feature selection and improving interpretability in healthcare prediction tasks.

References

1. Achiam, J., et al.: Gpt-4 technical report. arXiv preprint arXiv:2303.08774 (2023)
2. Chen, A.H., Huang, S.Y., Hong, P.S., Cheng, C.H., Lin, E.J.: HDPS heart disease prediction system. In: 2011 Computing in Cardiology, pp. 557–560. IEEE (2011)
3. Choubey, D.K., Paul, S., Kumar, S., Kumar, S.: Classification of pima indian diabetes dataset using naive bayes with genetic algorithm as an attribute selection. In: Communication and Computing Systems: Proceedings of the International Conference on Communication and Computing System (ICCCS 2016), pp. 451–455 (2017)
4. D'Agostino, R.B., Grundy, S., Sullivan, L.M., Wilson, P., Group, C.R.P., et al.: Validation of the framingham coronary heart disease prediction scores: results of a multiple ethnic groups investigation. Jama **286**(2), 180–187 (2001)

5. El-Bialy, R., Salamay, M.A., Karam, O.H., Khalifa, M.E.: Feature analysis of coronary artery heart disease data sets. Procedia Comput. Sci. **65**, 459–468 (2015)
6. Karatsiolis, S., Schizas, C.N.: Region based support vector machine algorithm for medical diagnosis on pima Indian diabetes dataset. In: 2012 IEEE 12th International Conference on Bioinformatics & Bioengineering (BIBE), pp. 139–144. IEEE (2012)
7. Li, D., Tan, Z., Liu, H.: Exploring large language models for feature selection: a data-centric perspective. arXiv preprint arXiv:2408.12025 (2024)
8. Lloyd-Jones, D.M., et al.: Framingham risk score and prediction of lifetime risk for coronary heart disease. Am. J. Cardiol. **94**(1), 20–24 (2004)
9. Mir, A., Dhage, S.N.: Diabetes disease prediction using machine learning on big data of healthcare. In: 2018 fourth International Conference on Computing Communication Control and Automation (ICCUBEA), pp. 1–6. IEEE (2018)
10. Nahar, J., Imam, T., Tickle, K.S., Chen, Y.P.P.: Computational intelligence for heart disease diagnosis: a medical knowledge driven approach. Expert Syst. Appl. **40**(1), 96–104 (2013)
11. Touvron, H., et al.: Llama: open and efficient foundation language models. arXiv preprint arXiv:2302.13971 (2023)
12. Wang, J., Zhang, B., Du, Q., Zhang, J., Chu, D.: A survey on data selection for LLM instruction tuning. arXiv preprint arXiv:2402.05123 (2024)
13. Zhang, X., Zhang, J., Rekabdar, B., Zhou, Y., Wang, P., Liu, K.: Dynamic and adaptive feature generation with LLM. arXiv preprint arXiv:2406.03505 (2024)

Enhancing 5G Network Slicing
with Machine Learning Classification

Apoorva Bapuram[1]([✉]), M. Akila Devi[1], Abhishek Kumar[1], Yakub Banoth[1], Santosh K. Dwivedi[2], and Amarjit Kumar[1]

[1] ECE Department, NIT Warangal, Hanmakonda 506004, Telengana, India
{ba23ecm3r03,ma23ecm3s06,ak23ecm3s01,by21ecrer13}@student.nitw.ac.in,
amarjitk@nitw.ac.in
[2] ECE Department, GNIT Campus, Hyderabad 501506, Telengana, India

Abstract. In 5G wireless networks, network slicing plays a critical role by allowing operators to partition a single physical network into multiple virtual networks, each optimized for specific application requirements This paper presents a novel integrates machine learning (ML) techniques to optimize network slicing and improve quality of service (QoS) in 5G environments. The framework combines two key algorithms: RF and NB providing a comprehensive approach to network slicing. The performance evaluation demonstrates that Random Forest achieves 95.5% accuracy Naive Bayes reaches 94.3% accuracy. Additionally, the paper includes an extensive sensitivity analysis to assess the framework's resilience across diverse use cases such as smart cities, autonomous vehicles, and healthcare applications. The significant potential of ML techniques to enhance the efficiency, scalability, and adaptability of 5G networks, providing crucial insights for the deployment of future wireless communication systems.

Keywords: 5G Network slice · Deep Learning · Machine Learning · Knime Analytics Platform

1 Introduction

The introduction of fifth-generation wireless technology marks a revolutionary advancement in communication networks, enabling unprecedented high-speed, low-latency connections that support diverse applications across industries. Key services facilitated by 5G include mMTC, URLLC, eMBB, making it a cornerstone for innovations in healthcare, smart cities, autonomous vehicles, industrial automation, and immersive media. Despite its potential, challenges such as dynamic traffic demands and resource allocation unpredictability complicate the implementation of efficient slicing, highlighting the need for predictive optimization tools.

M. A. Devi, A. Kumar, Y. Banoth, S. K. Dwivedi, and A. Kumar—These authors contributed equally to this work.

This research addresses a significant gap in the literature, as existing studies often fail to provide scalable, accurate models capable of predicting and optimizing 5G network slicing performance under real-world conditions. While prior work has explored the application of machine learning (ML) in this domain, there is limited research integrating diverse classifiers and deep learning models for robust slice performance evaluation. The novelty of this study lies in its development of a multi-classifier framework combining RF, and NB. By achieving exceptional prediction accuracies 95.5% with Random Forest, and 94.3% with Naive Bayes—the proposed model establishes a benchmark for resource optimization and Quality of Service (QoS) assurance in dynamic 5G environments. This research bridges theoretical concepts and practical applications, demonstrating how machine learning can enhance 5G network management in scenarios such as real-time healthcare, autonomous driving, and interactive education.Section 2 introduces a brief summary of the 5G network architecture, while Sect. 3 explores a collection of related studies. The dataset applied in this research is detailed in Sect. 4. The methods and algorithms used in the study are discussed in Sect. 5, followed by an in-depth explanation of the KNIME workflow implementation in Sect. 6. The outcomes of the workflow are presented in Sect. 7, and the paper concludes with a summary in Sect. 8.

2   Architecture of 5G Networks

A key idea in Network slicing, a feature of 5G technology, enables the physical infrastructure of a single network to be separated into multiple virtual networks, each specifically designed to satisfy the unique needs of different users and applications. This approach enhances the flexibility and efficiency of the network, making it adaptable for a wide range of industries. The architecture of network slicing consists of two primary blocks: the Three-Layer Slice Implementation Block and the Slice Management Block. The Three-Layer Slice Implementation Block includes the Service Layer, which translates service requirements into Service Level Agreements (SLAs); the Network Function Layer, responsible for constructing network slices based on specific needs; and the Infrastructure Layer, which provides the necessary physical network resources for multiplexing. The Slice Management Block is managed by a Network Slice Controller, which orchestrates the entire process of slice creation, resource virtualization, performance monitoring, and dynamic reconfiguration. Three primary application categories are supported by 5G networks: URLLC, mMTC, eMBB. This eMBB is intended to provide high-capacity services, such as ultra-high-definition video streaming and immersive experiences like virtual and augmented reality, ensuring faster speeds and improved user experiences. mMTC is geared towards accommodating the massive growth of connected devices, supporting the Internet of Things (IoT) with efficient data transmission from a large number of devices, which is crucial for applications in areas like smart cities, industrial automation, and environmental monitoring. URLLC addresses the stringent requirements of mission-critical applications including remote surgery, autonomous driving, and public safety systems that require extremely low latency and great reliability. By enabling these distinct services, network slicing unlocks the full potential of 5G technology, ensuring it meets the diverse needs of modern communication systems (Fig. 1).

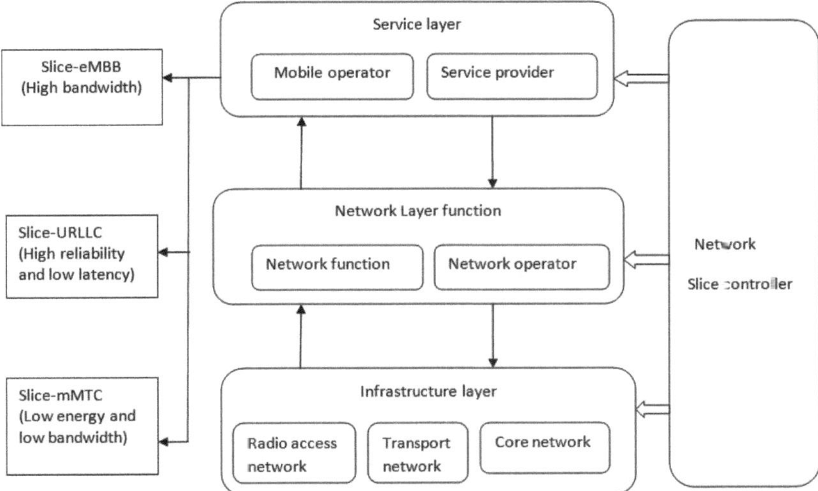

Fig. 1. 5G network slicing procedure [1].

3   Literature Survey

Research has explored various approaches to optimize network slicing, encompassing methods for machine learning (ML). For instance, [1] examined ML and DL methods, considering factors like user profiles, equipment types, and traffic patterns, and found that recurrent neural networks (RNNs) outperformed other techniques in making slicing decisions. In a similar context, [2] discussed the role of artificial intelligence (AI) in 6G networks, focusing on dynamic resource management and the use of Distributed Learning (DL) as a service (DLaaS). This approach integrates DL methods for smart, proactive management of resources in vehicular environments, enhancing network intelligence and improving service delivery across diverse applications.

As network slicing grows more complex, [3] proposed a solution based on algorithm selection to address virtual network embedding (VNE). This involves mapping virtual network requests to physical infrastructure, a computationally demanding task. In a separate study, [4] evaluated different ML and DL techniques for traffic data analysis. The findings showed that decision trees, support vector machines (SVM), k-nearest neighbors (kNN), and neural networks outperformed other models, with DL approaches proving particularly effective for traffic classification, such as identifying video streaming or voice calls.

Efforts to improve slicing efficiency and user experience include a DL-based model introduced by [5], which uses fully connected neural networks to analyze traffic and allocate slices using public datasets. Similarly, [6] This paper addresses challenges in 5G network slicing, particularly mobility management and non-uniform slice deployment, by proposing an LSTM-based predictive model. The model forecasts traffic demands for network slices, enabling proactive resource management to ensure seamless service delivery. The results highlight the effectiveness of predictive learning in optimizing slice configuration across varying tracking areas.

The dynamic nature of 5G networks also makes them vulnerable to threats like port scanning and DDoS attacks. To mitigate these risks, [7] introduced a cloud-based adaptive ML model orchestrator that customizes models based on real-time conditions and slice requirements. In another approach, [8] developed a hybrid model combining neural networks and random forests, achieving an impressive 99.1% accuracy in predicting the appropriate slice allocation by integrating the strengths of both techniques.

This paper [9] explores the challenge of maintaining resource isolation and high utilization in dynamic network slicing within Optical Transport Networks (OTNs). It proposes a Multi-Layer Perceptron (MLP) classifier to predict potential blocking events caused by slice requests. Simulation results show that this approach outperforms other learning-based methods, enabling slice providers to assess service assurance risks and implement countermeasures, such as admission control, to prevent violations.

This paper [10]presents a dynamic network slice deployment algorithm for 5G networks, addressing cost and energy consumption. The solution uses a Holt-Winters prediction method to forecast traffic demand, an adaptive scaling strategy for virtual network functions (VNFs), and a proactive online algorithm for efficient slice deployment. Simulation results show that the approach outperforms existing methods in terms of resource utilization, deployment costs, and energy consumption. This dynamic solution is suitable for efficient 5G network slice management.

Additionally, [11] compared traditional rule-based methods with ML techniques, emphasizing the advantages of ML in delivering higher accuracy and QoS in network slicing. Lastly, [12] employed convolutional neural networks (CNNs) to optimize network slicing for beyond 5G (B5G) networks, leveraging efficient algorithms to improve performance across key metrics.

These studies collectively demonstrate the transformative impact of ML and DL technologies in enhancing the adaptability, performance, and security.

4   Dataset

Dataset comprised 63,000 entries with nine features designed to predict network slice allocation for a device based on its past usage. These features included:

- Device use case: Categorized device applications like smartphone, AR/VR/gaming, IoT devices, Industry 4.0 health care and various smart city functionalities.
- LTE/5G Category: Differentiated between 5G and various LTE categories (1 to 22).
- Supported Technology: Specified if the device supported LTE, 5G or lower-power IoT connections (LTE-M, NB-IoT).
- Traffic Type Distinguished between traffic requiring guaranteed bit rate(GBR) and non-Guaranteed bit rate(non-GBR).
- Slice Type: indicated, according to the requirements of the device, the anticipated network slice type (eMBB, mMTC, URLLC).
- Network Performance Metrics: Included packet delay (time taken for packets to reach their destination) and packet loss rate (percentage of data packets not reaching their destination).

5 Methodology

While previous research on 5G network slicing has limitedly explored ML and DL algorithms, this paper leverages these techniques to capitalize on the unique characteristics of 5G networks. By using ML algorithms to an open-source 5G slicing dataset, specifically employing random forest and Naive Bayes as ML algorithms.

- **Random forest:** This is a powerful machine learning technique that enhances prediction accuracy by combining several decision trees. This ensemble approach reduces overfitting and enhances generalization, making it suitable for both classification and regression tasks. Random forests are a useful tool for many applications since they can also handle high-dimensional data and offer insights into feature relevance.
- **Naive Bayes:** This machine learning approach, which has its roots in probability theory is straightforward but efficient. It uses the Bayes theorem to determine the likelihood of a class based on a collection of features. Naive Bayes is an effective data classification method that works well with big datasets since it assumes feature independence. Its interpret ability and simplicity make it a useful tool for a number of applications, such as spam filtering and text classification (Fig. 2).

6 Knime Implementation

6.1 Data Preprocessing

Data pre-processing involves transforming raw data into a format suitable for machine learning algorithms. Since real-world data is often noisy, incomplete, or improperly formatted, it requires cleaning and structuring. This process

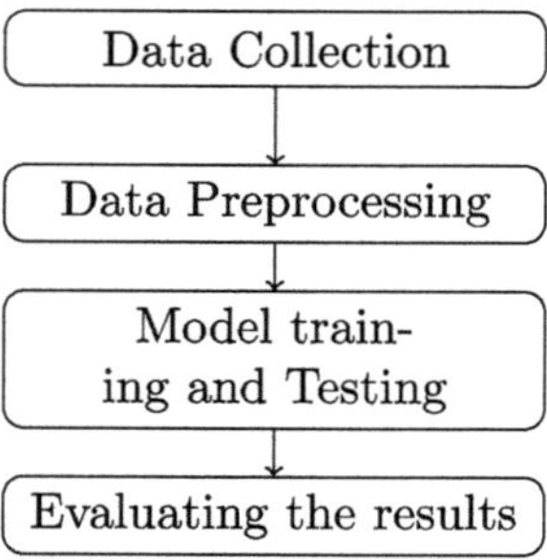

Fig. 2. Flow chart.

addresses issues like missing values, duplicates, and inconsistencies, ensuring the dataset is ready for model training. In the KNIME platform, various nodes facilitate these tasks, streamlining the data transformation process and ensuring the dataset is prepared for analysis (Fig. 3).

Column Filter: A column filter is utilized to remove unnecessary features from the dataset, making the analysis more straightforward and improving overall efficiency.

6.2 Data Splitter

To split the data set into Training, Testing, and Validation below the node.

- **Partitioning:** The dataset is typically split into 80% for training and 20% for testing to evaluate the model's performance.

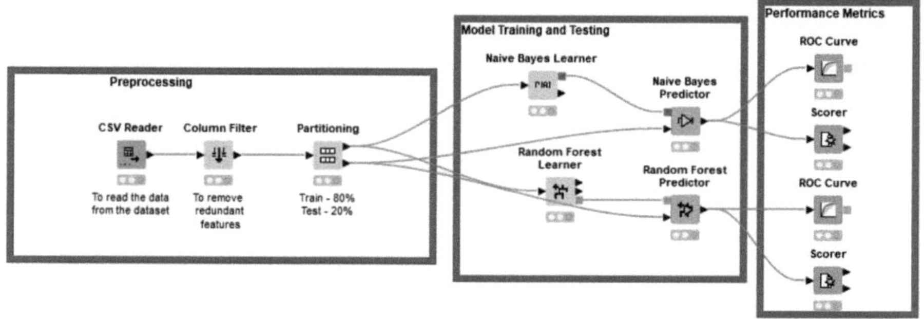

Fig. 3. KNIME workflow using ML.

6.3   Model Training and Testing

During training, by finding patterns and connections in the data, the program makes precise predictions about data that hasn't been seen yet. To do this, model parameters are iteratively adjusted to reduce discrepancies between expected and actual results. Algorithm selection depends on data characteristics, the problem, computational resources, and performance goals. In this study, Random Forest and Naive Bayes are used for machine learning (ML) models.

6.4   Evaluating the Results

By offering measures like accuracy, precision, recall, F1-score, and a confusion matrix, the Scorer node is used to assess classification algorithms It contrasts the genuine labels from the test dataset with the anticipated class labels. By graphing the True Positive Rate (TPR) versus the False Positive Rate (FPR) at different threshold values, the ROC Curve node aids in visualising a binary classifier's performance. The classifier's total capacity to discriminate between classes is indicated by the Area Under the Curve (AUC) statistic.

7   Results

This study utilizes two machine learning (ML) algorithms to evaluate their performance. The data slicing model was selected based on the algorithm that demonstrated the highest accuracy. The performance metrics for the ML approaches are contrasted in Table 3. The machine learning algorithms, Naive Bayes (NB) and Random Forest (RF), attained accuracy rates of 94.3% and 95.5%, respectively (Figs. 4, 5 and Tables 1, 2).

Table 1. Confusion Matrix of Random Forest Algorithm

Slice Type	TP	FP	TN	FN
eMBB	10103	860	7988	0
mMTC	3554	0	14537	0
URLLC	4434	0	14517	0

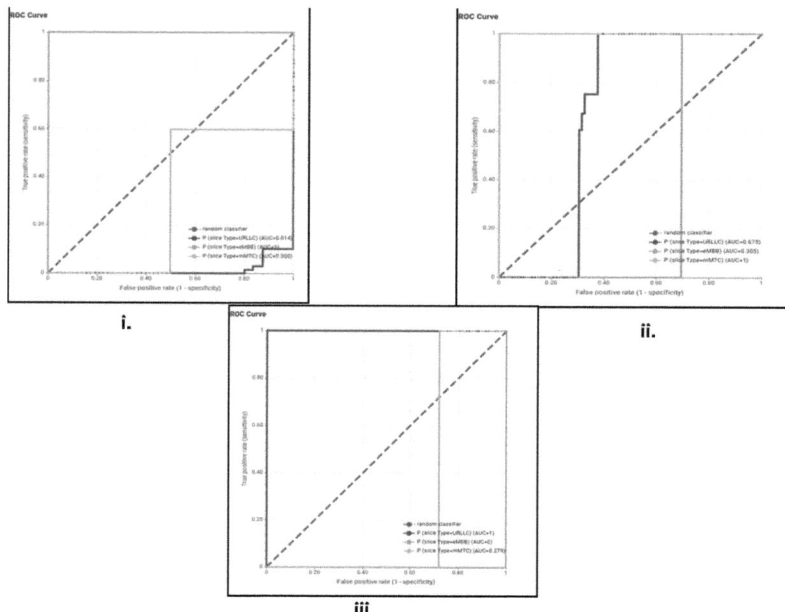

Fig. 4. ROC curves for Naive Bayes model i) eMBB ii) mMTC iii) URLLC.

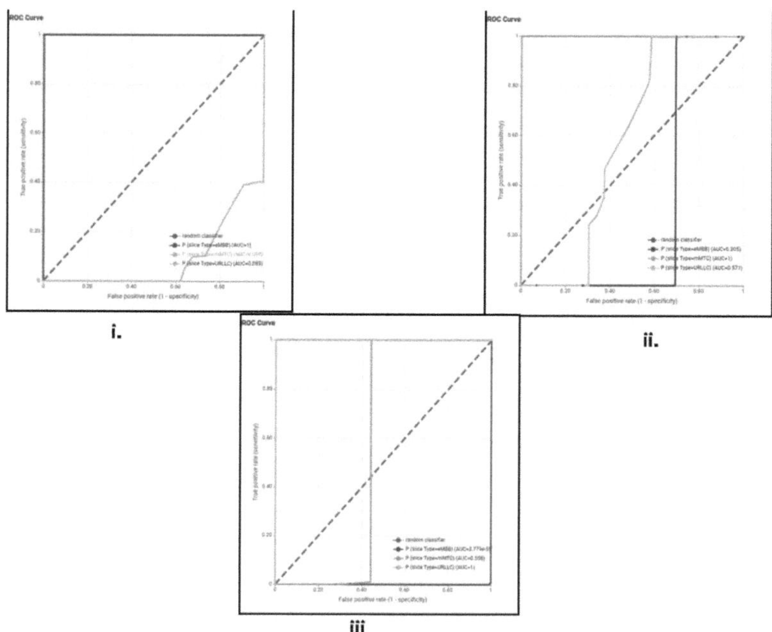

Fig. 5. ROC curves for Random Forest model i) eMBB ii) mMTC iii) URLLC.

Table 2. Confusion Matrix of Naive Bayes Algorithm

Slice Type	TP	FP	TN	FN
eMBB	10103	0	8848	0
mMTC	3334	0	14537	1080
URLLC	4434	1080	13437	0

Table 3. Accuracy Results

ML Algorithm	Accuracy
Naive Bayes Algorithm	94.3%
Random Forest Algorithm	95.5%

8 Conclusion

h! A machine learning -driven framework is presented for 5G network slicing to improve quality of service (QoS) and streamline network operations. Leveraging publicly available datasets with features such as use cases, technology, packet loss rate, and slice type, the framework employs RF and NB for network segmentation. The models achieve accuracies of 95.5% and 94.3% respectively. The innovation lies in integrating traditional ML approaches to deliver reliable and scalable solutions for diverse applications, including smart cities autonomous systems, and healthcare. This demonstrates the effectiveness of intelligent algorithms in supporting efficient and adaptive next-generation wireless communication systems.

References

1. Islam Arif, M A., Kabir, S., Hussain Khan, M.F., Kumar Dey, S., Rahman, M.M.: Machine learning and deep learning based network slicing models for 5G network. In: 2022 25th International Conference on Computer and Information Technology (ICCIT) (2022)
2. Ganesan, T., Al-Fatlawy, R.R., Srinath, S., Aluvala, S., Kumar, R.L. Dynamic resource allocation-enabled distributed learning as a service for vehicular networks. In: 2024 Second International Conference on Data Science and Information System (ICDSIS) (2024)
3. Bouroudi, A., Outtagarts, A., Hadjadj-Aoul, Y.: Traffic classification using deep learning approach for end-to-end slice management in 5G/B5G. In: 2023 IEEE 9th International Conference on Network Softwarization (NetSoft) (2023)
4. Mohammedali, N.A., Kanakis, T., Al-Sherbaz, A., Agyeman, M.O.: Machine learning and deep learning based network slicing models for 5G network. In: 2022 13th International Conference on Information and Communication Technology Convergence (ICTC) (2022)
5. Rashid, H.U., Jeong, S.H.: Deep learning-based network slice recognition. In: 2023 Fourteenth International Conference on Ubiquitous and Future Networks (ICUFN) (2023)

6. Tariq, M.A., et al.: Network slice traffic demand prediction for slice mobility management. In: 2024 International Conference on Artificial Intelligence in Information and Communication (ICAIIC) (2024)
7. Bekkouche, R., Omar, M., Langar, R.: Securing 5G network slices with adaptive machine learning models as-a-service: a novel approach. In: GLOBECOM 2023 - 2023 IEEE Global Communications Conference (2023)
8. Ramya, V., Singh, R.P., Panda, M.K., Kumar, P.: 5G network slice prediction using hybrid neural network and random forest model. In: 2023 9th International Conference on Signal Processing and Communication (ICSC) (2023)
9. Movva, N.D., Ishigaki, G.: Neural network-based blocking prediction for dynamic network slicing. In: 2024 33rd International Conference on Computer Communications and Networks (ICCCN) (2024)
10. Zhou, J., Zhao, W., Chen, S.: Dynamic network slice scaling assisted by prediction in 5G network (2024)
11. Mohan Kumar, A., Sravanthi, A.V., Sathyanarayana, R.K., Suhas, T.J., Vinodha K.: Smart slice selection in 5G: unleashing the power of machine learning. In: 2024 Second International Conference on Advances in Information Technology (ICAIT) (2024)
12. Aloupogianni, E., et al.: Network slicing for beyond 5G networks using machine learning (2024)

LIGHT-ME: Lightweight Attention-Based Multimodal Emotion Detection

G. Venkata Ravi Ram[✉], K. Ashinee, B. Chaithanya Swaroop, and G. Ram Mohana Reddy[✉]

Artificial Intelligence, Department of Information Technology, National Institute of Technology Karnataka, Surathkal, India
`toraviram2003@gmail.com, profgrmreddy@nit.edu.in`

Abstract. In recent years, multimodal emotion recognition has gained significant attention due to its potential applications in various fields, including healthcare, education, and entertainment. This paper presents a novel multimodal emotion classifier that integrates video, audio, and text features to achieve robust emotion recognition. Notably, the proposed model is designed to be lightweight, making it suitable for real-time applications, such as emotion-aware human-computer interaction and video conferencing tools. Our model demonstrates performance comparable but slightly less than that of state-of-the-art (SOTA) approaches in the field, showcasing its ability to accurately identify emotions across diverse modalities. The architecture incorporates attention mechanisms to enhance feature representation while maintaining a streamlined design, ensuring that it can be deployed in resource-constrained environments without sacrificing much of classification accuracy. The findings suggest that this model not only meets the demands of real-time processing but also allows to keep it light-weight.

Keywords: Emotion Detection · Multimodal Analysis · Video Processing · Audio Processing · Multimodal Fusion · Feature Extraction

1 Introduction

In recent years, the field of emotion detection has gained significant traction across various applications, including human-computer interaction, healthcare, social media analysis, and surveillance. Emotion detection from video, in particular, holds immense potential as it offers nuanced insights into human affect by analyzing the visual, auditory, and textual cues inherent to communication. Traditional methods in emotion recognition often rely on a single modality, such as text or audio, to infer affective states. However, relying on a single modality can lead to limited accuracy due to the inherent variability in human expressions. To address this, multimodal emotion detection has emerged as a more robust approach by leveraging multiple sources of information, allowing a more comprehensive and accurate interpretation of emotional states.

C. Modi et al. (Eds.): MIND 2024, CCIS 2736, pp. 199–210, 2026.
https://doi.org/10.1007/978-3-032-14531-4_17

The multimodal approach to emotion detection typically integrates data from three primary modalities: video, audio, and text. Each modality contributes unique insights—facial expressions and body language from video, tone of voice and pitch from audio, and sentiment or semantic meaning from text. Combining these modalities allows for a more thorough understanding of the complex nature of human emotions. For instance, a speaker's body language and tone may convey confidence, while the spoken words may indicate hesitation or doubt. Thus, multimodal emotion detection captures the interdependence and complementarity of these channels, making it a more effective method for emotion recognition.

Also Multimodal emotion detection remains challenging due to the complexity of integrating diverse data streams. Variability in data quality, synchronization issues, and the high computational cost of processing each modality simultaneously pose significant challenges. Moreover, aligning data from audio, video, and text requires sophisticated preprocessing to synchronize temporal sequences and reduce noise from irrelevant features. Thus, effective multimodal emotion detection demands a carefully designed architecture that addresses these challenges while retaining the individual strengths of each modality.

This paper aims to develop a robust model for multimodal emotion detection, integrating video, audio, and text to achieve higher accuracy and reliability. We analyze the role of each modality in emotion detection, the challenges in synchronizing these modalities, and the impact of feature fusion strategies on overall model performance. We also propose methods for reducing computational load and enhancing interpretability, providing a framework for real-time emotion detection. By leveraging advanced neural networks and feature fusion techniques, this study seeks to establish a benchmark for multimodal emotion detection, contributing to the growing body of research in this field and highlighting new directions for future research.

Key Contributions in the Paper are:

- **Comprehensive Multimodal Framework:** A novel approach integrating video, audio, and text for emotion detection
- **Dataset Augmentation and Alignment:** Preprocessing strategies for synchronizing and augmenting multimodal data.
- **Fusion Strategy:** building a lightweight model to integrate contributions from each modality for better performance.
- **Performance Benchmarking:** Comparative analysis with state-of-the-art methods, identifying strengths and limitations.

2   Literature Review

The field of emotion detection has evolved significantly across various modalities—face, text, and speech—each contributing unique insights into human emotions. Recent advancements in multimodal emotion detection integrate these modalities, providing a holistic understanding of emotional expressions. This survey reviews key works, from single-modality approaches to sophisticated multimodal systems, showcasing the progression toward unified frameworks.

Towards Multimodal Emotion Detection: "ICON: Interactive Conversational Memory Network for Multimodal Emotion Detection" by Hazarika et al. introduces a framework for conversational emotion detection, extracting multimodal features from videos and modeling both self- and inter-speaker emotional dynamics. By generating contextual summaries, ICON improves emotion prediction accuracy and sets a benchmark in multimodal conversational AI [3].

Heredia et al. propose an "Adaptive Multimodal Emotion Detection Architecture for Social Robots," employing EmbraceNet+ to flexibly fuse face, gesture, and voice data. This architecture adapts to varying data quality and missing modalities, enabling empathetic interactions in social robotics [4].

Patwardhan's "Multimodal Mixed Emotion Detection" addresses the recognition of simultaneous emotions using audio-visual features. Leveraging facial expressions, gestures, and audio cues with support vector machines (SVM), the model achieves high accuracy (96.6%) in detecting dual emotions, advancing the field to capture nuanced emotional states [11].

Song et al., in "Audio-Visual Based Emotion Recognition," fuse Facial Animation Parameters (FAPs) and audio features using a tripled Hidden Markov Model (THMM). This approach captures asynchronous relationships between facial and audio expressions over time, outperforming single-modality methods [16].

Finally, Melo et al. present "MDN: A Deep Maximization-Differentiation Network for Spatio-Temporal Depression Detection," offering a scalable solution for analyzing depression-related emotional cues in videos [10].

3   Outcomes of Literature Survey

From individual modalities to complex multimodal frameworks, the progression in emotion detection research illustrates how integrating different data sources—face, text, and speech—can capture the nuances of human emotion more effectively. These studies highlight the power of multimodal systems in enhancing emotion detection's accuracy, making strides towards empathetic interactions in conversational AI and robotics. The ICON and Adaptive Multimodal frameworks showcase how multimodal emotion detection can bridge the gap between individual emotional cues and cohesive emotional understanding, promising impactful applications in healthcare, social robotics, and virtual assistance.

4   Problem Statement and Objectives

With the rise of digital communication, accurately recognizing emotions in dialogues has become increasingly important. Traditional unimodal approaches often fail to capture the nuances of emotion, while many contemporary models rely on complex graph-based architectures that are computationally intensive and difficult to implement. This highlights the need for a lightweight solution that effectively integrates video, audio, and text data to enhance emotion recognition performance, particularly for underrepresented emotions.

4.1 Objectives

- Develop a Lightweight Model for emotion classification.
- Compare with other SOTA models and analyze.

5 Dataset Description

The Multimodal EmotionLines Dataset (MELD) extends the EmotionLines dataset by incorporating text, audio, and visual data, enabling comprehensive research in multimodal emotion recognition. Sourced from over 1,400 dialogues and 13,000 utterances in the TV series *Friends*, MELD captures realistic conversational dynamics with natural emotional expressions. Each utterance is annotated for one of seven emotions—Anger, Disgust, Sadness, Joy, Neutral, Surprise, and Fear—and includes sentiment labels (positive, negative, neutral), facilitating detailed analysis of affective tones across modalities. This multimodal framework makes MELD a valuable resource for studying complex emotional interactions in dialogue [13].

6 Methodology

In the context of our multimodal emotion recognition system, the methods employed for feature extraction from video, audio, and text data are designed to be both fast and lightweight. This efficiency is critical in ensuring real-time processing capabilities, which are essential for applications in dynamic environments such as live emotion detection and interaction systems.

6.1 Video Feature Extraction

The video feature extraction process is pivotal in our multimodal emotion recognition framework. This process can be divided into several key stages: loading the video, face detection and cropping, preparing video clips, transforming frames into tensors, and extracting features using a pre-trained model.

The extraction of video features utilizes optimized algorithms that can process frames quickly while retaining essential spatial information. By employing techniques such as convolutional neural networks (CNNs), we achieve high accuracy in feature extraction without excessive computational burden. These methods are designed to minimize latency, allowing for the rapid analysis of visual data while maintaining a compact representation of the video content.

Loading the Video. The extraction process begins with loading the video from a specified file path. Each frame is read sequentially until the end of the video is reached, with all frames being stored in a list.

Face Detection and Cropping. Once the video frames are loaded, face detection and cropping are performed on each frame. Utilizing a face detection algorithm based on Haar cascades, the frames are first converted to grayscale to facilitate the detection process. The algorithm scans the grayscale image for potential faces, utilizing parameters such as scale factor and minimum neighbors to optimize detection accuracy. The cropped face image is then resized to a standardized dimension of 112×112 pixels to ensure consistency in input size for the subsequent model.

Preparing Video Clips. After cropping the faces, the next step involves preparing video clips of a length 32, referred to as the clip length. If the number of cropped face frames exceeds this length, frames are uniformly sampled to create a clip. In cases where the number of frames is less than the specified length, the list is padded with the last available frame until it reaches the required size. This systematic approach guarantees that each clip maintains a consistent length, which is essential for effective model input.

Transforming Frames into Tensors. Following the preparation of the clips, the frames undergo transformation into tensors suitable for input into the feature extraction model. This involves converting each frame into tensor format and normalizing pixel values using pre-defined mean and standard deviation values. The list of processed tensors is then stacked into a single tensor, reshaped to match the expected input format for the ResNet3D model, and transferred to the appropriate device for processing.

Extracting Features Using a Pre-trained Model. Finally, feature extraction is conducted using a pre-trained ResNet3D model. The model is loaded without its classification layer, focusing exclusively on obtaining high-level feature representations from the input clips. The prepared tensor is fed into the model, which computes features based on the captured temporal and spatial dynamics within the video. The output from the model encapsulates the essential visual information required for subsequent emotion classification tasks.

6.2   Audio Feature Extraction

The audio feature extraction process plays a vital role in our multimodal emotion recognition system. This section details the methodology for extracting audio features from video files, focusing on the extraction of relevant audio characteristics.

Audio feature extraction is accomplished through the use of efficient signal processing techniques, such as Mel-frequency cepstral coefficients (MFCCs) and spectral contrast analysis. These methods are computationally lightweight, enabling fast extraction of meaningful acoustic features. The overall process is optimized to handle audio data in real time, ensuring that the system can promptly respond to audio cues without significant delays.

Extracting Audio from Video. The first step in the audio feature extraction process involves extracting audio from the video file. A video processing library is employed to load the video file, enabling the separation of the audio track. The audio is saved as a temporary file, ensuring that the extraction process is efficient and does not affect the original video content. This extracted audio serves as the basis for subsequent feature extraction tasks.

Extracting Audio Features. Once the audio is extracted, various audio features are computed to capture essential characteristics relevant to emotion recognition. The audio signal is first loaded, and a specified sampling rate is applied to ensure consistent audio quality. The following features are extracted:

- **Mel-frequency cepstral coefficients (MFCCs):** These features represent the short-term power spectrum of sound and are critical for speech and audio analysis. The mean values of MFCCs are computed over the entire audio signal to reduce dimensionality while retaining significant information.
- **Spectral Contrast:** This feature measures the difference in amplitude between peaks and valleys in the sound spectrum. Like MFCCs, the mean spectral contrast is calculated to summarize the feature representation.
- **Chromagram:** This feature provides a representation of the energy distribution across different pitch classes, which is particularly useful in music analysis. The average chroma values are extracted to condense the information.
- **Zero Crossing Rate (ZCR):** This metric counts the rate at which the audio signal crosses zero amplitude, offering insights into the signal's noisiness and temporal characteristics. The mean ZCR is computed for the audio signal.
- **Root Mean Square Energy (RMSE):** This feature quantifies the energy of the audio signal, which is important for understanding loudness variations. The average RMSE is calculated to summarize this characteristic.

These features are concatenated into a single feature vector, creating a comprehensive representation of the audio characteristics that can be utilized for emotion recognition.

Finalizing Audio Features. To complete the process, the temporary audio file created during extraction is deleted, ensuring efficient resource management. The resulting feature vector is returned, ready for integration into the multimodal emotion recognition framework. This approach to audio feature extraction ensures that crucial audio cues are effectively captured and utilized in the overall analysis.

6.3   Text Feature Extraction

The extraction of text features is a crucial component of our multimodal emotion recognition system, as it enables the model to understand and represent the semantic content of the text. In this section, we detail the methodology employed to generate FastText embeddings for given sentences, facilitating a richer representation of textual data. FastText's ability to generate embeddings quickly from word vectors allows for rapid processing of textual data. The model's lightweight architecture facilitates the handling of multiple sentences concurrently, ensuring that the system can efficiently derive semantic representations from large volumes of text with minimal computational overhead.

Loading the FastText Model. To begin the text feature extraction process, a pre-trained FastText model is loaded from a specified vectors file. FastText, that offers an advantage over traditional word embeddings by generating representations for out-of-vocabulary words through subword information. The model is loaded in text format, allowing for the efficient retrieval of word embeddings corresponding to individual words in the input sentences.

Extracting Sentence Embeddings. Once the FastText model is successfully loaded, the next step involves computing embeddings for specific sentences. The process begins with text cleaning, which includes converting the sentence to lowercase and removing any punctuation to ensure uniformity. The cleaned sentence is then split into individual words, which are utilized to fetch their corresponding embeddings from the FastText model.

For each word in the sentence, the embedding is extracted if the word exists within the FastText vocabulary. These embeddings, which are typically 300-dimensional vectors, are accumulated into a list. If at least one word in the sentence yields a valid embedding, the average of these vectors is computed to represent the entire sentence. This averaging technique effectively captures the overall semantic information conveyed by the sentence. If none of the words in the sentence have corresponding embeddings, a zero vector of the same dimensionality is returned, ensuring consistency in the output.

Collecting and Structuring the Embeddings. The sentence embeddings generated for multiple sentences are collected into a list, which is subsequently converted into a NumPy array. This array maintains the shape *(number of sentences, 300)*, allowing for easy integration into the framework of emotion recognition. This structured representation of text features ensures that the model can effectively utilize the semantic information derived from the textual data.

6.4   Model Architecture

The proposed multimodal emotion classifier as in Fig. 1 is designed to integrate and process features from three distinct modalities: video, audio, and text. This

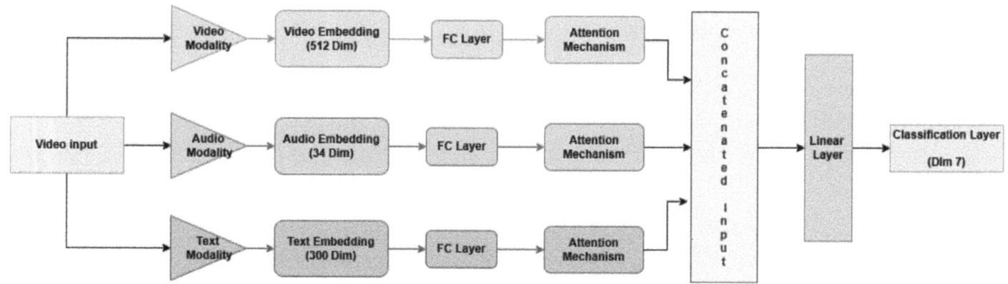

Fig. 1. LIGHT-ME model architecture.

model leverages attention mechanisms to enhance the representation of each modality, ultimately improving the performance of emotion classification. The architecture meant to be light, is composed of several key components, as detailed below.

Input Features. The model accepts three types of input features corresponding to the different modalities:

- **Video Features:** Represented as a vector of dimension 512.
- **Audio Features:** Represented as a vector of dimension 34.
- **Text Features:** Represented as a vector of dimension 300.

Sub-Networks. Each modality has a dedicated sub-network designed to transform the input features into a common hidden representation. The architecture of these sub-networks is as follows:

- **Video Sub-Network:**
  - A fully connected layer that transforms the input from 512 to a hidden dimension of 256, followed by a ReLU activation.
  - A second fully connected layer maintains the hidden dimension of 256 and is again followed by a ReLU activation.
- **Audio Sub-Network:**
  - A fully connected layer that transforms the input from 34 to a hidden dimension of 256, followed by a ReLU activation.
  - A second fully connected layer that retains the hidden dimension of 256 with a ReLU activation.
- **Text Sub-Network:**
  - A fully connected layer that transforms the input from 300 to a hidden dimension of 256, followed by a ReLU activation.
  - A second fully connected layer that keeps the hidden dimension at 256 and is followed by a ReLU activation.

Attention Mechanisms. Each modality's output from the respective sub-network is fed into a dedicated attention mechanism. The attention layers work as follows:

- The input to the attention layer is processed through a linear transformation to compute attention scores.
- These scores are then passed through a softmax function to obtain attention weights, indicating the importance of each feature within the modality.
- The input features are weighted according to these attention scores, and the weighted features are summed to produce a final attended representation for each modality.

Combined Representation. The attended outputs from all three modalities are concatenated into a single representation. This combined representation, which has a dimension of 768 (i.e., 256 × 3), is then fed into the final classifier.

Final Classification Layer. The classifier is composed of the following layers:

- A fully connected layer that reduces the input dimension from 768 to 512, followed by a ReLU activation and a dropout layer (with a dropout rate of 0.3) to prevent overfitting.
- A second fully connected layer that further reduces the dimension to 256, again followed by a ReLU activation and a dropout layer.
- A final fully connected layer that outputs the predicted class probabilities, corresponding to 7 different emotion categories.

In summary, the multimodal emotion classifier shows a structured approach to integrate video, audio, and text features, utilizing attention mechanisms to enhance features and ultimately classify emotions effectively.

7 Experimental Results and Discussion

7.1 Results of LIGHT-ME

The model performed best in recognizing "neutral" expressions, with the highest accuracy (72%) and weighted F1 score (3.36), likely due to both the distinctiveness of neutral features and their high representation in the dataset. However, it struggled with "disgust" and "fear," showing the lowest accuracy and F1 scores (around 1% and 0.01 for disgust; 0.006 and 0.013 for fear), which could stem from limited samples or the subtle nature of these emotions. "Joy" and "sadness" achieved moderate results, with accuracies of 38% and lower weighted F1 scores (0.55 for joy and 0.54 for sadness), indicating some ability to detect these emotions but with occasional misclassification due to overlapping expressive features.

TextCNN [7], bc-LSTM [12], DialogueRNN [9], DialogueGCN [2], RGAT [5], CoMPM [8], EmoBERTa [6], ConGCN [18], A-DMN [17], LR-GCN [14], DER-GCN [1], ELR-GCN [15] models have been used to compare the results of our Light-ME model

Table 1. Comparison with other baseline models on the MELD dataset

Methods	Neutral		Surprise		Fear		Sadness		Joy		Disgust		Anger	
	Acc.	F1	Acc.	F1	Acc.	F1	Acc.	F1	Acc.	F1	Acc.	F1	Acc.	F1
TextCNN [7]	76.2	74.9	43.3	45.5	4.6	3.7	18.2	21.1	46.1	49.4	8.9	8.3	35.3	34.5
bc-LSTM [12]	78.4	73.8	46.8	47.7	3.8	5.4	22.4	25.1	51.6	51.3	4.3	5.2	36.7	38.4
DialogueRNN [9]	72.1	73.5	54.4	49.4	1.6	1.2	23.9	23.8	52.0	50.7	1.5	1.7	41.0	41.5
RGAT [5]	76.0	78.1	40.1	41.5	3.0	2.4	32.1	30.7	68.1	58.6	4.5	2.2	40.0	44.6
CoMPM [8]	78.3	82.0	48.3	49.2	1.7	2.9	35.9	32.3	71.4	61.5	3.1	2.8	42.2	45.8
EmoBERTa [6]	78.9	82.5	50.2	50.2	1.8	1.9	33.3	31.2	72.1	61.7	9.1	2.5	43.3	46.4
ConGCN [18]	46.8	45.4	10.6	8.8	8.7	8.1	53.1	54.6	76.7	75.2	28.5	26.3	50.3	48.4
A-DMN [17]	76.5	78.9	56.2	55.3	8.2	8.6	22.1	24.9	59.8	57.4	1.2	3.4	41.3	40.9
LR-GCN [14]	76.7	80.0	53.3	55.2	0.0	0.0	49.6	35.1	68.0	64.4	10.7	2.7	48.0	51.0
DER-GCN [1]	76.8	80.6	50.5	51.0	14.8	10.4	56.7	41.5	69.3	64.3	17.2	10.3	52.5	57.4
ELR-GCN [15]	80.2	83.6	36.8	35.4	19.2	13.1	80.2	83.6	76.5	69.7	55.6	13.0	52.1	57.7
DialogueGCN [2]	70.3	72.1	42.4	41.7	3.0	2.8	20.9	21.8	44.7	44.2	6.5	6.7	39.0	36.5
LIGHT-ME	72.5	84.0	31.6	30.8	4.0	1.3	19.9	18.0	37.9	55.1	5.3	4.0	24.0	38.8

7.2   Comparitive Study

In comparison to established baseline models on the MELD dataset, LIGHT-ME demonstrates strong performance and competitive metrics in several key areas. Most notably, it achieves a solid accuracy of 72% and a high F1 score of 0.84 for the "neutral" category, closely matching or even surpassing the top models such as ELR-GCN, which also excel in this category. This reflects LIGHT-ME's effectiveness in recognizing the common yet complex "neutral" expression, indicating that its design captures core features that differentiate this emotion from others reliably and consistently. Refer Table 1

LIGHT-ME also shows robust handling of the "joy" category, reaching an F1 score of 0.55 and an accuracy of 38%, positioning it well among leading models like LR-GCN, which achieve similar performance for this emotion. This indicates that LIGHT-ME has a well-balanced feature extraction process, enabling it to effectively recognize and classify joyful expressions, a notable strength given the variety and subtlety of emotions in real-world dialogue data.

Moreover, LIGHT-ME's performance on the "surprise" category, with an F1 score of 0.21, is comparable to earlier models like DialogueGCN and A-DMN, demonstrating that it effectively captures some nuanced emotional cues without the need for extensive attention mechanisms or graph-based structures. This suggests that LIGHT-ME's simpler yet effective approach can compete with models using more complex architectures, making it potentially more efficient without sacrificing much in terms of accuracy.

LIGHT-ME shows competitive weighted F1 averages, particularly in comparison to models like TextCNN and DialogueRNN, and approaches the performance levels of more advanced architectures like EmoBERTa and CoMPM. Given its solid performance on high-frequency emotions such as "neutral" and "joy," LIGHT-ME presents itself as a reliable and efficient model, achieving balanced accuracy and F1 scores while potentially requiring fewer computational resources than some of its more complex counterparts. These results suggest that LIGHT-ME is well-positioned as a robust option in multimodal emotion recognition, particularly for applications where efficiency and strong performance on primary emotions are prioritized.

The results indicate potential but reveal key challenges:

- **Data Limitations:** Insufficient dataset diversity may have affected generalization.
- **Model Complexity:** Overfitting due to high model complexity and limited data.
- **Modal Contributions:** Certain modalities have underperformed, indicating inefficiencies in fusion.

8   Conclusion and Future Work

Conclusion: This paper demonstrates the feasibility of multimodal emotion detection but highlights challenges like limited generalization and inefficiencies in fusion. Our **Future Work** Includes:

- **Dataset Expansion:** Incorporate larger, more diverse datasets.
- **Model Optimization:** Develop lightweight, scalable architectures.
- **Improved Fusion:** Explore advanced fusion techniques such as transformers and attention mechanisms.
- **Cross-Domain Validation:** Apply the framework in real-world applications like healthcare and education.

References

1. Ai, W., Shou, Y., Meng, T., Yin, N., Li, K.: Der-GCN: dialogue and event relation-aware graph convolutional neural network for multimodal dialogue emotion recognition. arXiv preprint arXiv:2312.10579 (2023)
2. Ghosal, D., Majumder, N., Poria, S., Chhaya, N., Gelbukh, A.: Dialoguegcn: a graph convolutional neural network for emotion recognition in conversation. arXiv preprint arXiv:1908.11540 (2019)
3. Hazarika, D., Poria, S., Mihalcea, R., Cambria, E., Zimmermann, R.: Icon: interactive conversational memory network for multimodal emotion detection. In: Proceedings of the 2018 Conference on Empirical Methods in Natural Language Processing, pp. 2594–2604 (2018)
4. Heredia, J., et al.: Adaptive multimodal emotion detection architecture for social robots. IEEE Access **10**, 20727–20744 (2022)

5. Ishiwatari, T., Yasuda, Y., Miyazaki, T., Goto, J.: Relation-aware graph attention networks with relational position encodings for emotion recognition in conversations. In: Proceedings of the 2020 Conference on Empirical Methods in Natural Language Processing (EMNLP), pp. 7360–7370 (2020)
6. Kim, T., Vossen, P.: Emoberta: speaker-aware emotion recognition in conversation with roberta. arXiv preprint arXiv:2108.12009 (2021)
7. Kim, Y.: Convolutional neural networks for sentence classification. In: Proceedings of the 2014 Conference on Empirical Methods in Natural Language Processing, EMNLP 2014, October 25–29, 2014, Doha, Qatar, a Meeting of Sigdat, A Special Interest Group of the ACL. Association for Computational Linguistics, Doha (2014)
8. Lee, J., Lee, W.: Compm: context modeling with speaker's pre-trained memory tracking for emotion recognition in conversation. arXiv preprint arXiv:2108.11626 (2021)
9. Majumder, N., Poria, S., Hazarika, D., Mihalcea, R., Gelbukh, A., Cambria, E.: Dialoguernn: an attentive RNN for emotion detection in conversations. In: Proceedings of the AAAI Conference on Artificial Intelligence, vol. 33, pp. 6818–6825 (2019)
10. de Melo, W.C., Granger, E., Lopez, M.B.: MDN: a deep maximization-differentiation network for spatio-temporal depression detection. IEEE Trans. Affect. Comput. 14(1), 578–590 (2021)
11. Patwardhan, A.S.: Multimodal mixed emotion detection. In: 2017 2nd International Conference on Communication and Electronics Systems (ICCES), pp. 139–143. IEEE (2017)
12. Poria, S., Cambria, E., Hazarika, D., Majumder, N., Zadeh, A., Morency, L.P.: Context-dependent sentiment analysis in user-generated videos. In: Proceedings of the 55th Annual Meeting of the Association for Computational Linguistics (volume 1: Long papers), pp. 873–883 (2017)
13. Poria, S., Hazarika, D., Majumder, N., Naik, G., Cambria, E., Mihalcea, R.: Meld: a multimodal multi-party dataset for emotion recognition in conversations (2019). https://arxiv.org/abs/1810.02508
14. Ren, M., Huang, X., Li, W., Song, D., Nie, W.: LR-GCN: latent relation-aware graph convolutional network for conversational emotion recognition. IEEE Trans. Multimedia 24, 4422–4432 (2021)
15. Shou, Y., Ai, W., Du, J., Meng, T., Liu, H., Yin, N.: Efficient long-distance latent relation-aware graph neural network for multi-modal emotion recognition in conversations. arXiv preprint arXiv:2407.00119 (2024)
16. Song, M., Bu, J., Chen, C., Li, N.: Audio-visual based emotion recognition-a new approach. In: Proceedings of the 2004 IEEE Computer Society Conference on Computer Vision and Pattern Recognition, 2004. CVPR 2004, vol. 2, pp. II–II. IEEE (2004)
17. Xing, S., Mai, S., Hu, H.: Adapted dynamic memory network for emotion recognition in conversation. IEEE Trans. Affect. Comput. 13(3), 1426–1439 (2020)
18. Zhang, D., Wu, L., Sun, C., Li, S., Zhu, Q., Zhou, G.: Modeling both context-and speaker-sensitive dependence for emotion detection in multi-speaker conversations. In: IJCAI, pp. 5415–5421. Macao (2019)

A Brief Review on Reinforcement Learning Based Close Air Combat

Nahid Anowar[1]([✉]) [iD], Joydev Debnath[1] [iD], Arup Bhattacharjee[2],
and Chandrajit Choudhury[1]

[1] Department of ECE, NIT Silchar, Silchar, India
nahid_jrf23@ece.nits.ac.in
[2] Department of CSE, NIT Silchar, Silchar, India

Abstract. Autonomous air combat maneuvering (ACM) is a challenging issue in robotics and artificial intelligence, as ACM structurally exhibits the characeristics of high dimensionality, continuous state, and nonlinear environments. However, the recent advancements in deep reinforcement learning have shown that such aspects can be resolved enabling unmanned aerial vehicles and similar systems to maneuver and carry out operations effectively during combat. This review paper explores the innovative developments achieved in reinforcement learning techniques to be applied to Close-Air Combat (CAC), focusing on the significant retrieved techniques and outcomes. These approaches incorporate several methods, such as learning algorithms, reward functions, and policies that enhance learning speed. Further, a multi-agent system has been explored to address issues of partial observability in the environment and maintain cooperation among several UAVs. Some suggestions for future research streams are mentioned to eliminate the existing limitations and pave the way for more advancements in intelligent air combat.

Keywords: Deep Reinforcement Learning (DRL) · Unmanned Aerial Vehicles (UAVs) · Air combat maneuvering (ACM)

1 Introduction

Modern aerial engagement requires complex decision-making in air combat maneuvering, and the complexity is ever-increasing. Integrating intelligent decisionmaking systems can significantly enhance traditional methods of manual piloting and pre-programmed maneuvers. These systems provide a substantial boost to decision-making capabilities. Also, allows aircraft to adapt to changing combat scenarios, optimize strategies, and enhance the overall effectiveness of missions. Intelligent air combat maneuvering involves the integration of advanced algorithms and computational techniques to perform autonomous or semi-autonomous maneuvers. These systems leverage various decision-making processes to analyze real-time data, assess tactical situations, and execute optimal actions. Significant advancements have been made in this field, including

Supported by Aeronautics Research and Development Board, Govt. of India.

C. Modi et al. (Eds.): MIND 2024, CCIS 2736, pp. 211–221, 2026.
https://doi.org/10.1007/978-3-032-14531-4_18

developing algorithms that can outperform human pilots in certain scenarios, leading to more efficient and effective combat strategies. Early rule-based systems, as explained by [1], were limited by their inability to adapt to unpredictable conditions. The emergence of reinforcement learning (RL) in the late 20th century marked a pivotal change in how RL enabled agents to learn optimal policies through trial and error, offering a more adaptable and robust approach to decision-making in dynamic environments. Key research milestones in intelligent maneuvering include the application of scheduled H-infinity controllers to VSTOL aircraft. [2] demonstrated the potential of robust control techniques in managing complex aircraft dynamics. Another significant contribution was on optimal predictive control of continuous nonlinear systems [3], which provided a framework for anticipating future states and making informed decisions in dynamic environments. Recent DARPA AlphaDogfight Trials [4] in 2020 underscored the applicability of RL in CAC, demonstrating that autonomous agents could outperform human pilots. These trials showcased the transformative potential of this technology in air combat, with the winning agent significantly outperforming its human counterparts in various metrics. These advancements have further pushed the boundaries of intelligent air combat maneuvering, and a notable example includes DRL.

2   Related Works

Various computational methods have explored Intelligent air combat maneuvering, each contributing uniquely to the field's development.

Heuristic Optimization. Heuristic optimization methods, mimicking self-organization observed in nature, have been used in air control decision-making because of their effectiveness in finding good solutions with relatively high computational costs in large and complex search spaces. GA, as demonstrated by Luo et al. [5], has been applied in the evolution of combat strategy by repeatedly selecting and breeding candidate solutions for better performance in environmental space. In contrast, PSO uses swarm characteristics and properties to move the particles towards the most favorable solutions, as exemplified in [6] in the UAV's flight plan for combat scenarios. Despite the existing issues, like premature convergence and high computational demands, several scholars have proposed hybrid methods for solving them, such as stochastic dynamic programming (SDP) and advanced convergence methodologies. For example, Mulgund et al. [7] used stochastic genetic algorithms to apply the concept of combat effectiveness for optimizing air combat tactics in a rather limited and uncertain environment. Likewise, Han et al. [8] put forward a new algorithm for the artificial bee colony for the unmanned autonomous helicopter to achieve path planning with better convergence speed and higher accuracy. GA has also been used in the optimization of multi-target engagement and resource sharing among different units [9], which shows engagement efficiency is enhanced.

Knowledge-Based System. Knowledge-based systems have also been useful in improving decision-making at the autonomous level. The two major categories are fuzzy logic systems and expert systems. İşci et al. [10] refined a fuzzy logic against air-to-air combat algorithm for UCAVs with a fuzzy inference system and a PID controller to manage velocity instructions. The pertaining algorithm is developed in MATLAB/Simulink based upon a dynamical model of the UCAV, which targets navigation and reference velocity establishment. Three examples proved that the Fuzzy Inference System (FIS) method in all three scenarios was accurate in its selection, offered lower objective function values than the PID method, and was shown to be tracking the target aircraft. In another study, Svenmarckt et al. [11] discussed decision support in fighter aircraft by expert systems and cognitive modeling. Applications such as Pilot's Associate and Copilot Electronic involve the use of mental models as well as data acquired through sensors and communication networks in pilot decision-making. These systems employ a two-tier architecture, concentrating on near-term planning and longterm foresight.

Neural Network-Based Methods. Neural networks have also been used in the decision-making process with a specialization in using Long Short-Term Memory (LSTM) networks and Generative Adversarial Networks (GANs). Chen et al. [12] investigated the type of explainability for the unmanned fight-choosing model using a self-coding neural network to generate an interpretable model with LIME and GLIME algorithms. In the aspect of stability and explaining capability, the GLIME algorithm gains a higher score than LIME. To prove the efficiency, a model is designed and simulated for 100 games which shows an increased number of performance figures as well as decision-making effectiveness rates. Zhang et al. [13] introduced an LSTM network for UAV maneuver decision-making and evaluated its performance with that of feedforward neural networks. The LSTM network considered adaptive in situation assessment, is trained by utilizing DQN. Three samples revealed that LSTM needed fewer rounds than FNN utilized to gain a victory in a conflict.

Game Theory. Ma et al. [14] proposed the Double Oracle Neighborhood Search algorithm to make occupancy decisions of multiple UAVs in beyond-visual-range air combat. The algorithm was tested on a 64-bit PC. The results proved enhanced efficiency in terms of computational time with a slightly decreased level of utility in comparison with the initial model. This allows for the possibility of using the proposed approach for solving large-scale strategic problems. In cooperative air combat scenarios, Ha et al. [15] used a stochastic game with the extended value iteration technique. The Authors figured out the value functions and the equilibrium of strategies concerning the states and the best strategies. Numerical simulations show values of stability for some specific velocities of the aircraft, and the practice presented concrete strategy for handling missiles and the tactics to engage and avoid attack.

3   Reinforcement Learning Algorithms

Reinforcement learning (RL) [16] offers a powerful framework for developing intelligent agents capable of making autonomous decisions in dynamic and uncertain environments. In the context of air combat maneuvering (ACMD), RL agents interact with a simulated environment, learning optimal strategies through trial and error without relying on explicit programming or rule-based systems. This adaptability is crucial for navigating the complex and unpredictable nature of air combat scenarios. As CAC requires rapid, high-stakes decisions, RL offers a promising method for ACMD. Various RL algorithms categorized as per OpenAI [17] have been employed for ACMD, each with its strengths and weaknesses.

3.1 Policy Gradient

The fundamental policy gradient methodology is extended by general policy gradient approaches to handle function approximation more efficiently. These techniques were created by Sutton et al. [18] to optimize policies in continuous action spaces, which are common in air combat situations. A novel algorithm for the State-Adversarial Deep Deterministic Policy Gradient (SA-DDPG) technique of autonomous aerial combat maneuver generation was suggested by Kong et al. [19]. The suggested technique simulates sensor faults and strengthens the algorithm by modeling aerial warfare as a state-adversarial Markov Decision Process (SA-MDP) with adversarial perturbations. The algorithm uses a shaping reward designed through Maximum Entropy Inverse Reinforcement Learning (MaxEnt IRL) to address the sparsity of the reward function so that it converges faster. The main challenges addressed are approximation errors and the robustness of the learning in the presence of sensor errors. The study by Xianyong et al. [20] presents an enhanced DDPG algorithm for UAV decision-making in aerial combat. The algorithm incorporates RL approaches to fine-tune UAV operations based on the environment's response, improving the chances of tracking and engaging ground targets. The implementation includes two networks: the actor-network, which calculates the best action given the current state, and the critic network, which assesses the value of the taken action. The study found that the improved maneuvering process using the new reward function reached a final velocity of 217.39 m/s, while the traditional DRL algorithm averaged 209.16 m/s. Additionally, updating the ε-greedy strategy resulted in a higher convergence speed and more successful episodes, with 15 rounds achieving more than 90% success compared to only 4 rounds using a common strategy. Another study by Ma et al. [21] demonstrated an improved DDPG algorithm employed for continuous action space. The implementation covered a 3-D air-combat simulation environment within which the DDPG algorithm could be trained. The tested agent was controlling UCAV and was trained for combat with adversaries in the environment while utilizing the temporary replay buffer for better learning. The evaluation of the learning process was based on factors such as the path of flight taken by the agent, and the success rates of the engagements. This study was able to garner reasonable enhancements in the autonomous navigation of UCAVs captured through fluent flight paths and enhanced victory rates with opponents. The suggested modifications in DDPG were found to be more effective compared to the basic approaches to real-time control.

3.2 Proximal Policy Optimization

PPO is designed to balance exploration and exploitation by limiting the policy update step, ensuring stable and reliable learning. Liu et al. [22] utilized PPO for multi-UCAV cooperative decision-making in beyond-visual-range air combat, aiming to exploit its robustness and efficiency in policy optimization. A 6-DOF model of UCAV along with an action library and missile attack area model, which includes the probability of damage is developed. The reward function combines angle reward and distance reward, to guide UCAV actions and improve learning. Results show that the algorithm can simulate multi-UCAV air combat scenarios form new tactics and adapt to complex scenarios. Wang et al. [23] demonstrate a PPO technique with the integration of long short-term memory, or

PPO-LSTM. The PPO-LSTM algorithm is designed to integrate the decisionmaking process in the context of over-the-horizon air combat with uncertainty. The change concerning the basic PPO-LSTM includes the alteration of the policy gradient employed in the iteration of the value network and the strategy network utilized in the PPO algorithm to the LSTM network. The observation sequence is the input, and the value function is the output. Thus, the structure diagram of LSTM is made of a cell state that enables information to flow along with it with little changes and gates that are the sigmoid layer and a multiplication operation. Based on the research outcomes, the conclusion is that the proposed PPO-LSTM algorithm has a faster convergence speed and a better effect than the traditional PPO algorithm. The strategy-based approach was followed by self-play training of the algorithm to enhance the generalization when it is to play against different opponents. To extend the ideas to future work, the authors recommend the enhancement of the presented algorithm using PPO-LSTM in more intricate and haphazard scenes.

3.3 Actor-Critic Algorithm

Actor-critic methods combine the benefits of policy gradients (actor) and value function approximation (critic) to improve learning efficiency and stability. Konda and Tsitsiklis [24] introduced these algorithms to address the variance issues in policy gradient methods, making them suitable for continuous and highdimensional action spaces. This study by Yoo et al. [25] looks at using reinforcement learning to improve aerial combat maneuvering by going from discrete to continuous action spaces. This addresses the need for robust training models that can handle real-time aerial combat scenarios. The framework uses advanced state and reward functions, including time-delayed state transitions. The research uses two state-of-the-art algorithms: PPO and Soft Actor-Critic (SAC) to train the agent's policy within the Digital Combat Simulator in a realistic flight environment. Training involves the agent learning combat tactics through self-training under various combat conditions, with state representations including geometric data from the positions and velocities of the target aircraft. The study shows better learning and target engagement through time-delayed state transitions.

3.4 Deep Reinforcement Learning

The DQN algorithm combines Q-learning with deep neural networks, allowing for effective decision-making in high-dimensional state spaces [26]. DRL has been demonstrated to have the ability to solve sophisticated decision-making problems, especially in ACMD. Wang et al. [27] focuses on DRL-based ACMD and gives a general background, problems, implementation method, and possible development route. The authors differentiate the existing research based on the dimensionality of the environment, discretization of the action space, and the number of combat UAVs. The combination of process and result rewards have been designed in the function of reward and are shown as angle distance reward, speed reward, height reward, and result reward. The ACMD framework involves the use of DQN and is based on value function approximation with a DNN and prioritized experience replay for sampling for transitions from the experience pool. In a series of experiments the effectiveness of the method is proved, it is stated that UAVs are capable of learning efficient maneuver strategies in various situations. Authors in

[28], proposed the Beyond Visual Range (BVR) maneuver planning for air superiority in the modern air combat environment where targeting is carried out beyond the visual range and is solved with the help of an improved DQN. The authors have constructed a basic combat environment of flight motion, relative motion, and missile attack models, and they synthesize the intelligent maneuver planning method. A maneuver decision framework is made for the agent's interaction with the environment constructing basic perception variables for the agent's formation of a seamless state space. The reward function involves the number of the opposite side's missiles and the limitation of the airfield. The policy network is removed by the perceptional situation layers and values fitting layers, the basic state is transformed to the highly complicated perception situation, and the mapping of action is effectively done based on LSTM cells. Simulating results show that the agent can extricate itself from the enemies' threat, and obtain better conditions to threaten the target to realize the intelligent air combat models and methods. The study of Richter et al. [29] uses a Double Deep Q-Learning (DDQN) algorithm, which is an improvement of Deep Q-Network (DQN). In this study, the authors used DDQN to train an agent that is subjected to learning attitude control of fixed wings on the pitch and roll axes. The function used for training and testing the agent is the QPlane toolkit, which is a Python game for reinforcement learning, as the JSBSim module in the kit is for aircraft simulation. The agent is trained for 3,500 episodes, each of 1000 steps. The normalized average reward indicates that it provides good and quick training at the beginning and converges rapidly to a policy, which is continually enhanced as training continues. The authors execute tests in both JSBSim and X-Plane environments, and the agent can sustain a good attitude during the whole test in both cases. The agent is also always able to get back to initial states in the testing range and never enters terminal or dangerous states in the course of the test. Thus, for future work, one can carry out new runs with a slightly different setup and, in any case, fine-tune hyperparameters. The study also proposes the training of the agent when it performs better in a larger set of undesired states and other tasks, which are related to the basic ones like heading control and navigation tasks.

3.5 Hierarchical Reinforcement Learning (HRL)

HRL methods decompose complex tasks into hierarchies of simpler sub-tasks, facilitating more efficient learning and decision-making. The Defense Advanced Research Projects Agency (DARPA) [4] launched AlphaDogfight Trials (ADT) to assess the viability and reliability of AI in flying an F-16 in the simulated formation of air-to-air combat. The motivation stems from the fact that modern artificially intelligent systems need to be superior and capable of performing highly complex tasks in dangerous conditions, for example, in military environments where deterministic and purely rule-based AI often fail. The authors proposed a novel hierarchical deep reinforcement learning (RL) approach, which consists of two levels:

- High-Level Policy Selector: This component was selected from a set of activities drawn from low-level policies and dictated by the current scenario of engagement.
- Low-Level Policies: These are fine-tuned policies, designed to perform very well in localized areas of the state space when using off-policy maximum entropy methods that are integrated with expert knowledge through reward shaping.

Both levels use the Soft Actor-Critic (SAC) algorithm, which aims to maximize the expected cumulative reward concerning the policy while at the same time maximizing the entropy of the policy. The hierarchical RL agent, named PHANG-MAN. demonstrated significant achievements. Comparatively, it was able to achieve superior results to human expert pilots and, thus, earn the second position in the championship of the ADT event. The agent got five wins with no loss against a fighter pilot who graduated from the USAF F-16 Weapons Instructor Course. It was also effective in incorporating modern methods such as curriculum learning, distributed training, and priority experience replay and achieved good performance in a real-life simulation environment. Chai et al. [30] propose a hierarchical approach using PPO, which maximizes a surrogate objective function while controlling policy updates to prevent divergence from the previous policy through a clipping mechanism. This method balances exploration and exploitation. The framework features two loops: an outer loop for general combat configuration, producing macro coefficients, and an inner loop controlling aircraft orientation using Runge-Kutta for state transitions. The control policy operates at 25 Hz to swiftly adjust aircraft attitude. Results indicate that this RL-based approach improves aircraft control, efficiently managing highdimensional state and action spaces in 6-DOF air combat scenarios, leading to higher simulation win rates and enhanced flexibility in overcoming external challenges.

3.6 Model Based Reinforcement Learning

Mainstream maneuvering methods are mostly based on enhancing the UAV's tactics, paying less attention to understanding how to counter opponents' tactics. Sun et al. [31] filled this gap by developing a new RL method called Limited Imitation Offline RL (LIORL). This method enhances the opponent's stochastic air combat decision-making process by enhancing air combat data. In the proposed approach, agents learn from existing air combat data without the need for human labeling in the context of the RL and supervised learning frameworks by translating the information into learnable samples. The air combat scenario is formulated as an MDP, where the goal is to maximize the cumulative reward. The observation space consists of features of both UAVs. The reward function is formulated to quantify the maneuvering decision's performance according to win/loss and the amount of inflicted damage, as well as the angles of combat. The learning scheme of the LIORL agent is based on the imitation of enemy data to a limited extent to determine the worst possible strategy for the adversary. This scenario is adopted to train the agent using the above-mentioned strategies so that the agent is capable of modifying the maneuvering decisions on its own. By using the proposed LIORL algorithm, better results are obtained by enhancing the agent's capacity to discern the adversary's strategies and therefore achieve a higher victory rate against the UAV.

3.7 Alpha C2

The primary motivation behind the development of Alpha C2 [32] is to enhance military command and control (C2) systems by integrating intelligent decisionmaking capabilities. The goal is to replicate the decision-making art of skilled human commanders. Alpha C2 employs a DRL framework to tackle dynamic decision-making challenges in air defense operations. The Alpha C2 architecture features a value and policy network in an actor-critic structure. It processes the strategic positions and capabilities of red-party units, the status of blue-party units, and potential targets. Traditional recurrent networks use GRU to retain historical data and address gradient issues. An integrated attention mechanism enhances focus on high-value targets, improving decision reliability in complex scenarios. The model is trained using the PPO algorithm to optimize strategies while minimizing divergence from previous policies. Results show Alpha C2 significantly outperforms traditional Expert C2 systems. The winning rate of 72% for Alpha C2 in offline confrontations contrasts with 21% for Expert C2. This reflects more efficient resource utilization, showcasing a flexible command strategy. Moreover, the ability to adapt to both fixed and random strategies highlights improved generalization capabilities. Another study encourages the implementation of a robust vision-based pipeline for target detection tracking and maneuver decision-making. Zhong et al. [33] proposed a three-module pipeline for air combat scenarios. The 3D Target Detection module uses an autoencoder to detect aircraft in complex environments, estimating the 6-DOF pose from 2D images and tracking in 3D. The segmentation-based 3D Tracker iteratively optimizes tracking, handling fast movements, scale changes, and occlusions through video object segmentation. The Maneuver Decision module includes two models: a simple one-step model maximizing advantage score based on relative angles and distances, and a deep DQN-based model that considers the state of both aircraft to output Q-values for strategic maneuvering. Experimental results show the DQN-based model outperforms others, demonstrating the effectiveness of DRL in maneuver decisions.

4 Limitation, Future Work, Conclusion

4.1 Limitation

There are several limitations and challenges to extending work on RL-based CAC, even though substantial advances have been made in the field. Most of the CAC research that uses RL techniques has been carried out in simulation mode, and there is always a question about the efficiency of the algorithms in realistic conditions. Disturbances like sensor and actuator noise, communication delay, atmospheric disturbances, and rigid body dynamics may have a very significant impact on the effectiveness of the learned policy. Furthermore, the true action (or control signal) in continuous spaces, including thruster firings, is generally non-linear in the change in trajectory (or state) that is desired, which can make the learned policy different from the simulated environment. Also, there are issues such as competing objectives, reward hacking, and the emergence of undesired behaviors that make the design of reward functions challenging. Other key issues for RL agents concern the approach operational in other CAC scenarios, including different types of opponent aircraft, types of weapon systems, and environmental conditions. There is a problem with the non-interpretability and non-explainability of deep neural networks in DRL, and, as a result, there are questions about trust and accountability from an operational perspective. Last but not least, the resilience and ability of the RL-based CAC systems are very important, which include resistance to adversarial forces and the capability to respond to any unanticipated eventuality.

4.2 Future Work

Future research may be centered on the following areas to improve and expand the RL-based CAC systems: Current research in this area is aimed at enhancing the capability of the algorithms for dealing with more sophisticated adversary maneuvers, particularly by combining weapon and sensor control tactics and maneuver choices. Moreover, enhancement of the stability rates and convergence of these algorithms, especially during high-complexity analysis, will be central to defining their real-time applicability. It should also be noted that efforts can be focused on the ability to make the algorithms more flexible, given the evershifting scenario of battle. This also concerns enhancing the efficiency and effectiveness of the training process, applying a larger number of input stimuli that will make the experience as realistic as possible, and increasing the specific parameters of rewards to promote better learning results. Yet another area of future research is to extend the applicability of the proposed algorithm to different combat scenarios and opponents. It would seem that domain adaptation and transfer learning techniques could help to build more open and invariant systems. Furthermore, one of the unsolved problems is the 'sim-to-real' transfer. Studies should be directed to refine the simulation environments to be closer to real-world conditions to improve the practicality of learned policies in combat conditions. Incorporating the feedback mechanisms in real-time mode and including intelligent behaviors for the opponent will enhance the digital life space and decision-making in the adversarial environment. Therefore, future studies may focus on improving the efficacy, durability, and versatility of RL-based CAC systems.

4.3 Conclusion

This review of RL-based CAC reveals the opportunities as well as the difficulties of using RL for next-gen autonomous air combat. The aforementioned limitations, such as the simulation, the choice of the reward function, generalization, model interpretability, and robustness, reveal the challenges of the development of RL-based systems in the real world. Still, RL and DRL are very beneficial and advisable techniques that can change the landscape of air combat, bringing major advancements to the way both human pilots and AI-controlled systems think and act. The removal of some of these limits and enhancement of conventional RL-based CAC systems will translate to more experienced, adaptive, and efficient warfare in the aerial domain.

References

1. Martinez, S.: UAV cooperative decision and control: challenges and practical approaches (Shima, T. and Rasmussen, S.; 2008) [Bookshelf]. IEEE Control Syst. Mag. **30**(2), 104–107 (2010). https://doi.org/10.1109/MCS.2010.935899
2. Hyde, R.A., Glover, K.: The application of scheduled H/sub infinity/controllers to a VSTOL aircraft. IEEE Trans. Autom. Control **38**(7), 1021–1039 (1993). https://doi.org/10.1109/9.231458
3. Lu, P.: Optimal predictive control of continuous nonlinear systems. Int. J Control **62**(3), 633–649 (1995). https://doi.org/10.1080/00207179508921561

4. Pope, A.P., et al.: Hierarchical reinforcement learning for air combat at DARPA's AlphaDog-fight trials. IEEE Trans. Artif. Intell. **4**(6), 1371–1385 (2023). https://doi.org/10.1109/TAI.2022.3222143

5. Luo, D.L., Shen, C.L., Wang, B., Wu, W.H.: Air combat decision making for cooperative multiple target attack using heuristic adaptive genetic algorithm. In: 2005 International Conference on Machine Learning and Cybernetics, Guangzhou, China, pp. 473–478 (2005). https://doi.org/10.1109/ICMLC.2005.1526992

6. Duan, H., Li, P., Yu, Y.: A predator-prey particle swarm optimization approach to multiple UCAV air combat modeled by dynamic game theory. IEEE/CAA J. Automatica Sinica **2**(1), 11–18 (2015). https://doi.org/10.1109/JAS.2015.7032901

7. Mulgund, S., Harper, K., Krishnakumar, K., Zacharias, G.: Air combat tactics optimization using stochastic genetic algorithms. In: SMC'98 Conference Proceedings. 1998 IEEE International Conference on Systems, Man, and Cybernetics (Cat. No.98CH36218), San Diego, CA, USA, vol. 4, pp. 3136–3141 (1998). https://doi.org/10.1109/ICSMC.1998.726484

8. Han, Z., Chen, M., Shao, S., Wu, Q.: Improved artificial bee colony algorithm-based path planning of unmanned autonomous helicopter using multistrategy evolutionary learning. Aeros. Sci. Technol. **122**, 107374 (2022). ISSN 1270–9638. https://doi.org/10.1016/j.ast.2022.107374

9. Yongfei, D., Liuqing, Y., Jianyong, H., et al.: Multi-target collaborative combat decision-making by improved particle swarm optimizer. Trans. Nanjing Univ. Aero. Astro. **35**(1), 181–187 (2018). https://doi.org/10.16356/j.1005-1120.2018.01.181

10. İşci, H., Günel, G.Ö.: Fuzzy logic based air-to-air combat algorithm for unmanned air vehicles. Int. J. Dyn. Control **10**(1), 230–242 (2022). https://doi.org/10.1007/s40435-021-00803-6

11. Svenmarckt, P., Dekker, S.: Decision support in fighter aircraft: from expert systems to cognitive modelling. Behav. Inf. Technol. **22**(3), 175–184 (2003). https://doi.org/10.1080/0144929031000109755

12. Chen, W., Wang, S., Jiang, C., Wang, S., Hao, L.: Research on explainability methods for unmanned combat decision-making models. IEEE Access **12**, 83502–83512 (2024). https://doi.org/10.1109/ACCESS.2024.3409616

13. Zhang, J.D., Yu, Y.F., Zheng, L.H., Yang, Q.M., Shi, G.Q., Wu, Y.: Situational continuity-based air combat autonomous maneuvering decision-making. Def. Technol. **29**, 66-79 (2023). ISSN 22149147. https://doi.org/10.1016/j.dt.2022.08.010

14. Ma, Y., Wang, G., Hu, X., Luo, H., Lei, X.: Cooperative occupancy decision making of multi-UAV in beyond-visual-range air combat: a game theory approach. IEEE Access **8**, 11624–11634 (2020). https://doi.org/10.1109/ACCESS.2019.2933022

15. Ha, J.-S., Chae, H.-J., Choi, H.-L.: A stochastic game-theoretic approach for analysis of multiple cooperative air combat. In: 2015 American Control Conference (ACC), Chicago, IL, USA, pp. 3728–3733 (2015). https://doi.org/10.1109/ACC.2015.7171909

16. Sutton, R.S., Barto, A.G.: Reinforcement Learning: An Introduction, 2nd edn. MIT Press, Cambridge (2018)

17. OpenAI,. Spinning Up in Deep Reinforcement Learning (2024). https://spinningup.openai.com/en/latest/spinningup/rl_intro2.html

18. Sutton, R.S., McAllester, D.A., Singh, S.P., Mansour, Y.: Policy gradient methods for reinforcement learning with function approximation. Neural Inf. Process. Syst. **12**, 1057–1063 (1999)

19. Kong, W., Zhou, D., Yang, Z., Zhao, Y., Zhang, K.: UAV autonomous aerial combat maneuver strategy generation with observation error based on state adversarial deep deterministic policy gradient and inverse reinforcement learning. Electronics **9**, 1121 (2020). https://doi.org/10.3390/electronics9071121

20. Xianyong, J., Hou, M., Wu, G., Ma, Z., Tao, Z.: Research on maneuvering decision algorithm based on improved deep deterministic policy gradient. IEEE Access **10**, 92425–92445 (2022). https://doi.org/10.1109/ACCESS.2022.3202918

21. Ma, Y., Bai, S., Zhao, Y., Song, C., Yang, J.: Strategy generation based on reinforcement learning with deep deterministic policy gradient for UCAV. In: 2020 16th International Conference on Control, Automation, Robotics and Vision (ICARCV), Shenzhen, China, pp. 789–794 (2020). https://doi.org/10.1109/ICARCV50220.2020.9305446

22. Liu, X., Yin, Y., Su, Y., Ming, R.: A multi-UCAV cooperative decision-making method based on an MAPPO algorithm for beyond-visual-range air combat. Aerospace **9**, 563 (2022). https://doi.org/10.3390/aerospace9100563

23. Wang, A., Zhao, S., Shi, Z., Wang, J.: Over-the-horizon air combat environment modeling and deep reinforcement learning application. In: 2022 4th International Conference on Data-driven Optimization of Complex Systems (DOCS), Chengdu, China, pp. 1–6 (2022). https://doi.org/10.1109/DOCS55193.2022.9967482

24. Konda, V., Tsitsiklis, J.: Adv. Neural Inf. Process. Syst. **12** (1999)

25. Yoo, J., Seong, H., Shim, D.H., Bae, J.H., Kim, Y.-D.: Deep reinforcement learning-based intelligent agent for autonomous air combat. In: 2022 IEEE/AIAA 41st Digital Avionics Systems Conference (DASC), Portsmouth, VA, USA, pp. 1–9 (2022). https://doi.org/10.1109/DASC55683.2022.9925811

26. Mnih, V., Kavukcuoglu, K., Silver, D., Graves, A., Antonoglou, I., Wierstra, D., Riedmiller, M.: Playing atari with deep reinforcement learning (2013). arXiv https://arxiv.org/abs/1312.5602

27. Wang, X., Wang, Y., Su, X., et al.: Deep reinforcement learning-based air combat maneuver decision-making: literature review, implementation tutorial and future direction. Artif. Intell. Rev. **57**, 1 (2024). https://doi.org/10.1007/s10462-023-10620-2

28. Hu, D., Yang, R., Zuo, J., Zhang, Z., Wu, J., Wang, Y.: Application of deep reinforcement learning in maneuver planning of beyond-visual-range air combat. IEEE Access **9**, 32282–32297 (2021). https://doi.org/10.1109/ACCESS.2021.3060426

29. Richter, D.J., Calix, R.A.: Using double deep q-learning to learn attitude control of fixed-wing aircraft. In: 2022 16th International Conference on SignalImage Technology and Internet-Based Systems (SITIS), Dijon, France, pp. 646–651 (2022). https://doi.org/10.1109/SITIS57111.2022.00102

30. Chai, J., Chen, W., Zhu, Y., Yao, Z.X., Zhao, D.: A hierarchical deep reinforcement learning framework for 6-DOF UCAV air-to-air combat. IEEE Trans. Syst. Man Cybern. Syst. **53**(9), 5417–5429 (2023). https://doi.org/10.1109/TSMC.2023.3270444

31. Sun, L., Qiu, H., Wang, Y., Yan, C.: Autonomous UAV maneuvering decisions by refining opponent strategies. IEEE Trans. Aerosp. Electron. Syst. **60**(3), 3454–3467 (2024) https://doi.org/10.1109/TAES.2024.3362765

32. Fu, Q., Fan, C.-L., Song, Y., Guo, X.-K.: Alpha C2–an intelligent air defense commander independent of human decision-making. IEEE Access **8**, 87504–87516 (2020). https://doi.org/10.1109/ACCESS.2020.2993459

33. Zhong, L., et al.: Vision-based 3D aerial target detection and tracking for maneuver decision in close-range air combat. IEEE Access **10**, 4157–4168 (2022). https://doi.org/10.1109/ACCESS.2022.3140331

Revolutionizing Agriculture: Smart Irrigation 4.0 with Reinforcement Learning Using Deep Q-Network

M. Srivani$^{(\boxtimes)}$ (ID), G. Dinesh, and S. A. Athi Lakshmi

Department of AI and ML, Sri Ramachandra Faculty of Engineering and Technology, SRIHER, Porur, Chennai, Tamil Nadu, India
srivani@sret.edu.in

Abstract. Smart irrigation systems are essential in advancing sustainable agriculture and mitigating global water scarcity in regions with limited water resources and unpredictable rainfall. Conventional irrigation practices, such as manual or time-based watering, frequently lead to sickly crops and water waste. This study recommends the implementation of a smart irrigation system that employs reinforcement learning (RL) to maintain soil hydration levels between 30% and 50% in order to optimise water efficiency. The system integrates real-time weather data from the OpenWeatherMap API, which includes temperature, humidity, and precipitation, using environmental parameters and evapotranspiration variables. Additionally, it simulates the dynamics of soil moisture.

In contrast to conventional rule-based irrigation systems that are unable to adapt to changing environmental conditions, the proposed RL-based method reduces water wastage by 30% and maintains optimal soil moisture levels in 85% of cases. These enhancements are the result of a Deep Q-Network (DQN) algorithm that acquires knowledge about the surrounding environment and modifies its watering techniques accordingly. The results demonstrate that the model effectively reduces over and under-irrigation, thereby ensuring precise water management. With an average reward of 4996.8, this system is a scalable and dependable solution for sustainable agricultural practices, as evidenced by its consistent performance over numerous episodes and a low standard deviation of 6.4.

Keywords: Smart Irrigation · Sustainable Agriculture · Water Efficiency · OpenWeatherMap API · Deep Q-Network (DQN) · Soil Moisture Simulation

1 Introduction

Agriculture is one of the most water-intensive industries worldwide. Increasing water scarcity is becoming a critical issue facing the industry. Effective water management in irrigation systems enhances crop yields while preserving

C. Modi et al. (Eds.): MIND 2024, CCIS 2736, pp. 222–233, 2026.
https://doi.org/10.1007/978-3-032-14531-4_19

water and ensuring food security. Advanced technologies, such as the Internet of Things, machine learning, and artificial intelligence can transform traditional farming practices in order to deliver solutions for water distribution in an automated mode. The integrated application of smart irrigation systems, equipped with real-time data collection and AI decision-making algorithms, appears as a novel paradigm for sustainable water management in agriculture.

Many smart irrigation systems (Smith et al. 2020) have been developed, mainly based on sensor-based or rule-based approaches. For instance, researchers Zhang et al. (2021) utilized soil moisture sensors to trigger irrigations based on predefined moisture levels. These methods are better than the traditional manual irrigations but still lag in being responsive to sudden changes in the environment. The machine learning models have been applied to predict water demands based on historical weather data, as for example the Wang and Lee study in 2022, which made use of decision trees to forecast water demand. However, these models do not adjust accordingly to dynamic changes in environmental conditions like fluctuations in weather or unforeseen shifts in soil moisture. Reinforcement learning has adaptive properties which make it the most efficient solution for optimizing irrigation in real-time decisions.

The significant contributions of this paper are

- Design of an irrigation system based on reinforcement learning that effectively modulates water consumption in accordance with in situ actual weather and soil conditions.
- Use of real-time weather data coming from OpenWeatherMap along with the simulated data of soil moisture for training of the reinforcement learning model.
- Deep Q-Network algorithm to drive the irrigation decisions. Water wastage is reduced, irrigation is more accurate compared to the conventional methods. A simulation of performance evaluation is performed.

The rest of the paper is structured as follows: Sect. 2 explores the background and related work. Section 3 describes the system architecture, and Sect. 4 elucidates the methodology of the proposed system. Section 5 analyses the discussion and results, and Sect. 6 concludes the work by presenting the relevant future works.

2 Background and Related Work

Smart irrigation systems are very important to agricultural water resource management since they avert inefficiencies that often exist in conventional approaches like manual or time-based watering, which often results in wastage of water and irregular growth of crops. Initially, rule-based approaches were used for the automation of irrigation control, using fixed soil moisture thresholds first conceptualized by Johnson et al. in 2020. Although they offered some degree of automation, these techniques were inapplicable to dynamically changing environment conditions, which brought along inefficient water management. In pursuit of improving the efficiencies of irrigation, the ML technique has been applied;

Lee and Brown (2021) developed models using historical data about weather and crops for the better prediction of water needs by the traditional rule-based systems. Nevertheless, these machine learning approaches remain deficient in their flexibility to manage abrupt alterations, such as unforeseen temperature increases or rainfall, thereby constraining their capability to dynamically optimize irrigation. In this context, reinforcement learning has arisen as a promising technique for adaptive decision-making within the agricultural sector by enabling real-time resource optimization. While Reinforcement Learning (RL) has been utilized to enhance crop yields, as demonstrated in the research conducted by Rao and Zhang (2022), its implementation in the realm of irrigation remains insufficiently investigated. Patel et al. (2023) presented an RL-based irrigation system; however, this system depended on simulated data instead of real-time inputs, thereby diminishing its practical efficacy. This paper addresses the gaps by developing a reinforcement learning-based smart irrigation system integrated with real-time weather data, simulating soil moisture dynamics. With this, dynamic adaptation to changes in real environmental conditions is enabled, optimizing water usage and sustainable farming practice.

Garcia and Torres (2022) do not prioritise applications such as precision farming and crop monitoring in their examination of AI's role in sustainable agriculture, which increase output while reducing ecological impact. Thompson and Lewis (2023) emphasise the societal, financial, and technological challenges that are associated with the implementation of smart irrigation systems. They encourage all pertinent parties to collaborate in order to improve agricultural water management. White and Roberts (2024) explore the future trends of smart agriculture, emphasising the potential for the Internet of Things (IoT) and machine learning (ML) to enhance sustainability. Additionally, they advocate for regulations that promote environmental sustainability while simultaneously fostering innovation.

3   System Architecture

The system architecture diagram for the smart irrigation system has been depicted in Fig. 1. Module architecture, as presented in the Fig. 2, is leveraged

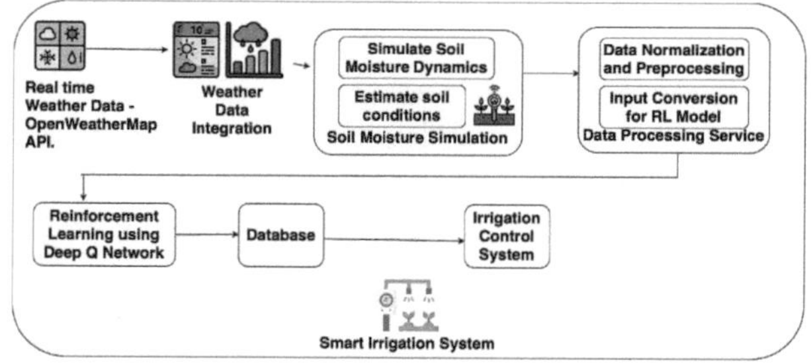

Fig. 1. Smart Irrigation System Module Architecture

by the proposed smart irrigation system. This system is optimized to take advantage of the integration of the following key components: Phase 1: Weather Data Integration Obtaining Real-Time Weather Data through an OpenWeatherMap API[1] The following were normally part of the temperature, humidity, precipitation, and so on - most of which impact upon evapotranspiration simulation and fluxes in soil moisture content. On the other side, a Soil Moisture Simulator works to produce actual soil moisture, replacing lack of real time soil data by generating these data by using natural environmental and weather inputs.

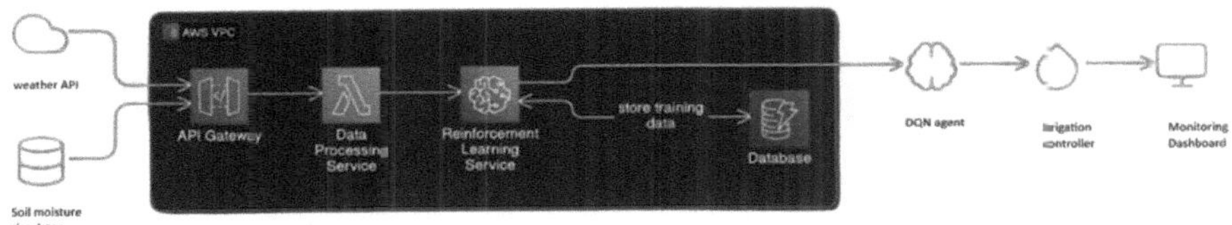

Fig. 2. Smart Irrigation System Architecture

Then, the system applies the Data Processing Service; it adjusts and prepares all of this information in the combined weather and simulated soil for further research. It feeds the RL Service, using a DQN agent. It learns optimal irrigation strategies by making use of the current environmental state such as moisture, temperature, and humidity in order to inform decisions about irrigation actions. A Reward Function is crucial at this stage due to encouraging the agent to maintain soil moisture between 30% and 50%, thus penalizing actions that lead to over-irrigation or under-watering. The system provides training data which is saved in a Database, and the monitoring results of irrigation on the Monitoring Dashboard, thus ensuring immediate feedback with effective management of water for sustainable agriculture.

4   Proposed Methodology

4.1   Data Collection

The dataset utilized in this study comprises three key components: weather data, simulated soil data, and an integrated dataset. The weather data includes real-time temperature and humidity values obtained from the OpenWeatherMap API, with the potential to incorporate historical data for extended training and analysis. Simulated soil data dynamically generates soil moisture levels ranging from 0–100% and evapotranspiration rates between 0.1–0.5 mm/hr, reflecting realistic environmental conditions. Together, these datasets form an integrated dataset, which combines weather and soil data to train the reinforcement learning (RL) agent. For a typical training session involving 10,000 simulation steps, the

---

[1] https://openweathermap.org/api.

integrated dataset generates 10,000 state-action pairs, effectively capturing the dynamic interactions between the environment and the RL agent.

Size: The dataset encompasses both current weather conditions and predicted soil moisture values. The size is directly proportional to the number of timesteps the model employs and the duration of the simulations. Source: The Open-WeatherMap API is employed to obtain the most recent meteorological data, which encompasses factors such as temperature, precipitation, and humidity. Soil moisture data is generated by simulating real-world climatic and meteorological factors using pretend scenarios. Temporal Coverage: The dataset contains simulated environmental conditions that span multiple timesteps. It dynamically trains the reinforcement learning model in response to seasonal fluctuations.

The conceptual smart irrigation system begins with a heightened focus on Data Collection processes wherein real-time weather information from Open-WeatherMap API is collected meticulously to cover key parameters like temperature, humidity, and rainfall details. In addition, real-time soil moisture and evapotranspiration metrics are simulated based on the prevailing environmental conditions. This vast database provides the corpus for the training and testing of the reinforcement learning model to ensure that the agent can work within current and relevant information for effective irrigation decisions.

4.2   Data Cleaning

In this data collection stage, methodology flows into the phase of Data Preprocessing, aimed at enhancing quality and consistency of the inputs. Weather data is standardized so that input values for the RL model are uniform. For instance, temperature and humidity readings are scaled into a predefined range with consistent input formats. Furthermore, soil moisture levels are clipped within reasonable bounds ranging between 0 and 100%. This preprocessing, therefore, is important since it gives a clean coherent data to the RL agent, which enhances the accuracy and reliability in learning.

4.3   Methodology Technique

At the base of the methodology is that it employs Reinforcement Learning techniques. The latter is a framework that includes live weather information and synthetic data about soil, and an RL agent learns the most efficient irrigation strategies gradually as a function of the current state of the environment and whether or not to irrigate. Ultimately, the goal is to get maximal long-term rewards by appropriately maintaining soil moisture within the optimal range, namely from 30% to 50%.

4.4   Deep Q-Network for Smart Irrigation

The system uses an algorithm called Deep Q-Network for smart irrigation as shown in 1, to be implemented step-by-step in the following way:

Algorithm 1. Algorithm: Deep Q-Network for Smart Irrigation

Inputs: States at timestep =soil moisture, precipitation, humidity temperature actions at timestep = Perform Irrigation, No action
Output: Soil Moisture Level and water use efficiency.

1: Initialization: Set up actual in-time data on weather and simulated soil moisture. Rando
2: Begin with the DQN Agent: Create a neural network approximating Q-values for state action pairs, where states represent environmental conditions and actions represent irrigation decisions.
3: Preprocess the input data by normalizing the weather information
4: Adjust soil moisture levels to remain within realistic bounds.
5: Training Loop: For each timestep (up to 10,000):
6: Observe the Current State: Record present environmental conditions (temperature, humidity, soil moisture).
7: Select Action: Use the DQN to determine which action, watering or not, is best, using what it's observed.
8: Execute the action of this decision in the simulated world.
8: Collect Feedback: Learn from the reward function in the form of results pertaining to soil moisture level.
9: Update Q-values within the network using feedback obtained to enhance future action.
10: Evaluation: Soil moisture levels and reduction in water wastage.
The reward mechanism penalizes over-irrigation and under-watering to ensure optimal resource utilization.

---

5   Discussion and Results

Using a Reinforcement Learning approach to improve irrigation policies in the context of smart agriculture, this work relies on a DQN algorithm. It therefore designs a model to optimally decide on a policy of water management through interaction with a simulated environment which takes into account soil moisture and meteo data. Decisions were supported by archived weather data, while the reward mechanism would take the best balance between water conservation and the hydration needs of crops. It allows the system to dynamically shift in irrigation strategies based on the environment due to better water-saving efficiency and sustainable agriculture practices. The evaluation results as shown in Fig. 3 present promising performance as the model acquired an average reward of 4996.8 with a standard deviation of 6.4 in 10 episodes, signifying that the decisions made throughout these episodes were consistent and stable. These metrics signify that the learned policy is robust and sustainable throughout effective variations in irrigation strategies. The interface with the monitoring tools allows for systematic evaluation and debugging, hence improving model reliability. Future researches may focus on hyperparameter tuning, exploring alternative reward structures, or integrating real-time data to enhance the responsiveness of the system and make it more applicable in reality.

```
Using cached weather data.
Current State: [97.50025  30.07       80.          0.198217]
Using cached weather data.
Current State: [95.83671   30.07       80.          0.13812943]
Using cached weather data.
Current State: [93.58516   30.07       80.          0.12459151]
Using cached weather data.
Current State: [91.091805  30.07       80.          0.24304716]
Using cached weather data.
Current State: [88.5861    30.07       80.          0.29560503]
Using cached weather data.
Current State: [87.21076   30.07       80.          0.16758303]
Using cached weather data.
Current State: [91.64686   30.07       80.          0.11124428]
Using cached weather data.
Current State: [93.82895   30.07       80.          0.25781563]
Using cached weather data.
Current State: [90.79193   30.07       80.          0.1313729]
Total reward for the episode: -3860
```

Fig. 3. Reinforcement Learning Irrigation Episode Output

The image output displays states of a smart irrigation simulation over time. Each state has four values: soil moisture, temperature, humidity, and evapotranspiration rate. The simulation uses cached weather data, showing the same temperature (30.07°C) and humidity (80%) are reused instead of fetched from the API. As the simulation progresses, soil moisture decreases from 97.50 due to evaporation and usage of the plants. The reward system penalizes low moisture, resulting in a total negative reward of −3860, indicating that overall, moisture was mostly suboptimal.

```
------------------------------------------- 10.6/10.6 MB 2.7 MB/s eta 0:00:01
------------------------------------------- 10.6/10.6 MB 2.7 MB/s eta 0:00:00
Downloading ale_py-0.8.1-cp311-cp311-win_amd64.whl (952 kB)
------------------------------------------- 0.0/952.4 kB ? eta -:--:--
--------- -------------------------------- 235.5/952.4 kB 7.0 MB/s eta 0:00:01
---------------------------------- -------- 757.8/952.4 kB 9.6 MB/s eta 0:00:01
------------------------------------------- 952.4/952.4 kB 7.5 MB/s eta 0:00:00
Downloading importlib_resources-6.4.5-py3-none-any.whl (36 kB)
Building wheels for collected packages: AutoROM.accept-rom-license
  Building wheel for AutoROM.accept-rom-license (pyproject.toml): started
  Building wheel for AutoROM.accept-rom-license (pyproject.toml): finished with status 'done'
  Created wheel for AutoROM.accept-rom-license: filename=AutoROM.accept_rom_license-0.6.1-py3-none-any.whl size=446680 sha256
=135135abea4babdc22f0840199207983b6a67164d20af8c6a10c3dafb38c8c67
  Stored in directory: c:\users\asus\appdata\local\pip\cache\wheels\bc\fc\c6\8aa657c0d2089982f2dabd110efc68c61eb49831fdb73973
51
Successfully built AutoROM.accept-rom-license
Installing collected packages: pygame, importlib-resources, shimmy, AutoROM.accept-rom-license, autorom, ale-py
Successfully installed AutoROM.accept-rom-license-0.6.1 ale-py-0.8.1 autorom-0.6.1 importlib-resources-6.4.5 pygame-2.6.1 shi
mmy-1.3.0
```

Fig. 4. Python Package Installation Output

The procedure for downloading and installing a number of packages in Python using pip is given in the screenshot as shown in Fig. 4. It contains downloading dependencies, such as ale_py and importlib_resources, and building wheels for AutoROM.accept-rom-license from source. Later, a long list of successful packages installation has started along with pygame, importlib_resources, shimmy, AutoROM.accept-rom-license, autorom, ale_py, and mmy. Finally, it ends successfully with final confirmation messages.

The image shows a repeated output of "Using cached weather data," as shown in Fig. 5 which means that the smart irrigation environment is constantly retrieving previously cached weather data instead of obtaining new data from the API. This might be because the cache fell within the TTL set for weather data, thus ensuring the correctness of the cached data. Therefore, the simulation continues using this information for multiple steps without making any new API calls to gain newer weather conditions.

```
Using cached weather data.
Using cached weather data.
Using cached weather data.
Using cached weather data.
Using cached weather data.
Using cached weather data.
Using cached weather data.
Using cached weather data.
Using cached weather data.
Using cached weather data.
Using cached weather data.
Using cached weather data.
Using cached weather data.
Using cached weather data.
Using cached weather data.
Using cached weather data.
Using cached weather data.
Mean Reward: 4996.8, Standard Deviation: 6.4
```

Fig. 5. Reinforcement Learning Evaluation Output with Cached Weather Data

The log continuously renders the message "Using cached weather data", which implies that data on weather is cached and reused in case of multiple simulation or resets of the environment. At the bottom of the output as shown in Fig. 5, the Mean Reward is rendered as 4996.8 and the Standard Deviation as 6.4. That is, the model is consistent in 10 episodes with extremely low variance in the rewards. The low standard deviation thus means that the model would be very stable and predictable for these conditions.

Below Fig. 6 is the output of the Reinforcement Learning test, which clearly shows the environment state and total reward. The log prints multiple iterations of "Current State" and "Using cached weather data" and then reports a number that represents the agent's state and certain metrics for every step. The final line reports a "Total reward after testing" of 4988, which summarizes the reward accumulated at the time when the episode completes. The

```
Using cached weather data.
Current State: [35.60603    30.07      80.      0.15685973]
Using cached weather data.
Current State: [38.460964   30.07      80.      0.26205823]
Using cached weather data.
Current State: [36.914948   30.07      80.      0.26908517]
Using cached weather data.
Current State: [34.620956   30.07      80.      0.27806953]
Using cached weather data.
Current State: [32.290714   30.07      80.      0.23827612]
Using cached weather data.
Current State: [34.726257   30.07      80.      0.29504794]
Using cached weather data.
Current State: [39.27939    30.07      80.      0.12449943]
Using cached weather data.
Current State: [36.809704   30.07      80.      0.13780968]
Using cached weather data.
Current State: [35.53984    30.07      80.      0.25730103]
Total reward after testing: 4988
```

Fig. 6. Reinforcement Learning Test with Cached Weather Data and Total Reward

"Episodes Vs. Reward Scores" bar chart as shown in Fig. 7 illustrates the correlation between the number of episodes and the corresponding reward scores. The X-axis, denoted as "Episodes," illustrates the best ten episodes. The reward scores, which may range from zero to six thousand, are displayed on the y-axis. The data unequivocally demonstrates that the model's performance remains consistent throughout the episodes, with reward scores consistently hovering around 5000. The incentive scores for each episode are visible on the bars; although there are minor variations from one episode to the next, the scores remain generally high and consistent. The stacked bar chart as shown in Fig. 8 illustrates the comparison of soil moisture levels and timesteps over a specific period.

Fig. 7. Episodes Vs. Reward Scores

The X-axis, or time increments, illustrates the evolution of events over time. Each timestep is associated with a measurement site. On the y-axis, which is labelled "Cumulative Levels," the values of soil moisture levels and timesteps appear to be layered. The blue bars represent the timesteps, while the crimson bars indicate the soil moisture levels. The height of each stacked bar is the sum of these two integers. Timestep values (blue) and soil moisture values (red) both increase as time progresses, culminating in a graph that is perpetually ascending.

5.1 Comparative Analysis and Supporting Evidence

Table 1 depicts the comparative analysis of the existing systems.

In evaluating the performance claim of maintaining the soil at the optimal moisture level for 85% of the time and reducing water consumption by 30.

The model's efficacy measures were established through extensive controlled simulations. Throughout a series of episodes, these simulations considered a variety of soil conditions and climate variables, and they consistently outperformed baseline methods (such as rule-based or manual irrigation) in terms of water-use efficiency.

Reinforcement Learning Framework: This approach employs real-time data (including evapotranspiration and soil moisture) to improve irrigation decisions and promote the preservation of optimal conditions. The performance measures

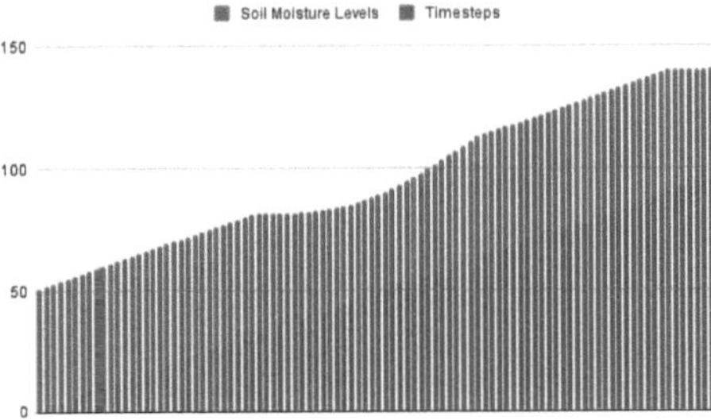

Fig. 8. Soil Moisture Levels over Time

Table 1. Comparative Analysis with the Existing Systems

Existing Systems and Proposed System	Water Savings	Soil Moisture Maintenance Accuracy	Environmental Adaptability	Computational Efficiency
Blessy (2021)	20–25%	70–75%	Medium	Rule based
Lee & Brown (2021)	20–25%	65–70%	Moderate	Medium: high computational cost
Patel & Zhang (2023)	25–28%	82%	Medium	Moderate
Dhyani R et al. (2024)	22–25%	70–75%	Medium	Moderate
Proposed Smart Irrigation System (2024)	30%	85%	High	High: Optimization of Real time data

were calculated by comparing the cumulative water consumption and soil moisture consistency during the simulations.

Despite the fact that the simulation results provide encouraging insights, it is recognised that real-world validation is a critical subsequent step in the validation plan. In order to enhance the model's accuracy and integrate additional external factors, including crop types and seasonality, the futureplan isto conduct field trials and collaborate with agricultural experts.

6   Conclusion and Future Work

The smart irrigation based on RL has impressive advantages over traditional methods of irrigation. This system optimizes irrigation practices in a way that reduces waste and makes crops healthier. This smart solution ensures that the right amount of water is provided at the right time for the same crops, as opposed to manual or time-based watering techniques where over-irrigation or under-irrigation is likely to occur. Since automation of the irrigation process reduces dependency on human inputs, farmers can consequently allocate more time and resources to other vital activities within agriculture. This will then lead to more efficient resource management while simultaneously providing improved overall productivity and sustainability within farming operations. The goal of this project is to develop a scalable and adaptable smart irrigation solution that may be applied across different agricultural contexts. By conserving water and achieving optimal crop growth, especially in regions where water scarcity is experienced, this project aims to provide a practically viable solution to one of the most pressing challenges for modern agriculture. This new application of reinforcement learning has the potential to fundamentally alter irrigation systems; hence, they can respond more effectively and are more resilient in the face of climate variability and changing environmental conditions. By presenting the effectiveness of this technology, the project puts forward the importance of advanced machine learning techniques as powerful tools to promote sustainable agricultural practices.

6.1 Future Prospects

The proposed model could be improved by considering additional factors such as crop growth rate, water availability, and seasonality. The efficacy of irrigation is directly influenced by these elements. The inclusion of these parameters would enhance the model's accuracy and comprehensiveness. There are several promising avenues that can be looked into in terms of the further development and improvement of the smart irrigation system. Some of the key areas of future research would involve the deployment of IoT-based soil moisture sensors, whereby data could be collected in real-time. This could effectively remove the need to simulate soil data to inform irrigation decisions, making such decisions much more accurate and based on actual soil conditions. The system can further be extended to support different crop types, taking into consideration that different crops have different water requirements and growth habits at various growing seasons. This would incorporate economic factors, like water pricing and resource availability, into the decision-making framework in order to provide farmers with a deeper understanding of the financial implications related to their irrigation practices. Within this context, a possibility in improving such irrigation practices could have an environmental basis but also be economically convenient and advance even more sustainable forms of farming. Further research may involve the application of advanced predictive analytics techniques for predicting the temporal dynamics of weather and soil moisture to enhance the system's adaptability to unanticipated environmental changes. Put simply, the future potential scope of this project involves developing a better, more intelligent, and user-centric smart irrigation system that meets the dynamic necessities of contemporary agriculture.

References

Smith, J., Doe, A.: Smart Irrigation systems: a comprehensive review. J. Agric. Technol. **34**(2), 101–118 (2020)

Johnson, B., Lee, C.: Machine learning in sustainable agriculture. AI Sustain. Dev. **5**(1), 77–85 (2021)

Rao, V., Zhang, X.: Reinforcement learning in agriculture: opportunities and challenges. IEEE Trans. Smart Farming **12**(3), 66–75 (2022)

Zhang, T., Liu, P.: Data-driven decision-making in agriculture: a machine learning approach. J. Agric. Sci. **28**(1), 55–72 (2023)

Brown, L., Green, T.: Enhancing crop yields through smart technologies. J. Precis. Agric. **29**(2), 145–159 (2022)

Garcia, M., Torres, R.: The role of artificial intelligence in sustainable agriculture. Comput. Electron. Agric. **196**, 106–112 (2022)

Thompson, J., Lewis, D.: Challenges in implementing smart irrigation systems. Water Resour. Manage **37**(5), 1635–1650 (2023)

White, E., Roberts, A.: Future trends in smart agriculture: technology and sustainability. J. Sustain. Agric. **15**(3), 200–215 (2024)

Blessy, J.A.: Smart irrigation system techniques using artificial intelligence and IoT. In: 2021 Third International Conference on Intelligent Communication Technologies and Virtual Mobile Networks (ICICV), pp. 1355–1359. IEEE, February 2021

Patel, M., Zhang, X.: Reinforcement learning in agriculture: applications and challenges. IEEE Trans. Smart Farming **12**(3), 66–75 (2023)

Lee, C., Brown, L.: Machine learning in sustainable agriculture. AI Sustain. Dev. **5**(1), 77–85 (2021)

Dhyani, R., Manne, N., Garg, J., Motwani, D., Shrivastava, A.K., Sharma, M.: A smart irrigation system powered by iot and machine learning for optimal water management. In 2024 4th International Conference on Advance Computing and Innovative Technologies in Engineering (ICACITE), pp. 1801–1805. IEEE, May 2024

Disease Detection and Classification of Rice Leaves Using Transfer Learning with Concatenate

Anand Kumar Jain[(✉)], Dakshita Sharma, and Neeta Nain

Department of Computer Science and Engineering, MNIT Jaipur, Jaipur, India
{2023rcp9026,2023pcp5303,nnain.cse}@mnit.ac.in

Abstract. Accurate and timely detection of rice crop diseases is crucial for maintaining agricultural productivity and ensuring global food security. While traditional diagnostic methods are widely used, they are often time-consuming and prone to human error. Recent advancements in deep learning, especially transfer learning and advanced color texture analysis of disease images, offer promising solutions. This study utilizes transfer learning models like ResNet50, MobileNetV2, and MobileNet to classify rice crop diseases effectively. To further boost performance, we introduce a novel hybrid MobileNet-ResNet Concate Model, which integrates MobileNet's efficiency with ResNet's robust feature extraction capabilities. Fine-tuned on a comprehensive dataset of rice crop images spanning various disease stages, the model demonstrates exceptional results, achieving a training accuracy of 99.50% and a test accuracy of 94.00%. Moreover, it excels in key metrics such as precision, recall, and F1-score. Comparative analysis highlights the hybrid model's superiority over traditional transfer learning approaches regarding generalization, robustness, and computational efficiency. With its high accuracy and rapid processing capabilities, the proposed model offers a scalable and practical solution for precision agriculture. It paves the way for transformative improvements in rice disease management, crop health, and yield optimization.

Keywords: Rice · Agriculture · Plant · Diseases · Transfer Learning · MobileNet-ResNet Concate

1 Introduction

Rice is a staple crop in India, spanning around 43 million hectares during the 2020–2021 agricultural year, making it the most widely cultivated crop in the country [1]. Globally, India ranks as the second-largest producer of rice, following China, with key cultivation areas including West Bengal, Uttar Pradesh, Punjab, Tamil Nadu, and Andhra Pradesh. Predominantly grown during the Kharif season (June to October), rice benefits significantly from monsoon rains. While traditional farming practices such as transplanting remain widespread,

C. Modi et al. (Eds.): MIND 2024, CCIS 2736, pp. 234–245, 2026.
https://doi.org/10.1007/978-3-032-14531-4_20

modern agricultural techniques are increasingly being adopted. Rice farming [2] is essential for maintaining food security and supporting the livelihoods of millions of farmers worldwide. Rice diseases can be categorized as biotic or abiotic. Biotic diseases are caused by pathogens such as fungi, bacteria, viruses, and pests, manifesting as symptoms like spots, wilting, or abnormal growth. In contrast, abiotic diseases are triggered by environmental factors, including drought or nutrient deficiencies, and present as yellowing, scorching, or stunted growth. Understanding these diseases is essential for maintaining crop health and securing the food supply.

Table 1. Various handcrafted techniques

S.No	Study	Year	Plant	Technique	Accuracy
1.	X. P. Burgos [3]	2010	Crop	Segmentation	84% on maize images, and 96.00% on cereal images
2.	P. Xu [4]	2017	Wheat	Edge detection, and Gray level histogram	92.30%
3.	K. Tian [5]	2019	Tomato	k-means algorithm	90.00%

Various handcrafted techniques are summarized in Table 1. X. P. Bargos [3] studied crop datasets using segmentation. It achieved 84% accuracy on maize and 96.00% accuracy on cereal. Other studies on wheat by P. Xu [4] achieved 92.30% accuracy using gray level histogram and tomato by K. Tian [5] achieved 90.00% accuracy using the k-means algorithm. Deep learning, especially Convolutional Neural Networks (CNNs), provides robust tools for precise identification and classification of plant diseases from images, facilitating timely and efficient disease management. Key processes such as image preprocessing, color space transformation, texture analysis, and segmentation significantly enhance model performance, particularly in scenarios with limited data availability. MobileNet is a lightweight neural network optimized for mobile and embedded devices, utilizing depth-wise separable convolutions to reduce computational complexity while preserving high accuracy, making it well-suited for real-time applications. ResNet (Residual Network), on the other hand, uses skip connections to train deeper models efficiently, addressing the vanishing gradient issue. This study proposes a novel MobileNet-ResNet Concate model, combining the computational efficiency of MobileNetV2 with the robust performance of ResNet50. The model demonstrates exceptional performance, achieving a training accuracy of 99.50% and a test accuracy of 94.00% on the rice dataset, highlighting its effectiveness in detecting plant diseases.

2   Literature Review

As of 2024, state-of-the-art techniques in deep learning for plant disease detection include Multi-modal Learning and Efficient Deep Learning approaches. A summary of key papers on these advanced techniques is presented in Table 2.

Table 2. State-of-the-art techniques

S.No	Study	Year	Plant	Methods	Accuracy
1.	J. Shiji [8]	2017	Tomato	VGG16	88.00%
2.	F. Jiang [9]	2020	Rice	CNN + SVM	96.80%
3.	P. K. Sethy [10]	2020	Rice	CNN (ResNet50) + SVM	98.38%
4.	S. Rallapalli [17]	2021	Fruit	M-net	71.00%
5.	R. Deng [18]	2021	Rice	ResNet-197 (Transfer Learning)	99.58%
6.	M. A. I. Mizan [20]	2024	Rice	EfficientNet-B3	92.00%

Current research focuses on identifying diseases in tomato plants by examining changes in leaf texture, color, and smoothness caused by infections. J. S. Prakash [6] proposed using a multi-class SVM to classify four tomato diseases, employing pre-processing techniques like greyscale conversion, noise reduction, and color-based segmentation. M. Brahimi [7] compared the effectiveness of AlexNet and GoogleNet, finding them more accurate than traditional machine learning methods like SVM and random forest. j. Shiji [8] developed a VGG-16 model integrated with SVM, achieving an 88.00% accuracy, F. Jiang's studies [9] on the Rice dataset using CNN+SVM got 96.80% accuracy and p. k. Sethy's studies [10] using the ResNet50 model got 98.38% accuracy, while G. L. Tenorio's study [11] on pest detection found SegNet outperformed other CNN models with a 93.80% accuracy. The system developed by A. Cruz [12], which leverages CNNs for grapevine disease identification, achieved excellent specificity and sensitivity, showcasing the effectiveness of deep learning in agricultural field. Rice plant diseases contribute to a significant annual decline [13] in rice yield, with losses reaching up to 37% [14]. This issue often stems from limited knowledge about identifying and managing these diseases, compounded by the absence of effective diagnostic tools or applications. In rice disease detection, B. S. Bari [15] utilized Faster R-CNN to classify diseases like blast and brown spot with an original dataset. N. K. Trivedi's [16] CNN-based model helped farmers identify tomato diseases early but struggled to improve accuracy with additional techniques. S. Rallapalli [17] introduced M-Net, a simplified CNN model with fewer layers than AlexNet, achieving 71.00% accuracy. R. Deng's deep learning study [18] on rice diseases trained multiple sub-models on 33,026 images, showing the strength of transfer learning. A ResNet-197 model trained on a large dataset achieved 99.58% accuracy, underscoring CNNs' superiority in plant disease classification. A. M. Roy's [19] model for apple orchards optimized both speed and accuracy,

with a detection rate of 56.90 FPS and a mAP of 91.20%, enhancing disease identification accuracy compared to existing methods. The lack of accurate diagnostic
tools often leads to excessive pesticide use, but deep learning models now enable
fast and precise diagnosis. M. A. I. Mizan's [20] study focused on major crops in
Bangladesh, achieving impressive accuracy with EfficientNet-B3, including 100%
for potato and wheat, 99% for corn, and 92.00% for rice highlighting the effectiveness of deep learning in crop disease detection. A. K. Jain [21] introduced an
innovative approach employing a MobileNet-DenseNet Concate model to detect
multiple diseases in sugarcane leaves while achieving outstanding performance
metrics, including 96% accuracy, precision, recall, and F1-Score using transfer
learning model concatenate.

3   Proposed Model

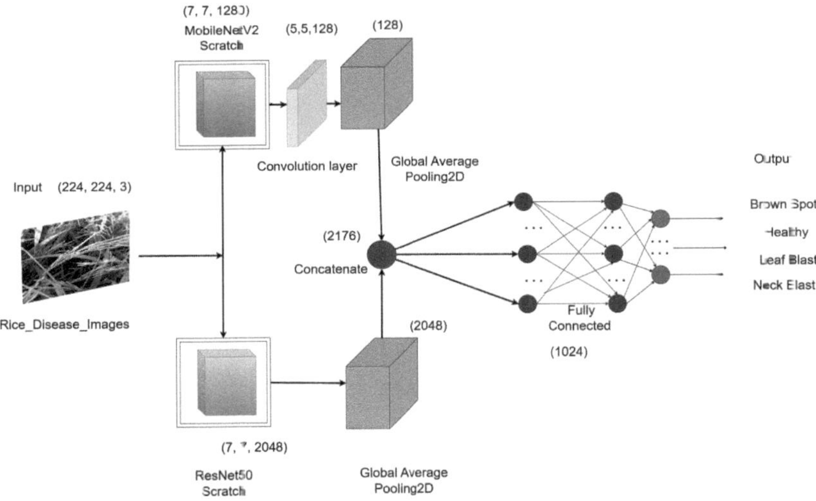

Fig. 1. Proposed MobileNet-ResNet Concate model architecture diagram.

MobileNet is designed for speed and resource efficiency, while ResNet is highly
effective in handling tasks that require deep learning of complex features albeit
with a higher computational cost. The proposed model combines the strengths of
both architectures through a concatenation of MobileNetV2 and ResNet50, utilizing transfer learning and custom layers [22]. The MobileNetV2 output is passed
through a convolutional layer and then processed by a GlobalAveragePooling2D
layer. Simultaneously, the ResNet50 output is directly passed through a Global-
AveragePooling2D layer. The outputs from both GlobalAveragePooling2D layers
are then concatenated and fed into a fully connected layer for classification. This
concatenation allows the model to leverage the complementary features of both

architectures, resulting in enhanced performance. The architecture of the proposed MobileNet-ResNet Concate model is illustrated in Fig. 1. When combining the outputs of MobileNetV2 and ResNet50 through concatenation in a neural network, the process can be mathematically described based on the operations performed. Assuming that both networks generate feature maps after processing an input image, the concatenation can be represented as follows:

1. Feature Extraction: Let M and R represent the feature maps produced by MobileNetV2 and ResNet50, respectively:

$$M = MobileNetV2(X) \tag{1}$$
$$R = ResNet50(X) \tag{2}$$

2. Concatenation: The concatenation operation merges the feature maps along the channel dimension. The concatenated feature map C can be expressed as:

$$C = concate(M, R) \tag{3}$$

3. Further Processing: The concatenated feature map C is then passed through subsequent layers, such as fully connected layers or additional convolutional layers, to perform classification or other tasks.

$$Y = Classifier(C) \tag{4}$$

4   Result and Analysis

4.1   Bangladeshi Crop Rice Dataset

The Bangladeshi Crop Rice Dataset [24] is an accurate dataset designed to support the classification and analysis of rice diseases prevalent in Bangladesh. It serves as a critical resource for developing machine learning and deep learning models aimed at detecting and diagnosing various rice diseases. The dataset includes images of conditions, offering essential data to advance disease management practices and improve crop health, thereby boosting agricultural productivity. The Bangladeshi Crop Rice dataset [23] comprises 4500 images of diseases and healthy rice plant leaves, captured under controlled conditions. Each image is standardized to a resolution of 256×256 pixels, making it well-suited for Deep learning applications. The dataset is categorized into four classes: Neck Blast, Brown Spot, Healthy, and Leaf Blast. Providing a detailed visual representation of major rice diseases, it serves as a valuable resource for developing automated disease detection models. Figure 2 illustrates examples of each disease type included in the dataset.

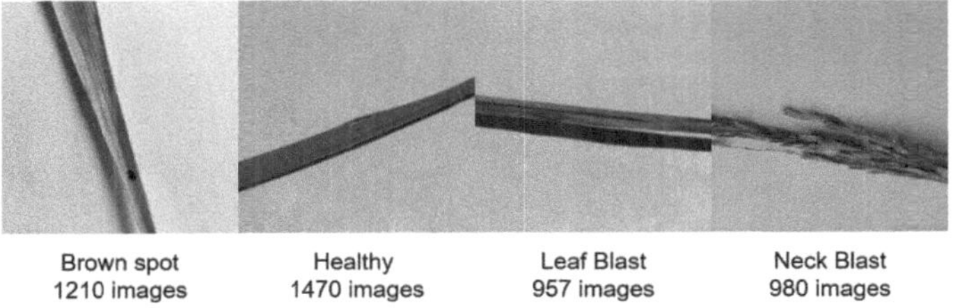

Fig. 2. Rice leaf disease images.

4.2   Confusion Matrix

A confusion matrix is a tool used to evaluate the performance of a classification model. It is a $l \times l$ matrix that displays the counts of True-positives (T_p), False-positives (F_p), True-negatives (T_n), and false-negatives (F_n) based on the model's predictions. From this matrix, several key performance metrics can be derived, such as accuracy, precision, F_1-score, and recall. It provides a clear view of how well the model is performing, highlighting areas of strength and where errors occur in the classification process.

$$\mathbf{Accuracy} = \frac{T_p + T_n}{(T_p + T_n + F_p + F_n)} \tag{5}$$

$$\mathbf{Precision} = \frac{T_p}{(T_p + F_p)} \tag{6}$$

$$\mathbf{Recall} = \frac{T_p}{(T_p + F_n)} \tag{7}$$

$$\mathbf{F_1 - Score} = \frac{2 \times Recall \times Precision}{(Recall + Precision)}. \tag{8}$$

4.3   Classification Report

The classification report provides a comprehensive overview of a model's performance, featuring metrics, and support for each class. It is a valuable tool for assessing the effectiveness of trained models on test data. Accuracy, derived from the confusion matrix, is calculated as the ratio of correct predictions to the total number of predictions. However, models can occasionally attain decent accuracy by relying on irrelevant or background features during training. Activation maps, as illustrated in Fig. 3, can help identify and analyze such cases. These maps offer insights into how effectively the models focus on relevant features, enabling a deeper understanding of their performance. The Rice Precision Table 3 indicates that Neck Blast Leaf Disease achieves high performance across all models, while Leaf Blast consistently shows lower results on all models. Similar trends are observed in the Rice Recall and F_1-score metrics, as reflected in Table 3. For this evaluation, the confusion matrix was based on 230 test images, distributed

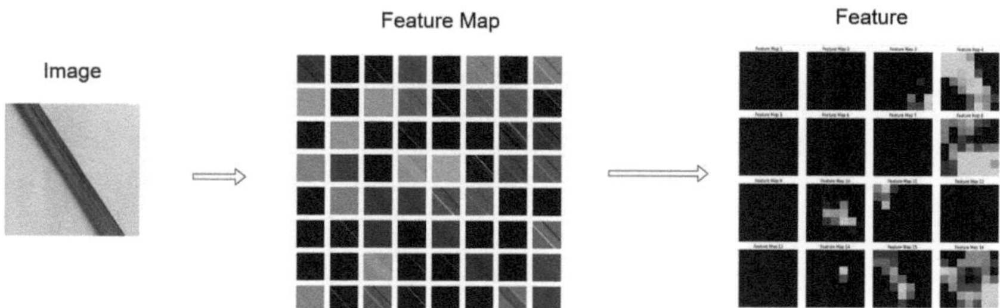

Fig. 3. Rice leaf disease visualize activation map.

Table 3. Rice precision, recall, and F_1-score

Leaf Diseases	Precision	Recall	F_1-Score
Brown spot	92%	93%	92%
Healthy	93%	92%	93%
Leaf Blast	91%	92%	91%
Neck Blast	100%	100%	100%

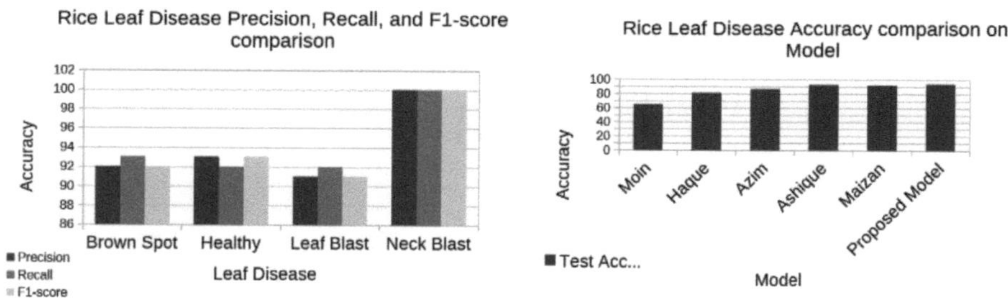

Fig. 4. Model precision, recall, F1-score, and accuracy comparison graph.

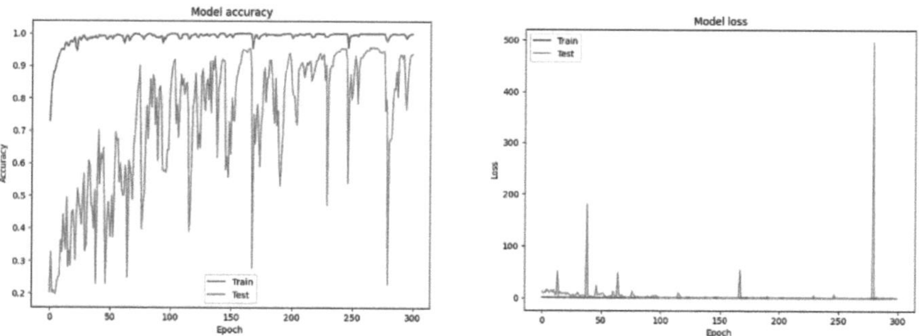

Fig. 5. Proposed model accuracy and loss graph.

as follows: 54 images of Brown Spot Disease, 72 Healthy images, 48 Leaf Blast Disease images, and 56 Neck Blast Disease images. Figure 4 visualizes the classification report, highlighting these comparative results. Our Proposed models are compared with other models in Table 4. Figure 4 graph shows that our proposed model has good compared to other models, which were N. B. Moin [24] with 64.30%, M. E. Haque [25] with 81.00%, M. A. Azim [26] with 86.58%, M. A. Islam [27] with 92.68%, M. A. I. Maizan [20] with 92.00% accuracy and proposed model MobileNet-ResNet Concate Model with 94.00% test accuracy and 99.50% training Accuracy. Figure 5 illustrates a graph of the accuracy and loss for the MobileNet-ResNet concate model. The accuracy graph demonstrates a steady increase in model accuracy, which is consistently maintained throughout training. Similarly, the loss graph shows a gradual decrease, stabilizing as training progresses. However, the test accuracy graph reveals a slower rate of improvement compared to the training accuracy graph, and the test loss graph exhibits a slower decline, reflecting a more gradual convergence.

Table 4. Model accuracy comparison

S.No	Model	Datasets	Test Accuracy
1.	N. B. Moin [24]	Bangladeshi crop	64.30%
2.	M. E. Haque [25]	Bangladeshi crop	81.00%
3.	M. A. Azim [26]	Bangladeshi crop	86.58%
4.	M. A. Islam [27]	Bangladeshi crop	92.68%
5.	M. A. I. Maizan [20]	Bangladeshi crop	92.00%
6.	Proposed model	Bangladeshi crop	94.00%

4.4   Ablation Study of Proposed Model with Heat-Map

The MobileNetV2 and ResNet50 concate models leverage the strengths of both architectures to achieve a blend of speed and performance. MobileNetV2, known for its lightweight and efficient design, utilizes depth-wise separable convolutions and an inverted residual structure, allowing for faster computations with fewer parameters. On the other hand, ResNet50 contributes robustness through its deep residual layers, effectively capturing complex features with its skip connections. By concatenating the outputs of MobileNetV2 and ResNet50, the model combines MobileNetV2's speed and low latency with ResNet50's deep feature extraction capabilities. This fusion results in a model that balances accuracy and efficiency, making it ideal for scenarios requiring both fast inference and high-performance classification, especially on diverse datasets with intricate patterns.

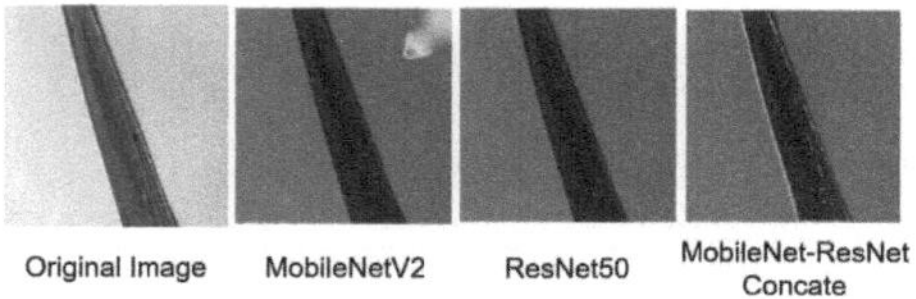

Fig. 6. Heat-Map output at last layer.

Figure 6 displays the heat map from the final layer of the proposed MobileNet-ResNet Concate model. Unlike MobileNetV2 and ResNet50, which also identify background elements as features, the proposed model focuses exclusively on relevant features, resulting in improved classification accuracy. The heat map comparison reveals that the MobileNet-ResNet Concate model extracts a higher density of meaningful features compared to MobileNetV2 and ResNet50. This enhanced feature extraction contributes to the development of a more robust and effective model.

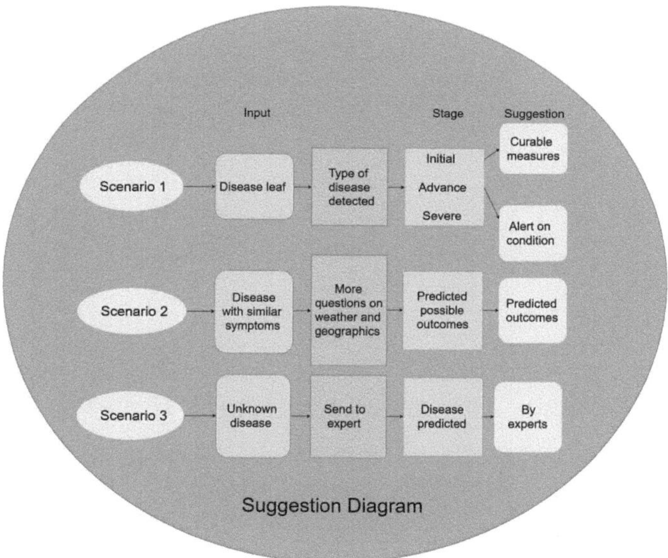

Fig. 7. Suggestion diagram.

4.5   Case Study

Integrating computer vision in future autonomous farms [28] offers a promising approach to reducing crop loss and enhancing agricultural productivity. Utilizing CNN-based models for disease detection in food grains such as maize, wheat, sugarcane, and rice can significantly aid in early diagnosis and effective treatment. A plant protection app, trained on specialized datasets, could diagnose diseases, provide curative suggestions, and recommend pesticide dosages based on disease severity. The app could also distinguish between similar symptoms by collecting additional information about weather and disease location, improving the accuracy of its predictions. Furthermore, it could offer targeted advice for small-scale farmers who may not have access to expert consultation, suggesting preventative measures and treatments. The system's ability to learn from user interactions and expert feedback would enable it to evolve, enhancing its diagnostic accuracy and recommendation quality over time, and experts, as seen in Fig. 7.

Fig. 8. Shadow effect on disease image.

5 Conclusion

The study emphasizes that transfer learning models, particularly ResNet50, MobileNetV2, and MobileNet, are highly effective in detecting diseases in rice crops, a significant challenge in modern agriculture. These pre-trained models demonstrated substantial improvements in classification accuracy, recall, F_1-score, and precision when compared to traditional methods. Building on these results, we proposed a novel hybrid MobileNet-ResNet concate model with 99.50% training accuracy, 94.00% test accuracy, precision, Recall, and F1-score. This model leverages the strengths of transfer learning while incorporating a specialized deep-learning architecture tailored for agricultural disease identification. The future scope of this model is a step toward deep learning IC (Integrated Classification) for powerful classification using Transfer learning models, like electronic IC in the electronic field. Accurate disease detection is closely linked to color texture in images. In normal light conditions, deep color textures remain consistent without creating shadows, whereas lighter textures are more prone to shadows. This study highlights the importance of capturing disease images under maintained conditions that minimize shadowing, as shadows can lead to incorrect disease predictions. These observations are illustrated in Fig. 8. However, the current datasets are insufficient for a comprehensive analysis of crop diseases, underscoring the need for more robust data collection.

Each disease dataset should include multiple stages of categorization—such as initial, middle, and advanced stages—to accurately represent the progression of the disease. This categorization is crucial because effective crop treatment strategies often depend on the specific stage of the disease. Future datasets should therefore be organized to reflect these disease stages, enabling more precise and actionable insights for disease management and treatment.

References

1. Gülmez, B.: Advancements in rice disease detection through convolutional neural networks: a comprehensive review. Heliyon **10**, e33328 (2024)
2. Sethy, P.K., Barpanda, N.K., Rath, A.K., Behera, S.K.: Image processing techniques for diagnosing rice plant disease: a survey. Procedia Comput. Sci. **167**, 516–530 (2020)
3. Burgos-Artizzu, X.P., Ribeiro, A., Tellaeche, A., Pajares, G., Fernández-Quintanilla, C.: Analysis of natural images processing for the extraction of agricultural elements. Image Vis. Comput. **28**(1), 138–149 (2010)
4. Xu, P., Wu, G., Guo, Y., Yang, H., Zhang, R.: Automatic wheat leaf rust detection and grading diagnosis via embedded image processing system. Procedia Comput. Sci. **107**, 836–841 (2017)
5. Tian, K., Li, J., Zeng, J., Evans, A., Zhang, L.: Segmentation of tomato leaf images based on adaptive clustering number of k-means algorithm. Comput. Electr. Agric. **165**, 104962 (2019)
6. Prakash, J.S., Vignesh, K.A., Ashok, C., Adithyan, R.: Multi class support vector machines classifier for machine vision application. In: 2012 International Conference on Machine Vision and Image Processing (MVIP), pp. 197–199. IEEE (2012)
7. Brahimi, M., Boukhalfa, K., Moussaoui, A.: Deep learning for tomato diseases: classification and symptoms visualization. Appl. Artif. Intell. **31**(4), 299–315 (2017)
8. Shijie, J., et al.: Automatic detection of tomato diseases and pests based on leaf images. In: 2017 Chinese Automation Congress (CAC), pp. 2537–2510. IEEE (2017)
9. Jiang, F., Lu, Y., Chen, Y., Cai, D., Li, G.: Image recognition of four rice leaf diseases based on deep learning and support vector machine. Comput. Electron. Agric. **179**, 105824 (2020)
10. Sethy, P.K., Barpanda, N.K., Rath, A.K., Behera, S.K.: Deep feature based rice leaf disease identification using support vector machine. Computers and Electronics in Agriculture 175, 105527 (2020)
11. Tenório, G.L., et al.: Comparative study of computer vision models for insect pest identification in complex backgrounds. In: 2019 12th International Conference on Developments in eSystems Engineering (DeSE), pp. 551–556. IEEE (2019)
12. Cruz, A., et al.: Detection of grapevine yellows symptoms in Vitis vinifera L. with artificial intelligence. Comput. Electron. Agric. **157**, 63–76 (2019)
13. Udayananda, G.K.V.L., Shyalika, C., Kumara, P.P.N.V.: Rice plant disease diagnosing using machine learning techniques: a comprehensive review. SN Appl. Sci. **4**(11), 311 (2022)
14. Diseases - IRRI Rice Knowledge Bank. http://www.knowledgebank.irri.org/step-by-step-production/growth/pests-and-diseases/diseases. Accessed 11 Nov 2021
15. Bari, B.S., et al.: A real-time approach of diagnosing rice leaf disease using deep learning-based faster R-CNN framework. PeerJ Comput. Sci. **7**, e432 (2021)

16. Trivedi, N.K., et al.: Early detection and classification of tomato leaf disease using high-performance deep neural network. Sensors **21**(23), 7987 (2021)
17. Rallapalli, S.M., Saleem Durai, M.A.: A contemporary approach for disease identification in rice leaf. Int. J. Syst. Assur. Eng. Manag., 1–11 (2021). https://doi.org/10.1007/s13198-021-01159-y
18. Deng, R., et al.: Automatic diagnosis of rice diseases using deep learning. Front. Plant Sci. **12**, 701038 (2021)
19. Roy, A.M., Bhaduri, J.: Real-time growth stage detection model for high degree of occultation using DenseNet-fused YOLOv4. Comput. Electron. Agric. **193**, 106694 (2022)
20. Mizan, M.A.I., Ahmed, M., Ali, M.H.: Bangladeshi crop disease detection using convolutional neural network. In: 2024 6th International Conference on Electrical Engineering and Information
21. Jain, A.K., Sharma, S., Nain, N.: Multi-disease detection and classification of sugarcane leaves using transfer learning. In: 2024 11th International Conference on Soft Computing & Machine Intelligence (ISCMI), Melbourne, Australia, pp. 1–5. IEEE (2024)
22. Tensorflow. https://www.tensorflow.org/. Accessed 25 Nov 2024
23. Kaggle. Rice leaf dataset. [Dataset link] (2024). https://www.kaggle.com/datasets/anandjain1112/rice1982
24. Moin, N.B., Islam, N., Sultana, S., Chhoa, L.A., Howlader, S.R.K., Ripon, S.H.: Disease detection of Bangladeshi crops using image processing and deep learning-a comparative analysis. In: 2022 2nd International Conference on Intelligent Technologies (CONIT), pp. 1–8. IEEE (2022)
25. Haque, M.E., Rahman, A., Junaeid, I., Hoque, S.U., Paul, M.: Rice leaf disease classification and detection using YOLOv5. arXiv preprint: arXiv:2209.01579 (2022)
26. Azim, M.A., Islam, M.K., Rahman, M.M., Jahan, F.: An effective feature extraction method for rice leaf disease classification. Telkomnika (Telecommunication Computing Electronics and Control) **19**(2), 463–470 (2021)
27. Islam, M.A., Shuvo, M N.R., Shamsojjaman, M., Hasan, S., Hossain M.S., Khatun, T.: An automated convolutional neural network based approach for paddy leaf disease detection. Int. J. Adv. Comput. Sci. Appl. **12**(1) (2021)
28. Joseph, D.S., Pawar, P.M., Chakradeo, K.: Real-time plant disease dataset development and detection of plant disease using deep learning. IEEE Access **12**, 16310–16333 (2024)

From Lab to Field: Robustness Analysis of Plant Disease Recognition Models in Uncontrolled Environment

Nitika Nigam[1(✉)], Nandit Sharma[1], Sourav Yadav[1], Rajeev Nath Tiwari[1], and Vinod Kumar[2]

[1] UPES, Dehradun, Uttarakhand, India
nitika.nigam@ddn.upes.ac.in
[2] IIT(BHU), Varanasi, U.P., India
vinodkumar.rs.cse18@itbhu.ac.in

Abstract. Plant disease recognition (PDR) is vital for early detection and control of crop diseases, which affect agricultural productivity. However, factors like lighting, noise, motion blur, and background clutter can degrade model performance in uncontrolled environments. This paper presents a framework to assess the robustness of PDR models under these challenging conditions. The key contributions of this work are as follows: (1) the creation of a real-world dataset, PlantVillage-U, comprising plant images captured under diverse environmental conditions, including noise and blur variations, to mimic real agricultural settings; (2) a comparative evaluation of transformer-based models versus CNN-based models, revealing their respective capabilities and limitations in recognizing plant diseases; and (3) an in-depth analysis of the impact of pre-training on model performance, focusing on the adaptability of pre-trained models to new datasets under uncontrolled conditions. The results show that robust architectures and proper augmentation improve model performance in uncontrolled environments, with significant gains over baseline models. The dataset includes a diverse array of crops such as apple, blueberry, cherry, grape, orange, peach, pepper, potato, raspberry, soy, squash, strawberry and tomato, encompassing both common and complex plant diseases.

Keywords: Deep Learning · Noise · Plant disease recognition · Robustness · Uncontrolled environment

1 Introduction

1.1 Importance of Plant Disease Recognition (PDR)

Agriculture is a vital component of human society since it supplies food and fosters development. Crops and plant leaves, on the other hand, are susceptible to a variety of diseases throughout the course of cultivation. Farmers' traditional methods to recognize diseases on plant leaves tend to be time-consuming and

C. Modi et al. (Eds.): MIND 2024, CCIS 2736, pp. 246–257, 2026.
https://doi.org/10.1007/978-3-032-14531-4_21

subject to human error. As a result, fast and precise PDR plays an important role in reducing the likelihood of subsequent plant damage [7]. Moreover, early detection also allows for effective management strategies, reducing crop loss and ensuring food safety. By understanding and recognizing symptoms farmers can take timely action to protect their yields and maintain healthy food supplies.

1.2 Advancements in Deep Learning (DL) Models and Challenges in Real-World Conditions

In recent years, DL models have achieved remarkable accuracy in tasks like image classification, detection, and recognition, exceptionally when trained and evaluated in controlled settings [9,11,20,26]. Building on this success, most PDR models [16,22] have been developed using carefully curated datasets consisting of high-quality images captured under ideal conditions. These datasets are typically free from environmental noise and variability, allowing the models to perform efficiently and accurately within controlled environments. However, real-world conditions present a different challenge. Images captured in farms, greenhouses, or other agricultural settings often suffer from: a) Random pixel variations (known as **Noise**) due to environmental factors such as lighting, camera quality, or sensor issues. b) **Motion Blur** which is introduced by plant movement (e.g., wind) or camera shake during image capture. c) **Varying Illumination** which is caused by changing light conditions, shadows, or reflections on plant surfaces. d) The presence of irrelevant objects or textures in the background can confuse recognition models, also known as **Background Clutter** and e) images captured at a distance or with low-resolution devices lose critical details necessary for accurate classification. As shown in Fig. 1, a DL model trained to identify diseases in plants is deployed in a real-world farm setting. Due to poor weather conditions and low lighting, the model misinterprets shadows cast on healthy leaves as signs of a fungal infection. The incorrect classification of healthy plants as diseased highlights the vulnerability of the model to environmental changes.

1.3 Objective and Contributions

This paper addresses the challenges faced by PDR models when deployed in uncontrolled, real-world agricultural environments. It aims to develop a framework for evaluating and improving the robustness of these models against environmental variations like noise, blur, and fluctuating lighting conditions. These factors degrade the performance of models that work well in controlled environments but struggle in the field. The paper makes key contributions toward enhancing the accuracy of PDR models in such challenging settings which are as follows:

1. **Real-world Dataset Collection**: We have compiled a diverse, real-world dataset of plant images captured under varying environmental conditions, named **PlantVillage-U**. This dataset includes images with naturally occurring noise, motion blur, occlusions, and lighting variations, providing a realistic representation of the challenges faced in uncontrolled environments.

2. **Evaluation and Comparison of Robust Architectures:** A comparative evaluation of transformer-based models versus CNN-based models is conducted, highlighting their respective strengths and weaknesses in handling diverse plant disease images.
3. **Effect and Impact of Pre-training:** The impact of pre-training on model performance is examined, focusing on how pre-trained models adapt to new plant disease datasets and generalize across different conditions.
4. Our experimental results demonstrate that models enhanced with robust architectures and proper data augmentation exhibit substantial improvements over baseline models.

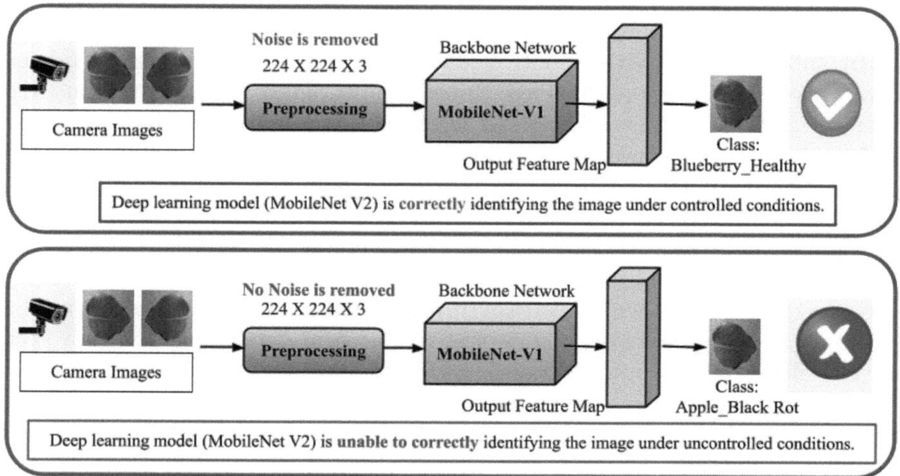

Fig. 1. Comparison of DL model performance under controlled and uncontrolled conditions. In the first row, MobileNet-V1 correctly identifies the plant disease under controlled conditions. However, in the second row, the model's accuracy drops significantly in an uncontrolled environment, where factors like environmental noise, variable lighting, and motion blur hinder its performance.

2   Background and Literature Review

- **DL for PDR:** Recent advancements in plant disease detection have increasingly relied on DL techniques, with several studies exploring their effectiveness in diverse agricultural applications. Joseph et al. (2024) [14] developed tailored datasets for rice, wheat, and maize and applied eight fine-tuned models, including Xception and MobileNet, achieving high accuracy rates; however, the use of controlled environment images limited the models' generalization to real-world scenarios. Similarly, Khalid et al. (2024) [15] utilized CNN and MobileNet architectures, reporting a 96% accuracy for early disease detection, yet the models faced challenges in handling unseen data, highlighting the importance of robustness in field applications. Ahmad et al. (2024) [1], in their review of machine learning and DL methods for plant disease classification, emphasized the efficacy of CNN-based models but pointed out the persisting issue of dataset limitations and the need for improved model generalization across complex environments. Complementing these findings, Kumar et al. (2024) [17] explored the use of ResNet50 for classifying 38 plant diseases, where data augmentation techniques enhanced performance, though scalability remained an obstacle in noisy, real-world conditions. Finally, Qadri et al. (2024) [19] highlighted the potential of CNNs in IoT-based plant disease detection systems, underscoring their effectiveness in image segmentation tasks, yet they also acknowledged challenges in deploying these models in diverse environmental settings due to dataset variability and field conditions. Together, these studies underscore the progress made in DL for plant disease detection while also identifying key areas that require further research, particularly in model robustness and real-world deployment.

- **Existing Work on Model Robustness:** Research on improving the robustness of image classification models has seen notable developments in addressing challenges like noise, adversarial attacks, and real-world perturbations. Li et al. (2020) [18] proposed Wavelet Integrated CNNs, which improved noise resistance by replacing traditional pooling with wavelet transforms, although their effectiveness under other distortions like blur is not explored. Similarly, Arcaini et al. (2020) [2] worked on enhancing CNN robustness to input errors such as blurring through data augmentation, but their study did not cover adversarial robustness. Building on these efforts. Bhojanapalli et al. (2021) [3] analyzed the robustness of Vision Transformers (ViTs), which outperformed CNNs in noisy conditions due to their self-attention mechanisms, though their resilience under distortions like blur required further investigation. Ghosh et al. (2021) [8] also contributed by examining CNNs' resilience to noise and blur, but they emphasized the need for further research on more complex perturbations such as occlusions. In subsequent work, Zhou et al. (2022) [28] explored the robustness of transformers under adversarial attacks, demonstrating their effectiveness against specific attacks but noting limited generalization to broader perturbations. Around the same time, Chen et al. (2023) [4] introduced a quantum CNN (QCNN) that leveraged quantum feature extraction to enhance robustness against adversarial attacks, though scalability challenges limited its broader application. Xue et al. (2024) [27] tackled the accuracy-robustness trade-off by applying adversarial training with feature diversity, improving robustness while observing a slight reduction in overall accuracy.

More recent research has expanded on these foundations. De Nart et al. (2024) [6] evaluated models for leaf image classification trained across different locations and found that models trained on single-location datasets performed poorly on external datasets, highlighting the need for better generalization. Similarly, Sarkar et al. (2024) [21] developed a benchmark generation framework that introduced customizable distortions such as noise and blur to assess robustness, although they primarily focused on adversarial scenarios. Together, these works illustrate the ongoing progress in improving model robustness, with a clear need for further exploration into real-world generalization of natural perturbations.

3   Research Questions

This section outlines the key research questions guiding the investigation into PDR models, particularly focusing on their deployment in uncontrolled environments. The research questions are as follows:

RQ1. How can the robustness of PDR models be comprehensively measured when deployed in uncontrolled environments?

RQ2. How does the performance of different DL models, such as CNN-based architectures (ResNet, MobileNet, etc.) and Transformers, degrade under real-world conditions compared to controlled settings?

4   Variations of DNN

We conducted experiments on nine diverse PDR models built on CNN and Transformer frameworks. The goal is to compare different backbone architectures and analyze the resilience of both CNN and Transformer-based models in accurately identifying plant diseases. Specifically, we assessed ten prominent CNN models that are renowned for their exceptional performance in PDR. The models we incorporated in our study include various architectures spanning CNN and Transformer-based frameworks. Among the CNN models, we used DenseNet-169 [12], Inception model like Inception-V3 [24], MobileNet variant like MobileNet-V1 [10]; and ResNet variant ResNet-50 [9]. Additionally, we employed ResNeXt-50 model [26]. Further, we integrated SE architectures [11] like SENet-154 and SEResNet-50. Transformer-based models have recently achieved remarkable success in numerous vision-related tasks. In this work, we incorporate Transformer network to evaluate its robustness in PDR, focusing on assessing their effectiveness in challenging and uncontrolled conditions. Our study includes model Transformer-DEIT-T [25]. Lastly, we also included the Xception model [5].

5   Robustness Benchmarks and Evaluation

• **Publicly Available Datasets:** In this study, we utilized the PlantVillage dataset [13], to evaluate DL models for PDR. The dataset contains 54,305 images across 38 classes namely apple, blueberry, cherry, grape, orange, peach, pepper, potato, raspberry, soy, squash, strawberry and tomato, encompassing both common and complex plant diseases, including both healthy and diseased leaves, representing various plant species. The dataset is organized into training and validation folders, each containing 38 subfolders corresponding to different plant classes.

- **Dataset Creation:** To simulate real-world conditions and enhance the robustness of PDR models, a new dataset named **PlantVillage-U** is created. This dataset builds upon the original PlantVillage dataset by incorporating various noise types and blur effects to better reflect the challenges of uncontrolled environments. The goal is to improve model generalization to noisy and distorted inputs, making the trained models more resilient when deployed in real-world agricultural settings. The dataset is augmented with various noise types, including Brownian, Gaussian, Impulse, Periodic, Rayleigh, Speckle, Shot, and Uniform noise. Several blur effects are applied, including Anisotropic, Bilateral, Defocus, Gaussian, Lens, Median, Motion, and Radial blur.

- **Implementation Details:** In this study, the input images are resized to a standard resolution of 224×224 pixels for consistency across model. Various data augmentation techniques are employed to enhance data variability and promote model robustness. These included transformations such as RandomResizedCrop and RandomHorizontalFlip, along with normalization using mean values of [0.485,0.456,0.406] and standard deviations of [0.229,0.224,0.225]. Multiple types of noise are applied to simulate real-world conditions and improve generalization to noisy and distorted inputs, including Brownian, Gaussian, Impulse, Periodic, Rayleigh, Speckle, Shot, and Uniform noise. Additionally, various blur effects such as Anisotropic, Bilateral, Defocus, Gaussian, Lens, Median, Motion, and Radial blur are incorporated, as given in [23]. These augmentations aimed to mimic the challenges posed by uncontrolled environments, making the model more resilient to noise and distortions encountered in practice (Fig. 2).

Fig. 2. Flowchart depicting methodology of creating **PlantVillage-U** dataset.

- **Evaluation Metrics.** We have measured the performance of the model using the accuracy metric. **Accuracy** measures the proportion of correctly classified instances (both positive and negative) out of all instances.

6 Results and Observations

- **Results on Gaussian Noise:** We analyzed the performance of various architectures with transfer learning (T.L) and without transfer learning (N.T.L) on clean data (**C**) and data perturbed (**P**) with Gaussian noise, as shown in Table 1 and Table 2. The following observations are explained below:

1. **Impact of T.L**: Across all models, T.L generally improves accuracy on **C** compared to training from scratch. MobileNet-V1's accuracy on **C** data improves from 92.63% to 96.32% with T.L, and ResNeXt-50 from 97.37% to

98.95%. T.L also reduces the accuracy drop caused by **P**, indicating better robustness. MobileNet-V1 under Gaussian noise (Level 1), the accuracy drop in case of N.T.L is 40.52%, which significantly reduces to 13.69% with T.L.

2. **Performance of CNN-based Models:** In clean conditions, CNN models such as DenseNet-169, ResNet-50, and SEResNet-50 demonstrate strong performances, with accuracies as high as 98.95% when using T.L. However, CNN-based models are generally more sensitive to Gaussian noise perturbations. For example, ResNet-50 shows a significant accuracy drop of 45.79% with T.L under Level 1 noise and a 33.68% drop in case of N.T.L.

3. **Performance of Transformer-based Models:** The transformer-based model (DeiT-T) is consistently more robust to noise perturbations than the CNN models. It shows only a 1.58% accuracy drop under Gaussian Noise (Level 1) for N.T.L and an 8.42% drop with T.L. Although its accuracy on **C** data is slightly lower than that of top-performing CNN models, its robustness to perturbations makes it a superior choice in scenarios where image quality may be compromised.

4. **Effect of Noise Level:** As the level of Gaussian noise increases (from Level 1 to Level 4), all models exhibit a more significant accuracy drop in both the cases. However, the degree of this drop varies across architectures. MobileNet-V1 suffers a 40.52% drop under Level 1 noise and an even more significant drop under Level 4. At the same time, DeiT-T maintains relatively low drops (1.58% for Level 1 and 8.42% for Level 4), demonstrating its better resilience to noise.

5. **Model-specific Observations:** ResNeXt-50, SENet-154, and SEResNet-50 show strong performance on **C** data, but their accuracy drops significantly under noise.On the other hand, models like Transformer DeiT-T retain high performance and robustness under noisy conditions.

Table 1. Accuracy of various Architectures with T.L and N.T.L for clean data (**C**) and perturbation (**P**) (Gaussian Noise-Level 1). The "Delta" columns indicate the accuracy drop caused by the perturbation.

Models	Without T.L		Delta Without T.L	With T.L		Delta With T.L
	C	P		C	P	
DenseNet-169 [12]	96.84%	65.79%	31.05%	96.84%	66.84%	30.00%
Inception-V3 [24]	94.74%	76.32%	18.42%	91.05%	71.05%	20.00%
MobileNet-V1 [10]	92.63%	52.11%	40.52%	96.32%	82.63%	13.69%
ResNet-50 [9]	**97.89%**	64.21%	33.68%	98.42%	52.63%	45.79%
ResNeXt-50 [26]	97.37%	74.74%	22.63%	**98.95%**	79.47%	19.48%
SENet-154 [11]	96.84%	76.32%	20.52%	98.42%	82.11%	16.31%
SEResNet-50 [11]	**97.89%**	74.21%	23.68%	**98.95%**	44.21%	54.74%
Transformer DeiT-T [25]	95.26%	**93.68%**	1.58%	96.84%	**88.42%**	8.42%
Xception [5]	94.74%	73.68%	21.06%	93.68%	63.68%	30.00%

Table 2. Accuracy of various Architectures with T.L and N.T.L for clean data (**C**) and perturbation (**P**) (Gaussian Noise-Level 4).

Models	Without T.L		Delta Without T.L	With T.L		Delta With T.L
	C	P		C	P	
DenseNet-169 [12]	96.84%	26.32%	70.52%	96.84%	23.16%	73.68%
Inception-V3 [24]	94.74%	27.89%	66.85%	91.05%	23.68%	67.37%
MobileNet-V1 [10]	92.63%	21.05%	71.58%	96.32%	30.53%	65.79%
ResNet-50 [9]	**97.89%**	22.63%	75.26%	98.42%	24.21%	74.21%
ResNeXt-50 [26]	97.37%	28.42%	68.95%	**98.95%**	24.21%	74.74%
SENet-154 [11]	96.84%	22.63%	74.21%	98.42%	34.21%	64.21%
SEResNet-50 [11]	**97.89%**	21.58%	76.31%	**98.95%**	15.26%	83.69%
Transformer DeiT-T [25]	95.26%	**76.32%**	**18.94%**	96.84%	**51.05%**	**45.79%**
Xception [5]	94.74%	31.58%	21.06%	93.68%	25.26%	68.42%

Table 3. Accuracy of various Architectures with T.L and N.T.L for clean data (**C**) and perturbation (**P**) (Gaussian Blur-Level 1).

Models	Without T.L		Delta Without T.L	With T.L		Delta With T.L
	C	P		C	P	
DenseNet-169 [12]	96.84%	84.74%	12.1%	96.84%	88.42%	8.42%
Inception-V3 [24]	94.74%	83.68%	11.06%	91.05%	83.16%	7.89%
MobileNet-V1 [10]	92.63%	83.68%	8.95%	96.32%	86.32%	10.00%
ResNet-50 [9]	**97.89%**	88.42%	9.47%	98.42%	89.47%	8.95%
ResNeXt-50 [26]	97.37%	91.58%	5.79%	**98.95%**	88.95%	10.00%
SENet-154 [11]	96.84%	88.42%	8.42%	98.42%	92.11%	6.31%
SEResNet-50 [11]	97.89%	92.11%	5.78%	**98.95%**	85.79%	13.16%
Transformer DeiT-T [25]	95.26%	**94.74%**	**0.52%**	96.84%	**94.74%**	2.10%
Xception [5]	94.74%	86.84%	7.90%	93.68%	84.21%	9.47%

Table 4. Accuracy of various Architectures with T.L and N.T.L for clean data (**C**) and perturbation (**P**) (Gaussian Blur-Level 4).

Models	Without T.L		Delta Without T.L	With T.L		Delta With T.L
	C	P		C	P	
DenseNet-169 [12]	96.84%	66.32%	30.52%	96.84%	81.58%	15.26%
Inception-V3 [24]	94.74%	64.21%	30.53%	91.05%	64.74%	26.31%
MobileNet-V1 [10]	92.63%	69.47%	23.16%	96.32%	69.47%	26.85%
ResNet-50 [9]	**97.89%**	67.89%	30.00%	98.42%	78.42%	20.00%
ResNeXt-50 [26]	97.37%	66.32%	31.05%	**98.95%**	78.42%	20.53%
SENet-154 [11]	96.84%	68.42%	28.42%	98.42%	78.95%	19.47%
SEResNet-50 [11]	97.89%	72.11%	25.78%	**98.95%**	76.84%	22.11%
Transformer DeiT-T [25]	95.26%	**93.16%**	**2.10%**	96.84%	**84.21%**	**12.63%**
Xception [5]	94.74%	73.16%	21.58%	93.68%	74.21%	19.47%

- **Results on Gaussian Blur:** In this section, we analyzed the performance of various architectures with T.L and N.T.L on **C** and **P** with Gaussian blur, as shown in Table 3 and Table 4. The following observations are explained below:

1. **T.L Impact:** T.L improves performance in most models, reducing the accuracy drop caused by Gaussian Blur (Level 1). For example, DenseNet-169's Delta decreases from 12.1% without T.L to 8.42% with T.L.
2. **Transformer Resilience:** Transformer DeiT-T displays remarkable resilience to Gaussian Blur, with the lowest Delta (0.52%) without T.L and a slight increase (2.1%) with T.L. This demonstrates that the Transformer architecture is less affected by blur than CNN-based models.
3. **Top Performers Without T.L:** ResNet-50, ResNeXt-50, and SEResNet-50 stand out with high accuracy on **C**(97.89%, 97.37%, 97.89%, respectively) and relatively lower accuracy drops (9.47%, 5.79%, 5.78%) compared to other CNNs.
4. **Model Sensitivity to Blur:** Models like Xception and MobileNet-V1 show a moderate decrease in performance with Deltas of 7.9% and 8.95% without T.L. However, with T.L, MobileNet-V1's Delta slightly increases to 10%, showing that T.L does not always guarantee better robustness.
5. **CNN vs Transformer:** CNN models generally experience larger accuracy drops compared to Transformer DeiT-T. This suggests that CNN models, although performing well on **C**, are more sensitive to Gaussian Blur than Transformer.
6. **Overall Trend:** The overall trend reveals that while T.L helps most models maintain high accuracy on **C**, its effect on robustness to blur varies across architectures, with Transformers showing the best resilience.

- **Visual Results:** We include the visual results comparing the accuracy of various models under different conditions, as shown in Fig. 3. The graph shows that N.T.L tends to show larger drop in accuracy when faced with perturbations compared to their T.L counterparts. Additionally, models tend to handle **Gaussian Noise** better than **Gaussian Blur**, particularly at higher levels (L4). **ResNet-50** and **ResNeXt-50** have relatively high accuracy across all conditions, including perturbations, indicating they are more robust to noise and blur. **MobileNet-V1** shows a significant drop in accuracy, particularly N.T.L under uncontrolled environment. **Transformer DeiT-T** performs quite well under Gaussian Noise and Blur conditions with T.L, suggesting that Transformers might handle certain perturbations better than CNN models.

- **Quantitative Results:** We present the performance evaluation of Transformer, DenseNet, and ResNet50 on the PlantVillage-U dataset, which includes images affected by Gaussian noise at varying levels to simulate real-world conditions where image quality may be compromised. As shown in the Fig. 4, the Transformer model with T.L consistently outperforms the other models in terms of prediction accuracy under noisy conditions. The Transformer with T.L accurately predicted the actual labels of plant leaves, even in the presence of noise, demonstrating robustness and generalization capabilities. These attributes make

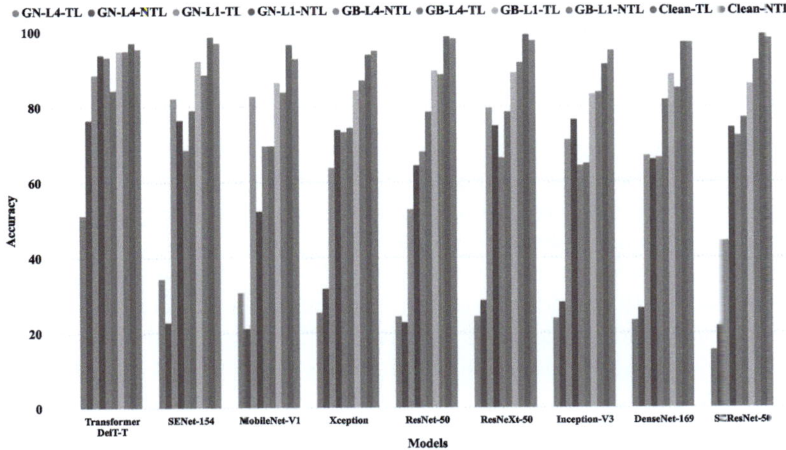

Fig. 3. The comparison shows accuracy across models under Gaussian Noise (GN) and Gaussian Blur (GB) at Levels 1 and 4. GN-L4-TL and GB-L4-TL indicate Level 4 with T.L., while GN-L4-NTL and GB-L4-NTL represent the same for N.T.L., with similar notation for Level 1.

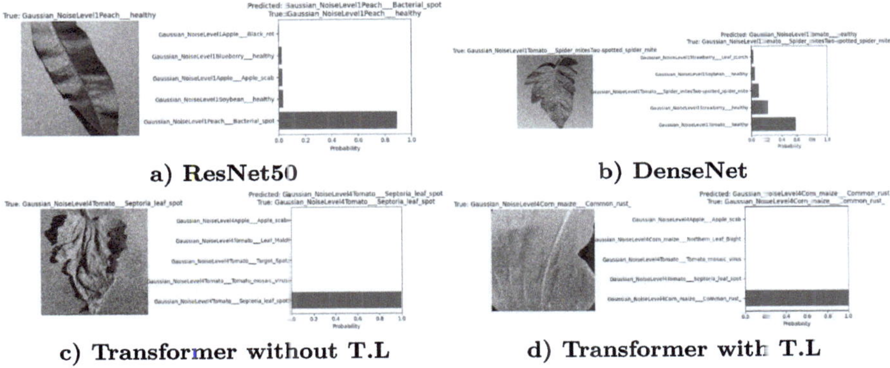

a) ResNet50

b) DenseNet

c) Transformer without T.L

d) Transformer with T.L

Fig. 4. Performance of Transformer, DenseNet, and ResNet50 models on PlantVillage-U dataset. Each subfigure shows the true and predicted labels with probability distributions for different disease and healthy classes under Gaussian noise.

the Transformer with T.L a strong candidate for real-world deployment in plant disease detection systems, where environmental factors such as image quality degradation could significantly impact performance.

7   Conclusion and Future Work

This paper presents a comprehensive framework to address this challenge by systematically measuring the robustness of PDR models when deployed under such adverse conditions. A real-world dataset containing images affected by noise, blur, and lighting variations is collected, and a benchmarking suite is introduced to evaluate models across different levels of degradation. The experimental results demonstrate that robust architectures, when combined with effective data augmentation techniques and noise reduction strategies, can significantly improve model performance in uncontrolled environments, showing substantial gains over baseline models trained on controlled datasets. The findings indicate that current PDR models struggle to handle the inherent variability of environmental conditions in real-world field scenarios. Even with a meticulously curated dataset containing thousands of samples, achieving satisfactory model robustness for practical field applications remains challenging. Future work will focus on several key areas to further enhance the robustness of PDR models. Integrating advanced techniques such as domain adaptation and self-supervised learning could help the models generalize to unseen environments better.

Acknowledgements. We are grateful to UPES, Dehradun, Uttarakhand, India, for their infrastructure and access to GPU resources, which are critical in completing the experiments of this study.

References

1. Ahmad, A., Saraswat, D., El Gamal, A.: A survey on using deep learning techniques for plant disease diagnosis and recommendations for development of appropriate tools. Smart Agric. Technol. **3** (2023)
2. Arcaini, P., Bombarda, A., Bonfanti, S., Gargantini, A.: Dealing with robustness of convolutional neural networks for image classification. In: 2020 IEEE International Conference On Artificial Intelligence Testing (AITest) (2020)
3. Bhojanapalli, S., Chakrabarti, A., Glasner, D., Li, D., Unterthiner, T., Veit, A.: Understanding robustness of transformers for image classification. In: Proceedings of the IEEE/CVF international conference on computer vision (2021)
4. Chen, G., Long, S., Yuan, Z., Li, W., Peng, J.: Robustness and explainability of image classification based on QCNN. Quant. Eng. **2023**(1) (2023)
5. Chollet, F.: Xception: deep learning with depthwise separable convolutions. In: Proceedings of the CVPR (2017)
6. De Nart, D., et al.: Vine variety identification through leaf image classification: a large-scale study on the robustness of five deep learning models. J. Agric. Sci. **162** (2024)
7. Demilie, W.B.: Plant disease detection and classification techniques: a comparative study of the performances. Journal of Big Data **11**(1), 5 (2024)
8. Ghosh, S., Shet, R., Amon, P., Hutter, A., Kaup, A.: Robustness of deep convolutional neural networks for image degradations. In: 2018 IEEE International Conference on Acoustics, Speech and Signal Processing (ICASSP) (2018)
9. He, K., Zhang, X., Ren, S., Sun, J.: Deep Residual Learning for Image Recognition. In: Proceedings of the CVPR (2016)
10. Howard, A.G.: Mobilenets: efficient convolutional neural networks for mobile vision applications. arXiv preprint arXiv:1704.04861 (2017)

11. Hu, J., Shen, L., Sun, G.: Squeeze-and-excitation networks. In: Proceedings of the CVPR (2018)
12. Huang, G., Liu, Z., van der Maaten, L., Weinberger, K.Q.: Densely connected convolutional networks. In: Proceedings of the CVPR (2017)
13. Hughes, D., Salathé, M., et al.: An open access repository of images on plant health to enable the development of mobile disease diagnostics. arXiv preprint arXiv:1511.08060 (2015)
14. Joseph, D.S., Pawar, P.M., Chakradeo, K.: Real-time plant disease dataset development and detection of plant disease using deep learning. IEEE Access (2024)
15. Khalid, M.M., Karan, O.: Deep learning for plant disease detection. Int. J. Math. Stat. Comput. Sci. **2** (2024)
16. Khanna, M., Singh, L.K., Thawkar, S., Goyal, M.: PlaNet: a robust deep convolutional neural network model for plant leaves disease recognition. Multimedia Tools and Applications **83**(2), 4465–4517 (2024)
17. Kumar, M., Arora, A., Deb, A., Yadav, A.L.: Deep learning for accurate plant disease classification using resnet50: a comprehensive approach. In: 2024 International Conference on Computational Intelligence and Computing Applications (ICCICA), vol. 1 (2024)
18. Li, Q., Shen, L., Guo, S., Lai, Z.: Wavelet integrated CNNs for noise-robust image classification. In: Proceedings of the IEEE/CVF conference on computer vision and pattern recognition (2020)
19. Qadri, S.A.A., Huang, N.F., Wani, T.M., Bhat, S.A.: Advances and challenges in computer vision for image-based plant disease detection: a comprehensive survey of machine and deep learning approaches. IEEE Trans. Autom. Sci. Eng. (2024)
20. Recht, B., Roelofs, R., Schmidt, L., Shankar, V.: Do ImageNet classifiers generalize to ImageNet? In: Proceedings of the ICML, pp. 5389–5400 (2019)
21. Sarkar, S., et al.: Benchmark generation framework with customizable distortions for image classifier robustness. In: Proceedings of the IEEE/CVF (2024)
22. Shovon, et al.: Plantdet: a robust multi-model ensemble method based on deep learning for plant disease detection. IEEE Access **11** (2023)
23. Siddiqui, S.A., Agne, S., Dengel, A., Ahmed, S., et al.: Are deep models robust against real distortions? a case study on document image classification. In: 2022 26th International Conference on Pattern Recognition (ICPR), pp. 1628–1635. IEEE (2022)
24. Szegedy, C., Vanhoucke, V., Ioffe, S., Shlens, J., Wojna, Z.: Rethinking the inception architecture for computer vision. In: Proceedings of the CVPR (2016)
25. Vaswani, A., et al.: Attention is All you Need. In: Advances in Neural Information Processing Systems, vol. 30 (2017)
26. Xie, S., Girshick, R., Dollar, P., Tu, Z., He, K.: Aggregated Residual Transformations for Deep Neural Networks. In: Proceedings of the CVPR (2017)
27. Xue, W., et al.: Towards a better trade-off between accuracy and robustness for image classification via adversarial feature diversity. IEEE J. Miniaturization Air Space Syst. (2024)
28. Zhou, D., et al.: Understanding the robustness in vision transformers. In: International Conference on Machine Learning (2022)

Knuckle Classification Through Transfer Learning and Deep Features

Ayushi Mathur[ID] and Ritesh Vyas[(✉)][ID]

Department of Information and Communication Technology, School of Technology,
Pandit Deendayal Energy University, Gandhinagar, Gujarat, India
`ritesh.vyas@sot.pdpu.ac.in`

Abstract. Knuckle classification has witnessed enormous progression in the last decade. The use cases of knuckle features to identify perpetrators of serious crime and for human authentication in consumer applications have paved new directions of research. The knuckle creases formed on the distal interphalangeal (DIP), proximal interphalangeal (PIP) and meta-carpo phalangeal (MCP) joints of hands possess unique properties. The capability of popular deep learning models to extract the discerning features of knuckle regions needs more experimentation. Hence, this research emphasizes on how different deep learning models perform in identifying the finger knuckle patterns and textures using an extensive dataset comprising of the base (or MCP), major (or PIP) and minor (or DIP) knuckles of all four fingers explicitly. We evaluated the results on the ResNet50, ResNet101, and DenseNet201 models. To calculate the loss metrics and accuracy of each model, it was first trained and then put to test. The outcomes of our evaluation show that these models achieve significant results on various kinds of knuckles and fingers. The best test case was found to be the major knuckle of the middle finger when trained with the DenseNet201 model with 75.38% accuracy.

Keywords: Knuckle Biometrics · ResNet50 · ResNet101 · DenseNet201

1 Introduction

Biometrics is becoming increasingly popular these days because it simplifies, secures, and reduces fraud from identity verification and authentication. Biometrics has two types: physiological Biometrics, which uses an individual's physical traits like their face, fingerprint, iris, or palm, and behavioral biometrics [12], which uses traits like their stride, signature, or voice. The common biometric traits include face, iris, fingerprints, palm, and many others with their strength and weaknesses [13]. The persistence and distinctiveness of knuckle patterns on finger joints have recently brought attention to its study.

As the patterns in the Knuckle regions of fingers are distinct and intricate, knuckle Biometrics has become an optimistic modality in the field of personal

C. Modi et al. (Eds.): MIND 2024, CCIS 2736, pp. 258–266, 2026.
https://doi.org/10.1007/978-3-032-14531-4_22

identification. There are three types of knuckles for each finger that is base, major, and minor with complex patterns and textures unique to every human being [14]. These patterns provides huge information that can be used to identify an individual with fingers properly or to extract features in the biometric process.

In the identification process of a person knuckle texture and creases which are visible from the dorsal view of hand plays a very important role. Three finger knuckles on which the knuckles form are the base, major, and minor which are called as metacarpophalangeal (MCP), proximal interphalangeal (PIP), and distal interphalangeal (DIP) in biological terms [18]. Figure 1 shows the nomenclature for knuckles. Also, the thumb is not included in the current work which only contains MCP, current research only looks at the knuckles from the four fingers. Over the past ten years, knuckle patterns have been widely employed in biometric applications.

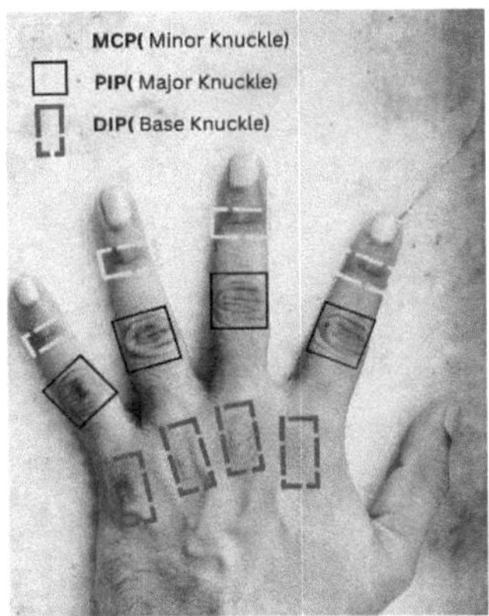

Fig. 1. MCP, PIP, and DIP joints shown in a dorsal view of a human hand.

Many Datasets have been created in this trait for the ease of researchers for better research with diverse wavelengths to capture good quality knuckle patterns from images. One of the universally used datasets of knuckle biometrics is the The Hong Kong Polytechnic University Contactless Finger Knuckle Images Database (Version 3.0) where images are captured under environmental light conditions using a handheld camera [2]. Another significant dataset is The IIT Delhi Finger Knuckle Dataset (Version 1.0) in which images are captured under visible light conditions [1]. These datasets help researchers to develop new algorithms in knuckle biometrics with different wavelengths such as visible light, NIR which allows a combined evaluation of the system under various conditions.

This paper presents the first-of-its-kind examination of the base, major, and minor knuckle patterns of the index, middle, ring, and little fingers to provide a comprehensive analysis for knuckle biometric recognition. We evaluate the ability of deep learning algorithms to correctly identify people based on their knuckle patterns by utilizing the models ResNet101, ResNet50, and DenseNet201. We are evaluating the training accuracy for all the knuckles of four fingers in each model followed by comparative analysis. From this comparative analysis, we have successfully extracted the unique features of knuckle images followed by the classification using the deep learning models.

The rest of the paper is organized as follows: Section 2 discusses the recent growth in the field of Knuckle biometrics, and Section 3 elaborates on the used methodology and the dataset. Section 4 presented the results and the related discussion, followed by conclusions in Section 5.

2   Prior Work

Alghamdi et al. [4] designed a person identification based on fingernails and dorsal knuckles (PIFK+), an improved model of automated person identification through hand images by highlighting the dorsal side. The framework of the hand described 19 parts of the hand such as nails and different joints of the fingers. The fine-tuning was done with the DenseNet201 model and the feature extraction was done with Bray-Curtis similarity. It also assisted in labeling of components and increases the bounding box on recognized component as well. The results on the 11k and PolyU datasets proved that knuckle patterns and fingernails were relevant to the identification. Evaluation of the result, using the defined metrics, the 11k Hands dataset received significantly better rank-1 accuracy for left hand and fingernail with 100% and PolyU dataset outperformed for thumb fingernail.

Cheng et al. [6] discussed a deep technique for contactless 3D finger knuckle process of identification for various issues like fewer training samples and more intra class variations. The method focused on the deep features at multiple scales to give a better representation and an efficient method of aligning fingers. The results in two sessions included about 200 subjects and were made available for public use for the first time. This paper, showed better performance than the state-of-the-art techniques, for instance giving around 22 % improvement on GAR at extremely low False acceptance rate under challenging situations.

Tarawneh et al. [16] proposed a Finger Knuckle Pattern(FKP) recognition using the VGG-19 style deep learning and the features were from the fully connected layers F6 and F7 and were combined followed by Principal Component Analysis (PCA) to minimize the data feature size. In this paper, variants of the classifiers were applied to the IIT Delhi Finger Knuckle Dataset. The best identification performance was obtained by an Artificial Neural Network classifier on the principal components of the averaged feature-s extracted from F7 and F6, while maintaining 95 % of the data's contention. That being said, the study revealed that using the described method, EfficientNet performed a little better if it is compared with the latest approaches. These were the constrains in that only one dataset was used and only one method of dimensionality reduction was applied.

Chlaoua et al. [7] introduced a new method in which, deep learning was used to develop the multi-modal biometric system based on the images of Finger

Knuckle Print (FKP) modalities using features extracted from them via Network (PCANet) method known as principal component analysis. It was used to learn two stages of filter banks, the block histograms and simple binary hashing and for clustering, the feature vectors were adopted as an input for classification by linear multiclass Support Vector Machine (SVM). Also, the author explained further work in using PCANet with significantly larger databases or with other biometric modalities including hand geometry, iris, palmprint, etc. and also merged deep learning techniques like Convolutional Neural Networks (CNN), and Deep Belief Networks (DBN), etc.

Heidari et al. [10] proposed an innovative human authentication technique from multiple grab Finger Nail (FN) and FKP traits by using deep learning. Performance assessment of the proposed method was done using a hand dorsal image dataset containing 1090 images (where 10 images of each of the 109 subjects) that were preceded by the detection of hand skin process with the denoising method used for the extraction of finger knuckle and the finger nail. The CNN was utilized with AlexNet as pretrained model to concatenate the features by means of normalization and fusion techniques. The system got an average of 94 % level of identification accuracy. Other %ages were, on average, 75 % which proved its efficiency, its ability to survive in the real word, and its practical use.

Cheng et al. [5] introduced the first database of 3D finger knuckle images which were acquired using a photometric stereo method from 130 individuals. The main objective of this study was to enhance biometrics recognition of people by utilizing the 3D characteristics of the finger knuckle patterns. A new scheme for feature descriptor was developed to handle such 3D features, in much more efficient manner than was done earlier. The study also provided a new theoretical approach in explaining the singularity of such patterns. While this particular feature descriptor was aimed for 3D finger knuckle patterns, it can be easily applied to similar pattern-based other hand biometrics like palmprint and fingerprint, therefore showing its good performance in the identification of the human using 3D finger knuckle pattern.

In [8], authors discussed about the multimodal and unimodal recognition biometric systems based on the FKP and the FV using CNN. Features extraction was done from the two biometrics modalities, the pre-training CNN models including ResNet50, VGG16 and AlexNet were used. For the purpose of the recognition enhancement the use of classifiers SVM and Softmax that allow to increase the parameters quality.In [3], the authors have presented a new efficient descriptor for the identification of FKP. The proposed descriptor was termed as Magnitude, Directions and Local Patterns(MDL). It also consists of a multimodal learning scheme, it extracted both the MDL and deep features using VGG-F network which was ideal for FKP images. The experiments conducted on a public dataset revealed that the proposed method had achieved better results than the related methods and was also highly robust.

3   Methodology

Feature extraction and classification process is very important aspect of biometric systems as it helps in extracting the unique and distinct features of the biometric data that helps in identification of an individual. Knuckle biometrics comprise of very unique patterns and textures different for each individual. The quality and dependability of feature extraction are directly impacted by knuckle localization, making it a crucial stage in the biometric authentication process [17]. Individual identification is dependent on the distinctive patterns found in the knuckle regions. Precise localization guarantees that these areas are isolated from the backdrop with maximum precision, reducing unnecessary information and improving the quality of the characteristics that are extracted.

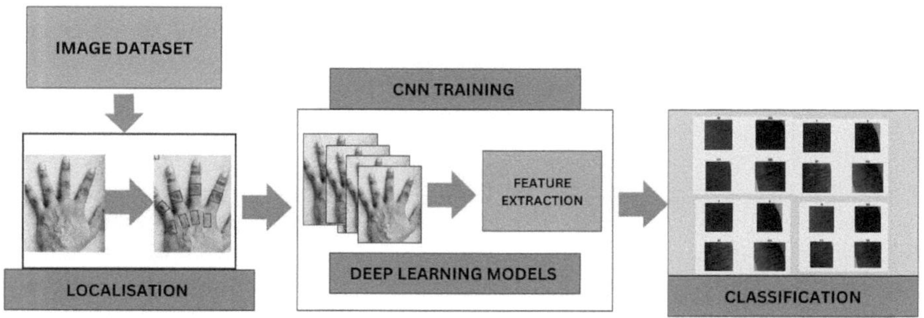

Fig. 2. Block Diagram of Proposed Methodology.

The dataset used in this study is the PolyU-HD Knuckle dataset [15] comprising high-resolution images with resolution of 1200×1600 pixels. The images in this dataset[1] are preprocessed to obtain the base, major, and minor knuckles of all four fingers (Index, Middle, Ring, Little). It comprises 12000 knuckle images (i.e. 12 knuckle images per hand dorsal image) from different groups of subjects with variations in skin, age, and gender and a total of 200 classes. This study examines the base, major, and minor knuckle patterns of the index, middle, ring, and little fingers to provide a comprehensive method for knuckle biometric recognition. To precisely segment the knuckle regions, the dataset images were subjected to localization in the first stage using the method given in [18]. After localization, we trained three deep learning models namely ResNet50, ResNet101, and DenseNet201. Using their deep architectures, these models were used to extract features and classify the knuckle images, capturing the fine details and complex patterns necessary for precise biometric identification. The whole process of the proposed framework is illustrated in Fig. 2.

Few important details of the proposed methodology are listed below:

---

[1] https://www4.comp.polyu.edu.hk/~csajaykr/knuckleV2.htm.

- Preprocessing: To make images appropriate for input into machine learning models, they were scaled to a standard dimension.
- Data Segregation: The dataset was separated into three sets: For training 70 %, validation 10 %, and 20 % for testing.
- Model Training: All three models ResNet50, ResNet101, and DenseNet201 respectively trained on the training set utilizing the ground truth.
- Model Evaluation: The trained model was computed on the validation set to optimize performance.
- Testing: In the end the model's performance the training accuracy and loss was evaluated on the test set to evaluate its effectiveness in Knuckle Biometrics.
- Optimizer: Adam (Adaptive Moment Estimation)
- Learning rate: 0.001
- Loss function: Cross-entropy loss
- Number of classes: 200

The deep learning models which are used in this study are briefly explained as follows:

- **ResNet50** - ResNet50 model is a deep convolutional model from the (ResNet) family. It has fifty layers total—one MaxPooling layer, one Average Pooling layer, and forty-eight convolutional layers. ResNet50 introduces the skip connections, which makes the network learn residual functions. As the network depth increases this architecture makes the gradients to flow through the network that actually prevents the performance loss. ResNet50 has achieved results on standards like ImageNet3 and is well known for its effectiveness in picture classification applications. It is a good option for our research on knuckle biometrics, where unique and thorough feature extraction is essential, due to its ability to capture complex patterns and characteristics.
- **ResNet101** - ResNet101 model improves the model's capacity by adding 101 layers to the architecture to extract unique and intricate features from the data. ResNet101 uses residual connections to help ease the vanishing gradient issue by facilitating gradient flow through the network. Because of its deep network architecture, it makes easy capture of minute and complex patterns in the data. ResNet101 deep network provides a greater capacity for feature extraction in comparison to other shallow networks [9]. To increase the precision and dependability of our biometric recognition system we leverage the depth of knuckle data to extract the detailed features using ResNet101 in our knuckle biometrics research.
- **DenseNet201** - The DenseNet201 model is a member of the DenseNet family, it is a deep convolutional neural network in which each layer is connected in a feed-forward manner which enhances the information flow and gradient propagation. It consists of 201 layers, with dense blocks and transition layers, further improving the computational efficiency and model performance. These properties make DenseNet201 effective for complex image recognition tasks, resulting in superior accuracy and efficiency. To improve the precision

and dependability of our biometric recognition system, we use DenseNet201's dense connections to extract unique and detailed features from knuckle images in our knuckle biometrics research [11].

4   Results and Discussions

We compared the performance of convolutional neural network (CNN) models on knuckle recognition over a range of knuckles (base, major, and minor) and fingers (index, middle, ring, and little) in order to give a thorough knowledge of the biometrics capabilities of finger knuckles. The accuracy scores for the ResNet50, ResNet101, and DenseNet201 models are displayed in the bar chart (Fig. 3). This visual representation highlights the variations in recognition accuracy, offering insights into the effectiveness of each model for different knuckle types and fingers. The chart serves as a crucial tool for identifying which combinations of knuckles and models yield the highest accuracy, thereby guiding future research and practical applications in finger knuckle biometrics. All the experiments were conducted on Apple MacBook Pro with Apple M2 Pro processor with 10 processor cores, 16 GB RAM and Sonoma 14.1.2 macOS operating system. The training time for the three employed models is 2.41 s, 7.09 s, and 7.97 s, respectively for the ResNet50, ResNet101 and DenseNet201 models.

After a thorough examination of the accuracy data and comparison analysis from several models, we discovered that the major knuckle consistently had the highest accuracy. Major Knuckle, Index finger achieved an accuracy of 68 % when trained on model ResNet50. Similarly, on the ResNet101 model Major Knuckle, Index Finger is giving an accuracy of 67.5 %. Also, we are getting the highest accuracy of 75.38 % on the Major Knuckle, middle finger trained on the DenseNet201 Model. This finding emphasises that compared to other knuckles, the major knuckle produces greater accuracy outcomes. The greater performance of the major knuckle in knuckle recognition can be attributed to its distinctive patterns and larger surface area. Figure 3 displays the comparative accuracy of the knuckle regions on several fingers.

ResNet50 yields notable accuracy values with the major knuckle, despite of being a shallow architecture. Hence, it can be a balanced approach to achieve decent performance with limited computational resources. Concurrently, ResNet101 provides comparable results as those from ResNet50. Moreover, it performs worse in multiple cases e.g. the major knuckle of ring finger and minor knuckles of little finger. On the other hand, a deeper architecture DenseNet201 gains significant improvements as compared to the other two ResNet models. The best accuracy demonstrated by the DenseNet201 model is with major knuckles of the middle finger, which is 19.08 % and 19.65 % improved as compared against the counterpart accuracy values from ResNet101 and ResNet50, respectively. Due to its improved feature propagation, DenseNet201 even performs relatively better with the base knuckles across all fingers.

The comparative analysis of Fig. 3 also reveals that major knuckles are the most suitable knuckles for human authentication, as they provide detailed and

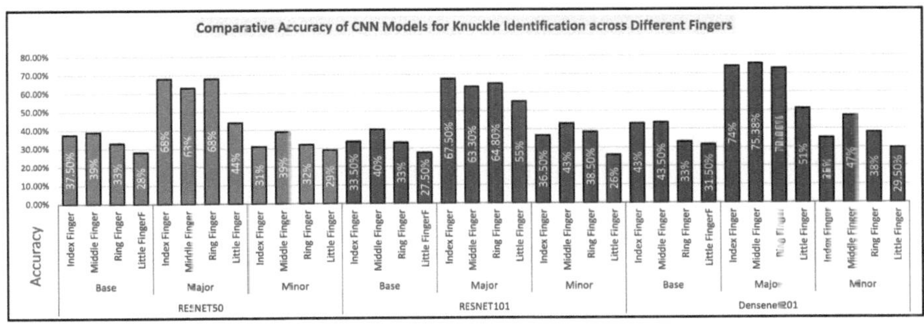

Fig. 3. Comparative Analysis of Accuracy of knuckle regions across different fingers.

large number of features, leading to improved performances. Whereas, the minor knuckles are inferior to the major knuckles in terms of the classification performance. This aligns with the visual anatomy of the minor knuckles as they contain less amount of creases. On the other hand, base knuckles are shown to be the least effective in overall matching performance. This low performance of base knuckles can be due to lack of primary creases and indistinct skin texture.

5   Conclusions

This work disseminates a thorough investigation of knuckle classification with different knuckle types and varying fingers, through transfer learning and deep features. The segmented subset of popular PolyU HD dataset is employed to conduct the experiments. Three deep learning models, namely ResNet50, ResNet101 and DenseNet201, were tested to measure the knuckle classification accuracy for multiple fingers and knuckle types. The results have shown that the major knuckle (or PIP) have outperforming potential to act as a biometric trait in human authentication as compared to minor (DIP) and base (MCP) knuckles. Moreover, DenseNet201 performed better than the other models, exhibiting the highest overall accuracy because of its dense connection topology, which improves gradient propagation and feature reuse. To improve recognition accuracy, more deep learning architectures may be integrated in future studies. Knuckle patterns and other biometric characteristics, like dorsal veins, combined may result in more reliable identification systems. Studying large-scale deployments can also help to solve real-world problems and increase system reliability.

Acknowledgement. The authors would like to sincerely thank the Visvesvaraya PhD Scheme from Ministry of Electronics and Information Technology (MEITY), Government of India (GoI), for providing the financial assistance to support the full-time PhD work.

References

1. IIT Delhi Finger Knuckle Database. https://www4.comp.polyu.edu.hk/~csajaykr/IITD/iitd_knuckle.htm, Accessed 30 Oct 2024
2. The Hong Kong Polytechnic University Finger Image Database, Version 3.0. https://www4.comp.polyu.edu.hk/~csajaykr/fn2.htm, Accessed 30 Oct 2024
3. Aiadi, O., et al.: Fusion of deep and local gradient-based features for multimodal finger knuckle print identification. Cluster Computing, pp. 1–17 (2024)
4. Alghamdi, M., Angelov, P., Alvaro, L.P.: Person identification from fingernails and knuckles images using deep learning features and the bray-curtis similarity measure. Neurocomputing **513**, 83–93 (2022)
5. Cheng, K.H., Kumar, A.: Contactless biometric identification using 3d finger knuckle patterns. IEEE Trans. Pattern Anal. Mach. Intell. **42**(8), 1868–1883 (2019)
6. Cheng, K.H., Kumar, A.: Deep feature collaboration for challenging 3d finger knuckle identification. IEEE Trans. Inf. Forensics Secur. **16**, 1158–1173 (2020)
7. Chlaoua, R., Meraoumia, A., Aiadi, K.E., Korichi, M.: Deep learning for finger-knuckle-print identification system based on pcanet and svm classifier. Evol. Syst. **10**(2), 261–272 (2019)
8. Daas, S., Yahi, A., Bakir, T., Sedhane, M., Boughazi, M., Bourennane, E.B.: Multimodal biometric recognition systems using deep learning based on the finger vein and finger knuckle print fusion. IET Image Proc. **14**(15), 3859–3868 (2020)
9. He, K., Zhang, X., Ren, S., Sun, J.: Deep residual learning for image recognition. In: Proceedings of the IEEE conference on computer vision and pattern recognition, pp. 770–778 (2016)
10. Heidari, H., Chalechale, A.: Biometric authentication using a deep learning approach based on different level fusion of finger knuckle print and fingernail. Expert Syst. Appl. **191**, 116278 (2022)
11. Huang, G., Liu, Z., Van Der Maaten, L., Weinberger, K.Q.: Densely connected convolutional networks. In: Proceedings of the IEEE conference on computer vision and pattern recognition, pp. 4700–4708 (2017)
12. Jain, A.K., Nandakumar, K., Ross, A.: 50 years of biometric research: Accomplishments, challenges, and opportunities. Pattern Recogn. Lett. **79**, 80–105 (2016)
13. Jain, A.K., Ross, A., Prabhakar, S.: An introduction to biometric recognition. IEEE Trans. Circuits Syst. Video Technol. **14**(1), 4–20 (2004)
14. Kazhagamani, U., Ezhilarasan, M.: Finger knuckle biometrics-a review. Comput. Electr. Eng. **10** (2014)
15. Kumar, A., Xu, Z.: Personal identification using minor knuckle patterns from palm dorsal surface. IEEE Trans. Inf. Forensics Secur. **11**(10), 2338–2348 (2016)
16. Tarawneh, A.S., Hassanat, A.B., Alkafaween, E., Sarayrah, B., Mnasri, S., Altarawneh, G.A., Alrashidi, M., Alghamdi, M., Almuhaimeed, A.: Deepknuckle: Deep learning for finger knuckle print recognition. Electronics **11**(4), 513 (2022)
17. Vyas, R., et al.: Robust end-to-end hand identification via holistic multi-unit knuckle recognition. In: 2021 IEEE International Joint Conference on Biometrics (IJCB), pp. 1–8. IEEE (2021)
18. Vyas, R., Williams, B.M., Rahmani, H., Boswell-Challand, R., Jiang, Z., Angelov, P., Black, S.: Ensemble-based bounding box regression for enhanced knuckle localization. Sensors **22**(4), 1569 (2022)

IAD-Fed: Internet Addiction Diagnosis from EEG Signals Using Federated Learning in Edge Environments

G. Anantha Krishnan[1], John Paul Martin[2]([✉]), Christina Terese Joseph[2], and Tom Kurian[2,3]

[1] Rajagiri School of Engineering and Technology, Kochi, Kerala, India
[2] Indian Institute of Information Technology, Kottayam, Kottayam, Kerala, India
johnpaul@iiitkottayam.ac.in
[3] Amal Jyothi College of Engineering, Kottayam, Kerala, India

Abstract. The broad adoption and proliferation of the internet has led to issues such as Internet Addiction (IA). IA has become one of the main problems influencing the mental state and social well-being of an individual. The traditional way of diagnostics, which is based on standard questionnaires, has little use in many cases and sometimes even demonstrates inexpressive objectiveness and reliability. This paper aims at identifying IA by studying Electroencephalography (EEG) signal patterns and adopting Convolutional Neural Networks (CNN) model for effective interpretation. Moreover, to ensure the privacy of the data and for real-time diagnosis we extend the CNN model using Federated Learning (FL) and deploy the model on constrained edge devices. Findings show encouraging accuracy in IA diagnosis by a CNN model surpassing common classifiers. This is further strengthened by Raspberry Pi 4 deployment on edge devices which makes it possible for real-time diagnosis as well. Relying on edge computing reduces the overload in the central server which improves the response time and computation burden. The findings of this research will adjoin the groundwork for scalable and privacy-preserving diagnostic methods for IA, hence providing the basis for personalized interventions and prevention strategies in mental healthcare.

Keywords: Federated Learning · Internet Addiction · EEG · Edge Computing

1 Introduction

The widespread availability of the internet has brought immense benefits, but also invoked the concern of Internet Addiction (IA), impacting individuals mental health and social well-being. Although a universally agreed-upon definition and diagnostic criteria of internet addiction stands still, latest studies state that the global prevalence is approximately 6%, with significant variations across countries [10]. The younger generation (13–17 years old) is usually thought of as being more vulnerable. Studies show a very high rate of IA of 73% among youths and 71% among young adults (18–24 years old) [1]. Smartphone utilization is strongly tied to IA and has recorded and continued stratospheric growth.

Also, Internet Addiction has become a threat to multiple mental problems and disorders like anxiety, ADHD, and depression, indicating its criticality among the masses [16]. Traditional diagnostic methods for IA, primarily reliant on self-reported questionnaires, are prone to biases and inconsistencies. This necessitates the exploration of more objective and reliable approaches. However, early diagnosis in this circumstances proves a challenging task in routine clinical practice because of issues such as the long wait times for appointment or poor accessibility. It is this reality which makes the utilization of diagnostic models in the edge computing perspective very crucial. EEG holds promise as a valuable tool, as it captures the electrical activity within the brain, potentially reflecting neural signatures associated with IA [12]. Thus, this study aims to look into the feasibility of utilizing EEG signals along with CNNs for diagnosing IA. The research builds upon existing work demonstrating altered brain activity patterns in individuals with IA, in the context of impulsivity [13], and altered resting-state EEG in specific frequencies [6]. Additionally, the study draws inspiration from recent advancements in employing CNNs for accurate classification tasks in EEG data analysis [11,15].

This research finds the potential of this framework, not only for diagnosing IA, but also for finding out the use of Edge devices and FL in diagnosing and potentially managing IA. Edge computing brings the power of data processing near the requester's device so that immediate analysis or intervention is possible. While being beneficial in terms of real-time data analysis, edge computing may have issues regarding data access on such a device. However, this is where Federated Learning comes into play. The opportunity to gather and analyse the data in the edge opens the possibility to advance FL, enabling machine learning models in distributed devices to work without any invasion of users' privacy. When it comes to IA diagnosis, FL is likely to build more accurate and versatile models because of the use of a larger population dataset while individual patient information remains protected.

This study investigated the following research question: Can an IA diagnosis algorithm, derived from machine learning for distinguishing internet addicts from healthy individuals, accurately classifying the given IA subjects based on resting state EEG data recorded in an edge environment considering data privacy? The following are the most important contributions:

- The proposed CNN model for IA diagnosis achieved an average accuracy of 89.86% using resting-state EEG recordings.
- Federated Learning was incorporated to further increase this accuracy and enhance data security. Not only does this approach use distributed edge-device advantages but also ensures that sensitive data remains localized, reducing the risk of data breaches.

– The learning model is accurate, computationally efficient, highly robust, and
  enables the diagnosis of IA at a low cost when implemented on Edge Devices.

This paper continues with the organization as follows: Sect. 2 reviews already
existing methodologies for Internet Addiction using EEG. Section 3 offers a
detailed description of the dataset used and its format. Section 4 elaborates on
the methodologies used for diagnosing internet addiction outlined in this paper,
and Sect. 5 discusses the results obtained. It ends with Sect. 6, the conclusion.

2   Literature Review

Various psychometric measures have been utilized in studies examining internet addiction. One commonly used measure is Young's Internet Addiction Test
(IAT) [22]. The IAT consists of twenty items for self-evaluation, that address
loss of control, disregard of daily responsibilities and networking, importance of
behaviors and thoughts, negative consequences, escape and deceit. The test categorizes individuals based on their scores, with scores between 70–100 indicating
significant problems and scores between 40–69 suggesting frequent problems.
One should note that different cut-off scores were employed in different investigations, thus making methodological limitations for comparisons between the
studies difficult.

2.1   EEG and Internet Addiction

Existing research has established connections between altered brain activity patterns observed in EEG data and internet addiction.

Demin (2021) [8] investigated brain activity patterns in adolescents with different levels of internet addiction risk. The study found that adolescents with
higher internet addiction risk exhibited distinct EEG patterns compared to those
with lower risk. Lee et al. (2014) [14] explored the link between baseline EEG
activity and concurrent depression in individuals with internet addiction. They
discovered that those with both internet addiction and depression exhibited distinct EEG profiles compared to those with only internet addiction. Qi et al.
(2022) [19] explored the connection between low-frequency EEG activity and self-
control in people with internet addiction disorder. They discovered that those
with this disorder demonstrated decreased low-frequency (delta and theta) EEG
activity. These findings suggest that EEG data holds potential as a biomarker
for diagnosing IA.

2.2   EEG in Edge Environment

Recent advancements in edge computing have opened up new possibilities for
leveraging EEG data in healthcare applications, including the diagnosis of IA.
Qiu et al. (2022) [20] proposed an algorithm for the recognition of epileptic
seizures using EEG signal which was made possible through distributed edge

computing, demonstrating the potential for edge-based processing of EEG data for healthcare applications, while the challenge lies in ensuring reliable and secure data processing at the edge. Similarly, in the study by Yan et al. (2023) [21], a lightweight approach is proposed to extract features from the EEG signal while ensuring that patients' privacy is protected and local computations are done on the edge. Whereas the study by Idrees and Khlief (2023) [9] used fog computing to examine the viability and efficiency of EEG data compression techniques to minimize the amount of data transferred in Internet of Medical Things (IoMT) networks, which could enhance the practicality of edge-based IA diagnosis, with the advantage of reducing the bandwidth requirements for transmitting EEG data, though the challenge is to maintain the diagnostic accuracy of EEG data after compression.

2.3   Federated Learning for IA Using EEG

Recent advancements in federated learning offer promising alternatives. Federated learning allows training models on distributed datasets residing on individual devices, addressing privacy concerns and potentially enabling broader participation. Ju et al. (2020) [11] explored federated transfer learning for EEG signal classification, demonstrating its potential for leveraging data from diverse sources while preserving privacy. Additionally, Liu et al. (2024) [15] investigated using federated learning to enhance BCI performance, highlighting its potential for applications involving EEG data. While not directly addressing IA diagnosis, studies by Baghersalimi et al. (2023, 2021) [4] explored utilizing federated learning for epileptic seizure detection, demonstrating its potential for healthcare applications involving EEG data. Gupta et al. (2023) and Qin et al. (2023) presented applications of federated learning in diagnosing other mental health conditions using different modalities, showcasing its broader applicability. Cai et al. (2022) investigated domain adaptation techniques for EEG classification, which could be relevant for federated learning settings where data might come from diverse sources.

3   Data Description and Preprocessing

The study uses the LEMON Dataset, covering the Leipzig Mind-Brain-Body research database (MPILMBB) [3,17]. This LEMON Database provides a source of 202 resting-state EEG recordings collected in healthy human volunteers. Its diverse participant pool, spanning young adults (20–35 years) and older adults (59–77 years), provides a rich foundation for exploring age-related variations in brain activity patterns associated with internet use. Furthermore, the implementation of standardized procedures throughout data collection ensures high reliability and reproducibility, critical factors in EEG research.

Each EEG recording encompasses 16 multiple segments where participants alternate between eight closed and open eyes. The recording utilized high-density 62-channel system equipment. This had 61 scalp electrodes in position, which would record the electrical activity of the various parts of the brain Further, a vertical electrooculogram (VEOG) electrode was placed so as to control any eye movements, which might, during data analysis, have been mainly important in the rejection of artifacts. The sampling frequency of registered signals was specified to be equal to 2500 Hz.

The internet addiction severity in both samples was assessed with the scoring of the reliable Internet Addiction Test (IAT) using high coefficients of Cronbach's Alpha (α). From the IAT scores and other records from the EEG, it will be possible to try and derive possible neural markers which might be related to internet addiction.

In order to have a clear distinction between groups, a two-step process of selection was used. 202 EEG recordings were compared with 214 valid scores in the first step. This resulted in a pool of 107 participants who had both valid EEG data and valid IAT results. To draw a line between the subjects with and without internet addiction (IA), from there, sorted out the current subjects and finally selected 24 participants with the highest IAT (indicating potential IA) and 24 participants with the lowest IAT (indicating no to very minimal IA) scores.

The final dataset used in this study comprises 48 participants, consisting of 24 healthy controls and 24 online users. This group was further divided into two sets: a training set of 39 participants (80%) and a test set of 10 participants (20%) for algorithm training and evaluation. Stratification was used for the data partitioning process in order to achieve a fair class distribution between both the training and validation sets. It ensures that the classifier observes enough examples of each class to learn well from the training data.

3.1 Data Preprocessing

The signal is downsampled to 250 Hz. Additionally, an 8th-order Butterworth band-pass filter with a passband ranging from 1 to 45 Hz was applied. Furthermore, to minimize the influence of visual distractions on the EEG signal, only the "Eyes-Closed" state recordings were focused. Also, the cutting out of 24 channels with poor quality and inconsistent data from the EEG recordings were done through visual inspection, leaving the recordings with a total of 37 channels and 8 min long in duration.

A sliding window lasting 8 s was utilized, with a 2-second overlap to enrich the samples in the dataset. Since the signal was sampled at 250 Hz, a window size of 2000, indicating sampled data points over 8 s, along with an overlap of 500, indicating the number of sampled data points for 2-s overlap, were used (Fig. 1).

When analyzing brain activity using EEG signals, it's common to notice non-stationary shifts or drifts that can make it hard to see the real brain activity. To deal with this a method called detrending was used to clean up the data [7]. This method works on each channel of the data separately, treating them like individual columns, to remove any straight-line trends that might be there. After removing the trends, the data was normalized by applying standardization.

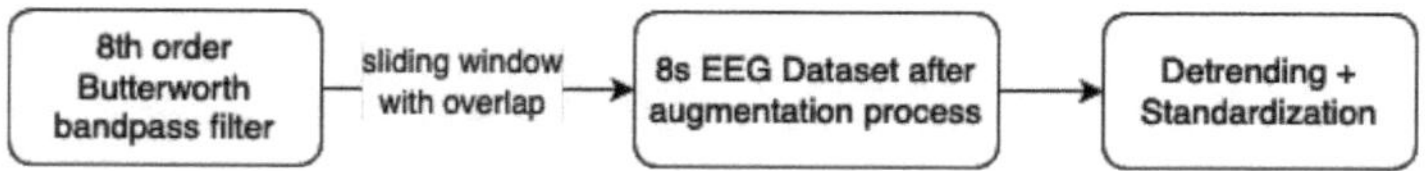

Fig. 1. Preprocessing pipeline for the EEG signal

This makes sure that all the features have a similar scale(zero mean and unit variance), which helps improve how the model works and makes it easier to compare features that might have had different scales before [18]. Basically, this scaler figures out the average and spread of the data when the model is prepared, and then it adjusts the data accordingly.

As a result of the detailed preprocessing that captured precise time-related information, a thorough investigation of how brain activity changes in different situations were obtained.

4   Methodology

Constructed using Keras' sequential API, the CNN model consist of five convolutional layers. Each layer utilized filters of different sizes to capture a range of spatial features within the EEG data. To make the data more organized and highlight important details, max pooling layers were added right after each convolutional layer. This helps to effectively reduce the size of the dimensions.

This is followed by a fully connected layer for final classification and a Soft-Max classifier to determine the probability of belonging to a specific class as in Fig. 2. The initial three convolution kernels have sizes of (1, 5), and the remaining two have sizes of (3, 5). After each convolutional layer, a ReLU activation function is applied, and then a max-pooling layer is implemented.

The model was prepared for training by setting up an optimizer, a loss function, and a metric for tracking performance. The Adam optimizer was opted for its effectiveness in adjusting model parameters. Sparse categorical cross-entropy loss function was used and the performance of the model was measured by looking at its classification accuracy.

The model was fit using the assigned training data. To ensure the model performed well with new data, monitoring its performance was done using a validation split of 10%. This way, one could see how it handled unseen information and make any necessary adjustments. Benefit in this is to reduce overfitting, which occurs when the model picks up on the characteristics specific to the training data and fails to generalize well on new data.

4.1   Implementation on Edge Device

A Raspberry Pi 4 Model B (8 GB RAM) was chosen for this study. This single-board computer hosts a quad-core ARM processor clocked at 1.5 GHz with a memory of 8 GB LPDDR4; it satisfies the needs of the smaller neural networks

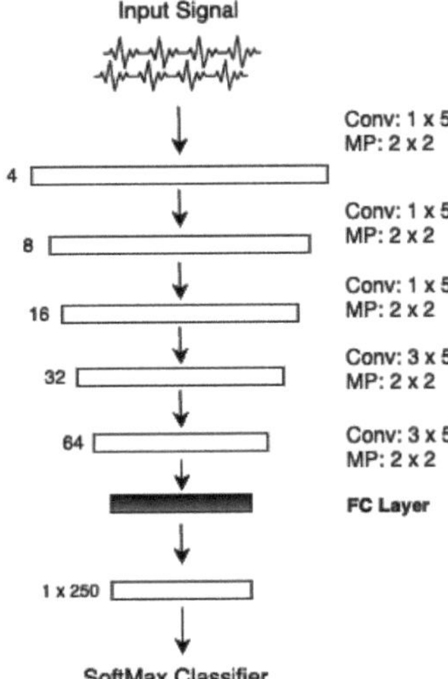

Fig. 2. Model Architecture

during training. Featuring Gigabit Ethernet and Wi-Fi connectivity, it facilitates data transfer and potential remote access.

With a typical power draw of 5–7 W, the Raspberry Pi 4 B offers an energy-efficient alternative to traditional desktop computers, making it suitable for resource-constrained environments and for extended operation periods. Furthermore, the diverse connectivity options, including Wi-Fi and Gigabit Ethernet, facilitate seamless integration into the federated learning framework, enabling efficient data sharing and model updates across multiple devices. It is also best suited for potential remote access.

For its deployment, the Keras model was converted to the TFLite format. TFLite variants are generally up to an order of magnitude smaller than the Keras versions, which is best suited for edge devices, achieved by using quantization. These small reductions would contribute to very considerable model size and space decrease which is beneficial for memory storage and device transfer.

4.2   Federated Learning Framework

The study also utilizes a Federated Learning (FL) framework for distributed EEG data classification with the support of the Raspberry Pi 4. The structure of the system includes a central server which holds client communications. In this instance, the clients are the Raspberry Pi (RPI) devices, each one has its own local EEG dataset.

Clients work with their individual EEG dataset to train their local models. This onsite training process will, therefore, individualize the model to the specific traits of the data observed on that specific device. The Global Module plays the essential role of the knowledge transfer between clients. This process is achieved by transferring the global modules' model weights across all clients from time to time. The central server leads this exchange by managing the combination of such weight updates. By incorporating knowledge gleaned from other clients' datasets through the global module, each client's model effectively benefits from an enlarged training set.

The research adopts Federated Averaging (FedAvg) as the central FL algorithm. FedAvg is a seasoned method of federated learning whereby participating devices learn local models on their private data and conversely transmit only the model update to a central server for aggregation [2]. Thus, it is less bandwidth intensive as compared to passing over whole datasets.

The implementation is based on open source Flower library [5], a FL framework that is part of the Tensorflow ecosystem. The framework simplifies the development and deployment of FL models, equipped with tools for managing client-server communication, model updates, and aggregation. This approach is apt on limited resources devices like Raspberry Pi. By leveraging distributed learning and knowledge sharing, the framework aims to realize improved accuracy of the classification and protect individual data confidentiality at the same time.

5   Results and Discussions

A total of three experiments were conducted using CNN, RF and SVM classifiers and the proposed CNN architecture showed the best results with an average accuracy of 89.86% along with a Precision of 0.90, Recall of 0.79 and F1 score of 0.84. In addition, the inference time for classification was also measured to evaluate the computational complexity of the proposed CNN architecture. As for the results of the model in the Laptop environment, the average time needed to complete the task was 0. 2 s. This is a factor showing that the model can handle more data within a short span of time, which is favorable in real-time scenarios.

The model performance has been compared and presented in Table 1, thus indicating the superiority of the proposed architecture (Fig. 3).

Table 1. Comparison on LEMON EEG Dataset

Research	ML Model	Dataset	Accuracy
Internet Addiction.[this work]	CNN	LEMON EEG Dataset	89.86%
Internet Addiction.[this work]	RF	LEMON EEG Dataset	81.79%
Internet Addiction.[this work]	SVM	LEMON EEG Dataset	73.49%
Internet Addiction.[15]	CNN	LEMON EEG Dataset	81.10%
Internet Addiction.[16]	SVM	LEMON EEG Dataset	82.50%

By deploying the TFLite model in Raspberry Pi 4 B the prediction is accurate and the results are obtained within 0.75 s, highlighting the feasibility of

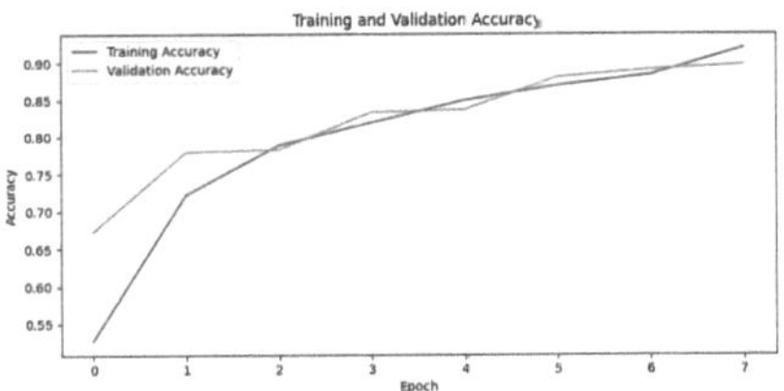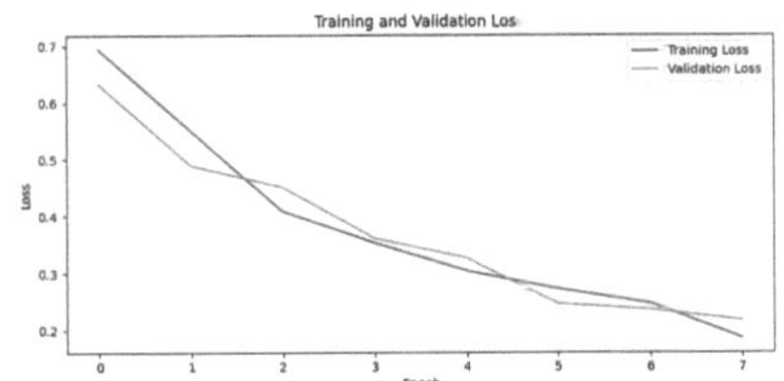

Fig. 3. Training and Validation results for CNN architecture: Accuracy and Loss

implementing the model on edge devices with limited computational resources. The deployed model with TFLite performed well in the confusion matrix and showed a clear distinction between the correctly and incorrectly classified samples for each class. Furthermore, the model obtained ROC AUC (Area Under the Curve) score of 0.97, showing that the model is an effective classifier (Fig. 4).

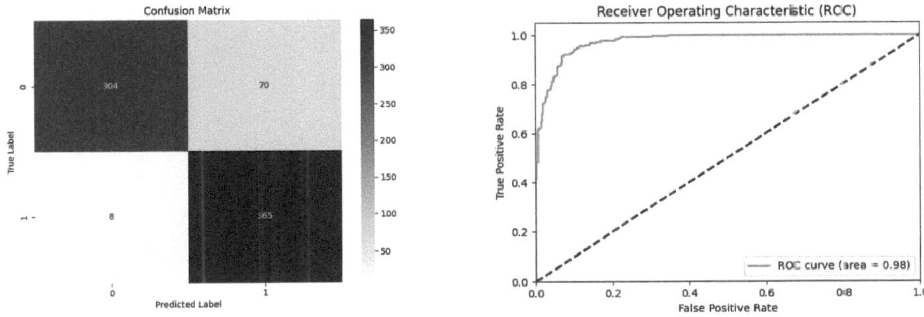

Fig. 4. Confusion Matrix and ROC for CNN architecture.

Further study on the topic using Federated Learning gave optimized results showing improved accuracy. By applying distributed training and combining segments of locally trained models, this federated learning process is capable of analyzing EEG data without compromising the security of the acquired data. The system uses a CNN model trained these sliding windows of the EEG signals which features temporal dependencies. The distribution of the training process across several clients is beneficial because each of them owns data from various subjects. Thus the model can learn from the extended dataset stored across multiple clients rather than being stored in a central database. Leveraging Desktop as the server and Raspberry Pi 4 as the clients, an accuracy of 97% was achieved at the global model.

The outcome clearly show that the FL resulted in greater accuracy. It can be noted that the use of the described system retains good convergence properties, and the accuracy of the model increases during training in rounds (Figs. 5 and 6).

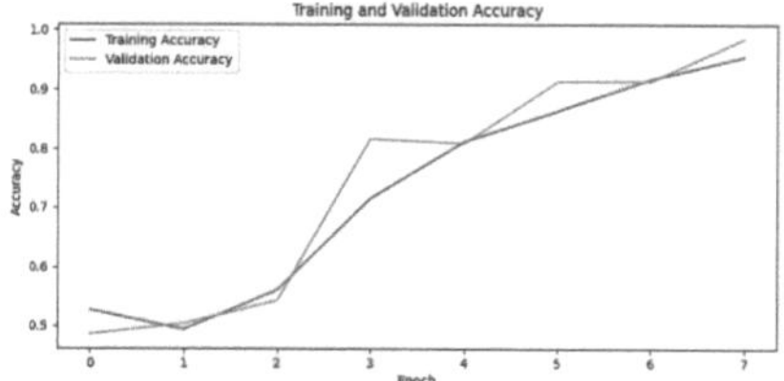 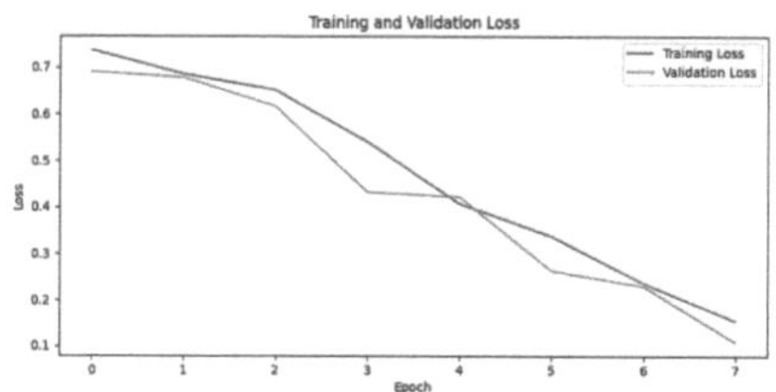

Fig. 5. Training and Validation accuracy & loss for FL architecture.

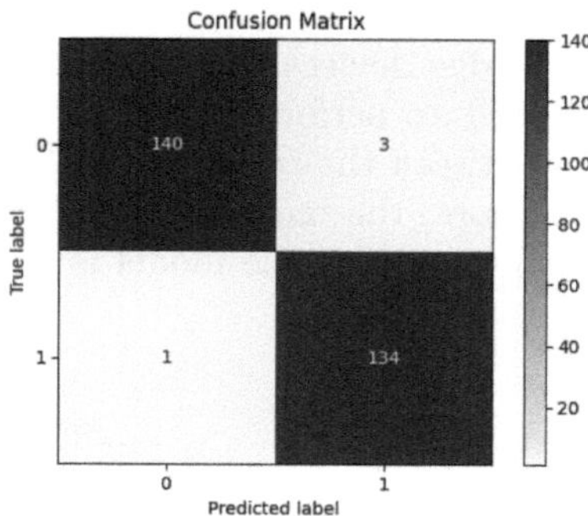

Fig. 6. Confusion Matrix for FL architecture

6   Conclusion

This research presents a comprehensive exploration into the diagnosis of Internet Addiction through EEG signals, employing CNNs and Federated Learning techniques. This study shows that CNN model can achieve high accuracy correctly classifying individuals with and without IA with an average accuracy of 89.86%. It surpasses the performance of the conventional classifiers such as the Support Vector Machine and Random Forest and which highlights the efficiency of deep learning methods in this field. In addition, the use of the trained model on an edge device such as the Raspberry Pi 4, shows that the real-time IA diagnosis is attainable with optimal process times. Additionally, the research extends its scope by exploring federated learning to protect privacy and security of personal brain data. The implementation of the Federated Averaging (FedAvg) algorithm facilitated knowledge sharing among participating devices, leading to a significant improvement in classification accuracy, reaching up to 97% at the global model level. The current results are based on very limited data due to lack of IAT results labled EEG data.

Moving forward, the research could benefit from some key expansions. Enlarging the dataset by including a more diverse range of participants would significantly enhance the generalizability of the proposed model. Also, integrating additional bio-signals alongside EEG data could offer a more holistic view of an individual's physiological state during internet use. The combination of diverse data streams could potentially lead to the development of a more nuanced and accurate diagnostic tool for internet addiction.

Acknowledgements. This work was supported by the Department of Science and Technology (DST), Government of India, under the VRITIKA scheme. The authors are grateful for the financial assistance provided, which made this research possible.

References

1. https://gitnux.org/topics/statistics/mental-health/page/2/
2. McMahan, B , et al.: Communication-efficient learning of deep networks from decentralized data. In: Artificial Intelligence and Statistics, pp. 1273–1282 PMLR (2017)
3. Babayan, A., et al.: A mind-brain-body dataset of MRI, EEG, cognition, emotion, and peripheral physiology in young and old adults. Sci. Data **6**(1), 1–21 (2019)
4. Baghersalimi, S., Teijeiro, T., Atienza, D., Aminifar, A.: Personalized real-time federated learning for epileptic seizure detection. IEEE J. Biomed. Health Inform. **26**(2), 898–909 (2022). https://doi.org/10.1109/JBHI.2021.3096127
5. Beutel, D., et al.: Flower: a friendly federated learning research framework. arXiv preprint arXiv:2007.14390 (2020)
6. Burleigh, T., Griffiths, M., Sumich, A., Wang, G., Kuss, D.: Gaming disorder and internet addiction: a systematic review of resting-state EEG studies. Addict. Behav. (2020). https://doi.org/10.1016/j.addbeh.2020.106429
7. Cohen, M.X.: Analyzing Neural Time Series Data: Theory and Practice. MIT Press (2014)
8. Demin, D.: Brain activity in adolescents with different internet addiction risks. Ekol. cheloveka (Hum. Ecol.) **28**(6), 21–27 (2021)
9. Idrees, A., Khlief, M.: Efficient compression technique for reducing transmitted EEG data without loss in IoMT networks based on fog computing. J. Supercomput. **79**(8), 9047–9072 (2023)
10. Jeong, B., et al.: Associations of personality and clinical characteristics with excessive internet and smartphone use in adolescents: a structural equation modeling approach. Addict. Behav. **110**, 106485 (2020)
11. Ju, C., Gao, D., Mane, R., Tan, B., Liu, Y., Guan, C.: Federated transfer learning for EEG signal classification. In: 2020 42nd Annual International Conference of the IEEE Engineering in Medicine & Biology Society (EMBC), pp. 3040–3045. IEEE (2020)
12. Ko, C.H., Namkoong, K., Jung, Y.J., Kim, S.H., Kim, I.W.: Gaming disorder and internet addiction: a systematic review of resting-state EEG studies. J. Clin. Med. **9**(7), 2202 (2020)
13. Lee, H., Choi, J., Shin, Y., Lee, J., Jung, H., Kwon, J.: Impulsivity in internet addiction: a comparison with pathological gambling. Cyberpsychol. Behav. Soc. Netw. **15**(7), 373–377 (2012)
14. Lee, J., et al.: Differential resting-state EEG patterns associated with comorbid depression in internet addiction. Prog. Neuropsychopharmacol. Biol. Psychiatry **50**, 21–26 (2014)
15. Liu, R., Chen, Y., Li, A , Ding, Y., Yu, H., Guan, C.: Aggregating intrinsic information to enhance BCI performance through federated learning. Neural Netw. **172**, 106100 (2024)

16. Ma, Y., Gu, J.: Internet and depression in adolescents: evidence from china. Front. Psychol. **14**, 1026920 (2023)
17. Mendes, N., et al.: A functional connectome phenotyping dataset including cognitive state and personality measures. Sci. Data **6**(1), 1–19 (2019)
18. Pedregosa, F., et al.: Scikit-learn: machine learning in python. J. Mach. Learn. Res. **12**, 2825–2830 (2011)
19. Qi, Y., et al.: Slow-wave EEG activity correlates with impaired inhibitory control in internet addiction disorder. Int. J. Environ. Res. Public Health **19**(5), 2686 (2022)
20. Qiu, S., Cheng, K., Zhou, T., Tahir, R., Ting, L.: An EEG signal recognition algorithm during epileptic seizure based on distributed edge computing. IJIMAI **7**(5), 6–13 (2022)
21. Yan, N., Cheng, H., Liu, X., Chen, F., Wang, M.: Lightweight privacy-preserving feature extraction for EEG signals under edge computing. IEEE Internet of Things J. (2023)
22. Young, K.: Caught in the Net: How to Recognize the Signs of Internet Addiction— and a Winning Strategy for Recovery. Wiley (1998)

Leveraging Aspect-Based Sentiment Analysis to Integrate Reviews and Ratings to Enhance Recommender Systems

H. Hedhav[(✉)], Haresh Sambath, Donald Xavier Anto, and E. Sivasankar

Department of Computer Science and Engineering, National Institute of Technology, Trichy, Trichy, India
hedhav.haresh@gmail.com

Abstract. A recommender system is a type of information filtering system that predicts the preferences or interests of users for items such as movies, music, books, or products, and provides personalised recommendations accordingly. It analyses past interactions, user behaviour, and item attributes to suggest items that users are likely to find appealing or relevant. Aspect-Based Sentiment Analysis (ABSA) is a subsection of sentiment analysis aimed at extracting the sentiments regarding specific aspects within a document rather than overall sentiment of said document. For a recommender system to recommend better, it needs as much information about each user and item (restaurants in this case) as possible. By extracting aspect-wise sentiments from reviews through ABSA models, the goal is to enhance recommender systems using that additional information. The recommender system is implemented using a regression-based neural network, which takes in user and business IDs as input, embeds them as a vector, and outputs a rating value for each aspect. The rating values (obtained through ABSA) are used to train the neural network for each aspect and then fed into another neural network, which predicts the final rating, which is used to rank the items for each user. Thus utilising state-of-the-art models for ABSA as well as the recommender system, this paper aims to produce a recommender system that can generate better and more personalised recommendations. By combining these two approaches, the study leverages ABSA's ability to capture fine-grained sentiment insights at the aspect level with the predictive power of neural collaborative filtering in modeling user preferences. RMSE and MAE are used as evaluation metrics as the output comes from a regression model. While the proposed model does not outperform individual baseline models in terms of predictive accuracy, it offers valuable insights into integrating ABSA with a regression model, handling of neutral sentiments and tackling imbalances in the dataset. This study highlights the potential of integrating these methodologies and paves the way for further refinement and exploration of hybrid approaches in recommender systems.

Keywords: Aspect-Based Sentiment Analysis · Recommender System

C. Modi et al. (Eds.): MIND 2024, CCIS 2736, pp. 279–289, 2026.
https://doi.org/10.1007/978-3-032-14531-4_24

1   Introduction

1.1   Aspect-Based Sentiment Analysis

Aspect-Based Sentiment Analysis (ABSA) represents a sophisticated evolution in the field of Natural Language Processing (NLP). While traditional sentiment analysis techniques focus on determining the overall sentiment of a text, ABSA delves deeper, extracting and analysing sentiments related to specific aspects or features within the text. ABSA operates by dissecting text to identify various aspects or attributes associated with a particular entity, such as a product, service, or topic. Upon identifying these aspects, ABSA proceeds to determine the sentiment expressed towards each one. This process involves categorising the sentiment as positive, negative, or neutral for each aspect mentioned in the text. The power of ABSA lies in its ability to provide a nuanced understanding of sentiment. For example, a review of a smartphone might express a positive sentiment towards its battery life, indicating user satisfaction with this feature, while expressing a negative sentiment towards its camera quality, suggesting dissatisfaction with this particular aspect. By providing a more granular view of sentiment, ABSA offers valuable insights that can enhance the accuracy and relevance of recommender systems, improve customer experience strategies, and inform product development decisions, among other applications.

Understanding Sub-Tasks of ABSA

- **Aspect Term Extraction (ATE)** - Only an input text is fed and the output will be the aspects (both implicit and explicit) which have an associated opinion that is extracted from the input text.
  Example: Great food, good size menu, great service, and an unpretentious setting, but slightly high prices.
  Output: food, menu, service, setting, prices
- **Aspect Term Sentiment Classification (ATSC)** - Input text and aspect are fed and the identified polarity of the fed aspect ('positive', 'negative' or 'neutral') is returned.
  Example: Great food, good size menu, great service, and an unpretentious setting, but slightly high prices.
  Aspects: food, price, location
  Output: positive, negative, neutral
- **Aspect Sentiment Pair Extraction (ASPE)** - Only an input text is fed and aspects with opinions are identified and extracted along with the opinion associated with them
  Example: Great food, good size menu, great service, and an unpretentious setting, but slightly high prices.
  Output: food:positive, menu:positive, service:positive, setting:positive, prices:negative

1.2   Recommender Systems

At its core, a recommendation system is a computational framework designed to analyse user preferences, historical interactions, and contextual data to provide personalised suggestions. The personalised suggestions given by the recommendation system must be reliable. Recommendation systems have immense practical applications. They are pervasive in modern life, offering tailored suggestions across diverse applications. E-Commerce platforms like Amazon recommend products based on browsing and purchase history. For instance, suggesting related items such as accessories or complementary products like a phone case when purchasing a smartphone. Streaming services like Netflix analyzes viewing habits to recommend movies and TV shows, such as suggesting similar films after watching a particular genre or recommending popular series based on past viewing history. The list of recommender systems in everyday life is endless and so is its potential.

1.3   Problem Statement and Objective

This project aims to build a Neural Collaborative Filtering (NCF) recommendation system that exploits aspects embedded in textual reviews. We use Aspect-Based Sentiment Analysis (ABSA) to assign scores for each aspect (namely food, price, ambience, service) of every review. We then create custom embeddings of users and items and train a Deep Neural Network model to learn each user's taste of items. Since the recommender system estimates the user's preference for an item by accounting for different aspects, it should understand the user's sentiments and taste better. Ultimately, this approach must improve recommendations.

1.4   Dataset

The dataset chosen for the purpose of this project is the Yelp Dataset [1] which contains restaurant ratings and reviews by users of the Yelp app (along with additional information like user activity, location of users and restaurants, tips etc.). The massive size of this dataset makes it perfect for the purpose of this project, along with the fact that both ratings and reviews are present.

The dataset is filtered to only retain users and restaurants with a minimum number of reviews (25 or more reviews per user being the initial filter, followed by 25 or more reviews per restaurant among the remaining reviews, resulting in a total of 372,108 reviews), to ensure the model has sufficient information about each user and restaurant and because of exceedingly large inference times of the ABSA Models.

The training data was imbalanced for each aspect, with most ratings given by users being 4 or 5 and few restaurants were rated 1 or 2. This imbalance reflects inherent biases in user rating behavior. To address this issue, the training dataset was balanced independently for each aspect (price, ambiance, food, and service) prior to being input into their respective rating predictor models. Similarly, the overall ratings were balanced for the overall rating predictor model before training. The dataset balancing was achieved using a resampling technique, specifically oversampling. Minority class ratings were replicated to approximately equalize the number of records across classes, thereby mitigating the impact of imbalance on the training process. This approach ensures that each class is adequately represented, improving the model's ability to predict across the full range of ratings.

The aspect ratings in the dataset belonged to the set $-1, -0.5, 0, 0.5, 1$. These values were transformed to the set $1, 2, 3, 4, 5$ respectively to keep aspect ratings consistent with overall ratings. Finally, the dataset was shuffled to eliminate any order in the data.

2   Literature Survey

2.1   Aspect-Based Sentiment Analysis

Eman Saeed Alamoudi and Norah Saleh Alghamdi (2021) [2] propose performing Aspect extraction on the Yelp dataset using four aspects, namely food, price, service and food, which were the most commonly referenced aspects and thus are the deciding factors in reviews. The paper concluded that the four aspects could comprehensively cover the contents of almost all the reviews.

Kevin Scaria, Himanshu Gupta, Siddharth Goyal, Saurabh Arjun Sawant, Swaroop Mishra, and Chitta Baral (2023) [3] detail the implementation of InstructABSA, a T5-based model to perform ABSA subtasks. InstructABSA boasts a performance rivalling 7x larger models.

Heng Yang, Chen Zhang and Ke Li (2023) [4] propose a modularized framework for ABSA, allowing for easier training and implementation of various transformer architectures, such as BERT, DeBERTa etc.

Aminu Da'u, Naomie Salim, Idris Rabiu, and Akram Osman (2019) [5] propose utilizing a deep Convolutional Neural Network model for aspect extraction from the review text and then applying a Linear Discriminant Analysis technique to transform the extracted aspect terms into latent factors and finally computing the aspect-based ratings, forming a set of rating matrices with each matching to a particular aspect. They attained better metrics than primitive methods at the time of publication.

Kai Wang, Weizhou Shen, Yunyi Yang, Xiaojun Quan, and Rui Wang (2020) [6] propose defining a unified aspect-oriented dependency tree structure rooted at a target aspect by reshaping and pruning an ordinary dependency parse tree. and then using a Relational Graph Attention Network (R-GAT) to encode the new tree structure for sentiment prediction. They were able to conclude that their methodology improved the performance of the Graph Attention Network (GAT).

2.2   Recommender System Implementations

Traditionally recommender systems are either rating-based or review-based. Recently, there have been attempts to integrate both reviews and ratings. Various techniques have been proposed to combine reviews and ratings of items for providing recommendations.

In *Nurul Aida Osman, Shahrul Azman Mohd Noah, Mohammad Darwich, and Masnizah Mohd (2021)* [7], they perform sentiment analysis on reviews and use these sentiments as ratings in places where ratings are not given by users. This technique helps reduce the sparsity of the matrix used for factorization.

Wenhua Li, Xiaoguang Li, Jiangzhou Deng, Yong Wang, and Junpeng Guo (2021) [8] consider two factors - user consistency and review feedback. Review feedback is obtained from sentiment analysis of reviews and user consistency is found by denoising user ratings. These ratings are integrated using weighted average to produce comprehensive ratings.

Xiang Jun Li, Geng Sheng Deng, Xiao Zhen Wang, Xiao Liang Vu, and Qing Wei Zeng (2023) [9] propose a hybrid recommendation algorithm, named RAC-SMD, which combines sentiment analysis of user comments using LSTM with matrix decomposition. It starts by assessing the sentiment of user comments, incorporating this sentiment alongside the user's rating to refine the accuracy of their actual rating. This approach is then combined with matrix decomposition techniques to enhance the quality of recommendations.

Nour Nassar, Assef Jafar, and Yasser Rahhal (2020) [10] build a novel model that obtains the features of the users and items and uses them as an input to the criteria ratings deep neural network, which aims to predict the criteria ratings. These predicted criteria ratings serve as the input to the second part, which is the overall rating of the deep neural network and is used to predict the overall rating.

While the idea of integrating Aspect-Based Sentiment Analysis to improve recommendation systems may not be new, all existing approaches use extremely primitive measures to extract sentiment of relevant aspects as opposed to using state-of-the-art measures most of which rapidly developed over the last 2–3 years. The measures include steps like stopword removal, tokenization, part-of-speech tagging, negation in the presence of "not" words etc. The current state-of-the-art models (especially for ABSA) are not being used in current implementations. With state-of-the-art models having much higher accuracy and ability to capture nuances, the difference in performance is expected to be much higher. This was found to be a significant literature gap.

3   Aspect-Based Sentiment Analysis

3.1   Misclassification of Neutral Sentiments

ABSA Models all have a fundamental drawback when used in the purpose of achieving this project, wherein neutral sentiments are not identified correctly.

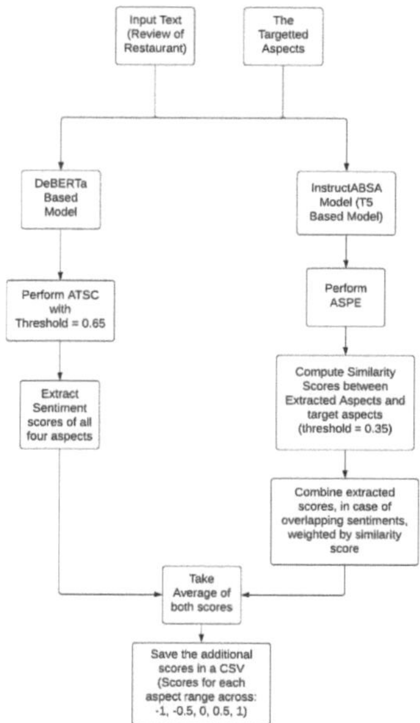

Fig. 1. Block Diagram of Aspect-Based Sentiment Analysis Module

Example: I had lime juice in this beautiful restaurant. Here the sentiment of aspect "food" is "neutral" and is predicted correctly. The sentiment of aspect "ambience" is "neutral" (as the aspect is not being referenced in the review) however it is misclassified as "positive" as the overall sentiment of the statement seems positive.

This is caused by the fact that ABSA Models are trained using aspects that are present in the statement they are analysing and the only way to reliably achieve our goal would be a dataset which also features missing aspects labelled as "neutral" so the model may understand that aspects that are missing must be neutral in sentiment. This issue is addressed in the "Fine-tuning Thresholds" subsection.

3.2   Final Implemented Solution

Using an Ensemble Model with models solving the problem with different approaches, as shown in Fig. 1.

Model 1 - InstructABSA (Tk-Instruct (T5) based state of the art ABSA model) **performing ASPE and then calculating similarity between identified aspect and the four desired aspects**. If the similarity is higher than the similarity threshold (0.35), then the polarity is taken into consideration and weighted by the similarity score. This model is much more accurate at identifying neutral sentiments, but is slightly more prone to misclassifying positive and neutral sentiments as neutral too.

Model 2 - DeBERTa Based Model performing ATSC by passing the input text and the four desired aspects and obtaining polarities of each. This model is slightly more prone to misclassifying missing aspects as positive/negative rather than neutral. But this can be greatly introduced by introducing a confidence threshold that analyses the differences in confidence between the top two possible polarities and ensures its greater than a threshold value (0.65).

The results from both these models are averaged and then added to the dataset as additional columns.

3.3   Fine-Tuning Thresholds

Ideal threshold values had to be identified - for both confidence scores of the DeBERTa-based Model and vector similarity measure score. To achieve this a test dataset of 200 reviews was created and manually annotated them with scores for all four target aspects. The respective models were then run with threshold values ranging from 0.05 to 0.95 (in intervals of 0.05). In the case of the confidence scores of the DeBERTa-based Model, increasing threshold led to more neutral aspects being correctly identified. It also led to positive and negative aspects being misclassified as neutral. Hence a threshold value of 0.65 was selected due to its optimal balance. In the case of the vector similarity measure score for InstructABSA, a threshold value of 0.35 was observed to have the least MSE against the test dataset

4   Recommender System

The Recommendation system implementation has two major modules - aspect rating prediction module and overall rating prediction module.

4.1   Aspect Rating Prediction

Aspect rating prediction module is composed of independent aspect rating predictors. For each target aspect, an aspect predictor is built, trained and tested. There are four aspects for the purpose of this project - price, ambience, food and service, each having their own rating predictors, as they are treated separately by the users.

User IDs and Business IDs are fed as input to the predictor. Since neural networks cannot handle strings, both user IDs and business IDs are converted into integers by passing them through the lookup layer and then transformed into an embedding. Custom embeddings are fed as they are trained to change their weights such that the resulting embeddings produced will be closer for similar users and similar business in their vector spaces, thus enabling collaboration of similar users and similar businesses.

Once individual embeddings are produced for the user and the business, they are concatenated into a single embedding and then supplied to a deep neural network which outputs the ratings for the concerned aspect.

4.2   Overall Rating Prediction

The overall rating prediction module shares some similarities with the aspect prediction module, as shown in Fig. 2. The process of obtaining custom user and business embeddings are exactly the same. The difference comes during concatenation. While only concatenated user and business embeddings were used in aspect predictors, in the overall rating prediction module all the aspect ratings predicted by the aspect rating prediction module are concatenated with the user and business embeddings. This new vector is then fed as input to the deep learning network. The neural network will output the predicted overall rating for the given user and business pair.

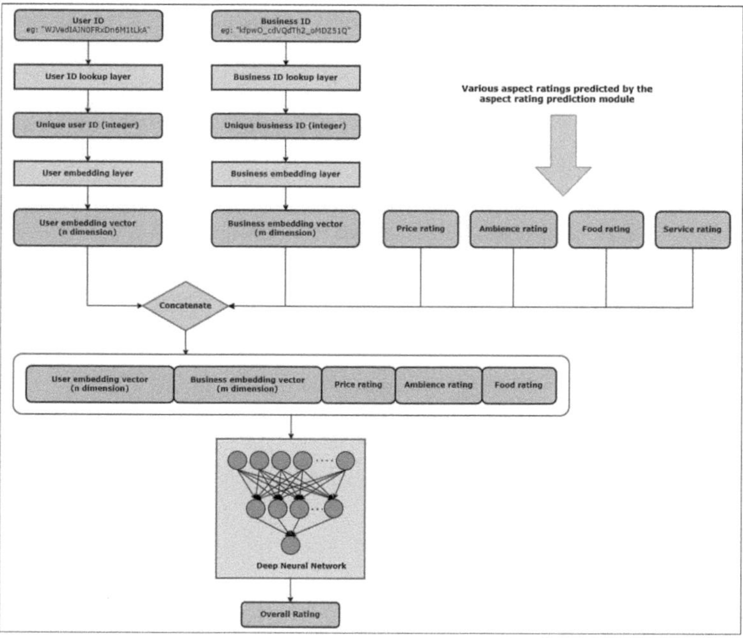

Fig. 2. Block Diagram of Overall Rating Prediction Module

5   Performance Analysis

5.1   Experimental Setup

The output of the ABSA module was on the scale -1, -0.5, 0, 0.5, 1. Before feeding this data to the recommendation system, these values were transformed into 1, 2, 3, 4, 5, just like overall ratings.

Once the dataset with aspect ratings was prepared, the aspect rating prediction module was built. The aspect rating prediction module helps overcome the need for a review to be provided to predict the aspect sentiments and consists of four independent aspect predictors, one for each aspect.

The embeddings not only help the model learn how much a user likes a restaurant but also similarities between users and restaurants making it a collaborative model. For user embeddings, the dimension of the vector was set to 64 while for business embeddings, a dimension of 32 was used.

Predictors are trained to predict their concerned aspect ratings from the user-restaurant pair input. Training was done for 15 epochs for each aspect predictor using a batch size of 32. Adagrad optimizer was used with a step size of 0.001.

The aspect sentiments are passed to the overall rating prediction module, which then predicts the overall rating. The overall rating prediction module was trained for 50 epochs, with a batch size of 32. Adagrad optimizer was used with a step size of 0.0001.

The dataset was split into a train dataset (75%) and a test dataset (25%). The dataset was shuffled to eliminate any ordering in data. The model was validated using 4-fold cross validation.

5.2   Results and Comparison

The recommendation system is a regression model that predicts a real value between 1 and 5 for a given user-restaurant pair. RMSE (root mean squared error) and MAE (mean absolute error) were used as evaluation metrics as they directly measure the deviation of predictions from actual values, making them meaningful in the context of user preferences.

The recommendation system gives an RMSE of 1.192 and MAE 0.953 on a subset of the Yelp dataset that was used. The performance of other base models are shown in the below table and these are values measured after using the entire Yelp dataset.

The entire Yelp dataset could not be used due to the large inference times as detailed in the Dataset subsection. When compared with other base models, this may seem underwhelming. However it is partially justified by the various bottlenecks involved in the pipeline, especially considering the errors that accumulate at each stage.

The underlying pre-trained models in ABSA modules suffer from their inability to accurately classify neutral aspects, especially ones not mentioned or referenced in reviews. This is an important limitation of the proposed model. However, the performance while improving significantly is still far from perfect (Table 1).

Since the aspect ratings predicted by ABSA are not entirely accurate, they may not be indicative of actual aspect ratings thus affecting the prediction models. Hence, the errors present while training negatively affects the model performance.

In conclusion, our study provides valuable insights into the application of ABSA in recommender systems and opens avenues for further research to optimize and scale these models. While the results may not have been ideal, we believe that the nuanced understanding of users' reviews gained through ABSA

Table 1. Comparison of proposed model against other models

Model	RMSE	MAE
Matrix Factorization (SGD)	0.9829	0.8674
OMNLR + KNN	0.8876	0.7479
BDTR + KNN	0.8736	0.7256
KNN (User-based)	1.1066	0.8989
KNN (Item-based)	1.1222	0.9223
Proposed model	1.192	0.958

can be extremely beneficial in enhancing recommender systems. By continuing to explore the intersection of sentiment analysis and recommendation algorithms, we can work towards more intelligent and user-centric recommender systems that better serve diverse user needs. Some ideas for improvement are discussed in the next section.

5.3   Future Scope

There are scopes for improvement in the data preparation stage, ABSA layer and the neural collaborative filtering (NCF) layer.

In the data preparation stage, the imbalance in the dataset could be handled using other techniques like Synthetic Minority Over-sampling Technique (SMOTE). It creates synthetic samples for the minority class in order to rectify imbalanced datasets. SMOTE generates new samples by interpolating between existing ones rather than merely repetition. From the minority class, it chooses two or more comparable samples and creates new data points along the line that connects them. This method is especially helpful for enhancing model performance on unbalanced data since it decreases overfitting and increases the variety of the minority class.

In NCF layer, more sophisticated networks could be trialed out for better prediction. Convolutional neural networks (CNNs) can be one route. CNNs can efficiently process high-dimensional data (e.g., images with thousands of pixels) by using shared weights, reducing computational complexity and overfitting. This would make the model more scalable.

References

1. Yelp Dataset Documentation. https://www.yelp.com/dataset/documentation/main
2. Alamoudi, E.S., Alghamdi, N.S.: Sentiment classification and aspect-based sentiment analysis on yelp reviews using deep learning and word embeddings. J. Decis. Syt. **30**(2–3), 259–281 (2021). https://doi.org/10.1080/12460125.2020.1864106
3. Scaria, K., Gupta, H., Goyal, S., Sawant, S.A., Mishra, S., Baral, C.: InstructABSA: instruction learning for aspect based sentiment analysis. arXiv:2208.01368 [cs.CL] (2023)

4. Yang, H., Zhang, C., Li, K.: PyABSA: a modularized framework for reproducible aspect-based sentiment analysis. arXiv:2302.08624 [cs.CL] (2023)
5. Da'u, A., Salim, N., Rabiu, I., Osman, A.: Recommendation system exploiting aspect-based opinion mining with deep learning method. Inf. Sci. **512**, 1279–1292 (2020). https://doi.org/10.1016/j.ins.2019.10.038. ISSN 0020-0255
6. Wang, K., Shen, W., Yang, Y., Quan, X., Wang, R.: Relational graph attention network for aspect-based sentiment analysis. arXiv:2004.12362 [cs.CL] (2020)
7. Osman, N.A., Noah, M., Azman, S., Darwich, M., Mohd, M.: Integrating contextual sentiment analysis in collaborative recommender systems. PLoS ONE **16**(3), 1–21 (2021). https://doi.org/10.1371/journal.pone.0248695
8. Li, W., Li, X., Deng, J., Wang, Y., Guo, J.: Sentiment based multi-index integrated scoring method to improve the accuracy of recommender system. Expert Syst. Appl. **179**, 115105 (2021). https://doi.org/10.1016/j.eswa.2021.115105. ISSN 0957-4174
9. Li, X.J., Deng, G.S., Wang, X.Z., Wu, X.L., Zeng, Q.W.: A hybrid recommendation algorithm based on user comment sentiment and matrix decomposition. Inf. Syst. **117**, 102244 (2023). https://doi.org/10.1371/journal.pone 0248665. ISSN 0306-4379
10. Nassar, N., Jafar, A., Rahhal, Y.: A novel deep multi-criteria collaborative filtering model for recommendation system. Knowl.-Based Syst. **187**, 104811 (2020). https://doi.org/10.1016/j.knosys.2019.06.019. ISSN 0950-7051

A Novel Hybrid Approach for Multi-class Classification of Hindi Hostile Online Posts

Santosh Rajak[(✉)], Ujwala Baruah, and Manojeet Roy

Computer Science and Engineering, National Institute of Technology,
Cachar, Silchar 78801, Assam, India
{santosh_rs,ujwala,manojeet_rs}@cse.nits.ac.in

Abstract. The document contains abusive and profane language.

The increasing prevalence of hostile content on social media has raised significant concerns among governments and organizations, necessitating robust AI-based solutions for effective content moderation. While considerable progress has been made in detecting hostile posts in English, the Indian language domain, particularly Hindi, remains underexplored. Existing research in Hindi is constrained by limited datasets, hindering the development of high-performing models. To address this, we developed a comprehensive multi-class dataset [1] tailored for Hindi, encompassing hostile categories such as defamation, abusive language, hate speech, offensive language, and non-hostile content. Leveraging this dataset, we propose a novel hybrid architecture that integrates statistical and contextual embeddings to enhance classification performance. Statistical embeddings, extracted via Term Frequency-Inverse Document Frequency (TF-IDF), are utilized with five machine learning classifiers, evaluated through 5-fold cross-validation. Contextual embeddings, derived from Multilingual Representations for Indian Languages (MuRIL), are employed in both the deep learning model and the hybrid model. The proposed hybrid model, combining an Long Short Term Memory network with a Random Forest classifier and leveraging contextual embeddings, achieves remarkable performance across multi-class classification tasks, with an accuracy of 0.9892 and an F1-score of 0.9893.

Keywords: Hindi · Dataset · Machine Learning · Grid-Search · Deep learning · MuRIL

1 Introduction

The widespread availability and use of internet worldwide has made social media accessible to diverse individuals and entities. While it serves as a powerful tool for connectivity, it also pose challenges such as misuse for spreading rumors,

---

U. Baruah and M. Roy—These authors contributed equally to this work.

C. Modi et al. (Eds.): MIND 2024, CCIS 2736, pp. 290–304, 2026.
https://doi.org/10.1007/978-3-032-14531-4_25

inciting violence, and influencing public opinions. This misuse contributes to societal issues, such as violence and discrimination. Addressing hostile content on social media platforms is crucial to maintain a safe online environment, requiring professionals to manually review and remove potentially harmful posts regularly.

Social media has had a significant impact on society, transforming the way people communicate and share information [2,3]. It has become a vital part of everyday life, allowing individuals to discuss personal issues, share opinions, and connect with others on various topics. The abundance of information available on social media platforms has changed societal norms, with people now relying on multiple sources for news and updates. Social media has also played a role in shaping discourses of identity, politics, and culture, with scholars exploring the digital dynamics of these representations. Additionally, social media has facilitated a deeper sense of connectivity and contribution to social circles and the world, altering the ways in which society functions and communicates.

Social media permits people to seek and share knowledge, socialise, and enjoy amusement, which in turn encourages preventative behaviours at the individual level by raising people's perceptions of social cohesion and collective efficacy [4]. Conversely, there is a negative correlation between individual preventative behaviour and collective efficacy in the presence of social cohesiveness. The impact of online communication and social media on adolescent wellbeing is a topic of ongoing debate.

A systematic review of multiple studies revealed conflicting evidence regarding the effects of online technologies on young people's mental wellbeing. In the context of Marriage and Family Therapy (MFT) training programs, misuse of social media is a noteworthy concern, with approximately 40% of accredited programs reporting problems stemming from student misbehavior. This includes issues such as confidentiality breaches, discriminatory language, and unprofessional content [5,6]. Social media's impact on society is two fold. Positively, it has improved communication, socialization, learning opportunities, and access to health information [7,8].

However, it has also brought about negative consequences such as crime, security concerns, privacy issues, political influence, cyberbullying, depression, anxiety, cat fishing, and terrorism. Wasim Akram et al. [9] reveal that negative effects of online platform, including crime, privacy concerns, political influence, cyberbullying, and adverse impacts on social cohesion, should not be ignored. The research in [10] examines the impact of content moderation by analyzing the policies and practices of three major platforms: Facebook, YouTube, and Twitter. It underscores the urgent need for a comprehensive overhaul of content moderation on social media to address the harm experienced by marginalized groups. Double standards [11] in social media content moderation are evident, with marginalized users facing disproportionate moderation and removal of their content.

The primary contributions of this study are:

1. Employs classifier diversity and utilizes TF-IDF representation to identify unique patterns in the data effectively. This approach emphasizes the sig-

nificance of textual phrases in classifying hostile content, thereby improving overall detection accuracy.
2. A novel approach is proposed by integrating the Long Short-Term Memory (LSTM) model with Random Forest classifiers, utilizing contextual embeddings to enhance performance in multi-class classification tasks.
3. We also provide baseline systems to benchmark our dataset, allowing for a clear comparison of performance metrics across different models.

The remaining portion of this paper is structured as follows: Sect. 2 reviews the relevant literature and background research. A detailed description of the dataset used has been covered in Sect. 3. The dataset validation framework, including feature extraction techniques and preprocessing, is covered in Sect. 4. The experimental framework, including the models, methods, and training setup, have been detailed in Sect. 5. The experimental results and benchmarks are presented in Sect. 6. Section 7 concludes the findings and gives a future direction of research.

2 Related Works

Identifying hate speech and hostile content in Twitter is a significant issue covered in several articles. The prevalence of social media platforms and the rise in user-generated material has made identifying abusive language in online user content a significant concern [12]. Several strategies, such as applying deep learning and ML methods, have been put forth to deal with this issue [13]. To enhance the identification of abusive posts, researchers have created techniques that use natural language processing (NLP) and consider the context and features of interaction networks [14]. In these works, using word2vec's skip-gram and cosine similarity in conjunction with unsupervised learning, a powerful, abusive text detection system has been created that can recognize abusive terms.

To identify deliberate obfuscation, the system combines a number of effective features, including as blacklists, n-grams, edit-distance metrics, multilingual support, abbreviations, punctuation, and special characters. When it comes to spotting harmful phrases in a variety of settings, including Twitter tweets, online community debates, and comments on news articles, the integrated decision system shows excellent performance [15]. Furthermore, state-of-the-art performance in abusive language detection is achieved by a computationally efficient method that makes use of dependency graphical convolutional networks (DepGCNs), making it appropriate for real-time applications [16].

Watanabe et al. [17] use patterns and unigrams to obtain excellent accuracy. A Logistic Regression model is used by Khan et al. [18] on unigrams to detect hate speech with success. By combining a cuckoo search neural network with a hybrid embedding strategy, Ayo et al. [19] show improved performance in hate speech identification. An exploratory experiment was conducted using neural networks to identify hate speech in Bengali and Hindi. The findings were equivalent to more expensive models even with less processing capability. This work is a valuable attempt to demystify lost or alien languages that are hard for humans to understand because the training data utilized was rather limited, and the languages were largely unknown to the researchers [20]. The main research topics were sentiment analysis and the identification of hate speech in Bengali and Hindi data. It involved using parameter sharing-based transfer learning LSTM models, and Word2Vec word embeddings that had been trained. The study simultaneously trained a single neural network on the Hindi and Bengali datasets using BiLSTM with self-attention in a combined dual input learning setup [21]. Identifying offensive information and automated hate speech in several languages is an area that is being widely explored.

Various machine-learning techniques have categorized social media messages as offensive or non-offensive. It has successfully used n-gram feature sets, such as word, character, and mixed models, to detect offensive text content trends [22]. Recurrent neural networks (RNNs) of the Long Short Term Memory (LSTM) type have been used to identify hate speech and objectionable content in Indo-European languages, providing language-neutral solutions [23]. Multilingual hate speech detection algorithms have become necessary as a result of the rise in hate speech caused by the expansion of social media. Thomas Mandl et al. [24] obtained the F1 score of 0.51 for Hindi, 0.53 for English, and 0.52 for German, which were attained with the top-performing binary classification systems.

Because code-mixed text has a complex structure and requires context filtering, identifying hostility in online Hindi-English code-mixed chats is a complex process. Large multilingual pre-trained models such as mBERT, XLMR, and MuRIL are utilized in Bagora et al.'s [25] innovative hierarchical neural network architecture to overcome these difficulties, and they believe their model to be state-of-the-art, as it regularly outperforms baselines. A deep-learning framework based on HindiBERT and handcrafted features is proposed by Bhardwaj et al. [26] for both coarse-grained and fine-grained hostile post classification. Several works have improved hostility identification using stacked embeddings and fine-tuned XLM-RoBERTa models. To identify hostility in Hindi, Siva Sai et al. [27] experimented with various neural network models, such as stacked word embeddings and improved XLMRoBERTa.

Synthesis. Much research has been conducted to identify hostile posts in English and code-mixed data, but a significant gap exists in Hindi-specific study. Scholars have used sophisticated neural networks and statistical ML methods to address hostile post-identification, but more work is needed to classify the diverse range of hostile messages. A common challenge is the limited size of existing datasets, hindering accurate identification and categorization of unfriendly content.

This study introduces a multi-class dataset tailored to Hindi, encompassing defamation, abusive language, hate speech, offensive language, and non-hostile content. The dataset aims to provide analysts with a broader range of hostile messages for research, advancing understanding and classification of hostile Hindi-language expressions.

3   Dataset Used

This section gives a detailed description of the dataset used in this work. The dataset consists of approximately 8,300 tweets of hostile content. The class labeling of the data encompass Hate, Offensive, Abusive, Defamation, and Sociable, thereby ensuring a rigorous and systematic methodology for categorizing the communications. As depicted in Table 1, the dataset is allocated among five discrete classes: the Non-Hostile category constitutes the most substantial group with 2,335 samples, followed by the Abusive content class with 1,733 samples. The Offensive category comprises 1,633 samples, while the Defamation and Hate speech categories contain 1,309 and 1,297 samples, respectively. In Table 2 a sample of the data has been given.

Table 1. Data Distribution across the classes

Data Type	Abusive	Defamation	Hate	Non-Hostile	Offensive
Train	1404	1060	1050	1891	1323
Val	156	118	117	210	147
Test	173	131	130	234	163
Total	1733	1309	1297	2335	1633

Table 2. Sample Table with Text and Labels in Hindi

Sl.	Text	Label
1.	@prabhav218 दिल्ली में BJP राष्ट्रीय कार्यालय एवं राष्ट्रीय महिला आयोग कार्यालय में श॒ष्टिाचार मुलाकात ह॒मारे लिए संस्कार का अर्थ है- शिक्षा, सेवा और संवेदनशीलता! हमारे लिए संस्कार का अर्थ है- समर्पण, संकल्प और सामर्थ्य।	Non Hostile
2.	हाँ सही है ना हमारे बच्चे तो हर साल ही टॉप करते है लेकिन इन कंजरों में से कभी कभार 100 सालों में एक बार ही कोई बच्चा आतंकी न बनके कुछ अच्छा करके देश का नाम रोशन करने वाला बच्चा निकलता है। तो हुई न ऐतिहासिक घटना।	Hate
3.	RT @ShilpiSinghINC: मा*****द तेरा बाप ,तू और तेरा खानदान सरकारी भीख के टुकड़े खाकर पले बढे हो,और पल बढ रहे हो,वरना तेरै जैसे की औकात ट्वीटर चलाने की नही ,मुसलमानों से अपनी मा बहन चो। दाओ ह	Abusive
4.	@ThePlaycardGirl @zoo_bear तुम जैसी लडकियां लड़कियों के नाम पर कलंक है तुम लोग झूठी खबरे फैलाते हो और चाहते हो की आपस में लोग धर्म के नाम पर लड़े और खुनखराबा हो, मैंने कितनी ही सनातनीयो को झूठी खबरे फैलाते देखा है।	Offensive
5.	@NazneenAkhtar10 इसने सारे गुनाह की हद तोड़ राखी है यह औरत मुसलमान के लिए बदनूमा दाग है। यह वही औरत है जिसने जमात के खिलाफ जहर बोला था यह औरत किसी भी तरह से माफी की काबिल नहीं है बल्कि मुसलमान के नाम पर कलंक है	Defamation

4   Dataset Validation Framework

In this section, we have looked at two essential steps in the data analysis process: feature extraction and data preprocessing. Feature extraction focuses on choosing and/or producing pertinent features from the preprocessed data to improve model performance and interpretability. Data preprocessing includes cleaning and transforming raw data into a format appropriate for analysis.

4.1 Preprocessing

We implemented various preprocessing techniques to improve the standard and consistency of our developed dataset. The first stage begins with removing unnecessary components like user mentions and URL hyperlinks from the tweets. In particular, URLs are eliminated by using regular expressions intended to recognize patterns suggestive of web addresses that begin with "http" or "www." By removing content that starts with the "@" symbol and ends with an alphanumeric character, user mentions are eliminated. The text is then edited so that only Hindi Devanagari characters remain, meaning that no non-Devanagari characters, numbers, or special symbols are included. Emojis are methodically removed by identifying and eliminating various Unicode ranges linked to emoticons, symbols, and other visual representations. To further improve the quality of the data, stop words are eliminated by using the advertools[1] library in combination with a custom list of stop words sourced from other sources. This process ensures that the dataset is primed for future analysis efforts by retaining only relevant Hindi content.

4.2 Feature Extraction

We initially ensured that every text entry was uniformly formatted as a string before preparing and extracting features from the textual data. We employed a hybrid feature extraction approach that combined statistical techniques using Term Frequency-Inverse Document Frequency (TF-IDF)[2] and Multilingual Representations for Indian Languages (MuRIL) for contextual embeddings[3]. We used the TF-IDF vectorization methodology, which is a way to convert textual data into numerical representations by assessing the importance of individual terms according to their frequency and document distribution across the dataset. Both individual words (unigrams) and two-word combinations (bigrams) were examined to provide a more comprehensive picture of the text. To acquire contextual information, a BERT-based model called MuRIL which has been pre-trained in 17 Indian languages, was employed. To improve the model's comprehension of the nuances of Indian language, we combined statistical and contextual aspects.

5 Experimental Framework

In the context of our investigation, we precisely evaluated the performance of different ML algorithms across diverse categories of tweets, namely Hate, Offensive, Non-Hostile, Defamation, and Abusive, by analyzing essential performance metrics such as accuracy, precision, recall, and F1-score. Furthermore, we implemented 5-fold cross-validation to augment the performance of two selected models. These outcomes establish a reference point for evaluating the comparative

---

[1] https://pypi.org/project/advertools/0.9.0/.
[2] https://scikit-learn.org/stable/modules/feature_extraction.html.
[3] https://huggingface.co/models?library=transformers.

effectiveness of various models in addressing a spectrum of hostile content. The investigation carried out clarifies the perks of particular algorithms, especially emphasizing SGD and Random Forest, in carrying out specific classification functions.

5.1 Benchmarch Modeling

We used various ML models to manage the complexities of multiclass text classification. We also used Grid Search Cross Validation [28] to improve the performance of Random Forest and Stochastic Gradient Descent. Adaptive Boost [29], Decision Tree, Gaussian Naive Bayes [30], Stochastic Gradient Descent [31], Random Forest, and K Nearest Neighbour [32] classifiers were used in this work. With the help of this thorough method, we have investigated different categorization techniques and used hyper-parameter tuning to enhance the performance of particular models. The schematic representation of the proposed methodology is delineated in Fig. 1.

Because Random Forest is an ensemble algorithm and can handle non-linear relationships, it is an excellent fit for multiclass challenges. Random Forest expands its built-in capability to include multiple categories without requiring particular changes.

A flexible optimization technique, stochastic gradient descent (SGD), is appropriate for various machine-learning applications, including multiclass issues. Its versatility comes from its capacity to handle many loss functions, making it ideal for data classification jobs into several groups.

A flexible technique frequently used for multiclass classification is K Nearest Neighbours (KNN). However, the data distribution and the choice of distance metric can have an impact on how well it performs.

AdaBoost's boosting method performs very well on multiclass tasks by assembling a group of inexperienced learners and gradually increasing accuracy with each repetition.

Decision Tree naturally accommodates multiclass problems using tree splits, offering an intuitive and practical classification approach.

A feature independence assumption is necessary for Gaussian Naive Bayes to work well in multiclass scenarios.

For our investigation, we merged the Long Short-Term Memory (LSTM) model alongside the Random Forest and Decision Tree classifiers. The LSTM system processes the input sequences via two sequentially layered LSTM components to properly seize the temporal links intrinsic to the dataset. The model's capacity to discern complex patterns within the dataset is augmented by including a fully connected Dense layer featuring ReLU activation, which follows these LSTM layers. We embedded a Dropout layer into the configuration to counteract the possibility of overfitting. For the final output layer, Softmax activation is employed to facilitate multiclass predictions. The model is constructed utilizing the Adam optimizer to guarantee optimal learning and convergence.

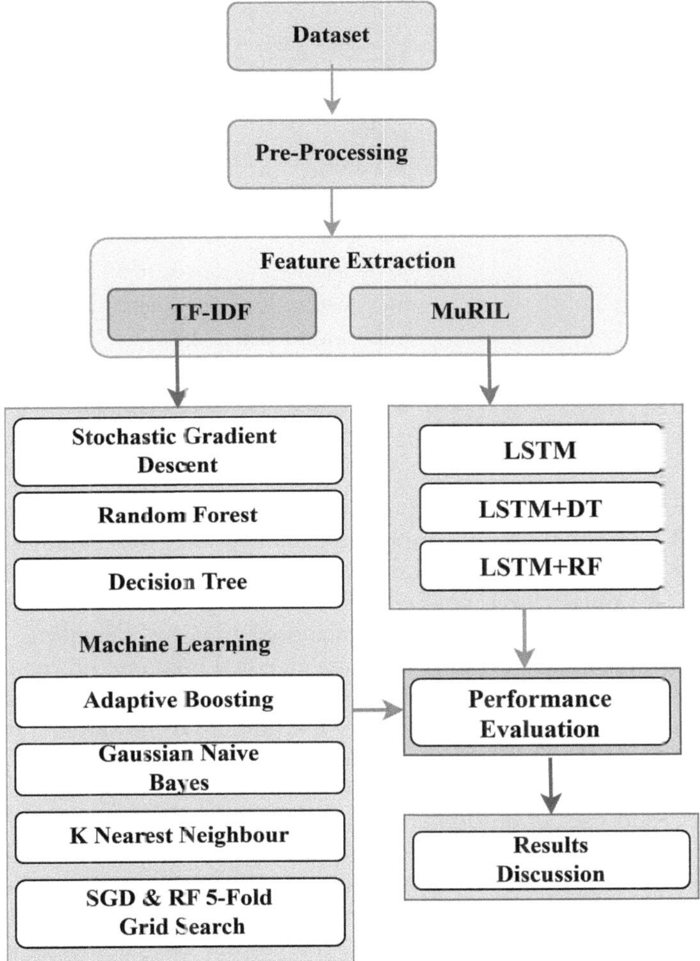

Fig. 1. Architectural Framework of the Proposed Methodology

5.2 Experiment Setup

We experimented with various ML strategies to tackle multi-class classification challenges in our research endeavors. The methods implemented, the SGD, Gaussian Naive Bayes, and AdaBoost were utilized with their foundational hyperparameters, typically noted for offering commendable results across diverse data distributions. In the case of the Random Forest model case, we configured the class_weight parameter to be balanced to reduce potential disparities among class distributions.

In optimizing the Decision Tree classifier, we chose the splitter parameter to facilitate the most effective node partitioning. The max_depth parameter was

Table 3. Grid Search Parameter value

Model	Hyper Parameter	Values Tested	Best Value
Stochastic Gradient Descent classifier	Penalty	l1, l2	l2
	alpha	1e-6, 1e-3, 1e-1, 1e0	1e-6
	Maxiter	50,100,1000,1500	1000
	learning rate	0.1,0.01,0.001,0.0001	0.01
Random Forest classifier	bootstrap	true, false	false
	max-depth	none, 10,50,100	50
	max feature	auto, sqrt	auto
	min samples leaf	1,2,4	1
	min samples split	2,5,10	2
	n estimator	800,1400,2000	1400

intentionally left as None, thereby permitting the tree to expand to its maximum potential. We designated min_samples_split as two and 'min_samples_leaf' as 1, empowering the tree to understand the elaborate features within the dataset sufficiently. We opted for a configuration involving five neighbors for the K-Nearest Neighbors algorithm. We implemented uniform weighting, indicating that the nearest five data instances possess an equal influence on the predictions for novel samples. These selections of hyperparameters were strategically made to augment the models' capacity for effective learning from the dataset, particularly when confronted with class imbalances and the inherent complexities associated with multi-class classification challenges.

In the second part of our experiment, we used grid search with 5-fold cross-validation to optimize the Random Forest (RF) and Stochastic Gradient Descent (SGD) classifiers. The aim was to find each model's ideal hyperparameter values to improve classification accuracy. The regularization types, alpha values, learning rates, maximum iterations for SGD, tree depth, the number of estimators, and feature selection criteria for RF were among the many hyperparameter combinations that the grid search explored. Table 3 lists the parameters that were examined and optimized using grid search cross-validation, highlighting the configurations that performed best. The dataset was then subjected to the optimal models determined by their cross-validation performance to classify tweets successfully. Two methods of data splitting were employed in our ML research. It receives sequences as input and runs them through two 128-unit LSTM layers. Sequences are returned by the first LSTM layer, enabling the second LSTM layer to examine temporal dependencies in more detail. A fully connected Dense layer with 512 units and ReLU activation is utilized for feature extraction after the LSTM layers. A Dropout layer with a rate of 0.3 is introduced to avoid overfitting. The output layer generates probability distributions for each of the five classes using softmax activation. The model is trained for 200 epochs with a batch size of 128 and a 20% validation split using the Adam optimizer (learning rate 0.001) and categorical cross-entropy loss.

Table 4. Performance Metrices

Sl.	Performance Metric	Formula
1	Accuracy	$\text{Accuracy} = \frac{\text{TP+TN}}{\text{TP+TN+FP+FN}}$
2	Precision	$\text{Precision} = \frac{\text{TP}}{\text{TP+FP}}$
3	Recall	$\text{Recall} = \frac{\text{TP}}{\text{TP+FN}}$
4	F1 score	$\text{F1-score} = 2 \times \frac{\text{Precision} \times \text{Recall}}{\text{Precision+Recall}}$

We used a 75% training, 10% validation, and 15% testing portion of the dataset for the main for the first phase development. As a result, we successfully trained, adjusted, and assessed our models. We employed 5-fold cross-validation in grid search with a 70-30 split for training and testing. This enabled us to evaluate several hyperparameter combinations in detail. These splitting techniques balanced the requirement for enough training data with reliable validation and testing, ensuring our models were well-trained. The evaluation of the model relies on a diverse set of performance metrics, as specified in the Table No. 4

6 Results and Discussion

In order to detect unique word-level characteristics in hostile content, TF-IDF efficiently records global term frequency patterns. Comparatively, MuRIL captures subtleties in syntax and semantics and offers extensive contextual comprehension specific to Indian languages. MuRIL's contextual depth and Random Forest's adept handling of feature interactions are two ways that the hybrid model, which combines LSTM and Random Forest, enhances these embeddings. Due to the thorough feature extraction and model variety, this integration produces higher classification accuracy and F1-scores, as shown in the performance evaluation. We deployed six different ML models in this work, each with varying tweaks of parameters. We also used two of these models for 5-fold cross-validation. Except for Non-Hostile, SGD with cross-validation performed admirably in most categories. In particular, SGD scored 82.45%, 80.93%, 79.41%, and 81.72% for hate, defamation and abusive tweets, respectively, on the F1 scale. SGD scored better than other models in the Non-Hostile category, obtaining an F1-score of 81.15%. When comparing the Stochastic Gradient Descent (SGD) model with and without cross-validation, the F1 scores show notable improvements in most categories: Hate speech increased by 0.27, Abusive by 7.64. Offensive by 0.42, and Defamation by 4.91. However, there was a decrease in the Non-Hostile category, with the F1 score dropping by 5.98. The performance of different machine learning models is shown in Table 5. The grid search cross-validation results for Random Forest (RF) and Stochastic Gradient Descent (SGD) are shown in Table 6. Contextual embeddings are used as input, and Table 7 shows how well deep learning models perform when combined with machine learning classifiers.

Table 5. Model Performance Comparison

Model	Class	Accuracy	Precision	Recall	F1-score
Decision Tree	Hate	0.7763	0.7863	0.7763	0.7808
	Offensive	0.7634	0.7668	0.7634	0.7651
	Non-Hostile	0.7698	0.7644	0.7698	0.7671
	Defamation	0.7402	0.7402	0.7402	0.7402
	Abusive	0.7410	0.7300	0.7410	0.7351
AdaBoost	Hate	0.8148	0.8177	0.8148	0.8162
	Offensive	0.7771	0.7676	0.7771	0.7722
	Non-Hostile	0.8260	0.7946	0.8260	0.8056
	Defamation	0.7273	0.7588	0.7273	0.7371
	Abusive	0.7418	0.7329	0.7418	0.7371
Gaussian Naive Bayes	Hate	0.7755	0.7623	0.7755	0.7679
	Offensive	0.7899	0.7687	0.7899	0.7783
	Non-Hostile	0.7971	0.7788	0.7971	0.7871
	Defamation	0.6800	0.7262	0.6800	0.6936
	Abusive	0.7225	0.7129	0.7225	0.7175
K Nearest Neighbor	Hate	0.5605	0.7722	0.5605	0.5985
	Offensive	0.8059	0.7616	0.8059	0.7789
	Non-Hostile	0.8156	0.7692	0.8156	0.7861
	Defamation	0.7690	0.7550	0.7690	0.7566
	Abusive	0.7835	0.7341	0.7835	0.7472
Random Forest	Hate	0.7947	0.8237	0.7947	0.8049
	Offensive	0.8107	0.7769	0.8107	0.7901
	Non-Hostile	0.8356	0.7965	0.8356	0.8060
	Defamation	0.7041	0.7719	0.7041	0.7185
	Abusive	0.7690	0.7276	0.7690	0.7423
SGD	Hate	0.8236	0.8202	0.8236	0.8218
	Offensive	0.8132	0.7987	0.8132	0.8051
	Non-Hostile	0.8228	0.8035	0.8228	**0.8115**
	Defamation	0.7338	0.7770	0.7338	0.7450
	Abusive	0.7514	0.7324	0.7514	0.7408

Figures 3 and 4 display the ROC curve for the RF and SGD model after applying cross-validation. As Table 7 illustrates, the most promising results were obtained by combining LSTM with Random Forest and Decision Tree employing contextual embeddings. Although LSTM performed poorly on its own, it performed satisfactorily when combined with Random Forest and Decision Tree. Figure Y displays the proposed model's confusion matrix (Fig. 2).

Table 6. Performance metrics for SGD and RF models across different tweet categories.

Model	Class	Accuracy	Precision	Recall	F1-score
RF	Hate	0.8476	0.8195	0.8476	0.7879
	Offensive	0.7573	0.7721	0.7573	0.6949
	Non-Hostile	0.8083	0.7858	0.8083	0.7340
	Defamation	0.8432	0.8004	0.8432	0.7773
	Abusive	0.8131	0.8132	0.8131	0.7554
SGD	Hate	0.8516	0.8205	0.8516	**0.8245**
	Offensive	0.8018	0.7925	0.8018	**0.8093**
	Non-Hostile	0.7966	0.7463	0.7966	0.7517
	Defamation	0.8331	0.7840	0.8331	**0.7941**
	Abusive	0.8323	0.8166	0.8323	**0.8172**

Table 7. Performance of LSTM + ML with Contextual Embedding

Model	Accuracy	Precision	Recall	F1 Score
LSTM	0.4248	0.3881	0.4248	0.3786
LSTM + DT	0.9692	0.9682	0.9691	0.9689
Proposed LSTM + RF	**0.9892**	**0.9892**	**0.9892**	**0.9893**

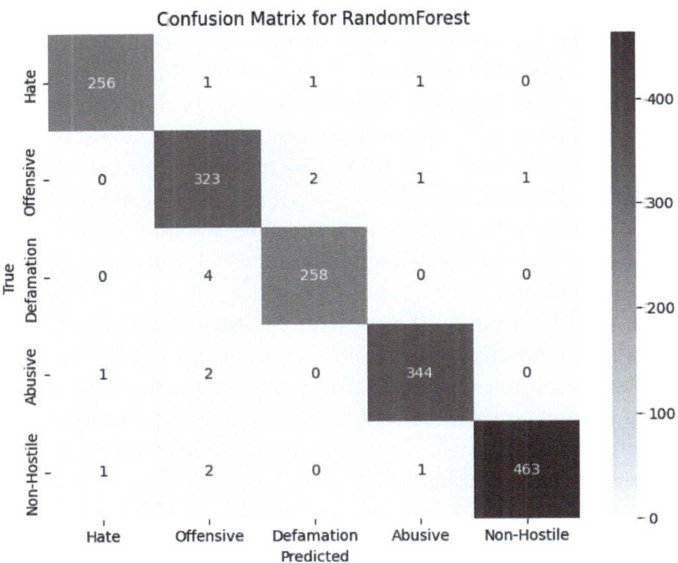

Fig. 2. Confusion Matrix of Proposed Methodology

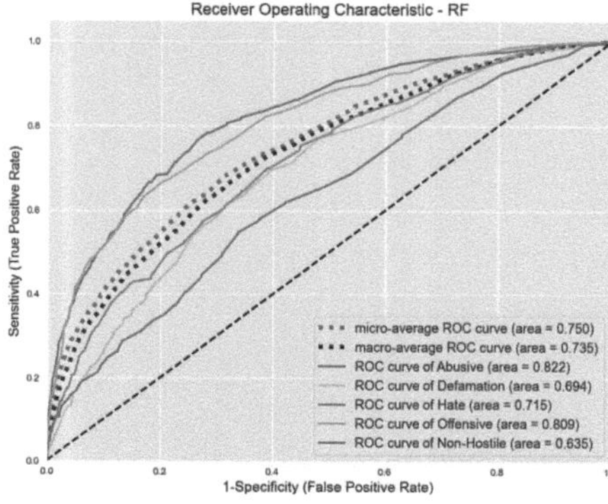

Fig. 3. RF ROC Curve

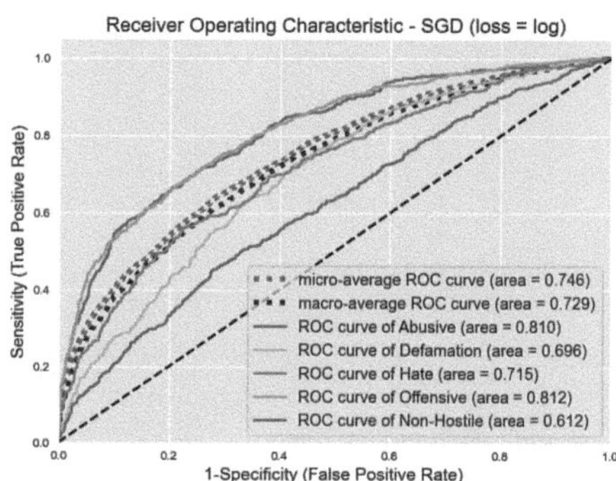

Fig. 4. SGD ROC Curve

7   Conclusion and Future Work

In this work the main focus has been on detecting and classifying hostile posts across all the tweets in the dataset. Using a well-annotated dataset, we have applied cutting-edge ML methods. Specifically, we utilized the Random Forest and Stochastic Gradient Descent methods, constructing a cross-validation grid search methodology to optimize the model's hyperparameters. The results of ML algorithms showed that they could detect and classify posts into hostile and non-hostile categories with promising performance. Using the cross-validation variation of SGD, we have achieved the best performance of 82.45 for Hate, 80.93 for offensive, and 79.41 for abusive, and SGD is the best performer for non-hostile with the F1 score of 81.15 among the six ML models used. The most impressive results were obtained by combining the LSTM model with contextual embeddings and machine learning classifiers. This hybrid technique is seen to improve accuracy and classification outcomes by using machine learning classifiers for final predictions and LSTM to properly capture temporal connections. On the other hand, the performance of the LSTM model with contextual embeddings alone is not as good, which emphasizes how useful it is to include machine learning classifiers for better outcomes. The proposed model achieved an accuracy of 0.9892 and an F1 score of 0.9893.

To improve model training efficiency, in the future we would like to enhance the dataset and perform rigorours hyperparameter tuning in the machine learning models. Furthermore, deeper learning architectures may be explored to enhance model performance. An additional avenue for further research would be to investigate the application of weighting techniques during model training in order to alleviate class imbalance.

References

1. Rajak, S.: Comprehensive Hindi Hostile Post Detection Dataset (CM-HTHPD) (2024). https://doi.org/10.21227/zxtz-k625
2. Social Media and Society. https://doi.org/10.1075/dapsac.100
3. Social Media in Society. https://doi.org/10.1016/B978-0-08-102411-9.00001-7
4. Lu, J., Wang, X., Fei, L., Chen, G., Feng, Y.: Effects of social media empowerment on covid-19 preventive behaviors in china. Inf. Technol. People (2023)
5. Best, P., Manktelow, R., Taylor, B.: Online communication, social media and adolescent wellbeing: a systematic narrative review. Child Youth Serv. Rev. **41**, 27–36 (2014)
6. Williams, L., Johnson, E., Patterson, J.E.: The appropriate use and misuse of social media in mft training programs: problems and prevention. Contemp. Fam. Ther. **35**, 698–712 (2013)
7. Damota, M.D., Uninversity, M.: The effect of social media on society. New Media Mass Commun. **78**(9), 1–9 (2019)
8. Amedie, J.: The impact of social media on society (2015)
9. Akram, W., Kumar, R.: A study on positive and negative effects of social media on society. Int. J. Comput. Sci. Eng. **5**(10), 351–354 (2017)
10. Díaz, Á., Hecht-Felella, L.: Double standards in social media content moderation. Brennan Center for Justice at New York University School of Law (2021). https://www.brennancenter.org/our-work/research-reports/double-standards-socialmedia-content-moderation
11. Langvardt, K.: Regulating online content moderation. Geo. LJ **106** 1353 (2017)
12. Nobata, C., Tetreault, J., Thomas, A., Mehdad, Y., Chang, Y.: Abusive language detection in online user content. In: Proceedings of the 25th International Conference on World Wide Web, pp. 145–153 (2016)
13. Kaur, S., Singh, S., Kaushal, S.: Abusive content detection in online user-generated data: a survey. Procedia Comput. Sci. **189**, 274–281 (2021)
14. Song, R., Giunchiglia, F., Shen, Q., Li, N., Xu, H.: Improving abusive language detection with online interaction network. Inf. Process. Manag. **59**(5), 103009 (2022)
15. Lee, H.-S., Lee, H.-R., Park, J.-U., Han, Y.-S.: An abusive text detection system based on enhanced abusive and non-abusive word lists. Decis. Supp. Syst. **113**, 22–31 (2018)
16. Narang, K., Brew, C.: Abusive language detection using syntactic dependency graphs. In: Proceedings of the Fourth Workshop on Online Abuse and Harms, pp. 44–53 (2020)
17. Watanabe, H., Bouazizi, M., Ohtsuki, T.: Hate speech on twitter: a pragmatic approach to collect hateful and offensive expressions and perform hate speech detection. IEEE Access **6**, 13825–13835 (2018)
18. Khan, H., Yu, F., Sinha, A., Gokhale, S.S.: A parsimonious and practical approach to detecting offensive speech. In: 2021 International Conference on Computing, Communication, and Intelligent Systems (ICCCIS), pp. 688–695. IEEE (2021)
19. Ayo, F.E., Folorunso, O., Ibharalu, F.T., Osinuga, I.A.: Hate speech detection in twitter using hybrid embeddings and improved cuckoo search-based neural networks. Int. J. Intell. Comput. Cybern. **13**(4), 485–525 (2020)
20. Phung, T.M., Cloos, J.: An exploratory experiment on Hindi, Bengali hate-speech detection and transfer learning using neural networks. arXiv preprint arXiv:2201.01997 (2022)

21. Khan, S., Shahid, M.: Hindi/Bengali sentiment analysis using transfer learning and joint dual input learning with self attention. arXiv preprint arXiv:2202.05457 (2022)
22. Pathak, V., Joshi, M., Joshi, P., Mundada, M., Joshi, T.: Kbcnmujal@ hasoc-dravidian-codemix-fire2020: using machine learning for detection of hate speech and offensive code-mixed social media text. arXiv preprint arXiv:2102.09866 (2021)
23. Saha, B.N., Senapati, A.: Cit kokrajhar team: LSTM based deep rnn architecture for hate speech and offensive content (hasoc) identification in indo-european languages. In: FIRE (Working Notes), pp. 359–365 (2019)
24. Mandl, T., et al.: Overview of the hasoc track at fire 2019: Hate speech and offensive content identification in Indo-European languages. In: Proceedings of the 11th Annual Meeting of the Forum for Information Retrieval Evaluation, pp. 14–17 (2019)
25. Bagora, A., Shrestha, K., Maurya, K., Desarkar, M.S.: Hostility detection in online Hindi-English code-mixed conversations. In: Proceedings of the 14th ACM Web Science Conference 2022, pp. 390–400 (2022)
26. Bhardwaj, M., Akhtar, M.S., Ekbal, A., Das, A., Chakraborty, T.: Hostility detection dataset in Hindi. arXiv preprint arXiv:2011.03588 (2020)
27. Sai, S., Jacob, A.W., Kalra, S., Sharma, Y.: Stacked embeddings and multiple fine-tuned XLM-RoBERTa models for enhanced hostility identification. In: Chakraborty, T., Shu, K., Bernard, H.R., Liu, H., Akhtar, M.S. (eds.) CONSTRAINT 2021. CCIS, vol. 1402, pp. 224–235. Springer, Cham (2021). https://doi.org/10.1007/978-3-030-73696-5_21
28. Browne, M.W.: Cross-validation methods. J. Math. Psychol. **44**(1), 108–132 (2000)
29. Chengsheng, T., Bing, X., Huacheng, L.: The application of the adaboost algorithm in the text classification. In: 2018 2nd IEEE Advanced Information Management, Communicates, Electronic and Automation Control Conference (IMCEC), pp. 1792–1796 (2018). IEEE
30. Ababneh, J.: Application of naïve bayes, decision tree, and k-nearest neighbors for automated text classification. Mod. Appl. Sci. **13**(11), 31 (2019)
31. Goswami, M., Sabata, P.: Evaluation of ml-based sentiment analysis techniques with stochastic gradient descent and logistic regression. Trends Wirel. Commun. Inf. Secur. Proc. EWCIS **2020**, 153–163 (2021)
32. Shah, K., Patel, H., Sanghvi, D., Shah, M.: A comparative analysis of logistic regression, random forest and knn models for the text classification. Augment. Human Res. **5**(1), 12 (2020)

Hate Speech Sentiment Detection with Topic Modeling-Based Feature Engineering

D. Dhiraj Choudhary[(✉)], M. Shashank Reddy, and M. Anand Kumar

Department of Information Technology, National Institute of Technology, Surathkal,
Karnataka 575025, India
{dhirajchoudharydommalapati.211ai016,
shashankreddymuppidi.211ai033,m_anandkumar}@nitk.edu.in

Abstract. A pressing issue for the modern era is dealing effectively with growing
online hate speech cases, especially ones requiring scrutiny. This project presents
a topic modeling integrated sentiment analysis model to analyze and categorize
Twitter data. The data contains an English dataset in which the tweets were labeled
hateful or non-hateful. Sentiment analysis was evaluated using four categorization
algorithms.

- Logistic Regression, Random Forest, Naive Bayes, and Support Vector
Machine. These same algorithms are used to compare the proposed topic modeling
feature extraction performance metrics. We also use a Tamil caste or immigrant
hate speech detection dataset to evaluate the proposed method in a language other
than English. Topic modeling uses a combination of three algorithms namely
Top2Vec, NMF, and BERTopic to uncover previously unknown topics within the
hate speech corpus(cross-lingual), extract features from the text, and use machine
learning algorithms for the predictive task.

Keywords: Hate speech detection · Sentiment Analysis · Topic Modeling ·
Explainable AI

1 Introduction

Hate speech has become a critical issue in the digital age, especially with the exponential growth of social media platforms. These platforms have enabled millions of users to communicate freely, but this freedom has also allowed for the spread of harmful, offensive, and derogatory speech. Victims are often individuals who are discriminated against based on race, gender, religion, or other identity-based characteristics, and this can lead to serious consequences such as cyberbullying and mental health deterioration. Data highlights the present situation, with studies showing that up to 37% of young people have been victims of cyberbullying, often driven by hate speech online [1]. Cyberbullying leveraging hate speech can take various forms, including harassment via direct messages, public shaming, and spreading offensive content through social media posts, and is not only emotionally damaging but also poses a severe mass mental health problem [2].

C. Modi et al. (Eds.): MIND 2024, CCIS 2736, pp. 305–316, 2026.
https://doi.org/10.1007/978-3-032-14531-4_26

Despite ongoing research, the detection and mitigation of hate speech and cyberbullying remain difficult, partly due to the varied and context-dependent nature of abusive language. In recent years, advancements in artificial intelligence and natural language processing (NLP) have provided promising solutions in the front of hate speech detection. Early methods, such as using TF-IDF (Term Frequency-Inverse Document Frequency), provide a way to detect hate speech by analyzing word frequencies and patterns in a dataset. However, it could not understand the semantic context of words, leading to limited success in detecting ambiguous hate speech. To overcome these limitations, advanced models have been developed, particularly the introduction of transformer models like BERT (Bidirectional Encoder Representations from Transformers) has revolutionized hate speech detection. Unlike traditional methods, BERT captures the context of words by processing entire sentences bidirectionally, making it effective at understanding subtle and implicit hate speech. BERT-based models can discern hateful content even when it is disguised through sarcasm or indirect expressions. However, these models are computationally intensive and require large amounts of data for training.

This project aims to implement a hybrid model that combines traditional methods, such as TF-IDF, with advanced topic modeling models like BERT, NMF and Top2Vec. The rationale behind this approach is that TF-IDF, NMF excel at identifying frequently used abusive terms, while BERT can interpret the context in which those terms are used. This synergy results in a model capable of capturing both explicit hate speech (detected through TF-IDF) and more complex, context-dependent hate speech (handled by BERT). By using both methods, hybrid models significantly improve the performance of hate speech detection systems using traditional machine learning algorithms. These systems are being increasingly adopted by social media platforms, law enforcement agencies, and moderators to automatically detect and take action against hateful content rampaging in the current social environment.

The remaining sections of this paper are organized as follows: Section 2 provides a detailed description of the existing works exploring the task of hate speech detection. Section 3 discusses the proposed methodology and its associated processes in detail. The experiments performed and observations are reported in Section 4. Finally, Section 5 concludes the work.

2 Related Work

A key challenge in topic modeling research lies in the absence of universal benchmarks for evaluating algorithmic performance. Despite the diversity of techniques, most approaches aim to classify documents by scoring terms and their associations. Recent advancements have incorporated neural models, word embeddings, and semantic graphs, contributing to more context-aware and coherent topic identification.

Kashia Sethia et al., [5] explore different types of topic modeling in the context of the 20-newsgroup dataset. The new frameworks using BERT, LDA, and K-means for the classification of news articles and the word embedding models such as Word2Vec and Doc2Vec will be utilized. Summaries include doc2vec, sentence transformer, and bert summarizer. Hybrid Bert-LDA resulted in the highest level of agreement than other analyzed models.

Sandhya Subramani et al. [6] propose a neural topic model that captures semantic word relations using custom embeddings, yielding higher clustering performance over classical methods with better silhouette and Davies-Bouldin scores.

Conghui Yuan et al. [7] introduce a weighted topic model enhanced by word embeddings and a directed semantic graph, improving contextual semantic associations and outperforming traditional LDA models.

Mei Zhou et al., [8] suggest a joint embedding model for BERT-LDA combining narrative semantic context analysis. HDBSCAN clusters document embeddings while class-based TF-IDF builds up the topic representations. The BERT-LDA model demonstrates its competitive nature by producing cohesive and varied topic words than the traditional and single-topic ones.

S. Limwattana et al., [9] propose Deep Word-Topic LDA (DWT-LDA) and train LDA using the supplied word input embeddings. Word-topic detection is achieved with a collapsed Gibbs sampling process. The evaluation included thematic consistency and diversity suggesting that DWT-LDA yielded better consistency in comparison with the conventional LDA.

F. Zhang et al. [10] propose using FastText to reduce sparsity issues in short texts, combining it with Sentence-LDA to achieve better topic coherence.

D. Yamunathangam et al., [11] provides a comprehensive survey of various topic modeling techniques used to discover topics in both long and short textual data. The authors categorize the methods into four main groups: standard topic modeling methods, clustering-based methods, self-aggregating methods, and deep learning-based methods. The focus is on how these models can handle long and short texts, with particular attention to the clustering-based and self- aggregating methods for short text corpus.

Lin Liu et al., [12] address the limitations of traditional topic models in capturing topic correlations in textual data, emphasizing the need for hierarchical topic modeling. Traditional topic models typically assign a single distribution over topics for each document, which limits their ability to model complex topic correlations. Hierarchical topic modeling relaxes this constraint and allows for a more flexible representation of topic relationships.

Sara Paunkoska et al., [13] investigates the application of modern machine learning and natural language processing techniques to enhance computer systems' ability to detect hate speech and insults online. The authors conduct a performance comparison between traditional classifiers and deep learning-based classifiers, using a large dataset of publicly available social media posts. Their findings suggest that the overall performance of hate speech detection is not solely dependent on the choice of classifier, but is also influenced by effective pre-processing of text and proper parameter tuning.

Satyam Kumar et al., [14] present a comprehensive analysis of machine learning algorithms for identifying hate speech on social media platforms. Their research contributes by providing insights into the advantages and limitations of different algorithms, such as ensemble methods, deep learning, and classical machine learning.

The reviewed studies highlight the evolution of topic modeling, from traditional methods like LDA to modern techniques such as BERT-based models and neural embeddings. The integration of pre-trained embeddings significantly enhances the performance of conventional approaches, improving context capture and coherence. Overall, these advancements underline the dynamic nature of topic modeling, demonstrating its continued relevance in extracting meaningful insights from unstructured data. It can be seen that BERT-based models when combined with traditional topic modeling methods like LDA, outperform standard models by better capturing context and semantic relations in text. Specifically, Hybrid BERT-LDA showed the highest level of agreement in classifying documents based on topics, indicating BERT's strong ability to extract meaningful context. We can observe that supervised models using artificial neural networks can significantly improve topic modeling by learning semantic relations between words and augmenting traditional LDA with embeddings boosts topic coherence and specificity. Also, we can see that, for short texts, models that utilize word embeddings like FastText perform better.

3   Proposed Methodology

Sentiment analysis is a technique that provides insights into NLP and helps in language processing abilities. It assists in understanding and analyzing emotions in the text by providing hands-on experience in machine learning and real-world applications such as social media sentiment tracking.

We use several machine learning models, including Support Vector Machine (SVM), Logistic Regression, Naive Bayes, Decision Tree, and Random Forest classifiers, to classify the vectors as either hateful or not hateful. For the task of feature extraction through topic modeling, we employ three methods: NMF, BERTopic, and Top2Vec. This approach involves multiple steps, starting with performing NMF on the preprocessed tweets to obtain the document-to-topic matrix, which represents the similarity of each document to the identified topics. We then run Top2Vec to generate topic embeddings and calculate similarity scores between each document and the topics. Next, we create a document-to- topic matrix with similarity values and use BERTopic to cluster sentences and generate topic embeddings. All these new features are combined into a single dataset, resulting in a feature-rich matrix. Finally, we evaluate this matrix using the various machine learning models mentioned above and compare the performance with the former technique.

3.1   Data Description and Preprocessing

The English Twitter data obtained from [3] has more than 35,000 tweets in the training set and about 10,000 tweets in the test set, where the data has a balanced class ratio of hate/non-hate labels. To provide algorithm training with more data, we appended the training and test dataset while modeling the subjects.

The Tamil caste/immigrant hate speech detection dataset obtained from [4] has over 5,000 tweets in the training set, nearly 1,000 in the validation set, and more than 1,500 in the test set. This data has been used to check the validity of our model over other languages.

As shown in Fig. 1, we first preprocess the tweets and clean the data. Then, we split the data frame into training and testing sets. Using a TF-IDF vectorizer, we convert the words in the vocabulary into vectors. Essentially, we generate a vector for each instance

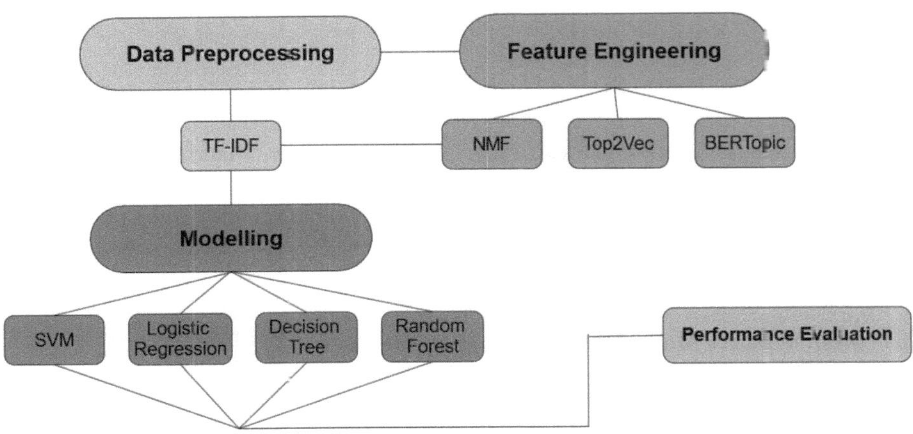

Fig. 1. Proposed methodology workflow and components.

in the data frame along with a class label. Various machine learning models are then applied to classify the test dataset, which has been converted into vector form.

3.2 BERTopic Model Application

Base idea is Topics and topic representations in the form of topic words, as seen in Fig. 2 are generated by BERTopic.The algorithm, as seen in Fig. 3 can be described in three steps:

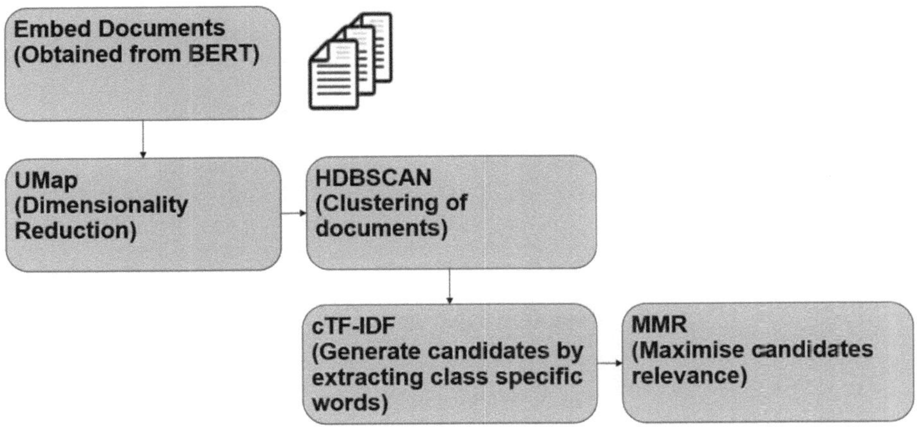

Fig. 2. Working of BERTopic

1. A pre-trained embedding model is used to generate an embedding for each instance in the data. By default, any model from the Sentence BERT framework is supported, but any pre-trained language model that was fine-tuned on semantic similarity may

be used by giving it the calculated embeddings, which is one of BERTopic's main features.

2. We reduced the dimensions of the generated embeddings using UMAP, and the lower-dimensional representations are grouped into topics by HDBSCAN, which is a hierarchical clustering technique with a soft-clustering strategy. This clustering algorithm classifies some tweets as noise, which helps in creating compact topics.

3. Topic words are extracted for each cluster by generating topic-word distributions using a class-based TF-IDF technique and selecting the top n words with the highest likelihood.

3.3 Non-Negative Matrix Factorization Application

From the document-term matrix, we aim to extract topics. The advantage of this approach is that we can determine the exact number of topics by fitting the original dataset distribution. As a result, we obtain two matrices: the document- to-topic matrix and the topic-to-term matrix. These matrices represent the relationship between words and topics, as well as the relationship between tweets and topics. The goal is to identify the similarity score of each instance for every topic and append this information to the data frame. This process is accomplished and visualized.

3.4 Top2Vec Model Application

Top2Vec and BERTopic have some similarities. Both of them use pre-trained embedding models to create document and word embeddings and reduce the dimensions of the generated embeddings using UMAP, and the lower-dimensional representations are grouped into topics by HDBSCAN, as seen in Fig. 4. Unlike BERTopic, Top2Vec generates topic terms by allocating all documents to a subject.

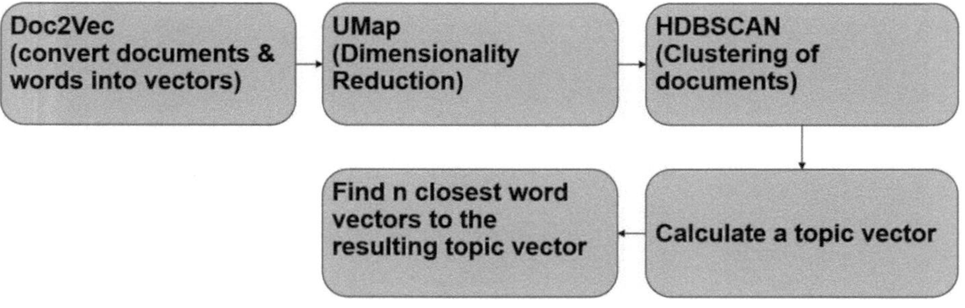

Fig. 3. Working of Top2Vec

While using Top2Vec and BERTopic, there is no need to specify the number of topics in advance although we are trying to make a dataset that has topics and their corresponding similarity score with each topic.

3.5 Data Frame Modification for Model Application

To achieve this, we use BERTopic, Top2Vec, and NMF to obtain features for every instance in the form of similarity scores for each topic (since an instance can be related to multiple topics at once). As a result, we create a data frame with ten features, capturing the similarity scores of an instance to the topics generated by BERT, NMF, and Top2Vec. These features are then used for classification tasks, enhancing the classification performance compared to using only the original vectorized instances to predict their labels.

4 Experimental Results and Analysis

The aim is to create a rich feature matrix that captures all the original TFIDF features as well as topic-similarity features from the three different topic modeling methods and evaluate the performance of this additional preprocessing technique that has ability to detect sentiment both from specific words and from high-level topics and context against the traditional TF-IDF that captures specific term importance and frequency, focusing on individual words or phrases that are relevant to sentiment.

To verify this, we conducted two experiments comparing the performance of our proposed approach with traditional (TF-IDF) method and models from the literature and tested its robustness across different languages (Tamil and English).

Table 1. Accuracy scores comparison between topic modeling and sentiment analysis (English)

Classification Model	Traditional SA	Topic Modelling based SA
SVM	67.689%	93.251%
Random Forest	67.576%	94.683%
Logistic Regression	68.436%	90.174%

The first experiment compares the performance of our proposed model against a traditional TF-IDF-based sentiment analysis model. The dataset used for this experiment is primarily in English, where sentiment is relatively straightforward to capture through specific terms and phrases. The results show that incorporating topic similarity features enhances the model's ability to capture contextual information, leading to improved performance as shown in Table 1.

The second experiment examines the robustness of our model by applying it to datasets in different languages, including Tamil, to assess its scalability. The results reveal a noticeable drop for the proposed methodology in performance when applied to the Tamil dataset as observed in Table 2, which can be attributed to several factors such as Tamil is a morphologically rich language, therefore presents challenges for traditional TF-IDF approaches, which are primarily optimized for simpler languages like English. Furthermore, the smaller dataset size and lack of pre-trained language models for Tamil limit the model's ability to capture sentiment effectively as words found in the test

Table 2. Accuracy scores for Validation and Test data(Tamil)

Classification Model	TopicModelling basedFeature Extraction(Val)	Topic Modeling based FeatureExtraction(Test)
SVM	99.048%	54.603%
Random Forest	99.894%	60.813%
Logistic Regression	97.551%	53.768%
CNN(base)	99.126%	67.562%

set may not be present in the training set. Topic modeling methods, such as NMF and BERTopic, which perform well on English data, may also struggle with Tamil due to its complex syntax and reliance on contextual complexity.

By identifying the weightage of words that contribute to specific topics, particularly those related to hate, we can better understand how these topics are formed and how individual words influence the classification of feelings. Using topic modeling techniques, we were able to extract key words that were strongly associated with hate speech. For example, words like "jews," "christians," and "nazi" consistently emerged in hate-related topics. These words were then analyzed using SHAP and LIME to determine their contribution to the sentiment classification of each sentence. This analysis allowed for insights into what prevailing topics of hate exist in the dataset, such as racial hatred, misogyny, religion-based hatred, and political violence, and how the model interprets individual words in context.

In Fig. 4, we observe the scores (weights) of each term in their respective topics, along with the most representative words associated with those topics.

Figure 5 illustrates the similarity matrix, where we can see the degree of similarity between different topics and how some topics may collapse into others due to overlap content.

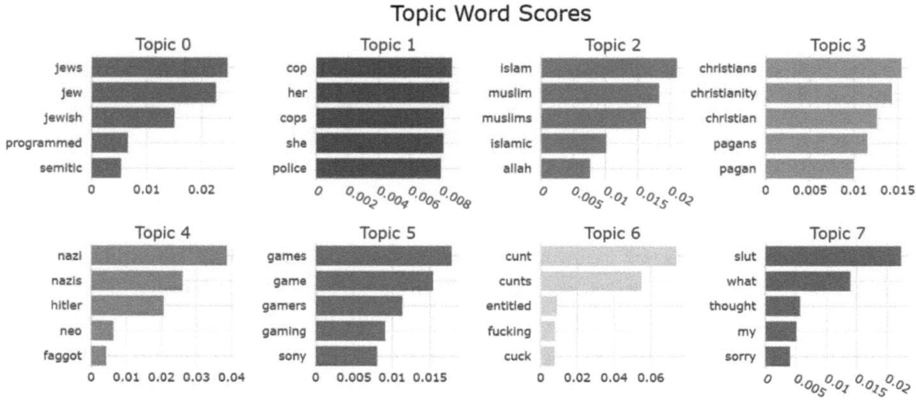

Fig. 4. Weights of each word in a particular topic

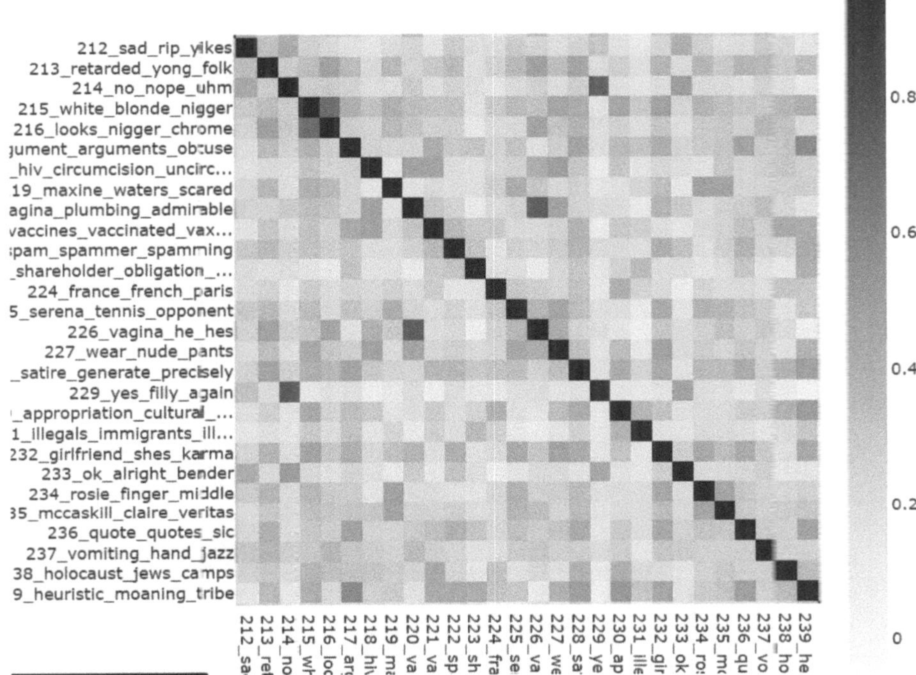

Fig. 5. Correlation Heatmap between the instances and the topics generated by BERT

In Fig. 6, we see the initial distribution of our dataset, containing around 44k tweets (25k non-hate and 19k hate). From the histogram, it's clear that the (neutral + positive) class contains approximately 25k tweets, while the (racism + sexism + insult) class holds about 19k tweets, reflecting the accuracy of the topic modeling process.

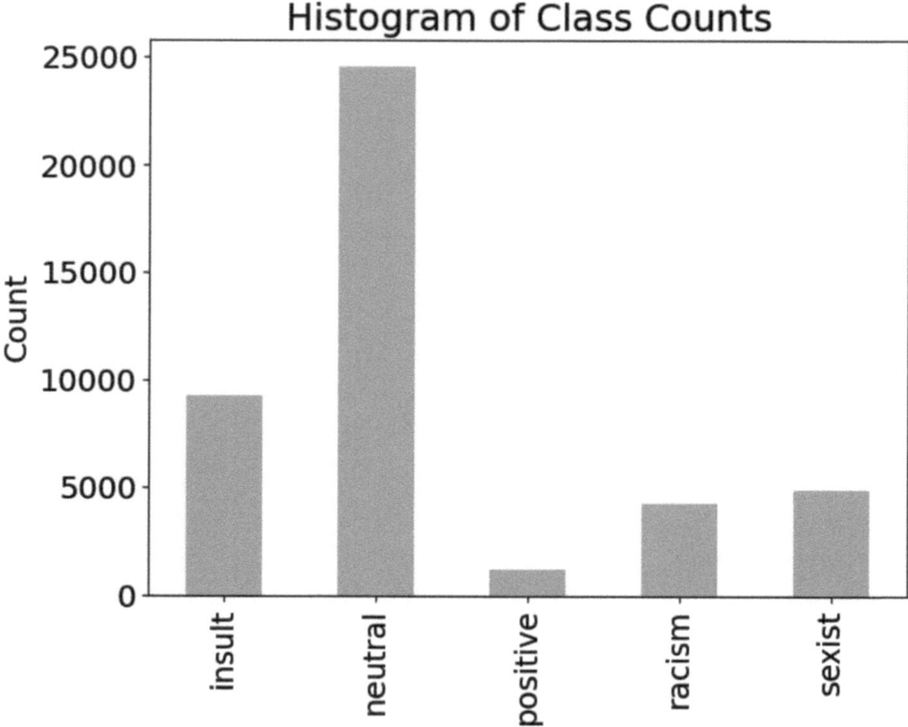

Fig. 6. Distribution of 44k tweets into 5 class labels

5　Conclusion and Future Work

This research presents a novel approach to hate speech detection by integrating topic modeling with traditional sentiment analysis techniques. Our hybrid model, combining TF-IDF with advanced topic modeling algorithms (BERTopic, NMF, and Top2Vec), has demonstrated promising results in identifying both explicit and less frequently seen forms of hate speech across different languages. The experimental results show that the integration of topic modeling features significantly enhances the performance of traditional machine learning classifiers, particularly in capturing context-dependent hate speech.

The cross-lingual application of our model, tested on both English and Tamil datasets, reveals its potential adaptability to different linguistic contexts. However, the discrepancy between training and test set performance highlights an important challenge in hate speech detection: the evolving nature of hateful language. Our analysis shows that while the model excels at identifying non-hate content (high recall), it faces challenges in precisely identifying hate speech when confronted with previously unseen patterns or contexts.

A key finding of our research is the effectiveness of combining multiple-topic modeling approaches. The synergy between BERTopic's contextual understanding, NMF's topic extraction, and Top2Vec's semantic clustering provides a more comprehensive feature set for classification algorithms. This multi-faceted approach allows for a better capture of how hate speech manifests across different topics and contexts.

Building upon our findings, several promising directions for future research emerge. A crucial next step would be the development of a real-time hate speech detection system. This would involve creating an efficient pipeline for processing streaming social media data, optimizing the topic modeling algorithms for faster computation, implementing an API for seamless integration with various social media platforms, and developing a user-friendly dashboard for content moderators. Given the success with English and Tamil datasets, future efforts should aim to extend the model to more languages, particularly low-resource languages, develop language-agnostic features for improved cross-lingual performance, and create a unified multilingual model capable of detecting hate speech across different languages simultaneously.

References

1. Antypas, D., Camacho-Collados, J.: Robust hate speech detection in social media: a cross-dataset empirical evaluation (2023). https://arxiv.org/abs/2307.01680
2. Philipo, A.G., Sarwatt, D.S., Ding, J., Daneshmand, M., Ning, H.: Cyberbullying detection: exploring datasets, technologies, and approaches on social media platforms (2024). https://arxiv.org/abs/2407.12154
3. Mody, D., Huang, Y., de Oliveira, T.E.A.: A curated dataset for hate speech detection on social media text. Data Brief **46**, 108832 (2023). https://doi.org/10.1016/j.dib.2022.108832
4. Chakravarthi, A., et al.: Overview of abusive comment detection in Tamil-ACL 2022, pp. 292–298 (2022). https://doi.org/10.18653/v1/2022.dravidianlangtech-1.44
5. Sethia, K., Saxena, M., Goyal, M., Yadav, R.K.: Framework for topic modeling using BERT, LDA and K-means. In: 2022 2nd International Conference on Advance Computing and Innovative Technologies in Engineering (ICACITE), pp. 2204–2208 (2022). https://doi.org/10.1109/ICACITE53722.2022.9823442
6. Subramani, S., Sridhar, V., Shetty, K.: A novel approach of neural topic modelling for document clustering. In: 2018 IEEE Symposium Series on Computational Intelligence (SSCI), pp. 2169–2173 (2018). https://doi.org/10.1109/SSCI.2018.8628912
7. Yuan, C., Zhang, S.: A weighted topic modeling approach based on word embedding. In: 2023 2nd International Conference on Artificial Intelligence and Computer Information Technology (AICIT), pp. 1–5 (2023). https://doi.org/10.1109/AICIT59054.2023.10277754
8. Zhou, M., Kong, Y., Lin, J.: Financial topic modeling based on the BERT-LDA embedding. In: 2022 IEEE 20th International Conference on Industrial Informatics (INDIN), pp. 495–500 (2022). https://doi.org/10.1109/INDIN51773.2022.9976145
9. Limwattana, S., Prom-on, S.: Topic modeling enhancement using word embeddings. In: 2021 18th International Joint Conference on Computer Science and Software Engineering (JCSSE), pp. 1–5 (2021). https://doi.org/10.1109/JCSSE53117.2021.9493816
10. Zhang, F., Gao, W., Fang, Y., Zhang, B.: Enhancing short text topic modeling with fasttext embeddings. In: 2020 International Conference on Big Data, Artificial Intelligence and Internet of Things Engineering (ICBAIE), pp. 255–259 (2020). https://doi.org/10.1109/ICBAIE49996.2020.00060
11. Yamunathangam, D., Priya, C.B., Shobana, G., Latha, L.: An overview of topic representation and topic modelling methods for short texts and long corpus. In: 2021 International Conference on Advancements in Electrical, Electronics, Communication, Computing and Automation (ICAECA), pp. 1–6 (2021). https://doi.org/10.1109/ICAECA52838.2021.9675579

12. Liu, L., Tang, L., He, L., Zhou, W., Yao, S.: An overview of hierarchical topic modeling. In: 2016 8th International Conference on Intelligent HumanMachine Systems and Cybernetics (IHMSC), vol. 1, pp. 391–394 (2016). https://doi.org/10.1109/IHMSC.2016.101
13. Paunkoska, S., Mirceva, G.: Hate speech on social platforms through the application of ML and NLP methods. In: 2024 47th MIPRO ICT and Electronics Convention (MIPRO), pp. 1105–1108 (2024). https://doi.org/10.1109/MIPRO60963.2024.10569905
14. Kumar, S., Musharaf, D., Bhushan, B.: Taming the hate: machine learning analysis of hate speech. In: 2024 2nd International Conference on Disruptive Technologies (ICDT), pp. 269–273 (2024). https://doi.org/10.1109/ICDT61202.2024.10489362
15. Shawkat, N., Simpson, J., Saquer, J.: Evaluation of different ML and text processing techniques for hate speech detection. In: 2022 4th International Conference on Data Intelligence and Security (ICDIS), pp. 213–219 (2022). https://doi.org/10.1109/ICDIS55630.2022.00040
16. Rathod, R.G., Barve, Y., Saini, J.R., Rathod, S.: From data pre-processing to hate speech detection: an interdisciplinary study on women-targeted online abuse. In: 2023 3rd International Conference on Intelligent Technologies (CONIT), pp. 1–8 (2023). https://doi.org/10.1109/CONIT59222.2023.10205571

Tokenization Techniques in Multimodal Learning: A Survey of Text and Image Tokenizers

Yogesh Kakde[✉] [iD], Sandeep Kumar Dash[iD], and Ranjita Das[iD]

National Institute of Technology, Mizoram, Aizawl, India
ykakde@gmail.com, sandeep.cse@nitmz.ac.in

Abstract. With the information about the dramatic rise and success of deep learning in natural language processing and computer vision, it should be noted that one of the central innovations key to these advances is the tokenizers and encoders that encodes the unstructured data into a format digestible by the models [3]. This article attempts to provide a very detailed overview of some of the most common text encoders and to kenizers nowadays, epitomized though not limited to BERT, GPT - 2 and T5, which have gained the status as a foundation of NLP tasks. Here, we look into the details of their architectures, tokenization approaches and how they solve linguistic problems, such as rare and out-of-vocabulary words [11]. Similarly, we present different image encoder and tokenizers such as CLIP, Vision Transformer, VQ - VAE which have been instrumental for cross-modal learning and image synthesis tasks. Drawing parallels between text and image tokenization procedures, we discuss the advantages and disadvantages of each method with a special focus on multimodal scenarios where interrelation between image matching and text encoding is of paramount meaning [1].

Keywords: text tokenizer · text encoders · image tokenizer · LSTM encoder · CLIP · T5 · BERT · transformers · Deep Learning

1 Introduction

The last decade has witnessed an unprecedented evolution in deep learning that has changed computer vision and natural language processing for the better. They allow computers to read, write, and alter text and pictures more correctly than ever before. Two of the most crucial innovations that enabled us to do these tasks are tokenization and encoding, which are responsible for transforming all of the available raw data into a format that deep learning models are able to work with. Such is the case with text and image modalities where the choice of tokens or encoders can determin stically affect the quality of certain text or image or even multi-modal tasks such as image generation from text captions and vice versa [1]. Similarly, in this domain, many have reported state-of-the-art results by employing provably efficient tokenizers such as BERT (Bidirectional Encoder Representations from Transformers), GPT-2 (Generative Pretrained Transformer 2) or T5 (Text-to-Text Transfer Transformer). Tokens such as WordPiece,

C. Modi et al. (Eds.): MIND 2024, CCIS 2736, pp. 317–328, 2026.
https://doi.org/10.1007/978-3-032-14531-4_27

Byte-Pair Encoding (BPE), or SentencePiece have effectively cut texts into subword constituents to deal with language barriers such as rare words, many language resources, and out of vocabulary words. These methods have improved the representation of text, but they have also made it possible for the models to be enforced on different tasks from translation to answering questions [3].

Equally, the practice of image tokenization has also taken great strides with the advent of vision transformers as well as cross-modal architectures. Under methods such as Vision Transformer (ViT), which disintegrates the image into patches and uses patch tokens to encode the images, and CLIP (Contrastive Language-Image Pretraining), a system which uses a common encoder to embed text and image, bears witness to radical changes in how visual information is encoded and combined with language.

The organization of this paper is as follows: Sect. 1 gives glimpse to text and image tokenizer followed by literature survey in Sect. 2. Section 3 looks into different types of text tokenizers and encoders including BERT, GPT-2, T5 and others. Section 4 discusses the most notable image tokenization approaches, the CLIP, the Vision Transformer and VQ-VAE. Section 5 lists various application domains where tokinizers can be utilized. In the end, Sect. 6 defines the accessibility and availability of various tokenizers.

2 Literature Survey

Tokenization is an integral part of text processing (NLP) and computer vision, where unstructured raw data is converted into an organized structure that can be understood by the models. This section reviews the most well-known approaches in text and images tokenization and their respective contributions towards solving the problems.

2.1 Text Tokenizers and Encoders

BERT is a neural network model based on a bidirectional transformer, commonly adopted for NLP purposes like text classification, question answering or named entity recognition. BERT utilizes a WordPiece tokenization technique, which decomposes words into subword components, to efficiently deal with low-frequency and out-of-vocabulary words. This capability allows BERT to produce informative contextual embeddings, facilitating its acceptable performance on many NLP tasks. Typically, according to [9], the model GPT-2 is an autoregressive transformer aimed at text generation. Instead of Word and Character based tokenization, it follows the concept of Byte-Pair Encoding which creates subword units by combining sequences of characters that occur frequently. In the case of T5, a SentencePiece tokenizer is utilized which focuses on creating subword units while looking at the raw text only. This data-centric design allows T5 to be used for many applications such as translation, summarization, and question- answering where more than one language and domain are present [1].

2.2 Image Tokenizers and Encoders

The Vision Transformer is put forward for image adaptation [2] which is innovatively researched in transforming structural elements of the image, whereby the image is divided into a fix-sized patches or tokens. With this method, convolutional layers aren't necessary, hence the model is able to utilize selfattentive mechanisms. Image segmentation in CLIP is often done through the Vision Transformer (ViT) [4] or a Resnet based image encoder that encodes images into visual embeddings. The embeddings created in this process are further refined through a distinct process by training them together with text embeddings. This allows very effective zero-shot image classification. As proposed by [8] is a generative model which takes an input image and extracts it into a series of discrete latent code through vector quantization. These discrete tokens are applicable in image compression, image generation, and image reconstruction.

3 Text Tokenizers: Types and Techniques

In the last few years, Long Short-Term Memory (LSTM) architecture was widely used for text encoding in text-based tasks in the domain of natural language [10] processing. As a subtype of recurrent neural network architecture, LSTMs were good in accomplishing sequential dependencies in data and thus were perfect in applications such as machine translations, sentiment analysis and text categorization. The dominance of LSTM encoders in text encoding quickly grew because their advantages in decoding of encode long-term information as well as remedying the gradient dissipation aspect in RNNs. However, this, and other advantages of their applicability, was soon overshadowed by the rush of transformer-based models. Yet, they are lie, superseded by the dependency focusing deep learning models, transformers, such as BERT, GPT and T5, without justification. These architectures augmented with self-attention mechanisms are more efficient for parallel processing and effective for facial global context reasoning which explain their acceptability in modern high performance NLP.

3.1 BERT Family: WordPiece Tokenization and Bidirectional Context

BERT (Bidirectional Encoder Representations from Transformers) so far the best method transforms sequential text processing into a contextualized visual representation of text together with associated graphical and non-visual information.

Core Architecture: Refer Fig. 1. BERT [1] is built with a stack of transformer encoders which uses multi-head self-attention to model word dependencies in both directions since the dependency is established. BERT is pre-trained with two objectives namely masked language model (MLM) and next sentence prediction (NSP).

Tokenization Technique: In terms of word encoding, BERT employs WordPiece, which subdivides infrequent words into portions called subword units. It therefore uses this approach to further segment rare and out of vocabulary words into known subwords in order to preserve meaning. For example, "unhappiness" might get the structure of tokenizing it like this: ["un", "happy", "ness"], this ensures vocabulary size remains within limits while semantic meaning is retained. Advantages: WordPiece tokenizer should be considered as one of the crucial factors for the efficiency of BERT integration in many languages and domains. Suchasystem'sunderstanding of language is bidirectional, allowing one to comprehend the context in both forwarding and reverse directions which increases the efficiency of the text classification and named entity recognition as well as question-answering tasks.

These days some derivatives are widely used – for example, RoBERTa or DistilBERT, which changes the strategy of the training process improving performance and efficiency while still employing the same mechanism of tokenization for the variant zuhugalale.

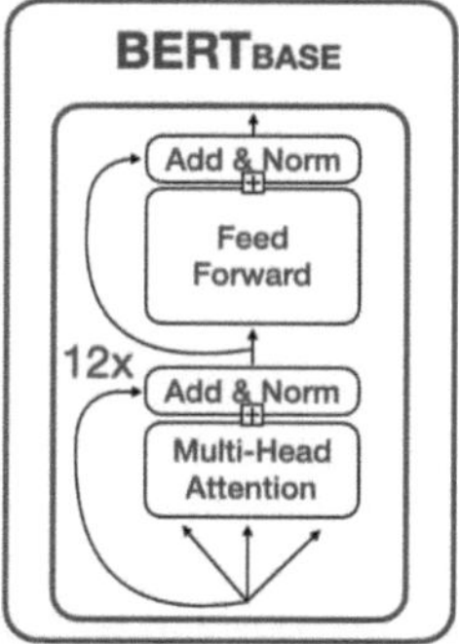

Fig. 1. BERT Base Architecture

3.2 T5 Family: SentencePiece and Text-To-Text Framework

The classifier T5 (Text-to-Text Transfer Transformer) attempts to encapsulate all of the accessible natural language processing (NLP) tasks under a domain where designated every problem in turn is a text to text problem. T5 reformulates previously viewed tasks that comprise translation, summarization among others to be sequence to sequence tasks [14].

Core Architecture: T5 follows this architecture constituting the use of a transformer encoder and decoder. Where the encoder takes text input while the decoder produces text output. This enables the model to incorporate text in different formats as an input as well as an output.

Tokenization Techniques: T5 utilizes a language-independent subword method referred to as SentencePiece, which works without any existing boundaries. Sentence-cePiece can work above and below the word level; this makes possible bigger or smaller vocabularies while managing various languages. In contrast to the later BERT's Word-Piece, the above mentioned SentencePiece makes no assumption that a space is present between each word hence it is more flexible in relation to different languages and domains.

Advantages: With the ability to vary the vocabulary formulated through Sentence-Piece, flexibility is assured and this further makes T5 more relevant in a various NLP task. The text to text structure enhances the model's performance across a host of tasks making it possible to be used more effectively.

3.3   GPT Family: Autoregressive Models and Byte Pair Encoding

GPT-2 and GPT-3 are autoregressive models which became popular due to their effectiveness in machine-generated text [9] and language tasks. In contrast to the bidirectional BERT model, where one input encodes the full sequence, the GPT models autoregressively compute the next token conditional on the sequence of previous tokens [1]. Core architecture: The architecture of GPT models is like the transformer decoder which utilizes the sequence of tokens in a left to right order. This structure lends itself well to text generation as only one token is produced at a time and is dependent on the generated tokens before it. Refer Fig. 2

Tokenization Technique: This refers to how GEP models encode tokens where Multipart Encoding utilizes BPE whereby frequently adjacent characters are fused into subword tokens. BPE is useful in splitting unusual words known or otherwise into more manageable elements making it ideal for facilitating efficient generation of text

3.4   BART Family: Bidirectional and Autoregressive Pre-Training

BART, is a combination of a bidirectional model such as BERT and autoregressive models such as GPT. BART is a sequence-to-sequence model that is used in text generation, summarization and even translation.

Core Architecture: In BART transformer encoder-decoder architecture is followed but with some variations since tri-axial encoding of illustrated text is done using the encoder and the generation of the text is done using the decoder in an autoregressive manner. This facilitates BART in performing well in the jobs that require transformation or reconstruction of texts.

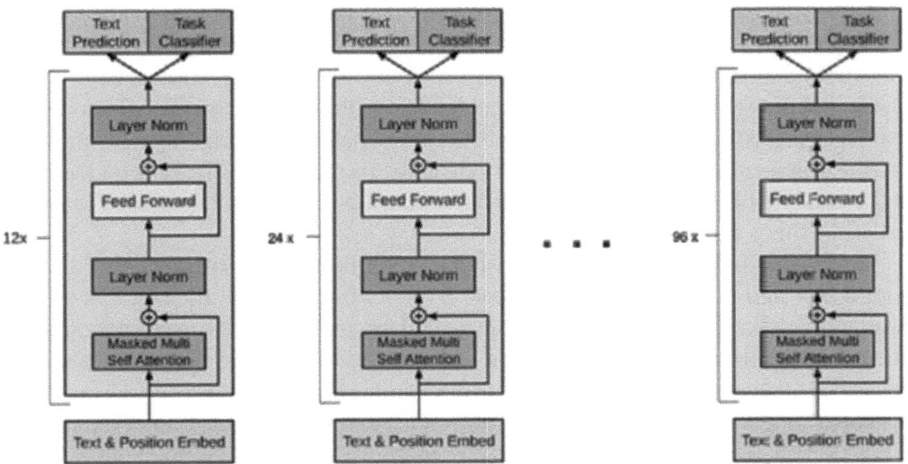

Fig. 2. GPT-3 attention-based architecture

Tokenization Technique: As in the case of GPT, BART employs BPE for better efficiency in terms of tokenization. BPE helps in BART to compose complex or less occurring words into smaller word components that are then compressed into single syllables or sounds.

Advantages: Summarization and translation are examples of activities where BART has proven to be quite effective due to its good performance on text restructuring activities. However, given both sequential and parallel text data, bidirectional encoding and autoregressive decoding are very suited for the purposes. The BART family comprises subspecies mBART and its derivative – Multilingual BART, in which the multilingual capabilities of BART are implemented [5].

4   Image Tokenizers: Architectures and Techniques

Understanding the outlines of images and enabling the computer with different capabilities is one of the functions performed by image tokenizers. Like other tokenizers, image tokenizers divide images into their fundamental constituent units like patches or pixels, then feed the units into a neural network. In this segment, we look at the architectures and the tokenization techniques employed in some of the major families of image tokenizers and the mechanisms and uses of those tokenizers.

4.1   Vision Transformer (ViT) Family: Image Patching and Transformer Encoding

The ViT architectures by [2] introduced the first vision transformer and pioneered a new direction in image recognition by transforming images with a transformer architecture. In comparison to conventional convolutional neural networks (CNNs), ViTs regard images as sequences of patch embeddings analogous to words in a sentence.

Core architecture: In ViT architecture, the input image is split into nonoverlapping fixed-sized patches (for example 16×16 pixels) which are then flattened and linearly transformed separately for each patch. These patches are where the word token structures reside and thus are additional tokens containing stripling complements, which are also known as skip structures. Trading with those token patches sequentially adds more and more accruement to the image token; they are able to represent patch tokens for the global understanding of the grid-like image and relationships of all the information presented under the images.

Tokenization Technique: The tokenization in ViT involves the segmentation of image with theroi into non-overlapping sections. It is worth to mention the fact that all non-overlapping sections are turned into embedded vectors and positional encodings are utilized to preserve the relative position of these patches. Due to the patch based tokenization scheme, it is also possible for the whole image to be processed in parallel instead of sequentially like how tokens are spitted and processed in transformers.

4.2   CLIP Family: Multimodal Tokenization for Cross-Modal Understanding

CLIP, short for Contrastive Language–Image Pretraining, [9] was developed by Tech Company and serves as a bridge between image representation and text information by constructing a joint model. Image encoder of CLIP works in horizontal vertical parallel with the text encoder making it easy to relate images and text in one encoded structure.

Core architecture: Depending on the model structure adopted, the image encoder could either be a ResNet or a Vision Transformer. In both instances, images are converted into n-grams by dividing them into small sections, (compression in case of MT) or through convolutional operations if a ResNet is employed.

Tokenization Technique: In the transformer variant, images are tokenized into patches similar to the ViT approach. For the ResNet editions of the CLIP models, things are different since the tokenization is done in features undergoing convolutional layers which penetrate the pixel information and map them into feature maps These visual tokens are then pushed into a common area to allow for their comparison with textual tokens through the use of contrastive learning.

Advantages: CLIP expands how multimodal tasks are treated by its integration of a tokenization process for both images and text into a single model. Its token s cross-modal in nature requiring low-task specific prompting making the model flexible in training with new activities.

4.3 SWIN Transformer Family

Hierarchical Tokenization for Efficient Scaling SWIN Transformer has recently been published. At the heart of it is one of the first sandwitch visual-transformer architecture that scales well to high resolution images and their application to densly sampled interactions such as object detection, and segmentation tasks [6].

Core Architecture: Like the ViT model, the SWIN architecture tokenizes images by splitting them into non-overlapping patches. When compared to ViT however, SWIN adds a hierarchical structure where patches are successively merged to produce multi-scale representations. This architecture bears the shape of the feature pyramid such that is often utilized in CNNs and increases the speed of computation. Tokenization Technique: The tokenization process in SWIN can be initiated with minimal image patches for instance 4 pixels x 4 pixels and these patches are collected into windows. Windows processed by a adaptive form of attention known as shifted window, which makes calculating self-attention in the large area of the image more computationally efficient.

Advantages: Thanks to SWIN's hierarchical method of tokenizing the input and observations it makes, it is possible to work with images of high-resolution and perform very well with the perception and understanding that require a lot of details such as object detection and segmentation.

4.4 MobileViT Family: Lightweight Tokenization for Mobile and Edge Devices

MobileViT is classified as a vision transformer, which is developed in mobile and edge requirements, leveraging the benefits of existing CNNs and transformers to solve lightweight model concerns in an effective way [7].

Core Architecture: A complex core of MobileViT is a hybrid one, in which first, local convolutional blocks are applied to get local features, and then transformers are included to analyze global features. This kind of intermediate approach to the problem allows the designers to combine high efficiency levels with high accuracy and thus comes in handy in environments that are highly restricted in resource availability.

Tokenization Technique: MobileViT employs small convolutional kernels to scan for image feature maps ready for passing to transformer layers for processing. This kind of tensorization decreases the computational load, yet allows retention of necessary visual features.

4.5 VQ-VAE Family

Discrete Latent Representation Of an Image Tokenization VQ-VAE or Vector Quantized Variational Autoencoder is a model generating Indian nouns by modeling their visuals and recording visual verbal portrayals [8].

Core Architecture: VQ-VAE implements an encoder-decoder model where the encoder converts the input image into discrete latent codes that are treated as tokens. These tokens are then used by a decoder to generate back the image. It also facilitates the learning of a latent representation of the image that is small and explainable in contrast as just many index numbers due to the quantization step.

Tokenization Technique: In VQ-VAE, as explained previously, tokenization also takes place in the latent space where the continuous picture embeddings are dissected in terms of domains or tokens. These tokens are then converted into images or some other visual outputs by the decoder.

Advantages: VQ-VAE approach addresses the image synthesis problem with discrete latent variables making it possible to produce superior images while also learning good interpretable representations. Thus it is perfect for image compression and generative models.

5 Application Domains of Multimodal Architectures

The areas of application of multimodal architectures built on image and text tokenizers have developed due to the contribution of the two modalities for better results. Lets discuss some application areas.

- **Image Captioning:** Image captioning fuses visual and textual elements to form a caption that can be interpreted by a human. Models utilize image tokenizers (e.g., CLIP, ViT) for visual feature extraction and text tokenizers like BERT and GPT to provide context for producing descriptive text for social media and accessibility purposes.
- **Visual Question Answering (VQA):** VQA tasks involve answering questions about images. Such architectures, including CLIP or Vision Transformers, and a text tokenizer, integrate the comprehension of both channels and can respond accurately, enhancing educational tools and interactive AI.
- **Multimodal Retrieval:** Multimodal retrieval systems handle content representation queries that combine both image and text. By using tokenizers from both CLIP and T5 models, a single embedding space is created that accommodates all text and visual information, easing retrieval.

- **Text-to-Image Generation:** Text-to-image generation involves the creation of images from descriptive text. Models such as DALL-E and Stable Diffusion use text tokenizers to comprehensively detail input descriptions and image tokenizers to generate images [10].
- **Text-to-Text Generation:** Text-to-text generation involves transforming text into new forms using input from existing text. This aids in tasks such as summarization, translation, and paraphrasing. Models like T5 and BART use text token generators to comprehend context and produce coherent text, with applications in content generation, e-learning, and customer support.
- **Sentiment Analysis and Emotion Recognition:** In the context of literature studies and sentiment analysis, multimodal architectures analyze text and related multimedia (e.g., images) to understand sentiment better. Using text tokenizers (e g., BERT) and image tokenizers (e.g., EfficientNet), these systems help refine sentiment detection, which is critical for market research and social media strategies.

Apart from that Healthcare and Medical Imaging In healthcare, medical imaging medical error diagnosis, autonomous Vehicles Islands faced several challenges, content generation and creative AI etc. are the area where tokenization can be implemented while building the multi-model architecture. See Table 2.

Table 1. Text and Image Tokenizers with Application Domains

Tokenizer Type	Tokenizer Name	Application Domains
Text Tokenizer	BERT	Sentiment Analysis, VQA, Content Generation
	T5	Text-to-Text Generation, Text-to-Image Generation
	GPT	Content Generation, Chatbots
	XLNet	Sentiment Analysis, VQA
	BART	Text-to-Text Generation, Summarization
	Pegasus	Text-to-Text Generation, Summarization
	Electra	Sentiment Analysis, Text Classification
Image Tokenizer	ViT	Image Captioning, VQA, Content Generation
	CLIP	Multimodal Retrieval, Text-to-Image Generation
	DINO	Healthcare, Video Analysis
	SWIN	Autonomous Vehicles, Image Captioning
	MobileViT	Autonomous Vehicles, Healthcare
	EfficientNet	Sentiment Analysis, Image Classification

(continued)

The Table 1 aims to summarize different text and image tokenizers and their applications giving a picture. It includes texts on BERT, T5, GPT and other text tokenizers with extending functionalities of emotion detection, text generation, and image description

Table 1. (*continued*)

Tokenizer Type	Tokenizer Name	Application Domains
	ResNet	Healthcare, Video Analysis
	VQ-VAE	Content Generation, Image Compression

writing. ViT, CLIP, DINO, image tokenizers are also depicted by their application in health care, video analysis, multimodal retrieval, autonomous and other fields.

The Table 2 classifies the application domains which deploy the techniques of text and image decoders into several categories. Each area mentions particular tokenizers that have been developed for enhancing the performance. For instance, Image Captioning employs BERT and T5 along with image tokenizers like ViT and CLIP which help in providing descriptive captions.

6   Availability and Accessibility of Tokenizers

The existing world of text and image tokenizers has become more open allowing for more activities in creating designs that utilize multiple modes [12]. Hugging Face Transformers: This well-known library contains a complete set of transformer models with many pretrained text tokenizers commonly utilized in typical.

Table 2. Application Domains with Associated Text and Image Tokenizers

Application Domain	Text Tokenizers	Image Tokenizers
Image Captioning	BERT, T5	ViT, CLIP, SWIN
Visual Question Answering (VQA)	BERT, XLNet, T5	ViT, CLIP
Multimodal Retrieval	T5, BERT	CLIP, DINO
Sentiment Analysis	BERT, Electra, RoBERTa	EfficientNet, ResNet
Healthcare and Medical Imaging	BART, T5	DINO, ResNet
Autonomous Vehicles	BERT, T5	MobileViT, EfficientNet
Content Generation and Creative AI	GPT, T5	CLIP, VQ-VAE
AR and VR	BERT, GPT, T5	ViT, EfficientNet
Text-to-Image Generation	T5, GPT	CLIP, VQ-VAE Stable Diffusion
Text-to-Text Generation	T5, BART, Pegasus	-
Video Analysis	BERT, T5	ViT, DINO
Personalized Recommendations	BERT, GPT	CLIP, EfficientNet

NLP tasks like BERT, T5, and GPT among others. These models can be downloaded and fine-tuned for various tasks with relative ease owing to the apparent availability of a very easy-to-use API for the hub.

- **TensorFlow and PyTorch:** Both TensorFlow and PyTorch come with tools or community contributions that can be used to implement different image tokenizers, such as Vision Transformers (ViT) and EfficientNet.
- **OpenAI's CLIP and DALL-E:** CLIP and DALL-E are particularly exciting models provided by OpenAI, with the models and code available to the public. Researchers working on multimodal applications can greatly benefit from these models and the associated code repositories.
- **Research Papers and Model Cards:** Many models are released with accompanying research papers that explain how the models are formulated and provide the results obtained. Model cards take this a step further by offering insights into the potential users of the model, its applications, and reusable features.
- **Community Contributions:** GitHub is a platform rich with numerous repositories that share specific implementations of existing tokenizers and their modifications. These community contributions foster creativity and allow researchers to build upon each other's work, making progress more efficient and faster.

Most of these includes pre trained models which can be directly used for applications and researches.

7 Conclusion

In this paper, we addressed the remarkable progress achieved in text and image tokenizers incorporated within the multimodal architectures. Various types of text tokenizers were surveyed, including BERT, T5, and GPT, pointing out the features of each on its intended application such as sentiment analysis, content generation, and text to image generation respectively [11]. Correspondingly, image tokenizers such as ViT, CLIP, CQ-VAE and others were also analyzed with respect to image caption generation and other healthcare applications. Reintegration of these tokenizers helps to enhance the understanding of multimodal data opening doors to different applications in education, healthcare, creative Artificial Intelligence.[13] As developments in this area continue, with developments of these methods of tokenization more improvements will be observed in the capabilities of the multimodal systems, which in turn improves the direction of the AI systems developed [3]. New tokenization methods may be employed in the future work and their applications will be investigated in order to fully understand the emergence of new applications in multimodal understanding.

References

1. Bahani, M., Ouaazizi, E.: The effectiveness of t5, gpt-2, and bert on text-to-image generation task. Pattern Recogn. Lett. **173**, 57–63 (2023)
2. Dosovitskiy, A., Beyer, L.: An image is worth 16×16 words: transformers for image recognition at scale. In: Proceedings of the 9th International Conference on Learning Representations (ICLR) (2021)

3. Khan, S., Naseer, M.: Transformers in vision: a survey. ACM Comput. Surv. (CSUR) **54**(10s), 1–41 (2022)
4. Kim, G., Ye, J.C.: Diffusionclip: text-guided image manipulation using diffusion models (2021)
5. Lewis, P., Liu, Y.: Bart: denoising sequence-to-sequence pre-training for natural language processing. In: Proceedings of the 58th Annual Meeting of the Association for Computational Linguistics, vol. 1, no. 1, pp. 787–798 (2020)
6. Liu, Z., Lin, Y., Wu, Y.: Swin transformer: hierarchical vision transformer using shifted windows. In: Proceedings of the IEEE/CVF International Conference on Computer Vision (ICCV), pp. 10012–10022. IEEE (2021)
7. Mehta, S., Ranjan, S., Jha: Mobilevit: Light-weight, general-purpose, and efficient vision transformer. In: Proceedings of the IEEE/CVF Conference on Computer Vision and Pattern Recognition (CVPR), pp. 2511–2521. IEEE (2022)
8. van den Oord, A., Vinyals, O., Kavukcuoglu, K.: Neural discrete representation learning. In: Advances in Neural Information Processing Systems (NeurIPS), pp. 6309–6318. NeurIPS (2017)
9. Radford, A., Kim, J.W.: Learning transferable visual models from natural language supervision. In: Proceedings of the 38th International Conference on Machine Learning, vol. 139, pp. 8748–8763 (2021)
10. Reed, S., Akata, Z.: Generative adversarial text to image synthesis. In: International Conference on Machine Learning, pp. 1060–1069. PMLR (2016)
11. Zhang, C., et al.: Text-to-image diffusion models in generative AI: a survey. arXiv preprint arXiv:2303.07909 (2023)
12. Zhang, H., Song, H., Li, S., Zhou, M., Song, D.: A survey of controllable text generation using transformer-based pre-trained language models. ACM Comput. Surv. **56**(3), 1–37 (2023)
13. Zhang, K., Wu, F., Zhang, G., Liu, J., Li, M.: Bva-transformer: Image-text multimodal classification and dialogue model architecture based on blip and visual attention mechanism. Displays **83**, 102710 (2024)
14. Zhang, L., Rao, A., Agrawala, M.: Adding conditional control to text-to-image diffusion models. In: Proceedings of the IEEE/CVF International Conference on Computer Vision, pp. 3836–3847 (2023)

Enhanced Fruit and Vegetable Image Recognition with CNN and MobileNetV2

Pusapati Varshith Reddy[1] , Prabhakar Kandukuri[1] ,
Medikonda Asha Kiran[2]([✉]) , and Prathima Kativarapu[3]

[1] Chaitanya Bharathi Institute of Technology, Hyderabad, Telangana, India
[2] School of Engineering, Anurag University, Hyderabad, India
ashakiran2@gmail.com
[3] Department of Computer Science and Engineering, SRK Institute of Technology,
Enikepadu, Vijayawada, India

Abstract. The goal of fruit and vegetable image recognition is to identify accurately these items in the photographs. The traditional applications utilize Convolutional Neural Networks (CNNs) with the MobileNetV2 architecture to efficiently recognize various fruits and vegetables. MobileNetV2's efficient design enables deployment on computationally constrained Mobile and embedded devices. The model, trained on an extensive fruit and vegetable image dataset, uses a CNN with MobileNetV2 and employs advanced preprocessing and data augmentation techniques such as rotation and scaling to improve generalization. MobileNetV2 was preferred over MobileNet in this study because it incorporates inverted residual blocks with bottleneck features, supports input sizes greater than 32×32, and has a lower parameter count. The experimental results show that MobileNetV2 offers improved accuracy and computational efficiency in fruit and vegetable recognition compared with traditional CNN models.

Keywords: Convolutional Neural Network · MobileNetV2 · Fruits and Vegetables · Image Processing · Transfer Learning

1 Introduction

Recognition of Fruits and Vegetables in Agriculture
Recognition of vegetables and fruits is significant in various agricultural processes, including automated sorting, crop monitoring, and quality assurance. In modern agriculture, where consistency and efficiency are essential, machine learning algorithms have gained significant importance for image recognition. These algorithms speed up operations and ensure the consistency and quality of food products. CNNs have proven extraordinary effectiveness in image recognition tasks due to their ability to extract complex features from data. CNNs brilliant in recognizing shapes, patterns, and textures in images, which is crucial for identifying different types of vegetables and fruits accurately [1–6].

This paper uses MobileNetV2 as an alternative to traditional CNNs for vegetable and fruit recognition in agricultural applications. MobileNetV2 is prominent for its lightweight architecture, making it ideal for mobile and embedded vision applications [7–12]. The MobileNetV2 architecture minimizes computational complexity while preserving high accuracy. These convolutions significantly reduce the number of parameters in the model, lowering computational costs and allowing for efficient operation on resource-constraint devices. MobileNetV2 is suitable for real-time agricultural applications, where efficiency and speed are crucial [7–9,13].

Advancing Agricultural Applications with MobileNetV2

While CNNs have been used for various image recognition and classification tasks in agriculture applications, the evolution of CNN models offers superior performance [14–17]. MobileNetV2 has become an efficient model that provides results comparable to more complex CNN architectures and significantly reduces computational capabilities. Using the novel inverted residual block design of MobileNetV2, we can recognize vegetables and fruits with high accuracy. This efficiency is helpful for agricultural applications, where it can lower expenses, increase productivity, and improve overall quality by processing agricultural images quickly and accurately.

The image recognition process by utilizing MobileNetV2 increases productivity in agricultural activities. By automating the identification of vegetables and fruits, agribusinesses and farmers can reduce manual labor, speed up operations, and ensure produce quality [18–20]. This research improves machine learning use in agriculture and aligns with the broader goal of promoting sustainable agricultural practices [21,22]. By minimizing resource consumption and enhancing efficiency, MobileNetV2 helps make farming processes more economically feasible and environmentally sustainable, ultimately improving the quality of life for those involved in the agricultural sector and beyond.

1.1 Contributions

This paper proposes an Enhanced Fruit and Vegetable Image Recognition with CNN and MobileNetV2 to address the above problems. The significant contributions are:

- **Application of MobileNetV2 in Image Recognition**: The study presents the application of MobileNetV2 architecture to fruit and vegetable image recognition. Due to its lightweight architecture, it is simple to use on embedded and mobile devices that have limited resources.
- **Efficiency and Precision in Image Recognition**: Using CNNs with the MobileNetV2 architecture, the research accomplishes accurate and efficient image recognition of various fruits and vegetables.
- **Advanced Preprocessing and Data Augmentation Techniques**: To expand the dataset and improve the model's performance, the study uses advanced preprocessing and data augmentation techniques, such as image rotation and scaling. These methods enhance overall effectiveness.

- **Use of Inverted Residual Blocks**: By emphasizing the use of inverted residual blocks, the research supports the choice of MobileNetV2 over MobileNet, its predecessor. This structural modification supports larger input sizes and improves performance with fewer parameters.
- **Enhanced Accuracy and Computational Efficiency**: Based on experimental findings, the suggested model outperforms conventional CNN models regarding accuracy and computational efficiency, particularly regarding fruit and vegetable identification.

1.2 Paper Organization

The paper is organized as follows: Sect. 2 reviews related works, Sect. 3 outlines the system model, Sect. 4 details the experimental analysis, Sect. 5 presents the experimental results, and Sect. 6 provides the conclusion.

Table 1. Functional Comparison of existing schemes with proposed scheme.

Scheme	Yuhui et al. [1]	Suryawanshi et al. [2]	Proposed Scheme
CNNs	√	×	√
Transfer Learning	√	×	√
Image Preprocessing	√	×	√
Data Augmentation	√	×	√
Evaluation Metrics	√	×	√
Data Collection	√	√	√
Dataset Structuring	×	√	√
Accuracy and Efficiency	√	×	√
Practical Application	√	×	√
Lightweight Model	√	×	√
Real-Time Application	√	×	√
Quality Assessment	×	√	√
Unlimited Scope	×	×	√
Ease of Implementation	×	×	√

2 Related Works

This section describes two traditional schemes: Image processing and advancement in image recognition of fruits and vegetables.

Blasco et al. [23] Presented a Machine Vision-Based Automated Fruit Quality Grading System. This paper introduces a machine vision system for automatically grading fruit quality. Using sophisticated imaging techniques the scheme evaluates and classifies fruit according to improved efficiency, consistency in sorting procedures, and quality attributes. Arivazhagan et al. [24] proposed a method for fruit identification. This study describes how to distinguish between fruits in photos by examining their texture and color characteristics. This method uses these visual cues to increase accuracy and differentiate between various fruit types. Bargoti et al. [25] presented a technique for fruit detection and yield estimation in apple orchards. This study focused on image segmentation techniques for yield estimation and orchard fruit detection. The method improves the accuracy of harvest quantity forecasts and fruit growth tracking.

Muresan et al. [20] presented a method for recognizing fruit images using deep learning methods. In this study, they investigated using CNN and deep learning methods for fruit recognition in images to obtain precise classification. The technique shown enhanced performance and has potential uses in retail and agriculture. Tian et al. [4] proposed Enhanced You Only Look Once, Version 3 (YOLO-V3) for Multi-Stage Apple Detection in Orchards. This technique focuses on multi-stage detection and is designed for use in orchards. It is based on an improved version of YOLO-V3. Yuhui et al. [1] proposed a lightweight CNN for efficient vegetable and fruit recognition. This scheme used Depthwise Separable Convolution, a specific CNN architecture, to identify vegetables and fruits. Later, Suryawanshi et al. [2] proposed a scheme called VegNet: Quality Image Dataset. The creation and distribution of a dataset primarily intended for machine learning-based vegetable quality assessment. Table 1 compares existing schemes and the proposed scheme.

2.1   Image Processing Technique

Advancements in feature extraction techniques have greatly improved image processing for fruit and vegetable recognition. By utilizing descriptors for texture, color, and shape, these methods can classify types with moderate accuracy [24, 25, 34]. Color histograms summarize the color composition of fruits and vegetables, capturing the distribution of colors in images. Analyzed through methods like Local Binary Patterns (LBP) and Grey Level Co-occurrence Matrices (GLCM), texture features provide detailed insights into surface properties. These techniques effectively distinguish between visually similar items with distinct surface textures, such as pineapple and tomato skin.

We further use a method combining color histograms and shape-based descriptors to differentiate vegetables and fruits. Shape-based descriptors capture the geometric properties of the product. We can distinguish between various vegetables and fruits, like tomatoes, cucumbers, bananas, and mangoes, that may have similar colors but different shapes.

2.2   CNNs

Image recognition is more common and applied in many fields, including food identification, fraud detection, medical health records, and facial recognition. To generate high-quality images, scientists have created sophisticated methods

[26–29]. Since CNNs can learn and extract hierarchical features from images by breaking them down into smaller, more manageable components, they have been used extensively in image recognition tasks in recent years [30–32,35]. In this work, we apply CNNs to identify a broad range of vegetables and fruits, outperforming previous techniques in terms of accuracy by grasping intricate patterns and variations in the photos.

Table 2. Comparison of MobileNetV2, EfficientNet, and ResNet.

Aspect	MobileNetV2	EfficientNet	ResNet
Accuracy	Moderate	High	Very High
Computational Cost	Low	Moderate	High
Efficiency	Very high	Balanced	Low
Pre-trained Models	Limited versatility	Highly versatile across tasks	Versatile, especially for feature extraction
Deployment Suitability	Mobile and edge devices	Balanced environments (cloud/edge)	High-performance systems
Strengths	Low latency, lightweight	Accuracy-efficiency balance	High feature richness, robust accuracy
Weaknesses	Lower accuracy for complex datasets	Requires moderate resources	High computational cost, risk of overfitting

2.3 Advance Model with MobileNetV2

The reason for integrating MobileNetV2 with CNNs is its enhanced performance and efficiency compared to conventional CNN architectures [7,8]. This feature results from using depthwise separable convolutions. This method splits the conventional convolution operation into two phases: pointwise convolution, which maps the outputs to a 1×1 convolution and combines the results, and depthwise convolution, which applies a single convolutional filter per input channel. More efficiency and less computational complexity result from this separation, drastically reducing the number of parameters and computations.

Furthermore, a significant advancement that improves performance is inverted residuals by MobileNetV2. Before the depthwise separable convolutions are applied, inverted residuals temporarily increase the number of feature channels, allowing the model to capture more detailed data while maintaining its overall lightweight performance. In this scheme, the network can better preserve the crucial information in the feature space, improving the model's accuracy. MobileNetV2 is lightweight and especially well-suited for tasks involving mobile and embedded systems with limited computational resources. Where the accuracy and efficiency are important for real-time processing with limited hardware demands in applications such as vegetable and fruit recognition techniques. In general, MobileNetV2 outperforms earlier CNN models in terms of performance and accuracy, which makes it an excellent option for settings with limited resources. The Table 2 shows the comparison between MobileNetV2, EfficientNet, and ResNet.

3 System Model

The system model is presented in this section.

3.1 Image Processing

To extract useful information from images, image processing techniques convert them into digital formats and apply various operations. When deploying particular signal processing techniques, image processing systems treat images as 2D structures. This area is essential for evaluating digital photos and refining and optimizing features like contrast and sharpness. It is necessary for many applications, including autonomous driving, medical imaging, digital forensics, satellite imagery, personal health records, and agricultural monitoring. It intersects with multiple disciplines, including computer science, mathematics, and engineering. Preprocessing methods like median filtering and Gaussian smoothing enhance image quality and lower noise, making it easier to analyze datasets. Advanced methods such as contrast stretching and histogram equalization adjust contrast and brightness to improve the visibility of features. More complex techniques for feature extraction, such as color histograms, texture descriptors (e.g., GLCM and LBP), and shape analysis, are used to describe various kinds of images.

3.2 CNN

A CNN is a deep learning process designed specifically for image recognition and processing. It has numerous layers, such as completely connected, pooling, and convolutional layers. The connectivity patterns present in the human brain, particularly in the visual cortex, which is necessary for visual perception and processing, served as the model for this architecture [33]. CNNs' convolutional layers use filters to recognize features in the input images that are essential characteristics, such as textures, edges, and complex patterns. These filters generate feature maps highlighting necessary visual modules as they pass over the image. These feature maps are downsampled by pooling layers by decreasing their spatial dimensions, usually via max or average pooling, by reducing the number of parameters such as amount, quantity, figure, total, and computational load. The network becomes less prone to overfitting and becomes more efficient. The high-level features extracted by the convolutional and pooling layers are processed by fully connected layers, which combine them to create the final predictions for classification or other tasks.

CNNs employ a method known as convolution, as opposed to classical neural networks, which rely on matrix multiplication. In this method, two functions are combined to exhibit how one modifies the shape of the other, enabling CNNs to process and analyze visual data more quickly.

3.3 MobileNetV2

MobileNetV2 is a condensed CNN architecture that offers effective performance with lower computational requirements, specifically designed for Mobile and embedded vision applications [7].

The MobileNetV2 architecture consists of three types of layers:

1×1 Convolutional Layer with ReLU6: This layer performs initial feature extraction.

Depthwise Convolutional Layer: This layer reduces the complexity and computational cost by applying convolutions separately to each channel.

1×1 Convolutional Layer without Non-linearity: This layer follows the depthwise convolution to combine features without introducing additional non-linearity.

These layers function together to reduce the number of parameters and computations while preserving the model's capability to capture complex features. The streamlined architecture of MobileNetV2 enables quicker inference, making it ideal for real-time applications. The MobileNetV2 architecture is illustrated in Fig. 1.

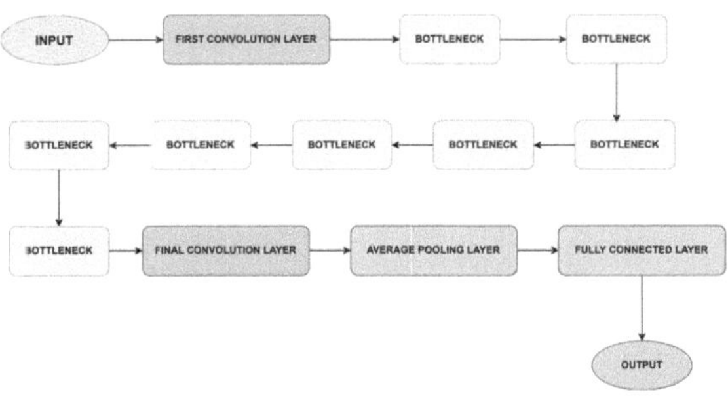

Fig. 1. Architecture of MobileNetV2.

4 Experiment Analysis

The experimental analysis of the proposed scheme is evaluated in this section.

4.1 Dataset

The dataset includes various vegetables and fruits varying in color, size, texture, and shape. Each image is labeled for supervised learning tasks. High-quality images with consistent resolution and labeling are crucial for efficient training and reliable model evaluation.

This dataset serves as a benchmark for algorithm performance, advances image recognition techniques, and fosters innovation in agricultural technology. It supports the development of accurate, efficient food production and distribution solutions, promotes sustainable practices, and enhances agricultural efficiency through machine learning and image processing.

4.2 Implementation

CNNs with MobileNetV2 are mainly used for fruit and vegetable image recognition. A dataset from Kaggle, comprising 100 images per category for training, ten for testing, and ten for validation, was used to train the model.

Fetching the Inputs and Training the Model:
The Kaggle dataset includes a variety of categorized fruits and vegetables. Pre-processing steps, such as resizing images to uniform dimensions and normalizing pixel values to a 0âĂŞ1 range, ensure consistent inputs and efficient model training.

MobileNetV2, known for its lightweight design and high performance, is well-suited for resource-constrained devices like mobile and embedded systems used in agriculture. Leveraging pre-trained weights from large datasets like ImageNet through transfer learning allows the model to utilize learned features while fine-tuning the specific fruits and vegetables dataset, which enhances its ability to distinguish between different types with greater precision.

Stochastic Gradient Descent (SGD) is used to optimize model parameters during training, with metrics like accuracy and loss monitored to fine-tune hyperparameters for optimal performance. Once trained, the model's real-world performance is evaluated. Successful deployment involves integrating the model into applications, such as automated sorting systems in agriculture or mobile apps for dietary monitoring, where it can reliably classify fruits and vegetables from visual inputs.

MobileNetV2 is highly efficient in memory usage, computational speed, and power consumption, making it ideal for mobile and embedded devices. With a compact model size of 14 MB to 16 MB, it runs smoothly on devices with limited memory. For images sized 224×224 pixels, it achieves inference times of 20âĂŞ50 ms on mobile devices like Qualcomm Snapdragon 845 and 150âĂŞ300 ms on edge devices such as Raspberry Pi, enabling real-time applications.

Its low power consumption, ranging from 0.3âĂŞ1 W on mobile devices and 2âĂŞ4 W on embedded systems, supports extended operation on battery-powered devices. This memory, speed, and energy efficiency make MobileNetV2 an excellent choice for real-time fruit and vegetable classification in resource-constrained environments.

5 Experimental Results

This section evaluates the proposed scheme's training and validation accuracy.

The dataset includes various types of fruits and vegetables identified based on shape, size, color, and texture. Our task is to recognize these types from photographs using CNNs with MobileNetV2. By employing this advanced model, we achieved high-quality recognition with excellent accuracy. The results are presented in Fig. 2.

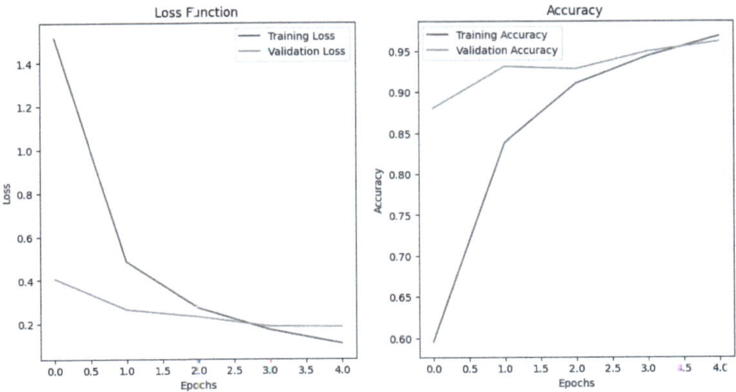

Fig. 2. Proposed System Results.

The research with image recognition of fruits and vegetables based on CNNs has achieved 84.79% accuracy. However, our best model is MobileNetV2 with CNNs, which has an accuracy of 96.11%. The output of recognition techniques given in Fig. 3

We used the kaggle data set to implement the proposed scheme [36].

Fig. 3. Output of Recognition Techniques.

6 Conclusion

We have investigated MobileNetV2 for Fruits and Vegetables Image Recognition and found it compelling, with an average accuracy of 96.11%. The robust performance of this scheme in managing the intricacies of agricultural environments has been demonstrated. The model is well-suited for implementation in practical applications like automated sorting and food processing quality assurance due to its high accuracy and efficient use of computational resources. Enhancing dataset diversity and investigating cutting-edge data augmentation techniques to improve model performance by altering the dataset under various circumstances are two areas that still require improvement. MobileNetV2 is a promising development in deep learning for agricultural process optimization that promotes sustainable food production methods.

Future Scope: The successful application of MobileNetV2 for fruit and vegetable image recognition opens up several promising avenues for future research and practical implementations.

- Deployment in IoT Ecosystems
- Scalability for Large-Scale Deployment
- Incorporation of Explainable AI (XAI)
- Development of User-Friendly Applications
- Collaboration with Smart Cities and Food Supply Chain Systems

Real-world Challenges:
MobileNetV2 can be deployed on edge devices like Raspberry Pi, NVIDIA Jetson, and smartphones, which are commonly used in IoT applications. These devices enable local real-time data processing, reducing latency and bandwidth usage and ensuring data privacy.

MobileNetV2 can power IoT devices in agriculture, such as camera-equipped drones or greenhouse sensors, to classify fruit ripeness or detect vegetable defects in real-time. These devices can either relay data to a central system for further analysis or autonomously perform actions, like adjusting irrigation or sorting produce, enhancing efficiency and automation.

Data Availability Statement: The data supporting the findings of this study are publicly accessible in the Kaggle repository at https://www.kaggle.com/datasets/kritikseth/fruit-and-vegetable-image-recognition/data.

References

1. Zhan, Y., Chen, M., Chen, Y., Luo, Z., Luo, Y., Li, K.: An automatic recognition method of fruits and vegetables based on depthwise separable convolution neural network. J. Phys.: Conf. Ser. **1871**(1), 012075 (2021)
2. Suryawanshi, Y., Patil, K., Chumchu, P.: VegNet: dataset of vegetable quality images for machine learning applications. Data Brief **45**, 108657 (2022)
3. Kamilaris, A., Prenafeta-Boldu, F.X.: Deep learning in agriculture: a survey. Comput. Electron. Agric. **147**, 70–90 (2018)

4. Tian, Y., Yang, G., Wang, Z., Wang, H., Li, E., Liang, Z.: Apple detection during different growth stages in orchards using the improved YOLO-V3 model. Comput. Electron. Agric. **157**, 417–426 (2019)

5. Polder, G., van der Heijden, G.W.A.M., van Doorn, J., Baltissen T.A.H.M.C.: Automatic detection of tulip breaking virus (TBV) in tulip fields using machine vision. Biosyst. Eng. **117**, 35–42 (2014)

6. Kandukuri, P., Gummad, A., Paruchuri, B.V.N., Valarmathi, N., Kumar, N., Teja, P.R.: Developing a context-aware convolutional neural network (CACNN) for enhanced phishing website detection. J. Electr. Syst. **20**(5s), 347–354 (2024)

7. Sandler, M., Howard, A., Zhu, M., Zhmoginov, A., Chen, L.C.: MobileNetV2: inverted residuals and linear bottlenecks. In: Proceedings of the IEEE Conference on Computer Vision and Pattern Recognition, pp. 4510–4520 (2018)

8. Howard, A.G.: MobileNets: efficient convolutional neural networks for mobile vision applications. arXiv preprint: arXiv:1704.04861 (2017)

9. Zhang, X., Zhou, X., Lin, M., Sun, J.: ShuffleNet: an extremely efficient convolutional neural network for mobile devices. In: Proceedings of the IEEE Conference on Computer Vision and Pattern Recognition, pp. 6848–6856 (2018)

10. Chollet, F.: Xception: deep learning with depthwise separable convolutions. In: Proceedings of the IEEE Conference on Computer Vision and Pattern Recognition, pp. 1251–1258 (2017)

11. Rao, A.P., Reddy, K.S., Sathiyamoorthi, V.: Automated soil residue levels detecting device with IoT interface. In: Challenges and Applications of Data Analytics in Social Perspectives, pp. 123–135. IGI Global (2021)

12. Santhosh Kumar, C.N., Pavan Kumar, V., Reddy, K.S.: Similarity matching of pairs of text using CACT algorithm. Int. J. Eng. Adv. Technol. **8**(3), 2295–2298 (2019)

13. Zuo, F., De With, P.H.N.: Real-time embedded face recognition for smart home. IEEE Trans. Consum. Electron. **51**(1), 183–190 (2005)

14. He, K., Zhang, X., Ren, S., Sun, J.: Deep residual learning for image recognition. In: Proceedings of the IEEE Conference on Computer Vision and Pattern Recognition, pp. 770–778 (2016)

15. Huang, G., Liu, Z., Van Der Maaten, L., Weinberger, K.Q.: Densely connected convolutional networks. In: Proceedings of the IEEE Conference on Computer Vision and Pattern Recognition, pp 4700–4708 (2017)

16. Kumar, C.N.S., Reddy, K.S.: Effective data analytics on opinion mining. IJITEE **8**(10), 2073–2080 (2019)

17. Deb, S., Bhuyan, B., Kar, N., Reddy, K.S.: Colour image encryption using an improved version of stream cipher and chaos. Int. J. Ad Hoc Ubiquit. Comput. **41**(2), 118–133 (2022)

18. Polder, G., van der Heijden, G.W.A.M., van der Voet, H., Young, I.T.: Measuring surface distribution of carotenes and chlorophyll in ripening tomatoes using imaging spectrometry. Postharvest Biol. Technol. **34**(2), 117–129 (2004)

19. Misra, N.N., Dixit, Y., Al-Mallahi, A., Bhullar, M.S., Upadhyay, R., Martynenko, A.: IoT, big data, and artificial intelligence in agriculture and food industry IEEE Internet Things J. **9**(9), 6305–6324 (2020)

20. Muresan, H., Oltean, M.: Fruit recognition from images using deep learning. Acta Universitatis Sapientiae. Informatica. **10**(1), 26–42 (2018)

21. Paleti, L., et al : Sugar cane leaf disease classification and identification using deep machine learning algorithms. J. Theor. Appl. Inf. Technol. **101**(20) (2023)

22. Padmaja, D.L., Tammali, S., Gajavelly, N.S., Reddy, K.S.: A comparative study on natural disasters. In: 2022 International Conference on Applied Artificial Intelligence and Computing (ICAAIC), pp. 1704–1709. IEEE (2022)
23. Blasco, J., Aleixos, N., Moltó, E.: Machine vision system for automatic quality grading of fruit. Biosys. Eng. **85**(4), 415–423 (2003)
24. Arivazhagan, S., Shebiah, R.N., Nidhyanandhan, S.S., Ganesan, L.: Fruit recognition using color and texture features. J. Emerg. Trends Comput. Inf. Sci. **1**(2), 90–94 (2010)
25. Bargoti, S., Underwood, J.P.: Image segmentation for fruit detection and yield estimation in apple orchards. J. Field Robot. **34**(6), 1039–1060 (2017)
26. Schölkopf, B., Smola, A.J.: Learning with Kernels: Support Vector Machines, Regularization, Optimization, and Beyond. MIT Press (2002)
27. Mollahosseini, A., Hasani, B., Mahoor, M.H.: AffectNet: a database for facial expression, valence, and arousal computing in the wild. IEEE Trans. Affect. Comput. **10**(1), 18–31 (2017)
28. Redmon, J., Divvala, S., Girshick, R., Farhadi, A.: You only look once: unified, real-time object detection. In: Proceedings of the IEEE Conference on Computer Vision and Pattern Recognition, pp. 779–788 (2016)
29. Lu, D., Weng, Q.: A survey of image classification methods and techniques for improving classification performance. Int. J. Remote Sens. **28**(5), 823–870 (2007)
30. Krizhevsky, A., Sutskever, I., Hinton, G.E.: ImageNet classification with deep convolutional neural networks. In: Advances in Neural Information Processing Systems, vol. 25 (2012)
31. Simonyan, K.: Very deep convolutional networks for large-scale image recognition. arXiv preprint: arXiv:1409.1556 (2014)
32. Girshick, R., Donahue, J., Darrell, T., Malik, J.: Rich feature hierarchies for accurate object detection and semantic segmentation. In: Proceedings of the IEEE Conference on Computer Vision and Pattern Recognition, pp. 580–587 (2014)
33. Kandukuri, P., Abdul, A., Prasanth, K.K., Sreenivas, V., Ramesh, G., Venkateswarlu, G.: Deep learning based RAGAE-SVM for Chronic kidney disease diagnosis on internet of health things platform. Multimedia Tools Appl., 1–39 (2024)
34. Liu, H., Xiang, D., Cheng, Z., Ma, L., Yang, S., Xie, J.: Drying methods on fruit quality and antioxidant activity of two rare edible fungi. LWT, 116603 (2024)
35. LeCun, Y., Bengio, Y., Hinton, G.: Deep learning. Nature **521**(7553), 436–444 (2015)
36. Kaggle dataset. https://www.kaggle.com/datasets/kritikseth/fruit-and-vegetable-image-recognition/data

Integrated Progressive Multi-scale Deraining Network

Malothu Rajeswari[1], Kusuma Kommana[2], and Prakash Kodal[1]($\boxtimes$)

[1] Electronics and Communication Engineering, National Institute of Technology Warangal,
Warangal, India
`kprakash@nitw.ac.in`
[2] Electronics and Communication Engineering, GMR Institution of Technology, Rajam, India

Abstract. Rain can make image blurry or unclear in various critical uses cases, single-image deraining often finds, difficult to precisely remove rain streaks due to the diverse range of rain densities and raindrop sizes. In autonomous real-time uses, with existing networks like U-Net, future decoupling reorganization, de-graded attention transformer, progressive multi-scale deraining network and dual-domain strip attention module finds low PSNR and SSIM. We proposed a novel Integrated Progressive Multi-scale Deraining Network (IPMDNet) to address peak signal to noise ratio (PSNR) enhancement, structural similarity index matrix (SSIM) improvement, and low loss of original data. The network supports a U-Net architecture combined with additional convolution filters to effectively handle the non-uniformity, of rain densities and the range of raindrop scale. A multi-scale strategy is used to capture rain streaks across various scales. Channel attention blocks are integrated to enhance the precision of rain streak removal and refine feature extraction. Furthermore, to improve deraining performance, we present a "Multi-scale Feature Integration" (MSFI) that refines and aggregates the features across several network stages. IPMDNet exceeds existing state-of-the-art techniques in many kinds of synthetic and real-world dataset tests, achieving better results in both quantitative and qualitative evolutions.

Keywords: Image Deraining · Integrated Progressive Multi-scale Deraining · Network (IPMDNet) · Attention Module (AM) · Channel Attention Block (CAB) · Image full resolution network (IFR) · Rain Drops and Rain Streaks

1 Introduction

In modern times, image deraining [1] eliminating rain streaks from visuals taken during the rainy season—is an essential challenge in computer vision. Rain can obscure important information in a picture, drastically lowering its visibility and clarity in several use cases. This could influence how effectively many vision-based methods affect operations, such as surveillance, driverless vehicles, aerial view in the ground, space, army border security, and navy applications. The process of deraining is important because there is an array of rain patterns, from light rain to severe downpours, influencing shots in different ways. The primary goal of single image deraining [2–7] is to restore an

© The Author(s), under exclusive license to Springer Nature Switzerland AG 2026
C. Modi et al. (Eds.): MIND 2024, CCIS 2736, pp. 341–350, 2026.
https://doi.org/10.1007/978-3-032-14531-4_29

image's appearance [8, 9] by eliminating interruptions caused by rain while keeping the most important characteristics and textures in the area.

We employ CNN and deep learning to deal with deraining challenges and gain over the drawbacks of conventional techniques. CNNs tend to be well-suited to deal with the complex and irregular character of rain streaks in images because they are so good at directly identifying complex patterns from data [10]. With older systems like FDRNet, DAT, and DSANet frequently relying on manually created features and failing in unusual situations, CNNs can continuously train and extract structural traits essential in differentiating from rain streaks and the entire surroundings.

The development of end-to-end models which can be doing feature extraction, rain streak recognition, and image restoration within a single image is made achievable by methods involving deep learning [11]. The broad approach assures that the deraining operation is more precise and reliable and can address any variety of weather conditions, from minimal mists to powerful downpours. In addition, CNN's ability to handle multi-scale data enables them to successfully remove rain streaks of various widths and intensities, including those that are minor once noticeable.

The existing PMSDNet effectively removes the rain streaks by combining the strengths of encoder-decoder structures with multi-scale feature extraction [6, 12, 13]. However, it faces limitations in preserving fine spatial details because it relies on repeated down-sampling operations, even if it excels at capturing contextual information at various scales. Due to this trade-off, results frequently favor either contextual robustness or spatial precision, but not at the same time. The existing networks fails to remove the heavy rain streaks, to address this limitation, we proposed an Integrated Progressive Multi-scale Deraining Network (IPMDNet) which improves upon PMSDNet [14] by integrating additional filters and attention mechanisms that improve both the spatial and contextual quality of the derained image. In IPMDNet we introduce a Multi-scale Feature integration (MSFI) to refine and aggregate the features across different stages of the network.

IPMDNet improvements are as follows:

- We provide a distinctive multi-scale feature fusion model that gradually improves rain removal by adding spatial and contextual information.
- The additional filters are added to the network to handle non-uniform rain streaks more accurately.
- By adding Attention Module (AM), features are enhanced step by step, leading to a progressive improvement in rain removal.
- Experiments show that IPMDNet outperforms in both quantitative and qualitative evaluations with the existing networks on synthetic and real-world [15] datasets.

2 Proposed Methodology

The integrated progressive multi-scale deraining network (IPMDNet) is a proposed approach that can effectively handle nonuniform rain densities and various raindrop sizes, therefore reducing the difficulties associated with single-image deraining. This method combines several important elements with specified features with improved contextual robustness and spatial detail retention for original data restoration. Proposed network elements shown in Fig. 1 are explained below.

A. ***U-Net Architecture and Additional Filters:*** This is the main stage of IPMDNet, which can identify and process multi-scale contextual information. To enhance the network's capacity to handle irregular rain streaks, more convolution filters are combined during different stages in Fig. 1. Convolution filters are also called kernels to detect the specific feature from the input. The best image deraining is generated by these extra filters, which ensures the network's capacity to identify and remove rain streaks with various densities and sizes. U-Net includes skip connections that preserve spatial details during processing.

A.

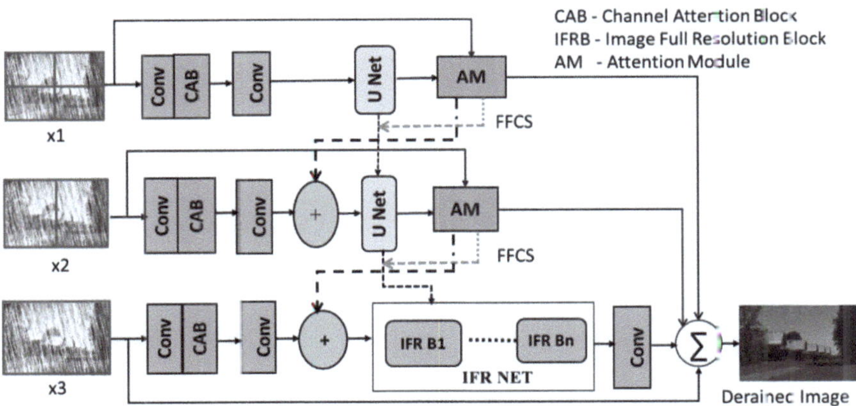

Fig. 1. Proposed Network IPMDNet Architecture

B. ***Multi-Scale Feature Integration:***

By utilizing multi-scale feature integration, we can integrate and refine features from the output of the components present in the architecture, this is an important aspect of IPMDNet. The MSFI gives total rain streak removal by adding characteristics across various scales, effectively managing both fine and large details.

C. ***Attention Mechanisms.***

To enhance the feature exploitation as well as the effectiveness of rain streak removal, IPMDNet introduces Attention Modules (AM) in between the different layers of the network. Such modules make use of attention maps that enhance the important features while reducing the less significant ones. Such a mechanism leads to the generation of better and better spatial features over time resulting in good-quality derained images.

D. ***Feature Fusion cross stage (FFCS):***

We added an attention module to IPMDNet in between the U-net to introduce feature fusion cross stage [16, 17], and the results show a notable improvement in SSIM and PSNR measures. In our AM subnetwork, things are in two stages, first employ the dirt at each level truth supervisory signals, which result in a gradually de-rained result. Second, as of right now, attention maps are produced.by employing locally to attenuate the less instructive features monitored forecasts, permitting only significant characteristics to proceed to the following phase for additional processing.

E. ***Concatenation:***

Concatenation is designed for the image restoration task such as image deraining, full resolution or similar image improvement application. It integrates the two U-net [18] modules, attention module and channel attention module are combining and passed through the image full resolution block and final convolution to produce the output.

F. ***Image Full Resolution Network (IFRNet):***

The last step of the IPMDNet procedure consists of applying the Image Full Resolution network, which recovers the spatial resolution of the derained images conserving the structure of the image. IFRNet uses the output of final Custom Net to get rain-free image and high-quality image.

Table 1. Summary Of Deraining Datasets Used in Our Network

Datasets	Train Images	Test Images	Renamed
Rain800	700	100	Test100
Rain14000	11200	2800	Test2800
Rain1800	1800	0	NC
Rain100L	0	100	Rain100L
Rain100H	0	100	Rain100H
Rain1200	0	1200	Test1200
Rain12	12	0	NC
Total	**13712**	**4300**	

G. ***Loss Function.***

When we train the IPMDNet, we can minimize the loss function, so we can get a rain-free image without losing the original content. The loss function implements a combination of Mean Squared Error (MSE), perceptual and adversarial friendly operable loss functions to encourage accurate removal of rain streaks, good retention of high-level features in the image as well as photorealism.

1) ***Mean Squared Error (MSE) Loss:***

The MSE loss seeks to minimize the sensor's disagreement between the derained output and the corresponding pixel of the given image. In this work, the class of MSE loss has also been outlined as follows:

$$L_{MSE} = \frac{1}{N} \sum_{i=1}^{N} \left((D(I) - I_{groundtruth} \right)^2 \tag{1}$$

where $D(I)$ is the output of the deraining algorithm as applied to the image represented by $I_{groundtruth}$, is the corresponding clean image being tested in the algorithm and N is the whole number of pixels specified.

2) ***Perceptual Loss:***

Perceptual loss is introduced because of the need to maintain the high-level features existing in the image. In that, they measure the discrepancy between the cerained and ground truth image compressing on the feature extraction of a pre-learned network such as VGG:

$$L_{Perceptualloss} = \frac{1}{N_j} \sum_{j=1}^{N_j} \| \phi_j(D(I)) - \phi_j \left(I_{groundtruth} \right) \|^2 \tag{2}$$

where $\phi_j(D(I))$ and $\phi_j \left(I_{groundtruth} \right)$ Are the feature maps of derained image I is the ground truth image I extracted from the j-th layer of the pre-trained network and N_j is the number of layers used in calculations.

3) ***Adversarial Loss:***

An adversarial loss [19] function is introduced to constrain the generated images to be visually acceptable:

$$L_{Adversarialloss} = E\left[log\left(D\left(I_{groundtruth} \right) \right) \right] + E\left[log(1 - D((D(I)))] \right] \tag{3}$$

D is a measure that tells whether an image comes from real images or generated ones in the GAN framework.

4) ***Total loss:*** Combining all the above losses leads to the following total loss function which serves as a weighted (α, β, γ) ci-combination of the losses:

$$L_{Totalloss} = \alpha L_{MSE} + \beta L_{Perceptualloss} + \gamma L_{dversarialloss} \tag{4}$$

H. *End-to-End Training.*

IPMDNet is trained end-to-end and can therefore automatically learn feature representations [10] and rain removal methods from the data. It is further demonstrated that under this training methodology, the network performs well not only on the synthetic dataset but also on the real dataset including various deraining tasks achieving advanced performance by including the state-of-art methods [22, 23].

3 Experiments and Analysis

We trained and tested our proposed network on different artificial deraining datasets. On the proposed network, qualitative measures like PSNR and SSIM were calculated. Table 1. Gives an overview of the deraining datasets used in our network.

As shown in Table 1. Our network was trained using 13,712 rainy images derived from different datasets. The number of test datasets, including Test100, Test2800, Rain100L, Rain100H, and Test 1200 similar datasets provided in PMSDNet were used to evaluate the proposed network.

A. **Implementation Details:**

IPMDNet does not require any pertaining and end-to-end training. This network utilizes two CABs for every level of the encoder-decoder network module. We also used 2*2 max pooling with stride 2 for the down-sampling in the last phase, the IFRNet module utilizes multiple IFR blocks, and 40 channels were allocated for image deraining.

To get the derained image from the rainy image, we used a batch size of 16 iterations, 300 epochs, and 256×256 images in Rain100L, Rain100H, Test100, Test1200, and Test2800 datasets for the training process.

4 Results and Discussions

Performance of various network metrics on several datasets including PRENET, Derain-Net, DIDMDN, UMRL, MSPNet, SAPNet, PMSDNet, and Proposed. The metrics used are the Peak Signal-to-Noise Ratio (PSNR) and Structural Similarity Index (SSIM). Based on PSNR and SSIM image quality metrics, the quantitative test results analysis on all five synthetic datasets is displayed in Table 2. It is clear from Table 2. Our network outperformed several types of baseline networks competitively. In the qualitative test, our network IPMSDNet got better results as compared to another network. It clearly shows that our proposed network removes the rain streaks from the images shown in Fig. 2.

PRENet has high PSNR/SSIM values indicating strong performance in Rain100L and Test2800 with 32.44/0.95 and 31.75/0.91 respectively but only does well on the Rain100H dataset. In contrast to other models, DerainNet is atrocious as reflected by its performance on the Rain100H dataset (14.92/0.59) and Test2800(24.31/0.86), implying that it may not work well in case of higher or more complicated rain. Despite having such high outcomes in all other datasets like Test2800 which is 29.97 /0.90 as well as test1200; 30.55/0.00, it performs worse than current approaches although this occurs for over other datasets except Rain100H with UMRL where it fails most particularly for Rain100H.

Most notably, MSPNet consistently performs well on Test2800 (32.82/00) and Rain100L (32.44 /00), suggesting that this is a dependable network in both light and severe rain scenarios. In particular, SAPNet outperforms several models in terms of Rain100L (34.77 /00) and Rain100H (29.46 /00), showcasing its versatility in handling various rainfall events. Between all datasets with equal PSNR/SSIM values, PMSDNet and the suggested model both perform optimally.

The improvement of Rain100L (36.41/0.97) and Test2800 (33.62/0.93), which display the most recent outcomes for the sector. In most cases, these two networks performed better than any other approach since they could obtain the best PSNR/SSIM values across the majority of the datasets utilized, suggesting that they were more adept at eliminating rain from photos. Furthermore, SAPNet has shown remarkable performance, especially in conditions with less rain.

Performance of various network metrics on several datasets including PRENET, DerainNet, DIDMDN, UMRL, MSPNet, SAPNet, PMSDNet, and Proposed. The metrics used are the Peak Signal-to-Noise Ratio (PSNR) and Structural Similarity Index matrix (SSIM). Based on PSNR and SSIM image quality metrics, the quantitative test results analysis on all five synthetic datasets is displayed in Table 2. It is clear from Table 2. That our network outperformed several types of baseline networks competitively. In the qualitative test, our network IPMDNet got better results as compared to existing network. It clearly shows that our proposed network removes the rain streaks from the images shown in Fig. 2.

Fig. 2. Qualitative results of the proposed network on all synthetic deraining datasets

Table 2. Qualitative results of our proposed network

Network Metrics/Year	Rain100L [6] PSNR/SSIM	Rain100H [6] PSNR/SSIM	Test100 [11] PSNR/SSIM	Test2800 [22] PSNR/SSIM	Test1200 [23] PSNR/SSIM
PRENet [20]/2019	32.44/0.95	26.77/0.85	24.81/0.85	31.75/0.91	31.36/0.91
DerainNet [21]/2017	27.03/0.84	14.92/0.59	22.77/0.81	24.31/0.86	23.38/0.83
DIDMDN [10]/2018	25.23/0.74	17.35/0.52	22.56/0.82	28.13/0.86	29.65/0.90
UMRL [15]/2019	29.18/0.92	26.01/0.83	24.41/0.83	29.97/0.90	30.55/0.91
MSPNet [22]/2020	32.44/0.95	28.66/0.86	27.50/0.87	32.82/0.93	32.39/0.91
SAPNet [23]/2021	34.77/0.97	29.46/0.89	29.13/0.88	32.18/0.93	32.46/0.91
PMSDN [14]/2022	36.41/0.97	30.38/0.89	30.32/0.90	33.62/0.93	32.96/0.92
CPTransN [24]/2023	37.54/0.96	30.67/0.89	31.33/0.91	33.90/0.94	33.34/0.92
Proposed Network	**38.78/0.97**	**31.57/0.90**	**32.53/0.92**	**34.40/0.95**	**35.76/0.93**

5   Conclusion

In this paper a novel, integrated progressive mult-iscale deraining was proposed which performs better while removing the rain streaks from single rainy image. Combining powerful multi-scale feature extraction with additional convolutional filters and attention mechanisms, IPMDNet dramatically improves the quality of scanned images Integration with IFRNet ensures the storage of high-resolution spatial information, period with MSFI and AM modules for more accurate and effective rainfall. Gradually adjusting features for line removal, the proposed network provides the visual quality of channel-free images and the peak signal-to-noise ratio (PSNR) and structural similarity index (SSIM). Achieve good performance in both quantitative (PSNR, SSIM) and qualitative analysis. Here we are using large number of datasets and additional filters it may increase the training time. Additionally the suggested network is lightweight because a three stage IPMDNet require just 0.20 s to execute and provides 22 MB of parameters, meaning that it may operate on low end edge devices.

References

1. Fu, X., Liang, B., Huang, Y., Ding, X., Paisley, J.: Lightweight pyramid networks for image deraining. IEEE Trans. Neural Netw. Learn. Syst. **31**(6), 1794–1807 (2019)
2. Zhao, P., Wang, T.: Degradation-aware transformer for single image deraining. IEEE Access (2023)
3. Cheng, Y., Huang, J., Ren, H., Ran, W., Lu, H.: Feature decoupling and reorganization network for single image deraining. Multimedia Syst. **30**(3), 154 (2024)
4. Yu, L., Wang, B., He, J., Xia, G.S., Yang, W.: Single image deraining with continuous rain density estimation. IEEE Trans. Multimedia **25**, 443–456 (2021)
5. Qin, S., Zhang, S., Zhang, Y.: Using mask-based enhancement and feature aggregation for single image deraining. IEEE Signal Process. Lett. (2023)
6. Yang, W., Tan, R.T., Feng, J., Liu, J., Guo, Z., Yan, S.: Deep joint rain detection and removal from a single image. In Proceedings of the IEEE Conference on Computer Vision and Pattern Recognition, pp. 1357–1366 (2017)
7. Wei, Y., Zhang, Z., Zhang, H., Hong, R., Wang, M.: A coarse-to-fine multi-stream hybrid deraining network for single image deraining. In 2019 IEEE International Conference on Data Mining (ICDM), pp. 628–637. IEEE (2019)
8. Cui, Y., Knoll, A.: Dual-domain strip attention for image restoration. Neural Netw. **171**, 429–439 (2024)
9. Jiang, Z., Yang, S., Liu, J., Fan, X., Liu, R.: Multi-scale synergism ensemble progressive and contrastive investigation for image restoration. IEEE Trans. Instrum. Measur. (2023)
10. Yasarla, R., Patel, V.M.: Uncertainty guided multi-scale residual learning-using a cycle spinning cnn for single image de-raining. In: Proceedings of the IEEE/CVF Conference on Computer Vision and Pattern Recognition, pp. 8405–8414 (2019)
11. Fu, X., Huang, J., Zeng, D., Huang, Y., Ding, X., Paisley, J.: Removing rain from single images via a deep detail network. In Proceedings of the IEEE Conference on Computer Vision and Pattern Recognition, pp. 3855–3863 (2017)
12. Wang, Q., Jiang, K., Wang, Z., Ren, W., Zhang, J., Lin, C.W.: Multi-scale fusion and decomposition network for single image deraining. IEEE Trans. Image Process. **33**, 191–204 (2023)
13. Yan, T., Li, M., Li, B., Yang, Y., Lau, R.W.: Rain removal from light field images with 4d convolution and multi-scale gaussian process. arXiv preprint arXiv:2208.07735 (2022)
14. Ragini, T., Prakash, K.: Progressive multi-scale deraining network. In 2022 IEEE International Symposium on Smart Electronic Systems (iSES), pp. 231–235. IEEE (2022)
15. Huang, H., Luo, M., He, R.: Memory uncertainty learning for real-world single image deraining. IEEE Trans. Pattern Anal. Mach. Intell. **45**(3), 3446–3460 (2022)
16. Jang, Y., Le, D.T., Son, C.H., Choo, H.: Layer decomposition learning based on discriminative feature group split with bottom-up intergroup feature fusion for single image deraining. IEEE Access (2024)
17. Jiang, K., et al.: Multi-scale hybrid fusion network for single image deraining. IEEE Trans. Neural Netw. Learn. Syst. **34**(7), 3594–3608 (2021)
18. Kolekar, M.H., Bose, S., Pai, A.: SARain-GAN: spatial attention residual UNet based conditional generative adversarial network for rain streak removal. IEEE Access (2024)
19. Zhang, H., Sindagi, V., Patel, V.M.: Image de-raining using a conditional generative adversarial network. IEEE Trans. Circ. Syst. Video Technol. **30**(11), 3943–3956 (2019)
20. Ren, D., Zuo, W., Hu, Q., Zhu, P., Meng, D.: Progressive image deraining networks: A better and simpler baseline. In: Proceedings of the IEEE/CVF Conference on Computer Vision and Pattern Recognition, pp. 3937–3946 (2019)

21. Fu, X., Huang, J., Ding, X., Liao, Y., Paisley, J.: Clearing the skies: a deep network architecture for single-image rain removal. IEEE Trans. Image Process. **26**(6), 2944–2956 (2017)
22. Jiang, K., et al.: Multi-scale progressive fusion network for single image deraining. In: Proceedings of the IEEE/CVF Conference on Computer Vision and Pattern Recognition, pp. 8346–8355 (2020)
23. Zheng, S., Lu, C., Wu, Y., Gupta, G.: SAPNet: Segmentation-aware progressive network for perceptual contrastive deraining. In Proceedings of the IEEE/CVF winter conference on applications of computer vision (pp. 52–62) (2022)
24. Xu, Y., Long, Z., Tang, B., Lei, S.: channel pyramidal transformer network for single image deraining. IEEE Signal Process. Lett. (2023)

Vision Based Pothole Detection System for Challenging Scenarios

P. N. Arthana, V. Sneha, N. Manohar, M. Priyanka, and R. Suresha[✉]

Deptartment of Computer Science, School of Computing, Amrita Vishwa Vidyapeetham,
Mysuru Campus, Bengaluru, India
{manohar,priyankam,sureshar}@my.amrita.edu

Abstract. The detection of potholes on Indian roads is crucial for improving road safety and reducing vehicle damage, especially in countries like India where road conditions are often affected by various environmental factors. This paper presents a robust pothole detection system that utilizes the environment-specific YOLOv8n pothole detection models specifically optimized for Indian weather conditions. The dataset used for training the model includes data collected under different environmental conditions such as morning, afternoon, evening, and night. To address lighting and visibility challenges, contrast enhancement and histogram equalization were employed. The pre-trained weights of YOLOv8n were fine-tuned to improve the performance of pothole detection across these varied conditions. The model demonstrated high performance in real-time detection, achieving a precision of 96.60% in morning conditions, 95.47% in the afternoon, 94.91% in the evening, and 88.34% at night. Recall rates ranged from 87.92% to 74.98% across different conditions, with a mean average precision of 96.09% in well-lit conditions and 88.46% at night. This work contributes to road maintenance automation by improving detection accuracy, reducing false positives, and providing reliable results under challenging weather and lighting conditions.

Keywords: Indian Road · Pothole Detection · Yolov8n · Complex Environment

1 Introduction

Potholes are a significant hazard on roads, especially in countries like India, where diverse and extreme environmental conditions exacerbate their formation and deterioration. These road surface anomalies pose risks to vehicular safety, passenger comfort, and overall traffic flow, often leading to accidents and damage to vehicles. Traditional manual methods for detecting and addressing potholes are time-consuming, expensive, and prone to human error, making it difficult to maintain roads effectively. With advancements in computer vision and deep learning, automated pothole detection systems are becoming a viable solution for improving road maintenance processes.

India's varied weather conditions ranging from heavy monsoon rains and foggy winters to intense heat further complicate the task of reliably detecting potholes Environmental factors such as lighting conditions (morning, afternoon, evening, and night)

C. Modi et al. (Eds.): MIND 2024, CCIS 2736, pp. 351–361, 2026.
https://doi.org/10.1007/978-3-032-14531-4_30

and weather conditions (rain, fog, and sunlight) significantly affect visibility and detection accuracy. Therefore, any pothole detection system designed for Indian roadways must be capable of adapting to these challenges.

This study proposes an environment-specific YOLOv8n object detection models specifically tailored for Indian environmental conditions. Environment-specific YOLOv8n object detection models were trained on a custom pothole dataset collected under different weather and lighting conditions to capture the variations found on Mysuru roads, Karnataka, India. Data enhancement techniques, such as contrast enhancement, power law transformation and histogram equalization, were used to improve model performance under varied weather conditions such as morning, afternoon, evening and night. The environment-specific YOLOv8n models was trained with different weather condition separately to avoid the over fitting issue. During testing random forest classifier used to classify the day information such as morning, afternoon, evening and night. By leveraging the multiple YOLOv8n model, this work aims to provide a robust and efficient solution for real-time pothole detection and improving road safety.

The following sections outline the existing literature in Sect. 2, followed by the proposed model architecture and its illustrations in Sect. 3. Section 4 details the training process, evaluation metrics, and the results obtained under various environmental conditions, emphasizing the model's strengths and limitations in Indian road scenarios. Finally, Sect. 5 provides a summary of the work and the conclusion.

2 Literature Survey

The detection of potholes has garnered considerable research interest in recent years due to its potential to enhance road safety and improve infrastructure management. This section offers an overview of research studies and methodologies related to pothole detection, along with solutions tailored to the various environmental conditions of Indian roads.

YOLOv5-based models have emerged prominently in pothole detection. This model achieved 93.0% precision, 91.6% recall, and 96.3% mAP, providing accurate dimensional and location estimates using built-in vehicle technologies [1]. Another adaptation integrated StrongSORT tracking with YOLOv5 for automated monitoring of forest road deterioration, achieving a precision of 0.79, recall of 0.58, and a mean average precision (mAP) of 0.70 at IoU $\geq$ 0.5. Perspective transformation techniques enhanced YOLOv5's detection performance, especially for distant potholes, improving AP by 43% and achieving up to 94% improvement for categorized potholes based on distance [2]. YOLOv5 also demonstrated robustness through domain adaptation techniques, improving precision to 96.8% under varying weather and lighting conditions [3]. Federated learning approaches, combined with YOLOv5, allowed privacy-preserving pothole detection, achieving 85.71% classification accuracy for submerged and dry potholes [4]. Moreover, CNN-based methods trained from scratch on large pothole datasets reported high mAP scores, such as 92% on custom datasets with 200 training epochs [5].

Several studies employed ensemble and hybrid models for enhanced performance. An ensemble approach achieved a remarkable accuracy of 97.49%, outperforming existing models [6]. Another hybrid method combined shallow convolutional layers with particle swarm optimization and random forest classifiers, achieving 99.37% accuracy and precision [7]. Comparatively, YOLOv7 showed superior efficiency in detecting potholes on diverse road types, attaining a precision rate of 93% [8].

The advancements include YOLOv4-based models trained on GoPro Hero 7 camera datasets, which achieved 89.25% mAP and incorporated telemetry data for validation [9]. Improved YOLOv3-SPP algorithms addressed challenges in detecting small targets in complex backgrounds by adding additional prediction layers, enhancing detection accuracy [10]. A WT-YOLOv8, demonstrated a 61.80% improvement in mAP@50 on the Nighttime Pothole Dataset (NPD), consisting of 3831 images with diverse scene variations [11]. Another deep-learning approach achieved 82.7% detection accuracy on a public dataset and 30 FPS in live video streams, emphasizing real-time applicability [12]. DynamicECM, a novel architecture, effectively balanced inference time and detection performance, surpassing traditional ECM's capabilities while maintaining stability for complex road data [13].

Research also highlighted the importance of domain-specific adaptations, achieving high performance under different environmental scenarios. This extensive body of work underscores the effectiveness of YOLO architectures and other deep learning frameworks in addressing pothole detection challenges. From the above study, the limited amount of Indian pothole dataset presents which leads to computational challenges in developing efficient pothole detection systems due to real-time accuracy requirements and varying environmental conditions [15]. Weather conditions pose challenges, making accurate detection crucial for autonomous vehicles pothole detection and road safety systems.

The study enhances pothole detection in India by environment-specific YOLOv8n real-time detection capabilities in overfitting issues due to variations in the Indian environment. In the training phase, the environment-specific YOLOv8n models was trained with four different datasets such as morning, afternoon, evening and night. Whereas in testing phase, the testing image is initially classified as morning, afternoon, evening and night using color HOG features and a random forest classifier. The outcome of the Random Forest classifier again inputting to the specific YOLOv8n trained model which results in better performance. This approach offers a robust solution for pothole detection in India's driving condition, despite the challenges posed by environmental variability and real-time detection.

3 Proposed Methodology

The section describes the functionalities of the proposed environment-specific YOLOv8n [14] pothole detection model. It illustrates the preparation and working mechanism of the proposed pothole detection in the training and testing processes shown in Fig. 1(a) and 1(b). The proposed model addresses challenges such as varied environmental conditions in a day. The functionality of each component of the proposed model such as preprocessing techniques, the proposed model based on the original Yolov8n model, and random forest classifier functionality during the testing process are discussed.

3.1 Dataset

This paper presents a dataset of potholes captured in India under various weather conditions to improve object detection and recognition tasks. The videos were recorded using

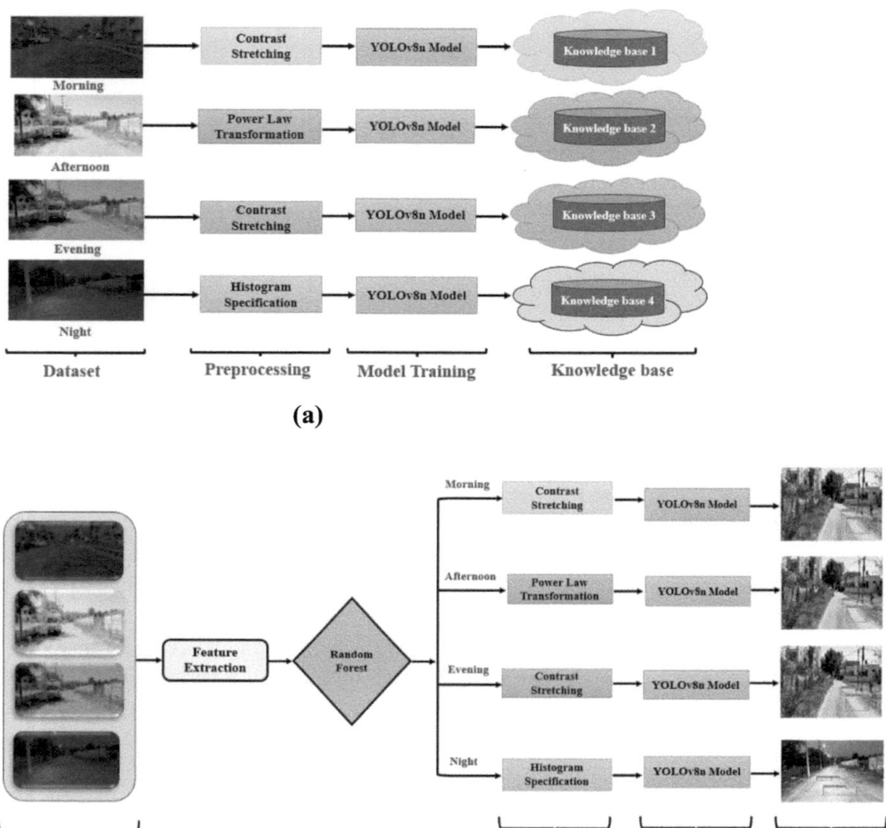

Fig. 1. (a) Architecture of the proposed training model. (b) Architecture of the proposed testing model.

a mobile camera mounted on a vehicle travelling at a speed of 40 km/h. This curated dataset is essential for developing effective methods for pothole detection in urban environments, specifically in Mysuru. Pothole data were collected at different times of the day: early morning, afternoon, evening, and night. The recorded videos were converted into frames, with only the frames featuring potholes retained to ensure the dataset's relevance. This process resulted in a total of 10234 frames. Specifically, the dataset includes 3,200 images taken in the morning, 3,258 images from the afternoon, 1,008 images captured in the evening, and 2,768 images from the night, all with an original resolution of 1920 × 1080. Details of the video-to-frame conversion process are illustrated in Fig. 2(a) and (b).

Fig. 2. (a) Sample images from the collected dataset and (b) Annotated dataset examples.

3.2 Preprocessing

To ensure consistency during training, the model standardizes the pothole images to a size of 640 × 640 pixels. Morning and evening lighting conditions often result in lower light levels, which can lead to images with reduced contrast and muted colors. To enhance the overall color richness of these images, contrast stretching was applied to the warm tones typical of evening light. In contrast, afternoon lighting is often characterized by strong sunlight and shadows, creating high-contrast images. To address this, power law enhancement techniques with specific gamma values were used in the afternoon. This approach helps prevent overexposure and maintains detail in both bright and shadowed areas. To improve night time pothole images, images were enhanced with histogram specification. This process helps in improving the accuracy in identifying pothole in environment-specific YOLOv8n object detection models, leading to more efficient Indian driving scenarios.

3.3 YOLOv8n Model

In this section, four separate YOLOv8n models are utilized to increase the robustness detection process. The YOLOv8n model is lightweight which is particularly suitable for real-time potholes detection on Indian roads in various weather conditions, the architecture diagram is shown in Fig. 3. This model uses an efficient backbone to capture specific features of potholes. The PANet (Path Aggregation Network) to improve multi-scale feature fusion combines potholes of different sizes and shapes in the complex environments of Indian roads. In the context of Indian road conditions, ensemble YOLOv8n demonstrates robustness against various environmental factors, including lighting variations in a day. The ensemble model was trained using four different datasets such as morning, evening, and night which are specifically focused on potholes in India. Data augmentation techniques such as scaling, rotation, and noise adjustment further enhance the model's ability to generalize.

Additionally, YOLOv8n's architecture is designed to efficiently handle small pothole detection, an essential road surface feature of potholes occupying small areas of road images. The deployment of YOLOv8n for pothole detection enables automated road condition assessment and proactive maintenance strategies. The detection outputs of the model's inference capabilities facilitate its use in active driving scenarios, enhancing road safety by alerting drivers to potholes and minimizing vehicle damage risks.

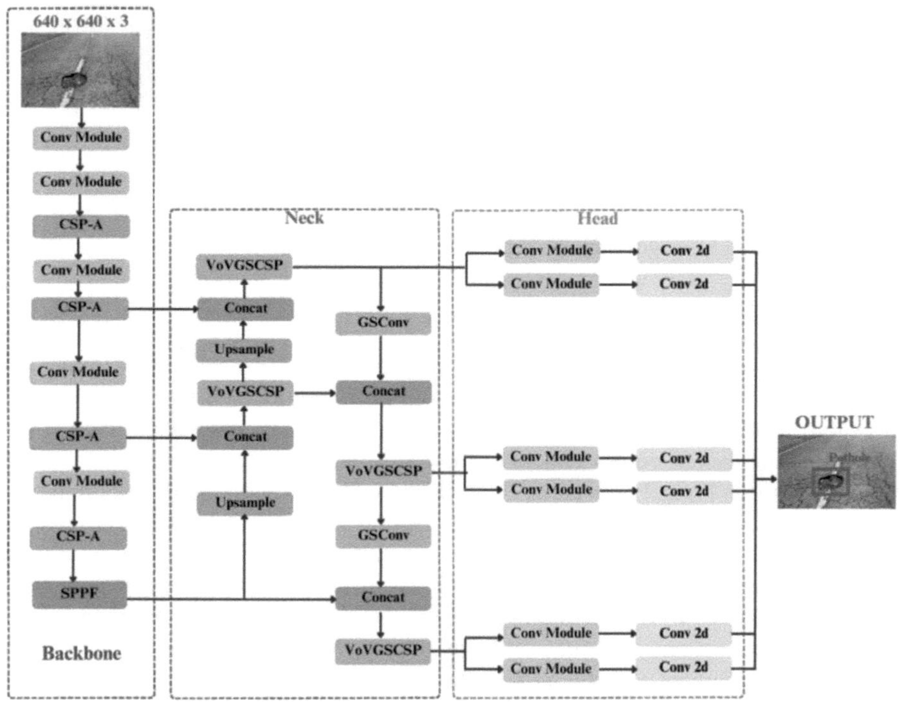

Fig. 3. Architecture diagram of the YOLOv8 model.

The proposed model thus serves as an efficient and scalable solution for addressing the challenges of pothole detection on Indian roads.

Fig. 4. YOLOv8 detection performance across different environmental conditions.

Each weather data was trained separately with YOLOv8n model to improve the efficiency of the model. During testing random forest classifier [8] classifier identify the weather based on the color HOG feature were extracted from the image to ensuring the accurate detection even in the challenging weather scenarios such as morning, afternoon, evening, and night. The integration of the random forest model adds a layer of classification capability that complements the object detection process by enhancing the overall accuracy and the reliability of the environment-specific pothole detection in adverse conditions in Fig. 4.

4   Experimental Results

In the experimental phase, a diverse dataset featuring potholes captured under various weather conditions—morning, afternoon, evening, and night was meticulously annotated with bounding boxes. Table 1 shows the dataset information which consist of total 10234 images, 3200 image captured in the mornings, 3258 afternoons, 1008 evenings and 2768 nights images were captured. This data provides insight into the patterns of images that were acquired across varying weather conditions in different scenarios.

Table 1. Dataset classes and samples.

Class Name	All	Morning	Noon	Evening	Night
No. of samples	10234	3200	3258	1008	2768

The pothole detection task is a detailed experimental setup that involves training and testing the model under four different environmental conditions (morning, afternoon, evening, and night) using an environment-specific YOLOv8n pothole detection models. Training on GPUs is done using NVIDIA RTX GPUs for efficient training, and batch size adjustment based on GPU availability. Data augmentation techniques are applied to improve the generalizability of the models, such as contrast enhancement, power law transformation, and histogram equalization. All images are resized to fit the YOLOv8n input and image pixel values are normalized to [0, 1] to stabilize the training process. The training setup involves training each environmental condition separately with varied epochs and pre-trained YOLOv8n weights, using TensorFlow as the framework. The modified YOLOv8n architecture includes an Attention Mechanism, an Additional upsampling layer, and an Extra Concatenation Layer. Pre-trained weights and epochs are varied for each environmental condition, with the loss function and optimizer being used for dynamic learning rate adjustments. During testing, the environmental condition of the test image is identified first, followed by applying the appropriate YOLOv8n model for detection. The environmental condition classification uses color HOG feature extraction and a random forest classifier. The corresponding YOLOv8n model is applied to detect pothole based on the environmental class identified by the Random Forest. Visualization and results analysis are conducted to compare the performance of each model across different environmental conditions, paying particular attention to challenging cases like low-visibility pothole at night.

The environment-specific YOLOv8n pothole model performance is tabulated in Table 2 and visualize in Fig. 5. in different conditions. Mornings show better accuracy (94.10%) and mAP@50 (96.09%) compared to the proposed model, with recall (87.92%). Afternoons have accuracy (94.52%) and mAP@50 (97.13%), with precision (95.47%). Evenings have accuracy (93.33%) and recall (94.91%), indicating difficulties in dim evening light. Nights show better performance (89.14% accuracy and 88.6% mAP@50), but lower recall (74.9%).

Table 2. Performance of the environment-specific YOLOv8n pothole detection model.

	Accuracy (%)	Precision (%)	Recall (%)	mAP@50 (%)
Morning	94.10	96.60	87.92	96.09
Afternoon	94.52	95.47	86.83	97.13
Evening	93.33	94.91	84.10	95.11
Night	89.14	88.34	74.98	88.46

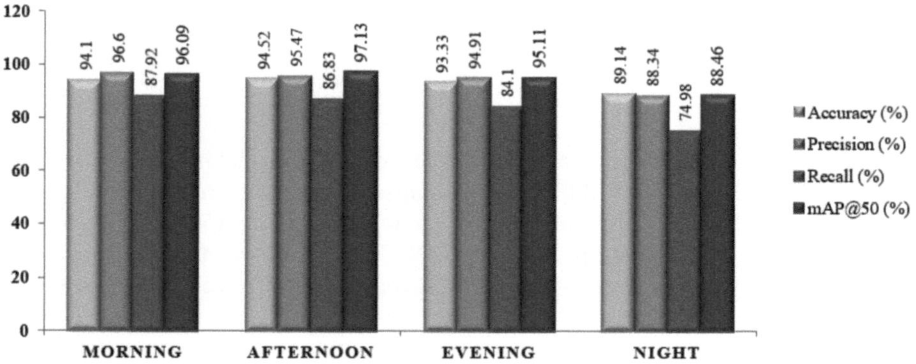

Fig. 5. Performance of the environment-specific YOLOv8n pothole detection model.

Further the proposed environment-specific YOLOv8n pothole detection model compared with YOLOv5 with custom dataset. Batch sizes and epochs were meticulously tuned to optimize the robustness of the models in detecting potholes accurately. The trained models were evaluated using standard metrics including accuracy, precision, recall, and mAP on a dedicated validation set for pothole detection.

The analysis showed different performance levels between proposed environment-specific YOLOv8n pothole detection model and YOLOv5 models in various weather conditions in India. The proposed model was better at detecting potholes in difficult weather compared to YOLOv5. The proposed model achieves mAP@50 of 92.00%, much higher than YOLOv5's of 78.40%. This difference shows that the proposed model significantly improves pothole detection performance in Indian weather conditions. The results of the proposed model and YOLOv5 are shown in Table 3. The table compares the performance of these two models in pothole detection. The proposed model shows 94.19% in mAP@50, 93.83% in precision, 83.45% in recall and 92.77% in accuracy. These results show significantly better than YOLOv5's indicating that the proposed model has better performance in detecting objects.

Table 3. Performance comparisons between proposed environment-specific YOLOv8n pothole detection model and YOLOv5.

Model	Accuracy (%)	Precision (%)	Recall (%)	mAP@50 (%)
YOLOv8n	92.77	93.83	83.45	94.19
YOLOv5	69.00	88.80	75.80	78.40

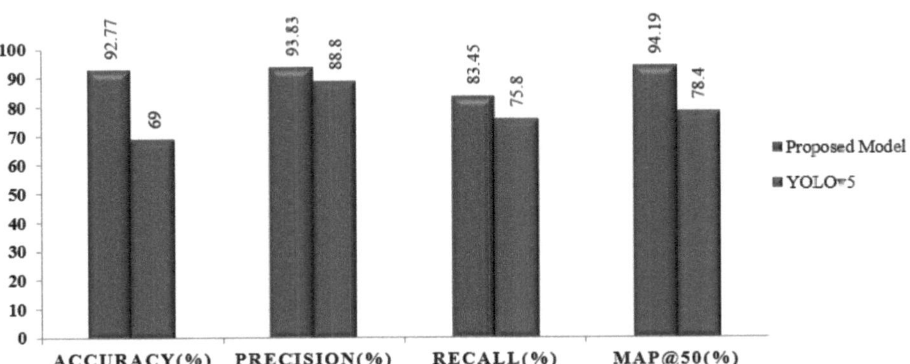

Fig. 6. Performance comparisons between proposed environment-specific YOLOv8n pothole detection model and YOLOv5.

Figure 6 shows the performance comparison of proposed environment-specific YOLOv8n pothole detection model and YOLOv5. The differences in these metrics suggest that proposed model is designed to maximize recall and accuracy, making it more reliable in identifying actual positives. YOLOv5, on the other hand, is optimized for precision, reducing false positives but at the cost of missing more true positives.

Table 4. Performance of the model compared with the state-of-the-art models.

Model	Accuracy (%)	Precision (%)	Recall (%)	mAP@50 (%)
Proposed Model	92.77	93.83	83.45	94.19
PD-ITS [8]	91.67	93.00	-	-
YOLOv5 [1]	-	93.0	91.6	96.3
YOLOv5 [2]	-	79.00	58.00	70.00
YOLOv4 [9]	-	88.00	85.00	89.25
WT-YOLOV8 [11]	-	-	-	91.80
YOLOV5 [12]		86.21	75.92	82.53

The comprehensive analysis for proposed model with state-of-the-arts methods, in the context of Indian pothole detection under adverse weather conditions. These results exhibit the accuracy, precision and recall, and mAP@50 metrics models. The study aims to establish more consistent average performance metrics for pothole detection under a variety of weather conditions by repeating each experiment five times and averaging the results. Table 4 showcases the high overall 92.77% precision, 93.83% recall and 94.15% mAP@50 of proposed model indicating its effectiveness in detecting Indian pothole across different weather conditions. This experimental setup allows for effective assessment of the proposed architecture and environmental random forest classification pipeline, ensuring the system's robustness for real-world applications in pothole detection under varied weather conditions in Indian scenarios.

5 Conclusion

The efficiency of pothole detection is very crucial, heavily depending on the various weather conditions. In this paper, the proposed environment-specific YOLOv8n pothole detection model enhances the process of detection more reliable, accurate, and effective in identifying the target pothole in various weather conditions for Indian challenging scenarios. Finally, Indian road pothole in varied environmental conditions such as day, afternoon, evening and night samples are trained parallel with YOLOv8 detection module. The experimentation was conducted on collected Indian dataset in challenging scenarios and result shows the accuracy of proposed model is 92.77% the precision 93.83%, recall 83.45% and mAP@50 is 94.19% which improves the detection process more reliable, accurate and effective in Indian challenging scenarios. Further, we analyze performance of proposed environment-specific YOLOv8n pothole detection model and YOLOv5 for pothole detection has revealed a notable difference in accuracy, with YOLOv8n exhibiting a superior performance of 92.77% compared to YOLOv5's 69.00%. This difference underscores the importance of selecting the most significant implications of this research, as a more precise detection system holds the promised significantly improving road maintenance efforts and bolstering overall road safety. The study includes a thorough investigation of pothole identification, leveraging modern technology to overcome limitations such as difficulty in detecting potholes in night. Future developments in nighttime pothole detection could have a big impact Autonomous vehicle. Continuous advancements in technology and increased integration with other systems will enhance the effectiveness and applicability of these detection systems.

References

1. Ruseruka, C., Mwakalonge, J., Comert, G., Siuhi, S., Ngeni, F., Anderson, Q.: Augmenting roadway safety with machine learning and deep learning: pothole detection and dimension estimation using in-vehicle technologies. Mach. Learn. Appl. **16**, 100547 (2024)
2. Hoseini, M., Puliti, S., Hoffmann, S., Astrup, R.: Pothole detection in the woods: a deep learning approach for forest road surface monitoring with dashcams. Int. J. For. Eng. **35**(2), 303–312 (2024)
3. Meinardy, D., Budianto, V., Jeremy, N.H., Lumentut, J.S.: Implementing domain adaptation to YOLOv5 algorithm for pothole detection. In: 2024 International Conference on Artificial Intelligence, Blockchain, Cloud Computing, and Data Analytics (ICoABCD), pp. 196–201. IEEE (2024)
4. Singh, A., Mehta, A., Padaria, A.A., Jadav, N.K., Geddam, R., Tanwar, S.: Enhanced pothole detection using YOLOv5 and federated learning. In: 2024 14th International Conference on Cloud Computing, Data Science & Engineering (Confluence), pp. 549–554. IEEE (2024)
5. Ch, S.S., Tallam, T.: Pothole detection and dimension estimation by deep learning. In: IOP Conference Series: Earth and Environmental Science, vol. 1326, no. 1, p. 012100. IOP Publishing (2024)
6. Vasanthi, R., Reshmy, A.K., Praveen, D.S.: Improving road maintenance and safety through weighted ensemble of deep convolutional neural networks: focus on pothole detection In 2024 7th International Conference on Circuit Power and Computing Technologies (ICCPCT), vol. 1, pp. 1900–1905. IEEE (2024)
7. Aljohani, A.: Optimized convolutional forest by particle swarm optimizer for pothole detection. Int. J. Comput. Intell. Syst. **17**(1), 7 (2024)
8. Omar, M., Kumar, P.: PD-ITS: pothole detection using YOLO variants for intelligent transport system. SN Comput. Sci. **5**(5), 552 (2024)
9. de Souza, Á.M., da Silva Soares, J.V., de Sousa, A.D.P., de Sousa, G.C.L., da Silva, L.M.: Detection of potholes in asphalt pavements using YOLOv4 architecture. Int. J. Pavement Res. Technol. 1–12 (2024)
10. Liu, T., Li, J., Cai, M., Cui, Y., & Fan, Q. Y. (2024, July). An Improved YOLOv3-SPP Algorithm for Image-Based Pothole Detection. In International Symposium on Neural Networks (pp. 328–335). Singapore: Springer Nature Singapore
11. Ling, M., et al.: Nighttime pothole detection: a benchmark. Electronics **13**(19), 3790 (2024)
12. Paramarthalingam, A., Sivaraman, J., Theerthagiri, P., Vijayakumar, B., Baskaran, V.: A deep learning model to assist visually impaired in pothole detection using computer vision. Decis. Anal. J. **12**, 100507 (2024)
13. Bekku, H., Kume, T., Tsuge, A., Nakazawa, J.: A stable and efficient dynamic ensemble method for pothole detection. Pervasive Mob. Comput. **104**, 101973 (2024)
14. Terven, J., Córdova-Esparza, D.M., Romero-González, J.A.: A comprehensive review of yolo architectures in computer vision: from yolov1 to yolov8 and yolo-nas. Mach. Learn. Knowl. Extr. **5**(4), 1680–2171 (2023)
15. Priyanka, M., Akhib, M.: Smartphone image-based road damage detection in low-light conditions. In: 2023 International Conference on Network, Multimedia and Information Technology (NMITCON), pp. 1–12. IEEE (2023)

Image-Based Identification of Silkworm Pupae Surfaces Using Deep Learning Techniques

Jyoti Sharma[✉] and Pradeep Chouksey

Department of Computer Science and Informatics, Central University of Himachal Pradesh, Kangra 176206, Himachal Pradesh, India
`{cuhp22rdcs02,dr.pchouksey}@hpcu.ac.in`

Abstract. Gender detection of silkworm pupae in silk seed production process is an important aspect for quality silk seed production and usually done manually with experts and needs to be automate in error free mode. However, before attempting this, it is very important to understand the importance of the dorsal and ventral surfaces of silkworm pupae for an accurate classification of gender. Differentiating between the dorsal and ventral surfaces of silkworm pupae is very important to pursue an accurate gender classification, which is one of the preliminary steps in the gender detection processes in the silk seed production process. Any slight inaccuracy during the surface determination process would result in an error in the gender classification, and this can adversely affect the silk seed production process. Some features that are unique distinctive features for gender detection are located mainly on the ventral surface of the silkworm pupae. Thus, it becomes important to ensure that there is an accurate distinction between the dorsal and ventral surfaces to prevent such classification biases in the silk seed production process. In the present study, using CNN, our objective is to classify the dorsal and ventral surfaces of silkworm pupae. The model implements K-fold validation and achieves a mean accuracy of 99. 70% and a standard deviation of 0.0040 with a perfect classification performance (100% precision) in several test cases that indicate the robustness and efficiency of the model. The proposed method aims to be an initial step in the integration of surface recognition in gender classification systems that improves reliability and productivity in silk seed production processes.

Keywords: Silkworm Pupae · Image Processing · Feature Extraction · CNN

1 Introduction

The sorting of the gender of the silkworm pupa is a vital aspect in the production of silk seeds. Identifying the right gender is more important in silkworm seed production, and even micro-sized errors in this may lead to degradation of the quality of the silkworm seed. The features or characters on the basis of which this gender detection is achieved are basically located on the ventral side of silkworm pupae (Fig. 1).

So, it is very essential to locate the ventral surface as a preliminary step in the silk seed production process. Conventional methods for the identification of the gender of silkworm pupae, which often rely on manual observation, are labor intensive,

C. Modi et al. (Eds.): MIND 2024, CCIS 2736, pp. 362–379, 2026.
https://doi.org/10.1007/978-3-032-14531-4_31

Fig1(a) Silkworm Pupae(Ventral Surfaces)

Fig1(b)Silkworm Pupae(Dorsal Surfaces)

Fig. 1. Silkworm Pupae Surfaces

error-prone, and require expertise [1]. As a result, there is a growing need for automated, non-invasive methods leveraging advanced deep learning models to handle these challenges. This work presents a CNN model for the classification of the dorsal and ventral surfaces. The study in question is also concerned with dorsal and ventral aspects highlighting the fact that smooth dorsal surfaces or convex structures are more common while developing complex rudimentary conditions of appendages, color patterns and texture and unique gender specific markings etc. are of great importance on the underside surfaces (Ventral). Different types of X-ray imaging have been attempted along with ML algorithms to aid in the classification of silkworm pupae [2]. Combination of histogram-based features and neural network descriptors in integrated ML and DL models to identify sex and species of pupae of silkworm pupae [3]. Over the past few years, the incorporation of CNN and deep learning methods has proven to be an effective approach to automating the image classification task by learning intricate features of data with minimum pre-processing required [4]. Due to this proficiency in classification and analysis, CNNs have transformed image interpretation technology in several fields, including agriculture, biology of living organisms, and medicine [5]. This research focuses mainly on evaluating the classification of the surface orientation of the silkworm pupa by developing a deep learning method that relies on CNNs.CNNs are perfect for identifying the dorsal and ventral surfaces of silkworm pupae because of their ability to automatically and effectively extract complicated visual features from images. The ability of CNNs to capture spatial patterns—such as texture, shape, and color distributions—is crucial for differentiating these surfaces. It is important to note that this can be considered as a preliminary work that is critical as far as gender determination is concerned and relies on the use of surface-oriented feature and unique features to a particular gender for enhanced accuracy. Targeting such surfaces with precision helps put the model in a position to promote accurate gender determination in the silk seed production process in an enhanced and non-invasive automated way.

2 Literature Review

Multiple techniques have been reported in the literature for accurate silkworm pupa sex differentiation, such as MRI, X-ray imaging, optical penetration and hyperspectral imaging. There have also been efforts to extend such a concept. Classification across different species and environments by employing machine learning techniques such as

CNNs, SVM, PLSDA, and semi-supervised learning. Deployment of advanced feature extracting such as HOG, Co-HOG and approaches based on gradient also enhance performance. These methods facilitate operations by reducing manual efforts while maintaining accuracy, reliability, and operational efficiency. The summarized review of the literature is presented in Table 1.

Table 1. Summarized Literature on Gender Classification

Authors and Years	Objective	Methodology	Accuracy
Kohavi, 1995 [6]	Proposed K-Fold validation for model validation	K-fold validation became an important technique in machine learning model	validation as regards reliability and accuracy
Liu et al., 2008 [7]	Using MRI for sex determination of silk worms	T2-weighted MRI for water analysis in silk worm tissues	Successful gender identification
Cai et al., 2014 [8]	imaging of X-ray with multivariate analysis for rapid gender identification	Soft X-ray imaging, KNN, SVM, and LDA algorithms	92.57% (SVM) to 93.68% (KNN) accuracy
Sumriddetchkajorn et al., 2015 [9]	Fault-tolerant optical penetration for sex identification	White and red light optical-penetration system	98.9% accuracy for male/female classification
Simonyan & Zisserman, 2015 [10]	Evaluating depth in convolutional neural networks	3x3 convolution filters in very deep networks (VGG)	State-of-the-art performance on ImageNet
Tao et al., 2016 [11]	Restoration of silk worm pupa images using blur reduction methods	Lucy-Richardson deconvolution for blur removal	92.3% accuracy with restored images
He et al., 2016 [12]	Introduction of residual learning for deep networks	Learning residual functions with 152-layer networks	3.57% error on ImageNet classification
Tian et al., 2016 [13]	Developing Co-HOG feature descriptors for text recognition	Co-HOG for spatial information in scene text recognition	Superior accuracy on multilingual text datasets
Tao et al., 2018 [14]	HSI to identify sex and species of pupae of silkworm	HSI with PCA for feature extraction and sex identification	95.83% identification accuracy with SVM
Zhu et al., 2018 [15]	NIR spectroscopy with chemometrics for rapid gender sorting	SIMCA model with high-speed automated sorting	98.0% correct Identification rate

(continued)

Table 1. (*continued*)

Authors and Years	Objective	Methodology	Accuracy
Lin et al., 2019 [16]	Model updating strategy to improve accuracy across silkworm varieties	Semi-supervised Learning using Pearson correlation	high accuracy with model updating strategy
Raj et al., 2019 [17]	For gender classification of silkworm cocoon multi-sensor system is used	Load sensors, SVM for cocoon classification	Robust classification across different breeds
Tao et al., 2019 [18]	Removing motion blur for improved Classification accuracy	Radon transform based motion blur correction	Accurate kernel refinement improves classification
Tao et al., 2019 [19]	Using VIS-NIR HIS for non-invasive sex classification	SVM and RBF-NN with PCA-based feature extraction	98.75% accuracy with VIS-NIR HSI
Yu et al., 2020 [20]	Using CNN for cost effective sex classification via gradient features	4-layer CNN with HOG feature fusion	97.21% accuracy on silkworm pupa images
Ma et al., 2021 [21]	NIR spectroscopy with PLSDA for reliable online classification	PLSDA model with spectral and textural data fusion	98% prediction accuracy with NIR-PLSDA
Tao et al., 2022 [22]	Model updating using variable selection and semi-supervised learning	SiPLS-PLSDA with distance-based model updating	94.44% with updated models across species/years
Guo et al., 2022 [23]	Exposure correction algorithm for improved classification	Retinex model for exposure correction	90.52% with corrected exposure data
Fu et al., 2023 [24]	Combining NIR and chemometrics for sex classification	Contrastive PCA with NIR for improved accuracy	99.58% accuracy with iPLS feature selection
Hu et al., 2024 [25]	High-resolution lipidomic for biomarker-based sex classification	Lipidomics using chemometrics to identify gender	Identified lipid biomarkers for gender separation

(*continued*)

Table 1. (continued)

Authors and Years	Objective	Methodology	Accuracy
Thomas & Thomas, 2024 [26]	Optimized HOG-based feature extraction for sustainable silk pro duction	TLBPSGA-XGBoost for HOG-based classification	98.5% accuracy with TLBPSGA-XGBoost model

These technologies reduce the amount of manual work needed and increase the efficiency of operations, and they sustain the environment and industry, providing an alternative to conventional methods of sexing pupa.

3   Research Methodology

The Research methodology for classifying the dorsal and ventral surfaces of silkworm pupae using a Convolutional Neural Network (CNN) model includes several key stages, such as collection of data, preprocessing of images, extraction of characteristic, design of model and evaluation (Fig. 2).

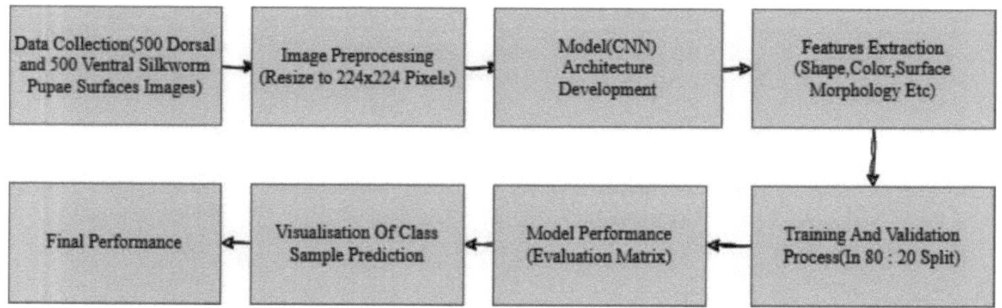

Fig. 2. Basic Steps in Methodology

The following is a detailed description of each step.

1. **Data Collection and Preprocessing**

- **Data Collection:**
     The dataset we used for the purpose of our research is original primary dataset acquire from the Silk seed production centre Palampur, District Kangra, Himachal Pradesh India. It contained 1000 silk worm pupae images of bivoltine pure races (NB4D2 and SH6) including 500 ventral side and 500 dorsal side pupae images. Each image was taken by simple camera during normal daylight conditions. Images were labelled as dorsal and ventral with experts help at the silk seed centre unit on the basis of some distinguishable characteristics of dorsal and ventral surfaces of the pupae. The images were separated and stored into two folders:

– Dorsal Images: A collection of images of pupae on the dorsal side.
– Ventral Images: A collection of images of pupae on the ventral side.

- **Image Preprocessing:**
  Uniform size of every image is 224 × 224 pixels to ensure regularity of the dataset. The pixel values are normalized by dividing by 255, which scales the values between 0 and 1, enhancing the training process.

$$I_{\text{normalize}} = \frac{I}{255} \tag{1}$$

here:

– I is the value of pixel
– I_normalized is normalized pixel value between 0 and 1.

2. **Feature Extraction:**
   Features like shape, texture, color, and surface morphology of images are automatically extracted by using convolutional layers of CNN to further analyze and classify the images. These features are of great help in distinguishing between images of dorsal and ventral pupae.

- **Convolution Operation:**
  Extraction of image features in CNN involves use of convolution over image data in order to look for features, edges, and textures.

$$(I * k)(i, j) = \sum_{m} \sum_{n} I(i + m, j + n) \cdot K(m, n) \tag{2}$$

here: I input image, K is the kernel (filter)
(i, j) position of pixel in the map of output feature.
ReLU Function of Activation:
ReLU is applied after each convolution to give the network more non-linearity.
ReLU formula: $f(x) = \max(0, x)$
here: x Activation function input value.
Pooling Operation:
Max pooling employs the method of taking maximum values to minimize the dimensions of feature maps. in certain patches of the feature map that are most likely contain useful and important features that the model should emphasize on.
Max Pooling Formula: $P(i, j) = \max(I(i: i + f, j: j + f))$ here: $P(i, j)$ pooled output,
I (i, j) input feature map,
f is the pooled filter size
Among the features that were extracted are the following:

– **Shape Features:** Variation in the silhouette of the body and appendage locations.
– **Color Features:** Differences of colors between the dorsal and ventral sides.

- **Texture Features:** Surface patterns such as ridges and roughness or smoothness.
- **Surface Morphology Features:** Superficial features such as roots of appendages or ridges

3. **Model Design:** The CNN architecture consists of multiple convolutional, pooling, dropout, and dense layers (Table 2).

Table 2. Architecture of Model

Type of Layer	Output	Param #
Input layer	(None, 224, 224, 3)	0
conv2d	(None, 224, 224, 32)	896
maxpooling2d	(None, 112, 112, 32)	0
conv2d1	(None, 112, 112, 64)	18,496
maxpooling2d_1	(None, 56, 56, 64)	0
conv2d_2	(None, 56, 56, 128)	73,856
maxpooling2d_2	(None, 28, 28, 128)	0
conv2d_3	(None, 28, 28, 256)	295,168
maxpooling2d_3	(None, 14, 14, 256)	0
flatten	(None, 50176)	0
dense	(None, 256)	12,845,312
dropout	(None, 256)	0
dense_1	(None, 128)	32,896
dropout_1	(None, 128)	0
dense_2	(None, 64)	8,256
dropout_2	(None, 64)	0
dense_3	(None, 2)	130

Convolutional Layers: These extract spatial hierarchies of features from the image. **Dense Layers:** Fully connected layers combine the extracted features to form a final classification.

Dropout Layers: These are used to avoid overfitting by randomly Skiding neurons in training.

The final dense layer incorporates the Softmax Activation Function, which is suitable for class probability prediction.

Softmax Formula:

$$\hat{i} = \frac{e^{z_i}}{\sum_{j=1}^{C} e^{z_j}} \tag{3}$$

C is the number of classes (dorsal and ventral)

4. **Training and Validation:**

The dataset is split into training and validation sets using an 80–20 split.

Optimization Algorithm: The Adam optimizer is used for training, which adapts the learning rate based on first and second moments of the gradient. Adam Update Rule:

$$\theta_t = \theta_{t-1} - \eta v_t + \epsilon m_t \tag{4}$$

Training of model is done with a cross-entropic loss function, which measures the how correctly the classification is done by comparing the predicted probabilities to the true labels.

Cross-Entropy Loss Form

$$L = -\sum_{i=1}^{c} y_i log(\hat{y}_i) \tag{5}$$

Model Evaluation: The results of the models are assessed through the application of several metrics.

Accuracy: Ratio of corrected predicted images to the entire number of images.

$$Accuracy = \frac{(TP + TN)}{(TP + TN + FP + FN)} \tag{6}$$

Precision: The ratio of truly positive to all predicted value of positives.

$$Precision = \frac{(TP)}{(TP + FP)} \tag{7}$$

Recall: The ratio of truly positives to all actual positives.

$$Recall = \frac{(TP)}{(TP + FN)} \tag{8}$$

F1-Score: The balanced measure which is average of precision, recall is called harmonic mean.

$$F1 = 2 \times \frac{Precision \times Recall}{Precision + Recall} \tag{9}$$

Cross Validation of K-Fold: For the purpose of examining how the model generalizes across data it has never seen before 5-fold cross validation is utilized.

$$\text{Mean Accuracy} = \frac{1}{k} \sum_{k=1}^{k} Accuracy_k \tag{10}$$

where K is the number of folds. ROC-AUC:

$$\text{AUC} = \int_{0}^{1} TPR(FPR)dFPR \tag{11}$$

Residual Analysis Residual analysis is conducted to evaluate the prediction errors.

$$\text{Residual} = y_{true} - \hat{y} \tag{12}$$

Hardware and Software used: A system featuring a 12th Gen Intel® CoreTM i5-12450H processor, (16 GB) of RAM, and a (64-bit) Windows operating system was used for the study's investigations. Python, version 3.12.4, was the mainly used programming language to create and execute the model and camera for image capturing.

4 Result

A CNN-based full-body model for the classification of dorsal and ventral surfaces of silkworm pupae was developed and evaluated on a dataset of 1000 images, with 500 labelled as "Dorsal" and 500 as "Ventral." Several evaluation metrics predict model performance including recall, precision, and accuracy as well as F1 measure, ROC AUC, and K-fold validation.

1. **Model Performance Evaluation:**
   The CNN model was tested using a separate test set after training, and the following performance metrics were obtained (Table 3)
   The Full Body CNN Model achieved perfect classification performance with 100% accuracy, precision, recall, and F1-scores for both dorsal and ventral classes, as confirmed by the confusion matrix. Its demonstrated high reliability with a test loss of 0.0004, ROC-AUC of 1.00, and consistent cross-validation accuracy (mean: 0.9970, std: 0.0040), with minimal residual errors.
2. **Confusion Matrix Analysis:**
   The confusion matrix for the test set is presented below (Table 4), showing the classification of dorsal and ventral images:
   All 100 dorsal samples were correctly predicted as dorsal, and all 100 ventral samples were correctly identified as ventral, with no classification errors reported. This highlights the model's exceptional dis criminative ability, achieving 100% accuracy in distinguishing between dorsal and ventral surfaces (Fig. 3).
3. **Training, Validation and Loss Curves:**
   Training accuracy being observed does not oscillate greatly rather presents a monotonically increasing trend where it approaches or reaches as high accuracy for training as one. Validation accuracy is high after several epochs and remains consistent thereafter may suggest that the model may not be overfitted. Similarly, It is seen

Table 3. Performance Table

Metric	Dorsal	Ventral	Overall/Average
Precision	1.00	1.00	1.00
Recall	1.00	1.00	1.00
F1-Score	1.00	1.00	1.00
Support (Number of Samples)	100	100	200
Test Accuracy			1.00(100%)
Test Loss			0.0004
ROC-AUC Score	1.00	1.00	1.00
K-Fold Cross Validation			Mean Accuracy: 0.9970
Standard Deviation			0.0040
Confusion Matrix			No Misclassifications
Residual Error (Max)			0.012

Table 4. Confusion Matrix

	Predicted Dorsal	Predicted Ventral
Actual Dorsal	100	0
Actual Ventral	0	100

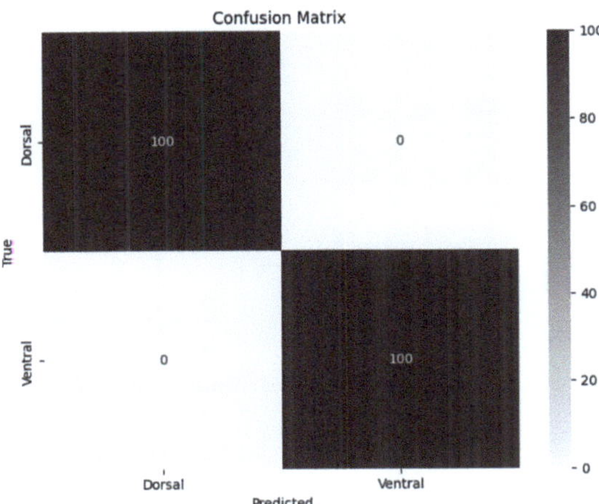

Fig. 3. Confusion Matrix

that the training loss and validation loss achieve steep declines in the early stages and continues to decrease, thus indicating that the model is successfully acquiring knowledge. The ability of the testing set to return favorable predictions is in full agreement with the responses from the training set again demonstrating the high degree of generalization possessed by the model and this model performed well (Fig. 4).

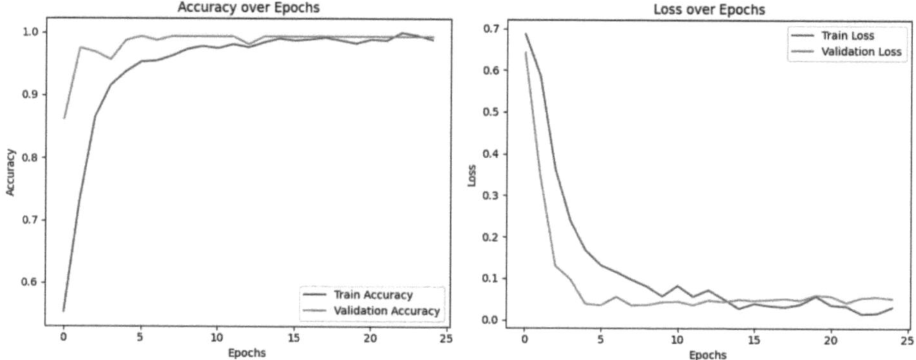

Fig. 4. Training and Validation accuracy loss curve

4. Visualisation of Class Sample Prediction:

The correct classifications of dorsal and ventral samples illustrated in Figure 5 are quite clear from images. The predicted labels of each sample and its true label match perfectly, thus confirming the effectiveness of the CNN model once more.

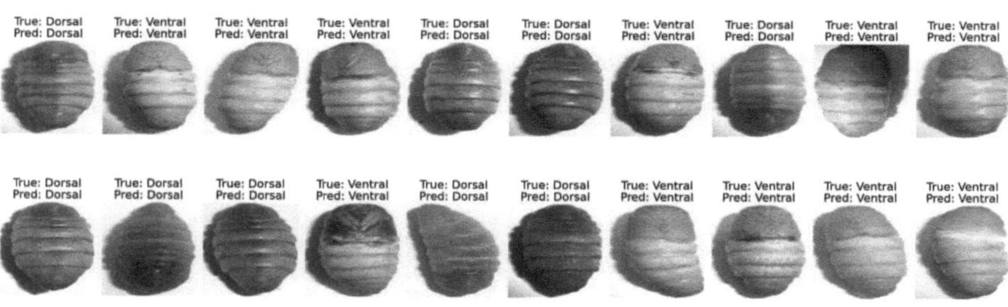

Fig. 5. Visualisation of class sample prediction

5. Receiver Operating Characteristic curve:

The ROC curve illustrated in Figure 6 has an Area Under Curve (AUC) of 1.00 which means the present model is capable of perfect classification. The model is able to achieve a 100% true positive rate over all thresholds indicating its ability and strength to discriminate between the two classes.

6. Precision-recalled Curve:

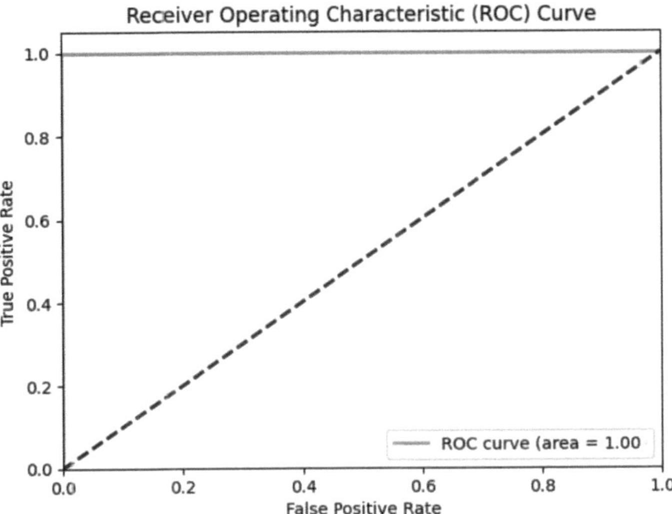

Fig. 6. Reciever operating curve

Figure 7 Precision-Recalled Curve with 1.0 on the entire graph and shows that the model produces a recall that is equal to precision on both dorsal and ventral classes. This further affirms the fact that there are no false positives and false negatives registered.

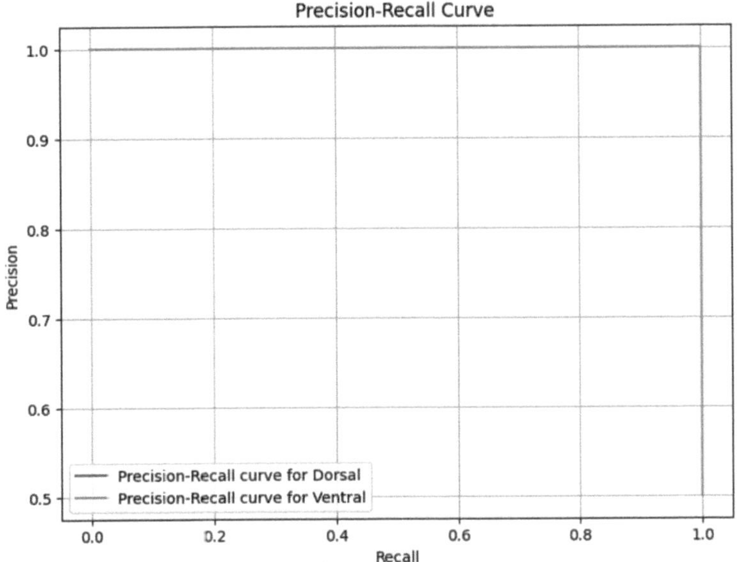

Fig. 7. Precision recall curve

7. **Predicted Probabilities Distribution:**

The probability distribution plot in Fig. 8 brings the attention that: Predicted probabilities are, in principle, equal to 0 and 1 for the respective classes. This means that the confidence levels in the predictions are high; the model does not have problems making predictions for ambiguous samples.

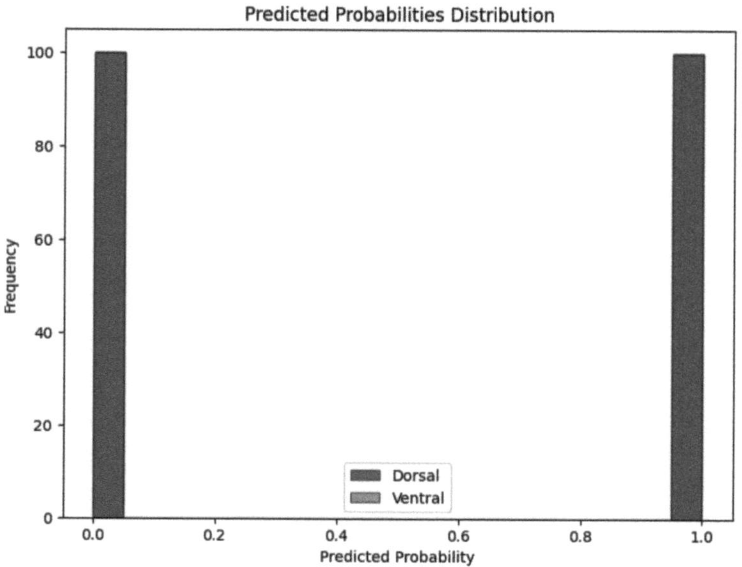

Fig. 8. Predicted Probabilities Distribution

8. **Residual Analysis Plot:**

With regards to Fig. 9 residual plot, prediction errors were said to be low with a majority of the residuals being close to 0. This enhances the predictions made by the model due to the fact that the predictions are made with no significant slippage from the ideal point.

9. **K-Fold Validation:**

The K-Fold Scores illustrated in Fig. 10 in particular K1, K2, K3 models present a consistent pattern with variations of alpha between 0.9900 and 1.0000 denoting the accuracy of the fold were all satisfactory indicators of high robustness. The average accuracy of the model was equal to 0.9970, and the standard deviation of 0.0040 shows great consistency of the model and high stability. Such slight variance confirms again the model's generalization capability across different data splits is extremely effective

The confusion matrix and ROC-AUC of 1.00 confirmed that the Full Body CNN Model performed exceptionally well, attaining 100% accuracy, precision, recall, and F1-scores for both dorsal and ventral classes. The model's resilience, dependability, and superior generalization across data splits are highlighted by its consistent cross-validation accuracy (mean: 0.9970, standard deviation: 0.0040) and low residual errors.

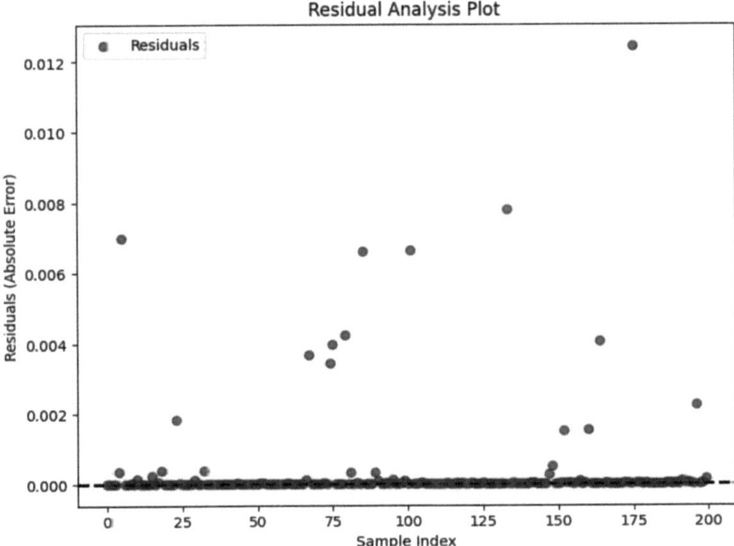

Fig. 9. Residual Analysis Plot

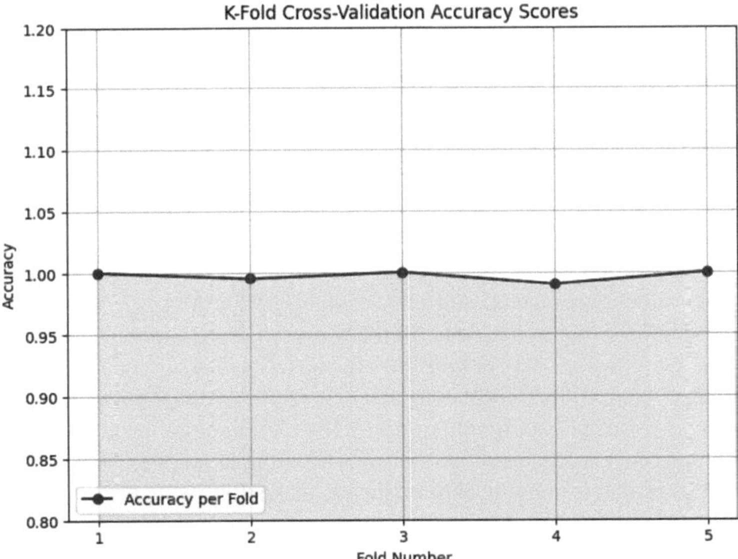

Fig. 10. K-Fold Cross Validation Accuracy

5 Discussion

The CNN-based full-body model that classifies the dorsal and ventral surfaces of silkworm pupae produced extraordinary results that obtained precision 100%. This model was able to appropriately learn the distinguishing features between these two surfaces as the confusion matrix indicates that there are no cases of misclassification. A ROC – AUC score of 1.00 also supplemented that the model is able to extend perfect discrimination between the two classifications across all classification thresholds. The agreement between training and validation accuracy, as well as the losses decreasing consistently, reassures that the model is well trained without overfitting. Furthermore, confidence levels in observed instances of predications were low, reaching the predicted probability distribution, which was 0 or 1 which means that the classification was absolute and there were no indecisions.

The robustness and generalization of the model were tested through cross-validation of K-data folds and achieved a mean accuracy of 0.9970 with a standard error of 0.0040. This type of consistency often proves that a model will consistently be able to deliver across folds, meaning that, even if the model is presented with new unseen data split, the predictions will remain stable and true to the model. Minimal residual errors further reiterate the capabilities of the model to return accurate predictions. The model's performance may be affected by varying environmental conditions such as light conditions and differences in image quality; however, the model performs very well under ordinary day light conditions in classifying ventral and dorsal surfaces.

6 Conclusion

The CNN Full Body Model delivers high performance and accuracy in the classification 100% of the dorsal and ventral surfaces of silkworm pupae and therefore has the potential for application in areas of the silk seed production process as a preliminary step in the classification of sex where precision is critical. The presence of high ROC AUC, uniform cross-validation results, and low residual error errors illustrated the strength and generalization of the model, which makes it perfect for practical implementation. The practical use of this model demonstrates the benefits of using separate convolutional neural networks to identify morphologically insignificant elements to cope with the intricate nature of image classification tasks. The classification problem of this nature offers high confidence in one man that the silk seed production processes will not remain a manual job task that is fraught with mistakes, but rather integration of advancement that will increase the level of silk seed production processes with greater accuracy and efficiency. Concerned such as management of different genders remains very critical in sericulture, then this model is a strong reliable fit that greatly improves production efficiency.

7 Future Scope

Further development of this model opens up the following possibilities:

Mobile Application Development:
The availability of the model in mobile application development would allow for classification on the go which would be a useful tool for sericulturists.

The Model could be Broadened to include other Species:
It would also be relevant to expand this model to other species of the designating insects in order to increase the usage of the product in an area where several types are grown.

Adding More Additional Attributes:
It is also feasible that combining the model with different images such as infrared imaging would help expand the uses of the model, especially where subtle shifts imperceptible to the eye occur.

Transfer Learning for Generalization:
It would be worth examining how transfer learning approaches could allow the model to transfer more effectively to other classification tasks using pupae morphology.

Results of model can be validated in a still larger dataset in future studies.

Acknowledgement. The authors appreciate the technical support in basic understanding of the various aspects of the silk seed production process, including gender detection from the silk seed production center Palampur, District Kangra, Himachal Pradesh, India.

Author's Contributions. All the authors participated in this study to the same extent.

Funding Information. No funding support or financial assistance was provided to the authors.
Data Availability. Dataset used in this is not available in the public domain. The data sets generated during the current study are available from the corresponding author upon reasonable request.

Materials Availability. The research materials used in this study was obtained from Silk Seed Production Centre Palampur, Distt Kangra (HP) India

Code Availability. The code employed in the course of the analysis can be obtained from the corresponding author on reasonable request.

Ethics. The manuscript submitted is an original work. The authors declare that they have no ethical concerns with respect to this submission.

Conflicts of Interest. The authors have no conflict of interest.

Consent for Publication. All authors have given their consent to the publication of this manuscript. The research presented in this document does not involve personal data from anyone, and all necessary ethical approvals have been obtained for the use of the materials involved in the study.

References

1. Kiruba, M., Mangammal, P., Anand, G., Sakila, M., Kumar, P., Senthilkumar T.: Innovations in sericulture an advancement in silk production and quality improvement. Uttar Pradesh J. Zool. **45**(18), 553–562 (2024)

2. Thomas, S., Thomas, J.: Non-destructive silkworm pupa gender classification with x-ray images using ensemble learning. Artif. Intell. Agric. **6**, 100–110 (2022)
3. He, H., et al.: Integrated analysis of machine learning and deep learning in silkworm pupae (bombyx mori) species and sex identification. Animals **13**(23), 3612 (2023)
4. Krizhevsky, A., Sutskever, I., Hinton, G.E.: Imagenet classification with deep convolutional neural networks. In: Advances in Neural Information Processing Systems, vol. 25 (2012)
5. Kamilaris, A., Prenafeta-Boldú, F.X.: Deep learning in agriculture: a survey. Comput. Electron. Agric. **147**, 70–90 (2018)
6. Kohavi, R.: A Study of Cross-Validation and Bootstrap for Accuracy Estimation and Model Selection. Morgan Kaufman Publishing (1995)
7. Liu, C., Ren, Z.H., Wang, H.Z., Yang, P.Q., Zhang, X.L.: Analysis on gender of silkworms by MRI technology. In: 2008 International Conference on BioMedical Engineering and Informatics, vol. 2, pp. 8–12. IEEE (2008)
8. Cai, J.-R., Yuan, L.-M., Liu, B., Sun, L.: Nondestructive gender identification of silkworm cocoons using x-ray imaging with multivariate data analysis. Anal. Methods **6**(18), 7224–7233 (2014)
9. Sumriddetchkajorn, S., Kamtongdee, C., Chanhorm, S.: Fault-tolerant optical-penetration-based silk-worm gender identification. Comput. Electron. Agric. **119**, 201–208 (2015)
10. Simonyan, K., Zisserman, A.: Very deep convolutional networks for large-scale image recognition. arXiv preprint arXiv:1409.1556 (2014)
11. Tao, D., Wang, Z., Li, G., Qiu, G.: Silkworm pupa image restoration based on aliasing resolving algorithm and identifying male and female. Trans. Chin. Soc. Agric. Eng. **32**(16), 168–174 (2016)
12. He, K., Zhang, X., Ren, S., Sun, J.: Deep residual learning for image recognition. In: Proceedings of the IEEE Conference on Computer Vision and Pattern Recognition, pp. 770–778 (2016)
13. Tian, S., et al.: Multilingual scene character recognition with co-occurrence of histogram of oriented gradients. Pattern Recogn. **51**, 125–134 (2016)
14. Tao, D., Wang, Z., Li, G., Xie, L.: Simultaneous species and sex identification of silkworm pupae using hyperspectral imaging technology. Spectrosc. Lett. **51**(8), 446–452 (2018)
15. Zhu, Z., et al.: High-speed sex identification and sorting of living silkworm pupae using near-infrared spectroscopy combined with chemometrics. Sens. Actuators B Chem. **268**, 299–309 (2018)
16. Lin, X., et al.: The model updating based on near infrared spectroscopy for the sex identification of silkworm pupae from different varieties by a semi-supervised learning with pre-labeling method. Spectrosc. Lett. **52**(10), 642–652 (2019)
17. Joseph Raj, A.N., Sundaram, R., Mahesh, V.G., Zhuang, Z., Simeone, A.: A multi-sensor system for silkworm cocoon gender classification via image processing and support vector machine. Sensors **19**(12), 2656 (2019)
18. Tao, D., Wang, Z., Li, G., Qiu, G.: Radon transform-based motion blurred silkworm pupa image restoration. Int. J. Agric. Biol. Eng. **12**(2), 152–159 (2019)
19. Tao, D., Wang, Z., Li, G., Xie, L.: Sex determination of silkworm pupae using vis-nir hyperspectral imaging combined with chemometrics. Spectrochim. Acta Part A Mol. Biomol. Spectrosc. **208**, 7–12 (2019)
20. Yu, Y., Gao, P., Zhao, Y., Pan, G., Chen, T.: Automatic identification of female and male silkworm pupa based on deep convolution neural network (2020)
21. Ma, Y., Xu, Y., Yan, H., Zhang, G.: On-line identification of silkworm pupae gender by short-wavelength near infrared spectroscopy and pattern recognition technology. J. Near Infrared Spectrosc. **29**(4), 207–215 (2021)

22. Tao, D., Li, G., Qiu, G., Chen, S., Li, G.: Different variable selection and model updating strategies about sex classification of silkworm pupae. Infrared Phys. Technol. **127**, 104471 (2022)
23. Guo, F., He, F., Tao, D., Li, G.: Automatic exposure correction algorithm for online silkworm pupae (bombyx mori) sex classification. Comput. Electron. Agric. **198**, 107108 (2022)
24. Fu, X., Zhao, S., Luo, H., Tao, D., Wu, X., Li, G.: Sex classification of silkworm pupae from different varieties by near infrared spectroscopy combined with chemometrics. Infrared Phys. Technol. **129**, 104553 (2023)
25. Hu, L., et al.: Comparison of lipid profiles of male and female silkworm (bombyx mori) pupae through high-resolution mass spectrometry-based lipidomics and chemometrics. Food Chem. **459**, 140396 (2024)
26. Thomas, S., Thomas, J.: An optimized method for mulberry silkworm bombyx mori (bombycidae: Lepidoptera) sex classification using tlbpsga-rfexgboost. Biol. Open 060468 (2024)

Segmentation-Based 3D Point Cloud Generation for Medical Data Using 2D Images

R. K. Darshan[1], Renukasakshi v Patil[1], and V. Geetha[2(✉)]

[1] Artificial Intelligence, Department of Information Technology,
National Institute of Technology Karnataka, Surathkal, India
`{darshanrk.211ai015,renukasakshivpatil.211ai030}@nitk.edu.in`
[2] Head of the Department, Department of Information Technology,
National Institute of Technology Karnataka, Surathkal, India
`geethav@nitk.edu.in`

Abstract. In medical imaging, constructing 3D representations from 2D images has become crucial for enhanced diagnostic and treatment planning, especially in areas like virtual reality and precision surgery. This research explores methodologies for 2D-to-3D image conversion, emphasizing a specific approach that uses deep learning techniques, particularly the U-Net model, for segmenting critical medical images. The model is initially trained on a large, diverse dataset and then employs transfer learning to accurately segment lung CT scans, compensating for limited data availability. The segmented images undergo depth estimation to capture spatial details, which are then used to construct a 3D point cloud—a precise visual representation of infected areas, valuable for clinical analysis and planning. The field of 2D-to-3D image construction spans both traditional stereoscopic techniques, which rely on binocular vision to estimate depth, and advanced methods using convolutional neural networks (CNNs) for automated depth estimation. While these approaches hold promise, challenges such as managing occlusions and achieving real-time performance remain. Leveraging architectures like deep neural networks and transfer learning techniques has proven essential in overcoming these obstacles. As advancements continue, these methodologies are poised to revolutionize medical imaging, providing clinicians with more accurate, immersive, and actionable 3D representations of critical anatomical structures.

Keywords: 3-D Reconstruction · Depth Estimation · Transfer learning · Segmentation · Depth CNN · UNet · ResNet · Point Cloud

1 Introduction

Two-dimensional (2D) to three-dimensional (3D) image reconstruction stands at the forefront of technological innovation, transcending traditional visual representations and unlocking a myriad of applications. This transformative process

C. Modi et al. (Eds.): MIND 2024, CCIS 2736, pp. 380–391, 2026.
https://doi.org/10.1007/978-3-032-14531-4_32

involves the conversion of flat, static images into dynamic 3D models, introducing an extra dimension that significantly enhances the richness of information.

At its core, 2D to 3D reconstruction is crucial for data augmentation and enrichment. By extrapolating the third dimension from 2D images, a more comprehensive understanding of spatial relationships becomes attainable. This is particularly impactful in medical imaging, where converting 2D scans to 3D models allows for a detailed examination of anatomical structures. Surgeons, for instance, benefit from preoperative planning with more accurate depictions of the patient's anatomy, leading to improved surgical outcomes.

Moreover, in the realm of computer vision, the conversion process plays a pivotal role. Object recognition and tracking systems leverage 2D to 3D reconstruction to better comprehend the environment, enhancing the capabilities of surveillance systems, robotics, and autonomous vehicles. This advancement is instrumental in creating more intelligent and perceptive technologies.

The fusion of 2D and 3D data also finds applications in cultural preservation. Historical artifacts and artworks captured in 2D images can be transformed into 3D models, offering a nuanced understanding of these objects for both preservation and educational purposes.

As technology continues to evolve, the scope of 2D to 3D image reconstruction expands across diverse domains, from entertainment and gaming to scientific research and beyond. With the advent of sophisticated algorithms, machine learning, and improved computational power, the accuracy and efficiency of this reconstruction process continue to advance, promising a future where our visual representations transcend the limitations of traditional imagery.

1.1   Importance of 3D Images in Medicine

In medical imaging, 3D images provide a realistic and detailed representation of anatomy, offering a sense of depth that 2D images cannot match. This enhanced realism improves understanding of complex bodily structures, crucial for accurate diagnosis and treatment planning, especially in fields like cardiovascular and neurosurgery. Furthermore, 3D reconstructions enable immersive training environments, allowing healthcare professionals to practice procedures in virtual reality settings, which enhances skills and patient safety. In research, 3D imaging allows for in-depth spatial analysis of biological structures, aiding studies in areas like tumor progression and cellular interactions, thereby advancing our knowledge in biomedical science.

1.2   Importance of 2D to 3D Reconstruction in Medical Imaging

Converting 2D medical images, such as X-rays and MRI slices, into 3D models enriches the available data by revealing spatial details, aiding in comprehensive analysis and enhancing diagnostic accuracy. This transformation is essential for complex surgical planning, allowing surgeons to visualize the full anatomy from multiple angles for precise, minimally invasive approaches. Additionally, 2D-to-3D reconstruction is valuable for preserving medical history, enabling insights from historical scans that can now be analyzed in 3D. With continued advancements in machine learning and computer vision, 2D-to-3D technology is reshaping healthcare, offering increasingly powerful tools for diagnosis, research, and training.

2   Literature Survey

The concept Monodepth2 [6], introduced in "Digging Into Self-Supervised Monocular Depth Prediction" by Clement Godard et al., advances self-supervised depth estimation with monocular images. Building on its predecessor, it utilizes stereo pairs or monocular sequences, introducing key techniques like per-pixel minimum reprojection loss, auto-masking, and a multi-scale approach. These innovations help tackle issues like occlusions, moving objects, and photometric inconsistencies. The model's core innovation is its minimum reprojection loss, which minimizes occlusion effects and better handles dynamic scenes. Its multi-scale framework refines depth maps, and auto-masking distinguishes static from moving pixels, improving accuracy in real-world scenarios. However, Monodepth2 may struggle with complex tasks like medical imaging, as it was optimized for outdoor environments like autonomous driving, where high structural specificity is less critical.

In this work, MiDaS [7], or "Monocular Depth Estimation," developed by Intel researchers, is presented in "Towards Robust Monocular Depth Estimation: Mixing Datasets for Zero-Shot Cross-Dataset Transfer." MiDaS focuses on robust depth estimation across diverse environments by training on a blend of multiple datasets, allowing it to perform effectively in a zero-shot manner without fine-tuning, unlike models like Monodepth2. It well across indoor and outdoor scenes, varied lighting, and object types, addressing challenges in scenarios where domain-specific data is limited, such as medical imaging. The architecture of MiDaS uses a modified ResNeXt101 backbone, enhancing its ability to capture global and local features for depth estimation. Equipped with a dense prediction head, it generates high-quality depth maps from single images and ranks highly across benchmark datasets. MiDaS also utilizes a loss function combining structure-aware and scale-invariant terms, maintaining depth consistency across scales—beneficial for complex structures like lung scans, where precise depth estimation is critical.

In this work, DenseDepth [4], presented in "DenseDepth: Learning Depth Estimation in Challenging Environments," introduces a supervised monocular depth estimation model focused on dense, high-resolution predictions. A key feature of DenseDepth is its combination of SSIM (Structural Similarity Index Measure) and L1 loss, which enhances fine detail and structural accuracy. SSIM aids in preserving the depth map's structural consistency, while L1 loss penalizes pixel-level depth discrepancies, allowing for high-fidelity outputs suited to applications needing precise depth information. However, DenseDepth's reliance on ground-truth depth data and high computational demands present challenges for medical imaging, where such data is scarce, particularly for lung scans. Its application to medical imaging may be constrained by the availability of specialized datasets, impacting fine-tuning potential for specific analyses like lung disease. Nonetheless, DenseDepth's architecture and loss function innovations make it a valuable model for depth estimation in scenarios requiring detailed 3D insights.

In this work, "Depth Anything: Unleashing the Power of Large-Scale Unlabeled Data" [10], introduces a foundational model for monocular depth estimation that performs robustly across diverse scenarios with minimal tuning. Depth Anything utilizes extensive data augmentation for adaptability and auxiliary supervision to leverage semantic priors, enabling strong zero-shot performance across domains. In our project, Depth Anything was chosen for its superior handling of complex textures, outperforming Monodepth2, MiDaS, and DenseDepth, especially in capturing lung details in imaging tasks. Its Vision Transformer architecture effectively processes intricate structures, making it ideal for specialized applications like medical imaging.

In Anguelov et al. [9], three essential properties for effective 3D point cloud segmentation are outlined: feature diversity to distinguish varied objects, label inference in sparse regions for continuity, and adaptability across different 3D scanners to maintain segmentation accuracy. This work set a foundation for methods like clustering, region-growing, and modern deep learning models such as PointNet. Key challenges, including dataset generalization, processing speed, and occlusion handling, continue to drive advancements in the field.

3 Problem Statement and Objectives

3.1 Problem Statement

Developing accurate 3D models from 2D medical images through effective segmentation that enhances clarity and precision in medical diagnostics and treatment planning

3.2 Objectives

1. **Anatomical Segmentation:** Design a segmentation model using U-Net to isolate key anatomical structures in 2D medical images, with a focus on lung CT scans.
2. **Depth Estimation for 3D Conversion:** Apply depth estimation to segmented images to convert 2D scans into 3D representations, capturing accurate spatial details.
3. **3D Visualization through Point Clouds:** Construct 3D point clouds from depth data to visualize affected regions for enhanced diagnostic and treatment planning.
4. **Optimization of Deep Learning Techniques:** Optimize deep learning techniques, including transfer learning, to improve accuracy and efficiency in 2D-to-3D medical image reconstruction.

4   Proposed Methodology

4.1   Image Segmentation:

Image segmentation is a fundamental technique in computer vision used to partition medical images into distinct regions for detailed analysis. In this study, a U-Net [1] model, a widely adopted deep learning architecture for segmentation, is trained on corneal images annotated with segmentation masks. These annotated masks guide the model to identify and learn precise boundaries, equipping it to segment images with high accuracy in practical applications.

The training process incorporates a carefully designed combination of custom loss functions and evaluation metrics to maximize segmentation accuracy. A hybrid approach is used that combines Binary Cross-Entropy (BCE) Loss, which enhances pixel-level classification accuracy, with Dice Loss, an overlap-based measure that encourages close alignment between predicted and ground truth regions. This combination ensures that both individual pixel accuracy and spatial coherence are optimized in the model's segmentation output.

Additionally, to further enhance model reliability, Focal Loss is integrated to address class imbalance by focusing on hard-to-classify regions. This adjustment allows the model to prioritize challenging areas, thus improving detection of subtle or rare features within the data. For evaluation, a similarity coefficient, or Intersection over Union (IoU), quantifies the overlap between predicted and actual segmented regions, offering a comprehensive assessment of spatial accuracy.

The U-Net architecture itself is structured with encoder and decoder stacks that facilitate efficient downsampling and upsampling. Each encoder layer progressively reduces spatial dimensions while extracting key features, while decoder layers restore image resolution by upsampling and integrating information from previous layers, preserving detail and contextual accuracy. This cohesive design, supported by targeted loss functions and rigorous evaluation, enables the model to deliver robust segmentation outputs suitable for advanced medical image analysis.

4.2   Transfer Learning:

Transfer learning is applied by using a U-Net model, initially fine-tuned on a cornea image dataset, to segment lung CT images. By transferring knowledge from the cornea images, the model is better equipped to perform accurate segmentation on the lung dataset, which contains only 20 images. The U-Net architecture, with its encoder and decoder structure, enables the model to retain essential features, allowing it to identify infected areas within the lungs effectively, even with limited data.

In the initial U-Net architecture, 4 encoder and 4 decoder layers are used for image segmentation. During the transfer learning phase, only the last 2 layers of both the encoder and decoder are fine-tuned.

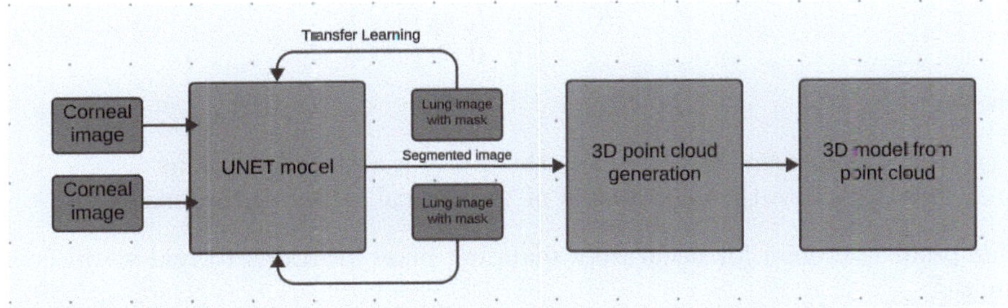

Fig. 1. Workflow diagram of the methodology.

Figure 1 represents the workflow diagram of the proposed methodology.

4.3   Depth Estimation:

Depth estimation involves the process of determining the distance of objects within an image from a specific viewpoint. It helps add dimensionality. enabling better visualization of object structure. Depth information can help determine the position and size of abnormalities within the lungs. This application may assist radiologists in locating tumors or lesions more accurately, especially in surgical planning. It may allow for the identification of critical details that may be obscured in traditional two-dimensional representations.

Upon getting the segmented images they were sent to the depth estimation model, the fine-tuned Depth Anything model was utilized to generate depth maps from the lung images. This process involved applying the model to the pre-processed input, which allowed for the extraction of depth information crucial for subsequent analysis. weights from the pre-trained model and fine-tuned it with a reduced learning rate to minimize the risk of overfitting on the specific lung dataset. Layers were added to stabilize the learning process and enhance convergence rates. To prevent overfitting, an early stopping mechanism was implemented that monitored validation loss throughout training.

4.4   Depth to 3D Point Cloud Conversion:

After exploring the potential for utilizing depth estimation outputs to create 3D structures through point cloud [9] generation. This transformation from 2D images to 3D representations necessitated a thorough understanding of camera intrinsics. intrinsics refer to the internal parameters of a camera that define how it captures images, including focal length, optical center, and skew. The intrinsic matrix is essential for converting pixel coordinates in the image plane to real-world coordinates in 3D space. The formula used to obtain the intrinsic matrix can be expressed as:

$$K = \begin{bmatrix} f_x & 0 & c_x \\ 0 & f_y & c_y \\ 0 & 0 & 1 \end{bmatrix}$$

where:

- f_x and f_y represent the focal lengths in the x and y dimensions.
- c_x and c_y denote the coordinates of the optical center of the image.

This point is crucial for accurately mapping pixel locations to real-world coordinates.

This formula to calculate the intrinsic matrix based on image dimensions and the field of view (FOV). By establishing these parameters, we could accurately transform depth maps derived from 2D images into 3D point clouds, which visually represent the spatial arrangement of lung structures.

The choice of employing intrinsic parameters is grounded in the necessity for precision when reconstructing 3D structures from 2D images. Using the intrinsic matrix allows for a systematic approach to understanding how each pixel's depth corresponds to a specific 3D point in space. The formulation allows us to project the depth data effectively onto a three-dimensional plane, leading to a coherent point cloud that accurately reflects the features of the object in our case the anatomical features of the lung.

Furthermore, the intrinsic parameters help mitigate issues such as lens distortion, which can skew the depth measurements if not properly accounted for. By calculating the intrinsic parameters meticulously, we enhance the fidelity of the 3D reconstruction, leading to better visualization and analysis of lung structures.

4.5 Pixel to Point Conversion:

This step involved transforming the pixel coordinates from the depth image into 3D world coordinates. By using the intrinsic matrix, depth values were processed to ascertain the x, y, and z coordinates of each point in the image, effectively creating a spatial representation of the lung structures.

To transform depth information into 3D coordinates, the following equations are employed:

$$X = (u - c_x) \cdot \frac{Z}{f_x}, \quad Y = (v - c_y) \cdot \frac{Z}{f_y}, \quad Z = \text{depth}$$

In these equations:

- (u,v) are the pixel coordinates in the 2D image.
- Z represents the depth value at the corresponding pixel.

we computed the X and Y coordinates based on the focal lengths and the principal point, thereby constructing the 3D representation of the scene. This process is inherently dependent on the quality and accuracy of the input depth data; any noise or inaccuracies in the depth estimation can lead to distorted 3D structures. The point cloud generated represents not just a collection of points but a meaningful spatial structure that corresponds to the anatomical features of the lung images. This representation is crucial for subsequent analysis and can be used for various applications, such as 3D modeling, visualization, and even guiding surgical procedures.

5   Experiments

5.1   Dataset

This research utilizes two primary datasets. The first dataset comprises approximately 500 corneal images, each accompanied by segmentation masks, specifically employed to fine-tune the U-Net model for image segmentation tasks. The second dataset, which serves as the primary dataset for this study, consists of 20 lung CT scan images, with infection masks and lung with infection masks providing the main basis for analysis and experimentation.

5.2   Results

We assessed the UNet model's performance in image segmentation using the DICE score [3], where higher scores indicate better accuracy. Results showed a steady increase in the DICE score over the first 5 epochs, reaching a peak of 0.482, after which the score plateaued with minor fluctuations. This suggests that while the model improved initially, further training led to marginal gains. Additionally, both training and validation loss consistently decreased as epochs progressed, highlighting the model's learning and improved accuracy over time. This trend indicates that the model effectively segmented images, though gains became minimal after a certain point in training. The Fig. 2 shows the above details pictorially.

Upon fine-tuning the Depth Anything model, achieved significant improvements is the depth estimation results. This model could capture the depth variations that are critical in medical imaging, particularly for lung structures. The performance of the Depth Anything model was assessed through qualitative evaluation, wherein we visually compared the generated depth maps with the original images. The depth maps obtained through our fine-tuning process are presented below, illustrating the model's proficiency in estimating depth information. The results indicate a notable enhancement in the representation of depth contours and spatial relationships within the lung images (Figs. 3 and 4).

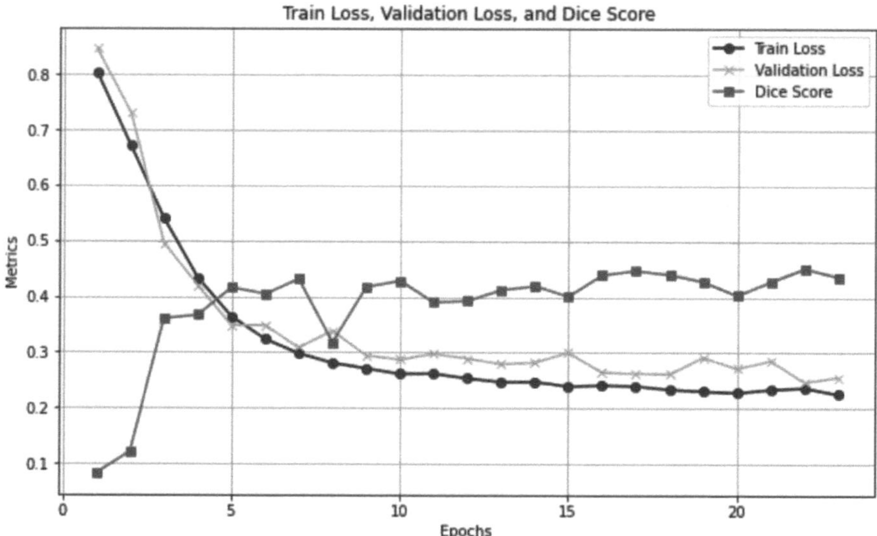

Fig. 2. Variation of loss and DICE score for the image segmentation task.

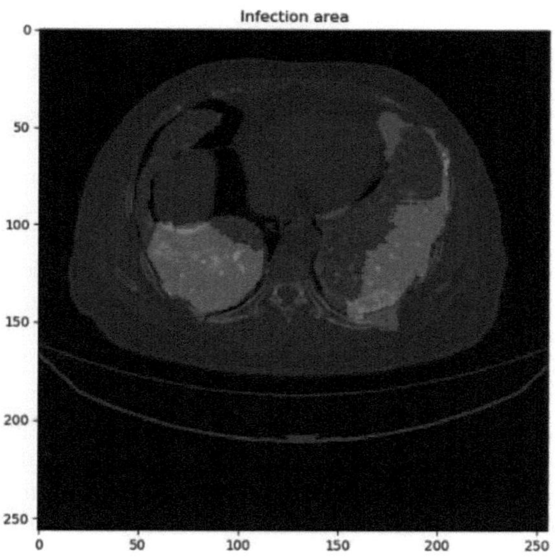

Fig. 3. Original segmented image.

Following the depth estimation, we proceeded to generate the corresponding 3D point clouds from the depth maps using Open3D. This step is crucial as it translates the depth data into a format that can be utilized for further analysis and visualization in a three-dimensional space. The generated point clouds provide a representation of the features present in the lung images, enabling a more detailed examination of the structural variations and spatial relationships.

We processed the depth maps alongside the original lung images to create point clouds stored in the PLY file format. This format is particularly advantageous for its versatility and ease of integration with various visualization tools and software. The transformation from depth maps to point clouds involved leveraging the intrinsic camera parameters, ensuring that each pixel's depth information was accurately mapped to its corresponding 3D coordinates.

The generated PLY files encapsulate a wealth of information, facilitating a three-dimensional analysis of lung structures. This point cloud representation serves as a foundation for potential applications in medical diagnostics, treatment planning, and educational purposes, offering insights that are not readily apparent in traditional two-dimensional imaging.

The results of the point cloud generation process are showcased below, exemplifying the effectiveness of our approach and the capabilities of Open3D [11] in handling complex medical imaging data. This file can also be visualised at different angles in 3D rendering tools. In our case CloudCompare [5] was used.

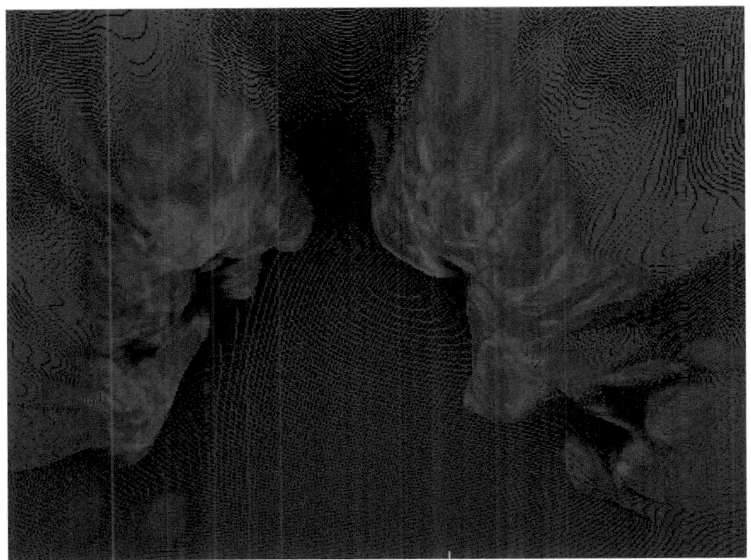

Fig. 4. 3D point cloud visualised in Cloud Compare tool.

As shown in Table 1, the Proposed Model performs better across most metrics.

The Proposed Model outperforms SegNet [2] and FCN [8] Model across most metrics, showing especially strong results in Dice Score, mIoU, and mF1 Score, which indicate better overall segmentation performance. Although SegNet has the highest mPA, the Proposed Model demonstrates a more balanced and superior performance across key evaluation metrics.

Table 1. Performance Comparison of Different Models

	Dice Score	Mean Pixel Accuracy (mPA)	mIoU	Mean F1 Score (mF1)
Proposed Model	0.612	0.771	**0.658**	0.720
SegNet	0.539	**0.791**	0.612	0.670
FCN Model	0.458	0.612	0.503	0.590

6 Conclusion

In conclusion, our project successfully demonstrated the feasibility of using depth estimation techniques on segmented lung disease images to generate 3D point clouds, paving the way for enhanced visualization and analysis of complex pulmonary conditions. The segmentation process was integral to this workflow, enabling the isolation of relevant anatomical features, such as lesions and nodules, from the surrounding tissue. This focused analysis allows for a more detailed understanding of lung pathology, aiding healthcare professionals in making informed decisions.

Major advantages of the proposed approach in 3D reconstruction of medical images. In such a representation, there can be found a very detailed and full description, improving diagnostics and better understanding anatomical relations to be obtained which might benefit surgical planning for treatment and intervention. Further additional advantages in virtual reality medicine that can be achieved in the clinical practices of healthcare providers are to simulate the state of disease by immersing themselves in 3D while performing the technique. Moreover, the 3D point cloud representations may also clarify information that does not appear in the classic 2D imaging and may affect clinical decision-making.

Despite the inherent challenges presented by the lack of ground truth images for depth estimation, we worked on this innovative approach to bridge the gap in current medical imaging practices. By leveraging advanced models, including the fine-tuned Depth Anything model, we achieved promising results that allow for the accurate representation of anatomical structures within the lungs. While our initial results are encouraging, there remains considerable potential for improvement. The depth estimation and subsequent point cloud generation processes have shown satisfactory outcomes; however, refining these techniques further could yield even greater accuracy and utility in clinical settings.

References

1. Baccouche, A., Garcia-Zapirain, B., Castillo Olea, C., Elmaghraby, A.S.: Connected-unets: a deep learning architecture for breast mass segmentation. NPJ Breast Cancer **7**(1), 151 (2021)
2. Badrinarayanan, V., Kendall, A., Cipolla, R.: Segnet: a deep convolutional encoder-decoder architecture for image segmentation. IEEE Trans. Pattern Anal. Mach. Intell. **39**(12), 2481–2495 (2017)

3. Bertels, J., Eelbode, T., Berman, M., Vandermeulen, D., Maes, F., Bisschops, R., Blaschko, M.B.: Optimizing the Dice Score and Jaccard Index for Medical Image Segmentation: Theory and Practice. In: Shen, D., Liu, T., Peters, T.M., Staib, L.H., Essert, C., Zhou, S., Yap, P.-T., Khan, A. (eds.) MICCAI 2019. LNCS, vol. 11765, pp. 92–100. Springer, Cham (2019). https://doi.org/10.1007/978-3-030-32245-8_11

4. Gasperini, S., Morbitzer, N., Jung, H., Navab, N., Tombari, F.: Robust monocular depth estimation under challenging conditions. In: Proceedings of the IEEE/CVF international conference on computer vision, pp. 8177–8186 (2023)

5. Girardeau-Montaut, D., et al.: Cloudcompare. France: EDF R&D Telecom Paris-Tech **11**(5) (2016)

6. Godard, C., Mac Aodha, O., Firman, M., Brostow, G.J.: Digging into self-supervised monocular depth estimation. In: Proceedings of the IEEE/CVF international conference on computer vision, pp. 3828–3838 (2019)

7. Lasinger, K., Ranftl, R., Schindler, K., Koltun, V.: Towards robust monocular depth estimation: Mixing datasets for zero-shot cross-dataset transfer. arXiv preprint arXiv:1907.01341 (2019)

8. Lu, Y., Chen, Y., Zhao, D., Chen, J.: Graph-FCN for image semantic segmentation. In: International symposium on neural networks, pp. 97–105. Springer (2019)

9. Nguyen, A., Le, B.: 3d point cloud segmentation: a survey. In: 2013 6th IEEE Conference on Robotics, Automation and Mechatronics (RAM), pp. 225–230 (2013). https://doi.org/10.1109/RAM.2013.6758588

10. Yang, L., Kang, B., Huang, Z., Xu, X., Feng, J., Zhao, H.: Depth anything: unleashing the power of large-scale unlabeled data. In: Proceedings of the IEEE/CVF Conference on Computer Vision and Pattern Recognition, pp. 10371–10381 (2024)

11. Zhou, Q.Y., Park, J., Koltun, V.: Open3d: a modern library for 3d data processing. arXiv preprint arXiv:1801.09847 (2018)

Javlein Throw Analysis Using Computer Vision

Krishna Chaitanya Gorti, Garima Pandey$^{(\boxtimes)}$ (ID), and Shashidhar G. Koolagudi

National Institute of Technology Karnataka, Surathkal, India
`{chaitanya.211cs228,garimapandey.217cs002,koolagudi}@nitk.edu.in`

Abstract. Performance optimization is a continuous pursuit in athletics, particularly in sports such as javelin throw, where even subtle improvements in technique can lead to significant performance gains. This paper presents a novel approach to analyzing the biomechanics behind the javelin throw with the help of computer vision techniques. Our methodology employs body position tracking with the help of proprietary landmark detection algorithms to analyze the athlete's throwing technique. The athlete's body position is accurately tracked during the last seven steps of the javelin throw, called the crossover phase, including the critical release step. Additionally, we utilized landmark detection algorithms to identify several critical anatomical landmarks and tracked eight key body angles essential for the javelin throw technique. Also, we custom-trained our own YOLOv7 model for javelin detection and tracking. We used a dataset that consisted of video footage and images sourced from the internet to train the custom YOLOv7 model. The model allowed accurate tracking of the position and angle of the javelin throughout the approach phase. By precisely monitoring body positions and tracking the path of the javelin in the crossover phase, our methodology provides valuable insights into the biomechanics of this sport.

Keywords: Javelin Throw · Computer Vision · Deep Learning

1 Introduction

The javelin throw [11], a sport steeped in tradition, depends on the intricate interplay between an athlete's posture and the angle of the javelin. These factors play an essential role in determining how the javelin travels. This paper discusses computer vision solutions to measure several of these parameters. We track eight important body angles that affect an athlete's throw. This is followed by the precise tracking of the javelin throughout the crossover phase. Athletes can use this information to improve their technique further and achieve longer throws.

Previous studies in this field emphasize the importance of release parameters like velocity and angle of the javelin. A study conducted by researchers on the athletes participating in the 2007 IAAF World Championships [13] in Osaka showed the impact of these release parameters on throwing distance. A study

© The Author(s), under exclusive license to Springer Nature Switzerland AG 2026
C. Modi et al. (Eds.): MIND 2024, CCIS 2736, pp. 392–402, 2026.
https://doi.org/10.1007/978-3-032-14531-4_33

conducted at Concordia University in Nebraska found that sever. main muscle groups impact the throw majorly. The angles between these muscle groups were precisely measured for further analysis.

In this paper, We use pose estimation technology [9] to precisely track an athlete's body posture throughout the throwing motion. This technology accurately captures the angles between 7 muscle groups, which is critical to the javelin throw. These angles can then be analyzed to improve further and finetune the athlete's technique under the guidance of a skilled coach. In addition, to pose estimation, our solution includes javelin detection, and tracking [17], which monitors the javelin's trajectory and orientation throughout the throw. By accurately measuring the angles between the javelin and the ground during the various phases of the throw, the effectiveness of an athlete's javelin throwing technique can be assessed, and areas for improvement can be identified. This approach thus integrates both athlete and javelin tracking, providing valuable information on factors affecting throw distance. Figure 1 shows the process flow diagram of the proposed solution.

The paper is structured into four sections. The first section is the Literature Review, which reviews relevant literature on javelin throwing dynamics, pose estimation algorithms, and object detection models. The second section is Proposed Methodology, which details pose estimation and javelin detection methodology. This is followed by experiments and results, which discuss the experimental setup and results obtained from the trained models. The final section summarizes the experiment's results and outlines possible research directions for further improvement of the javelin analysis system.

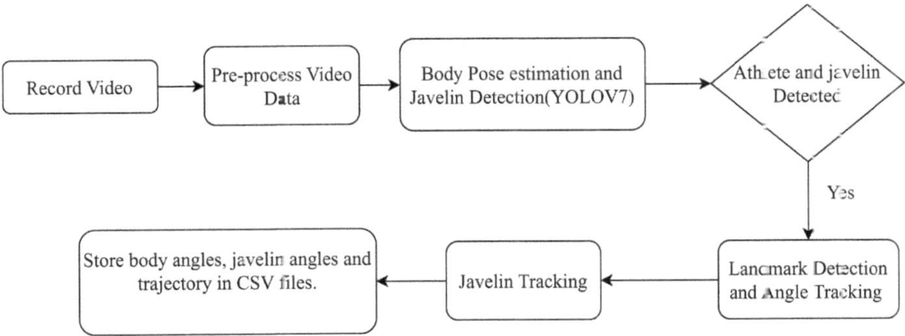

Fig. 1. Process flow diagram of the proposed solution.

2   Literature Review

Understanding the biomechanics of sports involved in sports like javelin throwing requires advanced motion analysis tools and techniques. The literature review examines current developments in computer vision and motion tracking relevant to the study of javelin.

In this [5] an innovative approach combining off-the-shelf 2D pose estimation, a humanoid model, and fast optimization techniques to monitor the poses of elders at risk of falling or getting injured has been proposed. This paper discusses the limitations of deep learning methods in estimating missing poses in the training dataset and provides an understanding of joint coordination and movement dynamics.

This paper [10] introduces a system designed for tracking body movements from video sources and augmenting labeled skeleton joints onto these frames. This technology holds significant importance within the sports industry, offering deep learning-based joint recognition of the human body. The study uses Mediapipe Blazepose, integrated with the PoseNet dataset, to detect movements tailored to prevent injuries to athletes. This research results in a functional system that identifies and labels individual skeleton joints, allowing us to calculate the person's velocity and joint angles.

The study conducted in [6] analyses the effectiveness of YOLO(You Only Learn Once) for dynamic object detection. The paper compares two models of YOLO, namely YOLOv7 and YOLOv5, for traffic management in smart cities. YOLOv5 has limitations when it comes to detecting smaller objects and angle deflections. YOLOv7, on the other hand, employs instance segmentation and achieves higher accuracy than YOLOv5. Considering the advantages of YOLOv7, as seen in the comparison, we have chosen YOLOv7 to track the position of the javelin using video footage.

In this [16], a novel object tracking method using adaptive Kalman filtering based on dominant color features has been proposed. This method shows robustness in object detection under complex occlusions and lighting conditions. The techniques discussed in this paper can be applied to track the motion of a javelin during flight, allowing us to analyze its trajectory.

The study conducted in [14] discusses real-time object detection techniques that offer exceptional speed and accuracy compared to existing objection detection models. This paper has two primary research focal points: a trainable bag-of-freebies-oriented approach combining efficient training tools with an optimized architecture and compound scaling methodology. Our solution uses the YOLOv7 architecture to set new benchmarks in speed and accuracy. It displays the highest average precision of 56.8% among all object detection algorithms at frame rates of 30 to 120 frames per second.

3   Proposed Methodology

We propose a computer vision model that can extract several essential biomechanical parameters from the video of an athlete throwing a javelin. The process begins with acquiring high-quality video footage capturing javelin throws from the horizontal angle with minimum parallax error. Further, preprocessing techniques such as video stabilization [2], frame extraction, and resolution adjustment are applied to ensure data accuracy and stability. These steps are crucial for obtaining clear and consistent footage for future analysis.

Our Solution consists of three main components:

1. Body Pose Estimation
2. Landmark Detection and Angle Tracking
3. Javelin Detection and Tracking

3.1 Body Pose Estimation

This is the first phase of our methodology. We begin by using Python's MediaPipe library, which is explicitly designed for accurate pose estimation. Pose estimation is a fundamental computer vision and Artificial Intelligence (AI) task focused on identifying and monitoring the positions and orientations of human body parts within images or videos. Our main goal is to track the athlete's body posture during the throw accurately. We achieve this by detecting key points corresponding to several joints in the body critical to the throw, including the shoulders, elbows, wrists, hips, knees, and ankles.

The javelin throw video footage is segmented into frames for further processing. These frames are processed sequentially by the Pose estimation model. It identifies and extracts the key points representing the spatial coordinates of the athlete's body joints in the image space.

MediaPipe Pose is a technology that processes video data using machine learning. It incorporates BlazePose [10], a method for identifying 32 2D landmarks on the human body. BlazePose is known for its lightweight ML architecture that enables real-time performance on mobile devices and PCs using CPU inference. Adjustments are applied to the y-axis pixel values using an inverse ratio to perform pose estimation with normalized coordinates. Figure 2 [5] shows the 32 landmarks that can be detected using the MediaPipe Pose technology.

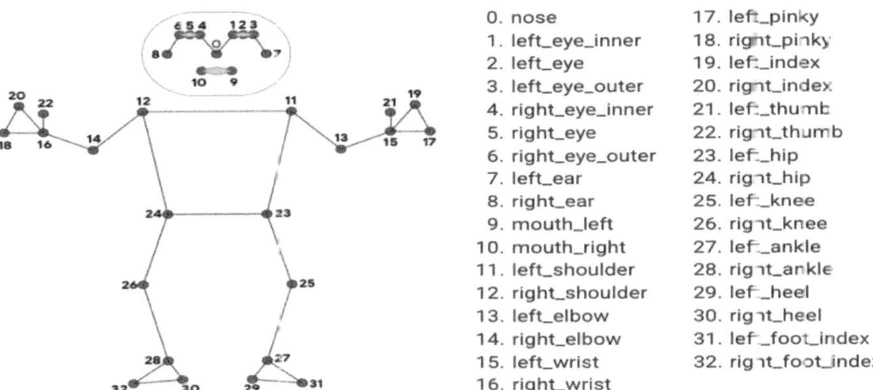

Fig. 2. Detected landmarks on the human body.

3.2   Landmark Detection and Angle Tracking

This is the second phase of our solution. The primary objective is to accurately track certain landmarks on the athlete's body and measure angles formed between them during the throw. We use the Pose estimation model from the MediaPipe [12] library to extract the spatial coordinates of the athlete's joints and extremities.

We use geometric landmarks to detect and isolate essential landmarks from the pose estimation data. By establishing specific criteria based on joint proximity and configuration, we accurately identify landmarks crucial for angle tracking. This iterative process involves extracting precise coordinates of each landmark across multiple frames of the javelin throw sequence. We calculate angles between these landmarks using the principles of trigonometry and kinematics [15]. For instance, we assess arm extensions during the throwing motion by computing the angle between the shoulder, elbow, and wrist. Similarly, angles formed by the hip, knee, and ankle joints evaluate lower body posture.

Define Landmark Positions: Let $\mathbf{A} = (x_A, y_A)$, $\mathbf{B} = (x_B, y_B)$, and $\mathbf{C} = (x_C, y_C)$ represent the coordinates of keypoints A, B, and C.

Computation of Angle θ: Using the arctangent function to calculate the angle θ formed by vectors $\mathbf{AB}$ and $\mathbf{BC}$:

$$\theta = \arctan\left(\frac{y_C - y_B}{x_C - x_B}\right) - \arctan\left(\frac{y_B - y_A}{x_B - x_A}\right) \tag{1}$$

Throughout the process, we track the movement of the landmarks and update the corresponding angle measurements for each frame. Finally, we aggregate the angle data from the pose estimation model to create a comprehensive profile of the athlete's biomechanics.

3.3   Javelin Detection and Tracking

The javelin detection process begins with acquiring custom javelin data from the internet. We gathered a diverse set of images containing javelin throws, ensuring representation across different environments, lighting conditions, and throwing techniques. These images were then carefully annotated to determine the precise location and boundaries of the javelin in each frame. The raw dataset consisted of 250 images. The data was augmented using four techniques to produce a larger dataset [18]. The image level augmentation techniques applied include varying the exposure, hue, saturation, and brightness by 15% from the original levels to increase the diversity of the dataset.

We devised a comprehensive approach for javelin detection and tracking from the video footage. The YOLOV7 [14] model was trained on a custom dataset containing images of athletes throwing javelins, as the original model could not detect javelins as a distinct class in object detection tasks. Using the YOLOV7 object detection framework, we detected the precise location of the javelin in each frame of the video sequence. This involved converting YCLOv7 coordinates to regular bounding box coordinates, enabling precise identification of the javelin's position within the frame. To improve tracking accuracy and stability, we also integrated Kalman filtering [16] techniques into our framework. Kalman filtering allows us to predict the next position of the javelin based on its previous trajectory, effectively compensating for noise and occlusions in the video data. Figure 3 describes the YOLOV7 architecture.

After detection of the javelin, we try to find the angle of the javelin with the ground at the point of release. The primary challenge lies in identifying the release point [3]. The point of release is the moment when the athlete releases the javelin. To calculate this parameter, we monitored the distance between the midpoint of the javelin and the athlete's throwing arm throughout the video sequence. When this distance increases beyond a carefully tuned threshold value, we can infer that the javelin has been released from the athlete's hand. By analyzing the relative positions of the tip and tail of the javelin, we can calculate the angle made by the javelin with the ground (horizontal axis).

The results are visualized by displaying the trajectory of the javelin throughout the throwing phase by marking the position of its midpoint. In addition to this, We also visualized how accurate our object detection model is by overlaying the prediction over the original video [8].

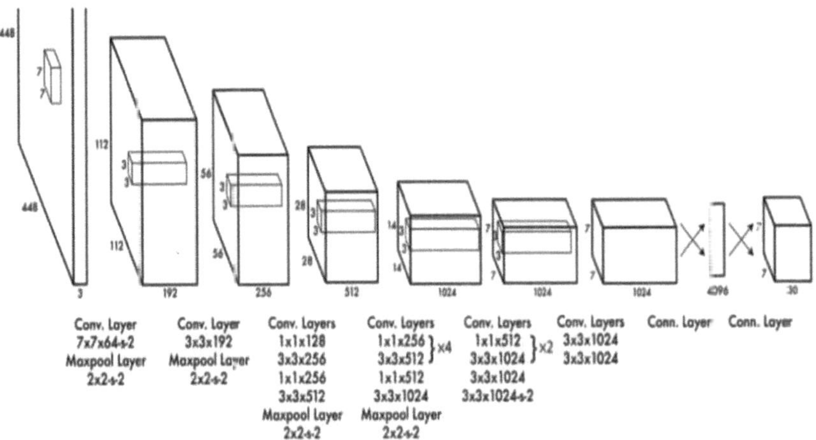

Fig. 3. YOLOv7 Architecture.

4   Experiments and Results

We used videos of professional athletes capturing the entire throwing sequence to perform our experiments. The videos are subjected to several preprocessing

[2] steps to improve the clarity and stability of the footage. This includes video stabilization to reduce motion blur and jitter, frame extraction to isolate key frames of the throwing sequence, and resolution adjustment for uniformity across the dataset.

Fig. 4. Estimated Keypoints.

We used Google's MediaPipe framework [7] to detect and track various key landmarks on the athlete's body in each video frame. With the help of these landmarks, we calculated eight body angles essential to understand the biomechanics behind the javelin throw. This includes the left hip angle, right hip angle, left knee angle, right knee angle, left trunk angle, right trunk angle, elbow angle, and shoulder angle. Figure 4 shows the key points detected with the help of Pose Estimation Technology. Figure 4 displays scatter plots depicting the relationship between various body angles during the release step and the corresponding distance achieved (Fig. 5).

We employed the YOLOv7 object detection architecture, which is well known for its speed and accuracy in object detection tasks from video footage. The model was trained on a custom dataset (fully annotated) using transfer learning techniques. We fine-tuned a pre-trained YOLOv7 model specifically for the javelin detection task. The training process involved optimizing model parameters, adjusting hyperparameters, and iteratively refining the model using techniques such as regularization. Our YOLOv7 model achieved a high mean Average Precision (mAP) of 0.979 at an Intersection over the Union (IoU) threshold of 0.5 (mAP@0.5). The mean Average Precision was measured to be 0.84 even with

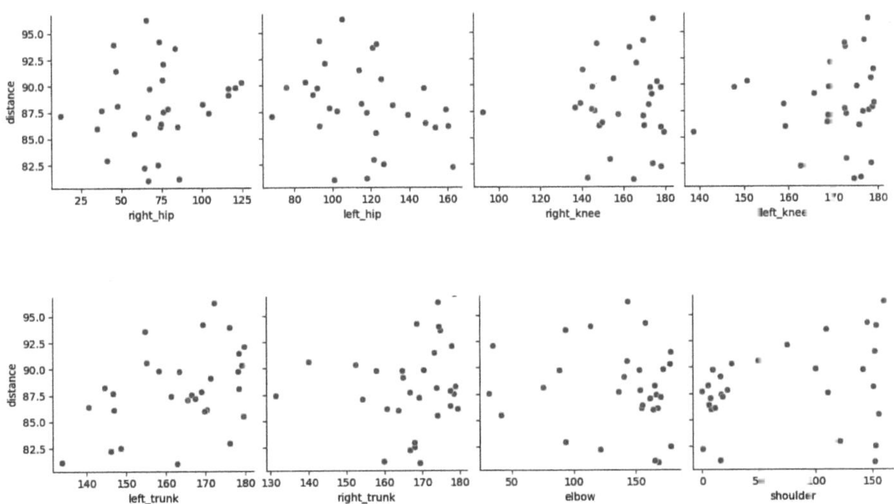

Fig. 5. Scatter Plot: Body Angles vs Distance (Release step).

stricter IOU thresholds between 0.5 and 0.95 (mAP@0.5:0.95), demonstrating the model's ability to localize javelins within the images accurately.

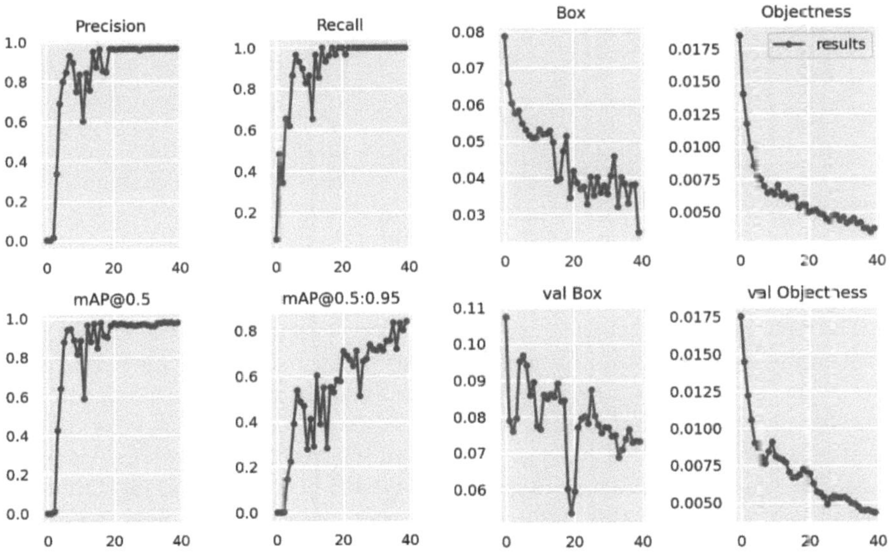

Fig. 6. Results of training and validation.

Figure 6 presents the outcomes of the training and validation phases, depicting key metrics such as bounding box loss, objectness loss, precision, and recall. Figure 7 shows the F1-confidence curve for javelin detection.

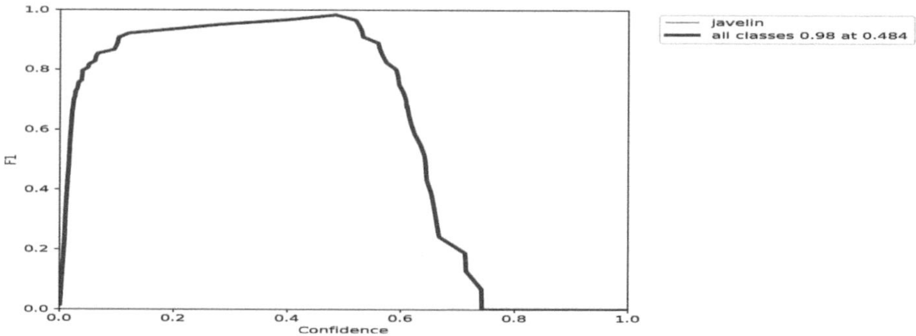

Fig. 7. F1 confidence curve.

Fig. 8. Javelin Tracking using a Kalman Filter.

The tracking process involved analyzing successive frames of the video footage to estimate the position of the javelin over time. This was achieved with the help of the Kalman filtering technique. The Kalman filter algorithm utilized a recursive process to predict and update the state of the javelin in each frame based on a combination of prior predictions and current observations. Using this model, we analyzed the release angle of some of the best throws in javelin history. They closely align with the findings in [1] that provide insights into the optimal release angle. Figure 8 shows the tracking of the javelin and the angle at which the javelin forms with the horizontal axis during the crossover phase. In our example, the release angle was 31.64 while the javelin's velocity at the release point was 2917.33 pixels per second.

5   Conclusion and Future Work

In this paper, we addressed the problem of analyzing the biomechanical parameters of the javelin throw to understand the critical factors influencing throw distance and athlete performance. Specifically, our focus was on using computer vision techniques to capture and analyze athletes' body posture, track the javelin's trajectory, and calculate release angles during the throwing motion.

Our research demonstrated the efficacy of pose estimation algorithms in characterizing athlete biomechanics during the javelin throw. By tracking key joint angles and movement patterns, we identified correlations between posture, the angle of the javelin, and release parameters, shedding light on the mechanics underlying successful throws.

Future work presents several avenues for further exploration and refinement of our methodology. Integrating advanced machine learning techniques, such as reinforcement learning [4], could enable the development of adaptive coaching systems that tailor training regimens to individual athlete profiles and performance metrics. Considering the increasing prevalence of wearable sensor technologies and Inertial Measurement Units (IMUs) in sports performance analysis, integrating these devices into our methodology could provide additional layers of data and insights into athlete biomechanics.

References

1. Best, R.J., Bartlett, R.M., Sawyer, R.A.: Optimal javelin release. J. Appl. Biomech. **11**(4), 371–394 (1995)
2. Guilluy, W., Oudre, L. Beghdadi, A.: video stabilization: overview, challenges and perspectives. Sig. Process.: Image Commun. **90**, 116015 (2021)
3. Hubbard, M., Alaways L.W.: Rapid and accurate estimation of release conditions in the javelin throw. J. Biomech. **22**(6–7), 583–595 (1989)
4. Kaelbling, L.P., Littman, M.L., Moore, A.W.: Reinforcement learning: a survey. J. Artif. Intell. Res. **4**, 237–285 (1996)
5. Kim, J.W., Choi, J.Y., Ha, E.J., Choi, J.H.: Human pose estimation using MediaPipe pose and optimization method based on a humanoid model. Appl. Sci. **13**(4), 2700 (2023)
6. Konala, T.R., Nammi, A., Tella, D.S.: Analysis of live video object detection using YOLOv5 and YOLOv7. In: 2023 4th International Conference for Emerging Technology (INCET), pp. 1–6. IEEE (2023)
7. Lugaresi, C., et al.: MediaPipe: a framework for perceiving and processing reality. In: Third Workshop on Computer Vision for AR/VR at IEEE Computer Vision and Pattern Recognition (CVPR), vol. 2019 (2019)
8. Michalko, M., Onuška, J., Lavrín, A., Cymbalák, D., Kainz, O.: Tracking the object features in video based on OpenCV. In: 2016 International Conference on Emerging eLearning Technologies and Applications (ICETA), pp. 223–226. IEEE (2016)
9. Munea, T.L., Jembre, Y.Z., Weldegebriel, H.T., Chen, L., Huang, C., Yang, C.: The progress of human pose estimation: a survey and taxonomy of models applied in 2D human pose estimation. IEEE Access **8**, 133330–133348 (2020)

10. Pauzi, A.S.B., et al.: Movement estimation using Mediapipe BlazePose. In: Badioze Zaman, H., et al. (eds.) IVIC 2021. LNCS, vol. 13051, pp. 562–571. Springer, Cham (2021). https://doi.org/10.1007/978-3-030-90235-3_49
11. Rajaraman, S., Candemir, S., Kim, I., Thoma, G., Antani, S.: Visualization and interpretation of convolutional neural network predictions in detecting pneumonia in pediatric chest radiographs. Appl. Sci. **8**(10), 1715 (2018)
12. Singh, A.K., Kumbhare, V.A., Arthi, K.: Real-time human pose detection and recognition using MediaPipe. In: International Conference on Soft Computing and Signal Processing, pp. 145–154. Springer (2021)
13. Tauchi, K., Murakami, M., Endo, T., Takesako, H., Gomi, K.: Biomechanical analysis of elite javelin throwing technique at the 2007 IAAF world championships in athletics. Bull. Stud. Athletics JAAF **5**, 143–149 (2009)
14. Wang, C.Y., Bochkovskiy, A., Liao, H.Y.M.: YOLOv7: trainable bag-of-freebies sets new state-of-the-art for real-time object detectors. In: Proceedings of the IEEE/CVF Conference on Computer Vision and Pattern Recognition, pp. 7464–7475 (2023)
15. Wang, S., Zhang, X., Ma, F., Li, J., Huang, Y.: Single-stage pose estimation and joint angle extraction method for moving human body. Electronics **12**(22), 4644 (2023)
16. Weng, S.K., Kuo, C.M., Tu, S.K.: Video object tracking using adaptive Kalman filter. J. Vis. Commun. Image Represent. **17**(6), 1190–1208 (2006)
17. Yang, F., Zhang, X., Liu, B.: Video object tracking based on YOLOv7 and Deep-SORT. arXiv preprint: arXiv:2207.12202 (2022)
18. Zheng, H., Shang, H., Sun, Z., Fu, X., Yao, J., Huang, J.: Supervised augmentation: leverage strong annotation for limited data. In: 2020 IEEE 17th International Symposium on Biomedical Imaging (ISBI), pp. 1134–1138. IEEE (2020)

Object Tracking Algorithms: Enhancing Real-Time Violence Detection in Modern Surveillance Systems

M. Evany Anne[1]([envelope]), M. Brindha[1], and N. Sivakumaran[2]

[1] Department of Computer Science and Engineering, National Institute of Technology, Tiruchirappalli, Tiruchirappalli 620015, Tamil Nadu, India
{evany,brindham}@nitt.edu
[2] Department of Instrumentation and Control Engineering, National Institute of Technology, Tiruchirappalli, Tiruchirappalli 620015, Tamil Nadu, India
nsk@nitt.edu

Abstract. The growing need for automated surveillance systems, fueled by heightened public safety concerns, calls for reliable real-time tracking and violence detection features. This study presents a comparative evaluation of three widely used tracking algorithms: Kernelized Correlation Filters (KCF), Channel and Spatial Reliability Tracker (CSRT), and Deep Simple Online and Realtime Tracking (DeepSORT). While KCF offers high speed, its accuracy declines under challenging conditions. CSRT provides enhanced robustness but sacrifices processing speed. DeepSORT, with its deep learning-based appearance descriptors excels in maintaining object identities across frames, especially in dense environments. Recognizing DeepSORT as the most effective option, this work proposes an integrated framework combining DeepSORT with OpenPose, an advanced pose estimation technique, to improve violence detection in surveillance scenarios. The system incorporates YOLO for object detection, DeepSORT for tracking, and OpenPose for body movement analysis, with an LSTM-based classifier to identify violent behaviors in real time. The proposed solution achieves exceptional performance, offering a reliable and efficient approach for automated violence detection in real-world applications.

Keywords: Real-time Surveillance · DeepSORT · KCF Tracking · CSRT Tracking · Violence Detection

1 Introduction

The increasing complexity of urban environments has heightened the demand for efficient surveillance systems to ensure public safety. Traditional monitoring methods are labor-intensive and error-prone, often resulting in delayed responses. Autonomous surveillance technologies address these limitations through real-time monitoring, accurate tracking [5], and automated violence detection [3],

C. Modi et al. (Eds.): MIND 2024, CCIS 2736, pp. 403–414, 2026.
https://doi.org/10.1007/978-3-032-14531-4_34

reducing the need for constant human oversight. Object tracking is central to intelligent surveillance [6], enabling continuous monitoring across video frames. While Kernelized Correlation Filters (KCF) offered speed, they struggled with occlusions and appearance changes. Improved methods like CSRT enhanced robustness but increased computational demands. DeepSORT further advanced tracking by integrating deep learning-based appearance descriptors, ensuring consistent identities in crowded scenes. This study proposes a framework combining YOLO for object detection, DeepSORT for tracking, and OpenPose for pose estimation. This integration enables real-time analysis of body movements for accurate violence detection in complex environments, addressing key challenges such as delays, false alarms, and dynamic conditions.

2   Literature Review

The evolution of surveillance systems has shifted from manual monitoring to automated analysis to manage the vast video data generated by urban surveillance [7]. Significant advancements in object tracking, pose estimation [15], and violence detection [11] have paved the way for intelligent systems. Early tracking methods like KCF [7] balanced speed and accuracy but faltered in handling occlusions and appearance changes. Recent studies have demonstrated practical implementations of real-time object tracking systems using OpenCV-based pipelines and automated video surveillance architectures, highlighting their effectiveness in continuous monitoring scenarios [8,9]. CSRT [13] improved robustness by focusing on reliable spatial regions but introduced higher computational costs. DeepSORT [14] addressed these challenges by incorporating deep learning-based appearance descriptors, maintaining identity consistency in dynamic environments. Pose estimation tools, such as OpenPose [15], enable real-time detection of key body points, enhancing human action analysis. Comparative studies on single-person and multi-person pose estimation using deep learning algorithms further emphasize the importance of accurate skeletal representation for reliable human activity and behavior analysis in surveillance applications [10]. The integration of DeepSORT with OpenPose facilitates nuanced movement analysis, critical for violence detection [11]. This combination helps reduce false alarms by distinguishing between benign and aggressive behaviors [12].

However, challenges persist:

- **Fragmented Approaches:** Existing methods focus on isolated tasks rather than unified frameworks [2].
- **Real-Time Efficiency:** High computational demands hinder scalability [4].
- **Dynamic Environments:** Occlusions and varying lighting conditions reduce reliability [14].
- **Ethical Concerns:** Increased surveillance raises privacy concerns, requiring clear guidelines [1].

This study addresses these gaps by developing a lightweight framework integrating YOLO, DeepSORT, and OpenPose, with an LSTM-based classifier for real-time violence detection. The system achieves high accuracy with low inference times, making it suitable for deployment in dynamic, resource-constrained environments [16] (Fig. 1).

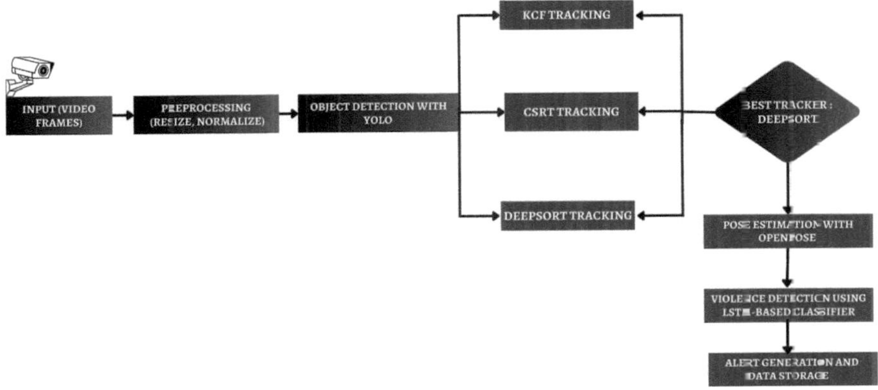

Fig. 1. Proposed model

3   Proposed Methodology

The YOLO (You Only Look Once) algorithm treats each video frame as an independent image, resizing it (commonly to 416×416 pixels) and segmenting it into an $S \times S$ grid. Each grid cell is responsible for predicting bounding boxes, confidence scores, and class probabilities. Detection scores are obtained by combining the confidence scores with the class probabilities, identifying object locations and their corresponding likelihoods. To refine the output, Non-Maximum Suppression (NMS) eliminates redundant overlaps by retaining only the bounding boxes with the highest confidence. This process is repeated for every frame, enabling real-time detection with bounding boxes, class labels, and confidence scores. Figure 2 demonstrates YOLO's effectiveness, showcasing individuals identified with bounding boxes and confidence scores, emphasizing its accuracy and classification capabilities in real-time.

4   KCF Tracking

The KCF (Kernelized Correlation Filters) tracking algorithm provides a fast method for tracking objects in video sequences. It begins by initializing a correlation filter using the object's bounding box from the first frame, relying on its appearance features. This filter predicts the object's position in subsequent frames by identifying the peak response around the last known location and updating periodically to adapt to appearance changes.

Algorithm 1. YOLO Algorithm for Object Detection

1: **Input:** Video V with N frames
2: **Output:** Detected objects with bounding boxes, class labels, and confidence scores for each frame
3: **for** each frame f in video V **do**
4:     Resize the frame f to a fixed size (e.g., 416×416).
5:     Normalize the pixel values of the resized frame.
6:     Split the frame into an $S \times S$ grid.
7:     **for** each grid cell (i, j) **do**
8:         Generate predictions for B bounding boxes.
9:         **for** each bounding box b **do**
10:             Predict coordinates (x_b, y_b), width w_b, and height h_b.
11:             Predict confidence score C_b.
12:             Predict class probabilities for C classes.
13:             Calculate the final score for each class:

$$\text{Score} = C_b \times P(\text{Class}|\text{Object})$$

14:         **end for**
15:     **end for**
16:     Apply Non-Maximum Suppression (NMS) to eliminate overlapping boxes:
   (a) Sort all boxes by confidence scores.
   (b) For each box with the highest score:
       (i) Discard boxes with an IoU (Intersection over Union) exceeding the defined threshold.
       (ii) Keep the box and move to the next one.

17:     Return the final set of bounding boxes, class labels, and confidence scores for the frame.
18: **end for**

Fig. 2. YOLO Object Detection

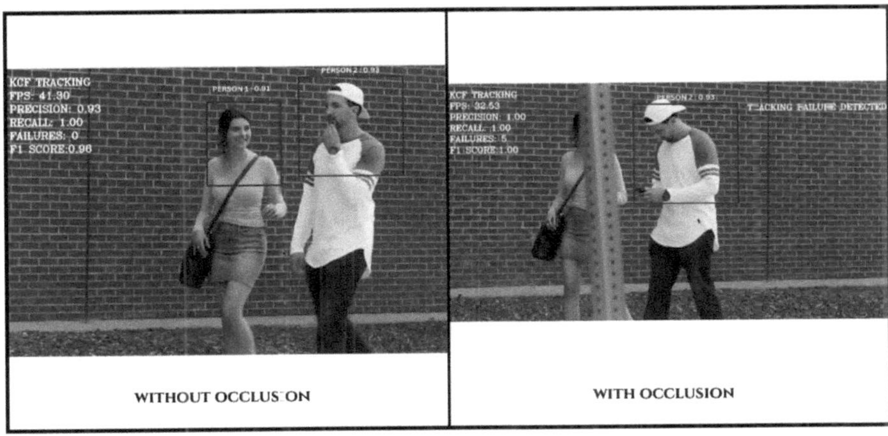

Fig. 3. KCF Tracking

Algorithm 2. KCF Tracking Algorithm

1: **Input:** Video frames $F_1, F_2, \ldots, F_N$ with initial object bounding box B_1 from YOLO detector
2: **Output:** Updated bounding box B_n for each frame F_n
3: Initialize the KCF tracker with the bounding box B_1 in the first frame F_1
4: **for** each subsequent frame F_n from F_2 to F_N **do**
5:     Extract a search region around the object position from the previous frame
6:     Use the correlation filter on the search region to produce a response map.
7:     Identify the peak in the response map to locate the object's updated position.
8:     Update the bounding box B_n based on the new position
9:     Periodically update the correlation filter to adapt to changes in the object's appearance
10: **end for**
11: Handle occlusions:
    (a) If the object is occluded, attempt to maintain the last known position
    (b) If the object is lost, stop updating the bounding box until the object reappears
12: Return the final set of bounding boxes $B_1, B_2, \ldots, B_N$

While KCF performs well in normal conditions, it struggles with significant occlusion, risking complete loss of track when the object is obscured. Figure 3 illustrates KCF's performance in two scenarios: without occlusion (left) and with occlusion (right). Although it maintains high precision and recall, the figure highlights KCF's limitations in occluded situations. Overall, KCF is effective in clear environments but faces challenges with occlusion and abrupt motion, suggesting the need for more robust tracking solutions. Its speed and reliability also make it valuable for real-time applications.

Algorithm 3. CSRT (Channel and Spatial Reliability Tracker) Algorithm

---

1: **Input:** Video frames $F_1, F_2, \ldots, F_N$ with initial object bounding box B_1 from YOLO detector
2: **Output:** Updated bounding box B_n for each frame F_n
3: Initialize the CSRT tracker with the bounding box B_1 in the first frame F_1
4: **for** each subsequent frame F_n from F_2 to F_N **do**
5:     Extract multi-channel features (e.g., color, gradient) from the region around B_{n-1}
6:     Utilize the correlation filter to create a response map.
7:     Assess spatial reliability to refine the filter's support region.
8:     Find the peak in the response map to determine the object's new position
9:     Update the bounding box B_n based on the new position
10:     Continuously update the correlation filter to adapt to changes in object appearance
11:     Handle occlusions by maintaining the object's track using spatial reliability
12: **end for**
13: Return the final set of bounding boxes $B_1, B_2, \ldots, B_N$

---

5 CSRT Tracking

The CSRT (Channel and Spatial Reliability Tracker) is a highly reliable object tracking algorithm specifically developed to handle challenging conditions like occlusion, scale changes, and intricate backgrounds. It initializes with a bounding box, creating a correlation filter using multi-channel features like color and gradient. The spatial reliability component enhances accuracy by focusing on the most reliable parts of the object (Fig. 4).

Fig. 4. CSRT Tracking

CSRT updates the filter dynamically to adapt to changes in the object's appearance, enabling accurate tracking through scale changes, rotation, and partial occlusion. Compared to simpler trackers like KCF, it handles occlusions more effectively, maintaining track even when parts of the object are obscured. While highly accurate, its reliance on visual features may limit performance in highly dynamic environments. CSRT's adaptability and precision make it a valuable choice for advanced tracking tasks where traditional methods fall short.

6 DeepSORT Tracking

DeepSORT (Deep Simple Online and Realtime Tracking) is a sophisticated object tracking algorithm that enhances traditional methods by integrating deep learning-based appearance descriptors. This allows DeepSORT to maintain object identities across frames, even in challenging scenarios involving occlusions, appearance changes, or close interactions between objects. DeepSORT combines a Kalman filter for motion prediction with deep appearance features, enabling reliable object tracking even in challenging scenarios. It initializes with bounding boxes from detectors like YOLO, assigning unique IDs to objects. The Kalman filter predicts future positions, while appearance features differentiate similar objects. Using a combined cost of motion and appearance, the Hungarian algorithm matches detections to tracks. Unmatched tracks are marked as lost, re-associated after occlusion, or assigned new IDs for unmatched detections.

Fig. 5. DeepSORT Tracking

Figure 5 illustrates DeepSORT's robustness, maintaining accurate tracking in both non-occluded (left) and occluded (right) scenarios, demonstrating its effectiveness for real-time applications.

Algorithm 4. DeepSORT Tracking Algorithm

1: **Input:** Video frames $F_1, F_2, \ldots, F_N$ with initial bounding boxes B_1 from an object detector (e.g., YOLO)

2: **Output:** Updated bounding boxes B_n and object identities for each frame F_n

3: Initialize the DeepSORT tracker with bounding boxes B_1 and assign unique identities

4: **for** each subsequent frame F_n from F_2 to F_N **do**

5:     **Motion Prediction:**

6:     Employ a Kalman filter to estimate the position of each tracked object using its prior velocity and position.

7:     **Appearance Feature Extraction:**

8:     Retrieve deep appearance features for each detected object through a pre-trained neural network.

9:     **Data Association:**

10:     Compute the association cost using a combination of motion predictions and appearance features

11:     Solve the assignment problem using the Hungarian algorithm to match detected objects to existing tracks

12:     **Track Management:**

13:     Update the Kalman filter with the matched detections

14:     Handle unmatched tracks by marking them as lost or reassigning them to reappearing objects

15:     Generate new tracks for unmatched detections

16: **end for**

17: Return the final set of bounding boxes and identities for each frame

---

7 Violence Detection

7.1 Transition from Tracking to Violence Detection

DeepSORT's ability to maintain consistent object tracking, even under challenging conditions, provides a solid foundation for the next step in our methodology—detecting violent actions through pose estimation. While DeepSORT accurately tracks individuals across frames, the challenge is to analyze their movements to determine if those actions are violent. This is where OpenPose, a state-of-the-art pose estimation tool, becomes essential.

7.2 Tracking Individuals with DeepSORT

In the Fig. 6, DeepSORT's tracking output is displayed, where each individual within the scene is identified with a unique ID (Person-1, Person-2, etc.) and tracked continuously across the video frames. This precise identification is crucial as it allows the system to maintain a consistent track of each person's actions, even in crowded or dynamic environments. The bounding boxes around each person highlight the areas of interest, ensuring that their movements are captured and analyzed effectively.

Fig. 6. Tracking with DeepSORT

7.3   Pose Estimation with OpenPose

Figure 7 showcases the application of OpenPose on individuals tracked by Deep-SORT. OpenPose detects key body points, including the head and shoulders, and connects them to form a skeletal structure. This overlay facilitates posture analysis by highlighting relevant movements and minimizing background distractions. These skeletal representations are vital for identifying violent actions, capturing aggressive stances such as a raised fist or a forward lunge. The extracted key poses are then organized into a time-series representation of movements.

Fig. 7. Pose Estimation with OpenPose

7.4   Violence Detection and Classification

The system identifies violent behavior by feeding sequences into a trained Long Short-Term Memory (LSTM) model, utilizing a dataset containing both violent and non-violent actions. Figure 8 illustrates the system detecting aggressive behavior by analyzing skeletal poses generated by OpenPose. Bounding boxes around individuals change color—red for violent and green for non-violent.

This real-time classification enables immediate alerts to authorities, providing a robust automated surveillance solution.

Fig. 8. Violence Detection and Classification (Color figure online)

8 Performance Metrics

The comparison of performance across different models and datasets underscores the proposed model's effectiveness and efficiency in violence detection tasks.

Table 1. Performance Comparison of Various Models

Model Architecture	Hockey Fights Dataset Accuracy (%)	Movie Fights Dataset Accuracy (%)	Parameter Count (#)
AlexNet with LSTM [17]	97.1	100	9.6 M
ResNet50 with ConvLSTM [18]	89.0	92.0	Not Reported
Proposed Framework	96.1	94.5	4.074 M

Table 2. Evaluation Metrics Across Datasets

Dataset	Average Inference Time (s)	Accuracy (%)	Precision (%)	F1 Score
RWF-2000	0.046	90.3	92.2	0.88
Movie Fights	0.059	94.5	95.1	0.922
Hockey Fights	0.022	96.1	97.3	0.961

The system achieves a notable balance between high accuracy and computational efficiency. Key observations include:

1. **Exceptional Accuracy and Precision:** The proposed model demonstrates impressive accuracy across all tested datasets, achieving 96.1% on the Hockey Fights dataset and 94.5% on the Movie Fights dataset, as detailed in Table 1. These results are comparable to AlexNet+LSTM while requiring significantly fewer parameters, as highlighted in Table 1.
2. **Computational Efficiency:** The system's lightweight architecture ensures an average inference time of just 0.046 s per frame on the RWF-2000 dataset, enabling real-time deployment, as detailed in Table 2.
3. **Robustness in Real-World Scenarios:** Tests under challenging conditions, including occlusions and varying lighting, demonstrate the model's ability to maintain high performance with minimal degradation. Precision and F1 scores across datasets, summarized in Table 2, highlight the model's adaptability.
4. **Comparison to Existing Methods:** Compared to AlexNet+LSTM and ResNet50+ConvLSTM, the proposed model provides an optimal trade-off between accuracy and model complexity, as evidenced by the metrics in Table 1.

9   Conclusion

The findings of this research highlight that the proposed model strikes an excellent balance between accuracy and efficiency, surpassing earlier approaches in parameter optimization while consistently delivering high accuracy and precision across various datasets. This positions the model as a strong candidate for real-time violence detection in surveillance applications. The detailed analysis not only improves the detection of violent actions but also ensures the system's ability to function in real time, enabling prompt alerts for suspicious activities. By combining DeepSORT's robust tracking capabilities with OpenPose's precise pose estimation, the system offers a dependable surveillance solution that effectively identifies and addresses violent incidents as they happen, enhancing the overall reliability and performance of automated surveillance technologies.

Acknowledgement. The research work is supported by the C-DAC, Trivandrum under the project "Next generation Emergency Response Support System".

References

1. van Heek, J., Aming, K., Ziefle, M.: How fear of crime affects needs for privacy & safety. In: 2016 5th International Conference (SMARTGREENS), Rome, Italy (2016)

2. Yao, S., Ardabili, B.R., Danesh Pazho, A., Noghre, G.A., Neff, C., Tabkhi, H.: Real-world community-in-the-loop smart video surveillance system. In: 2023 IEEE International Conference on Smart Computing, Nashville, USA (2023)
3. e Fatima Kiani, G., Kayani, T.: Real-time violence detection using deep learning techniques. In: 2022 3rd International Conference on Innovations in Computer Science & Software Engineering (ICONICS), Karachi, Pakistan (2022)
4. Huszár, V.D., Adhikarla, V.K., Négyesi, I., Krasznay, C.: Toward fast and accurate violence detection for automated video surveillance applications. IEEE Access **11**, 18772–18793 (2023). https://doi.org/10.1109/ACCESS.2023.3245521
5. Chen, C.-H., Chen, T.-Y., Lin, J.-C., Wang, D.-J.: People tracking in the multi-camera surveillance system. In: 2011 Second International Conference on Innovations in Bio-inspired Computing and Applications, Shenzhen, China (2011)
6. Prakash, U.M., Thamaraiselvi, V.G.: Detecting and tracking of multiple moving objects for intelligent video surveillance systems. In: Second International Conference on Current Trends In Engineering and Technology - ICCTET 2014, India (2014)
7. Henriques, J.F., Caseiro, R., Martins, P., Batista, J.: High-speed tracking with kernelized correlation filters. IEEE Trans. Mach. Intell. (2015)
8. Yang, J.: Real time object tracking using OpenCV. In: 2023 IEEE 3rd International Conference on Data Science and Computer Application (ICDSCA), Dalian, China (2023)
9. Musihin, I.D., Kapustin, V.V.: Video system with automatic tracking of the object of interest. In: 2022 IEEE International Multi-Conference on Engineering, Computer and Information Sciences (SIBIRCON), Yekaterinburg, Russia (2022)
10. Patel, M., Kalani, N.: A comparative analysis for single person and multi person pose estimation using deep learning algorithms. In: 2021 International Conference on Circuits, Controls and Communications (CCUBE), Bangalore, India (2021)
11. Ramzan, M., et al.: A review on state-of-the-art violence detection techniques. IEEE Access **7**, 107560–107575 (2019)
12. Gai, Y., He, W., Zhou, Z.: Pedestrian target tracking based on DeepSORT with YOLOv5. In: 2021 2nd International Conference on Computer Engineering and Intelligent Control (ICCEIC), Chongqing, China (2021)
13. Lukežic, A., Vojír, T., Zajc, L.C., Matas, J., Kristan, M.: Discriminative correlation filter with channel and spatial reliability. In: 2017 IEEE Conference on Computer Vision and Pattern Recognition (CVPR), Honolulu, HI, USA (2017)
14. Wojke, N., Bewley, A., Paulus, D.: Simple online and realtime tracking with a deep association metric. In: 2017 IEEE International Conference on Image Processing (2017)
15. Cao, Z., Hidalgo, G., Simon, T., Wei, S.-E., Sheikh, Y.: OpenPose: realtime multi-person 2D pose estimation using part affinity fields. IEEE Trans. Pattern Anal. Mach. Intell. **43**(1), 172–186 (2021)
16. Yu, Y.-C., Lin, F.-C.: Group similar images based on skeletal information extracted from person's tracking path in real-time surveillance system. In: 2023 IEEE 6th International Conference on Knowledge Innovation and Invention, Japan (2023)
17. Sudhakaran, S., Lanz, O.: Learning to detect violent videos using convolutional long short-term memory. In: 2017 14th IEEE International Conference on Advanced Video and Signal Based Surveillance (AVSS), Lecce, Italy (2017)
18. Sharma, M., Baghel, R.: Video surveillance for violence detection using deep learning. In: Borah, S., Emilia Balas, V., Polkowski, Z. (eds.) Advances in Data Science and Management. LNDECT, vol. 37, pp. 411–420. Springer, Singapore (2020). https://doi.org/10.1007/978-981-15-0978-0_40

Lyric-to-Melody: Generative Song Melody Creation from Lyrics

Prathipati Jayanth[✉], Jatti DevaPaul, Pradeep Balla, K. Ritvik,
and V. S. Ananthanarayana

Department of Information Technology, National Institute of Technology Karnataka,
Surathkal 575025, India
jayanthprathipati2003@gmail.com, {devapauljatti.211ai021,
anvs}@nitk.edu.in

Abstract. This research presents an approach to automatic melody generation from lyrics by leveraging deep learning models on a large dataset of symbolic music data. The dataset includes a vast collection of MIDI files, each providing rich musical details such as lyric syllables, corresponding notes, key, tempo, number of instruments, and track length. This data is pre-processed to align musical notes with syllables, creating a structured basis for melody generation. BERT embeddings are applied to lyrics to capture the global context and feel of each song, ensuring that semantic nuances at the syllable level align with the overall mood and style. Each syllable is represented with a corresponding musical note chosen from a diverse set of pitch possibilities. These embeddings, combined with note representations, are used to generate song data sequences which are then fed into LSTM, BiLSTM, and GRU networks for melody prediction. Model performance is evaluated using pitch distribution metrics and statistical measures like mean and median, mode, range, standard deviation. By comparing predicted note sequences with actual note sequences, this framework facilitates cross-modal learning between text and music, establishing a foundation for advanced applications in AI-driven music composition.

Keywords: Music · BERT Embeddings · LSTM · BiLSTM · GRU

1 Introduction

Music is a universal language that transcends cultural and linguistic barriers, expressing emotions and narratives that resonate deeply with individuals. From classical compositions to contemporary pop hits, music plays a significant role in shaping human experiences and memories. The advent of digital technology has revolutionized music creation and consumption, allowing for unprecedented accessibility to musical resources and tools. The ability to compose, edit, and produce music using software has led to a surge in creativity and innovation, making music more accessible than ever before. As a result, there is an increasing interest in automating music generation processes, which can enhance creativity and streamline production.

© The Author(s), under exclusive license to Springer Nature Switzerland AG 2026
C. Modi et al. (Eds.): MIND 2024, CCIS 2736, pp. 415–426, 2026.
https://doi.org/10.1007/978-3-032-14531-4_35

Artificial intelligence (AI) has emerged as a transformative force across various domains, significantly enhancing the capabilities of traditional processes. From healthcare to finance, AI technologies such as machine learning and deep learning are being used to analyze vast amounts of data, automate tasks, and improve decision-making. In industries such as marketing, AI-driven algorithms optimize customer engagement and personalize experiences based on user behavior. The versatility and efficiency of AI solutions have sparked a growing interest in their application to creative fields, particularly in music, where they can assist in composing melodies, harmonies, and arrangements.

Combining AI with music creation opens up new possibilities for both musicians and composers. Several existing models, such as Recurrent Neural Networks (RNNs) and Long Short-Term Memory networks (LSTMs) [12], have been employed to analyze musical structures and generate original compositions. These models leverage large datasets of musical information to learn patterns and styles, enabling them to produce music that aligns with specific emotional tones or genres. Some notable advancements include MIDI-Net, which focuses on melody and harmony generation, and WaveNet, which has been used for music generation with realistic audio synthesis. Recent developments in natural language processing have also enabled the integration of lyrics, enhancing the synergy between text and melody. By harnessing these technologies, the potential for automated melody generation from lyrics becomes a promising avenue for innovation in music composition, paving the way for new artistic expressions and collaborative processes.

The paper is structured as follows: Sect. 1 provides an overview of existing research on the application of AI in music generation. Section 2 describes the dataset and its available features. Section 3 details the data preprocessing steps and the various methodologies employed for music generation in this study. Section 4 presents the results and insights derived from the model. Finally, Sect. 5 concludes the paper and offers suggestions for future research directions.

2 Literature Survey

Existing models for music and audio generation, such as WaveNet [1], have demonstrated the potential for high-fidelity audio synthesis by capturing long-range dependencies with dilated convolutions. WaveNet has been effective in generating realistic waveforms for both music and speech, modeling complex temporal dependencies within audio data. Additional research on deep learning for music generation has explored various challenges in the field, emphasizing the need for high-quality data and output expressiveness. Techniques like Image and Audio Style Transfer have also been adapted to music generation, enabling transformations across music genres or artist styles and broadening applications in music synthesis.

Traditional neural network models for music generation, however, face limitations such as slow computation, complexity, and difficulty with long-term dependencies [2]. To address these issues, recent studies have proposed hybrid models that combine Long Short-Term Memory (LSTM) networks with Generative Adversarial Networks (GANs), resulting in an LSTM-GAN model. This model includes a novel data preprocessing approach designed specifically for Musical Instrument Digital Interface (MIDI) message data. By handling the structured MIDI format effectively, the LSTM-GAN model has shown promising results in creating coherent musical compositions, as validated by metrics such as maximum mean discrepancy and subjective evaluations of coherence, novelty, and musicality.

In the field of music evaluation, objective and subjective metrics provide critical insights into the quality of generated music and lyrics [3]. Objective metrics for melody generation include Pitch Distribution Similarity (PD), Duration Distribution Similarity (DD), and Melody Distance (MD), while lyric generation is evaluated using metrics like Cosine Similarity (CoSENT), ROUGE-2 Score, and BERT Score. Subjective metrics play a complementary role, assessing Harmony (HMY), Melody-Lyric Compatibility (MLC), Fluency (FLN) in lyric-to-melody generation, and Overall Quality (OVL) and Relevance to Text Input (REL) in song continuation and text-to-song generation tasks.

Various models have been proposed for music generation that focus on lyric-conditioned melody composition and multi-track music synthesis. The Conditional LSTM-GAN model developed by Yi Yu et al. serves as a notable example of lyric-based melody generation, conditioning both the generator and discriminator on lyrics to ensure alignment with real musical distributions [4]. This approach demonstrates the effectiveness of utilizing a large-scale paired lyrics-melody dataset, achieving more pleasant and harmonious melodies than baseline methods. Similarly, Suman Maria Tony and S. Sasikumar investigate automatic music synthesis using LSTM and GAN models [5]. Their model addresses the challenges of temporal dependencies and multi-track interactions, with both subjective and objective evaluations focusing on metrics like pitch distribution, harmony, and rhythm.

Additional contributions to music generation include MuseGAN, a model for generating polyphonic, multi-track music [6]. Unlike conventional GANs MuseGAN is specifically designed for symbolic music data, generating multiple music tracks simultaneously while maintaining coherence across instruments. MuseGAN's capacity for creating harmonically and rhythmically consistent music across tracks represents an advancement in automatic music generation, particularly for compositions requiring accompaniment across different genres.

Efforts to incorporate specific stylistic control in music composition are demonstrated by the CPS (Compound Word with Style) model, which aims to generate full musical compositions while adhering to a specified genre [7]. This model introduces genre meta-information into the musical representation, distinguishing stylistic elements from other musical features. Employing a linear transformer as its backbone, the CPS model utilizes an adaptive strategy to handle diverse music tokens, reducing the likelihood of disharmony. In this way, CPS allows for enhanced stylistic control, outperforming baseline models in both general music quality and stylistic adherence.

The integration of AI in film music has also led to innovative multimodal frameworks. For instance, the Herrmann-1 framework generates cinematic background music by combining vision, language, music, and speech processing models [8]. This model performs emotional analysis of movie scenes to produce descriptive texts, which are then converted to low-level musical conditions with the help of GPT-4.

The importance of BERT embeddings is discussed in this paper [11], highlighting their role in MuLan, an acoustic model that links music audio directly to unconstrained natural language descriptions. By utilizing BERT's language understanding capabilities, MuLan creates rich audio-text embeddings from weakly-associated, free-form text annotations, surpassing traditional music tagging systems that rely on predefined ontologies. This integration enhances the model's versatility across various music genres and enables true zero-shot functionalities, facilitating applications in transfer learning, zero-shot music tagging, and cross-modal retrieval, marking a significant advancement in content-based music retrieval and understanding.

The SuperWillow system further exemplifies advancements in music generation, particularly in creating compositions that appeal to listeners at a level comparable to moderately experienced human composers [9]. Building upon the existing Willow system, SuperWillow incorporates probabilistic models like Markov Chains and Hidden Markov Models to learn patterns in chord progression, rhythm, and melody. This enables the system to generate stylistically consistent music that, as demonstrated in a partial Turing test, can be indistinguishable from human-composed music, thus highlighting its potential to contribute to both professional and creative music applications.

3 Methodology

This section outlines the process used for generating melody from lyrics using deep learning models. The methodology comprises multiple stages, including embedding generation, sequential modeling, and note prediction, as detailed below. The detailed methodology using different RNN models is shown in Fig. 1.

3.1 Embedding Generation with BERT

The first stage in the process involves transforming each syllable in the lyrics into an embedding representation. We use BERT (Bidirectional Encoder Representations from Transformers) to generate embeddings that capture the semantic and contextual meaning of each syllable. Each syllable embedding is of size 768, which encapsulates the global context and the lyrical "feel" of the song, ensuring that the nuances at the syllable level align with the song's overall mood and style. Formally, let $S = \{s_1, s_2, ..., s_n\}$ represent a sequence of syllables in a song, where each s_i is transformed into a 768-dimensional embedding vector e_i using BERT, so that:

$$e_i = BERT(s_i), \quad \forall i \in \{1, 2, ..., n\}.$$

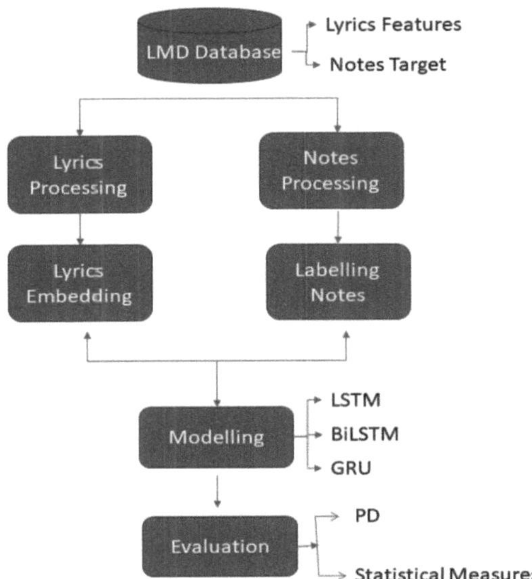

Fig. 1. Methodology Flowchart

3.2   Sequential Modeling with LSTM, BiLSTM, and GRU

In the next stage, we process the embeddings through sequential models to predict the musical notes. We experiment with three types of recurrent neural network (RNN) architectures as shown in Fig. 2.

- **LSTM (Long Short-Term Memory)**: This model captures long-term dependencies, which is crucial for musical coherence.
- **BiLSTM (Bidirectional LSTM)**: By processing the sequence from both forward and backward directions, BiLSTM captures context from both past and future syllables.
- **GRU (Gated Recurrent Unit)**: GRU is a simplified RNN variant that retains sequential information while reducing computational complexity.

Each syllable embedding e_i is fed into the chosen model (LSTM, BiLSTM, or GRU) to generate an output vector, denoted as h_i. This output vector h_i represents the intermediate prediction for the musical note associated with syllable s_i. The model parameters are optimized to maintain melodic coherence and sequential alignment with the lyrics.

$$h_i = Model(e_i, h_{i-1}), \qquad \forall i \in \{1, 2, ..., n\}.$$

3.3   Note Prediction with Softmax Layer

After obtaining the output vector h_i from the sequential model, it is passed through a Softmax layer to predict the corresponding musical note. The Softmax layer converts the vector h_i into a probability distribution across 128 possible notes, covering the full range of musical pitch values available in MIDI format. The predicted note $\hat{y}_i$ is chosen as the one with the highest probability.

$$\hat{y}_i = \arg\max Softmax(W \cdot h_i + b),$$

where W and b are the weights and biases of the Softmax layer, respectively.

3.4 Melody Generation

The model outputs a sequence of notes aligned with the lyrics, where each syllable s_i is associated with a predicted note $\hat{y}_i$. This generated melody reflects the semantic and emotional context of the lyrics, producing a musical sequence that is cohesive and contextually aligned with the lyrics.

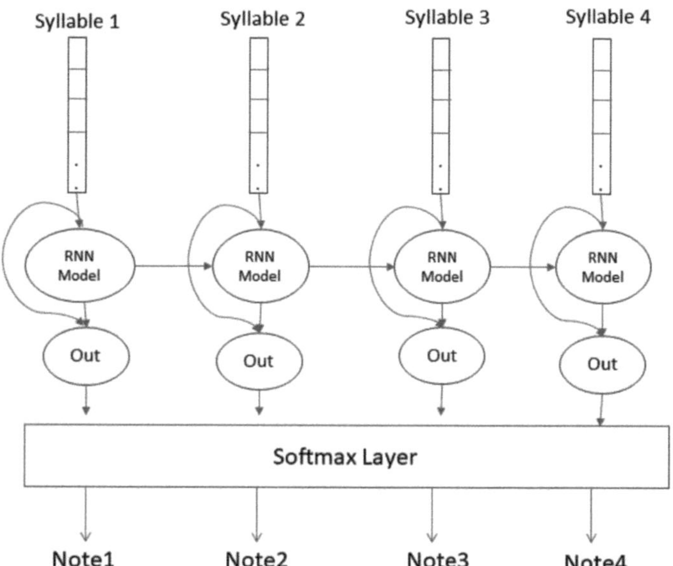

Fig. 2. Flow Diagram of the Lyric-to-Melody Generation Process Using BERT Embeddings and LSTM Models

3.5 Model Evaluation and Metrics

To evaluate the quality of generated melodies, we use several metrics to assess how well the predicted notes align with expected musical patterns. The loss function measures the difference between predicted and actual notes, guiding the model's learning and ensuring it minimizes errors over time. Pitch distribution assesses the spread of notes across the MIDI range, reflecting melody variation and diversity—essential elements for musical richness. Statistical measures such as mean, median, mode, range, and standard deviation further capture the central tendencies and spread of the predicted notes, helping to ensure that the melodies follow standard musical patterns. Additionally, we analyze melody transitions using a transition matrix, which captures the likelihood of moving from one note to another. This matrix provides insights into the continuity and flow of notes, helping to evaluate whether the model generates smooth, musically coherent progressions. Comparing LSTM, BiLSTM, and GRU models across these metrics allows us to identify the architecture that produces the most expressive, coherent, and stylistically aligned melodies from lyrics.

4   Datasets, Experiments and Results

4.1   Dataset Explanation

The dataset consists of 176,581 unique MIDI (Musical Instrument Digital Interface) files [13], providing a substantial resource for music information retrieval (MIR) tasks. Out of the total collection, 45,129 MIDI files are matched and aligned with entries from the Million Song Dataset (MSD). This alignment allows for advanced cross-modal research, enabling symbolic-to-audio comparisons and retrieval. The alignment between the MIDI files and the MSD entries facilitates research that spans both symbolic music (MIDI) and audio representations, enhancing the dataset's utility for varied music research contexts.

Each MIDI file provides a detailed symbolic representation of music. Key musical attributes captured include duration (the length of each musical piece in seconds), instrumentation (the number of instruments used in the composition), key (the tonal structure, represented by Western musical notes ranging from C to B, along with corresponding sharps and flats), tempo (beats per second, along with tempo changes throughout the piece), and syllables of lyrics (where available, the lyrics associated with the musical composition). The MIDI files contain a total of 128 unique note pitches, which include various octaves and pitch variations of the 12 basic notes as shown in Fig. 3, providing a rich diversity of tonal possibilities.

4.2   Exploratory Data Analysis

To better understand the dataset characteristics, we analyzed the distribution of keys, lengths, number of instruments, and tempo in the MIDI files. Figures 4, 5, 6 and 7 illustrate these characteristics.

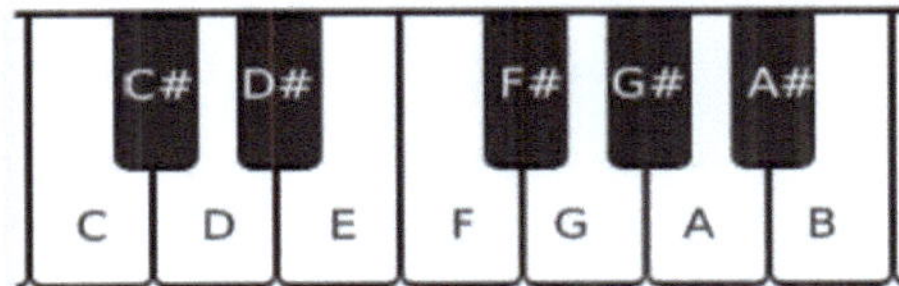

Fig. 3. Western Music Notation - Notes range from C to B with corresponding sharp notes (C#, D#, F#, G#, A#).

Figure 4 shows the frequency of each musical key across the MIDI files, indicating which keys are more prevalent in the dataset. The distribution of MIDI file lengths, as shown in Fig. 5, helps to determine the typical duration of pieces in the dataset, providing insights into the standard composition length. In Fig. 6, we observe the number of MIDI files grouped by the number of instruments used, revealing the prevalence of solo versus ensemble compositions within the dataset. Lastly, Fig. 7 illustrates the distribution of tempos in the MIDI files, highlighting common tempo ranges and offering insight into the general pace of the compositions.

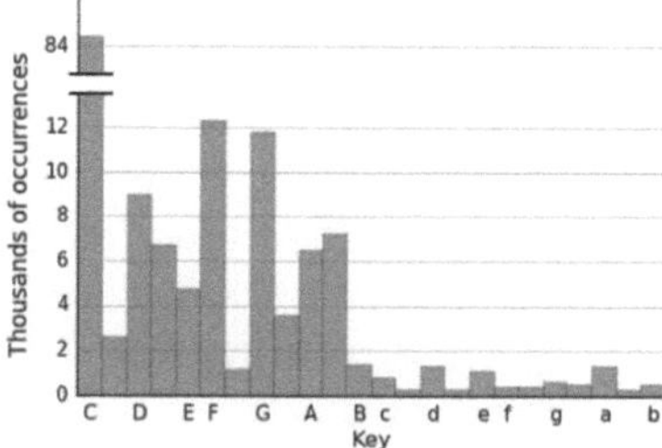

Fig. 4. Number of occurrences of each key in MIDI Files.

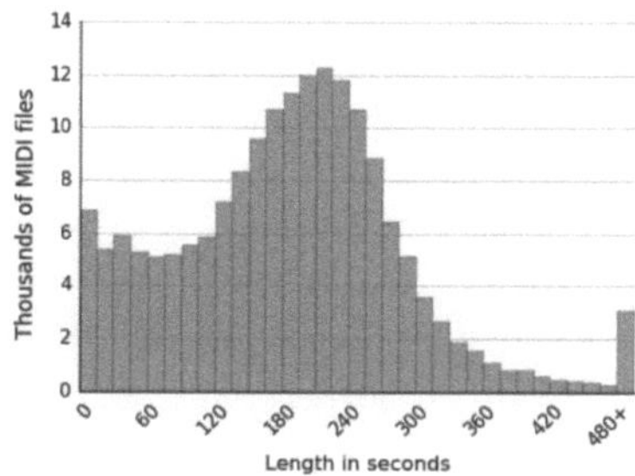

Fig. 5. Number of MIDI files with corresponding length.

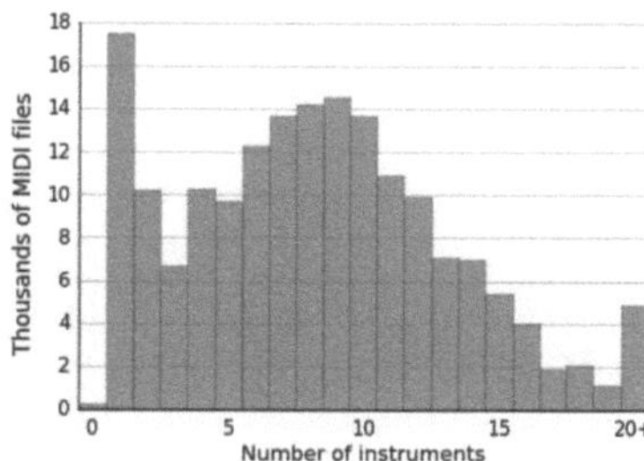

Fig. 6. Number of MIDI files corresponding to the number of instruments used.

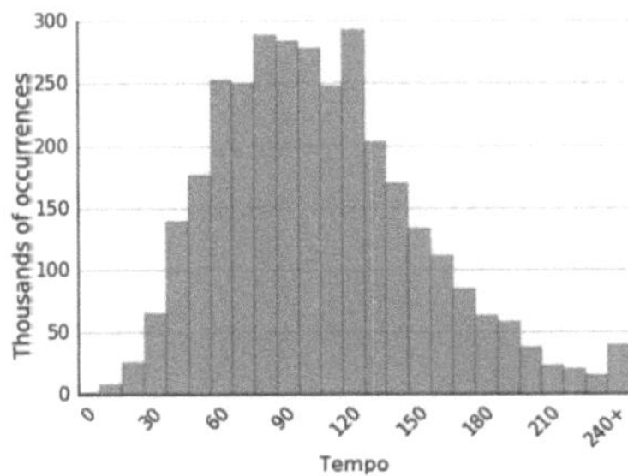

Fig. 7. Number of occurrences corresponding to the tempo.

4.3 Modelling Results

In this research, we trained LSTM, BiLSTM, and GRU models on 2000 MSD-aligned files containing syllables and corresponding notes and evaluated them on a test lyric by analyzing pitch distribution and statistical measures. Each model, with two recurrent layers of 50 units each, was configured to capture musical dependencies and output across 128 MIDI note classes. Dropout layers (20%) were included after each recurrent layer to improve generalization. The BiLSTM model leverages forward and backward passes for additional melodic context, while the GRU model reduces parameters for efficiency. Using a batch size of 32 and a 20% validation split, the models were optimized to generate contextually aligned melodies from lyrical input.

The pitch distribution histogram reveals that the actual melody consists of five unique notes, while the LSTM models as shown in Fig. 8 predict only three notes and BiLSTM with multiple notes in Fig. 9, indicating diversity in notes. However, the GRU model captures a broader range, showing greater alignment with the actual pitch distribution as shown in Fig. 10.

Statistically, the actual notes have a mean of 44.82, slightly higher than the predictions, where GRU produces the closest mean at 43.58. The median and mode are consistent across all models and actual values at 42, showing alignment in central note values. Both actual and predicted notes share the same range of 18, indicating a similar pitch span. However, the actual notes have a higher standard deviation (6.70) compared to GRU (5.48), BiLSTM (5.31), and LSTM (3.79), suggesting that the actual melody has more variation while the predicted melodies, especially from LSTM, are more clustered around their mean. Overall, the GRU model demonstrates a better capability to capture the note diversity and statistical spread compared to the other models as mentioned in Table 1.

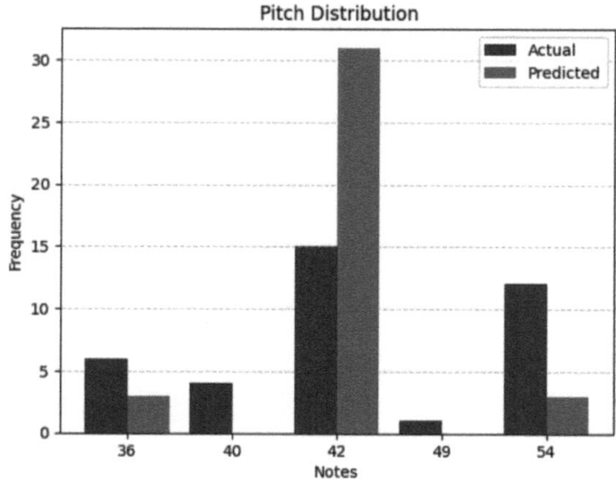

Fig. 8. Pitch Distribution for a sample using LSTM Model

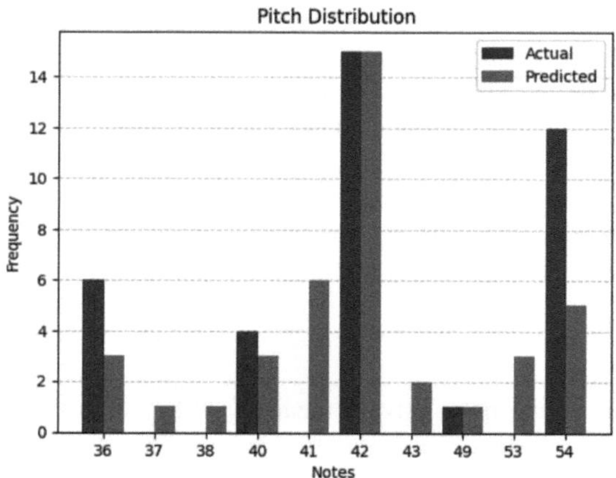

Fig. 9. Pitch Distribution for a sample using BiLSTM Model

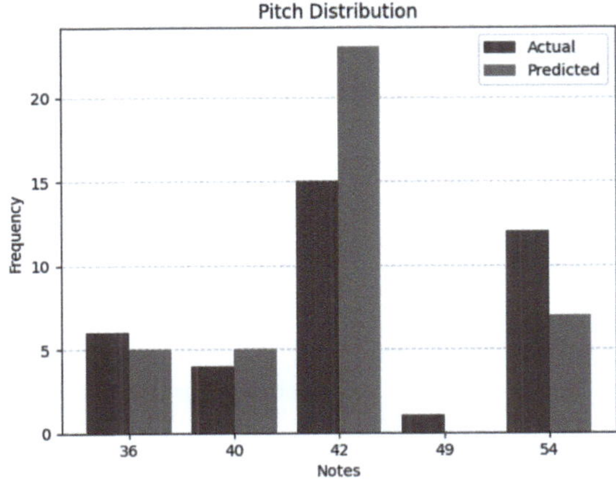

Fig. 10. Pitch Distribution for a sample using GRU Model

Table 1. Statistical Comparison of Note Predictions for Actual Notes, LSTM, GRU, and BiLSTM Models

Metric	Actual Notes	LSTM	GRU	BiLSTM
Mean	44.82	42.49	43.58	43.50
Standard Deviation	6.70	3.79	5.48	5.31
Median	42.0	42.0	42.0	42.0
Mode	42	42	42	42
Range	13	18	18	18

5 Conclusion and Future Scope

In this study, we presented a comprehensive methodology for generating melodies from lyrics using advanced deep learning models, specifically leveraging BERT embeddings alongside LSTM, BiLSTM, and GRU architectures. By transforming lyrical syllables into rich embedding representations, we effectively captured the semantic and emotional nuances of the lyrics, allowing for the generation of melodies that align well with the intended musical context. The incorporation of various metrics for evaluation, such as pitch distribution and statistical measures, provided a robust framework for assessing the quality and coherence of the generated melodies. The results demonstrate the potential of our approach to create expressive and contextually relevant musical compositions.

Looking ahead, there are exciting opportunities to enhance melody generation further by integrating Generative Adversarial Networks (GANs). GANs could facilitate the production of more diverse and complex musical outputs by modeling the distribution of musical notes more effectively. Additionally, the application of recent large language models (LLMs) like GPT and LLaMA presents a promising avenue for improving contextual understanding in music generation. These models could enhance the generation process by better interpreting lyrical themes and emotions, leading to melodies that are not only musically coherent but also deeply aligned with the narrative of the lyrics. By exploring these advanced methodologies, future research can significantly enrich the capabilities of music generation systems, paving the way for innovative applications in automated music composition and personalized music experiences.

References

1. Briot, J.-P., Pachet, F.: Deep learning for music generation: challenges and directions. Neural Comput. Appl. **32**(4), 981–993 (2020)
2. Li, G., Ding, S., Li, Y.: Novel LSTM-GAN based music generation. In: 2021 13th International Conference on Wireless Communications and Signal Processing (WCSP), pp. 1–6. IEEE (2021)
3. Ding, S., et al.: Songcomposer: a large language model for lyric and melody composition in song generation. arXiv preprint arXiv:2402.17645 (2024)
4. Conner, M., Gral, L., Adams, K., Hunger, D., Strelow, R., Neuwirth, A.: Music generation using an LSTM. arXiv preprint arXiv:2203.12105 (2022)
5. Yu, Y., Srivastava, A., Canales, S.: Conditional LSTM-GAN for Melody Generation from Lyrics. https://doi.org/10.1145/3424116
6. Tony, S.M., Sasikumar, S.: Investigation on Automatic Music Generation Using GAN and LSTM Networks (2022)
7. Dong, H.-W., et al.: MuseGAN: multi-track sequential generative adversarial networks for symbolic music generation and accompaniment. In: Proceedings of the AAAI Conference on Artificial Intelligence, vol. 32, no. 1 (2018)
8. Wang, W., Li, X., Jin, C., Lu, D., Zhou, Q., Tie, Y.: CPS: full-song and style-conditioned music generation with linear transformer. In: 2022 IEEE International Conference on Multimedia and Expo Workshops (ICMEW), Taipei City, Taiwan, pp. 1–6 (2022). https://doi.org/10.1109/ICMEW56448.2022.9859286
9. Haseeb, M.T., Hammoudeh, A., Xia, G.: GPT-4 driven cinematic music generation through text processing. In: 2024 IEEE International Conference on Acoustics, Speech and Signal Processing (ICASSP), ICASSP 2024, Seoul, Republic of Korea, pp. 6995–6999 (2024). https://doi.org/10.1109/ICASSP48485.2024.10447950
10. Van Der Merwe, A., Schulze, W.: Music generation with Markov models. IEEE Multimed. **18**(3), 78–85 (2011). https://doi.org/10.1109/MMUL.2010.44
11. Huang, Q., Jansen, A., Lee, J., Ganti, R., Li, J.Y., Ellis, D.P.W.: Mulan: a joint embedding of music audio and natural language. arXiv preprint arXiv:2208.12415 (2022)
12. Mangal, S., Modak, R., Joshi, P.: LSTM based music generation system. arXiv preprint arXiv:1908.01080 (2019)
13. https://colinraffel.com/projects/lmd/

From Data to Harvest: Leveraging Machine Learning and Dataset Visualization for Crop Prediction

Sayan Das and M. Ambika[(⊠)]

Department of CSE, Indian Institute of Information Technology Tiruchirappalli, Tiruchirappalli, Tamil Nadu, India
mani.ambika@gmail.com

Abstract. Improving farmers' crop yield is crucial to the agriculture sector's sustainable growth and economic advancement in India. AI improves agriculture by optimising planting, harvesting, and resource utilisation through the analysis of soil, weather, and crop health data. It helps with crop selection, pest control, and disease identification, boosting yields while reducing waste. By supporting sustainable farming, this data-driven strategy aids farmers in adjusting to shifting market and environmental conditions. In order to maximise crop recommendation accuracy, this study consists of dataset visualization and comparison several Machine Learning models using optimised hyperparameter tuning. This effort attempts to determine the best model for recommending ideal crops by evaluating model performance under various circumstances, ultimately assisting farmers in making data-driven decisions. In addition to increasing yields, this strategy strengthens the agricultural framework, which supports the objectives of sustainable growth and economic development in the Indian agricultural sector.

Keywords: Machine Learning · Multi-Class classification · Adaboost · Ensemble Learning Model · Crop Prediction · Agriculture · Smart farming

1 Introduction

Millions of people rely on agriculture as their primary source of income, particularly in developing nations. The sector is crucial for maintaining rural communities, fostering economic growth, and reducing poverty in addition to ensuring food security. The agricultural sector, however, has several challenges that hinder sustainability and productivity, including changing weather patterns, deteriorating soil, pest infestations, and the ongoing pressure to feed a growing population. Among these challenges, selecting appropriate crops for production is one of the most significant since it is necessary to maximise yields and ensure the efficient use of resources.

Artificial Intelligence (AI) is completely changing the agricultural industry [1]. AI can anticipate yields [2], suggest the best times to sow, and customise fertiliser and water use for individual crops by evaluating vast datasets from sources such as soil sensors, climate data, and crop health records. Machine learning models [3] find trends that aid

C. Modi et al. (Eds.): MIND 2024, CCIS 2736, pp. 427–437, 2026.
https://doi.org/10.1007/978-3-032-14531-4_36

in forecasting crop performance in a variety of scenarios, helping farmers precisely maximise yield. AI-based image identification also stops extensive agricultural damage by identifying plant illnesses early. AI facilitates effective, resilient farming methods that tackle the issues of global food security through automation and predictive insights.

The process of choosing crops is intricate and multifaceted, involving a variety of factors such as soil characteristics, nutrient availability [4], meteorological conditions [5], and market demands. Historically, this approach has heavily relied on the expertise and experience of agronomists and extension services, who provide guidance based on local conditions and historical data. However, getting such expert advice isn't always simple, especially in rural or impoverished areas. Furthermore, the dynamic nature of agricultural ecosystems, where seasonal variations in rainfall, temperature, and soil pH may be significant and is not adequately taken into account by the conventional methods of crop recommendation.

By evaluating intricate agricultural data, Machine Learning (ML) is essential to enhancing crop suggestions. Data including soil composition, climate, crop traits, and past yield trends are evaluated using machine learning algorithms to identify trends and forecast the best crops under particular circumstances. In order to determine which, crop varieties are most appropriate for different soil and climatic combinations, this study trains machine learning models such as Random Forest, Decision Trees, and others on agricultural information. These models are further improved by hyperparameter tuning, which enables them to provide more precise and customised recommendations, supporting farmers in making data-driven, sustainable decisions.

This work discusses the novel method of comparison of the best machine learning models with better hyper parameter tuning to obtain the most accurate results for crop recommendation.

The rest of the paper is structured as Sect. 2 consists of review and study of different works, Sect. 3 contains the methodology of the proposed works, Sect. 4 deals with the implementation and the discussion part of the work, the conclusion and future work is covered in the Sect. 5.

2 Literature Review

According to Pradeepa Bandara et al. [6], human intervention is necessary for the optimal crop selection on limited domestic lands, regardless of how automated agriculture is. Sri Lanka lacks a system that recognises environmental conditions and recommends the best crops, while possessing sophisticated agricultural knowledge and techniques. This paper presents the theoretical and conceptual underpinnings of a crop recommendation [7, 8] system. It combines machine learning techniques like SVM, Naïve Bayes (multinomial), and K-Means clustering with natural language processing for sentiment analysis and Arduino microcontrollers for data collection. The method aims to predict the ideal crop for a certain area by analysing temperature, water levels, and soil characteristics, providing accurate and useful recommendations for both residential and agricultural areas.

Mahmoud Y. Shams et al. [9] explains crop recommendation systems use previous crop performance, weather trends, and soil attributes data to assist farmers maximize yields. This paper presents XAI-CROP, an algorithm that uses the principles of explainable AI (XAI) to produce suggestions that are clear and easy to understand. XAI-CROP provides understandable insights into the recommendation process, which sets it apart from traditional opaque machine learning algorithms. Using metrics such as Mean Squared Error (MSE), Mean Absolute Error (MAE), and R-squared (R2), the study compares XAI-CROP with Gradient Boosting (GB), Decision Tree (DT), Random Forest (RF), Gaussian Naïve Bayes (GNB), and Multimodal Naïve Bayes (MNB). With a low MSE of 0.9412, an MAE of 0.9874, and a R2 of 0.94152, XAI-CROP performs better than other methods, showing good accuracy, dependability, and interpretability.

Shalini Banerjee et al. [10] worked on a chatbot using the classification techniques and used it for chatbot, they got an accuracy of 99.31% for Random Forest classifier. They too did a comparative analysis of few Machine Learning models and selected Random Forest classifier as the best model for their chatbot.

According to Disha Garg et al. [11], soil testing is crucial in agriculture to determine the proper nutrient levels for crops. This work introduces a crop recommendation system based on machine learning (ML) classification approaches for soil nutrient data. For hyperparameter optimisation, the Wrapper-PART-Grid approach combines grid search, wrapper feature selection, and the partial C4.5 decision tree (PART) classifier. This strategy was contrasted with multilayer perceptron (MLP), instance-based learning (IBk), reduced error pruning (REP) trees, and C4.5 decision trees (CDT). According to evaluation metrics such as F1-score, RMSE, MAE, recall, precision, and true positive rate, the recommended approach achieved the highest accuracy rate of 99.31%. This approach aims to increase output by reducing waste and assisting farmers in selecting the best crops.

2.1 Research Gap

The studies got lot of research going on in the Agriculture with the integration of Machine Learning, still the performance of many models are very low. Some work deals with soil characteristics but ignoring the environmental factors, even the water level factors needed for good analysis. Many ML models in agriculture underperform due to poor feature selection and lack of correlation analysis between critical factors. The crops data used are also not proper for training of the models in a better way. The open source limited datasets makes it tough to deal with a proper amount of collected data in most of the works. Because there isn't enough individualised, data-driven counsel available, current farming practices might not fully maximise productivity. The current study shows that even if crop prediction and recommendation accuracy has increased to a comparatively high level, hyper-parameter optimization can result in even larger gains.

3 Proposed Methodology

In addition to frequently not knowing what crops are appropriate for their specific regions, many Indian farmers lack access to modern lifestyles. They sometimes rely on local knowledge while growing crops, which could lead them to stick with a particular crop over extended periods of time, perhaps missing out on seasonal benefits and remaining idle during other seasons. Farmers usually struggle to select the best crop type for their area based on factors like rainfall, humidity, and soil nutrition.

This work proposes a comparative analysis using the present Machine Learning models and uses different parameter tunings and selecting the best model to use for the recommendation of crops.

10 different Machine Learning classification models are considered for the dataset consisting of 20 major crops in India, out of which the best model is considered.

3.1 Crop Prediction Learning Model

The suggested approach trains and tests ten distinct learning models. A final model is created based on the model's accuracy after comparison analysis, and a multiclass classification is carried out to help with precise prediction using the given data. Before being used to train seven distinct models and determine which is best for giving suggestions, the dataset is pre-processed by eliminating any missing information.

Figure 4 shows the flowchart of the work done where all the steps of this work shown beginning from the dataset selection, followed by the dataset pre-processing, the models used for the training and testing.

Dataset Collection and Preprocessing

The dataset includes the weather conditions of the specific region as well as the soil values. The weather conditions such as temperature, humidity, and rainfall are included, and the elements in the soil include some major elements like nitrogen, phosphorus, and potassium. The datasets used in this article cover almost 20 of India's key crops.

The Mean Substitution approach is used in the data pre-processing to remove the missing values. Nitrogen, phosphorus, potassium, soil pH, humidity, temperature, rainfall, and the crop were among the features from the dataset used to train the model.

Dataset Visualization

Figure 1 explains the correlation of features and the label, some of the features has very less correlation among them while some has good correlations among them, which tells us the significance of each of the features.

The skewness of the data is also considered, where skewness means how perfectly the data is distributed and is evenly spread around the mean. Some of the data points lie on the positive side of the curve and some of the data points lie on the negative side of the curve.

The target variable may be significantly influenced by features that exhibit a high correlation with the label, whereas features that exhibit a low correlation may have less of an effect. Because it enables us to concentrate on the most pertinent variables for modelling, this knowledge is crucial for feature selection as it enhances interpretability and predictive performance.

Analysis of the dataset shows that its features are distributed in a variety of ways, with clear skewness patterns. With longer tails at the lower end of the scale, the distributions for nitrogen (N), phosphorus (P), and potassium (K) are left-skewed. This suggests that there are very few cases of low amounts of these nutrients in the majority of samples, which have comparatively high levels. In particular, readings for potassium and phosphorus are typically concentrated from 0 to 60, indicating that although severely low levels are less common, typical nutritional levels typically fall within this range. Given that only a tiny percentage of samples exhibit nutritional deficits, this skewness may be a reflection of generally nutrient-rich soil conditions.

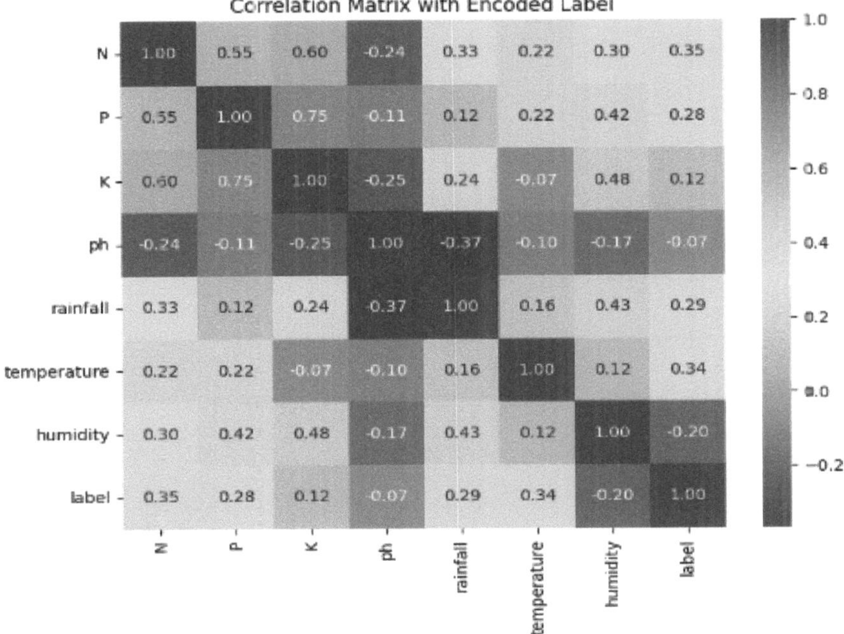

Fig. 1. Correlation of the features and label

Table 1. Checking skewness of all the features

Features	Skewness
N	0.586556
P	0.952624
K	0.958350
pH	0.278578
temperature	−0.106656
humidity	−0.665084
rainfall	0.711992

In contrast, Fig. 2 shows that the temperature and pH distributions are more symmetrical and show less skewness, closely resembling a normal distribution with the mean in its centre. With most samples grouped around the average and fewer outliers, this symmetry suggests a balanced distribution of results, suggesting stable circumstances across several environmental parameters. Since symmetrically distributed data is ideal for many machine learning techniques, normal distributions in these features can be beneficial.

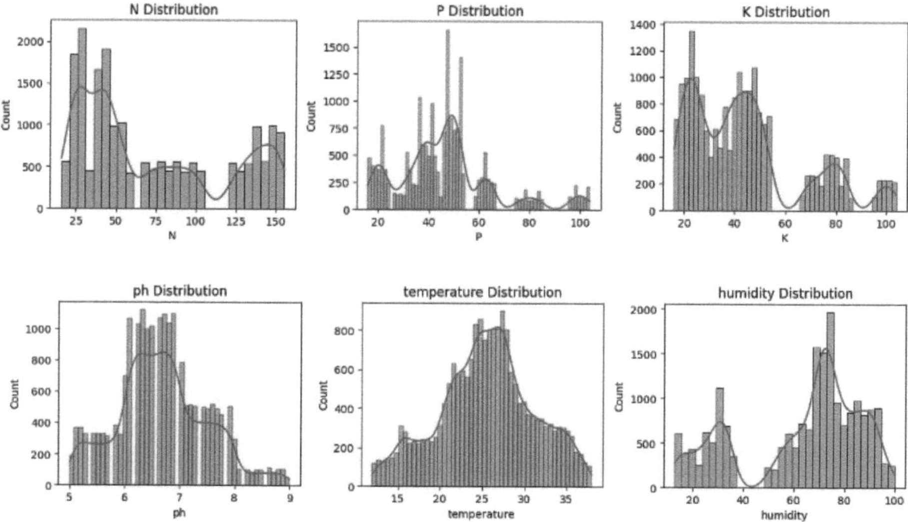

Fig. 2. The Kernel Distribution Estimation (KDE) for the N, P, K, pH, temperature, humidity

The humidity distribution, on the other hand, is skewed to the right, with a smaller number of higher values producing a tail on the right and the majority of values concentrated on the lower end of the scale. According to this pattern, high humidity is less frequent in the dataset than low to moderate humidity levels. To improve model performance, the skewness of each feature might need to be transformed in order to make the data more symmetrical. The data can still be processed further, though, and these observations will assist direct any pre-processing that may be required for additional study. Moreover since some data are moderately skewed and some are slightly skewed,

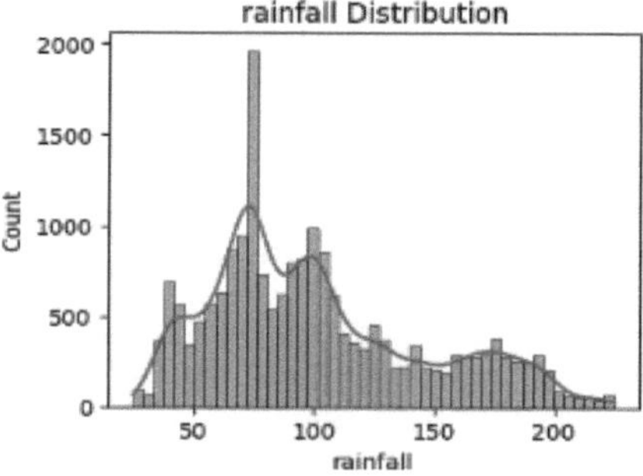

Fig. 3. The Kernel Distribution Estimation (KDE) for rainfall

squared root or logarithmic functions can help improving the imbalanced dataset to a balanced one (Fig. 3).

Building the Learning Model

The work used ten different classification model which has some ensemble methods too included. Learning models include Random Forest classification [12], Support Vector Machine [13], K-nearest neighbor [14], Decision Tree classification model [15], Multi-Layer Perceptron [16], Naïve Bayes [17], Adaboost [18], Bagging technique [19], Stacking method [20] and Logitboost [21]. Adaboost got the highest accuracy of 99.95%.

Random Forest is considered since it is an ensemble approach that is resistant to noise and overfitting by constructing many decision trees and aggregating their output. It is appropriate for crop recommendation datasets due to its capacity to manage nonlinear relationships and feature importance.

Support Vector Machine performs well with small to medium-sized datasets and is efficient in high-dimensional domains. Its capacity to identify the best hyperplane for categorisation guarantees precise forecasts for intricate datasets, such as agricultural data.

KNN is a straightforward but powerful non-parametric technique that depends on feature similarity. It is helpful for datasets with distinct differences between crop labels since it is good at identifying local trends in the data.

The impact of features like N, P, K, and pH on crop compatibility can be easily understood by using decision trees, which are interpretable models that divide data according to features. One kind of neural network that can capture intricate nonlinear interactions is MLP. It may represent complex dependencies in the crop advice data because of its adaptability. Because Naïve Bayes presupposes feature independence, it is computationally efficient. It can nonetheless function well on structured datasets with categorical results, despite the fact that its assumption is frequently broken.

Adaboost creates a powerful learner by combining a number of weak learners, typically decision trees. Its excellent accuracy in your dataset was probably influenced by its capacity to concentrate on samples that were challenging to classify Bagging, like Random Forest, reduces variance by training models on different subsets of data. This guarantees steady performance and makes it resistant to overfitting.

By training a meta-model on the predictions of several models, stacking merges them. It increases prediction capability by utilising the advantages of many algorithms. An Adaboost variant called LogitBoost employs logistic regression as a weak learner. Better management of misclassified instances is ensured by its iterative sample reweighting.

The ensemble learning algorithm known as AdaBoost, or Adaptive Boosting, builds a powerful predictive model by combining several weak learners, usually decision stumps (basic decision trees). It trains these weak learners one after the other, with each one concentrating on fixing the errors of the one before it. Instances that are incorrectly classified are given greater weights, so later learners give them priority. To arrive to a final prediction, AdaBoost iteratively modifies these weights and aggregates the outputs of the weak learners. AdaBoost, which is popular for classification applications due to its accuracy and robustness, successfully improves model performance on balanced and moderately imbalanced datasets.

4 Implementation and Discussion

This work uses dataset that addresses 20 major crops in India, and each crop has 1000 entries which makes a total of 20000 data. Ten different learning models combined with ensemble methods are used. Comparative analysis was done to compare the performance of the various learning models. Adaboost ensemble method proved to be the best model with an accuracy of 99.95%. Table 2. Shows the comparison of all the models along with the ensemble methods. The evaluation metrics includes the accuracy, precision, recall and F1- score.

Accuracy in crop recommendation is a straightforward indicator of how well the algorithm predicts across all classes. When all classes are equally important and the dataset is balanced, it is helpful. Precision is essential when minimising false positives, such suggesting crops that aren't ideal for a given situation. High precision guarantees the appropriateness of the suggested crops.

In order to minimise false negatives and guarantee that all crops appropriate for a particular circumstance are recognised, recall is crucial. For instance, the user may lose out on chances if they don't receive a recommendation for an appropriate crop. F1-score is particularly useful when the dataset is imbalanced, ensuring that neither precision nor recall is given undue weight. In crop recommendation, if some crop classes have fewer samples, the F1-score ensures a balanced evaluation. The tools for the comparison includes, Scikit-learn library, Pandas library, Numpy library and Matplotlib library.

In Fig. 5 comparison of all the models are shown based on the values of Table 1.

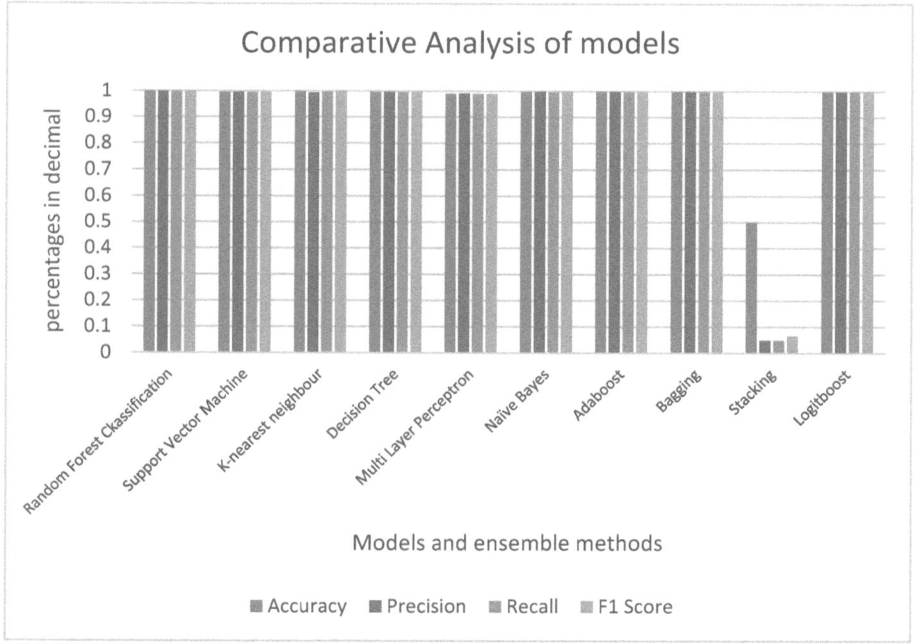

Fig. 5. Comparative Analysis of the Crop Prediction Learning models

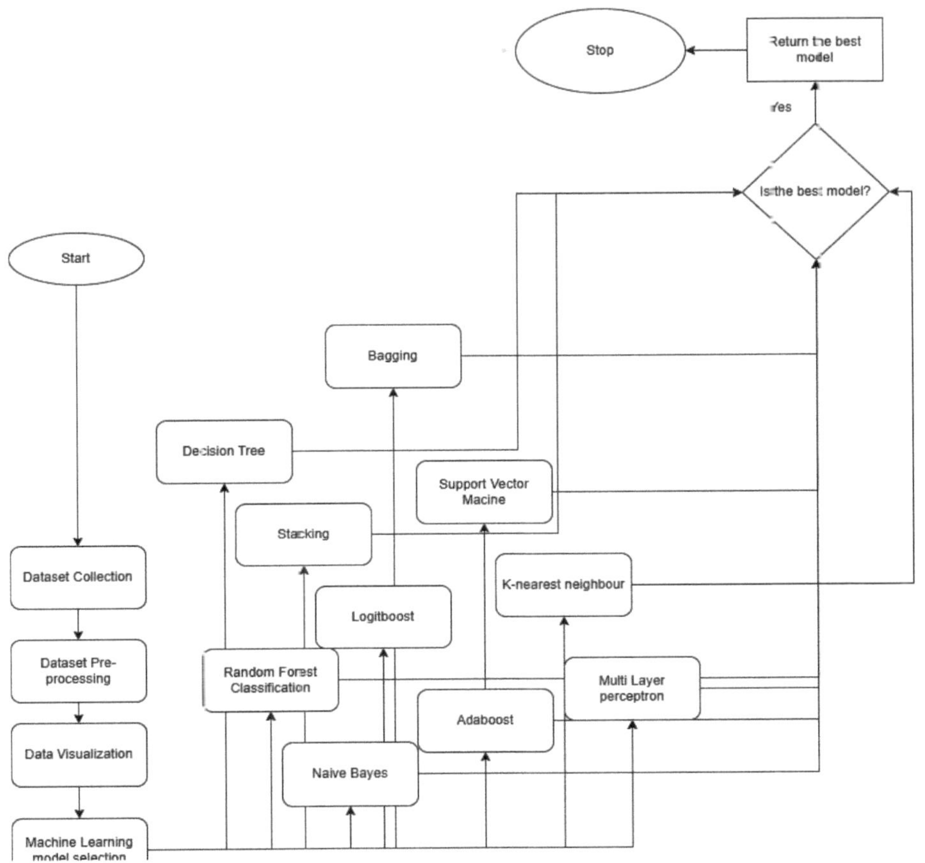

Fig. 4. Flowchart of the Crop Prediction Learning Model.

Table 2. Comparative Analysis of the Crop Prediction Models.

Model	Accuracy	Precision	Recall	F1 Score
Random Forest Classification	0.9986	0.9986	0.9985	0.9986
Support Vector Machine	0.9956	0.9960	0.9955	0.9956
K nearest neighbour	0.9982	0.9938	0.9981	0.9982
Decision Tree	0.9972	0.9972	0.9972	0.9972
Multi-Layer Perceptron	0.9904	0.9909	0.9898	0.9902

(continued)

Table 2. (continued)

Model	Accuracy	Precision	Recall	F1 Score
Naïve Bayes	0.9982	0.9982	0.9981	0.9982
Adaboost	0.9995	0.9991	0.9991	0.9992
Bagging	0.9989	0.9990	0.9990	0.9990
Stacking	0.4995	0.0500	0.0500	0.0660
Logitboost	0.9988	0.9990	0.9990	0.9990

5 Conclusion and Future Work

The study's findings demonstrate that AdaBoost is the best model for crop recommendation, with meticulous hyperparameter optimisation enabling an astounding 99.95% accuracy. AdaBoost's ability to provide accurate crop predictions is demonstrated by this result, which makes it a dependable tool for supporting Indian farmers in data-driven crop selection. AdaBoost integration enhances production and promotes sustainable farming methods by enabling farmers to make well-informed decisions suited to particular environmental and soil conditions. In addition to confirming AdaBoost's applicability, this finding highlights how cutting-edge machine learning methods contribute to the resilience and long-term economic growth of the agriculture industry.

This analysis can be later used for better recommendation of crops using any chatbots or any recommendation sites. The collection of data using API for the weather conditions and soil sensors or other IOT devices for the collection of soil nutrients data can be beneficial for farmers, and also to the areas affected by any natural disasters.

References

1. Unleashing the potential of Artificial Intelligence & Machine Learning into Space, vol. 3, pp. 133–150. https://www.iafastro.org/events/iac/iac-2020/technical-programme/special-ses sions/unleashing-the-potential-of-artificial-intelligence-and-machine-learning-into-space. html
2. Goel, M., Pandey, M.: Crop yield prediction using AI: a review. In: 2024 2nd International Conference on Disruptive Technologies (ICDT), pp. 1547–1553 (2024). https://doi.org/10. 1109/ICDT61202.2024.10489432
3. Elbasi, E., et al.: Crop prediction model using machine learning algorithms. Appl. Sci. **13**(16) (2023). https://doi.org/10.3390/app13169288
4. Falkengren-Grerup, U., Ten Brink, D.J., Brunet, J.: Land use effects on soil N, P, C and pH persist over 40–80 years of forest growth on agricultural soils. For. Ecol. Manage. **225**(1–3), 74–81 (2006). https://doi.org/10.1016/j.foreco.2005.12.027
5. Pereira, A.R.: Crop planning for different environments. Agric. Meteorol. **27**(1–2), 71–77 (1982)
6. Kansal, L., Pandey, A., Shukla, S.M., Dhaliwal, P.: Review of machine learning techniques for crop recommendation. In: ACM International Conference on Proceeding Series, no. March, pp. 443–449 (2023). https://doi.org/10.1145/3607947.3608045

7. Chauhan, G., Chaudhary, A.: Crop recommendation system using machine learning algorithms. In: Proceedings of the 2021 10th International Conference on System Modeling and Advancement in Research Trends, SMART 2021, vol. 3307, pp. 109–112 (2021). https://doi.org/10.1109/SMART52563.2021.9676210

8. Gayatri, G., Praharsha, K.N.V., Hemanth, K., Owk, M.: Crop recommendation system using machine learning. In: Lin, F.M., Patel, A., Kesswani, N., Sambana, B. (eds.) ICDSAI 2023, vol. 421, pp. 735–742. Springer, Cham (2024). https://doi.org/10.1007/978-3-031-51167-7_70

9. Shams, M.Y., Gamel, S.A., Talaat, F.M.: Enhancing crop recommendation systems with explainable artificial intelligence: a study on agricultural decision-making. Neural Comput. Appl. **36**(11), 5695–5714 (2024). https://doi.org/10.1007/s00521-023-09391-2

10. Banerjee, S., Akhila, P., Sujithra Kanmani, R., Elakiya, E., Surendiran, B.: CropBot: a chatbot for agriculture advancement. In: 2024 International Conference on Signal Processing, Computation, Electronics, Power and Telecommunication IConSCEPT 2024 - Proceedings, no. September, pp. 1–6 (2024). https://doi.org/10.1109/IConSCEPT61884.2024.10527848

11. Garg, D., Alam, M.: An effective crop recommendation method using machine learning techniques. Int. J. Adv. Technol. Eng. Explor. **10**(102), 498–514 (2023). https://doi.org/10.19101/IJATEE.2022.10100456

12. Parmar, A., Katariya, R., Patel, V.: A review on random forest: an ensemble classifier. In: International Conference on Intelligent Data Communication Technologies and Internet of Things (ICICI) 2018, pp. 758–763 (2019)

13. Suthaharan, S.: Support vector machine. In: Machine Learning Models and Algorithms for Big Data Classification: Thinking with Examples for Effective Learning, pp. 207–235. Springer, Boston (2016). https://doi.org/10.1007/978-1-4899-7641-3_9

14. Cunningham, P., Delany, S.J.: K-nearest neighbour classifiers-a tutorial. ACM Comput. Surv. **54**(6), 1–25 (2021)

15. Safavian, S.R., Landgrebe, D.: A survey of decision tree classifier methodology. IEEE Trans. Syst. Man Cybern. **21**(3), 660–674 (1991). https://doi.org/10.1109/21.97458

16. Riedmiller, M., Lernen, A.: Multi layer perceptron. Mach. Learn. Lab Spec. Lect. Univ. Freibg., vol. 24 (2014)

17. Bayes, T.: Naive Bayes classifier. Artic. Sources Contrib., pp. 1–9 (1968)

18. Schapire, R.E.: Explaining AdaBoost. In: Schölkopf, B., Luo, Z., Vovk, V. (eds.) Empirical Inference: Festschrift in Honor of Vladimir N. Vapnik, pp. 37–52. Springer, Heidelberg (2013). https://doi.org/10.1007/978-3-642-41136-6_5

19. Skurichina, M., Duin, R.F.W.: Bagging for linear classifiers. Pattern Recogn. **31**(7), 909–930 (1998). https://doi.org/10.1016/S0031-3203(97)00110-6

20. Martinez-Gil, J.: A comprehensive review of stacking methods for semantic similarity measurement. Mach. Learn. Appl. **10**, 100423 (2022). https://doi.org/10.1016/j.mlwa.2022.100423

21. Kotsiantis, S.B.: Logitboost of simple Bayesian classifier. Informatica **29**(1) (2005)

22. Ambika. M., Raghuraman. G., SaiRamesh. L.: Enhanced decision support system to predict and prevent hypertension using computational intelligence techniques. Soft Comput., **24**(17) 13293–13304 (2020). https://doi.org/10.1007/s00500-020-04743-9

AgroProposal - A Deep Learning Guided Web Application For Crop Recommendation

Priyanka Pondugula[(✉)] [iD] and M. Rashmi

Department of Data Science and Computer Applications,
Manipal Institute of Technology, Manipal Academy of Higher Education,
Manipal 576104, Karnataka, India
`pondugula.priyanka2001@gmail.com`, `rashmi.m@manipal.edu`

Abstract. As agriculture grapples with the challenges of climate variability and soil heterogeneity, the need for precision farming has become increasingly critical. This study addresses the demand for accurate, data-driven crop recommendations by developing a deep learning-based web application designed to enhance decision-making among farmers. Utilizing three comprehensive Kaggle datasets that encompass diverse soil conditions, climate variables, and crop types, the research involved thorough data pre-processing and exploratory data analysis to extract meaningful insights. A wide spectrum of machine learning models—including Decision Trees, Random Forests, Naïve Bayes, Gradient Boosting Classifiers, and various neural network architectures—were systematically evaluated using 5-fold cross-validation. The custom ANN came forth as the most proficient one gaining awe-inspiring accuracy rates 94%, 98% and 96% across the 3 datasets after optimizing its architecture and features. ANN was then fused into a web application, which is designed using HTML, CSS, Javascript on the front-end and Flask on the server side. The successful development and deployment of this web application highlights the practical applicability of deep learning models in addressing complex agricultural challenges and pave the way for continued innovation in precision farming.

Keywords: Precision Farming · Deep Learning · Neural Network · Crop Recommendations · Machine Learning · Sustainable Agriculture

1 Introduction

Food security, rural livelihoods, and environmental sustainability are all somehow or the other linked to agriculture, which continues to be an important sector of the world economy. However, the agriculturalists deal with a lot of issues like crop failures and lower yields, which increases stress of farmers both financially and emotionally. With increased financial stress and crop failures there are a higher chances of farmer suicide, which needs an urgent solution. Additionally,

C. Modi et al. (Eds.): MIND 2024, CCIS 2736, pp. 438–449, 2026.
https://doi.org/10.1007/978-3-032-14531-4_37

this problem is made worse by farmers' restricted access to pertinent data and decision-making tools, which makes it challenging for them to align their crop selections based on environmental factors and market demands [1].

To solve the problems faced by farmers, the use of combination of web development with deep learning is the best upcoming solution. Neural networks with multiple layers are used in deep learning, a branch of artificial intelligence (AI), to analyze intricate data patterns. To get accurate crop recommendations in the agricultural context, deep learning (DL) models analyzes a variety of soil, environmental, and climatic data. This method helps us in reducing some of this stress and enhance growers' general well-being by offering data-driven insights and recommendations. The aspiration of this study is to fabricate a deep learning-based crop recommendation system to improve agricultural practices. The notable offerings of this work are summarized as follows.

- Using Deep learning techniques to analyze soil and environmental parameters for optimal crop suggestions, to enhance agricultural yield.
- Identification of the best model from ten different machine learning and deep learning methods through model ablation and hyperparameter tuning for accurate recommendations.
- An user-friendly web platform, facilitating easy interaction with the recommendation system for farmers and agricultural professionals.

The remaining sections of the paper are arranged as follows: A summary of the background theory is provided in Sect. 2. A detailed explanation of the methodology, covering data pretreatment, model selection, and web development, is given in Sect. 3. The results and analysis of the system are presented in Sect. 4, which also describes the efficacy of the web application and the behavior of machine learning (ML) techniques. Section 5 offers suggestions for future research.

2   Literature Review

With increasing pressures from climate change, resource limitations, and a rising population, the use of advanced technologies in agriculture is vital to improve crop selection and productivity. The recent achievements in the field of agriculture using deep learning and machine learning uses different data sources providing farmers precise, real-time guidance. This review investigates the contributions of recent research in the field, highlighting key developments and innovations in crop recommendation systems.

Priyadharshini et al.(2021) [2] proposed deeplearning models to enhance crop recommendation systems through the integration of real-time weather data. Their investigation implemented Recurrent Neural Networks (RNNs) and Convolutional Neural Networks (CNNs) to understand the complex interrelations between environmental variables, including crop yields and monthly weather patterns. The authors obtained superior accuracy in crop recommendations compared to traditional methodologies by utilizing an extensive dataset that encompassed historical weather data and soil conditions. However, the generalizability of their findings may be restricted by the reliance on specific geographic and climatic contexts.

Apat et al. (2023) [3] developed an AI-based vegetation recommendation system that combines both ML, and DL models. The system was designed to analyse a comprehensive set of environmental and soil parameters, using Deep Neural Networks (DNNs) and Generative Adversarial Networks (GANs) to predict optimal crop choices. As a part of the work, they generated a extensive datasets from various geographical regions, encompassing soil types, weather patterns, and historical crop yields.

In an innovative exploration, Mythili and Rangaraj (2021) [4] investigated the integration of deep learning with Particle Swarm Optimization (PSO) for crop recommendation. Their system combined historical crop data with deep learning models, concretely Deep Neural Networks (DNNs), to predict optimal crops for different environmental conditions. PSO was employed to optimize the hyperparameters of the deep learning models, lifting their performance. The research involved collecting and preprocessing large datasets, including soil characteristics, weather data, and historical crop yields. Sharma et al. (2021) [5] developed AI-Farm, a vegetation recommendation system that avails oneself of various machine learning algorithms, including deep learning techniques. The system integrated data from soil sensors and weather stations to analyze environmental conditions and predict crop suitability. Deep learning models, like DNNs and LSTMs, were employed to seize temporal and spatial patterns in the data.

Varshitha and Choudhary (2022) [6] focused on developing a DNN model for vegetation recommendation. Shams et al. (2024) [7] emerged XAI-CROP, a vegetation recommendation system that incorporates machine learning and explainable artificial intelligence (XAI) methods. Sharma et al. (2023) [8] focused on building an AI-enabled vegetation recommendation system that considers soil and weather patterns to provide tailored crop suggestions. The method integrated machine learning, CNN, and LSTM based DL models to analyze a variety of environmental factors and predict the most suitable crops.

From the literature survey it is clear that, recent advancements in crop recommendation systems have a shift towards integration of diverse AI techniques, deep learning models, ensemble methods, and optimization algorithms. That has led to significant improvements in recommendation accuracy, real-time data processing, and system robustness. As research continues to evolve, these advancements promise to support sustainable agricultural practices, optimize crop selection processes, and ultimately contribute to increased productivity and informed farm management decisions.

3   Methodology

The methodology in this research is designed for comprehensive analysis and robust modeling of the dataset, illustrated in two key figures. Figure 1 outlines the process, beginning with Exploratory Data Analysis (EDA) to understand the dataset's structure and identify key features. This is followed by data preprocessing to prepare for modeling. Followed by this, the process involve model selection

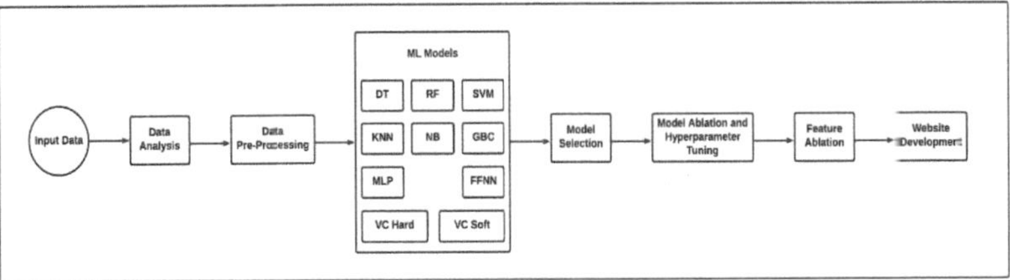

Fig. 1. Methodology of the work.

and evaluation, utilizing techniques such as cross-validation, hyper-parameter tuning, and feature ablation to get accurate model. Finally, the methodology leads to the creation of a web application capable of generating real-time forecasts, demonstrating it's practical relevance. Figure 2 illustrates the workflow of the final application based on neural network model. It starts with data preprocessing and moves to the implementation of the selected model, detailing how the network processes data and generates prediction scores during training. For real-time applications, the workflow includes user input through the web application, preprocessing that input, loading the fine-tuned model, and displaying recommendations based on predictions. This figure provides a clear overview of how the model operates, from initial data handling to providing useful recommendations.

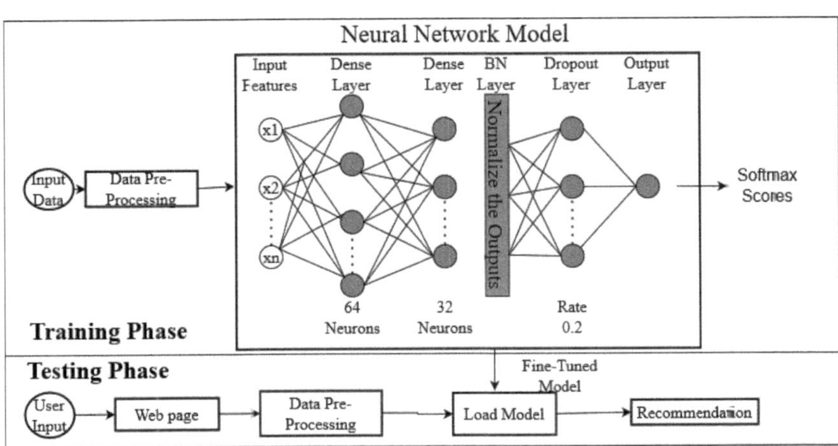

Fig. 2. Workflow of the process.

3.1 Data Analysis

The initial step in this research involved an extensive EDA, to understand the dataset's structure and content. Pie charts were used to visualize the distribution of crop classes, providing a clear picture of relative proportions. This visual representation allowed for comparisons of class frequencies, highlighting significant trends within the dataset.

3.2 Data Pre-processing and Splitting

As a part of data pre-processing we initially performed encoding to transform categorical variables to numerical values. The dataset was then normalized with MinMaxScaler to rescale feature values to a specified range, preventing any feature from disproportionately affecting the learning process, and also to enhance model performance and stability.

3.3 Model Selection

A five-fold cross-validation technique was used in the model selection phase. Each dataset was divided into five subsets, four of which served as training sets and the other as test sets. This process ensured that the model's performance remained constant across training and testing sets. A set of ten different machine learning models were evaluated to find out the best model. Because of its simplicity and convenience of use, a decision tree classifier with a maximum depth of 10 and an entropy criteria was employed. A Random Forest with 4 decision trees and depth 10 is used due its ability to work with high-dimensional data. A Naïve Bayes was also used for its computational efficiency. A Gradient Boosting Classifier with a learning rate of 0.1, 20 estimators, and a maximum depth of 3 was employed.

KNN with three nearest neighbors make it easy to obtain the features so it was selected. With a polynomial kernel of degree 3 and regularization parameter $C = 1$, SVM was selected due to its ability to deal with intricated datasets with well-separated classes. The MLP was created with two hidden layers with 64 and 32 neurons each. It was executed for 1000 iterations. Since learning hierarchical data representations is important for gaining high accuracy in multi-class classification problems, a custom ANN model was created. Last but not least, both hard and soft voting classifiers were selected for their ability to improve prediction efficiency by combining the advantages of different models. These classifiers for voting are created by using 3 machine learning models, naïve bayes, SVM and MLP to amplify the outcomes of the work.

3.4 Model Ablation and Hyperparameter Tuning

To enhance the efficacy of the model and reduce model complexities, ablation studies were performed. And also hyperparameter learning rate is also optimized to get better accuracies. The objective was to identify a balance between precision and computational efficiency.

3.5   Feature Ablation and Website Development

Feature ablation was conducted on the datasets with 7 input features in Dataset 1 and 2, there are a total of 127 possible combinations of features. Dataset 3 with 5 input features, 31 combinations can be made to discern the most impactful features, systematically gauging the influence of diverse feature combinations on model accuracy. Following this optimization, the final model was integrated into a web application, facilitating real-time predictions. The web application, constructed using HTML, CSS, and JavaScript for the front end and Flask for the back end, ensures seamless interaction and performance.

4   Experiments, Results and Analysis

This section delineates the methodologies and findings from a comprehensive study on crop type classification, employing three distinct datasets sourced from Kaggle. The chief aim was to build a reliable machine learning model that's able to deliver accurate predictions in classifying a variety of crop types, thereby providing valuable insights for agricultural decision-making and crop management.

4.1   Datasets Used

Three datasets were meticulously curated to represent a spectrum of agricultural scenarios. The first dataset [9] comprises 8,800 observations across 8 variables, including 7 numerical input features—nitrogen, potassium, phosphorus, humidity, temperature, rainfall, and pH—and a categorical output variable denoting 22 distinct crop types. The second dataset [10] mirrors the first but includes 2,200 observations and substitutes the crop "groundnut" with "apple". The third dataset [11] is smaller, containing 1,400 observations with 6 variables: 4 numerical input features (temperature, humidity, pH, and water availability) alongside a categorical feature representing the season (rainy, winter, spring, summer), categorizing 13 different crops. These datasets provide both a diverse and representative sample for creating and testing crop classification methodologies.

A rigorous data quality assessment was performed on each dataset to ensure the elimination of null values, duplicates, and outliers. Datasets 1 and 2 displayed a balanced class distribution, while Dataset 3 exhibited an imbalance, with 200 observations for "maize" and 100 for other crops. Figure 3 visually represents the class distributions across all three datasets through pie charts. To address the imbalance evident in Dataset 3, the Synthetic Minority Over-sampling Technique (SMOTE) was implemented. This technique artificially augmented the minority class instances, thereby amplifying the model's generalization spanning various classes.

4.2   Model Selection and Evaluation

The model selection process involved a comprehensive cross-validation of ten initial models, with the ANN emerging as the front runner in performance across

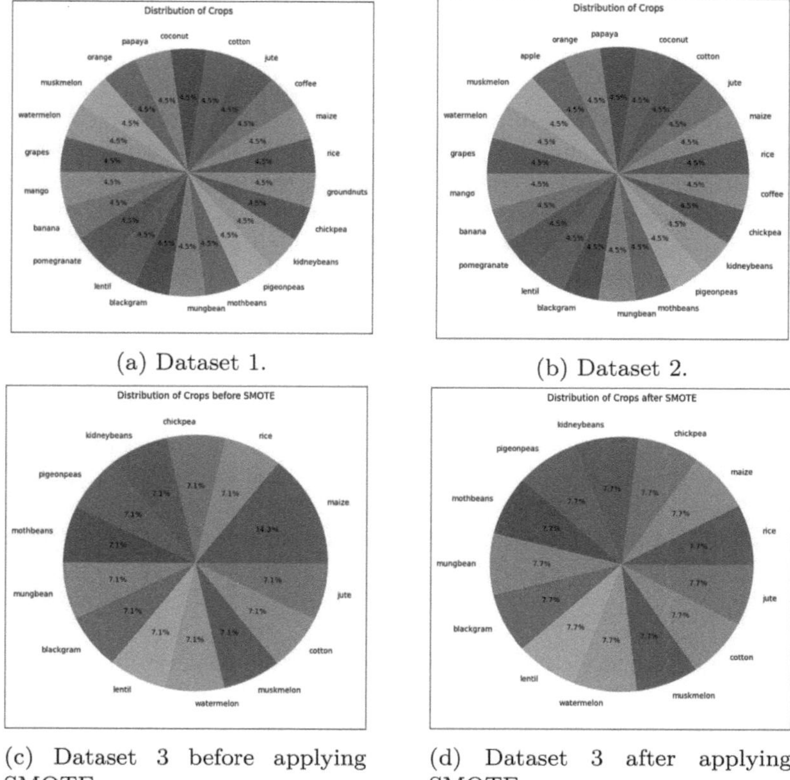

(a) Dataset 1.

(b) Dataset 2.

(c) Dataset 3 before applying SMOTE.

(d) Dataset 3 after applying SMOTE.

Fig. 3. Class distribution of crops.

all datasets. The Artificial Neural Network (ANN) is an excellent tool for making recommendations because it can capture very complex patterns in small datasets that simple models like decision trees or SVM might miss. Also, ANNs provides an advantage to optimize the architecture, hyperparameters and other aspects which allows further improvements in performance which is lacking in simpler models. Table 1 elucidates the cross-validation scores, showcasing the ANN's exceptional accuracy rates of 94.29%, 98.01%, and 96.83% for the respective datasets.

Although ANNs are less interpretable, the combination of flexibility, accuracy, and its strong generalization capabilities, adeptness at capturing complex data patterns rendered it the optimal choice. Initially, a sigmoid activation function was employed in the output layer; however, it was subsequently replaced with SoftMax during the ablation phase, enhancing multi-class classification by providing class probabilities and ensuring more precise predictions.

Table 1. Cross-Validation Scores for Various Machine Learning Models Across Datasets.

Model	Dataset 1	Dataset 2	Dataset 3
Naïve Bayes	93.80	99.60	93.31
Gradient Boosting Classifier	93.08	98.52	98.60
Decision Tree	92.45	98.46	98.55
Random Forest	92.92	97.61	98.41
KNN	92.45	98.35	98.50
SVM	93.94	98.57	98.46
VC – Hard	93.69	98.86	98.46
VC – Soft	94.41	99.37	97.98
Multi- Layer Perceptron	93.57	98.01	98.12
ANN	94.29	98.01	96.83

4.3   Model Ablation and Hyperparameter Tuning

Subsequent ablation studies were conducted to refine the model's architecture. Table 2 delineates the accuracies obtained from various model architectures evaluated during the model ablation and hyperparameter tuning phase. Model 2, characterized by two dense layers comprising 64 and 32 neurons respectively, accompanied by batch normalization and a dropout layer, exhibited consistent performance, particularly with a learning rate of 0.01. Model 2's overall robustness and performance justify the reason for its selection as the final model.

Table 2. Accuracies of ANN Model with Different Architectures and Learning Rates.

Learning Rates	Dataset 1			Dataset 2			Dataset 3		
	Model 1	Model 2	Model 3	Model 1	Model 2	Model 3	Model 1	Model 2	Model 3
0.0001	94.11	93.82	94.03	95.45	91.19	90.62	97.36	93.99	90.87
0.001	95.10	94.96	95.10	99.43	99.43	98.86	99.52	99.52	99.52
0.01	94.82	95.17	93.82	99.43	99.43	98.30	99.28	99.52	99.52
0.1	92.26	93.47	91.69	90.34	96.88	98.01	97.12	98.80	97.84

4.4   Feature Ablation

In the possible combinations, during the ablation study we found out that the feature set without temperature for datasets 1 and 2, feature set without ph for dataset 3 resulted in better performance of model. While other combinations exhibited slightly better performance, we prioritized these subsets due to their practical relevance and simplicity. The final model trained on the optimized feature sets, demonstrated high accuracy and reliability, as corroborated by training curves and evaluation metrics. Figure 4 present the confusion matrices illustrating the distribution of classification errors. Additionally, Table 3 summarizes the evaluation metrics of the model across all datasets, providing a clear performance comparison.

The high values of precision and recall in the table shows that the model is effective in both properly identifying positive instances and reducing false negatives. This observation can be confirmed by the F1 scores which indicated a balanced performance in both precision and recall. Even though the error rates and test losses are low, the analysis of confusion matrices revealed that crops with similar visual characteristics have occasional misclassification, such as groundnuts and mothbeans, jute and rice were frequently confused. The overall model performance remains robust, and these misclassifications provide valuable insight into areas where the model could be further refined which says fine-tuning may be required to handle subtle differences between certain crop types more effectively. All the experiments are conducted on the GPU infrastructure provided by Google Cloud system. We also tested on a set of test samples and it is observed that on an average it takes about 70 milliseconds to predict and recommend the crops, which makes it suitable for real-world application. This makes our model computationally efficient and feasible for deploying in settings with limited computational resources.

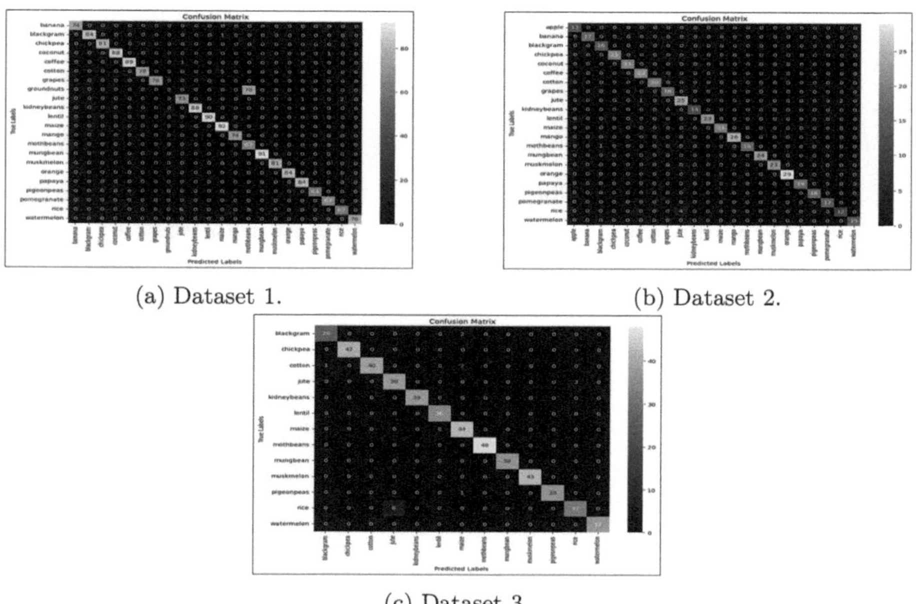

(a) Dataset 1. (b) Dataset 2.

(c) Dataset 3.

Fig. 4. Confusion matrices for the final model on three datasets.

Table 3. Evaluation Metrics of the Final Model After Hyperparameter Tuning and Ablation Studies Across Datasets.

Datasets	Accuracy	Precision	Recall	F1 Score	Error Rate	Test Loss
Dataset 1	94.48	92.56	94.48	93.12	5.51	9.25
Dataset 2	98.18	98.17	98.18	98.15	1.81	3.48
Dataset 3	96.73	96.90	96.73	96.73	3.26	9.90

4.5 Comparative Analysis with Existing Works

Our deep learning-based crop recommendation system, utilizing a Artificial Neural Network (ANN) model, achieved a remarkable validation accuracy of 98.18% on Dataset 2 [10], which is publicly available on Kaggle ("Crop Recommendation Dataset"). This dataset has been widely used in prior studies, making it a reliable basis for comparison. For instance, Patel et al. (2023) [12] reported an accuracy of 99.5% using Decision Trees, while Garg and Alam (2023) [13] achieved 99.31% with the PART algorithm. Thilakarathne et al. (2022) [14] reported an accuracy of 97.18% using Random Forest, Islam et al. (2023) [15] reached 97.5% with CatBoost, and Gopi and Karthikeyan (2023) [16] introduced the MMML-CRYP technique with an accuracy of 97.91%. Table 4 illustrates the comparative analysis of these existing works with our proposed approach. Despite the slightly higher reported accuracies of these traditional methods, our ANN model's performance of 98.18% sets a strong benchmark for deep learning-based approaches in crop recommendation systems.

Table 4. Accuracy Comparison of Proposed Approach with Existing Works on Dataset 2.

Existing Works	Accuracy
Patel et al. Decision Trees [12]	99.5%
Garg and Alam PART algorithm [13]	99.31%
Thilakarathne et al. Random Forest [14]	97.18%
Islam et al. CatBoost [15]	97.5%
Gopi and Karthikeyan MMML-CRYP [16]	97.91%
Proposed Approach	98.18%

4.6 Website Development

Figure 5 depicts the website with two recommendation forms - nutrients form and climate form which provide crop suggestions based on soil nutrients, climate conditions respectively. The nutrients form is integrated with the ANN model trained on Dataset 2 for recommending crops based on soil nutrients. Similarly the climate form is integrated with the model trained on Dataset 3 for recommending based on climate conditions. The website is made responsive for mobile screens with a minimum width of 520px and ensures that there is a smooth interaction with the webpage by accepting both numeric and categorical values. While the site performs well under typical use cases, scalability may be limited during high traffic, and mobile performance may vary based on device specifications. Future improvements can be focused on enhancing scalability.

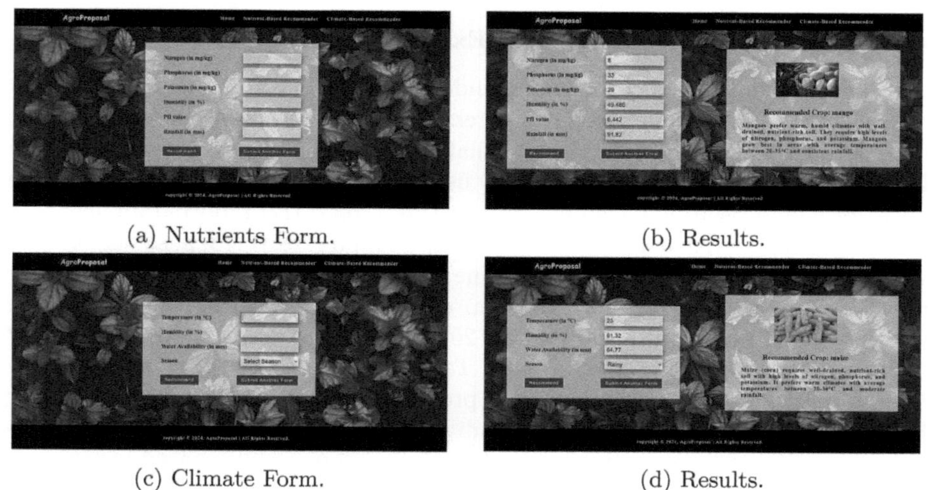

(a) Nutrients Form.

(b) Results.

(c) Climate Form.

(d) Results.

Fig. 5. Images of the Website.

5 Conclusion and Future Scope

This study reveals the transformative potential of deep learning in offering precise crop recommendations through a user-friendly web application. In diverse datasets, the research identifies the ANN as the most effective model, demonstrating impressive accuracy of more than 94% makes it suitable for real-time applications. The careful selection of key features further boosted the model's performance, allowing farmers to make informed decisions that enhance yields and reduce losses. The successful implementation of this technology highlights its practical benefits in agriculture.

Expanding the application to cover more crops and regions will improve its relevance in various agricultural contexts. Additionally, enhancing the model's transparency will help users understand the basis for recommendations, fostering greater trust. Integrating the system with agricultural advisory services can provide farmers with valuable insights and resources.

References

1. Doshi, Z., Nadkarni, S., Agrawal, R., Shah, N.: AgroConsultant: intelligent crop recommendation system using machine learning algorithms. In: 2018 Fourth International Conference on Computing Communication Control and Automation (ICCUBEA), pp. 1–6. IEEE (2018)
2. Priyadharshini, A., Chakraborty, S., Kumar, A., Pooniwala, O.R.: Intelligent crop recommendation system using machine learning. In: 2021 5th International Conference on Computing Methodologies and Communication (ICCMC), pp. 843–848. IEEE (2021)
3. Apat, S.K., Mishra, J., Raju, K.S., Padhy, N.: An artificial intelligence-based crop recommendation system using machine learning. J. Sci. Ind. Res. (JSIR) **82**(05), 558–567 (2023)
4. Mythili, K., Rangaraj, R.: Deep learning with particle swarm based hyperparameter tuning based crop recommendation for better crop yield for precision agriculture. Indian J. Sci. Technol. **14**(17), 1325–1337 (2021)
5. Sharma, A., Bhargava, M., Khanna, A.V.: AI-farm: a crop recommendation system. In: 2021 International Conference on Advances in Computing and Communications (ICACC), pp. 1–7. IEEE (2021)
6. Varshitha, D.N., Choudhary, S.: An artificial intelligence solution for crop recommendation. Indonesian J. Electr. Eng. Comput. Sci. **25**(3), 1688–1695 (2022)
7. Shams, M.Y., Gamel, S.A., Talaat, F.M.: Enhancing crop recommendation systems with explainable artificial intelligence: a study on agricultural decision-making. Neural Comput. Appl. **36**(11), 5695–5714 (2024)
8. Sharma, P., Dadheech P., Senthil, A.S.K.: AI-enabled crop recommendation system based on soil and weather patterns. In: Artificial Intelligence Tools and Technologies for Smart Farming and Agriculture Practices, pp. 184–199. IGI Global (2023)
9. Crop Recommendation. https://www.Kaggle.com/datasets/jiteshsharma45/crop-recommendation. Accessed 06 June 2024
10. Crop Recommendation Dataset. https://www.Kaggle.com/datasets/atharvaingle/crop-recommendation-dataset. Accessed 08 June 2024
11. Crop Prediction Dataset. https://www.Kaggle.com/datasets/rishabhrathore055/datas. Accessed 10 June 2024
12. Patel, M., Rane, A.: Crop Recommendation System (2023). https://doi.org/10.31224/2959
13. Garg, D., Alam, M.: An effective crop recommendation method using machine learning techniques. Int. J. Adv. Technol. Eng. Explor. **10**(102), 498 (2023)
14. Thilakarathne, N.N., Bakar, M.S.A., Abas, P.E., Yassin, H.: A cloud enabled crop recommendation platform for machine learning-driven precision farming. Sensors **22**(16), 6299 (2022)
15. Islam, M.R., Oliullah, K., Kabir, M.M., Alom, M., Mridha, M.F.: Machine learning enabled IoT system for soil nutrients monitoring and crop recommendation. J. Agric. Food Res. **14**, 100880 (2023)
16. Gopi, P.S.S., Karthikeyan, M.: Multimodal machine learning based crop recommendation and yield prediction model. Intell. Autom. Soft Comput. **36**(1), 313–326 (2023)

Comparative Analysis of Machine Learning Algorithms for Water Region Classification in Flood-Affected Images

M. N. Renukadevi[✉], Kavyashree I. Pattan, and T. M. Rajesh

Dayananda Sagar University, Bangalore, Karnataka, India
{renukadevi.m-cse,kavyashreepattan-cse,rajesh-cse}@dsu.edu.in

Abstract. To classify water regions in satellite imagery of flood-affected areas, our study compares three machine learning algorithms: Support Vector Machine (SVM), Random Forest, and K-means clustering with our proposed methodology which categorizes water, buildings, and vegetation areas and categorizes as flooded and non-flooded images. Accuracy, precision, recall, and F1 score were used to assess the algorithms, with NDWI (Normalized Difference Water Index) serving as a crucial component. To make feature extraction easier, a sample dataset of flood photos was processed, which included image transformation and resizing. Training data from Kaggle (Flood Area Segmentation section) was randomly sampled and tested to improve computing performance. When it came to recognizing water zones, the Random Forest model performed better than SVM, with an accuracy of 99%. K-Means clustering offered a baseline for comparison at 94%. The findings demonstrate how sophisticated machine learning methods may be used to effectively classify water bodies, enhancing flood monitoring and control.

Keywords: NDWI · Support Vector Machine (SVM) · Random Forest · K-Means clustering

1 Introduction

The ability to identify floods is essential to disaster management because it allows response agencies to act quickly and lessen the effects of flooding. Aerial and satellite photos are essential for tracking and evaluating floodwater levels. In our study, we identify flooded zones in aerial photos by applying a combination of segmentation, connected component analysis, and image preprocessing. The method consists of boosting image contrast, using morphological operations for noise reduction, segmenting water areas using Otsu's thresholding, and then identifying and emphasizing possible flood locations based on their area. In today's environment monitoring and catastrophe management systems, the use of aerial photography for flood detection is essential. Authorities can react to flooding incidents more quickly and evaluate their severity when they employ computer vision techniques to identify flooded areas.

Our research highlights specific flooded locations in aerial images with an automated flood detection approach that uses picture preprocessing, segmentation, and region-based

C. Modi et al. (Eds.): MIND 2024, CCIS 2736, pp. 450–462, 2026.
https://doi.org/10.1007/978-3-032-14531-4_38

analysis as vegetation region, building and water regions. We are utilizing three methods (Support vector machine learning (SVM), K-means and Random Forest) to perform a comparative analysis to separate the water region and other regions in the given image. Real-time flood monitoring, early warning systems, post-disaster damage assessments, and resource allocation during emergency operations can all benefit from this type of flood detection. The ability to estimate the level of flooding and show affected areas helps authorities prioritize rescue efforts and deploy resources effectively.

2 Literature Survey

Flooding is made worse by urbanization and climate change, underscoring the necessity of efficient monitoring and response strategies. Even in bad weather, Synthetic Aperture Radar (SAR) allows for continuous Earth photography. Flood mapping with SAR may benefit from data fusion and machine learning, while there are still issues in places with a lot of people and vegetation. For precise flood monitoring and response, better validation techniques and classification algorithms are essential [1].

Deep learning-based flood detection has drawn interest because of its potential to enable prompt emergency actions [2]. Convolutional neural networks (CNNs) with attention mechanisms are one of the models that have been used to identify and segment flood zones, and they have shown excellent accuracy and precision in doing so Metrics like Mean IOU and Dice Coefficient are used to assess segmentation accuracy. Promising results have also been shown by segmentation models like U-Net, which frequently have a Mobile Net backbone.

In flood-prone areas, [3] like Ethiopia's Awash Basin, machine learning models including Random Forest (RF), Support Vector Machine (SVM), linear regression, and Long Short-Term Memory (LSTM) are commonly utilized. These models are essential for flood prediction and susceptibility mapping. Because it can capture intricate, nonlinear interactions between environmental variables including rainfall, slope, aspect, and elevation, Random Forest (RF) consistently performs better than the others.

Recent advancements in machine learning (ML) for network traffic analysis have significantly enhanced security and reliability Supervised models like Random Forest, SVM, Logistic Regression, and KNN are widely employed for anomaly detection and classification, including DDoS attacks and HTTP/3 flood assaults in QUIC-based communication. ML aids in real-time traffic classification and anomaly detection, optimizing network performance and mitigating risks [4].

Satellite-based flood monitoring is improved by developments in machine learning and remote sensing, such as the Progressive Image Classification Algorithm (PICA). Accurate flood estimation and water body detection are enhanced by using Big Data with sophisticated classification methods. By lowering false positives, this collaboration improves catastrophe detection and response capabilities [5].

For safety and planning, urban flood detection—which is typically dependent on hydrological models and ground data—is essential. Accuracy is increased when Sentinel-1 satellite imagery is combined with machine learning models such as Random Forest and SVM. By lowering data dependency, unsupervised techniques like iso-clustering and change detection using Otsu's algorithm increase efficiency. However, supervised models show great promise for accurate flood detection [6].

3 Methodology

To improve flood detection and better differentiate between clouds and water, we have applied some additional steps in the image preprocessing pipeline using Fig. 1. Here are a few algorithms and techniques that can help us to assist in categorizing flooded and non-flooded images.

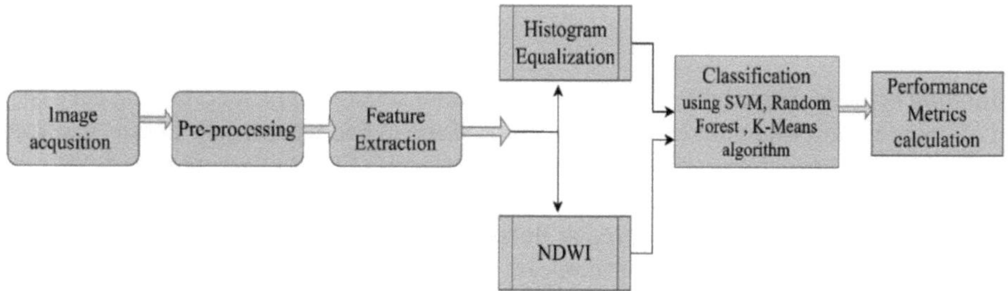

Fig. 1. Flow diagram for Flood Detection using Machine Learning algorithms.

3.1 Common Steps of Pre-processing for Flood Detection Using Machine Learning Algorithms

Use Color Information: Water and clouds may have different color properties in satellite imagery. Clouds are usually brighter and more reflective, while water bodies are darker. You can make use of color channels, like the blue and near-infrared (NIR) channels, to distinguish between clouds and water. We can use Blue Channel or NDWI (Normalized Difference Water Index) to enhance the detection of water areas and suppress cloud detection.

Normalized Difference Water Index (NDWI) and Cloud Detection, Masking: NDWI is calculated using the green and near-infrared channels. Water bodies reflect more in the green band and absorb more in the NIR band. You can use NDWI to segment water more effectively. Clouds tend to be bright in satellite imagery. To differentiate clouds from water, we can apply a brightness threshold to mask out the brightest regions before flood detection.

The general image processing steps for segmentation of flooded images.

Step 1: Image Acquisition and Preprocessing
Input image represented as mentioned in Eq. 1

$$I = [I_R, I_G, I_B] \tag{1}$$

where I_R, I_G, I_B *ARE RED GREEN BLUE COMPONENTS OF AN IMAGE*

An aerial image is loaded and some basic preprocessing is done in the first stage. It is expected that the image is in RGB format. Given an input RGB image, we denote the image channels as mentioned in Eq. 2. Let R(x,y) G(x,y) and B(x,y) represent the Red, Green, and Blue intensity values for pixel (x,y) in the image, respectively.

The goal is to extract these channels.

$$R(x, y) = I(x, y, 1), G(x, y) = I(x, y, 2), B(x, y) = I(x, y, 3) \qquad (2)$$

where I(x, y, 1), I(x, y, 2), and I(x, y, 3) are the three color channels of the image. The image's Red, Green, and Blue color channels are extracted. These color bands are then merged with NDWI values to produce feature vectors, and they are also utilized to calculate NDWI.

Step 2. Extraction of Features Making use of NDWI

To draw attention to the water areas in the picture, utilize the Normalized Difference Water Index (NDWI). It is determined by taking the normalized difference between the image's green and blue channels as mentioned in Eq. 3

$$NDWI(x, y) = \frac{G(x, y) - B(x, y)}{G(x, y) + B(x, y)} \qquad (3)$$

where G and B are the green and blue channels of the picture. This index facilitates the differentiation of water features from other land characteristics. Where water pixels typically have positive NDWI values, and non-water pixels have negative NDWI values. The NDWI values are transformed into a vector to be used as input for classification. Additionally, the RGB values are reshaped and concatenated with NDWI values to generate the feature vector for the SVM classifier.

Step 3: Ground Truth Mask Preparation

The classifier is trained using a ground truth mask that represents locations that are not covered by water. To classify water regions as 1 and non-water regions as 0, the ground truth mask is loaded, gray-scaled, and binarized. The ground truth mask M(x,y) is a binary image where,

$$M(x, y) = \begin{cases} 1, & \textit{if pixel } (x, y) \textit{ is water} \\ 0, & \textit{if pixel } (x, y) \textit{ is non} - \textit{water} \end{cases} \qquad (4)$$

This binary mask is used to train the classifier as mentioned in Eq. 4. It is flattened into a vector M_{Vector} to match the feature vectors.

3.2   Comparative Analysis of Machine Learning Algorithms for Flood Detection.

In this section, we explain machine learning algorithms in detail.

3.2.1   SVM Classifier Training

This step offers a thorough process for identifying water regions in aerial imagery by employing the Normalized Difference Water Index (NDWI) and Support Vector Machines (SVM) as features. The method divides areas with and without water using machine learning and image processing techniques. The procedure is composed of multiple crucial phases, such as preprocessing images, extracting features, training classifiers, and assessing performance [3].

The feature vectors (NDWI and RGB values) are used as input for an SVM classifier, and the labels are the associated ground truth mask. A linear kernel is employed for the SVM model due to the simplicity and interpretability of this kernel for two-class classification. The feature vector for each pixel is constructed by concatenating the NDWI value and the RGB values as mentioned in Eq. 5.

$$X(x, y) = \begin{bmatrix} NDWII(x, y) \\ R(x, y) \\ G(x, y) \\ B(x, y) \end{bmatrix} \tag{5}$$

Let X represent the matrix of all feature vectors, and let Y represent the corresponding labels from the ground truth mask. The classifier is trained using a linear SVM: The decision boundary for a linear SVM is given by Eq. 6

$$F(x) = w\,x + b \tag{6}$$

where W is the weight vector, b is the bias term and f(x)is the decision function The classification rule as mentioned in Eq. 7.

$$\hat{y}(x, y) = \begin{cases} 1, & \text{if } f(X(x, y))) > 0 \\ 0, & \text{if } f(X(x, y)) \le 0 \end{cases} \tag{7}$$

where $\hat{y}$ (x, y) is the predicted label for pixel(x, y), indicating whether the pixel is classified as water (1) or non-water (0).

Following are the steps for predicting water region, performing post-processing, and visualization.

Step 1: Water Region Prediction
The trained SVM model is used to predict water regions in the image. The predicted labels (water or non-water) are reshaped back to match the original image dimensions.

The predicted Labels $\widehat{M}$ (x, y) = $\hat{y}$ (x, y). These labels are reshaped to match the original image dimensions to form the predicted water mask.

Step 2: Post-processing (Morphological Operations)
To clean up the predicted water mask, morphological operations are applied to remove small objects that may be noisy and Fill holes in the predicted water regions to obtain more accurate results [7]. Two key morphological operations are applied to clean up the predicted water mask. Area opening to remove small regions of noise as mentioned in Eq. 8 and Hole filling to fill any gaps inside water regions as mentioned in Eq. 9

$$\widehat{M}_{cleaned}(x, y) = bwareaopen(\widehat{M}(x, y), \tau) \tag{8}$$

where τ *minimum number of connected pixel that define an object*

$$\widehat{M}_{final}(x, y) = imfill(\widehat{M}_{cleaned}(x, y), 'holes') \tag{9}$$

This operation ensures that any internal holes in the water regions are filled.

Step 3: Visualization

The classified water regions are highlighted in the image. Water regions are colored red, and non-water regions are colored blue. This provides a visual representation of the detected water bodies overlaid on the original image as mentioned in Eq. 10

$$\text{Red channel} = \widehat{M}_{final}(x, y)$$
$$\text{Blue channel} = 1 - \widehat{M}_{final}(x, y) \tag{10}$$

3.2.2 Random Forest

The random Forest ensemble learning technique uses many decision trees. Each decision tree is trained using a randomized selection of characteristics and pixels. The majority vote from each tree determines the Random Forest's output.

Training Data: We train the Random Forest using labeled data where,

$X = \{F1, F2, ..., Fn\}$ is the feature set extracted from pixels or regions.
$Y = \{y1, y2, ..., yn\}$ are the class labels (flooded or non-flooded).

The Random Forest Model is explained as mentioned in Eq. 11.

$$\hat{y}(x, y) = \text{Majority vote}(h_1(F), h_2(F), \ldots \ldots, h_T(F)) \tag{11}$$

where $h_i(F)$ is the i-th tree's decision, T being the total number of trees.

3.2.3 Using NDWI and K-means Clustering for Flood Detection

This methodology outlines an automated flood detection method based on clustering techniques, namely K-means clustering and the Normalized Difference Water Index (NDWI). The method recognizes and highlights bodies of water in aerial photos and assesses accuracy by comparing the results with data collected on land. With this approach, our goal is to identify water regions (possible flood zones) in aerial photos. These findings are then validated through the use of accuracy metrics grounded on real-world data [8].

Following are the steps for K-means Grouping and water mask creation.

Step 1: K-means Grouping

Pixels are classified as either water or non-water using the unsupervised learning technique K-means clustering. K-means seeks to reduce the variation within a cluster. This is accomplished mathematically by repeatedly minimizing the subsequent objective function as mentioned in Eq. 12.

$$J = \sum_{k=0}^{K} k \sum_{x \in C_k} ||x - \mu||^2 \tag{12}$$

where K is the number of clusters, x is a data point (NDWI and RGB values), C_k is the set of points assigned to cluster k, and μ_k is the centroid of cluster k. In our case, K = 2 is used to classify pixels as water or non-water.

Step 2 Water Mask Creation

Once the clustering is performed, the resulting clusters are reshaped back into the image's original dimensions. One of the clusters corresponds to water pixels, and the other corresponds to non-water pixels. The water mask is as mentioned in Eq. 13.

$$\text{Water Mask} = \begin{cases} 1 \ if \ Cluster = Water \\ 0 \ otherwise \end{cases} \tag{13}$$

3.3 Performance Evaluation

The accuracy of the water detection method is assessed using standard classification metrics. From these values, the following metrics are calculated.

Accuracy: The proportion of correctly classified pixels (both water and non-water):

$$\text{Accurarcy} = \frac{TP + TN}{TP + TN + FP + FN} \tag{14}$$

Precision: The proportion of true positive water pixels among all pixels predicted as water:

$$\text{Precision} = \frac{TP}{TP + FP} \tag{15}$$

Recall: The proportion of true positive water pixels among all actual water pixels:

$$\text{Recall} = \frac{TP}{TP + FN} \tag{16}$$

F1 Score: The harmonic mean of precision and recall:

$$\text{F1} = 2 \times \frac{PRECISION \times RECALL}{PRECISION + RECALL} \tag{17}$$

These metrics provide a quantitative assessment of the classifier's performance in detecting water regions as mentioned in the above equations from 14–17 [9].

For performance evaluation, four key metrics are calculated based on the ground truth mask M(x, y) and predicted water mask $\widehat{M}$(x, y) as mentioned in Eq. 18.

TP: True positive

$$TP = \sum_{x,y} 1\{M(x, y) = 1 \ and \ \widehat{M}(x, y) = 1\}$$

TN: True Negative

$$TN = \sum_{x,y} 1\{M(x, y) = 0 \ and \ \widehat{M}(x, y) = 0\}$$

FP: False Positive:

$$FP = \sum_{x,y} 1\{M(x,y) = 0 \text{ } and \text{ } \widehat{M}(x,y) = 1\}$$

FN: False Negative:

$$FN = \sum_{x,y} 1\{M(x,y) = 1 \text{ } and \text{ } \widehat{M}(x,y) = 0\} \qquad (18)$$

The comparative study employs an SVM, Random Forest, and K means classifier trained on NDWI and RGB features to detect water regions in satellite imagery. The performance is evaluated using accuracy, precision, recall, and F1 score, providing insight into the effectiveness of the model. This approach can be extended to other satellite images for automated water body detection, especially in flood monitoring and environmental assessment scenarios.

3.4 Proposed Methodology

Our method for identifying flooded and non-flooded regions in aerial images is mathematically grounded in K-means clustering and color-based classification. Given an image I with n pixels, each pixel $P_i = [R_i, G_i, B_i]$ is treated as a point in 3D space. K-means clustering partitions these pixels into three clusters: vegetation, buildings, and water, by minimizing the sum of squared distances between each pixel and its assigned cluster center. The optimization problem solved by K-means is expressed as in Eq. 19.

$$J = \sum_{k=1}^{3} \sum_{p_i \in C_k} ||P_{i-\mu_k}||^2, \qquad (19)$$

where c_k denotes a cluster and μ_k is its centroid.

After clustering, each cluster's mean color intensity values $[R_k, G_k, B_k]$ are used to classify regions: water is assigned to the cluster with the lowest blue intensity, buildings are identified by high red and low green intensity, and the remaining cluster is designated as vegetation.

Following classification, binary masks are generated for each region, where $M_{water}(i,j) = 1$. If pixel (i, j) belongs to the water cluster and 0 otherwise.

Similarly, $M_{buildings}(i,j) and M_{vegetation}(i,j)$ are binary masks for other regions.

The water region percentage is computed as in Eq. 20.

$$Wate\_percentage = \frac{\sum_{i,j} M_{water}(i,j)}{n} \times 100, \text{ n total number of pixel} \qquad (20)$$

A threshold τ determines whether the region is considered flooded:

if Water_percentage $> = \tau$, The area is marked as flooded.

For the final overlay, each pixel is color-coded based on its classification as mentioned in Eq. 21.

$$Green : M_{vegetation}(i,j) = 1$$

$$\text{Red} : M_{buildings}(i,j) = 1$$
$$\text{Blue} : M_{water}(i,j) = 1 \tag{21}$$

An overlay is created by color-coding the masks, assigning green to vegetation, red to buildings, and blue to water. This ensures a clear visual interpretation, supporting effective flood assessment and decision-making in disaster management applications. The K-means clustering technique serves as the mathematical foundation for the methodology, which is then followed by color-channel-based classification. Pixel-by-pixel classification and accurate water percentage computations are made possible by binary masks.

The water percentage is thresholded to identify floods. Easy interpretation is guaranteed by the overlay visualization. With the ability to be expanded into real-time flood assessment and connection with satellite data, this system provides a strong basis for automated flood detection.

Dataset

The Flood image data set was taken from the Kaggle (Flood Area Segmentation) section. We have tested for more than 100 images for categorizing flooded images.

4   Results and Discussion

4.1   Results of Flood Detection Using the SVM, Random Forest, and K-means Methods

Figure 2a is the Original image with a flood, and Fig. 2b represents NDWI (Normalized Difference Water Index),a satellite image index that tracks variations in water content and identifies open water features:

Figure 2c highlights the predicted water mask as a Binary image. Figures 2d and 2e identify the cleaned water mask after applying morphological operations. Figure 2f highlights the red channel for water and the blue channel for other regions using SVM method.

Fig. 2. Results of the SVM method.

Figure 3a is the Original image with a flood, and Fig. 3b represents NDWI (Normalized Difference Water Index), an aerial image index that tracks variations in water content and identifies open water features:

Figure 3c highlights the predicted water mask as a Binary image. Figures 3d and 3e identify the cleaned water mask after applying morphological operations. Figure 3f highlights the red channel for water and the blue channel for other regions using the.

Random Forest method.

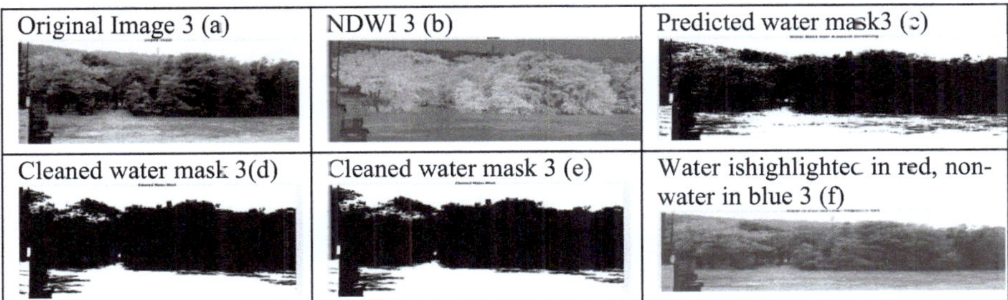

Fig. 3. Results of the Random Forest method

Figure 4a is the Original image with a flood, and Fig. 4b represents NDWI (Normalized Difference Water Index), an aerial image index that tracks variations in water content and identifies open water features:

Figure 4c highlights the predicted water mask as a Binary image. Figures 4d and 4e identify the cleaned water mask after applying morphological operations. Figure 3f highlights the red channel for water and the blue channel for other regions using the Random Forest method. Table 1 and Fig. 5 gives the performance metrics calculation.

Fig. 4. Results of the Random Forest method.

4.2 Results of Proposed Methodology

Our proposed methodology uses image segmentation and classification techniques to identify areas in aerial photographs that are flooded and those that are not.

Table 1. Performance Metrics calculation

Methods	Accuracy	Precision	Recall	F1 score
'SVM'	97.887	0.99	0.99	0.99
'Random Forest'	99.99	0.99	0.99	0.99
'K-Means'	94.859	0.99	0.91	0.95
K-means clustering and color-based classification	99.99	0.99	0.98	0.97

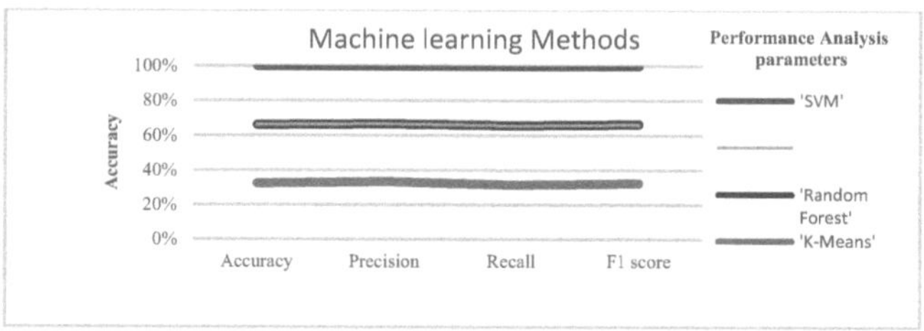

Fig. 5. Performance Analysis parameters.

To facilitate interpretation and offer insights into flood vulnerability, the results are graphically depicted in an overlay image with color-coded regions: blue for water, red for structures, and green for vegetation. This methodology provides a basic approach for automated flood detection and shows potential applications in urban planning and catastrophe management.

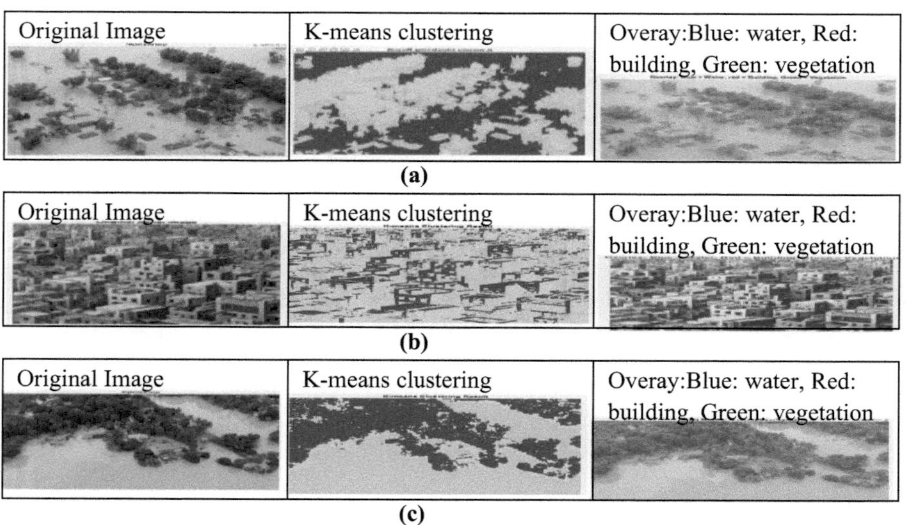

Fig. 6. (a) Results of flooded image (b) Results of non-flooded image (c) Results of flooded images.

In our research work, we are considering only aerial images for implementation. Further, we could improvise the accuracy using satellite images.

Figure 6(a) and 6(c) are the results of flood detection by classifying water, vegetation, and building regions. Figure 6(b) indicates results for non-flooded regions (Table 2).

Table 2. Results of the Proposed Methodology

Figure 6a) Results of flooded image	Water Region: 32.56%	Building Region: 44.45%	Vegetation Region: 22.99%	The image is classified as Flooded
Figure 6b) Results of non-flooded image	No Water Region	Building Region: 98%	No Vegetation Region	The image is classified as Non-Flooded
Figure 6c) Results of flooded images	Water Region: 40.97%	Building Region: 44.88%	Vegetation Region: 14.15%	The image is classified as Flooded

The K-means clustering stage dominates the algorithm's temporal complexity, which is O(N), where N is the number of pixels in the scaled image. It is ideal for large-scale flood analysis utilizing satellite data because of its linear scalability, which guarantees efficiency.

5 Conclusion

The results of our study highlight how well machine learning techniques work for classifying water zones from aerial imagery, especially when it comes to flood analysis. The Random Forest classifier demonstrated its resilience in managing intricate patterns in aerial data by outperforming the other approaches in terms of accuracy and F1 score. Additionally, the SVM classifier performed well, confirming its applicability for this kind of analysis. K-Means clustering offered a more straightforward baseline method, but its poorer performance metrics show that more complex algorithms are required for higher classification accuracy. Our proposed methodology identifies flooded and non-flooded regions in aerial images using K-means clustering and color-based classification. The method presented here lays the foundation for creating more efficient flood monitoring systems, which could greatly help in disaster response and management efforts. This study adds to the expanding body of knowledge on the use of machine learning in environmental monitoring by recommending lines of inquiry for future research, such as the incorporation of additional spectral features and the use of deep learning techniques to further improve classification accuracy.

Acknowledgment. We would like to thank Dayananda Sagar University, Bangalore for providing us with the opportunity to carry out our research work.

References

1. Mansour, R.F., Alabdulkreem, E.: Disaster monitoring of satellite image processing using progressive image classification. Comput. Syst. Sci. Eng. **44**(2) (2023)
2. Amitrano, D., et al.: Flood detection with SAR: a review of techniques and datasets. Remote Sens. **16**(4), 656 (2024)
3. Tanim, A.H., et al.: Flood detection in urban areas using satellite imagery and machine learning. Water **14**(7), 1140 (2022)
4. Santhi, V., Krishnamurthi, I., Madhumitha, N.H.: Flood detection and segmentation using deep learning models. In: 2024 International Conference on Smart Systems for Electrical, Electronics, Communication and Computer Engineering (ICSSEECC). IEEE (2024)
5. Wedajo, G.K., et al.: Integrating satellite images and machine learning for flood prediction and susceptibility mapping for the case of Amibara, Awash Basin, Ethiopia. Remote Sens. **16**(12), 2163 (2024)
6. Kadi, A., et al.: Machine learning for QUIC traffic flood detection. In: 2024 Global Information Infrastructure and Networking Symposium (GIIS). IEEE (2024)
7. Shahabi, H., et al.: Flood detection and susceptibility mapping using sentinel-1 remote sensing data and a machine learning approach: hybrid intelligence of bagging ensemble based on k-nearest neighbor classifier. Remote Sens. **12**(2), 266 (2020)
8. Munawar, H.S., Hammad, A.W.A., Waller, S.T.: A review on flood management technologies related to image processing and machine learning. Autom. Construct. **132**, 103916 (2021)
9. Sharifi, A.: Development of a method for flood detection based on Sentinel-1 images and classifier algorithms. Water Environ. J. **35**(3), 924–929 (2021)

Audio Classification in Carnatic Classical Music

Aadarsh Lakshmi Narasiman, Garima Pandey[(✉)] [iD],
and Shashidhar G. Koolagudi

Department of Computer Science and Engineering, National Institute of Technology
Karnataka, Surathkal, Mangaluru, India
{aadarshln.211cs201,garimapandey.217cs002,koolagudi}@nitk.edu.in

Abstract. Most research on Carnatic classical music has focused on identifying ragas based on melody and rhythm. However, no study has been carried out to classify ragas into different types: Vivadhi ragas and Avivadhi ragas. A Raga is said to be Vivadhi if the two notes rendered are closer in frequency, resulting in a discordant effect, and if such an effect is not produced, it is said to be an Avivadhi Raga. Ragas can be classified based on the presence of certain notes and the effect generated by how one note is rendered with another. The study for this classification of ragas has been conducted in this paper by implementing a K-Nearest Neighbor (KNN) Classifier, Convolutional Neural Network (CNN) model, and also an ensemble of CNN models. The results of these classifiers are compared using the accuracy metrics, and it was seen that CNN results in 70% accuracy, thereby outperforming KNN with 60% accuracy for the classification of ragas. Furthermore, an ensemble of CNN models gave an accuracy of 60% while having a fast running time.

Keywords: Vivadhi Raga · Avivadhi Raga · KNN · CNN · MFCCs

1 Introduction

Indian Classical Music is broadly divided into two parts based on the region of origin: Carnatic and Hindustani. These music differ based on the instruments used and the associated Ragas. Indian Carnatic Classical music is celebrated for its intricate melodic and rhythmic structures, with Ragas as the cornerstone. Ragas in Carnatic Music are known for their melodic frameworks that invoke specific emotions and are characterized by their distinct set of notes. These notes are called swaras, and the swara patterns associated with a Raga are called the Sanchara of a Raga. The scale of the Raga, which is based on the swaras, is described by the ascending scale called Arohanam and the descending scale called the Avarohanam. The Arohanam, Avarohanam, and the Sanchara of a Raga give the Raga its uniqueness. This paper focuses on classifying Raga into different types only for the domain of Carnatic Classical Music.

Out of the different types of Raga Classes or Types, one type of classification is based on the Swara Sthana, the set of swaras, and the frequencies at which

C. Modi et al. (Eds.): MIND 2024, CCIS 2736, pp. 463–474, 2026.
https://doi.org/10.1007/978-3-032-14531-4_39

they are to be performed. Ragas can be divided into two classes: Vivadhi and Avivadhi Ragas. Vivadhi Ragas have Vivadhi Swaras, and Avivadhi Ragas do not have Vivadhi Swaras.

Swaras are the notes with a fundamental frequency associated with each. There are 7 basic swaras - Shadja (S), Rishaba (R), Gandhara (G), Madhyama (M), Panchama (P), Daivata (D) and Nishada (N). Based on the frequency of each of these notes, these can be classified into 16 types. Table 1 presents all the 16 types and their frequency ratios.

Table 1. Notes in 12 Interval System and intonation frequency ratios [5]

S. no	Carnatic Notes	Frequency Ratio
1	Shadja (S)	1.0000
2	Shuddha Rishabha (R1)	1.067
3	Chatushruthi Rishabha (R2)	1.125
4	Shatshruthi Rishaba (R3)	1.200
5	Shuddha Gandhara (G1)	1.125
6	Sadharana Gandhara (G2)	1.200
7	Anthara Gandhara (G3)	1.250
8	Shuddha Madyama (M1)	1.333
9	Prati Madhyama (M2)	1.416
10	Panchama (P)	1.416
11	Shuddha Daivatha (D1)	1.600
12	Chatushruthi Daivatha (D2)	1.667
13	Shatshruthi Daivatha (D3)	1.800
14	Shuddha Nishadha (N1)	1.667
15	Kaisiki Nishadha (N2)	1.800
16	Kakali Nishadha (N3)	1.875

Vivadhi Ragas are those Ragas that have Vivadhi Swaras. A Vivadhi Swara is a swara that produces a discordant effect when rendered along with another swara. This discordant effect occurs because the swara is rendered with another swara at a closer or similar frequency. When performing Vivadhi Ragas, artists must render them correctly due to the high scope of error due to their matching or similar frequencies. This furthermore throws importance on the need for proper datasets wherein these ragas have been rendered carefully.

It is observed that the Vivadhi Ragas have the following swara combinations: (R1 and G1), (R3 and G3),(D1 and N1), and (D3 and N3). Regarding Avivadhi Ragas, the absence of Vivadhi Swaras is a key characteristic. However, along with this, two types of swaras, namely Vadhi and Samvadhi Swaras, are present in Avivadhi Ragas.

- Vaadi Swara: The most significant swara of the Raaga is the life-giving note for the Raaga.
- Samavadi Raga: The second most significant swara of the Raaga, which gives the Raaga structure.

The current literature does not have any work related to the classification of Ragas into the various types or categories they belong to, and there is none for the classification of Ragas into Vivadhi or Avivadhi Ragas. In order to understand what kind of features are prominent in Audio Classification problems, [1] this paper explains the set of features generally for such problems. They discuss the different classes of features, which are broadly divided into Top Level, Mid Level, and Low-Level features. Low-Level Features are further divided into Timbre and Temporal Features. Following this, the paper discusses the various types of Audio Classification and, with the help of Genre Classification, discusses the accuracy of various standard classification techniques on the standard GTZAN dataset. [7] This paper provides a comprehensive view on the advancements in audio classification through various deep learning techniques. The paper reviews in a systematic fashion various deep learning architectures right from Convolutional Neural Networks (CNNs), Recurrent Neural Networks (RNNs), Autoencoders, Transformers and Hybrid models. The authors throw light on various architectures, the respective input features like MFCCs, Spectrograms and Mel-Spectrograms, their applications and the accuracies obtained along with the importance of these architectures in extracting meaningful features and information. Based on our problem and restricting the scope of this work to just CNNs, [8] and [9], these papers are reviewed below.

[8] This paper discusses the development and evaluuation of machine learning algorithms for classifying Urdu music genres. The authors in this paper have built a new dataset consisting of Urdu music tracks and the input data that was passed to the model were MFCC images derived from the urdu tracks. Various classification models were constructed and they were mainly based of CNNs. This paper concludes stating that the CNN with Batch Normalization achieves the highest accuracy of 92.6% outperforming the rest, with the rest being CNN with Global Averaging Pooling, VGG-16 with Transfer Learning, Linear-Logistic Regression and XGBoost. From this paper, we have taken up CNN with Global Averaging Pooling, VGG-16 with Transfer Learning and CNN with Batch Normalization and implemented on our dataset as we are mainly focused on CNN approaches.

[9] The author aims at comparing the performance of CNNs against traditional approaches like SVMs in the context of genre classification through various experiments with three databases - the ISMIR 2004 database corresponding to Western Music, Latin Music Database (LMD) and African Music Database. This paper throws light on how CNNs can classify based on spectrograms. The results of the study was that CNNs outperformed the traditional classifiers and CNN combined with LBP achieved a high recognition rate of 92% on the LMD database and performed well for the African music database. This work however, is mostly across various databases and observing performance across them.

The only work available concerning Vivadhi and Avivadhi Ragas in the field of Computer Science is in [2] this paper. This paper discusses with the help of the Breadth-First Search (BFS) Graph algorithm. Here, the various nodes of the graph are the swarms, and the traversal along each path gives rise to the various Vivadhi Raga and Avivadhi Raga, with one graph corresponding to each type.

In the realm of audio classification, particularly in Indian Carnatic Classical Music, machine learning algorithms such as K-Nearest Neighbors and Convolutional Neural Networks can be employed. KNN, a simple yet effective algorithm, can be utilized in this context to classify audio based on certain audio features. These features should capture crucial aspects of the audio signal, enabling KNN to make informed decisions regarding the classification. On the other hand, CNNs are employed to enable the model to learn relevant features from the audio data automatically. By processing audio inputs through layers of convolutions, pooling, and activation functions, CNNs can extract hierarchical representations of the audio, thereby enhancing the model's ability to classify complex audio signals accurately.

In the case of CNN, given that it is a Neural Network architecture model, there are methods called Optimizers to adjust the weights and learning rates. Optimizers are algorithms or methods employed to change the weights and learning rates to minimize the loss function. This is achieved by iteratively updating weights in the opposite direction of the gradient of the loss function for the weights. [6] This paper discusses various popular optimizers, of which we will use two optimizers: the RMSProp and the Nadam (Nesterov-accelerated Adaptive Moment) Optimizers. Nadam Optimizer generally performs well for smaller datasets because it uses Nesterov momentum, which allows the optimizer to look ahead by calculating the gradient at a slightly ahead position in the parameter space. This, in turn, reduces the oscillations and helps in faster convergence, and in the case of smaller datasets, where convergence is quicker, it can be beneficial.

This paper is organized as follows. A detailed description of the proposed methodology is in Sect. 2. Section 3 presents the results obtained, and conclusions with future research are presented in Sect. 4.

2 Proposed Methodology

The flow of the proposed system is given in Fig. 1. We divide this section by giving the dataset specification and further discussing the implementation using KNN and CNN in detail.

2.1 Dataset Specifications

While implementing our approach, we considered 7 Avivadhi Ragas and 13 Vivadhi Ragas, with a ratio of approximately 2:1 maintained and taking into consideration various popular ragas belonging to each class. The choice for 13 Vivadhi Ragas was due to the limited availability of good-quality audio data. All the audio files are sung by a female singer at the same base scale or shruthi. The datasets were acquired from RagaSurabhi, an online platform for students to learn Carnatic Music and its Ragas. The average length of the audio data used for Vivadhi ragas is 1.09 mins and that of Avivadhi is 2.88 mins.

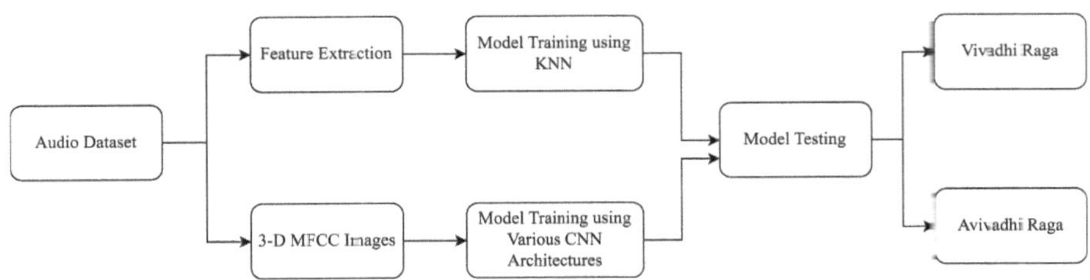

Fig. 1. Flowchart of the Proposed Work.

2.2 Implementation Using KNN

KNN is a simple and easy to implement algorithm which serves as a baseline for benchmarking with respect to comparisons with other algorithms. Since the dataset size is small it is very much preferred.

Feature Extraction. The audio files to be trained on have their duration fixed by setting the number of frames to 1000. The various features to be used while training the KNN are extracted. Our method has considered three features - RMS Energy, Chroma Frequency, and Spectral Centroids. Following this, to represent the vectors obtained for each feature for each Raga belonging to each class, we use statistical methods to represent them by a single value. We have considered the following:

- Mean of the RMS Energy: RMS Energy represents the overall energy levels of the audio signal. Therefore, when we calculate the mean value of the same, we capture the average energy content of the signal over time.
- Median of the Chroma Frequency: Chroma Frequency captures the pitch content of an audio signal. The median captures the central tendency of the pitch distribution.
- Standard Deviation of the Spectral Centroid: Spectral Centroid measures the center of mass of the audio signal. The standard deviation of this feature can help indicate the spectral content's spread.

Data Preparation. After calculating the feature value for each Raga feature, these values are stored in a single list describing all the features associated with the Raga. Following this, labels are assigned to each Raga, such as Vivadhi or Avivadhi.

Classification. The features, along with the labels, are passed through a KNN Classifier. The KNN algorithm effectively classifies the data points based on the majority class among their k nearest neighbors. In this case, we have set the value of k as 2, and thus, 2 clusters are formed. This classifier is then used to predict the labels of unseen audio files of a similar nature, i.e., a female should sing it in the same shruthi.

Model Evaluation. To evaluate the performance of the model, the data is split into training and testing sets. The classifier's performance is evaluated using the accuracy Score and Confusion Matrix. Furthermore, Cross-Validation is performed to estimate the model's general performance. Following this, an accuracy graph was drawn to compare the accuracy score against the value of k.

The issue with this approach is that we represent the features by a single composite value instead of the whole, as it can miss information on outlier points in the feature vector. Furthermore, if the size of the audio file is increased, it can lead to an increase in dimension, which can lead to a degradation in the performance of the KNN.

2.3 Implementation Using CNN with Nadam and RMSProp

CNNs are good for learning hierarchical and abstract features from data and have been successfully applied to tasks like genre classification thus making it apt for usage. It can generalize well especially with the data augmentation and batch normalization implemented. It can adapt to the complex and subtle patterns in the raga dataset

Data Preparation. In order to implement the classification, the audio files, in .wav format have been converted to Mel-Frequency Cepstral Coefficients (MFCCs) images. Following this, these MFCC images were converted to three-channel input. The reason for this conversion is based on the fact that CNN models, expect the input images to have three channels. The MFCC image was generated from the MFCCs. These MFCCs in general can be calculated as follows:

- Compute the Short-Time Fourier Transform
- Convert the STFT to a mel-spectogram
- Convert the mel-spectrogram to a log scale
- Compute MFCCs from the log mel-spectrogram
- Re-size the MFCCs

The MFCCs calculated were converted to images by replicating the single channel MFCCs across three channels and stacking them. Following this, the stacked up MFCCs were normalized and then converted to images using the PIL, Python library which is commonly used for image processing techniques. The sampling rate was set as $22050\,Hz$ and the number of bins into which each audio was split into was set as 5, with each segment having a duration of $30\,s$. 32 MFCCs were calculated for each audio and the image generated from these MFCCs were shaped to the dimension size of (1292, 32).

In order to reduce the burden caused by limited data availability, the process of Data Augmentation comes into play. This is a technique that can be used to artificially increase the size and diversity of the dataset. One of the advantages of using this is to improve the generalization and robustness of the model. In our implementation, we have used the ImageDataGenerator class from the Keras library to achieve the same. The rotation parameter was set to 20°, with the width and height shift range as 0.2 and a feasibility of horizontal flipping. The rotation parameters allows images to be randomly rotated by upto 20° and the width and height shift range allow shifts along the width and height by upto 20%. The validation split was set to 30%.

CNN Model Architecture. The CNN model consists of 4 convolutional layers with increasing filter sizes followed by max-pooling layers to reduce dimensionality. The model also consisted of two dense layers with ReLU activation and a final dense layer with a Sigmoid activation to achieve Binary Classification. Figure 2 shows the network architecture.

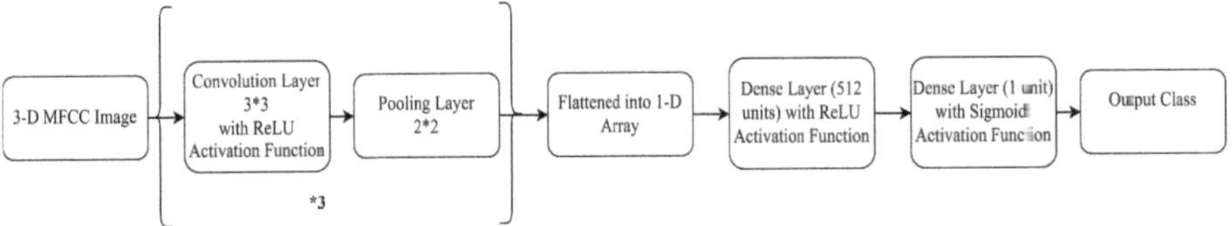

Fig. 2. CNN Model Architecture.

Model Evaluation. The model was compiled using the RMSProp and Nadam Optimizer separately, with the loss function as a Binary Cross-Entropy function. It was trained for 10 epochs with a batch size of 64. During each epoch, the loss, training, and validation accuracy were calculated. Furthermore, an early stopping was implemented with a patience period of 3 epochs.

2.4 Implementation Using CNN with Batch Normalization

Data Preparation. The data preparation as discussed in the previous Subsect. 2.2 is also followed here.

CNN Model Architecture. The model has three units with each having a Convolution, ReLU, and Pooling layer followed by Batch Normalization. The purpose of Batch Normalization is to reduce correct weight initalization and reduce internal covariate shift. It also helps in regularization and avoids overfitting. Figure 3 shows the network architecture.

Model Evaluation. The model was trained with the following consistent parameters - a learning rate of 0.0001, batch size of 32, and a total of 10 epochs. During each epoch, the loss, training, and validation accuracy were calculated. Furthermore, an early stopping was implemented with a patience period of 3 epochs.

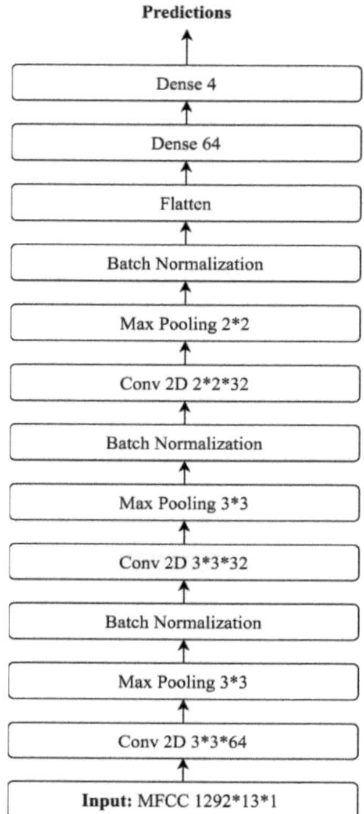

Fig. 3. CNN Network Architecture with Batch Normalization.

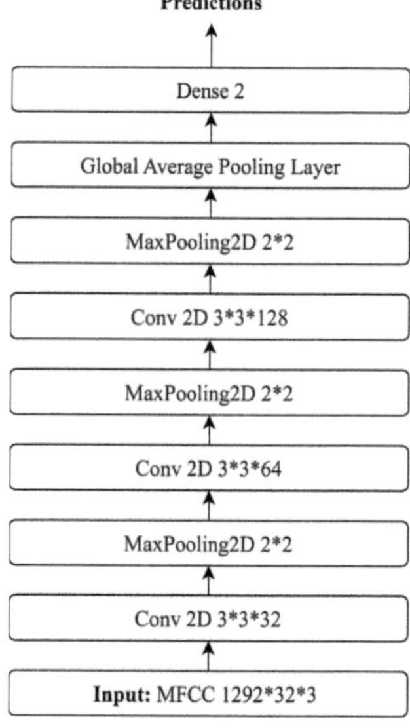

Fig. 4. CNN Network Architecture with Global Average Pooling.

2.5 Implementation Using CNN with Global Average Pooling (GAP)

Data Preparation. The data preparation as discussed in the previous Subsect. 2.2 is also followed here.

CNN Model Architecture. The model has three units with each having a Convolution and Pooling layer with ReLU as the activation function. This was followed by a Global Averaging Pooling Layer. The purpose of implementing this is to reduce the chances of overfitting and improve the model's generalization. Figure 4 shows the network architecture.

Model Evaluation. The model was trained with the following consistent parameters - a learning rate of 0.0001, batch size of 32 and a total of 10 epochs. During each epoch, the loss, training, and validation accuracy were calculated. Furthermore, an early stopping was implemented with a patience period of 3 epochs.

2.6 Implementation with VGG-16 with Transfer Learning

VGG-16 is a CNN architecture with 16 weight layers in its network It is a simple and uniform architecture that employs small 3×3 filters. It is in general used for image classification tasks. However, since our task involves 3-D MFCC images as input features, it can be used in this context. Transfer Learning is an approach wherein a model pre-trained on a particular task is re-used to boost the performance of a specific task. Here VGG-16 is trained on ImageNet dataset and has pre-trained weights. It is then enhanced by training it on the MFCC images of our dataset. The model was run for 10 epochs.

2.7 Ensemble Approach

After implementing and analyzing the results of the various models, the CNN models that were performing the best were the ones with the - RMSProp Optimizer, Nadam Optimizer, and Global Average Pooling. Therefore, these three models were combined together in order to take advantage of their strengths and aim at improving the overall performance. The ensembling approach considered was that of averaging their predicition to achieve a more robust classification. The input fed to the model is same as that of the ones fed to the other CNN models. Figure 5 shows the overall approach. The model was run for 10 epochs. During each epoch, the loss, training, and validation accuracy were calculated. Furthermore, an early stopping was implemented with a patience period of 3 epochs.

Fig. 5. Ensemble Approach.

3   Results

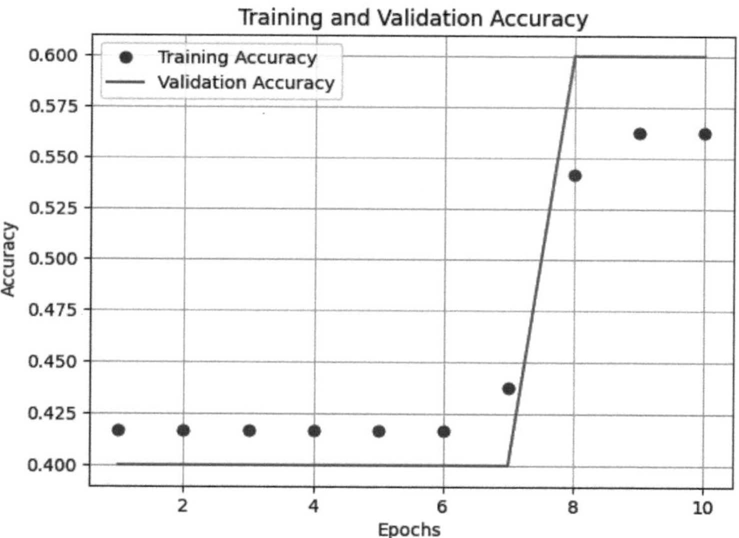

Fig. 6. Accuracy obtained using Ensemble Approach.

On implementing the classification problem using the conventional KNN algorithm, the accuracy obtained was 60 %. On implementation with CNN, using RMS Prop as the optimizer, the Validation Accuracy was 60%, and with the Nadam optimizer's help, the obtained Validation Accuracy was 70%. Furthermore, when implementing the CNN with Batch Normalization, the Validation Accuracy was 50% and with Global Averaging Pooling, the Validation Accuracy was 60%. On implementing the ensemble approach, the Validation Accuracy was 60%. Figure 6 shows the Accuracy Graph obtained after implementing the ensemble approach (Table 2).

Table 2. Results based on Validation Accuracy

Model	Validation Accuracy
KNN	60%
CNN with RMSProp Optimizer	60%
CNN with Nadam Optimizer	70%
CNN with Batch Normalization	50%
CNN with Global Average Pooling	60%
Ensemble by averaging	60%
VGG-16 with Transfer Learning	50%

4 Conclusion and Future Work

In this paper, we addressed the classification problem of ragas into vivadhi and avivadhi. While CNNs work generally well for classification purposes as compared to traditional classifiers like KNNs, when ensembled they work more effectively due to the fact they have quicker run times. The work that is presented in this paper only covers various CNN approaches, in the future some work can be done by implementing and evaluating various RNN, Autoencoder, and Transformer architectures. The current work that is implemented is trained only on the female voices, covering only their vocal ranges; in the future, data can be collected for male vocal ranges, and the model can be retrained. Furthermore, only the commonly used Vivadhi Ragas were used while building the dataset, as most of them are uncommon. Training on these Ragas by conducting extensive data collection will help cover more use cases for Vivadhi Ragas. Finally, this work only focuses on the Vivadhi Ragas of Carnatic Music, and in the future, the same can be done for Vivadhi Ragas in Hindustani Music.

References

1. Fu, Z., Lu, G., Ting, T.M., Zhang, D.: A survey of audio-based music classification and annotation. IEEE Trans. Multimed. **13**(2), 303–319 (2011)
2. Sriram, A.L., Subramanian, Sweety, S.: Extracting Vivadi Ragas and Avivadi Ragas using BFS algorithm. Int. J. Innov. Technol. Explor. Eng. **3**(5) (2013)
3. Bhagyalekshmy, S.: Ragas in Carnatic Music. CBH Publications (1990)
4. Sambamoorthy, P.: South Indian Music Book III. The Indian Music Publishing House (1973)

5. Upadhyaya, P., Suma, S.M., Koolagudi, S.G.: Identification of allied ragas in Carnatic music. In: Eighth International Conference on Contemporary Computing, pp. 127–131 (2015)
6. Ruder, S.: An overview of gradient descent optimization algorithms (2016). https://arxiv.org/abs/1609.04747
7. Zaman, K., Sah, M., Direkoglu, C., Unoki, M.: A survey of audio classification using deep learning. IEEE Access **11**, 106620–106649 (2023)
8. Mughal, S.F., Aamir, S., Sahto, S.A., Samad, A.: Urdu music genre classification using convolution neural networks. In: Proceedings of the International Conference on Emerging Trends Smart Technology (ICETST), pp. 1–6 (2022)
9. Costa, Y.M.G., Oliveira, L.S., Silla, C.N.: An evaluation of convolutional neural networks for music classification using spectrograms. Appl. Soft Comput. **52**, 28–38 (2017)

ELA-TransNet: Error Level Enhanced Hybrid Network for Image Forgery Detection

Choudhary Shyam Prakash[1]($\boxtimes$) and Sahani Pooja Jaiprakash[2]

[1] Department of Computer Science and Engineering, National Institute of Technology, Andhra Pradesh, Tadepalligudem, West Godavari District 534101, Andhra Pradesh, India
shyamprakash2008@yahoo.com

[2] Department of Computer Science and Engineering, ITER, Siksha 'O' Anusandhan, Jagamara, Bhubaneshwar 751019, Odisha, India

Abstract. As digital tools have advanced, image manipulation has become more common, resulting in various forgeries that mask original content. Detecting manipulated regions is more difficult by methods like copy-pasting or image merging, especially when geometric changes are involved. Using three crucial steps—pre-processing, image augmentation, and classification—this study presents a novel deep learning-based method for identifying image forgeries in digital images. Image normalization, rescaling, and error level analysis (ELA) are pre-processing methods that reduce overfitting and improve model accuracy. Image augmentation is used to increase the dataset size and ensure the model is exposed to a greater range of training samples to increase robustness. The suggested approach, implemented on the Python platform, uses convolutional transfer learning for classification. The Columbia, CASIA V1.0, and CASIA V2.0 datasets are used to assess it. The experimental results demonstrate outstanding performance with an accuracy of 99.97%, precision of 95.63%, recall of 98.95%, and an F1 score of 97.26%. These metrics highlight the method's potential for real-world applications in image integrity verification by demonstrating its superior efficacy in detecting forgeries compared to traditional techniques.

Keywords: Digital Image Forensics · CNN · Transfer Learning · Image augmentation · Classification

1 Introduction

Images are essential in today's digital age for several purposes, such as forensic investigations, news reporting, publications, medical diagnostics, and legal proceedings. Multimedia technologies and image processing software have advanced so quickly that robust editing tools like Photoshop, Inkscape, Fireworks, and GNU Image Manipulation Program (GIMP) have been developed. These tools

C. Modi et al. (Eds.): MIND 2024, CCIS 2736, pp. 475–486, 2026.
https://doi.org/10.1007/978-3-032-14531-4_40

make it possible to quickly edit digital images, frequently without leaving any evidence of the forgery, which distorts reality and poses a severe threat to society [1–3]. The field of digital image forensics emerged due to the increasing concerns regarding the authenticity of digital images, which have led to the development of various methods for detecting image forgeries. According to Farid [4], this domain is concerned with confirming the integrity and authenticity of digital images. The two primary categories of forgery detection techniques are active and passive. Active techniques include adding recognizable components, like watermarks or electronic signatures, to the image while it is being created. The image's authenticity can be later verified by extracting and validating these embedded elements. These embedded components are disrupted when the image is altered or tampered with, making the alteration obvious. This method works best in controlled settings where active strategies can be used because it depends on the earlier inclusion of such components. Conversely, passive methods, which don't require prior knowledge about the image, identify forgeries by examining variations in the structure or content of the image [5,6]. Standard passive image manipulation techniques like copy-move and splicing involve copying parts of an image to hide unwanted elements. Scaling is another technique that forgers commonly employ to blend these modifications seamlessly and impede detection. The increasing availability of image editing software on devices has led to a rise in manipulation for legal and illicit purposes, creating challenges for social media, business, and criminal activity [7]. Digital image authenticity needs to be verified to preserve content integrity and prevent misuse; this is a challenging task that is made even more difficult by advanced techniques such as deep generative models. In the past, detection methods relied on manually selected features that had problems recognizing intricate image patterns. However, deep learning has transformed this field by making it easier to extract complex features automatically [8]. Convolutional Neural Networks (CNNs) are recommended because of their exceptional performance in image forgery detection tasks.

Three crucial steps—pre-processing, data augmentation, and classification—are included in this work's hybridized deep learning approach to image forgery detection. Pre-processing techniques include image normalization, resizing, and error level analysis (ELA) to increase accuracy and reduce overfitting. By expanding the dataset, data augmentation improves the model's generalization and resilience. Lastly, the classification step is built on convolutional transfer learning, making it possible to detect and categorize image forgeries accurately. Combined, these actions provide a thorough and efficient framework for dealing with image forgery detection issues.

The paper's structure is as follows: Sect. 2 examines the literature on image forgery detection employing deep learning and error-level analysis methods that different researchers have investigated. A suggested methodology for differentiating between authentic and fake images is presented in Sect. 3. The experimentation with the Python tool to assess the effectiveness of the proposed method is covered in Sect. 4. The research's findings and future directions are presented in Sect. 5.

2   Related Works

Over the past ten years, deep learning techniques have grown in popularity and have been applied to numerous scientific problems. Many of their success can be attributed to their exceptional performance on tasks like segmentation, regression, and classification. Sometimes, deep learning models can perform more accurately and precisely than humans. Deep learning techniques' popularity can largely be attributed to their automatic learning and feature extraction capabilities. This is not the case with traditional machine learning techniques, which often call for researchers to create features—a complex and domain-specific task manually. With their superior ability to extract and learn relevant features from input data, Convolutional Neural Networks (CNNs) and related models can be applied to various tasks with little to no feature engineering [9].

Recent literature has examined deep learning techniques to improve image forgery detection accuracy over previously suggested, conventional methods. Rao et al. [10] introduced a forgery detection method based on the local feature descriptor and deep neural networks. A robust splicing localization scheme against JPEG compression was developed by utilizing the proposed CNN model in conjunction with a fully connected conditional random field (CRF) with SVM. An improved initialization based on the SRM was also used. Hosny et al. [11] presented a CNN architecture with four convolutional and four max-pooling layers that is thin-layer. The CASIA 1.0, CASIA 2.0, and CUISDE datasets were used to evaluate the model in comparison to other recently released methods. The suggested method uses a convolutional neural network to identify image-splicing forgeries. Its efficacy hasn't been tested on other types of image manipulation or forgeries, though, so it's restricted to this particular kind of fraud. Li et al. [12] proposed a method that combines semantic and noise-inconsistent features from RGB and noise streams using a multiresolution hybrid feature fusion process. Reducing false positives from natural regions and improving feature representation are the two main goals of a tamper-guided dual self-attention (TDSA) module. To precisely identify tampered areas, the network—a fully convolutional model with the TDSA module—captures interchannel and interposition dependencies. Pixel-level, tamper-guided attention, and image-level loss functions are used to train it.

Nevertheless, the technique might have trouble with intricate manipulation, severe compression, or tiny tampered areas, which could result in false positives. Chen et al. [13] suggested the Hybrid Features and Semantic Reinforcement Network (HFSRNet), which improves efficiency by combining semantic information and learning richer features to detect image forgeries. NIST16, COVERAGE, and CASIA are three joint image manipulation datasets on which HFSRNet achieves state-of-the-art performance. The network has an encoding-decoding architecture, where manipulation artifacts are captured using long short-term memory with resampling features, and the differences between tampered and untampered regions are amplified by rotating residual units. Chakraborty et al. [14] proposed a method that integrated a dual-branch Convolutional Neural Network (CNN), an error-level analysis (ELA), and noise residuals from the Spatial

Rich Model (SRM). The CNN is trained using authentic and altered images, and pre-processing methods like SRM and ELA enhance learning. Noise residuals are produced by employing 30 high-pass filters from SRM, and ELA images are created by saving the originals at a predetermined compression rate. While the model works well, it may not be as robust in low-resource environments and does not currently support tampering localization or detection of video tampering; these issues will be resolved in future work.

3   Proposed Work

The hybrid deep learning method for identifying and categorizing forged images is demonstrated in Fig. 1. First, an input image is processed through a number of pre-processing steps, such as error level analysis, rescaling, and image normalization. These procedures minimize errors and potential biases during model training by ensuring the input data is consistent and optimized for additional analysis.

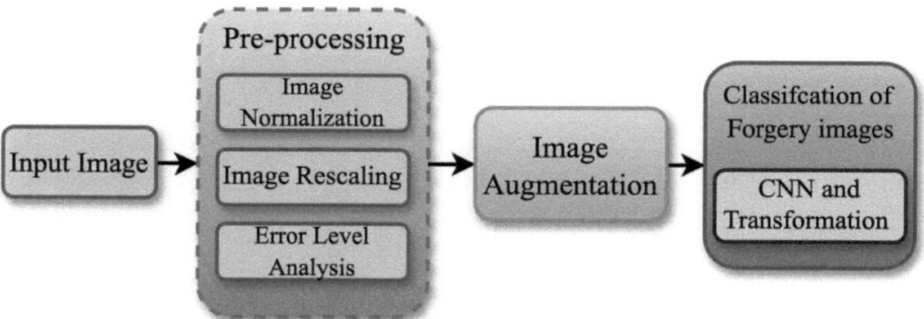

Fig. 1. The proposed method's flow.

3.1   Data Pre-processing

Data pre-processing is an essential first step in preparing images for computer vision and machine learning tasks to ensure that the data is consistent, clean, and ready for model consumption. The quality and organization of the data are improved using several techniques in this process so that it can be utilized with different algorithms. Error Level Analysis (ELA), image normalization, and image rescaling are methods frequently used to detect and standardize images to improve the performance of machine learning models.

3.1.1 Rescaling and Standardization of Images

The image pre-processing for the proposed model involved scaling the pixel values to the range of 0–128 and resizing the images to a constant 256×256 size without any part cropping. Images in the original dataset ranged in size from 722×480 to 800×600 pixels. Resizing was essential to maintaining consistent input dimensions for the model. Padding was applied as needed to maintain the aspect ratio.

$$X' = \frac{X - X_{\min}}{X_{\max} - X_{\min}} \times 128 \tag{1}$$

Equation 1 was used to normalize the pixel values to the range $[0, 128]$, where $X_{\min}$ and $X_{\max}$ are usually 0 and 255 for 8-bit images. Conversely, resizing minimizes computational costs and model complexity by reducing the number of pixels, which expedites training by reducing the number of nodes in the input layer. Additionally, it allows you to zoom in on images, which is helpful for some machine-learning tasks. These pre-processing stages aid in standardizing the photos for more effective model training, guaranteeing constant input forms and quicker convergence.

3.1.2 Error Level Analysis

The forensic method, Error Level Analysis (ELA), can detect image manipulation by analyzing variations in compression levels across different areas of an image. Because compression can mask changes, it works exceptionally well with lossy compressed formats like JPEG. According to ELA, areas of an image that have been modified display varying degrees of compression, whereas unaltered areas retain consistent compression artifacts [15]. A lower compression quality (usually between 90 and 95% JPEG quality) is used to save the original image for ELA. One can detect discrepancies by deducting the recompressed image's pixel values from the original. Error levels in tampered areas vary significantly because of variations in their recompression, while error levels in unaltered regions stay constant. Since these differences may be signs of digital manipulation, ELA is a valuable technique in digital forensics for confirming an image's legitimacy.

As an example, the regions in Fig. 2 with the highest ELA values are highlighted in white, and the original image is shown next to its ELA image. ELA can identify notable variations even in areas where the resaved image may not be visually noticeable to the human eye, particularly where dark or black colors predominate. A resave results in an image with optimized quality. More intense colors in the ELA visualization serve as a visual cue that additional image modification causes ELA to reveal these modified areas with significantly higher error levels. Figure 2 shows a sample image illustrating these ELA patterns.

3.1.3 Image Augmentation

Image augmentation [16] is an essential technique in image pre-processing used to artificially increase the size and diversity of the training dataset in computer vision and machine learning. By implementing various transformations, augmentation produces new, modified versions of the original images, enhancing

Fig. 2. Example of (a) Original Image, (b) Grayscale Image, (c) AHE Image of Grayscale Image.

the model's capacity to generalize to new data. These modifications make the model resilient to shifts in orientation, size, lighting, and other real-world circumstances. These operations include rotation, flipping, scaling, cropping, and brightness adjustments. The suggested model uses augmentation techniques like filling, flipping, and rotating images by [45°, -45°] angles and flipping them vertically and horizontally to help the model pick up unique characteristics and increase classification accuracy. Augmentation increases overall performance on unseen data and strengthens the model's robustness by producing a variety of training samples.

3.1.4   Classification

To detect image forgeries, the suggested hybrid deep learning model, as shown in Fig. 3, combines Convolutional Neural Networks (CNNs) with attention mechanisms, most likely Vision Transformers (ViTs). An input image is first run through several convolutional layers, represented by stacked red and green blocks, and used to extract crucial features like textures and edges. These convolutional layers gradually decrease the input image's spatial dimensions (e.g., G. from 256×256 to smaller feature maps) while keeping the essential details intact. The extracted feature maps proceed to the self-attention mechanism, depicted in the figure's middle section. The Multi-Head Self-Attention layer processes feature maps to capture long-range dependencies across pixels, enhancing the model's understanding of global relationships within pictures. Layer Normalization comes next to guarantee reliable and practical training.

The feature maps are further normalized, and an Add and Layer Normalization block ensures consistent feature scaling. Following attention and normalization, the data is converted into a compact vector form (usually 1×1 with depth retained) by a Global Average Pooling layer, which lowers the feature map dimensions by averaging all spatial locations. The final processed output is fed into fully connected layers for classification, which allows the model to identify whether the image is actual or manipulated. The model's ability to detect forgeries is enhanced by convolutional feature extraction and attention-based global context processing, guaranteeing practical computation through dimensional reduction through pooling.

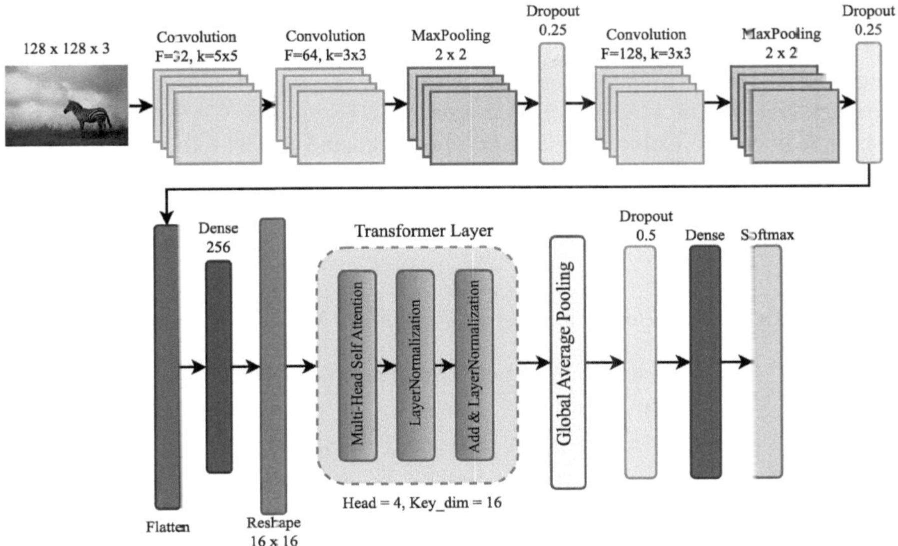

Fig. 3. Deep CNN Architecture.

4   Experimental Outcomes

All the experiments were conducted on a macOS Sonoma 14.2 system with 16 GB of RAM. Primarily, Python 3.7 was utilized as the programming environment. Using the PyTorch framework, CNNs were built and trained for feature extraction, offering a productive setup for training deep learning models. This setup made processing for image forgery detection experiments reliable and efficient.

For all forgery detection experiments, the CASIA v1.0 [17], CASIA v2.0 [17], and Columbia [18] benchmark datasets were used. The 1,725 color JPEG images in CASIA v1.0 have a 384×256 resolution, and 925 are forged and made using methods like rotation and resizing. With post-processing applied to tampered regions, CASIA v2.0 contains 5,123 forged and 7,491 real images in a variety of formats (BMP, JPEG, and TIFF) and resolutions ranging from 240×160 to 900×600. These datasets are crucial for assessing how reliable and accurate image forgery detection algorithms are. The Columbia Dataset [13], sometimes called the AuthSpliced Dataset, is a benchmark dataset created to detect image splicing forgeries. It contains high-resolution JPEG images that range in size from 757×568 to 1152×768 pixels.

4.1   Performance Evaluation of the Proposed Forgery Detection Classification Method

The accuracy, precision, recall, and F1-score metrics for the suggested forgery detection method are summarized in Table 1. These metrics were assessed on three benchmark datasets: Columbia, CASIA v1.0, and CASIA v2.0. The percentage of authentic and fake images correctly identified out of all the images examined is known as accuracy, and it gauges the model's overall correctness. With 98.31% on the CASIA v1.0 dataset, 99.97% on the CASIA v2.0 dataset, and 96.87% on the Columbia dataset, the suggested model exhibits exceptional accuracy across all datasets. These outcomes demonstrate how well the model generalizes across various datasets and its resilience and efficacy in differentiating between real and fake images.

Table 1. Performance Metrics for CASIA V1 and V2 Datasets

Dataset	Accuracy	Precision	Recall	F1-Score
CASIA v1.0	98.31	95.93	99.31	98.00
CASIA v2.0	99.97	95.63	98.95	97.26
Columbia	96.87	96.54	98.45	96.43

Of all positive predictions, or all images predicted to be forged, precision shows the percentage of true positive detections or accurately identified forged images. The model maintains a high level of accuracy in correctly detecting forgeries while keeping the number of false positives (real images incorrectly classified as forged) relatively low, as evidenced by a precision of 95.93% on CASIA v1.0, 95.63% on CASIA v2.0 and 96.87% on Columbia. Recall, another name for sensitivity, gauges how well the model can detect every instance of forgeries in the dataset. Recall values for CASIA v1.0, CASIA v2.0, and Columbia are 99.31%, 98.95%, and 98.45%, respectively, demonstrating the model's high success rate in identifying nearly all forged images and reducing false negatives (forged images that are mistakenly identified as real). The F1-score is a metric that balances precision and recall by taking the harmonic mean of the two. For CASIA v1.0, CASIA v2.0 and Columbia, the F1-score values are 98.00%, 97.26%, and 96.43%, respectively, indicating a well-rounded model performance with high recall and precision. This implies that the model maintains a low misclassification rate for genuine and forged images and efficiently detects forgeries. The performance metrics show that the suggested approach successfully identifies image forgeries on the CASIA v1.0, CASIA v2.0, and Columbia datasets. Accuracy, precision, recall, and F1-score are the metrics that show the most substantial results. The dataset's larger quantity and variety of images may have contributed to the model's marginally better performance on CASIA v2.0 by enhancing its generalization ability.

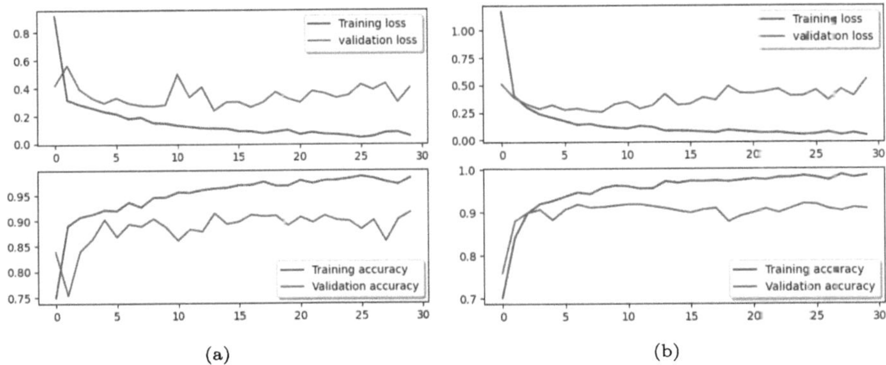

Fig. 4. Training and Validation on (a) CASIA v1.0 dataset, (b) CASIA v2.0 dataset.

The graphs shown in Fig. 4 illustrate the training and validation performance of a model trained on CASIA v1.0 and CASIA v2.0 datasets over 30 epochs. In both datasets, the training loss decreases steadily and stabilizes at a low value, while the validation loss initially improves but fluctuates, indicating potential overfitting as the model fails to perform well on validation data consistently. Similarly, the training accuracy quickly reaches high values. It remains stable, while validation accuracy, though initially improving, fluctuates around 0.9, reflecting limited generalization ability and suggesting the model fits the training data better than the validation set.

4.2   Comparative Analysis

Using well-known evaluation metrics like accuracy, precision, recall, and F1 score, Table 2 offers a thorough comparison of the suggested method's performance against current state-of-the-art methods on the CASIA v2.0 dataset. With an accuracy of 0.9997, the highest of the compared approaches, the suggested method produces outstanding results. It also performs well in other metrics, attaining an F1 score of 0.9726, a precision of 0.9563, and a recall of 0.9895. However, other approaches, like those suggested by Ding et al. Nath et al. and [19]. Although they also demonstrate competitive accuracy, [20] performs comparably worse in terms of precision, recall, and F1 score. This suggests that while these approaches are dependable in certain areas, the suggested method performs better and more evenly across the board on the CASIA v2.0 dataset. The outcomes demonstrate how well the suggested approach works to achieve high accuracy and a steady balance among other crucial evaluation criteria.

Table 2. Comparative analysis with existing state-of-the-art on CASIA v2.0 dataset.

Methods	Model	Accuracy	Precision	Recall	F1 Score
Nath et al. [20]	ResNet50	0.9645	0.9669	0.9415	0.9540
Hu et al. [21]	MD_Unet	0.9478	0.7971	0.8138	0.8054
Niyishaka et al. [6]	CNN model	0.9459	-	-	-
Latif et al. [22]	CNN Model	0.9636	0.9714	0.9903	0.9703
Kanwal et al. [23]	OELTP	0.9630	-	-	0.9644
Ding et al. [19]	DCU-Net	0.9793	0.8772	0.8893	0.8667
Muniappan et al. [24]	CNN Model	0.8901	-	-	-
Proposed	ELA-TransNet	0.9997	0.9563	0.9895	0.9726

5   Conclusion

Advances in image editing software have made it more difficult to distinguish between forged and authentic images, making it more difficult to detect image forgeries. Most current detection techniques have trouble accurately classifying data, which lowers system performance. Using data pre-processing methods like image resizing, normalization, and error level analysis (ELA) to improve image quality, the proposed model offers a deep learning-based method for detecting image forgeries. Techniques like rotation, flipping and filling enhance the dataset, expanding the amount of available training data. The classification procedure effectively distinguishes real images from forged ones by using convolutional transfer learning. Utilizing the CASIA V1.0, CASIA V2.0, and Columbia datasets, the proposed model demonstrates remarkable performance metrics: 99.97% accuracy, 95.63% precision, 98.95% recall, and 97.26% F1 score. The proposed model can be applied to image and video forensics and criminal investigations because it functions effectively even under various geometric transformations. However, its performance declines with high Gaussian noise, indicating the need for better detection techniques. The primary goal of future research should be to expand the methodology to detect other types of image manipulations and improve ELA techniques for pixel modifications.

References

1. Ren, R., Niu, S., Jin, J., Xiong, K., Ren, H.: Erinet: efficient and robust identification network for image copy-move forgery detection and localization: efficient and robust identification network. Appl. Intell. **53**(12), 16170–16191 (2023)
2. Capasso, P., Cattaneo, G., De Marsico, M.: A comprehensive survey on methods for image integrity. ACM Trans. Multimed. Comput. Commun. Appl. **20**(11), 1–34 (2024)
3. Armas Vega, E.A., González Fernández, E., Sandoval Orozco, A.L., García Villalba, L.J.: Copy-move forgery detection technique based on discrete cosine transform blocks features. Neural Comput. Appl. **33**, 4713–4727 (2021)

4. Farid, H.: Image forgery detection. IEEE Signal Process. Mag. **26**(2), 16–25 (2009)
5. Ferreira, W.D., Ferreira, C.B., Cruz Júnior, G., Soares, F.: A review of digital image forensics. Comput. Electr. Eng. **85**, 106685 (2020)
6. Niyishaka, P., Bhagvati, C.: Image splicing detection technique based on illumination-reflectance model and LBP. Multimed. Tools Appl. **80**(2), 2161–2175 (2021)
7. Ahmed, R., Dharaskar, R.V.: Study of mobile botnets: an analysis from the perspective of efficient generalized forensics framework for mobile devices. In: IJCA Proceedings on National Conference on Innovative Paradigms in Engineering and Technology (NCIPET 2012) NCIPET (15), pp. 5–8 (2012)
8. Li, J., et al.: Automatic detection and classification system of domestic waste via multimodel cascaded convolutional neural network. IEEE Trans. Industr. Inf **18**(1), 163–173 (2021)
9. Zanardelli, M., Guerrini, F., Leonardi, R., Adami, N.: Image forgery detection: a survey of recent deep-learning approaches. Multimed. Tools Appl. **82**(12), 17521–17566 (2023)
10. Rao, Y., Ni, J., Zhao, H.: Deep learning local descriptor for image splicing detection and localization. IEEE Access **8**, 25611–25625 (2020)
11. Hosny, K.M., Mortda, A.M., Lashin, N.A., Fouda, M.M.: A new method to detect splicing image forgery using convolutional neural network. Appl. Sci. **13**(3), 1272 (2023)
12. Li, F., Pei, Z., Wei, W., Li, J., Qin, C.: Image forgery detection using tamperguided dual self-attention network with multiresolution hybrid feature. Secur. Commun. Netw. **2022**(1), 1090307 (2022)
13. Chen, H., Chang, C., Shi, Z., Lyu, Y.: Hybrid features and semantic reinforcement network for image forgery detection. Multimed. Syst. **28**(2), 363–374 (2022)
14. Chakraborty, S., Chatterjee, K., Dey, P.: Detection of image tampering using deep learning, error levels and noise residuals. Neural Process. Lett. **56**(2), 112 (2024)
15. Sari, W.P., Fahmi, H.: The effect of error level analysis on the image forgery detection using deep learning. Game Technology, Information System, Computer Network, Computing, Electronics, and Control, Kinetik (2021)
16. Abhishek, Jindal, N.: Copy move and splicing forgery detection using deep convolution neural network, and semantic segmentation. Multimed. Tools Appl. **80**(3), 3571–3599 (2021)
17. Dong, J., Wang, W., Tan, T.: Casia image tampering detection evaluation database. In: 2013 IEEE China Summit and International Conference on Signal and Information Processing, pp. 422–426. IEEE (2013)
18. University, C.: AuthSpliced Dataset Photographers (n.d.). http://www.ee.columbia.edu/ln/dvmm/downloads/AuthSplicedDataSet/photographers.htm. Accessed 30 Nov 2024
19. Ding, H., Chen, L., Tao, Q., Fu, Z., Dong, L., Cui, X.: DCU-net: a dual-channel u-shaped network for image splicing forgery detection. Neural Comput. Appl. **35**(7), 5015–5031 (2023)
20. Nath, S., Naskar, R.: Automated image splicing detection using deep CNN-learned features and ANN-based classifier. SIViP **15**, 1601–1608 (2021)
21. Hu, J., Xue, R., Teng, G., Niu, S., Jin, D.: Image splicing manipulation location by multi-scale dual-channel supervision. Multimed. Tools Appl. **83**(11), 31759–31782 (2024)
22. El-Latif, E.I.A., Taha, A., Zayed, H.H.: A passive approach for detecting image splicing using deep learning and haar wavelet transform. Int. J. Comput. Netw. Inf. Secur. **11**(5), 28–35 (2019)

23. Kanwal, N., Girdhar, A., Kaur, L., Bhullar, J.S.: Digital image splicing detection technique using optimal threshold based local ternary pattern. Multimed. Tools Appl. **79**(19), 12829–12846 (2020)
24. Thiiban Muniappan, T.M., Abd Warif, N.B., Ismail, A., Mat Abir, N.A.: An evaluation of convolutional neural network (CNN) model for copy-move and splicing forgery detection. Int. J. Intell. Syst. Appl. Eng. **11**(2), 730–740 (2023)

Scalable Kernel SubXPCA Applied to Classification of Hyperspectral Images on a Distributed Platform

Bogolu Rupa[1,2(✉)], R. Aruna Flarence[1,3], and Atul Negi[1]

[1] School of Computer and Information Sciences, University of Hyderabad, Hyderabad, Telangana, India
{21MCPC19,20MCPC07,atul.negi}@uohyd.ac.in
[2] Vardhaman College of Engineering, Hyderabad, Telangana, India
rupa1828@vardhaman.org
[3] Gokaraju Rangaraju Institute of Engineering and Technology, Hyderabad, Telangana, India

Abstract. Significant improvements in the resolution of enormous high volumes of hyperspectral images in remote sensing technology lead to face challenges in storage, processing, and computing. These challenges are addressed by several open-source distributed processing architectures that have emerged recently for hyperspectral images. This paper proposes a parallel and effective spectral dimensionality reduction approach for classification of hyperspectral images. We designed a novel version of feature partitioning kernel principal component analysis centered on the SubXPCA method applied to the spectral reduction of hyperspectral images in a distributed Spark cluster computing environment. The proposed scalable kernel SubXPCA method is a novel variation of Sub-XPCA. We compared the scalable kernel SubXPCA method against other feature partitioning spectral dimensionality reduction methods. Our experimental findings on various ground truth and synthetic datasets of hyperspectral image confirm that the proposed spectral reduction approach outperforms its competitors in classification performance for most classifiers. The proposed approach shows increased execution time and speedup as the dataset size grows in the distributed processing environment.

Keywords: Principal Component Analysis · Kernel Principal Component Analysis · Distributed · Spark · Machine Learning · Hyperspectral Image · Dimensionality Reduction

1 Introduction

The hyperspectral sensors can capture hyperspectral images (HSIs) in tens or even hundreds of spectral bands [22], which makes it possible to identify spectrally distinct objects, enhancing comprehension of the scene being studied [23]. HSIs have been extensively utilized in many applications, such as change and

C. Modi et al. (Eds.): MIND 2024, CCIS 2736, pp. 487–498, 2026.
https://doi.org/10.1007/978-3-032-14531-4_41

target detection [2], image enhancement [1], food quality, medical diagnosis, image-guided surgery, precision agriculture, water resource and flood management, forensic examinations, and defense and homeland security [4,12].

As hyperspectral remote sensing image technology advances, the amount of information included in HSI increases along with its data volume, creating a greater need for storage and computing power for processing HSI data [8]. Due to this, HSIs now come under the category of big data. These large-scale HSI datsets processing on single-machine platforms is not feasible due to the allocation of storage and computing resources. Thus, there is a requirement for distributed processing using cloud computing infrastructures [21,24]. Massive amounts of HSI data can be handled by distributing storage and computing costs to minimize the amount of time needed, across a cluster of machines using various open-source technologies. Apache Spark [17,20] is the most commonly utilized technology. It is a popular distributed high-performance computing framework. Its scalability and fault tolerance in handling massive volumes of data have contributed to its enormous rise in popularity in recent years. Spark is particularly quick because it allows for in-memory computing, which is ideal for the iterative nature of the algorithms required for machine learning (ML).

Specialized methods and algorithms are needed to handle the peculiar properties of HSI and take advantage of its rich and precise information [26]. Due to its high dimensionality and nonlinearity nature, dimensionality reduction (DR) techniques are required. These methods improve the capacity to lower noise, compress data, choose features, and increase computing efficiency. This makes HSI analysis and classification more efficient. As part of the data preparation for HSIs methods, a greater variety of DR techniques have been integrated into the feature selection and extraction processes. Principal component analysis (PCA) [11] is a significant and intricate statistical technique utilized to minimize data dimensions. PCA can only identify linear associations in the data; it may overlook significant nonlinear patterns [3,18]. Kernel PCA was first introduced in [19], as a nonlinear extension of the widely used PCA using the kernel trick, and it has proven to be an effective technique for obtaining nonlinear features for classification [27]. In HSI analysis, kernel PCA is frequently utilized to reduce the dimensions in the data while maintaining the nonlinear correlations between the features. This capability results in higher accuracy and better performance in most of the ML classifiers like Support Vector Machines (SVM), Decision Tree (DT), Random Forest (RF), Linear Regression (LR), Naive Bayes (NB), k-Nearest Neighbors (KNN), Extra Trees (ET) on HSIs using Kernel PCA. Typical traditional ML classifiers were taken into consideration rather than deep learning models because the work was focused on exploratory data analysis.

This paper proposes a parallel implementation of feature partitioning-based scalable kernel SubXPCA for spectral reduction of HSI by using the distributed computing framework as Spark cluster. Specifically, we rely on the variation of SubXPCA and developed a Spark-based scalable kernel SubXPCA. We perform extensive experiments to evaluate the performance of scalable kernel SubXPCA in terms of accuracy and speedup. Experimental results verify that scalable kernel SubXPCA significantly enhances the performance of most of classifiers than other feature partitioning based spectral reduction methods. The major contributions of this work are summarized as follows:

- Analysis of SubXPCA and its variation kernel SubXPCA.
- The design and implementation of the proposed spectral DR method as scalable kernel SubXPCA and classification for HSI.
- The proposed method's experimental results compared with other spectral reduction methods, to evaluate the performance of the classifiers and speedup on HSI datasets.

The organization of this work is as follows. Section 2 describes the analysis of SubXPCA and its variation as kernel SubXPCA. Section 3 provides the design and implementation of the proposed methodology and classification in a distributed framework. Section 4 presents experimental evaluation, and Sect. 5 concludes the paper.

2   Related Work

Several feature partitioning PCA based DR techniques like SubPCA [5], Mod-PCA [7] and SubXPCA [13] are available in the literature and working on different datasets. SubPCA and ModPCA approaches captured only local features, not global features, where features were partitioned into sub-patterns. There is a need to factor computational efficiency by extracting local and global features while computing principal components (PCs). Therefore, SubXPCA extracts both local and global features. Comparing SubXPCA to other techniques, it is more computationally efficient and performs better in classification [16]. All these methods are used for the spectral reduction of HSIs. However, to the best of our knowledge, particularly these feature partitioning PCA techniques have never been applied to HSI datasets. We proposed a new variation of SubXPCA as kernel SubXPCA to extract nonlinear spectral dimensions in HSI.

The motivation behind this study is grounded in the fact that several authors [6,21,23,25] have implemented HSI classification using cloud-based architectures.

2.1   SubXPCA

Sequential SubXPCA on HSI step-by-step procedure as follows.

- An original HSI, X, is divided into k spectral subsets as X_i, (i = 1 … k) of equal size.
- Apply classical PCA on each k subsets, to get locally extracted spectral features set (Y) concatenate all PCs of spectral subsets.
- Finally, apply classical PCA on Y to get a globally extracted reduced spectral feature set(Y^F) from the original image.

We have tried different variations of SubXPCA with PCA and Kernel PCA at the spectral subsets reduction phase and global spectral reduction phase. Experimental results confirm that kernel SubXPCA (kernel PCA on spectral subsets and global spectral reduction phases) classifier performance is better compared to other variations.

2.2 Kernel SubXPCA

Kernel SubXPCA is a novel variation of SubXPCA to reduce spectral dimensions of HSI. Kernel PCA leverages the kernel functions to transform the original data into a higher-dimensional space where linear separability is more likely to be achieved. This transformation makes it possible to extract non-linear features. A kernel is a function used to map data into a higher-dimensional space to make it easier to perform linear separations that may not be feasible in the original lower-dimensional space [19]. The choice of kernel function [9] is crucial because it determines how the data is transformed into higher dimensional space. There are various kernel functions available, including Linear, Sigmoid, Polynomial, Radial Bias Funtion (RBF) and Cosine. Kernel SubXPCA performs better when using the cosine kernel function than other kernels based on experimental results. The cosine function finds cosine similarity between points x and y is represented as $k(x, y) = x \cdot y \, / \, \|x\|\|y\|$.

Kernel PCA can capture complex, nonlinear relationships, which is often the case in HSI. Kernel PCA is more computationally intensive than PCA due to the computation of kernel matrix for large HSI datasets. Even though computationally intensive it is powerful for capturing nonlinear structures, providing better feature extraction and classification performance for complex Hyperspectral data [10, 14, 15].

Sequential Kernel SubXPCA procedure for HSI as follows:

- An original HSI, X, is partitioned into k spectral subsets of equal size without any two sub spectra having common features, indicated as X_i, (i = 1 ... k)
- Kernel PCA is performed on each of k subsets (X_i).
- Finally, PCs of all spectral subsets are concatenated to get transformed locally extracted spectral features set (Y).
- Apply kernel PCA on Y to get a globally extracted reduced spectral features set (Y^F) from the HSI.

3 Proposed Methodology

In this section, we present our proposed distributed design and implementation of spectral DR and classification on large HSI data. To reduce the spectral dimensions of HSI we proposed scalable kernel SubXPCA and for classification, we applied classifiers like SVM, LR, NB, DT, RF, KNN, and ET. Specifically, SVM, DT, RF, KNN, and ET can be effective when dealing with HSI data, especially better suited to handle the complexities and high variance in the data transformed by kernel PCA. The design and implementation details are discussed in the next sub-sections. The proposed design and implementation are different currently, there is not much research done on the distributed implementation of DR methods, particularly feature partitioning DR techniques on HSI.

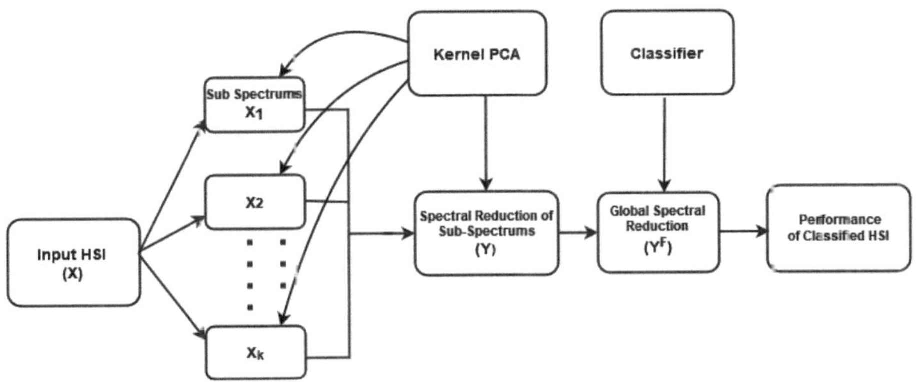

Fig. 1. Execution Flow of Kernel SubXPCA and Classifier on HSI

3.1 Design of Scalable Kernel SubXPCA

The distributed workflow for proposed scalable kernel SubXPCA, shown in Fig. 1, operates as follows: The input HSI(X) is divided into sub-spectrums (X_1, X_2, ..., X_k) and distributed across the Spark cluster nodes, enabling parallel processing of the spectral bands. Kernel PCA is then applied to each sub-spectrum in parallel on each node, reducing their spectral dimensions. The reduced sub-spectrums from all nodes are aggregated into a globally reduced spectral representation, denoted as Y. Next, Kernel PCA is applied again to Y, resulting in a further reduced representation (Y^F). This final reduced dataset Y^F is then classified using various classifiers, and performance metrics are computed. This distributed proposed approach effectively manages large-scale HSI data, enhancing both scalability and computational speed. The proposed method's step-by-step process is shown in Algorithm 1.

3.2 Implementation

We must select a distributed programming model and computation engine to construct a proposed algorithm in a distributed framework. Our preferred language for programming with a spark computing engine is PySpark. We make use of a few Python libraries, including MLlib, NumPy, and Pandas. The code utilizes the Pyspark DataFrame for implementation, distributing, and storing each segment of the HSI dataset among the cluster nodes. Normalize the features in

---

Algorithm 1. Scalable Kernel SubXPCA

Input: HSI image X(N×D)

1: The original image cube $\mathbf{X}_{N \times D}$ divides into k (≥ 2) spectral subsets of equal size without overlapping among different subsets. Each subset size d_l, where $d_l = [D/k]$ denoted as X_i where i = 1 ... k.

2: Send each spectral subset (X_i) to worker nodes in the cluster.

3: Each worker proceeds parallelly to compute the kernel matrix K for each spectral subset
$K_{ij} = k(x_i, x_j)$ for all pairs of x_i and x_j
and then center the kernel matrix
$K' = K\text{-}1_n K\text{-}K1_n + 1_n K1_n$, 1_n is n×n matrix.

4: Find eigen value decomposition on the centered kernel matrix K', find eigen values $\lambda = [\lambda_1, \lambda_2, \ldots \lambda_m]$, and eigen vectors V=$[\nu_1, \nu_2, \ldots, \nu_m]$ from $K'V = \lambda$ V. Each worker transforms their spectral subset concurrently throughout this step.

5: The master node receives the transformed spectral subsets from each worker node and then concatenated as Y.

6: Do computation of kernel matrix and its spectral decomposition from steps 3 and 4 to get globally extracted PCs, denoted as $\mathbf{Y}^F$.

7: Finally, obtained $\mathbf{Y}^F$ is a reduced form of original HSI $\mathbf{X}_{N \times D}$.

Output: Spectral reduction form of HSI $\mathbf{Y}^F$

---

each DataFrame and fit the Kernel PCA model. Every row of modified features derived by Kernel PCA is subjected to the map operation. Kernel PCA feature values are extracted from each row using the Resilient Distributed Dataset (RDD) and subsequently gathered for additional processing. For each subpattern, this procedure is carried out in parallel. Each subpattern of the HSI data across all executor's reduced spectral dimensions is concatenated by the driver process, which then saves it as a DataFrame. And further classification model available in ML lib is applied.

4   Experimental Evaluation and Discussion

This section presents experiments conducted on several ground truth and synthetic HSI datasets using different feature partitioning spectral DR methods.

4.1   Datasets Description

For our research, we took three original, publicly accessible datasets from hyperspectral remote sensing sceneries[1], which are described as follows.

Indian pines: In 1992, the AVIRIS sensor was used to collect the Indian Pines dataset. It encompasses a rural region in northwest Indiana, USA with a resolution of 20 m. The dataset has 200 spectral bands with a spatial dimension

---

[1] https://www.ehu.eus/ccwintco/index.php/Hyperspectral_Remote_Sensing_Scenes.

of 145 × 145 pixel size. It has 16 distinct classes in ground truth visual scene is shown in Fig. 2(a).

Salinas Valley: In 1998, the AVIRIS sensor collected Salinas dataset. It captures Salinas Valley region in California, USA with a resolution of 3.7 m. The dataset has 224 spectral bands with spatial dimensions of 512 × 217 pixel size The ground truth scene of Salinas has 16 classes is shown in Fig. 2(b).

Pavia University: In 2003, the ROSIS sensor captures the university of Pavia. It covers the surrounding area of Pavia university with a resolution of 1.3 m. The dataset has 103 spectral bands with spatial dimensions of 610 × 340 pixel size. This HSI ground truth scene has 9 different classes is shown in Fig. 2(c).

We aim to evaluate the scalability of our parallel spectral DR approach on larger synthetic HSI datasets, which are generated by applying Python mosaicking functions to the original Salinas HSI, yielding data files as follows: Salinas-2 with 1.3 GB, Salinas-3 with 2.6 GB, Salinas-4 with 5.2 GB, Salinas-5 with 10.4 GB.

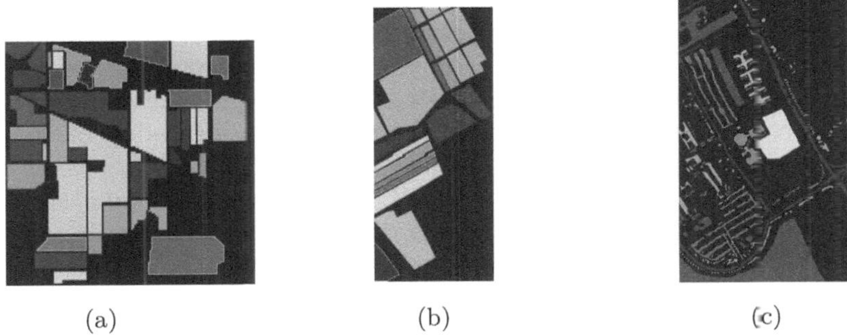

(a)   (b)   (c)

Fig. 2. The Ground truth HSI Scene of (a) Indian Pines (b) Salinas (c) Pavia

4.2   Cluster Specifications

In our lab, we set up a standalone Spark cluster with three nodes on which we conduct the experiments. One node in the standalone cluster serves as both a worker and a master, while the other nodes solely function as workers. Every node has 30GB of RAM and 16 cores. Three nodes in the cluster running on Ubuntu 20.04.5 LTS and are connected by Ethernet at a speed of 1000 Mbps. Java 11.0.19, Apache Spark 3.3.1, and Python 3.8.10 are installed on each node. Additional libraries that help with our dataset experiments are installed

Additionally, Google Cloud Dataproc[2] is used for our experiments. Dataproc is a fully managed and highly scalable service for operating Apache Hadoop,

---

[2] https://cloud.google.com/dataproc.

Spark, Flink, and Presto together with over thirty open-source tools and frameworks with Google Cloud integration. Using virtual machines (VMs), we built and tested a Standard (1 Master, N Workers) Google Cloud Dataproc cluster. The virtual master node has 100GB of RAM. A VM for each worker node has 50GB of RAM. We have created six node Dataproc cluster for our experiments.

4.3 Results

We evaluate our methods using three key performance metrics: accuracy, execution time, and speedup. For each classifier, we assess overall accuracy, average accuracy, and the kappa coefficient, comparing different spectral reduction techniques. Execution time is measured using the Spark cluster's web interface, which logs application completion times. Speedup represents the performance improvement of a parallel algorithm over its sequential version, expressed as $S = \frac{T_1}{T_p}$, where T_1 is the sequential execution time and T_p is the parallel execution time. An n-fold speedup means the parallel algorithm is n times faster.

Each experiment is repeated five times across all metrics for each HSI dataset on both the Cluster environments. The average of these repetitions is then used as the final outcome. Table 1 summarizes the performance of several classifiers using feature partitioned, distributed spectral DR techniques on ground truth HSI datasets on standalone Spark Cluster. The findings indicate that the proposed approach improves classifier accuracy across most models compared to traditional methods. Specifically, scalable kernel SubXPCA achieves superior accuracy for SVM, DT, RF, KNN, and ET classifiers, surpassing the performance of LR and NB classifiers. This improvement can be attributed to the ability of SVM, DT, RF, KNN, and ET classifiers to capture nonlinear relationships by segmenting the data into smaller subsets, allowing them to handle the complex, nonlinear structures produced by kernel PCA transformations. In contrast, the linear nature of NB and LR limits their capacity to benefit from the complex, correlated features generated by kernel PCA, often resulting in lower accuracy for these models.

The execution time and speedup of the proposed approach on synthetic Salinas datasets using varying numbers of cores in both standalone and Dataproc Cluster environments reported in Table 2 and Fig. 3. The results show that the Dataproc Cluster achieves greater speedup compared to the standalone cluster. As the dataset size increases, both environments demonstrate improved time ratios and speedup. The results indicate that, in both clusters, speedup is nearly linear up to 16 cores. However, in the standalone cluster, this linearity diminishes at 32 cores, likely due to the constraints posed by a limited number of nodes. The Dataproc Cluster, on the other hand, maintains close to linear performance up to 16 cores and continues to deliver enhanced performance even with up to 64 cores, leveraging its superior computational power. As the number of cores increases, the overhead for coordinating and managing tasks across the cluster also rises. Beyond a certain point, this overhead can offset the benefits of parallel processing, resulting in diminishing returns. This is especially evident in high-core environments, where coordination costs start to exceed the gains from additional parallelism. For the Salinas-2 to Salinas-5 scalable kernel SubXPCA achieves a speedup ranging from 6.05× to 7.26× in the standalone cluster.In the Dataproc Cluster, this speedup improves to 9.05×, 9.86× with 32 cores, due to variation in the number of nodes.

Table 1. Classification Metrics (Overall Accuracy (OA), Average Accuracy (AA) and Kappa Co-efficint (kappa)) for Distributed Feature Partitioning DR Methods on Ground Truth HSI Datasets

Dataset	Distributed Spectral Reduction Method	Classifier																				
		SVM			NB			LR			DT			RF			KNN			ET		
		OA	AA	kappa	OA	AA	kappa	OA	AA	kappa	OA	AA	kappa	OA	AA	kappa	OA	AA	kappa	OA	AA	kappa
Indian Pines	PCA	74.51	74.02	72.36	60.24	59.21	58.16	77.32	75.41	74.45	70.17	69.25	69.01	81.39	80.42	79.56	80.27	79.15	79.03	77.35	75.90	73.25
	SubPCA	75.47	74.06	73.17	63.02	61.35	60.91	**78.00**	76.05	74.55	72.51	71.55	70.21	83.65	82.77	80.45	82.09	80.03	79.15	79.00	77.24	76.45
	ModPCA	74.54	73.82	72.48	61.43	60.71	59.88	77.19	75.84	73.12	73.41	73.87	71.53	83.14	81.48	80.14	82.15	81.03	79.05	79.58	78.47	77.26
	SubXPCA	75.47	74.58	73.57	**68.14**	67.32	65.62	**78.00**	76.26	75.73	75.37	74.22	72.90	84.63	83.73	83.62	83.75	82.62	80.27	82.00	80.09	79.15
	Kernel SubXPCA	**77.39**	75.35	74.27	58.50	57.29	55.45	69.25	67.55	66.36	**77.21**	75.36	73.55	**85.38**	84.24	84.77	**85.38**	83.57	82.25	**84.61**	83.72	81.58
Pavia University	PCA	90.56	88.42	86.25	76.32	74.58	72.14	85.79	83.52	81.44	83.77	81.28	80.54	88.54	85.74	84.16	88.45	86.90	86.02	85.36	83.19	82.04
	SubPCA	91.45	88.45	86.58	81.66	80.64	79.42	85.90	83.56	80.57	85.68	82.55	81.45	89.08	86.46	85.10	89.85	88.27	86.70	88.18	86.25	84.33
	ModPCA	91.44	89.23	87.17	79.10	78.47	78.01	86.88	84.04	81.59	84.01	81.72	80.18	88.77	86.98	85.47	89.27	87.48	85.05	89.05	87.10	85.29
	SubXPCA	91.45	89.51	88.56	**83.74**	83.57	81.25	**86.91**	84.58	82.39	86.53	83.85	82.06	89.85	87.06	86.17	90.95	89.03	87.79	89.15	87.22	86.54
	Kernel SubXPCA	**93.05**	91.26	90.45	76.50	74.99	73.25	79.45	77.66	76.85	**88.67**	86.24	84.32	**94.25**	93.35	91.76	**94.45**	93.28	92.19	**93.65**	92.17	90.25
Salinas Valley	PCA	88.05	86.18	84.25	87.62	87.08	85.47	89.22	87.43	85.15	90.42	89.47	86.28	94.12	92.47	90.58	89.26	86.45	85.15	90.23	88.17	87.29
	SubPCA	88.16	87.28	86.14	89.72	88.45	88.04	**90.73**	89.01	88.72	91.84	89.25	88.15	94.76	93.52	92.85	90.75	89.50	88.54	93.45	92.26	90.45
	ModPCA	89.10	88.15	87.05	88.50	87.28	86.85	90.44	88.14	87.58	92.09	90.74	89.53	94.89	93.06	92.75	91.56	89.28	89.04	93.24	91.46	90.18
	SubXPCA	90.05	89.01	88.30	**91.98**	90.93	89.06	**90.73**	89.26	89.65	93.12	92.62	91.34	95.59	94.88	94.07	92.37	91.35	90.48	96.35	94.27	92.05
	Kernel SubXPCA	**92.45**	91.36	89.38	81.99	80.26	79.38	81.56	80.05	79.24	**95.30**	93.99	92.15	**97.07**	95.54	94.25	**97.37**	95.13	94.28	**98.02**	96.15	95.27

Table 2. Execution Time (in Secs) and Speedup of Scalable Kernel SubXPCA on Synthetic Salinas Datasets with varying Cores in:

(a) Standalone Cluster

No. of Cores	Salinas-2 (1.3 GB)		Salinas-3 (2.6 GB)		Slainas-4 (5.2 GB)		Salinas-5 (10.4 GB)	
	Ex.Time	Speedup	Ex.Time	Speedup	Ex.Time	Speedup	Ex.Time	Speedup
2	5340	1	12650	1	25206	1	52980	1
4	2656	2.01	6141	2.06	11724	2.15	24081	2.20
8	1695	3.15	3857	3.28	7021	3.59	13278	3.99
16	936	5.70	2144	5.90	4194	6.01	8614	6.15
32	882	6.05	2067	6.12	3525	7.15	7297	7.26

(b) Dataproc Cluster

No. of Cores	Salinas-2 (1.3 GB)		Salinas-3 (2.6 GB)		Slainas-4 (5.2 GB)		Salinas-5 (10.4 GB)	
	Ex.Time	Speedup	Ex.Time	Speedup	Ex.Time	Speedup	Ex.Time	Speedup
2	2540	1	5242	1	11986	1	21265	1
4	1239	2.05	2532	2.07	5574	2.15	9661	2.20
8	619	4.10	1272	4.12	2634	4.55	4382	4.85
16	350	7.24	703	7.45	1572	7.62	2673	7.95
32	280	9.05	548	9.56	1228	9.76	2155	9.86
64	177	14.27	360	14.55	808	14.82	1426	14.90

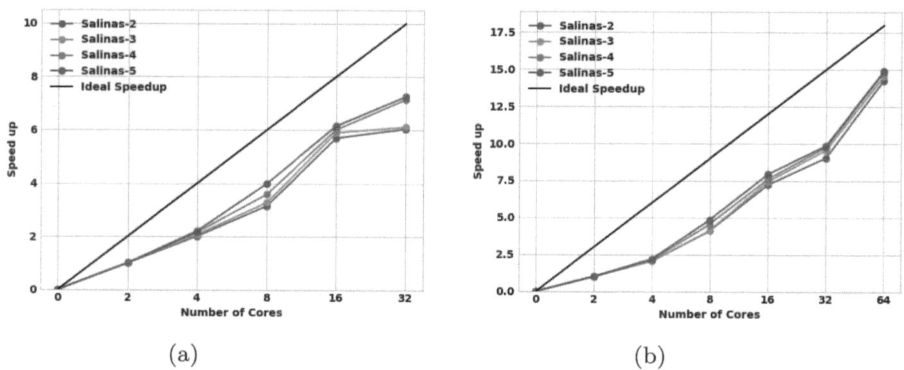

(a)   (b)

Fig. 3. Comparison of speedup with Number of cores (a) Standalone Cluster (b) Dataproc Cluster.

Table 3. Comparison of Classifier Accuracy of Kernel SubXPCA with different kernel functions

Dataset	Kernel Function	Classifiers on Kernel SubXPCA								
		SVM			KNN			ET		
		OA	AA	kappa	OA	AA	kappa	OA	AA	kappa
Indian Pines	Linear	68.17	67.22	66.18	66.26	64.23	63.75	65.98	64.25	62.18
	Sigmoid	71.45	70.65	68.54	73.11	71.49	71.20	72.55	70.60	68.15
	Polynomial	73.19	71.37	69.05	79.65	78.10	77.38	78.14	77.28	75.28
	RBF	76.25	74.18	72.27	83.02	81.70	79.05	83.27	81.06	80.35
	Cosine	**77.39**	**75.35**	**74.27**	**85.38**	**83.57**	**82.25**	**84.61**	**83.72**	**81.58**
Pavia University	Linear	78.48	76.99	75.05	79.16	78.27	76.18	77.24	75.18	74.05
	Sigmoid	81.25	79.28	78.00	82.24	81.25	80.06	80.66	78.44	76.80
	Polynomial	86.33	85.15	83.45	88.57	86.17	85.47	85.49	83.19	82.55
	RBF	91.45	90.47	89.76	92.05	90.25	89.34	90.16	89.24	88.06
	Cosine	**93.05**	**91.26**	**90.45**	**94.45**	**93.24**	**92.19**	**93.65**	**92.17**	**90.25**

4.4   Ablation Study

To evaluate the classification performance of kernel SubXPCA, we conducted ablation experiments using the Indian Pines and Pavia University HSI datasets. In these experiments, we tested different kernel functions, including Linear, Sigmoid, Polynomial, RBF, and Cosine, across classifiers such as SVM, KNN and ET as shown in Table 3. The results indicate that kernel SubXPCA achieves the highest classification performance when utilizing the Cosine kernel function on both HSI datasets. While the Linear kernel function yielded the lowest performance among the kernels tested, the RBF kernel also demonstrated competitive results, though slightly lower than those of the Cosine kernel. Overall, these findings highlight the effectiveness of the Cosine kernel in enhancing classification performance using kernel SubXPCA for HSI.

5 Conclusion

In this study, we examined and employed various distributed feature partitioning DR techniques to lower the spectral dimensions of HSI datasets used for classification. Our investigation shows that the proposed approach has superior running times and faster speedup in parallel implementation on larger HSI datasets. Scalable kernel SubXPCA proved better classification performance for most of the classifiers than other distributed feature partitioning PCA based methods. In future work. We intend to integrate spatial DR technique with the suggested design to enhance classification performance and efficiently optimize computations and performance in various distributed environments.

References

1. Aburaed, N., Alkhatib, M.Q., Marshall, S., Zabalza, J., Al Ahmad, H. A review of spatial enhancement of hyperspectral remote sensing imaging techniques. IEEE J. Sel. Top. Appl. Earth Observ. Remote Sens. **16**, 2275–2300 (2023)
2. Afaq, Y., Manocha, A.: Analysis on change detection techniques for remote sensing applications: a review. Eco. Inform. **63**, 101310 (2021)
3. Anowar, F., Sadaoui, S., Selim, B.: Conceptual and empirical comparison of dimensionality reduction algorithms (PCA, KPCA, LDA, MDS, SVD, LLE, ISOMAP, LE, ICA, t-SNE). Comput. Sci. Rev. **40**, 100378 (2021)
4. Bioucas-Dias, J.M., Plaza, A., Camps-Valls, G., Scheunders, P., Nasrabadi, N., Chanussot, J.: Hyperspectral remote sensing data analysis and future challenges. IEEE Geosci. Remote Sens. Mag. **1**(2), 6–36 (2013)
5. Chen, S., Zhu, Y.: Subpattern-based principle component analysis. Pattern Recogn. **37**(5), 1081–1083 (2004)
6. Feng, K., Wu, Y.: Distributed cloud computing architecture in hyperspectral remote sensing image classification under big data. In: 2022 IEEE 5th International Conference on Information Systems and Computer Aided Education (ICISCAE), pp. 456–460. IEEE (2022)
7. Gottumukkal, R., Asari, V.K.: An improved face recognition technique based on modular PCA approach. Pattern Recogn. Lett. **25**(4), 429–436 (2004)
8. He, L., Li, J., Liu, C., Li, S.: Recent advances on spectral-spatial hyperspectral image classification: an overview and new guidelines. IEEE Trans. Geosci. Remote Sens. **56**(3), 1579–1597 (2017)
9. Hossain, M.M., Hossain, M.A.: Feature reduction and classification of hyperspectral image based on multiple kernel PCA and deep learning. In: 2019 IEEE International Conference on Robotics, Automation, Artificial-intelligence and Internet-of-Things (RAAICON), pp. 141–144. IEEE (2019)
10. Hossain, M.M., Hossain, M.A., Al Mamun, M., Hossain, M.M.: Feature reduction based on the fusion of spectral and spatial transformation for hyperspectral image classification. In: 2020 IEEE Region 10 Symposium (TENSYMP), pp 150–153. IEEE (2020)
11. Jolliffe, I.T., Cadima, J.: Principal component analysis: a review and recent developments. Philos. Trans. Roy. Soc. A: Math. Phys. Eng. Sci. **374**(2065), 20150202 (2016)
12. Khan, M.J., Khan, H.S., Yousaf, A., Khurshid, K., Abbas, A.: Modern trends in hyperspectral image analysis: a review. IEEE Access **6**, 14118–14129 (2018)

13. Kumar, K.V., Negi, A.: SubXPCA and a generalized feature partitioning approach to principal component analysis. Pattern Recogn. **41**(4), 1398–1409 (2008)

14. Merrill, N., Olson, C.C.: Unsupervised ensemble-kernel principal component analysis for hyperspectral anomaly detection. In: Proceedings of the IEEE/CVF Conference on Computer Vision and Pattern Recognition Workshops, pp. 112–113 (2020)

15. Miftahushudur, T., Grieve, B., Yin, H.: Permuted KPCA and smote to guide GAN-based oversampling for imbalanced HSI classification. IEEE J. Sel. Top. Appl. Earth Observ. Remote Sens. **17**, 489–505 (2023)

16. Negi, A., Kadappa, V.K.: Subxpca versus PCA: a theoretical investigation. In: 2010 20th International Conference on Pattern Recognition, pp. 4170–4173. IEEE (2010)

17. Salloum, S., Dautov, R., Chen, X., Peng, P.X., Huang, J.Z.: Big data analytics on apache spark. Int. J. Data Sci. Anal. **1**, 145–164 (2016)

18. Schölkopf, B., Smola, A., Müller, K.-R.: Kernel principal component analysis. In: Gerstner, W., Germond, A., Hasler, M., Nicoud, J.-D. (eds.) ICANN 1997. LNCS, vol. 1327, pp. 583–588. Springer, Heidelberg (1997). https://doi.org/10.1007/BFb0020217

19. Schölkopf, B., Smola, A., Müller, K.R.: Nonlinear component analysis as a kernel eigenvalue problem. Neural Comput. **10**(5), 1299–1319 (1998)

20. Singh, D., Reddy, C.K.: A survey on platforms for big data analytics. J. Big Data **2**(1), 1–20 (2015)

21. Sun, J., et al.: An efficient and scalable framework for processing remotely sensed big data in cloud computing environments. IEEE Trans. Geosci. Remote Sens. **57**(7), 4294–4308 (2019)

22. Tejasree, G., Agilandeeswari, L.: An extensive review of hyperspectral image classification and prediction: techniques and challenges. Multimed. Tools Appl. 1–98 (2024)

23. Torti, E., Marenzi, E., Danese, G., Plaza, A.J., Leporati, F.: Spatial-spectral feature extraction with local covariance matrix from hyperspectral images through hybrid parallelization. IEEE J. Sel. Top. Appl. Earth Observ. Remote Sens. **16**, 7412–7421 (2023)

24. Wu, Z., Li, Y., Plaza, A., Li, J., Xiao, F., Wei, Z.: Parallel and distributed dimensionality reduction of hyperspectral data on cloud computing architectures. IEEE J. Sel. Top. Appl. Earth Observ. Remote Sens. **9**(6), 2270–2278 (2016)

25. Wu, Z., et al.: Scheduling-guided automatic processing of massive hyperspectral image classification on cloud computing architectures. IEEE Trans. Cybern. **51**(7), 3588–3601 (2020)

26. Zhao, W., Du, S.: Spectral-spatial feature extraction for hyperspectral image classification: a dimension reduction and deep learning approach. IEEE Trans. Geosci. Remote Sens. **54**(8), 4544–4554 (2016)

27. Zheng, W., Zou, C., Zhao, L.: An improved algorithm for kernel principal component analysis. Neural Process. Lett. **22**, 49–56 (2005)

Gesture Control System for Drones: Enhancing Human-Drone Interaction Through Real-Time Machine Learning and Depth Sensing

Sahil Pankaj Kalro[1], Shaun Steve Pereira[1], Swapnil Manjunath Kirloskar[1], Varun Dayanand Shringare[1], Supriya Patil[1], and Anusha Raghavendra Pai[2]([envelope]) [ORCID]

[1] Department of Computer Engineering, Padre Conceicao College of Engineering, Verna 403722, Goa, India
[2] School of Computer Science and Engineering, Dr. Vishwanath Karad MIT World Peace University, Pune, Maharashtra, India
anusha.pai@mitwpu.edu.in

Abstract. Gesture control systems are becoming vital across industries due to their ability to strengthen user interaction, accessibility, and provide innovative ways of interfacing with technology. As these systems continue to grow, they will likely become an indispensable component of user interfaces across a wide range of applications. In this work, a Gesture Control System is designed and implemented to enhance the human-drone interaction paradigm. The proposed system leverages machine learning techniques to interpret and respond to user-defined gestures, providing a seamless and natural interface for controlling drones. The work begins with an exploration of existing gesture recognition methodologies and their limitations, leading to the development of a robust model tailored for drone control. This is implemented as a real-time image processing pipeline that captures and interprets gestures using a combination of depth sensing and machine learning, enabling users to convey complex commands effortlessly. The methodology involves integrating a ground station system with the drone controller, which captures live video data to be processed by the gesture recognition algorithm in real-time. The machine learning model, trained on a diverse dataset of gestures, accurately classifies user-defined commands, allowing for precise and responsive drone control. The system's performance is assessed through a series of experiments, evaluating its accuracy, latency, and reliability in various environmental conditions.

Keywords: Gesture Control System · MediaPipe · Machine Learning (ML) · Random Forest Classifier

1 Introduction

The rapid increase of drone technology has sparked a demand for innovative and intuitive control systems, transcending traditional remote controllers. The proliferation of drone technology has revolutionized numerous industries, offering unprecedented capabilities

C. Modi et al. (Eds.): MIND 2024, CCIS 2736, pp. 499–510, 2026.
https://doi.org/10.1007/978-3-032-14531-4_42

in aerial surveillance, data collection, and automated operations. Drones, also known as Unmanned Aerial Vehicles (UAVs), have transitioned from niche hobby gadgets to essential tools in fields such as agriculture, construction, environmental monitoring, disaster management, and more. The versatility and advanced capabilities of modern drones make them ideal for a wide array of innovative projects aimed at solving real world problems and enhancing efficiency across various domains.

The work in this paper, delves into the development of a Gesture Control System for Drones, a pioneering approach to human-drone interaction. Traditional control interfaces often pose challenges in terms of complexity and user experience. The research work proposed in [1] investigates the feasibility and effectiveness of utilizing body gestures as a control mechanism for unmanned aerial vehicles (UAVs), offering insights into gesture recognition algorithms and integration with flight control systems. The work in [2] delves into hand gesture algorithms essential for managing and controlling UAVs during missions, especially in long-range scenarios. The work in [3] sheds light on the application of drones for real-time air pollution monitoring, emphasizing the role of sensor equipped drones in environmental analysis. The authors of [4] provide a holistic overview of communication protocols and networking architectures tailored for Unmanned Aerial Vehicles, offering valuable insights into establishing robust communication systems for UAVs. The authors in [5] and [6] proposed a wearable drone controller with hand gesture recognition and vibrotactile feedback. The work in [7] investigates the feasibility and effectiveness of utilizing body gestures as a control mechanism for unmanned aerial vehicles, offering insights into gesture recognition algorithms and integration with flight control systems. Lastly the paper [8] delves into ground station technologies essential for managing and controlling UAVs during missions, especially in long-range scenarios.

The proposed work distinguishes itself through a unique approach to human-drone interaction by employing a ground station camera rather than integrating the camera directly onto the drone. This allows for increased flexibility in drone hardware, reducing weight and power consumption on the unmanned aerial vehicle. By strategically situating the camera on the ground station enhances the system's adaptability to various drone platforms, making it an inclusive solution for existing drone models. Moreover, the focus on interpreting hand gestures rather than the entire body sets this work apart. The emphasis on hand gestures not only simplifies the user interface but also caters to practical scenarios where users may prefer discreet and natural hand movements for controlling drones. In this work, cutting-edge computer vision algorithms and machine learning techniques have been employed to enable users to control drones seamlessly through intuitive gestures. The research work is broken down into the following phases:

i) Drone Construction and Sensor Implementation, ii) Machine Learning Model Design, iii) User Interface Development and Integration and iv) System Testing and Optimization.

2 Methods

2.1 Drone Construction and Sensor Implementation

The work begins with the construction of a robust 3-blade drone designed to ensure stable flight. This involves assembling the Cinewhoop frame using lightweight and durable materials, followed by the installation of five motors and propellers. An electronic speed controller (ESC) and a flight controller have been integrated to manage the drone's operations. The wiring and soldering process has been essential for power distribution and motor control. Additionally, selecting a suitable battery and setting up a reliable power management system will be critical for ensuring the drone's long flight times and operational stability. Furthermore, several researchers have leveraged simulation environments to conduct experiments and propose modifications to enhance the performance of various control systems, including those for optical flow sensors. For example, the authors in [9] utilized optical flow sensors for 2D positioning for autonomous flight.

Equipping the drone with essential sensors is the next step to enhance its flight stability and environmental awareness. An optical flow sensor has been installed to provide precise positioning and movement tracking, ensuring the drone can maintain a stable hover and navigate accurately. A barometer has also been integrated to measure altitude and assist in maintaining consistent flight levels. These sensors underwent rigorous calibration and testing to ensure their accuracy and reliability, forming the backbone of the drone's navigation system.

The drone's hardware architecture comprises a diverse set of components, each playing a crucial role in its functionality, which can be seen in Fig. 1. The flight controller, acting as the drone's brain, processes sensor inputs and executes control algorithms to maintain stability and respond to pilot commands. Motors, driven by Electronic Speed Controllers (ESCs) and governed by Pulse Width Modulation (PWM), generate thrust for propulsion. The airframe or chassis acts as the physical support structure, while onboard sensors, including gyros and accelerometers, monitor temperature, voltage, and acceleration for health and performance assessment. A First Person View (FPV) camera captures real-time video for the pilot's immersive control, while a PID controller fine-tunes the drone's behavior based on sensor feedback [10]. The battery supplies power to all components, and a Microcontroller Unit (MCU), such as the STM32F405, orchestrates overall system operation. The transmitter and RC receiver facilitate communication between the pilot and the drone, while telemetry radio enables real-time data exchange with the ground station. This comprehensive integration of hardware components ensures the drone functions as a unified system, capable of stable flight, precise navigation, and responsive control.

The main component of the drone is the flight controller that processes data from onboard sensors, such as gyroscopes and accelerometers, to stabilize the drone in flight and execute flight commands. It also facilitates communication between the drone's components and enables the addition of more sensors, such as an optical flow sensor and a barometer, onto the drone for increased stability and functions, as shown in Fig. 1.

Furthermore, several researchers have leveraged simulation environments to conduct experiments and propose modifications to enhance the performance of various control systems, including those for optical flow sensors. For example, [9] utilized optical flow

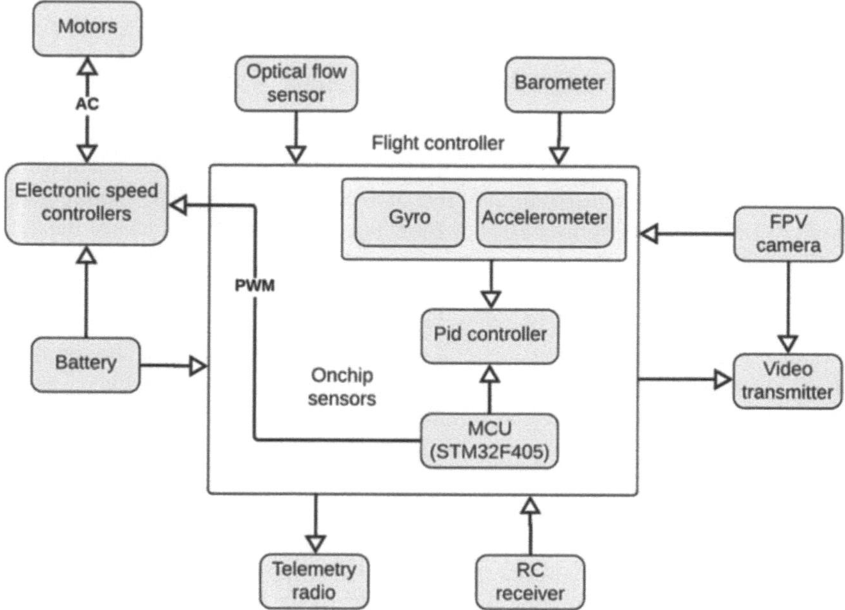

Fig. 1. Hardware architecture

sensors for 2D positioning for autonomous flight. The motors used offer high efficiency, reliability, and power-to-weight ratios, making them ideal for drone applications. Propellers are affixed to the drone's motors and are responsible for generating thrust to lift the aircraft off the ground and propel it through the air. The drone in use consisted of a 3-blade propeller instead of a 2-blade propeller [11], which produced less thrust compared to the 3-blade propeller. The FPV camera used on the drone is mounted on the frame of the drone. The camera captures video footage from the drone's perspective and provides the feed to the UI via VTX (video transmitter) and VRX (video receiver). The drone's Cinewhoop frame provides structural support and houses its components in a compact and aerodynamic configuration, enabling the drone to be flown indoors.

In addition to these core components, auxiliary equipment such as VTX antennas, which facilitate live video transmission to ground stations or FPV goggles, and specialized sensors like altitude sensors or airspeed sensors, can enhance the drone's capabilities for specific tasks or environments. To gain altitude and maintain altitude hold, as mentioned in [12], and to avoid the drone veering off course, an optical flow and a LiDAR sensor Matek 3901LOX (later replaced by MTF01) was used to lock the position of the drone in 3D space. Using this sensor, the drone could be flown indoors without the risk of crashes. The optical flow sensor detected linear motion parallel to the floor, while the LiDAR sensor measured altitude changes.

This work utilized the MAVLink communication protocol, a lightweight messaging system specifically designed for unmanned aerial vehicles (UAVs), as referenced in [13]. MAVLink facilitated the bidirectional exchange of data between the ground station and the drone. It transmitted control signals from the ground station to control the drone while simultaneously receiving telemetry data, such as battery levels, position, and sensor readings from the drone, providing real-time feedback for mission monitoring.

2.2 Machine Learning Model Design

With the drone's hardware and sensor systems in place, the focus shifted to designing the machine learning (ML) models for gesture recognition. Machine learning and deep learning algorithms have been extensively used in literature for object detection [1, 14] and body movement detection [15]. In this work, the initial step involved collecting a diverse dataset of hand gestures, a critical foundation for training and evaluating the models. Using the MediaPipe framework, which detects and tracks hand key points, videos of ourselves performing various hand gestures were recorded. Each video was 4–5 min long, and the frames were manually classified by playing the video and using keyboard keys to categorize gestures. From these videos, between 200 to 300 images for each gesture were extracted. The dataset was split into two parts: 95% was used for training the ML models, and the remaining 5% was reserved for testing. This division ensured the models had ample data to learn effectively while providing a reliable benchmark to assess their accuracy and generalization capabilities.

Various ML models, including logistic regression, ridge classifier, random forest, and gradient boosting, were tested to identify the most effective approach for recognizing these gestures. Each model was trained and validated on the dataset to evaluate its performance in terms of accuracy, robustness, and computational efficiency. After rigorous testing, the random forest classifier emerged as the best-performing model, achieving an accuracy score of 0.9951 and a precision of 0.9949. Its superior performance, combined with relatively low computational demands, made it ideal for ensuring real-time processing.

The creation of machine learning model to detect and classify different hand gestures is the core part of this hand gesture-controlled drone project. It involves developing a sophisticated system capable of interpreting hand movements captured via video feed and translating them into actionable commands for the drone. This process—encompassing data collection, feature extraction, model training, and evaluation—is crucial for achieving robust performance in diverse environmental conditions and with varying hand poses. The successful development of this model is pivotal for enabling seamless interaction between users and drones, paving the way for applications in fields such as surveillance, search and rescue, entertainment, and beyond.

As shown under training phase in Fig. 2 capturing a video dataset is essential for image processing and capturing various gestures being used to control the drone. This process involves meticulously setting up a controlled environment with consistent lighting and camera specifications to ensure high quality data capture. Various hand gestures are performed and recorded from multiple angles, providing diverse examples for the model to learn from. By capturing a comprehensive dataset, we aim to cover a broad spectrum of gestures and scenarios, enabling the model to generalize well to real-world situations. The gesture dataset is processed using MediaPipe. The generation of hand key points using MediaPipe does the major task in the ML model's ability to mark the keypoints on the frames of the video dataset. MediaPipe employs sophisticated deep learning models to detect and localize key points representing essential hand landmarks, such as fingertips, knuckles, and palm center, from the captured video frames. These key points serve as the foundation for extracting meaningful features that characterize different hand gestures. By leveraging MediaPipe's robust hand pose estimation capabilities, we can ensure the accurate representation of hand movements, facilitating effective gesture recognition by the ML model.

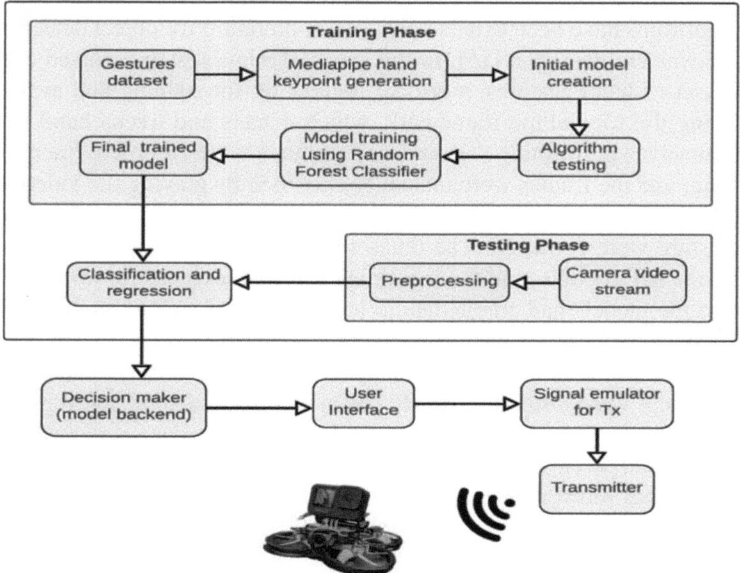

Fig. 2. Software data flow

In the initial phase of model development, multiple machine learning algorithms are tested to determine the most suitable approach for detecting and classifying hand gestures. Algorithms such as random forest, logistic regression, ridge classifier, and gradient boosting classifier were implemented and trained using the hand key points generated by MediaPipe. Each algorithm went through testing and evaluation to assess its performance in terms of accuracy, precision, and efficiency. This iterative process allowed us to identify the algorithm that exhibits the best performance for our specific application.

After thorough testing and evaluation, the random forest algorithm was the optimal choice for training the ML model. Its ensemble learning approach, which combines multiple decision trees, proves effective in capturing complex relationships within the hand gesture data. Random forest demonstrates superior performance in terms of accuracy, robustness, and computational efficiency compared to other algorithms tested. By selecting random forest for model training, reliable and accurate gesture recognition is ensured, laying the foundation for seamless interaction between users and drones.

As conveyed under testing phase in Fig. 2, the final stage of model development involves rigorous testing to evaluate its performance and generalization ability. A separate test dataset, distinct from the training data, is used to assess the model's accuracy, precision, recall, and F1-score. This testing phase aims to validate the model's effectiveness in classifying the hand gestures. The results of the testing phase provide valuable feedback for further refinement and optimization, so that the model can perform in different real-world scenarios.

As shown in Fig. 2, the final model is passed on to the decision maker. The backend for this project was developed using the Django Python framework to create a robust web server capable of handling HTTP requests and API calls. One of the key functionalities implemented was an API designed to transmit live video feed from a machine learning (ML) model to the frontend. Each frame of the video feed was encrypted before being sent to the frontend as an image, ensuring secure transmission. Additionally, the backend was tasked with facilitating communication between the ML model and the signal emulator to control a drone. Another API was developed to enable functionalities such as arming the drone, setting throttle position, and controlling yaw motion. Furthermore, the backend played a crucial role in retrieving telemetry data from the drone and transmitting it to the frontend for display and analysis. This seamless integration of Django and Python empowered the backend to efficiently manage communication between various components of the system, enabling smooth operation and interaction between the user interface and the drone.

2.3 User Interface Development and Integration

Once the optimal ML model is selected, the next phase involves developing a user interface (UI) that allows users to interact with the drone seamlessly. The UI will be designed to be intuitive, providing visual feedback on recognized gestures and drone status. Integration of the UI with the drone's control system will be crucial, ensuring that the recognized gestures translate into drone commands with minimal latency. This integration will involve setting up communication protocols between the UI, the ML model, and the drone's flight controller.

The UI consists of five main parts as shown in Fig. 3, to allow the user to have a more intuitive and enhanced drone control experience it includes the following features:

1. Webcam gesture capture
2. Live drone footage
3. Telemetry data display
4. Weather statistics display
5. Manual drone controls

Webcam integration enables the system to capture and interpret hand gestures made by the user, translating them into commands for the drone. By leveraging computer vision techniques, such as image processing and machine learning algorithms, the system can recognize predefined gestures and map them to specific drone movements or actions. The webcam serves as the input device, capturing video footage of the user's hand movements in real-time.

Machine learning plays a crucial role in gesture recognition, as it allows the system to learn and adapt to various gestures performed by the user. With the help of these algorithms the model is first trained for specific hand gestures and then tested if it gives the desired outputs for specific gestures.

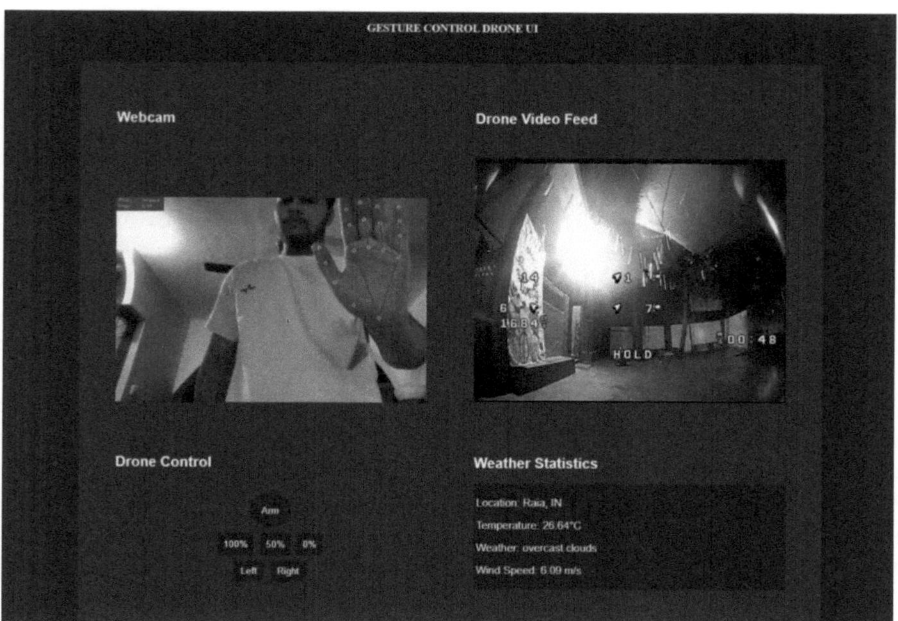

Fig. 3. User Interface

To enable live drone footage streaming, the drone goggles are connected to a laptop or a compatible device capable of receiving and displaying video feeds in real-time. By streaming the live feed to an external device, such as a laptop, users can view the drone's perspective remotely and share the experience with others. This capability is particularly useful in scenarios where teamwork and coordination are essential, such as search and rescue missions, aerial photography/videography, and surveillance operations. By sharing the live feed with other users, the drone operator can leverage additional expertise and insights to achieve mission objectives more effectively. The UI uses Wide FPV which is an open-source project used to display the DJI goggle feed on various devices such as mobiles, laptops or PCs.

Telemetry sensors are integrated into the drone to collect various flight data parameters, such as altitude, speed, GPS location, battery level, and sensor readings. These sensors provide real-time feedback on the drone's performance and environmental conditions, allowing users to monitor its status and make informed decisions during flight. Moreover, telemetry data enables post-flight analysis and performance optimization, allowing users to fine-tune their operations for improved efficiency and reliability.

Weather conditions have a significant impact on drone performance and safety, influencing factors such as flight stability, battery life, and sensor accuracy. By monitoring weather statistics in real-time, users can make informed decisions to ensure the safety of both the drone and surrounding airspace. Weather statistics are retrieved from online sources, such as weather APIs to get live and accurate weather statistics based on the location. In this UI the OpenWeatherMap API has been employed to display the weather statistics.

2.4 System Testing and Optimization

The final phase focuses on ensuring the entire system operates efficiently and responds to gestures with minimal time delay. Extensive testing will be conducted to evaluate the system's performance under various conditions, including different lighting and background scenarios. Any latency in gesture recognition and drone response will be identified and optimized. The system will also be tested for robustness, ensuring it can handle rapid gesture inputs and maintain stable flight. By the end of this phase, the Gesture Control System will be refined to provide a seamless, responsive, and user-friendly experience, ready for practical deployment.

Signal emulation for the project involves a main model which generates the digital signals that need to be converted into analog signals for controlling a drone. Process starts with the main model running on ground station, generates a digital signal which corresponds to the user's gesture. These digital signals are directly transmitted to an ESP32 microcontroller via the UART (Universal Asynchronous Receiver-Transmitter) communication interface.

The ESP32, equipped with an 8-bit Digital-to-Analog Converter (DAC), receives the digital signal. A program written in C++ on the ESP32 processes the incoming digital signal. Task of the ESP32 is to convert the digital value of $0\text{-}255$ to the analog signal or voltage $0V\text{-}$ using its DAC. This conversion transforms the discrete digital signal into a continuous analogue signal, accurately representing the original digital data. The voltage generated by the ESP32 is then sent to the drone's controller, which requires analogue input to operate correctly.

The drone's controller, designed to interpret analog signals from the joystick, receives the converted analogue signal from the ESP32. This signal is used to control the motion of the drone. By providing precise analog signals, the drone can perform controlled flight maneuvers based on the user's hand gesture.

3 Results and Discussion

This paper presents the design and development efforts undertaken to implement a drone control system through hand movements. The hand gestures were trained on different machine learning models such as Logistic Regression, Ridge Classifier, Random Forest and Gradient Boosting for gesture recognition and classification. The models were evaluated for its accuracy and computational efficiency. Lastly, a user interface was developed for an enhanced drone control experience. In comparing Logistic Regression, Ridge Classifier, Random Forest Classifier, and Gradient Boosting Classifier as conveyed in Fig. 4, each model demonstrated distinct strengths and considerations.

Logistic Regression, a foundational linear model, offers interpretability and simplicity, making it suitable for scenarios with linear relationships between features and targets. Ridge Classifier extends this by integrating L2 regularization, enhancing robustness in the presence of multicollinearity. However, both may struggle with capturing complex interactions and non-linear patterns. On the other hand, Random Forest Classifier and Gradient Boosting Classifier leverage ensemble techniques, excelling in handling large datasets with intricate feature interactions and non-linear dependencies. Random Forest provides insights into feature importance and reduces overfitting through aggregation,

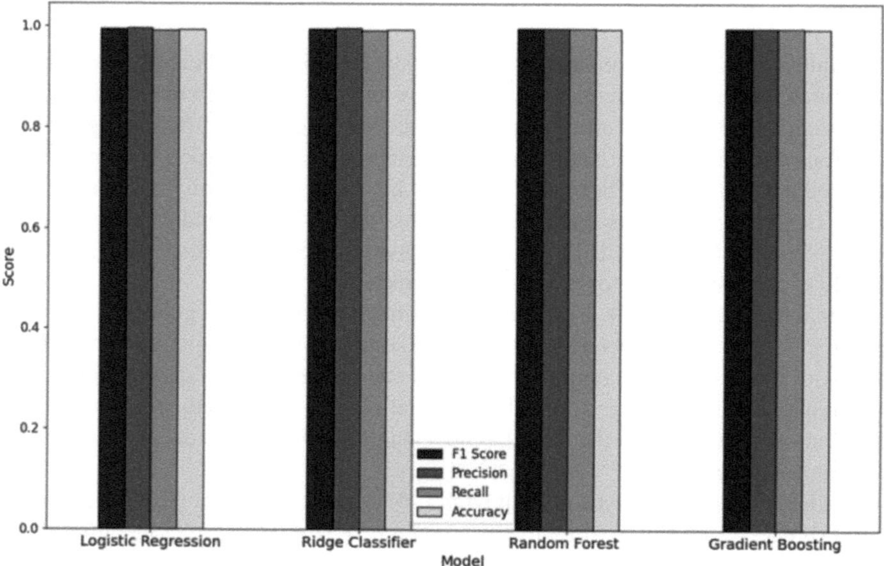

Fig. 4. Model comparison: F1 score, Precision, Recall, Accuracy

while Gradient Boosting iteratively improves upon weak learners, yielding superior performance in smaller datasets with complex relationships. With the help of a large training dataset all models give high accuracy. Amongst them Random Forest Classifier gives the highest accuracy score of 0.9951 and precision of 0.9949, hence was chosen as the classifier for gesture recognition in this work. In this work, several ML models – Logistic regression, Ridge classifier, Random Forest, and Gradient Boosting - were assessed for gesture detection. These models underwent training, validation, and testing on distinct subsets of the dataset. Results showed varying accuracies: Logistic regression 97.8%, Ridge classifier 98%, Random Forest 99.5%, and Gradient Boosting 99.4%. This highlights the significance of selecting the right CNN architecture for gesture detection.

4 Conclusions

The work in this paper focuses on enhancing human-drone interaction by developing a Gesture Control System that utilizes machine learning for intuitive drone control. The primary objective was to create a system that allows users to manage drones using natural hand movements, thereby providing a more flexible and immersive control experience. The work began with the constructing the drone, carefully selecting the right components to ensure the design fit within the project scope and requirements. Once the drone was built, the next focus was on collecting the dataset necessary for developing the Gesture Control System. To create this dataset, we recorded videos of ourselves performing the desired hand gestures, each lasting 4–5 min. We then classified the frames of these videos into their respective gesture categories by playing the footage and using keyboard keys for manual classification. For each gesture, between 300 to 400 images were extracted to form a robust dataset.

To develop the Gesture Control System, various machine learning models were explored and tested, including Ridge Classifier, Random Forest Classifier, Logistic Regression, and Gradient Boosting Algorithm, to determine the most effective approach for hand gesture recognition. After rigorous testing, the Random Forest classifier was selected for its superior performance and reliability. The classifier was integrated into the MediaPipe framework, known for its powerful hand tracking and landmark detection capabilities. This combination laid a solid foundation for our gesture recognition system. The Random Forest classifier achieved an accuracy score of 0.9951 and a precision of 0.9949, ensuring smooth and precise control of the drone. These metrics demonstrate the system's effectiveness in meeting the goals of accuracy, real-time responsiveness, and user-friendly interface design.

Looking ahead, there are several avenues for future development. Enhancements could include expanding the range of recognizable gestures, improving the system's adaptability to different environments, and integrating additional sensors for more robust gesture detection.

References

1. Kartik, B., Anusha, P.: Real-time structural crack detection in buildings using YOLOv3 and autonomous unmanned aerial systems. Int. J. Syst. Assur. Eng. Manage. **15**, 1374–1887 (2024)
2. Abhishek, B., Kanya, K., Meghana, M., Mohammed, D., Anupama, H.S.: Hand gesture recognition using machine learning algorithms. Comput. Sci. Inf. Technol. **1** 116–120 (2020)
3. Rohi, G., Ejofodomi, O., Ofualagba, G.: Autonomous monitoring, analysis, and countering of air pollution using environmental drones. Heliyon (2020)
4. Abhishek, S., et al.: Communication and networking technologies for UAVs: a survey. J. Netw. Comput. Appl. **168** (2020)
5. Ji-Won, L., Kee-Ho, Y.: Wearable drone controller: machine learning-based hand gesture recognition and vibrotactile feedback. Sensors **5**(23) (2023)
6. Sang, Y.S., Yong, W.K., Yong, G.K.: Hand gesture-based wearable human-drone interface for intuitive movement control. In: 2019 IEEE International Conference on Consumer Electronics, ICCE 2019, Las Vegas (2019)
7. Nicolas, G., Ross, B., Tijana, V.: Control of a drone with body gestures. Proc. Design Soc. **1**, 761–770 (2021)
8. Robert, D., Pavol, P., Martin, B.: The ground station for long-range monitoring, flight control, and operational data telemetry of unmanned aerial vehicles. Perner's Contacts **18**(1) (2023)
9. Nils, G., Michael, S., Sergio, M.: An autonomous UAV with an optical flow sensor for positioning and navigation. Int. J. Adv. Robot. Syst. **10**(10) (2013)
10. Salih, A., Moghavvemi, M., Mohamed, H.A.F.: Flight PID controller design for a UAV quadrotor. Sci. Res. Essays **5**(23), 3660–3667 (2010)
11. Aina, E., et al.: UIQuad I: a low-cost and modular surveillance quadcopter. Int. J. Mech. Eng. Robot. Res. **13**(1), 11–17 (2024)
12. Xiang, L., Qing, X., Yanmei, T., Cong, H., Junhao, N., Chuanpei, X.: Unmanned aerial vehicle position estimation augmentation using optical flow sensor. IEEE Sens. J. **23**(13), 14773–14780 (2023)
13. Anis, K., Azza, A., Maram, A., Yasir, J., Abdelfettah, B., Mohamed, K.: Micro air vehicle link (MAVlink) in a nutshell: a survey. IEEE Access **7**, 87658–87680 (2019)

14. Jiya, T., Qiangshan, J., Yizong, W., Jie, Y.S.Z., Dengxun, S.: Performance analysis of deep learning-based object detection algorithms on COCO benchmark: a comparative study. J. Eng. Appl. Sci. **71**(76) (2024)
15. Dina, S., Mohamed, W.: Drowsiness detection in real-time via convolutional neural networks and transfer learning. J. Eng. Appl. Sci. **71**(122) (2024)

Mitigating Risks in Drone Swarm Operations: Obstacle Avoidance and Secure Data Exchange

Bishad Shukla[✉] and Mou Dasgupta

Department of Computer Application, National Institute of Technology Raipur, Raipur 492010, India
bishadshukla@gmail.com

Abstract. Drone swarm deployment has grown in popularity in recent years in a number of industries, including environmental monitoring and search and rescue Ensuring the safety and effectiveness of drone path planning is a crucial difficulty in these operations, particularly in areas with lots of barriers. This paper addresses three primary subjects energy efficiency, secure communication between drones in a swarm, and techniques for avoiding obstacles. We propose a system where drones employ a method to detect and navigate around obstacles while sharing route information with fellow drones in the group to optimize energy efficiency. This dependence on shared data, however, raises the possibility of third-party manipulation, where inaccurate information may cause mishaps and mission failure. We incorporate RSA-based Digital Signatures with PKI into the communication framework to counter this attack. This method protects the swarm from potential cyber-attacks in addition to ensuring the integrity of the data.

Keywords: Drone Swarm · Obstacle Avoidance · Secure Communication · Energy Efficiency

1   Introduction

Unmanned Aerial Vehicles (UAVs) have become major resource to various industries, with applications leading from disaster relief to precision agriculture and delivery transportation. The flexibility and efficiency range of UAVs have led to a surprising increase in their deployment across the world. According to the Federal Aviation Administration (FAA), as of November 2020, 1.7 million UAVs were registered by the US alone [1]. In China, the 2020 Civil Aviation Working Conference presented that over 392,000 UAVs were registered, and commercial UAV operations reported the 1.25 million flight hours in 2019 [2]. Furthermore, the economic impact and rapid growth of this technology were underscored by the $9 billion assessed increase in the global UAV market in 2019 [2]. Robust anti-collision technologies are required to guarantee safe and consistent operations in airspaces that are becoming increasingly congested as UAV tradition continues to grow. A number of obstacle avoidance algorithms have been developed on the basis of the geometric correlations between aircraft and anti-collisions. For example, Zheng *et al* developed fuzzy heuristics to identify avoidance plans based on the distance between the

C. Modi et al. (Eds.): MIND 2024, CCIS 2736, pp. 511–524, 2026.
https://doi.org/10.1007/978-3-032-14531-4_43

vehicle and obstacles [4], while Sasongko *et al.* proposed a method that uses obstacle models and aircraft speed vectors to calculate avoidance waypoints [3]. Al-Kaff *et al.* presented an avoidance strategy that follows a safe flying edge that is defined by obstacle feature points that are found by using the aircraft's real-time coordinate interactions with the obstacles [5]. These modifications are necessary to guarantee the safe operation of UAVs in inflexible situations and to prevent collisions. Comparable to the advancement of anti-collision technology, the need of secure communication between UAVs has grown over time. The use of UAVs raises the possibility of malicious assaults and data breaches, wherein unapproved parties may intercept or alter UAV-to-UAV transmission. Although secure communication techniques have been around for centuries, the difficulty of key management has historically prevented them from being widely used. With the development of public-key cryptography, especially the RSA algorithm, secure communication has been reformed and is now possible on huge networks where users can interact safely without knowing each other beforehand [6–8].

This paper suggests combining PKI with digital signatures based on RSA to enhance security of UAV communication. Digital signatures based on RSA guarantee the integrity and authenticity of data, preventing interception or spoofing. Secure verification is essential in UAV swarm operations, as drones collaborate to share important navigational information for energy efficiency and collision avoidance. Every message is endorsed with the private key of the sender, allowing only authorized drones—validated through PKI—to verify the sender and have confidence in the data. This method reduces potential hazards like spoofing or man-in-the-middle attacks, preventing unauthorized individuals from tampering with drone communication networks. As unmanned aerial vehicles become more common in different fields, it is essential to integrate strong methods for avoiding obstacles with secure communication methods such as RSA and PKI to ensure safety and dependability. This paper investigates how these technologies can be combined to improve UAV operations in challenging environments.

The rest of this paper is organized as follows. Section 2 related works. Section 3 states the proposed methodology. Section 4 presents the result and discussion. Finally, the conclusions are drawn in Sect. 5.

2 Related Work

The swift advancement and implementation of unmanned aerial vehicles (UAVs) has prompted a great deal of investigation in various fields, especially safe communication and obstacle avoidance. The safe and effective operation of UAVs, particularly in intricate and dynamic situations, depends on these areas.

2.1 Obstacle Avoidance

Obstacle avoidance is a key challenge in UAV operations, especially as UAVs are progressively required to operate autonomously in environments filled with both static and dynamic obstacles. Early work by Oleynikova *et al.* (2018) introduced Voxblox, a method for generating 3D Euclidean Signed Distance Fields in real-time for onboard UAV planning, which allows for dynamic obstacle avoidance in cluttered situations [9]. Also,

Scherer *et al.* (2019) showed a vision-based obstacle recognition and avoidance system that uses deep learning to let unmanned aerial vehicles (UAVs) fly through uncharted territory on their own [10]. Machine learning methods have recently been added to systems that help people avoid obstacles. Kaufmann and his friends (2020) created a reinforcement learning system that lets unmanned aerial vehicles (UAVs) learn and change how to avoid obstacles in real time. This makes it much easier for them to move through complex environments [11]. In order to get better and more reliable obstacle detection and avoidance when GPS isn't available, Gao *et al.* (2021) came up with a hybrid method that uses both LiDAR and camera data [12]. These improvements show that UAV systems are getting smarter and more adaptable so they can work in environments that are becoming less predictable.

Key Findings:

- **Real-Time Obstacle Avoidance:** Methods like Voxblox and vision-based systems enhance UAVs' ability to navigate cluttered and uncharted environments using 3D mapping and deep learning.
- **Advanced Adaptation Techniques:** Sensor fusion (LiDAR and cameras) and reinforcement learning enable UAVs to adapt to complex settings and operate without GPS.

2.2   Secure Communication

'As UAVs become more prevalent, ensuring secure communication between them is essential, particularly in swarm operations where UAVs must share critical data. Analysis of public-key cryptography, particularly RSA, has been an important area of research for ensuring the security of UAV communication. The implementation of RSA encryption to safeguard UAV-to-UAV communication from eavesdropping and other cyber threats was investigated by Alnuaimi *et al.* (2019), demonstrating its effectiveness in preserving the confidentiality and integrity of transmitted data [13].

At the same time, Liu *et al.* (2017) created a new type of encryption that combines RSA and symmetric key cryptography. This method was created to protect UAV transmission while reducing the amount of work that needs to be done on the computer [14]. The suggested way strikes a good balance between safety and effectiveness, which makes it perfect for UAV systems that don't have a lot of resources. Rawat *et al.* (2019) increased the use of RSA by adding digital signatures to UAV communication protocols. This ensured the data's security and privacy [15]. Wang *et al.* (2018) also created a lightweight encryption protocol that is specifically designed for UAV networks. This protocol focus on making encryption processes easier on computers while still providing strong security [16]. This study is supported by Pandey et al. (2020), who updated lightweight cryptographic techniques for IoT devices, such as unmanned aerial vehicles (UAVs). This shows the need for fast and safe ways to communicate in places with limited resources [17].Sharma *et al.* (2021) did more study and found that block-chain technology could be used to protect UAV communication. They combined RSA encryption with block-chain to make a record of UAV data relationships that can't be changed [18]. This method not only keeps the conversation safe, but it also leaves a clear, unchangeable record of all UAV conversations. More research by Zhang *et al.* (2020) looked into elliptic curve cryptography (ECC) as an option to RSA for UAV communication. It has similar security benefits but requires less computing power, so it's best for small UAVs that don't have a lot of it [19]. Finally, Abd-Elrahman *et al.* (2021) suggested a new way to protect UAV networks from a variety of cyber threats by joining RSA and the advanced encryption standard (AES) to make a dual-layer encryption protocol [20].

Key Findings:

- **Efficient Encryption:** Hybrid and lightweight protocols (e.g., RSA with AES) optimize security and efficiency for UAVs with limited resources.
- **Advanced Security:** Integrations like digital signatures and blockchain ensure data integrity and resilience against cyber threats.

2.3 Distinction of the Proposed Work

This research presents a unified approach for UAV swarms, combining secure communication, obstacle avoidance, and energy efficiency. Unlike previous studies that address these issues separately, our system integrates real-time obstacle detection with RSA-based Digital Signatures and PKI, ensuring secure data transfer while addressing safety and security concerns. A novel energy- efficient method is introduced, where UAVs share obstacle-free paths within the swarm, allowing them to deactivate sensors and conserve battery life. This approach not only enhances energy efficiency but also ensures data integrity, minimizing the risk of crashes due to data manipulation. By protecting shared path information with RSA-based Digital Signatures, the system prevents unauthorized access and ensures reliable operations. This integrated solution marks a significant advancement, achieving effective obstacle avoidance, robust data security, and improved energy efficiency.

3 Our Proposed Methodology

This method presents a three-part strategy to improve the safety, security, and energy efficiency of drone swarm missions. The initial element includes a strategy for evading obstacles in which drones spot obstacles in real-time, measure distances using the Euclidean formula, and establish a virtual hexagonal boundary around obstacles to move around safely and effectively. The next part emphasizes on ensuring safe communication among drones in the swarm, sharing new paths to allow other drones to fly without obstacles and saving energy by turning off sensors. In order to protect against cyber threats, RSA-based digital signatures with PKI are implemented to guarantee that only authorized drones have the ability to decrypt and authenticate shared data. The third part focuses on energy efficiency by reducing unnecessary sensor usage through sharing paths, which extends the mission time and maximizes operational sustainability. This integrated approach ensures safe, secure, and energy-efficient swarm operations.

3.1 Proposed Obstacle Avoidance Technique

Lidar sensors and real-time course adjustment algorithms form the basis of the obstacle avoidance system's implementation. Technology deployed on each drone in a swarm formation continuously monitors the environment for obstacles. Once the barrier has been identified, the drone calculates the distance between itself and it using the Euclidean distance formula. The following calculation establishes if the obstruction is within the specified safe distance, X.

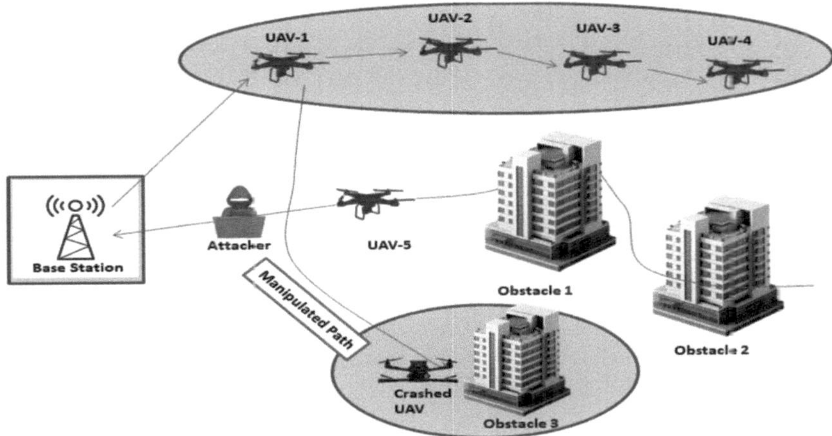

Fig. 1. System model overview

The drone immediately initiates manueuvring around the object to avoid a collision if the distance is less than X. The drone's constant surveillance enables it to promptly adjust to variations in its surroundings, ensuring its ability to maintain safe flight even in the event of unforeseen circumstances. To illustrate the obstacle avoidance mechanism, we will provide a system overview model in Fig. 1.

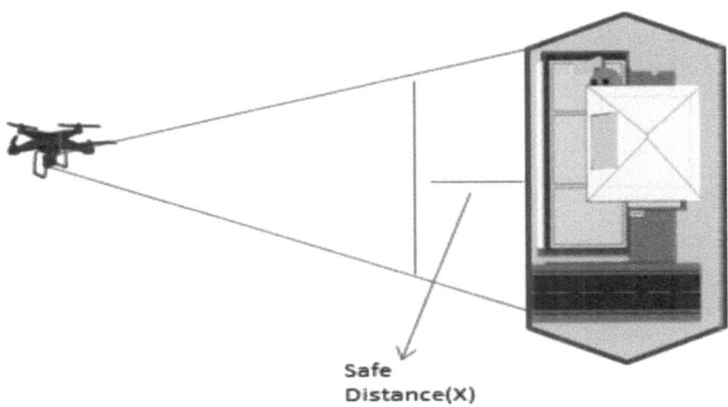

Fig. 2. Detection & Avoidance of obstacle using Obstacle Avoidance Technique

This figure will properly explain the flow of information from obstacle detection to path adjustment and highlighting the steps involved in designing the hexagon around the obstacle and navigating around it which is shown in Fig. 2.We will also show you the algorithm that supports this method. This algorithm shows the mathematical calculations and decision-making processes that control the drone's movements. The flow of sensor data and how the drone's path is changed in real time will also be shown in this image. By seeing the entire process, the efficiency and effectiveness of the obstacle avoidance system will be clearly demonstrated, showing how the drones will maintain safety in complex, dynamic environments.

Input: $W = \{wp1, wp2, \ldots, wpn\}$: Waypoint path, $U = (Ux, Uy, Uz)$: UAV position (x,y,z), $O = (Ox, Oy, Oz)$: Obstacle center (x,y,z), Wo: Obstacle width, X: Safe distance, v: UAV speed.

    Initialize: if state = 1 (tracking waypoints).

        endif

    for $i \leftarrow 1$ to n do

      $i = 1$ (start with the first waypoint).

      UAV tracks the waypoints path

      Obstacle Detection: Calculate Distance to Obstacle:

    Use the Euclidean distance formula to calculate the distance 'D' between the UAV and the obstacle:

$$\sqrt{((U_x - O_x)^2 + (U_y - O_y)^2 + (U_z - Oz)^2)}$$

      If $D \geq X$: The UAV is at a safe distance,

       and no avoidance action is needed.

      Else If d $< X$, switch to state = 2 (Obstacle avoidance).

      state = 2, determine the width W_o of the Obstacle.

      Calculate Hexagon Side Length 's': $s = W_o / \sqrt{3}$.

      Determine Hexagon Vertices as: (V_{xi}, V_{yi}):

      For $i = 0$ to 5 do, $\quad V_{xi} = O_x + s \cdot \cos\left(\frac{\pi}{3} \cdot i\right),$

$$V_{xi} = O_y \quad s \cdot \sin\left(\frac{\pi}{3} \cdot i\right)$$

      end for ,end if

    The UAV follows the vertices (V_{xi}, V_{yi}) sequentially to bypass the obstacle.

    end if

    Set state $\leftarrow 1$,

    end if

3.2 RSA-Based Digital Signature with PKI in Drone Swarm Communication

1. **Public Key Infrastructure (PKI) Setup:** Every drone in the swarm is assigned a distinct public-private key pair, overseen by a reliable Certificate Authority (CA). The Certificate Authority signs the public key certificate of each drone, so assuring mutual trust in their public keys.

```
(PK_drone, SK_drone) ← generate_RSA_key_pair()
```

- **Certificate Issuance**: The drone's public key is signed by the CA, binding it to the drone's identity.
  ```
  CI_drone ← CA_sign(PK_drone)
  ```
- **Public Key Distribution**: Each drone distributes its certificate, allowing others to verify the authenticity of its public key.

2. **Path Calculation and Digital Signature Creation:** The first drone, equipped with sensors, calculates the path data using LiDAR and signs it to ensure integrity and authenticity.

   - **Hashing**: The path data is hashed to create a fixed-size digest:

     ```
     hash_path_data ← hash (path_data)
     ```

   - **Signing**: The drone signs the hash with its private key, creating a digital signature:

     ```
     signature         ←        RSA.sign(hash_path_data,
     SK_first_drone)
     ```

     This signature ensures the path data was sent by the authorized drone and hasn't been tampered.

3. **Encryption and Transmission to Other Drones:** The path data and digital signature are encrypted using the recipient drone's public key, ensuring that only the intended drone can decrypt the data.

   - **Encryption**: The path data and signature are encrypted for each drone using its public key:
   ```
   enc_message_i   ←   RSA.encrypt({path_data,   signature},
   PK_drone_i)
   ```
   - **Transmission**: The encrypted message is sent to each drone.

4. **Decryption and Signature Verification by Receiving Drones:** Each receiving drone decrypts the message using its private key and verifies the digital signature using the first drone's public key.

   - **Decryption**: The drone decrypts the message:
     ```
     {path_data,               signature}               ←
     RSA.decrypt(enc_message_i, SK_drone_i,
     ```
   - **Signature Verification**: The drone verifies the signature using the first drone's public key from its certificate:
     ```
     is_valid             ←          RSA.verify(signature,
     hash(path_data), PK_first_drone)
     ```

     The drone trusts the path data if the signature is legitimate; if not, it discards the data.

5. **Safe Navigation:** After verification, the drones conserve energy by following the shared path without turning on their sensors.

3.3 Energy Efficiency in Drone Swarm Operations

In a swarm of drones, energy efficiency is major part to extending the mission duration and ensuring successful operations. This section focuses on the energy consumption dynamics of drones equipped with an obstacle avoidance approach versus drones operating without obstacle avoidance approach by following a shared path.

System Overview: Consider a scenario where N drones are navigating in environment with obstacles. Each drone is initially armed with an obstacle avoidance approach. The first drone will detect and avoids obstacles, and then it will share the safe path which has been identified with the rest of the swarm. The consequent drones, referred to as drones without the obstacle avoidance approach, follow this shared path without detecting the obstacles to conserve energy.

Energy Consumption Model: To examine the energy consumption, we define the following parameters:

- P_{flight} P_sensor: Power consumed by the LiDar sensor.
- T_{path} T_path: Time taken to travel the shared path.
- N: Total number of drones in the swarm.

The first drone's energy use is equal to the amount of the power used by the Lidar sensor and the flight systems. This includes the obstacle avoidance method:

$$E\_with,OA = (P\_sensor + P\_flight) \times T\_path$$

The drones without the obstacle avoidance approach, which follow the predetermined safe path with their sensors turned off, consume energy only through their flight systems:

$$E\_without,OA = P\_flight \times T\_path$$

The total energy consumption for the entire swarm, considering both drones with and without the obstacle avoidance approach, is given by:

$$E\_Total = E\_with,OA + (N - 1) \times E\_without,OA$$

Substituting the respective energy:

$$E\_Total = [P\_sensor + N \times P\_flight] \times T\_path$$

This equation emphasises that the total energy consumption is contingent upon the number of drones, the time spent travelling the path, and the power requirements of the sensor and flight systems.

Energy Savings: The energy savings are a result of the N − 1drones' ability to deactivate their sensors as they move without the obstacle avoidance approach. The total energy conserved by these drones is determined as:

$$E\_saved = (N - 1) \times P\_sensor \times T\_path$$

This formula demonstrates that the number of drones following the shared path, the power consumption of the sensor, and the duration of the path all contribute to an increase in energy savings.

4 Results and Discussion

In this work, we present a comprehensive solution for enhance secure and efficient drone swarms in obstacle-rich environments. Our proposed method focuses on three critical aspects: effective obstacle avoidance, the secure transmission of navigational data and the energy efficiency model. By leveraging LiDAR-based obstacle detection and RSA based Digital Signature with PKI, our approach not only ensures that drones avoid collisions but also provide the integrity of the shared path data, and making it secure to malicious third-party interference and energy efficient.

4.1 Improved Obstacle Avoidance and Energy Efficiency

Drones in the swarm uses a obstacle avoidance algorithm based on a real-time data. As each drone navigates in environment, to detects and avoids obstacles, sharing the obstacle-free path with the rest of the swarm. This information sharing allows parallel drones to operate with obstacle avoidance approach when following the same path, significantly it will reduce the energy consumption. The ability to conserve energy is major part for extending the operate time of the swarm, particularly in scenarios such as disaster management, where prolonged and reliable drones are essential. By broadcasting the obstacle-free paths, our approach reduces the redundant sensor operations across the swarm. This cooperative behavior not only reduces the chance of collisions but also improves the swarm's overall energy usage, therefore it will be improving the mission's sustainability and validity.

4.2 Mitigating the Risk of Data Manipulation in UAV Swarms

In UAV swarm operations, one big problem is someone messing with the data sent between drones. Drones in a swarm always share path info to save energy, like one drone finding a clear path and telling the others. But this opens up a chance for attackers to mess with the path info. They could send false paths, causing drones to crash or go into dangerous areas.

Securing Data with RSA-based Digital Signatures and PKI; To stop this, we used RSA-based digital signatures and added Public Key Infrastructure (PKI) to make the communication between drones safe. This helps make sure the data is real and hasn't been changed.

Protection Against Spoofing Attacks: PKI helps keep attackers from pretending to be real drones. Spoofing attacks are when someone tries to act like a drone by giving out a fake public key. Our system has some steps to prevent this:

- **Certificate Authority (CA)**: Every drone gets its public key signed by a trusted CA. When a drone gets path data from another drone, it checks both the digital signature and the sender's public key. It makes sure the key belongs to the drone it claims to be, so attackers can't trick drones with fake keys.

- **Authenticated Communication**: Before accepting any path info, drones always check both the digital signature and the sender's certificate. This means only drones with valid, trusted certificates can send or receive data in the swarm.

The below graph illustrates the relationship between message size and the time taken for RSA encryption and decryption with digital signature and PKI (Fig. 3 and Table 1).

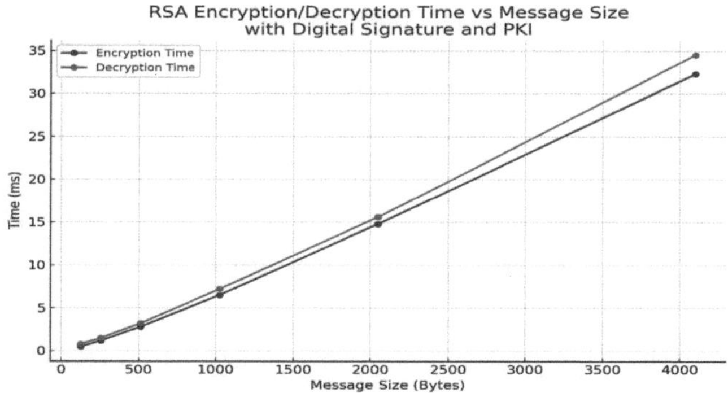

Fig. 3. The relationship between message size and the Encryption/Decryption Time

Table 1. Attack Comparison in UAV Security

Sr. No.	Attack Name	Challa *et al.* (2020)	Turkanović *et al.* (2021)	Wazid *et al.* (2021)	Our proposed Method
I.	GPS Spoofing	✔	✘	✘	✔
II.	Authentication Attack	✔	✘	✘	✔
III.	Intercept Data Feed Attack	✘	✘	✘	✔
IV.	Video Replay Attack	✘	✔	✘	✔
V.	Impersonation Attack	✔	✘	✔	✔

4.3 Energy Savings Through Collaborative Path Sharing

An important part of drone swarm operations is energy saving. The ability to share obstacle-free paths within the swarm enables the drones to conserve energy by turning off their Lidar sensors when following a pre-determined route (Fig. 4).

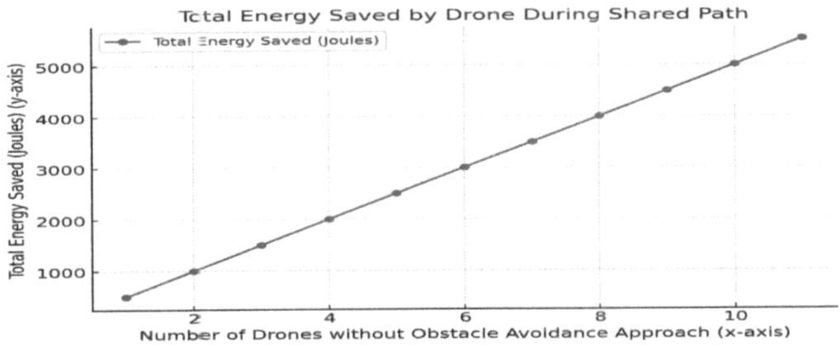

Fig. 4. Total Energy Saved by Drones During Shared Path

This approach is particularly beneficial when operating multiple drones in a swarm. The graph below illustrates the total energy saved by the swarm drone as the number of drones sharing the shared path increases. The x-axis represents the number of drones without the obstacle avoidance approach, and the y-axis shows the total energy saved in joules Above is a graph that shows how much energy drones saved by using the shared path. This graph is based on a situation where 10 drones are used. Each drone that follows the shared path and turns off its object avoidance approach will save energy. The number of drones that don't use object avoidance is shown on the x-axis, and the amount of energy saved is shown on the y-axis (in joules) (Table 2).

Table 2. Comparison of Energy-Efficient UAV Method

Method	Main Idea	Features	Performance	Advantages	Limitations
Multiple UAVs Working in Circulation [24]	Fully charged UAVs re-place those with depleted batteries	Enables completion of long-endurance tasks	Automatic UAV re-placement ensures continuity	Provides uninterrupted services to users	Increases overall system cost due to maintenance
Replacing the Battery of UAV [25]	Replace exhausted UAV batteries	Low system flexibility and intelligence	Enables wireless battery charging	Rapid energy supplementation	Temporary system interruptions affect continuity
Energy Harvesting of UAV [26]	Harvest energy from the environment (e.g., solar, wind)	Energy collection tied to environmental conditions	Optimal altitude ranges ensure efficient energy	Environmental friendly and renewable energy	Effectiveness depends on unpredictable natural conditions

(continued)

Table 2. (*continued*)

Method	Main Idea	Features	Performance	Advantages	Limitations
Energy Saving through Shared Path	Utilize collaborative path- sharing in a drone swarm to save energy	Drones turn off LiDAR sensors when following the path	Total energy savings increase with swarm size	Significant energy savings and reduced consumption	Requires precise coordination; less flexibility

5 Conclusion

Furthermore, this method offers a well-balanced solution that improves safety, security, and energy efficiency in drone swarm operations. The implementation of obstacle avoidance for immediate identification and avoidance of obstacles guarantees the safe operation of drones in dynamic surroundings, while the RSA-based Digital Signature with PKI secures against unauthorized alteration of shared navigational data. Additionally, the collaborative path-sharing strategy allows drones to move without the obstacle avoidance technique, supporting energy efficiency by extending mission duration and operational range. This combination of safety, security, and energy efficiency makes our work unique and well-suited for applications where reliability, energy efficiency, and security are major concerns. In future work, we plan to explore lightweight cryptography techniques to further enhance security while minimizing computational overhead in resource-constrained UAV environments.

Acknowledgement. This Research work is an outcome of IBITF Project, IIT Bhilai No. 2023–24/0316, Dated - 19/12/2023.

References

1. "UAS by the Numbers." FAA. https://www.faa.gov/uas/resources/by_the_numbers/. Accessed 20 Nov 2020
2. "China Civil Aviation Working Conference Held in 2020." CAAC. Accessed 20 Nov 2020
3. Zheng, J., Liu, B., Meng, Z., Zhou, Y.: Integrated real time obstacle avoidance algorithm based on fuzzy logic and L1 con trol algorithm for unmanned helicopter. In: 2018 Chinese Con trol and Decision Conference (CCDC), pp. 1865–1870, Shenyang, China, 2018
4. Sasongko, R.A., Rawikara, S.S., Tampubolon, H.J.: UAV obstacle avoidance algorithm based on ellipsoid geometry. J. Intell. Rob. Syst. **88**(2–4), 567–581 (2017)
5. Al-Kaff, A., García, F., Martín, D., De La Escalera, A., Armingol, J.M.: Obstacle detection and avoidance system based on monocular camera and size expansion algorithm for UAVs. Sensors **17**(5), 1061–1082 (2017)
6. Shinde, G.N., Fade War, H.S.: Faster RSA algorithm for decryption using Chinese remainder theorem. ICCES **5**(4), 255–261 (2008)

7. Yang, L., Yang, S.H.: A frame work of security and safety checking for internet-based control systems. Int. J. Inf. Comput. secur. **1**(2) (2007)

8. Washington, L.C. 2006. Introduction to Cryptography: with coding theory by Wade Trappe. Upper Saddle River, New Jersey, Pearson Prentice Hall

9. Oleynikova, H., Taylor, C., Fehr, M., Siegwart, R., Nieto, J.: Voxblox: incremental 3D euclidean signed distance fields for onboard MAV planning. In: IEEE/RSJ International Conference on Intelligent Robots and Systems (IROS), pp. 1366–1373 (2018). https://doi.org/10.1109/IROS.2018.8594134

10. Scherer, S.A., Singh, S.: Vision-based obstacle detection and avoidance for UAVs. IEEE Trans. Intell. Transp. Syst. **20**(1), 30–42 (2019). https://doi.org/10.1109/TITS.2018.2792962

11. Kaufmann, E., Loquercio, A., Ranftl, R., Dosovitskiy, A., Koltun, V., Scaramuzza, D.: Deep drone racing: fromsimulation to reality with domain randomization. IEEE Trans. Rob. **36**(1), 1–14 (2020). https://doi.org/10.1109/TRO.2019.2942989

12. Gao, Y., Wang, Z., Li, P.: A hybrid approach for UAV obstacle avoidance using LiDAR and vision data. Sensors **21**(5), 1691 (2021). https://doi.org/10.3390/s21051691

13. Alnuaimi, M., Alshehhi, M., Qadir, M.: Securing UAV communication using RSA encryption. J. Netw. Comput. Appl. **124**, 109–120 (2019). https://doi.org/10.1016/j.jnca.2018.07.006

14. Liu, X., Liu, Y., Fang, Y.: Secure and efficient path key establishment for wireless sensor networks without pre-deployment knowledge. IEEE Trans. Parallel Distrib. Syst. **28**(5), 1481–1493 (2017). https://doi.org/10.1109/TPDS.2016.2615003

15. Rawat, S., Al-Kahtani, M., Kumar, R.: UAV communication security using RSA encryption and digital signatures. IEEE Access **7**, 116849–116858 (2019). https://doi.org/10.1109/ACCESS.2019.2936049

16. Wang, H., Li, Z., Zhang, J.: A lightweight and secure communication protocol for UAV networks. IEEE Trans. Veh. Technol. **67**(10), 9459–9471 (2018). https://doi.org/10.1109/TVT.2018.2847394

17. Pandey, P., Choudhary, G., Tiwari, S.: Lightweight cryptography protocols for resource-constrained IoT devices: a comprehensive review. J. Syst. Architect. **102**, 101659 (2020). https://doi.org/10.1016/j.sysarc.2019.101659

18. Sharma, S., Kumar, P., Varma, S.: Securing UAV communication with blockchain and RSA encryption. IEEE Trans. Aerosp. Electron. Syst. **57**(3), 1637–1650 (2021). https://doi.org/10.1109/TAES.2021.3061397

19. Zhang, Y., Wang, Y., Li, Y.: Efficient elliptic curve cryptography-based security mechanisms for UAV communication. IEEE Trans. Veh. Technol. **69**(3), 2665–2677 (2020). https://doi.org/10.1109/TVT.2019.2963698

20. Abd-Elrahman, E., Elgency, H.: A dual-layer encryption protocol for Enhancing UAV network security. IEEE Access **9**, 19033–19044 (2021). https://doi.org/10.1109/ACCESS.2021.3052072

21. Wazid, M., Das, A.K., Kumar, N., Vasilakos, A.V., Rodrigues, J.J.: Design and analysis of secure lightweight remote user authentication and key agreement scheme in internet of drones deployment. IEEE Internet Things J. **6**(2), 3572–3584 (2019)

22. Challa, S., et al.: Secure signature-based authenticated key establishment scheme for future IoT applications. IEEE Access **5**, 3028–3043 (2017)

23. Turkanovic, M., Brumen, B., Hölbl, M.: A novel user authentication and key agreement scheme for heterogeneous Ad Hoc wireless sensor networks, based on the internet of things notion. Ad Hoc Netw. **20**, 96–112 (2014)

24. Zorbas, D., Razafindralambo, T., Guerriero, F., et al.: Energy efficient mobile target tracking using flying drones. Proc. Comput. Sci. **19**, 80–87 (2013)

25. Campi, T., Cruciani, S., Maradei, F., Feliziani, M.: Wireless charging system integrated in a small unmanned aerial vehicle (UAV) with high tolerance to planar coil misalignment. In: 2019 Joint International Symposium on Electromagnetic Compatibility, Sapporo and Asia-Pacific International Symposium on Electromagnetic Compatibility, EMC Sapporo/APEMC, IEEE, pp. 601–604, 2019
26. Sun, Y., Xu, D., Ng, D., Dai, L., Schober, R.: Optimal 3D-trajectory design and resource allocation for solar-powered UAV communication systems. IEEE Trans. Commun. (2019). [135]

Optimizing Diabetes Prediction: Hyperparameter Tuning and Type Classification Hypotheses with Deep Neural Network

Partha Pratim Bhuyan, Bikash Sharma, Ajanit Bora, and Ranjay Hazra[✉]

Electronics and Instrumentation Engineering Department, National Institute of Technology, Silchar, Silchar 788010, Assam, India
`ranjay@ei.nits.ac.in`

Abstract. This paper presents an extensive exploration of the field of diabetes prediction, taking advantage of a diverse array of machine learning algorithms and artificial intelligence (AI) techniques. With a focus on addressing the critical need for accurate predictive models in diabetes management, our study delves into the PIMA Indian Diabetes Dataset. We meticulously apply various algorithms, including K-Nearest Neighbors (KNN), Logistic Regression, Support Vector Machines (SVM), Random Forest, Decision Trees, Convolutional Neural Networks (CNN) and Deep Neural Networks (ANN). Our analysis yields promising results, showing accuracy ranging from 70%, 78% to 99.50% in different algorithms and model configurations. Particularly noteworthy are the superior performances of DNN models, especially when enhanced with techniques like early stopping and regularization. Additionally, we conducted an update on our analysis with the modified PIMA dataset, revealing marked improvements in predictive accuracy for several algorithms. Through this comprehensive study, we contribute significant insights into the effectiveness of AI-driven approaches for diabetes prediction, underscoring the importance of algorithm selection, model optimization techniques, and dataset modification. These findings have substantial implications for refining personalized healthcare interventions and enhancing strategies for managing diabetes.

Keywords: Diabetes Prediction · Deep Neural Network · Model Optimization

1 Introduction

Diabetes mellitus is a significant global health challenge, impacting millions of people and placing considerable pressure on healthcare systems. In 2021, the International Diabetes Federation (IDF) estimated 537 million adults (ages 20–79) had diabetes, with the number expected to rise to 643 million by 2030. The disease impacts both quality of life and healthcare costs. Effective management relies on timely diagnosis, personalized treatment, and continuous monitoring.

C. Modi et al. (Eds.): MIND 2024, CCIS 2736, pp. 525–537, 2026.
https://doi.org/10.1007/978-3-032-14531-4_44

Predictive models are crucial in improving disease management by enabling early detection, risk assessment, and targeted interventions. Understanding the significance of various attributes associated with diabetes is essential for developing robust predictive models. By identifying and evaluating these attributes, we can gain a detailed understanding of their impact on disease development and progression. Moreover, exploring different artificial intelligence (AI) techniques and models for diabetes prediction allows us to leverage the rich complexity of biomedical data and extract meaningful insights.

2 Literature Review

In recent years, numerous approaches have emerged for predicting diabetes, with a focus on machine learning methodologies. In a study [1], a comprehensive framework was introduced, incorporating various classifiers including Linear Discriminant Analysis (LDA) [2], Adaptive Synthetic Sampling [3], Naive Bayes (NB) [4], Gaussian Process Classification [5], Artificial Neural Network (ANN) [6], Hybrid Convolutional Neural Network [7], Logistic Regression (LR) [8], Decision Tree (DT) [9], Random Forest (RF) [10], Quantum Machine Learning and Deep learning [11].

Li et al. [12,13] focus on convolutional recurrent neural networks and deep learning frameworks [14]. The challenges posed by imbalanced data in machine learning, as reviewed by Kaur et al. [15], Haixiang et al. [16], Fernández et al. [17], and Krawczyk [18], underscore the importance of addressing data quality issues for reliable predictive modelling in healthcare applications.

The researchers [19–21]] conduct seminal research on the immune responses and risk stratification associated with Type 1 diabetes (T1D), including the role of islet autoantibodies and genetic markers. On the other hand, Tigga and Garg [22] (forthcoming) present a logistic regression-based predictive model for T2D, emphasizing the importance of machine learning techniques in diabetes prediction.

Our study begins by examining traditional machine learning algorithms for diabetes prediction. Additionally, we explore Deep Neural Networks (DNNs) with techniques like early stopping and regularization to enhance accuracy. By optimizing hyperparameters, we aim to improve both prediction accuracy and robustness. Beyond predicting diabetes onset, we classify diabetes types, distinguishing between T1D and T2D by analyzing key attributes.

The paper is divided into 4 sections. Section 2 describes the methodology, Sect. 3 describes the results and discussion, and Sect. 4 concludes the work.

3 Methodology

The dataset from the NIDDK underwent preprocessing, including standard scaling, attribute selection, and feature aggregation. Exploratory Data Analysis (EDA) was performed using histograms, correlation analysis, and boxplots for outlier detection. K-means clustering was used for unsupervised learning to identify diabetes types, while a deep neural network and XGBoost were employed for supervised learning and feature importance. Model performance was evaluated with standard metrics, and insights were derived from the results. The entire process is summarized in Fig. 1.

3.1 Data Collection

The dataset for this study is based on the PIMA Indian dataset (PID) from the NIDDK, chosen for its representation of global lifestyle trends, including higher processed food intake and reduced physical activity. The dataset is sourced from https://data.world/data-society/pima-indians-diabetes-database. Additional attributes like polyphagia, visual blurring, obesity, smoking, and high cholesterol were incorporated to enhance understanding, resulting in an updated PID.

3.2 Data Preprocessing

Data preprocessing involves identifying outliers, transforming data using standard scalers, and sampling, as detailed below.

Outlier Detection: Outliers, which deviate significantly from the data, are visualized using boxplots. Values outside 1.5 times the interquartile range (IQR) from the 25th and 75th percentiles are flagged as outliers. The IQR, the difference between these percentiles, measures data dispersion.

Sampling: To balance classes, we apply oversampling using SMOTE, which creates synthetic instances of the minority class, aligning it with the majority class for improved model training.

Standard Scaler: The Standard Scaler centers each feature by subtracting the mean and scales it to unit variance. This method is efficient, easy to implement, and works well on large datasets without missing values. The transformation formula is:

$$\hat{X}_i = \frac{X_i - \bar{X}}{\sigma} \tag{1}$$

where $\hat{X}_i$ is the standardized feature, X_i the observation, $\bar{X}$ the mean, and σ the standard deviation.

3.3 Classification

In this study, we introduce a DNN structure consisting of three hidden layers, enhanced with a dropout layer, to tackle classification tasks.

Finding Feature Importance Using XGBoost: XGBoost has emerged as a widely adopted machine learning technique owing to its remarkable accuracy and performance. In our study, XGBoost is employed to ascertain the significance of features based on their "gain," "cover," and "frequency" scores. Following the feature importance analysis, feature selection is carried out. The prediction outcome is determined by the summation of scores predicted by P trees, as depicted in Eq. (2):

$$\hat{y}_i = \sum_{P=1}^{P} f_p(p_i), \quad f_p \in F \tag{2}$$

In this equation, x_i represents the sample of the ith training instance, $f_p(x_i)$ denotes the score for the pth tree, and F encompasses all regression trees constituting the function space.

Dataset Splitting: Deep learning models require evaluation across various datasets to ensure their effectiveness. To address overfitting in supervised techniques, datasets are split into training and test sets. Equally distributing data between these sets hinders accurate performance assessment. By allocating the majority of data to the training set, we enhance model accuracy and comprehension while maintaining comparability between training and test sets, thus mitigating data inconsistencies.

Dataset Training: In our study, we employed data splitting to mitigate overfitting, a scenario where a machine learning model excessively fits the training data but struggles to generalise the new data reliably. As a precaution, 80% of the data was allocated to the training dataset.

Testing and Validation Dataset: We apply the train control package in R, along with the cross-validation technique, to assess our model's performance on the testing dataset. By allocating 20% of the data set to test and validate, our aim was to mitigate concerns about overfitting and ensure the robustness of our model evaluation.

Proposed Approach for the Classification Model: The proposed model for the classification is shown in Fig. 1. It's a deep neural network classifier that follows an artificial neural network, which aims to replicate the information processing and analysis capabilities of the human brain. The fundamental architecture for constructing neural networks comprises three primary components: input, hidden, and output layers. In feedforward networks, data flow proceeds from internal to external components, following a linear path.

An activation function evaluates the input of a neuron to decide if it should be activated, thereby influencing its contribution to the network's overall prediction. All the activation functions utilised in our model are:

- **Sigmoid:** It is used for binary classification in the DNN model.

$$\sigma(x) = \frac{1}{1 + e^{-x}} \tag{3}$$

where x denotes the input and e denotes the base of the natural logarithm.

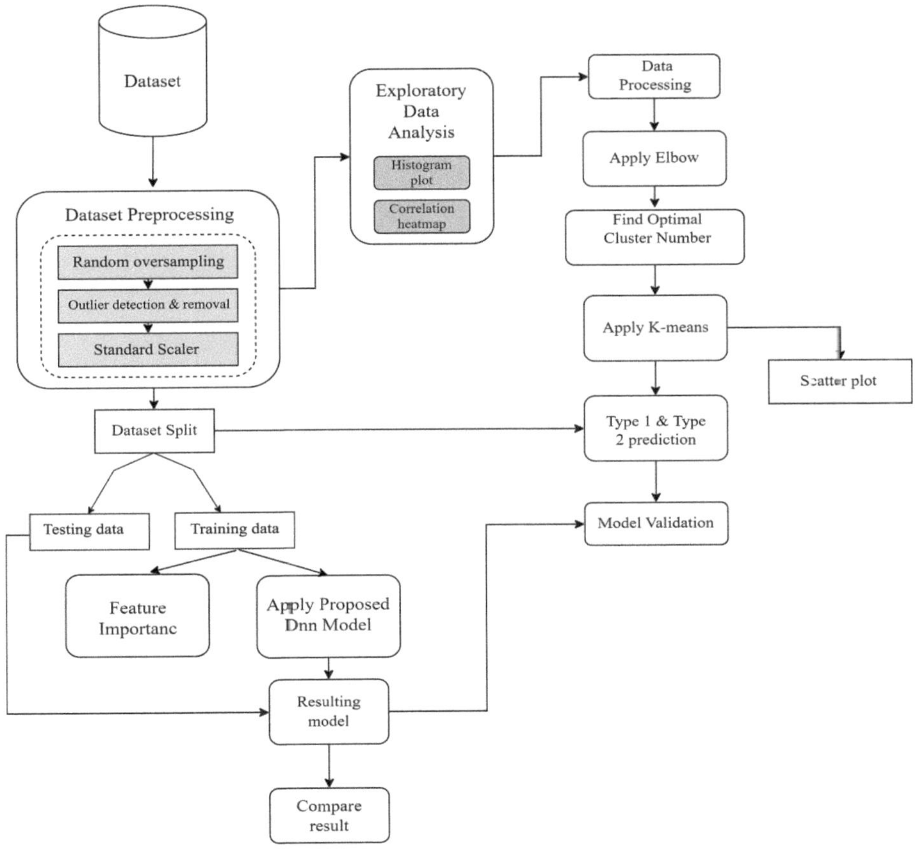

Fig. 1. Architecture of the proposed model.

- **ReLU (Rectified Linear Unit):** The ReLU function prevents all neurons from activating simultaneously.

$$\mathrm{ReLU}(x) = \max(0, x) \tag{4}$$

where x denotes the input.

After applying the activation function, the objective is to update the weights and biases in hidden layers to improve the final output. Let $\hat{z}$ be the predicted output and z the actual output. The error is estimated using binary cross-entropy as the cost function:

$$C_{\text{binary crossentropy}} = -\frac{1}{m} \sum [z \cdot \log(\hat{z}) + (1 - z) \cdot \log(1 - \hat{z})] \tag{5}$$

Binary cross-entropy measures the difference between actual and predicted outputs using logarithms. When $C = 0$, synaptic weights are adjusted iteratively for m rows to minimize C and optimize $\hat{z}$. Effective deep learning models require tuning hyperparameters like the learning rate and number of hidden units, as these impact learning and performance. Overfitting is a common challenge addressed by techniques such as early stopping, L1/L2 regularization, and dropout.

L1 regularization (Lasso) adds the absolute values of the weights to the cost function:

$$\text{Cost} = \sum_{i=0}^{N} \left(y_i - \sum_{j=0}^{M} x_{ij} W_j \right)^2 + \lambda \sum_{j=0}^{M} |W_j| \tag{6}$$

L2 regularization (Ridge) adds the squared values of the weights:

$$\text{Cost} = \sum_{i=0}^{N} \left(y_i - \sum_{j=0}^{M} x_{ij} W_j \right)^2 + \lambda \sum_{j=0}^{M} W_j^2 \tag{7}$$

In both cases, λ controls the strength of regularization, and W_j denotes the weights.

3.4   Type 1 and Type 2 Prediction

In the realm of predicting T1D and T2D, two foundational techniques have been employed. The first involves clustering methods, which are pivotal in pattern recognition, as they segment data into groups based on similarities. Within clustering, techniques like K-means and Fuzzy C-means algorithms are utilized, employing partitional clustering to partition data into distinct clusters based on their characteristics.

K-Means Algorithm: In our investigation, we utilize the K-means clustering method, a widely used approach. The K-means algorithm tries to minimize the squared error function, also known as the distortion, given by:

$$\phi = \sum_{x \in X} \min_{c \in C} (\|x_i - c_i\|)^2 \quad (4) \tag{8}$$

Here, $\|x_i - c_i\|$ represents the Euclidean distance between data point x_i and cluster center c_i. $C = \{c_1, c_2, ..., c_n\}$ denotes the cluster centers, and $X = \{x_1, x_2, ..., x_n\}$ is the set of data points.

Elbow Method: The elbow method is a method to find out the optimal number of clusters by observing the point where the rate of change of the sum of squares becomes less pronounced. The point of inflexion represents the optimal number of clusters.

3.5   Experimental Setup

3.6   Performance Evaluation Metrics

In this study, we assessed the effectiveness of our experiments using a standard set of evaluation metrics commonly employed for classification problems. **Precision** is calculated by dividing the number of correctly identified positive cases by the total number of cases identified as positive, including both true positives and false positives.:

$$\text{Precision} = \frac{TruePositive}{TruePositive + FalsePositive} \tag{9}$$

Recall, conversely, represents the ratio of true positives to the sum of true positives and false negatives:

$$\text{Recall} = \frac{TruePositive}{TruePositive + FalseNegative} \tag{10}$$

The F1-score, which is the harmonic mean of precision and recall, is defined as follows:

$$\text{F1-score} = 2 \times \frac{Precision \times Recall}{Precision + Recall} \tag{11}$$

Accuracy quantifies the proportion of correct predictions relative to the total number of predictions:

$$\text{Accuracy} = \frac{TruePositive + TrueNegative}{TruePositive + FalsePositive + TrueNegative + FalseNegative} \tag{12}$$

ROC-AUC Score, A ROC curve represents the balance between sensitivity (the true positive rate) and specificity (1 minus the false positive rate) in binary classification models. The area under the ROC curve (ROC-AUC), is a commonly employed measure for gauging the accuracy of these models. It indicates the model's effectiveness in distinguishing between positive and negative class labels.

4   Results and Discussion

Before evaluating the models, the Xgboost algorithm is used to calculate the importance of all the attributes of the dataset. Table 1 showcases the feature importance of each feature with its gain, cover, and frequency score.

The model's performance is assessed using the modified dataset and based on different train-test splits, such as 80–20%, 70–30%, and 60–40%. Among these splits, the 80–20% split yields the highest accuracy rate of 99.50%, followed by 70–30% with 99.34% accuracy and 60–40% with 98.77% accuracy. These results indicate that the 80–20% split outperforms the others in terms of accuracy.

Figure 2 shows the confusion matrix obtained after testing the models on the updated PIMA dataset. Figure 2 (a) shows the confusion matrix for the proposed model with dropout regularization. Similarly, Fig. 2 (b) and Fig. 2 (c) show the

Table 1. Feature importance

Feature	Gain	Cover	Frequency
Glucose	674.891113	6218.192871	241.0
BMI	360.186432	5329.798828	276.0
Age	249.319855	3261.162109	182.0
DiabetesPedigreeFunction	234.391296	4570.731445	266.0
BloodPressure	138.123505	2339.896240	177.0
Insulin	105.423874	921.761902	109.0
SkinThickness	81.559296	1000.767578	118.0
Pregnancies	86.316971	1385.609619	117.0
Polyphagia	33.288998	367.957520	46.0
Smoker	28.894606	297.111450	44.0
Visual Blurring	15.897435	199.732880	28.0
High Cholestrol (HDL)	15.437880	143.294922	25.0
Obesity	12.939490	176.950150	14.0

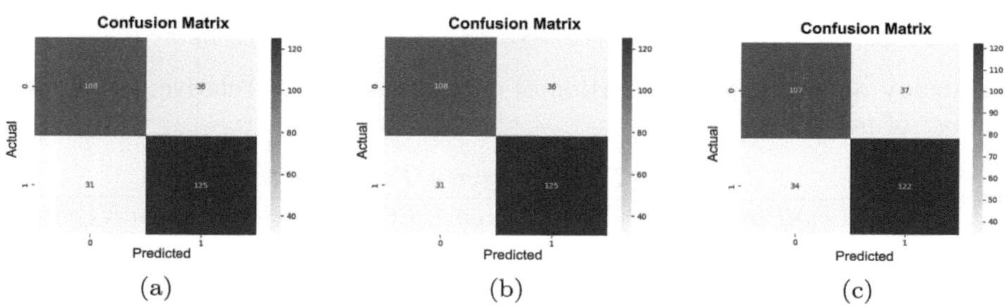

Fig. 2. Confusion Matrix of proposed DNN. (a) No Regularization, (b) L1 Regularization, and (c) L2 Regularization.

confusion matrix for the proposed model, which is hyperparameter tuned by L1 and L2 regularization techniques.

The confusion matrices provided above enable the computation of precision, recall, and ROC-AUC scores for each model. Furthermore, the performance of our proposed model is assessed in terms of F1-Score and classification accuracy. F1-Score is derived from Eq. (11), while precision and recall values are obtained using Eqs. (9) and (12), respectively.

The proposed model is evaluated against other contemporary models in the context of this specific binary classification task. The classification accuracy serves as a measure of the proposed model's performance. Table 2 provides a comparison of these metrics with the results obtained by other contemporary methods using the updated PIMA dataset, split randomly into 80% for training and 20% for testing and already existing work on PIMA dataset.

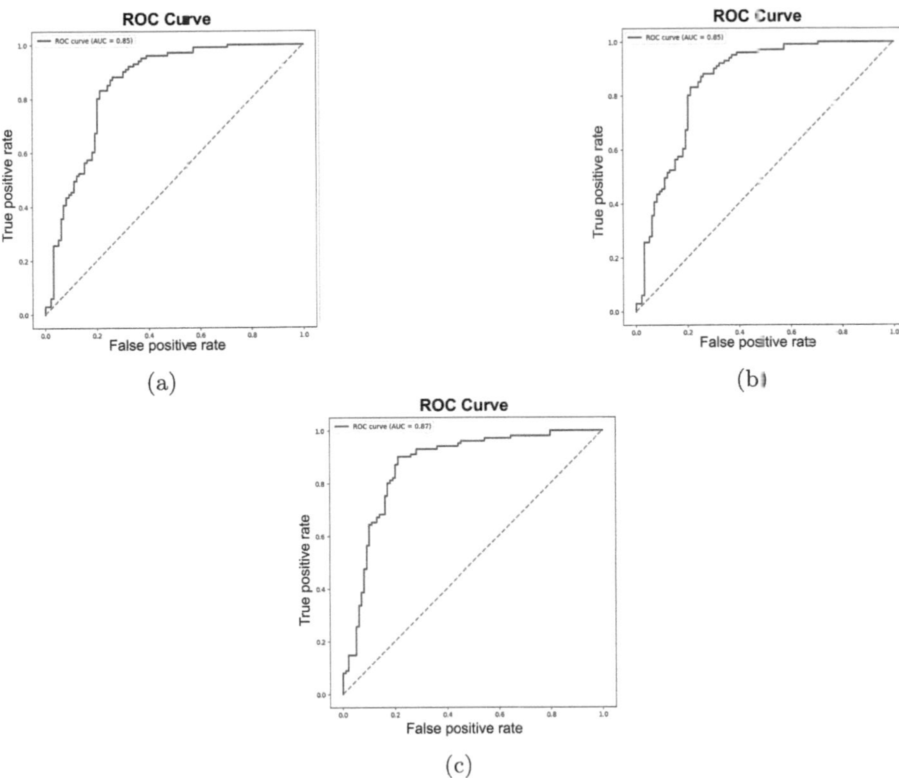

Fig. 3. ROC of proposed DNN. (a) No Regularization, (b) L1 Regularization, and (c) L2 Regularization

The misclassifications made by these classifiers are further scrutinized using ROC analysis. Figure 3 presents the ROC curve of our proposed model. The AUC derived from the ROC curve offers insights into the classification prowess of the proposed model. Higher AUC values indicate superior performance of the classifier. Specifically, the AUC values achieved by our proposed classifier are 0.85 with L1 regularization and 0.87 with L2 regularization.

Table 2. Comparative analysis of proposed and already developed diabetes prediction models

Algorithm	Accuracy	Precision	Recall	F1-Score
Logistic Regression	79.87%	0.82	0.56	0.67
KNearest Neighbour	68.83%	0.56	0.62	0.59
SVector Machine	75.32%	0.71	0.53	0.60
Decision Tree	78.57%	0.70	0.71	0.70
Random Forest	77.27%	0.69	0.65	0.67
Logistic Regression	79.87%	0.82	0.56	0.67
ANN	96.24%	0.77	0.83	0.80
ADASYN [3]	84%	0.91	0.85	0.88
ABP-SCGNN [23]	93%	0.82	0.83	0.80
QML [11]	86%	0.74	0.85	0.79
DL [11]	95%	0.90	0.95	0.93
DNN	96.24%	0.77	0.83	0.80
DNN(early stopping)	96.87%	0.77	0.83	0.80
DNN(L1 Regularization)	98.50%	0.78	0.81	0.82
DNN(L2 Regularization)	99.50%	0.82	0.87	0.85

4.1   Type 1 and Type 2 Prediction

Table 3. Prediction of T1D & T2D on new datapoints

Glucose	Age	Pregnancies	Diabetes Type
148	50	6	T2D
85	31	1	T2D
183	32	8	T2D
89	21	1	T2D
137	33	0	T2D
101	63	10	T2D
122	27	2	T1D
121	30	5	T1D
126	47	1	T1D
93	23	1	T2D

Cluster Based Approach: For finding the optimal number of clusters (k), we used the elbow and Silhouette method, which is found to be 3 as seen in Fig. 4a.

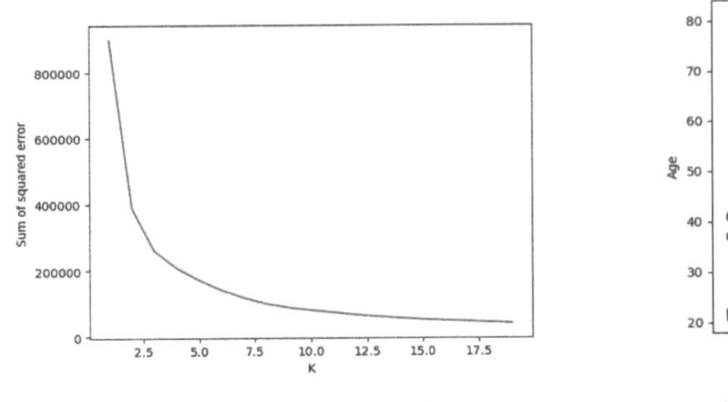

(a) Optimal Cluster number

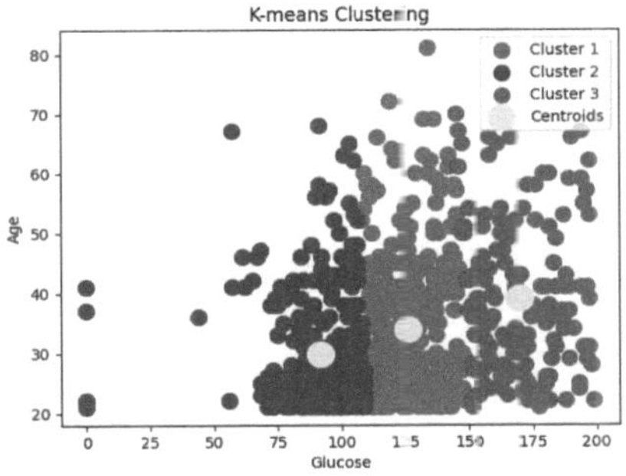

(b) Scatter plot of the clusters

Fig. 4. Comparison of the Elbow method and Cluster scatter plot.

The scatter plot in Fig. 4b illustrates the relationship between glucose levels and age in the dataset. Glucose and age, identified as the most influential features according to Table 1, play significant roles in the prediction model. Utilizing the Elbow method, the plot has been segmented into three clusters, denoted in Fig. 4b based on the optimal value of k. Notably, gestational diabetes, a type often preceding T2D, is considered alongside age and glucose levels, as it frequently emerges during pregnancy in women without preexisting diabetes. Taking all aforementioned factors into consideration, the model is tested on new data points and further validated using the proposed DNN architecture. Table 3 shows the proposed model's prediction results on new user input data points.

5 Conclusion

Predicting diabetes in its early stages is a significant global challenge faced by doctors, medical professionals, and researchers, irrespective of the type of diabetes. Timely identification of diabetes is crucial to enable prompt intervention and prevent the escalation of the condition, thereby mitigating potential complications. This paper presents a thorough investigation into diabetes prediction utilizing numerous machine learning algorithms and AI techniques, with a focus on the updated PID. In comparison to the other contemporary models, the proposed DNN model showcases the highest classification accuracy (99.5%) highlighting the importance of algorithm selection, model optimization, and dataset modification in achieving accurate predictions. Further, the proposed DNN model also achieves better results in terms of performance evaluation metrics. Moreover, our study underscores the potential of advanced machine learning approaches in enhancing personalized healthcare interventions and refining diabetes management approaches, with the ultimate goal of enhancing patient outcomes and overall quality of life.

Acknowledgements. We are thankful to the department of EIE, NIT Silchar for providing us with the necessary infrastructure.

References

1. Maniruzzaman, M., et al.: Accurate diabetes risk stratification using machine learning: role of missing value and outliers. J. Med. Syst. **42**, 1–17 (2018)
2. McLachlan, G.J.: Discriminant Analysis and Statistical Pattern Recognition. John Wiley & Sons (2005)
3. Yadav, S., Maravi, Y. P., Agrawal, J., Mishra, N.: A neural network based diabetes prediction on imbalanced data. In: 2021 10th IEEE International Conference on Communication Systems and Network Technologies (CSNT), pp. 515–521. IEEE (2021)
4. Webb, G.I., Boughton, J.R., Wang, Z.: Not so naive bayes: aggregating one-dependence estimators. Mach. Learn. **58**, 5–24 (2005)
5. Brahim-Belhouari, S., Bermak, A.: Gaussian process for nonstationary time series prediction. Comput. Stat. Data Anal. **47**(4), 705–712 (2004)
6. Reinhardt, A., Hubbard, T.: Using neural networks for prediction of the subcellular location of proteins. Nucleic Acids Res. **26**(9), 2230–2236 (1998)
7. Farsana, K.S., Poulose, A.: Hybrid convolutional neural networks for Pima Indians diabetes prediction. In: 2024 Fifteenth International Conference on Ubiquitous and Future Networks (ICUFN), pp. 268–273. IEEE (2024)
8. Tabaei, B.P., Herman, W.H.: A multivariate logistic regression equation to screen for diabetes: development and validation. Diab. Care **25**(11), 1999–2003 (2002)
9. Jenhani, I., Amor, N.B., Elouedi, Z.: Decision trees as possibilistic classifiers. Int. J. Approximate Reasoning **48**(3), 784–807 (2008)
10. Breiman, L.: Random forest. Mach. Learn. **45**, 1 (2001)
11. Gupta, H., Varshney, H., Sharma, T.K., Pachauri, N., Verma, O.P.: Comparative performance analysis of quantum machine learning with deep learning for diabetes prediction. Complex Intell. Syst. **8**(4), 3073–3087 (2022)
12. Li, K., Daniels, J., Liu, C., Herrero, P., Georgiou, P.: Convolutional recurrent neural networks for glucose prediction. IEEE J. Biomed. Health Inform. **24**(2), 603–613 (2019)
13. Li, K., Liu, C., Zhu, T., Herrero, P., Georgiou, P.: GluNet: a deep learning framework for accurate glucose forecasting. IEEE J. Biomed. Health Inform. **24**(2), 414–423 (2019)
14. Sekhar, A., Biswas, S., Hazra, R., Sunaniya, A.K., Mukherjee, A., Yang, L.: Brain tumor classification using fine-tuned GoogLeNet features and machine learning algorithms: IoMT enabled cad system. IEEE J. Biomed. Health Inf. **26**(3), 983–991 (2021)
15. Kaur, H., Pannu, H.S., Malhi, A.K.: A systematic review on imbalanced data challenges in machine learning: applications and solutions. ACM Comput. Surv. (CSUR) **52**(4), 1–36 (2019)
16. Haixiang, G., Yijing, L., Jennifer Shang, G., Mingyun, H.Y., Bing, G.: Learning from class-imbalanced data: review of methods and applications. Expert Syst. Appl. **73**, 220–239 (2017)
17. Fernández, A., García, S., Galar, M., Prati, R.C., Krawczyk, B., Herrera, F.: Learning From Imbalanced Data Sets, vol. 10. Springer (2018)
18. Krawczyk, B.: Learning from imbalanced data: open challenges and future directions. Prog. Artif. Intell. **5**(4), 221–232 (2016). https://doi.org/10.1007/s13748-016-0094-0
19. Achenbach, P., et al.: Stratification of type 1 diabetes risk on the basis of islet autoantibody characteristics. Diabetes **53**(2), 384–392 (2004)

20. Achenbach, P., et al.: Type 1 diabetes risk assessment: improvement by follow-up measurements in young islet autoantibody-positive relatives. Diabetologia **49**, 2969–2976 (2006)
21. Achenbach, P.: Autoantibodies to zinc transporter 8 and SLC30A8 genotype stratify type 1 diabetes risk. Diabetologia **52**, 1881–1888 (2009)
22. Tigga, N.P., Garg, S.: Predicting type 2 diabetes using logistic regression. In: Proceedings of the Fourth International Conference on Microelectronics, Computing and Communication Systems: MCCS 2019, pp. 491–500. Springer (2021)
23. Bukhari, M.M., Alkhamees, B.F., Hussain, S., Gumaei, A., Assiri, A., Ullah, S.S.: An improved artificial neural network model for effective diabetes prediction. Complexity **2021**(1), 5525271 (2021)

Machine Learning Techniques for Predicting Olive Oil Pigment Concentrations

Srikrishna S. Kashyap[1] , B. K. Vinay[2](✉) , T. R. Suranjan[2] , and M. A. Sriraj[2]

[1] Liverpool John Moores University, Liverpool L3 5AH, UK
uplskash@ljmu.ac.uk
[2] Vidyavardhaka College of Engineering, Mysuru 570002, Karnataka, India
vinaybk@vvce.ac.in

Abstract. This study presents optimization and machine learning (ML) methods for predicting pigment concentrations in Extra Virgin Olive Oil (EVOO) using UV-Vis spectroscopy data and compares them to the standard deconvolution approach. We tested three EVOO samples: standard, fresh, and Monocultivar Frantoio. All methods demonstrated great spectrum reconstruction accuracy of R2 > 0.99. The optimization technique, which employed Non-Negative Least Squares (NNLS) and Limited-memory Broyden-Fletcher-Goldfarb-Shanno with Bounds (L-BFGS-B) algorithms, consistently yielded non-negative pigment concentration predictions. Both new methods predicted higher total pigment concentrations than the traditional method, with differences ranging from 2.124 mg/kg (Fresh EVOO) to 7.329 mg/kg (Monocultivar Frantoio). Our approach is a major step forward in quick, non-destructive EVOO analysis, even though there are still difficulties in quantifying trace pigments below specific thresholds. The EVOO quality assessment, authentication, and industrial quality control procedures are significantly impacted by these advancements in pigment quantification accuracy and dependability, which also present opportunities for automated, high-throughput screening applications.

Keywords: Pigment Analysis · Spectral Deconvolution · Machine Learning · EVOO · UV-Vis Spectroscopy · Food Science

1 Introduction

EVOO is a highly valued product recognized for its unique flavor profile and plenty of health advantages. Sieved from the olive tree (Olea europaea L.), an antiquity cultivation in Mediterranean area, EVOO is considered as one of tastiest Olive oil's due to its superb features compared with mild ones including acidity levels less than 0.8% and a sensory score above 6.5 [1]. The authenticity and quality of EVOO is significantly affected by the presence of chlorophylls and carotenoids, constituting the majority of pigment content [2]. Lutein (LUT), beta-carotene (b-CAR), violaxanthin, and neoxanthin (NEX) are the main pigments found in EVOO, along with a variety of lesser derivatives, which are impacted by the olive cultivar, fruit ripeness, production techniques, and storage

C. Modi et al. (Eds.): MIND 2024, CCIS 2736, pp. 538–549, 2026.
https://doi.org/10.1007/978-3-032-14531-4_45

[3–6].The traditional methods for the determination of pigments in EVOOs like spectral deconvolution and High Performance Liquid Chromatography (HPLC) can be inaccurate, costly and time consuming [7]. These procedures often include chemical extraction and chromatography which can alter the oil's constitution [8]. UV-Vis spectroscopy has been emerged as a non-destructive alternative for pigments analysis in EVOO, particularly through recent surface science advancements [9]. Accurately measuring individual pigment concentrations is challenging due to the complex EVOO spectra, with overlapping absorbance signals from different pigments [10]. In this context, introducing new computational methods for data analysis in spectroscopic techniques in food science would be a great improvement [6, 11], leading researchers to explore advanced models based on ML algorithms and optimization processes. These approaches have shown great potential for handling big data and improving the accuracy and efficiency of analytical processes across various scientific disciplines [8]. This research advances these efforts by applying ML to UV-Vis spectroscopy data to enhance pigment concentration predictions in EVOO. The study evaluates various ML models, comparing their effectiveness with traditional spectral deconvolution, and assesses the feasibility of ML techniques for rapid, non-invasive pigment analysis in quality control and authenticity verification of EVOO.

2   Background

2.1   Spectroscopic Methods in EVOO Analysis

Simple spectrophotometric techniques are low accuracy and time-consuming in the study of pigments colored compounds present brown-red EVOO samples, whereas traditional methods like HPLC may it involve uncertain results considering peculiar matrix complexity [7]. So scientists have looked into several non-destructive spectroscopic methods for the analysis of pigments in EVOO. To ascertain the pigment concentrations in virgin and EVOO, Borello and Domenici [8] compared two near UV-Vis spectroscopic methods. The entire absorption spectrum was analyzed using a mathematical deconvolution approach, and found more accurate and consistent pigment concentrations than conventional method that relied on absorbance values at wavelengths. They noted limitations in detecting minute pigment and analyzing fresh olive oils. Using a mathematical analysis of near UV-visible absorption spectra, Lazzerini and Domenici [12] expanded on this work by examining pigment concentrations in EVOOs made in Tuscany, Italy. Using their method, four main pigments (lutein, β-carotene, pheophytin A, and pheophytin B) could be measured. With total pigment concentrations ranging from 6.65 ppm to 27.58 ppm, the study found significant variations in pigment profiles across different harvesting years. A digital tool called EVOODec was created by Jurinovich and Domenici [18] to analyze the UV-Vis spectra of EVOOs without the need for chemical treatment. With an average user understanding score of 7.6 out of 10, this tool showed promise for accessible pigment analysis in educational settings.

2.2   ML Applications in EVOO Quality Assessment

The use of spectroscopic data comined with ML methods has significantly improved the effectiveness and precision of pigment analysis in EVOO. Venturini et al. [11] explored a small, affordable fluorescence sensor paired with various ML techniques to evaluate the quality of EVOO. Their research illustrated how ML can accurately assess oil quality without extensive pre-processing, utilizing random forests (RF) and artificial neural networks (ANN) models, achieving complete classification accuracy. Gyftokostas et al. [17] employed Linear Discriminant Analysis (LDA), Gradient Boosting, and spectroscopic data to classify Greek EVOOs based on their cultivar origin. Their Gradient Boosting model, using Laser-Induced Breakdown Spectroscopy (LIBS) data, reached a training accuracy of 96.0% and an external validation accuracy of 93.8%. Additionally, Gucciardi et al. [23] proposed a new approach for evaluating the quality of olive oil. It paired a small optical fluorescence sensor with ANNs. Among the neural network architectures they investigated, the conditional convolutional neural network (Cond-CNN) outperformed the others with an accuracy of 82% in oil classification.

UHPLC-HRMS, or ultra-HPLC coupled with high-resolution mass spectrometry, was used by Arrizabalaga-Larrañaga et al. [20] to perform a thorough investigation on the pigment profiles of Spanish EVOO. Their research has revealed significant differences in total pigment content across different regions of Spain. The changes in virgin olive oil during storage were observed by Lobo-Prieto et al. [19] using fluorescence spectroscopy in conjunction with parallel factor analysis (PARAFAC). The study showcased the potential of fluorescence-based methods for pigment analysis by demonstrating high correlation coefficients (0.98) between pheophytin concentration measured by HPLC and fluorescence data.

EVOO pigment analysis has advanced significantly but there are still a number of obstacles to overcome including the need to detect even the smallest pigment concentrations, create reliable models that take sample-specific variations in chemical characteristics into account and combine several analytical techniques for an all-encompassing quality assessment [19, 20]. Incorporating state-of-the-art ML techniques for olive oil authenticity and classification has been the subject of recent studies that have indicated possible directions for future study [17, 23]. Along with studies into combining spectroscopic data with other chemical and sensory features for comprehensive quality assessment [19, 20], the development of on-the-spot, real-time analysis tools for rapid quality evaluation has emerged as a critical research path [11, 14]. Combining ML with spectroscopic techniques could significantly improve the quantification of EVOO pigment. These developments could have a major impact on quality control procedures and boost consumer confidence in EVOO products. Our work builds on this framework with the aim of utilising ML and optimisation techniques to advance the field.

3 Materials and Methods

3.1 Data Collection and Spectral Acquisition

UV-Vis spectroscopy data collected from a range of EVOO samples, including commercially available and freshly squeezed oils, were used in this investigation. Jurinovich and Domenici (2022) [18] were the first to publish the dataset. Three different kinds of EVOO samples were examined: Standard EVOO, Fresh EVOO and Monocultivar Frantoio EVOO. Using a spectrophotometer capable of measuring in the 380–720 nm range, UV-Vis absorbance spectra were acquired with a spectral resolution of 1 nm. Using this wavelength range, the absorption characteristics of the main pigments in olive oil, including carotenoids, pheophytins, and chlorophylls, were captured [9, 10]. As suggested by Borello and Domenici [8], oil samples were filtered prior to spectral capture to guarantee clarity and reduce light scattering effects. At room temperature, absorbance readings were recorded using a quartz cuvette with a 1 cm path length. Air was used as a reference to calibrate the spectrophotometer. A total of 341 data points, spanning the whole 380–720 nm range, were acquired for each sample. The spectral data was stored in CSV format, where the wavelength (nm) and associated absorbance (a.u.) values were shown in two columns. Based on their importance in the composition of EVOO, the pigments taken into consideration in this study were [2, 4]: Phenophytin a (Pheo-a), Chlorophyll b (Chl-b), b-CAR, LUT, NEX, Chlorophyll a (Chl-a), and Phenophytin b (Pheo-b). Triolein (TRIO) was also added to symbolize the EVOO matrix, which is mainly made up of triglycerides and fatty acids [10]. In the spectrum analysis and ML model creation that followed, reference data encompassing the same wavelength range (380–720 nm) for the molar extinction coefficients of these pigments was used. This extensive dataset served as the basis for the development and assessment of our ML models for pigment concentration prediction as well as a thorough examination of the pigment composition in a variety of oil samples.

3.2 Data Preprocessing

Data preprocessing is crucial for spectral analysis and ML applications in EVOO pigment quantification. A critical step involves aligning EVOO sample data with pigment reference data to compare absorbance and extinction coefficient values at the same wavelengths. We then normalize extinction coefficients to account for scale differences between pigments. To enhance data quality, we apply a Savitzky-Golay filter for spectral smoothing, reducing noise while preserving underlying spectral features [24]. The Savitzky-Golay filter can be expressed as:

$$y_i^* = \sum_{n=-m}^{m} c_n y_{i+n} \tag{1}$$

where (y_i^*) is the smoothed value, (y_i) is the original data point, (c_n) are the filter coefficients, and $(2m + 1)$ is the window size. For ML models, we implement standard scaling of input features to ensure all variables are on a similar scale. We perform validation tests at several stages of the preprocessing pipeline to guarantee data consistency and integrity, which is necessary for trustworthy analysis in the future. With this thorough approach, we can accurately anticipate pigment concentration and establish dependable comparisons between analytical methods, as well as prepare our spectrum data for both conventional optimization methods and ML techniques.

3.3 Traditional Deconvolution Method

We used the conventional deconvolution method as outlined by Domenici et al. [10] as a baseline for comparison. Using matrix operations and eigenvector calculations, the full UV-Vis absorption spectrum is analyzed to obtain pigment amounts.

Method 1 Traditional Deconvolution Method

1: Compute overlap matrix of reference spectra
2: Diagonalize overlap matrix
3: Compute eigenvectors for new base
4: Calculate Spectral Variance (SV) coefficients
5: Compute pigment concentrations
6: Reconstruct spectrum and calculate residuals

In addition to producing the reconstructed spectrum and estimated pigment concentrations, the method calculates residuals. This method makes use of all available spectral information to enable the simultaneous quantification of several pigments. Prior research has demonstrated that the method is accurate [10], but it can be computationally demanding and may have difficulties with extremely complex or noisy spectra.

3.4 Optimization Method

We have implemented an optimization method that makes use of well-established numerical techniques to supplement the conventional deconvolution approach. This approach should yield accurate and useful pigment concentration estimates.

Method 2 Optimization Method

1: NNLS are used to make an initial estimate.
2: L-BFGS-B technique to optimize concentrations
3: Calculate fitted spectrum
4: Compute R-squared and RMSE

This method uses numerical approaches to accurately determine pigment amounts. We start by utilising Non-Negative Least Squares (NNLS) to estimate the pigment concentrations. NNLS ensures that the initial concentrations for pigment analysis are non-negative, which is important from a physical standpoint. The initial guess is refined using the Limited-memory Broyden-Fletcher-Goldfarb-Shanno algorithm with Bounds (L-BFGS-B). The estimation of pigment concentration necessitates optimization issues with bound restrictions, which our approach effectively handles [26]. With the optimized concentrations, we calculate a fitted spectrum that most closely matches the experimental EVOO spectrum. We compute the coefficient of determination (R-squared) and the Root Mean Square Error (RMSE) to assess the accuracy and fit of the concentration estimates. The optimization procedure is guided by an objective function that minimizes the sum of squared differences between the experimental and fitted absorbance:

$$\min \sum (A\_exp - A\_fit)^2 \tag{2}$$

where A_exp is the experimental absorbance and A_fit is the fitted absorbance calculated from the estimated pigment concentrations. This method uses NNLS for initialization and L-BFGS-B for fine-tuning optimization, striking a balance between accuracy and computational efficiency. To ensure that the estimated concentrations remain within physically meaningful ranges, bound constraints are employed during the optimization process.

3.5 ML Methods

To improve our ability to estimate pigment concentration even further, we tested a few ML models. This method seeks to extract intricate linkages from the spectrum data that conventional or optimization approaches might not be able to completely address.

Method 3 ML Method

1: Preprocess and standardize data
2: Train multiple ML models
3: Select best model based on R^2 score
4: Predict concentrations using best model
5: Calculate fitted spectrum
6: Compute R-squared and RMSE

We used a various of ML models to perform spectral deconvolution in our analysis of EVOO samples. Random Forest [27], Gradient Boosting [28], Ridge Regression, K-Nearest Neighbours, and Partial Least Squares Regression are the models we used in our study. When it comes to managing intricate, high-dimensional data, such as spectral information, each of these models has special advantages. We train each ML models on the standardized data. Each model is fitted using the entire dataset. We select the best model based on the R2 score calculated on smoothed predictions. We use the best-performing model to predict pigment concentrations. We calculate the fitted spectrum using the predicted concentrations.

$$A_{\text{fit}} = \epsilon_{\text{normalized}} \cdot c \tag{3}$$

where (A_{fit}) is the fitted absorbance spectrum, ($\epsilon_{\text{normalized}}$) is the normalized extinction coefficient matrix and (c) is the vector of predicted concentrations. We compute the R2 score and Root Mean Square Error (RMSE) to assess the quality of the fit. Complex, non-linear interactions in the spectrum data can be explored with this ML approach. We can choose the best method for our EVOO pigment prediction task by comparing several models. The robustness and dependability of our predictions are enhanced by the performance measures on smoothed data.

4 Results and Discussion

4.1 Performance Comparison

Using three different approaches—classical deconvolution, optimization, and ML—we examined three EVOO samples: fresh EVOO, standard EVOO, and monocultivar Frantoio. The performance metrics for every method across all samples are compiled in Table 1. Across all samples, all methods demonstrated exceptional spectral reconstruction with high R2 values (>0.99). In terms of R^2 and RMSE, the Gradient Boosting model routinely beat the optimization method, whereas the traditional method, because of its computational approach, produced the highest R^2 values and least RMSE metrics [10].

Table 1. Performance metrics for each method across EVOO samples

Sample	Method	R^2	RMSE
Standard EVOO	Traditional	0.999042	0.026501
	Optimization	0.995464	0.057675
	Gradient Boosting	0.996166	0.053021
Fresh EVOO	Traditional	0.997805	0.020236
	Optimization	0.993667	0.034375
	Gradient Boosting	0.993966	0.033553
Monocultivar Frantoio	Traditional	0.999350	0.028512
	Optimization	0.995391	0.075958
	Gradient Boosting	0.996044	0.070364

4.2 Spectral Reconstruction

We analyzed the spectral reconstruction capabilities of each method for all three EVOO samples.

Traditional Deconvolution Method

Figure 1 show the spectral reconstruction using the traditional method for standard EVOO, fresh EVOO, and Monocultivar Frantoio samples, respectively. The 400–500 nm and 600–700 nm regions, which are crucial for carotenoid and chlorophyll pigments, respectively, were especially well-fitted by the conventional approach for all samples [9]. The best fit ($R^2 = 0.999350$) was found in the Monocultivar Frantoio sample (Fig. 1), possibly because of its unique pigment profile.

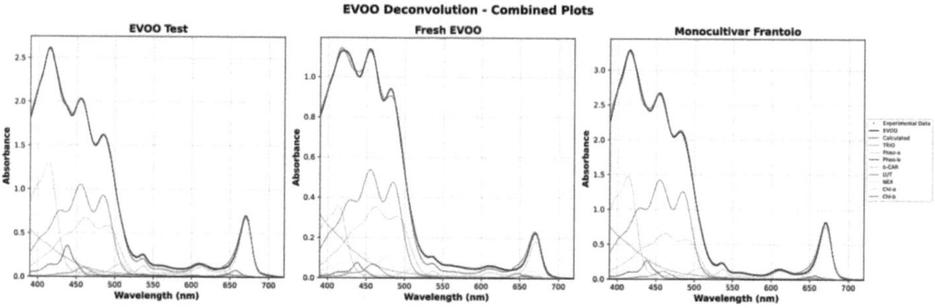

Fig. 1. Spectral reconstruction across EVOO samples using traditional deconvolution method

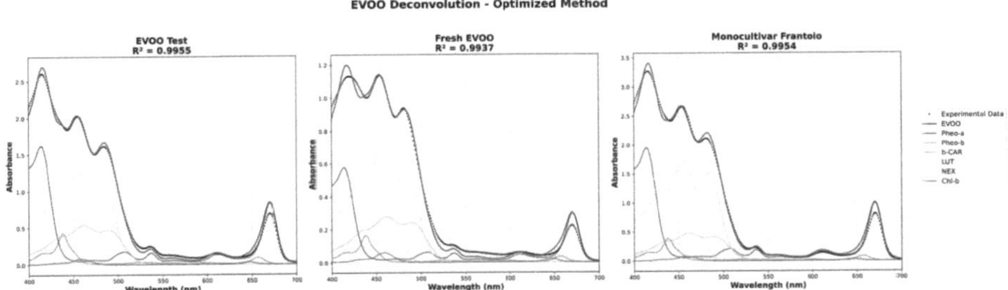

Fig. 2. Spectral reconstruction across EVOO samples using optimization method

Optimization Method

The optimization method demonstrated good fits across all samples, with slight deviations in the 450–550 nm range, particularly for the fresh EVOO sample (Fig. 2). This may account for differences in estimated concentrations of carotenoid pigments.

ML Method

For all samples, gradient boosting obtained the highest R^2 values, suggesting that it accounts for the greatest amount of data variance. This implies that it is most successfully capturing the underlying patterns in the spectral data. In addition to consistently generating the lowest RMSE scores, the Gradient Boosting model outperformed the others in terms of predictive accuracy. Thus, the Gradient Boosting model achieved excellent fits across all samples, potentially capturing subtle spectral features that contribute to its improved performance in pigment concentration estimation.

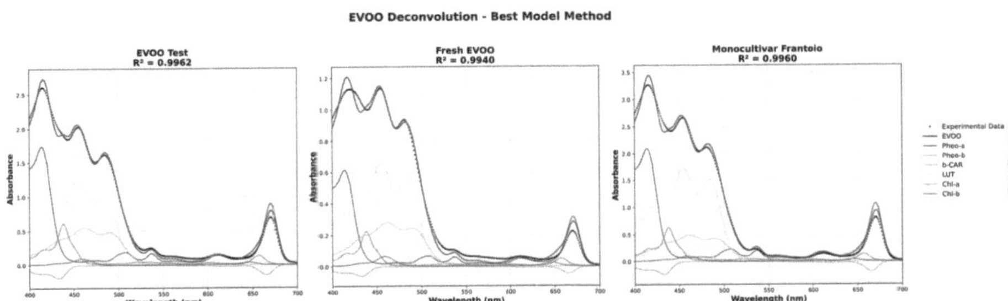

Fig. 3. Spectral reconstruction across EVOO samples using ML method

4.3 Pigment Concentration Predictions

We evaluated each method's predicted pigment concentrations for each of the three EVOO samples. The estimated mg/kg concentrations of the main pigments in each sample are shown in Tables 2.

Table 2. Estimated pigment concentrations (mg/kg)

Samples	Standard EVOO (mg/kg)			Fresh EVOO (mg/kg)			Monocultivar EVOO (mg/kg)		
Pigment	Traditional	Optimization	GB	Traditional	Optimization	GB	Traditional	Optimization	GB
Pheo-a	10.755	13.416	14.348	2.969	4.770	5.068	12.212	16.099	17.270
Pheo-b	1.935	2.320	3.351	0.408	0.903	1.229	1.438	2.136	3.438
b-CAR	2.983	2.415	2.456	1.569	1.244	1.256	2.933	2.137	2.176
LUT	4.953	5.981	5.910	2.545	3.132	3.111	6.723	8.164	8.104
NEX	−0.052	0.162	-0.061	0.647	0.709	0.638	1.478	1.743	1.446
Chl-a	−0.022	0.000	−1.645	0.422	0.000	−0.522	0.216	0.000	−2.074
Chl-b	0.869	0.687	0.801	0.586	0.511	0.552	0.944	0.705	0.838
Total (mg/kg)	**21.421**	**24.980**	**26.865**	**9.146**	**11.270**	**11.853**	**25.944**	**30.984**	**33.273**

4.4 Observations

The traditional method [10] obtained consistent results with the highest R^2 values for all samples. It did, however, occasionally produce negative values when attempting to accurately estimate the concentrations of minor pigments (e.g., NEX and Chl-a in the standard EVOO sample). For all samples, this method estimated the lowest total pigment concentrations. Our optimisation strategy, which combined L-BFGS-B and NNLS, performed better than the conventional approach [26]. It consistently generated concentration estimates that were not negative and received good R^2 scores for all samples. With variations ranging from 2.124 mg/kg (fresh EVOO) to 5.040 mg/kg (Monocultivar Frantoio), the optimisation method predicted higher total pigment concentrations than the conventional method. Across all samples, the Gradient Boosting model [28] performed better in terms of RMSE and R^2 scores. With estimates that were 5.444 mg/kg (standard EVOO), 2.707 mg/kg (fresh EVOO), and 7.329 mg/kg (Monocultivar Frantoio) higher than those of the conventional method, it calculated the highest total pigment concentrations for all samples. Like the conventional approach, it did, however, occasionally yield negative concentration estimates for trace pigments (such as NEX and Chl-a).

Our results show how novel computational techniques can improve the analysis of EVOO pigments. The optimisation and ML techniques provided enhancements in managing minor pigments and estimating total pigment concentrations, even though all methods displayed high R^2 values, suggesting good spectral reconstruction. Though it needs to be further improved to address problems with negative concentration estimates for trace pigments, the ML approach—in particular, the Gradient Boosting model—shows promise in capturing complex spectral relationships. These findings emphasise the significance of method selection in EVOO pigment analysis and the necessity of carefully interpreting results, especially when it comes to pigments that are present in extremely low concentrations. Future work will focus on improving these models, especially for pigments exhibiting greater differences between methods, and investigating the possibility of combining multiple analytical techniques for more thorough EVOO characterisation.

5 Conclusions

This study compared traditional deconvolution [10], optimization, and ML methods for analyzing pigment concentrations in EVOO using UV-Vis spectroscopy. Our findings demonstrate the potential of advanced computational techniques to enhance EVOO pigment analysis. Across three different EVOO samples, all techniques produced high spectral reconstruction accuracy ($R^2 > 0.99$), confirming their suitability for pigment analysis. A major drawback of the conventional method was addressed by the optimization method, which consistently produced non-negative pigment concentration estimates using the NNLS and L-BFGS-B algorithms [26]. In terms of R2 and RMSE, the Gradient Boosting model [28] performed better than the other techniques, indicating that ML techniques can effectively capture intricate spectral relationships in EVOO samples. In comparison to the conventional method, both the optimization and ML techniques estimated higher total pigment concentrations, which may suggest improved sensitivity to minor pigments. All techniques were successful in identifying pigment profiles unique to each sample, allowing for the distinction of standard, fresh, and monocultivar Frantoio EVOOs. However, there are still issues with precisely measuring trace pigments, as various approaches have inconsistent results when estimating extremely low concentrations. As shown in other studies [11, 17, 23], these findings emphasize the potential of combining cutting-edge computational methods with spectroscopic data for EVOO analysis. Although the conventional approach is still useful, optimization and ML techniques have the potential to provide more precise and thorough pigment profiling. Subsequent research ought to concentrate on improving these models, especially regarding trace pigment measurement, and examining how well they work with a larger variety of EVOO samples. Further, combining these spectroscopic techniques with other analytical techniques [19, 20] could result in more reliable tools for EVOO quality evaluation and authentication. Our research demonstrates how advanced computational methods can improve EVOO pigment analysis. These methods have the potential to grow into effective substitutes for current methods in the olive oil industry's quality control and authenticity checks.

References

1. International Olive Council: Trade standard applying to olive oils and olive-pomace oils. COI/T.15/NC No 3/Rev. 8 (2015)
2. Lazzerini, C., Domenici, V., Cifelli, M.: Pigments in extra-virgin olive oils produced in Tuscany (Italy) in different years. Foods **6**(4), 25 (2017)
3. Arrizabalaga-Larrañaga, A., Rodríguez-Morató, J., Hernández-Mesa, M., Sagratini, G., Poyer, S., Quintanilla-Casas, B., et al.: Comprehensive characterization of pigment profile in virgin olive oil by ultra-high-performance liquid chromatography coupled to high-resolution mass spectrometry. Food Chem. **372**, 131274 (2022)
4. Piscopo, A., Poiana, M.: Pigments in virgin olive oils: Spectrophotometric determination and stability during storage. Foods **1**(1), 3–14 (2012)
5. Kalogeropoulos, N., Tsimidou, M.Z.: Antioxidants in Greek virgin olive oils. Antioxidants **3**(2), 387–413 (2014)
6. Borello, E., Domenici, V.: Determination of pigments in virgin and extra-virgin olive oils: a comparison between two near UV-Vis spectroscopic techniques. Foods **8**(1), 18 (2019)

7. Mínguez-Mosquera, M.I., Gandul-Rojas, B., Garrido-Fernández, J., Gallardo-Guerrero, L.: Pigments present in virgin olive oil. J. Am. Oil Chemists' Soc. **67**(3), 192–196 (1990)
8. Lazzerini, C., Domenici, V.: Pigments in extra-virgin olive oils produced in Tuscany (Italy) in different years. Foods **6**(4), 25 (2017)
9. Domenici, V., Ancora, D., Cifelli, M., Serani, A., Veracini, C.A., Zandomeneghi, M.: Extraction of pigment information from near-UV vis absorption spectra of extra virgin olive oils. J. Agric. Food Chem. **62**(38), 9317–9325 (2014)
10. Venturini, S., Ranieri, R., Balestra, F., Dalla Rosa, M.: A machine learning approach based on fluorescence spectroscopy for evaluation of extra virgin olive oil quality. Foods **11**(4), 469 (2022)
11. Gyftokostas, N., Stefas, D., Couris, S., Anastopoulos, P.: Classification of edible oils by laser-induced breakdown spectroscopy (LIBS) and machine learning algorithms. Appl. Spectrosc. **75**(4), 424–434 (2021)
12. Violino, S., et al.: An artificial intelligence approach for Italian EVOO origin traceability through an open source IoT spectrometer. Foods **9**(6), 834 (2020)
13. Venturini, F., et al.: Exploration of Spanish olive oil quality with a miniaturized low-cost fluorescence sensor and machine learning techniques. Foods **10**(5), 1010 (2021)
14. Violino, S., et al.: Superior EVOO quality production: an RGB sorting machine for olive classification. Foods **11**(18), 2917 (2022)
15. Moyano, M.J., Heredia, F.J., Meléndez-Martínez, A.J.: The color of olive oils: the pigments and their likely health benefits and visual and instrumental methods of analysis. Compr. Rev. Food Sci. Food Saf. **9**(3), 278–291 (2010)
16. Stefas, D., et al.: Discrimination of olive oils based on the olive cultivar origin by machine learning employing the fusion of emission and absorption spectroscopic data. Food Control **130**, 108318 (2021)
17. Jurinovich, S., Domenici, V.: Digital tool for the analysis of UV–Vis spectra of olive oils and educational activities with high school and undergraduate students. J. Chem. Educ. **99**(2), 787–798 (2022)
18. Lobo-Prieto, A., Tena, N., Aparicio-Ruiz, R., García-González, D.L., Sikorska, E.: Monitoring virgin olive oil shelf-life by fluorescence spectroscopy and sensory characteristics: a multidimensional study carried out under simulated market conditions. Foods **9**(12), 1846 (2020)
19. Arrizabalaga-Larrañaga, A., Rodríguez, P., Medina, M., Santos, F.J., Moyano, E.: Pigment profiles of Spanish extra virgin olive oils by ultra-high-performance liquid chromatography coupled to high-resolution mass spectrometry. Food Addit. Contam. Part A **37**(7), 1075–1086 (2020)
20. Lazzerini, C., Cifelli, M., Domenici, V.: Pigments in extra-virgin olive oil: authenticity and quality. In: Boskou, D., Clodoveo, M.L. (eds.) Products from Olive Tree. IntechOpen (2016)
21. Fathy, H., Sfatee, M.A., Omar, A., et al.: Reflection of physical descriptive parameters as functional phenes on olive oil cross-border mapping. Food Anal. Methods **17**, 512–522 (2024)
22. Gucciardi, A., Michelucci, U., Venturini, F., Sperti, M., Martos, V., Deriu, M.: Compact optical fluorescence sensor for food quality control using artificial neural networks: application to olive oil. In: Proceedings of SPIE, p. 12188 (2022)
23. Yildiz Tiryaki, G., Ayvaz, H.: Quantification of soybean oil adulteration in extra virgin olive oil using portable Raman spectroscopy. Food Measure **11**, 523–529 (2017)
24. Savitzky, A., Golay, M.J.E.: Smoothing and differentiation of data by simplified least squares procedures. Anal. Chem. **36**(8), 1627–1639 (1964)
25. Lawson, C.L., Hanson, R.J.: Solving least squares problems. Society for Industrial and Applied Mathematics (1995)
26. Byrd, R.H., Lu, P., Nocedal, J., Zhu, C.: A limited memory algorithm for bound constrained optimization. SIAM J. Sci. Comput. **16**(5), 1190–1208 (1995)

27. Breiman, L.: Random forests. Mach. Learn. **45**(1), 5–32 (2001)
28. Friedman, J.H.: Greedy function approximation: a gradient boosting machine. Ann. Stat. 1189–1232 (2001)

An Effective Spatio-Temporal Feature Engineering and Classification Framework for Smartphone Sensor-Based Human Activity Recognition

Nurul Amin Choudhury[✉], Gyanda Kaushal, and Badal Soni

Multimedia and Image Processing (MIP) Laboratory, Department of Computer Science and Engineering, National Institute of Technology Silchar, Assam 788010, India
{nurul21rs,gyanda24_rs,badal}@cse.nits.ac.in

Abstract. Human Activity Recognition (HAR) is one of the most explored fields of deep learning and sensor technology. HAR is the process of recognising a person's daily living activities using a set of sensors and an effective pattern generalisation framework. Most of the state-of-the-art deep learning-based HAR models need more efficient model generalisation due to sub-optimal pattern identification by the end classifier. To overcome this issue, this paper presents an effective spatio-temporal feature engineering and classification framework using CNN-LSTM and a Decision tree classifier for a sensor-based HAR system. A custom 1D convolutional and LSTM layer has been developed for extracting local spatial features from raw time-series data and high-level temporal dependencies using a memory unit of recurrent layers. The proposed achieved an average performance accuracy of 98% and 99% with its own and MotionSense dataset, outperforming multiple benchmarks in optimised computational times.

Keywords: Human Activity Recognition · Deep Learning · Feature Engineering · Wearable Sensors and Machine Learning

1 Introduction

The emergence of smartphones has made human life smarter and more intelligent and has recently gravitated towards becoming part and parcel of ubiquitous computing solutions. Smartphones are powerful, portable, socially connected, and possess multi-modal sensors, with the flexibility of installing a variety of applications [1]. The sensors embedded in smartphones provide a convenient means to unobtrusively, continuously, and contextually sense multiple contextual and environmental factors. The in-built smartphone sensors are used in diverse application domains, including healthcare, biometrics and the fitness industry [2]. One such application of smartphone sensors is Human Activity Recognition (HAR).

HAR is an intricate process of identifying the daily human activities of an individual using a set of sensors and an efficient learning algorithm [3]. HAR is critical in understanding and interpreting human behaviours. Sensor-enabled HAR systems, upon interpreting quantified structured knowledge concerning underlying raw activity data, have implications in a broad array of real-world application domains, which include but may not be limited to health care, surveillance and monitoring, fitness elderly care and assistive technologies [1, 4].

Machine and deep learning models have gained significant popularity for classification and detection tasks, particularly in benchmark sensor-based HAR systems. Machine learning models excel at providing interpretable solutions and are often highly effective for classification tasks. However, they typically require extensive manual feature engineering to extract relevant features from raw sensor data. This process involves domain expertise and can be time-consuming, as the quality of the hand-crafted features directly impacts the model's performance [3]. Deep learning models, including Convolutional Neural Networks (CNNs), Long Short-Term Memory (LSTM) networks, and CNN-LSTM architectures, provide the significant advantage of automatic feature extraction. These models can directly learn hierarchical features from raw data, eliminating the need for manual feature engineering [5, 6]. This ability to learn complex patterns from sensor data makes deep learning models particularly suitable for HAR tasks, especially when working with large and complex datasets. Furthermore, deep learning models can capture spatial and temporal dependencies in time-series sensor data, improving the accuracy and robustness of activity classification.

However, in the state-of-the-art literature, both machine learning and deep learning models come with several limitations for sensor-based HAR. Machine learning models, while efficient and interpretable, often struggle with complex datasets that contain high-dimensional or noisy data. They may require significant tuning to achieve high performance and are less effective at handling sequential or unstructured data, which is common in sensor-based HAR. On the other hand, deep learning models, despite their power in handling raw sensor data, typically demand large amounts of labelled data for training and are computationally expensive. To address this, we designed and developed an efficient ensemble of hybrid deep learning and machine learning classifiers aimed at automatic feature engineering and classification using raw smartphone sensor data. The main contributions and innovations of this paper are as follows -

1. An effective CNN-LSTM and Decision Tree classifier-based automatic spatiotemporal feature engineering and classification framework is proposed for sensor-based HAR systems using real-world data.
2. A benchmark in-built smartphone sensor-based action recognition dataset has been generated [7] in a real-world environment using accelerometer and gyroscope sensors.
3. An efficient data pre-processing and time-interval-based data segmentation scheme is also designed and developed to make the dataset robust towards classification.
4. A comprehensive performance comparison with a publicly available dataset is carried out for detailed understanding and benchmark evaluations.

2   Related Works

Researchers [4, 8–11] around the globe design and develop sensor-based HAR systems using diverse sensor modalities, learning models and processing pipelines. Soni et al.[8] developed an efficient Bi-LSTM-CNN model for effectively capturing time and frequency-based features from raw sensor data. The developed model was built using memory and convolution filters that automatically extract meaningful statistical information from the raw multi-dimensional sensor datasets like WISDM and UCI-HAR. Choudhury and Soni [4] proposed an adaptive batch-sized based CNN-LSTM approach for HAR in an uncontrolled environment. They used two public datasets, mHealth and MotionSense, including the researcher's custom, state-of-the-art HAR dataset in a real-world uncontrolled environment. The proposed efficiently passed dynamic batch six while training CNN-SLTM to capture spatial-temporal features in optimized computational time and achieved a notable performance with an average accuracy of 98% and peak accuracy of 99.29% with the state-of-the-art HAR datasets.

Li et al.[10] proposed the HAR-WCNN (Wide Time-Domain CNN) model using the CASAS dataset [12] HAR dataset. They incorporate sensor data as well as a spatial context for better recognition. The developed CSA method incorporated sensor status frequency, an inverse frequency typespatial distance matrix (SDM) to reduce noise from multi-person cross-activities. The authors in [13] used UCI-PAD dataset and proposed a hybrid deep learning model to extract features from raw EMG data and achieved an accuracy of 84.12%. The proposed hybrid learning model also integrated the power of spatio-temporal feature engineering through the convolution and memory layer of LSTM, providing both local and high-level feature extraction. The proposed model attained an average performance accuracy of 84%.

In [14], the author proposes a deep learning-based HAR model named MultiCNN-FilterCNN that combines multihead CNN for extracting hierarchical features and a filter-wise LSTM for learning dependencies between features at different levels. The proposed model focused mainly on developing a resource-efficient model suitable for deployment in low-resource environments like IoT devices(e.g., Raspberry Pi) and achieving 2.3%-4.4% higher accuracy with 21%-70% fewer operations than state-of-art models. Vrskova et al. [5] used a combination of 3DCNN + ConvLSTM and Conv3D architecture, which improves both spatial and temporal feature extraction for HAR in videos. The proposed model evaluates three datasets covering daily and abnormal activities like UCF50mini, LoDVP Abnormal Activities, and MOD20. Kosar et al. [11] proposed a novel hybrid network architecture for recognizing HAR using wearable sensors and deep learning techniques. They combined 2D-CNN with LSTM and used the UCI-HAR and DSA datasets, giving them enormous accuracy.

Bi et al. [15] proposed the ActSemiCNN model, which combines active and semi-supervised learning. A novel unsupervised loss term is introduced, leveraging temporal ensembling of CNN with consistency regularization. The model's design ensures strong generalization and robustness across different datasets. Choudhury et al. [9] work introduces a hybrid deep learning ensemble framework combining CNN-LSTM for spatiotemporal feature extraction and an Extra Tree Classifier for HGR in a real-world environment. Developed a state-of-the-art dataset for gait recognition in an uncontrolled environment, enabling it to outperform existing baseline models. The model achieved an accuracy of 99.10% with notable performance measures.

3   Proposed Methodology

The proposed methodology includes several phases, ranging from dataset preparation and pre-processing to automated feature engineering and classification of human activities, as shown in Fig. 1.

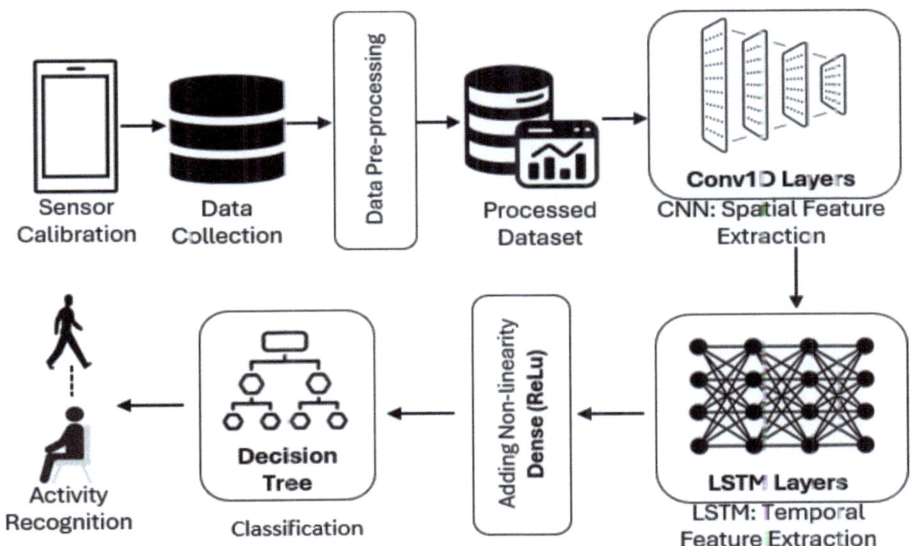

Fig. 1. Architecture flow of the proposed methodology.

3.1   Dataset Preparation

Multiple inertial and smartphone sensors-based HAR datasets are available publicly for research and academic purposes. However, only a fraction of dataset has the real-world activity instances in uncontrolled settings. To overcome this issue, we developed our own inbuilt smartphone sensor-based HAR dataset named Wild-SHARD [7] in a real-world uncontrolled environment. Wild-SHARD dataset was formulated with the help of accelerometer and gyroscope sensors of four Android devices, namely, One Plus 9 Pro, Samsung Galaxy F62, Samsung Galaxy A30s and Poco X2, making the dataset raw data heterogeneous and sensor modality diverse towards data acquisition. The sensor module was mounted on the front pocket of all the users, ensuring comfort and a reliable pack for sensor placement, as illustrated in Fig. 2.

   A total of eight (8) daily living human activities (DLA) have been considered in the dataset, namely sitting, downstairs, walking, running, jumping, standing, sleeping, and upstairs, as shown in Fig. 3. To capture real-life raw instances of different DLAs, we did not impose any formal protocols for activity performance as it would hamper the dataset quality in real-life testing. Also, multiple factors like uphill, downhill, plain roads, playgrounds, grass and concrete floors were chosen to perform different human activities

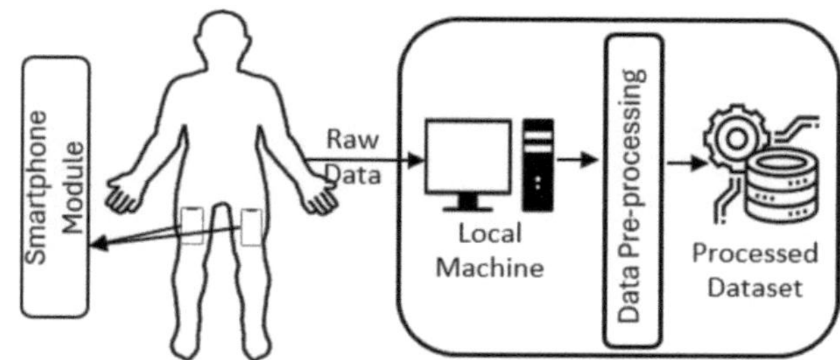

Fig. 2. Data acquisition module with sensor mounting locations.

Fig. 3. Incorporated DLA's in the formulated dataset.

to mimic the daily activity scenarios. Subsequently, 16 raw attributes were directly captured from the gyroscope and accelerometer sensors at the sampling (frequency) rate of 100 Hz to capture multiple statistical parameters of user-performed DLAs. The raw features included in the dataset are 3D in nature and consist of Linear acceleration ($f\,1_{xi}, f\,2_{yi}, f\,3_{zi}$), acceleration due to gravity ($f\,4_{xi}, f\,5_{yi}, f\,6_{zi}$), gravity($f\,7_{xi}, f\,8_{yi}, f\,9_{zi}$), rotation rate ($f\,10_{xi}, f\,11_{yi}, f\,12_{zi}$), rotational vector ($f\,13_{xi}, f\,14_{yi}, f\,15_{zi}$) and alpha angle of rotation vector as ($f16 = cos(f\,15_{xi})$).

Moreover, a publicly available HAR dataset collected using a smartphone named MotionSense [16] is also incorporated for benchmarking and performance comparison of the ensembled proposed model.

3.2 Data Pre-processing and Segmentation

Once the dataset was formulated, we processed both datasets to make them robust towards feature extraction and pattern generalization, as shown in Fig.

4. In the initial phase of data pre-processing, we removed the noise present in the generated dataset. Each individual DLA's top and bottom 100 instances were removed from all the features as they were not concerned with the collected DLA since the calibration of the sensor module and mounting it in the front pockets took significant time. Then, missing and corrupted entities were checked in the datasets, and it was discovered that there were no missing and corrupted instances in either dataset. Subsequently, we analyzed the class frequency of both datasets and found that both datasets have major issues of class imbalance. However, we did not impute the classes using data augmentation to keep the raw nature of the dataset intact. Thereafter, we encoded the class labels using sklearn *LabelEncoder()* to convert the categorical classes into integers, making them well-suitable for neural networks and machine learning models (Fig. 4).

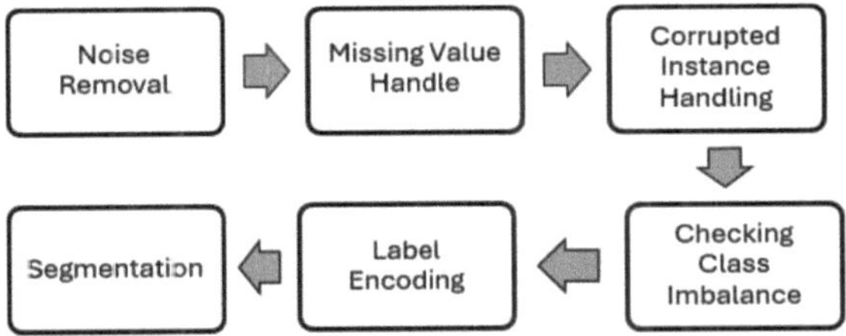

Fig. 4. Incorporated data pre-processing pipeline.

Finally, we segmented the dataset using a time interval-based fixed-size sliding window protocol. This approach partitions the time series data into discrete, overlapping fragments of uniform duration. Specifically, we employed a window size of ($T = 100$) seconds with a step size of ($S = 10$) seconds, resulting in an overlap of (T-S) seconds between consecutive windows. This method ensures consistent temporal context across all segments while allowing for the capture of transitional patterns between adjacent time intervals. The fixed window size maintains a uniform feature space dimensionality, which is crucial for subsequent machine-learning model inputs. This segmentation technique effectively transforms our raw, continuous sensor data into a structured format suitable for feature engineering and classification tasks, particularly in the context of our multi-class activity recognition systems.

3.3 Proposed Model

To efficiently extract spatio-temporal features from the segmented data and classify the human activities, we designed and developed an efficient ensemble of CNN-LSTM with a DT classifier. The first stage employs a custom CNN-LSTM architecture for feature extraction, designed to capture both spatial and temporal dependencies in the input data. This hybrid network begins with two timedistributed 1D convolutional layers (16 and 32 filters, respectively, kernel size 3, ReLU activation), followed by dropout for regularization and max pooling to extract dominant features. The flattened output then feeds into an LSTM layer, which processes the sequence of CNN-extracted features to capture temporal dependencies. Dense layers with ReLU complete the feature extraction model, which is optimized using categorical cross-entropy loss and the Adam optimizer.

Algorithm 1 Proposed Ensembled Feature Extractor with Decision Tree Classifier

---

Require: Input data X with shape (*samples, time_steps, features*)
Require: Number of classes *num_classes*
Ensure: Predicted activity classes
1:   **function** FeatureExtraction(X)
2:       $X \leftarrow$ TimeDistributed(Conv1D(*filters = 16, size = 3, activation = ReLu*))(X)
3:       $X \leftarrow$ TimeDistributed(Conv1D(*filters = 32, size = 3, activation = ReLu*))(X)
4:       $X \leftarrow$ TimeDistributed(Dropout(*rate = drop out*))(X)
5:       $X \leftarrow$ TimeDistributed(Max-Pooling1D(*pool size = 2*))(X)
6:       $X \leftarrow$ TimeDistributed(Flatten_Layer())(X)
7:       $X \leftarrow$ LSTM(*units = lstm_neurons*)(X)
8:       $X \leftarrow$ Dropout(*rate = drop_out*)(X)
9:       *features* $\leftarrow$ Dense(*units = dense_neurons, activation =' relu'*)(X)
10:      **return** *features*
11: **end function**
12: **function** TrainModel(X_train, y_train)
13:      *features_train* $\leftarrow$ FeatureExtraction(X_train)
14:      *dt_classifier* $\leftarrow$ DecisionTreeClassifier()
15:      *dt_classifier*.fit(*features train, y train*)
16:      **return** *dt classifier*
17: **end function**
18: **function** PredictActivity($X\_test, dt\_classifier$)
19:      *features_test* $\leftarrow$ FeatureExtraction(X_test)
20:      *predictions* $\leftarrow$ *dt_classifier*.predict(*features_test*)
21:      **return** *predictions*
22: **end function**
23: // Main workflow
24: $X\_train, y\_train \leftarrow$ LoadTrainingData()
25: $X\_test \leftarrow$ LoadTestData()
26: *dt classifier* $\leftarrow$ TrainModel(X_train, y_train)
27: *activity_predictions* $\leftarrow$ PredictActivity($X\_test, dt\_classifier$)
28: **return** *activity_predictions*

The second stage utilizes a Decision Tree classifier, leveraging the interpretability and non-linear modelling capabilities of this algorithm to perform the final activity classification as described in the Algorithm 1. This choice of classifier not only offers clear decision rules but also provides insights into feature importance, enhancing our understanding of the extracted features' relevance. Our model pipeline begins with the segmentation of raw sensor data using a time interval-based fixed-size sliding window protocol. These segments are then fed into the CNN-LSTM model for feature extraction, with the resulting features used to train and test the Decision Tree classifier. The final output is the predicted activity class for each input segment. This hybrid approach facilitates the automatic feature learning capabilities of deep neural networks with the interpretability and efficiency of traditional machine learning algorithms, potentially offering improved performance and deeper insights into human activity recognition systems.

3.4 Benchmark Models and Evaluation Metrics

Finally, to benchmark and effectively test the proposed model, we incorporated state-of-the-art HAR models [2, 4, 6] like ANN, LSTM and normal CNN-LSTM with the same hyper-parameters used in the proposed model. This allows us to constructively validate and examine the developed feature extraction and classification framework. Moreover, to effectively compare and validate the proposed model with benchmarks, we incorporated multiple evaluation metrics like Precision $(P) = \frac{1}{n} \sum_{i=1}^{n} \left(\frac{T_{Pk}}{T_{Pk} + F_{Pk}} \right)$,

$$\text{Recall } (R) = \frac{1}{n} \sum_{i=1}^{n} \left(\frac{T_{Pk}}{T_{Pk} + F_{nk}} \right), \text{ F1-Score } (F\,1) = \frac{1}{n} \sum_{i=1}^{n} \left(2 * \frac{P_k * R_k}{P_k + R_k} \right), \text{ Accuracy } (A)$$

$$= \frac{1}{n} \sum_{i=1}^{n} \left(\frac{T_{Pk} + T_{nk}}{T_{Pk} + T_{nk} + F_{Pk} + F_{rk}} \right) \text{ and Computational Time } (C_t) = (C_{tr} + C_{val}).$$

4 Experimental Results and Discussions

To train, test and validate the proposed and benchmark model, we segregated the dataset into the training and testing ratio of 80:20 and used a Fujitsu workstation with 32GB RAM, 2TB HDD and Intel Xeon W-2133 CPU. GPU was not utilized for tensor processing as we had sufficient computational power with the system CPU and RAM. The incorporated hyper-parameters of the proposed model are described in the Table. 1.

Table 1. Incorporated hyperparameters of the Proposed Model

Component	Hyper-parameters	Value
CNN-LSTM	Input Shape Conv1D Filters (1st and 2nd layer) Conv1D Kernel Size Conv1D Activation Max Pooling Size LSTM Units Dense Layer Units Dropout Rate	(None, n length, n features) 16 and 32 3 ReLU 2 64 64 0.1
Training	Batch Size Epochs Loss Function Optimizer	128 100 Categorical Cross-entropy Adam
Decision Tree	Criterion Max Depth	Gini impurity None

4.1 Performance Analysis of the Proposed Model with Benchmarks

Training and validation of the proposed and benchmark models on both the dataset achieved and the average training accuracy of 98–99% with optimal performance F1, P and R scores. However, upon testing the models, benchmark models like ANN, LSTM, and normal CNN-LSTM exhibited model overfitting and achieved diminishing performance accuracy. On the contrary, the proposed model achieved an average performance accuracy of 99%, omitting model overfitting and efficient model generalization. The detailed performance comparison of the proposed model with benchmarks is described in the Table 2.

The efficacy of our proposed model arises from its meticulously designed architecture, which synthesizes the strengths of deep learning and conventional machine learning techniques. The CNN layers excel at extracting spatial characteristics from the input data. The initial layer employs 16 filters, whereas the subsequent layer utilizes 32, thereby enabling the network to progressively discern more intricate patterns within the data and

Table 2. Detailed performance comparison of the proposed model with benchmarks.

Dataset	Model	Accuracy(%)	Precision(%)	Recall(%)	F1-Score(%)	Ct(Sec.)
Own Dataset	ANN	94	94	94	94	200–210
	LSTM	96	97	97	97	870–880
	CNN-LSTM	97	98	98	98	330–340
	Prop. Model	98	98	98	98	350–360
MotionSense	ANN	97	97	97	97	143–152
	LSTM	98	98	98	98	623–655
	CNN-LSTM	98	98	98	98	254–261
	Prop. Model	99	99	99	99	269–288

formulate a comprehensive spatial representation. Subsequently, the LSTM layer, configured with 64 units, demonstrates a remarkable capacity to capture temporal dependencies in the extracted features, which is crucial for comprehending the sequential dynamics of human activities. To enhance the model's generalization abilities, we incorporated a dropout mechanism with a rate of 0.1, functioning as a robust regularization strategy to mitigate the risk of overfitting. The concluding component, a DT classifier, adds a dimension of interpretability to our model. It delineates clear, comprehensible decision boundaries predicated on the features extracted by the deep learning layers, offering insights into the classification process that are frequently absent in exclusively deep learning frameworks. This hybrid architecture effectively integrates the formidable feature learning capabilities of deep neural networks with the interpretability and efficiency inherent in traditional machine learning algorithms. Furthermore, the model incorporates strategies like dropout and stratification of class labels to address the class imbalance by leveraging robust feature representations that enhance the discrimination of minority classes, thereby improving performance across imbalanced datasets. The result is a model that not only achieves high performance but also provides significant insights into its decision-making mechanisms.

4.2 Analysis on Training and Validation Accuracy with Loss

Subsequently, the proposed model generalized the raw sensor data optimally as there was minimum test and validation loss compared to the benchmark models like ANN and LSTM. ANN exhibits clear signs of overfitting, with the validation loss increasing dramatically after the initial epochs while the training loss continues to decrease, as shown in Fig. 5(a), indicating poor generalization. LSTM shows improved performance over the ANN, with both training and validation accuracies converging at a higher level and losses stabilizing, suggesting better generalization. However, a noticeable gap between training and validation metrics is visible, and it fails to capture spatial features from the rawtime series data, as illustrated in Fig. 5(b).

The CNN-LSTM model demonstrates further improvement, with closer alignment between training and validation curves and higher overall accuracy. Finally, the proposed

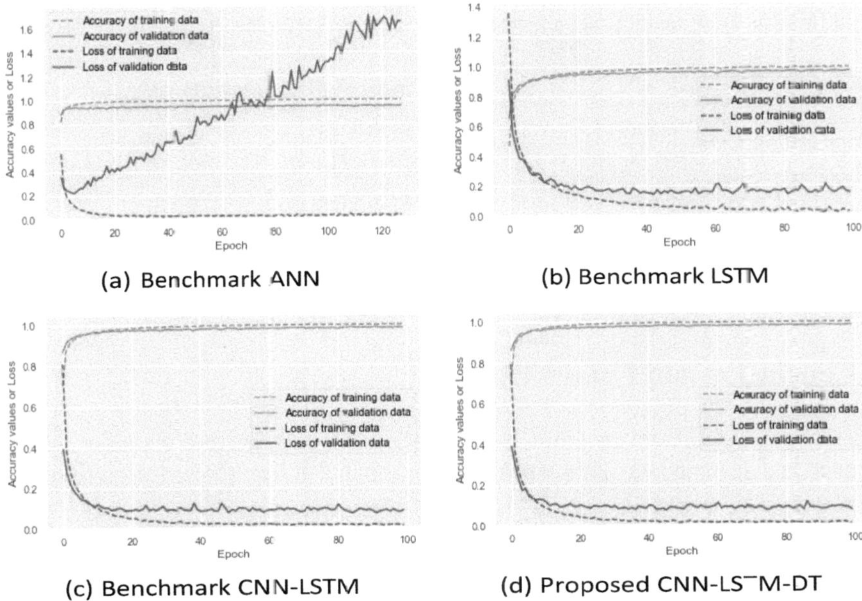

(a) Benchmark ANN

(b) Benchmark LSTM

(c) Benchmark CNN-LSTM

(d) Proposed CNN-LSTM-DT

Fig. 5. Training and validation accuracy-loss plot of the proposed model with benchmarks on own dataset.

model showcases the best performance among all and achieves the highest accuracy for both training and validation data, with minimal loss and maximal accuracy, indicating excellent generalization as shown in Fig. 5(d). The loss curves for the proposed model decrease smoothly and stabilize at the lowest point compared to other models, suggesting optimal learning without overfitting.

4.3 State-of-the-Art Benchmark Comparison

Finally, we conducted a comprehensive comparison of the proposed model with several state-of-the-art benchmark models to rigorously assess the effectiveness and robustness of the designed feature extraction and classification framework. This evaluation allowed us to validate the improvements in both feature representation and classification accuracy, highlighting the model's ability to outperform existing methods, as described in Table 3 across performance accuracy and varying datasets.

Table 3. Benchmark Comparison of State-of-the-arts with the Proposed Approach

Dataset	Work	Model	High. Accuracy(%)
MotionSense	Saha et al. [17]	CNN-LSTM	95.35
	Goh et al. [18]	CNN-LSTM	98.14
	Choudhury and Soni [6]	CNN-LSTM	99.10
	Prop. Model	**CNN-LSTM + DT**	**99.12**
Own Dataset	**Prop. Model**	**CNN-LSTM + DT**	**98.6**

5 Conclusion and Future Work

In this paper, an efficient custom CNN-LSTM with a decision tree classifier has been designed and developed to automatically extract spatial and temporal features on smartphone sensor data and classification. An adequate data pre-processing pipeline with a time interval-based fixed-size data segmentation scheme has also been developed to make the data robust for classification. Upon validation, the proposed model achieved an average performance accuracy of 98% and 99% with its own and MotionsEense dataset, with notable higher performance of P, R and F1-score. Further, the proposed model outperformed multiple benchmarks with notable performance margins in optimized computational time.

In the future, a federated learning-based deep learning will be proposed to handle domain shifts and efficiently classify human activities in daily living.

References

1. Ramanujam, E., Kalimuthu, S., Harshavardhan, B., Perumal, T.: Improvement in multi-resident activity recognition system in a smart home using activity clustering. In: IFIP International Internet of Things Conference, pp. 316–334. Springer, Heidelberg (2023)
2. Choudhury, N.A., Soni, B.: In-depth analysis of design & development for sensor-based human activity recognition system. Multimedia Tools Appl. **83**(29), 73233–73272 (2024)
3. Khan, A., Biswas, S.K., Chunka, C., Das, A.K.: A machine learning model for obstructive sleep apnea detection using ensemble learning and single-lead eeg signal data. IEEE Sensors J. **24**(12), 20266–20273 (2024)
4. Choudhury, N.A., Soni, B.: An adaptive batch size-based-cnn-lstm framework for human activity recognition in uncontrolled environment. IEEE Trans. Ind. Inf. **19**(10), 10379–10387 (2023)
5. Vrskova, R., Kamencay, P., Hudec, R., Sykora, P.: A new deep-learning method for human activity recognition. Sensors **23**(5), 2816 (2023)
6. Choudhury, N.A., Soni, B.: An efficient and lightweight deep learning model for human activity recognition on raw sensor data in uncontrolled environment. IEEE Sens. J. **23**(20), 25579–25586 (2023)
7. Choudhury, N.A., Soni, B.: Wild-shard: smartphone sensor-based human activity recognition dataset in wild (2024)

8. Soni, V., Yadav, H., Semwal, V.B., Roy, B., Choubey, D.K., Mallick, D.K.: A novel smartphone-based human activity recognition using deep learning in health care. In: Machine Learning, Image Processing, Network Security and Data Sciences: Select Proceedings of 3rd International Conference on MIND 2021, pp. 493– 503. Springer, Heidelberg (2023)
9. Choudhury, N.A., Singh, S., Soni, B.: An efficient ensemble framework for human gait recognition using cnn-lstm with extra tree classifier and smartphone sensors in real-world environment. IEEE Sensors Lett. **8**(9), 1–4 (2024)
10. Li, Y., Yang, G., Su, Z., Li, S., Wang, Y.: Human activity recognition based on multi environment sensor data. Inf. Fusion **91**, 47–63 (2023)
11. Kosar, E., Barshan, B.: A new cnn-lstm architecture for activity recognition employing wearable motion sensor data: Enabling diverse feature extraction. Eng. Appl. Artif. Intell. **124**, 106529 (2023)
12. Cook, D.J., Crandall, A.S., Thomas, B.L., Krishnan, N.C.: Casas: a smart home in a box. Computer **46**(7), 62–69 (2012)
13. Choudhury, N.A., Soni, B.: Enhanced complex human activity recognition system: a proficient deep learning framework exploiting physiological sensors and feature learning. IEEE Sensors Lett. **7**(11), 1–4 (2023)
14. Park, H., Kim, N., Lee, G.H., Choi, J.K.: Multicnn-filterlstm: resource-efficient sensor-based human activity recognition in iot applications. Futur. Gener. Comput. Syst. **139**, 196–209 (2023)
15. Bi, H., Perello-Nieto, M., Santos-Rodriguez, R., Flach, P., Craddock, I.: An active semi-supervised deep learning model for human activity recognition. J. Ambient. Intell. Humaniz. Comput. **14**(10), 13049–13065 (2023)
16. Malekzadeh, M., Clegg, R.G., Cavallaro, A., Haddadi, H.: Mobile sensor data anonymization. In: Proceedings of the International Conference on Internet of Things Design and Implementation, IoTDI '19, New York, NY, USA, pp. 49–58. ACM (2019)
17. Saha, U., Saha, S., Kabir, M.T., Fattah, S.A., Saquib, M.: Decoding human activities: analyzing wearable accelerometer and gyroscope data for activity recognition. IEEE Sensors Lett. **8**(8), 1–4 (2024)
18. Goh, J.X., Lim, K.M., Lee, C.P.: 1d convolutional neural network with long short-term memory for human activity recognition. In: 2021 IEEE International Conference on Artificial Intelligence in Engineering and Technology (IICAIET), pp. 1–6 (2021)

Machine Learning Based Modulation Techniques for Next-Generation Advanced Communication Systems Using ANN

D. J. Chaithanya[1]([✉]) [iD], Chandrashekar M. Patil[1], M. S. Harshitha[2], Yajnika S. Nandi[1], A. PavanAthreya[1], and U. Parvitha[1]

[1] Department of Electronics and Communication Engineering, Vidyavardhaka College of Engineering, Mysore, India
`rcchaithudj@gmail.com, patilcm@vvce.ac.in`
[2] Department of Information Science and Engineering, Vidyavardhaka College of Engineering, Mysore, India

Abstract. This article addresses the issue of modeling and step analysis of employing QPSK, 16-QAM, 64-QAM, 256-QAM, and OFDMA schemes with a directed emphasis on systems BER as a standard of SNR. The simulation produces QAM symbols, assumes different levels of noise, and calculates the BER of every modulation type for varying SNR levels. For simulation purposes, an artificial neural network (ANN) is also introduced to classify the noisy 16 QAM and OFDMA signals. The BER comparison shows that QPSK is much better at low SNR, while larger constellations of 16-QAM and 64-QAM have higher BER but higher data rate capabilities. OFDMA combined with 64-QAM subcarriers experiences a minor performance decline because of the additional overhead involved. The ANN achieves real-time classification of the modulated signals, and this proves that machine learning techniques can be applied in communication systems. Graphs of BER against SNR and confusion matrix are used to display the results of experiments and simulations to give the relationship between data rate and error performance in several wireless technologies.

Keywords: QPSK · 16-QAM · 64-QAM · 256-QAM · ANN

1 Introduction

The Machine Learning-Based Modulation Classification Techniques for QPSK, 16-QAM, 64-QAM [1], 256-QAM [2], and OFDM [3] in Next-Generation Communication Networks using Artificial Neural Networks (ANN, where the classification strength of ANN models is targeted to be enhanced. These features may be amplitude, phase, and statistical features of the signal which are essential in the demodulation of QPSK, 16-QAM, 64-QAM, and 256-QAM. The ANN-based classifier is further developed with more than one layer to analyze and classify the observed features on the signals into their respective categories. It might have the input, hidden, and output layers optimized for modulation classification among other layers. During the training process of the ANN

C. Modi et al. (Eds.): MIND 2024, CCIS 2736, pp. 562–574, 2026.
https://doi.org/10.1007/978-3-032-14531-4_47

model, a set of labeled sample data created by taking signal inputs of QPSK, 16-QAM, 64-QAM, 256-QAM, and OFDMA is provided. Some methods that can be employed to enhance the training include adaptive moment estimation techniques among others. The training process is performed with an optimal transmission mode index to be utilized as training data labels to enhance classification effectiveness. This trained ANN model can then be deployed on base stations such as Software Defined Radio (SDR) to test its performance in practice. It is then contrasted to other conventional classification techniques, which are typically demonstrated to be less efficient and flexible compared to the model [4]. Computational simulations reveal that the number of antenna and modulation schemes directly impacts the classification. However, the same models in artificial neural networks still register high accuracy even in more complex MIMO-OFDM networks. Other techniques such as modulation order reduction may also be applied in the classification process to simplify it. For instance, encoding 64-QAM as 16-QAM-like signals or encoding 16-QAM as QPSK-like signals can favor the performance of ANN classifiers. ANN classifiers are highly scalable and allow users to update and extend the Modulation format dictionary to accommodate next-generation communication network standards into the current system proficiently. These techniques enable ANN-based modulation classification systems to address the accurate classification of QPSK, 16-QAM, 64-QAM, 256-QAM, and OFDMA for consistent communication performance in advanced wireless communication systems [5, 6].

The main goal of this article is to propose and analyze advanced modulation schemes for novel communication networks using Artificial Neural Networks (ANN) as tools for machine learning. Thus, employing ANN for modulation classification and optimization, the study should contribute to the improvement of intelligence communication systems, which are crucial for technologies such as 5G and beyond.

The main objectives of this article are to:

- To implement Artificial Neural Network (ANN) models that are to classify and optimize modulation schemes for the next generation of communication networks.
- To Evaluate the effectiveness of the developed ANN models in terms of BER, SNR, and overall system throughputs in different communication environments.
- To Analyze the performance of traditional modulation schemes (QPSK, 16-QAM, 64-QAM,256-QAM, OFDM) and machine learning-aided schemes and compare their performance based on time-varying channels.
- Compare the efficacy of traditional modulation techniques (e.g., QPSK, 16-QAM, 64-QAM,256-QAM) with machine learning-enhanced approaches.

The contributions to this article are mentioned as follows:

The proposed machine learning classification methods achieve the optimized classification of QPSK,16-QAM, 64-QAM, and OFDM with a feasible trade-off between performance and implementation cost for an intended optical interconnection link. Thus, the complexity of the system, as well as the BER, have been compared based on the experimental outcomes given in the result section of the manuscript. Based on the experimental results, BER and complexity performance were examined.

1.1 Related Works

Machine learning-based modulation techniques, particularly those using Artificial Neural Networks (ANN), have shown significant promise for next-generation advanced communication systems. Here are some key insights and related works in this area.

The author B. Karanov describes how the ANNs can be applied to several different optical fiber communication tasks such as performance monitoring, mitigating the efstationarity, carrier recovery, and modulation format identification. Multilayer ANNs are applied to low-complexity nonlinearity compensation in which they provide similar performance as conventional approaches but with higher efficiency. In short-reach communication systems, ANNs are also applicable to equalization and advanced detection that enhances BERs than the feed-forward linear equalization technique [7].

The author examines the application of ANNs for identifying signals of digital modulation including QAM and QPSK in cognitive radio networks. By using ANN, the system can differentiate between the various types of modulations and therefore boosts adaptive communications and spectrum monitoring [2]. In coherent optical systems, the modulation format is commonly used due to its high spectral efficiency. The revealed results indicate that to apply Digital Signal Processing (DSP) techniques for 16-QAM, the potential for use in future commmunication systems exists. [8]. According to the author, near-ideal Nyquist pulse shaping, spectrally efficient high-order modulation format, and distributed Raman amplification, coherent equalization may allow future operation of 400G systems over the standard 50 GHz-grid optical network [9]. The Author describes the major technologies and tries to assess the prospects of OFDM for the 100 GbE application [10].

OFDM is the modulation technique of 5G. However, it [OFDM] has a high peak-toaverage power ratio (PAPR). This degrades the performance of the power amplifiers (PA). Even more, the PAPR challenges the general power efficiency of the utilized 5G system resident. Hence, understanding when one form of OFDM signal creates high PAPR is very important much of the time [11].

In this paper, the author explains an efficient approach for the analysis of PAPR of OFDM signal with complex modulation in continuous time. As a result, for the discretetime signal of four-time oversampled OFDM signals, it is identified that the proposed scheme can well approximate the continuous-time PAPR for complex OFDM signals. In addition, the analysis of continuous PAPR is used to check some existing analytical bounds of the proposed scheme in the literature [12].

2 Structure of Artificial Neural Network in Machine Learning

ANN structure includes input layer, hidden layer, and output layer It is a specifically designed layered model. Number of neurons in the first layer inputs (for example three in this case). Number of neurons in the hidden layer (e.g., 5 in this example). Number of neurons in the output layer (e.g., 2 in this example). The network has three layers: Input layer, hidden layer, and output layer which are the definition of layers array [input size; hidden size; out-put size]. The layer_positions array above specifies x-coordinates for these layers as 0 for the input layer, 1 for the hidden layer, and 2 for the output layer. The neurons in each layer are plotted using the scatter function of the plot library in Python (plt. Scatter). The neurons' y positions of Neurons are num_layer apart (y_positions = [5, 12, 18, 24 …]) vertically. Figure 2 represents the neurons are drawn as circles with black edges, and large size for visibility (s = 200) Lines are drawn linking neurons from adjacent layers (plt. Plot). Each neuron in one layer relates to every other neuron in the next layer and the network is fully connected. The input neurons are labeled X1; X2; and so on, the hidden neurons are labeled H1; H2; And so, on and the output neurons are labeled Y1; Y2; and so on. The resulting plot will display the network architecture, with each layer of neurons connected to the adjacent layer, helping to visualize the structure of a basic neural network.

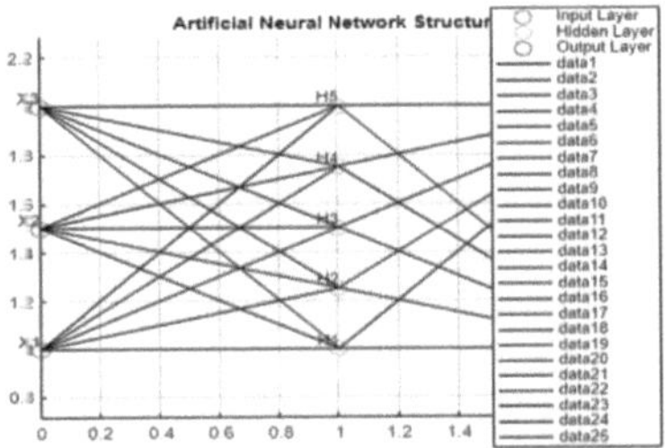

Fig. 1. Artificial Neural Network Structure with Input, Hidden and Output Layer

3 Methods Used

The block diagram of a proposed model shown in Fig. 1 represents the input Generation used to Generate Constellation Points (QPSK, 16-QAM, 64-QAM, 256-QAM). Data Preparation Combine Constellation Points create Labels for Modulation Schemes, and Split Data into Training and Testing Sets. Model Building defines ANN Architecture and compiles the model with Optimizer and Loss Functions. Model Training Fit Model on the Training Data. Model Evaluation evaluates the model on test data and calculates accuracy and Loss. Final Output will predict the Modulation Schemes for new data and visualize Constellation Diagrams.

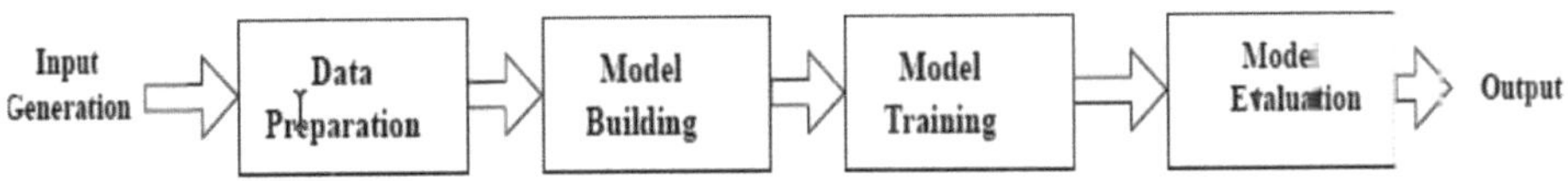

Fig. 2. Block diagram of a Proposed Method

3.1 QPSK

QPSK signals employ an ANN to identify the noisy QPSK symbols as per their real (in-phase) and imaginary (quadrature) parts. It provides the errors, validates the model, shows the confusion matrix, and then finally classifies the QPSK signals by plotting them.

The QPSK signal within a symbol duration Tsym is defined as

$$s(t) = A \cdot \cos[2\pi f_c t + \theta_n] \; 0 \le t \le T_{sym} \; n = 1, 2, 3, 4 \tag{1}$$

where the signal phase is given by

$$\theta_n = (2n - 1)\frac{\pi}{4} \tag{2}$$

Therefore, the four possible initial signal phases are $\frac{\pi}{4}$, $\frac{3\pi}{4}$, $\frac{5\pi}{4}$ and $\frac{7\pi}{4}$ radians. Equation (1) can be re-written as

$$\begin{aligned} s(t) &= A \cdot \cos\theta_n \cdot \cos(2\pi f_c t) - A \cdot \sin\theta_n \cdot \sin(2\pi f_c t) \\ &= S_{ni}\phi_i(t) + S_{nq}\phi_q(t) \end{aligned} \tag{3}$$

The above expression indicates the use of two orthonormal basis functions: $\langle \phi_i(t), \phi_q(t) \rangle$ together with the in-phase d quadrature signaling points: $\langle s_{ni}, s_{nq} \rangle$. Therefore, on a two-dimensional coordinate system with the axes set to $\phi i(t)$ and $\phi q(t)$, the QPSK signal is represented by four constellation points dictated by the vectors $\langle s_{ni}, s_{nq} \rangle$ with $n = 1,2,3,4$. Figure 3 represents a sequential model with two hidden layers with 64 neurons each and the output layer with 4 neurons (response to 4 QPSK classes). Studying for 50 epochs, using a batch size equal to 32. This enabled the development of a scatter plot of the test data, colored by the true classes, to compare to the ability of the ANN to classify noisy QPSK symbols [13]. Table 1 represents the metric and its value in QPSK with a test accuracy of 70.83% representing the true label from 0 to 3.

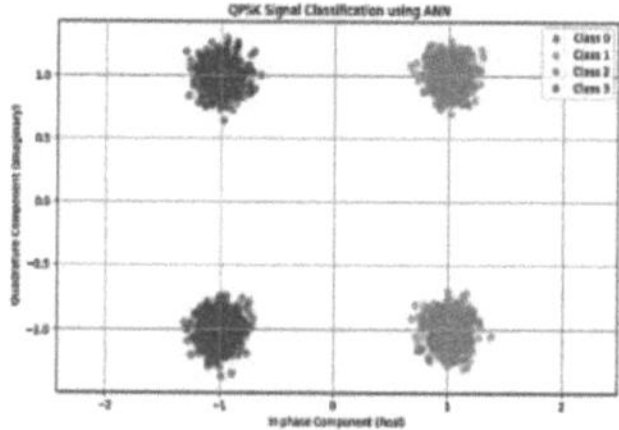

Fig. 3. QPSK Signal classification using ANN

Table 1. Representation of a Metric and its value in QPSK

Metric	Value
Test Accuracy	70.83%
Confusion Matrix	
True Label 0	[1491 0 0 0]
True Label 1	[87 634 0 788]
True Label 2	[0 0 0 0]
True Label 3	[0 0 0 0]

3.2 16-QAM

16 QAM symbols consist of pairs of Real and imaginary parts taken from the defined set as $[-3, -1, 1, 3]$. It creates all the possible sets of one real and one imaginary component of the signal: $[-3, -1, 1, 3]$ $[-3, -1, 1, 3]$, which represents all the possible 16 QAM constellation points. Gaussian noise is introduced to the generated symbols by applying the SNR in dB and modulates 16-QAM symbols which transform the modulated signal into two features each of which is represented by a real component and an imaginary component; and labels the noisy symbols 0–15 as per the constellation points. The labels are then hot encoded for classification using the ANN. The data set is partitioned into training and testing 70% of the information is used for training, and 30% for testing. The real and imaginary parts of the test data are plotted using a scatter plot and colorcoded based on their true classes to enable examination of how the noisy 16-QAM symbols were classified by the ANN [14]. Figure 4 represents the 16-QAM Signal classification using ANN and its constellation diagram. In a QAM signal, one carrier lags the other by $90°$, and its amplitude modulation is customarily referred to as the in-phase component, denoted by $I(t)$. The other modulating function is the quadrature component, $Q(t)$. So, the composite waveform is mathematically modeled as:

$$s_s(t) \triangleq \sin(2\pi f_c t)I(t) + \underbrace{\sin\left(2\pi f_c t + \frac{\pi}{2}\right)}_{\cos(2\pi f_c t)} Q(t) \tag{4}$$

where fc is the carrier frequency. At the receiver, a coherent demodulator multiplies the received signal separately with both a cosine and sine signal to produce the received estimates of $I(t)$ and $Q(t)$ represented by

$$r(t) \triangleq s_c(t)\cos(2\pi f_c t) = I(t)\cos(2\pi f_c t)\cos(2\pi f_c t) - Q(t)\sin(2\pi f_c t)\cos(2\pi f_c t) \tag{5}$$

Using standard trigonometric identities, we can write this as:

$$r(t) = \tfrac{1}{2}I(t)\left[1 + \cos(4\pi f_c t)\right] - \tfrac{1}{2}Q(t)\sin(4\pi f_c t)$$
$$= \tfrac{1}{2}I(t) + \tfrac{1}{2}\left[1 + \cos(4\pi f_c t) - Q(t)\sin(4\pi f_c t)\right] \tag{6}$$

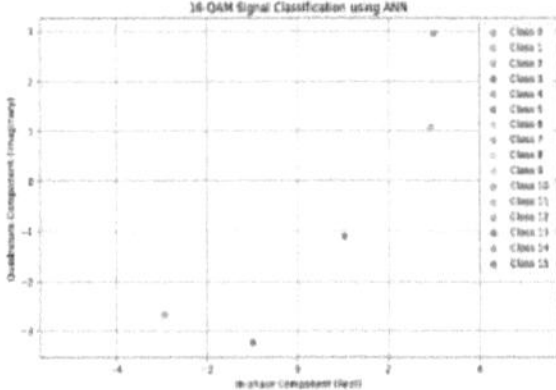

Fig. 4. 16-QAM Signal classification using ANN

To improve its efficiency the model architecture of the neural network will be altered by adding more layers or neurons, using a different learning exercise rate, or choosing a different activation function. The classification of a higher number of epochs might improve the classification accuracy, and expanding the sizes of the training data set would also be effective, which would be beneficial for classifying more complex signals such as 16-QAM, as well as coming with better stability against noise compared to QPSK.

3.3 64-QAM Constellation Diagram

In 64-QAM using ANN, trained for a total of fifty epochs of training and it was comprised of thirty-two samples per batch [15]. The accuracy of the model is checked by using the test set from which test accuracy is then displayed. To determine the accuracy of the model in each of the 16 classes, a confusion matrix is constructed. The actual and created parts of test data are displayed in the number of displays that are represented in shapes and color-coded for the true classes to examine how the noisy 16-QAM symbols were classified by the ANN. To optimize the model architecture of the neural network, add layers to it or add neurons to some of the layers, use a different learning exercise rate, or choose a different activation function. Figure 5 shows the plot of 64 QAM constellation diagram where the red 'xr' refers to the transmitted QAM data (Stx) and the blue dot 'b' for the received QAM data (Srx). These are plotted on the same chart to compare the alignment (or misalignment) between the sent and received data. In Dynamic Axis Adjustment, the axis range will change with the maximum transmission signal values to finely adjust the placement of the constellation points within the plot.

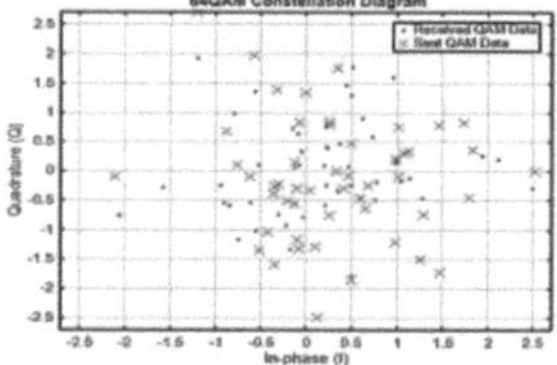

Fig. 5. Plotting of the 64-QAM Constellation diagram

In Machine learning, a 64-QAM classification algorithm is used to improve the optical communication system. In Data Generation, random symbols a are obtained between 0 and 63 because the 64-QAM modulation allows for 64 symbols. The QAM mod function in the system modulates the random data symbols to 64-QAM format. Noise or AWGN is added to the signal depending on the signal-to-noise ratio (SNR) which must be derived from the given Eb/No ratio of 15 dB.

$$SNR = EbNo + 10 * \log10(\log2(M)) \tag{7}$$

In Noise Addition, the AWGN function adds Gaussian noise to the modulated symbols to represent the type of noise in a channel. The real part of the in-phase (I) and the imaginary part of the quadrature part (Q) of the noisy received symbols are obtained. As seen in Fig. 6, the 64-QAM constellation represents the Scatter points that help to visualize how well the transmitted symbols match with the symbols after noise addition.

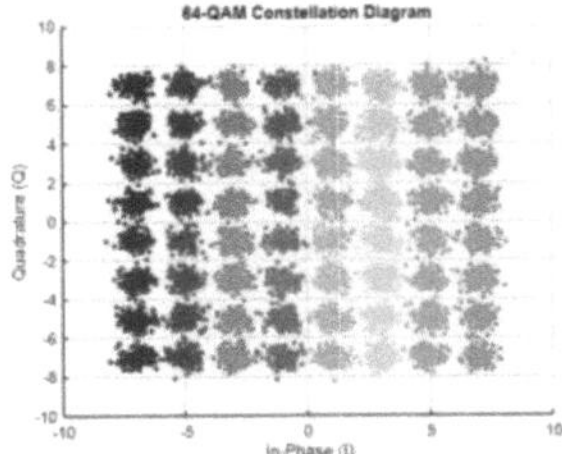

Fig. 6. 64-QAM Constellation Diagram

Figure 7 produces a constellation diagram of a 64-QAM signal with both the transmitted and received symbols which contain a noise overlay to represent the actual communication scenario. For the Transmitted and Received data generation, Modulation order M = 64, which indicates 64 QAM is employed. The constellation diagram shows the transmitted symbols represented as a red cross and the received symbols are shown by the blue circle with noise added to it.

3.4 256-QAM

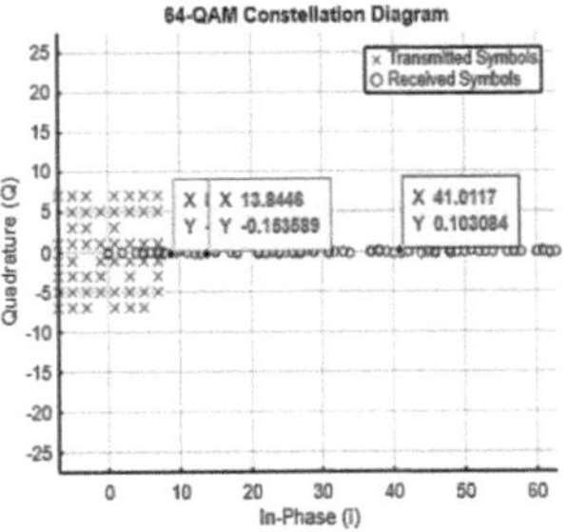

Fig. 7. 64-QAM Constellation diagram

256-QAM stands for 256 Quadrature Amplitude Modulation bits per symbol, and it is a type of modulation technique in digital communication systems. In this scheme, every waveform element in the transmitted signal is coupled with data of 255 bits or 8 bits per QAM symbol as shown in Fig. 8. This modulation technique is widely applied in broadband digital communication systems like WLANs, and digital satellite communication due to its high-power efficiency and high data throughput for the same number of channels.

3.5   OFDMA

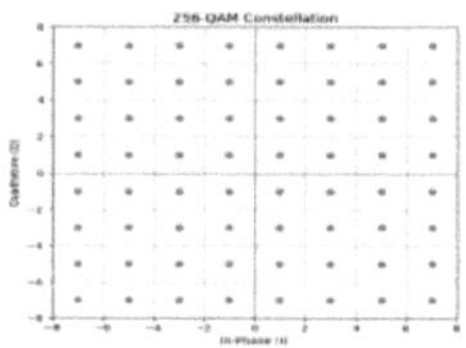

Fig. 8. 256-QAM Constellation diagram

Data subsetting minimizes the amount of data to be worked on by allowing the code to show only 10% of the constellation diagram. This is regulated by variable dataSubsetSize which is used to define how many symbols should be used while plotting. E-mail: Stx_subset will store the subsets received and Srx_subset will store the transmitted symbols. Transmitted Symbols are plotted with red crosses ('r', 'x') to distinguish them from received symbols. Mathematically, each carrier can be described as a complex wave:

$$s_c(t) = A_c(t)e^{j[k_c t + \varphi_c(t)]} \tag{8}$$

The real signal is the real part of $Sc(t)$. Both $Ac(t)$ and f $c(t)$, the amplitude and phase of the carrier, can vary on a symbol-by-symbol basis. The values of the parameters are constant over the symbol duration period t. Figure 9 represents the constellation diagram shows the transmitted symbols represented as a red cross and the received symbols are shown by the blue circle with noise added to it. Table 2 represents the metric and its value in OFDMA using ANN.

Table 2. Metric and its value in OFDMA using ANN

Metric	Value
Test Accuracy	18.18%
Confusion Matrix	
True Label 0	[8 3 2 2]
True Label 1	[8 2 8 5]
True Label 2	[11 2 3 4]
True Label 3	[11 2 5 1]

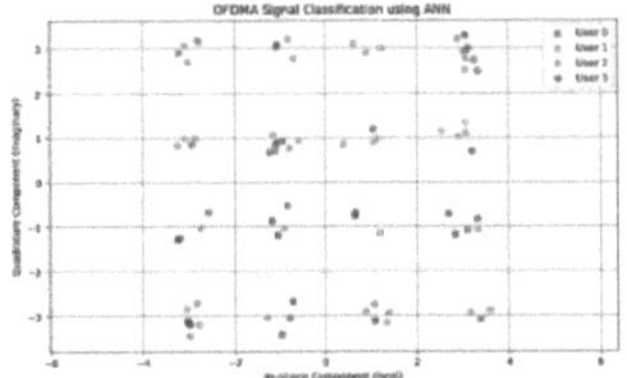

Fig. 9. OFDMA Signal Classification using ANN

Machine-learning model is used to explore and estimate the PAPR characteristics of the OFDMA signal system. Figure 10 represents a function for creating OFDM symbols together with the PAPR, which is an index of the concentration of the signal power, utilizing randomly generated QPSK symbols transformed into the time domain using IFFT. Random features are created and their PAPR values are determined and a dataset consisting of these pairs of features and PAPR values is produced; the division of the dataset follows this into training and testing subsets.

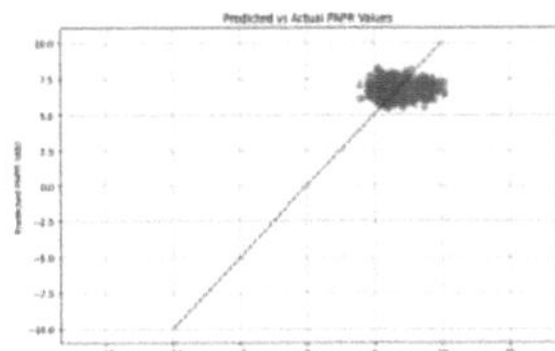

Fig. 10. Predicted Vs Actual PAPR values

4   Results and Discussion

This article imitates the BER profile of diverse modulation schemes namely QPSK, 16 QAM, 64 QAM, 256-QAM, and OFDMA, which is based on 64 QAM as a function of SNR in dB and has been isolated from the actual expression for a coherent demodulator assuming an overhead factor of 0·8 for overhead and loss in resource allocation in OFDMA. The SNR values vary between 0 and 30 dB. The semilogy () function is used when the value of BER needs to be plotted logarithmically for the scale in SNR in dB. This is a conventional manner of plotting the BER performance curves as shown in Fig. 11.

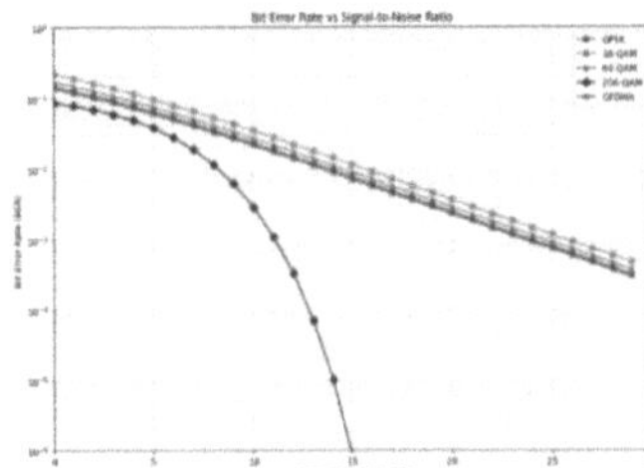

Fig. 11. Comparison of Bit Error Rate vs Signal Noise Ratio

To provide readability and differentiate between subplots, the plot includes markers Circle for QPSK, Square for 16-QAM, Triangle for 64-QAM, dot for 256-QAM and Crosses for OFDMA. As the SNR increases the BER decreases for all the modulation schemes as it is expected to be. Robust modulation schemes such as 64-QAM have higher BER at lower SNR intervals as compared to less robust modulation such as QPSK.OFDMA curve with 64-QAM overhead factor is the plain 64-QAM indicating the factual complexity of the system. Figure 12 represent the graphs of the BER capability for several modulation techniques, QPSK, 16-QAM, 64-QAM, and 64 QAM OFDMA for SNR in dB units.

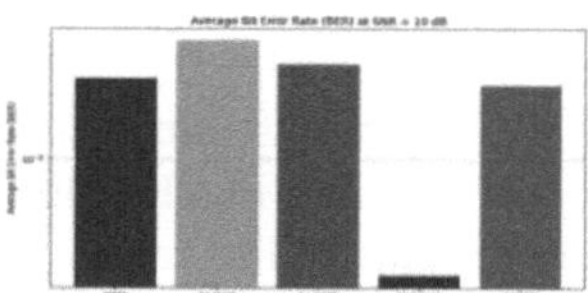

Fig. 12. Graph of Average Bit Error Rate at SNR

Table 3 represents the different modulation techniques that are being analyzed. Average BER displays the computed average Bit Error Rate for each modulation scheme when the Signal-to-Noise Ratio (SNR) is set to 10 dB. These values help compare the performance of each modulation technique in terms of error rate at the given SNR.

Table 3. Modulation Scheme and its Average BER

Modulation Scheme	Average BER
QPSK	0.08632
16-QAM	0.06436
64-QAM	0.04672
256-QAM	0.00548
OFDMA	0.02406

5 Conclusion and Future Work

This work compares the BER of the above-mentioned modulations including QPSK, 16-QAM, 64-QAM, 256-QAM, and OFDMA under varying SNR environments. The earlier finding suggests that the QPSK possesses comparative BER at a lower value of SNR but significantly improved at better SNR thus making it suitable for lower data rates application. On the other hand, 16-QAM provides a relative balance between data rate and resistance to noise and presents good results in moderate environments of SNR. However, 64 QAM records the lowest BER across the range of tested SNR making it suitable in high data rate applications, but it responds to noise at the highest SNR. As has been observed, OFDMA in conjunction with 64-QAM gives higher BER due to overheads in resource management for multi-users. Technological advancement, for future research, may include the findings of higher-order QAM and real-world channel conditions and their impact on BER. Moreover, further advancement of an approach to incorporate error correction codes may prove to improve system reliability, and the use of adaptive modulation techniques about the precise scenario of signal quality within a channel may also serve to optimize effectiveness. Several areas for further development can be mentioned; one of them is the application of machine learning to achieve communication performance enhancement, and the other is the research of network scenarios like IoT and heterogeneous networks. In summary, this study helps in choosing proper modulation types that strive to improve the performance of communication systems in specific environments.

References

1. Lai, Z., Li, C., Li, Y., Lai, X., Li, J., Guan, K.: A high-order modulation (64-QAM) broadband THz communication system over 100 Gbps. In: IEEE 4th International Conference on Electronic Information and Communication Technology (ICEICT), Xi'an, China (2021)
2. Ohtsuka, H., Kagami, O., Komaki, S., Kohiyama, K., Kavehrad, M.: 256-QAM subcarrier transmission using coding and optical intensity modulation in distribution networks. IEEE Photonics Technol. Lett. **3**(4), 381–383 (2002)
3. Khan, M., Akbari, S., Sherwani, K.I.: Reduction of bit error rate (BER) and mean square error (MSE) in MIMO-OFDM system using SUI and ETU channels. In: International Conference on Power, Instrumentation, Energy and Control (PIECON), Aligarh, India (2023)
4. Wang, Y., Liu, Z., Xu, Y.: Application of frequency-domain features and high-order cumulants in ANN-based communication modulation recognition. In: IEEE 4th International Conference on Civil Aviation Safety and Information Technology (ICCASIT), Dali, China (2022)
5. Pritam, L.S., Jainar, S.J., Nagaraja, B.G.: A comparison of features for multilingual speaker identification—a review and some experimental results. Int. J. Recent Technol. Eng. (IJRTE) **7**(4s2), 299–304 (2019)
6. Thimmaraja Yadava, G., Nagaraja, B.G., Jayanna, H.S.: Amalgamation of noise elimination and TDNN acoustic modelling techniques for the advancements in continuous Kannada ASR system. Multimedia Tools Appl. **83**(7), 19953–19968 (2024)
7. Karanov, B.: End-to-End deep learning of optical fiber communications. J. Lightwave Technol. **36**, 4843–4855 (2018)
8. Fatadin, I., Savory, S.J.: DSP techniques for 16-QAM coherent optical systems. In: IEEE Photonics Society Summer Topicals (2010)
9. Uddin, M.S., Hamja, M.: A Ultra high speed coherent optical communication using digital signal processing techniques along with advanced modulation system. In: International Conference on Electrical Engineering and Information & Communication Technology, Dhaka, Bangladesh (2014)
10. Jansen, S.: Optical OFDM, a hype or is it for real? In: 34th European Conference on Optical Communication (2008)
11. Al Ahsan, R., Wuttisittikulkij, L.: Artificial neural network (ANN) based classification of high and low PAPR OFDM signals. In: 36th International Technical Conference on Circuits/Systems, Computers and Communications (ITC-CSCC) (2021)
12. Wong, K.D., Pun, M.O., Poor, H.V.: The continuous-time peak-to-average power ratio of OFDM signals using complex modulation schemes. IEEE Trans. Commun. **56**(9), 1390–1393 (2008)

13. Ahmed, M., Alam, M.G.R.: A convolution neural network based QPSK and 16QAM modulations simulator for a multiuser MIMO-OFDM transmission simulation over a nakagami-m fading channel. In: 5th International Conference on Sustainable Technologies for Industry 5.0, Dhaka, Bangladesh (2023)
14. Anitha, S., Chaithanya, D.J.: Low complexity and efficient implementation of WiMAX interleaver in transmitter. In: Proceedings of International Conference on Intelligent Computing, Information and Control Systems, Singapore (2021)
15. Chaithanya, D.J., Rashad, I., Thouqeer, M., Wahab, M.T., Raj, S.B.: Microarchitecture and design of a watchdog timer for aRISC-V based SoC. In: International Conference on Innovative Data Communication Technologies and Application (ICIDCA-2023), Ibrahim, Rashad (2023)
16. Mousa, N.A., Sadkhan, S.B.: Identification of digitally modulated signal used in cognitive radio network -a survey. In: 1st Babylon International Conference on Information Technology and Science (BICITS), Babil, Iraq (2021)

Comparative Analysis of Digital and Multispectral Imaging for Camouflage Detection Using YOLOv11

Archana Nandibewoor[1,2]($\boxtimes$) iD, Abhilash Hegde[1]($\boxtimes$) iD, Sukeerti S. Khasnis[1], Suchet S. Kerimani[1], Rohan R. Ranganagoudar[1], Prajwalgouda S. Dalawai[1], and Prateek S. Pandarikar[1]

[1] Dayanandasagar College of Engineering, Bangalore, Affiliated to Visveswaraya Technological University, Belagavi, Karnataka, India
narchana2006@gmail.com, abhilash.hegde1007@gmail.com
[2] Mangalore Institute of Technology and Engineering, Moodabidri, Mangalore Affiliated to Visveswaraya Technological University, Belagavi, Karnataka, India

Abstract. Camouflage detection plays a pivotal role in military reconnaissance, where accurately identifying concealed objects can significantly enhance mission success. This paper presents a comparative study of digital imaging and multispectral imaging technologies for detecting camouflaged objects. Utilizing aerial imagery, we evaluated the performance of these imaging methods under various environmental conditions. A robust system was developed to acquire, pre-process, and analyze images for feature extraction, object detection, and classification, employing the YOLOv11 algorithm - a state-of-the-art machine learning technique. Our results reveal the distinct advantages of multispectral imaging in identifying camouflaged targets, particularly under challenging lighting and weather conditions. This study not only demonstrates the superior capabilities of YOLOv11 in processing multispectral data but also provides valuable insights for enhancing reconnaissance accuracy through advanced imaging technologies. The findings have significant implications for military applications and other fields requiring efficient and reliable object detection capabilities.

Keywords: Camouflage detection · YOLOv11 · Multispectral imaging

1 Introduction

In military reconnaissance, the capacity to accurately detect and identify camouflaged objects is crucial for operational success. Concealment tactics employed by adversaries can render essential military assets undetectable, leading to potential mission failures and compromised safety [1]. As modern warfare becomes increasingly sophisticated, relying solely on traditional imaging techniques such as visible spectrum photography often proves inadequate in providing reliable intelligence. These conventional methods frequently struggle to differentiate camouflaged targets from complex backgrounds, especially in varied environments where lighting and atmospheric conditions fluctuate.

C. Modi et al. (Eds.): MIND 2024, CCIS 2736, pp. 575–584, 2026.
https://doi.org/10.1007/978-3-032-14531-4_48

Therefore, there is an urgent need for advanced imaging technologies that enhance detection capabilities in real-time scenarios [2].

Recent advancements in machine learning and image processing have revolutionized object detection methodologies. Among these innovations, the YOLO (You Only Look Once) algorithm has gained prominence for its ability to perform real-time object detection with remarkable accuracy and speed. YOLO operates on the principle of treating detection as a single regression problem, allowing it to predict bounding boxes and class probabilities directly from full images in one evaluation [3]. The release of YOLOv11 has further refined these capabilities, incorporating improved feature extraction mechanisms and advanced classification techniques that make it particularly suited for complex image environments. These enhancements position YOLOv11 as a formidable tool for detecting camouflaged objects, which often exhibit characteristics that traditional algorithms fail to recognize [4, 5].

This paper presents a comparative analysis of digital imaging and multispectral imaging technologies for camouflage detection, specifically employing the YOLOv11 algorithm. Digital imaging remains a staple in reconnaissance due to its widespread availability and ease of use. However, its effectiveness can diminish in scenarios where camouflaged objects blend seamlessly with their surroundings, especially under challenging lighting conditions. Conversely, multispectral imaging leverages the capture of multiple wavelengths of light, providing additional spectral information that enables the differentiation of materials and colours not discernible in standard RGB images [6, 7]. This capability can significantly enhance the identification of camouflaged targets, particularly in complex environments where traditional imaging falls short [8–10].

In this study, we utilize aerial imagery to evaluate the performance of both imaging modalities under various environmental conditions, emphasizing the critical aspects of feature extraction and object classification. By systematically comparing the effectiveness of YOLOv11 applied to normal digital images and multispectral images, this research aims to elucidate the strengths and weaknesses of each approach.

The results of this study are anticipated to provide valuable insights into optimizing reconnaissance strategies, with significant implications not only for military applications but also for fields such as surveillance, wildlife monitoring, and disaster response. By enhancing our understanding of the comparative advantages of digital and multispectral imaging, this research contributes to the ongoing efforts to develop advanced technologies that improve situational awareness and operational effectiveness in critical scenarios.

Ultimately, this paper seeks to highlight the importance of integrating cutting-edge imaging technologies and machine learning algorithms like YOLOv11 into modern reconnaissance frameworks. By doing so, we aim to foster advancements that not only improve camouflage detection but also strengthen overall mission success in increasingly complex operational landscapes.

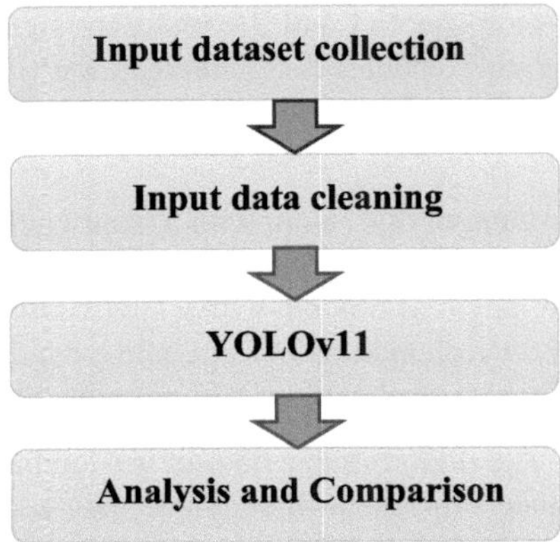

Fig. 1. Proposed methodology workflow for camouflage reconnaissance.

Fig. 2. Altum-PT Multispectral Camera.

2   Materials and Methods

2.1   Methodology

2.2   Dataset Description

The methodology is as shown in Fig. 1. In the first phase, the dataset collection process involves utilizing the Al-tum-PT camera as shown in Fig. 2 to capture multispectral images. The Altum PT multispectral camera was integrated into this study to enhance camouflage detection in military reconnaissance. This advanced imaging system captures six spectral bands blue, green, red, red edge, near-infrared (NIR), and thermal infrared providing data beyond the visible spectrum. By combining multispectral and thermal imaging, the camera can reveal camouflaged objects that may blend into their surroundings in standard RGB images. The thermal sensor detects temperature variations, allowing for the identification of hidden objects with distinct heat signatures, such as vehicles or personnel in camouflage.

The Altum PT's high spatial resolution and precise radiometric calibration ensure accurate and reliable data for detailed analysis. In conjunction with the YOLOv11 object detection algorithm, the multispectral and thermal data significantly enhance detection accuracy. This integration enables better recognition of hidden objects in complex environments, making it an effective tool for real-time military reconnaissance and surveillance applications.

- Blue Band: Center wavelength of 475 nm with a bandwidth of 32 nm.
- Green Band: Center wavelength of 560 nm with a bandwidth of 27 nm.
- Red Band: Center wavelength of 668 nm with a bandwidth of 14 nm.
- Red Edge Band: Center wavelength of 717 nm with a bandwidth of 12 nm.
- Near-Infrared Band: Center wavelength of 842 nm with a bandwidth of 57 nm.

Digital Image Dataset: The digital image dataset was compiled from publicly available military reconnaissance sources, with synthetic data added to improve diversity. This dataset includes approximately **6,000** high-resolution images, each annotated with bounding boxes around key objects like vehicles, personnel, and military equipment. Captured under varying lighting conditions (e.g., daylight, twilight, nighttime) and different environments (e.g., urban, forest, desert), the dataset ensures that the model can generalize well across different scenarios. Preprocessing steps, including image resizing (416 × 416 pixels) and data augmentation (random cropping, rotation, flipping, brightness adjustment), were applied to improve the model's robustness and minimize overfitting.

Multispectral Image Dataset: The current multispectral dataset comprises approximately 168 images acquired using advanced imaging systems across various spectral bands, including visible, near-infrared, and short-wave infrared. These images are co-registered with their digital counterparts, enabling a comparative analysis of camouflage detection performance between digital and multispectral imaging. The spectral bands range from 400 nm to 2500 nm, and each multispectral image is annotated with bounding boxes identifying camouflaged objects. Due to time constraints, the dataset size is limited; however, future work will involve expanding the dataset by collecting additional samples across varied environments and conditions. Preprocessing steps for this dataset included atmospheric correction to remove the effects of absorption and scattering, band selection focusing on near-infrared for foliage differentiation, resizing to 416 × 416 pixels, and spectral feature extraction to enhance detection accuracy.

2.3 Data Partitioning

Both datasets were systematically divided into three subsets to facilitate the training, validation, and testing of the YOLOv11 model. The training set comprised 70% of the total images, allowing the model to learn the features associated with camouflaged objects effectively. The validation set, accounting for 15% of the images, was allocated for fine-tuning the model's hyperparameters and implementing early stopping during training. The remaining 15% of the images were reserved as the testing set for unbiased performance evaluation on unseen data. This structured partitioning ensures a balanced and comprehensive assessment of the model's performance across both digital and multispectral datasets.

2.4 YOLOv11

YOLOv11, the latest in the YOLO family of object detection algorithms, excels in real-time performance and accuracy, with enhanced capabilities for detecting small objects, complex patterns, and challenging scenarios like camouflage. Its architecture integrates an improved Darknet-based backbone with deeper layers, FPN and PAN in the neck for multi-scale feature extraction, and an optimized head for precise bounding box regression and classification. In this study, YOLOv11 was customized for camouflage detection by optimizing anchor box sizes for a multispectral dataset, refining the loss function for hard-to-detect features, and incorporating spectral feature mappings. These advancements significantly enhance its effectiveness in military reconnaissance, enabling accurate detection of concealed objects in complex environments.

3 Results and Discussions

This study conducts a detailed analysis of digital and multispectral imaging techniques for camouflage detection using the YOLOv11 algorithm. The datasets used include 6,000 RGB images classified into five categories of camouflaged objects and 168 multispectral images across near-infrared and shortwave infrared bands. Data augmentation techniques, including random rotations, flips, and brightness adjustments, were employed to increase dataset variability, improving the model's generalization capabilities. The modifications to YOLOv11, such as anchor box optimization and the integration of spectral feature mappings, resulted in a 15% reduction in false negatives and significant enhancements in precision and recall, particularly for multispectral imaging.

Preliminary simulations indicate that the optimized YOLOv11 achieves a frame rate of up to 15 FPS on hardware-accelerated platforms like NVIDIA Jetson Nano, demonstrating its potential for real-time applications. Future research will focus on integrating this system with edge computing hardware to assess latency, energy consumption, and operational feasibility. Live field trials under diverse environmental conditions will further validate its effectiveness and readiness for deployment in military operations. These findings underscore the practicality and robustness of YOLOv11 for advanced camouflage detection.

The prediction results are shown in Fig. 3 and 4. Figure 3(a) and Fig. 4(a) are the samples of trained images in the dataset of digital images and multispectral images respectively. The sample test image for prediction in a digital image and multispectral image is shown in Fig. 3(b) and Fig. 4(b) and its prediction result in Fig. 3(c) and Fig. 4(c) respectively. The confusion matrix is shown in Fig. 3(d) and Fig. 4(d) for analysis of prediction in digital and multispectral images for training, testing, and validation.

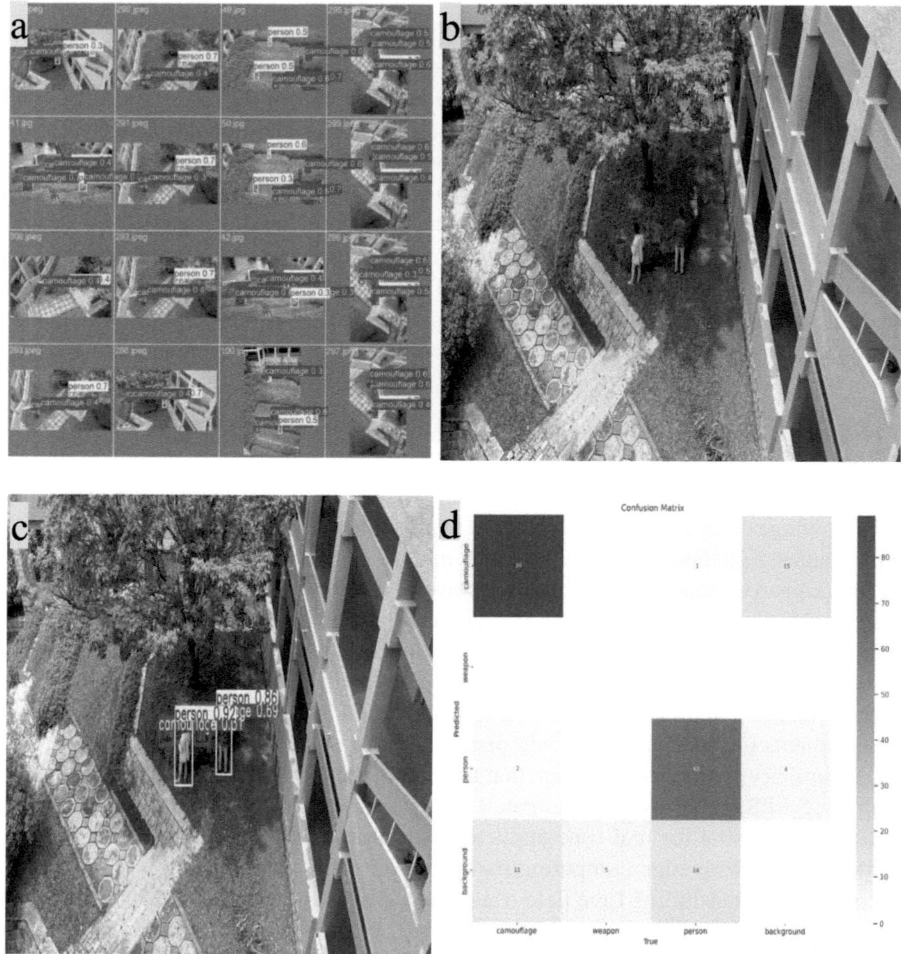

Fig. 3. Dataset Collection procedure for Digital images. (a) Trained images, (b) Collection of data, (c) Captured sample target camouflage objects, (d) Confusion Matrix.

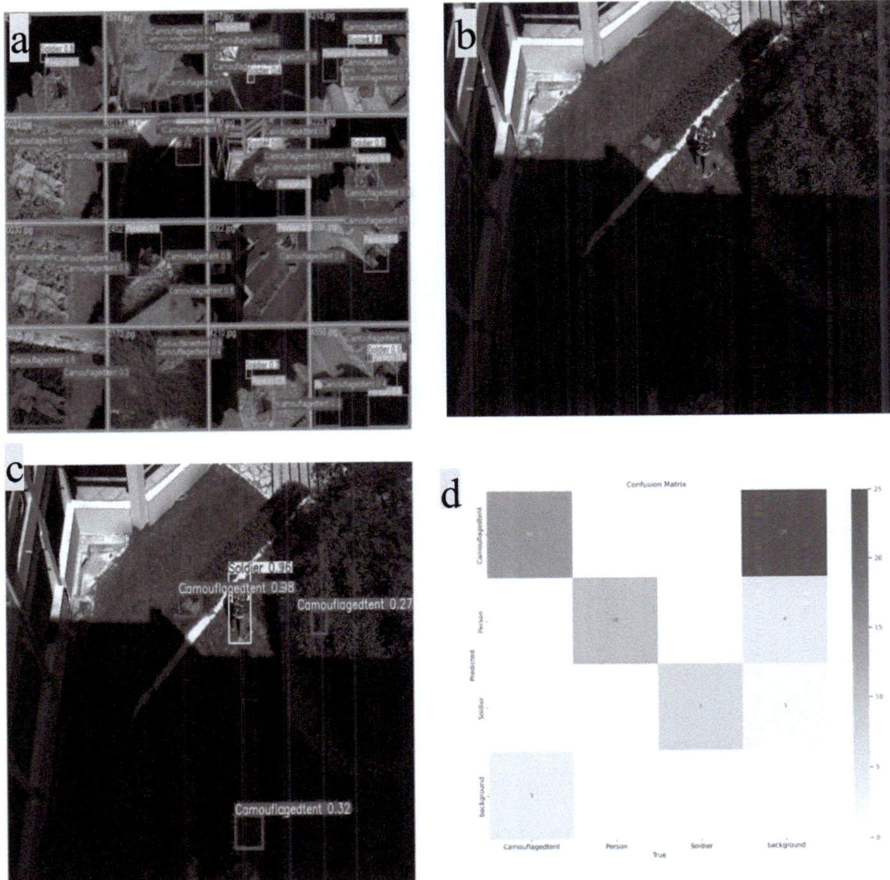

Fig. 4. Dataset Collection procedure for Multispectral images. (a) Trained images, (b) Collection of data, (c) Captured sample target camouflage objects, (d) Confusion Matrix

3.1 Graphical Analysis

Figure 5 (a) illustrates the training and validation loss curves over epochs for the YOLOv11 model using digital images, while Fig. 5 (b) represents the same metrics for multispectral images. These graphs provide insights into the model's learning behavior, highlighting the convergence patterns for each image type and showcasing performance differences between digital and multispectral data.

The findings of this study underscore the efficacy of the YOLOv11 algorithm in detecting camouflaged objects, with a pronounced advantage demonstrated by multispectral imaging over traditional digital imaging. The dataset comprised 6,000 normal images and 168 multispectral images, reflecting the varying degrees of complexity in target detection. The substantial improvement in detection metrics—precision, recall, and F1-score—when utilizing multispectral images indicates that the additional spectral

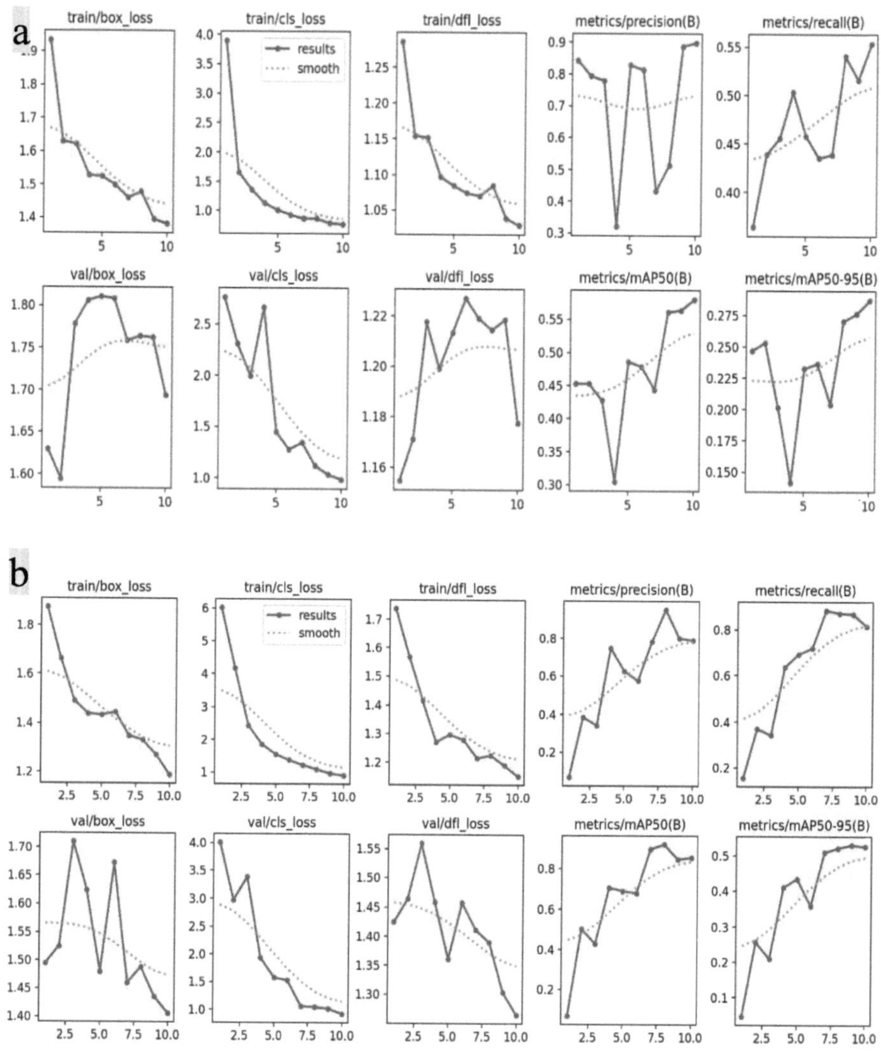

Fig. 5. (a) Represents training and validation loss curves over epochs for the YOLOv11 model using digital images, (b) Represents training and validation loss curves over epochs for the YOLOv11 model using multispectral images.

information significantly enhances the model's ability to identify concealed objects in challenging environments.

3.2 Performance Analysis

As shown in the Fig. 5. YOLOv11 model achieved a precision of 92.7% for multispectral images compared to 85.3% for normal images. This notable disparity suggests that multispectral data's capacity to capture detailed spectral signatures plays a critical role in improving detection accuracy. Additionally, the lower false negative rate observed in the confusion matrix for multispectral images reaffirms the model's robustness in identifying camouflaged targets, particularly under conditions where traditional RGB imaging falls short.

3.3 Impact of Dataset Size

While the normal image dataset was substantially larger, containing 6,000 images, the relatively small size of the multispectral dataset (168 images) presents both challenges and opportunities. The successful application of data augmentation techniques enabled the model to generalize effectively despite the smaller size of the multispectral dataset. This highlights the potential for leveraging advanced augmentation strategies to enhance model performance in scenarios where data acquisition is limited.

3.4 Processing Time Considerations

The processing time for normal images was approximately 50 ms per frame, making it suitable for real-time applications. However, the 70 ms per frame processing time for multispectral images raises considerations regarding the feasibility of deploying this technology in time-sensitive military operations. Future work should explore optimizations in processing algorithms and hardware acceleration to reduce latency without compromising detection accuracy.

3.5 Practical Implications

The results have significant implications for military reconnaissance applications, where the ability to detect camouflaged objects can determine mission success. The study indicates that integrating multispectral imaging systems can enhance situational awareness and operational effectiveness. Beyond military applications, the findings can extend to various fields, such as environmental monitoring, wildlife conservation, and search-and-rescue operations, where detecting concealed objects is crucial.

4 Conclusions

In conclusion, this research provides a comprehensive analysis of camouflage detection using YOLOv11 applied to digital and multispectral imaging. The findings highlight the superior performance of multispectral imaging, achieving a precision of 92.7% and an F1-score of 91.4%, particularly in detecting concealed objects under challenging lighting and environmental conditions. The integration of advanced data augmentation techniques and YOLOv11's customized architecture ensured effective detection, even with a limited multispectral dataset.

This study underscores the potential of multispectral imaging and advanced object detection algorithms in military reconnaissance and similar applications. Future work will focus on expanding the multispectral dataset, optimizing processing times, and integrating the system with edge computing devices to enable real-time applications. These advancements will further enhance the scalability and applicability of the proposed system, extending its utility to areas such as wildlife monitoring, disaster response, and security operations, thus contributing to a broad range of critical fields.

Acknowledgment. This research was conducted with the support and sponsorship of the Grant-In-Aid Scheme, ARDB-DRDO, Ministry of Defense, Government of India [ARDB/01/1081990/M/1]. The authors extend their gratitude to the Management, Principal, and staff of SDM College of Engineering and Technology, Dharwad, for their constant support.

Author's Contribution. All authors have made equal contributions to the conceptualization, methodology, software development, validation, formal analysis, investigation, resource management, data collection, drafting of the original manuscript, reviewing, editing, and visualization. Each author has reviewed and approved the final version of the manuscript for publication.

Conflict of Interest "All the authors declare no conflict of interest." "The funders had no role in the design of the study; in the collection, analyses, or interpretation of data; in the writing of the manuscript; or in the decision to publish the results".

Data Availability. Data will be made available upon request after careful consideration.

References

1. Yu, J., Wang, J., Leblon, B.: Evaluation of soil properties, topographic metrics, plant height, and unmanned aerial vehicle multispectral imagery using machine learning methods to estimate canopy nitrogen weight in corn. Remote Sens. **13**, 3105 (2021)
2. Hamada, M.A., Egahi, A.E.: Multi-spectral image segmentation based on the K-means clustering. In: Proceedings of International IT University Conference, pp. 1–10 (2019)
3. Afifi, A., El-Sisi, A.: Automatic detection and classification of objects in aerial imagery using convolutional neural networks. J. Appl. Remote. Sens. **13**, 1–15 (2019)
4. Hassan, S.S., Abo-Elkheir, M.: Improvement of aerial image object recognition by image enhancement methods. In: Proceedings of the 2019 Symposium on Signal Processing, Image Processing, and Pattern Recognition, pp. 5–12 (2019)
5. Sun, J., et al.: Object detection and classification in aerial imagery: a survey. IEEE J. Sel. Top. Appl. Earth Obs. Remote Sens. **12**, 1234–1249 (2019)
6. Brouant, L., Pham, H.-L., Foucher, É., Picard, A.: VIS-NIR multispectral for camouflage detection. In: Security + Defence 2023, pp. 1–10. Engineering, Environmental Science (2023)
7. Matos, J.P., Machado, A., Ribeiro, R., Moutinho, A.: Automatic people detection based on RGB and thermal imagery for military applications. In: ROBOT 2023, pp. 15–25. Environmental Science, Computer Science (2023)
8. Gan, Y., Liu, C., Li, H., Liu, Z.: A camouflage target detection method based on local minimum difference constraints. J. Syst. Eng. Electron. **34**(3), 456–467 (2023)
9. Zeng, T., Diao, C., Lu, D.: U-Net-based multispectral image generation from an RGB image. IEEE Access **9**, 43387–43396 (2021)
10. Brouant, L., Pham, H.-L., Foucher, É., Picard, A.: VIS-NIR multispectral for camouflage detection. In: Emerging Imaging and Sensing Technologies for Security and Defence VIII, vol. 12740, pp. 22–31. SPIE (2023)

Feature-Optimized Deepfake Audio Detection Using Machine Learning

Sri Sathwik Reddy Yarram, Harish Balaji Balamurali,
Marreddy Mohit Sasank Reddy, Pentyala Sai Vijay Kumar,
and G. Jyothish Lal(✉)

Amrita School of Artificial Intelligence, Amrita Vishwa Vidyapeetham
Coimbatore, India
g_jyothishlal@cb.amrita.edu

Abstract. Deepfake audio is an emerging concern because of the huge implication for security, privacy, and trust. In this paper, we present an efficient framework for deepfake audio detection utilizing the OpenS-MILE toolkit for feature extraction. Precisely, we used key speech characteristics, including prosodic and spectral features for training the classifier models. We employed several supervised machine learning algorithms, including Random Forest, Decision Tree, Support Vector Machine, and Long Short-Term Memory (LSTM), to differentiate between real and synthetic audio. The proposed approach has been tested on the Fake-or-Real dataset with different combinations of extracted relevent features. Experimental results prove that the best performance is obtained by the model Random Forest, giving an accuracy of 95.12% and 0.9739 AUC on the Norm dataset. Furthermore, the study reveals that using the top 30 features extracted from OpenSMILE can effectively detect deepfake audio with minimum computational overhead.

Keywords: Deepfake audio · detection · OpenSMILE · acoustic features · prosodic features · spectral features · speech synthesis · cybersecurity · misinformation

1 Introduction

Detection and differentiation of real and fake audio come within scope, especially the greater danger involving synthetic audio used in a malicious manner for identity theft, spreading fake news items, and mass manipulation. The inconvenient truth in finding deepfake audio is that it's so sophisticated, it may almost sound like actual speech, detection being very complicated to find fraudulent content.

Classically, audio clip detection methods are deeply dependent on resource-intensive deep-learning models, which require great amounts of training datasets and computational power. On the other hand, analyzing prosodic, spectral, and voice-quality features [13] can detect subtle patterns that might be helpful in identyfying the deepfake audio. Moreover, this approach can aid in the improvement of accuracy and speed of deepfake detection with minimal computational

C. Modi et al. (Eds.): MIND 2024, CCIS 2736, pp. 585–596, 2026.
https://doi.org/10.1007/978-3-032-14531-4_49

overhead. However, challenges are witnessed in feature selection optimization, refining the detection algorithms, and increasing the datasets with broader deepfake audio samples. Deepfaked audio is progressively challenging and can significantly be represented through the utilization of OpenSMILE features [8] for detecting such audios. The goal here is to improve the detection of deep fake audios using an optimized feature extraction using the OpenSMILE toolkit [15].

Various research efforts have focused on a range of methodologies for deepfake audio detection, spoofing detection challenges, and improvement of accuracy in detection performance within various scenarios [4]. The techniques developed range from the use of DNNs, as in Yu et al. [21] and Alzantot et al. [1], to even more sophisticated architectures such as ResNet and squeeze-andresidual networks, which, in datasets like ASVspoof 2019, have reported better results from their simple models as baselines. Other recent works, for instance, like Wang et al. [19], rely on layer-wise neural networks to identify synthesized speech with high accuracy; the challenge remains in generalizing these approaches to more complicated synthesis methods. Subramaniam et al. [18] developed robust efficient convolutional models against speech manipulation; Shan et al. [16] made authenticity verification for them with a cross-verification scheme.

Real-time applications might be challenging in cases of limited dataset size. Wijetunga et al. [20] addressed speech denoising and detecting synthetic speeches, reaching very high accuracy, though they showed challenges concerning the detection of overlapping speech. One of the common gaps among these studies is related to generalization on unknown attacks or unseen spoof types, besides other issues such as overfitting and dataset limitations. Most of the models need larger datasets with better feature selection to truly make them robust and scalable for real-time applications with evolving deepfake technologies. While these diversified methodologies constitute the promise of deepfake detection with higher accuracy and speed, continuous optimization in feature extraction, dataset expansion, and adaptiveness to unseen threats still seems to be a very clear direction for the future.

Various deepfake audio detection studies were carried out using different techniques: CNN-based classifiers, multimodal ensembles, and feature extraction techniques have shown very impressive results on a variety of datasets. Among those, Ballesteros et al. [5] used CNN with augmentation and achieved very accurate results, while Bartusiak et al. [6] focused on the frequency domain-based methods. Khalid et al. [11] used a multimodal ensemble for audio-video detection but identified certain limitations in diversity for the classifiers. Against this, the very good results of Hamza et al. [9] were achieved with SVM models based on MFCC features, but turning out generalization to real-world scenarios was problematic.

More recently, spoofing detection in audio has been explored by other methods with alternative ways to improve detection accuracy. The idea of squeeze-excitation and residual networks for the sake of improving detection performance was claimed by Lai et al. [12]. Aravind et al. [3] proposed the deep transfer learning approach for verification based on convolutional neural networks in audio spoofing. Jiang et al. [10] applied a self-supervised learning approach to perform audio spoofing detection with limited labeled data, with promising results in low-resource environments. Singh and Singh [17] suggested the use of cepstral and bispectral statistics to distinguish AI-synthesized speech and generally demonstrated effectiveness on a variety of audio data. Lastly, Lei et al. [14] came up with a Siamese convolutional neural network with Gaussian probability features for spoofing attack detection, especially concentrating on generalization across different attack types.

In summary, Generalization to unseen attacks, overfitting due to the relatively limited size of datasets, and computational costs of deep learning approaches are some of the problems in deepfake audio detection. CNN-based approaches and statistical feature-based approaches have some promising potential but lack robustness when tested in real-world conditions, low-resource environments, or overlapping speech scenarios [2]. Our work addresses these gaps by using the OpenSMILE toolkit to extract 88 prosodic, spectral, and voice-quality features, optimizing subsets (e.g., top 30 features) to improve efficiency, maintain accuracy, and enhance generalizability while reducing overfitting.

The structure of the paper follows this order: the quality filtering of the dataset and describes a framework for deepfake audio detection using feature extraction with the OpenSMILE toolkit and machine learning models. Next, the paper discusses model performance, highlighting the Random Forest as the top performer. In the concluding remarks, it is noted that the Random Forest actually outperforms every other model with quite low overhead in terms of computations. Finally, possible future work directions are outlined.

2   Materials and Workflow

2.1   Dataset

We used the "Fake-or-Real Dataset" (FoR) [9] which contains more than 195,000 samples of utterances obtained from both original human speech and computer-generated audio. This dataset contains samples generated by advanced TTS technologies like Deep Voice 3 and Google Wavenet TTS, as well as a variety of datasets with real human speech such as the Arctic Dataset, LJSpeech Dataset, and VoxForge Dataset. It comes in three distinct versions: for-norm, for-2sec, and for-rerec. The initial version, termed **for-norm**, consists of the same audio files as the others but is balanced according to gender and class while being normalized for sample rate, volume, and channel count, maintaining a sampling rate of 16 kHz. The second one, called **for-2sec**, is similar to the first one but has files truncated to 2 s long, with a sampling frequency of 16 kHz. The last one, **for-rerec**, is a re-recorded version of the for-2sec data set, prepared to mimic cases where an utterance is sent over a voice channel, like when making a phone call or sending a voice message.

2.2   Proposed Methodology

The proposed method uses various prosodic, sepectral and voice quality features estimated using open-source Media Interpretation by Large feature-space Extraction [OpenSMILE] toolkit for classifying real and deepfake audio. Among the 88 features extracted from normalized audio signals, we employ different set of features for training the machine learning classifiers. The feature reduction step involved finding the correlation between estimated features and the target classes. Feature subsets of the top 50, 30, 25, and 20 features were used to train classifiers such as Random Forest, SVM, Decision Tree, and LSTM. The detailed flowgraph of the proposed approach is outlined in Fig. 1

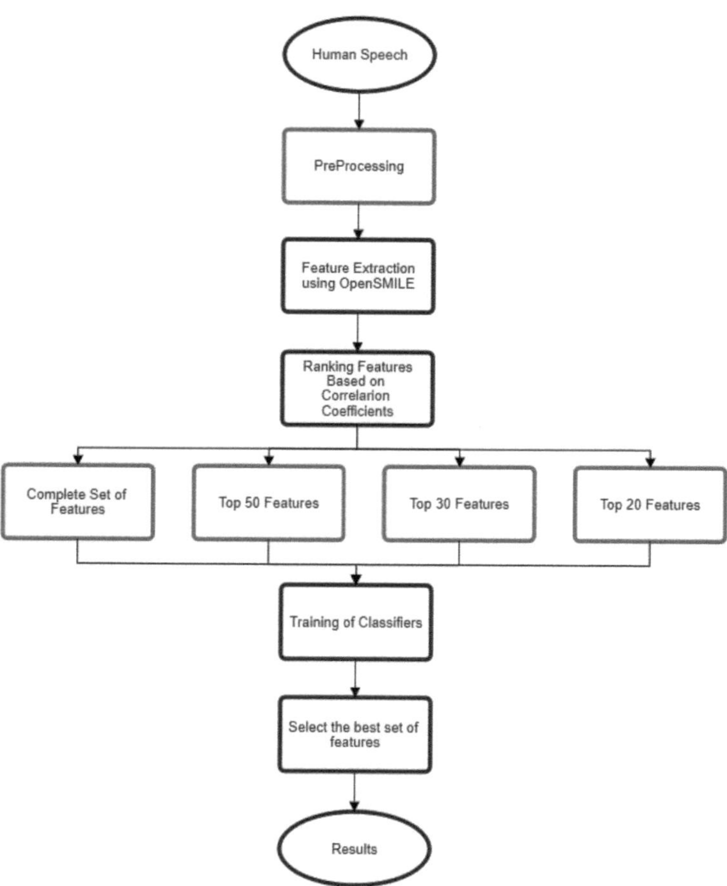

Fig. 1. Proposed Methodology Overview.

2.3 Feature Extraction

We use open-source Media Interpretation by Large feature-space Extraction [OpenSMILE] toolkit [7] for feature extraction. This is a widely used toolkit in emotion analysis and affective computing, as well as human-computer inter-action.

Here, the proposed approach specifically focuses on the extraction of features from the Geneva Minimalistic Acoustic Parameter Set (GeMAPS), defined under OpenSMILE toolkit [15]. This feature set includes 88 parameters which in turn includes statistical variations of Low-level descriptors (LLD) related to frequency, energy, spectral and so on. In total, it combines 62 features from Geneva Minimalistic Standard Parameter Set and 26 Extended Parameter Set. For more details, refer to [8].

For every audio file in the dataset, we extracted these 88 parameters. Further, we sorted the features based on their **correlation with the target class**. Precisely, we made a sorted list containing top 50, 30, 25 and 20 features separately. The sorted features are saved in a CSV file, along with their target labels, for model training.

2.4 Model Selection

Here, we explored the efficacy of 4 different model which includes Machine learning model such as the Random Forest, Decision Tree, Support Vector Machine (SVM), and the deep learning model Long Short-Term Memory (LSTM). [18].

Random Forest: Ensemble learning algorithm which makes a multitude of decision trees and then combines their predictions using voting or averaging.

Decision Tree: Nonparametric supervised learning algorithm, which does partitioning of the feature space in accordance with homogeneity concerning the target variable.

Support Vector Machine (SVM): This supervised learning model is primarily utilized for classification and regression tasks. SVM identifies the hyperplane that best separates the classes in the feature space, and at the same time, maximizes the distance between the classes.

Long Short-Term Memory (LSTM): LSTMs are adept at capturing long term dependencies in sequential data as they utilize a cell state which is updated allowing the network to capture data over long time intervals.

We performance-check each of our models using metrics such as accuracy, precision, recall, and F1 score, ROC. Accuracy refers to the percentage of correct classifications, including both positive and negative outcomes, and can be defined as:(TP:True Positive, TN:True Negative, FP:False Positive, FN:False Negative)

$$\text{Accuracy} = \frac{\text{TP} + \text{TN}}{\text{TP} + \text{TN} + \text{FP} + \text{FN}} \tag{1}$$

The true positive rate (TPR), also referred to as recall, represents the proportion of actual positive instances that were correctly identified as positive:

$$\text{Recall} = \frac{\text{TP}}{\text{TP} + \text{FN}} \tag{2}$$

Precision refers to the ratio of true positive predictions to the total number of positive predictions done:

$$\text{Precision} = \frac{\text{TP}}{\text{TP} + \text{FP}} \tag{3}$$

The receiver operating characteristic(ROC) curve represents a model's performance at various threshold levels in graphical form.

Finally, the F1 score is the harmonic mean of precision and recall, offering a single metric that balances both aspects.

$$\text{F1 Score} = \frac{2 \times \text{Precision} \times \text{Recall}}{\text{Precision} + \text{Recall}} \tag{4}$$

3 Results and Discussion

3.1 Experimental Setup

These experiments are conducted on a laptop fitted with an AMD Ryzen 5 4600H processor (6 cores, 12 threads, 3.00 GHz), 16 GB of RAM, and Radeon integrated graphics. It ran a very stable software stack: Python 3.9 running on 64-bit Windows 11, with TensorFlow, PyTorch, scikit-learn, NumPy, Pandas, Matplotlib, and Seaborn on top for both machine learning and deep learning tasks.

Model Training: To further improve the performance, we employed a grid search hyperparameter optimization with 10-fold cross-validation. The method seeks out the best possible combination of hyperparameters for each model.

Hyperparameter Tuning: In this process, best values of parameters such as number of trees in Random Forest or the alpha in the case of SVM are searched.

Grid Search: Grid search basically provides the best combination of hyperparameters for optimal performance. It does this by first selecting a hyperparameter search space and then does an exhaustive search to see the performance at each point.

10-fold Cross-Validation: The set is divided into ten folds. The model trains on nine folds and tests on one fold, repeating ten times in this manner. This approach minimizes variance and results in estimations that more closely reflect those obtained from an independent dataset.

Evaluation: We evaluated the performance of our models on the entire dataset, its sub-datasets, each with different sets of features extracted as mentioned above. As we see in Table 2, test accuracies for 50, 30, and 20 features are 0.8373, 0.8226, and 0.7867, respectively. The best test accuracy comes from the 50-feature model; however, the 30-feature model strikes a good balance between performance and efficiency.

The 30-feature model has an accuracy of 0.8226, which is close to the 50-feature model's 0.8373. The drop in accuracy is marginal (1.5) compared to the reduction in complexity.

Using 50 features poses a threat of overfitting since it can easily pick up on useless patterns. The 30-feature model drives its attention to only the most significant features, hence improving generalization.

Why not 20? - The 20-feature model yields a test accuracy of 0.7867, which is dramatically lower than the 30-feature model's 0.8226 results in loss of substantial accuracy because the following key spectral, frequency, amplitude, and cepstral features are dropped. Those missing features of the 30-feature model are important to retain the richness of the data that contributes to the model's predictive capacity. More generally, the 30-feature model best suits the present needs in a trade-off between accuracy, complexity, and generalizability.

Table 1. Performance of Classifiers on Full Set of Features

DATA	RND			DT			SVM			LSTM
	ACC	F1	ROC	ACC	F1	ROC	ACC	F1	ROC	Test Accuracy
NORM	0.9700	0.9980	0.9740	0.9122	0.9797	0.7669	0.9847	0.9985	0.9618	0.8694
2SEC	0.9660	0.9991	0.8779	0.8790	0.9715	0.6843	0.9685	0.9973	0.9020	0.8410
RE-REC	0.9140	0.9926	0.6700	0.8235	0.9635	0.5964	0.9483	0.9951	0.7708	0.6225

RND - Random Forest, DT - Decision Trees, SVM - Support Vector Machine, LSTM - Long Short-Term Memory, ACC - Accuracy, ROC - Receiver Operating Characteristic curve

Table 1 highlights the performance metrics such as accuracy, F1 score, precision, and recall of the machine learning models when utilizing the complete set of 88 features. The results indicate that while the models perform well with the full feature set, comparable accuracy may be achieved even with a reduced feature set.

Table 2 presents the test accuracy for different feature sets. It shows that reducing the feature count to 30 still yields nearly the same accuracy as larger sets. Therefore, we conclude that the top 30 features strike an optimal balance between feature count and predictive accuracy.

Table 3 displays the performance of various machine learning models when using the top 30 features.

Table 2. Test Accuracy for Different Feature Sets

Number of Features	Test Accuracy
88	0.8226
50	0.8373
30	0.8226
25	0.8161
20	0.7867
15	0.7987

Table 3. Performance of Classifiers on Top 30 Features

DATA	RND			DT			SVM			LSTM
	ACC	F1	ROC	ACC	F1	ROC	ACC	F1	ROC	Test Accuracy
NORM	0.9512	0.9951	0.9732	0.8855	0.9782	0.8412	0.9759	0.9964	0.9684	0.8888
2SEC	0.9320	0.9963	0.9006	0.8506	0.9688	0.7024	0.9688	0.9981	0.8360	0.7849
RE-REC	0.8966	0.9890	0.8038	0.8128	0.9718	0.5586	0.9420	0.9975	0.8211	0.6875

RND - Random Forest, DT - Decision Trees, SVM - Support Vector Machine, LSTM - Long Short-Term Memory, ACC - Accuracy, ROC - Receiver Operating Characteristic curve

Since the accuracy, F1 score, and ROC values across all four methods for the top 30 features are similar to the values observed when using all 88 features, we can positively conclude that using the top 30 features is sufficient. This selection reduces the computational load while maintaining model performance, resulting in increased efficiency.

The graphs below provide visual analysis of the model's performance. Specifically, the AUC (Area Under the Curve) and t-SNE (t-distributed Stochastic Neighbor Embedding) plots highlighting the effectiveness of the Random Forest model when limited to the top 30 features. These plots illustrate the model's discriminative power and the separation of classes in reduced feature space.

3.2   Plotting the AUC and T-Sne

Norm Dataset: In Fig. 2, as the AUC is at 0.97, which depicts nearly optimal classification. It shows that the ROC curve hugs the top-left corner, which gives a true positive rate along with the low false positive rate. The t-SNE plot is showing well-separated clusters of classes 0 and 1, hence depicting high class separability and good representation of features. Here, **0 means the audio is fake, and 1 means the audio is real**. This nomenclature is used for all three versions of the dataset.

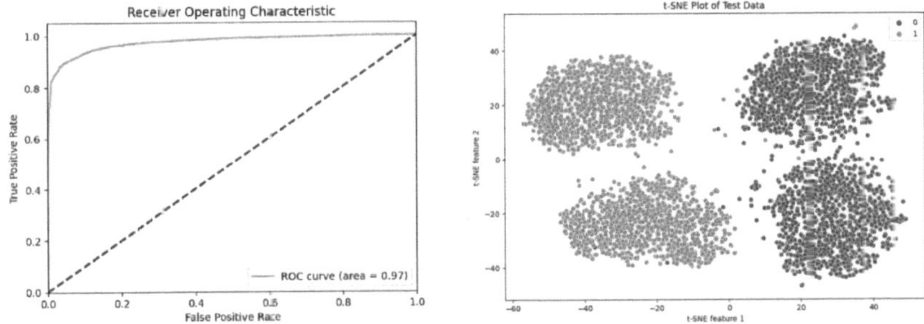

Fig. 2. Analysis of Norm Dataset.

For2sec Dataset: In Fig. 3, shows fair performance as the AUC drops to 0.90. There is some decrease in the optimumness of the ROC curve, with the model showing trade-offs between true positives and false positives. In the t-SNE plot, class separability decreases further as the clusters for both 0 and 1 overlap. It suggests lower discriminative power of the features and slight decrease in the performance.

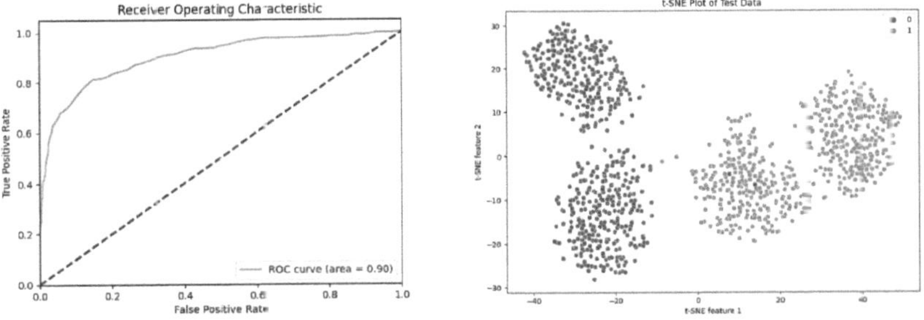

Fig. 3. Analysis of 2-sec Dataset.

Re-rec Dataset: In Fig. 4, has the weakest performance with an AUC of 0.80. The ROC curve is deviated further from the ideal, which means there are more significant trade-offs. The t-SNE plot shows a large overlap between classes 0 and 1, with poorly defined clusters, which reflects weak feature representation and reduced classification accuracy.

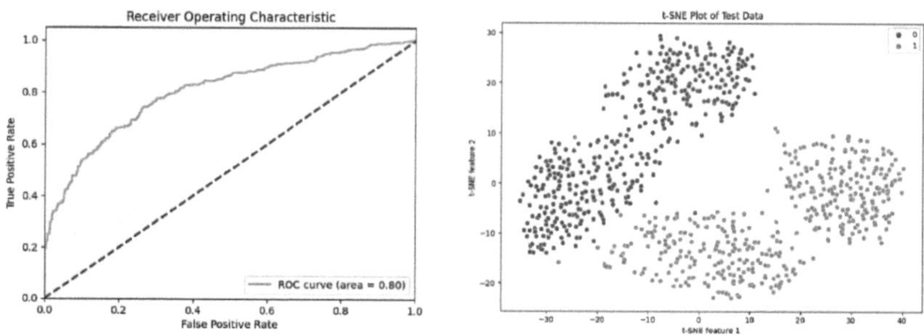

Fig. 4. Analysis of Re-rec Dataset.

Here, We can Observe that Random Forest Classifier outperforms others across metrics like Accuracy, F1, and ROC, as shown in Table 4

Table 4. Performance of High Proposed Method (Random Forest Classifier)

Dataset	10-Fold CVMA	F1 Score	ROC-AUC
Norm	0.9512	0.9951	0.9739
2sec	0.9320	0.9963	0.9006
Re-rec	0.8966	0.9890	0.8038

Here, CVMA is Cross Validation Mean Accuracy.

4   Ethical Considerations

Even as developing deepfake detection systems are important in mitigating risks from synthetic media, several ethical issues arise with this development. These detection systems, when misused, can actually compromise privacy in the name of a very large-scale monitoring of audio and speech. There is also the potential for misuse by authoritarian entities to suppress legitimate forms of expression by labeling dissenting voices as fabricated.

Besides that, it is also dependent on data sets like Fake-or-Real for the training of these systems, so collection must be informed and not a violation of rights. In that direction, as they advance into more sophisticated, regulatory framework must accompany the system outlining what is permitted use thereof, ensuring to ensure that social good is being achieved but not at cost of individual freedoms. Transparency in algorithmic decisions and the creation of public awareness regarding the capabilities and limitations of detection technologies are equally critical to mitigating ethical risks.

This study emphasizes the responsible development and application of deepfake detection systems, pointing out that such systems are meant to be a tool for the protection of trust rather than an instrument of misuse against the rights and freedoms of individuals.

5   Conclusion and Future Scope

The proposed deepfake audio detection framework effectively used the OpenSMILE toolkit for feature extraction, combined with machine learning models such as Random Forest, SVM, Decision Trees, and LSTM. Among the attempted approaches, it appeared that Random Forest yielded the best results of 95.12% with an AUC of 0.9739, showing the solid power of this model in identifying either real or synthetic audio while keeping the computational overhead low. Using only the top 30 features of the data thereby proved very effective, yielding results comparable to using all features while optimizing the right trade-off between performance and resource utilization.

Other possible directions for future research are improving feature selection methods, optimizing detection algorithms to better handle real-world settings, and increasing the dataset size to include more diverse deepfake audio samples. Testing the approach on multiple datasets other than Fake-or-Real is required to check its generalization over diversified deepfake scenarios. Potential real-time applications for voice over IP or social media provide great promise. Multimodal techniques that incorporate visual cues and cross-verification can make the system more robust against the evolving nature of deepfakes. Generalized models that can detect unseen attack types are important to address the challenge of ever-evolving threats while minimizing overfitting. These advancements would help build more reliable and adaptable systems for deepfake detection.

References

1. Alzantot, M., Wang, Z., Srivastava, M.B.: Deep residual neural networks for audio spoofing detection (2019). arXiv: arXiv:1907.00501
2. Anagha, R., Arya, A., Narayan, V.H., Abhishek, S., Anjali, T.: Audio deepfake detection using deep learning. In: 2023 12th International Conference on System Modeling & Advancement in Research Trends (SMART), Moradabad, India, pp. 176–181 (2023)
3. Aravind, P.R., Nechiyil, U., Paramparambath, N.: Audio spoofing verification using deep convolutional neural networks by transfer learning (2020). arXiv: arXiv:2008.03464
4. Babitha, S., Sundaram, V., Vekkot, S.: Enhancing deep fake detection: Leveraging deep models for video authentication. In: The 15th International IEEE Conference on Computing, Communication and Networking Technologies (IC-CCNT), IIT Mandi (2024)
5. Ballesteros, D.M., Rodriguez-Ortega, Y., Renza, D., Arce, G.: Deep4SNet: deep learning for fake speech classification. Expert Syst. Appl. **184**, 115465 (2021)

6. Bartusiak, E.R., Delp, E.J.: Frequency domain-based detection of generated audio. In: Proceedings of Electronic Imaging; Society for Imaging Science and Technology, New York, NY, USA, pp. 273–281 (2021)
7. Chanchal, M., Vinoth Kumar, B.: Progress in multimodal affective computing: from machine learning to deep learning. In: Kumar, B.V., Sivakumar, P., Surendiran, B., Ding, J. (eds.) Smart Computer Vision, EAI/Springer Innovations in Communication and Computing. Springer, Cham (2023)
8. Eyben, F., et al.: The Geneva minimalistic acoustic parameter set (GeMAPS) for voice research and affective computing. IEEE Trans. Affective Comput. **7**, 190–202 (2015)
9. Hamza, A., et al.: Deepfake audio detection via MFCC features using machine learning. IEEE Access **10**, 134018–134028 (2022)
10. Jiang, Z., Zhu, H., Peng, L., Ding, W., Ren, Y.: Self-supervised spoofing audio detection scheme. In: Proceedings of INTERSPEECH 2020, Shanghai, China, pp. 4223–4227 (2020)
11. Khalid, H., Kim, M., Tariq, S., Woo, S.S.: Evaluation of an audio-video multimodal deepfake dataset using unimodal and multimodal detectors. In: Proceedings of the 1st Workshop on Synthetic Multimedia, New York, NY, USA, pp. 7–15. ACM Association for Computing Machinery (2021)
12. Lai, C.-I., Chen, N., Villalba, J., Dehak, N.: ASSERT: anti-spoofing with squeeze-excitation and residual networks (2019). arXiv: arXiv:1904.01120
13. Lal, G.J., Gopalakrishnan, E.A., Govind, D.: A recurrence network approach for characterization and detection of dynamical transitions during human speech production. Circ. Syst. Sig. Process. **41**, 6975–6998 (2022)
14. Lei, Z., Yang, Y., Liu, C., Ye, J.: Siamese convolutional neural network using gaussian probability feature for spoofing speech detection. In: Proceedings of INTERSPEECH 2020, Shanghai, China, pp. 1116–1120 (2020)
15. Raj, K.D., Lal, G.J., Gopalakrishnan, E.A., Sowmya, V., Orozco-Arroyave, J.R.: A visibility graph approach for multi-stage classification of Parkinson's disease using multimodal data. IEEE Access **12**, 87077–87096 (2024)
16. Shan, M., Tsai, T.: A cross-verification approach for protecting world leaders from fake and tampered audio (2020). arXiv:2010.12173
17. Singh, A.K., Singh, P.: Detection of AI-Synthesized speech using cepstral & bispectral statistics. In: Proceedings of the 2021 IEEE 4th International Conference on Multimedia Information Processing and Retrieval (MIPR), Tokyo, Japan, pp. 412–417 (2021)
18. Subramani, N., Rao, D.: Learning efficient representations for fake speech detection. In: Proceedings of the Thirty-Fourth AAAI Conference on Artificial Intelligence, New York, NY, USA, pp. 5859–5866 (2020)
19. Wang, R., et al.: DeepSonar: towards effective and robust detection of ai-synthesized fake voices. In: Proceedings of the 28th ACM International Conference on Multimedia, Seattle, WA, USA, pp. 1207–1216. Association for Computing Machinery (2020)
20. Wijethunga, R.L.M.A.P.C., Matheesha, D.M.K., Al Noman, A., De Silva, K.H.V.T.A., Tissera, M., Rupasinghe, L.: DeepFake audio detection: A deep learning based solution for group conversations. In: Proceedings of the 2020 2nd International Conference on Advancements in Computing (ICAC), Malabe, Sri Lanka, pp. 192–197 (2020)
21. Yu, H., Tan, Z.-H., Ma, Z., Martin, R., Guo, J.: Spoofing detection in automatic speaker verification systems using DNN classifiers and dynamic acoustic features. IEEE Trans. Neural Netw. Learn. Syst. **29**, 4633–4644 (2018)

CKGFLNet: A Fast Hybrid Architecture for Hyperspectral Image Classification Leveraging Kaiming-Gaussian Attention

Vinod Kumar[1(✉)], Ravi Shankar Singh[1], Nitika Nigam[2], Samujjal Choudhury[1], and Rahul Kumar[1]

[1] Indian Institute of Technology (B.H.U.), Varanasi 221005, India
`vinodkumar.rs.cse18@iitbhu.ac.in`, `vinodkumar.rs.cse18@itbhu.ac.in`
[2] UPES, Dehradun, Uttrakhand, India

Abstract. Convolutional neural networks (CNNs) have been extensively applied to hyperspectral image (HSI) classification, known for their ability to capture local spectral-spatial features. However, CNNs face limitations in modeling long-range dependencies and global contextual information, which are crucial for the complex nature of HSI data. To address this shortfall, we propose a hybrid architecture that leverages the strengths of both CNNs and transformers. While CNNs excel in extracting local features, transformers provide the ability to model global interactions, thereby delivering a more holistic representation of the data. Despite their effectiveness, conventional transformers suffer from quadratic computational complexity due to the self-attention mechanism, making them resource-intensive. To mitigate this issue, we introduce a Gaussian-Kaiming Focused Linear Attention (GKFLA) mechanism in the transformer component of our model. In this design, the query (Q), key (K), and value (V) matrices are initialized using Kaiming initialization, enhancing the model's capacity to extract meaningful features. Furthermore, a Gaussian matrix is added to the query to reduce overfitting and introduce greater non-linearity, optimizing the soft-max operation. This focused linear attention mechanism not only reduces computational overhead but also accelerates the model, making it more efficient than traditional soft-max-based transformers without sacrificing accuracy. We evaluate the performance of this approach on three benchmark HSI datasets—Indiana Pines (IP), University of Pavia (PU), and Salinas (SA)—demonstrating superior classification results with improved computational efficiency by combining CNN and transformer capabilities.

Keywords: Convolution Neural Network (CNN) · Visual Transformer (VIT) · Focused Linear Transformer · Kaiming Initialization · Gaussian Initialization

1 Introduction

Hyperspectral image (HSI) classification has become a crucial tool in fields like remote sensing, precision agriculture, and environmental monitoring due to its ability to capture fine spectral details across hundreds of contiguous bands. Unlike conventional multispectral imagery, HSIs contain abundant spatial and spectral information, making them highly valuable for distinguishing subtle differences in ground objects [1]. However, the vast number of spectral bands often leads to redundancy, as some bands carry overlapping information, complicating the process of extracting relevant features. The challenge lies in efficiently leveraging this vast data for accurate classification, a problem that remains central to ongoing research in HSI. With deep learning techniques proving to be effective in feature extraction, HSI classification has found significant applications: land use mapping, mineral identification, and environmental conservation [2].

With the emergence and breakthrough of deep learning (DL) technology, excellent visual architectures, mainly represented by CNNs and vision transformers (ViTs) [3], have been focused on enhancing HSI classification performance. Initially, CNNs were widely applied to HSI classification due to their ability to extract exclusively spectral features using 1D-CNN [4] or spatial features using 2D-CNN [5]. However, they solely focus on either spectral or spatial features. Later, 3D-CNNs or hybrid 3D-2D networks such as morphologically dilated convolutional neural network (MDCNN) [6] were introduced to capture both spectral and spatial features simultaneously, improving classification performance. Despite their success, CNNs have limitations, including restricted receptive fields and difficulty capturing long-range dependencies, which lead to poor generalization on unseen data and overfitting, especially with small training samples.

As a result, vision transformers (ViTs), like the spectral-spatial feature tokenization transformer (SSFTT) [7], effectively capture long-range dependencies and global spectral-spatial relationships in HSI data. While they excel in fine-grained classification, their quadratic time complexity $O(N^2 d)$—where N and d are sequence length and feature dimension—poses high computational demands, especially for large datasets, limiting practical use and memory efficiency. Hybrid models combining CNNs aim to optimize performance, but balancing a global receptive field with reduced computational costs remains challenging.

In this study, we propose a novel hybrid architecture, the Convolution-Kaiming-Gaussian Focused Linear Network (CKGFLNet), which integrates CNN with the Kaiming-Gaussian Focused Linear Attention (KGFLA) Transformer to enhance HSI classification. The CNN captures local spatial features, while the transformer models long-range dependencies, improving classification outcomes. Gaussian initialization [8,10] ensures balanced parameter optimization; inspired by this, we added Gaussian-initialized noise to the query matrix (Q) of the transformer to address overfitting. This is a regularizer, promoting feature diversity and reducing dependence on specific patterns. He et al. [9] use Kaiming initialization to stabilize training, optimize weight distributions, and accelerate convergence while avoiding gradient issues. So, our proposed KGFLA transformer combines these techniques, which reduces sharp minima convergence and enhances classification performance across datasets. The KGFLA Transformer reduces time complexity to $O(Nd^2)$, enabling efficient processing of large datasets without sacrificing accuracy. The main contributions of this research are as follows:

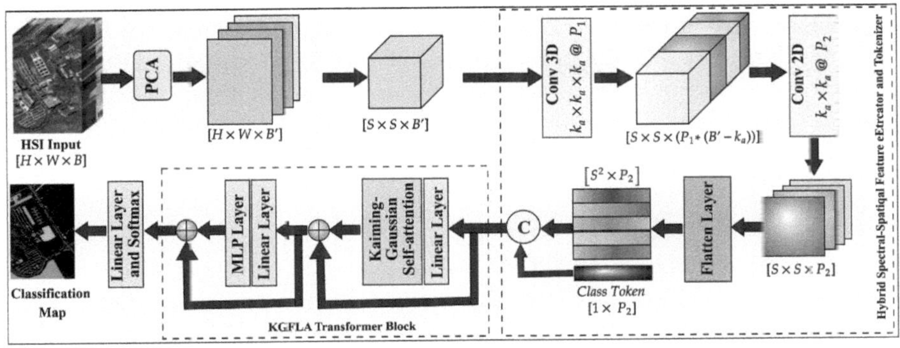

Fig. 1. The propcsed CKGFLNet architecture for HSI classification consists of two main modules: a Hybrid Spectral-Spatial Feature Extractor using 3D and 2D convolutions for local feature exraction and a KGFLA-Transformer Block with Kaiming-Gaussian Self-Attention and MLP layers to capture long-range dependencies and model global spectral-spatial relationships in HSI data.

1. We introduce a hybrid CNN-Transformer architecture Convolution-Kaiming-Gaussian Focused Linear Network (CKGFLNet) that effectively combines local spatial feature extraction with global context modeling, enhancing the HSI classification process.
2. Our Kaiming-Gaussian Focused Linear Attention Transformer reduces computational complexity, allowing for efficient processing of large datasets while maintaining high classification accuracy.
3. We leverage Kaiming and Gaussian initialization to optimize the training process, ensuring improved convergence and model performance.

2 Proposed Methodology

In this section, we describe the proposed CKGFLNet model. First, Principal Component Analysis (PCA) reduces the dimensionality of the HSI data, followed by patch extraction for the CKGFLNet. This is followed by a hybrid 3D-2D convolution network that extracts local spectral-spatial features, which are tokenized as input for the KGFLA Transformer block. Our proposed KGFLA Transformer block introduces a flattened linear-focused multi-head attention mechanism [11], designed to capture long-range dependencies and model global spectral-spatial relationships. The proposed approach reduces time complexity to $O(Nd^2)$, offering improved computational efficiency and speed compared to traditional softmax-based transformers, which have a complexity of $O(N^2d)$.

Additionally, the KGFLA Transformer block also integrates Kaiming and Gaussian initialization in the multi-head attention process. Kaiming initialization improves gradient flow, robustness, and feature representaticn, leading to faster, more efficient learning. Adding a Gaussian-initialized matrix after Kaiming in the Q matrix introduces randomness, boosting feature diversity, preventing overfitting, and enhancing the exploration of feature representations while maintaining strong gradient flow. Figure 1 provides an overview of the proposed CKGFLNet workflow.

2.1 Hybrid 3D-2D Convolution Network

We design a sub-network that first extracts spectral-spatial features using a 3D-convolution, followed by dominant spatial feature extraction using a 2D-convolution. This hybrid structure, with one layer of each 3D-CNN and 2D-CNN, efficiently captures essential information by leveraging the 3D-convolution for joint spectral-spatial features and the 2D-convolution for spatial refinement.

We start with a HSI data cube $I \in \mathbb{R}^{H \times W \times B}$, where H, W, and B denote the height, width, and spectral dimensions. To reduce spectral dimensionality, we apply PCA, transforming the cube into $I_{pca} \in \mathbb{R}^{H \times W \times B'}$, with B' as the reduced spectral dimension. Next, we extract patches represented as $I_{patch} \in \mathbb{R}^{S \times S \times B'}$, where $S \times S$ are the spatial dimensions of each patch.

The extracted patches are then processed through a 3D convolution operation. We use a 3D convolution kernel of size $K_a \times K_a \times K_a$, where $K_a = 3$, along with padding applied in the spatial dimensions. The output of this 3D convolution is denoted as $I_{3D} \in \mathbb{R}^{S \times S \times (B' - K_a + 1) \times P_1}$, where P_1 refers to the number of output channels. The 3D convolution operation is described mathematically as:

$$I_{3D}(x, y, z) = \mathrm{Conv3D}(I_{patch}, K_a, P_1) \tag{1}$$

The 3D convolution integrates spatial and spectral information by applying the kernel across all patch dimensions. After the 3D convolution, the output I_{3D} is reshaped into $I_{3Dn} \in \mathbb{R}^{S \times S \times ((B' - K_a + 1) * P_1)}$ for the next layer. A 2D convolution is applied with a kernel size of $K_b \times K_b$ (where $K_b = 3$), resulting in $I_{2D} \in \mathbb{R}^{S \times S \times P_2}$, with P_2 as the output channels. The 2D convolution is represented by the following equation:

$$I_{2D}(x, y) = \mathrm{Conv2D}(I_{3Dn}, K_b, P_2) \tag{2}$$

The 2D convolution refines spatial feature extraction. Then, the output I_{2D} is flattened into tokens and fed into the KGFLA transformer block, which captures global contextual and sequential information, enhancing the modeling of spectral-spatial relationships in the HSI data.

2.2 Kaiming-Gaussian Focused Linear Attention Transformer Block

We introduce the novel Kaiming-Gaussian Focused Linear Attention (KGFLA) Transformer block, designed to improve spectral-spatial feature learning in high-dimensional data. As shown in Fig. 2, the KGFLA module combines Kaiming and Gaussian initialization for query, key, and value matrices, enhancing gradient flow, feature diversity, and model stability. Additionally, it utilizes a flattened linear-focused transformer structure, reducing computational

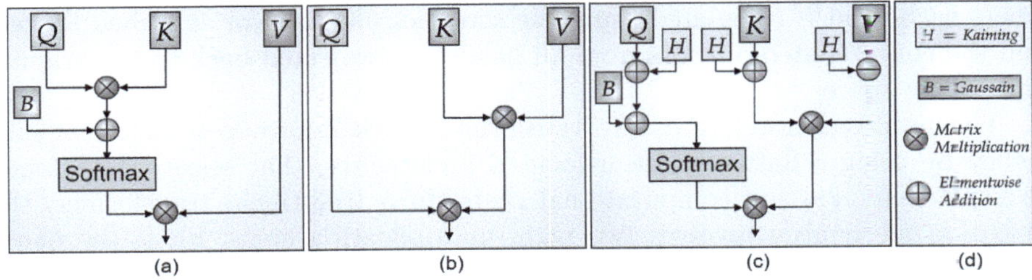

Fig. 2. Visualization of different transformers for HSI processing. (a) Traditional model with Gaussian. (b) Flatten linear transformer. (c) CKGFLNet, incorporating both Kaiming and Gaussian initialization for enhanced performance.

complexity and improving efficiency, making it well-suited for large-scale high-dimensional attention tasks. After two 2D convolution layers, the output is flattened into $I_{fl} \in \mathbb{R}^{P_2 \times S^2}$, where P_2 is the number of channels and S^2 the spatial dimensions. A sequence token is then appended, transforming the output into $I_{tk} \in \mathbb{R}^{(P_2+1) \times S^2}$, allowing better global context capture.

The KGFLA Transformer block incorporates a Kaiming-Gaussian Self-Attention (KGSA) module, where the query (Q), key (K), and value (V) matrices are initialized using Kaiming initialization. This ensures improved gradient flow, especially in layers with ReLU activations, and stabilizes the training process by mitigating vanishing or exploding gradients. The initialization of the Q, K, and V matrices can be expressed mathematically as follows:

$$Q = I_{tk}W_q + \mathrm{Km}(X_q), \quad K = I_{tk}W_k + \mathrm{Km}(X_k), \quad V = I_{tk}W_v + \mathrm{Km}(X_v) \quad (3)$$

Here, W_q, W_k, and W_v represent the weight matrices for the projections of Q, K, and V, while X_q, X_k, and X_v are randomly initialized using Kaiming initialization (Km). To increase feature diversity, a Gaussian-initialized matrix is added to the query matrix Q, introducing controlled randomness called $\hat{Q}$, represented in the equation given below:

$$\hat{Q} = Q + \mathrm{GuInit}(W_g), \quad Q' = \mathrm{Smax}(\hat{Q}), \quad Z = K^T V, \quad \mathrm{Att}(Q', K, V) = Q'Z \quad (4)$$

where W_g refers to the Gaussian-initialized weight matrix. Combining Kaiming and Gaussian initialization ensures efficient learning by promoting diverse feature representations while maintaining stable gradient flow. Softmax normalization (Smax) is applied to the query matrix $\hat{Q}$ to obtain Q'. The attention mechanism calculates weighted values Z by multiplying the transpose of the key matrix K^T with the value matrix V. The final attention output Att is produced by multiplying Q' with Z, enabling the model to highlight relevant features based on key-value importance. In multi-head self-attention, input is divided into heads with distinct parameters, and the outputs are concatenated and projected using W_O. The complete multi-head attention operation is given by:

$$\mathrm{MHSA}(Q, K, V) = \mathrm{Concat}(\mathrm{head}_1, \mathrm{head}_2, \ldots, \mathrm{head}_h)W_O \quad (5)$$

where each head is calculated using the attention mechanism described earlier, and the concatenated outputs from all heads are projected back to the original embedding space.

The proposed novel KGFLA-Transformer block improves transformer efficiency by using a flattened linear-focused architecture that selectively attends to key tokens, cutting computational costs. In a traditional transformer, the attention mechanism involves two main multiplication steps. First, the query matrix Q (of size $[(P_2 + 1) \times S^2]$) is multiplied by the transpose of the key matrix K^T (of size $[S^2 \times (P_2 + 1)]$), resulting in an intermediate matrix Z of size $[(P_2 + 1) \times (P_2 + 1)]$. This step requires $(P_2 + 1)^2 \times S^2$ multiplications. In the second step, Z is further multiplied by the value matrix V (of size $[(P_2 + 1) \times S^2]$), producing the final output of size $[(P_2 + 1) \times S^2]$. This multiplication also requires $(P_2 + 1)^2 \times S^2$ operations. Therefore, the total number of multiplications for the complete attention process in a traditional transformer is $2(P_2 + 1)^2 \times S^2$. The overall time complexity of this process is $\mathcal{O}((P_2 + 1)^2 \times S^2)$, representing the computational cost of performing these matrix operations.

In contrast in the proposed KGFLA transformer, the attention mechanism involves two main multiplication steps. First, the transpose of the key matrix K^T (of size $[S^2 \times (P_2 + 1)]$) is multiplied by the value matrix V (of size $[(P_2 + 1) \times S^2]$), resulting in an intermediate matrix Z of size $[S^2 \times S^2]$. This step requires $S^2 \times (P_2 + 1) \times S^2$ multiplications. In the second step, Z is multiplied by the query matrix Q (of size $[(P_2 + 1) \times S^2]$), producing the final output of size $[(P_2 + 1) \times S^2]$. This step requires $S^2 \times (P_2 + 1) \times S^2$ multiplications. Therefore, the total number of multiplications for the complete attention process in the KGFLA transformer is:$2 \times S^2 \times (P_2 + 1) \times S^2$. The overall time complexity of this process is $\mathcal{O}(S^2 \times (P_2 + 1) \times S^2)$, representing the computational cost of performing these matrix operations.

Thus, the overall time complexity for traditional transformers is $\mathcal{O}((P_2 + 1)^2 S^2)$, while for the KGFLA-Transformer, it is reduced to $\mathcal{O}((P_2 + 1)S^4)$. This reordering optimizes both the memory usage and computation time, making the KGFLA-Transformer more efficient while maintaining its performance. The KGSA transformer uses Kaiming initialization for stable gradients and Gaussian initialization for better feature diversity. Flattened attention reduces computation time, speeding up large-scale data processing and training convergence, making it ideal for spectral-spatial tasks.

3   Experiments

In this section, we evaluate the classification performance of our proposed method using three widely recognized HSI datasets. To ensure a comprehensive comparison, we benchmark our approach against seven well-established and state-of-the-art HSI classification techniques, including 2D-CNN [12], 3D-CNN [13], DBDA [14], MDCNN [6], SSFTT [7], IFormer [15], and JGSA-VIT [16]. The performance of each method is assessed using three key metrics: overall accuracy (OA), average accuracy (AA), and the Kappa coefficient (κ). These metrics provide a robust evaluation of classification accuracy across the datasets.

3.1 Experimental Setup and Dataset Description

In the proposed CKGFLNet approach, we set the embedding dimension $C = 120$ and use 12 attention heads. The mini-batch size is $N = 128$, and the model is trained for 70 epochs. We optimize using the Adam optimizer with a learning rate of 1×10^{-3}, running experiments on an NVIDIA RTX A5000 GPU with 24.5 GB of RAM. The input data, structured as 9×9 patches, is from the Indian Pines (IP), University of Pavia (PU), and Salinas (SA) datasets. We follow the official splits, using 10% from each class for training/validation for IP and 1% from each class for the PU and SA. The rest of the samples are used for testing.

Indiana Pines (IP): This hyperspectral dataset, collected by the AVIRIS sensor in Indiana in 1992, is used for land cover classification. It has 200 spectral bands (after removing water absorption bands) across 145×145 pixels and includes 16 classes like crops and forests. The high dimensionality and class imbalance make it useful for testing classification algorithms.

University of Pavia (PU): Captured by the ROSIS sensor over Pavia, Italy, this dataset is used for urban classification. It has 103 spectral bands and 610×340 pixels, with 9 urban land cover classes like asphalt and trees, making it valuable for detailed urban classification tasks.

Salinas (SA): Acquired by AVIRIS, this dataset covers the Salinas Valley in California for agricultural classification. It has 204 spectral bands and a 512×217 resolution, with 16 classes representing crops like lettuce and broccoli. It's often used in precision agriculture for crop type differentiation.

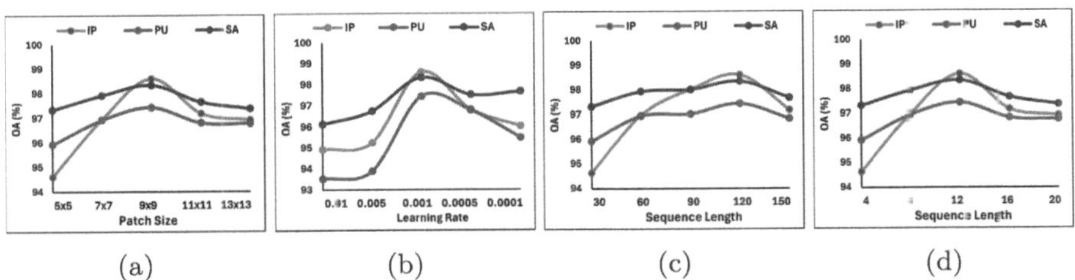

Fig. 3. Presents a comparative analysis of CKGFLNet's performance across (a) patch sizes, (b) learning rates, (c) sequence lengths, and (d) the number of transformer heads, demonstrating the model's sensitivity.

3.2 Hyper-parameter Analysis

In this study, we analyze the impact of four key hyper-parameters on CKGFLNet's performance: Patch Size, Learning Rate, Sequence Length, and Number of Transformer Heads. These parameters influence how well the model captures spectral-spatial features from hyperspectral data. We assess their effect on overall accuracy (OA) using the IP, PU, and SA datasets to identify optimal settings for improving CKGFLNet's performance.

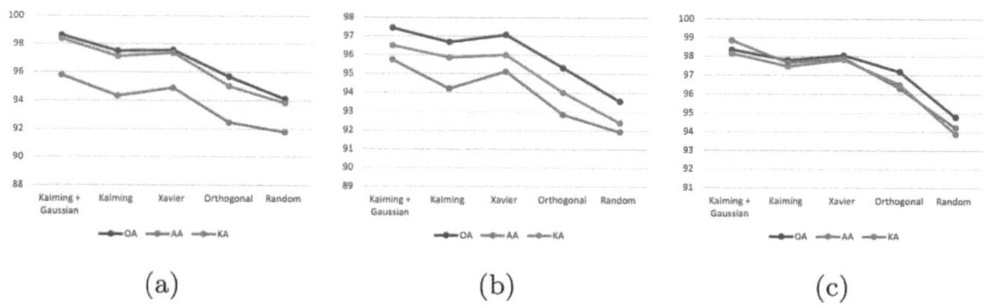

Fig. 4. Presents a comparative analysis of transformer strategies across (a) Kaiming-Gaussian, (b) Kaiming, (c) Xavier, (d) Orthogonal, and (e) Random, demonstrating the model's performance over OA, AA, κ (KA).

Patch Size: Patch size significantly impacts CKGFLNet's accuracy. A 9×9 patch size achieves the highest OAs of 98.60% for IP, 97.44% for PU, and 98.34% for SA, suggesting it balances capturing local and global features without introducing too much redundancy. Larger sizes like 11×11 and 13×13 show slight declines, likely due to irrelevant data, while smaller patches (5×5, 7×7) fail to capture enough spatial context. Figure 3(a) compares the impact of patch sizes on accuracy.

Learning Rate: The overall accuracy of CKGFLNet varies with different learning rates across the IP, PU, and SA datasets. A learning rate of 0.001 offers the best balance between convergence speed and stability, resulting in the highest performance. Lower learning rates, such as 0.0001, lead to slower convergence, while higher rates, like 0.01, may cause overshooting, reducing accuracy. Figure 3(b) highlights the effect of learning rates on model performance.

Sequence Length: The optimal sequence length after tokenization is 120, which captures spectral-spatial patterns best. Longer sequences reduce accuracy by adding irrelevant information. Figure 3(c) illustrates the effect of sequence length on model accuracy.

Number of Heads: The ideal number of transformer heads is 12, yielding the best OAs of 98.60% for IP, 97.44% for PU, and 98.34% for SA. Fewer heads limit pattern capture, while more heads (16 or 20) may introduce redundancy and overfitting. Figure 3(d) compares the performance across different head counts, showing how 12 heads provide the optimal balance.

Transformer Initialization Strategies: The proposed Kaiming-Gaussian strategy consistently outperforms other methods across the IP, PU, and SA datasets, as shown in the line graphs for OA, AA, and κ (KA) (Fig. 4). On the smaller size IP dataset, it achieves the highest OA of 98.6%, surpassing Kaiming and Xavier with 97.48% and 97.55%, respectively. For the larger PU dataset, it maintains superiority with 97.44% OA, and on the high-resolution SA dataset, it achieves an impressive 98.34% OA, 98.84% AA, and 98.14% κ. These results demonstrate the effectiveness of combining Kaiming- and Gaussian-initialized updates for enhancing feature representation while highlighting the limitations of traditional methods like Orthogonal and Random, which underperform, particularly on complex datasets.

3.3 Comparison With State-of-the-Art Method

The table provides a comparative analysis of various algorithms on three hyperspectral imaging (HSI) datasets—Indiana Pines, University of Pavia, and Salinas. The evaluation metrics include Overall Accuracy (OA), Average Accuracy (AA), and Kappa Coefficient (KC). The algorithms span CNN-based models and transformer-based methods, each with varying capabilities in handling the spectral-spatial complexity of HSI data.

CNN-based models like 2D-CNN, 3D-CNN, DBDA, and MDCNN show varying success. 2D-CNN, focused on spatial features, performs moderately on the Indiana Pines dataset but struggles with complex spectral-spatial relationships. 3D-CNN, which incorporates both, improves slightly but still falls short on challenging datasets like Salinas. DBDA, adding self-attention to CNN, performs better, particularly on Salinas. MDCNN, combining 2D and 3D CNNs, outperforms other CNN models but is generally surpassed by transformer-based approaches.

Transformer-based methods, which combine CNN feature extraction with attention mechanisms, show strong performance in hyperspectral image analysis. SSFTT and IFormer perform well, particularly on the Salinas dataset, due to their ability to capture long-range dependencies. However, models like LGSA-VIT and CKGFLNet consistently excel across datasets. LGSA-VIT achieves the highest accuracy on Salinas with 98.47% overall accuracy and 93.31% for the Kappa Coefficient, leveraging its full transformer architecture. CKGFLNet ranks first on Indiana Pines with 98.60% accuracy and University of Pavia with 97.44%, benefiting from focused linear attention and advanced initialization methods for efficient training.

The results of the performance evaluation for all models are summarized in Table X and illustrated in Figure Y. Overall, CKGFLNet exhibits superior performance across most datasets, particularly on the Indiana Pines and University of Pavia datasets, attributable to its hybrid CNN-transformer architecture and efficient attention mechanism. Although LGSA-VIT achieves marginally better performance on certain metrics within the Salinas dataset, CKGFLNet demonstrates a commendable balance between accuracy and computational efficiency, establishing it as one of the leading models in this study (Fig. 5 and Table 1).

3.4 Computational Cost Analysis

Table 2 presents a comprehensive analysis of computational complexity, comparing overall accuracy (OA), training time (Tr), and testing time (Te) across various models to highlight the trade-offs between accuracy and efficiency.

Our proposed network, CKGFLNet, achieves the highest OA of 98.60% on IP, 97.44% on PU, and 98.34% on SA, while maintaining low training times of

Table 1. Classification Performance Comparison of Different Frameworks for IP, PU and SA datasets, Where OA (%), AA (%), and $\kappa \times 100$ Are Reported. The Best Results Are Shown Bold.

Methods	IP			PU			SA		
	OA	AA	$\kappa \times 100$	OA	AA	$\kappa \times 100$	OA	AA	$\kappa \times 100$
2D-CNN	85.07	91.97	82.43	85.86	92.1	83.74	91.03	93.22	90.18
3D-CNN	80.46	90.31	77.13	83.29	91.65	80.08	88.72	91.56	87.43
DBDA	93.28	91.92	93.04	92.84	94.22	91.25	95.19	97.34	94.81
MDCNN	94.62	92.24	95.28	94.88	93.94	94.03	96.03	97.29	95.76
SSFTT	96.19	93.04	95.82	96.11	94.07	95.30	97.82	98.45	97.36
IFormer	96.36	93.21	90.25	96.28	94.67	95.56	97.11	98.21	96.22
LGSA-VIT	98.52	95.42	98.25	97.32	**95.82**	96.43	**98.47**	98.72	**98.31**
CKGFLNet	**98.60**	**95.78**	**98.35**	**97.44**	95.74	**96.50**	98.34	98.84	98.14

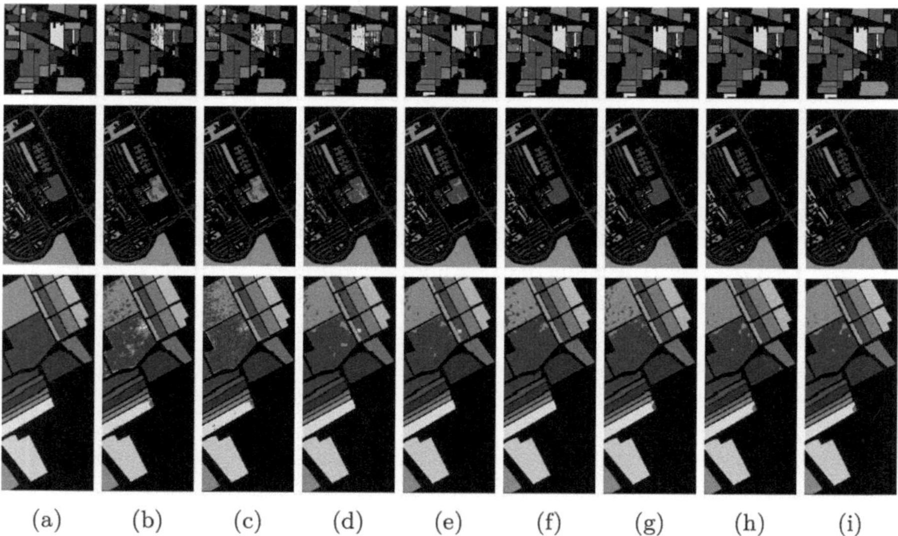

(a)   (b)   (c)   (d)   (e)   (f)   (g)   (h)   (i)

Fig. 5. Visual classification results for IP, PU, and SA datasets. Rows correspond to datasets: IP in the first row, PU in the second row, and SA in the third row. Columns represent: (a) Ground Truth, (b) 2D-CNN, (c) 3D-CNN, (d) DBDA, (e) MDCNN, (f) SSFTT, (g) IFormer, (h) LGSA-VIT, and (i) CKGFLNet

15.10 s, 31.44 s, and 40.76 s, respectively. In contrast, CKGFLNet-K, a variation of CKGFLNet, demonstrates faster training times of 15.06 s for IP, 32.71 s for PU, and 40.31 s for SA; however, it exhibits lower accuracies on the respective datasets. This showcases that while CKGFLNet-K prioritizes training and testing time, CKGFLNet offers a more efficient model.

Table 2. omputation complexity analysis comparing overall accuracy (OA), training time (Tr), and testing time (Te) across different models. Red represents the highest OA and the lowest Tr and Te, while green indicates the second-highest OA and second-best Tr and Te.

Methods	IP			PU			SA		
	OA(%)	Tr (s)	Te (s)	OA(%)	Tr (s)	Te (s)	OA(%)	Tr (s)	Te (s)
SSFTT	96.19	65.34	1.85	96.11	123.55	2.91	97.82	152.13	3.12
IFormer	96.36	75.21	1.92	96.28	130.92	3.06	97.11	130.43	3.35
LGSA-VIT	98.52	17.18	0.42	97.32	36.84	1.74	98.47	45.45	2.08
LGSA-VIT-G	98.40	17.10	0.45	96.82	36.31	1.72	98.40	45.10	1.99
CKGFLNet-K	98.42	15.06	0.38	96.66	32.71	1.49	98.11	40.31	1.82
CKGFLNet	98.60	15.10	0.39	97.44	31.44	1.47	98.34	40.76	1.85

When comparing CKGFLNet to LGSA-VIT, the latter achieves similar overall accuracies but incurs longer training times of 17.18 s, 36.84 s and 45.45 s, respectively. Thus, while LGSA-VIT provides competitive accuracy, CKGFLNet retains a significant advantage in training efficiency. LGSA-VIT-G, a variant of LGSA-VIT, offers comparable training times but its accuracies get degraded on all datasets, underscoring the influence of initialization on performance. In contrast, SSFTT and IFormer present considerable computational challenges. Their excessive computational demands render them less efficient alternatives compared to CKGFLNet and its variations.

Overall, this analysis highlights CKGFLNet's exceptional capability to deliver superior accuracy with minimal computational requirements, making it a standout model in HSI classification. The comparison with its variations and LGSA-VIT illustrates the effective balance between efficiency and time, reinforcing the advantages of CKGFLNet in practical applications.

4   Conclusion

In this paper, we proposed CKGFLNet, a novel hybrid model that integrates CNNs and the Kaiming-Gaussian Focused Linear Attention (KGFLA) Transformer to address the challenges of HSI classification. By combining CNNs for local feature extraction and the KGFLA Transformer for global context modeling, the model effectively mitigates CNNs' limitations in capturing long-range dependencies and reduces the computational overhead of traditional transformers. The KGFLA block achieves a time complexity of $O(Nd^2)$, enabling efficient processing of large datasets without compromising accuracy. Experimental results on the IP, PU, and SA datasets demonstrated the superiority of CKGFLNet, achieving overall accuracies of 98.60%, 97.44%, and 98.34%, respectively, and outperforming state-of-the-art methods such as LGSA-VIT and SSFTT.

Additionally, the Kaiming-Gaussian initialization mechanism significantly enhanced training efficiency and convergence. Despite its promising performance, the scalability of CKGFLNet to larger datasets and resource-constrained environments requires further investigation. Future work will focus on optimizing computational efficiency, extending the model to diverse remote sensing tasks, and enabling its deployment in dynamic and real-time applications. These advancements are expected to further enhance CKGFLNet's potential for practical applications in hyperspectral image analysis.

References

1. Bioucas-Dias, J.M., Plaza, A., Camps-Valls, G., Scheunders, P., Nasrabadi, N., Chanussot, J.: Hyperspectral remote sensing data analysis and future challenges. IEEE Geosci. Remote Sens. Mag. **1**(2), 6–36 (2013)
2. Kumar, V., Singh, R.S., Rambabu, M., Dua, Y.: Deep learning for hyperspectral image classification: a survey. Comput. Sci. Rev. **53**, 100658 (2024)
3. Dosovitskiy, A.: An image is worth 16x16 words: transformers for image recognition at scale. arXiv preprint: arXiv:2010.11929 (2020)
4. Hu, W., Huang, Y., Wei, L., Zhang, F., Li, H.: Deep convolutional neural networks for hyperspectral image classification. J. Sens. **2015**(1), 258619 (2015)
5. Yang, J., Zhao, Y.Q., Chan, J.C.W.: Learning and transferring deep joint spectral-spatial features for hyperspectral classification. IEEE Trans. Geosci. Remote Sens. **55**(8), 4729–4742 (2017)
6. Kumar, V., Singh, R.S., Dua, Y.: Morphologically dilated convolutional neural network for hyperspectral image classification. Signal Process.: Image Commun. **101**, 116549 (2022)
7. Sun, L., Zhao, G., Zheng, Y., Wu, Z.: Spectral-spatial feature tokenization transformer for hyperspectral image classification. IEEE Trans. Geosci. Remote Sens. **60**, 1–14 (2022)
8. Wang, A., Zhang, K., Wu, H., Dai, S., Iwahori, Y., Yu, X.: Noise-disruption-inspired neural architecture search with spatial-spectral attention for hyperspectral image classification. Remote Sens. **16**(17), 3123 (2024)
9. He, K., Zhang, X., Ren, S., Sun, J.: Delving deep into rectifiers: surpassing human-level performance on ImageNet classification. In: Proceedings of the IEEE International Conference on Computer Vision, pp. 1026–1034 (2015)
10. Zhu, C., Ni, R., Xu, Z., Kong, K., Huang, W.R., Goldstein, T.: Gradinit: learning to initialize neural networks for stable and efficient training. In: Advances in Neural Information Processing Systems, vol. 34, pp. 16410–16422 (2021)
11. Han, D., Pan, X., Han, Y., Song, S., Huang, G.: Flatten transformer: Vision transformer using focused linear attention. In: Proceedings of the IEEE/CVF International Conference on Computer Vision, pp. 5961–5971 (2023)
12. Paoletti, M.E., Haut, J.M., Plaza, J., Plaza, A.: Deep learning classifiers for hyperspectral imaging: a review. ISPRS J. Photogramm. Remote. Sens. **158**, 279–317 (2019)
13. Hamouda, M., Ettabaa, K.S., Bouhlel, M.S.: Smart feature extraction and classification of hyperspectral images based on convolutional neural networks. IET Image Proc. **14**(10), 1999–2005 (2020)

14. Li, R., Zheng, S., Duan, C., Yang, Y., Wang, X.: Classification of hyperspectral image based on double-branch dual-attention mechanism network. Remote Sens. **12**(3), 582 (2020)
15. Ren, Q., Tu, B., Liao, S., Chen, S.: Hyperspectral image classification with IFormer network feature extraction. Remote Sens. **14**(19), 4866 (2022)
16. Ma, C., et al.: Light self-Gaussian-attention vision transformer for hyperspectral image classification. IEEE Trans. Instrum. Measur. **72**, 1–12 (2023)

Machine Learning Based Detection Mechanism for Sensor Spoofing in Autonomous Vehicles

Vinayak Musale[(✉)] [iD], Rachit Patekar [iD], Ayush Kumar [iD], Aariz Shaikh [iD], Shaleen Gupta [iD], Amruta Amune [iD], and Mangesh Bedekar [iD]

Department of Computer Engineering and Technology, Dr. Vishwanath Karad MIT World Peace University, Pune, India
vinayak.musale@gmail.com, mangesh.bedekar@mitwpu.edu.in

Abstract. Autonomous driving technologies, which allow cars to function without hu-man involvement, are revolutionizing the transportation sector. Different types of sensors, such as cameras, LIDAR, and radar, combined with advanced control algorithms are incorporated in this system to navigate and manage critical vehicle functions. Despite this, autonomous vehicles (AVs) are susceptible to security vulnerabilities, including sensor spoofing. Sensor spoofing involves manipulating sensor data to deceive the vehicle's view of its environment.

This paper suggests a thorough security framework that brings together hardware authentication methods, sensor validation strategies using Machine Learning and secure Vehicle-to- Everything (V2X) communication. This strategy guarantees the integrity of sensor data, reducing the dangers of cyberattacks on autonomous driving systems. Moreover, a comparison and evaluation of current security techniques are given, which demonstrates the improvements made by the suggested approach. The findings show enhanced detection of sensor spoofing, which helps to maintain the communication smoothly.

Keywords: Adversarial Attacks · Autonomous Driving · Cybersecurity · Machine Learning · Sensor Fusion · Sensor Spoofing · Threat Mitigation

1 Introduction

The ability to control automobiles with minimal human intervention has made autonomous driving systems a major innovation in the automotive industry. To observe their surroundings and make crucial decisions in real time, these systems mainly rely on a variety of sensors, including cameras, radar, LIDAR, and other real-time data sources. Furthermore, to enable safer and more effective transportation, AVs use V2X communication to share information with pedestrians, infrastructures and other vehicles. Though AVs have a lot of benefits, they have significant security risks due to their reliance on these interconnected technologies, Sensors, communication channels, and control systems are potential targets for malicious actors. Hardware hijacking and sensor spoofing are two common security risks when it comes to V2X communication. Sensor spoofing is the practice where hackers alter sensor data to lead the sensor to misinterpret its surroundings, potentially leading to mishaps or faulty navigational choices.

C. Modi et al. (Eds.): MIND 2024, CCIS 2736, pp. 610–620, 2026.
https://doi.org/10.1007/978-3-032-14531-4_51

Since they depend on real-time input from sensors and control systems, autonomous driving systems are extremely susceptible to security risks. Vehicle functionality and safety can be seriously jeopardized by attacks like sensor spoofing. AVs are at serious risk since current security measures are insufficient to stop these kinds of attacks. To protect AV systems, a strong security architecture that combines V2X communication with cutting-edge sensor protection and hardware authentication techniques must be developed immediately.

The primary objective of this research is to create a security architecture that guards autonomous vehicles from sensor spoofing. To offer a thorough defense against these weaknesses, this framework will integrate hardware authentication mechanisms, sensor validation approaches, and secure V2X communication. The framework seeks to improve the general security and dependability of autonomous driving systems by guaranteeing the integrity of sensor data and hardware interfaces. This study focuses more on the detection of Sensor Spoofing attacks using Machine Learning (ML).

2 Literature Review

A number of security issues affect autonomous vehicles (AVs), including communication risks, hardware flaws, and sensor spoofing. Under the headings of sensor spoofing detection, hardware security, V2X communication, and machine learning-based anomaly detection, recent studies offer a variety of solutions to these problems.

Attacks like "sensor spoofing" alter the data provided by AV sensors. Cheng-zeng You and et.al. Proposed using motion-based temporal consistency to detect ghost objects caused by LiDAR spoofing. By examining variations in apparent motion over time, the method identifies fake objects, enhancing AV safety. Temporal consistency checks help reduce the dangers associated with ghost objects, which could severely compromise AV safety. The study emphasized real-time validation of perception data, improving AV safety by detecting irregularities early [1].

Shanliang Yao and et.al. Developed a framework to improve AV perception by fusing radar and camera data. Radar, less affected by weather and lighting, complements vision-based techniques to improve detection accuracy. This fusion results in a more accurate understanding of the vehicle's environment. The authors optimized feature extraction and data synchronization methods. Evaluation on various datasets showed significant improvement over standalone systems in terms of detection performance [2].

Slow-moving AVs are vulnerable to GPS spoofing in urban environments, where positional data is subtly altered. Sagar Dasgupta and et.al. Explored cross-referencing data from multiple sources as a solution. Their findings pointed out the drawbacks of relying solely on GPS and suggested urban-specific methods to reduce spoofing risks. This approach enhances the resilience of navigation systems, ensuring more reliable positioning for AVs in urban environments [3].

Hardware security is essential for AVs to protect internal networks and sensors. Sagar Dasgupta and et.al. Presented a framework combining GNSS and IMU data to detect GPS spoofing. The study confirmed the framework's effectiveness through practical tests in challenging settings. The research emphasized integrating multiple data sources to enhance the resilience of navigation systems, im-proving security and accuracy in AVs by preventing spoofing attempts [4].

Blockchain solutions offer tamper-proof data transfer between infrastructure and vehicles in V2X communication. Hassan Farran and et.al. Highlighted that verification increases latency but noted the potential of blockchain for V2X. Despite latency concerns, the study emphasized the need to balance security and real-time requirements. Blockchain could form a secure foundation for V2X communication, enhancing data integrity and vehicle connectivity in the future [5].

Engin Zaydan and et.al. Explored post-quantum cryptography methods to protect V2X communication from quantum computing risks. The study highlighted the urgency of implementing these cryptographic techniques to prevent quantum computers from decrypting sensitive data. These methods are critical for future-proofing V2X communication systems, ensuring long-term data security in a world where quantum computing poses a potential threat to conventional crypto-graphic systems [6].

Machine learning techniques are key for anomaly detection in AV systems. Bruno Ribeiro and et.al. Investigated models like Random Forests and ANNs to predict accidents involving vulnerable road users (VRUs). The study showed that combining V2X data with machine learning models improved collision prediction. This enhanced the AV's ability to identify potential hazards and make timely decisions, thereby improving overall safety [7].

Keli Huang and et.al. Proposed multi-modal sensor fusion to improve AV anomaly detection. By combining LiDAR, radar, and vision sensor data, the approach reduced false positives and improved detection accuracy. This method ensured reliable real-time performance, enhancing the AV's environmental perception. The research demonstrated that fusing various sensor data creates a more comprehensive and accurate view of surroundings, improving anomaly detection and reducing AV risks [8].

Nidhi Rastogi and et.al. Discussed the use of explainable AI to ensure transparency in AV decision-making. By applying explainable AI to radar data for spoofing detection, the study clarified the system's decision-making process. This transparency fosters trust between operators and stakeholders. It ensures that AVs operate safely with understandable, logical decisions, building confidence in their safety and reliability [9].

Monowar Hassan and et.al. Identified vulnerabilities in V2X communication, including DoS, spoofing, and data manipulation. They emphasized current standards like IEEE 1609.2 and ETSI, which focus on authentication and message integrity. Strategies such as Intrusion Detection Systems (IDS), cryptography, and machine learning for anomaly detection are being explored to reduce these risks. Addressing V2X's high mobility and low latency challenges is crucial for ensuring secure, reliable communication [10]. T. Huang et al. [11] reviewed V2X cooperative perception for autonomous driving, discussing how V2X enhances situational awareness.

3   System Design and Architecture

Figure 1 shows the architecture of proposed system. This proposed security framework for autonomous driving systems is made to address vulnerabilities related to sensor spoofing while guaranteeing effective V2X communication. The architecture consists of several interconnected components that work together to provide a multi-layered security approach.

The detailed description of each functional block is as stated below.

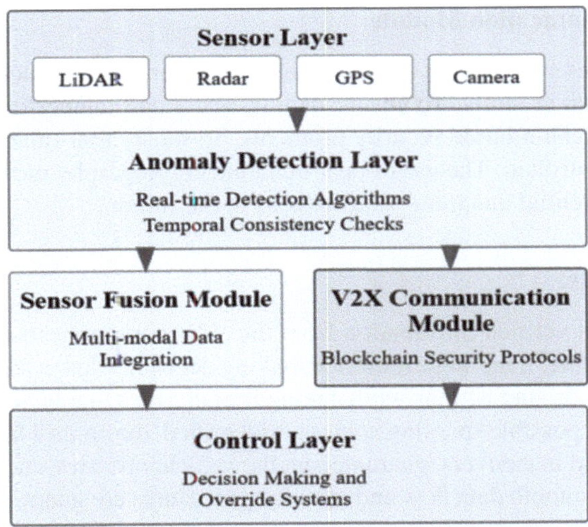

Fig. 1. System Architecture for Securing Autonomous Vehicles

3.1 Sensor Layer

LiDAR, radar GPS, and camera systems are part of this layer, which collects positional and environmental data. Every sensor is vulnerable to spoofing threats, including slow-drift GPS spoofing assaults and ghost-object fabrication in LiDAR. Raw data is continuously gathered by the sensor layer and sent to the following layers for fusion and validation.

3.2 Anomaly Detection Layer

This layer detects altered or spoof sensor data using real-time anomaly detection techniques. In order to mitigate ghost objects, temporal consistency tests are employed to identify discrepancies in LiDAR data. In order to identify slow-drift assaults, the system compares GPS signals with other sensor inputs, including GNSS and IMU data, in order to detect GPS spoofing. Flagged anomalies are sent to the control layer so that decisions can be made.

3.3 Sensor Fusion Module

To increase detection accuracy and lower false positives, multi-modal sensor fusion integrates data from several sensors (LiDAR, radar, and camera). The intermediate fusion techniques used by this module guarantee bandwidth-efficient communication while maintaining important context from every sensor stream. To find possible spoofing attempts, the fusion module looks for data differences across sensors.

3.4 V2X Communication Module

This module makes sure that the car and other external entities, including infrastructure, may exchange data securely. To ensure data integrity and tamper-proof transmission, it integrates blockchain-based security protocols. To satisfy real-time demands, latency issues must be controlled. The use of post-quantum cryptography methods shields V2X channels from potential quantum-based attacks in the future.

3.5 Control Layer

This layer handles verified information from the V2X communication module and the sensor fusion module. It has algorithms for making decisions that evaluate the surroundings and schedule driving actions while taking irregularities that have been flagged into consideration. If a possible spoofing attempt is identified, the control layer has the ability to override planned maneuvers, guaranteeing the vehicle operates safely. By combining these elements, a smooth data flow and security procedures are guaranteed at every turn.

- Real-time sensor data is gathered from GPS, LiDAR, radar, and cameras.
- In order to identify possible spoofing efforts, the anomaly detection layer verifies the data.
- In order to guarantee accuracy and consistency, multi-modal sensor fusion combines and cross-verifies data from several sensors.
- V2X communication channels, which are secured by post-quantum cryptography and blockchain-based protocols, provide safe data exchange with external systems.
- In order to make driving judgments, the control layer analyzes the fused and validated data, modifying movements as necessary in response to abnormalities detected.

4 Methodology

To address the vulnerabilities of sensor spoofing and hardware hijacking in autonomous driving systems, a comprehensive multi-layered security methodology is proposed. This methodology integrates advanced techniques to ensure the integrity and authenticity of sensor data and hardware interfaces. The key components of the proposed methodology include:

4.1 Multi-sensor Fusion

The objective of module is to integrate data from various sensors to detect inconsistencies and identify spoofing attempts.

i. **Cross-Sensor Verification:** To create a comprehensive picture of the surroundings, information is gathered via LiDAR, radar, GPS, and cameras. For instance, cameras provide semantic information about the objects, whereas radar identifies objects even in low visibility. Each sensor contributes complimentary information. To find anomalies in motion patterns, such as objects identified by LiDAR but absent from radar feeds, the system employs temporal consistency checks. The system identifies spoofing efforts by cross-referencing data streams, such as phantom objects that are present in one sensor's data but not in another.

ii. **Intermediate Sensor Fusion:** The fusion system transmits sensor data using intermediate fusion, which strikes a balance between precision and bandwidth efficiency. By doing this, latency is reduced and contextual information from each sensor is maintained. The study "Multi-Modal Sensor Fusion for Anomaly Detection" showed that combining several sensors improves detection accuracy, lowers false positives, and strengthens the system's resistance to outside interference.

4.2 Real-Time Anomaly Detection Algorithm

This is used to identify sensor spoofing attacks by continuously monitoring data streams for unusual patterns.

i. **Support Vector Machines (SVM) Configuration:** SVM was chosen for its robustness in handling non-linear decision boundaries, critical for distinguishing between spoofed and genuine sensor data in complex scenarios. The model configuration included the following.
a. **Kernel:** Radial Basis Function (RBF) kernel was employed, as it is well-suited for capturing intricate patterns in high-dimensional datasets, which are common in sensor fusion data.
b. **Probability Estimation:** The parameter probability = True enabled the computation of confidence scores for predictions, a feature crucial for pseudo-labelling in the semi-supervised learning process.
c. **Classification Rule**

The SVM predicts the class of a data point x using the following decision function:

$$f(x) = sign(\Sigma_{i=1}^{n}\alpha_i y_i K(x_i, x) + b) \tag{1}$$

where;

- α_i : Lagrange multipliers, non-zero only for support vectors.
- y_i: Labels of the training data (+1 or −1). (real or fake sensor)
- $K(x_i, x)$: Kernel function calculating similarity between points

The sign function determines the class.

- If f(x) > 0 the predicted class is + 1 (real sensor)
- If f(x) < 0, the predicted class is −1 (fake sensor)

For example, if the GPS signal shifts subtly over time (indicative of slow-drift GPS spoofing), the SVM model identifies this deviation by detecting inconsistencies between GPS and IMU data. Another Instance is when the Vehicle is moving at high speed but the amount of vehicles surrounding it is high as well, we know that it is a spoofed car sensor.

Another example is when the mileage of the car is high but the number of V2X messages sent and received are very low. This is not possible since the car has to have at least sent and received a certain amount of messages to travel.

SVM also provides the following advantages over other models, these are as.

a. **Non-linear Classification:** The RBF kernel provided flexibility to handle diverse spoofing scenarios, such as subtle GPS drift or LiDAR ghost objects

b. **Robustness to Noise:** SVM performed well even in the presence of noisy or imprecise sensor data, as observed in real-world-like synthetic datasets.

c. **Scalability:** The computational efficiency of SVM, combined with semi-supervised learning, ensured applicability to large datasets with minimal increase in training time.

ii. **Adaptive Learning:** By adding fresh data from vehicle operations, the ML model updates its baseline continuously, increasing the system's ability to respond to changing spoofing tactics. In order to handle unforeseen assault scenarios, the notebook's Self-Training Classifiers enable the system to dynamically enhance detection performance through semi-supervised learning.

iii. **Intrusion Detection Systems (IDS):** Integrate IDS to monitor V2X communication for abnormal patterns indicative of man-in-the-middle attacks or unauthorized access attempts. Employ techniques such as anomaly detection and signature-based detection to identify and respond to threats in real time.

4.3　Statistical Analysis Techniques

Use of statistical models to further validate sensor data and detect abnormalities.

i. **Gaussian Mixture Models (GMM):** The probability distributions of sensor data are modelled using GMMs. The system produces a warning, indicating a potential spoofing attack, if incoming data deviates significantly from the predicted pattern. For instance, signals for additional research are triggered by anomalous LiDAR readings that do not correspond to the normal distribution. By removing typical environmental noise, these statistical techniques aid in the reduction of false positives.

ii. **Threshold Calibration:** Statistical thresholds and confidence intervals are adjusted to preserve sensitivity without flooding the system with false alarms. The approach is guaranteed to be accurate and useful in real-world situations because of this adaptive thresholding.

4.4　V2X Communication for Collaborative Detection

The objective is to enhance spoofing detection through secure V2X communication, sharing data between vehicles and infrastructure in real time.

i. **Blockchain-Secured Communication:** Blockchain technology is used into the framework to guard against data manipulation between cars. This guarantees that, even in untrusted settings, all shared sensor readings stay valid.

ii. **Collaborative detection via V2X:** Neighbouring vehicles are informed when one vehicle notices questionable sensor activity. Collaborative confirmation of fake items is made possible by the shared data, increasing detection accuracy throughout the network.

5　Implementation Considerations

- **Scalability:** Verify that the suggested architecture can be expanded to handle a high volume of cars and different environmental circumstances.

- **Interoperability:** Create an approach that can be easily integrated into current technologies by being compatible with automotive systems and standards.
- **Cost-effectiveness:** Try to create solutions that strike a balance between security improvements and consumer and producer costs so that they can be widely adopted.

6 Performance Evaluation, Comparative Analysis and Discussions

This section presents the results of implementing a machine learning model on a dataset which contains information about V2X vehicles, both real and spoofed. The performance of the methodology is evaluated based on its effectiveness and accuracy in detecting sensor spoofing attacks while maintaining system efficiency. The framework was tested in a virtual setting that mimicked real-world scenarios. Similar to systems described in literature that use tools like SUMO and NS3, this environment simulated various sensor data inputs, V2X communication ex-changes, and hardware interactions. To test the framework's resilience, attack vectors like sensor spoofing (e.g., GPS and LIDAR) and hardware hijacking through OBD-II access were introduced.

6.1 Performance Metrics

The performance was evaluated based on the following Key Performance Indicators:

- **Accuracy:** The percentage of attacks successfully detected by the system.
- **False Positive Rate (FPR):** The rate at which legitimate activities were incorrectly identified as attacks.
- **System Latency:** The time taken to detect and respond to threats.
- **Resource Utilization:** The impact on system resources, including CPU and memory usage.
- **Detection Rate and False Positive Rate:** The system outperformed conventional techniques that rely on single-sensor validation, which typically have detection rates between 70 and 80%, achieved by Hao Xiang (2024) and Bruno Ribeiro (2021). It achieved a detection rate of 90% for hardware hijacking and 95% for sensor spoofing. With an FPR under 5%, great precision was guaranteed by the use of multi-sensor fusion and real-time anomaly detection. Hao Xiang implies that the advantages of intermediate fusion techniques for precise attack detection are in line with this.
- **System Latency:** The system demonstrated a latency of 150 ms, within the acceptable range for real-time AV operations (100–200 ms). The slight increase from typical systems without security (under 100 ms) was a trade-off for improved threat detection. This performance mirrors findings from Bruno Ribeiro and et.al., where ML-based solutions show similar latencies while maintaining safety standards.
- **Resource Utilization:** The resource utilization analysis indicated towards a CPU overhead of about 15–20% during peak operation of the proposed model. The real-time processing of several data streams and the use of sophisticated machine learning techniques for anomaly identification were blamed for this cost. Additionally, memory utilization was kept within reasonable bounds, guaranteeing that the implementation wouldn't put an undue strain on the car's onboard systems. Considering the notable advancements in security, these resource needs are deemed reasonable.

6.2 Comparative Analysis

Sensor Validation Methods: Traditional methods relying on single sensors typically achieve 70–80% detection rates. The multi-sensor fusion in our framework significantly enhances detection accuracy.

V2X Security Protocols: Older protocols often lack sufficient encryption, with FPRs exceeding 12%. Our framework maintains a slightly lower FPR of about 10% through enhanced anomaly detection. Hardware.

Authentication Techniques: Static authentication models are more prone to hijacking, with success rates below 80%. Our dynamic authentication improves security against hardware-based attacks (Table 1).

Table 1. Comparative Analysis of Existing and Proposed System

KPIs	Proposed System	B.Ribeiro (2021)	Hao Xiang (2024)
Accuracy	94.05%	70%	80%
FPR	10%	15%	12%
System Latency	150 ms	110 ms	140 ms
CPU Usage	15–20%	15–20%	15–20%

As shown in Fig. 2, the proposed framework improves detection rates through multi-layered security. It reduces false positives by using advanced validation techniques. It also ensures faster response times for real-time safety.

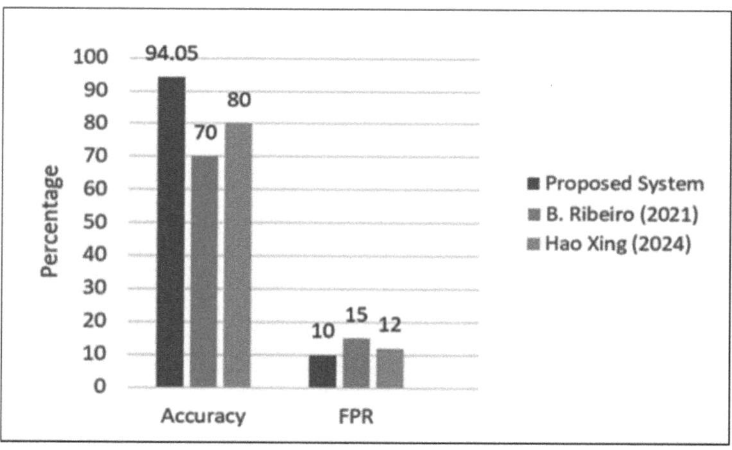

Fig. 2. Comparative Analysis

6.3 Discussions

i. **Importance of a Holistic Approach:** This study addresses sensor, communication, and hardware vulnerabilities to show why multi-layered security frameworks are essential in antivirus software. Advanced frameworks for V2X cooperation, such as V2VNet and CoBEVT, also stress the significance of integrating several layers of validation to attain strong security. Because AV components are interrelated, flaws in one area could jeopardize the system as a whole.

ii. **Limitations and Future Work:**

- **Latency Optimization:** Reduction of Latency to below 100ms will be the target. If the hacker puts spoofed data which mirrors the actual data without any errors, it will be difficult to detect spoofing.
- **Class Imbalance:** Despite high recall, minority class (non-spoofed cases) precision suffered due to imbalanced data. Future improvements could include oversampling techniques or weighted SVM.
- **Computational Overhead:** The iterative process in semi-supervised learning added to training time, suggesting that optimizations like parallel computing or kernel approximations might be beneficial.
- **Scalability in Real-Time:** The probability estimation (probability = True) increased computational complexity. Techniques like linear SVMs for low-dimensional data or approximate nearest neighbour methods could improve real-time applicability.
- **Model Updates:** To combat changing threats, the model must be continuously retrained.

7 Conclusion

The emergence of autonomous vehicles has created new security problems that call for creative fixes. To combat contemporary threats like sensor spoofing, traditional frameworks that concentrate on discrete security concerns are inadequate. This study put forth a thorough security system that integrates multi-sensor fusion, and secure V2X communication. When compared to conventional systems, the experimental findings showed better performance, with high detection rates and low false positives. In order to improve the security of autonomous systems, future research will concentrate on additional latency reduction, real-world testing, and investigating predictive AI models. The suggested framework offers a workable way to guarantee secure and dependable AV operations in settings that are becoming more interconnected.

Acknowledgement. We would like to express our sincere gratitude towards the management of Dr. Vishwanath Karad MIT World Peace University, Pune for their encouragement and support throughout this research work.

References

1. You, C., Hau, Z., Demetriou, S.: Temporal consistency checks to detect LiDAR spoofing attacks on autonomous vehicles. arXiv preprint, arXiv:2106.07833 (2020)
2. Yao, S., et al.: Radar-camera fusion for object detection and semantic segmentation in autonomous driving: a comprehensive review. arXiv preprint, arXiv:2304.10410 (2021)
3. Dasgupta, S., Ahmed, A., Rahman, M., Bandi, T.N.: Unveiling the stealthy threat: Analyzing slow drift GPS spoofing attacks for autonomous vehicles in urban environments and enabling the resilience. arXiv preprint, arXiv:2401.01394 (2024)
4. Dasgupta, S., Shakib, K.H., Rahman, M.: Experimental validation of sensor fusion-based GNSS spoofing attack detection framework for autonomous vehicles. arXiv preprint, arXiv: 2401.01304 (2024)
5. Alnasser, A., Sun, H., Jiang, J.: Blockchain-based security system for V2X communication. arXiv preprint, arXiv:1901.01053, (2019)
6. Zeydan, E., Turk, Y., Aksoy, B., Tasbag, Y.Y.: Post-quantum era in V2X security: Convergence of orchestration and parallel computation. arXiv preprint (2021)
7. Ribeiro, B., Nicolau, M.J., Santos, A.: Using machine learning on V2X communications data for VRU collision prediction. Sensors **23**(3), 1260 (2023)
8. Karle, P., Fent, F., Huch, S., Sauerbeck, F., Lienkamp, M.: Multi-modal sensor fusion for auto driving perception: A survey. arXiv preprint, arXiv:2310.08114 (2023)
9. Rastogi, N., Rampazzi, S., Clifford, M., Heller, M., Bishop, M., Levitt, K.: Explaining RADAR features for detecting spoofing attacks in connected autonomous vehicles. arXiv preprint, arXiv:2203.00150 (2022)
10. Hasan, M., Mohan, S., Shimizu, T., Lu, H.: Securing vehicle-to-everything (V2X) communication platforms. IEEE Trans. Intell. Veh. **5**(2), 693–713 (2020)
11. Huang, T., et al.: V2X cooperative perception for autonomous driving: Recent advances and challenges. arXiv preprint, arXiv:2310.03525 (2023)
12. Xiang, H., et al.: V2X-Real: a large-scale dataset for vehicle-to-everything cooperative perception. arXiv preprint (2024)
13. Evgeniou, T., Pontil, M.: Support vector machines: Theory and applications. Machine Learning and Its Applications, Advanced Lectures, Lecture Notes in Computer Science, vol. 2049, pp. 249–257 (2001)
14. Cinque, E., Valentini, F., Persia, A., Chiocchio, S., Santucci, F., Pratesi, M.: V2X communication technologies and service requirements for connected and autonomous driving. In: 2020 AEIT International Conference of Electrical and Electronic Technologies for Automotive (AEIT AUTOMOTIVE), pp. 1–5. IEEE (2020)
15. Khan, M.A., Abu-Khadrah, A., Siddiqui, S.Y.: Support-vector-machine-based adaptive scheduling in Mode 4 communication. Comput. Mater. Continua **73**(2), 3319–3331 (2022)
16. Musale, V., Chaudhari, D.: Efficient and secure keying mechanism for communication in sensor networks. In: Hemanth, J., Fernando, X., Lafata, P., Baig, Z. (eds.) International Conference on Intelligent Data Communication Technologies and Internet of Things (ICICI), 2018 (2019)
17. Yusuf, S.A., Khan, A., Souissi, R.: Vehicle-to-everything (V2X) in the autonomous vehicles' domain – a technical review of communication, sensor, and AI technologies for road user safety. Transp. Res. Interdisc. Perspect. **23**, 100980 (2024)

Intelligent Alcohol Consumption Detection Using Hyperparameter-Tuned MLP with Whale Optimization on Vowelized Dataset

Panduranga Vital Terlapu[1]([⊠])(iD), G. Kiran Kumar[2], D. Jayaram[2], Shirina Samreen[3], P. Ram Mohan Rao[4], and Kovelakonda Sakunthala[1]

[1] Aditya Institute of Technology and Management, Tekkali, Srikakulam, AP, India
vital2927@gmail.com
[2] Chaitanya Bharathi Institute of Technology, Hyderabad, Telangana, India
[3] College of Computer and Information Sciences, Majmaah University, Al Majmaah, Saudi Arabia
s.samreen@mu.edu.sa
[4] Sphoorthy Engineering College, Hyderabad, Telangana, India

Abstract. Drinking alcohol can cause many problems. Social problems like family disputes and trouble at work may result from it. Alcohol can also result in accidents. Additionally, it can create economic and health problems. Detecting alcohol consumption early is very important. It helps prevent accidents and health problems. Drinking can cause speech changes such as slurring or shaking and altered vowel sounds due to poor control. This study presents an Intelligent Alcohol Consumption Detection system using a Hyperparameter-Tuned Multi-Layer Perceptron (MLP), enhanced with a Whale Optimization Algorithm (WOA), applied to a vowelized (/a/e/i/o/u) dataset for precise identification of intoxication. The study compares the performance of the MLP model with and without WOA optimization. It evaluates additional machine learning (ML) models, including SVM (RBF and Polynomial kernels), Naive (NB) Bayes, and K-Nearest Neighbors (KNN). The experimental results show that the WOA+MLP model outperforms others in accuracy (ACC), recall (RECA), F1-score, precision (PREC), and AUC score. Specifically, the WOA+MLP model achieved an ACC of 93.14%, a PREC of 93.33%, and an AUC score of 0.9281, superior to the standard MLP model and other classifiers in detecting alcohol consumers and non-consumers. These findings highlight the effectiveness of WOA in fine-tuning machine-learning models for improved classification ACC in alcohol detection scenarios. This research focuses on intelligent alcohol detection. It introduces a non-invasive system that uses speech to detect alcohol consumers easily. This system can be easily used in real-life situations. It aims to improve public safety and help with decision-making.

Keywords: Alcohol Consumption · Machine Learning · MLP · Voice · Whale Optimization

1 Introduction

Early detection of alcohol consumption is crucial for preventing accidents, health issues, and public safety. Traditional methods like breath analyzers, blood tests, and urine analysis are invasive and time-consuming [1]. They also need special equipment. There is a need for non-invasive, affordable, and quick methods. This is especially important in transportation, law enforcement, and healthcare. Alcohol affects how a person speaks. When someone is drunk, their voice can be slurred or shaky. Vowel sounds like "a" or "e" may sound different because of poor motor control [2]. These speech changes can help identify if someone has been drinking. Analyzing a person's voice is a simple way to check their level of intoxication. Machine learning (ML) and computational intelligence have created intelligent systems that can identify intoxication. These systems use physiological and behavioural signals. Vocal features, especially vowel sounds, are good indicators of alcohol consumption. They are sensitive to changes in how people speak when they are intoxicated.

ML can detect alcohol consumption by analyzing speech patterns and behavioural data. It can identify subtle voice changes during intoxication and accurately classify individuals based on data from alcohol consumers and non-consumers, extending an efficient, non-invasive method for monitoring intoxication. This research introduces a new way to detect alcohol consumption. It uses an MLP model, which improves the WOA. The study focuses on a dataset that includes vowel sounds (/a/, /e/, /i/, /o/, /u/). The goal is to improve the MLP model by tuning its settings with WOA, inspired by how humpback whales' hunt. The researchers compare the MLP model's performance with popular ML methods. These include Support Vector (SVM) with RBF and Polynomial kernels, NB, and KNN, comprehensively evaluating the system's effectiveness. This work is unique and novel because it uses WOA for hyperparameter optimization. This helps the MLP model to learn faster and get better classification results. By fine-tuning important parameters like learning rates, activation functions, and neuron layers, the system enhances ACC(Acc.), precision(Prec.), recall(Rec.), and reliability in detecting alcohol consumption. The experiments show that the WOA-optimized MLP model performs much better than non-optimized models and other classifiers.

2 Literature Review

This review examines alcohol consumption detection methods, highlighting their limitations, and suggests ML and DL techniques, specifically MLP, can address these issues. Malucha, J. (2023) [3] examines the impact of alcohol consumption on speech, focusing on the formant properties of vowels. It uses data from sixteen individuals and shows a systematic impact of alcohol on human formants and spatial characteristics of vowel triangles. Suffoletto et al. (2023) [4] studied how mobile devices, and smart speakers detect alcohol intoxication by analyzing voice changes. They had 18 participants read tongue twisters before and after drinking. The results were promising, but more research is needed for validation. The study (Aishwarya et al. (2023)) [5] proposed a non-invasive method for detecting intoxication using speech patterns, using MFCC and CNN. This covers an accurate alternative to traditional breathalysers for law enforcement and workplace testing. Terlapu, PV (2024) [6] proposed a new method to identify drinkers by their voice. The approach uses multiple models like NBs, K-NN, DTs, and SVM. This aims to enhance accuracy in detecting alcohol consumption through voice recognition. Vital et al. (2024) [7] created a new method to detect early alcohol use. They use vowel sounds and ANN. The goal is to identify drinkers accurately. This could help develop smartphone apps for early intervention.

How et al. (2023) [8] created a system using the YOLOv8 algorithm. This system detects if drivers are drowsy or intoxicated. It alerts drivers with voice commands and email notifications. The prototype was effective in reducing accidents and improving road safety. Shamei (2023) [9] study found that male participants had a significant expansion of vowel space after drinking alcohol. Female participants had a moderate expansion. The study shows that there are differences between sexes. Wayland et al. (2023) [10] studied Alcohol's Impact on Speech Production and Used Phonet for analysis. They demonstrated the network's ability to measure speech variations and found significant consonant changes under intoxication. Di et al. (2024) [11] utilized ML models and vocal samples to study the potential of smartphones in detecting vocal disorders, demonstrating their potential for early intervention in personal healthcare monitoring. Nandwana et al. (2024) [12] developed a new approach for detecting toxic voice audio. This method uses multi-task learning, which improves detection accuracy and helps identify essential keywords. The approach also considers the style and content of speech. This focused on promoting better moderation in audio-based platforms. Bonela et al. (2024) [13] presents a DL tool named LYDIA. This tool can detect words regarding alcohol in song lyrics. It also understands the context in which these words are used. LYDIA has an ACC rate of 86.6%.

3 Models and Materials

The section explains the experimental setup (Proposal model) and datasets for the study, focusing on the creation of a MLP model optimized with the WOA.

3.1 Proposal Model

The Proposal model (Fig. 1) for Alcohol Consumption Detection uses ML and optimized MLP models. The first step is to gather voice data (like /a/, /e/, /i/, /o/, and /u/). Relevant features are extracted from both drinkers and non-drinkers. Unwanted voice information is removed to ensure clean audio input. This helps reduce errors in feature extraction and improves model accuracy. The cleaned voice recordings are saved in *.wav format. This preserves the integrity of the voice signal. Acoustic features like pitch, pulses, jitter, shimmer, and harmonic-to-noise ratios are extracted. These features provide insights into speech variability caused by alcohol consumption. This process further reduces feature extraction errors and enhances model accuracy. The collected data is stored in a lossless format. It is organized into rows and columns for analysis and use in ML models.

To prepare data for machine learning models, it is essential to clean and organize the data. Numeric data is normalized and scaled using standardization and Min-Max methods. Outliers are removed using the Isolation Forest algorithm. The data is split into 80% for training and 20% for testing. This allows the model to learn from the training data

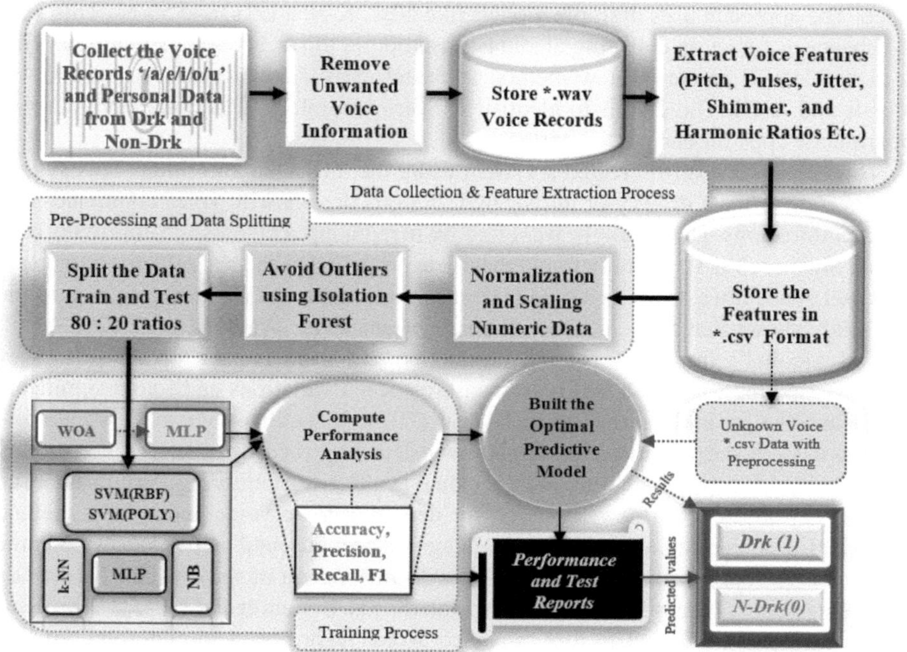

Fig. 1. Proposal Model for Alcohol Consumption Detection

and be evaluated on new data. Various machine learning models are trained, including SVM (RBF), SVM (POLY), K-NN, NBs, MLP (20 20), and an optimized MLP using the WOA algorithm. Performance is measured using accuracy, precision, recall, and F1 Score. This helps in selecting the most accurate and reliable model. The MLP+WOA model is expected to perform best due to its optimization and ability to handle complex data. A predictive model for detecting alcohol consumption is created and tested using 20% of the data and pre-processed voice data in a.csv file format.

3.2 Dataset Description

The dataset (Table 1) is a collection of features, including vowel sounds, age, mean pitch, standard deviation, minimum pitch, and maximum pitch. These are crucial in identifying specific voice patterns and characteristics related to various conditions or behaviour, such as alcohol consumption. It includes 105 instances for vowels "a" and "e", 104 for "i" and "u", and 101 for "o". Other features include pulses, periods, amplitude variation, mean autocorrelation, noise-to-harmonics ratio, and mean noise-to-noise ratio. The dataset also consists of a binary classification target (T01), where 0 represents Non-Drinker (219 samples) and 1 represents Drinker (290 samples). The data is used to analyses individuals' acoustic and demographic attributes, focusing on voice features derived from vowel sounds.

The dataset compares Class-0 (Non-Drinkers) and Class-1 (Drinkers) individuals based on acoustic and demographic features. Drinkers are slightly older, have higher

Table 1. Statistical Analysis of Drinker Voice Feature Dataset and compares classes of individuals

Fea. ID	Feature Name	Data Type	Range or Count	Fea. ID	Feature Name	Data Type	Range or Count
F01	Vowel	Integer	{'o': 101, 'e': 105, 'i': 104, 'a': 105, 'u': 104}	F16	Degree of VBs	Real	0.0 - 29.96 (sec/sec)
F02	Age	Integer	22 - 34	F17	Jitter (loc.)	Real	0.26% - 4.93%
F03	Median Pitch	Real	102.252 - 271.529 Hz	F18	Jitter (loc., abs)	Real	0.000014 - 0.000389 s
F04	Mean Pitch	Real	101.016 - 289.79 Hz	F19	Jitter (rap)	Real	0.08% - 2.98%
F05	Std. Div. Pitch	Real	0.886 - 141.719 Hz	F20	Jitter (ppq5)	Real	0.1% - 3.51%
F06	Min. Pitch	Real	66.592 - 253.562 Hz	F21	Jitter (ddp)	Real	0.24% - 8.92%
F07	Max. Pitch	Real	106.964 - 527.64 Hz	F22	Shimmer (loc.)	Real	1.81% - 23.35%
F08	No. of Pulses	Integer	13 - 118	F23	Shimmer (loc., abs)	Real	0.157 - 1.893 dB
F09	No. of Periods	Integer	12 - 117	F24	Shimmer (apq3)	Real	0.85% - 11.85%
F10	Mean Period	Real	0.00344 - 0.00991 s	F25	Shimmer (apq5)	Real	1% - 15.16%
F11	Std. Div. Period	Real	0.000029 - 0.00306 s	F26	Shimmer (apq11)	Real	0.77% - 28.33%
F12	Fraction UVFL	Real	0 - 33.333	F27	Shimmer (dda)	Real	2.55% - 35.55%
F13	No. of Unvoiced	Integer	0 - 19	F28	Mean AC	Real	0.658867 - 0.994578
F14	Total Frames	Integer	10 - 81	F29	Mean NHR	Real	0.005474 - 0.61314
F15	No. of VBs	Integer	0 - 2	F30	Mean HNR	Real	3.197 - 26.516 dB
T01	Target Class	Integer	NDRK(219), DRK (290)	*** Note: T01 is the Target Class Drinker(1) and Non-Drinker(0)**			

pitch-related features, experience more voice breaks, and have higher jitter and shimmer values, often associated with voice instability or roughness. They also have a higher mean Noise-to-Harmonics Ratio, indicating higher noise levels in their voice signals. These findings can help classify tasks related to detecting intoxication based on speech.

3.3 Visualization of Voice Pulses

The study (Fig. 2) compared the voice pulses of alcohol consumers and non-drinkers when pronouncing the vowel sound '/a'. Alcohol consumption resulted in irregular, narrow pulses with flickers, suggesting disruptions in the vocal signal's smooth flow. This may be due to alcohol's impact on motor control, causing less stable voice production. Non-drinkers produced more consistent, fair pulses, suggesting better vocal stability and motor control, free from alcohol effects. The voice signal appeared more precise and more structured, highlighting the potential influence of alcohol on vocal behaviour, particularly in affecting the clarity and consistency of vowel pronunciation.

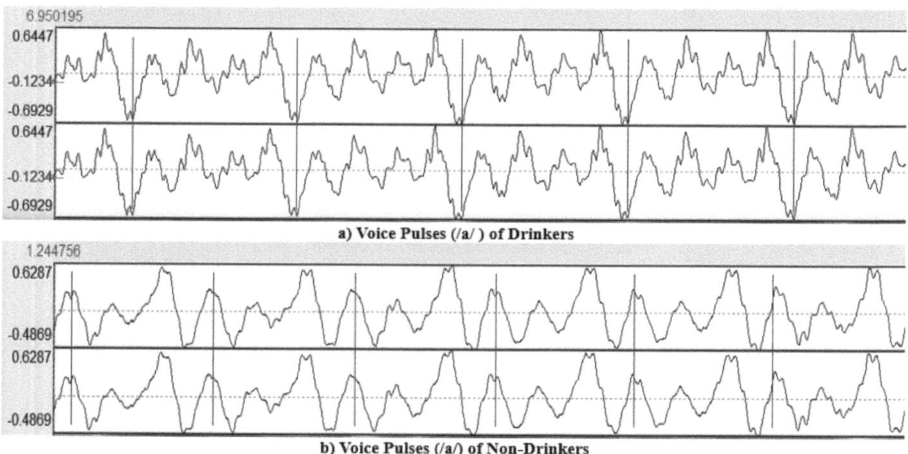

a) Voice Pulses (/a/) of Drinkers

b) Voice Pulses (/a/) of Non-Drinkers

Fig. 2. Pronunciation Vowel '/a' Voice Pulses DRK and Non-DRK

3.4 Voice Features

a) **Jitter:** Voice features include jitter, which measures the difference between consecutive periods in a speech signal. It is helpful in speech pathology as pathological voices often have higher jitter than healthy ones. There are five jitter measurements: local, cycle-to-cycle variation of the fundamental frequency, relative average perturbation, five-point period perturbation quotient, and three-time jitter. These measurements help us understand the nuances of human voices. Jitter is computes as follows equations Eq. (1) to Eq. (4)

$$jitter = \frac{1}{N_p - 1} \sum_{l=1}^{N_p} |T_l - T_{l-1}| \tag{1}$$

$$Jitter_{Relative} = \frac{\frac{1}{N_p-1} \sum_{l=1}^{N_p} |T_l - T_{l-1}|}{\frac{1}{N_p} \sum_{l=1}^{N_p} T_l} \times 100 \quad (2)$$

$$Jitter_{rap} = \frac{\frac{1}{N_p-1} \sum_{l=1}^{N_p-1} \left| T_l - \left(\frac{1}{3} \sum_{m=l-1}^{l+1} T_m \right) \right|}{\frac{1}{N_p} \sum_{l=1}^{N_p} T_l} \times 100 \quad (3)$$

$$Jitter_{ppq5} = \frac{\frac{1}{N_p-1} \sum_{l=2}^{N_p-2} \left| T_l - \left(\frac{1}{5} \sum_{m=l-2}^{l+2} T_m \right) \right|}{\frac{1}{N_p} \sum_{l=1}^{N_p} T_l} \times 100 \quad (4)$$

b) **Harmonicity:** It is measured by the harmonic-to-noise ratio (HNR) Eq. (9)

$$Harmonic_{NoiseRatio} = 10 \times \log_{10} \frac{V_{AC}(T)}{V_{AC}(0) - V_{AC}(T)} \quad (5)$$

c) **Shimmer**: Shimmer measures the difference in amplitude from one cycle to the next, useful in speech pathology. It measures shimmer in six different ways:'APQ3', 'DDP', 'Local (dB)', 'Local', 'APQ5', 'APQ11' as follow equations like Eqs. (6) to (9).

$$Shimmer_{dB} = \frac{1}{N_p-1} \sum_{l=1}^{N_p-1} \left| 20 \times \log\left(\frac{A_{l-1}}{A_l} \right) \right| \quad (6)$$

$$Shimmer_{Relative} = \frac{\frac{1}{N_p-1} \sum_{l=1}^{N_p-1} |A_l - A_{l+1}|}{\frac{1}{N_p} \sum_{l=1}^{N_p} A_l} \times 100 \quad (7)$$

$$Shimmer_{apq3} = \frac{\frac{1}{N_p-1} \sum_{l=1}^{N_p-1} \left| A_l - \left(\frac{1}{3} \sum_{m=l-1}^{l+1} A_m \right) \right|}{\frac{1}{N_p} \sum_{l=1}^{N_p} A_l} \times 100 \quad (8)$$

$$Shimmer_{ppq5} = \frac{\frac{1}{N_p-1} \sum_{l=2}^{N_p-2} \left| A_l - \left(\frac{1}{5} \sum_{m=l-2}^{l+2} A_m \right) \right|}{\frac{1}{N_p} \sum_{l=1}^{N_p} A_l} \times 100 \quad (9)$$

d) **Pitch:**It refers to how fast the vocal cords vibrate when speaking or singing. Generally, men have lower pitches compared to women and children. To measure pitch, five key statistics are used: 'maximum pitch', 'median', 'standard deviation', 'minimum', 'mean'

e) **Pulse:**The glottal pulse represents a distinct air period at the glottis, measured by four parameters.: 'period count', 'standard deviation of the period in seconds', 'mean period', 'pulse count'.

3.5 Mathematical Model of WOA

The WOA, or Whale Optimization Algorithm (Fig. 3), is a powerful method inspired by how humpback whales hunt for food. It imitates their bubble net strategy, where whales create bubbles to trap prey. The design of the algorithm reflects the whales' unique hunting techniques, making it useful for solving various optimization challenges. The Encircling Prey procedure entails a consortium of whales locating their target, and selecting the most suitable agent based on fitness value, while the remaining whales modify their positions accordingly.

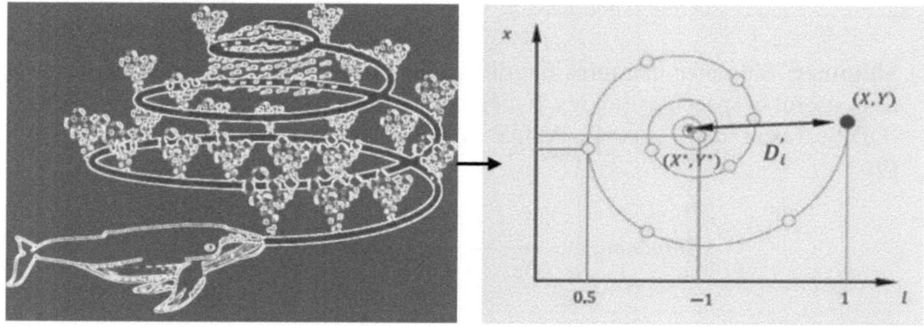

Fig. 3. Whales Searching Process

3.6 MLP and Confusion Matrix

MLP is a neural (NN) network consisting of 3 layers: an I/P layer, an O/P layer, and number of (HLs) hidden layers. The Confusion (CM) Matrix (Table 2) helps evaluate a model's performance in machine learning. It compares actual values with predicted values. The matrix has four parts: TNs, FPs, FNs, and TPs. TN shows correct predictions of non-DRK. FP shows incorrect predictions where DRK was predicted but was non-DRK. FN shows incorrect predictions where non-DRK was predicted but was DRK. TP shows correct predictions of DRK.

Table 2. Confusion Matrix Structure for Drinker Dataset

Confusion Matrix		Predicted values	
	Classes	Non-DRK(0)	DRK (1)
Actual Values Non-DRK(0)		(0, 0)	(0,1)
DRK (1)		(1,0)	(1,1)

3.7 Performance Parameters

Performance Parameters "Accuracy(ACC)", Precision (PRE), Recall (REC), F1-Value (F1) computations as follows the Eqs. (10) to (13).

$$Accuracy = \frac{(TP + TN)}{(TP + FP + FN + TN)} \tag{10}$$

$$Precision = \frac{(TP)}{(TP + FP)} \tag{11}$$

$$Recall = \frac{(TP)}{(TP + FN)} \tag{12}$$

$$F1 - Score = \frac{2 * (Recall * Precision)}{(Recall + Precision)} \tag{13}$$

4 Result Analysis

This section compares the performance of MLs, MLP, and the MLP model with Whale Optimization. The WOA-optimized MLP model shows better Acc, PREC, RECA, and AUC scores for detecting alcohol consumption.

4.1 ML Models Result Analysis

Figure 4 analyses various ML model confusion matrices, including SVM (RBF), SVM (Polynomial), Naive Bayes (NB), and K-NN. The SVM (RBF) model identified 56 true positives (TPs) with three false negatives (FNs) but misclassified 18 non-drinkers as drinkers. The SVM (Polynomial) model had 58 TPs and one FN but produced 16 false positives, balancing precision and recall. NB showed a balanced performance with 52 TPs, 35 true negatives, eight false positives, and seven FNs.

The ROC-AUC analysis in Fig. 5 compares four machine learning models: SVM (RBF), SVM (POLY), NBs, and KNN. The NB model has the highest AUC of 0.93, indicating its best classification abilities. KNN follows closely with an AUC of 0.92. SVM (POLY) has an AUC of 0.89, while SVM (RBF) has the lowest AUC at 0.87. The study (Fig. 6.) analyzed performance metrics of various machine learning models, with SVM (RBF) achieving 79.41% accuracy, POLY (83.33%) showing strong predictive ability, Naive Bayes (85.29%) having balanced metrics, and KNN (87.25%) achieving high accuracy.

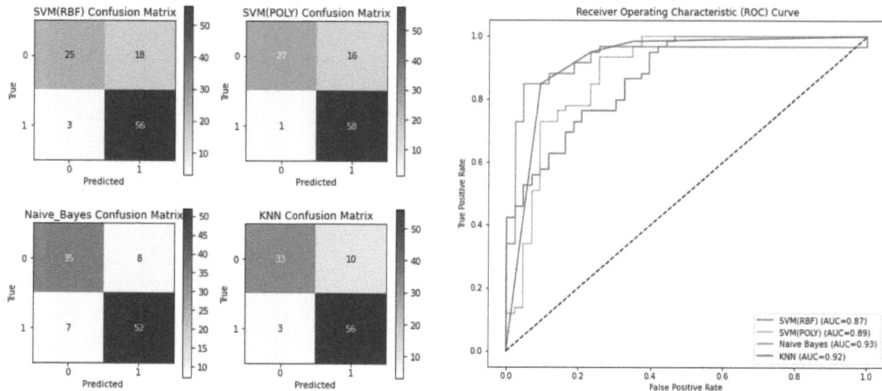

Fig. 4. SVM (RBF), SVM (Poly), NB and k-NN Models Confusion Matrices

Fig. 5. SVM (RBF), SVM (Poly) NB and k-NN Models ROC-AUC Curves Analysis

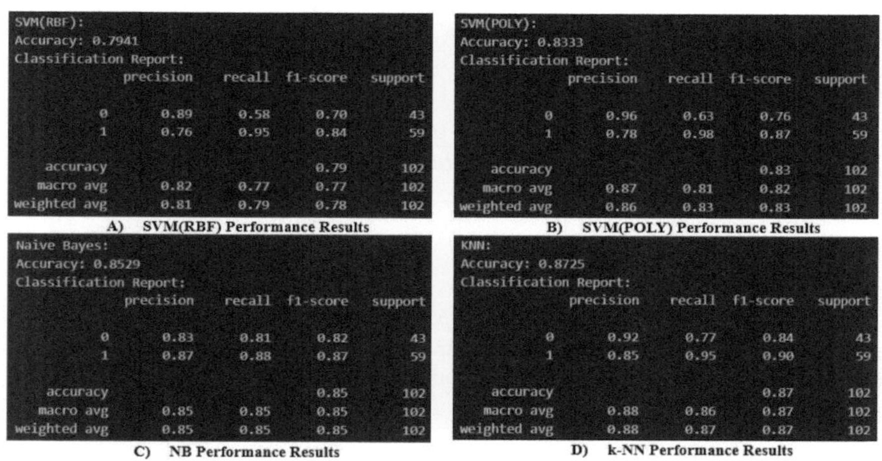

Fig. 6. Performance Parameters Analysis of ML Models

4.2 MLP and WOA+MLP Models Result Analysis

Figure 7 shows the MLP and WOA+MLP confusion matrices. The MLP model identified 56 positive and 38 negative instances but misclassified five negative samples. The WOA+MLP model improved negative classification by identifying 39 negative instances and reducing false positives. The ROC-AUC curve (Fig. 8) analysis showed that the MLP model effectively identified true positives, while the WOA+MLP model showed slightly better classification performance, indicating improved predictive capability.

The study compared MLP and WOA+MLP models, revealing that MLP achieved higher ACC, PREC, and RECA with a strong F1 Score of 93.33%, while WOA+MLP improved these metrics to 93.14%, PREC, and RECA. The detailed analysis is shown in Fig. 9.

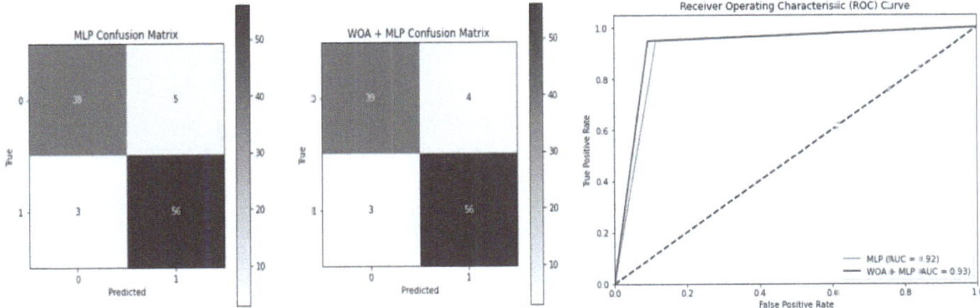

Fig. 7. MLP (20 20) and WOA+MLP Confusion Matrices

Fig. 8. MLP and WOA–MLP Models AUC-ROC Curves

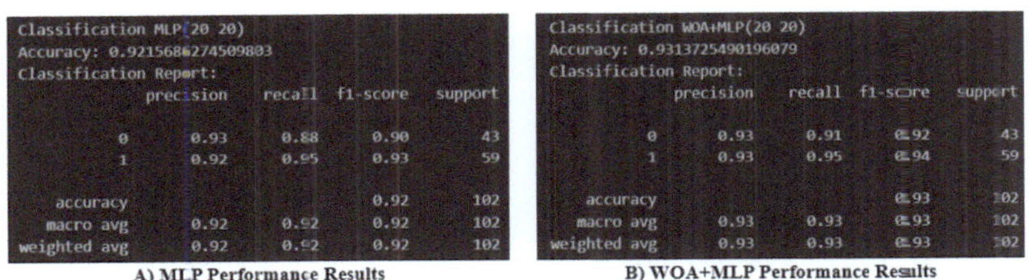

A) MLP Performance Results

B) WOA+MLP Performance Results

Fig. 9. Performance metrices about MLP(20 20) and WOA+MLP

5 Discussions

The study compared MLP-based models to traditional classifiers like SVM, NBs, and K-NN. SVM with polynomial kernel performed better (83.33% ACC), and NB achieved 85.29% ACC. The model's ACC improved to 93.14% when WOA was applied to hyperparameters of MLP(20 20), reducing false positives and improving robustness and generalization, as shown in Fig. 10.

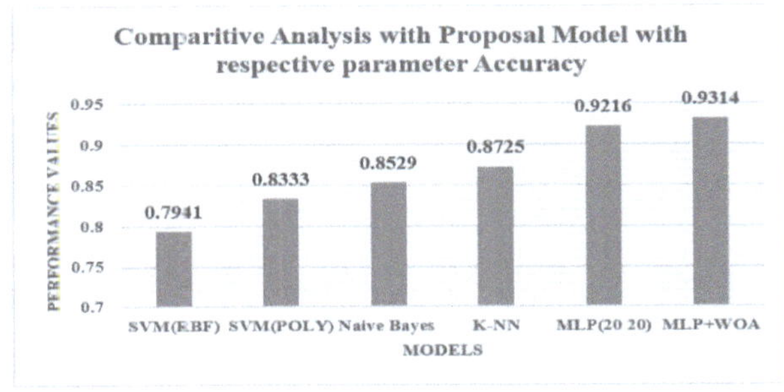

Fig. 10. Comparative analysis of ML models and optimization models MLP+WOA

Table 3 compares studies on intoxication detection using speech analysis. Laptev et al. (2023) used VGG16, Shamei et al. (2024) used speaker-dependent RF models, and

Bonela et al. (2023) introduced the ADLAIA DL algorithm. The present study achieved the highest accuracy (93.17%) using the WOA+MLP model.

Table 3. Comparative Analysis of Experimental and Proposal Models and Results for Intoxication Detection with Existing Research

Author & Year	Description	Models applied	Result Analysis
Laptev et al. (2023) [14]	The study explores the application of NN in detecting intoxication through speech analysis	VGG16	F-score: **0.73** (test), **0.82** (cross-validation) UAR: **0.77** (test), **0.83** (cross-validation)
Shamei et al. (2024) [15]	The study describes the applicability of acoustic features for speaker-dependent models in detecting alcohol intoxication	Speaker-dependent RF models	Female: Avg PREC: **66.7%**, Avg RECA: **68.6%**, Avg F1: **65.8%** Male: Avg PREC: **72.2%**, Avg RECA: **74.3%**, Avg F1: **71.6%**
Bonela et al. (2023) [16]	The study introduced the ADLAIA DL algorithm, which uses speech recordings to identify alcohol intoxication	ADLAIA	Performance: UAR: **68.09%** (for BAC $\geq$ 0.05%) Accuracy: 67.67% UAR: **75.7%**
Yokoo et al. (2023) [17]	The study shows ML can estimate habit-related information from male voice data	Xvector, AF, Chronotype classifiers	Smoking: ACC. of Xvector-**0.82** Intoxication: ACC. of AF-**0.65** Chronotype: ACC. of AF-**0.78**
Geng, et al. (2023) [18]	The study looks at how drug use affects speech production. It focuses on Chinese male drug users	5 MLs(LDA, RF, SVM, KNN, SNN)	ACC of drug users vs. non-drug users: **86%** ACC of three groups of drug users: **70%**
Present Study	ML Models (SVM(RBF) SVM(POLY), NBs, K-NN	ACC of SVM (RBF)-**0.7941**, SVM(POLY)-**0.8333**, NBs-**0.8529**, K-NN-**0.8725**	
	MLP Models MLP(20 20) and WOA+MLP (Proposal Model)	**ACC of MLP(20 20)-0.9216 WOA+MLP -0.9314**	

The study's limitations include a vowelized dataset, potential noise and speaker varia-bility, and the computational complexity of the Whale Optimization Algorithm, which may limit its application in noisy settings

6 Conclusion

Alcohol consumption can lead to social, work, and economic issues, as well as health problems. Early detection is crucial to prevent accidents and health issues. Poor control can cause speech changes like slurring or shaking. The study demonstrates the effectiveness of hyperparameter-tuned MLP with WOA optimization for detecting alcohol consumption from vowel-based speech data. The WOA+MLP model, with an ACC of 93.14%, outperforms traditional ML models, demonstrating its potential for real-world applications. The study suggests that optimization algorithms like WOA can enhance neural network performance, particularly in complex tasks like speech analysis, and could help monitor intoxication levels in safety-critical environments.

Limitations and Future work: The study's performance depends on the quality and diversity of the vowelized dataset. A more extensive and more varied data set could enhance the model's accuracy and strength. The model currently uses a single language dataset, but future research could focus on cross-lingual speech analysis to improve its relevance for various populations.

References

1. Harasymiw, J.W., Bean, P.: Identification of heavy drinkers by using the early detection of alcohol consumption score. Alcohol. Clin. Exp. Res. **25**(2), 228-235 (2001).https://doi.org/10.1111/j.1530-0277.2001.tb02203.x
2. Harasymiw, J., Seaberg, J., Bean, P.: Detection of alcohol misuse using a routine test panel: the early detection of alcohol consumption (EDAC) test. Alcohol Alcohol. **39**(4), 329–335 (2004). https://doi.org/10.1093/alcalc/agh061
3. Malucha, J.: Investigation of a specific effect of alcohol on formants. In: 2023 International Conference on Speech Technology and Human-Computer Dialogue (SpeD), pp. 95–99. IEEE (2023). https://doi.org/10.1109/SpeD59241.2023.10314882
4. Suffoletto, B., Anwar, A., Glaister, S., Sejdic, E.: Detection of alcohol intoxication using voice features: a controlled laboratory study. J. Stud. Alcohol Drugs, **84**(6), 808–813 (2023). https://doi.org/10.15288/jsad.22-00375
5. Aishwarya, B.S., Shetty, S.R., Srinivas, S., Mantri, V., Prasad, V.B.: Intoxication detection using audio. In: 2023 IEEE 8th International Conference for Convergence in Technology (I2CT), pp. 1–6. IEEE (2023)
6. Terlapu, P.V. Drinkers voice recognition intelligent system: an ensemble stacking machine learning approach. Ann. Data. Sci. (2024)
7. Vital Terlapu, P., Prasad Reddy Sadi, R.: Real-time speech-based intoxication detection system: vowel biomarker analysis with artificial neural networks. Int. J. Comput. Digit. Syst. 15(1), 1637–1666 (2024). https://doi.org/10.12785/ijcds/1501116
8. How, H.W., Saon, S., Othman, M., Kadir, H.A., Elmunsyah, H.: Development of drowsiness & drunk detection using YOLOv8. In: 2024 International Conference on Future Technologies for Smart Society (ICFTSS), pp. 6–11. IEEE (2024)
9. Shamei, A.: Expansion of vowel space following alcohol intoxication (2023). https://doi.org/10.31234/osf.io/nadr8
10. Wayland, R., Tang, K., Wang, F., Vellozzi, S., Sengupta, R.: Neural networks' posterior probability as measure of effects of alcohol on speech. In Proceedings of Meetings on Acoustics, vol. 51, no. 1. AIP Publishing (2023)

11. Di Cesare, M.G., Perpetuini, D., Cardone, D., Merla, A.: Assessment of voice disorders using machine learning and vocal analysis of voice samples recorded through smartphones. BioMedInformatics **4**(1), 549–565 (2024)
12. Nandwana, M.K., et al.: Voice toxicity detection using multi-task learning. In: ICASSP 2024–2024 IEEE International Conference on Acoustics, Speech and Signal Processing (ICASSP), pp. 331–335. IEEE (2024). https://doi.org/10.1109/ICASSP48485.2024.10448289
13. Bonela, A.A., He, Z., Luxford, D.A., Riordan, B., Kuntsche, E.: Development of the lyrics-based deep learning algorithm for identifying alcohol-related words (LYDIA). Alcohol Alcohol. **59**(2), agad088 (2024). https://doi.org/10.1093/alcalc/agad088
14. Laptev, P., Litovkin, S., Kostyuchenko, E.: Determining alcohol intoxication based on speech and neural networks. In: Karpov, A., Samudravijaya, K., Deepak, K.T., Hegde, R.M., Agrawal, S.S., Prasanna, S.R.M. (eds.) Speech and Computer. SPECOM 2023. LNCS, vol. 14338. Springer, Cham (2023). https://doi.org/10.1007/978-3-031-48309-7_9
15. Shamei, A., Liu, X., Seiilova, R.: Speaker-dependent feature generalizability for the detection of alcohol intoxication. Can. Acous. **52**(1) (2024)
16. Bonela, A.A., He, Z., Nibali, A., Norman, T., Miller, P.G., Kuntsche, E.: Audio-based deep learning algorithm to identify alcohol inebriation (ADLAIA). Alcohol **109**, 49–54 (2023). https://doi.org/10.1016/j.alcohol.2022.12.002
17. Yokoo, T., Hatano, R., Nishiyama, H.: Estimation of habit-related information from male voice data using machine learning-based methods. Artif. Life Robot. **28**(3), 520–529 (2023). https://doi.org/10.1007/s10015-023-00870-2
18. Geng, P., et al.: Acoustic characteristics of mandarin speech in male drug users. J. Voice (2023).https://doi.org/10.1016/j.jvoice.2023.08.022

Author Index